Tectonics and Magmatism in Turkey and the Surrounding Area

Geological Society Special Publications

Series Editors

A. J. HARTLEY

R. E. HOLDSWORTH

A. C. MORTON

M. S. STOKER

Special Publication reviewing procedures

The Society makes every effort to ensure that the scientific and production quality of its books matches that of its journals. Since 1997, all book proposals have been refereed by specialist reviewers as well as by the Society's Publications Committee. If the referees identify weaknesses in the proposal, these must be addressed before the proposal is accepted.

Once the book is accepted, the Society has a team of series editors (listed above) who ensure that the volume editors follow strict guidelines on refereeing and quality control. We insist that individual papers can only be accepted after satisfactory review by two independent referees. The questions on the review forms are similar to those for *Journal of the Geological Society*. The referees' forms and comments must be available to the Society's series editors on request.

Although many of the books result from meetings, the editors are expected to commission papers that were not presented at the meeting to ensure that the book provides a balanced coverage of the subject. Being accepted for presentation at the meeting does not guarantee inclusion in the book.

Geological Society Special Publications are included in the ISI Science Citation Index, but they do not have an impact factor, the latter being applicable only to journals.

More information about submitting a proposal and producing a Special Publication can be found on the Society's web site: www.geolsoc.org.uk.

GEOLOGICAL SOCIETY SPECIAL PUBLICATION NO. 173

Tectonics and Magmatism in Turkey and the Surrounding Area

EDITED BY

ERDIN BOZKURT
Geological Engineering Department, Middle Eastern Technical University, Turkey

JOHN A. WINCHESTER
Department of Earth Sciences, University of Keele, UK

JOHN D. A. PIPER
Department of Earth Sciences, University of Liverpool, UK

2000
Published by
The Geological Society
London

THE GEOLOGICAL SOCIETY

The Geological Society of London was founded in 1807 and is the oldest geological society in the world. It received its Royal Charter in 1825 for the purpose of 'investigating the mineral structure of the Earth' and is now Britain's national society for geology.

Both a learned society and a professional body, the Geological Society is recognized by the Department of Trade and Industry (DTI) as the chartering authority for geoscience, able to award Chartered Geologist status upon appropriately qualified Fellows. The Society has a membership of 8600, of whom about 1500 live outside the UK.

Fellowship of the Society is open to persons holding a recognized honours degree in geology or a cognate subject and who have at least two years' relevant postgraduate experience, or not less than six years' relevant experience in geology or a cognate subject. A Fellow with a minimum of five years' relevant postgraduate experience in the practice of geology may apply for chartered status. Successful applicants are entitled to use the designatory postnominal CGeol (Chartered Geologist). Fellows of the Society may use the letters FGS. Other grades of membership are available to members not yet qualifying for Fellowship.

The Society has its own Publishing House based in Bath, UK. It produces the Society's international journals, books and maps, and is the European distributor for publications of the American Association of Petroleum Geologists (AAPG), the Society for Sedimentary Geology (SEPM) and the Geological Society of America (GSA). Members of the Society can buy books at considerable discounts. The Publishing House has an online bookshop (*http://bookshop.geolsoc.org.uk*).

Further information on Society membership may be obtained from the Membership Services Manager, The Geological Society, Burlington House, Piccadilly, London W1V 0JU (Email: *enquiries@geolsoc.org.uk;* tel: +44 (0)171 434 9944).

The Society's Web Site can be found at *http://www.geolsoc.org.uk/*. The Society is a Registered Charity, number 210161.

Published by The Geological Society from:
The Geological Society Publishing House
Unit 7, Brassmill Enterprise Centre
Brassmill Lane
Bath BA1 3JN, UK

(*Orders*: Tel. +44 (0)1225 445046
Fax +44 (0)1225 442836)
Online bookshop: *http://bookshop.geolsoc.org.uk*

First published 2000

The publishers make no representation, express or implied, with regard to the accuracy of the information contained in this book and cannot accept any legal responsibility for any errors or omissions that may be made.

© The Geological Society of London 2000. All rights reserved. No reproduction, copy or transmission of this publication may be made without written permission. No paragraph of this publication may be reproduced, copied or transmitted save with the provisions of the Copyright Licensing Agency, 90 Tottenham Court Road, London W1P 9HE. Users registered with the Copyright Clearance Center, 27 Congress Street, Salem, MA 01970, USA: the item-fee code for this publication is 0305-8719/00/$15.00.

British Library Cataloguing in Publication Data
A catalogue record for this book is available from the British Library.

ISBN 1-86239-064-9
ISSN 0305-8719

Typeset by Type Study, Scarborough, UK
Printed by Arrowsmiths, Bristol, UK.

Distributors
USA
AAPG Bookstore
PO Box 979
Tulsa
OK 74101-0979
USA
Orders: Tel. +1 918 584-2555
Fax +1 918 560-2652
Email *bookstore@aapg.org*

Australia
Australian Mineral Foundation Bookshop
63 Conyngham Street
Glenside
South Australia 5065
Australia
Orders: Tel. +61 88 379-0444
Fax +61 88 379-4634
Email *bookshop@amf.com.au*

India
Affiliated East-West Press PVT Ltd
G-1/16 Ansari Road, Daryaganj,
New Delhi 110 002
India
Orders: Tel. +91 11 327-9113
Fax +91 11 326-0538
Email *affiliat@nda.vsnl.net.in*

Japan
Kanda Book Trading Co.
Cityhouse Tama 204
Tsurumaki 1-3-10
Tama-shi
Tokyo 206-0034
Japan
Orders: Tel. +81 (0)423 57-7650
Fax +81 (0)423 57-7651

Contents

Introduction — vii

Tethyan evolution

STAMPFLI, G. M. Tethyan oceans — 1

OKAY, A. İ. Was the Late Triassic orogeny in Turkey caused by the collision of an oceanic plateau? — 25

ROBERTSON, A. H. F. & PICKETT, E. A. Palaezoic–Early Tertiary Tethyan evolution of mélanges, rift and passive margin units in the Karaburun Peninsula (western Turkey) and Chios Island (Greece) — 43

ALTINER, D., ÖZCAN-ALTINER, S. & KOÇYIĞIT, A. Late Permian foraminiferal biofacies belts in Turkey: palaeogeographic and tectonic implications — 83

ROBERTSON, A. H. F. Mesozoic–Tertiary tectonic–sedimentary evolution of a south Tethyan oceanic basin and its margins in southern Turkey — 97

GÖNCÜOĞLU, M. C., TURHAN, N., ŞENTÜRK, K., ÖZCAN, A., UYSAL, Ş. & YALINIZ, M. K. A geotraverse across northwestern Turkey: tectonic units of the Central Sakarya region and their tectonic evolution — 139

FARINACCI, A., FIORENTINO, A. & RIDOLFI, V. Aspects of Jurassic radiolarite sedimentation in a ramp setting following the 'mid-Late Jurassic discontinuity', Barla Dağ area, Western Taurus, Turkey — 163

YILMAZ, A., ADAMIA, S., CHABUKIANI, A., CHKHOTUA, T., ERDOĞAN, K., TUZCU, S. & KARABIYIKOĞLU, M. F. Structural correlation of the southern Transcaucasus (Georgia)–eastern Pontides (Turkey) — 171

Neotethyan ophiolites

FLOYD, P. A., GÖNCÜOĞLU, M. C., WINCHESTER, J. A. & YALINIZ, M. K. Geochemical character and tectonic environment of Neotethyan ophiolitic fragments and metabasites in the Central Anatolian Crystalline Complex, Turkey — 183

YALINIZ, K. M., FLOYD, P. A. & GÖNCÜOĞLU, M. C. Geochemistry of volcanic rocks from the Çiçekdağ Ophiolite, Central Anatolia, Turkey, and their inferred tectonic setting within the northern branch of the Neotethyan Ocean — 203

PARLAK, O., HÖCK, V. & DELALOYE, M. Suprasubduction zone origin of the Pozantı–Karsantı Ophiolite (southern Turkey) deduced from whole-rock and mineral chemistry of the gabbroic cumulates — 219

Post-Tethyan basin evolution

KAZMIN, V. G., SCHREIDER, A. A. & BULYCHEV, A. A. Early stages of evolution of the Black Sea — 235

GÖRÜR, N., ÇAĞATAY, N., SAKINÇ, M., AKKÖK, R., TCHAPALYGA, A. & NATALIN, B. Neogene Paratethyan succession in Turkey and its implications for the palaeogeography of the Eastern Paratethys — 251

KARABIYIKOĞLU, M. F., ÇINER, A., MONOD, O., DEYNOUX, M., TUZCU, S. & ÖRÇEN, S. Tectonosedimentary evolution of the Miocene Manavgat Basin, western Taurides, Turkey — 271

KAYMAKÇI, N., WHITE, S. H. & VAN DIJK, P. M. Palaeostress inversion in a multiphase deformed area: kinematic and structural evolution of the Çankırı Basin (central Turkey), Part 1 – northern area   295

Neotectonics
BURCHFIEL, C. B., NAKOV, R., TZANKOV, T. & ROYDEN, L. H. Cenozoic extension in Bulgaria and northern Greece: the northern part of the Aegean extensional regime   325

YILMAZ, Y., GENÇ, Ş. C., GÜRER, F., BOZCU, M., YILMAZ, K., KARACIK, Z., ALTUNKAYNAK, Ş. & ELMAS, A. When did the western Anatolian grabens begin to develop?   353

BOZKURT, E. Timing of extension on the Büyük Menderes Graben, western Turkey, and its tectonic implications   385

KOÇYİĞİT, A., ÜNAY, E. & SARAÇ, G. Episodic graben formation and extensional neotectonic regime in west Central Anatolia and the Isparta Angle: a case study in the Akşehir-Afyon Graben, Turkey   405

TATAR, O., PIPER, J. D. A. & GÜRSOY, H. Palaeomagnetic study of the Erciyes sector of the Ecemiş Fault Zone: neotectonic deformation in the southeastern part of the Anatolian Block   423

Igneous activity
BOZTUĞ, D. S–I–A-type intrusive associations: geodynamic significance of synchronism between metamorphism and magmatism in Central Anatolia, Turkey   441

ARGER, J., MITCHELL, J. & WESTAWAY, R. W. C. Neogene and Quaternary volcanism of southeastern Turkey   459

YURTMEN, S., ROWBOTHAM, G., İŞLER, F. & FLOYD, P. A. Petrogenesis of basalts from southern Turkey: the Plio-Quaternary volcanism to the north of İskenderun Gulf   489

Index   513

It is recommended that reference to all or part of this book should be made in one of the following ways:

BOZKURT, B., WINCHESTER, J. A. & PIPER, J. D. A. (eds) 2000. *Tectonics and Magmatism in Turkey and the Surrounding Area.* Geological Society, London, Special Publications, **173**.

STAMPFLI, G. M. 2000. Tethyan oceans. *In*: BOZKURT, B., WINCHESTER, J. A. & PIPER, J. D. A. (eds) *Tectonics and Magmatism in Turkey and the Surrounding Area.* Geological Society, London, Special Publications, **173**, 1–23.

Introduction

Tethyan evolution

Turkey is sited at the collisional boundary between Gondwana in the south and Laurasia in the north and its geological history records the suturing of a succession of continental fragments. The Tethyan ocean, which existed between Laurasia and Gondwana, was not a single continuous oceanic plate, but rather comprised variable-sized continental fragments throughout its history (Fig. 1). These rifted from the Gondwana margin and, as the rifts widened, created oceans (mainly described as Prototethys, Palaeotethys and Neotethys in the literature), then subsequently collided with Laurasia so that these oceans sequentially closed. The present tectonic regime follows closure of the Neotethyan ocean. Although Prototethys has traditionally been regarded as a Late Proterozoic and/or Early Palaeozoic ocean, Palaeotethys as a Palaeozoic ocean and Neotethys as a Mesozoic–Early Tertiary ocean, the views expressed in this volume show that there is no common agreement. Many alternative models have been proposed for their evolution and they may, indeed, have overlapped in time. The models proposed below differ in subduction polarity, timing of ocean basin opening and closure, and in the location of suture zones. Figure 1 is a simplified tectonic map showing the location of the main Tethyan sutures and neighbouring major continental blocks in Turkey and its surrounding area.

The first section on Tethyan evolution opens with a broad review of Tethyan ocean development by **Stampfli**. Following definitions of the main Tethyan oceans and a brief literature review, Stampfli presents palaeocontinental reconstructions for key stages in the evolution of Prototethys, Palaeotethys, Neotethys, the Variscan, Eocimmerian and marginal oceans during the Palaeozoic and Mesozoic. His models also cover the Alpine Tethys, the Central Atlantic and the Vardar, North Atlantic and Valais oceans. After describing the Tethys sutures in the eastern Mediterranean area and presenting a tectonic map showing the present location of these sutures, Stampfli concludes that the Cretaceous-aged coloured mélanges in Turkey and Iran, now located within the Eurasian margin to the north of the Neotethys active margin, are separated from the Neotethyan suture: this would imply that the İzmir–Ankara–Erzincan Suture in Turkey (Fig. 1) is not the Neotethyan suture. In this scheme the İzmir–Ankara–Erzincan Suture represents a Jurassic back-arc oceanic basin opened along the complex pre-existing Karakaya–Palaeotethyan suture zone following the northward subduction of Neotethys since the Late Triassic.

Okay describes the E–W trending latest Triassic Cimmeride orogen in northern Turkey (box 2 in Fig. 2). He proposes that the Cimmeride deformation and metamorphism were caused by collision and partial accretion of an Early–Middle Triassic oceanic plateau (Nilüfer unit) to the active southern continental margin of Laurasia, with the Karakaya Complex interpreted as a Palaeotethyan subduction–accretion–collision complex. The age of the Cimmeride deformation in the Sakarya Zone (Fig. 1) is palaeontologically constrained between the latest Norian and Hettangian (215–200 Ma), compatible with ^{40}Ar–^{39}Ar phengite cooling ages of 214–192 Ma from eclogites and blueschists in the Nilüfer unit. Thick overlying Upper Triassic arkosic sandstone sequences containing extensive olistostromes of Permian and Carboniferous limestone (Hodul Unit), are interpreted to have formed during collision in foredeep basins in front of south-verging Hercynian continental thrust sheets of the Laurasian margin. Okay views the İzmir–Ankara–Erzincan Suture (Fig. 1) as representing both the Palaeotethyan and Neotethyan sutures and concludes that the Neotethys opened as a separate ocean during the Early Triassic.

Robertson & Pickett discuss Palaeozoic–Early Tertiary tectonic evolution of part of the Tethyan ocean, based on evidence from the Karaburun Peninsula (western Turkey) and nearby Chios Island (Greece) (Fig. 1; box 3 in Fig. 2), located near to the northern margin of the Anatolide–Tauride Platform (Taurides). Here, an exceptionally intact and unmetamorphosed tectono-sedimentary sequence forms a microcosm of the tectonic history of both Palaeotethys and Neotethys. A kilometre-thick mélange containing mainly Silurian–Carboniferous exotic blocks within a highly sheared matrix of turbidites, pelagic carbonates and channellized conglomerates is interpreted as an Upper Carboniferous–Lower Permian subduction/accretion complex that developed near the southern margin of Palaeotethys during the collision of a passive margin with a trench. Unconformably overlying Lower Triassic

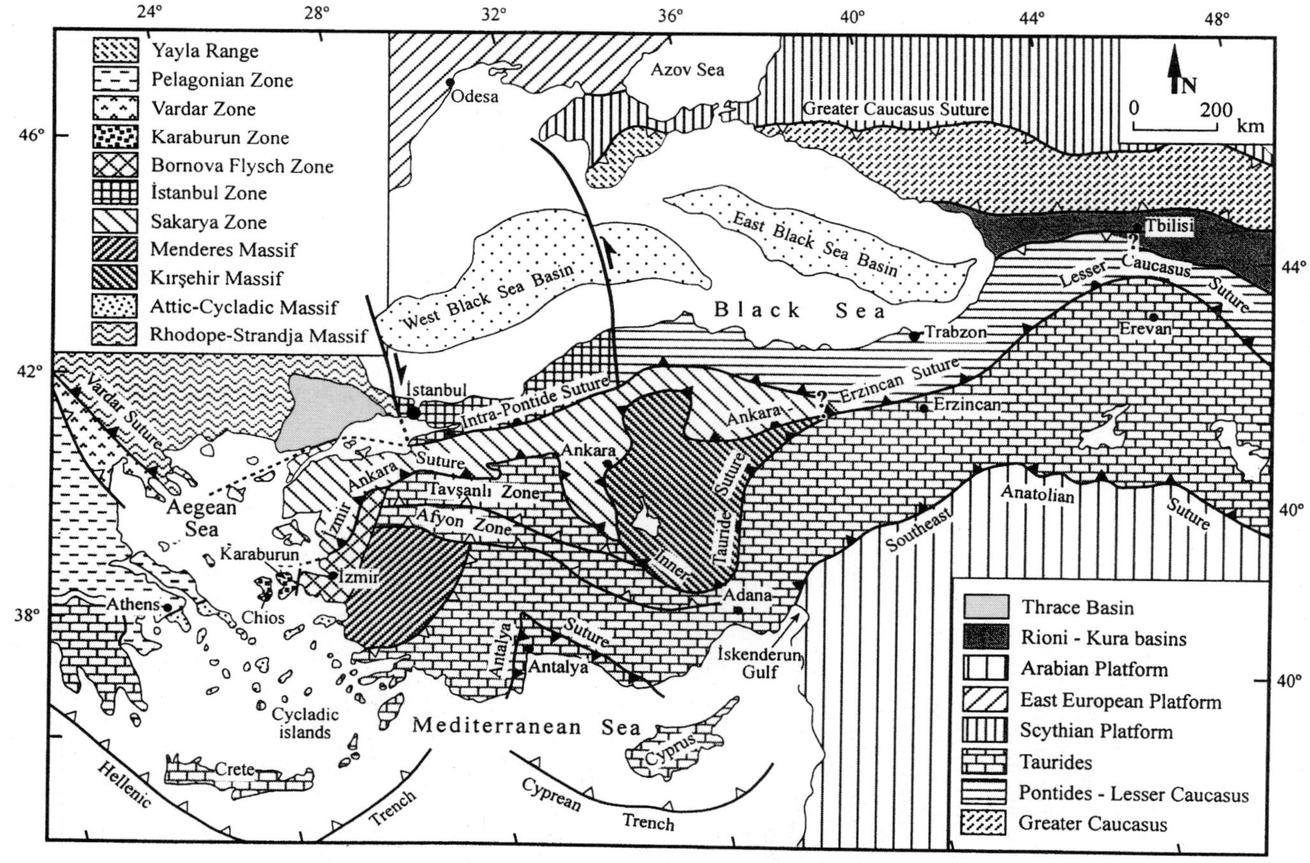

Fig. 1. Simplified tectonic map showing the location of the main Tethyan sutures and neighbouring major continental blocks in Turkey and its surrounding area, modified from Okay & Tüysüz (1999), with acknowledgements also to Stampfli (2000), Robertson (2000), Robertson & Pickett (2000), Altıner et al. (2000), A. Yılmaz et al. (2000). Heavy lines with filled triangles show sutures; polarity is indicated by the tip of triangles. Heavy lines with open triangles indicate thrust belts with triangles pointing in the direction of vergence. Heavy lines with half arrows, bounding the İstanbul Zone along its western and eastern margins, are the Western Black Sea Fault and West Crimean Fault, respectively. Arrows show the relative movement along these faults.

basinal successions record Early Triassic rifting of a northerly Neotethyan ocean. Subsequently, during Middle–Late Triassic time, the rift basin overstepped a subsiding shallow-water (Karaburun) platform bordering the northern Neotethys. Uplift associated with Cimmerian emplacement of the Karakaya Complex further north is recorded by a brief hiatus followed by deposition of deltaic sediments during latest Triassic–Early Jurassic time. During Campanian–Maastrichtian times, after prolonged passive margin subsidence the Karaburun carbonate platform underwent flexural uplift and erosion recording initial closure of the northerly Neotethys. The platform then collapsed into a foredeep during Maastrichtian–Danian times. During final stages of continental collision in the Early Tertiary, the mélange and the unconformably overlying rift and platform units were deformed and locally interleaved. Robertson & Pickett conclude by interpreting the Karaburun–Chios Mesozoic platform as part of the southern margin of a Neotethyan ocean basin bordering the Anatolide–Tauride Platform and the Menderes Massif to the south.

Altıner et al. describe the Late Permian palaeogeography and tectonic evolution of Turkey by analysing characteristics of the Late Permian carbonate platform and foraminiferal biofacies belts (box 4 in Fig. 2). The platform is reconstructed by assembling the Upper Permian outcrops from different, but juxtaposed, Triassic and Cretaceous–Tertiary tectonic units. Upper Permian marine carbonates occur in contrasting southern and northern biofacies belts. The Southern Biofacies Belt includes low-energy inner platform deposits of the Anatolide–Tauride Platform and the Arabian Platform (Fig. 1), while the highly deformed and fragmented Northern Biofacies Belt includes the Upper Permian of the Karakaya Orogen and outer platform and platform margin deposits of the Anatolide–Tauride Platform. Upper Permian blocks in the Karakaya Orogen display similar palaeontological and biofacies characteristics to the outer platform or platform margin deposits of the Taurides; they

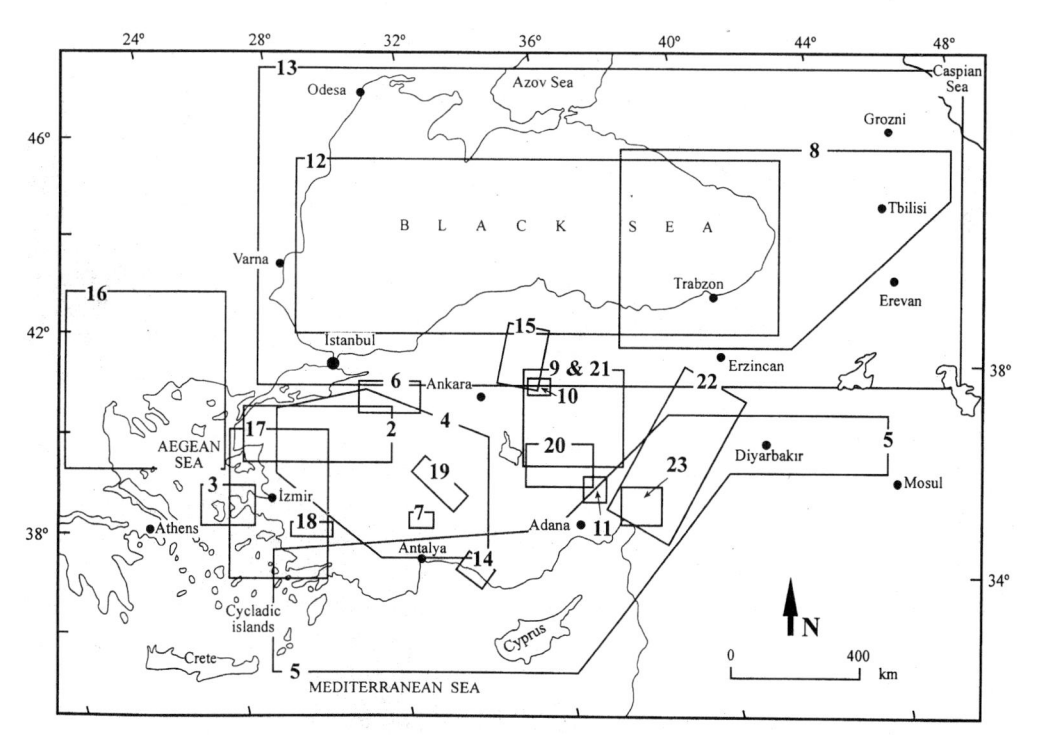

Fig. 2. Location map showing areas described in the papers in this volume. 2, Okay; 3, Robertson & Pickett; 4, Altıner et al.; 5, Robertson; 6, Göncüoğlu et al.; 7, Farinacci et al.; 8, A. Yılmaz et al.; 9, Floyd et al.; 10, Yalınız et al.; 11, Parlak et al.; 12, Kazmin et al.; 13, Görür et al.; 14, Karabıyıkoğlu et al.; 15, Kaymakçı et al.; 16, Burchfiel et al.; 17, Y. Yılmaz et al.; 18, Bozkurt; 19, Koçyiğit et al.; 20, Tatar et al.; 21, Boztuğ; 22, Arger et al.; 23, Yurtmen et al. Geographical location of the paper by Stampfli is not shown since it covers a broad review of Tethyan evolution.

represent the northernmost extent of this carbonate platform. The absence of transgressive Upper Permian deposits resting unconformably on pre-Permian basement of the Sakarya Continent strongly suggests that the carbonate platform was facing a trough or basin to the north. This would have separated the Late Permian carbonate platform in the south from the basement of the future Sakarya Continent in the north. Altıner *et al.* therefore reject a Gondwanan origin for the Sakarya Continent.

Robertson focuses on the Mesozoic–Tertiary tectonic evolution of the Tethyan ocean (Neotethyan), based on geological and geophysical information from southern Turkey and offshore areas of the easternmost Mediterranean (box 5 in Fig. 2). This region is dominated by the rifting, spreading and closure of several Neotethyan oceanic basins, including the inner Tauride basin formerly located north of the present Tauride Mountains. This opened in the Early Triassic (see also Robertson & Pickett, this volume) and, following closure in latest Cretaceous–Early Tertiary times, gave rise to the regionally extensive Lycian and Beyşehir–Hoyran–Hadim nappes. Further south, the Tauride Mountains originated as continental fragments which rifted from Gondwana in Late Permian–Early Triassic time to form the northern margin of a separate southern Neotethyan oceanic basin. Middle–Late Triassic spreading in this basin was followed, during Jurassic–Early Cretaceous time, by construction of Tauride-fringing carbonate platforms. By the late Early Cretaceous, regional convergence of the African and Eurasian Plates induced closure of both the northern and southern Neotethyan oceanic basins, activating north-dipping subduction zones. Southward 'roll-back' of pre-existing cold and dense oceanic lithosphere initiated genesis of supra-subduction type ophiolites, which were initially obducted on to the northern margin of the Tauride carbonate platforms in latest Cretaceous time. Thrusting to their present, more southerly positions in the Late Eocene (in the east) to Late Miocene (in the west) resulted from final suturing of the northern Neotethyan ocean.

The southern Neotethyan ocean was tectonically disrupted in latest Cretaceous time with emplacement of ophiolites and mélange onto the Arabian Platform in the east (e.g. Koçali ophiolite) and thrusting and strike-slip displacement of continental margin and ophiolite units in the west (Antalya Complex). Prolonged and episodic closure history in SE Turkey culminated in renewed arc volcanism, extensive subduction–accretion (Maden Complex) and thrusting to form the Southeast Anatolian Suture in the Eocene (Fig. 1); younger mélange occurrences (Misis-Andırın Mountains) suggest that subduction locally persisted until the Late Oligocene–Early Miocene. Further west (north of the Mediterranean Sea) an oceanic seaway between the eastern and western Anatolide–Tauride platforms (Isparta Angle) closed in the Late Palaeocene–Early Eocene, while further south, the southernmost oceanic basin preserved within an embayment of the North Africa/Arabian margin only began to experience collision-related deformation during Plio-Quaternary times. In southern Turkey, bordering the Mediterranean Sea, the emplaced Neotethyan units are unconformably overlain by Miocene (to Pliocene) basins which resulted from regional southward-directed crustal loading as convergence of Africa and Eurasia continued. They may also have been influenced by the initiation of a north-dipping subduction zone located south of Cyprus because the Antalya and Adana-Cilicia basins represent areas of crustal extension behind this zone since the Late Miocene. During Plio-Quaternary times continuing regional convergence was accommodated by left-lateral strike-slip faulting along the South Anatolian Transform Fault which delineated the southern margin of the Anatolian Plate during westward 'tectonic escape'. Robertson concludes that whereas SE Turkey today records a post-collisional setting, the easternmost Mediterranean records only incipient collision: this makes it ideal for the study of diachronous collisional processes.

Göncüoğlu *et al.* describe the stratigraphy and tectonic relationships of south-verging structural units across the root-zone of the Alpine İzmir–Ankara–Erzincan Suture around the central Sakarya area of NW Turkey (Fig. 1; box 6 in Fig. 2). In the northern unit (Central Sakarya Terrane), a Triassic rift basin assemblage which unconformably overlies a basement comprising a Variscan ensimatic arc complex and a fore-arc–trench complex, is itself unconformably overlain by a Liassic–Upper Cretaceous platform sequence. The middle tectonic unit (Central Sakarya Ophiolitic Complex) comprises a partly subducted Late Cretaceous accretionary complex. The southern tectonic unit includes a basement of Variscan metamorphic rocks formed in a back-arc setting, overlain by a Triassic–Lower Cretaceous succession interpreted as the passive continental margin of the Anatolide–Tauride Platform. Basinal deepening allowed deposition of a thick synorogenic flysch sequence subsequently influenced by high pressure/low temperature metamorphism. Göncüoğlu et al.'s evolutionary model contrasts with the others in this volume (e.g. Stampfli) in both the palaeogeographic distribution of the main plates during the pre-Alpine and Alpine period and in the proposed subduction polarity of the oceanic material.

Farinacci *et al.* reconstruct the basinal setting of the Tethyan Jurassic radiolarite deposits from the Barla Dağ area (western Taurides) (box 7 in Fig. 2). The main Kimmeridgian radiolarite sedimentation occurred within or just below the wave base in a storm-influenced carbonate ramp environment and followed uplift coeval with an Early Bajocian–Kimmeridgian discontinuity known as the 'main gap', a hiatus that lasted about 25 Ma. Pre-existing carbonate platforms were converted into ramps by block faulting which produced widespread Late Pliensbachian submergence. Change in the depositional bathymetry of some radiolarites suggests replacement of a deep basin by ramps. Farinacci *et al.* conclude that extensive shallow seas and a relatively narrow deep oceanic realm separated the Eurasian and African Plates in the western Tethys.

A. Yılmaz *et al*. describe and correlate Upper Cretaceous–Tertiary units of the Eastern Pontides (NE Turkey) and Transcaucasus (Georgia) (Fig. 1) which belong to the same geological belt and represent the former active margin of the Eurasian Plate (box 8 in Fig. 2). The Eastern Pontides, interpreted as the product of interference between a spreading ridge and the subduction zone during Late Jurassic–Cretaceous time, is divided into three structural units. From north to south, these are (1) the Southern Black Sea coast–Adjara Trialeti unit, (2) the Artvin–Bolnisi unit and (3) the imbricated Bayburt–Karabakh unit. They represent, respectively, a juvenile Santonian–Campanian back-arc, a south-facing arc formed mainly during Liassic–Campanian time, and a Malm–Campanian fore-arc basin of an active continental margin. To the south, these tectonic units are bordered by the Ankara–Erzincan–Lesser Caucasus suture, comprising ophiolites, mélanges and an ensimatic arc association produced during final closure of the Neotethyan ocean. By contrast, Jurassic–Early Cretaceous rift-related sediments indicate that the western part of the Eastern Pontides was a passive continental margin. This progressive change from east to west along the Eastern Pontides is explained by the progressive interaction between a spreading ridge and subduction zone, which ceased before the Maastrichtian. During the Middle Eocene renewed rifting resulted in the formation of new basins, some of which closed during Oligocene–Early Miocene times whereas others, such as the Black and Caspian Seas, survive as relict basins.

Neotethyan ophiolites

When combined with other geological data, geochemical studies, particularly those using immobile elements from basic extrusive igneous rocks, are useful for identifying tectonic settings of units within orogenic belts by comparison of their compositions with rocks from modern tectonic settings. Turkey is characterized by two major E–W trending ophiolite belts (ancient suture zones) that record closure and destruction of the Neotethyan oceans: these comprise the northern Neotethys (the İzmir–Ankara–Erzincan Suture) located between the Sakarya Continent in the north and the Anatolide–Tauride Platform in the south, and the southern Neotethys (Southeast Anatolian Suture) that formerly separated Gondwana (Arabian Platform) in the south from the Anatolide–Tauride Platform in the north (Fig. 1).

In two successive papers **Floyd *et al.*** and **Yalınız *et al*.** describe the petrogenesis and tectonic evolution of little-known Central Anatolian ophiolites and related rocks (boxes 9 and 10 in Fig. 2, respectively). These are found as allochthonous slices tectonically emplaced onto the Kırşehir Massif (Fig. 1) that represents the passive margin of the Tauride–Anatolide Platform. In both papers the ophiolitic rocks are considered to be remnants of suprasubduction zone oceanic crust formed by intra-oceanic subduction within the İzmir–Ankara–Erzincan ocean during the early Late Cretaceous. In the former paper, the metabasic rock associations are interpreted mainly as remnants of a tholeiitic arc and an adjacent back-arc basin with MORB-like compositions. The latter study details the petrology and tectonic setting of a dismembered example of the same ophiolitic assemblage which occurs as a huge tectonic slice in the northern Kırşehir Massif. For this the authors propose a two-stage emplacement model involving earlier obduction of MORB/OIB-type volcanic rocks and accretionary prism assemblages of the İzmir–Ankara–Erzincan oceanic plate on to the passive margin of the Anatolide–Tauride Platform, followed by the formation of a new, fore-arc type oceanic crust. The latter was emplaced southwards during the Late Cretaceous on to the Kırşehir Massif.

The Pozantı–Karsantı Ophiolite, one of the largest Late Cretaceous ophiolites in this region, crops out in the eastern Tauride belt of southern

Turkey and is described by **Parlak** *et al.* (box 11 in Fig. 2). It covers an area of approximately 1300 km^2 between the sinistral Ecemiş Fault Zone to the west and the sinistral East Anatolian Fault Zone to the east. The whole-rock and mineral chemistry of gabbro-norites from the ophiolite indicate production in a suprasubduction zone tectonic setting related to north-dipping subduction of the northern Neotethyan ocean by the beginning of the Late Cretaceous. Intra-oceanic subduction induced formation of the metamorphic sole and the generation of dyke swarms. Parlak *et al.* conclude that the ophiolite continued to accrete mélange and was finally obducted over the Anatolide–Tauride Platform during the Late Cretaceous or Early Palaeocene.

Post-Tethyan basin evolution

Following Late Cretaceous–Tertiary closure of the Neotethys ocean by collision of dispersed pieces of Gondwana with Eurasia, several post-orogenic Neogene basins of various sizes were developed. The third group of papers presents detailed description of basin-fill and basin-bounding structures and provides important evidence about the tectono-sedimentary and palaeoceanographic evolution of these basins.

Kazmin *et al.* report that the Black Sea comprises western and eastern sub-basins (Fig. 1; box 12 in Fig. 2), which mainly opened in the Eocene. Two depocentres in the western Black Sea are interpreted as products of early back-arc extension which formed in Barremian–Albian time north of the Pontide arc. By contrast, the central and eastern part of the basin opened as an inter-arc basin induced by the rifting of the Late Cretaceous arc. Following compression in the Eastern Pontides, the eastern Black Sea basin opened mainly in the Middle Eocene. Simultaneous opening of the central-eastern part of the Western Black Sea Basin and the Eastern Black Sea Basin (Fig. 1) is attributed to southward drift of the Pontides and clockwise rotation of the Andrusov Rise.

Görür *et al.* discuss Parathethyan evolution (box 13 in Fig. 2). Neogene basins of varying sizes formed after the Late Cretaceous to Tertiary collision of Gondwana with Eurasia. Paratethys, defined as an E–W trending land-locked basin extending from the Rhone Valley in the west to the Aral Sea in the east, forms one of them. Its isolation from marine realms such as the Mediterranean Sea caused drastic changes in palaeo-oceanographic conditions. From the distribution of these changes Paratethys is subdivided into three parts: Western, Central and Eastern. The Eastern Paratethys (EP) covers areas of the Black, Caspian and Aral Seas. Görür *et al.* detail the tectono-sedimentary and palaeo-oceanographic history of the EP between the Tarkhanian (Middle Miocene) and Cimmerian (Pliocene) and describe the Neogene marginal succession in the southern Black Sea coast and the Marmara regions of Turkey, supported by palaeogeographic maps. In the Tarkhian, the southern margin of the EP was a carbonate platform which emerged during the Late Tarkhian to Early Chokrakian. Isolation of the basin during the Karaganian was followed by marine conditions which prevailed until the Early Konkian when the EP was connected to the Indo-Pacific Ocean. Ensuing brackish conditions were followed by the widespread Early Sarmatian transgression, after which the EP was again isolated during Middle–Late Sarmatian. During the Pontian the EP was connected to the Marmara and NE Aegean regions but the link with the Mediterranean via the Marmara region did not form until the Late Akchaglylian. In this respect, the model by Görür *et al.* contradicts previous claims that the Marmara region formed a link between the Mediterranean and the Paratethys during most of the Middle Miocene.

In the Manavgat Basin, on the eastern flank of the 'Isparta Angle' in the western Taurides (box 14 in Fig. 2) **Karabıyıkoğlu** *et al.* show that Miocene basin fill unconformably overlies Mesozoic rocks which had been imbricated and overthrust by the Antalya Nappes and Alanya Massif metamorphic rocks during the Eocene. Irregularly distributed Burdigalian–Lower Langhian coarse clastic rocks (fluvial/alluvial fan and fan delta complexes) prograded into a shelf area filling the pre-existing topography. They are overlain by transgressive Langhian reefal shelf carbonates that onlap fan delta sediments and record a sharp rise in relative sea-level together with a decrease in sediment supply. Syn-sedimentary block faulting resulted in fragmentation and sudden deepening of the carbonate shelf during which Upper Langhian to Serravalian breccias, debris flows and hemipelagic sediments characterized by slumps and rock falls/slides were deposited. From the Tortonian until the Messinian, sedimentation was largely controlled by progressive uplift of the hinterland as shown by the rapid passage from high density currents and debris flows to turbulent coarse-grained fan deltas. These sediments were later folded during end-Miocene N–S compression, and unconformably overlain by undeformed Pliocene fluvial conglomerates. Three distinct episodes in the evolution of the

Manavgat Basin are distinguished as an Early Miocene fan-delta deposition, followed by Late Burdigalian to Langhian reef limestones and finally thick turbidites. Karabıyıkoğlu *et al.* conclude by suggesting that the Manavgat Basin and the northern part of the Adana Basin display similar evolution and might have been connected.

Kaymakçı *et al.* report that the Çankırı Basin (Central Anatolia; box 15 in Fig. 2), located where the Sakarya Continent became attached to the Pontides and the Kırşehir Massif collided and sutured along the İzmir–Ankara–Erzincan Suture (Fig. 1), experienced post-Middle Miocene (*c.* 9.7 Ma) deformation during extrusion-related transcurrent motions along the North Anatolian and East Anatolian fault zones. The main structures moulding the Ω-shaped Çankırı Basin, which contains more than 4 km of Upper Cretaceous to Plio-Quaternary sediments, are thrust faults defining its western and northern rims and a belt of NNE-striking folds marking its eastern margin. In the south, the basin fill onlaps on to the Kırşehir Massif. Other major structures affecting the basin are the dextral Kızılırmak and Sungurlu fault zones (KFZ and SFZ, respectively), which are splays of the North Anatolian Fault Zone. In making a kinematic and structural analysis of these structures by applying palaeostress inversion studies using fault slip data from four sub-areas, Kaymakçı *et al.* recognize four deformational phases and construct the palaeostress configuration for each. The first two (pre-Late Palaeocene and Late Palaeocene to Aquitanian) phases were characterized by thrusting and folding during the final northward subduction of Neotethys beneath the Pontides along a roughly E–W trending trench. The authors suggest that oblique transpression occurred and propose that subduction had a dextral strike-slip component. The Ω-shape of the basin is attributed to a 30° and 50° clockwise rotation along the western and eastern margins respectively during Eocene to Oligocene times, which resulted when collision of a promontory of the Kırşehir Massif indented the Sakarya Continent (Fig. 1). The third phase was a Burdigalian to pre-Tortonian (20.5 Ma to 9.7 Ma) extensional deformation, driven by gravitational collapse of the orogen following collision and further convergence of the Sakarya Continent and the Kırşehir Massif. During this phase, compression was replaced by extension and multidirectional normal faults were formed. In the final phase, linked to regional strike-slip deformation between the post-Middle Miocene (Tortonian, 9.7 Ma) and the present, most pre-existing structures were reactivated along inherited planes of weakness. The western margin, dominated by a pre-existing thrust fault belt, was reactivated into a zone of sinistral transpression as the conjugate of the KFZ and SFZ.

Neotectonics

As the 1999 earthquakes remind us, Turkey is located on the seismically active 'Mediterranean

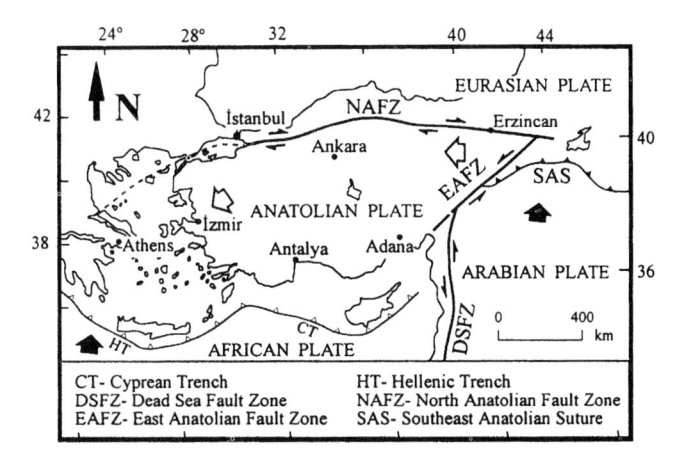

Fig. 3. Simplifed neotectonic map of Turkey showing its major structures. Heavy lines with half arrows are strike-slip faults with arrows showing the relative movement sense. The heavy line with filled triangles shows a major fold and thrust belt (Southeast Anatolian Suture): small triangles indicate direction of vergence. The heavy line with open triangles indicates an active subduction zone, its polarity indicated by the tip of small triangles. Bold filled arrows indicate relative movement direction of African, Arabian and Eurasian Plates; open arrows, relative motion of Anatolian Plate.

Earthquake Belt'. The Turkish section is outlined by three major structures (Fig. 3). The first of these is the Hellenic–Cyprus Trench, a convergent plate boundary between the African Plate in the south and the Anatolian Plate in the north. The African Plate is descending down the trench towards the NNE beneath the Anatolian Plate. The other two major structures are the dextral North Anatolian and sinistral East Anatolian fault zones (Fig. 3). Along these intracontinental strike-slip fault zones, the Anatolian Plate is being extruded towards the WSW between the converging Eurasian and Arabian Plates. The western half of the Anatolian Plate is dominated by N–S directed extension and consequent E–W, NE- and NW-trending horst and graben structures.

Western Anatolia is a part of the Aegean extensional system, embracing a large area that also includes much of Greece, Macedonia, Bulgaria and Albania. The origin and age of extension in the Aegean have long been debated and the papers contained in the Neotectonics section provide new evidence that contributes to a better understanding of this complex area. Following a broad review of extensional tectonics (Burchfiel *et al.* and Y. Yılmaz *et al.),* more geographically focused studies are described by Bozkurt, Koçyiğit *et al.* and Tatar *et al.*

Evidence from the southern Balkan Peninsula and the northern part of the Aegean extensional system is presented by **Burchfiel *et al.***, who review Middle Miocene to Recent tectonic evolution of Bulgaria and northern Greece (box 16 in Fig. 2). They suggest that Late Eocene to Early Miocene arc-normal extension continued contemporaneously with convergence in Greece, FYR Macedonia, Bulgaria and Turkey and was induced by crustal weakening due to magmatic and radiogenic heating of thickened crust following final closure of the Vardar–İzmir–Ankara Zone (Fig. 1) by northward subduction.

In Early or Middle to early Late Miocene time (26–21 Ma), major regional lithospheric extension occurred along NW-trending structures oblique to an older magmatic arc. A second phase of NE-directed extension produced low-angle detachment faults and was accompanied by a short period of coeval compression. Extension then migrated northward into SW Bulgaria at *c.* 16 Ma. Burchfiel *et al.* interpret this extension in terms of roll-back of the Hellenic Trench, which is also expressed by southward migration of the Hellenic volcanic arc. They speculate that this may also have occurred in FYR Macedonia and eastern Albania. N–S extension along E–W striking faults in central Bulgaria began at *c.* 9 Ma and extended westward, with decreasing magnitude, into SW Bulgaria and FYR Macedonia in the Quaternary, cutting across older NW-trending grabens. This continued extension is ascribed either to trench roll-back along the southern part of the Hellenic subduction system or to local anticlockwise rotation of NW Anatolia relative to part of Eurasia, including NW Greece and Albania (western Hellenides). In the Late Pliocene a widespread major erosion surface referred to as the 'sub-Quaternary surface' developed and is marked by an angular unconformity or disconformity between Upper Pliocene and Lower Quaternary strata. The presence of this surface high in the mountains (e.g. Rhodopian Mountains) demonstrates significant Quaternary displacements along normal faults associated with N–S extension. During the Late Pliocene (*c.* 3–4 Ma), deformation in SW Bulgaria and northern Greece was expressed by continued NE–SW to N–S extension and associated NE- to E–W striking dextral strike-slip faults; these functioned as transfer faults between areas of extension. This deformation is thought to have resulted from propagation of the dextral North Anatolian Fault Zone into the northern Aegean and formation of parallel faults to the north. In addition to these two different tectonic regimes, a third phase of E–W extension related to continued trench roll-back along the northern part of the Hellenic subduction system prevailed in western FYR Macedonia and eastern Albania. This, in turn, suggests that there were three areas with diffuse boundaries characterized by different styles of extensional tectonism in the southern Balkan region during the last 4 Ma: (1) N–S extension in central Bulgaria; (2) coupled strike-slip and NE–SW extension in SW Bulgaria, northern Greece and central FYR Macedonia; and (3) E–W extension in western FYR Macedonia and eastern Albania.

Based on evidence from seismicity and GPS studies, Burchfiel *et al.* propose that the active deformation in northern Greece, SW Bulgaria and FYR Macedonia is Late Quaternary N–S extension, and that dextral strike-slip movement on the North Anatolian Fault Zone must have begun at about 4 Ma. Finally they conclude that mountainous topography in the southern Balkan region results from Miocene to Recent extension with different causes involving a complex interplay between the Hellenic Trench, westward escape of Anatolia, and N–S extension and rotation of Anatolia (Fig. 3).

Y. Yılmaz *et al.* use new data to explain the

timing and mechanism of the western Anatolian graben system, and distinguish five major stages in the tectonic evolution of western Anatolia (box 17 in Fig. 2). The first stage is the Late Cretaceous–pre-Miocene pre-graben stage. Late Cretaceous to pre-Middle Eocene collision between the Sakarya Continent and Anatolide–Tauride Platform along the İzmir–Ankara–Erzincan Suture was followed by compression. This produced N- and S-directed thrusting which continued until the Late Eocene–Oligocene in the Pontides and the Late Miocene in the Taurides. Eocene crustal thickening and synchronous HT/M-HP metamorphism (main Menderes metamorphism) in the Menderes Massif (Fig. 1) was associated with widespread upper mantle and crustal melting. By the Early Miocene uplift and exhumation of the Menderes Massif had already occurred along low-angle thrusts, back thrusts and the associated normal faults recognized extensively in the massif.

In the second stage during the Early Miocene, N–S trending grabens were initially formed in an E–W extensional regime. The graben-bounding faults, which are strike-slip faults with considerable dip-slip movement, form conjugate pairs possibly developed during a N–S compression which is also indicated by development of gentle E–W trending folds and local reversed faults in the Lower Miocene successions. The Kale-Tavas Basin, initiated during the Chattian (Late Oligocene), earlier than the other N–S grabens, and previously regarded as a molasse basin with respect to the Menderes Massif, is interpreted here as a piggyback basin situated above the southerly transported Lycian Nappes. The authors also note the lack of stratigraphic contact between the Kale-Tavas Basin and the Menderes Massif and suggest that this part of the Menderes Massif remained buried beneath the Lycian Nappes throughout the Late Oligocene.

N–S extension in western Anatolia began during the Late Miocene (third stage, early N–S extension). During this stage, major N- and S-facing breakaway faults (low-angle detachment faults) were formed to bound the southern and northern flanks of the Bozdağ horst in the central Menderes Massif. Along these detachments the footwall high-grade metamorphic rocks and Miocene pre-tectonic granites of the Menderes Massif were progressively deformed, uplifted and juxtaposed against Upper Miocene continental red beds on the hanging wall. The faults remained active during the Late Miocene, later than previously considered (Early Miocene). Further away from the Bozdağ horst, in the hanging wall of the detachment faults, Upper Miocene sediment deposition was controlled by developing N–S trending cross-grabens. The graben-bounding cross-faults represent reactivated faults that controlled Early Miocene E–W extension. They also suggest that E–W trending normal faults and associated grabens, which initially began to develop during the Late Miocene, were linked to the extrusion of 9–6 Ma alkaline basaltic lavas. N–S extension ceased at the end of the Late Miocene and uplift produced a major erosion surface marked by an angular unconformity or disconformity above Upper Miocene–Lower Pliocene strata (fourth stage). The presence of this erosion surface high in the mountains demonstrates that N–S extension was rejuvenated (fifth stage, a later stage of N–S extension) and that significant Plio-Quaternary displacements occurred along these normal faults. Late Miocene structures were cut and displaced during this phase. This second, Plio-Quaternary, phase of N–S extension produced the existing E–W trending grabens in western Anatolia. The Lower–Middle Miocene fluvial-lacustrine sediments have no genetic relationships to these grabens, as recently suggested. Y. Yılmaz et al. conclude that the timing of westward escape of Anatolia along its boundary faults was synchronous with, and may be responsible for, rejuvenation of N–S extension and development of the neotectonic extensional regime in western Anatolia.

This paper also distinguishes two magmatic episodes: (1) an Oligocene–Early Miocene high-K calc-alkaline hybrid magmatism that is late/post collisional with respect to Tethyan convergence; (2) Late Miocene–Pliocene alkaline continental rift-related volcanism. The non-volcanic period (14–10 Ma) between these two phases corresponds to the time of transition from N–S compression to N–S extension in Western Anatolia. This period has been evaluated as late orogenic extension following excessive crustal thickening.

In two successive papers **Bozkurt** and **Koçyiğit** *et al.* describe new structural and stratigraphic evidence for episodic two-stage graben formation in two case studies from the Büyük Menderes Graben in western Anatolia and the Akşehir–Afyon Graben in west Central Anatolia, respectively (boxes 18 and 19 in Fig. 2, respectively). The basin fill in both grabens consists of two major sequences: deformed Miocene fluvio-lacustrine sediments overlain unconformably by undeformed, nearly horizontal Plio-Quaternary sediments. The older infill is folded and thrust faulted in the Akşehir-Afyon Graben while it is back-tilted northward and locally folded in the Büyük Menderes Graben. Both grabens exhibit evidence for two-stage extension where an initial phase of extension related to orogenic collapse of the overthickened crust, which followed Late Palaeogene collision across the Neotethyan ocean, was superseded by later and steeper normal faults during the Pliocene. The deformation of older basin fill is attributed to a short phase of compression resulting from a probable variation in kinematics of the Eurasian and African Plates in the Late Miocene, a time which also corresponds to a major break in sedimentation and magmatism, and a regional folding event, across many western Anatolian basins. Bozkurt further suggests that the Miocene sediments were deposited on the hanging wall of the normal fault(s) and that the metamorphic rocks of the Menderes Massif in the footwall were deformed, mylonitized and progressively exhumed. The second, neotectonic phase of extension, was triggered by the initiation of strike-slip movement along the North Anatolian and East Anatolian fault zones during the Pliocene and is attributed to the westward tectonic escape of the Anatolian Plate along these structures. The fault controlling the early phase of extension in the Büyük Menderes Graben may have been reactivated during the second phase. Koçyiğit et al. conclude that the Akşehir Fault is an oblique-slip normal fault forming part of the current extensional regime of west Central Anatolia and the Isparta Angle region; this contrasts with previous interpretations which interpreted it as a reverse fault belonging to a compressional neotectonic regime. Bozkurt further suggests that the basal Miocene red clastic rocks cannot be regarded as passive graben fill. Because the initiation of movement on younger faults bounding the present graben floor is constrained to c. 1 Ma, the age of the Büyük Menderes Graben is Pliocene, younger than previously considered (Early–Middle Miocene). He concludes that western Anatolia is an example of a region that experienced two modes of extension: 'core-complex mode' and 'wide-rift mode', reflecting significant changes in the tectonic setting of western Anatolia which can be attributed to orogenic collapse followed by tectonic escape.

The eastern and central parts of the Anatolian Plate are dominated by active, intracontinental dextral and sinistral strike-slip faults. **Tatar et al.** report a palaeomagnetic study from the Erciyes sector of the sinistral Ecemiş (or Central Anatolian) Fault Zone and comment on neotectonic deformation in the SE part of the Anatolian Plate (box 20 in Fig. 2). They also summarize palaeomagnetic evidence for neotectonic deformation across a broad zone extending for at least 300 km between the sinistral East Anatolian and dextral North Anatolian fault zones. Their palaeomagnetic study of young (1–2 Ma) lava flows across the Ecemiş Fault Zone identifies block rotations in this part of Anatolia of c. 10° counterclockwise during the last 1 million years. Between the East Anatolian Fault Zone in the south and the North Anatolian Fault Zone in the north, the degree of counterclockwise rotation during the tectonic escape within the last 2–3 Ma diminishes from c. 25° in the east to c. 10° in the SW. This reflects a transition from highly strained to less strained crust as the width of the Anatolian Plate confined between the Arabian–Eurasian pincer broadens to the west.

Igneous activity

Although volcanism in Turkey is currently quiescent, there is abundant evidence that magmatism has been associated with all stages in its tectonic evolution. Studies of both intrusive and extrusive rocks, their geochemistry and the relationship between deformation and their age of emplacement therefore provide vital additional information about the progressive tectonic evolution of the area.

Boztuğ describes the mineralogy and whole-rock major and trace element geochemistry of intrusive associations in the Kyrşehir Massif (Fig. 1; box 21 in Fig. 2). He evaluates the geodynamic significance of these data in the context of the Late Cretaceous synchronicity of these collision-related granitoids with metamorphism in the massif. The intrusive associations record differences in geological setting and are classified into three groups: (1) syn-collisional, S-type peraluminous two-mica leucogranites; (2) post-collisional, I-type metaluminous hybrid monzonites; and (3) post-collisional and within-plate, A-type alkaline rocks including monzonites and syenites. Boztuğ suggests that metamorphism and magmatism were synchronous during Late Cretaceous Anatolide–Tauride Platform and Pontide collision along the İzmir–Ankara–Erzincan Suture Zone (Fig. 1). The metamorphism inverted the passive margin of the Anatolian Plate during collision, accounting for a decrease in metamorphic grade from north to south. Subsequent magmatism along the Anatolian passive margin is manifested by successive episodes of syn-collisional peraluminous, post-collisional calc-alkaline hybrid and post-collisional within-plate alkaline pulses.

Arger et al. present evidence for two episodes of basaltic magmatism in southeastern Turkey (box 22 in Fig. 2) at c. 19–15 Ma and c. 2.3–0.6 Ma based on K–Ar dating. Each episode produced olivine–titanaugite basalts in both the Anatolian Plate and the Arabian Plate which are difficult to classify using any conventional model. Because Miocene magmatism predated the onset of the modern strike-slip regime in eastern Turkey, but the Plio-Quaternary magmatism did not, there is no obvious tectonic explanation for the timing or chemistry of this volcanicity. The authors therefore propose that both episodes, together with associated crustal thickening and uplift, resulted from inflow of plastic lower crust from adjoining regions. Thus, although this region has remained in a plate boundary zone for tens of millions of years, volcanism has no direct relationship to local plate motions. They suggest that both episodes of volcanism are the result of loading effects induced by glacial sea-level variations which caused net flow of lower crust from beneath the offshore shelf to the land; this could have been contemporary with Early–Middle Miocene moderate glaciation of Antarctica and more intense lowland glaciation of the northern hemisphere which began around 2.5 Ma.

Finally, in a study of some of the most recent volcanic rocks in the area, and rocks which can be most closely associated with the present tectonic regime, **Yurtmen et al.** describe petrographical and geochemical characteristics of Plio-Quaternary volcanicity represented by small scoria cones and associated basanite and alkali-olivine basalt lavas north of İskenderun Gulf (Southern Turkey) (Fig. 1; box 23 in Fig. 2). These volcanic rocks lie along the active sinistral NE–SW trending Karataş–Osmaniye Fault Zone (KOFZ) which forms part of the modern Anatolide–African plate boundary and the southern Neotethys suture. The main exposures are concentrated at the intersection of these two structures. The chemistry of the alkaline lavas resembles ocean island basalts (OIB) and intra-continental plate basalts, with a magmatic source in the asthenosphere similar to OIB. This source has HIMU character and is regarded as a mixture of depleted mantle with a plume component; it is classified as one of the mantle end-members for young extension-related alkaline basalts. Based on the similarity of geochemical characteristics of the İskenderun Gulf volcanics with OIB, Yurtmen et al. emphasize the importance of extension-related alkali basalts and subduction-related basalts although a tectonic interpretation is precluded by absence of local extension, subduction or mantle plume activity in this region.

The thematic set of papers in this volume have been selected from papers presented at the *Third International Turkish Geology Symposium*, held at the Middle East Technical University (METU), Ankara during September 1998. This meeting was sponsored by the Middle East Technical University, the Scientific and Technical Research Council of Turkey (TÜBİTAK), the American Association of Petroleum Geologists (AAPG) and a range of industrial sponsors, including the Turkish Petroleum Corporation (TPAO), BP Exploration, Etibank, Perenko, Rio Tur Madencilik A.Ş. (Rio Tinto), Perenko, Cominco and Arco. The editors would like to thank all reviewers, the Organizing Committee, the staff and students at METU who helped to ensure that the conference ran smoothly. Facilities supplied by the departments of Geological Engineering at METU, Earth Sciences at Keele University and Earth Sciences at Liverpool University during preparation of this volume are gratefully acknowledged. Thanks are due to Dr R. E. Holdsworth (Series Editor) for his continuous encouragement, help and comments during the preparation of this volume and to the Geological Society Publishing House, particularly to Joanna Cooke for her editorial work and Angharad Hills for her continuous help at every stage of this volume.

References

ALTINER, D., ÖZKAN-ALTINER, S. & KOÇYİĞİT, A. 2000. Late Permian biofacies belts in Turkey: palaeogeographic and tectonic implications. *This volume*.

OKAY, A. İ. & TÜYSÜZ, O. 1999. Tethyan sutures of northern Turkey. *In*: DURAND, B., JOLIVET, L., HORVÁTH, F. & SÉRANNE, M. (eds) *The Mediterranean Basins: Tertiary Extension within the Alpine Orogen.* Geological Society, London, Special Publications, **156**, 475–515.

ROBERTSON, A. H. F. 2000. Mesozoic–Tertiary tectonic-sedimentary evolution of south-Tethyan oceanic basins and its margins in southern Turkey. *This volume*.

ROBERTSON, A. H. F. & PICKETT, E. A. 2000. Palaeozoic–Early Tertiary Tethyan evolution of mélanges, rift and passive margin units in the Karaburun Peninsula (W Turkey) and Chios Island (Greece). *This volume*.

STAMPFLİ, G. 2000. Tethyan oceans. *This volume*.

YILMAZ, A., ADAMIA, S., CHABUKIANI, A., CHKHOTUA, T., ERDOĞAN, K., TUZCU, S. & KARABIYİKOĞLU, M. 2000. Structural correlation of the southern Transcaucasus (Georgia)-eastern Pontides (Turkey). *This volume*.

Tethyan oceans

GÉRARD M. STAMPFLI

Institut de Géologie et Paléontologie, Université de Lausanne, BFSH2-CH 1015 Lausanne, Switzerland (e-mail: gerard.stampfli@igp.unil.ch)

Abstract: Diachronous subsidence patterns of Tethyan margins since the Early Palaeozoic provide constraints for paleocontinental reconstructions and the opening of disappeared oceans. Palaeotethys opening can be placed from Ordovician to Silurian times and corresponds to the detachment of a ribbon-like Hun Superterrane along the Gondwanan margin. Neotethys opening took place from Late Carboniferous to late Early Permian from Australia to the eastern Mediterranean area. This opening corresponds to the drifting of the Cimmerian superterrane and the final closing of Palaeotethys in Middle Triassic times. Northward subduction of Palaeotethys triggered the opening of back-arc oceans along the Eurasian margin from Austria to the Pamirs. The fate of these Permo-Triassic marginal basins is quite different from areas to area. Some closed during the Eocimmerian collisional event (Karakaya, Agh-Darband), others (Meliata) stayed open and their delayed subduction induced the opening of younger back-arc oceans (Vardar, Black Sea). The subduction of the Neotethys mid-ocean ridge was certainly responsible for a major change in the Jurassic plate tectonics. The Central Atlantic ocean opened in Early Jurassic time and extended eastwards into the Alpine Tethys in an attempt to link up with the Eurasian back-arc oceans. When these marginal basins started to close the Atlantic system had to find another way, and started to open southwards and northwards, slowly replacing the Tethyan ocean by mountain belts.

There is still some confusion about what Tethys existed at what time (e.g. Şengör 1985). A consensus exists, however, regarding the presence of a mainly Palaeozoic ocean north of the Cimmerian continent(s): the Palaeotethys, a younger Late Palaeozoic–Mesozoic ocean located south of this continent – the Neotethys – and finally a Middle Jurassic ocean – the Alpine Tethys (Favre & Stampfli 1992; Stampfli & Marchant 1997), an extension of the Central Atlantic, which broke through the Pangea supercontinent. These three oceanic realms form the Tethyan domain *s.l.* extending from Morocco to the Far East (Şengör & Hsü 1984).

The subsidence history of these oceans to support this group's proposed paleocontinental reconstructions is discussed here. These reconstructions have been done in the frame of the IGCP 369 project and the EUROPROBE-PANCARDI project to serve as a basis for discussion. Through the ongoing process of data collection these reconstructions have evolved and will, it is hoped, evolve further to give a larger consensus about their validity.

These reconstructions are presently displayed on the website in Lausanne (*www.sst.unil.ch*), focusing mainly on the western Tethyan realm and the Alpine domain. The arguments which led to the present state of these reconstruction are found in Stampfli *et al.* (1998 *a, b*, 2000) and Stampfli & Mosar (1999). Regarding the Alpine domain *s.str.*, the reader is referred to Stampfli (1993) and Stampfli & Marchant (1997), which discuss the opening of the Piemont and Valais Oceans. The Late Variscan evolution of the western Tethyan realm is discussed in Stampfli (1996), a review paper with a large reference list about the southern Variscan domains.

Some definitions

The first geodynamically correct definition of the main Tethyan oceans, based on extensive field work in the Middle East, was given by Stöcklin (1974). He recognized a Late Palaeozoic?–Triassic oceanic realm cutting through the epi-Baikalian (Pan-African–Gondwanan) Platform and separating the Iranian Plate from Arabia – which he called Neotethys – and another older oceanic realm separating the Iranian epi-Baikalian (Panafrican) domain from the Variscan Turan domain to the north – which he called Palaeotethys.

Following this proposal, an investigation of the eastern Alborz Range was begun (Stampfli 1978), effectively defining it as a potential southern margin of Stöcklin's (1974) Palaeotethys ocean. The opening of this Palaeozoic ocean was placed in Silurian time. At the same time, the ophiolites of Mashhad were

From: BOZKURT, E., WINCHESTER, J. A. & PIPER, J. D. A. (eds) *Tectonics and Magmatism in Turkey and the Surrounding Area.* Geological Society, London, Special Publications, 173, 1–23. 1-86239-064-9/00/$15.00
© The Geological Society of London 2000.

recognized as most likely pertaining to the Palaeotethys suture [see the review of Ruttner (1993) concerning these ophiolites].

The drifting of the Irano–Afghan Block from Gondwana to Laurasia was then clearly recognized and constrained by the evolution of the microflora of the Iranian Block from a Gondwanan affinity in Carboniferous time (Coquel et al. 1977; Chateauneuf & Stampfli 1979) to a Eurasian affinity in Late Triassic time (Corsin & Stampfli 1977). The Eocimmerian Orogeny was also defined in Iran at that time, as a result of the closing Palaeotethys and Middle Triassic collision of the Iranian Block with the Eurasian Turan Block (Stampfli 1978).

This concept was later extended further west (Turkey) and east (Tibet, Far East) by Şengör (1979, 1984); who defined the Cimmerian Block as a ribbon-like microcontinent separating Neotethys from Palaeotethys (Şengör & Hsü 1984), he also defined, at the same time, the Cimmerian deformation as non-Hercynian or post-Hercynian. Şengör's definition of Palaeo- and Neotethys (e.g. Şengör 1989) is similar to Stöcklin's (1974), with a major deviation which became clearer with time – Şengör viewed the opening of the Neotethys as the spreading of a back-arc ocean. This proposal implied that the Gondwana margin was an active margin in Permo-Carboniferous times and that margin would then belong to the Variscan domain s.l. [the Podataksasi zone of Şengör (1990, 1991)]. This assumption was based on an erroneous interpretation of the uplift and erosion of the Neotethys rift shoulders of northern India, Oman, Iran and Turkey as proof of Variscan deformation within the epi-Baikalian (Pan-African) domain (e.g. Oman – Michard 1982; Şengör 1990: India – Fuchs 1982; Bagati 1990: Turkey – Demirtaşlı 1984). The geometry of this Permian unconformity, its age compared to the rifting period, the geochemistry of associated basalts, the sedimentary record and the geodynamic context, imply a synrift thermal uplift and not a contemporaneous orogenic event, as clearly demonstrated in all these areas by thorough field work done in the last ten years (e.g. Mann & Hanna 1990; Pillevuit 1993; Vannay 1993; Garzanti et al. 1994, 1996; Pillevuit et al. 1997). Similarly, Ricou (1974) and Braud (1987) never spoke of Variscan deformation or metamorphism concerning the Sanandaj–Sirjan Zone (the Iranian part of the Podataksasi Zone); they regarded the metamorphic rocks of this region as a retrogressed epi-Baikalian basement [this is also indicated on the 1: 1 000 000 map of Iran (Huber & Eftekhar-Nezhad 1978)]. Lower Permian limestones and volcanics rest on this Precambrian basement; they can be regarded as syn- to post-rift deposits of the northern margin of Neotethys. The development of greenschist facies metamorphic conditions in these areas may be related to the Permian Neotethyan rifting phase, or to younger intrusive events (Berberian & Berberian 1981) when these regions became part of the northern Neotethyan active margin.

Stöcklin's (1974) original definition of Palaeotethys is therefore correct; it separates the Variscan domain from the epi-Baikalian (Pan-African) domain and its closure in Triassic times produced the Eocimmerian tectonic event which is always found south of the Variscan domain. The complete Triassic closure of Palaeotethys on an Iranian transect was proven later on by paleomagnetic studies (e.g. Schmidt & Soffel 1984; Lemaire 1997; Soffel & Förster 1984), confirming the conclusions reached previously based on floral and microfloral distribution (e.g. Corsin & Stampfli 1977).

As proposed by Şengör (1979), the Cimmerian orogenesis was of collage type and never produced a large mountain belt. This can be explained by the presence of intra-oceanic arcs, back-arc marginal seas and oceanic plateaus, located between the Cimmerian and Eurasian Plates (see below), which strongly reduced the effects of crustal thickening. In many cases the Palaeotethys suture zone was used for the opening of Jurassic marginal oceans during the subduction of the Neotethys (e.g. Caspian Sea, İzmir–Ankara Suture), complicating somewhat the image one can reconstruct of the former geometry of the Palaeotethyan margins.

The reconstructions

Presented here are a set of maps, together with subsidence curves, for key times in the Palaeozoic and Mesozoic. The plate reconstructions were computed by the GMAP (Geographic Mapping and Palaeoreconstruction Package) program developed by Torsvik & Smethurst (1994). These maps are based mainly on a review of the following articles and books:

- palaeomagnetics – Embleton (1984), Klitgord & Schouten (1986), Rowley & Lottes (1988), Van der Voo (1993) and Powell & Li (1994);
- palaeoreconstructions – Zonenshain et al. (1985), Ziegler (1988b), Hutchison (1989), Ziegler (1990), Zonenshain et al. (1990), Baillie et al. (1994), Khain (1994a), Niocaill & Smethurst (1994), Şengör & Natal'in (1996), Stampfli (1996) and Torsvik & Eide (1998).

Most of the references concerning the palaeozoic evolution of the Alpine *s.l.* region have been taken from recent compilations, e.g.:

- Società-Geologica-Italiana (1979);
- IGCP project 5 – Flügel *et al.* (1987) and Sassi & Zanferrari (1989);
- IGCP project 276 – Baud *et al.* (1991*b*) and Carmignani & Sassi (1992);
- von Raumer & Neubauer (1993);
- IGCP project 369 (started in September 1994) – see web site *www.sst.unil.ch* or *www.geomin.unibo.it/orgv/igcp/igcp.htm*;
- PANCARDI project (a Europrobe project started in 1994) – see web site *www.geofys.uu.se/eprobe/*.

A *Pangaea A* fit is used since the Late Carboniferous (Stampfli 1996). The *Pangaea B* concept (Irving 1977) could be applied for the Early Carboniferous period, it would then slowly grades into a *Pangaea A* position towards the end of the Carboniferous. A Triassic *Pangaea B*, proposed by some palaeomagnetic studies (Muttoni *et al.* 1996; Lemaire 1997; Torq 1997), is certainly not supported by geological evidence, mainly regarding the 3000 km dextral strike-slip motion during the Triassic supposedly passing through Morocco. The feature generally used to transform the *Pangaea B in A* is the Tizi-n-Test Fault Zone of the High Atlas, but this has proven not to be a dextral but a sinistral Tertiary shear zone, and only of local importance (Jenny 1983). Also, the lasting marine sedimentation and calc-alkaline volcanism in southern Europe until the Late Carboniferous–Early Permian [see the review of field data in Stampfli (1996)] favour a *Pangaea A* model, because the *Pangaea B* model would imply a total closure of Palaeotethys up to the Caucasus before the Permian, but Late Carboniferous granites (De Bono 1998), or sedimentary sequences, of Greece (e.g. Phyllite–quartzite Group: Krahl *et al.* 1983, 1986; Krahl 1992) are not affected by Variscan metamorphism.

Prototethys

This as yet little known oceanic realm bordering Gondwana on its North African to Australian side in Late Proterozoic and/or Early Palaeozoic time will not formally be defined here. It could be characterized by the deposition of the 'Sinian' sedimentary cycle in many areas located in the vicinity of this ocean, as shown in Fig. 1 (Morocco, Arabia, Iran, India, China). Most of these areas are also affected by Pan-African deformation, followed by the deposition of a new cycle of sedimentation usually starting in Cambrian times. Was the Prototethys a mainly Late Proterozoic ocean or a mainly Cambrian ocean? This is still an open question. This group's reconstructions suggest that Baltica–Siberia could have drifted away from Gondwana, opening the Prototethys in Early Palaeozoic time, a model also proposed by Torsvik & Eide (1998).

The Early Palaeozoic subsidence curves from Iran and India (Fig. 2; curves 2 and 3), possibly associated to the Prototethys thermal subsidence, could also be interpreted as resulting from the formation of a flexural foreland basin in view of the accelerating subsidence. This, together with other arguments presently in review, led this group to propose the accretion of an arc to the Prototethyan margin in Ordovician times. Followed here the idea of Şengör & Natal'in (1996) concerning the development at that time of a large intra-oceanic arc complex south of Siberia, the Kipchak Arc now forming the Kazakhstan Plate. This arc was extended to the south of Baltica and included in it were all the Alpine basement elements which comprise Ordovician granites, sometimes associated with remnants of oceanic crust or even eclogites of that age (von Raumer *et al.* 1993, 1998). The Rheic ocean is then viewed as the back-arc ocean located between Baltica and this Panalpine Arc. The Mauretanian ocean would open at the same time, possibly also as a back-arc basin due to the drifting away of Avalonia from the west coast of Africa. In view of its obduction onto the Baltica passive margin, the Iapetus ocean is represented here as a suprasubduction zone ocean opening at the expense of an older ocean.

Palaeotethys

The gentle docking of the 'Pan-Alpine' Ordovician arc was immediately followed by the southward subduction of the fast-spreading Rheic ocean. After subduction of its mid-ocean ridge, slab roll-back affected the remnant Rheic ocean and triggered the opening of Palaeotethys (Fig. 3). The ribbon-like continent being drifted away from Gondwana is therefore a composite terrane that this group term the Hun Superterrane [it contains most of the areas devastated by Attila!; see Stampfli (1996) and von Raumer *et al.* (1998)]. This Hun Terrane includes all the fragments accreted to Europe during the Variscan cycle and it extended eastwards to the Karakum (Turan) and Tarim areas, and possibly to the north China Block.

Fig. 1. Early Ordovician reconstruction, 490 Ma. The drifting of Baltica (including Ta, Taymir) and Siberia took place either from the south American side of Gondwana or the Indian side of Gondwana. Palaeomagnetic data do not provide enough constraints to decide on this. An intra-oceanic arc extends from Siberia to the south of Baltica [Ki, Kipchak Arc of Şengör and Natal'in (1996); and the Hun Cordillera terranes). The opening of the associated back-arc oceans resulted in the seafloor spreading of the Rheic and Khanti-Mansi Oceans at the expense of Prototethys. The drifting of Avalonia (E-Av, W-Av) off the coast of western Africa probably opened the Rheic–Mauretanian Ocean at the same time. In Early Palaeozoic Gondwana is bordered by the following blocks. *Hun Cordillera terranes*: Early Palaeozoic active margin of the Hun composite terrane from west to east: OM, Ossa–Morena; Ch, Channel terrane; Sx, Saxo–Thuringian; Is, Istanbul; Po, Pontides; Li, Ligerian; Md, Moldanubian; MS, Moravo–Silesicum; He, Helvetic; sA, south Alpine; Pe, Penninic; AA, Austro–Alpine; Cr, Carpathian; Tn, north Tarim. *Hun Gondwana terranes*: blocks forming the northern margin of Palaeotethys, from west to east: Ib, Iberic; Ar, Armorica; Mo, Moesia; Ct, Cantabria; Aq, Aquitaine; Al, Alboran; Ia, intra-Alpine (Adria, Carnic, Austro–Carpathian); DH, Dinaric–Hellenic; Kr, Karakum–Turan; Pa, Pamirs; Ts, south Tarim; Qa, Qantang. *The Cimmerian terrane*: blocks forming the southern margin of Palaeotethys that were detached during the Late Permian opening of Neotethys, from west to east: Ap, Apulia *s.str.*; HT, Hellenides–western Taurides externides; Me, Menderes–Taurus; Ss, Sanandaj–Sirjan; Al, Alborz; Lt, Lut–Central Iran; Af, central Afghanistan; sT, south Tibet; SM, Sibu Masu. *Anamian blocks*: defining the future northern and southern branch of Paleotethys: nT, north Tibet; IC, Indochina and Borneo; sC, south China. Numbers 1–4 refer to the position of the subsidence curves of Fig. 2.

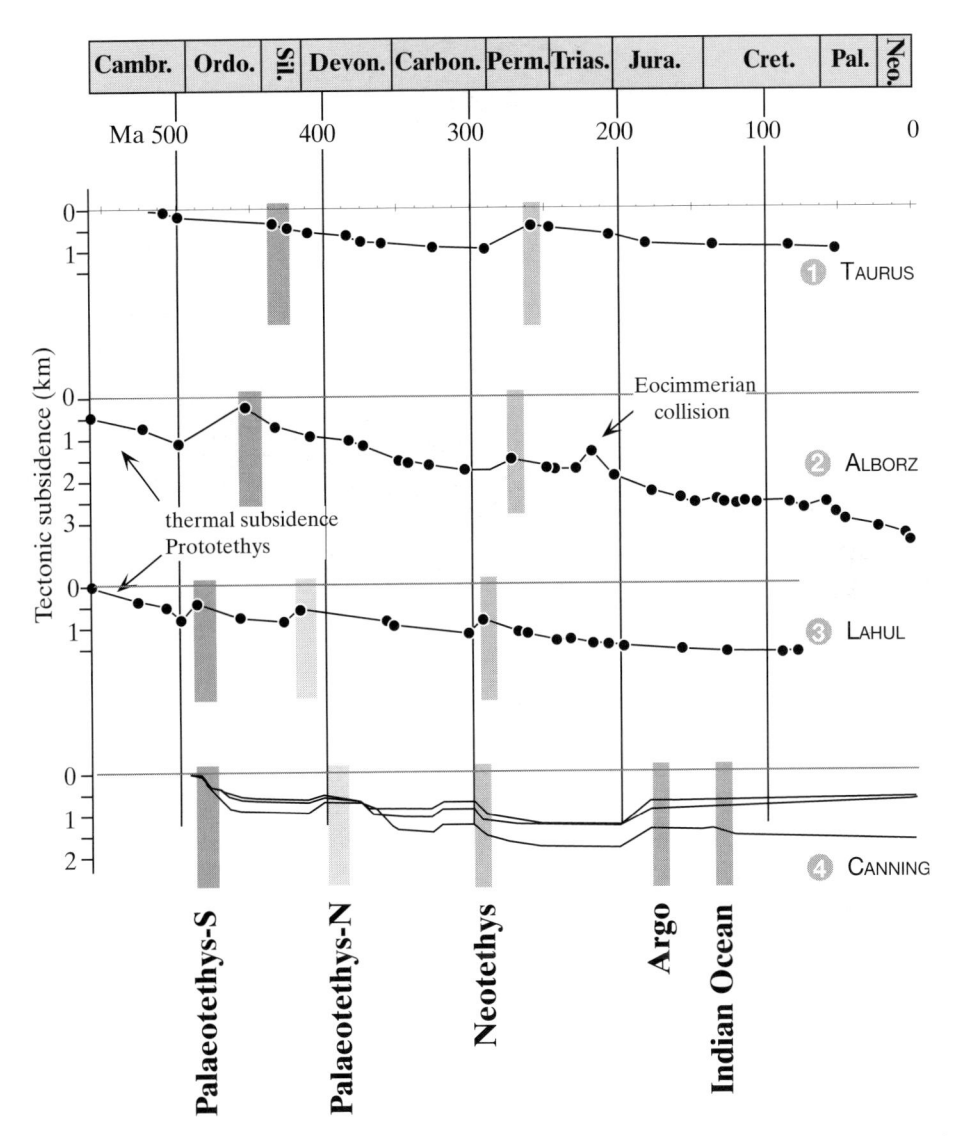

Fig. 2. Geohistory of the northern and southern margins of the Neotethys (southern margin of Palaeotethys) since the Cambrian. The stratigraphic data were collected for: curve 3 near the Neotethyan margin shoulder of northwest India, *c.* 100 km south of the Indus Suture Line in High Lahul (Vannay 1993); for curve 2 from the southeastern border of the Caspian Sea (Eastern Alborz Belt) near the Palaeotethyan margin shoulder in the Iranian Cimmerian Block (Stampfli 1978); and for curve 1 in the Central Taurus in Turkey, between the Palaeotethyan suture and the East Mediterranean oceanic realms (Demirtaşlı 1984, 1989) (see palaeogeographic position in Figs 1 and 3–5). The data for the Canning Basin are from AGSO (1995) and the timescale is from Gradstein & Ogg (1996).

Subsidence patterns of selected areas along the Palaeotethyan margins show that the thermal subsidence of this ocean was diachronous, starting in the Early Ordovician from the east (Australia–India) to Early Silurian in the west (Turkey). From faunal (e.g. Robardet *et al.* 1994) and palaeomagnetic data, it appears that Avalonia (east and west, including the Brabant Terrane) drifted away from Gondwana before the Hun Terranes. In contrast, the Armorican Terrane remained close to Gondwana until the Ordovician (Perroud *et al.* 1984), when it accelerated its journey to collide with Laurussia in the Devonian. Therefore, Armorica *s.l.*

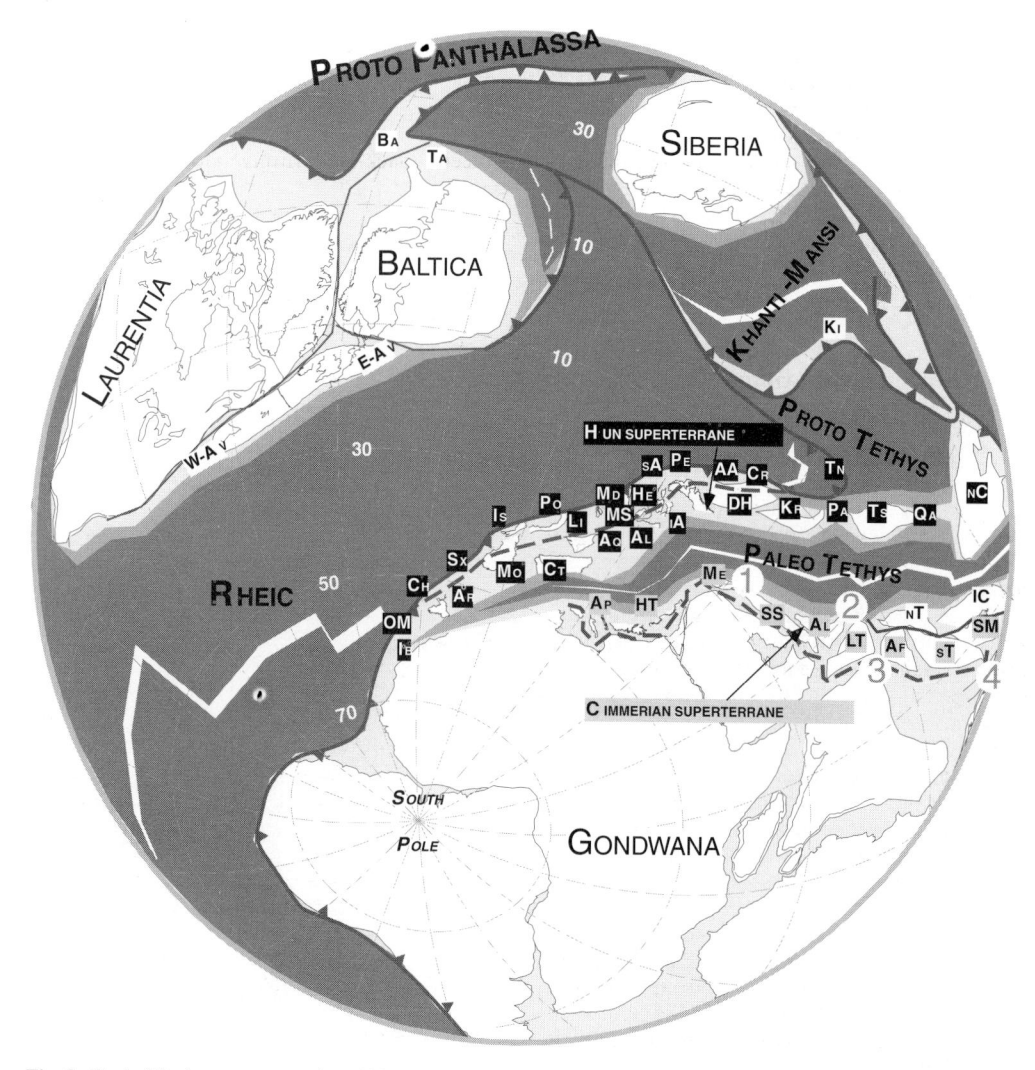

Fig. 3. Early Silurian reconstruction, 435 Ma. Baltica and Avalonia have been accreted to Laurentia–Barentsia (Ba) and the Pan-Alpine Arc accreted to Gondwana. The opening of Palaeotethys separated the Hun Superterrane from Gondwana. This opening took place in a context of diachronous back-arc spreading and slab roll-back of the Rheic Ocean. See Fig. 1 for the legend of the terranes. 1–4, Position of the subsidence curves of Fig. 2.

(Brioveria) is included in the Hun Superterrane and is placed north of North Africa based on palaeomagnetic data (see Torsvik & Eide 1998), which show a separation of Armorica from Gondwana not before Early Silurian.

Therefore, the Hun Superterrane is spread over a relatively large palaeolatitudinal area (from 60° south to the equator); large changes of facies are expected between Armorica and terranes now found in the Alps (e.g. Carnic and Austroalpine domain). For terranes located within the tropical zone (Alps, Spain, southern France) it can be shown that they present a very similar stratigraphic evolution to that of the Gondwana margin in Iran (Alborz) or Turkey (Taurus) from the Silurian to the Carboniferous; they are interpreted as representing the northern Palaeotethyan margin (Stampfli 1996). During the Carboniferous, terranes derived from the Hun Superterrane developed pelagic sedimentation followed by flysch deposits before being accreted to the Avalonia–Baltica margin. On the contrary, carbonate platform sedimentation continued along the Gondwana margin until the Namurian–Moscovian in North Africa (Lys 1986; Vachard *et al.* 1991), with pelagic sedimentation persisting until Permian time in the Sicanian basin located north of Sicily (Catalano *et al.* 1988, 1992; Kozur 1990). East of a Palaeo-Apulian promontory [southern Greece, Turkey, Iran; see Vai (1991)], the carbonate platform lasted until Early–Middle Triassic. This shows a diachronous closure of Palaeotethys from Moscovian to Early Triassic time from Morocco to Greece, along what was certainly a very oblique convergence zone.

The Variscan and Eocimmerian event and the marginal oceans

In Europe, the 'Variscides collisional processes' are generally regarded as extending from the Early Devonian to the Late Carboniferous, and the 'Tethyan cycle' (opening of the Alpine Tethys–Central Atlantic system) as not starting before mid-Triassic times. An apparent lack of major tectonic events during the Permian and Triassic southwestern Europe, or in the Appalachian domain, documents the welding of Gondwana with Laurasia to form the Permian Pangaea. But the Variscan domain extends over the whole Alpine area and even further in the Dinarides and Hellenides, northern Turkey and the Caucasus. It also extends in time, with deformations becoming younger, possibly grading into Eocimmerian (Triassic) deformations southwards and eastwards. As shown in Fig. 4, in the Late Carboniferous, the Palaeotethyan domain was not fully closed in southeast Europe and even persisted till the Early Triassic in the Hellenides [phyllite–quartzite Group; Krahl *et al.* (1983, 1986), Krahl (1992) and Stampfli *et al.* (2000)], and Middle Triassic times further East (e.g. Alborz Chain in Iran) (Fig. 5).

Stampfli *et al.* (1991) and Stampfli (1996) discussed this diachronous closure of the large Palaeotethys ocean, insisting on the likely development of back-arc oceans or basins within the Permo-Triassic Eurasian margin. From the Palaeo-Apulian promontory eastwards it is quite clear that an Eocimmerian domain of deformation is found just south of a relatively undeformed Variscan domain, which is represented by a Late Carboniferous–Early Permian arc and clastic sedimentation of Verrucano type, mainly affected by extension. In contrast, the Cimmerian deformations are accompanied by Triassic flysch, mélanges and volcanics (Stampfli *et al.* 1995, 1998*a*) and even collisional-type intrusive events (Reischmann 1998), marking the closing of either Palaeotethys or the marginal oceans.

As shown by the plate reconstructions (Fig. 4), the Permian margin of southeast Europe is of a transform type and little subduction took place along that margin at that time. However, the slab roll-back of Palaeotethys rapidly induced 'back-arc' rifting along the whole margin after Late Permian times. East of the palaeo-Apulian promontory, this back-arc rifting graded into seafloor spreading of the Maliak–Meliata marginal ocean (Kozur 1991). The western end of the Meliata Rift (Southern Alps–Ivrea) aborted in Late Permian, whilst its eastern part (Maliak–Meliata–Dobrogea) spread [e.g. Early Triassic mid-ocean ridge basalt (MORB) pillow lava of N-Dobrogea; Niculitel formation: Cioflica *et al.* (1980), Seghedi *et al.* (1990) and Nicolae & Seghedi (1996)]. This Early Triassic seafloor spreading is accompanied by a marked thermal subsidence along the Pelagonian northern margin (Fig. 7, curve 6). At Meliata, the oceanic series are not older than mid-Triassic (Kozur 1991), but these oceanic remnants represent the accretion–obduction of the Meliata Ridge during its subduction, implying an older age for the onset of seafloor spreading.

In the Dinaro–Hellenide domain (Fig. 6), the Late Carboniferous arc (Pelagonia) collided directly with the Gondwanan margin (Stampfli *et al.* 1995, 1998*a*; Vavassis *et al.* 1997). On the contrary, the Karakaya domain in northern Turkey (Şengör *et al.* 1980; Okay & Mostler 1994) is a complex Cimmerian deformation zone representing first the closure of Palaeotethys between an oceanic plateau (Nilüfer Formation) and the Gondwana margin, then the closing of the Karakaya Back-arc and its northward thrusting on the Variscan Sakarya margin of Eurasia (see Okay 2000). Therefore, the Karakaya and Küre back-arc sequences (Ustaömer & Robertson 1994, 1997, 1999) are not remnants of the Palaeotethys *s.str.* (see Kozur 1997*b*) but they represent a Marianatype marginal ocean developing south of the Sakarya margin. During the Cimmerian collision event, the Late Permian–Early Triassic Karakaya–Küre back-arc ocean subducted southwards (Şengör *et al.* 1980; Tüysüz 1990; Pickett & Robertson 1996), but the former subduction of Palaeotethys was to the north, under Eurasia, as clearly demonstrated in Iran (Stampfli 1978; Ruttner 1993), Afghanistan (Boulin 1988) and in the Pontides (see Ustaömer & Robertson 1999).

Late Permian–Middle Triassic back-arc basins are also known eastwards in the Caucasus (Nikishin *et al.* 2000), northeast Iran [e.g. the Agh–Darband sequence; Baud & Stampfli (1989) and Baud *et al.* (1991*a*)], northern

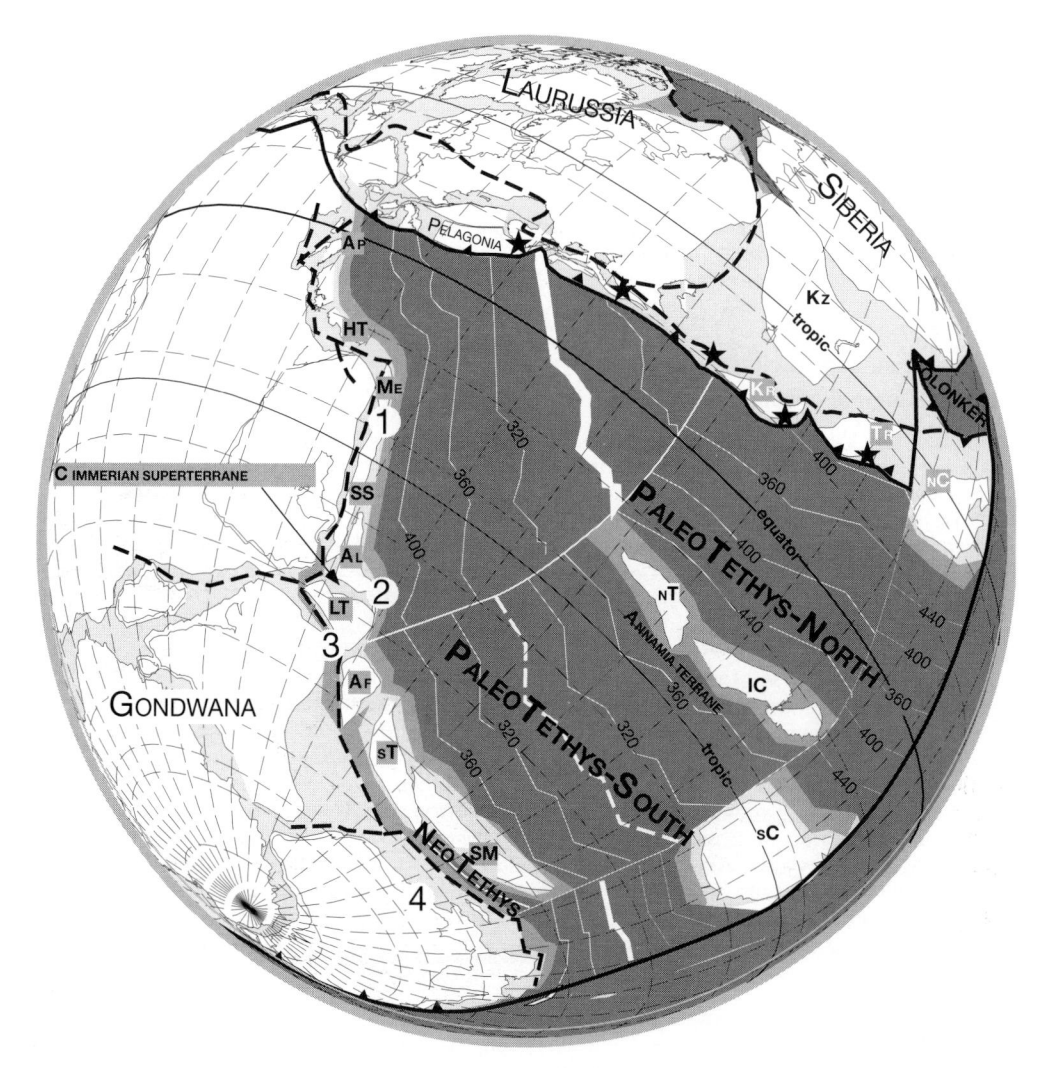

Fig. 4. Carboniferous–Permian boundary reconstruction, 290 Ma. As the Palaeotethys mid-ocean ridge is moving eastward, slab roll-back of the western Palaeotethys induces the collapse of the Variscan Orogen in southern Europe. This led eventually to the opening of back-arc rifts in the active Eurasian margin (see Fig. 5). See Fig. 1 for legend. Kz (Kazakhstan) is derived from the Kipchak Arc. 1–4, Position of the subsidence curves of Fig. 2.

Afghanistan (Boulin 1988) and in the Pamirs (Khain 1994a, b; Leven 1995). Due to the collision of the Cimmerian blocks with the Eurasian margin (e.g. Stampfli et al. 1991; Alavi et al. 1997; Fig. 6), these back-arc basins disappeared during the Late Triassic. Along the Cimmerian orogen, development of a carbonate platform resumed in the Norian or Liassic, marking the end of this orogenic cycle. The Upper Late Triassic–Liassic deposits usually start with continental clastics derived from the Cimmerian elevations – they are found on both sides of the suture in Turkey – where they rest unconformably on older sequences [Karakaya or Sakarya basement in the Pontides – Koçyiğit (1987) and Altıner et al. (1991): Triassic or Palaeozoic strata in the Taurus – Monod & Akay (1984)]. In Iran they are represented by the Shemshak Formation which also extends northwards on the Turan Plate (Stampfli 1978).

For the opening and oceanization of the Meliata Rift, the main geodynamic factor used is the thermal subsidence, which affected the whole Austro-Alpine domain and other internal

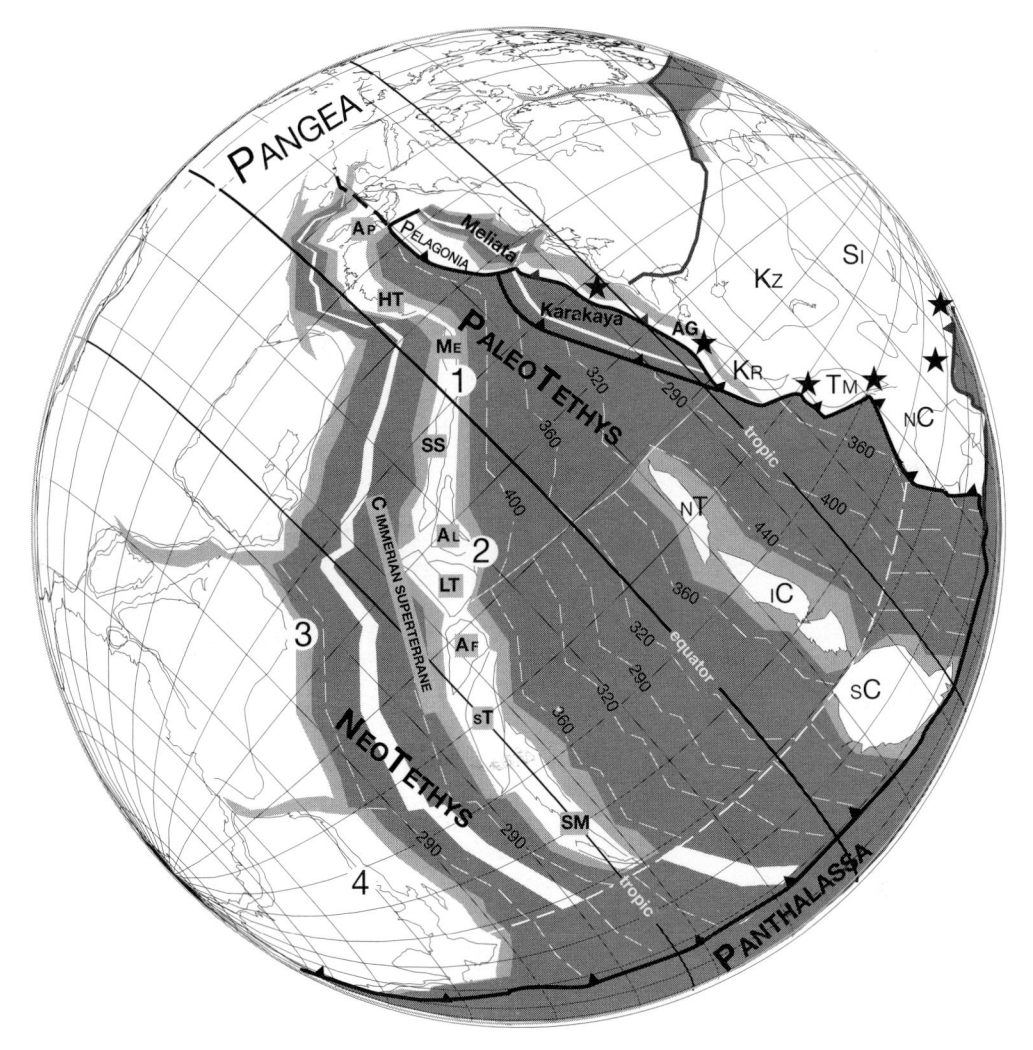

Fig. 5. Permian–Triassic boundary reconstruction, 248 Ma. The slab roll-back of the whole Palaeotethys induced the opening of back-arc oceans in the active Eurasian margin [Meliata, Karakaya (Ka) and Agh-Darband (Ag)] and the strong slab-pull force is opening the Neotethys Ocean. This opening is separating the Cimmerian Superterrane from Gondwana. See Fig. 1 for legend. 1–4, Position of the subsidence curves of Fig. 2.

parts of the Alps since the Permian–Triassic boundary, implying a rifting phase just before that time (Fig. 7). This subsidence induced the deposition of a more or less complete Triassic sequence, usually conformable on an Upper Permian clastic sequence, presenting a large diversity of facies and thicknesses locally approaching 3–4 km (e.g. Bernoulli 1981; Haas *et al.* 1995). This diversity is also linked to the counter-effect of the Eocimmerian deformation which induced local inversion or flexure of pre-existing basins, mainly in the southern margin of Meliata, and in the aborted part of the rift (e.g. subsidence curves 7 and 8 in Fig. 7). This Middle Triassic tectonic pulse is well established in the Dolomites where it is also accompanied by the emplacement of diapirs (Castellarin *et al.* 1996).

Without the presence of the Maliak-Meliata Ocean it would be difficult to explain the development of such a large-scale carbonate platform and marginal sequences before the opening of the Atlantic–Alpine Tethys system in the Early–Middle Jurassic. As can be seen from figure 7, this opening had no effect on all these peri-Apulian regions.

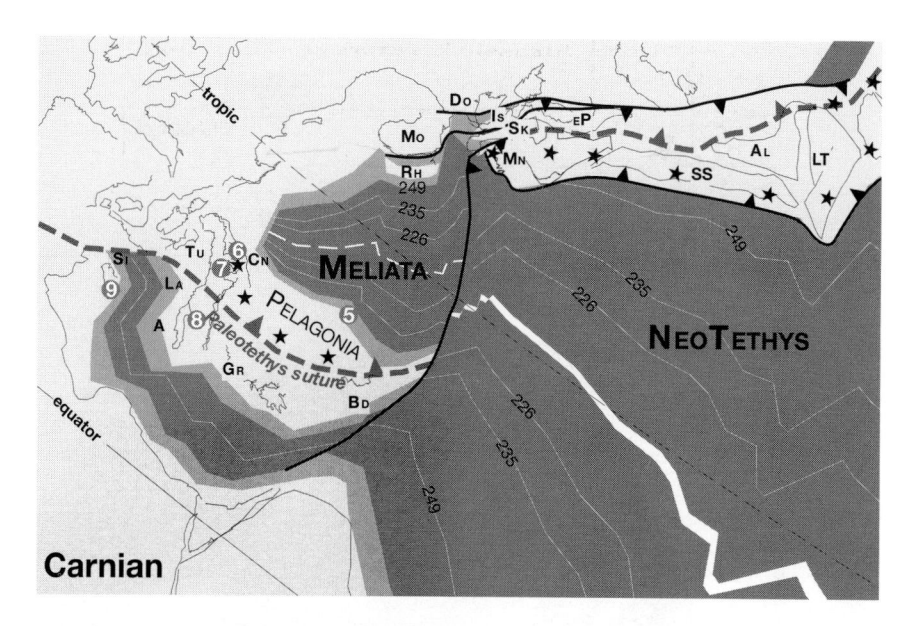

Fig. 6. Carnian reconstruction of the western Tethyan area. AL, Alborz; Ap, Apulia s.str.; Bd, Bey Dağları; Cn, Carnic Alps; Do, Dobrogea; eP, east Pontides; GR, autochthonous of Greece; Is, Istanbul; La, Lagonegro; LT, Lut–Tabas; Mn, Menderes, Taurus; Mo, Moesia; Rh, Rhodope; Si, Sicannian Basin; Sk, Sakarya; SS, Sanandaj–Sirjan; Tu, Tuscan Nappes. ★ Magmatic arc activity north of Neotethys and post-collisional volcanism south of Meliata. The Palaeotethys Suture is represented by a dashed line. 5–9, Position of the subsidence curves of Fig. 7.

Neotethys

Although the geodynamic evolution of the Neotethys Ocean is now relatively well constrained, mainly through the recent findings of well-dated Wordian MORB in Oman (Pillevuit 1993; Pillevuit *et al.* 1997), its relationship with the East Mediterranean Basin is more debatable. The Neotethys separated the Cimmerian microcontinent(s) from Gondwana between the latest Palaeozoic and the earliest Tertiary, following a diachronous opening from east to west (Figs 2 and 5). Depending on the authors, the East Mediterranean–Ionian Sea basin is regarded as opening in the Late Palaeozoic (Vai 1994) or as late as the Cretaceous (e.g. Dercourt *et al.* 1985, 1993). Most people would regard this ocean as opening in Late Triassic or Early Jurassic (e.g. Robertson & Woodcock 1980; Garfunkel & Derin 1984; Şengör *et al.* 1984; Finetti 1985) and therefore possibly being related to the Alpine Tethys–Atlantic opening. However, as shown in Fig. 7, the Alpine Tethys opening has no tectonic or thermal effect on areas located around the Ionian Sea, although a Jurassic extensional phase is clearly recognized along the Levant Transform margin (Garfunkel 1998; Stampfli *et al.* 2000).

A new interpretation showing that the East Mediterranean domain corresponded to an oceanic basin since the Late Palaeozoic has been proposed by Stampfli (1989). Subsequently, new plate tectonic reconstructions considering this basin as part of the Neotethyan oceanic system have been developed (Stampfli *et al.* 1991, 2000; Stampfli & Pillevuit 1993) (Figs 5 and 6). This is supported by: (1) geophysical characteristics of the Ionian Sea and East Mediterranean Basin (isostatic equilibrium, seismic velocities, elastic thickness), excluding an age of the seafloor younger than Early Jurassic; (2) subsidence patterns of areas such as the Sinai margin, the Tunisian Jeffara Rift, and Sicily and Apulia s.str. (Stampfli *et al.* 2000) (Fig. 7), confirming a Late Permian onset of thermal subsidence for the East Mediterranean and Ionian Sea basins and the absence of younger thermal events; (3) Triassic MORB found in Cyprus in the Mamonia complex (Malpas *et al.* 1993), certainly derived from the East Mediterranean sea-floor, although a more exotic nature of these basalts (from the Neotethys) cannot be excluded; and (4) Upper Permian Hallstatt-type pelagic limestone, similar to those found in Oman where they sometimes rest directly on MORB (Pillevuit 1993;

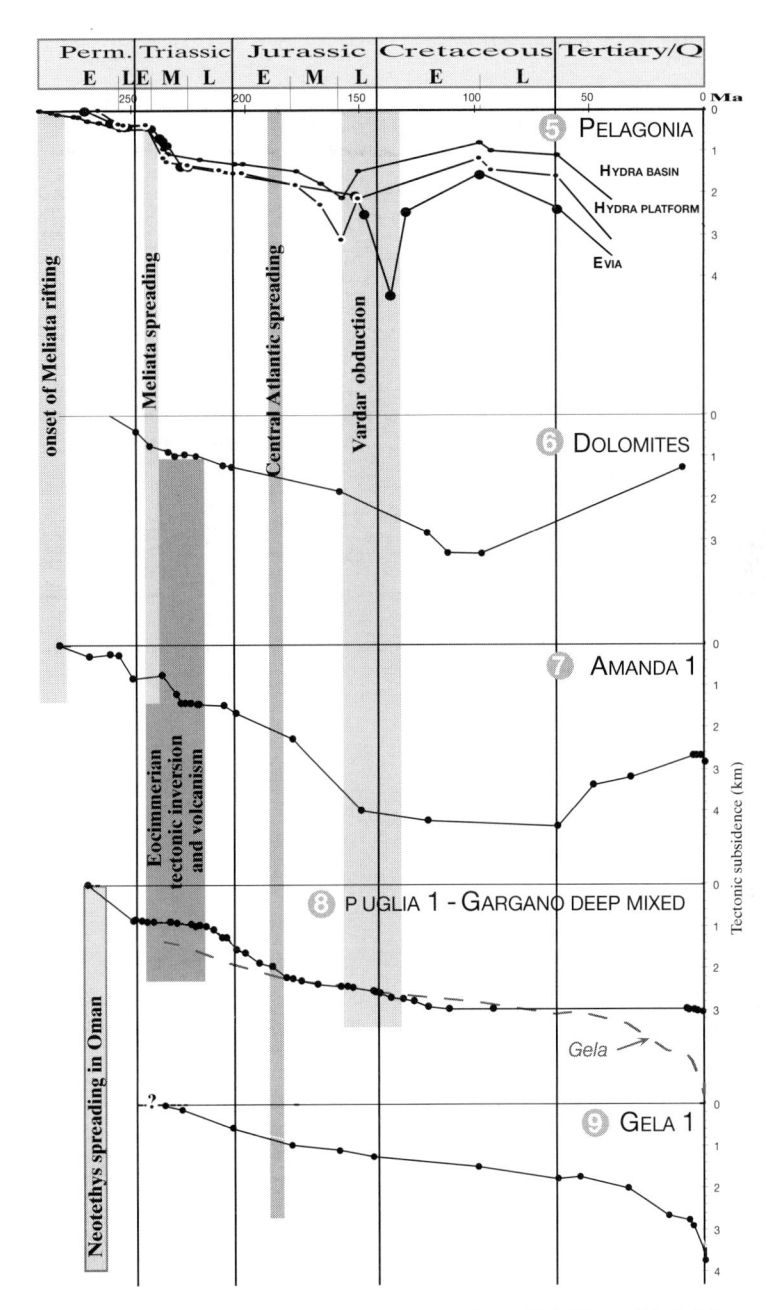

Fig. 7. Circum–Apulia tectonic subsidence curves modified from Stampfli & Mosar (1999). The curves for Pelagonia are from De Bono (1998) and the timescale is from Gradstein & Ogg (1996).

Niko *et al.* 1996), have also been reported from the Sosio Complex (Kozur 1995) in Sicily. This Late Permian pelagic macrofauna presents affinities with both Oman and Timor, and implies a Late Permian direct deep-water connection of the East Mediterranean Basin with the Neotethys.

The Alpine Tethys, the Central Atlantic and the Vardar

Field work in the Canary Islands and in Morocco (Favre *et al.* 1991; Favre & Stampfli 1992; Steiner *et al.* 1998) has allowed the onset of seafloor spreading to be dated as Toarcian in the northern part of the central Atlantic. Similar subsidence patterns between this region and the Lombardian Basin (Figs 9 and 11) led to the proposal of a direct connection between these two areas (Fig. 8). The Lombardian Basin aborted (Bertotti *et al.* 1993) as it could not link up with the Meliata Ocean whose already cold oceanic lithosphere was rheologically unbreakable relative to surrounding continental areas. Therefore, the Alpine Tethys rift opened along the Meliata northern margin separating the future Austro–Carpathian domain from Europe. Thermal subsidence and spreading started in Aalenian time in the west (Fig. 9) [the Briançonnais margin of Stampfli & Marchant (1997) and Stampfli *et al.* (1998*b*)] and Bajocian time eastward [the Austro–Alpine margin of Froitzheim & Manatschal (1996) and Bill *et al.* (1997)]. The Alpine Tethys was linked to the Eurasian back-arc basins located further east through the Moesian–Dobrogean Transform.

A transform ocean linked up the central Atlantic and the Alpine Tethys delayed seafloor spreading and thermal subsidence in this Maghrebide Ocean is well exemplified by the subsidence curve from the Rif area (Favre & Stampfli 1992; Favre 1995) (Fig. 9).

The rotation of Africa relative to Europe after the Late Triassic induced the subduction of the Meliata Ocean under the young Neotethys oceanic crust. This intra-oceanic subduction gave birth to the Vardar Ocean, which had totally replaced the Meliata Ocean by the end of the Jurassic. The Vardar Ocean obducted southward onto the Pelagonian margin in the Late Jurassic (Fig. 7), then subducted northward under Moesia, finally opening the Black Sea in the Late Cretaceous, and it represents the third generation of back-arc opening in that region.

The North Atlantic and the Valais Ocean

Since the Early Cretaceous (Fig. 10), there has been no possibility for the Atlantic mid-ocean ridge to link up eastwards with another ocean. A last attempt was made through the opening of the Valais and Biscay Oceans, but by the end of the Santonian the break-up between North America and Greenland had taken place and then, in Campanian time, the Biscay Ocean aborted (Ziegler 1988*a, b*). Closing of the Valais Ocean took place during the opening

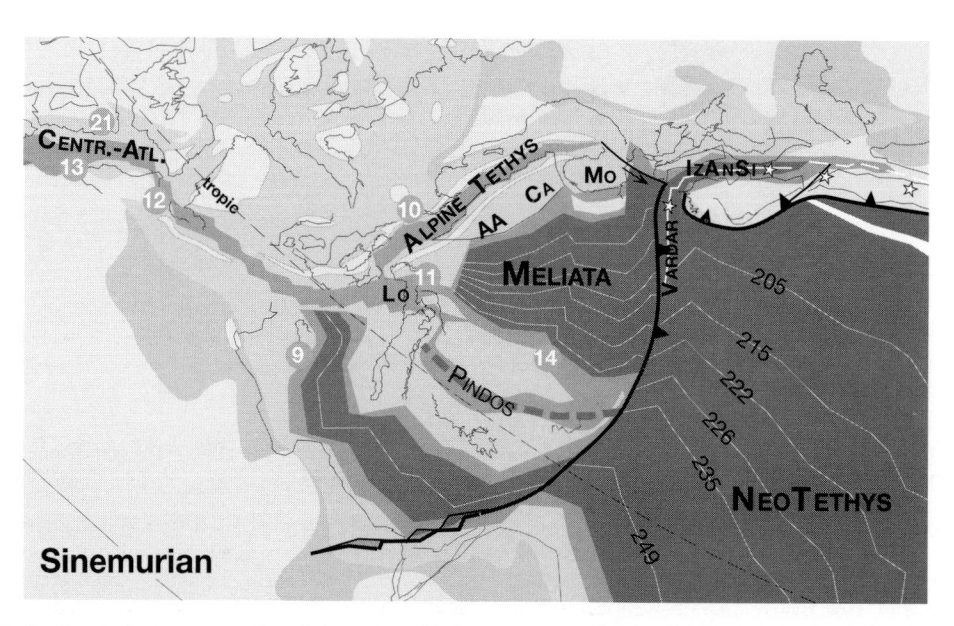

Fig. 8. Sinemurian reconstruction of the western Tethyan area. AA, Austro–Alpine; CA, Inner Carpathian; IzAnSi, İzmir–Ankara Ocean; MO, Moesia; Va, Vardar. ★, Magmatic arc activity. Palaeogeography modified after Ziegler 1988*a*. 9–21, Position of the subsidence curves of Figs 7, 9 and 11.

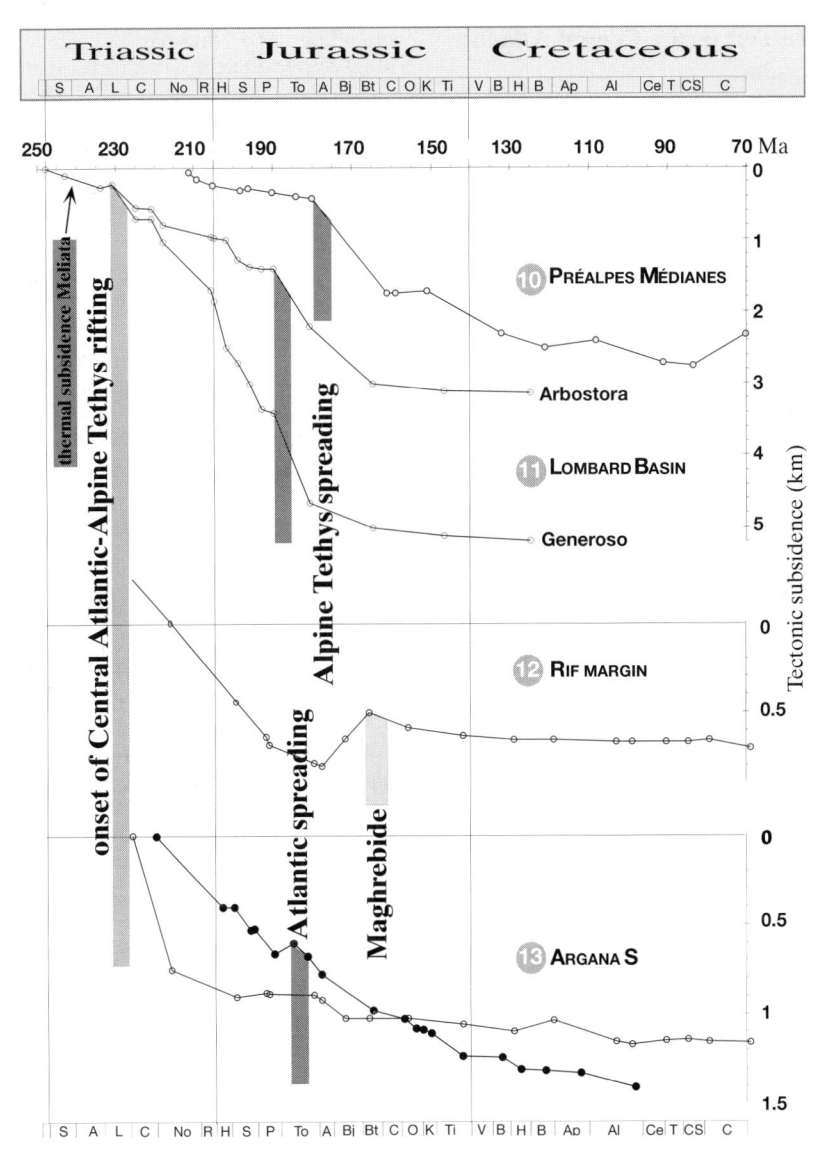

Fig. 9. Subsidence pattern of the Central Atlantic, Alpine Tethys system. Curve 10 (Préalpes Médianes Basin) is from Mosar *et al.* (1996); curves 11 (Lombard Basin) are modified from Greber *et al.* (1997); curve 12 (Rif) is from Favre (1995); and curves 13 have been calculated from the data of Ambroggi (1961), Adams *et al.* (1980) and Du Dresnay (1988). The timescale is from Gradstein & Ogg (1996).

of the Gulf of Biscay and subduction of the Alpine Tethys certainly started at the same time (Stampfli & Marchant 1997; Stampfli *et al.* 1998*b*).

Therefore, spreading of the Valais Ocean was short lived (Fig. 11). Its closure during the opening of the Biscay Ocean was only partial (there was no collision of the Briançonnais Peninsula with the Helvetic margin in the Late Cretaceous), but this event marks the onset of the inversion phase of pre-existing fault-bounded basins in many areas in Europe (Ziegler 1990).

Figures 10 and 11 show how different parts of southwest Europe or Iberia have been affected by these diverse rifting phases: the interpretation of these curves, however, requires a good knowledge of their initial position, some areas being mainly thermally influenced whereas others are more affected by faulting

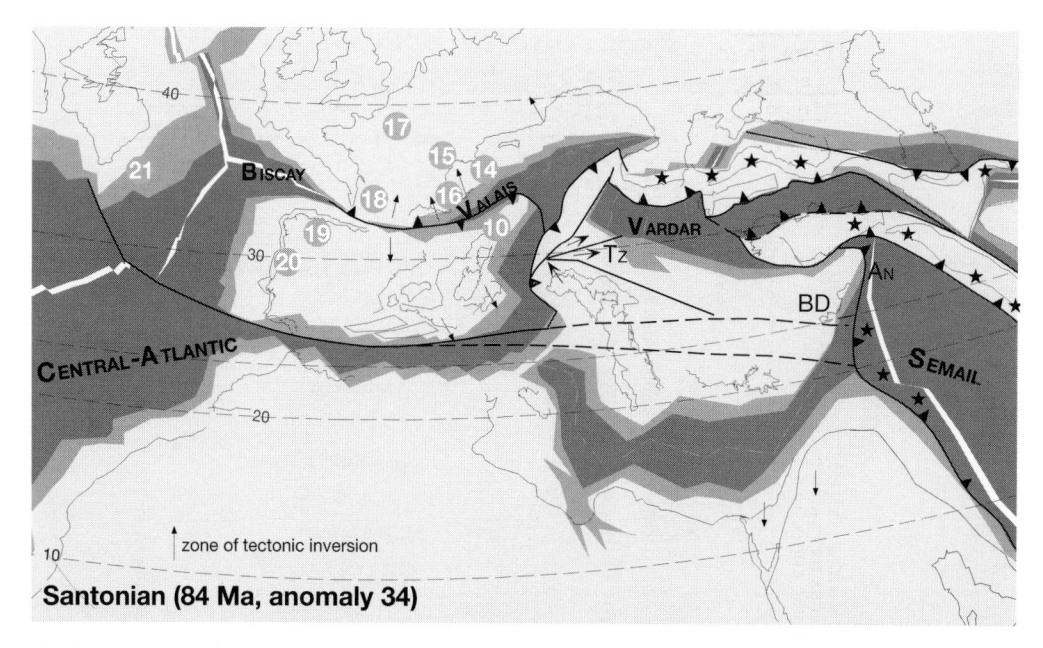

Fig. 10 Santonian reconstruction of the western Tethyan area. 10–21, Position of the subsidence curves of Fig. 11.

and extension of the upper crust. Generally speaking, all the areas shown on these figures have been affected by the opening of the North Atlantic and Valais Ocean system in Late Jurassic times.

Discussion – Tethys sutures in Turkey

Figure 12 shows a simplified tectonic map of the Aegean–Mediterranean area with the major sutures indicated as broad lines. The present back-stop of the East Mediterranean subduction zone (e.g. Cyprus, Crete) is regarded as the western extension of the Neotethys suture as found along the peri-Arabian ophiolitic zone (Ricou 1971) in Syria (e.g. Delaune-Mayere 1984), southern Turkey (e.g. Michard et al. 1984), Iraq and Iran (e.g. Ricou 1974; Braud 1987), and Oman (e.g. Bernoulli & Weissert 1987; Robertson & Searle 1990; Pillevuit et al. 1997). This obvious connection actually implies that the Neotethys ocean does continue into the East Mediterranean Basin, as discussed above, and provides a direct link between the pelagic Permian deposits from Sicily and their equivalent in Oman.

Actually, the Neotethys suture is composite and it contains the remnants of the southern passive margin of Neotethys (Hawasina Basin in Oman), as well as remnants of Cretaceous ophiolites (Semail Nappe in Oman) which obducted onto that margin. These peri-Arabian ophiolites correspond to the obduction of a younger intra-oceanic Cretaceous offspring of Neotethys (the Semail Ocean). Therefore, the northern margin of that Semail Ocean is the former northern active margin of Neotethys. Around the Arabic promontory (e.g. Maden Complex; Aktaş & Robertson 1984) and in Cyprus (Robertson & Xenophontos 1993), this younger ocean closed during the Palaeogene. In Oman this ocean is not yet closed (Sea of Oman) and is still subducting under the Makran active margin.

As the East Mediterranean subduction zone (Aegean active margin) is certainly not older than Miocene, it does not represent the western continuation of the Neotethys active margin. In our structural scheme, a possible extension of this Late Cretaceous–Palaeogene active margin westward under the Lycian Nappes, linking up the Antalya Suture with the Axios Vardar Zone, is proposed. This connection is necessary to provide a western limit to the Greater Apulian Block (Apulia; autochthonous and para-autochthonous units of the Dinarides and Hellenides, Bey Dağları Block). This limit corresponds to the former transform that separated Neotethys from Meliata (Figs 6 and 8).

In this scheme, the İzmir–Ankara Suture represents a Jurassic back-arc ocean related to

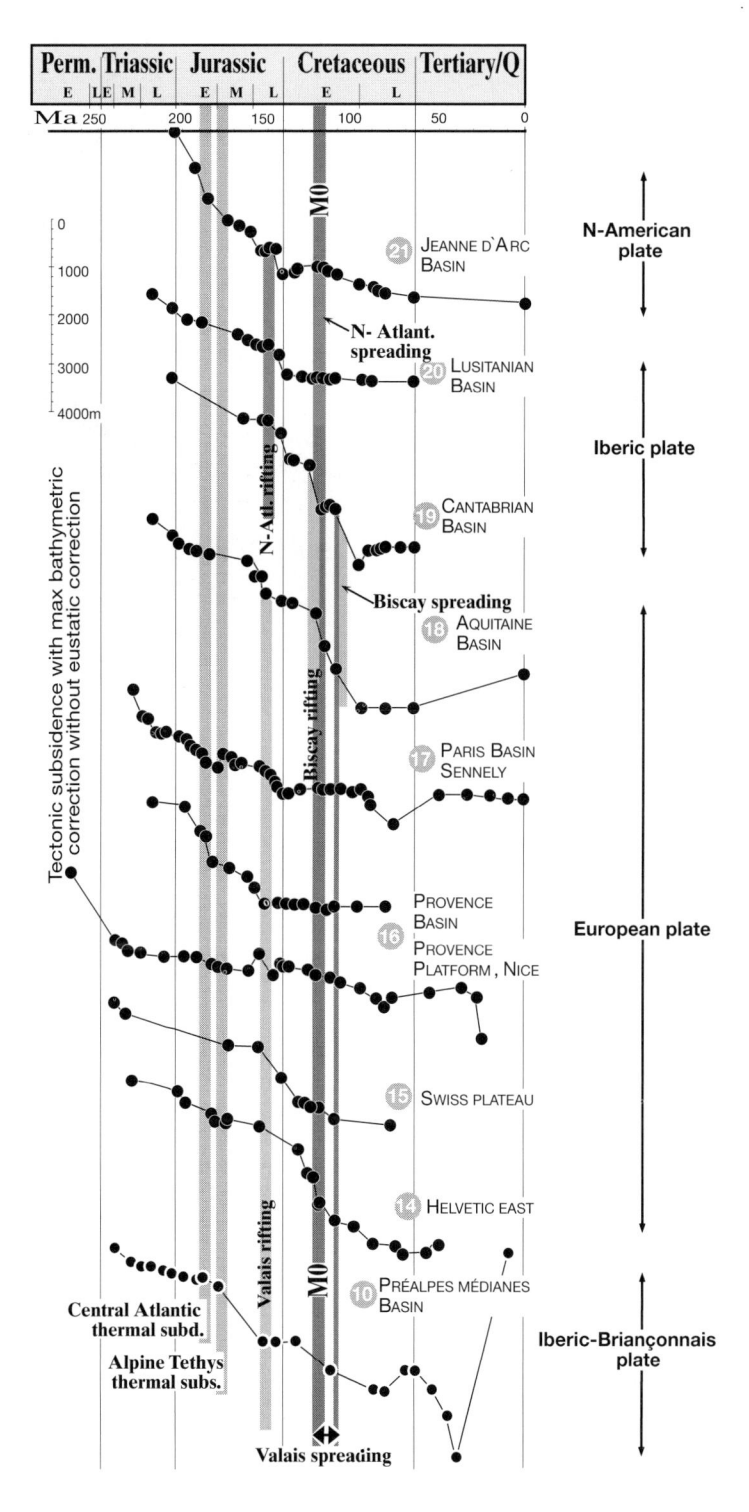

Fig. 11. Circum Iberia subsidence curves, modified from Borel (1998), curves 14 and 15 are modified from Wildi *et al.* (1989). The timescale is from Gradstein & Ogg (1996).

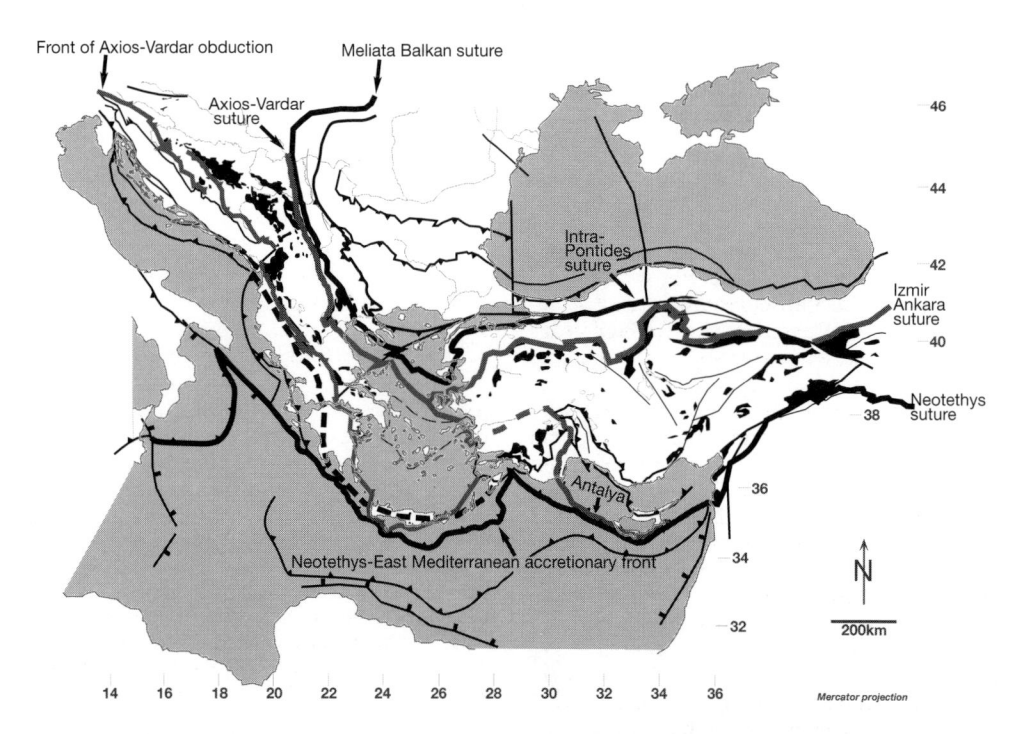

Fig. 12. Tectonic map of the eastern Mediterranean area showing major suture zones. Black areas represent major ophiolitic outcrops. A more complete coloured map can be found at this group's website (*www-sst.unil.ch*).

the northward subduction of Neotethys since Late Triassic, this back-arc basin opened within the complex Karakaya–Palaeotethys Suture Zone. It followed the collapse of the Eocimmerian Orogen and reseparated the Cimmerian Tauric–Menderes Block from Eurasia. This suture, characterized by its 'coloured mélanges' (Gansser 1960), can be followed up to the Iranian border. In Turkey, closure of the İzmir–Ankara Ocean postdates the Late Cretaceous ophiolite obduction found on its southern passive margin (Gutnic *et al.* 1979; Okay *et al.* 1996; Demirel & Kozlu 1998; Collins & Robertson 1998). Development of a new accretionary wedge along its northern margin is placed in the Middle Campanian and the final closure in the Eocene (e.g. Norman 1984; Koçyiğit 1991). Similar coloured mélanges are also found in Iran where they grade into Palaeogene flysch, followed by the main Eocene unconformity; they separate the Sanandaj–Sirjan Block from the central and north Iranian blocks (Stöcklin 1968, 1974, 1977, 1981). Therefore, the Cretaceous coloured mélanges in Turkey and Iran are clearly disconnected from the Neotethyan suture, and are located within the Eurasian margin to the north of the Neotethys active margin. In that sense, the İzmir–Ankara Suture is not the Neotethys suture.

The intra-Pontides (Okay *et al.* 1996), Meliata–Balkan Suture is of Cretaceous age and corresponds to the collision of a Vardarian arc with the Rhodope–Moesian Plate inducing the folding and northward thrusting of the Balkanide units (Georgiev *et al.* 1997; Tari *et al.* 1997). The Axios–Vardar Suture is the final Late Cretaceous–Palaeocene suture of the Vardar. The latter obducted onto the Pelagonian margin in Late Jurassic times – the front of this obduction is shown in Fig. 12. The Axios–Vardar Palaeogene Suture was somehow connected to the İzmir–Ankara Suture of similar age, although, the two domains were separated by the former Meliata–Neotethys transform.

In Fig. 12, the transported Palaeotethys suture in the Dinaro–Hellenides Zone is shown by a dashed line following the Budva–Pindos Zone under which it is hidden (De Bono 1998). The Palaeotethys Suture Zone in Turkey is possibly found in Chios and the Karaburun Peninsula (Kozur 1997*a*); further east it would be located in the vicinity of the İzmir–Ankara Suture Zone. Following Late Triassic or Jurassic opening of the latter, Palaeotethyan elements could have been dispersed on both sides of the İzmir–Ankara Rift. So far, most Palaeotethyan elements have been described from the northern side of that suture.

Conclusions

Palaeocontinental reconstructions show that a large oceanic space (Palaeotethys) remained open south of the Variscides until Late Palaeozoic time. The Devonian–mid-Carboniferous collisional processes in Europe were related to accretion of the Hun Superterrane to Avalonia–Baltica. This terrane presents strong affinities with the Palaeotethyan passive margin sequences found, for example, in northern Iran (Alborz). The latter is regarded as representing the southern margin of this ocean, the Hun Superterrane the northern margin. This northern margin was separated from Gondwana in Ordovician–Silurian times. After the accretion of the Hun Superterrane to Europe, between the Late Devonian and Early Carboniferous, subduction jumped south into the Palaeotethys. Northward subduction of the Palaeotethys is responsible for Upper Carboniferous calc-alkaline intrusions and volcanism found everywhere in the Variscan Alpine domain. Closure of the Palaeotethys was achieved after the Namurian north of Africa but was diachronous going eastwards. East of a Palaeo-Apulian promontory, subduction continued into the Permian and generated the opening of the Maliak–Meliata and Karakaya-Küre marginal Oceans.

Concomitant Late Permian opening of the marginal Meliata Ocean (within the Eurasian margin) and Neotethys (within the northern Gondwana margin) accelerated the closure of the Palaeotethys in the Dinaro–Hellenide region. Late Permian–Lower Triassic mélanges found in Greece point to a final closure of this Palaeozoic ocean at that time (Eocimmerian event). In Turkey, the collision of the Cimmerian terranes with the Eurasian margin was more complex due to the presence of large oceanic plateaux and Mariana-type back-arc basins between the two domains.

The Late Triassic–Early Jurassic intra-oceanic subduction of the Meliata Ocean generated the Vardar marginal Ocean, obducted in Late Jurassic onto the Dinaro–Hellenic area. Its subsequent northeast directed subduction generated the collision of an intra-oceanic arc with the Austro–Carpathian and Balkanide areas in late Early Cretaceous times.

Figure 12 is a first attempt at showing the present-day location of the sutures of the above described oceans in the eastern Mediterranean area. It clearly shows that the Neotethys suture s.str. is located to the south of Turkey. The İzmir–Ankara Suture is not the Neotethys suture, it is the suture of a marginal ocean located within the Neotethyan active margin.

In a southeast transect of Europe, it can be seen how the Variscan Orogen evolved into Early and Late Cimmerian deformations. The rather clear situation found in the Appalachians cannot be extrapolated much further than western Iberia, where Laurentia and Gondwana collided in the Carboniferous–Early Permian. In the rest of Europe, this collision never happened as such. There was a collision between the Eurasian active margin and terranes derived from Gondwana. One has to wait for the anticlockwise rotation of Africa in the Cretaceous to see a collision between Europe and Africa, giving birth to the Alpine Orogen.

I would like to acknowledge helpful input from G. Borel, C. Steiner and A. De Bono for the subsidence curves for the Alpine and Hellenide regions, and Jon Mosar for the elaboration of the palaeoreconstructions. This is a contribution to the FNRS project 2000-53646.98 'Geodynamics of Tethyan margins'. A. H. F. Robertson and A. İ. Okay reviewed this manuscript, I thank them for their constructive remarks and for sharing their knowledge of the Mediterranean area with me.

References

ADAMS, E., AGER, D. V. & HARDING, A. G. 1980. Géologie de la région d'Imouzzer des Ida-ou-Tanane (Haut Atlas occidental). *Notes Service géologique Maroc*, **41/285**, 59–80.

AGSO 1995. Preservation of old accumulations – key to Canning hydrocarbon exploration. *AGSO Research Newsletter*, **22**, 5–7.

AKTAŞ, G. & ROBERTSON, A. H. F. 1984. The Maden complex, SE Turkey: evolution of a Neotethyan active margin. *In*: DIXON, J. E. & ROBERTSON, A. H. F. (eds) *The Geological Evolution of the Eastern Mediterranean*, Geological Society, London, Special Publications, **17**, 375–402.

ALAVI, M., VAZIRI, H., SEYED-EMAMI, K. & LASEMI, Y. 1997. The Triassic and associated rocks of the Nakhlak and Aghdarband areas in central and northeastern Iran as remnants of the southern Turanian active continental margin. *Geological Society of America Bulletin*, **109**, 1563–1575.

ALTINER, D., KOÇYIĞIT, A., FARINACCI, A., NICOSIA, U. & CONTI, M. A. 1991. Jurassic, Lower Cretaceous stratigraphy and palaeogeographic evolution of the southern part of north-western Anatolia. *Geologica Romana*, **28**, 13–80.

AMBROGGI, R. 1961. *Etude géologique du versant méridional du Haut Atlas occidental et de la plaine du Souss.* Notes et Mémoires du Service géologique du Maroc.

BAGATI, T. N. 1990. Lithostratigraphy and facies variations in Spiti Basin (Tethys) Himachal Pradesh, India. *Journal of Himalayan Geology*, **1**, 35–47.

BAILLIE, P. W., POWEL, C. M., LI, Z. X. & RYALL, A. M. 1994. The tectonic framework of western Australia's Neoproterozoic to recent sedimentary basins. *In*: PURCELL, P. G. R. R. (ed.) *The Sedimentary Basins of Western Australia*. PESA, 45–62.

BAUD, A. & STAMPFLI, G. 1989. Tectonogenesis and evolution of a segment of the Cimmerides: the volcano–sedimentary Triassic of Aghdarban (Kopet–Dagh, North-East Iran). *In*: ŞENGÖR, A. M. C. (eds) *Tectonic Evolution of the Tethyan region*. Kluwer, 265–275.

——, STAMPFLI, G. & STEEN, D. 1991a. The Triassic Aghdarband Group: volcanism and geological evolution. *In*: RUTTNER, A. W. (ed.) *The Triassic of Aghdarband (AqDarband), NE-Iran, and its Pre-Triassic Frame*. Abhandlungen der Geologisches Bundes-Anstalt in Wien, **38**, 125–137.

——, THÉLIN, P. & STAMPFLI, G. 1991b. *Palaeozoic Geodynamic Domains and their Alpidic Evolution in the Tethys*. Mémoires de Géologie, Lausanne.

BERBERIAN, F. & BERBERIAN, M. 1981. Tectono–plutonic episodes in Iran. *In*: GUPTA, H. K. & DELANY, F. M. (eds) *Zagros, Hindu Kush, Himalaya, Geodynamic Evolution*. American Geophysical Union, Geodynamic Series, **3**, 5–32.

BERNOULLI, D. 1981. Ancient continental margins of the Tethyan Ocean. *In*: BALLY, A. W., WATTS, A. B., GROW, J. A., MANSPEIZER, W., BERNOULLI, D., SCHREIBER, C. & HUNT, J. M. (eds) *Geology of Passive Continental Margins; History, Structure and Sedimentologic Record (with Special Emphasis on the Atlantic Margin)*. AAPG, Education Course Note Series, **19**, 5/1–5/36.

—— & WEISSERT, H. 1987. The Upper Hawasina Nappes in the Central Oman Mountains: stratigraphy, palinspastics and sequence of nappes emplacements. *Geodinamica Acta*, **1/1**, 47–58.

BERTOTTI, G., PICOTTI, V., BERNOULLI, D. & CASTELLARING, A. 1993. From rifting to drifting: tectonic evolution of the South-Alpine upper-crust from the Triassic to the Early Cretaceous. *Sedimentary Geology*, **86**, 53–76.

BILL, M., BUSSY, F., COSCA, M., MASSON, H. & HUNZIKER, J. C. 1997. High precision U–Pb and $^{40}Ar/^{39}Ar$ dating of an Alpine ophiolite (Gets nappe, French Alps). *Eclogae geologicæ Helvetiae*, **90**, 43–54.

BOREL, G. 1998. *Dynamique de l'extension mésozoique du domaine briançonnais: les Préalpes médianes au Lias*. PhD Thesis, Université Lausanne.

BOULIN, J. 1988. Hercynian and Eocimmerian events in Afghanistan and adjoining regions. *Tectonophysics*, **148**, 253–278.

BRAUD, J. 1987. *La suture du Zagros au niveau de Kermanshah (Kurdistan Iranien), reconstitution paléogéographique, évolution géodynamique magmatique et structurale*. Thèse, Université Paris-Sud.

CARMIGNANI, L. & SASSI, F. P. 1992. Contribution to the geology of Italy with special regard to the Palaeozoic basements. *In*: IGCP project 276 Newsletter: Siena.

CASTELLARIN, A., SELLI, L., PICOTTI, V. & CANTELLI, L. 1996. La tettonica Medio Triassica e il diapirismo nelle Dolomiti occidentali. *In*: *Geologia delle dolomiti, San Cassiamo*. 25–46.

CATALANO, R., DI STEFANO, P. & KOZUR, H. 1988. New results in the Permian and Triassic stratigraphy of western Sicily with special reference to the section at Torrente San Calogero, SW of Pietra di Salome (Sosio valley). *Società geologica Italiana, Atti del 74 Congresso Nazionale*, **Vol. A**, 119–125

——, —— & —— 1992. New data on Permian and Triassic stratigraphy of western Sicily. *Neues Jahrbuch für Geologie und Paläontologie, Abhandlungen*, **184**, 25–61.

CHANNELL, J. E. T, MCCABE, C. & WOODCOCK, N. H. 1993. Palaeomagnetic study of Llandovery (Lower Silurian) Red Beds in NW England. *Geophysical Journal International*, **115**, 1085–1094.

CHATEAUNEUF, J. J. & STAMPFLI, G. 1979. Preliminary report on Permian palynology of Iran. *Proceedings of the International Palynological Conference*, Lucknow, **2**, 186–198.

CIOFLICA, G., LUPU, M., NICOLAE, I. & VLAD, S. 1980. Alpine ophiolites of Romania: tectonic setting, magmatism and metallogenesis. *Annale Institul Geologie Geofizica*, **56**, 79–96.

COLLINS A. S. & ROBERTSON, A. H. F. 1998. Processes of Late Cretaceous to Late Miocene episodic thrust-sheet translation in the Lycian Taurides, SW Turkey. *Journal of the Geological Society, London*, **155**, 759–772.

COQUEL, R., LOBOZIAK, S., STAMPFLI, G. & STAMPFLI-VUILLE, B. 1977. Palynologie du Dévonien supérieur et du Carbonifère inférieur dans l'Elburz oriental (Iran Nord-Est). *Revue de Micropaléontologie*, **20**, 59–71.

CORSIN, P. & STAMPFLI, G. M. 1977. La formation de Shemshak dans l'Elbourz oriental (Iran). Flore–stratigraphie–paléogéographie. *Géobios*, **10**, 509–571.

DE BONO, A. 1998. *Pelagonian margins in Central Evia Island (Greece). Stratigraphy and geodynamic evolution*. PhD Thesis, Université Lausanne.

DELAUNE-MAYERE, A. 1984. Evolution of a passive continental margin: Baler–Basit (NW Syria). *In*: DIXON, J. E. & ROBERTSON, A. H. F. (eds) *The Geological Evolution of the Eastern Mediterranean*, Geological Society, London, Special Publications, **17**, 151–159.

DEMIREL, I. H. & KOZLU, H. 1998. Evaluation of burial history, thermal maturity and source-rock assessment of the Upper Palaeozoic succession of the eastern Taurus region, southern Turkey. *Marine and Petroleum Geology*, **14**, 867–877.

DEMIRTAŞLI, E. 1984. Stratigraphic evidence of Variscan and early Alpine tectonics in southern Turkey. *In*: DIXON, J. E. & ROBERTSON, A. H. F.

(eds) *The Geological Evolution of the Eastern Mediterranean*. Geological Society, London, Special Publications, **17**, 129–146.
—— 1989. Stratigraphic correlation forms of Turkey. *In*: SASSI, F. P. & ZANFERRARI, A. (eds) *Pre-Variscan and Variscan events in the Alpine–Mediterranean Belts, Stratigraphic Correlation forms*. Rendic. della Società geologica Italiana, Società Geologica Italiana, Roma, **12–2**, 183–211.
DERCOURT, J., RICOU, L. E. & VRIELINCK, B. 1993. *Atlas Tethys, Palaeoenvironmental Maps*. Gauthier-Villars.
——, ZONENSHAIN, L. P., RICOU, L. C. & AL., E. 1985. Présentation des 9 cartes paléogéographiques au 1/20 000 000 s'étendant de l'Atlantique au Pamir pour la période du Lias à l'Actuel. *Bulletin de la Société Géologique de France*, **8**, 637–652.
DU DRESNAY, R. 1988. Répartition des dépôts carbonatés du Lias inférieur et moyen le long de la côte atlantique du Maroc: conséquences sur la paléogéographie de l'Atlantique naissant. *Journal of African Earth Sciences*, **7**, 385–396.
EMBLETON, B. J. J. 1984. Continental palaeomagnetism. *In*: VEEVERS, J. J. (ed.) *Phanerozoic Earth History of Australia*. Clarendon Press, 11–38.
FAVRE, P. 1995. Analyse quantitative du rifting et de la relaxation thermique de la partie occidentale de la marge transformante nord-africaine: le Rif externe (Maroc). *Geodinamica Acta*, **8**, 59–81.
—— & STAMPFLI, G. 1992. From rifting to passive margin: the Red Sea, the central Atlantic and the Alpine Tethys as examples. *Tectonophysics*, **215**, 69–97.
——, —— & WILDI, W. 1991. Jurassic sedimentary record and tectonic evolution of the northwestern corner of Africa. *Palaeogeography, Palaeoecology, Palaeoclimatology*, **87**, 53–73.
FINETTI, I. 1985. Structure and evolution of the Central Mediterranean (Pelagian and Ionian Seas). *In*: STANLEY, D. J. & WEZEL, F. C. (eds) *Geological Evolution of the Mediterranean Basin*. Springer, 215–230.
FLÜGEL, H. W., SASSI, F. P. & GRECULA, P. 1987. Pre-Variscan and Variscan events in the Alpine Mediterranean mountain belts. *In*: Alfa. Bratislava.
FROITZHEIM, N. & MANATSCHAL, G. 1996. Kinematics of Jurassic rifting, mantle exhumation, and passive-margin formation in the Austroalpine and Penninic nappes (Eastern Switzerland). *Geological Society of America Bulletin*, **108**, 1120–1133.
FUCHS, G. 1982. The geology of the Pin valley in Spiti, H.P, India. *Jahrbuch der Geologischen Bundesanstalt (Wien)*, **124**, 325–359.
GANSSER, A. 1960. Ausseralpine Ophiolithprobleme. *Eclogae geologicae Helvetiae*, **52/2**, 659–680.
GARFUNKEL, Z. 1998. Constrains on the origin and history of the Eastern Mediterranean basin. *Tectonophysics*, **298**, 5–35.
—— & DERBIN, B. 1984. Permian–early Mesozoic tectonism and continental margin formation in Israel and its implications for the history of the Eastern Mediterranean. *In*: DIXON, J. E. & ROBERTSON, A. H. F. (eds) *The Geological Evolution of the Eastern Mediterranean*. Geological Society, London, Special Publications, **17**, 187–201.
GARZANTI, E., ANGIOLINI, L. & SCIUNNACH, D. 1996. The mid Carboniferous to lowermost Permian succesion of Spiti (Po Group and Ganmachidam Formation; Tethys Himalaya, northern India): Gondwana glaciation and rifting of Neo-Tethys. *Geodinamica Acta*, **9**, 78–100,
——, NICORA, A., TINTORI, A., SCIUNNACH, D. & ANGIOLINI, L. 1994. Late Palaeozoic stratigraphy and petrography of the Thini Chu Group (Manang, Central Nepal): sedimentary record of Gondwana glaciation and rifting of Neotethys. *Rivista italiana di Palaeontologia e Stratigrafia*, **100**, 155–194.
GEORGIEV, G., STEFANESCU, M. & GABOR, T. 1997. Cimmeride thrust-fold structures beneath the Moesian Platform. *Proceedings of the EAGE 59th Conference and Exhibition, Geneva*, Abstract, 510.
GRADSTEIN, F. M. & OGG, J. G. 1996. A Phanerozoic time scale. *Episodes*, **19**, 1–2.
GREBER, E., LEU, W., BERNOULLI, D., SCHUMACHER, M. E. & WYSS, R. 1997. Hydrocarbon provinces in the Swiss Southern Alps – a gas geochemistry and basin modelling study. *Marine Petroleum Geology*, **14**, 3–25.
GUTNIC, M., MONDO, O., POISSON, A. & DUMONT, J.-F. 1979. Géologie des Taurus occidentales (Turquie). *Mémoire de la Société Géologique de France, Nouvelle série*, **58**, 1–112.
HAAS, J., KOVÀCS, S., KRYSTYN, L. & LEIN, R. 1995. Significance of Late Permian–Triassic facies zones in terrane reconstructions in the Alpine–North Pannonian domain. *Tectonophysics*, **242**, 19–40.
HUBER, H. & EFTEKHAR-NEZHAD, J. 1978. *Geological Map of Iran, 1:1 000 000*. NIOC/GSI Tehran, six sheets.
HUTCHISON, C. S. 1989. *Geological Evolution of South-East Asia. Oxford Monographs on Geology and Geophysics*. Clarendon Press.
IRVING, E. 1977. Drift of the major continental blocks since the Devonian. *Nature*, **270**, 304–309.
JENNY, J. 1983. Les décrochement de l'Atlas de Demnat (Haut Atlas central, Maroc): prolongation orientale de la Zone de décrochement du Tizi-n-Test et la clef de la compréhension de la tectonique atlasique. *Eclogae geologicae Helvetiae*, **76**, 243–251.
KHAIN, V. E. 1994a. *Geology of Northern Eurasia*. Gebrüder Borntraeger.
—— 1994b. Geology of Northern Eurasia (Ex-USSR). *Beiträge zur regionalen Geologie der Erde*, **24**, 346–385.
KLITGORD, K. D. & SCHOUTEN, H. S. 1986. Plate kinematics of the Central Atlantic. *In*: VOGT, P. R. & TUCHOLTE, B. E. (eds) *The Western North Atlantic Region. The Geology of North America*, **Volume M**, 351–378.
KOÇYIĞIT, A. 1987. Hasanoğlan (Ankara) yöresinin

tectono-stratigrafisi: Karakaya orojenik kuşağın ınevrimi. *Hacettepe University Earth Sciences*, **14**, 269–293 [in Turkish with English abstract].

—— 1991. An example of an accretionary forearc basin from Central Anatolia and its implications for the history of subduction of Neo-Tethys in Turkey. *Geological Society of America Bulletin*, **103**, 22–36.

KOZUR, H. 1990. Deep-water Permian in Sicily and its possible connection with the Himalaya–Tibet region. *Proceedings of the 5th Himalaya–Tibet–Karakorum Workshop Abstracts*, Milano.

—— 1991. The evolution of the Hallstatt ocean and its significance for the early evolution of the Eastern Alps and western Carpathians. *In*: CHANNELL, J. E. T., WINTERER, E. L. & JANSA, L. F. (eds) *Palaeogeography and Paleoceanography of Tethys*. Palaeogeography, Palaeoclimatology, Palaeoecology, **87**, 109–135.

—— 1995. First evidence of Middle Permian Ammonitico Rosso and further new stratigraphic results in the Permian and Triassic of the Sosio Valley area, Western Sicily. *Proceedings of the 1st Croatian Geological Congress, Opatija. Zbornik radova Proceedings*, **1**, 307–310.

—— 1997a. First discovery of Muellerisphaerida (inc. sedis) and Eoalbaillella (Radiolaria) in Turkey and the age of the siliciclastic sequence (clastic series) in Karaburun peninsula. *Freiberger Forschungsheft*, **C466**, 33–59.

—— 1997b. Pelagic Permian and Triassic of the western Tethys and its palaeogeographic and stratigraphic significance. *Berg und Hüttenmännischer Tag, Freiburg, Abstracts*, 21–25.

KRAHL, J. 1992. The young Palaeozoic and Triassic Tethyan rocks in the external Hellenides on Crete, the northern border of Gondwana. *Proceedings of the 8th Congress of the Geological Society of Greece*, 59–60.

——, KAUFMANN, G., KOZUR, H., RICHTER, D., FÖRSTER, O. & HEINRITZI, F. 1983. Neue Daten zur Biostratigraphie und zur tektonischen Lagerung der Phyllit-Gruppe und der Trypali-Gruppe auf der Insel Kreta (Griechenland). *Geologische Rundschau*, **72**, 1147–1166.

——, ——, RICHTER, D. *ET AL.* 1986. Neue Fossilfunde in der Phyllit-Gruppe Ostkretas (Griechenland). *Zeitschrift der deutschen geologischen Gesellschaft*, **137**, 523–536.

LEMAIRE, M.-M. 1997. *Les relations du bloc d'Ustyurt avec l'Iran et l'Eurasie d'après les données paléomagnétiques de formations permiennes à jurassiques de la plaque de Touran et de la plaque scythe.* PhD Thesis, Université Louis Pasteur, Strasbourg.

LEVEN, E. JA. 1995. Permian and Triassic of the Rushan–Pshart zone (Pamir). *Rivista italiana di Palaeontologia e Stratigrafia*, **101**, 3–16.

LYS, M. 1986. *Biostratigraphie du Carbonifère et du Permien en Mésogée. Etudes micropaléontologique et paléobiogéographique.* Doctorat d'Etat, Université Paris-Sud Orsay.

MALPAS, J., CALON, T. & SQUIRES, G. 1993. The development of a late Cretaceous microplate suture zone in SW Cyprus. *In*: PRICHARD, H. M., ALABASTER, T., HARRIS, N. B. W. & NEARY, C. R. (eds) *Magmatic Processes and Plate Tectonics*. Geological Society, London, Special Publications, **76**, 177–195.

MANN, A. & HANNA, S. S. 1990. The tectonic evolution of pre-Permian rocks, Central and South-eastern Oman mountains. *In*: ROBERTSON, A. H. F., SEARLE, M. P. & RIES, A. (eds) *The Geology and Tectonics of the Oman Region*. Geological Society, London, Special Publications, **49**, 307–325.

MICHARD, A. 1982. Contribution à l'étude de la marge nord du Gondwana: une chaîne plissée paléozoique, vraisemblablement hercynienne en Oman. *Comptes Rendus de l'Académie des Sciences de Paris, Sér.* **II**, **295**, 1031–1036.

——, WHITECHURCH, H., RICOU, L.-E., MONTIGNY, R. & YAZĞAN, E. 1984. Tauric subduction (Malatya–Elaziğ) and its bearing on tectonics of the Tethyan realm in Turkey. *In*: DIXON, J. E. & ROBERTSON, A. H. F. (eds) *The Geological Evolution of the Eastern Mediterranean*, Geological Society, London, Special Publications, **17**, 361–373.

MONOD, O. & AKAY, E. 1984. Evidence for a Late Triassic–Early Jurassic orogenic event in the Taurides. *In*: DIXON, J. E. & ROBERTSON, A. H. F. (eds) *The Geological Evolution of the Eastern Mediterranean*. Geological Society, London, Special Publications, **17**, 113–122.

MOSAR, J., STAMPFLI, G. M. & GIROD, F. 1996. Western Préalpes médianes: timing and structure. A review. *Eclogae geologicae Helvetiae*, **89**, 389–425.

MUTTONI, G., KENT, D. V. & CHANNELL, J. E. T. 1996. Evolution of Pangea: palaeomagnetic constraints from the Southern Alps, Italy. *Earth and Planetary Sciences Letters*, **140**, 97–112.

NICOLAE, I. & SEGHEDI, A. 1996. Lower Triassic basic dyke swarm in North Dobrogea. *Romanian Journal of Petrology*, **77**, 31–40.

NIKISHIN, A. M., CLOETINGH, S., BRUNET, M.-F., STEPHENSON, R. A., BOLOTOV, S. N. & ERSHOV, A. V. 2000. Scythian platform, Caucasus and Black Sea regions: Mesozoic–Cenozoic tectonic history and dynamics. *In*: CRASQUIN, S. & BARRIER, E. (eds), *Bulletin du Muséum National d'Histoire Naturelle, Paris, Peri-Tethys*, Memoire, **3**.

NIKO, S., PILLEVUIT, A. & NISHIDA, T. 1996. Early Late Permian (Wordian) non-ammonoid cephalopods from the Hamrat Duru Group, central Oman Mountains. *Transactions and Proceedings of the Palaeontological Society of Japan*, **183**, 522–527.

NIOCAILL, C. M. & SMETHURST, M. 1994. Palaeozoic palaeogeography of Laurentia and its margins: a reassessment of palaeomagnetic data. *Geophysical Journal International*, **116**, 715–725.

NORMAN, T. N. 1984. The role of the Ankara Mélange in the development of Anatolia (Turkey). *In*: DIXON, J. E. & ROBERTSON, A. H. F. (eds) *The Geological Evolution of the Eastern*

Mediterranean. Geological Society, London, Special Publications, **17**, 441–447.

OKAY, A. İ. 2000. Was the Late Triassic orogeny in Turkey caused by collision of an oceanic plateau? This volume.

—— & MOSTLER, H. 1994. Carboniferous and Permian radiolarite blocks from the Karakaya complex in Northwest Turkey. *Turkish Journal of Earth Sciences*, **3**, 23–28.

——, SATIR, M., MALUSKI, H., SIYAKO, M., MONIE, P., METZGER, R. & AKYÜZ, S. 1996. Palaeo- and Neo-Tethyan events in northwestern Turkey: geologic and geochronologic constraints. *In*: YIN, A. & HARRISON, T. M. (eds) *The Tectonic Evolution of Asia*. Cambridge University Press, 420–441.

PERROUD, H., VAN DER VOO, R. & BONHOMMET, N. 1984. Palaeozoic evolution of the Armorica plate on the basis of Palaeomagnetic data. *Geology*, **12**, 579–582.

PICKETT, E. A. & ROBERTSON, A. H. F. 1996. Formation of the late Palaeozoic–Early Mesozoic Karakaya complex and related ophiolites in NW Turkey by Palaeotethyan subduction–accretion. *Journal of the Geological Society, London*, **153**, 995–1009.

PILLEVUIT, A. 1993. *Les blocs exotiques du Sultanat d'Oman, évolution paléogéographique d'une marge passive flexurale*. Mémoires de Géologie (Lausanne), **17**.

——, MARCOUX, J., STAMPFLI, G. & BAUD, A. 1997. The Oman Exotics: a key to the understanding of the Neotethyan geodynamic evolution. *Geodinamica Acta*, **10**, 209–238.

POWELL, C. M. & LI, Z. X. 1994. *Reconstruction of the Panthalassan Margin of Gondwanaland*. Geological Society of America Memoirs, **184**, 5–9.

REISCHMANN, T. 1998. Pre-Alpine origin of tectonic units from the metamorphic complex of Naxos, Greece, identified by single Zircon Pb/Pb dating. *Proceedings of the 8th International Congress of the Geological Society of Greece*, Patras, Greece. **XXXII/3**, 101–111.

RICOU, L. E. 1971. Le croissant ophiolitique péri-arabe: une ceinture de nappes mises en place au crétacé supérieur. *Revue de Géographie Physique et de Géologie Dynamique*, **13–4**, 327–349.

—— 1974. *L'étude géologique de la région de Neyriz (Zagros Iranien)*. Thèse, Université Orsay, Paris-Sud.

ROBARDET, M., VERNIERS, J., FEIST, R. & PARIS, F. 1994. Le Paléozoique anté-varisque de France, contexte paléogéographique et géodynamique. *Géologie de la France*, **3**, 3–31.

ROBERTSON, A. H. F. & SEARLE, M. P. (eds) 1990. *The Geology and Tectonics of the Oman Region*. Geological Society, London, Special Publications, **49**, 797–834.

—— & WOODCOCK, N. H. 1980. Tectonic setting of the Troodos Massif in the East Mediterranean. *In*: PANAYIOTOU, A. (ed.) *Ophiolites*. Proceedings of the International Ophiolite Symposium, 36–49.

—— & XENOPHONTOS, C. 1993. Development of concepts concerning the Troodos ophiolite and adjacent units in Cyprus. *In*: PRICHARD, H. M., ALABASTER, T., HARRIS, N. B. & NEARY, C. R. (eds), *Magmatic Processes and Plate Tectonics*. Geological Society, London, Special Publications, **76**, 85–119.

ROWLEY, D. B. & LOTTES, A. L. 1988. Plate–kinematic reconstructions of the North Atlantic and Arctic: Late Jurassic to Present. *Tectonophysics*, **155**, 73–120.

RUTTNER, A. W. 1993. Southern borderland of Triassic Laurasia in NE Iran. *Geologische Rundschau*, **82**, 110–120.

SASSI, F. P. & ZANFERRARI, A. 1989. *Pre-Variscan and Variscan Events in the Alpine–Mediterranean Belts, Stratigraphic Correlation Forms*. Rendicotta della Società Geologica Italiana, Societa Geologica Italiana, Roma.

SCHMIDT, K. & SOFFEL, H. 1984. Mesozoic geological events in the Central-East-Iran and their relation to palaeomagnetic results. *Neues Jahrbuch für Geologie und Paläontologie, Abhandlungen*, **168**, 173–181.

SEGHEDI, I., SZAKACS, A. & BALTRES, A. 1990. Relationships between sedimentary deposits and eruptive rocks in the Consul unit (North Dobrogea) – implications on tectonic interpretations. *Dari de Seama ale Sedintelor Institut Geologie Geofizica*, **74**, 125–136.

ŞENGÖR, A. M. C. 1979. Mid-Mezozoic closure of Permo-Triassic Tethys and its implications. *Nature*, **279**, 590–593.

—— 1984. *The Cimmeride Orogenic System and the Tectonics of Eurasia*. Geological Society of America, Special Paper, **195**.

—— 1985. The story of Tethys: how many wives did Okeanos have. *Episodes*, **8**, 3–12.

—— 1989. The Tethyside orogenic system: an introduction. *In*: ŞENGÖR, A. M. C. (ed.) *Tectonic Evolution of the Tethyan Region*. Kluwer, 1–22.

—— 1990. A new model for the late Palaeozoic–Mesozoic tectonic evolution of Iran and implications on Oman. *In*: ROBERTSON, A. H. F. & SEARLE, M. P. (eds) *The Geology and Tectonics of the Oman Region*. Geological Society, London, Special Publications, **49**, 797–834.

—— 1991. Late Palaeozoic and Mesozoic tectonic evolution of the middle eastern Tethysides: implications for the Palaeozoic geodynamics of the Tethyan realms. *In*: BAUD, A., THÉLIN, P. & STAMPFLI, G. M. (eds) *Palaeozoic Geodynamic Domains and their Alpidic Evolution in the Tethys*. Mémoires de Géologie Lausanne, **10**, 111–150.

—— & HSÜ, K. J. 1984. The Cimmerides of Eastern Asia: history of the Eastern end of Palaeo-Tethys. *Mémoires de la Société Géologique de France*, **47**, 139–167.

—— & NATAL'IN, B. A. 1996. Palaeotectonics of Asia: fragments of a synthesis. *In*: YIN, A. & HARRISON, M. (eds) *The Tectonic Evolution of Asia*. Cambridge University Press, 443–486.

——, YILMAZ, Y. & KETİN, İ. 1980. Remnants of a pre–Late Jurassic ocean in the northern Turkey: fragments of Permian-Triassic Palaeo-Tethys?

Geological Society of America Bulletin, **91**, 599–609.

——, —— & SUNGURLU, O. 1984. Tectonics of the Mediterranean Cimmerides: nature and evolution of the western termination of Palaeo-Tethys. *In*: DIXON, J. E. & ROBERTSON, A. H. F. (eds) *The Geological Evolution of the Eastern Mediterranean*. Geological Society, London, Special Publications, **17**, 77–112.

SOCIETÀ-GEOLOGICA-ITALIANA 1979. *Palaeozoico e basamento in Italia aggionamenti e problemi*. S-G-I, Roma, 464.

SOFFEL, H. C. & FÖRSTER, H. G. 1984. Polar wander path of the Central-East-Iran Microplate, including new results. *Neues Jahrbuch für Geologie und Paläontologie, Abhandlungen*, **168**, 165–172.

STAMPFLI, G. 1978. *Etude géologique generale de l'Elbourz oriental au sud de Gonbad-e-Qabus (Iran NE)*. PhD. Thesis (1868), Université Genève.

—— 1989. Late Palaeozoic evolution of the eastern Mediterranean region. IGCP 276 Palaeozoic geodynamic domains and their alpidic evolution in the Tethys, Lausanne, Switzerland, Short course, I–II.

—— 1993. Le Briançonnais, terrain exotique dans les Alpes? *Eclogae geologicae Helvetiae*, **86**, 1–45.

—— 1996. The Intra-Alpine terrain: a Palaeotethyan remnant in the Alpine Variscides. *Eclogae geologicae Helvetiae*, **89**, 13–42.

—— & MARCHANT, R. H. 1997. Geodynamic evolution of the Tethyan margins of the Western Alps. *In*: PFIFFNER, O. A., LEHNER, P., HEITZMAN, P. Z., MUELLER, S. & STECK, A. (eds) *Deep Structure of the Swiss Alps – Results from NRP 20*. Birkhaüser AG, 223–239.

—— & MOSAR, J. 1999. The making and becoming of Apulia. *Mémoires Sécience Géologie (University of Padova)*. Special Volume, 3rd Workshop on Alpine Geology, **51/1**, 141–154.

—— & PILLEVUIT, A. 1993. An alternative Permo-Triassic reconstruction of the kinematics of the Tethyan realm. *In*: DERCOURT, J., RICOU, L.-E. & VRIELINCK, B. (eds) *Atlas Tethys Palaeoenvironmental Maps. Explanatory Notes*. Gauthier-Villars, 55–62.

——, DE BONO, A. & VAVASIS, I. 1995. Pelagonian basement and cover in Euboea (Greece). *Terra Nova*, **7**, 180.

——, MARCOUX, J. & BAUD, A. 1991. Tethyan margins in space and time. *In*: CHANNELL, J. E. T., WINTERER, E. L. & JANSA, L. F. (eds) *Palaeogeography and Palaeoceanography of Tethys*. Palaeogeography, Palaeoclimatology, Palaeoecology, Elsevier, **87**, 373–410.

——, MOSAR, J., DE BONO, A. & VAVASSIS, I. 1998*a*. Late Palaeozoic, Early Mesozoic plate tectonic of the western Tethys. *Proceedings of the 8th International Congress of the Geological Society of Greece*, Patra, **32**, 113–120.

——, ——, FAVRE, P., PILLEVUIT, A. & VANNAY, J.-C. 2000. Permo-Mesozoic evolution of the western Tethyan realm: the Neotethys/East-Mediter- ranean connection. *In*: CAVAZZA, W., ROBERTSON, A. H. F. R. & ZIEGLER, P. A. (eds) *Peritethyan Rift/Wrench Basins and Passive Margins, IGCP 369*. Bulletin du Muséum National d'Histoire Naturelle, Paris, in press.

——, ——, MARCHANT, R., MARQUER, D., BAUDIN, T. & BOREL, G. 1998*b*. Subduction and obduction processes in the western Alps. *In*: VAUCHEZ, A. & MEISSNER, R. (eds) *Continents and their Mantle Roots*. Tectonophysics, **296**, 159–204.

STEINER, C. W., HOBSON, A., FAVRE, P., STAMPFLI, G. M. & HERNANDEZ, J. 1998. The Mesozoic sequence of Fuerteventura (Canary islands): witness of an Early to Middle Jurassic seafloor spreading in the Central Atlantic. *Geological Society of America Bulletin*, **110**, 1304–1317.

STÖCKLIN, J. 1968. Structural history and tectonics of Iran: a review. *AAPG Bulletin*, **52**, 1229–1258.

—— 1974. Possible ancient continental margin in Iran. *In*: BURK, C. A. & DRAKE, C. L. (eds) *The Geology of Continental Margins* Springer, 873–887.

—— 1977. Structural correlation of the Alpine ranges betwen Iran and Central Asia. *Mémoires hors série de la Société géologique de France*, **8**, 333–353.

—— 1981. A brief report on geodynamics in Iran. *In*: GUPTA, H. K. & DELANY, F. M. (eds), *Zagros, Hindu-Kush, Himalaya, Geodynamic Evolution*. American Geophysical Union and Geological Society of America, **3**, 70–74.

TARI, G., DICEA, O., FAULFERSON, J., GEORGIEV, G., BYRNE, P. & MOUNTNEY, N. P., 1997. Cimmerian and Alpine stratigraphy and structural evolution of the Moesian platform. *In*: ROBINSON, A. G. (ed.) *Regional and Petroleum Geology of the Black-Sea and Surrounding Region*. AAPG Memoirs, **68**, 63–90.

TORQ, F. 1997. *Evolution et destruction de la Pangée du Carbonifère au Jurassique. Dérive des pôles magnétiques et étude de la dipolarité du champ*. Thèse, Université Paris VII.

TORSVIK, H. & EIDE, E. A. 1998. Phanerozoic palaeogeography and geodynamics with Atlantic details. NGU Report No. **98.001**.

—— & SMETHURST, M. A. 1994. *Geographic mapping and palaeoreconstruction package (GMAP)*. Software description and examples of palaeoreconstructions (version GMAP for Windows v.1.0).

TÜYSÜZ, O. 1990. Tectonic evolution of a part of the Tethys orogenic collage: the Kargi Massif, northern Turkey. *Tectonics*, **9**, 141–160.

USTAÖMER, T. & ROBERTSON, A. H. F. 1994. Late Palaeozoic marginal basin and subduction–accretion: the Palaeotethyan Küre complex, Central Pontides, northern Turkey. *Journal of the Geological Society, London*, **151**, 291–305.

—— & —— 1997. Tectonic–Sedimentary evolution of the North-Tethyan margin in the central Pontides of northern Turkey. *In*: ROBINSON, A. G. (ed.) *Regional and Petroleum Geology of the Black-Sea and Surrounding Region*. AAPG Memoirs, **68**, 255–290.

—— & —— 1999. Geochemical evidence used to test

alternative plate tectonics models for pre-Upper Jurassic (Palaeotethyan) units in the Central Pontides, N. Turkey. *In*: BOZKURT, E. & ROWBOTHAM, G. *Aspects of Geology of Turkey II.* Geological Journal, in press.

VACHARD, D., BEAUCHAMP, J. & TOURANI, A. 1991. Le Carbonifère inférieur du Haut-Atlas de Marakesh (Maroc): facies, microfossiles et traces fossiles. *Géologie Méditerranéenne*, **18**, 3–19.

VAI, G. B. 1991. Palaeozoic strike-slip rift pulses and Palaeogeography in the circum Mediterranean Tethyan realm. *Palaeogeography, Palaeoclimatology, Palaeoecology*, **87**, 223–252.

—— 1994. Crustal evolution and basement elements in the Italian area: palaeogeography and characterisation. *Bollettino di Geofisica Theorica ed Applicata*, **36**, 411–434.

VAN DER VOO, R. 1993. *Palaeomagnetism of the Atlantic, Tethys and Iapetus Oceans.* Cambridge University Press.

VANNAY, J. C. 1993. *Géologie des chaînes du Haut-Himalaya et du Pir Panjal au Haut-Lahul (NW Himalaya) paléogéographie et tectonique (PhD). Mémoires Géologie (Lausanne)*, **16**, 148.

VAVASSIS, I., DE BONO, A. & STAMPFLI, G. M. 1997. New stratigraphic data on the Pelagonian pre-Jurassic units of Evvia island (Greece). *Terra Abstract*, **9**, 160.

VON RAUMER, J. F. & NEUBAUER, F. (eds) 1993. *Pre-Mesozoic Geology in the Alps.* Springer.

——, STAMPFLI, G. M. & MOSAR, J. 1998. From Gondwana to Pangaea – an Alpine point of view. *In: Schriften der A. Wegener Stiftung 98/2/Giesen*, Terra Nostra, 154–156.

——, MÉNOT, R. P., ABRECHT, J. & BIINO, G. 1993. The pre Alpine evolution of the external massifs. *In*: VON RAUMER, J. F. & NEUBAUER, F. (EDS) *Pre-Mesozoic Geology in the Alps.* Springer, 221–240.

WILDI, W., FUNK, H. P., LOUP, B., AMATO, E. & HUGGENBERGER, P. 1989. Mesozoic subsidence history of the European marginal shelves of alpine Tethys (Helvetic realm, Swiss Plateau and Jura). *Eclogae geologicae Helvetiae*, **82**, 817–840.

ZIEGLER, P. A. 1988a. *Evolution of the Arctic–North Atlantic and the Western Tethys.* AAPG Memoirs, **43**.

—— 1988b. Post-Hercynian plate reorganisation in the Tethys and Arctic–North Atlantic domains. *In*: MANSPEIZER, W. (eds) *Triassic–Jurassic rifting. Developments in Geotectonics.* Elsevier, **22**, 711–755.

—— 1990. *Geological Atlas of Western and Central Europe.* 2nd Edition. Shell International Petroleum Mij.

ZONENSHAIN, L. P., KUZMIN, M. I. & KONONOV, M. V. 1985. Absolute reconstructions of the Palaeozoic oceans. *Earth and Planetary Science Letters*, **74**, 103–116.

——, —— & NATAPOV, L. M. 1990. *Geology of the USSR: A Plate Tectonic Synthesis.* AGU Geodynamic Series Monograph, **21**.

Was the Late Triassic orogeny in Turkey caused by the collision of an oceanic plateau?

ARAL İ. OKAY

İstanbul Teknik Üniversitesi, Avrasya Yerbilimleri Enstitüsü, Ayazağa, TR-80626 İstanbul, Turkey (e-mail: okay@itu.edu.tr)

Abstract: A belt of Late Triassic deformation and metamorphism (Cimmeride Orogeny) extends east–west for 1100 km in northern Turkey. It is proposed that this was caused by the collision and partial accretion of an Early–Middle Triassic oceanic plateau with the southern continental margin of Laurasia. The upper part of this oceanic plateau is recognized as a thick Lower–Middle Triassic metabasite–marble–phyllite complex, named the Nilüfer Unit, which covers an area of 120 000 km^2 with an estimated volume of mafic rocks of 2×10^5 km^3. The mafic sequence, which has thin stratigraphic intercalations of hemipelagic limestone and shale, shows consistent within-plate geochemical signatures. The Nilüfer Unit has undergone a high-pressure greenschist facies metamorphism, but also includes tectonic slices of eclogite and blueschist with latest Triassic isotopic ages, produced during the attempted subduction of the plateau. The short period for the orogeny (< 15 Ma; Norian–Hettangian) is further evidence for the oceanic plateau origin of the Cimmeride Orogeny. The accretion of the Nilüfer Plateau produced strong uplift and compressional deformation in the hanging wall. A large and thick clastic wedge, fed from the granitic basement of the Laurasia, represented by a thick Upper Triassic arkosic sandstone sequence in northwest Turkey, engulfed the subduction zone and the Nilüfer Plateau.

An east–west trending belt of latest Triassic deformation and regional metamorphism extends for over 1100 km in northern Turkey. The Early Mesozoic deformation (but not the regional metamorphism) was known previously (Şengör 1979; Bergougnan & Fourquin 1982) and was referred to as the Cimmeride deformation (Şengör et al. 1984). The Cimmeride deformation was ascribed to the closure of the Palaeotethys ocean following the collision of a Cimmerian continental sliver with the southern margin of Laurasia (Şengör 1979; Şengör et al. 1984). Here, an alternative explanation, involving the collision and partial accretion of an oceanic plateau to the southern margin of Laurasia, is proposed for the origin of the latest Triassic deformation and metamorphism in northern Turkey.

A tectonic map of Turkey and the surrounding region is shown in Fig. 1. During the Palaeozoic and Mesozoic, the various continental blocks that make up present-day Turkey were situated on the continental margins of the Tethys Ocean. The Pontides, which comprise the Strandja, İstanbul and Sakarya Zones, show Laurasian stratigraphic affinities, while the Anatolide–Tauride Block and the Kırşehir Massif are tectonically and stratigraphically related to Gondwana (Şengör & Yılmaz 1981; Okay et al. 1996; Okay & Tüysüz 2000). The İstanbul Zone is a continental fragment, which was translated south from the Odessa Shelf with the Cretaceous opening of the oceanic West Black Sea Basin (Fig. 1; Okay et al. 1994). Its stratigraphy is similar to that of the Scythian and Moesian platforms, with a fully developed Palaeozoic sedimentary sequence unconformably overlain by Triassic and younger sedimentary rocks (Haas 1968; Dean et al. 1997; Görür et al. 1997). In the İstanbul Zone, a weak latest Triassic deformation is marked by an unconformity between the Norian siliciclastic turbidites and the overlying Upper Cretaceous carbonates. The Strandja Zone consists of a Late Hercynian metamorphic and granitic basement unconformably overlain by Lower Triassic–Middle Jurassic sedimentary rocks (Chatalov 1988; Okay et al. 1996). The Anatolide–Tauride Block and the Kırşehir Massif are also devoid of Triassic metamorphism, and of any significant Triassic deformation. Several well studied Lower Mesozoic stratigraphic sections in the Taurides, including those in the Bornova Flysch Zone (Erdoğan et al. 1990) and in the central Taurides (Gutnic et al. 1979; Özgül 1997), show a continuous transition between Triassic and Jurassic with no evidence of an intervening deformation phase. The pre-Jurassic thrusting, described by Monod & Akay (1984) from a small locality in the central Taurides, is as yet of unknown significance. Late Triassic deformation and regional metamorphism in Turkey are predominantly found in the Sakarya Zone, which will form the main subject of this paper.

From: BOZKURT, E., WINCHESTER, J. A. & PIPER, J. D. A. (eds) *Tectonics and Magmatism in Turkey and the Surrounding Area.* Geological Society, London, Special Publications, **173**, 25–41. 1-86239-064-9/00/$15.00
© The Geological Society of London 2000.

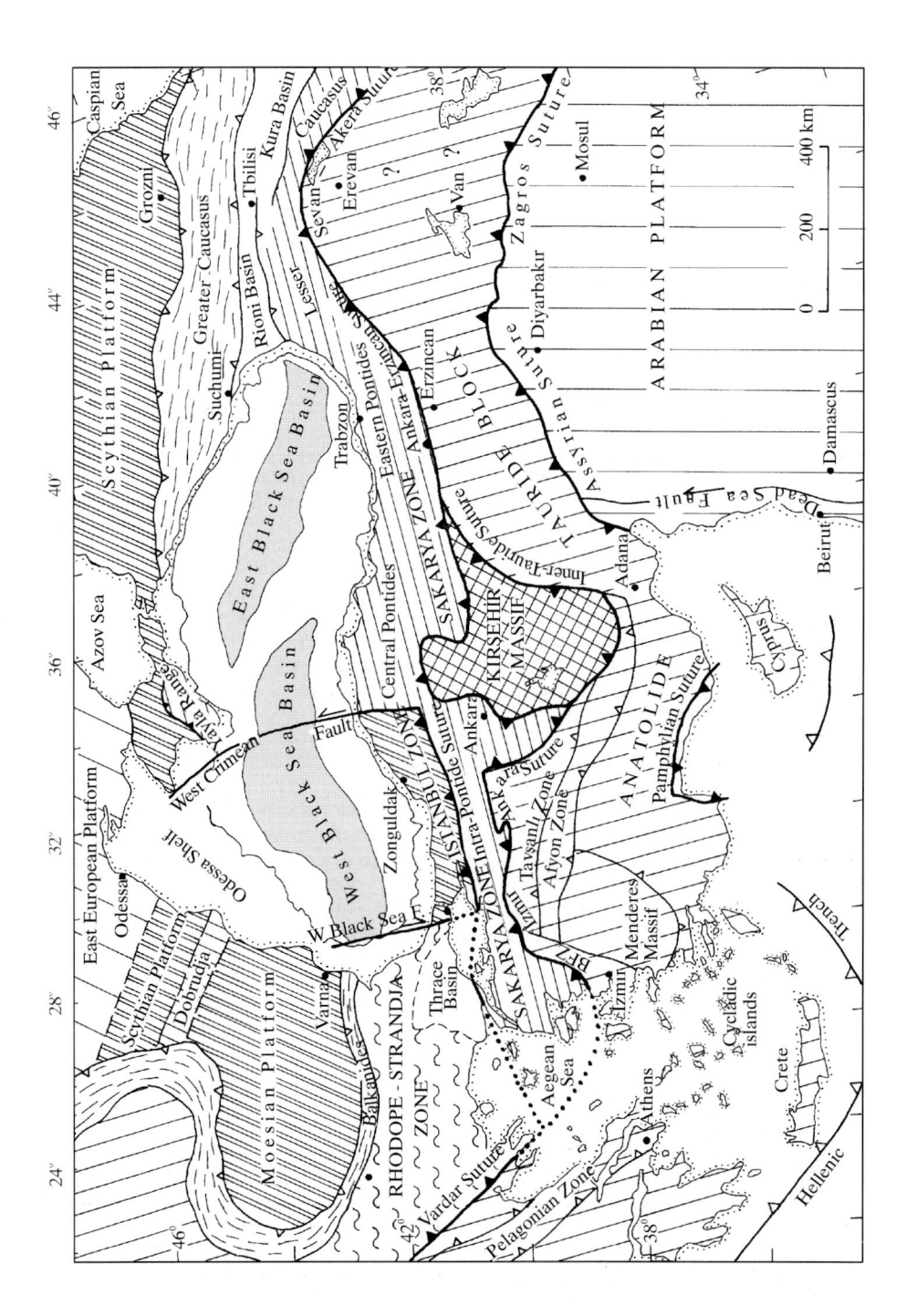

Fig. 1. Tectonic map of Turkey and the surrounding region [after Okay & Tüysüz (1999)].

The Sakarya Zone – a Hercynian unit with Palaeotethyan assemblages

The Sakarya Zone is an elongate continental fragment, 1500 km long and c. 90 km wide, which extends from the Biga Peninsula in the west to the Lesser Caucasus in the east (Figs 1 and 2). The İzmir-Ankara Suture forms its southern contact with the Anatolide–Tauride Block. The Sakarya Zone and the Anatolide–Tauride Block exhibit different Palaeozoic and Mesozoic stratigraphies and were amalgamated into a single continental unit during the latest Cretaceous–Palaeocene continental collision (Şengör & Yılmaz 1981; Okay & Tüysüz 1999). The Sakarya Zone is in contact with the İstanbul Zone in the northwest along the Intra-Pontide Suture of Early Eocene age (Okay et al. 1994), while in the northeast it is bounded by the oceanic East Black Sea Basin (Fig. 1).

A distinctive stratigraphic feature of the Sakarya Zone is a regional earliest Jurassic unconformity. Strongly deformed sedimentary, metamorphic and magmatic rocks of Devonian–Late Triassic age are unconformably overlain by the Lower–Middle Jurassic conglomerates and sandstones. The pre-Jurassic rocks, which are intermittently exposed along the 1500 km length of the Sakarya Zone (Figs 2 and 3), can be grouped into two categories: Hercynian continental units and Palaeotethyan subduction–accretion–collision complexes. As discussed below, there is still some controversy on the tectonic setting and correlation of the pre-Jurassic subduction–accretion complexes in the Sakarya Zone.

Tectonic interpretation of the pre-Jurassic subduction–accretion complexes

Bingöl et al. (1975) defined the Karakaya Formation from the Biga Peninsula in northwest Turkey as a heterogeneous, slightly metamorphosed Triassic unit of feldspathic sandstone, quartzite, conglomerate, siltstone, basalt, mudstone and radiolarian chert. They pointed out its wide extent throughout the Pontides, from the Biga Peninsula to the Ankara region, and interpreted the depositional environment of the Karakaya Formation as an Early Triassic intracontinental basin (Bingöl et al. 1975). A completely new and revolutionary interpretation of the Karakaya Complex as a Carboniferous–Triassic subduction–accretion complex was proposed by Tekeli (1981). Şengör et al. (1984) followed this tectonic interpretation, but separated the pre-Jurassic subduction–accretion complexes in the central Pontides from the Karakaya Complex, ascribing the former to the subduction of the Palaeotethys and the latter to that of a small Permian–Triassic back-arc basin of the Palaeotethys. This led to an artificial subdivision of the Cimmeride Orogen into the Karakaya Orogen in northwest Turkey and the Palaeotethyan Orogen in the central Pontides. The discovery of Carboniferous and Permian radiolarian cherts (Kozur & Kaya 1994; Kozur 1997; Okay & Mostler 1994), Triassic eclogites (Okay & Monié 1997) and blueschists (Monod et al. 1996) in the Karakaya Complex, as well as more precise characterization of its deformational and stratigraphic features (Okay et al. 1991, 1996; Pickett & Robertson 1996), confirmed Tekeli's (1981) interpretation of the Karakaya Complex as a subduction–accretion complex, formed through the subduction of an ocean as old as Carboniferous. The pre-Jurassic subduction–accretion complexes in the central Pontides are lithologically and temporally similar to those further west and east in the Sakarya Zone. Furthermore, there is no continental unit that separates the pre-Jurassic subduction–accretion complexes of the central Pontides from those further west or east. The İstanbul Zone, which was claimed to separate the pre-Jurassic subduction–accretion complexes (Ustaömer & Robertson 1993), is now known to have reached its present position in the Cretaceous or even later (Okay et al. 1994). Therefore, in this paper, all the pre-Jurassic subduction–accretion complexes in the Sakarya Zone are regarded as having formed during the subduction of the Palaeotethys and are collectively referred to as the Karakaya Complex, as initially intended by Tekeli (1981).

Palaeotethyan subduction–accretion–collision complexes in the Sakarya Zone

Outcrops of the Palaeotethyan subduction–accretion–collision complexes occur throughout the Sakarya Zone beneath the Jurassic and younger cover rocks (Tekeli 1981; Figs 2 and 3). They comprise two main tectonostratigraphic units (Fig. 4). At the base there is a thick sequence of metabasite, marble and phyllite of Triassic age, which is overlain by strongly deformed, but generally unmetamorphosed, clastic and mafic volcanic sequences of Late Palaeozoic–Triassic age. The clastic and mafic volcanic sequences can be further subdivided into those which formed during the

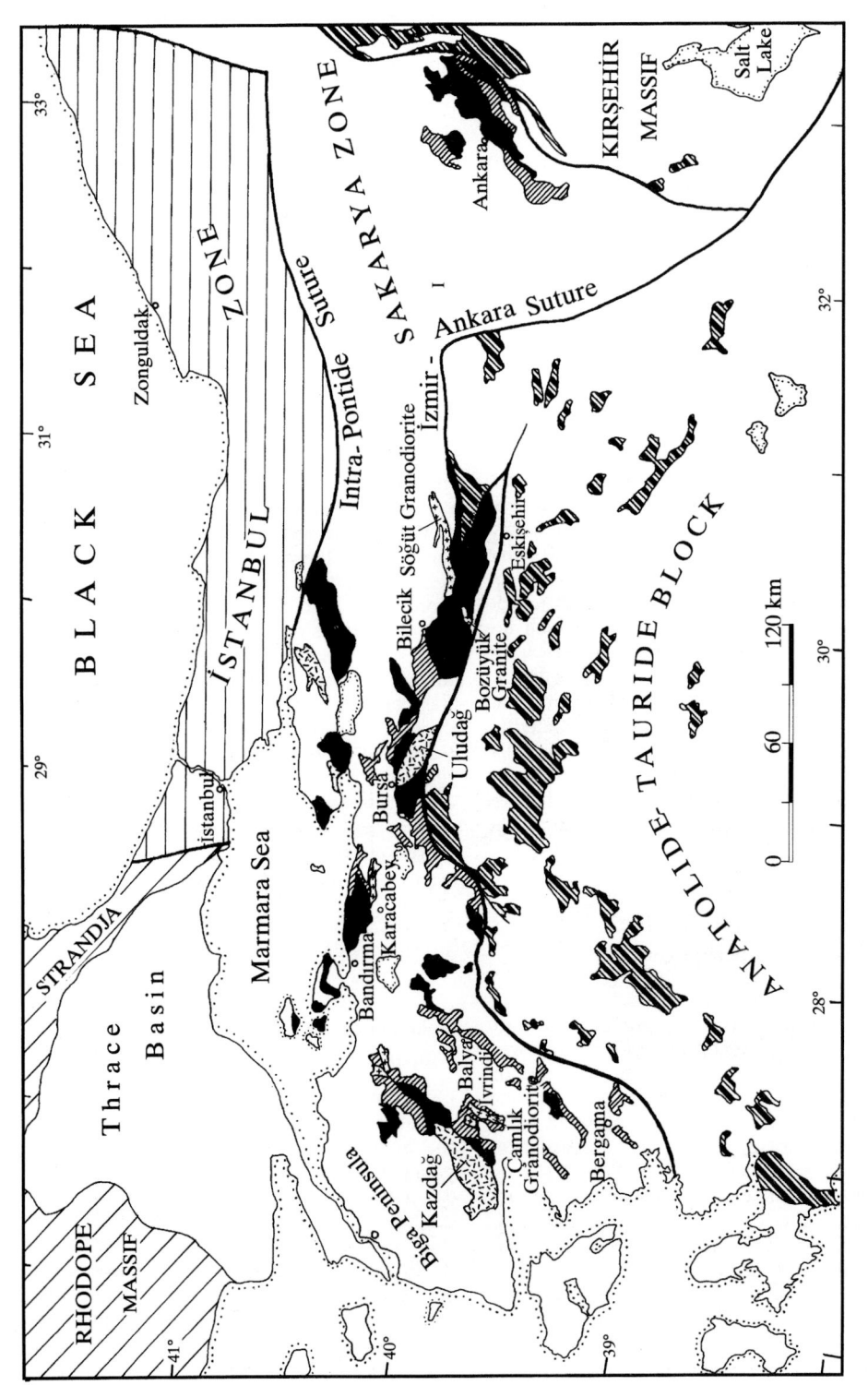

Fig. 2. Distribution of the pre-Jurassic outcrops in the western part of the Sakarya Zone and the Neotethyan subduction–accretion complexes and ophiolite in the Anatolide–Tauride Block. For key see Fig. 3.

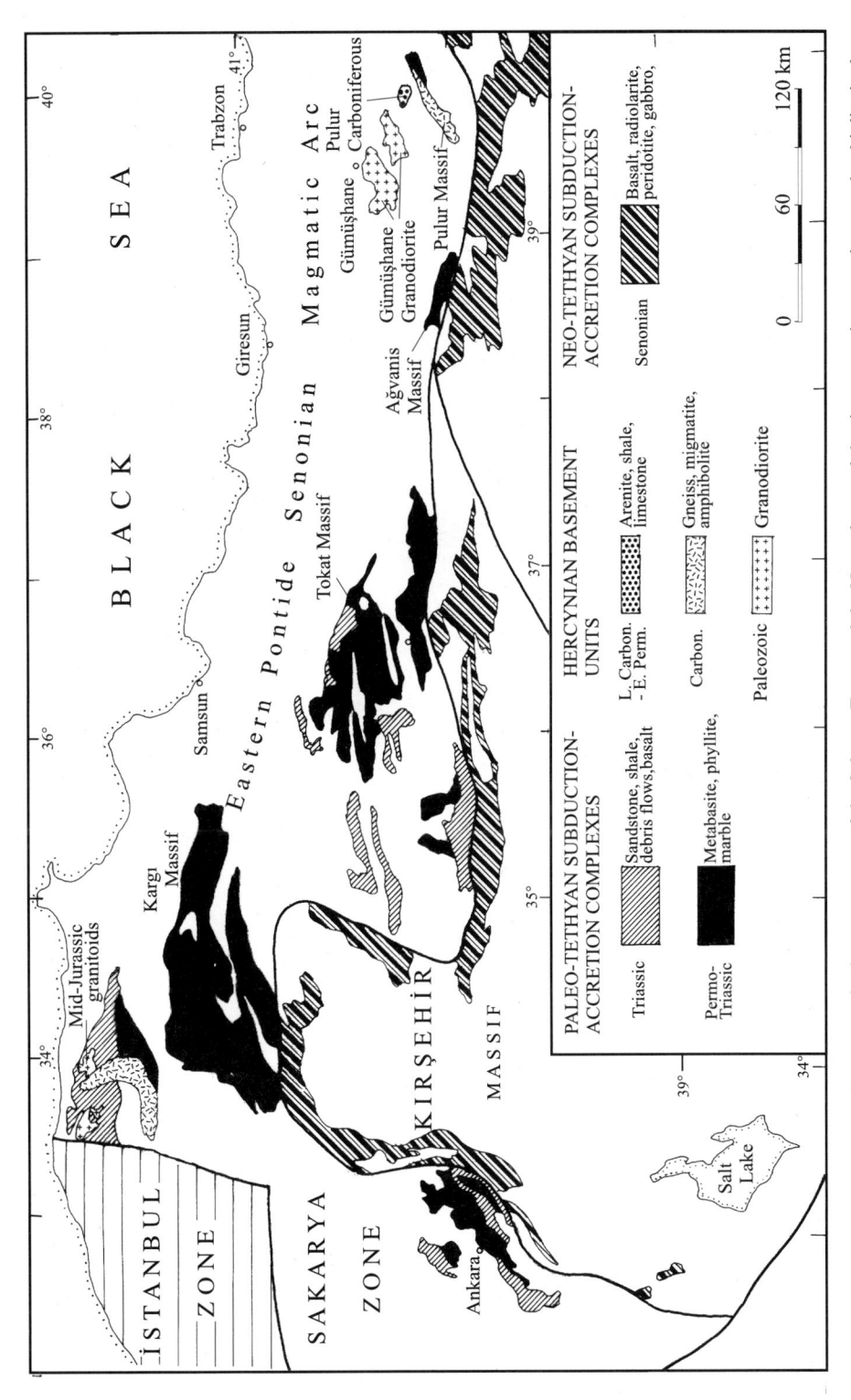

Fig. 3. Distribution of the pre-Jurassic outcrops in the eastern part of the Sakarya Zone, and the Neotethyan subduction–accretion complexes and ophiolite in the Anatolide–Tauride Block.

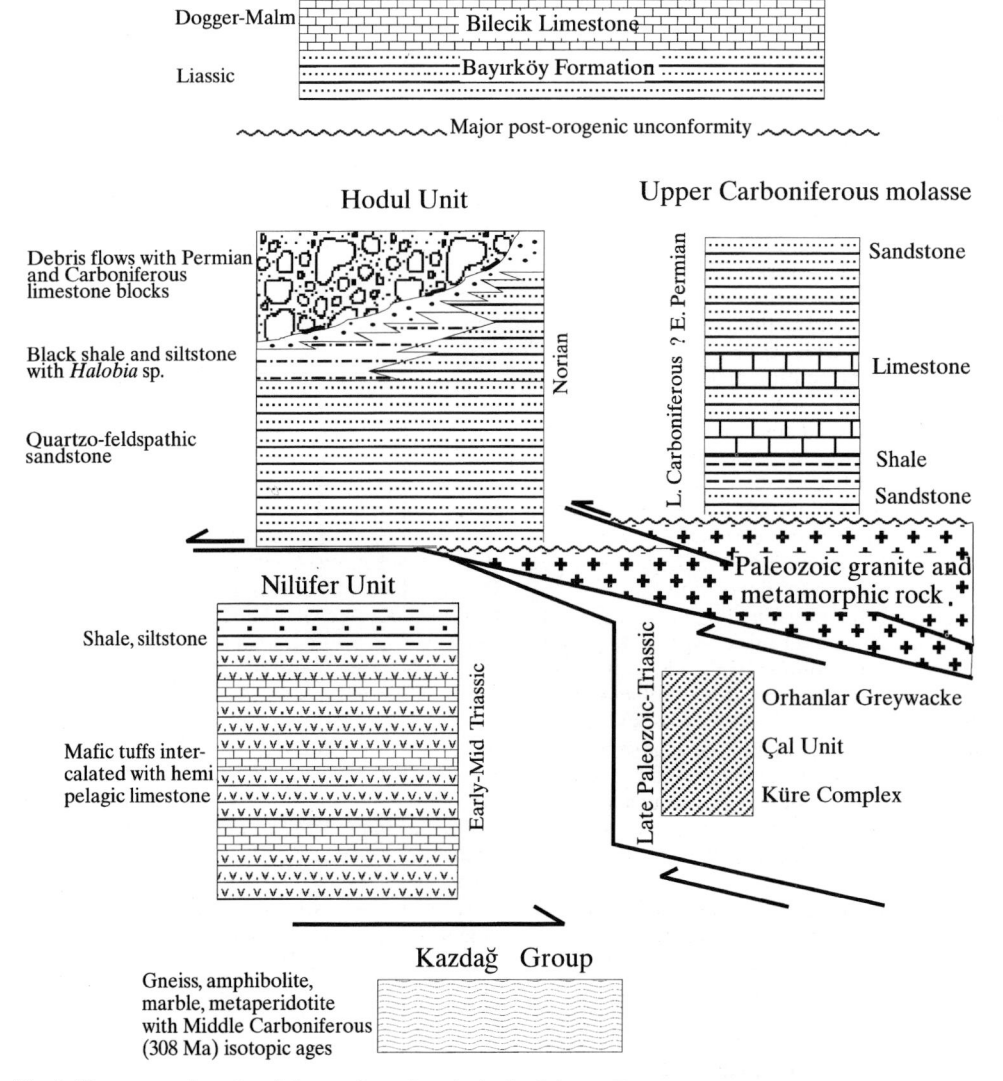

Fig. 4. Tectonostratigraphy of the pre-Jurassic units in the Sakarya Zone.

subduction–accretion of the Palaeotethys and those which formed during the collision of the oceanic plateau.

The Nilüfer Unit – a Triassic oceanic plateau?

At the base of the Karakaya Complex there is a strongly deformed metabasite–marble–phyllite unit of Triassic age, over 7 km in structural thickness. In the western part of the Sakarya Zone, between the Biga Peninsula and Bursa, this assemblage was mapped as the Nilüfer Unit (Okay et al. 1991, 1996; Leven & Okay 1996; Pickett & Robertson 1996). In northwest Turkey, the Nilüfer Unit is found wherever the base of the Triassic and older clastic and volcanic rocks are exposed (Fig. 2), suggesting that it forms a continuous layer at depth. Further east, the Nilüfer Unit is described under different names east of Bursa (Genç & Yılmaz 1995), north of Eskişehir (Monod et al. 1996; Monod & Okay 1999), around Ankara (Koçyiğit 1987, 1991; Akyürek et al. 1988), in the central Pontides (Tüysüz 1990; Ustaömer & Robertson 1994), in the Tokat (Yılmaz et al. 1997) and in the Agvanis massifs (Okay 1984)

(Figs 2 and 3). In all these localities the Nilüfer Unit forms the lowermost stratigraphic unit, suggesting stratal continuity at depth. The correlation of this metabasite–marble–phyllite unit across 1100 km in the Sakarya Zone is based on the present author's geological work in northwest Turkey (Okay et al. 1991, 1996; Leven & Okay 1996), north of Eskişehir (Monod et al. 1996; Monod & Okay 1999), in the Agvanis (Okay 1984) and in the Pulur massifs (Okay 1996; Okay & Şahintürk 1997), as well as on field reconnaissance in the Ankara region, in the central Pontides and in the Tokat Massif.

The Nilüfer Unit consists dominantly of metabasites, representing metamorphosed fine-grained mafic tuffs with rare pyroclastic flows and pillow lavas. These mafic rocks, which constitute about c. 80% of the sequence, are intercalated with carbonate layers, 0.5–200 m thick. The fine lamination, devoid of bioturbation, observed in many carbonate horizons, as well as the thin to medium bedding, suggests a pelagic environment of deposition. In the metabasites there are also phyllite horizons, up to several tens of metres thick. Apart from these rock types, the Nilüfer Unit also comprises minor (< 2%) metacherts, lenses of ultramafic rock and gabbro < 1 km long. Coarse-grained, continent-derived sediments, such as sandstones or conglomerates, are conspicuously absent in the Nilüfer Unit. In some regions, such as northeast of Kazdağ, the metabasites are overlain by phyllites and intercalated marbles, several hundred metres thick.

The uppermost part of the mafic volcanic sequence north of Bergama has been dated by conodonts in the intercalated limestones as Middle Triassic (Anisian–Ladinian boundary; Kaya & Mostler 1992), while conodonts from the lower parts of the Nilüfer Unit in the type section south of Bursa indicate an Early Triassic age (Kozur, pers. comm.). Thus, the mafic volcanic rocks of the Nilüfer Unit appear to have been generated in a relatively short period between 245 and 240 Ma. The trace element geochemistry of the mafic volcanic rocks of the Nilüfer Unit, from various regions in northwest Turkey, was studied by Pickett (1994), who showed that they are non-alkalic and non-orogenic in character and exhibit remarkably consistent geochemical characteristics. On basalt discrimination diagrams and multi-element plots they indicate a within-plate setting, which is supported by the analysis of relict igneous clinopyroxenes from the metabasites (Pickett 1994; Pickett & Robertson 1996). Triassic metabasites from the Ankara region (Çapan & Floyd 1985; Floyd 1993) and those from the Kargı Massif in the central Pontides (Doğan 1990) also exhibit trace element contents and ratios typical of within-plate oceanic island basalts.

The Nilüfer Unit is strongly deformed and in many areas shows the features of a broken formation. The early deformation is characterized by layer-parallel stretching, which resulted in the boudinage of the carbonate horizons. The stretching was followed by folding and shearing at high angles to the layering. Foliation is well marked in the mafic tuffs and phyllites, while the coarse-grained pyroclastic flows retain most of the igneous texture. Structural thicknesses > 7 km, as measured south of Bursa, are due to internal thrusting, as well as folding, although, because of the absence of marker horizons in the Nilüfer Unit, these internal thrusts are difficult to map. An exception occurs north of Eskişehir, where a tectonic slice of the Nilüfer Unit, 4 km thick and 25 km long, differentiated because of its blueschist facies metamorphism, tectonically underlies another slice of the Nilüfer Unit showing only greenschist facies metamorphism (Monod & Okay 1999).

The Nilüfer Unit commonly shows high-pressure greenschist facies metamorphism with the common mineral paragenesis of albite + chlorite + actinolite/barroisite + epidote in the metabasites. Rare, iron-rich metacherts in the Nilüfer Unit contain a sodic amphibole + quartz + epidote assemblage (Okay et al. 1996). Blueschist and eclogite facies rocks occur as exotic tectonic blocks and slices within the greenschist facies sequence. A small eclogite lens of garnet + glaucophane + omphacite + epidote + phengite occurs east of Bandırma on the southern coast of the Marmara Sea (Fig. 2; Okay & Monié 1997). A larger slice of the Nilüfer Unit, showing blueschist to eclogite facies metamorphism, is found north of Eskişehir (Monod et al. 1996; Monod & Okay 1999). Ar–Ar dating of phengites from both regions of high-pressure rocks, separated by 230 km, has given similar latest Triassic–earliest Jurassic ages [208–201 Ma in Bandırma and 214–192 Ma in Eskişehir; Monod et al. (1996) and Okay & Monié (1997)]. The greenschist facies metamorphism in the Nilüfer Unit postdates the high pressure–low temperature (HP–LT) metamorphism, and hence must also be of latest Triassic age. The regional metamorphism is further constrained stratigraphically as Late Triassic by the Middle Triassic depositional age of the Nilüfer Unit and the unconformable cover of the Liassic sandstones.

The Nilüfer Unit is overlain tectonically by two distinctive rock types. In many regions strongly deformed, but generally unmetamorphosed, Triassic clastic rocks of the Karakaya Complex lie over the Nilüfer Unit. These contacts have either been interpreted as tectonic (Okay *et al.* 1991, 1996) or as stratigraphic (Akyürek & Soysal 1983). In many localities, such as east of Bandırma and south of Bursa, the contacts are sheared and folded, although the consistent superposition of the Upper Triassic clastic rocks over the Nilüfer Unit suggests a sheared stratigraphic contact between the two units. In parts of northwest Turkey, such as in the Bilecik region, north of Karacabey, around Bozüyük (Fig. 2), the Nilüfer Unit is tectonically overlain by Palaeozoic granites (Yılmaz 1981; Genç & Yılmaz 1995).

The tectonic base of the Nilüfer Unit is exposed in two Miocene core complexes in the Kazdağ and Uludağ Ranges (Fig. 2). In both regions, Miocene and younger normal faults form the contact between the metabasites of the Nilüfer Unit and the underlying gneisses, amphibolites and marbles with Carboniferous zircon ages (Okay *et al.* 1996).

As discussed above, the scattered outcrops of the Nilüfer Unit are most probably connected at depth, and the Nilüfer Unit forms a blanket cover beneath Triassic and younger strata throughout most of the Sakarya Zone (Figs 2 and 3), and thus has an areal distribution of *c.* 120 000 km^2. Taking a minimum vertical thickness for the Nilüfer Unit as 2 km and considering that 80% of the sequence is made up of mafic magmatic rocks, the volume of basalt produced during the Early–Middle Triassic is estimated as 2×10^5 km^3. This is a large igneous province comparable in volume with the Columbia River basalts, the North Atlantic volcanic province, or the Deccan traps (Coffin & Eldholm 1994).

The Nilüfer Unit has been interpreted either as an ensimatic intra-arc to forearc sequence (Okay 1984; Okay *et al.* 1991, 1996), or as an oceanic seamount (Pickett & Robertson 1996). The thick and laterally extensive volcaniclastic sequences intercalated with sedimentary rocks, which form the bulk of the Nilüfer Unit, are characteristic of the sedimentary basins flanking active island arcs (e.g. Dickinson & Seely 1979). However, the geochemistry of the Nilüfer Unit does not indicate an arc affinity. On the other hand, a seamount origin for the Nilüfer Unit is unlikely as the seamounts have a maximum diameter of *c.* 35 km (e.g. Dominguez *et al.* 1998), while the Nilüfer Unit is more than 30 times this size. A third hypothesis for the origin of the Nilüfer Unit, suggested here, is that it represents the upper parts of a Triassic oceanic plateau, which was accreted to the Laurasian active continental margin during the Late Triassic.

Oceanic plateaux cover large areas of the present day oceanic basins (e.g. Ontong Java is 1.86×10^6 km^2, more than twice the size of Turkey), and are characterized by mafic crustal thicknesses > 10 km (Saunders *et al.* 1996; Gladczenko *et al.* 1997; Kerr *et al.* 1998). Because of their great crustal thicknesses and buoyancy, oceanic plateaux are less readily subducted than normal oceanic crust and are potentially preserved in the geological record (Burke *et al.* 1978; Ben-Avraham *et al.* 1981). Former oceanic plateaux have been described from Japan (e.g. Kimura *et al.* 1994), western North America (e.g. Ben-Avraham *et al.* 1981; Richards *et al.* 1991) and the Caribbean region (e.g. Kerr *et al.* 1998). During the subduction of the oceanic plateaux, it is usually the upper mafic volcanic layers that are detached from their plutonic substratum and accreted to the active continental margin (e.g. Tejada *et al.* 1996).

Evidence favouring an oceanic plateau origin of the Nilüfer Unit are as follows: (1) exclusive mafic magmatism associated with the absence of a known stratigraphic basement and lack of continent-derived detritus in the Nilüfer Unit suggest formation in an oceanic environment away from the continents; (2) intercalation of mafic rocks with hemipelagic limestone and shale is unusual for oceanic ridge magmatism, although it is to be expected in an off-axis intra-oceanic volcanism; (3) the areal distribution of the Nilüfer Unit and the volume of mafic magma produced is comparable with oceanic plateaux; (4) the geochemistry of the mafic rocks in the Nilüfer Unit show a consistent within-plate signature; (5) the presence of mafic eclogites and blueschists in the Nilüfer Unit implies association with an oceanic subduction zone.

The size of the Triassic oceanic plateau can be estimated from the outcrop distribution of the Nilüfer Unit. In the west outcrops of the Nilüfer Unit end abruptly in the Biga Peninsula; neither the equivalents of the Nilüfer Unit nor the Late Triassic deformation have been described from the mainland Greece. The easternmost extensive outcrop of the Nilüfer Unit is in the Agvanis Massif (Okay 1984); further east the Hercynian basement units are predominant, although small slivers of Nilüfer Unit have been described from the Pulur Massif (Fig. 3; Okay 1996). Between these two limits is an east–west length of *c.* 1100 km for the Nilüfer oceanic plateau. A minimum north–south length of *c.* 90 km for the

Nilüfer Plateau is indicated by the present-day width of the Sakarya Zone.

Subduction-related Palaeotethyan clastic and magmatic rocks

Subduction-related clastic and volcanic sequences include those accreted to the continental margin prior to the collision of the oceanic plateau, while the sequences deposited during the Late Triassic collision are designated as collision related. Differentiation of subduction- and collision-related Palaeotethyan sequences are often problematic, as subduction-related sequences are also involved in the collision. Nevertheless, such a distinction can be made in northwest Turkey, based on the age and lithology of the rocks. The subduction-related sequences are Late Palaeozoic–Triassic in age and are dominantly pelagic, while the collisional sequences are Late Triassic in age and also comprise shallow-marine series. The subduction-related sequences in northwest Turkey are the Orhanlar Greywacke and the Çal Unit, and in the central Pontides the Küre Complex. In addition to these formations there is a suite of Middle Jurassic granitoids in the northern part of the central Pontides, which may be related to the final phases of subduction of the Palaeotethys (Yılmaz & Boztuğ 1986).

The Orhanlar Greywacke. The Orhanlar Greywacke consists of homogeneous, grey siltstones and sandstones with a rich clay matrix (Brinkmann 1971; Okay et al. 1991, 1996). The greywackes are composed of very poorly sorted angular quartz, plagioclase, opaque, lydite, radiolarian chert, basalt and phyllite clasts in an argillaceous matrix. The Orhanlar Greywacke has undergone strong layer-parallel extension which has destroyed most of the bedding. In the type area north of Balya, the Orhanlar Greywacke contains small (< 1 m) olistoliths of dark Lower Carboniferous (Visean and Serpukhovian) limestone and is overlain by undeformed Liassic sandstones and siltstones (Leven & Okay 1996). South of Bursa, the Orhanlar Greywacke lies with a sheared stratigraphic contact over the Nilüfer Unit. It could represent trench fill clastics recycling material from the accretionary complex as well as from the arc.

The Çal Unit. Most of the Çal Unit consists of debris and grain flows with Upper Permian limestone and mafic clasts in a volcanic or volcano-clastic matrix (Okay et al. 1991, 1996; Leven & Okay 1996). There are also mafic pyroclastic flows, calciturbidites, pelagic limestones and radiolarian cherts in the sequence. Calciturbidites are made up of transported Upper Permian limestone grains. The radiolarian cherts, initially dated as Early Permian (Sakmarian–Artinskian) (Okay & Mostler 1994), are now redetermined as Late Permian (Dorashamian) (Kozur 1997). Associated with these rock types there are Anisian limestones, several hundred of metres thick. Like the other Karakaya units, the Çal Unit has been strongly deformed, largely destroying stratal continuity. It may represent an oceanic seamount, which was accreted to the Laurasian margin during the Middle Triassic (Leven & Okay 1996).

The Küre Complex. The Küre Complex, which crops out in the northern part of the central Pontides, consists mainly of dark shales intercalated with siltstones and fine-grained sandstones of pre-Middle Jurassic age. It includes tectonic slices of mafic pillow lava, gabbro and serpentinite (Ustaömer & Robertson 1994; Aydın et al. 1995), and represents a Franciscan-type subduction–accretion complex.

Kastamonu granitoids. These medium-sized plutons, < 20 km across, intrude the Upper Palaeozoic–Upper Triassic subduction–accretion complexes in the northern part of the central Pontides (Fig. 3), and are unconformably overlain by undated conglomerates and Upper Jurassic (Oxfordian) shallow-marine limestones (Yılmaz & Boztuğ 1986). They range from granodiorite to quartz–monzonite and show calc-alkaline geochemical features (Boztuğ et al. 1984). Yılmaz & Boztuğ (1986) related the formation of the granitoids to the northward subduction of the Palaeotethys.

Palaeotethyan clastic rocks formed during the collision of the oceanic plateau

The Hodul Unit

The Hodul Unit comprises collision-related clastic sequences in the western part of the Sakarya Zone, extending from the Biga Peninsula eastwards to the region north of Eskişehir (Okay et al. 1991, 1996). They form an easily recognizable unit with thick sequences of white arkosic sandstones, and extensive olistostromes with Permian and Carboniferous limestone blocks. The sequence is strongly deformed with anastomosing shear zones which have destroyed stratal continuity over distances as short as a few hundred metres. The clastic sequence ranges from coarse-grained and thickly bedded arkosic sandstones with pebbles of granite to distal turbidites with thinly bedded arkosic sandstones with shale intercalations (Okay et al. 1991). At several places in northwest Turkey the sequence is dated through macrofossils as Norian (Okay et al. 1991, 1996; Leven & Okay 1996). In regions near the İzmir–Ankara Suture, the arkosic sandstones pass up to greywackes and siltstones with exotic olistoliths of mafic volcanic rock and limestone (Okay et al. 1991, 1996; Fig. 2). The size of the limestone blocks in the Norian olistostromes ranges up to 1 km. A detailed palaeontological study of the foraminifera in the limestone blocks has shown the presence of almost all stages from Middle Carboniferous (Bashkirian) to uppermost Permian (Dorashamian) (Leven & Okay 1996). There are also rare blocks of pelagic limestone and radiolarite, which are dated by conodonts as Middle Carboniferous (Bashkirian) (Okay & Mostler 1994). The total thickness of the Hodul sequence is > 2 km: a more precise estimate is not possible because of the strong tectonism.

The arkosic nature of the clastic rocks, as well as granitic pebbles in the sandstones, indicate a granitic source area for the Hodul Unit. In only one locality in the Sakarya Zone, in the area west of Ivrindi, the arkosic sandstones of the Hodul Unit can be observed to lie transgressively over the Palaeozoic Çamlık Granodiorite (Fig. 2). In other regions, e.g. east of Bandırma, the Hodul Unit lies with a shear zone contact over the Nilüfer Unit. The Hodul Unit is overlain unconformably by undeformed shallow-water Liassic sandstones and siltstones.

Hercynian continental units in the Sakarya Zone

The Hercynian continental units of the Sakarya Zone consist of three different rock assemblages: Carboniferous high-grade metamorphic rocks, Palaeozoic granites and a latest Carboniferous molasse sequence.

Carboniferous high-grade metamorphic rocks

These consist mainly of gneiss, migmatite, amphibolite and marble with rare meta-ultramafic rocks. They crop out mainly in the cores of two Tertiary core complexes in the Kazdağ and Uludağ Ranges (Fig. 2), and also in a Tertiary thrust slice in the Pulur Massif in the eastern Pontides (Fig. 3). The metamorphic grade is amphibolite to granulite facies with cordierite + sillimanite + garnet + biotite subassemblages in the Kazdağ and Pulur regions (Okay 1996; Okay et al. 1996). The high-grade metamorphism is of Late Carboniferous age. Gneisses from the Kazdağ region, dated by the single-zircon stepwise Pb-evaporation technique, have yielded an age of 308 ± 16 Ma (Moscovian; Okay et al. 1996), while those from the Pulur region gave Sm–Nd and Rb–Sr ages between 303 and 322 Ma (Moscovian–Bashkirian; Topuz et al. 1997).

In both the Kazdağ and Uludağ regions, the Carboniferous high-grade metamorphic rocks are tectonically overlain by the Palaeotethyan subduction–accretion complexes, while in the Pulur Massif the metamorphic rocks are unconformably overlain by Lower–Middle Jurassic limestones (Okay & Şahintürk 1997).

Palaeozoic granodiorites

These granitoid rocks form several isolated bodies, up to 30 km across, in the Sakarya Zone (Figs 3 and 4). They are generally hornblende–biotite granodiorites and are locally strongly deformed (Yılmaz 1976, 1981; Servais 1982; Bergougnan 1987; Okay et al. 1991). The few available isotopic ages range from Devonian to Permian. The Çamlık Granodiorite in the Biga Peninsula has yielded a zircon age of 399 ± 13 Ma (Devonian) using a single-zircon stepwise evaporation technique (Okay et al. 1996); the Söğüt Granodiorite gave a K–Ar biotite age of 272 ± 3 Ma (Early Permian; Çogulu & Krummenacher 1967) and a total Pb age of 290 Ma (latest Carboniferous; Çogulu et al. 1965); the Gümüşhane Granodiorite in the eastern Pontides has yielded a well-defined Rb–Sr isochron age of 360 ± 2 Ma (earliest Carboniferous; Bergougnan 1987).

Most of the Palaeozoic granodiorites are unconformably overlain by Lower Jurassic continental to shallow-marine clastic rocks. Only the Çamlık Granodiorite in the Biga Peninsula also has an unconformable cover of Upper Triassic (Norian) arkosic sandstones (Okay et al. 1991, 1996; Fig. 2). Some of the granodiorites, like the Bozüyük and Söğüt bodies, have narrow contact aureoles, although the country rock into which these plutons initially intruded are very poorly preserved. In the few localities, where the Palaeozoic granodiorites are in contact with the Palaeotethyan subduction–accretion complexes (Fig. 2), the granodiorites are seen to be thrust over the Nilüfer Unit. A mylonite belt > 1000 m thick separates the Söğüt Granodiorite from the underlying Triassic metabasite–phyllite–marble unit (Yılmaz 1981).

Latest Carboniferous molasse

This coherent shallow-marine to continental sequence crops out only in the eastern Pontides (Ketin 1951). The lower part of the sequence, > 1100 m in thickness, consists of an intercalation of pebbly sandstone, quartzite, shale with thin coal horizons and shallow-marine limestone with fusulinids; this is conformably overlain by continental arkosic red sandstones, > 1000 m in thickness. Sandstones throughout the sequence contain clasts of acidic plutonic and volcanic rocks. Fusulinids from the limestones indicate a latest Carboniferous age (Late Kasimovian–Early Gzelian) for the lower part of the sequence: the overlying continental series may possibly extend into the Early Permian (Okay & Leven 1996; Okay & Şahintürk 1997). Although the base of the Upper Carboniferous sequence is not exposed, it probably unconformably overlies the nearby Carboniferous granites and high-grade metamorphic rocks in the Pulur and Gümüşhane regions (Fig. 3). Similar Upper Carboniferous sequences overlying metamorphic and granitic rocks are exposed in the Lesser Caucasus (Khain 1975), and can be compared with the Rotliegende Formation of Hercynian Europe.

Timing of the Cimmeride Orogeny in the Sakarya Zone

The age of the Cimmeride deformation in the Sakarya Zone can be constrained between the age of the youngest deformed strata of the Karakaya Complex and the oldest unconformably overlying undeformed beds. The age of the deformed olistostromes in the Hodul Unit in northwest Turkey is established at several localities as Norian. This is based on macrofauna, such as *Halobia suessi* (Mojsisovics), *Pinacoceras postparma* (Mojsisovics) and *Pseudocardioceras acutum* (Mojsisovics) from the Balya region (Leven & Okay 1996), and *Halobia styriaca* (Mojsisovics) from north of Bursa (Erk 1942). A more precise Middle–Late Norian age has been obtained from the Hodul Unit south of Ivrindi, based on the forms *Zugmayerella* sp., *Anadontophora* cf. *griesbachi* (Bittner), *Amonotis*(?) sp. and *Gonionautilus securis* (Dittmar) (Leven & Okay 1996).

The age of the oldest unconformably overlying undeformed series in northwest Turkey (Bayırköy Formation) is firmly established as Sinemurian, based on ammonites, brachiopods and foraminifera (Alkaya 1982; Altıner *et al.* 1991). The Sinemurian sequence extends upwards semi-continuously to the Late Cretaceous, with no significant deformation until the Senonian. This brackets the age of the Cimmeride Orogeny between the latest Norian and Hettangian (215–200 Ma). Only in the central Pontides, where the undeformed sequence overlying the subduction–accretion units is of Middle Jurassic age, is the Cimmeride deformation probably slightly later (Tüysüz 1993).

Ar–Ar dating of phengites from the eclogite in the Nilüfer Unit east of Bandırma gave a narrow metamorphic age between 208 and 201 Ma (Okay & Monié 1997). A similar isotopic age range was obtained (214–192 Ma) from phengites from the blueschists north of Eskişehir (Monod *et al.* 1996). The eclogite was metamorphosed at a temperature of *c.* 480°C, which is close to the closure temperature for Ar in phengite [generally taken to be in the 350–400°C range (Okay & Monié 1997)]. Therefore, these isotopic ages, which correspond to the Late Norian–Early Liassic (Hettangian) (Gradstein *et al.* 1994), reflect an early stage of cooling and are compatible with the palaeontological ages of the deformation.

The Cimmeride Orogeny in northern Turkey is remarkable for its relatively brief deformation and metamorphism, and in this respect contrasts strongly with orogenies like the Alpide or the Himalayan, which have life spans > 50 Ma.

The duration of the Palaeotethyan subduction and the age of the subducting oceanic crust

The duration of subduction is best given by the age range of the magmatic arc and the forearc sequence: in the Sakarya Zone neither are recognized. In their absence, the Late Triassic age of the high pressure metamorphism provides a minimum duration for the subduction. The age range of the subducting oceanic crust can be deduced from the age of the pelagic sediments in the accretionary complexes. Middle Carboniferous (Bashkirian) and Upper Permian radiolaria are described from chert blocks in the Karakaya Complex (Kozur & Kaya 1994; Okay & Mostler 1994; Kozur 1997), suggesting subduction of an oceanic crust at least as old as Middle Carboniferous. This is compatible with palaeogeographic reconstructions which indicate the continued presence of a large oceanic area north of Gondwana in the Carboniferous and Permian (e.g. Smith *et al.* 1981; Stampfli 1996).

Origin of the Carboniferous and Permian limestone blocks in the Upper Triassic olistostromes

Voluminous olistostromes with Carboniferous and Permian neritic limestone blocks occur in the upper parts of the Upper Triassic arkosic clastic sequences in northwest Turkey (Okay *et al.* 1991, 1996; Leven & Okay 1996). The origin of these exotic blocks, e.g. whether they were initially deposited on the southern or northern margin of the Palaeotethys and how they were incorporated into the Hodul Unit, cannot be satisfactorily answered because of apparently conflicting evidence. Data which argue for a southerly origin of the Carboniferous and Permian limestone blocks are as follows. (1) The Triassic olistostromes form a belt, 15–25 km wide and 280 km long, immediately northwest of the İzmir–Ankara Suture (Fig. 2). The density and the size of the exotic limestone blocks decrease northward and westward away from the İzmir–Ankara Suture. These observations suggest that the blocks in the olistostromes were derived from a carbonate platform that lay in the present southeast, e.g. in the direction of the Anatolide–Tauride Block. (2) There is a general westward increase in the age of the limestone blocks from Midian–Dzhulfian immediately adjacent to the İzmir–Ankara Suture to Murgabian–Midian further northwest (Leven & Okay 1996). Assuming that the top of the carbonate platform was eroding first, this again argues for a southward derivation of the blocks. (3) No marine Permian deposits are known along the Laurasian margin in the western Palaeotethys. In Bulgaria, the Permian is represented by terrigenous clastic and volcanic rocks (e.g. Yanev 1992), and, in the İstanbul Zone, Permian deposits are largely absent and the Early Triassic and possibly latest Permian consist of red beds and basic lavas. In contrast, Permian and Carboniferous deposits in the Anatolide–Tauride Block are generally marine carbonates (e.g. Monod 1977; Argyriadis 1978; Altıner 1983). On the other hand, Permian fusulinid assemblages from the exotic limestone blocks in the Hodul Unit are faunistically different to those in the Anatolide–Tauride Block,

and are similar to the fusulinid assemblages found in the northern margin of the eastern Palaeotethys in Afghanistan, the Pamirs and China (Leven & Okay 1996).

Assuming that the exotic Permo-Carboniferous limestone blocks in the Hodul Unit were deposited in the northern outer shelf of the Anatolide–Tauride Block, then a narrow sliver of this outer shelf must have rifted, possibly during the Early Triassic mafic magmatism, from the bulk of the Anatolide–Tauride Block. This Cimmerian continental sliver must have been translated northward and must have eventually been incorporated into the Upper Triassic olistostromes (Okay *et al.* 1996). However, a major problem with this tectonic scenario is that nowhere in the Sakarya Zone is there evidence for a Gondwanan continental basement, which must have underlain the Permo-Carboniferous limestones.

A model for the Cimmeride orogeny in Turkey

A basic tenet of the model presented here is that the Cimmeride deformation in Turkey is caused by the collision and partial accretion of an oceanic plateau with the Laurasian active continental margin during the Late Triassic. Accretion of an oceanic plateau to the Laurasian margin would imply a northward dipping subduction zone during the Late Triassic, as suggested by Robertson & Dixon (1984), Ricou (1994) and Stampfli *et al.* (1991) and others. The evidence in favour for the oceanic plateau origin of the Nilüfer Unit has been cited above. The alternative model for the Cimmeride deformation, involving the collision of a Cimmerian continent with the Laurasian margin (Şengör 1979; Şengör *et al.* 1984), is difficult to maintain as no coherent Cimmerian continent can be defined in northern Turkey. The pre-Jurassic rocks in the Sakarya Zone are either subduction–accretion complexes or fragments of a Hercynian continental sequence belonging to Laurasia. Furthermore, the comparatively short period of deformation and metamorphism in northern Turkey (< 15 Ma) would be incompatible with a continent–continent collision but would be explicable through the docking and partial accretion of a oceanic plateau to the Laurasian margin. It is also striking that neither the Late Triassic deformation nor the presence of the Nilüfer Unit have been documented from mainland Greece or from the Lesser Caucasus, illustrating a causal relation between the two.

Figure 5 shows two schematic cross-sections

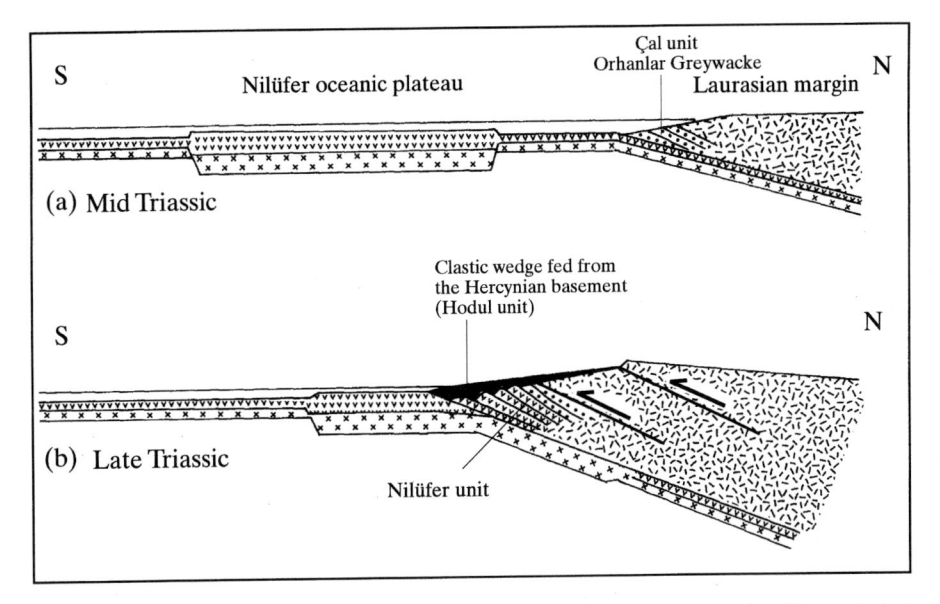

Fig. 5. Schematic cross-sections illustrating the tectonic evolution of the southern margin of Laurasia during: (**a**) Middle Triassic; and (**b**) Late Triassic.

illustrating the tectonic evolution of the Nilüfer oceanic Plateau. During the Early–Middle Triassic, the Nilüfer oceanic Plateau was generated during a voluminous intraplate mafic magmatism. This appears to have been part of a large Triassic mafic magmatic pulse in the Tethyan region. Triassic mafic tuffs, > 1000 m in thickness, are described from the allochthons in the Taurides (Huğlu Unit; Monod 1977; Gökdeniz 1981; Özgül 1997) and they are also widespread in the Triassic sections in Greece. The geochemistry of some Triassic mafic volcanic rocks in the eastern Mediterranean have a plume signature (Dixon & Robertson 1999) suggesting that the Triassic mafic magmatism was plume related. It is also possible, especially if a southerly origin is assumed for the exotic Permo-Carboniferous limestone blocks, that part of the outer shelf of the Anatolide–Tauride Block was also rifted during the Early Triassic magmatism.

In the Late Triassic, the northward movement of the Palaeotethyan plate brought the Nilüfer oceanic plateau near the Laurasian active margin. At this time the Laurasian margin was characterized by a Hercynian granitic and metamorphic basement with isolated molassic basins of Late Carboniferous–Permian age, and a thin cover of Triassic continental to shallow-marine sediments, e.g. as observed in Bulgaria or in the İstanbul Zone (e.g. Ganev 1974; Chatalov 1991). Subduction–accretion units of the Karakaya Complex, such as the Orhanlar Greywacke or the Çal Unit, were probably already part of the accretionary complex at the southern margin of the Laurasia (Fig. 5a). The collision of the Nilüfer oceanic Plateau with the active margin occurred during the Late Triassic (Norian), as indicated by the age of the syncollisional clastic sequences and the age of the HP–LT metamorphism. The volcanic edifice of the oceanic plateau was detached from its plutonic substratum and formed a thick accretionary wedge at the southern margin of Laurasia, represented by the Nilüfer Unit (Fig. 5b). The attempted subduction of a major oceanic plateau resulted in uplift and severe compressive deformation of the overlying active continental margin (e.g. Ben-Avraham et al. 1981; Cloos 1993). The arc and forearc sequences were probably largely eroded during the latest Triassic, and the Hercynian crystalline basement was internally sliced. A thick clastic wedge fed from the south vergent Hercynian thrust slices engulfed the subduction zone (Fig. 5b). This part of the model explains the unusual position of the arkosic sandstones of the Hodul Unit overlying the Nilüfer Unit, as well as the Carboniferous granites. The presence of Hercynian basement units both above and below the Nilüfer Unit suggests interdigitation of the two during the Late Triassic.

Conclusions

It is here proposed that the Late Triassic Cimmeride deformation and metamorphism in northern Turkey was caused by the collision and accretion of an Early–Middle Triassic oceanic plateau to the southern margin of Laurasia. The upper volcanic edifice of the oceanic plateau is preserved in northern Turkey as a very thick and extensive metabasite–marble–phyllite unit. Oceanic plateaux are common in the present-day Pacific Ocean and former examples have been described from Japan, western North America and the Caribbean region. This is apparently the first possible example of an ancient oceanic plateau from the Tethys Ocean.

An implication of the model presented here is that the İzmir–Ankara–Erzincan Suture represents both the Palaeo- and Neotethyan sutures. A suture can be defined as the site of a former ocean basin separating two continental plates (e.g. Moores 1981). It is expressed as a major tectonic line that forms a profound stratigraphic, structural, metamorphic and magmatic divide. The Araso Schuppen Zone, parts of the Insubric Line in the Western Alps (Trümpy 1975) and the Indus–Tsangpo Fault Zone in the Himalaya (e.g. Le Fort 1989) are examples of major sutures. The İzmir–Ankara–Erzincan Suture in Turkey is another example of a latest Cretaceous–Palaeocene suture across which stratigraphic, structural and metamorphic correlations are not possible (Şengör & Yılmaz 1981; Okay & Tüysüz 1999). However, as the Cimmeride Orogeny in Turkey was an accretionary rather than collisional orogeny, a Palaeotethyan suture as distinct from the Neotethyan suture does not exist. The Palaeotethyan suture shown on the maps (e.g. Şengör *et al.* 1984) does not correspond to a mappable tectonic line in the field, and does not form any stratigraphic, structural, metamorphic or magmatic boundary. Palaeotethyan subduction–accretion complexes, as well as the Hercynian continental units, occur on both sides of the supposed 'Palaeotethyan' suture line.

An important problem is the spatial and temporal relation between the Palaeo- and Neotethys. Accretionary complexes, consisting mainly of basalt, radiolarian chert, pelagic shale and serpentinite, occur widely to the south of the İzmir–Ankara Suture (Figs 2 and 3). Dating of radiolarian cherts from these accretionary complexes have yielded only Norian and younger ages, giving the duration of the İzmir–Ankara Neotethyan Ocean (Bragin & Tekin 1996; Bozkurt *et al.* 1997). The absence of Palaeozoic pelagic sedimentary rocks in the Neotethyan accretionary complexes reinforces the generally accepted view that the Neotethys opened as a separate ocean during the Triassic. It is possible that the narrow continental sliver with the Permo-Carboniferous limestone carapace, which may have rifted off from the northern margin of the Anatolide–Tauride Block during the Early Triassic, was responsible for the opening of the Neotethys and the closing of the Palaeotethys during its northward drift (Okay *et al.* 1996).

My fieldwork in northern Turkey over many years was generously supported by the Mineral Research and Exploration Institute of Turkey (MTA), the Turkish Petroleum Corporation (TPAO) and, in the last few years, by a grant from The Scientific and Research Council of Turkey (TÜBİTAK)-Glotek. Discussions with O. Monod, N. Okay, G. Stampfli and O. Tüysüz on various aspects of the Palaeotethyan problems are gratefully acknowledged. I also thank D. Altıner and G. Stampfli for critical reviews of this paper.

References

AKYÜREK, B. & SOYSAL, Y. 1983. Basic geological features of the region south of the Biga peninsula (Savaştepe–Kırkağaç–Bergama–Ayvalık). *Mineral Research and Exploration Institute of Turkey (MTA) Bulletin*, **95/96**, 1–13 [in Turkish with English abstract].

——, AKBAŞ, B. & DAGER, Z. 1988. *Çankırı-E16 Quadrangle, 1:100 000 Scale Geological Map and Explanatory Text.* Mineral Research and Exploration Institute of Turkey (MTA) Publications [in Turkish with English abstract].

ALKAYA, F. 1982. Taxonomic revision of the Lower Jurassic (Liassic) Phylloceratids of northern Turkey (Part I). *Geological Society of Turkey Bulletin*, **25**, 31–40 [in Turkish with English abstract].

ALTINER D. 1983. Upper Permian foraminiferal biostratigraphy in some localities of the Taurus Belt. *In*: TEKELİ, O. & GÖNCÜOĞLU, M. C. (eds) *Geology of the Taurus Belt*. Proceedings of the International Tauride Symposium. Mineral Research and Exploration Institute of Turkey (MTA) Publications, 255–268.

——, KOÇYİĞİT, A., FARINACCI, A., NICOSIA, U. & CONTI, M. A. 1991. Jurassic, Lower Cretaceous stratigraphy and palaeogeographic evolution of the southern part of north-western Anatolia. *Geologica Romana*, **28**, 13–80.

ARGYRIADIS, I. 1978. *Le Permien Alpino-Méditerranéen à la charnière entre l'Hercynien et l'Alpin*. Thèse, l'Université de Paris Sud, Centre d'Orsay.

AYDIN, M., DEMİR, O., ÖZÇELİK, Y., TERZİOĞLU, N. & SATIR, M. 1995. A geological revision of İnebolu, Devrekani, Ağlı and Küre areas: new observations in Palaeotethys–Neotethys sedimentary successions. *In*: ERLER, A., ERCAN, T., BİNGÖL, E. & ÖRÇEN, S. (eds) *Geology of the Black Sea*

Region. Proceedings of the International Symposium on the Geology of the Black Sea Region, Mineral Research and Exploration Institute of Turkey (MTA) Publications, 33–38.

BEN-AVRAHAM, Z., NUR, A., JONES, D. & COX, A. 1981. Continental accretion: from oceanic plateaus to allochthonous terranes. *Science*, **213**, 47–54.

BERGOUGNAN, H. 1987. *Etudes géologiques dans l'est-Anatolien*. PhD Thesis, University Pierre et Marie Curie, Paris, France.

—— & FOURQUIN, C. 1982. Remnants of a pre-Late Jurassic ocean in northern Turkey: fragments of Permian–Triassic Neotethys? Discussion. *Geological Society of America Bulletin*, **93**, 929–936.

BİNGÖL, E., AKYÜREK, B. & KORKMAZER, B. 1975. Geology of the Biga peninsula and some characteristics of the Karakaya blocky series. *Proceedings of the 50th Anniversary of Turkish Republic Earth Sciences Congress*, Mineral Research and Exploration Institute of Turkey Publications, 70–75 [in Turkish with English abstract].

BOZKURT, E., HOLDSWORTH, B. K. & KOÇYİĞİT, A. 1997. Implications of Jurassic chert identified in the Tokat Complex, northern Turkey. *Geological Magazine*, **134**, 91–97.

BOZTUĞ, D., DEBON, F., LE FORT, P. & YILMAZ, O. 1984. Geochemical characteristics of some plutons from the Kastamonu granitoid belt (Northern Anatolia, Turkey). *Schweizerische Mineral-ogische und Petrographische Mitteilungen*, **64**, 389–404.

BRAGIN, N. Y. & TEKİN, U. K. 1996. Age of radiolarian chert blocks from the Senonian ophiolitic mélange (Ankara, Turkey). *Island Arc*, **5**, 114–122.

BRINKMANN, R. 1971. Jungpaläozoikum und alteres Mesozoikum in NW-Anatolien. *Mineral Research and Exploration Institute of Turkey (MTA) Bulletin*, **76**, 56–67.

BURKE, K., FOX, P. J. & ŞENGÖR, A. M. C. 1978. Buoyant ocean floor and the origin of the Caribbean. *Journal of Geophysical Research*, **83**, 3949–3954.

ÇAPAN, U. Z. & FLOYD, P. A. 1985. Geochemical and petrographic features of metabasalts within units of Ankara melange, Turkey. *Ofioliti*, **10**, 3–18.

CHATALOV, G. A. 1988. Recent developments in the geology of the Strandzha Zone in Bulgaria. *Bulletin of the Technical University of İstanbul*, **41**, 433–465.

—— 1991. Triassic in Bulgaria – a review. *Bulletin of the Technical University of İstanbul*, **44**, 103–135.

CLOOS, M. 1993. Lithospheric buoyancy and collisional orogenesis: subduction of oceanic plateaus, continental margins, island arcs, spreading ridges, and seamounts. *Geological Society of America Bulletin*, **105**, 715–737.

COFFIN, M. F. & ELDHOLM, O. 1994. Large igneous provinces: crustal structure, dimensions, and external consequences. *Reviews of Geophysics*, **32**, 1–36.

ÇOĞULU, E. & KRUMMENACHER, D. 1967. Problèmes géochronométriques dans le partie NW de l'Anatolie Centrale (Turquie). *Schweizerische Mineralogische und Petrographische Mitteilungen*, **47**, 825–831.

——, DELALOYE, M. & CHESSEX, R. 1965. Sur l'age de quelques roches plutoniques acides dans la région d'Eskişehir, Turquie. *Archives des Sciences, Genève*, **18**, 692–699.

DEAN, W. T., MARTIN, F., MONOD, O., DEMİR, O., RICKARDS, A. B., BOLTYNCK, P. & BOZDOĞAN, N. 1997. Lower Palaeozoic stratigraphy, Karadere–Zirze area, central Pontides, northern Turkey. *In:* GÖNCÜOĞLU, M. C. & DERMAN, A. S. (eds) *Early Palaeozoic Evolution in NW Gondwana*. Turkish Association of Petroleum Geologists, Special Publications, **3**, 32–38.

DICKINSON, W. R. & SEELY, D. R. 1979. Structure and stratigraphy of the fore-arc regions. *AAPG Bulletin*, **63**, 2–31.

DIXON, J. E. & ROBERTSON, A. H. F. 1999. Are multiple plumes implicated in the Triassic break-up of the Gondwanan margin in the Eastern Mediterranean region. *Journal of Conference Abstracts*, **4**, 314.

DOĞAN, M. 1990. *Petrologic, magmatic and tectonic modelling of the Palaeotethyan Ophiolites from the Pontides*. PhD Thesis, İstanbul University [in Turkish with English abstract].

DOMINGUEZ, S., LALLEMAND, S. E., MALAVIEILLE, J. & VON HUENE, R. 1998. Upper plate deformation associated with seamount subduction. *Tectonophysics*, **293**, 207–224.

ERDOĞAN, B., ALTINER, D., GÜNGÖR, T. & ÖZER, S. 1990. Stratigraphy of the Karaburun Peninsula. *Mineral Research and Exploration Institute of Turkey (MTA) Bulletin*, **111**, 1–24 [in Turkish with English abstract].

ERK, A. S. 1942. *Etude géologique de la région entre Gemlik et Bursa (Turquie)*. Mineral Research and Exploration Institute of Turkey (MTA), Special Publications, Seri B, **9**.

FLOYD, P. A. 1993. Geochemical discrimination and petrogenesis of alkalic basalt sequences in part of the Ankara mélange, central Turkey. *Journal of the Geological Society, London*, **150**, 541–550.

GANEV, M. 1974. Stand der Kenntnisse über die Stratigraphie der Trias Bulgariens. *In:* ZAPFE, H. (ed.) *Die Stratigraphie der alpin–mediterranean Trias*. Springer, 93–96.

GENÇ, S. C. & YILMAZ, Y. 1995. Evolution of the Triassic continental margin, northwest Anatolia. *Tectonophysics*, **243**, 193–207.

GLADCZENKO, T. Z., COFFIN, M. F. & ELDHOLM, O. 1997. Crustal structure of the Ontong Java Plateau: modelling of new gravity and existing seismic data. *Journal of Geophysical Research*, **102**, 22 711–22 729.

GÖKDENİZ, S. 1981. *Recherches géologiques dans les Taurus occidentales entre Karaman et Ermenek, Turquie. Le series a 'tuffites vertes' Triassiques*. Thèse, l'Université de Paris Sud, Centre d'Orsay.

GÖRÜR., N., MONOD, O., OKAY, A. İ *ET AL.* 1997. Palaeogeographic and tectonic position of the Carboniferous rocks of the western Pontides (Turkey) in the frame of the Variscan belt.

Bulletin de la Société Géologique de France, **168**, 197–205.

GRADSTEIN, F. M., AGTERBERG, F. P., OGG, J. G., HARDENBOL, J., VAN VEEN, P., THUERRY, J. & HUANG, Z. 1994. A Mesozoic time scale. *Journal of Geophysical Research*, **99**, 24 051–24 074.

GUTNIC, M., MONOD, O., POISSON, A. & DUMONT, J. F. 1979. Géologie des Taurides Occidentales (Turquie). *Mémoires de la Société Géologique de France*, **137**, 1–112.

HAAS, W. 1968. Das Alt-Paläozoikum von Bithynien (Nordwest Türkei). *Neues Jahrbuch für Geologie und Paläontologie, Abhandlungen*, **131**, 178–242.

KAYA, O. & MOSTLER, H. 1992. A Middle Triassic age for low-grade greenschist facies metamorphic sequence in Bergama (İzmir), western Turkey: the first palaeontological age assignment and structural–stratigraphic implications. *Newsletter for Stratigraphy*, **26**, 1–17.

KETİN, İ. 1951. Über die Geologie der Gegend von Bayburt in Nordost-Anatolien. *İstanbul University, Faculty of Science Bulletin, Seri B*, **16**, 113–127.

KERR, A. C., TARNEY, J., NIVIA, A., MARRINER, G. & SAUNDERS, A. D. 1998. The internal structure of oceanic plateaus: inferences from obducted Cretaceous terranes in western Columbia and the Caribbean. *Tectonophysics*, **292**, 173–188.

KHAIN, V. E. 1975. Structure and main stages in the tectono–magmatic development of the Caucasus: an attempt at geodynamic interpretation. *American Journal of Science*, **275-A**, 131–156.

KIMURA, G., SAKAKIBARA, M. & OKAMURA, M. 1994. Plumes in central Panthalassa? Deductions from accreted oceanic fragments in Japan. *Tectonics*, **13**, 905–916.

KOÇYİĞİT, A. 1987 Tectonostratigraphy of the region of Hasanoğlan (Ankara): evolution of the Karakaya orogenic belt (in Turkish). *Hacettepe University Earth Sciences*, **14**, 269–294 [in Turkish with English abstract].

—— 1991. An example of an accretionary forearc basin from northern Central Anatolia and its implications for the history of subduction of Neotethys in Turkey. *Geological Society of America Bulletin*, **103**, 22–36.

KOZUR, H. 1997. Pelagic Permian and Triassic of the western Tethys and its palaeogeographic and stratigraphic significance. *Abstracts, XLVIII. Berg- und Hüttenmannischer Tag, Technische Universitat Bergakademie Freiberg*, 21–25.

—— & KAYA, O. 1994. First evidence of pelagic Late Permian conodonts from NW Turkey. *Neues Jahrbuch für Geologie und Paläontologie Monatshefte*, 339–347.

LEVEN, E. JA. & OKAY, A. İ. 1996. Foraminifera from the exotic Permo-Carboniferous limestone blocks in the Karakaya Complex, northwest Turkey. *Rivista Italiana Palaeontologia e Stratigrafia*, **102**, 139–174.

LE FORT, P. 1989. The Himalayan orogenic segment. *In*: ŞENGÖR, A. M. C. (ed.) *Tectonic Evolution of the Tethyan Region*. Kluwer, 289–386.

MONOD, O. 1977. *Recherches géologiques dans le taurus occidental au sud de Beyşehir (Turquie)*. Thèse, l'Université de Paris Sud, Centre d'Orsay.

—— & AKAY, E. 1984. Evidence for a Late Triassic–Early Jurassic orogenic event in the Taurides, *In*: DIXON, J. E. & ROBERTSON, A. H. F. (eds) *The Geological Evolution of the Eastern Mediterranean*. Geological Society, London, Special Publications, **17**, 113–122.

—— & OKAY, A. İ. 1999. Late Triassic Palaeotethyan subduction: evidence from Triassic blueschists. *Journal of Conference Abstracts*, **4**, 315–316.

——, ——, MALUSKI, H., MONIÉ, P. & AKKÖK, R. 1996. Schistes bleus du Trias supérieur en Turquie du NW: comment s'est fermée la Palaeo-Téthys? *16th Réunion des Sciences de la Terre*, 10–12 April 1996, Orléans, France, Abstracts, 43.

MOORES, E. M. 1981. Ancient suture zones within continents. *Science*, **213**, 41–46.

OKAY, A. İ. 1984. The geology of the Agvanis metamorphic rocks and neighbouring formations. *Mineral Research Exploration Institute of Turkey (MTA) Bulletin*, **99/100**, 16–36.

—— 1996. Granulite facies gneisses from the Pulur region, Eastern Pontides. *Turkish Journal of Earth Sciences*, **5**, 55–61.

—— & LEVEN, E. JA. 1996. Stratigraphy and palaeontology of the Upper Palaeozoic sequence in the Pulur (Bayburt) region, Eastern Pontides. *Turkish Journal of Earth Sciences*, **5**, 145–155.

—— & MONIÉ, P. 1997. Early Mesozoic subduction in the Eastern Mediterranean: evidence from Triassic eclogite in northwest Turkey. *Geology*, **25**, 595–598.

—— & MOSTLER, H. 1994. Carboniferous and Permian radiolarite blocks in the Karakaya Complex in northwest Turkey. *Turkish Journal of Earth Sciences*, **3**, 23–28.

—— & ŞAHİNTÜRK, Ö. 1997. Geology of the Eastern Pontides. *In*: ROBINSON, A. (ed.) *Regional and Petroleum Geology of the Black Sea and Surrounding Region*. AAPG Memoir, **68**, 291–311.

—— & TÜYSÜZ, O. 1999. Tethyan sutures of northern Turkey. *In*: DURAND, B., JOLIVET, L., HORVÁTH, F & SÉRANNE, M. (eds) *The Mediterranean Basin: Tertiary Extension within the Alpine Orogen*. Geological Society, London, Special Publications, **156**, 475–515.

——, ŞENGÖR, A. M. C. & GÖRÜR, N. 1994. Kinematic history of the opening the Black Sea and its effect on the surrounding regions. *Geology*, **22**, 267–270.

——, SİYAKO, M. & BÜRKAN, K. A. 1991. Geology and tectonic evolution of the Biga peninsula, northwestern Turkey. *Bulletin of the Technical University of İstanbul*, **44**, 91–256.

——, SATIR, M., MALUSKI, H., SİYAKO, M., MONIÉ, P., METZGER, R. & AKYÜZ, S. 1996. Palaeo- and Neotethyan events in northwest Turkey: geological and geochronological constraints. *In*: YIN, A. & HARRISON, M. (eds) *Tectonics of Asia*. Cambridge University Press, 420–441.

ÖZGÜL, N. 1997. Stratigraphy of the tectono–stratigraphic units in the region Bozkır–Hadım–

Takent (northern central Taurides). *Mineral Research and Exploration Institute of Turkey (MTA) Bulletin*, **119**, 113–174 [in Turkish with English abstract].

PICKETT, E. 1994. *Tectonic evolution of the Palaeotethys Ocean in northwest Turkey*. PhD Thesis, University of Edinburgh.

—— & ROBERTSON, A. H. F. 1996. Formation of the Late Palaeozoic–Early Mesozoic Karakaya Complex and related ophiolites in NW Turkey by Palaeotethyan subduction–accretion. *Journal of the Geological Society, London*, **153**, 995–1009.

RICHARDS, M. A., JONES, D. L., DUNCAN, R. A. & PAOLA, D. J. 1991. A mantle plume initiation model for the Wrangellia flood basalt and other ocean plateaus. *Science*, **254**, 263–267.

RICOU, L. E. 1994. Tethys reconstructed: plates, continental fragments and their boundaries since 260 Ma from Central America to south-eastern Asia. *Geodinamica Acta*, **7**, 169–218.

ROBERTSON, A. H. F. & DIXON, J. E. 1984. Introduction: aspects of the geological evolution of the Eastern Mediterranean. *In*: DIXON, J. E. & ROBERTSON, A. H. F. (eds) *The Geological Evolution of the Eastern Mediterranean*. Geological Society, London, Special Publications, **17**, 1–74.

SAUNDERS, A. D., TARNEY, J., KERR, A. C. & KENT, R. W. 1996. The formation and fate of large oceanic igneous provinces. *Lithos*, **37**, 81–95.

ŞENGÖR, A. M. C. 1979. Mid-Mesozoic closure of Permo-Triassic Tethys and its implications. *Nature*, **279**, 590–593.

—— & YILMAZ, Y. 1981. Tethyan evolution of Turkey: a plate tectonic approach. *Tectonophysics*, **75**, 181–241.

——, —— & SUNGURLU, O. 1984. Tectonics of the Mediterranean Cimmerides: nature and evolution of the western termination of Palaeo-Tethys. *In*: DIXON, J. E. & ROBERTSON, A. H. F. (eds) *The Geological Evolution of the Eastern Mediterranean*. Geological Society, London, Special Publications, **17**, 77–112.

SERVAIS, M. 1982. *Collision et suture téthysienne en Anatolie Centrale, (tude structurale et métamorphique (HP-BT) de la zone nord Kütahya*. PhD Thesis, Université de Paris-Sud, Centre d'Orsay.

SMITH, A. G., HURLEY, A. M. & BRIDEN, J. C. 1981. *Phanerozoic Palaeocontinental World Maps*. Cambridge University Press.

STAMPFLI, G. 1996. The Intra-Alpine terrain: a Palaeotethyan remnant in the Alpine Variscides. *Eclogae Geologica Helvetica*, **89**, 13–42.

——, MARCOUX, J. & BAUD, A. 1991. Tethyan margins in space and time. *Palaeogeography, Palaeoclimatology, Palaeoecology*, **87**, 373–409.

TEJADA, M. L. G., MAHONEY, J. J., DUNCAN, R. A. & HAWKINS, M. P. 1996. Age and geochemistry of basement and alkalic rocks of Malaita and Santa Isabel, Solomon Islands, southern margin of Ontong Java Plateau. *Journal of Petrology*, **37**, 361–394.

TEKELİ, O. 1981. Subduction complex of pre-Jurassic age, northern Anatolia, Turkey. *Geology*, **9**, 68–72.

TOPUZ, G., SATIR, M., ALTHERR, R. & SADIKLAR, M. B. 1997. Thermobarometrische und geochronologische Untersuchungen zur Metamorphosegeschichte Granat- und Cordieritführender Gneise des Pulur Massivs, NE Türkei. *Deutsche Mineralogische Vereinigung*, **Abstracts**, 45.

TRÜMPY, R. 1975. Penninic–Austroalpine boundary in the Swiss Alps: a presumed former continental margin and its problems. *American Journal of Science*, **275-A**, 209–238.

TÜYSÜZ, O. 1990. Tectonic evolution of a part of the Tethyside orogenic collage: the Kargı Massif, northern Turkey. *Tectonics*, **9**, 141–160.

—— 1993. A geo-traverse from the Black Sea to the central Anatolia: tectonic evolution of northern Neotethys. *Turkish Association of Petroleum Geologists Bulletin*, **5**, 1–33 [in Turkish with English abstract].

USTAÖMER, T. & ROBERTSON, A. H. F. 1993. A Late Palaeozoic–Early Mesozoic marginal basin along the active southern continental margin of Eurasia: evidence from the Central Pontides (Turkey) and adjacent regions. *Geological Journal*, **28**, 219–238.

—— & —— 1994. Late Palaeozoic marginal basin and subduction–accretion: the Palaeotethyan Küre Complex, Central Pontides, northern Turkey. *Journal of the Geological Society, London*, **151**, 291–305.

YANEV, S. N. 1992. The Permian in Northern Bulgaria. I. Formal lithostratigraphy related to the Lower Permian. *Geologica Balcanica*, **22**, 3–27.

YILMAZ, O. & BOZTUĞ, D. 1986. Kastamonu granitoid belt of northern Turkey: first arc plutonism product related to the subduction of the Palaeotethys. *Geology*, **14**, 179–183.

YILMAZ, Y. 1976. Geology of the Gümüşhane granite (petrography). *İstanbul University, Faculty of Science Bulletin, Seri B*, **39**, 157–172 [in Turkish with English abstract].

—— 1981. Tectonic evolution of the southern margin of the Sakarya continent. *İstanbul University Earth Sciences*, **1**, 33–52 [in Turkish with English abstract].

——, SERDAR, H. S., GENÇ, C. *ET AL.* 1997. The geology and evolution of the Tokat Massif, south-central Pontides, Turkey. *International Geological Review*, **39**, 365–382.

Palaeozoic–Early Tertiary Tethyan evolution of mélanges, rift and passive margin units in the Karaburun Peninsula (western Turkey) and Chios Island (Greece)

ALASTAIR H. F. ROBERTSON & ELIZABETH A. PICKETT[1]

Department of Geology and Geophysics, University of Edinburgh, Grant Institute, West Mains Road, Edinburgh EH9 3JW, UK (e-mail: Alastair.Robertson@glg.ed.ac.uk)
[1] *Present address: British Geological Survey, Murchison House, West Mains Road, Edinburgh EH9 3LA, UK*

Abstract: Similar Palaeozoic–Lower Tertiary units in the northern Karaburun Peninsula (western Turkey) and the adjacent island of Chios (Greece) provide insights into the tectonic evolution of the Tethyan Ocean near the junction between the Taurides/Anatolides and Hellenides. The northwest Karaburun Peninsula is dominated by a highly sheared mélange (*c.* > 2 km thick) with Silurian–Carboniferous blocks (up to hundreds of metres in size) of neritic and pelagic limestone, black ribbon chert, shale, extrusives and volcanogenic sediments. The blocks are set in a matrix of siliciclastic turbidites, pelagic carbonates and channelized conglomerates. Northern Chios is similarly composed of mélange (*c.* 3–4 km thick) with limestone blocks dated as Silurian–Carboniferous, black ribbon chert, shale and volcanics within predominantly siliciclastic sediments, including local conglomerates.

Both mélanges are unconformably overlain by Lower Triassic basinal successions (with sheared basal contacts), including terrigenous clastics, pelagic carbonate, radiolarite, lava and volcanogenic sediments, interpreted as rift successions. In both areas, successions shallow upwards into extensive Mesozoic carbonate platform facies, similar to those widely developed in the Taurides and Hellenides. On the Karaburun Peninsula, deposition was punctuated by emergence and localized deltaic siliciclastic deposition in latest Triassic–earliest Jurassic time. The Karaburun carbonate platform later emerged, eroded, then subsided in Campanian–Maastrichtian time; it then collapsed in the Maastrichtian–earliest Tertiary and was overlain by another mélange (Bornova Mélange), with blocks of both local platform-derived and accreted Mesozoic oceanic lithologies, and was finally thrust imbricated during continental collision. On Chios, a Lower Jurassic carbonate platform succession is overthrust by an exotic Carboniferous–Lower Jurassic mixed shallow-water carbonate and siliciclastic unit.

Most tectonic hypotheses for the mélange are problematic. They include formation as Palaeozoic basinal 'olistostromes', as an ideal Late Palaeozoic subduction–accretion complex basin, as an Early Triassic rift, or as entirely tectonic mélange. The mélange are seen here as the end-product of a combination of Late Palaeozoic southward(?) subduction–accretion (culminating in trench–microcontinent collision?), Early Triassic rifting and latest Cretaceous–Early Tertiary subduction/collision. Regional comparisons suggest that initial mélange formation took place in Late Carboniferous–Early Permian time. Subsequent Early Triassic rifting was associated with siliciclastic, calcareous and radiolarian deposition, and andesitic volcanism. The Triassic rift was overlain by a subsiding passive margin adjacent to a northerly Neotethyan oceanic basin from Middle Triassic–Late Cretaceous. This ocean closed in the Late Cretaceous–Early Tertiary, resulting in collapse of the passive margin, subduction–accretion and further mélange formation in a foredeep. Continental collision resulted in further deformation of the Palaeozoic mélange, thrust imbrication of the Mesozoic platform and shearing at its base.

The Tethys orogenic belt in the eastern Mediterranean is a superb natural laboratory for the study of tectonic processes and orogenic assembly, in part because of its diversity and the fact that it is still an active collisional area. The last decade and a half have seen immense strides in understanding this complex region. Prior to this, a widespread, if not universal, view was that there was only one (early) Mesozoic Tethyan ocean (Neotethys) in the eastern Mediterranean region (Dercourt *et al.* 1986) and that tectonic variety mainly resulted from long-distance thrusting over the North African foreland to the south [e.g. Ricou *et al.* (1984) for Turkey; Papanikolaou (1984*a*) for Greece]. Today, it is generally accepted that the eastern

Mediterranean region developed as a suite of microcontinental fragments and intervening small ocean basins (Şengör et al. 1984; Robertson & Dixon 1984; Dercourt et al. 1992), with each area requiring study in its own right. Formerly, a strongly argued view was that the eastern Mediterranean Sea is underlain by the North African foreland (Hirsch 1984), whereas today most workers see the eastern Mediterranean as an oceanic basin dating from the Triassic (Robertson & Woodcock 1980; Garfunkel & Derin 1984; Garfunkel 1998; Robertson 1998), or Late Permian (Stampfli et al. 1991, 2000; Stampfli 2000). Then, the main, Palaeozoic Tethys (Palaeotethys) was viewed as being located to the north of Turkey in the Pontide mountain range and beyond (Şengör et al. 1984; Tüysüz 1990), whereas today it is widely believed to lie further south, within Anatolia (Dercourt et al. 1992; Ustaömer & Robertson 1993, 1997).

There is currently a renewed interest in the Late Palaeozoic–Early Mesozoic geology and tectonics of central western Turkey, since this area includes evidence of both Palaeozoic and Mesozoic oceanic units. An initial problem concerns current terminology of Tethyan basins. Şengör et al. (1984) defines 'Palaeotethys' as a discrete oceanic basin whose remnants are located in the north, near the Eurasian continental margin (in the Pontides). 'Neotethyan' oceanic basins opened as back-arc basins above a south dipping Palaeotethyan subduction zone. By contrast, Stampfli et al. (2000) define 'Palaeotethys' as an ocean basin further south, that subducted northwards opening marginal basins to the north (i.e. Vardar or Meliata ocean). 'Neotethys', in turn, opened further south by rifting of 'Cimmerian' fragments from Gondwana. In the east (Oman, Himalayas), this oceanic basin opened in Late Permian time, but evidence from the eastern Mediterranean region (e.g. Levant, southern Turkey) favours Triassic–Early Jurassic opening (Garfunkel 1998; Robertson 1998).

A third alternative is that 'Palaeotethys' remained partly open and evolved into 'Neotethys' comparable to, for example, the modern southwest Pacific region (Robertson & Dixon 1984). Currently, there is a need to test alternative Tethyan tectonic models, using well-constrained data from critical areas. It is, therefore, a mistake to adopt a genetic terminology inherent in any particular tectonic model. To avoid this danger, Palaeotethys is defined here as simply an essentially Palaeozoic–Early Mesozoic ocean and Neotethys as an essentially Mesozoic–Early Tertiary ocean, wherever located and however formed (Robertson et al. 1996).

Within the Turkish region, it has been argued that Palaeotethys was mainly subducted northwards beneath Eurasia (Ustaömer & Robertson 1993, 1997; Robinson et al. 1995). Alternatively, Palaeotethys was subducted southwards beneath the then leading edge of North Africa, represented by the Antolide–Tauride Platform (Okay et al. 1996). Within the intervening area, evidence of apparent Early Mesozoic southward subduction was reported within the Permo-Triassic Karakaya Complex (Okay et al. 1996; Pickett & Robertson 1996), but it remains unclear where this unit was then located relative to North Africa or Eurasia. Recently, Okay (2000) suggests that Triassic subduction involving the Karakaya Complex was northwards, as proposed for the Central Pontide region (Ustaömer & Robertson 1993, 1997).

The focus here is on the evolution of Palaeotethys and its relationship to Neotethys as defined above. Evidence will be considered for the evolution of Palaeotethyan and Neotethyan units located in the vicinity of the northern margin of the Mesozoic Anatolide–Tauride Platform, based on evidence from two critical areas of western Turkey and Greece where, exceptionally, unmetamorphosed units are well exposed. In contrast, further east, the northern margin of Anatolide–Tauride Platform is mainly metamorphic, including blueschists related to closure of the northerly Neotethyan basin in Late Mesozoic–Early Tertiary time (Okay et al. 1996) and, therefore, the nature of any underlying Palaeotethyan units remains largely obscure.

Within the Karaburun Peninsula (western Turkey) and on the adjacent Greek island of Chios (Fig. 1), unmetamorphosed and relatively well-dated Palaeotethyan and Neotethyan units are well exposed and accessible for study. These two areas include important mélange units that have, in the past, been interpreted in very different ways, in terms of either subduction–accretion or rift processes. (The term 'mélange' is used here as a purely descriptive, non-genetic field term for a pervasively mixed unit.) In addition, Mesozoic carbonate platform units and a separate unit of Late Mesozoic mélange are also present.

Presented here are new results for Palaeozoic–Lower Tertiary lithologies exposed on the Karaburun Peninsula and Chios in the light of existing data, especially recent biostratigraphic results. Despite their proximity, the two areas have largely been studied separately over the last half century, but taken together they provide important insights into the tectonic evolution of a critical area linking the Taurides

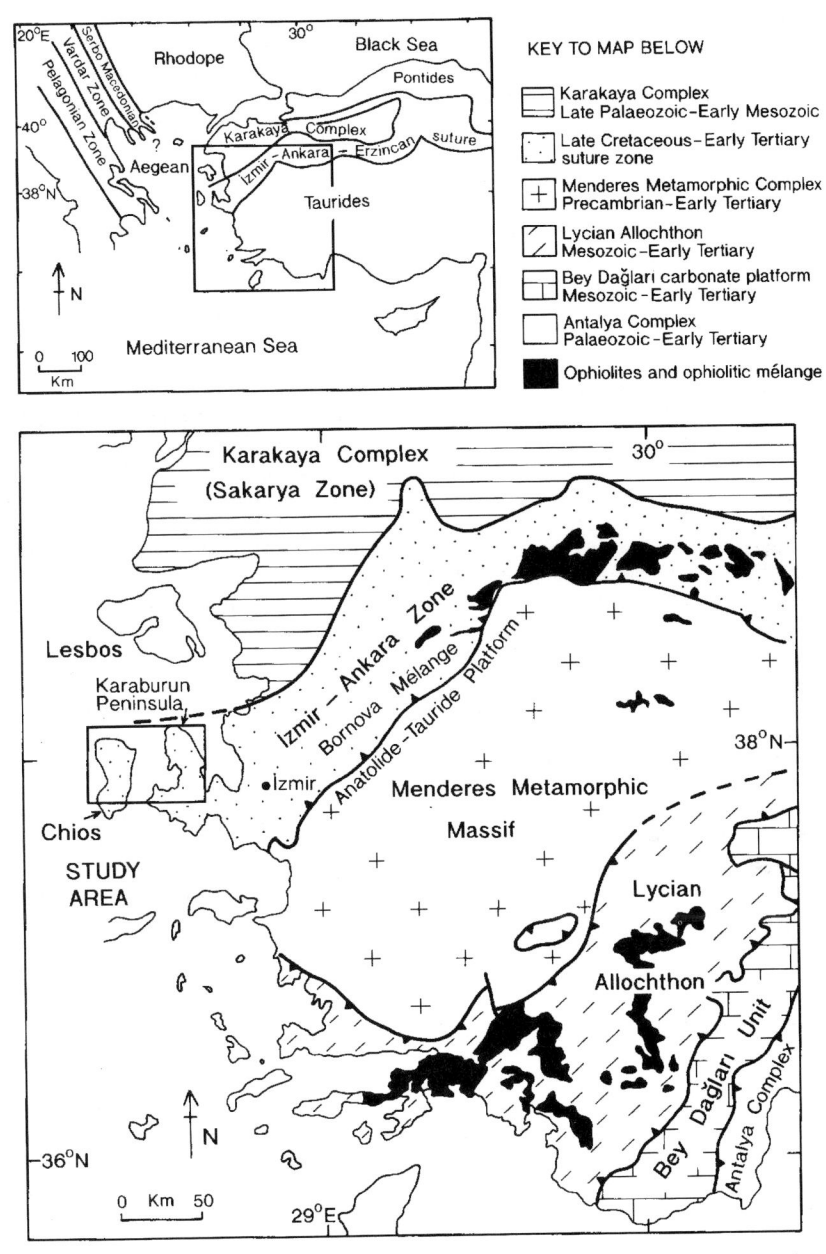

Fig. 1. Regional tectonic map showing the setting of the Karaburun Peninsula (Turkey) and Chios (Greece). Note also the location of the Palaeotethyan Karakaya Complex to the north, and the Neotethyan İzmir–Ankara–Erzincan (Suture) Zone and the Menderes Metamorphic Massif to the south.

and Hellenides. It will be concluded that the Karaburun–Chios mélanges represent the best known unmetamorphosed Palaeotethyan subduction–accretion complex in the eastern Mediterranean region, overlain by a rifted Mesozoic continental margin related to opening of a northerly Neotethyan oceanic basin.

Regional geological setting

The island of Chios and the Karaburun Peninsula lie at the western end of the Mesozoic–Early Tertiary (i.e. Neotethyan) İzmir–Ankara–Erzincan Suture Zone (İAESZ) that separates the metamorphic Menderes Massif to the south and related units further east from Upper Palaeozoic–Lower Mesozoic units to the north, notably the Karakaya Complex (Fig. 1). A zone of ophiolites and ophiolitic mélanges (forming the İAESZ) trending across northern Turkey was formed by closure of a Mesozoic ocean basin ('Northern Neotethys') during latest Mesozoic–Early Tertiary time (Şengör & Yılmaz 1981). To the north, the Karakaya Complex is interpreted as an accretionary wedge formed by subduction of Palaeotethyan oceanic lithosphere (e.g. Tekeli 1981; Okay et al. 1991; Pickett & Robertson 1996). Further south, the Menderes Massif and related units are interpreted as continental fragments rifted from North Africa (Gondwana) during Early Mesozoic time (Robertson et al. 1991), or Late Permian time (Stampfli et al. 1991; Stampfli 2000), to open a Neotethyan oceanic basin to the south.

The sequences on Karaburun and Chios begin with kilometres thick mélange units, including blocks and dismembered thrust sheets of Palaeozoic age and matrix sediments. In both areas the mélange is overlain by Lower Triassic basinal successions, including volcanogenic units, followed by thick Mesozoic carbonate platform units. In addition, on Chios, the above stratigraphy is structurally overlain by a large thrust sheet of Upper Palaeozoic–Lower Mesozoic (predominantly Permian) basinal to shallow-marine sediments.

The Karaburun Peninsula

Previous work

Classical work is summarized by Erdoğan et al. (1990) for Karaburun and Besenecker et al. (1968) for Chios. Early workers treated both the Karaburun and Chios units as entirely stratiform successions (e.g. Ktenas 1925) in which, for Karaburun (Fig. 2), Ordovician–Silurian (Holl 1966) and Devonian (Sözen 1973) corals and brachiopods, and Carboniferous fusulinids (Philippson 1911; Brinkmann et al. 1972) were recognized. 'Olistoliths' associated with lenses of Triassic limestone were reported in Karaburun by Konuk (1977).

Within what is termed here the Karaburun Mélange (new name), Erdoğan et al. (1990) recognized the Denizgiren Group [name introduced by Ktenas (1925)] which comprises two units of inferred Early Triassic age, the Karareis Formation [name introduced by Konuk (1977)] and the Gerence Formation, which Konuk (1977) regarded as laterally equivalent and intergradational. He interpreted the 'Karareis Formation' as a > 2 km thick Lower Triassic succession of mudstones and sandstones, mafic volcanics, black cherts and limestones, including 'olistoliths' of Carboniferous–Lower Triassic limestones. Erdoğan et al. (1990) reported the presence of deformed lenses of thinly bedded limestones and thin-shelled bivalves, associated with Triassic foraminifera, interbedded with green and black chert, and turbiditic sandstone. 'Olistostromal horizons', 5–10 m thick, set in a sandy or muddy matrix, with subrounded to rounded clasts yielded Silurian–Early Devonian to Late Carboniferous ages (Erdoğan, pers. comm.). Erdoğan et al.'s (1990) mapping also confirmed the presence of a laterally discontinuous sheet of Middle Carboniferous limestone (Ktenas 1925; Brinkmann et al. 1972) beneath a Mesozoic platform succession in the southwest (Alandere Formation). Erdoğan et al. (1990) interpreted what is termed here the Karaburun Mélange as a Lower Triassic rift-related succession in an unstable 'trench-type' setting adjacent to a developing Triassic carbonate platform.

Kozur (1995, 1997) gave an exhaustive critique of earlier work and went on to report new radiolarian, conodont, ostracod and Muellerphaerid (Incertae sedis) age data that confirm the presence of Silurian, Devonian and Carboniferous units. Lower units, including black cherts, contain Silurian–Upper Devonian radiolarians. A widespread and structurally higher 'turbidite–olistostrome' unit contains black radiolarites of Silurian–Devonian age and limestone 'olistoliths' with conodonts of Early Silurian (Wenlockian) and Late Devonian age. Silurian limestone olistoliths locally contain 'inclusions' of 'keratophyric tuff' associated with fragmental volcaniclastics (< 100 m thick near Ildır; Fig. 2). Kozur (1995, 1997) assumed a Silurian age for the volcaniclastics, although Erdoğan (pers. comm.) interprets these as inclusions within Lower Triassic 'olistostromes'; Middle Devonian is apparently absent. Kozur (1995, 1997) confirmed a Middle Carboniferous (Bashkirian) age for limestones in the upper part of the mélange in the southwest (Alandere Formation) and, elsewhere, he extracted Lower Visean radiolarians from black chert. Kozur (1995, 1997) supports Erdoğan et al.'s (1990) view that the Carboniferous is unconformably overlain by Lower Triassic basinal facies. Lithological contrasts and changes in vitrinite reflectance reported by Kozur (1995, 1997) suggest that Palaeozoic [conodont alteration index (CAI 4–5)] and Triassic (CAI 1.5) units are locally intersliced, especially in the east close to

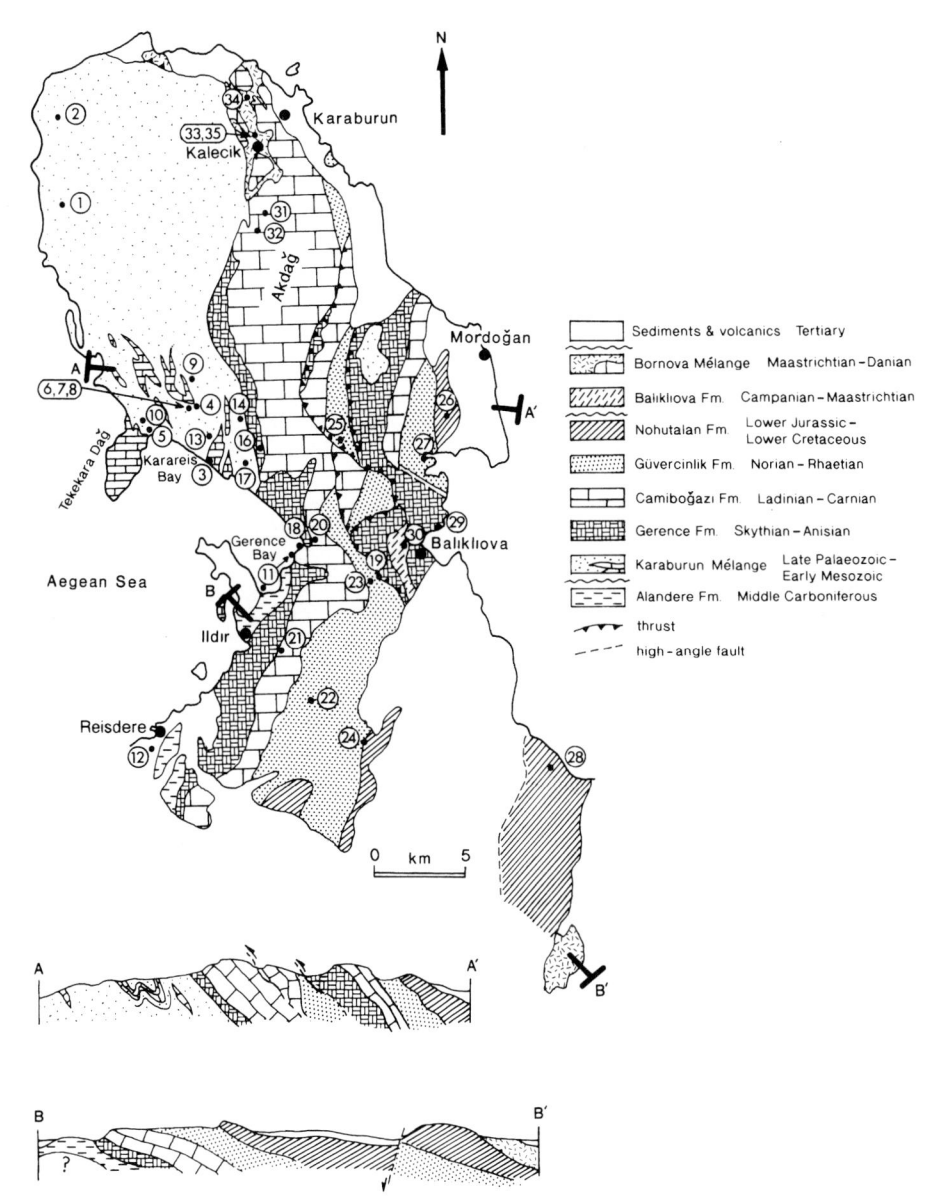

Fig. 2. Geological map and cross-sections of the northern part of the Karaburun Peninsula, showing the main tectonostratigraphic units and large-scale structure. Places mentioned in the text are numbered. Modified after Erdoğan et al. (1990).

the overlying Mesozoic platform succession, in contrast to the depositional interfingering inferred by Erdoğan et al. (1990).

Stratigraphy

Kozur's (1995, 1997) palaeontological results indicate that the northwestern Karaburun Peninsula cannot be considered entirely as a Lower Triassic stratiform succession (with sedimentary olistoliths), as envisaged by Erdoğan et al. (1990). Therefore, use of the terms Karareis Formation and Denizgiren Group are discontinued here and the term *Karaburun Mélange* is introduced. However, the term *Gerence Formation* remains valid for the overlying Lower

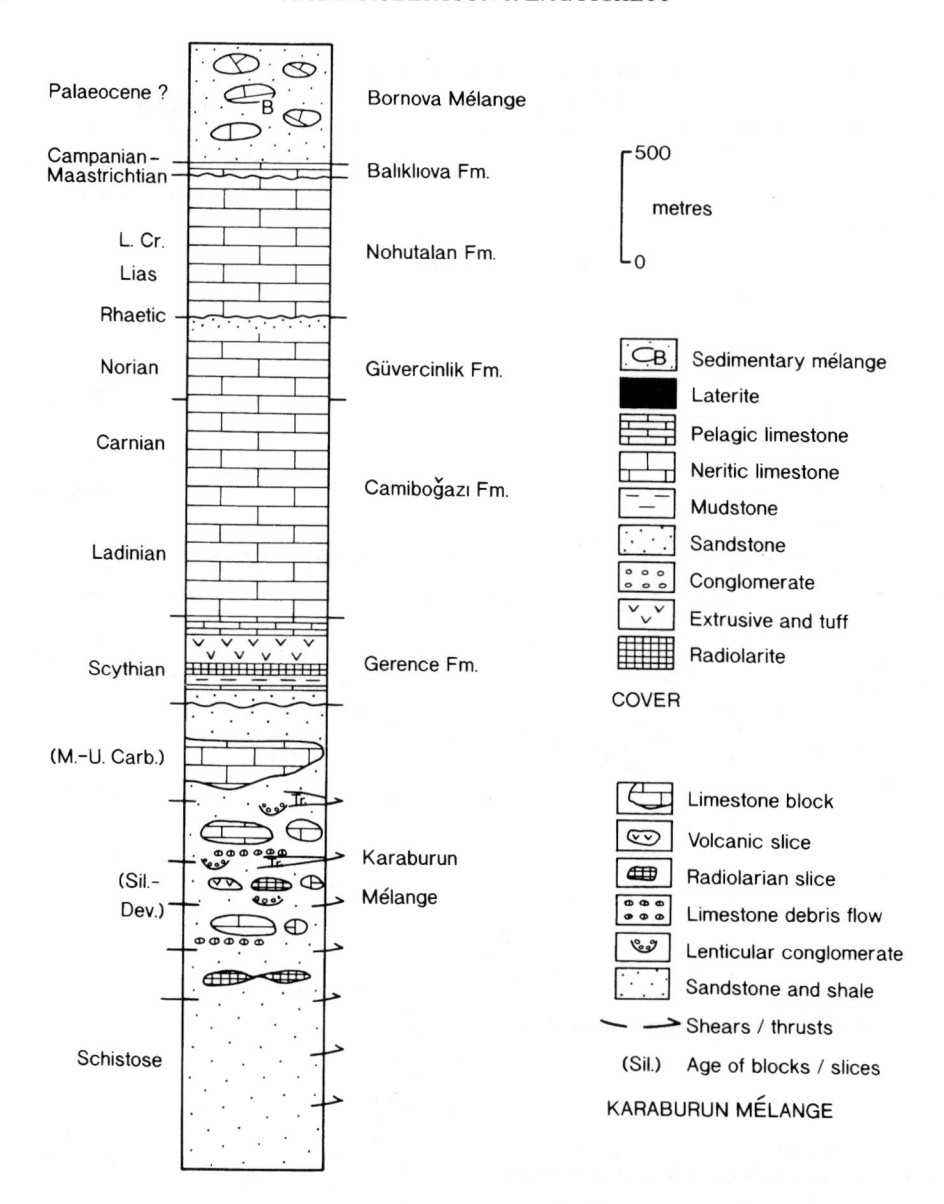

Fig. 3. Composite log of the Karaburun Peninsula succession. Data sources specified in the text.

Triassic succession at the base of the Mesozoic platform succession in the south, and this is correlated with intact volcanic–sedimentary units further north [uppermost part of the Karareis Formation of Erdoğan et al. 1990)]. Above the Karaburun Mélange, a thick, relatively intact, Mesozoic (Triassic–Upper Cretaceous) succession has long been recognized (Philippson 1911; Ktenas 1925; Kalafatcıoğlu 1961; Brinkmann et al. 1972). The Lower Triassic part of the succession was divided into seven formations by Brinkmann et al. (1972) but these are now best regarded as facies variants within a single Lower Triassic unit, i.e. the Gerence Formation (150–500 m thick). Therefore, Erdoğan et al.'s (1990) stratigraphy for the overlying Mesozoic platform succession (Fig. 3) and adjacent uppermost Cretaceous–Lower Tertiary mélange (i.e. the *Bornova Mélange*; Fig. 4) is retained.

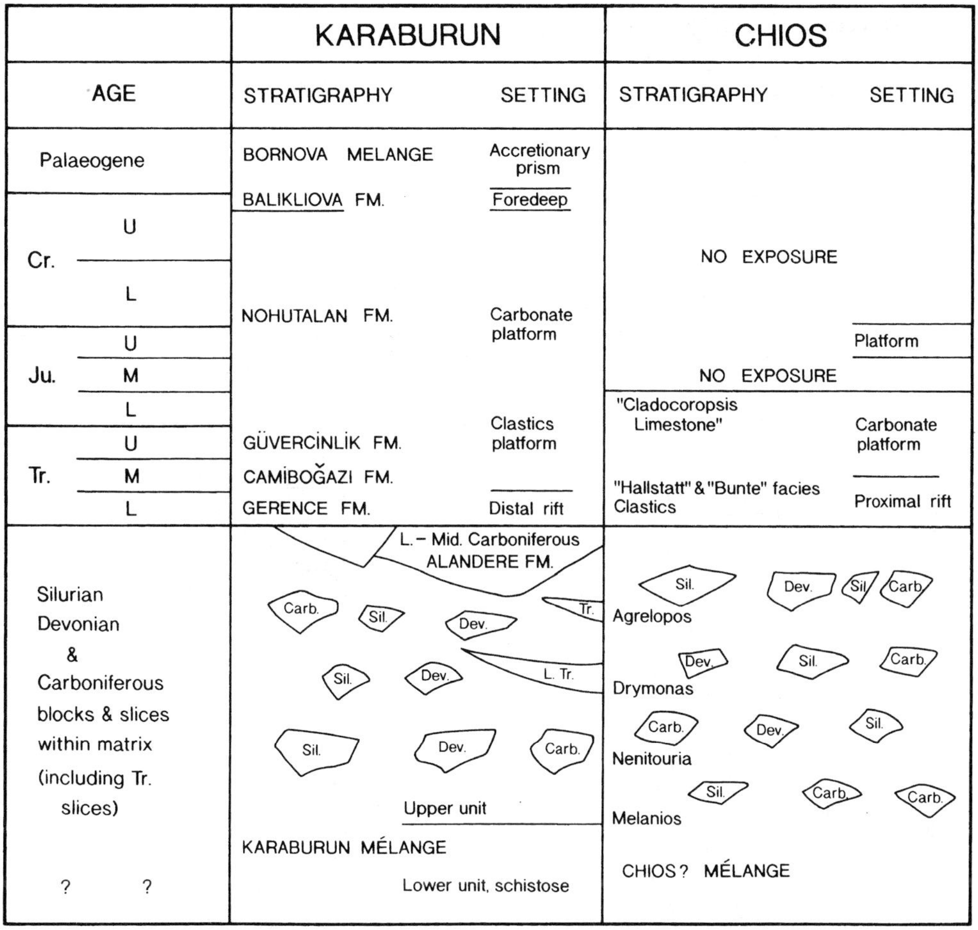

AGE	KARABURUN STRATIGRAPHY	SETTING	CHIOS STRATIGRAPHY	SETTING
Palaeogene	BORNOVA MELANGE	Accretionary prism		
	BALIKLIOVA FM.	Foredeep		
Cr. U			NO EXPOSURE	
Cr. L	NOHUTALAN FM.	Carbonate platform		
Ju. U				Platform
Ju. M			NO EXPOSURE	
Ju. L			"Cladocoropsis Limestone"	Carbonate platform
Tr. U	GÜVERCİNLİK FM.	Clastics platform		
Tr. M	CAMİBOĞAZI FM.		"Hallstatt" & "Bunte" facies Clastics	Proximal rift
Tr. L	GERENCE FM.	Distal rift		
Silurian Devonian & Carboniferous blocks & slices within matrix (including Tr. slices) ? ?	L.- Mid. Carboniferous ALANDERE FM. ... Upper unit KARABURUN MÉLANGE Lower unit, schistose		Agrelopos Drymonas Nenitouria Melanios CHIOS? MÉLANGE	

Fig. 4. Summary of tectonostratigraphy of Karaburun and Chios showing proposed correlations.

The Karaburun Mélange

The reconnaissance mapping from this study confirms that, beneath local Neogene volcanic-related units, the whole of the northwestern Karaburun Peninsula beneath the Mesozoic platform (Fig. 6a) is a mélange, with little or no lateral or vertical stratal continuity. The mélange is divisible into two units, based on lithology and state of metamorphism (Fig. 4).

Lower unit. A structurally lower, undated, schistose shale–sandstone unit of greenschist facies metamorphic grade is exposed in the northwestern coastal area (Fig. 2), extending up to *c.* 5 km inland (e.g. Fig. 2, 1 and 2). This unit, estimated as up to 500 m thick, is equivalent to the Ordovician(?) sericite schist of Kozur (1995, 1997; his Küçükbahçe Formation). The sediments strike nearly north–south and are mainly inclined to the east at 40–60°, although dips are locally variable. Lithologies are mainly sheared sandstone turbidites and shales, with little chert or exotic blocks. The unit is cut by veins and pods of metasomatic quartz up to tens of centimetres thick, not seen in structurally overlying units.

Upper unit. This encompasses the remainder of the outcrop (> 1.5 km thick) beneath the Mesozoic platform successions. It is a mélange comprising two main components: (1) competent blocks and disrupted thrust sheets of neritic and pelagic limestone, black ribbon radiolarite (lydite) and extrusive rocks; and (2) a sheared matrix of shale–sandstones and rare conglomerate.

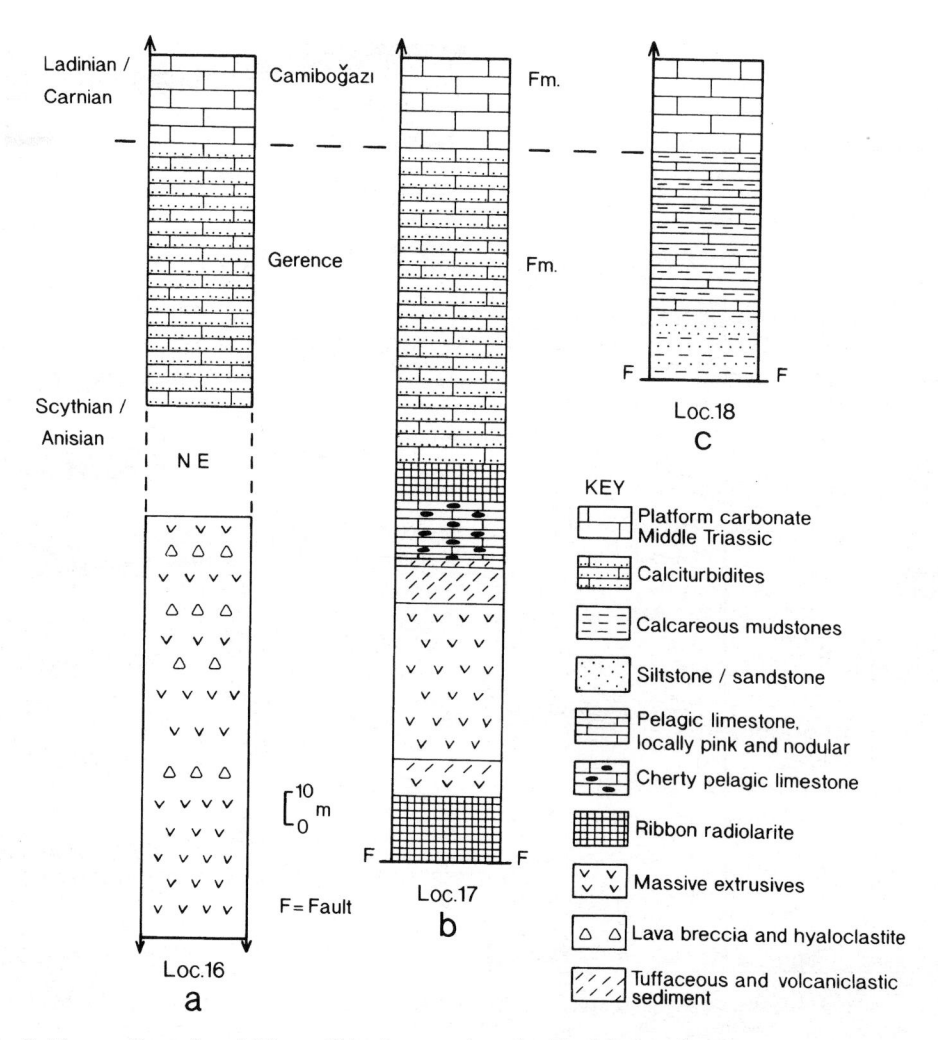

Fig. 5. Measured logs of partial Lower Triassic successions. See Fig. 2 for locations.

Limestone. Limestones mainly form disrupted sheets and blocks up to 600 m long by several hundreds of metres thick. The largest disrupted limestone sheets commonly dip at low to moderate angles (< 30°), discordant to the generally higher angle local tectonic fabric in the mélange. Smaller limestone blocks (metres to tens of metres in size) are commonly steeply inclined, concordant with the fabric of the adjacent mélange. The limestones are commonly internally sheared and tectonically brecciated with tension gashes and stylolites (Fig. 2.3 and 4). The margins of some individual blocks and sheets degenerate into angular jigsaw puzzle-type breccias mixed with argillaceous carbonate (Fig. 2.5). Within this marginal zone, limestone clasts are veined and brecciated, whereas the enclosing matrix sediment is undeformed. Elsewhere (Fig. 2.6), matrix sediments at the contact with tectonic inclusions are brecciated, veined and sheared. Excellent examples of disrupted blocks and sheets of limestone (up to 500 m long by 80 m wide), together with smaller, metre size blocks, are well exposed inland from Tekekara Dağ (e.g. Fig. 2.4, 6–8). Limestones in this area were dated as Silurian and Devonian by Kozur (1995), and as Carboniferous and Early Triassic by Erdoğan (1990*a*, *b*). A common facies is well-bedded, muddy, buff weathering pelagic limestone, locally pinkish and nodular (e.g. Fig. 2.6). Some blocks are composed of red, thin-bedded pelagic limestone, with nodules and lenticles of black replacement chert (e.g. Fig. 2.9).

A peninsula north of Karareis Bay (Tekekara

Fig. 6. Field photographs of Karaburun units. (**a**) Triassic carbonate platform (pale upper) overlying the Karaburun Mélange (dark lower), northwest Karaburun Peninsula. (**b**) Redeposited Middle Carboniferous limestone; coastal section north of Ildır. (**c**) Redeposited limestones showing layer-parallel extension; coastal section north of Ildır. (**d**) Phacoidal blocks of lava and sandstone in a sheared shaly matrix; coastal section north end of Karareis Bay. (**e**) Channelized conglomerate with clasts including black chert; coastal section north end of Karareis Bay. (**f**) Limestone breccia at top of Lias platform succession; southwest of Mordoğan (Fig. 2.25). Note angular clasts – probably a tectonic breccia.

Dağ; Fig. 2), mapped as Triassic limestone by Erdoğan (1990a, b), is composed of numerous large (tens to hundreds of metres) limestone blocks, dated as Midddle Carboniferous–Early Permian(?) by Garrasi & Weitchat (1968), with poorly exposed matrix sediments between individual blocks and sheets (e.g. Fig. 2.5). Disrupted thrust sheets and blocks of well-bedded pink limestone, with shaly intercalations, contain numerous well-preserved pisoliths, oncolites and ooids, with sparry overgrowths. Some oncolites are fragmented and largely replaced by dolomite rhombs. Stylolites and late-stage spar veins are also observed. In addition, directly north of Tekekara Dağ, a disrupted sheet of well-bedded calciturbidites passes stratigraphically upwards into thinly bedded dark grey ribbon chert, similar to units exposed within the Lower Triassic Gerence Formation (as redefined here; Fig. 2.10). North of Ildır, a block of Upper Silurian limestone includes siliceous tuff, and nearby siliceous tuffs (at Ildır Fatih College) are reported to contain limestone intercalations with Silurian or Devonian ostracods (Kozur 1995).

The coastal section between the southern end of Gerence Bay and Reisdere–Ildır and south of Ildır (Fig. 2) includes widespread Lower–Middle Carboniferous limestones, including the Serpukhovian–Bashkirian aged Alandere Formation (> 300 m thick) of Erdoğan et al. (1990). Dating by Kozur (1995, 1997) suggests that the more basinal facies are Early Carboniferous, (his Döşemealtı Formation), whereas shallow-water limestones are Middle Carboniferous (Serpukhovian and Bashkirian, i.e. latest Visean) in age. However, no lower boundary is exposed and the unit is only patchily exposed in the south. Where well exposed along the coast at the southern end of Gerence Bay (Fig. 2.11), the succession is mostly gently inclined but very tectonically disrupted, locally folded, inverted and cut by numerous faults and shear zones with spar-filled tension gashes. In places, the limestones are broken into tectonic blocks and breccia, with a buff silty matrix, but with little or no lateral stratal continuity. Even intact units show layer-parallel extension (Fig. 6c). Associated turbiditic sandstone and mudstone are tectonically juxtaposed with limestone and do not form part of a single intact succession. Meso-scale isoclinal folds locally verge northwest or north-northwest (e.g. Fig. 2.11).

The following Lower Carboniferous lithofacies are depositionally intercalated where well exposed in coastal sections, but cannot be placed in stratigraphic order because of tectonic disruption. (1) Medium-bedded to massive, dark grey limestones, locally very fossiliferous (including crinoids, bivalves, brachiopods, corals and bioclasts), interbedded with dark grey to black, finely laminated wackestones (Fig. 2.11 and Fig. 6b), gastropods with rare quartzite and calcareous sandstone clasts. The interbedded laminates contain minor black replacement chert and are very bituminous, giving off a fetid odour when split. (2) Medium- to thick-bedded massive calcarenite, with minor replacement chert, locally channelized down into facies (1) above. (3) Thin- and medium-bedded calcareous sandstone and quartzose sandstone showing sedimentary structures (e.g. grading and micro-cross-lamination) indicative of turbiditic deposition, and locally with small dark chert(?) and carbonate fragments (e.g. Fig. 2.11). (4) Buff coloured mudstones (up to 0.6 m thick; Fig. 2.11) are interbedded with the above limestones and sandstones.

The redeposited Lower Carboniferous limestones (e.g. calciturbidites; Fig. 2.11) are packstone with recrystallized fragments of shells, algae, crinoids, echinoderms, gastropods, molluscs, benthic foraminifera, micritic grains, ooids and scattered quartz grains set in a coarse spar cement. Interbedded sandstone is composed of angular to subrounded grains of plutonic quartz, quartzite, limestone bioclasts, crystalline limestone, folded micaschist (with strain-slip cleavage), green chlorite, feldspathic lava, chert, black siltstone, muscovite and plagioclase. In contrast, non-redeposited massive limestones (e.g. near Ildır) include packstone with numerous quite well-rounded oolitically coated grains of quartz, quartzite, rare schist, reddish recrystallized chert, micaceous sandstone and altered volcanics, together with common benthic foraminifera, pisoliths, algae and ooids.

Further southwest (c. 7 km southwest of Ildır, near Reisdere; Fig. 2.12), thick-bedded to massive neritic crinoidal limestones are rather poorly exposed beneath Neogene volcanic-related units. The limestone is locally a breccia with a buff muddy matrix and is cut by small fissures filled with pink carbonate.

Black chert. The chert is ribbon radiolarite, with beds of massive chert (< 20 cm thick), separated by dark grey shaly partings, from which Kozur (1995) extracted Silurian–Lower Devonian and Lower Carboniferous radiolaria. Sandstone, or other, interbeds were not observed within the ribbon chert. The chert comprises densely packed radiolarians, partly replaced and infilled by chalcedonic quartz, set in a matrix of microcrystalline quartz and fine terrigenous silt with fine quartz and sericite. In the structurally lower parts of the mélange, the black chert is commonly very highly sheared and intersliced with coarse turbiditic sandstone (e.g. Fig. 2.3). In places, contorted chert passes laterally into tectonic chert breccia, with angular clasts < 0.6 m in size (e.g. Fig. 2.3), set in a sandstone matrix. The chert is also commonly deformed into medium-scale chevron folds (Fig. 2.13).

From its brittle, competent style of deformation, it is clear that the chert was fully lithified before being tectonically incorporated into an incompetent matrix. The chert becomes more abundant in structurally higher levels of the mélange in the east (e.g. İdecik Çeşme; Fig. 2.14), where it forms sheared, disrupted units up to several tens of metres thick and is laterally continuous up to several hundreds of metres.

Volcanic rocks. Volcanic rocks are observed in two main areas. Along the coast on the northern side of Karareis Bay (Fig. 2.3 and 13), altered intermediate–basic extrusives are present as subrounded elongate blocks of aphyric, or feldspar–phyric, massive to locally pillowed lava. These lithologies are spatially associated with blocks and lenticles (< 10 m in size) of chevron-folded black and grey chert, set in a matrix of very highly sheared shale and turbiditic sandstone. Further south, near Ildır (Fig. 2), Kozur (1995) reported silicic volcaniclastic sediments (keratophyric tuffs) associated with Silurian limestone and interpreted the volcanics as Silurian in age, although Erdoğan (pers. comm.) regards these as thrust intercalations of Triassic volcanics.

Sandstone–shale matrix. Sandstone and shale form the matrix of the Karaburun Mélange. The sandstones are clearly turbidites, based on occurrences of grading, cross-lamination, convolute lamination, flute casts, shale rip-up clasts, slumping and synsedimentary deformation (e.g. Fig. 2.3). Dips are variable but are generally steeply inclined towards the east. Successions are mainly the right way up, implying generally eastward facing, but are locally inverted. Shales are commonly highly sheared with an 'argile scagliosa' (i.e. scaly clay) texture (e.g. Fig. 2.13 and Fig. 6d). Medium-grained sandstones contain plutonic quartz, folded mica schist, amygdaloidal basalt, muscovite, plagioclase, minor blue chlorite, black siltstone and common grains of basic extrusive igneous rock.

Three types of matrix are observed at different structural levels of the mélange. First, the metamorphosed sandstone (lower unit) is very sheared, with a schistosity picked out by muscovite. Large quartz grains show incipient fusion with a fine-grained quartzose matrix. Sandstones in the mid part of the mélange are strongly intersheared with black ribbon chert and local volcanics. Some sandstones in the higher part of the mélange, although sliced, are more laterally continuous (up to hundreds of metres) and include rare pelagic limestones and calciturbidites of Early Triassic age, based on foraminiferal occurrences (Konuk 1977; Erdoğan *et al.* 1990).

There are also occasional thicker beds (< 3 m) of pebblestones (e.g. Fig. 2.3 and 13) and rare channelized conglomerates, up to several tens of metres thick and several hundreds of metres long (e.g. Fig. 2.8 and 10). Bedding in the conglomerates varies from near vertical to subhorizontal (e.g. Fig. 2.8). Superficially, these conglomerates appear to be blocks but on closer inspection are seen to pass laterally and vertically into typical siliciclastic turbidites. The clasts within them are mainly subangular to angular (to locally well rounded), composed of black chert, or siliceous siltstone, quartzite and schist (e.g. Fig. 2.3), set in a muddy matrix (Fig. 6e). One conglomeratic unit (Fig. 2.8) is a debris flow deposit mainly composed of subrounded black chert clasts, up to 0.25 m in size, set in a silty matrix. Many of the medium-bedded, and thinner, sandstones are strongly sheared and show classic phacoidal textures (e.g. Fig. 2.3).

Dating of the matrix is problematic for two reasons: (1) Kozur (1995, 1997) dates the 'matrix' based mainly on intercalations of lydite (black chert) and pelagic limestone; however, in this study, the cherts and limestones were seen to be thrust bounded (layer-parallel shears) and, thus, do not date the actual mélange matrix; (2) Lower Triassic 'matrix sediments' identified by Erdoğan *et al.* (1990) may be thrust into the mélange, as suggested by contrasting (lower) conodont maturity indices in these sediments (Kozur 1995, 1997).

Interpretation of the Karaburun Mélange

The Karaburun Peninsula is interpreted as a mélange of Palaeozoic exotic blocks, mainly limestone, radiolarian chert, extrusives and volcanogenic sediments set in a matrix of mainly siliciclastic turbidites and shale. An original succession can be inferred by comparison of individual blocks. Silurian and Devonian limestone blocks record a pelagic, basinal setting, with input from an adjacent continental margin made up of mainly quartzo–feldspathic rocks. The succession included subordinate silicic-intermediate composition volcanics and volcaniclastics dated as Silurian by Kozur (1995, 1997), although some of the volcanics could be Early Triassic in age and sheared into the mélange.

Lower–Middle Carboniferous limestones are mainly redeposited facies and comprise calciturbidites and quartzose turbidites within finely laminated, deep-water, organic-rich calcareous muds. Kozur (1995, 1997) reported radiolaria in this Lower Carboniferous succession. The limestone is strongly folded and brecciated, and does not represent a single intact succession, but is instead a highly disrupted thrust sheet and associated blocks within the Karaburun Mélange. The presence of Middle Carboniferous shallow-water carbonates (Alandere Formation), possibly locally extending to the Lower Permian (Garrasi & Weitschat 1968), indicates input from an adjacent carbonate platform. As depositional contacts between deep- and shallow-water facies are rarely preserved, it is possible that the older basinal and younger neritic carbonates represent blocks derived from carbonate platform, slope and basinal facies. In addition, the black cherts could have been underlain by oceanic crust that is not preserved.

The siliciclastic matrix sediments are mainly turbidites, probably deposited in a deep-marine setting. Some sandstone turbidites (lithic sandstones and greywackes) are interbedded with the Middle Carboniferous basinal succession and others with Lower Triassic pelagic carbonates, but most units are bound by sheared contacts and remain undated. The lenticular conglomerates are interpreted as remnants of channels, probably in a deep subaqueous setting. The main source of larger clasts was lithified black chert and black siltstone, similar to block lithologies within the mélange.

Mesozoic platform successions of the Karaburun Mélange

The Karaburun Mélange is overlain by a succession dominated by Mesozoic platform carbonates, up to 3 km thick, forming the Akdağ Mountains of the central northern Karaburun Peninsula (Philippson 1911; Brinkmann et al. 1972). The succession, as preserved in the lowest exposed structural unit and an overlying thrust sheet, begins with basinal pelagic carbonates, siliceous and volcaniclastic sediments with local volcanics and Mn deposits (Gerence Formation: 150–500 m thick), and passes up into Middle Triassic (Anisian)–Upper Cretaceous platform carbonates, albeit with discontinuities (Fig. 4). Here, the Early Triassic, Late Triassic–Early Jurassic and Late Cretaceous intervals are focused on, which are critical to tectonic interpretation of the region.

Lower Triassic basinal facies. The succession in the Lower Triassic Gerence Formation is well exposed in the lowest structural unit, seen along the coast of Gerence Bay (Fig. 2). In the northwest, Lower Triassic facies beneath the Mesozoic carbonate platform, including volcanics, were placed in the uppermost part of the 'Karareis Formation' by Erdoğan et al. (1990), but are here included within the Gerence Formation.

Erdoğan et al. (1990) reported that the Lower Triassic Gerence Formation unconformably overlies the Middle Carboniferous Alandere Formation, 2 km northeast of Ildır (Fig. 2). Karstic weathering and ferruginous oxidation beneath the Lower Triassic limestone indicate subaerial exposure. Elsewhere (east of Reisdere), the Alandere Formation is depositionally overlain by thinly bedded limestones with intraformational breccias and chert intercalations (Erdoğan et al. 1990). Kozur (1995) reported a contact on the east side of Gerence Bay (Fig. 2) where karstified Upper Carboniferous limestones are overlain by a basal conglomerate with limestone pebbles in an argillaceous matrix, overlain by dolomite, then Scythian limestone. North of Gerence Bay, the contact between Palaeozoic lithologies of the mélange (e.g. black chert) and an overlying intact Lower Triassic volcano–sedimentary succession was observed to be strongly sheared and folded (e.g. Fig. 2.14; İdecik Çeşme). In summary, the contact between the mélange and the overlying Lower Triassic platform succession is interpreted as a sheared unconformity.

In contrast to the Karaburun Mélange beneath, several lines of evidence indicate that the Gerence Formation is an intact stratigraphical unit: (1) bedding is mainly gently inclined parallel to the overlying intact Middle Triassic platform carbonate succession; (2) exposures show considerable lateral stratigraphical and structural continuity, especially in the south; (3) the unit is less deformed by shearing than the underlying Karaburun Mélange, except near the contact in the north, where the succession is strongly folded, faulted and sheared, with local phacoids; and (4) fossil evidence indicates upward younging (Erdoğan et al. 1990).

The Karaburun Mélange is locally overlain (e.g. Fig. 2.16), with a poorly exposed sheared contact, by lava breccias and hyaloclastite, passing upwards (after an interval of no exposure) into a thickening- and coarsening-upward succession of calciturbidites, then into a Middle Triassic carbonate platform succession (Fig. 5a). The calciturbidites contain small subrounded, partly recrystallized, fragments of micrite, shells (including *Daonella*), echinoderms and algae in a micritic matrix packed with calcified radiolarians and scattered pelagic bivalve shells. Further south (Fig. 2.17) a relatively coherent succession begins with ribbon radiolarite, followed by plagioclase- and pyroxene–phyric andesite, green 'tuff', andesitic 'tuffaceous' conglomerate, cherty limestone (grey-buff with replacement chert) and red and black radiolarite, locally manganiferous (MnO up to 51.61%; Pickett 1994), overlain by well-bedded, dark grey, steeply dipping, graded calciturbidites (in beds up to 1 m thick) and fissile shale. Alternating laminae of radiolarian-rich limestone and shelly packstone were observed in thin section. Folds, with north–south axes, are locally west facing. Along the coast of Gerence Bay (Fig. 2.18) the succession is too tectonically disrupted to measure. Thin-bedded limestones and calcareous shales (< 15 m) locally pass upwards into pink, nodular, ammonite-bearing pelagic limestones, interbedded with reddish calcareous shales, locally containing lenticles of replacement chert. There are also intercalations of medium–thick-bedded turbiditic sandstones, including fine, medium and locally coarse facies (limestone conglomerates) with thin-bedded black chert and cherty limestone. Elsewhere in the vicinity (Fig. 2.20) the succession is more coherent, with thinly bedded, nodular, concretionary, pelagic limestone and buff calcareous mudstones (Ladinian?), overlain by thick-bedded Middle Triassic carbonates. A comparable succession, including pink pelagic limestone, is seen in the higher thrust sheet further east (Fig. 2.19).

Correlation of the above local successions (Fig. 7) indicates the following composite succession: (1) pelagic limestones (locally of Hallstatt type) and shales, (2) ribbon radiolarites, lavas and volcanogenic sediments; and (3) thickening- and coarsening-upward siliceous calciturbidites (Late Anisian; Brinkmann et al. 1972). There is marked lateral facies variation, with lavas and radiolarites being more extensive in the northwest. The succession passes conformably upwards into Middle Triassic (Upper Anisian) platform carbonates [Camiboğazı Formation: Erdoğan et al. (1990)]. In addition, within the higher thrust sheet in the east (4 km north of Balıklıova; Fig. 2), Erdoğan et al. (1990) reported massive reef limestones, up to 50 m thick, with coral, bryozoans and algae, again passing upward into Middle Triassic limestone.

Vesicular volcanics from the upper part of the Gerence Formation were analysed by X-ray fluorescence (XRF) (Table 1a) and found to range from andesitic to dacitic composition. The samples are associated with red chert and are overlain by Lower Triassic nodular dolomite–mudstone and bedded limestone from the road section north of Ildır. On mid-ocean ridge basalt (MORB)-normalized diagrams (Fig. 8a), the extrusives display a distinctive calc-alkaline pattern, with large-ion lithophile (LIL) element enrichment and high field strength (HFS) element depletion relative to MORB. Nb is depleted relative to Ce but is enriched with respect to MORB; Ce is very enriched relative to MORB and Ti is depleted. With the exception of Cr, the other HFS elements, i.e. P, Zr, Y and Sc, are near to average MORB composition (Pickett 1994).

Radiolarian chert horizons within the Lower Triassic succession in the above area (Fig. 2.16) locally contain ore-grade material and were formerly mined for Mn (Erdoğan pers. comm.).

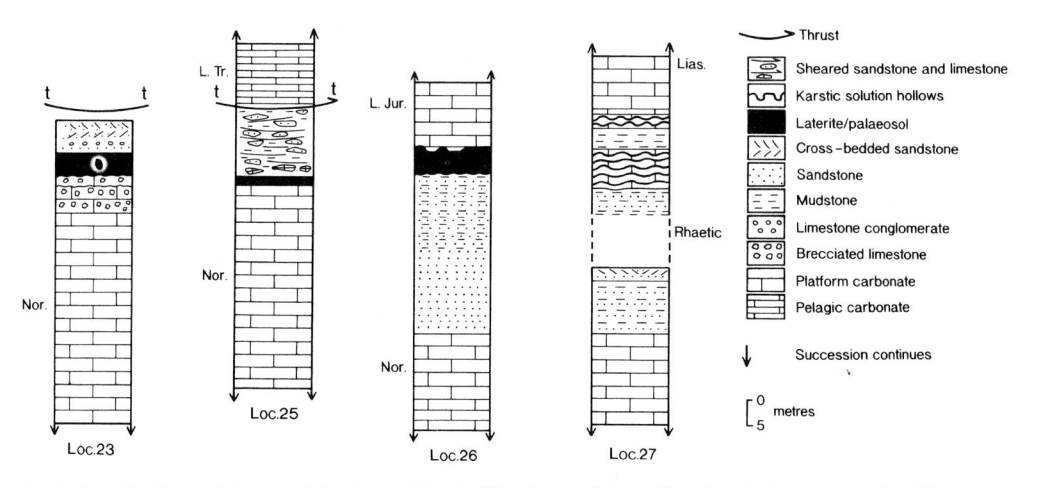

Fig. 7. Detailed logs of the top of the latest Triassic–Liassic transition within the platform succession. The quartzose sandstone intercalations are dated as Rhaetic (Erdoğan et al. 1990). For locations see Fig. 2.

XRF analysis of five samples of manganiferous chert yielded MnO concentrations of 0.89–51.61%; samples contained low values of Fe_2O_3 (< 2.6%), Al_2O_3 (< 6%) and CaO (< 1.7%). Trace element values are low with the exception of Ba and Sr, which correlate with high MnO values (Pickett 1994). The significance of these results is that they indicate that volcanism was associated with extensive hydrothermal activity (see below).

The Middle Triassic carbonate platform unit. The Lower Triassic Gerence Formation is conformably overlain by the Middle Triassic (Ladinian–Carnian) Camiboğazı Formation (500–> 1000 m thick), representing a thick, competent unit that forms much of the mountainous interior of the Karaburun Peninsula (Brinkmann *et al.* 1972; Fig. 2). An exposed transition between the two formations was reported locally in the south by Erdoğan *et al.* (1990), e.g. along the Ildır–Barbaros road (Fig. 2.21). The Camiboğazı Formation is dominated by little deformed, massive to thick-bedded, shallow-water platform carbonates, with algae, benthic foraminifera, ostracods, crinoids and gastropods. Mapping by Erdoğan *et al.* (1990) shows that the succession is repeated within two discrete thrust sheets above an exposed lower unit (that could also be allochthonous).

The Upper Triassic–Lower Jurassic disrupted platform succession. The Middle Triassic platform passes conformably upwards into dolomitic shallow-water carbonates of Late Triassic (Norian–Rhaetic) age (Güvercinlik Formation), exposed in each of the three tectonic units, especially in the lowermost one, widely exposed in the south (Erdoğan *et al.* 1990; Fig. 2.22 – the Ildır–Barbaros road). The carbonates include algal stromatolites, wackestones with megalodonts (large bivalves) and laminated dolomite. Neritic carbonates of the Güvercinlik Formation are also intercalated

Table 1. *Selected major- and trace-element analyses of extrusive rocks [from Pickett (1994)]*

(**a**) *Volcanics from the Karaburun Peninsula*

	06/10/92 2a	06/10/92 2d	06/10/92 2e
SiO_2	61.14	73.46	62.93
Al_2O_3	14.62	13.44	13.82
Fe_2O_3	6.10	2.72	7.67
MgO	0.98	0.64	0.85
CaO	5.27	2.19	4.58
Na_2O	2.59	2.97	1.42
K_2O	2.746	2.180	2.828
TiO_2	0.786	0.711	0.743
MnO	0.085	0.016	0.106
P_2O_5	0.165	0.148	0.143
LOI	5.68	2.07	4.86
Total	100.15	100.55	99.94
Nb	10.5	9.4	9.3
Zr	131.8	115.7	119.7
Y	24.3	21.2	22.5
Sr	164.3	300.4	171.3
Rb	87.3	67.1	94.3
Th	7.3	5.5	7.6
Pb	10.8	8.9	10.2
Zn	82.2	85.0	88.5
Cu	14.9	5.2	16.1
Ni	7.4	3.3	6.8
Cr	15.7	12.8	15.8
Nd	29.6	23.4	22.6
Ce	61.4	55.9	50.4
La	22.9	22.6	22.4
V	207.3	196.4	224.9
Ba	400.9	413.5	538.7
Sc	28.2	26.8	32.0

(**b**) *Volcanics from blocks in the Chios Mélange*

	C37	C40D	C45C
SiO_2	52.58	58.70	53.58
Al_2O_3	14.62	12.74	15.43
Fe_2O_3	8.15	8.83	8.26
MgO	9.05	4.70	5.51
CaO	6.35	4.33	3.30
Na_2O	4.15	3.34	5.02
K_2O	0.160	0.297	2.478
TiO_2	0.659	0.650	2.698
MnO	0.161	0.135	0.098
P_2O_5	0.137	0.136	0.491
LOI	4.97	6.47	3.62
Total	100.42	100.33	100.48
Nb	7.3	4.3	39.4
Zr	72.9	54.7	284.2
Y	19.5	14.7	32.5
Sr	85.3	253.7	372.3
Th	2.7	0.6	65.5
Pb	9.9	4.8	1.5
Zn	75.2	78.3	78.8
Cu	61.7	41.7	58.5
Ni	220.2	16.1	31.6
Cr	795.6	60.4	29.4
Nd	16.9	8.5	42.4
Ce	36.1	22.4	93.8
La	17.5	13.1	49.3
V	218.3	313.1	195.2
Ba	115.1	185.1	751.3
Sc	35.4	46.4	19.5

Analyses obtained by XRF method of Fitton & Dunlop (1985). LOI, loss on ignition.

with laterally discontinuous terrigenous clastic sediments of Norian–Rhaetic age, as follows (Fig. 7).

In the south, within the lowermost tectonic unit (east of Ildır; Fig. 2.22), local successions include c. 5 m of red mudstones, fine–medium-grained quartzose sandstones and limestones, with intraformational carbonate clasts. Individual clastic intervals, up to 1.4 m thick, are composed of fine–medium-grained amalgamated units. Occasional thicker intervals (up to 3 m thick) are medium-grained, well-sorted, quartzose sandstone. These relatively incompetent intercalations are sheared and folded, with small-scale duplex structures. To the north (west of Balıklıova; Fig. 2.23), thick-bedded Norian limestone is overlain by several metres of laterite, in turn overlain by sheared mudstone, sandstone, pebblestone and phacoids of limestone, beneath an overriding thrust sheet of Lower Triassic lithologies (Gerence Formation). A thrust plane cuts down-section towards the north where the Upper Triassic Güvercinlik Formation is tectonically removed.

Further south, in the same tectonic unit (near Barbaros; Fig. 2.24), the Lias is again shallow-water carbonate. However, the Rhaetic–Liassic transition is marked by medium–thick-bedded calcarenites, with subordinate limestone conglomerate, up to 2 m thick. Clasts are subrounded to rounded and mainly composed of limestone, with some black chert (< 2 cm in size). The calcarenites comprise well-rounded reworked neritic carbonate, quartz, schist and chert, set in a micritic matrix, partly recrystallized to dolomite.

In the northeast (Fig. 2.25), Erdoğan *et al.* (1990) mapped a lower thrust sheet in which the succession terminates in the Norian–Rhaetic Platform carbonates (Güvercinlik Formation). Up-section, steeply dipping thick-bedded Upper Triassic platform carbonates turn pink and become increasingly brecciated (Fig. 6f). Interstices between angular clasts are composed of pink ferruginous oxide sediment that becomes more abundant upwards. Several metres of sedimentary limestone breccia follow, with subangular clasts, then a reddish palaeosol, 3–5 m thick with caliche followed by c. 2 m of cross-bedded quartzose sandstone and

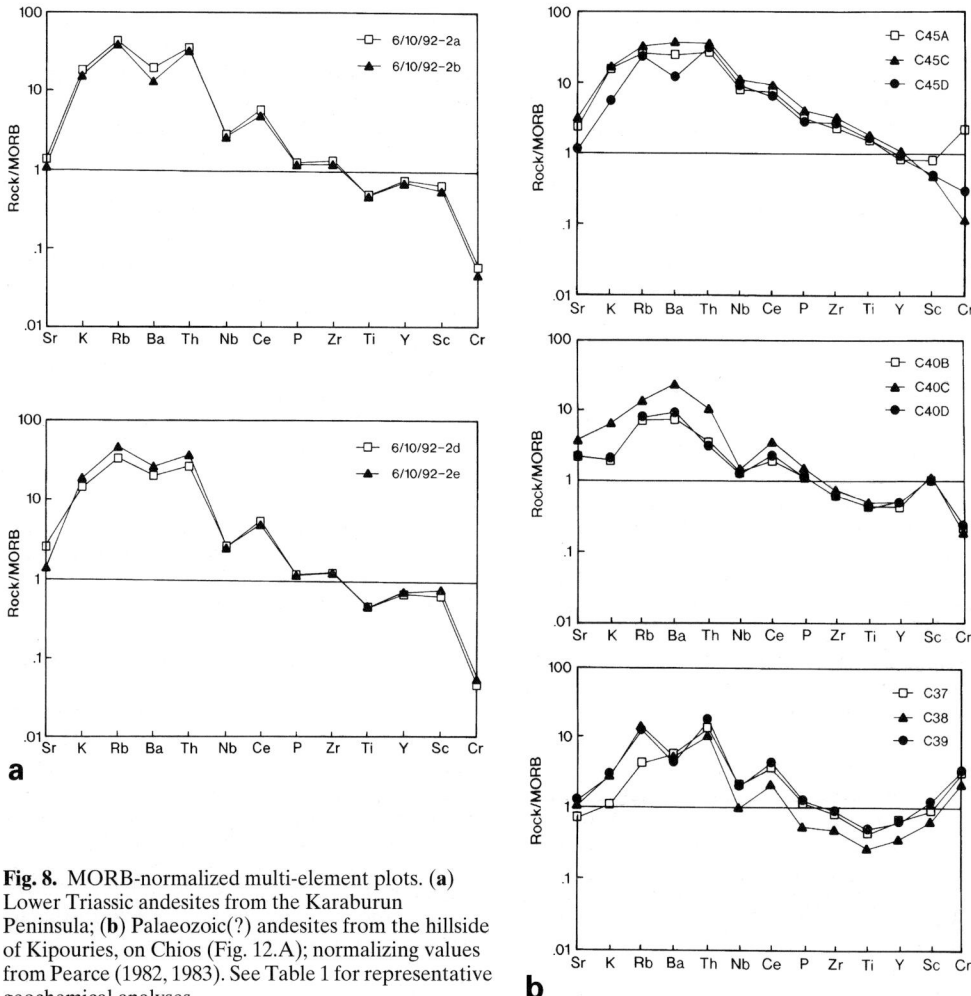

Fig. 8. MORB-normalized multi-element plots. (**a**) Lower Triassic andesites from the Karaburun Peninsula; (**b**) Palaeozoic(?) andesites from the hillside of Kipouries, on Chios (Fig. 12.A); normalizing values from Pearce (1982, 1983). See Table 1 for representative geochemical analyses.

conglomerate. A transition between tectonic breccia and proximal sedimentary breccia is seen locally. The sandstones are texturally and chemically mature lithic sandstone with subangular to subrounded grains of quartz, quartzite, radiolarian chert (variably recrystallized), together with subordinate mica schist, dark siltstone and aphyric basalt. The overlying lateritic sandstones include red radiolarite and other mainly subrounded to well-rounded grains in a ferruginous, calcareous matrix.

This intermediate-level tectonic unit is mapped as being directly overlain by the Lower Triassic (Gerence Formation) of the uppermost thrust sheet, in turn passing depositionally upwards into the Norian–Rhaetic Güvercinlik Formation (west of 26 in Fig. 2 and southwest of Mordoğan). The top of the Upper Triassic succession passes depositionally upwards into c. 30 m of soft-weathering calcareous mudstone and sandstone. Sandstone beds are thin–to medium-bedded (locally well cemented), preserving excellent way-up criteria (grading, channelized base of beds). These sandstones are composed of detrital bioclasts (e.g. algal micrite), reworked radiolarian chert, green (chloritic) altered lava, minor plutonic quartz and polycrystalline quartz, interbedded with ferruginous siltstone. The sandstone is depositionally overlain by several metres of red laterite, interbedded with thin-bedded sandstone. Stratigraphically underlying sandstones are cut by contemporaneous karstic solution hollows (several metres in diameter), infilled with red lateritic sediment. Above this, the Lower Jurassic carbonate platform succession is initially rubbly and pinkish, and then passes upwards into white, nearly massive, thick-bedded carbonates.

The continuation of this Lower Jurassic succession is very well exposed along the coast further southeast (Erdoğan et al. 1990; Fig. 2.27). This succession is dominated by classic peritidal stromatolitic carbonates (largely dolomitic), with excellent bird's-eye structure, well-preserved megalodonts and cryptalgal lamination. Minor reddish rubbly zones indicate periodical emergence. Towards the top of the succession, the limestone contains megalodonts and red stylolitic seams. A conformable Norian–Early Jurassic transition is well exposed. The lower part of the Lower Jurassic succession contains prominent siliciclastic intervals. Traced upwards, neritic carbonates become rubbly, then pass into red laterite, sandstone, mudstone, buff-brown coarse sandstone with large-scale cross-bedding, conglomeratic traction deposits, FeO concretions, abundant carbonized woody material and ferruginous mudstone intraclasts. The conglomeratic layers contain angular clasts of black chert and white quartz up to 3 cm in size. The sandstones consist mainly of highly strained, subrounded to angular quartz grains, displaying strong undulose extinction. Rare schist and devitrified volcanic grains are also present. The succession passes upwards into poorly exposed sandstone and mudstone, followed by a return to neritic carbonates, initially with thin (1 m thick) rubbly, red lateritic intercalations, marking the base of the 'Cladocoropsis limestones' of Brinkmann et al. (1972).

Jurassic–Cretaceous platform deposition. During the Lower Jurassic–Lower Cretaceous, carbonate platform deposition resumed (Nohutalan Formation) but successions are relatively thin (several hundred metres). As seen in a southerly exposure (Fig. 2.26), the Liassic–Lower Cretaceous succession (Nohutalan Formation; > 500 m) comprises monotonous thick-bedded limestones, with local cherty nodules and reef limestones in the upper part of the succession, together with a hiatus in the Late Jurassic and lateritic horizons in the Albian (Erdoğan et al. 1990).

Late Cretaceous demise of carbonate platform. Detailed mapping by Erdoğan et al. (1990) demonstrated the existence of an low-angle unconformity at the base of an Upper Cretaceous succession in the east (Balıklıova area; Fig. 2.29). In places, the Upper Cretaceous is mapped as resting directly on the Lower Triassic (Gerence Formation). A comparable unconformity is reported within a large block forming part of the Bornova Mélange on the mainland further east (Okay & Siyako 1993). Observations during this study revealed that the Upper Cretaceous succession is very disrupted and partly forms detached blocks within Maastrichtian turbidites. One outcrop of Upper Cretaceous neritic limestone, exposed directly west of Balıklıova old village (Fig. 2.30), shows a transition from well-bedded Upper Cretaceous limestone, with pinkish calcarenitic and pebbly intercalations, to 35 m of pinkish nodular pelagic limestones with *Globotruncana*, of Campanian–Maastrichtian age (Balıklıova Formation). These limestones, in turn, pass into mudstones and medium-bedded siliciclastic turbidites, dated as Maastrichtian–Danian (up to 300 m). On the mainland, further east (Savaştepe area), onset of pelagic carbonate deposition apparently occurred earlier, suggesting progressive collapse of the platform from east to west (Okay & Siyako 1993).

The Bornova Mélange. In the northeastern Karaburun Peninsula (near Kalecik; Fig. 2.31–33 and 35), the Mesozoic Karaburun Platform has disintegrated into blocks up to several kilometres in size, mainly composed of Lower Triassic limestone. These are set within sheared, medium–thin-bedded, partly lithified, lithoclastic sandstone turbidites and mudstones, locally interbedded with thin-bedded chalky pelagic limestones, with Campanian planktonic foraminifera and laminated chalky calciturbidites. Near their edges, blocks of competent Karaburun Platform rocks are sheared, veined and brecciated. For example, in quarries west of Balıklıova (Fig. 2.19), Upper Cretaceous neritic limestone is fragmented into blocks, tens of metres in size, embedded in calcareous turbidites. Quartzose sandstone beds, up to 0.45 m thick, are interbedded with reddish and brownish calcareous mudstone. Turbiditic sandstones locally abut directly against disrupted Upper Cretaceous limestone.

The mélange matrix includes numerous polymict debris flows, with clasts of red radiolarite, quartzite, pink pelagic limestone, basic extrusives (up to 5 m in size), serpentinite and other ophiolite-derived rocks (e.g. Fig. 2.34). Most clasts are < 30 m in size and commonly highly sheared. Where unoxidized in new road cuttings, the mudstone matrix is black and rich in fine plant material. The sandstones are grey and often phacoidally deformed. Shear fabrics in the matrix are commonly relatively gently inclined. The sandstones comprise angular to subangular clasts of plutonic quartz, metamorphic quartz, quartzite, radiolarian chert (in various states of recrystallization), mica schist, quartzose and micaceous siltstone, rare chloritized lava, biotite and muscovite set in a calcite spar matrix. In the northwest, the basal thrust of the Bornova Mélange cuts downwards into the Karaburun Mélange (Fig. 2.34) and blocks of black chert and Upper Palaeozoic (Lower Devonian)–Lower Triassic limestone (Erdoğan et al. 1990) are incorporated into the mélange.

Interpretation of Mesozoic Karaburun carbonate platform and overlying mélange

In the southwest, the Karaburun Mélange is unconformably overlain by a Lower Triassic basinal succession (Gerence Formation). The Middle Carboniferous succession, dismembered as a thrust sheet (Alandere Formation) within the mélange, was emergent and karstified before being overlain by Lower Triassic facies. Elsewhere, the contact between the mélange and the overlying (parautochthonous) Lower Triassic succession is strongly sheared. Above this, an upward passage over a short interval (tens of metres) into pelagic limestones, red radiolarian cherts and andesitic volcanics, suggestive of rapid deepening during the Scythian, is followed by overall shallowing upwards. The relative abundance of volcanics and radiolarian cherts increases northwards, suggesting the existence of deeper water and more distal conditions in this direction. Elsewhere, reef carbonates in the easterly (higher) thrust might represent a carbonate build-up on a horst. Localized manganiferous cherts are indicative of hydrothermal activity in a bathyal environment, probably associated with submarine volcanism (e.g. Robertson & Degnan 1998). A similar origin was also proposed for Mn-rich deposits above Triassic lavas in the Antalya Complex in southwest Turkey and for other Neotethyan Mn-cherts (Robertson 1981).

The implied subduction-influenced character of the volcanics must be treated with caution, as inferred Silurian volcanics from the Chios Mélange (see below) show a similar subduction influence and might thus reflect magmatism influenced by subduction in the Palaeozoic or even earlier, rather than coeval with Early Triassic subduction.

The overlying Middle Triassic (Ladinian–Carnian) platform limestones record tectonic subsidence (Camiboğazı Formation). The overlying Norian–Rhaetic interval (Güvercinlik Formation) displays characteristic features of intertidal–supratidal flats (e.g. birds-eyes, cryptalgal laminae) and back-reef lagoons, typical of the fenestral, dolomitic limestones of the Alpine Triassic 'Loferites' (Fischer 1964). Dolomitic micrite layers could represent tidal–channel levee deposits, with the dolomite probably being penecontemporaneous. Reef carbonates are absent in the successions studied.

The platform periodically became emergent, leading to development of laterite, palaeosols and karst. The local lenses of red sandstone may represent channel fills in a fluvio–deltaic setting. An abundance of woody plant material indicates that the emergent source landmass was densely vegetated. The high textural and chemical maturity of the sands is suggestive of recycling of quartzose metamorphic rocks, cherts and siltstone, presumably from the subjacent Karaburun Mélange, but input was solely from the underlying Triassic succession in one area studied (Fig. 2.25). The fact that platform carbonate beneath laterite and sandstone at one locality (Fig. 2.25) is brecciated, with clasts being reworked in an overlying conglomerate, suggests that the emergence was related to contemporaneous faulting. At this locality, sedimentary structures are suggestive of gravity deposition within a small tectonically controlled basin.

Two alternative explanations for the earliest Jurassic emergence and clastic deposition are: (1) rejuvenated crustal extension causing block uplift and erosion; and (2) compression (or transpression) inverting older basement rift faults. Similar uppermost Triassic–lowest Jurassic clastics occur throughout the northern margin of the Tauride–Anatolide Platform (e.g. Anamas Dağ) and have been related to a 'Cimmerian' compressional event with reported evidence of thrusting (Monod & Akay 1984). This latest Triassic–earliest Jurassic event was approximately coeval with deformation of the Karakaya Complex further north and the clastics of Karaburun (and Chios, see below); it could therefore represent a far-field echo of tectonic emplacement of the Karakaya Complex.

Shallow-water platform conditions resumed in the Early Jurassic–Early Cretaceous (Nohutalan Formation), with dolomites, patch reefs and hiatuses related to relative sea-level change. The mapped unconformity beneath the Upper Cretaceous (Campanian–Maastrichtian) succession (Balıklıova Formation) is then interpreted as the result of uplift and erosion related to transition from a passive margin to a thrust belt. A similar transition to a foredeep occurred, e.g. in Oman prior to overthrusting by nappes in the Late Cretaceous (Robertson 1987). In Oman, the erosional unconformity locally reached the Permian of the platform succession. A similar deepening-upward transition from neritic to pelagic carbonate is seen in Karaburun.

During the Maastrichtian–Danian, the Karaburun carbonate platform broke into blocks incorporated into the Bornova Mélange. The Bornova Mélange represents a sedimentary mélange of Maastrichtian–Danian age, with blocks of mainly Mesozoic Neotethyan lithologies, including abundant radiolarian chert, set in a basinal siliciclastic turbiditic matrix. The radiolarian chert and basalt clasts are assumed to have been accreted during subduction of oceanic crust before being incorporated into the Bornova Mélange. There are also extensively studied blocks of Upper Silurian–Lower Devonian limestones with coral near Kalecik (Gusic *et al.* 1984). These blocks may be recycled from the Palaeozoic Karaburun Mélange.

Erdoğan *et al.* (1990) mapped the Karaburun carbonate Platform as two main, tectonically repeated, thrust sheets above a lower unit (that could itself be allochthonous). Large-scale cutting downwards in the stratigraphy towards the west, combined with mainly west and northwest vergent folding, especially within the incompetent Rhaetic sandstones and mudstones, suggest that tectonic transport was towards the west with respect to present geographical coordinates. This thrusting was coeval with, or postdated, formation of the Bornova Mélange in latest Cretaceous–Palaeocene time. In places, thrusting exploited lithological contrasts at the base of the Lower Triassic coherent succession (Gerence Formation) and also incompetent Rhaetic clastic intercalations, in contrast to very competent neritic limestones above and below.

The island of Chios

The island of Chios (Fig. 9) consists of an underlying 'autochthonous' unit and an overthrust 'allochthonous' unit (Herget & Roth 1968; Besenecker *et al.* 1968, 1971). The lower unit is a deformed Palaeozoic thick unit, > 3–4 km, with 'olistoliths' of Devonian, Silurian and Carboniferous limestones up to hundreds of metres in size (i.e. the *Chios Mélange*; Table 1). During this study, detached blocks, principally limestone, black chert and volcanics, were identified within a matrix of siliciclastic turbidite, occasional carbonate debris flows, rare calciturbidites and channelized conglomerates. Early workers interpreted this unit as a Palaeozoic layer-cake succession, with Lower–Middle Silurian, Lower Devonian and Lower Carboniferous lithologies (Ktenas 1921, 1923a, b; Paekelmann 1939; Teller 1980; Besenecker *et al.* 1968, 1971; Kauffmann 1969). The mélange is unconformably overlain (with a locally sheared contact) by Lower Triassic radiolarian chert, basinal siliciclastic and volcanogenic rocks which pass upwards into a Lower Mesozoic carbonate platform succession. The overthrust exotic unit in the northeast of the island consists of Permian shallow-water carbonates, overlain by Lower Jurassic shallow-water terrigenous and calcareous sediments (Fig. 10). Permian lithologies are not known elsewhere on the island.

The Chios Mélange

This is dominated by deformed quartzose sandstone and shale, with blocks of limestone, volcanic rocks and chert set within a sheared mainly siliciclastic matrix (Papanikolaou & Sideris 1983). The mélange forms an eastward dipping stack of folded sandstone and shale, with four (named) major horizons of large blocks of Palaeozoic limestone and volcanics (Fig. 11), interpreted as sedimentary olistoliths with olistostromes by Papanikolaou & Sideris (1983). Well-preserved sedimentary way-up structures within the clastic sequence allow the recognition of predominantly west verging folds, locally displaying overturned limbs. Folded shales interbedded with the sandstone also display asymmetry towards the west. Intense shearing is also commonly observed, giving rise to classic phacoidal fabrics seen in tectonic mélange and slice complexes.

Blocks within the mélange. Limestone blocks (metres to kilometres in size) within the mélange range in age from Silurian to Early Carboniferous, with a predominance of older (i.e. Silurian–Devonian) limestone blocks in the uppermost mélange unit (e.g. at Agrelopos) and younger (i.e. Carboniferous) blocks in the lowest unit (e.g. at Melanios; Papanikolaou & Sideris 1983; Fig. 12). The uppermost mélange unit is well exposed near Kipouries (Fig. 12.A), where it contains numerous limestone blocks many tens of metres across, in a quartzitic conglomerate matrix. Further north (near Kampia; Figs 12.B and C and Fig. 13), limestone up to 500 m thick forms numerous blocks for several kilometres along-strike, together with numerous additional blocks (e.g. volcanics), set in a sheared siliciclastic matrix (Agrelopos Limestone and Kefala Volcanics; Fig. 16a).

Lithologies within the limestone blocks include limestone breccia (Fig. 16e) with fine-grained, marly matrix material, grey and pink micrite, marly limestone and limestone conglomerates. One block from the Agrelopos unit (2 km southwest of Kipouries; Fig. 12.A), consists of crinoidal, coralline and oolitic fragments with laminated micrite interstices, possibly representing lithified reef talus. Limestone blocks are commonly mantled by up to 10 m of limestone breccia. Laterite-filled fissures, up to 3 m wide, cut the limestone and surrounding limestone breccias, but not adjacent quartzitic conglomerates. Seams and patches of fine-grained marly material cut across laterite fissures, but terminate against the quartzitic conglomerate. The contact between quartzitic and limestone conglomerate is marked by a ferruginous crust.

A large limestone block *c.* 1 km southwest of Kampia (Fig. 12.C), dated as Silurian–Early Devonian (Papanikolaou & Sideris 1983), is locally karstified and cut by laterite-filled

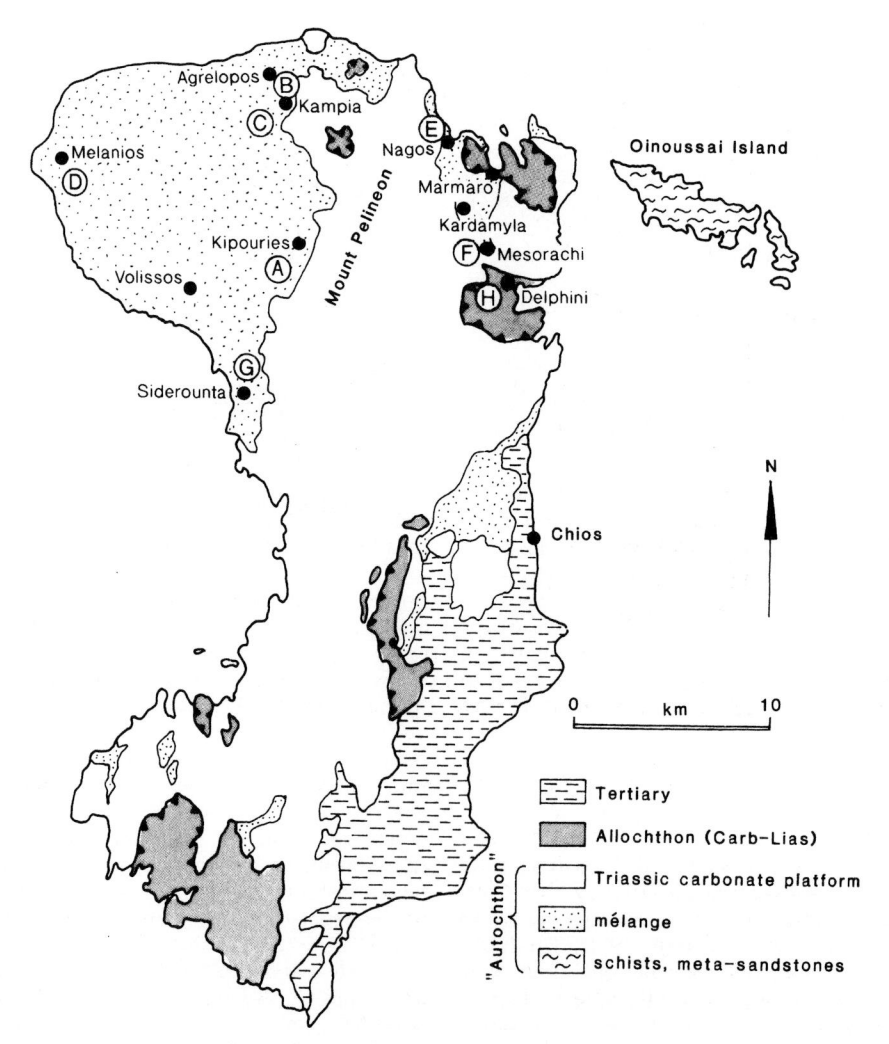

Fig. 9. Geological map of Chios modified after Besenecker *et al.* (1968). Note the location of the Chios Mélange in the northwest of the island; also the occurrence of klippen of Carboniferous–Liassic shallow-water sedimentary rocks (allochthon). Locations mentioned in the text are indicated.

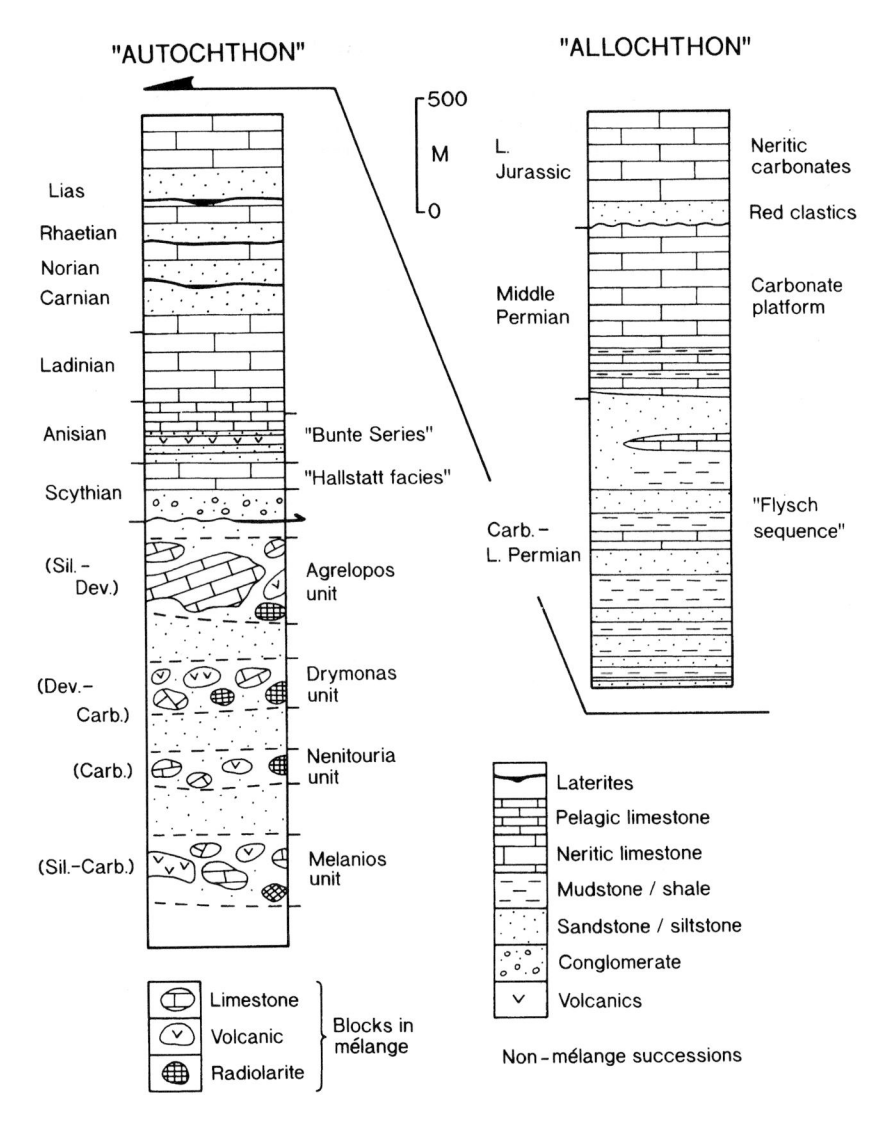

Fig. 10. Schematic stratigraphical successions through the relatively autochthonous lower unit and the allochthonous upper unit on Chios. Modified after Papanikolaou & Sideris (1992).

fissures (Figs 14 and 15). A laterite layer up to 10 m thick marks the base of the limestone block and is the site of an old galena and sphalerite mine (Papanikolaou & Sideris 1992). Grading and load structures within interbedded sandstones and shales beneath this limestone–laterite block indicate that the sequence is overturned. Thin-bedded redeposited carbonate horizons and smaller limestone blocks (up to 10 m across), also containing laterite fissures, occur within conglomerates surrounding this large limestone block. Elsewhere (e.g. near Agrelopos), palaeokarst and galena–sphalerite mineralization is associated with large limestone blocks.

Volcanogenic blocks, some > 50 m in size, are present in each of the four zones of exotic blocks and are especially well exposed in the structurally higher part of the mélange, near Kipouries (Fig. 12.A and Fig. 14). Successions within such blocks are commonly volcaniclastics, with dark green angular clasts up to 10 cm in size, rich in clinopyroxene phenocrysts (up to 3 mm across) and vesicles (0.5–6 mm across). A lithologically similar, but a larger, volcanic block west of Kampia (Fig. 12.C and Fig. 13) preserves a

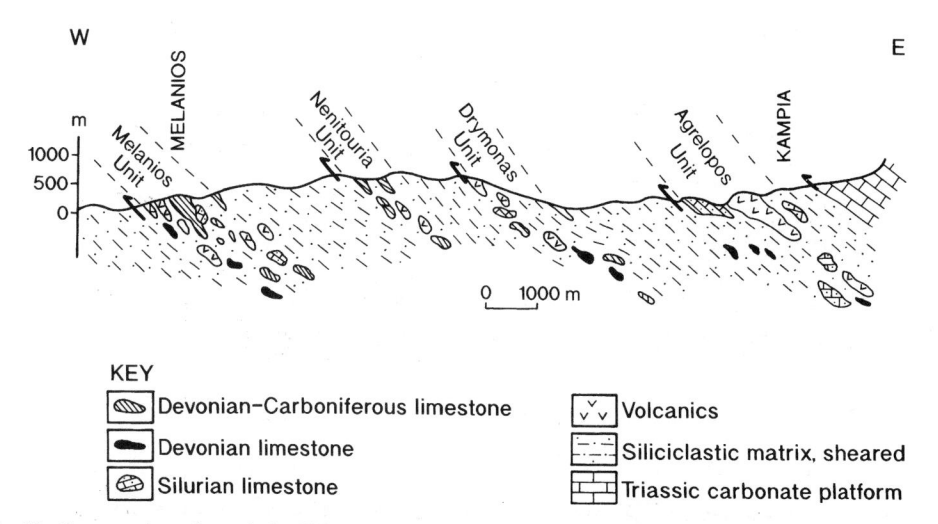

Fig. 11. Cross-sections through the Chios Mélange. Redrawn after Papanikolaou & Sideris (1992). Note the presence of four main zones with numerous blocks of Palaeozoic limestone and associated volcanics.

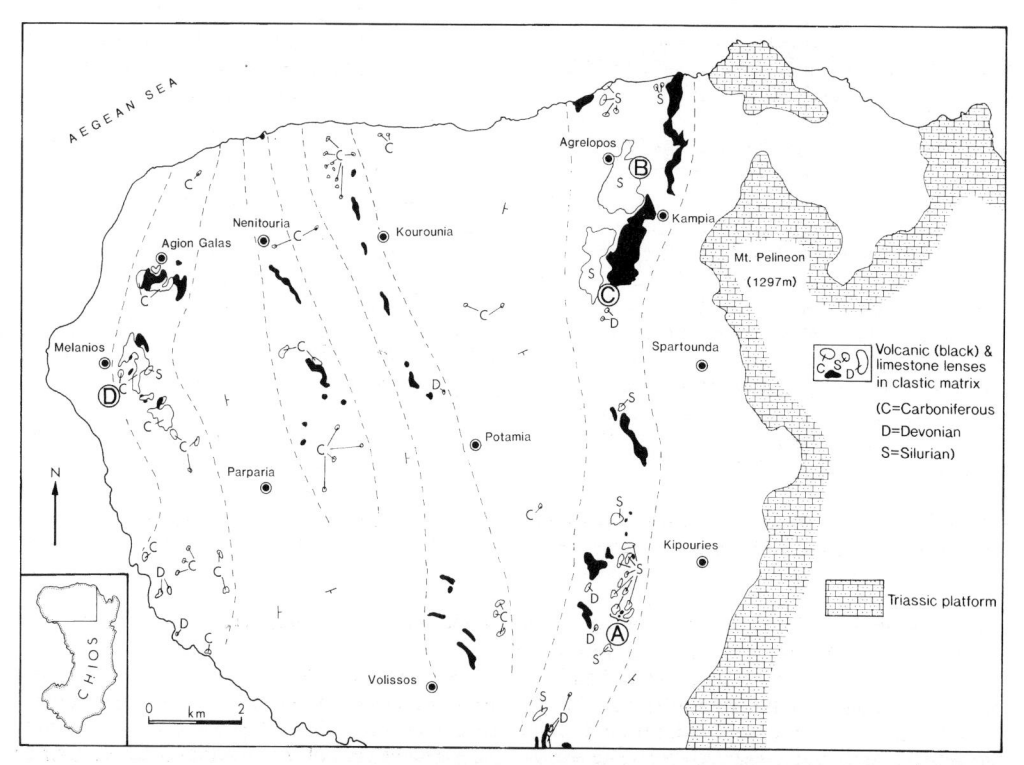

Fig. 12. Geological map of the Chios Mélange and the overlying Lower Mesozoic platform succession, showing the ages determined for the exotic blocks and locations mentioned in the text. Map modified after Besenecker *et al.* (1971) and Papanikolaou & Sideris (1992).

> 200 m thick pyroclastic succession, including volcaniclastics and a massive flow or sills. Blocks of lava breccia ('agglomerate') are also observed in the lowermost mélange horizon, *c.* 1 km southeast of Melanios village (Fig. 12.D).

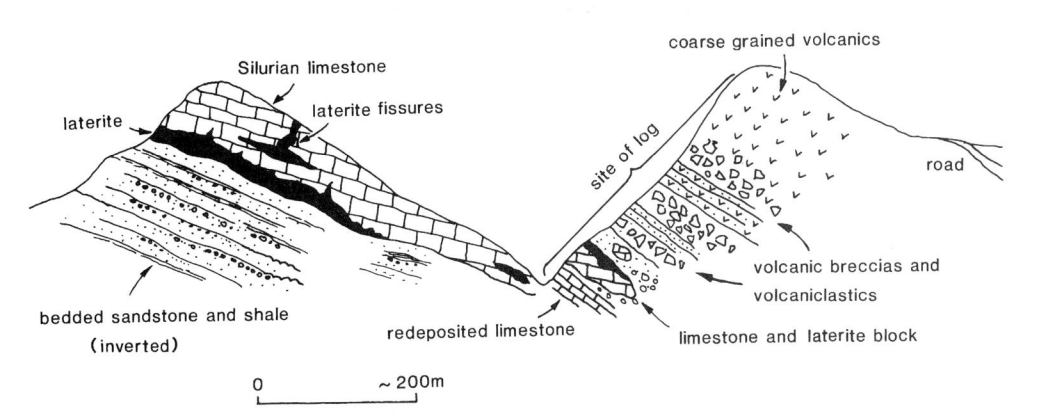

Fig. 13. Sketch section through the upper part of the mélange (Agrelopos unit, southwest of Kampia; Fig. 12, C). The Silurian limestones form a large block. Note that the associated sandstone and shale are inverted, suggesting the presence of large-scale folding within the Chios Mélange.

Eleven samples from three large volcanic blocks from the hillside near Kipouries (Fig. 12.A) were analysed by XRF and shown to be predominantly basaltic andesite to andesitic in composition (Table 1b; Pickett 1994). The data from one block, when plotted on MORB-normalized trace element plots, exhibit 'humped' traces, showing considerable enrichment in LIL elements and a gradual slope down to Y and Sc (Fig. 8b, samples C45A–D). By contrast, the samples from a neighbouring block display a sharp trough in Nb and depletion of Zr, Ti, Y and Cr relative to MORB (samples C40B–D). Samples from a third block also display relatively low Nb with depletion of several (HFS) elements relative to MORB (samples C37–C39).

Chert. Blocks of sheared black chert up to many metres across (Fig. 16b), without associated terrigenous sediment, are exposed within the Chios Mélange, mainly, in northeast Chios, near Nagos Bay (Fig. 9.E). The chert is typically in tectonic contact with sheared turbiditic sandstone and shale. Bedded chert is boudinaged, chevron folded and highly sheared.

Terrigenous matrix. Many of the limestone, chert and volcanic blocks are enveloped within sheared turbiditic sandstone and shale, although contact relations are rarely well exposed (Fig. 16d). Sedimentary structures observed in the clastic sediment include normal grading, lamination, cross-lamination, load casts, grooves, prods, flute casts and plant fragments. Matrix sandstone from the vicinity of large blocks of limestone and lava in the upper mélange unit (Agrelopos unit, southwest of Kipouries; Fig. 9.A) is very poorly sorted lithic sandstone with mainly subrounded to, locally, well-rounded grains of quartz, polycrystalline quartz, basic–silicic volcanics with rare micaschist, quartzose sandstone and siltstone, cut by a spaced fracture cleavage but without recrystallization.

Quartzitic conglomerates appear to form a matrix to the large limestone and volcanic blocks in some areas (e.g. Siderounta area; Figs 9 and 17), although it is possible that some of the conglomerate outcrops may also represent blocks. The quartzitic conglomerates are clast supported and comprise rounded cobbles, up to 15 cm across, composed of arkosic sandstone, acidic volcanic rocks, grey quartzite, white vein quartz and micaceous quartzites (Fig. 16c). The contacts between the limestone blocks and the conglomeratic matrix are relatively diffuse, with intervening zones of brecciated limestone and calcareous conglomerate (up to several metres wide). Polymict conglomerates containing sandstone, chert, limestone, green volcanic rocks and reworked laterite are also well exposed near Kampia (Fig. 12.B and C). Rounded cobbles of quartzite, volcanics and chert there are identical to those near Kipouries, and clearly appear to form a matrix of large limestone and volcanic blocks. The coarse quartzitic conglomerate mélange matrix appears to be confined to horizons with large limestone and volcanic blocks. Between the horizons of large limestone and volcanic blocks, the succession is predominantly bedded sandstone and shale with well-preserved sedimentary structures. Chromite grains were identified within these sediments (Stattegger 1984), suggesting derivation from ultramafic ophiolitic rocks not exposed in the area. Predominantly southward palaeoflow is indicated by flute casts near Potamia (Fig. 12), supported by evidence of north–south groove marks. In contrast, ripples and grooves from low levels of the mélange near Volissos (Fig. 12) indicate east–west (bimodal) palaeocurrents (Pickett 1994).

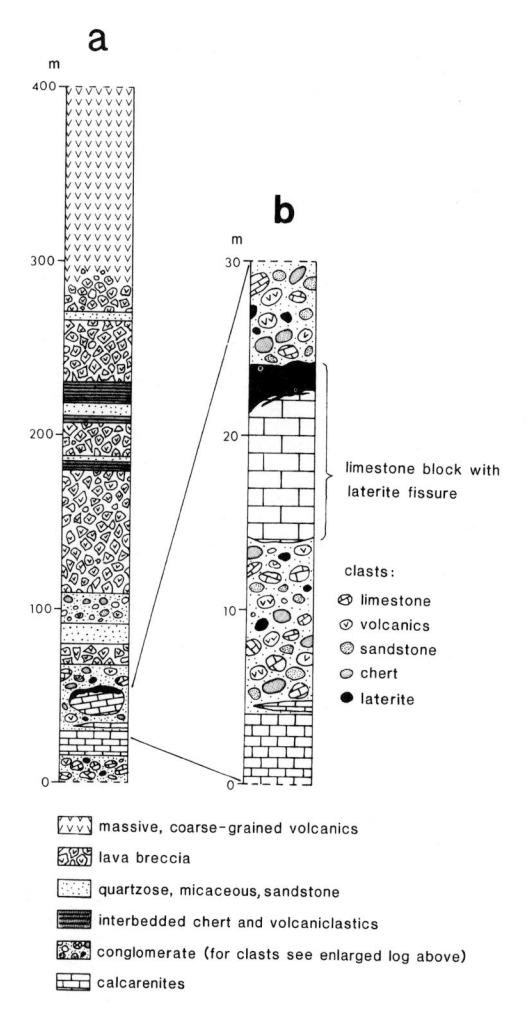

Fig. 14. (a) Logs through the upper part of mélange (Agrelopos unit, southwest of Kampia; Fig. 12.C); overall succession through mainly volcaniclastic unit. (b) Detailed log through an interval of heterogeneous blocks. See Fig. 13 for location of logs.

Interpretation of Chios units

The Chios Mélange comprises blocks of limestone, extrusives, possible intrusives (sills), volcaniclastics and black chert set in a turbiditic siliciclastic matrix (Papanikolaou & Sideris 1983). The dominance of Carboniferous limestone blocks in the lower part of the mélange but Silurian blocks higher in the unit, show that the mélange cannot simply be a tectonically disrupted stratigraphic succession. Instead, many features typical of subduction–accretion complexes are exhibited, including layer-parallel extension competent blocks and radiolarian cherts within a mainly pelagic turbiditic incompetent clastic matrix. The limestone blocks originated in a shallow-water setting, including coralgal reefs, whereas others are pelagic.

For the blocks near Kipouries (Fig. 12.A), the following sequence of events is inferred. Firstly, reef talus was lithified and subaerially exposed, resulting in karstification, brecciation and laterite formation. Later, this entire unit was detached as a coherent block. Break-up of the edges of the block led to fracturing and infilling by marly material. The block was finally emplaced in the mélange, together with volcanic blocks and quartzitic conglomerate. The thick clast-supported conglomerates with well-rounded clasts reflect deposition in a shallow-marine, or fluvial, setting – possibly within a major channel – without evidence of interbedded gravity deposits. The conglomerates are, therefore, likely to have become incorporated during initial detachment and subsidence of blocks from a parent carbonate platform. The limestone and volcanic blocks were detached and mixed with conglomerate during formation of the mélange. A problem is the timing of formation, as the matrix remains undated.

Pe-Piper & Kotopouli (1994) suggested that the volcanics represent a relatively intact Palaeozoic sequence that they correlated with pre-Middle Permian extrusives on the island of Lesbos (Hecht 1972; Katsikatsos et al. 1982), where sparse volcanics are exposed. The relative Nb depletion in two of the blocks analysed during this study is suggestive of a subduction component (Fig. 8b), but whether this is contemporaneous is unknown. In contrast, a sample from one block displays a typical 'humped' pattern of within-plate basalt (Pearce 1982, 1983). The volcanics might thus have originated in more than one tectonic setting, a view supported by Pe-Piper & Kotopouli (1994) who analysed a larger number of volcanic rocks from the structural upper part of the mélange in which they also identified both subduction

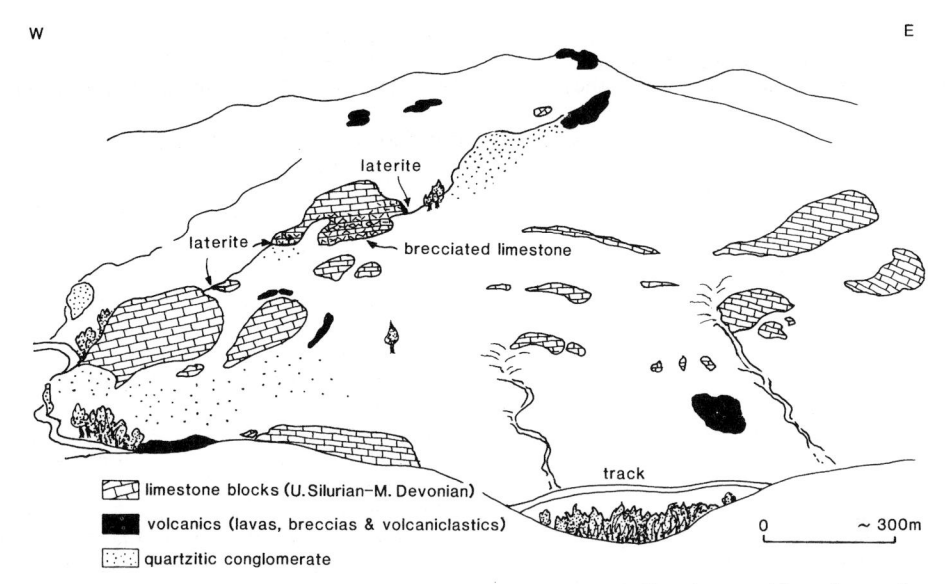

Fig. 15. Field sketch (oblique view of hillside) of the upper mélange unit (Agrelopos unit) southwest of Kipouries (Fig. 12.A). Note the blocks of limestone and volcanics up to several hundred metres in size, in a matrix of mainly quartzitic conglomerate.

and within-plate components. In summary, the Chios Mélange represents a once intact Silurian–Carboniferous carbonate platform succession with Silurian volcanics that was accreted as blocks within a mélange that includes (undated) radiolarite and a terrigenous matrix, locally with detrital chromite grains.

The Mesozoic carbonate platform of Chios

Northern Chios is dominated by Mount Pelineon, which represents a north–south trending Triassic–Lower Jurassic carbonate platform at its thickest extent. Successions were described in detail by Fryssalakis (1985) and only a brief summary is given here to allow comparison with Karaburun.

Clastic rocks at the base of the platform are represented by a Scythian–Anisian basinal succession in the Mesorachi area (Figs 9 and 10). The mélange grades upwards through sheared dark grey-green mudstone into intraformational conglomerate, with chert fragments within green-grey and red cherty mudstone. This conglomerate is interbedded with fine-grained sandstone and mudstone, and then passes into massive and thick-bedded clast-supported conglomerates containing abundant rounded white quartz clasts (1 mm to 3 cm across). Other clast lithologies include sandstone, black chert, green volcanic rocks, green mudstone and conglomerate, up to 12 cm across. The conglomerate is also interbedded with red and green mudstone and medium–coarse-grained sandstone with traces of cross-bedding. The conglomerates are laterally discontinuous and grade into sandstone over a few metres. A sandstone from within conglomerate near the base of the platform succession (at Mesorachi; Fig. 9.F) is poorly sorted lithic sandstone composed of moderately well-rounded to well-rounded grains of metamorphic quartzite, together with subordinate veined microcrystalline chert, micaschist, rare quartzose siltstone and sandstone, and scattered folded muscovite laths.

The conglomerates become matrix supported upwards and pass into medium-grained sandstone, then into sheared green mudstone with grey calcarenite lenses. The calcarenite then turns pinkish and dolomitic, with red mudstone partings, locally nodular (Lower Anisian Hallstatt Facies: Besenecker *et al.* 1968; Asserto *et al.* 1979; Fryssalakis 1985). This then locally passes into intraformational conglomerate ('Bunte Series'), dated as Early Anisian on the basis of conodonts (Kauffmann 1969). The conglomerate is reddish and contains angular dolomite fragments in a muddy matrix. It is overlain by red shales and cherts with bands of silicic conglomerate and pale tuff which, in turn, pass into nodular limestone containing gastropods.

Elsewhere, Triassic silicic conglomerates, very similar to those of the Mesorachi region,

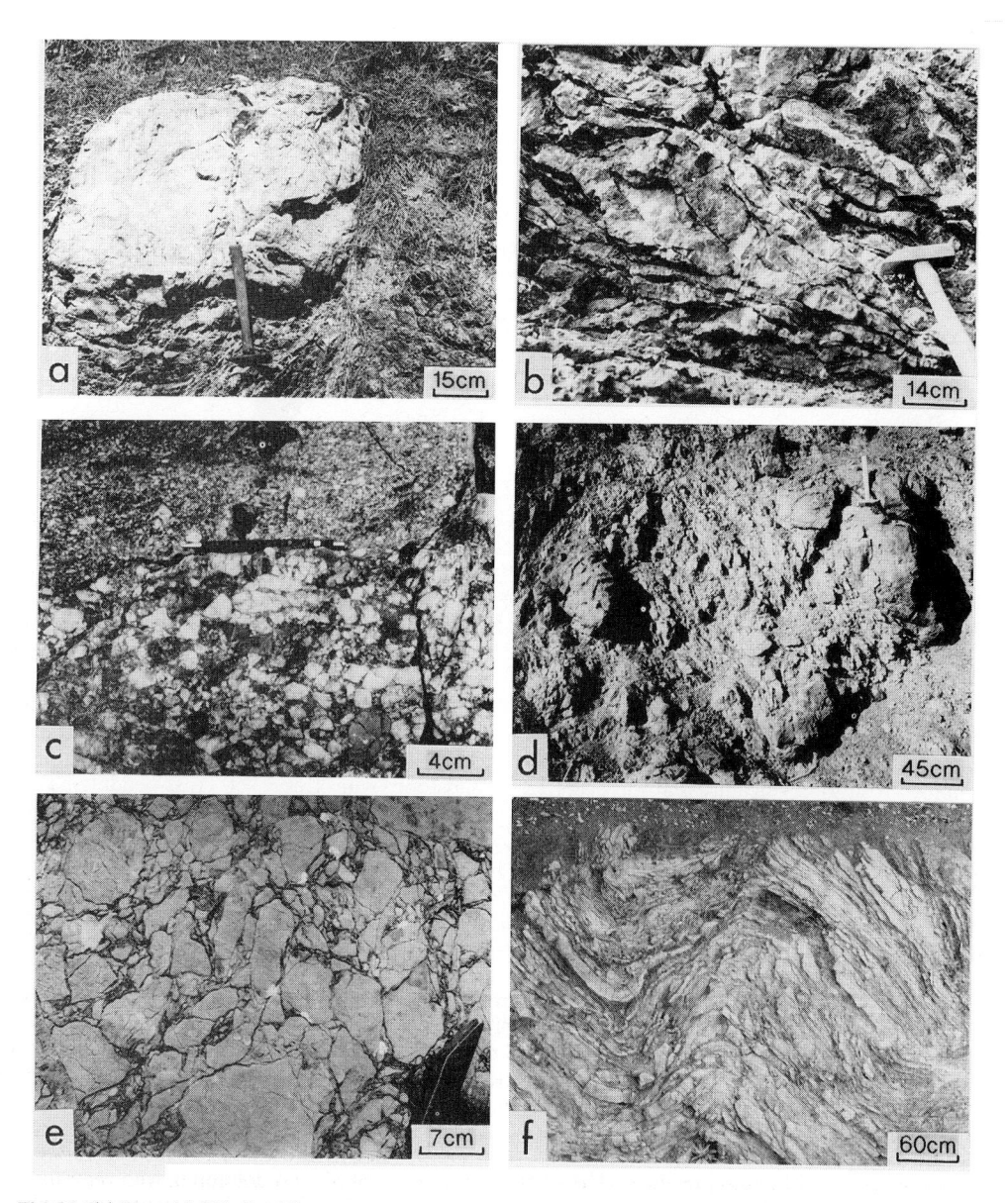

Fig. 16. (**a**) Phacoidal block of limestone in a sheared shaly matrix (Agrelopos unit), Sgourou Nero area. (**b**) Duplex structure is black ribbon chert ('lydite'). (**c**) Quartzose conglomerate, north of Siderounta. (**d**) Chacoidal structure within sandstone and shale; road north of Volissos. (**e**) Limestone breccia associated with limestone exotic; Sgourou Nero area. (**f**) Folded turbidites ('wildflysch') west of Delphini. Carboniferous–Upper Permian, upper nappe.

are also exposed near Kipouries (Fig. 12.A). Conglomerates containing fining-upwards sequences and sharp bases are interbedded with purple-grey sandstone. The conglomerate is predominantly made up of clasts of opaque white quartz (forming *c.* 80% of clasts) with minor amounts of limestone, laterite and black chert clasts. At the top of the logged section, the conglomerate contains coarser bands. Thin-section study reveals the presence of abundant lithic grains, including devitrified porphyritic andesite–dacite, glassy basalt, fine-grained sandstone, quartzite, mica schist and granodiorite. The grains are surrounded by dark red to

The upper nappe ('Chios Allochthon') is made up of mixed siliciclastic–carbonate sediments of Carboniferous–Early Permian age (Kauffmann 1969; Fig. 16f), passing up into a Middle Permian neritic succession with fusulines, algae, corals and brachiopods (Baud *et al.* 1991; Migiros & Sideris 1992). In turn, overlying transgressive Lower Liassic clastics (conglomerates, sandstones and siltstones), pass up into Jurassic neritic limestones and dolomites *c.* 300 m thick. South of the Mesorachi succession, in the east, near Delphini (Fig. 9), Triassic limestone of the lower units is tectonically overlain by highly folded lithologies of the overlying thrust sheet. Fold geometry in the Carboniferous 'flysch' and overall thrust geometry is reported to indicate compression from the northeast–southwest (Papanikalaou & Sideris 1983), whereas local structural measurements during this study, near Nagos (Fig. 9.E) and west of Delphini (Fig. 9.H), indicate emplacement from the southeast to the northwest. The Triassic limestone near the basal contact is very folded and appears to be locally overturned. Box folds in the limestone are common and may have been caused by back-thrusting and pop-up as the limestone was detached along the contact with the underlying mélange.

Interpretation of the Chios Platform and overthrust unit

The Triassic platform of Chios displays a well-preserved basal section which indicates that shallow-water carbonates became established above a terrigenous clastic basement (Fryssalakis 1985). Although sheared, the basal clastics appear to be a transitional unit between the Chios Mélange below and the overlying Triassic succession (e.g. at Mesorachi; Fig. 9.F). The probable explanation is that clastic sediment was eroded from the mélange during subaerial exposure (Late Permian–basal Triassic?) and that this was later sheared (during Early Tertiary collision?) beneath the more competent Mesozoic platform succession. After initial marine transgression, pronounced deepening upwards is suggested by the presence of overlying Lower Triassic cherts and red mudstones. The presence of intraformational chert breccia indicates reworking of previously lithified chert. Also, fining-upward sequences within terrigenous clastics are interpreted as coarse-grained, relatively proximal turbidites (e.g. near Kipouries). The clasts within the conglomerates and sandstones reflect both sedimentary and volcanic sources. Common opaque white

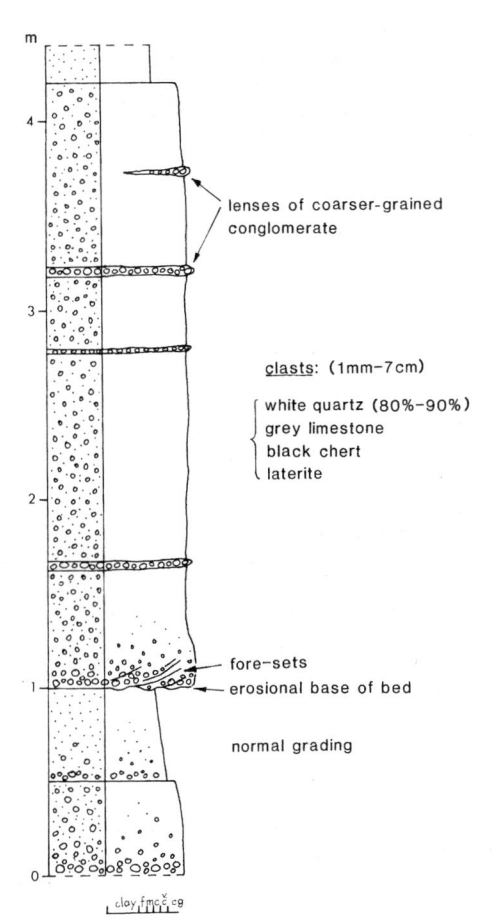

Fig. 17. Log of quartzose conglomerates north of Siderounta (Fig. 9.G). Note also the clast types. These conglomerates are interpreted as large channels within the matrix of the Chios Mélange.

opaque ferruginous material which also appears as seams within some of the grains.

The Anisian, mixed 'Bunte unit' is overlain by massive dolomites and algal limestones of Ladinian age, marking the establishment of a carbonate platform, > 1000 m thick. The overlying Carnian interval includes clastic deposits (Fryssalakis 1985). The earliest Jurassic is marked by emergence and non-marine clastic deposition. Above, the Early Jurassic interval includes at least three emersion horizons, characterized by intercalations of conglomerate and sandstone, together with bauxitic horizons (Papanikolaou & Sideris 1992). Upper Jurassic limestones are restricted to an isolated outcrop of Cladocoropsis limestone (Besenecker *et al.* 1968). Unconformably overlying Palaeogene conglomerates are seen locally in southeastern Chios.

quartz may have originated as vein quartz. Sedimentary structures are abundant and, at Mesorachi (Fig. 9.F) and north of Siderounta (Fig. 9.G), indicate southward directed flow in present day coordinates, parallel to the trend of the platform. However, more detailed studies suggest an intricate local palaeogeography (Fryssalakis 1985). Dolomites and algal limestones at the base of the overlying Ladinian carbonate platform succession indicate establishment of very shallow-marine conditions, followed by emergence in the latest Carnian. During Carnian–Norian time a relatively stable carbonate platform developed, followed by important emergence and clastic deposition in earliest Jurassic time. Summarizing, the Triassic–Early Jurassic was marked by overall subsidence, punctuated by emergence that was probably tectonically controlled. In contrast, the upper nappe records a Carboniferous–Lower Permian succession that shows sedimentary features suggestive of deposition from turbidity or storm-influenced currents. The succession then shallowed up into a Middle Permian shallow-water carbonate platform. This is unconformably overlain by Lower Jurassic clastic sequences passing into a Jurassic carbonate platform.

The lower ('autochthonous') and upper 'allochthonous' units of Chios differ strongly in their Palaeozoic history. For example, the exotic blocks in the Chios Mélange were clearly not derived from the upper unit platform, as Permian blocks are absent; Triassic is absent from the upper unit. The Permian of the upper nappes is comparable with the Permian of the Karakaya Complex further north (Fig. 2). However, no lower flysch succession is known there and the Lower Jurassic cover is pelagic (Bilecik Limestone) rather than neritic as in the upper nappe (Okay *et al.* 1991; Pickett & Robertson 1996). The upper nappe can also be compared with the autochthonous basement of Lesbos Island to the north, which includes Carboniferous siliciclastics and Permian neritic carbonates, although Triassic lithologies are absent from the upper Lesbos nappe.

Correlation of the Chios and Karaburun units

The Karaburun and Chios Mélanges are very similar and can be correlated (Table 1); Silurian–Carboniferous limestone blocks are abundant in both mélanges. Tectonically dismembered thrust sheets, or broken formation, rather than isolated mélange blocks are present in both areas. Scattered exposures of the Alandere Formation, south of Ildır, may mark a horizon of Carboniferous limestone blocks, as observed on Chios (e.g. at Agrelopos). Both mélange units contain similar siliciclastic sediments, black chert, black siltstone, limestone and intermediate basic composition volcanics. The lower schistose unit of Karaburun is not exposed on Chios but is apparently similar to metamorphics on the intervening island of Oinoussai (Ktenas 1925; Fig. 9).

Taken together, the exotic blocks of Karaburun and Chios can be restored as Silurian, Devonian–Upper Carboniferous (Lower Permian?) successions. Terrigenous intercalations (e.g. Lower Carboniferous of Karaburun) suggest derivation from a metamorphic (?Pan-African) basement. A Silurian age for at least some volcanism is suggested by the close spatial association of volcanics with large Silurian blocks in the Agrelopos unit near Kampia (Fig. 12). Also, a Silurian age is inferred in Karaburun, where tuffs are associated with dated Silurian limestone (e.g. Ildır, Fig. 2; Kozur 1995, 1997). The mainly subduction-influenced chemical signature might then signify Silurian subduction-related volcanism that was either contemporaneous or inherited from some even earlier (?Pan-African) magmatic event. The presence of within-plate-type extrusives in both Karaburun and Chios could relate to accretion of seamounts, removed from the chemical influence of subduction. The possible Silurian setting was a Palaeotethyan rift, followed by establishment of a subsiding carbonate margin, passing laterally into slope (calciturbites) and basinal (radiolarian) facies, presumably overlying Palaeotethyan oceanic crust. The presence of laterites and karstic weathering (e.g. some Chios blocks) points to periodical emergence and weathering in a warm, humid climate. In general, the Palaeozoic units could record a Palaeotethyan passive margin succession, e.g. related to a microcontinent within the Palaeotethyan ocean, if not the main continental margin.

The structural style of both the Chios and Karaburun mélanges is similar; fold emergence was mainly towards the west in both. Blocks were possibly detached along zones of weakness represented by the laterites. A relatively proximal setting for break-up with respect to a terrigenous source area is suggested by the coarse siliciclastic conglomerates in the upper part of the Chios Mélange (Agrelopos unit, near Kipouries). The Karaburun Mélange was strongly deformed after deposition of the Mesozoic platform succession, presumably in Early Tertiary time. The platform was imbricated,

Triassic cover rocks were sliced into the mélange, and the contact between the mélange and Mesozoic platform was strongly sheared. Inferred Tertiary deformation was less intense on Chios where the Triassic platform cover on the mélange, although sheared, is more intact. The upper nappe was presumably emplaced in the Early Tertiary, after loss (?erosion) of the upper part of a Mesozoic platform succession equivalent to the complete Mesozoic platform succession of Karaburun.

Both the Karaburun and Chios Mélanges are overlain by Lower Triassic basinal successions, including mudstones, radiolarites and volcanogenic lithologies, whereas mainly tuffaceous intercalations are exposed on Chios. Volcanics are exposed in northwestern Karaburun. The greater abundance of quartzose conglomerates, sandstones and intraformational conglomerates on Chios, compared to more pelagic *Daonella* limestones and Ammonitico Rosso in western Karaburun, suggests a relatively proximal setting for the former area. The Late Triassic–Early Jurassic intervals are similar in both areas, with platform carbonate subsidence punctuated by emergence and siliciclastic deposition. On Chios, shallow-water carbonates are preserved only until the Early Jurassic, but in Karaburun they extend to the latest Cretaceous. Both areas underwent west vergent folding and thrusting, which is dated as post-Maastrichtian in Karaburun.

Alternative tectonic models for Chios and Karaburun

The origin of the Karaburun and Chios Mélanges is clearly complex and a number of alternative tectonic scenarios can be considered.

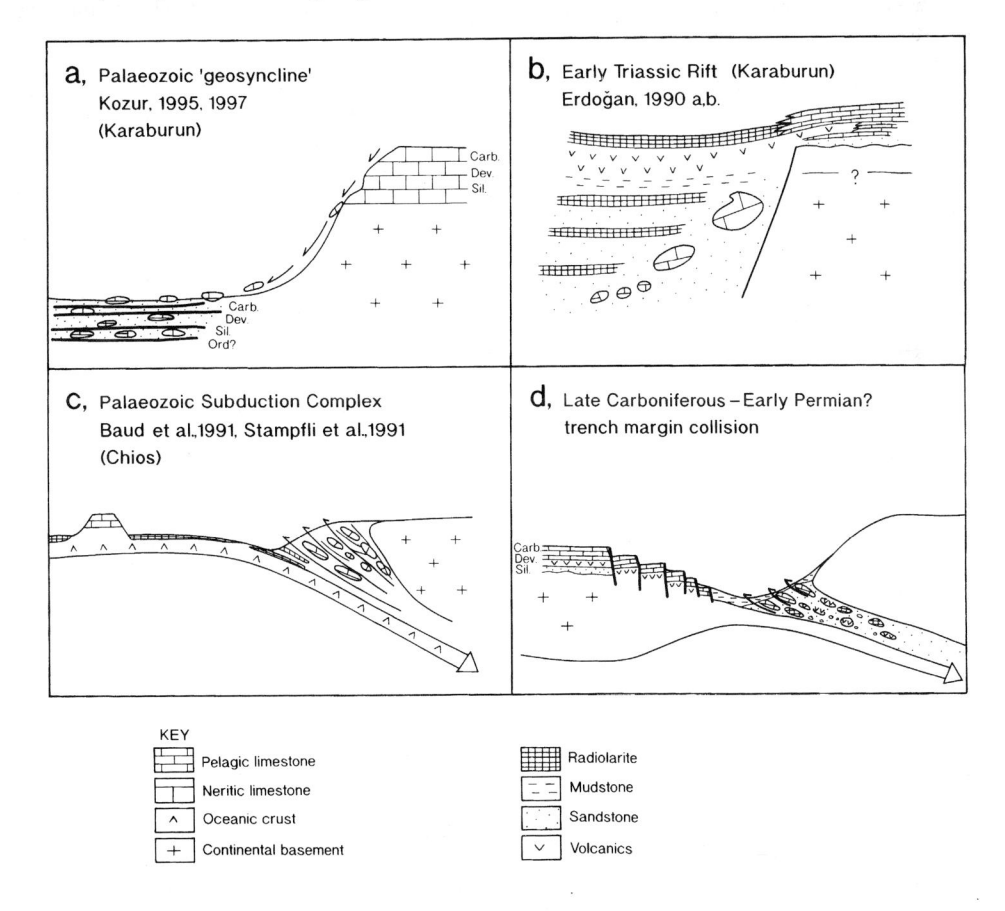

Fig. 18. Some alternative tectonic interpretations of the Karaburun and Chios Mélanges: (**a**) as a layer-cake succession of Ordovician(?)–Lower Permian(?) olistostromes (Kozur 1995); (**b**) as a highly deformed Triassic rift unit; (**c**) as a Palaeotethyan (i.e. Permo-Triassic) accretionary complex; (**d**) as an accretionary complex involving collision of a trench with a platform (microcontinent?) – favoured model.

Olistostromes within a long-lived deep-water basin (Fig. 18a)

Kozur (1995) interpreted the northwest Karaburun Peninsula as an intact, but deformed, layer-cake succession of basinal siliciclastics, extrusives and black cherts (lydites), together with olistoliths of Silurian–Devonian age. The blocks were periodically emplaced by gravity sliding throughout this immense time period. The relatively deformed and thermally mature nature of the Palaeozoic lithologies relates to 'Caledonian' or 'Hercynian' metamorphism in Kozur's view, and correlates with the metamorphic succession of the Menderes Massif (Dürr *et al.* 1978, 1995; Akkök 1983; Erdoğan & Güngör 1992, 1993). More recently, Kozur (1997) envisaged an origin of the mélange as an accretionary complex related to southward subduction of an ocean between Gondwana and Eurasia. He recognizes subduction-related olistostromes of numerous ages and, thus, infers persistence of subduction between Late Silurian and Early Carboniferous (punctuated by Early Devonian 'collision' to explain the apparent absence of Middle Devonian). On the other hand, slope facies, including debris-flow deposits, could relate to a passive carbonate margin slope, unrelated to subduction in this model.

There are a number of problems with the above olistostromal origin: (1) there is little evidence that blocks were emplaced into contemporaneous Palaeozoic deep-sea sediments; e.g. the Silurian–Devonian radiolarian cherts (basinal) are not interbedded with olistoliths (margin derived); (2) black cherts on both Chios and Karaburun commonly exhibit sheared contacts with associated siliciclastic turbidites, suggesting the cherts were first lithified and only later tectonically juxtaposed; (3) some matrix sediments on Chios were reported to be Triassic in age (Papanikolaou 1984*b*), although this remains to be confirmed, and, in Karaburun, matrix sandstones are locally interbedded with Early Triassic deep-water limestone (Erdoğan, pers. comm.); (4) on Chios, Carboniferous limestone blocks mainly occur structurally lower in the unit than Silurian limestone blocks (Papanikolaou & Sideris 1983), ruling out a simple layer-cake-type succession in which block lithologies were formed and successively emplaced as olistoliths; (5) detailed logging by the present authors shows that successions within Palaeozoic units are more deformed than suggested by Kozur (1995, 1997). Persistence of an unstable slope setting, in a single tectonic regime, for > 150 Ma is unlikely.

Rift setting (Fig. 18b)

The mélange and the overlying basinal successions, including the volcanics of Karaburun, were interpreted as a Triassic rift sequence (Erdoğan *et al.* 1990). In this model, Upper Palaeozoic limestone and chert blocks slid from the marginal to axial zones of a rift. The general absence of Permian blocks might suggest that related rift-shoulder uplift and erosion took place at this time. The siliciclastic turbidites would represent synrift, continentally derived turbidites, together with local channelized conglomerates. The Lower Triassic basinal cover sediments would then record the upper levels of the rift succession, explaining why no basal conglomerate is widely exposed and why Lower Triassic deep-sea radiolarian sediments occur directly above this mélange.

The main problems with this model are: (1) the thickness, estimated as 3–4 km on Chios and > 2 km on Karaburun, is high for a synrift sequence with no exposed footwall; (2) large (up to hundreds of metres) blocks are present at various levels in the mélange and there is no coherent overall stratigraphy (e.g. fining-upwards succession); (3) a rift origin alone does not explain the common evidence of intense shearing, folding and thrusting of the mélange, in contrast to the overlying Mesozoic platform succession of both Chios and Karaburun; (4) the contact between the mélange and the Lower Triassic successions varies locally from a sheared sedimentary gradation on Chios to a shear zone, or unconformity, in Karaburun, thus complicating an interpretation of these two as parts of a contiguous rift-related unit. The presence of a sheared unconformity, seen on Chios and Karaburun at the base of the Mesozoic successions, suggests that the underlying mélange is pre-Triassic in age; (5) comparable rift-related successions of Late Permian–Middle Triassic age, including siliciclastic turbidites, calciturbidites, radiolarian cherts, pelagic limestones and volcanics, occur widely elsewhere in southwestern Turkey; e.g. in the Lycian Nappes (Graciansky 1972; Collins & Robertson 1998, 2000) and the Antalya Complex (Gutnic *et al.* 1979; Robertson 1993). However, these rift successions are all coherent lithostratigraphical units with, at most, only minor detached and related debris flow units, suggesting that rifting alone is not a viable process to form the Chios and Karaburun Mélanges.

Subduction–accretion setting (Fig. 18c and d)

The mélanges could be interpreted as preserved remnants of an accretionary complex, developed in response to steady-state subduction of Palaeotethyan oceanic crust beneath an active continental margin, as previously suggested for Chios by Baud *et al.* (1991). Stampfli *et al.* (1991) interpreted the mélange as the imbricated outer slope of a Palaeotethyan accretionary complex. Pe-Piper & Kotopouli (1994) also interpreted the Chios Mélange sequence as an accretionary wedge and proposed that, along with Lesbos, it represents the southwestern extent of the Karakaya Complex, in pre-Middle Permian age. The presence of two magma types (i.e. a low-Ti type and a high-Ti–high-Nb type) were taken to indicate formation in a Palaeotethyan back-arc basin which Pe-Piper & Kotopouli (1994) correlated with the Karakaya Complex.

In the subduction–accretion model (Fig. 18c), the radiolarian cherts would represent accreted oceanic sediments, whereas the volcanics and limestone blocks might represent fragments of oceanic crust, or oceanic seamounts overlain by carbonate platforms; the terrigenous turbidites would be trench-fill, or trench-slope, basin sediments derived from the adjacent continental margin.

This above subduction model is also problematic. (1) There is little evidence of oceanic-derived material within the Karaburun or Chios Mélanges (e.g. MORB-type extrusives and related cherts), although occurrence of chromite grains of possible ultramafic ophiolitic origin has been reported from the matrix of the Chios Mélange (Stattegger 1984). (2) Typical accretionary successions are not observed (i.e. MORB lavas, overlain by cherts, then trench-type turbidites). (3) The successions within limestone blocks locally include intercalations of siliciclastic sediments (e.g. in Lower Carboniferous of Karaburun), suggesting a continental margin rather than oceanic origin. According to Pe-Piper & Kotopouli (1994), some of the igneous units on Chios intrude terrigenous sediments and contain rare quartzitic xenoliths, again suggesting continental affinities. Recent work supports an origin of the Karakaya Complex as an accretionary mélange; however, the extrusives present are of within-plate or MORB type and are of Triassic age (Pickett & Robertson 1996), unlike the volcanics of inferred Late Palaeozoic age in the Chios Mélange.

Subduction complexes can, however, show great structural and lithological variation in different areas and settings. One alternative subduction-related origin might involve derivation of blocks from the landward trench margin; i.e. from an adjacent continental margin with a subduction décollement located beneath the present erosion level for Chios and Karaburun. In this case, oceanic crust was subducted and not preserved, other than for derived chromite grains. However, it is unlikely that the entire kilometres thick mélange could form in this way.

A more plausible alternative is that steady-state subduction took place beneath a Palaeotethyan continental margin, accreting radiolarian chert from above the floor of Palaeozoic ocean crust that was subducted. This process continued until a continental margin unit, or microcontinent, collided with the trench, disrupting the succession into blocks and broken formation set within a matrix of contemporaneous deep-water siliciclastic sediments (Fig. 18d). The four main mélange bands on Chios, rich in exotic blocks, could be interpreted as accretion of individual slices which underwent intense layer-parallel extension (boudinage) within a subduction-trench setting (during underplating to the landward wall of the trench?). A similar process was inferred in the Central Pontides, where the leading edge of a Permian carbonate platform broke up as it collided with a trench (Ustaömer & Robertson 1997); other comparable Tethyan examples (e.g. in Greece and Yugoslavia) were summarized by Robertson (1994). A modern eastern Mediterranean example of this process is the collision of the Eratosthenes Seamount, a microcontinental fragment capped by a carbonate platform, with the Cyprus active margin to the north (Robertson 1998).

Origin as tectonic mélange

The mélange could also be interpreted as entirely tectonic mélange, created, e.g., during collisional tectonics. In this model, a Palaeozoic–Lower Triassic layer-cake succession of competent limestones and cherts, and incompetent sandstones and shales, was later strongly sheared to produce tectonic blocks ('phacoids') of limestone, chert and volcanics in a sheared sandstone–shale matrix. Such shearing might, in principle, have taken place *prior to*, or *following*, deposition of the overlying Mesozoic platform succession, i.e. pre-Early Triassic or post-latest Cretaceous. The presence of relatively unaltered (CAI 1–1.5) pelagic sediments within more metamorphosed Karaburun Mélange is evidence of pervasive tectonic slicing (Kozur

1995, 1997), as also indicated by the sheared contacts between the mélange and the overlying Mesozoic platform succession.

Problems with a purely tectonic origin are: (1) how matrix sediments were introduced; (2) the fact that some of the limestone blocks on Chios preserve relations compatible with superficial gravity emplacement (i.e. marginal sedimentary breccias); (3) the very deformed state of the mélange relative to the overlying Mesozoic platform successions.

Comparison with adjacent areas

The possibility of a pre-Triassic origin for the mélange is also suggested by comparison with similar, but more metamorphosed, mélange located along the northern margin of the Tauride–Anatolide Platform in the Kühtahya–Bolkardağ region, > 200 km east of the study area. Here, Silurian–Devonian limestones are cut by diabase dykes and were reported to be unconformably overlain by a mélange composed of limestone, also volcanics and volcanogenic rocks of intermediate–acidic composition (Özcan *et al.* 1988). Blocks include siltstone, sandstone, black chert, recrystallized limestone, crinoidal limestone and coralline limestone. The mélange is unconformably overlain by Lower Triassic red clastics, marking the base of a Mesozoic carbonate platform succession; ophiolitic rocks are absent. Recent biostratigraphical studies indicate the presence of Upper Silurian–Lower Ordovician limestones (similar to the Kalecik Limestone of northeast Karaburun) and Lower Permian limestones (comparable with the Tekekara Dağ Limestone of northwest Karaburun). The mélange in the Kühtahya–Bolkardağ region is unconformably overlain by shallow-marine Scythian sediments at the base of a Mesozoic Anatolide carbonate platform succession (Kozur 1997). It is, therefore, possible that Palaeotethyan mélange underlies the northern margin of the Anatolide–Tauride Platform (and equivalent preserved in the Lycian Nappes) for hundreds of kilometres alongstrike. Göncüoğlu *et al.* (2000) report that, in their view, in places, metamorphosed carbonate platform units correlated with the Anatolides (i.e. southerly origin) are underlain (with a normal contact) by conglomerates cut by acid and basic dykes related to a Late Palaeozoic magmatic arc. Furthermore, comparable Palaeotethyan-related mélange may be present further west on the island of Evvia (Greece) and in Attica on the mainland (Baud & Papanikalaou 1981; Baud *et al.* 1991; Stampfli, pers. comm.).

Working hypothesis: Upper Palaeozoic subduction complex reworked by Triassic rifting and Early Tertiary collisional deformation

The interpretation that best fits the available evidence is that the Chios and Karaburun Mélange's are the end-products of a multistage history that involved *Early Permian(?) subduction–accretion, Triassic rifting* and *Early Tertiary collisional processes.*

The subduction–accretion phase

A period of subduction within Palaeotethys culminated in collision of a continental margin or microcontinent with a trench, resulting in accretion of a Silurian–Carboniferous (to Lower Permian?) platform and marginal units as blocks and thrust slices within a matrix of deep-water, trench-type siliciclastic deposits. Since the youngest confirmed blocks in the mélange are Gzhelian (Kozur 1997), the age of the mélange accretion could range from Late Carboniferous to Early Permian.

Where was this accretionary prism located within Tethys? One alternative is that it might relate to a subduction zone dipping southwards beneath, what later became, the Mesozoic Anatolide–Tauride Platform. Southward subduction of Palaeotethys is inherent in regional tectonic models of, e.g. Şengör *et al.* (1984), Okay *et al.* (1996), Pickett & Robertson (1996) and Göncüoğlu *et al.* (2000). Alternatively, the mélange might relate to northward subduction beneath the southern margin of Eurasia, as inferred from studies of the Central Pontides (Ustaömer & Robertson 1993, 1997, 1999; Stampfli 2000; Stampfli *et al.* 2000). The main point is that the mélanges are associated with the Anatolide–Tauride Platform, suggesting a south Tethyan (Gondwanian) rather than north Tethyan (Eurasian) affinity. During the Late Palaeozoic, the Anatolide–Tauride block was attached to Gondwana in most tectonic reconstructions [e.g. see Stampfli 2000)]. Thus, for the mélanges to relate to northward subduction–accretion beneath Eurasia, the entire Palaeotethyan Ocean would have to be subducted, juxtaposing Eurasian and Gondwanian units prior to Early Triassic time. One other possibility is that the mélanges relate to northward Palaeotethyan subduction close to the Gondwana margin, e.g. beneath an intra-oceanic unit (oceanic arc or microcontinent). However, the simplest hypothesis is that the mélanges relate to southward subduction of

Palaeotethys beneath the Gondwanian margin during Late Carboniferous–Early Permian? time. In this model, northward subduction was, however, in progress beneath Eurasia during Triassic time (Robertson *et al.* 1999).

The Early Mesozoic rift phase (Fig. 19)

Where exactly was the Early Triassic Karaburun–Chios Rift located? The intact Triassic–Jurassic Karaburun–Chios succession escaped the major latest Triassic ('Cimmerian')

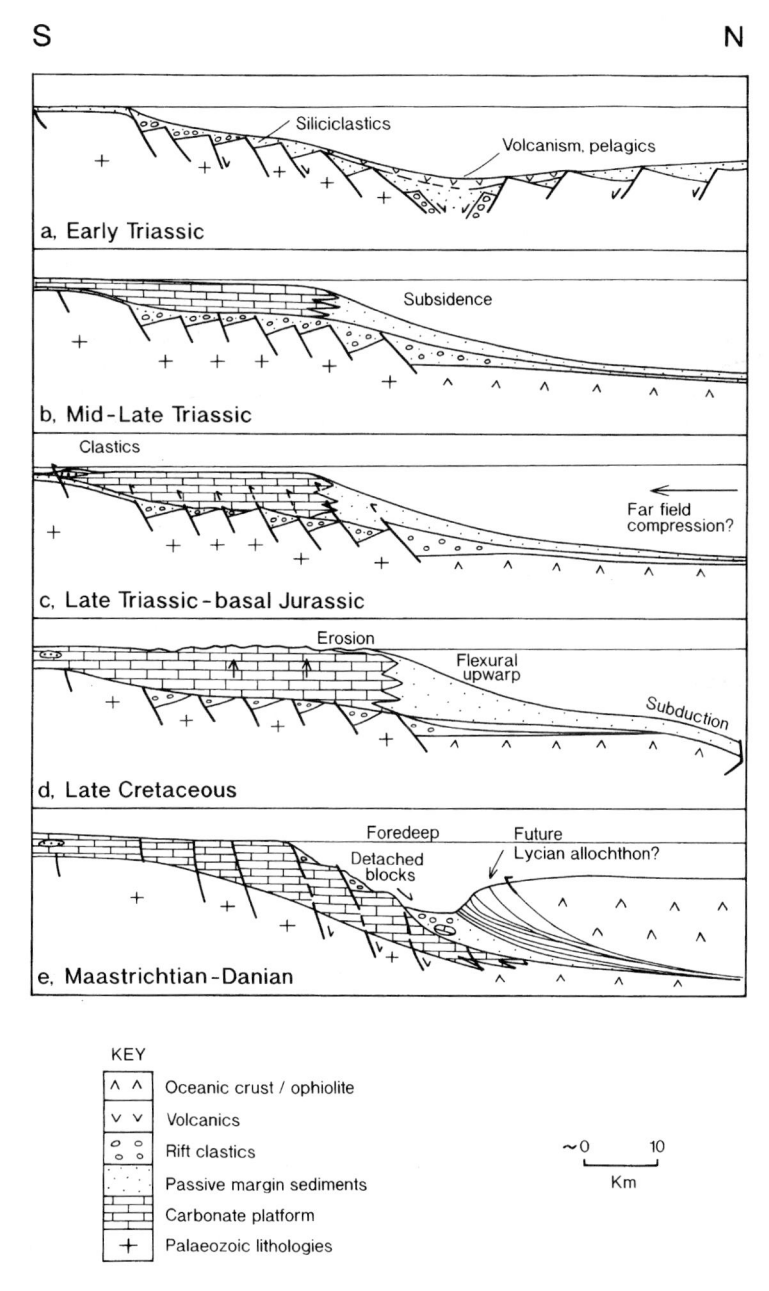

Fig. 19. Sketch sections showing the proposed Mesozoic–Early Tertiary tectonic evolution of Karaburun and Chios related to Triassic rifting and development of a subsiding passive margin (Jurassic–Cretaceous) emplaced in stages related to later collisional deformation (Late Cretaceous–Early Tertiary).

deformation that affected the Pontides (e.g. Karakaya Complex), favouring a southerly (near Gondwana) setting. The Karaburun Platform records an Early Triassic rifting, followed by continuous passive margin conditions until Late Cretaceous, briefly interrupted in the latest Triassic–earliest Jurassic. The Middle–Late Triassic subsiding carbonate platform is assumed to have bordered a developing Neotethyan oceanic basin, located to the north of the Anatolide–Tauride Platform (northerly Neotethys). This contrasts with Erdoğan et al. (1990), who believed that spreading was delayed until the Late Cretaceous.

During the Maastrichtian–Danian the platform strongly subsided as a foredeep ahead of an advancing thrust load (Bornova Mélange). The edge of the Karaburun carbonate Platform disintegrated, forming blocks within mélange. In the north, the underlying Karaburun units were locally reworked as blocks. In addition, oceanic material (radiolarian chert, basic extrusives and serpentinite) were emplaced from the leading edge of the overthrusting load into a foredeep. This load is assumed to have been oceanic crust, emplaced during suturing of the northerly Neotethyan oceanic basin. The oceanic-derived clasts were first accreted, then reworked into the foredeep, mainly by gravity processes. Subsequently, thrusting in the direction of tectonic transport (present west) detached the carbonate platform along zones of rheological weakness, namely the base of the well-bedded Lower Triassic basinal succession (Gerence Formation) and the overlying Rhaetic sandstones and shales, to form several large thrust sheets in Karaburun. An Upper Palaeozoic–Lower Mesozoic thrust sheet was also emplaced on Chios and may represent the leading edge of an allochthonous slab that includes Lesbos Island to the north. As continental collision intensified, the thrusting continued to propagate downwards into underlying Triassic rift units, shearing and dislocating the contact with the Lower Triassic basinal successions.

One apparent problem is the inferred westward or northwestward directions of thrusting and folding in Karaburun and Chios. However, palaeomagnetic studies show that this region was later tectonically rotated. In Karaburun, data from eight sites indicate consistently clockwise rotations of c. 44° post-11 Ma (Kissel et al. 1989). Restoring such rotations indicates that pre-Neogene thrust directions were more towards the southwest. Any westward component of emplacement might relate to transpressional deformation in this area. A smaller, more variable, data set from Chios (15 Ma andesites and 16 Ma rhyolites) suggests mean anticlockwise rotations of 25° although results vary locally and more data are needed (Kondopoulou et al. 1993, 1998). Also, the possibility of any pre-11 Ma rotations is unknown.

Accepting a southerly (i.e. Gondwana-related) origin of Karaburun–Chios in the Mesozoic, two local alternatives (Fig. 20) can be proposed. Firstly, Mesozoic Chios and Karaburun formed part of the northern margin of the Anatolide–Tauride Platform. However, these units escaped pervasive metamorphism, including high-pressure metamorphism (Harris

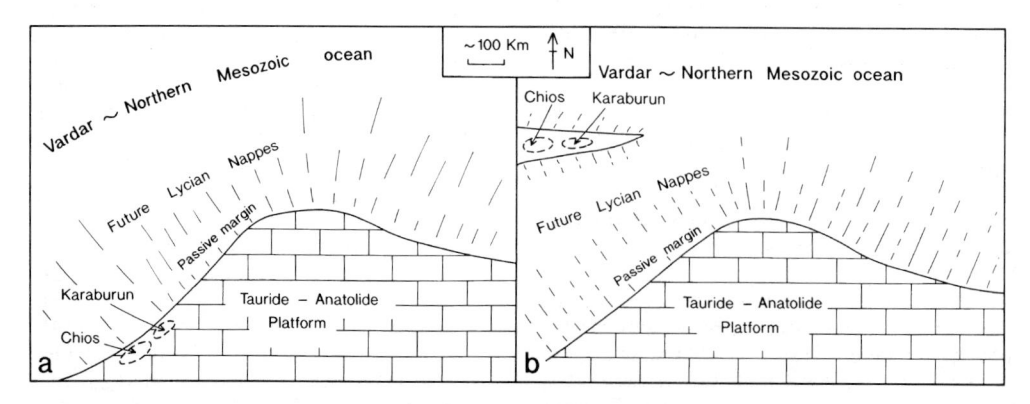

Fig. 20. Plate tectonic sketch maps showing alternative possible settings of the Karaburun–Chios unit: (**a**) as an embayment to the west of a promontory of the Anatolide–Tauride Platform, including the Menderes Massif; (**b**) as a separate small microcontinent located within Neotethys further north. (**a**) Is favoured – see text for explanation.

et al. 1994; Okay & Kelley 1994) during Late Cretaceous (Turonian–Coniacian) subduction, as observed in the Anatolide Platform units further east (*c.* 30° longitude; Fig. 1). A possible explanation for this difference is that the area to the east formed a north facing promontory that absorbed collision, whereas the Karaburun–Chios area to the west was in the 'lee' of this collision and escaped metamorphism (Okay & Siyako 1993). This also provides an explanation for the presence of Lower Carboniferous basinal facies (Kozur *et al.* 2000) at the base of the most proximally derived Lycian Nappe (e.g. Karadere; Collins & Robertson 1999). Okay *et al.* (1996) further suggest that the area to the west, including Karaburun–Chios, was bordered by a transform margin and thus shielded from collision. A transform fault also figures in plate tectonic reconstructions of Stampfli *et al.* (2000) for this region of Tethys.

A second alternative is that Karaburun–Chios represents a small crustal fragment (microcontinent) that was rifted from the Anatolide–Tauride Platform in the Early Triassic and was then located near the southern margin of the Mesozoic oceanic basin (Fig. 20b). Such a setting would explain the presently high structural (i.e. northerly) position of Karaburun–Chios within the İAESZ (Fig. 2). Such a fragment would have been located > 350 km north of the Anatolide–Tauride Platform to accommodate the restored minimum width of the Lycian Nappes that root along the northern margin of the platform (Okay 1989; Collins & Robertson 1999). One problem with this model is that there is no record of the expected passive margins bordering a Karaburun–Chios microcontinent, either to the north or south.

On balance, the present authors prefer the Anatolide–Tauride margin model (Fig. 20a). In this, the Lycian Nappes root to the north of the Karaburun–Chios Platform. The Lycian Nappes were overthrust southwards over this platform in the Late Cretaceous as a transitory load which finally came to rest further south. During initial collision in the latest Cretaceous, the distal passive margin of the Karaburun–Chios Platform and ophiolites (Lycian Peridotite) were detached and thrust southwards over the Menderes continental margin (Anatolide–Tauride Platform). As collision intensified in the Early Tertiary, the basal thrust cut downwards into the foreland, slicing off the leading edge of the Mesozoic platform that then disintegrated into blocks within the Maastrichtian–Palaeocene Bornova Mélange, leaving intact platform units only in the west, in Karaburun and Chios. The Karaburun and Chios Mélange units beneath the Mesozoic platforms were extensively resheared during this collisional phase. Emplacement over the Menderes Massif was complete by Early Eocene time according to Okay & Siyako (1993) or, in an alternative view, only after Late Eocene time according to Erdoğan *et al.* (1990).

Conclusions

- The mélange units of the Karaburun Peninsula (Turkey) and the island of Chios (Greece) are interpreted as an Upper Palaeozoic (Upper Carboniferous–Lower Permian?) subduction–accretion complex, developed near the south margin of a Palaeotethyan oceanic basin. This was followed by rifting and passive margin evolution of a northerly Neotethyan oceanic basin that was sutured in Early Tertiary time.
- Conclusions for Karaburun are:
  - The Karaburun Peninsula exposes two related tectonostratigraphic units. The lower of these, the Karaburun Mélange, is dominated by Silurian–Upper Carboniferous exotic blocks of neritic and pelagic carbonate, black chert and subordinate volcanogenic units. The blocks are set in a sheared matrix of incompetent siliciclastic turbidites, including localized channelized conglomerates.
  - Silurian–Upper Carboniferous carbonates accumulated in platforms, slope and basinal settings, floored, or bordered by continental crust in view of associated (interbedded) terrigenous sediments.
  - An overlying Lower Triassic succession, composed of interbedded pelagic carbonates, turbiditic sandstones, red ribbon radiolarites, pink Ammonitico Rosso, basic volcanics and volcanogenic sediments, is interpreted as a rift basin related to the opening of a younger, Neotethyan oceanic basin.
  - Overlying Middle–Upper Triassic successions of Karaburun record a subsiding shallow-water platform bordering a northerly Neotethyan oceanic basin.
  - During latest Triassic–earliest Jurassic, the carbonate platform emerged and was overlain by thin, deltaic deposits, locally derived from underlying Triassic rift units and basement. The possible driving mechanism was far-field stresses related to a collisional event within Tethys further north (possibly 'Cimmerian' emplacement of the Karakaya Complex).

- During the Late Cretaceous (Campanian–Maastrichtian) the Karaburun Platform underwent flexural uplift, followed by thrust loading related to the initial stages of convergence and continental collision.
- During the Maastrichtian–Danian, blocks were detached from an advancing thrust load and spalled into a foredeep (Bornova Mélange), together with previously accreted 'Neotethyan' oceanic lithologies (e.g. radiolarian chert, pelagic limestone, basalt, serpentinite).
- The Mesozoic carbonate platform was deformed into two thrust slices above a lower unit (also allochthonous?) by southwest vergent compression (restored co-ordinates) after Maastrichtian time.
- An original depositional contact between the Karaburun Mélange and overlying Lower Triassic rift-related succession was sheared, with local thrust intercalation of mélange during latest Cretaceous–Early Tertiary convergence or collision.
- Conclusions for Chios are:
  - Two main tectonostratigraphic units are present; the lower Chios Mélange, overlain with a sheared, but locally unconformable, contact by a Lower Triassic basinal succession, then by an Upper Triassic–Lower Jurassic carbonate platform. An upper exotic thrust sheet of Permian–Jurassic mixed shallow-water carbonate–siliciclastic unit was emplaced locally northwestwards (present coordinates) over a Mesozoic carbonate platform unit. The platform may correlate with the Permian of Lesbos Island.
  - The Chios Mélange comprises a c. 5 km thick unit of terrigenous shale, sandstone and conglomerate with blocks of limestones (Silurian–Carboniferous), black chert, lava and volcanogenic sediments. Local units restore as a carbonate platform (with laterites) that disintegrated into blocks mantled by coarse siliciclastic sediments.
- The Chios and Karaburun Mélanges and overlying platform successions can be correlated, with the Triassic units of Chios recording a more rift proximal setting than Karaburun.
- >The Pre-Triassic mélange is the end-product of a series of tectonic–sedimentary events. These began with initial subduction–accretion in the Late Palaeozoic (Late Carboniferous–Early Permian?), culminating in collision of a passive margin, or microcontinent with a trench. Secondly, in Early Triassic time, the resulting suture rifted giving rise to siliceous sediments, minor calcareous gravity deposits and basic–intermediate composition volcanics. The rift was then overstepped by a subsiding passive margin adjacent to a Neotethyan oceanic basin. A hiatus and siliciclastic deposition in the latest Triassic–earliest Jurassic may reflect a 'Cimmerian' collision further north. During the Late Cretaceous plate convergence, the Karaburun Platform underwent flexural uplift and erosion, then collapsed to form a foredeep (Bornova Mélange) in Maastrichtian–Danian time. Finally, the Palaeozoic mélange was intersliced with formerly overlying Triassic rift-related units during Early Tertiary continental collision. The original depositional contact between the Karaburun and Chios Mélanges and overlying Mesozoic platform units was strongly sheared during this time.
- Taking account of the regional tectonic setting and Neogene to present-day tectonic rotations, the Karaburun–Chios Mesozoic Platform is interpreted as part of the southerly margin of a Mesozoic, northerly Neotethyan, oceanic basin bordering the Tauride–Anatolide Platform and the Menderes Metamorphic Massif to the south.

AHFR's fieldwork was carried out with support of a grant from the Carnegie Trust for the Scottish Universities. EAP acknowledges support from NERC during a research studentship at the University of Edinburgh. We thank B. Erdoğan for introducing us to the Karaburun Peninsula and more recent discussion, and D. Papanikolaou, G. Migiros and Ch. Sideris for insights gained during a field excursion they led to Chios and Lesbos in 1992 as part of the 6th Congress of the Geological Society of Greece; H. Kozur also contributed to the discussion. We thank I. Sharp for assistance in the field and A. İ. Okay for help with our work on the Karakaya Complex in northwest Turkey. We are also grateful to İ. Özgenç for his help with sending samples to Edinburgh. D. Baty assisted with drafting some of the diagrams. The manuscript benefited from comments by A. İ. Okay and G. Stampfli.

References

Akkök, R. 1983. Structural and metamorphic evolution of the northern part of the Menderes Massif: new data from the Derbent area, and their implication for the tectonics of the massif. *Journal of Geology*, **91**, 342–350.

Asserto, R., Jacobshagen, V., Kauffmann, G. & Nicora, A. 1979. The Scythian/Anisian boundary in Chios, Greece. *Rivista Italiano di Paleontogia e Stratigraphica*, **85**, 715–736.

Baud, A. & Papanikolaou, D. 1981. Olistoliths and flysch facies at Permo–Triassic transition series from Attica, Eastern Greece. *Workshop meeting for IGCP projects: Triassic of the Tethys Realm – Project, No. 4 and Permo–Triassic Boundary in Tethys Realm–Project*, **106**, 33–34.

——, JENNY, C., PAPANIKOLAOU, D., SIDERIS, CH. & STAMPFLI, G. 1991. New observations on Permian stratigraphy in Greece and geodynamic interpretation. *Bulletin of the Geological Society of Greece*, **25**, 187–206.

BESENECKER, H., DÜRR, S., HERGET, G., KAUFFMANN, G., LUDTKE, G., ROTH, W. & TIETZE, K.-W. 1971. *1: 50 000 Geological Map of Chios (2 Sheets: Northern and Southern)*. Athens.

——, DÜRR, S., HERGET, G. ET AL. 1968. Geologie von Chios (Agais). *Geologica et Palaeontologica*, **2**, 121–150.

BRINKMANN, R., FLÜGEL, E., JACOBSHAGEN, LECHNER, H., RENDEL, B. & RICK, P. 1972. Trias, Jura und Kreide der Halbinsel Karabrun (West-Anatolien). *Geologica et Palaeontologica*, **5**, 139–150.

COLLINS, A. & ROBERTSON, A. H. F. 1998. Processes of Late Cretaceous to Late Miocene episodic thrust sheet translation in the Lycian Taurides, SW Turkey. *Journal of the Geological Society, London*, **155**, 759–772.

—— & —— 1999. Evolution of the Lycian Allochthon, western Turkey as a north-facing Late Palaeozoic to Mesozoic rift and passive continental margin. *In*: BOZKURT, E. & ROWBOTHAM, G. (eds) *Advances in Turkish Geology, Part 1: Tethyan evolution and fluvial to marine sedimentation*. Geological Journal, in press.

DERCOURT, J., RICOU, L. E. & VRIELYNCK, B (eds) 1992. *Atlas Tethys Palaeoenvironmental Maps*. Beicip-Franlab.

——, ZONENSHAIN, L. P., RICOU, L. E. ET AL. 1986. Geological evolution of the Tethys belt from the Atlantic to the Pamirs since the Lias. *Tectonophysics*, **123**, 241–315.

DÜRR, S., ALTHERR, R., KELLER, J., OKRUSCH, M. & SEIDEL, E. 1978. The median Aegean crystalline belt: stratigraphy, metamorphism. *In*: CLOSS, H. & SCHMIDT, K. (eds) *Alps, Apennines and Hellenides*. Schweizerbart, 455–476.

——, DORA, Ö. O., CANDAN, O., ÖZER, S. & GÜNGÖR, T. 1995. *Stratigraphy and Tectonics of the Menderes Massif*. International Earth Science Colloqium on Aegean region, 1995, Excursion Guide, 1–26.

ERDOĞAN, B. 1990a. Tectonic relations between İzmir–Ankara Zone and Karaburun Belt. *Mineral Research and Exploration Institute of Turkey (MTA) Bulletin*, **110**, 1–15.

—— 1990b. Stratigraphy and tectonic evolution of the İzmir–Ankara Zone between İzmir and Seferihisar. *Turkish Association of Petroleum Geologists Bulletin*, **2**, 1–20.

—— & GÜNGÖR, T. 1992. Stratigraphy and tectonic evolution of the northern margin of the Menderes Massif. *Turkish Association of Petroleum Geologists Bulletin*, **4**, 9–34.

—— & —— 1993. Stratigraphy and tectonic evolution of the northern margin of the Menderes Massif. *Bulletin of the Geological Society of Greece*, **28**, 269–286.

——, ALTINER, D., GÜNGÖR, T. & ÖZER, S. 1990. Stratigraphy of Karaburun Peninsula. *Mineral Research and Exploration Institute of Turkey (MTA) Bulletin*, **111**, 1–20.

FISCHER, A. G. 1964. The Lofer cyclothems of the Alpine Triassic. *In*: MERRIAM, D. F. (ed.) *Symposium on Cyclic Sedimentation*. Bulletin of the Geological Survey of Kansas, **169**, 107–149.

FITTON, J. G. & DUNLOP, H. M. 1985. The Cameroon line, West Africa and its bearing on the origin of oceanic and continental alkali basalts. *Earth and Planetary Science Letters*, **72**, 23–38.

FRYSSALAKIS, G. 1985. Le Trias de l'île de Chios (Mer Egée, Grèce). *Evolution paleogeographique, modalités de la sédimentation, implications paléostructurales*. PhD Thesis, Université de Paris VI.

GARFUNKEL, Z. 1998. Constraints on the origin and history of the Eastern Mediterranean basin. *Tectonophysics*, **298**, 5–37.

—— & DERIN, B. 1984. Permian–Early Mesozoic tectonism and continental margin formation on Israel and its implications for the history of the eastern Mediterranean. *In*: DIXON, J. E. & ROBERTSON, A. H. F. (eds) *The Geological Evolution of the Eastern Mediterranean*. Geological Society, London, Special Publications, **17**, 187–201.

GARRASI, C. & WEITSCHAT, W. 1968. *Geologie von Nordwest-Karaburun (West-anatolische Kuste)*. Hamburg, 177–184 [unpublished manuscript].

GRACIANSKY, P. C. 1972. *Récherches géologiques dans les Taurides Lycien Occidental*. DSc Thesis, Université de Paris-Sud.

GÖNCÜOĞLU, C. M., TURHAN, N., ŞENTÜRK, K., ÖZCAN, A. & UYSAL, S. 1998. Tectonic units of the central Sakarya region: a geotraverse across the İzmir–Ankara Suture in NW Anatolia, Turkey. *Third International Turkish Geology Symposium, METU – Ankara, Abstracts*, 26.

——, ——, ——, —— & YALINIZ M. K. 2000. A geotraverse across northwestern Turkey: tectonic units of the Central Sakarya region and their tectonic evolution. *This volume*.

GUSIC, I., WOHLFEIL, H. & WOHLFEIL, K. 1984. Zur Alterssellung und Fazies des Kalkes von Kalecik (Devon) und der Akdağ-Series (Trias) im nordöstlichen Teil von Karaburun (westl. İzmir, Türkei). *Neus Jahrbuch für Geologie und Paläontologie, Abhandlunger*, **167**, 375–404.

GUTNIC, M., MONOD, O. & POISSON, A. 1979. Géologie des Taurides occidentales (Türquie). *Mémoire de la Societé Géologique de France*, **137**, 1–112.

HARRIS, N. B. W., KELEY, S. P. & OKAY, A. İ. 1994. Post-collision magmatism and tectonics of northwest Turkey. *Contributions to Mineralogy and Petrology*, **117**, 241–252.

HECHT, J. 1972. Geological map. 'Plomari-Mytilini' sheet, 1: 50 000, I.G.M.E. Athens.

HERGET, G. & ROTH, W. 1968. Genese und Alterssellung von Vorkommen der Sb–W–Hg-

Formation in der Türkei und auf Chios, Griechenland. *Bayrische Academie der Wissenschaften, mathematisch–naturwissenschaftliche Klausur*, **127**, 118.

HIRSCH, F., 1984. The Arabian sub-plate during the Mesozoic. *In*: DIXON, J. E. & ROBERTSON, A. H. F. (eds) *Geological Evolution of the Eastern Mediterranean*. Geological Society, London, Special Publication, **17**, 217–224.

HOLL, R. 1966. Genese und Altersstellung von Vorkommen der Sb–W–Hg–Fo in der Türkei und der Insel Chios (Griechenland). *Bayrische Academie der Wissenschaften, mathematisch–naturwissenschaftliche Klausur*, **127**, 118.

KALAFATCIOĞLU, A. 1961. A geological study in the Karaburun peninsula. *Mineral Research and Exploration Institute of Turkey (MTA) Bulletin*, **56**, 40–49.

KATSIKATSOS, G., MATARANGAS, D., MIGIROS, G. & TRIANTAPHYLLIS, M. 1982. *The geological structure of Lesvos island*. IGME Report.

KAUFFMANN, G. 1969. *Die geologie von Nordost-Chios (Agais)*. Dissertation, University of Marburg.

KISSEL, C., LAJ, C., POISSON, A. & SIMEAKIS, K. 1989. A pattern of block rotations in Central Aegean. *In*: KISSEL, C. & LAJ, C. (eds) *Palaeomagnetic Rotations and Continental Deformation*. Kluwer, 115–129.

KONDOPOULOU, D., LECI, V. & SIMEAKIS, C. 1993. Palaeomagnetic study of the Tertiary volcanics in Chios Island, Greece. *Proceedings of the 2nd Congress Hellenic Geophysical Union*, Florina, 5–7 May, 72–84.

——, DE BONIS, L., KOUFOS, G. & SEN, S. 1998. Palaeomagnetic data and biostratigraphy of the Middle Miocene vertebrate locality of Thymiana (Chios Island, Greece). *Bulletin of the Geological Society of Greece*, **31**, 625–635.

KONUK, T. 1977. Bornova filişinin yaşı hakkında. *Ege University, Faculty of Science Bulletin, Seri B*, **1**, 65–74 [in Turkish with English abstract].

KOZUR, H. 1995. New stratigraphic results on the Palaeozoic of the Western parts of the Karaburun Peninsula, Western Turkey. *In*: PIŞKIN, O., ERGÜN, M., SAVAŞÇIN, M. Y. & TARCAN, G. (eds) *Proceedings of International Earth Sciences Colloquium on the Aegean Region*, **1**, 289–308

—— 1997. The age of the siliclastic series ('Karaeis Formation') of the Western Karaburun Peninsula Western Turkey. *In*: SZANIAWSKI (ed.) *Proceedings of the Sixth European Conodont Symposium (ECOSVI), Palaeontologia Polonica*, **58**, 171–189.

——, ŞENEL, M. & TEKİN, K 2000. First evidence of Hercynian Lower Carboniferous flyschoid deepwater sediments of the Lycian nappes, southwestern Turkey. *Geologia Croatica*, in press.

KTENAS, K. 1921. Sur la découverte du Devonian a l'île de Chio (mer Egée). *Compte Rendus Sommaires Société Géologique de France*, 131–132.

—— 1923a. Sur le Carbonifere de l'île de Chio (mer Egée). *Compte Rendus Sommaires Société Géologique de France*, 146–148.

—— 1923b. Sur la decouverte d'un horizon a *Productus cora* a l'île de Chio (mer Egée). *Compte Rendus Sommaires Société Géologique de France*, 206–207.

—— 1925. Contribution a l'étude de la presqu'ile d'Erythree (Asie mineure). *Annales de la Faculté de Science, Athens*, 57–112.

KUNT, H. 1994. *Petrology and geochemistry of Kadınhanı area (Konya), Central Turkey*. PhD Thesis, University of Glasgow.

MIGIROS, P. & SIDERIS, CH. (eds) 1992. *Guide Book for Post-congress Field Trip to Chios–Lesvos*. Fifth meeting of the International Geological Correlation Project, Project 276, Athens, 20–27.

MONOD, O. & AKAY, E. 1984. Evidence for an Upper Triassic–Early Jurassic orogenic event in the Taurides. *In*: DIXON, J. E. & ROBERTSON, A. H. F. (eds) *Geological Evolution of the Eastern Mediterranean*. Geological Society, London, Special Publications, **17**, 113–128.

OKAY, A. İ. 1989. Geology of the Menderes Massif and the Lycian Nappes south of Denizli, western Taurides. *Contributions to Mineralogy and Petrology*, **45**, 289–316.

—— 2000. Was the Late Triassic orogeny in Turkey caused by the collision of an oceanic plateau? *This volume*.

—— & KELLEY, S. 1994. Jadeite and chloritoid schists from northwest Turkey: tectonic setting, petrology and geochronology. *Journal of Metamorphic Geology*, **12**, 455–466.

—— & SİYAKO, M. 1993. The new position of the İzmir-Ankara Neo-Tethyan suture between İzmir and Balıkesir. *In*: TURGUT, S. (ed.) *Tectonics and Hydrocarbon Potential of Anatolia, Proceedings of the Ozan Sungurlu Symposium*. Turkish Association of Petroleum Geologists Publications, 333–354.

——, —— & BÜRKAN, K. A. 1991. Geology and tectonic evolution of the Biga Peninsula, northwest Turkey. *Bulletin of İstanbul Technical University, İstanbul*, **44**, 191–256.

——, SATIR, M., MALUSKI, H., SİYAKO, M., METZGER, R. & AKYÜZ, S. 1996. Palaeo- and Neo-Tethyan events in northwest Turkey: geologic and geochronologic constraints. *In*: YIN, A. & HARRISON, T. M. (eds) *The Tectonic Evolution of Asia*. Cambridge University Press, 420–441.

ÖZCAN, A., GÖNCÜOĞLU, M. C., TURAN, N., UYSAL ŞENTÜRK, K. & IŞIK, A. 1988. Late Palaeozoic evolution of the Kühtahya–Bolkardağ Belt. *METU Journal of Pure and Applied Science*, **21**, 211–220.

PAECKELMANN, W. 1939. Ergebnisse einer Reise nach der Insel Chios. *Zeitschrift der deutschen geologischen Gesellschaft*, **91**, 341–376.

PAPANIKOLAOU, D. J. 1984a. The three metamorphic belts of the Hellenides: a review and a kinematic interpretation. *In*: DIXON, J. E. & ROBERTSON, A. H. F. (eds) *The Geological Evolution of the Eastern Mediterranean*. Geological Society, London, Special Publications, **17**, 551–562.

—— 1984b. *Introduction to the geology of Greece: the pre-Alpine units*. International Geological Correlation Project, 5, Correlation of Prevariscan and Variscan Events of the Alpine–Mediterranean Mountain Belt. Field meeting in Greece, 17–23 September, Athens.

—— & SIDERIS, CH. 1983. Le Paléozoique de l'Autochthone de Chios: une formation à blocs de type wildflysch d'age Permien (pre parte). *Compte Rendus de l'Académie de Science de Paris*, **297**, 603–606.

—— & —— 1992. *In:* MIGIROS, P. & SIDERIS, CH. (eds) *Guide Book for Post-Congress Field Trip to Chios–Lesvos*. Fifth meeting of the International Geological Correlation Project, Project 276, Athens.

PEARCE, J. A. 1982. Trace element characteristics of lavas from destructive plate boundaries. *In:* THORPE, R. S. (ed.) *Andesite*. Wiley, 528–548.

—— 1983. Role of sub-continental lithosphere in magma genesis at active continental margins. *In:* HAWKSWORTH, C. J. & NORRY, M. J. (eds) *Continental Basalts and Mantle Xenoliths*. Shiva Geology Series, 230–249.

PE-PIPER, G. & KOTOPOULI, C. N. 1994. Palaeozoic volcanic rocks of Chios, Greece: Record of a Palaeotethyan suture. *Neues Jahrbuch für Mineralogie Monatshefte, Stuttgart*, **1**, 23–39.

PHILIPPSON, A. 1911. Reisen und Forschiungen im westlichen Kleinasien, II: Ionien und das westliche Lydien. *Petermanns Mitteilungen, Erganzungsheft*, **172**, 1–100.

PICKETT, E. A. 1994. *Tectonic evolution of the Palaeotethys ocean in NW Turkey*. PhD Thesis, University of Edinburgh.

—— & ROBERTSON, A. H. F. 1996. Formation of the Late Palaeozoic–Early Tertiary Karakaya Complex and related ophiolites in NW Turkey by Palaeotethyan subduction–accretion. *Journal of the Geological Society, London*, **153**, 995–1009.

RICOU, L.-E., MARCOUX, J. & WHITECHURCH, H. 1984. The Mesozoic organisation of the Taurides: one or several oceanic basins? *In:* DIXON, J. E. & ROBERTSON, A. H. F. (eds) *The Geological Evolution of the Eastern Mediterranean*. Geological Society, London, Special Publications, **17**, 349–360.

ROBERTSON, A. H. F. 1981. Metallogenesis on a Mesozoic passive continental margin, Antalya Complex, southwest Turkey. *Earth and Planetary Science Letters*, **54**, 323–345.

—— 1987. The transition from a passive margin to an Upper Cretaceous foreland basin related to ophiolite emplacement in the Oman Mountains. *Geological Society of America Bulletin*, **99**, 633–653.

—— 1993. Mesozoic–Tertiary sedimentary and tectonic evolution of Neotethyan carbonate platforms, margins and small ocean basins in the Antalya Complex, southwest Turkey. *Special Publication of the International Association of Sedimentologists*, **20**, 415–456.

—— 1994. Role of the tectonic facies concept in orogenic analysis and its application to Tethys in the Eastern Mediterranean region. *Earth-Science Reviews*, **37**, 139–213.

—— 1998. Tectonic significance of the Eratosthenes Seamount: a continental fragment in the process of collision with a subduction zone in the eastern Mediterranean (Ocean Drilling Program Leg 160). *Tectonophysics*, **298**, 63–82.

—— & DEGNAN, P. J. 1998. Significance of modern and ancient oceanic Mn-rich hydrothermal sediments, exemplified by Jurassic Mn-cherts from southern Greece. *In:* MILLS, R. A. & HARRISON, K. (eds) *Modern Ocean Floor Processes and the Geological Record*. Geological Society, London, Special Publications, **148**, 217–240.

—— & DIXON, J. E. 1984. Introduction: aspects of the geological evolution of the Eastern Mediterranean. *In:* DIXON, J. E. & ROBERTSON, A. H. F. (eds) *The Geological Evolution of the Eastern Mediterranean*. Geological Society, London, Special Publications, **17**, 1–74.

—— & WOODCOCK, N. H. 1980. Tectonic setting of the Troodos massif in the east Mediterranean. *In:* PANAYIOTOU, A. (ed.) *Ophiolites*. Proceedings of International Symposium – 1979, Cyprus. Cyprus Geological Survey Department, 36–49.

——, PICKETT, E. & USTAÖMER, T. 1999. Inter-relationships between Palaeotethys and Neotethys in the Eastern Mediterranean: possible role of changing subduction polarity. *EUG 10 Abstracts*, Cambridge Publications, **4**, 315.

——, CLIFT, P. D, DEGNAN, P. J. & JONES, G. 1991. Palaeogeographic and palaeotectonic evolution of the Eastern Mediterranean Neotethys. *Palaeogeography, Palaeoclimatology, Palaeoecology*, **87**, 289–343.

——, ——, BROWN, S. *ET AL.* 1996. Alternative tectonic models for the Late Palaeozoic–Early Tertiary development of Tethys in the Eastern Mediterranean region. *In:* MORRIS, A. & TARLING, D. H. (eds) *Palaeomagnetism and Tectonics of the Mediterranean Region*. Geological Society London, Special Publications, **105**, 239–263.

ROBINSON, A. G., BANKS, C. J., RUTHERFORD, M. M. & HIRST, J. P. P. 1995. Stratigraphic and structural development of the Eastern Pontides, Turkey. *Journal of the Geological Society, London*, **152**, 861–872.

ŞENGÖR, A. M. C. & YILMAZ, Y. 1981. Tethyan evolution of Turkey: a plate tectonic approach. *Tectonophysics*, **75**, 181–241.

——, —— & KETİN, İ. 1980. Remnants of a pre–Late Jurassic ocean in northern Turkey: fragments of Permian–Triassic Palaeotethys? *Geological Society of America Bulletin*, **91**, 599–609.

——, —— & SUNGURLU, O. 1984. Tectonics of the Mediterranean Cimmerides: nature and evolution of the western termination of Palaeo-Tethys. *In:* DIXON, J. E. & ROBERTSON, A. H. F. (eds) *The Geological Evolution of the Eastern Mediterranean*. Geological Society, London, Special Publications, **17**, 77–112.

SÖZEN, A. 1973. *Geologische Untersuchungen zur Genese der Ziinnober–Lagerstatte Kalecik/Karaburun (Türkei)*. Inaugural dissertation, München.

STAMPFLI, G. M. 2000. Tethyan oceans. *This volume*.

——, MARCOUX, J. & BAUD, A. 1991. Tethyan margins

in space and time. *Palaeogeography, Palaeoclimatology and Palaeoecology*, **87**, 373–409.

——, MOSAR, J., FAURE, P., PILLEVUIT, A. & VANNAY, J.-C. 2000. Permo-Mesozoic evolution of the western Tethys realm: the Neotethys East Mediterranean basin connection. *In:* ZIEGLER, P., CAVAZZA, W. & ROBERTSON, A. H. F. (eds) *Peritethyan Rift/Wrench Basins and Passive Margins, IGCP 369*. Bulletin du Muséum National d'Histoire Naturelle, Paris, in press.

STATTEGGER, K. 1984. Chromspinell im klastischen Karbon von Chios, Aegais (Vorbericht). *Anzeiger der Osterreichischen Akademie der Wissenschaften, Mathematisch–Naturwissenschaftlichen Klasse*, **129**, 1–2.

TEKELİ, O. 1981 Subduction complex of pre-Jurassic age, northern Anatolia, Turkey. *Geology*, **9**, 68–72.

TELLER, R. 1980. Geologische Beobachtungen auf der Insel Chios. *Denkschrift Academie der Wissenschaften Wien, mathematisch–naturwissenschaftliche Klausur*, **40**, 340–356.

TÜYSÜZ, O. 1990. Tectonic evolution of part of the Tethyside orogenic collage: the Kargı Massif, northern Turkey. *Tectonics*, **5**, 1–33.

USTAÖMER, T. & ROBERTSON, A. H. F. 1993. Late Palaeozoic–Early Mesozoic marginal basins along the active southern continental margin of Eurasia: evidence from the Central Pontides (Turkey) and adjacent regions. *Geological Journal*, **28**, 219–238.

—— & —— 1997. Tectonic–sedimentary evolution of the North Tethyan margin in the Central Pontides of Northern Turkey. *In:* ROBINSON, A. G. (ed.) *Regional and Petroleum Geology of the Black Sea and Surrounding Region*. AAPG Memoirs, **68**, 255–290.

—— & —— 1999. Geochemical evidence used to test alternative plate tectonic models for pre-Late Jurassic (Palaeotethyan) units in the Central Pontides, N Turkey. *In*: BOZKURT, E. & ROWBOTHAM, G. (eds) *Advances in Turkish Geology, Part 1*: Tethyan evolution and fluvial to marine sedimentation. Geological Journal, in press.

Late Permian foraminiferal biofacies belts in Turkey: palaeogeographic and tectonic implications

DEMİR ALTINER,[1] SEVİNÇ ÖZKAN-ALTINER[1] & ALİ KOÇYİĞİT[2]

[1]*Marine Micropalaeontology Research Unit, Department of Geological Engineering, Middle East Technical University, TR-06531 Ankara, Turkey*
(e-mail: demir@metu.edu.tr)
[2]*Tectonic Research Unit, Department of Geological Engineering, Middle East Technical University, TR-06531 Ankara, Turkey*

Abstract: Upper Permian marine carbonates are distinguished in two contrasting biofacies belts in Turkey. The Southern Biofacies Belt, represented by low-energy inner platform deposits of the Tauride Belt and the Arabian Platform, is rich in algae and smaller foraminifera but poor in fusulines. The Kubergandian and Murgabian stages are missing, although the rest of the Upper Permian consists of monotonous, shallow-marine carbonate deposits. The extremely tectonised and fragmented Northern Biofacies Belt includes the Upper Permian of the Karakaya Orogen and outer platform or platform margin deposits of the Tauride Belt. These deposits are rich in parachomata-bearing fusulines comprising *Cancellina, Verbeekina, Afghanella, Sumatrina, Neoschwagerina* and *Yabeina*. The reconstructed biostratigraphic scheme indicates that all Upper Permian stages (Kubergandian–Dorashamian) are present.

The lateral continuity of the two biofacies belts is detected by the presence of tongues of the Northern Biofacies Belt pinching out in the Southern Biofacies Belt. Upper Permian blocks in the Karakaya Orogen display similar palaeontologic and biofacies characteristics, with the outer platform or platform margin deposits of the Taurides constituting the northernmost extension of the carbonate platform. This platform was probably facing a basin or a trough to the north. The lack of any transgressive Upper Permian deposits resting unconformably on the pre-Permian basement of the Sakarya Continent strongly suggests that such a basin was located between the Late Permian carbonate platform in the south and the basement rocks of the future Sakarya Continent in the north.

As in the case of many regional studies, the Late Permian palaeogeographic and tectonic evolution of Turkey can be understood if the configuration of carbonate platforms, troughs and basins, and areas of non-deposition, erosion or continental deposition are clearly defined. Since the 1980s models have been proposed to explain the geological events within this critical time interval (Şengör & Yılmaz 1981; Şengör *et al.* 1984; Robertson & Dixon 1984; Stampfli *et al.* 1991; Dercourt *et al.* 1993; Okay *et al.* 1996). These models have assumed a wedge-shaped Palaeotethys embayment in the eastern Mediterranean and discussed the position of the Karakaya Basin with respect to the Palaeotethys. They have also discussed early rifting of a southerly Neotethys and depicted areas of continental and marine deposition. Reconstructions were highly simplified and mainly based on regional tectonic interpretations rather on local stratigraphic and facies data.

The Late Permian palaeogeography and tectonics are contributed to in this study by analysing the biofacies characteristics of the Late Permian carbonate platform. This platform is reconstructed by assembling the Upper Permian outcrops distributed in different tectonic units of Triassic and Cretaceous to Tertiary age. In conclusion, a possible Late Permian configuration for this important part of Turkey is speculated on.

Late Permian foraminiferal biofacies belts in Turkey

The Southern Biofacies Belt

In Turkey there are undifferentiated metamorphic rock belts including the Upper Permian and many outcrops which are not yet properly studied. However, the known Upper Permian platform carbonates are mainly present in two biofacies belts, namely the Southern and Northern Biofacies Belts (Fig. 1). The Southern Biofacies Belt, characterized by low-energy inner platform deposits, is recognized both in the Taurides and the Arabian Platform. The succession is dominated by carbonates with intercalations of quartz arenitic sandstones and it

From: BOZKURT, E., WINCHESTER, J. A. & PIPER, J. D. A. (eds) *Tectonics and Magmatism in Turkey and the Surrounding Area.* Geological Society, London, Special Publications, **173**, 83–96. 1-86239-064-9/00/$15.00
© The Geological Society of London 2000.

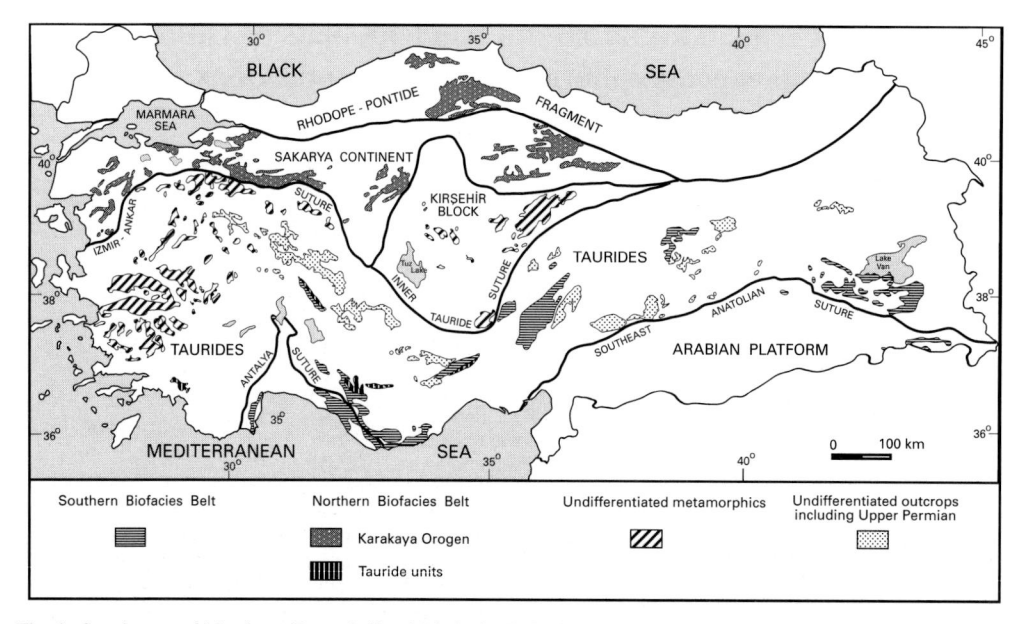

Fig. 1. Southern and Northern Foraminiferal Biofacies Belts in Turkey. Neotethyan sutures are modified from Şengör (1984).

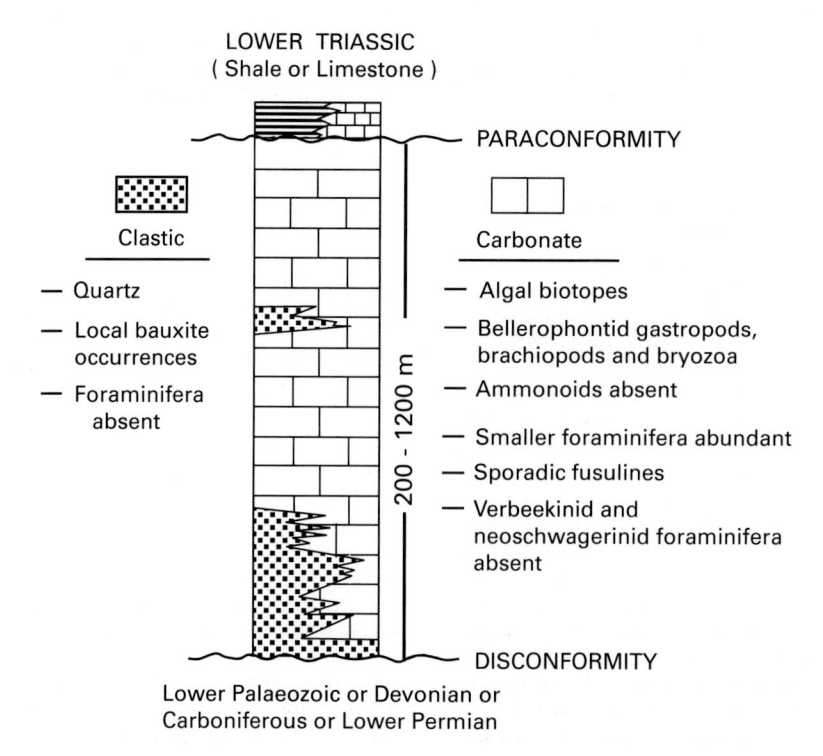

Fig. 2. Simplified columnar section of the Upper Permian displaying palaeontological characteristics of the Southern Biofacies Belt.

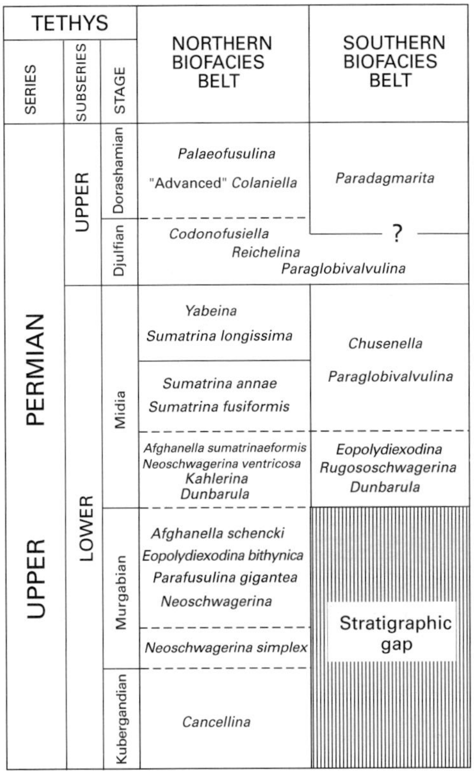

TETHYS			NORTHERN BIOFACIES BELT	SOUTHERN BIOFACIES BELT
SERIES	SUBSERIES	STAGE		
PERMIAN	UPPER	Dorashamian	Palaeofusulina "Advanced" Colaniella	Paradagmarita
		Djulfian	Codonofusiella Reichelina	? Paraglobivalvulina
	LOWER	Midia	Yabeina Sumatrina longissima	Chusenella
			Sumatrina annae Sumatrina fusiformis	Paraglobivalvulina
			Afghanella sumatrinaeformis Neoschwagerina ventricosa Kahlerina Dunbarula	Eopolydiexodina Rugososchwagerina Dunbarula
		Murgabian	Afghanella schencki Eopolydiexodina bithynica Parafusulina gigantea Neoschwagerina	Stratigraphic gap
			Neoschwagerina simplex	
		Kubergandian	Cancellina	

Fig. 3. Foraminiferal biostratigraphy in the Northern and Southern Biofacies Belts. Tethys scale is after Leven (1992).

includes local bauxite occurrences (Fig. 2). It overlies older Palaeozoic units unconformably and is overlain paraconformably by Lower Triassic rock units (Özgül 1976, 1984, 1997; Monod 1977; Argyriadis 1978; Lys & Marcoux 1978; Altıner 1981, 1984, 1997; Köylüoğlu & Altıner 1989).

The belt is rich in gymnocodiacean and dasycladacean algae. Bellerophontid gastropods, productid brachiopods and Bryozoa are abundant, but ammonoids are not recorded. Smaller foraminifera, particularly hemigordiopsids, are abundant. Among fusulines, schwagerinids are sporadic, and verbeekinid and neoschwagerinid forms are practically absent (Altıner 1984, 1997; Köylüoğlu & Altıner 1989). The Kubergandian and Murgabian stages are missing in the successions of the Southern Biofacies Belt (Fig. 3), although the rest of the Upper Permian, including the Midian, Djulfian and Dorashamian stages, is represented by monotonous shallow-marine and micritic carbonate deposits.

The palaeogeographic distribution of the important Late Permian foraminiferal taxon, *Paradagmarita*, is confined to this belt and helps to depict the belt in the western Tethys (Fig. 4). This genus extends over southern Turkey, including the Taurides and the Arabian Platform (Monod 1977; Lys & Marcoux 1978; Altıner 1981, 1984, 1997; Zaninetti et al. 1981; Şengör et al. 1988; Köylüoğlu & Altıner 1989), south Iran (Argyriadis & Lys 1977; Argyriadis 1978), Saudi Arabia (Okla 1994) and Oman (Montenat et al. 1976). This belt is also recognized in eastern Asia. The genus *Shanita*, typically confined to the Southern Biofacies Belt in Turkey (Altıner & Zaninetti 1977; Altıner 1981, 1984; Zaninetti et al. 1982), south Iran (Baghbani 1988) and Oman (Montenat et al. 1976), also occurs in western Yunnan (Sheng & He 1983), Burma and Thailand (Brönnimann et al. 1978; Whittaker et al. 1979; Zaninetti et al. 1979; Şengör et al. 1988). It corresponds to the Sibumasu Terrane of Metcalfe (1988) whose position was interpreted to be marginal to Gondwanaland in the Late Palaeozoic. West of Turkey, the Southern Biofacies Belt is probably present in Greece but not yet documented. Further west, the Upper Permian outcrops of western Serbia (Pantic 1969), the southern Alps (Neri & Pasini 1985; Noé 1987) and north Hungary (Haas et al. 1986; Bérczi-Makk 1992) display palaeontological characters similar to the Southern Biofacies Belt.

The Northern Biofacies Belt

The extremely tectonised and fragmented Northern Biofacies Belt consists of detached Upper Permian carbonate blocks floating in the Triassic Karakaya Orogen in north Turkey (Şengör & Yılmaz 1981; Şengör et al. 1984; Koçyiğit 1987) and the Upper Permian outer platform or platform margin deposits of the allochthonous Bolkar Dağı unit of Özgül (1976) in the Taurides (Fig. 1). The tectonic slices, including the Upper Permian in the Lycian Nappes [De Graciansky 1972; see also Collins & Robertson (1998)], are also interpreted to be equivalent to the Bolkar Dağı Unit of Özgül (1976).

The biologic evidence used to group the Upper Permian exposures in the Karakaya Orogen and the Bolkar Dağı Unit under one uniform biofacies belt are well defined and numerous (Table 1). Verbeekinid and neoschwagerinid populations, practically absent in the Southern Biofacies Belt, are well diversified and abundant in both the Karakaya Orogen and the Bolkar Dağı Unit (Skinner 1969; Demirtaşlı

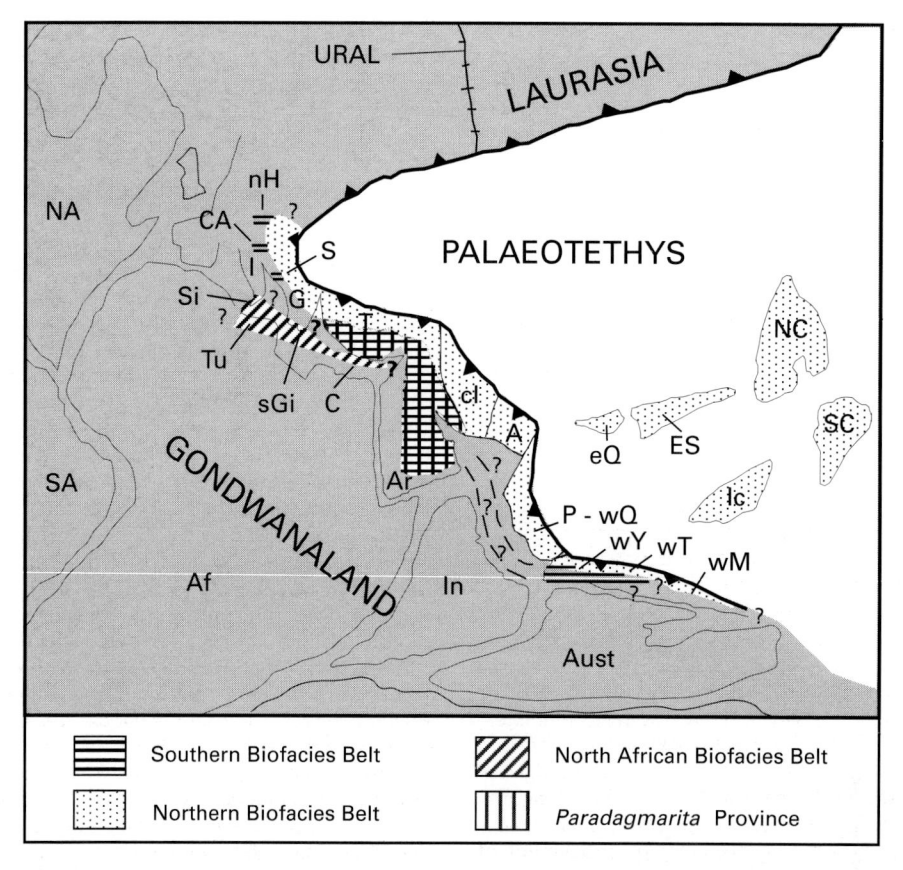

Fig. 4. Highly schematic Late Permian reconstruction displaying the foraminiferal biofacies belts. The base map has been modified and redrawn from Şengör et al. (1988). Note that the position of the Sakarya basement has been left in the Turkish blocks (T) side, as in the original publication, although this paper rejects the Gondwana origin of the future Sakarya Continent in its conclusions. A, Afghan block; Af, Africa; Ar, Arabia; Aust, Australia; C, Cyprus; CA, Calcareous Alps; cI, central Iran; ES, Emei Shan; eQ, east Qangtang block; G, Greece; I, Italy; Ic, Indo-China; nH, north Hungary; P–wQ, Pamir–west Qangtang block; SA, South America; SC, South China; sGi, southern Greece islands; Si, Sicily; T, Turkish blocks; Tu, Tunisia; wM, western Malaya; wT, western Thailand; wY, western Yunnan.

et al. 1984; Leven & Okay 1996; Özgül 1997; Altıner 1997). Schwagerinids, ozawainellids and schubertellids are highly diversified in contrast to their sporadic occurrences in the Southern Biofacies Belt. Among smaller foraminifera, Lasiodiscidae, Colaniellidae, Endotebidae and Abadehellidae occur typically (Table 1), and in some families, like Biseriamminidae, different steps in evolution seem to be confined either to the northern or southern belt (Altıner 1997). The genera *Paradagmarita* and *Louisettita* are confined to the southern belt, *Globivalvulina* sp. A and Genus A described recently by Altıner (1997), occur characteristically in the northern belt (Table 1). The most characteristic porcelaneous foraminifera in the northern belt is the rock-forming *Hemigordiopsis* and in the southern belt the presence of the pillared genus *Shanita*. The reconstructed biostratigraphic scheme indicates that all Upper Permian stages (Kubergandian–Dorashamian) are present in the Northern Biofacies Belt (Fig. 3).

The Northern Biofacies Belt, with its typical foraminiferal provinces (*Neoschwagerina, Yabeina, Lepidolina, Palaeofusulina, Colaniella*, etc.; Şengör et al. 1988), extends from Turkey through the Crimea, Transcaucasia, central and northern Iran, Afghanistan, the Pamirs to eastern Asia, including Indo-China, South China, North China and Japan (Fig. 4). West of Turkey, the equivalents are found also in Yugoslavia (Dinarides) and in the northern Calcareous

Table 1. *Comparisons of the diagnostic foraminiferal taxa of the Northern and Southern Biofacies Belts*

Northern Biofacies Belt	Southern Biofacies Belt
Verbeekinid and neoschwagerinid foraminifera	–
Schwagerinids, ozawainellids and schubertellids highly diversified	Schwagerinids, ozawainellids and schubertellids
Lasiodiscidae (*Lasiodiscus*)	–
Colaniellidae (*Colaniella*)	–
Biseriamminidae (*Globivalvulina* sp. A, Genus A)	Biseriamminidae (*Paradagmarita, Louisettita*)
Endotebidae (*Endoteba*)	–
Abadehellidae (*Abadehella*)	–
Hemigordiopsidae (*Hemigordiopsis* in rock-forming abundance)	Hemigordiopsidae (*Shanita*)

Alps (Şengör *et al.* 1988). This belt, characterized by outer platform to platform margin deposits, lies north of the Southern Biofacies Belt with inner platform deposits extending from north Hungary through south Turkey and Oman to Thailand. One broad conclusion that can be arrived at here is that the Late Permian carbonate platform extended as a linear belt from central Europe through Turkey to the Far East, and consisted of two laterally changing biofacies belts. This vast platform was therefore facing an ocean to the north and, in the light of other studies (Şengör & Yılmaz 1981; Şengör *et al.* 1984; Robertson & Dixon 1984), this ocean is identified as the wedge-shaped embayment of Palaeotethys.

In the western Tethys, further to the south of the Southern Biofacies Belt, there occurs another realm whose biologic characteristics are similar to those of the Northern Biofacies Belt. This belt extends from Tunisia (Glintzboeckel & Rabaté 1964; Skinnner & Wilde 1967; Gargouri & Vachard 1988; Vachard & Razgallah 1993) through Sicily (Skinner & Wilde 1966; Flügel *et al.* 1991), South Italy (Lagonegro, Monte Facito; Ciarapica *et al.* 1986; Panzanelli-Fratoni *et al.* 1987; Vachard & Miconnet 1989), southern Greece and the southern Greek islands (Nakazawa *et al.* 1975; Papanikolau & Baud 1982; Baud *et al.* 1991; Grant *et al.* 1991; Vachard *et al.* 1993, 1995; Altıner & Özkan-Altıner 1998), and finally to Cyprus (Reichel 1946; Nestell & Pronina 1997). Although it is not within the scope of this paper, this domain is provisionally named the 'North African Biofacies Belt'. Its spatial distribution is remarkable because it coincides, at least partially, with the Permian rift basin of Robertson *et al.* (1991) and areas of early rifting of the southern Neotethys (Stampfli *et al.* 1991; Dercourt *et al.* 1993).

Stratigraphic evidence for the assembly of Late Permian biofacies belts

Upper Permian blocks in the Karakaya Orogen

Controversial interpretations and observations made on the Triassic Orogen include a subduction–accretion complex (Tekeli 1981; Okay *et al.* 1991, 1996; Pickett & Robertson 1996), a deformed rock sequence from the back-arc basin of the Palaeotethys Ocean (Şengör & Yılmaz 1981; Şengör *et al.* 1984; Yılmaz 1990; Genç & Yılmaz 1995), and a Triassic rift-basin succession (Koçyiğit 1987; Altıner & Koçyiğit 1993) in northern Turkey. One of the most intriguing subjects is the provenance of Upper Permian, as well as other Upper Palaeozoic, blocks floating in the Karakaya Belt. Since Erk (1942), who was not aware of the tectonic setting of the Upper Permian units in the Bursa region, many authors have reported the blocky nature of the Upper Permian in northern Turkey (Bingöl *et al.* 1975; Şengör & Yılmaz 1981; Yılmaz 1981, 1990; Şengör *et al.* 1984; Koçyiğit 1987; Okay *et al.* 1991, 1996; Altıner & Koçyiğit 1993; Genç & Yılmaz 1995). However, opinions about the provenance of the blocks differ. Some authors, based on Altınlı (1975), who proposed that the Derbent Limestone was a transgressive unit over the basement rocks of the Sakarya region, suggested that the Karakaya Basin opened in the Upper Permian carbonate platform and interpreted the carbonate blocks as rock fall or mass-flow deposits originating from the basement (Şengör & Yılmaz 1981; Yılmaz 1981, 1990; Genç & Yılmaz 1995). Others, however, proposed a southerly Gondwana origin for the Upper Permian blocks (Koçyiğit 1987; Altıner & Koçyiğit 1993;

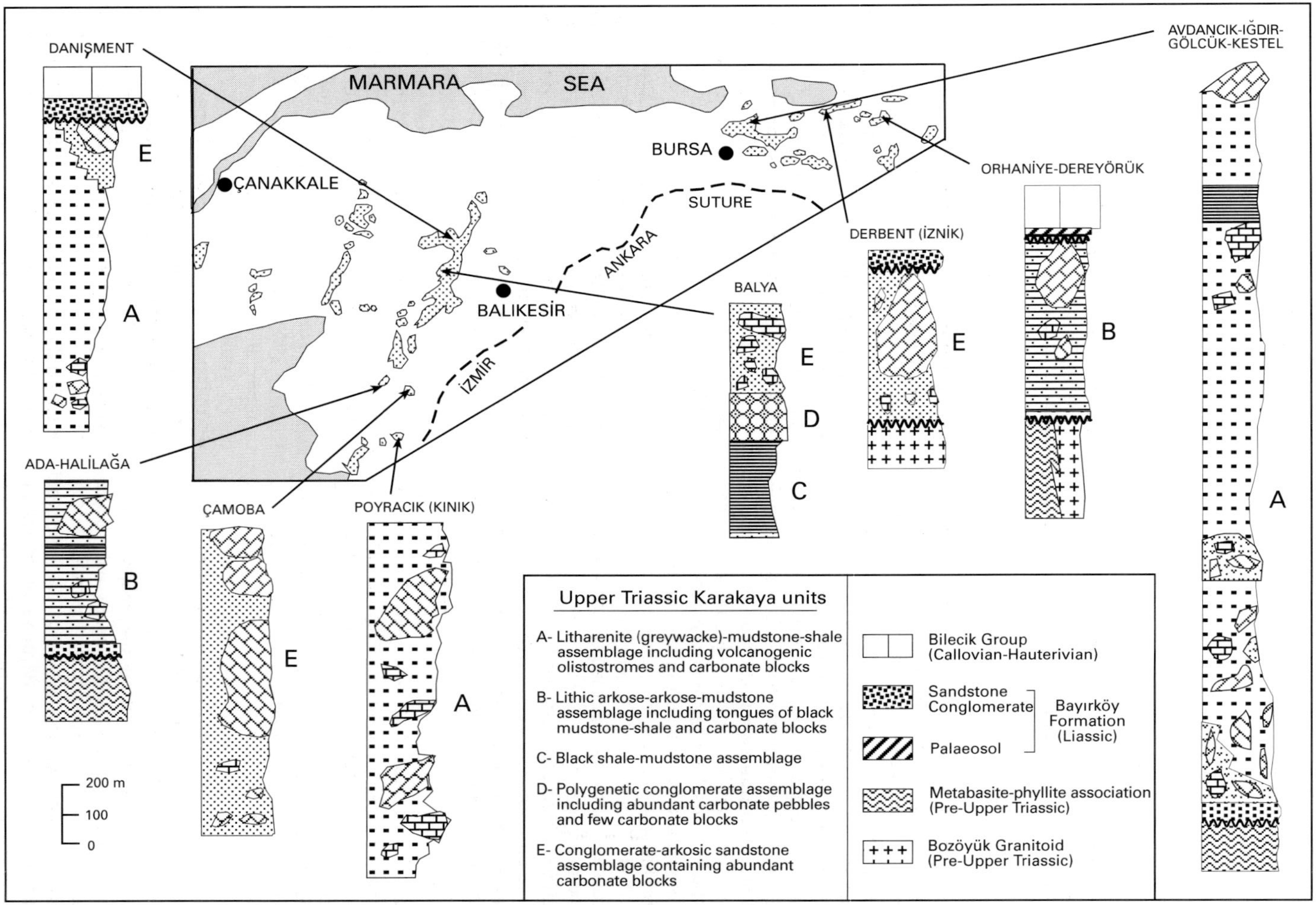

Fig. 5. Upper Triassic siliciclastic and blocky sequences of the Karakaya Orogen in the Balıkesir and Bursa regions.

Leven & Okay 1996; Okay et al. 1996). Among these authors, Okay et al. (1996) assumed a southward subduction of Palaeotethys which resulted in back-arc rifting in the Anatolide–Tauride Platform, where carbonate sedimentation prevailed during the Late Permian. The back-arc rift apparently developed into the Neotethyan Vardar Ocean in the Middle Triassic. The Anatolide–Tauride continental sliver, with its cap of Upper Permian carbonates, was obducted in the Late Triassic onto the Sakarya basement, shedding Upper Permian blocks into the siliciclastic deposits laid down contemporaneously on the Sakarya zone.

In order to test the Upper Permian stratigraphic interpretations and conclusions about the provenance of the carbonate blocks in the Karakaya units, the present authors recently studied the Bursa and Balıkesir regions in northwestern Anatolia, comprising the type locality of the Derbent Limestone of Altınlı (1975) and the more recently recognized Upper Permian Çamoba Formation in the Balıkesir region (Akyürek & Soysal 1983) (Fig. 5). Siliciclastic unmetamorphosed rock successions in the Bursa and Balıkesir regions are basically characterized by five lithologic entities, each of which could be named as a formal rock-stratigraphic unit.

Unit A. This unit is mainly composed of a microconglomeratic litharenite, litharenite (greywacke), mudstone and shale assemblage, including a variety of carbonate blocks ranging in age from Carboniferous (Visean) to Middle Triassic (Ladinian) and volcanogenic olistostromes (Fig. 5). Below the regional Liassic transgression, which has been dated as Hettangian at the earliest (Altıner et al. 1991), this rock assemblage transgressively overlies the pre-Late Triassic metabasite–phyllite rock association in the Avdancık–Iğdır–Gölcük–Kestel section in the Bursa region. The presence of the pelagic lamellibranch *Halobia* has previously been reported by Erk (1942) and Kaya (1991) from the same region, and blocks of Ladinian age in the studied sections suggest a Late Triassic age for the unit.

Unit B. This succession is transgressive over the pre-Triassic metabasite–phyllite rock association which was intruded by a granitoid body, most probably belonging to Bozüyük Granitoid of pre-Late Triassic age (Orta Sakarya Granite; Yılmaz 1981). This lithologic entity consists of conglomerates with pebbles derived from the underlying rock assemblage and a lithic arkose to arkosic sandstone and mudstone assemblage. The successions also include tongues of a black shale–mudstone assemblage (C-type lithologic entity; Fig. 5), as in the Ada–Halilağa area (Balıkesir region), and completely detached carbonate blocks of Permian and Triassic age. As this unit underlies the red palaeosol level of the Liassic Bayırköy Formation in the Orhaniye–Dereyörük area (Altıner et al. 1991), and because it contains carbonate blocks of Ladino-Carnian age and tongues of a black shale–mudstone unit of Norian age, a Late Triassic age is assigned to it.

Unit C. This succession recognized in the Balya area has been previously reported by Altıner et al. (1991) from the Edremit–Halılar area as the Bağcağız Formation of the Halılar Group (Krushensky et al. 1980) and by Okay et al. (1991, 1996) as a distinct succession in their Hodul Unit. Essentially represented by a thin-shelled, bivalve-bearing, black shale–mudstone succession, including thin siltstone intercalations, this unit does not contain any carbonate blocks or olistostromes in its matrix and has been attributed to the Norian, based on fossils identified in the Balya area (Leven & Okay 1996).

Unit D. This unit is principally composed of polygenetic conglomerates containing abundant limestone and subordinate radiolarian chert, granite and sandstone pebbles and a few limestone blocks. Its presence above a sharp contact with the black shale–mudstone rock assemblage indicates an abrupt relief change in the Karakaya Orogen in the Norian, at least in the Balya area.

Unit E. This lithologic entity, probably reflecting the youngest depositional period within the Karakaya Orogen, is characterized by conglomerates composed mostly of granite pebbles, arkosic sandstones and a variety of carbonate blocks, sometimes attaining sizes in the order of kilometres. The unit overlies the Bozüyük Granitoid unconformably in the Derbent (İznik) area of the Bursa region, although its stratigraphic position above the greywacke-dominated Danişment section [Orhanlar Greywacke of Brinkmann (1971, 1976); also see Okay et al. (1991)] is doubtful. This rock assemblage is widely distributed and corresponds to the Hodul Unit of Okay et al. (1991, 1996). It lies below the regional Liassic unconformity and is older than the Hettangian. Although a certain diachronism could be expected in regional studies, a Norian–Rhaetian age can be assigned

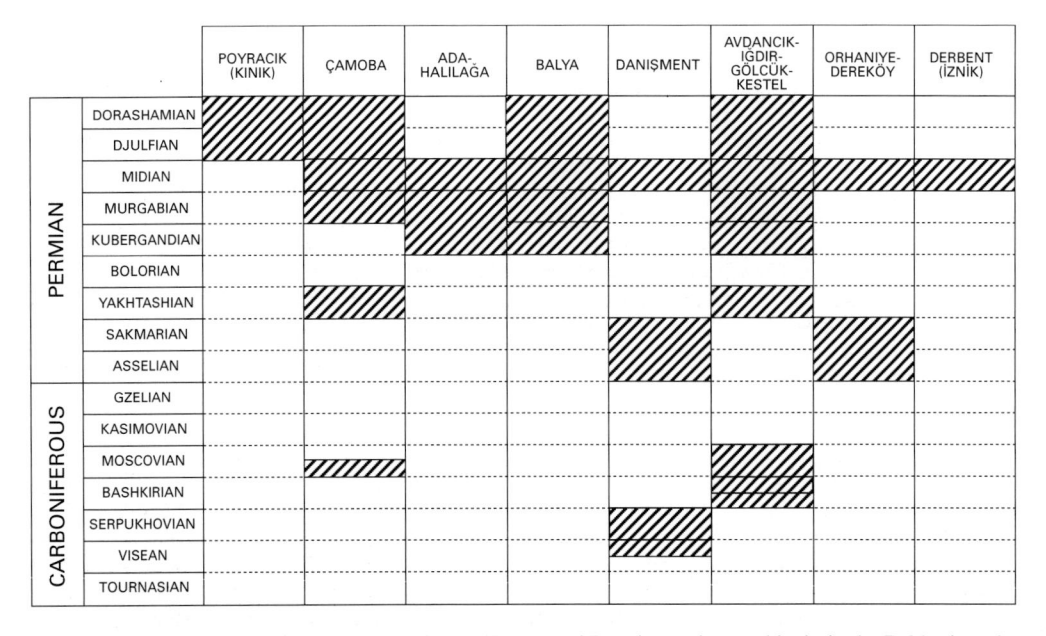

Fig. 6. Chronostratigraphic distributions of Carboniferous and Permian carbonate blocks in the Balıkesir and Bursa regions (northwestern Turkey).

to it, since it overlies the fossiliferous black shale–mudstone association of Norian age in the Balya area. Two of the study localities comprise the type areas of the Çamoba Formation (Akyürek & Soysal 1983) and Derbent Limestone (Altınlı 1975) (Fig. 5), both of which are considered as transgressive Upper Permian units over the basement rocks. This group's field observations show that the limestone outcrops of Late Permian age are totally chaotic and set in a siliciclastic matrix of arkosic sandstones in both sections. The limestone outcrops along the sections never have chronostratigraphic continuity, disappear laterally in short distances and are sometimes found juxtaposed with blocks of older ages.

In summary, Upper Palaeozoic (Carboniferous and Permian) carbonate exposures in unmetamorphosed siliciclastic rock assemblages of the Karakaya Orogen are chaotic and are blocks (olistoliths), olistostromes or tectonic slice complexes, transported into the sedimentation area during the Late Triassic. A northerly provenance, including the Moesia, İstanbul and Strandja Zones, is unlikely because these regions were above sea level and did not accumulate any marine carbonate deposits (Okay et al. 1996). Hence, the likely provenance is from the south. The northern margin of a carbonate platform of Gondwana origin is a possible depositional site for the Upper Permian blocks which complete the configuration of the platform to the north.

Further evidence for the southern origin of these chaotic blocks in the Karakaya units is the wide range of the ages of carbonate blocks from Visean to the latest Permian (Fig. 6). It is considered here that the time of accumulation of the various Carboniferous and Permian carbonate blocks is rather short, since they are identified only in Upper Triassic Karakaya units. This suggests a single source for carbonate blocks, including those of Late Permian age. The intense Hercynian deformation present in the Sakarya basement rocks is dated as Bashkirian–Moscovian (Okay et al. 1996) and almost excludes the possibility of derivation of Carboniferous (Visean–Moscovian) limestone blocks from the Sakarya basement. Hence, the most probable provenance of the Carboniferous and Permian blocks remains the Gondwana side because the chronostratigraphic order of blocks shows the best fit with the Carboniferous and Permian stratigraphy of the Taurus carbonate platform (Monod 1977; Altıner 1981, 1984; Özgül 1997).

Lateral continuity of biofacies belts in the Taurides

A difficult task in reconstructing the palaeogeography of the Taurides is to correlate

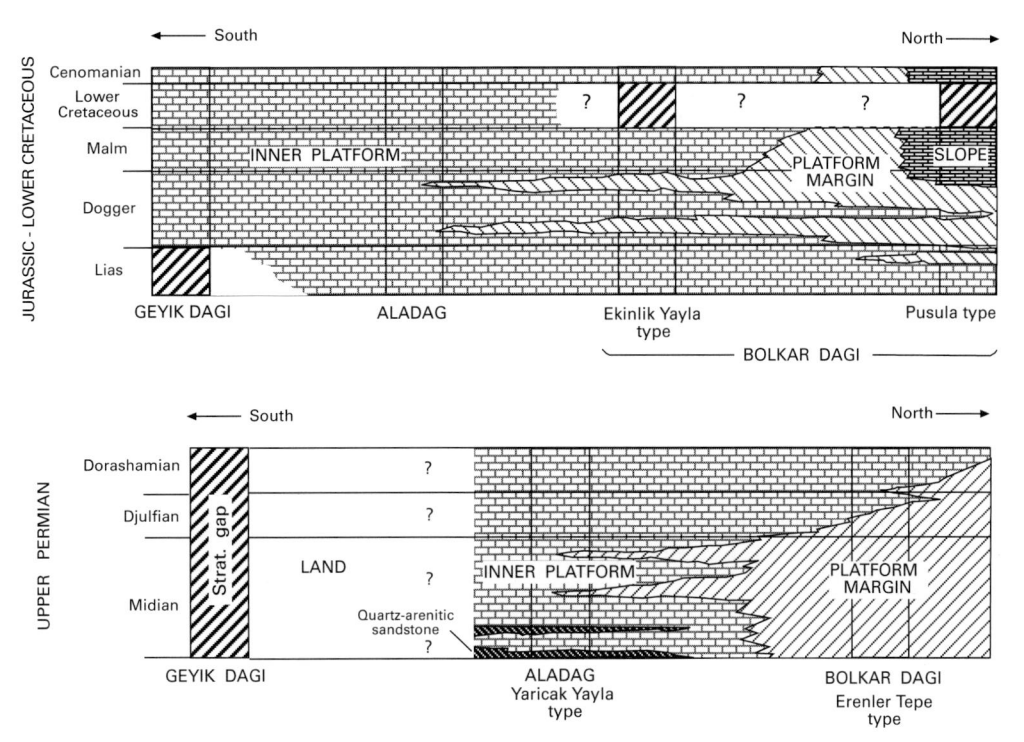

Fig. 7. Correlation and interpretation of the Jurassic–Lower Cretaceous and the Upper Permian facies in the Geyik Dağı, Aladağ and Bolkar Dağı Units (Bozkır–Hadım–Taşkent region, central Taurides). This figure is based on the present authors' palaeontological analyses of stratigraphic sections described in Özgül (1997).

carbonate facies exposed in different, but juxtaposed, tectonic units. Although the relative positions of tectonic units and their facies types are more or less understood following studies of tectonic polarity of nappes (Ricou et al. 1975; Şengör & Yılmaz 1981; Özgül 1984), direct physical evidence proving the lateral continuity of units is absent or poorly known.

In three of the juxtaposed tectonic units, the Geyik Dağı, Aladağ and Bolkar Dağı Units (Özgül 1976), two fundamental carbonate depositional episodes spanning the Late Permian and Jurassic–Early Cretaceous intervals are correlated and interpreted in Fig. 7, using chronostratigraphic and biofacies data from Özgül (1997), who investigated the stratigraphy of the Bozkır–Hadım–Taşkent region in the central Taurides. On these sections the Mesozoic and Upper Permian carbonates are reconstructed from south to north, depending on their facies characteristics and relative tectonic positions in the outcrop belts. The best fit occurs in the Geyik Dağı–Aladağ–Bolkar Dağı organization in both intervals. The Mesozoic reconstruction is provided by analysis of three distinct facies belts, algal and foraminiferal micritic limestones (inner platform), bioclastic limestones of high-energy type (platform margin) and calciturbiditic limestones with pelagic organisms (slope). Two distinct tongues pinching out in the Dogger of the Aladağ Unit are direct stratigraphic evidence for lateral continuity of facies and the tectonic units which contain them. More or less the same configuration is depicted for the Upper Permian facies (Fig. 7). Although to the south the Geyik Dağı Unit is characterized by a large stratigraphic gap corresponding to the Late Permian, to the north the Upper Permian of the Aladağ and Bolkar Dağı Units, represented by Southern and Northern Biofacies Belts, respectively, occurs in laterally changing inner platform and platform margin deposits. Two tongues of the Northern Biofacies Belt contain neoschwagerinid fusulines and pinch out in the Midian strata of the Southern Biofacies Belt. These tongues link the two biofacies belts and indicate times of maximum transgression during the Late Permian and the maximum aerial distribution of the Midian sediments in the Northern Biofacies Belt.

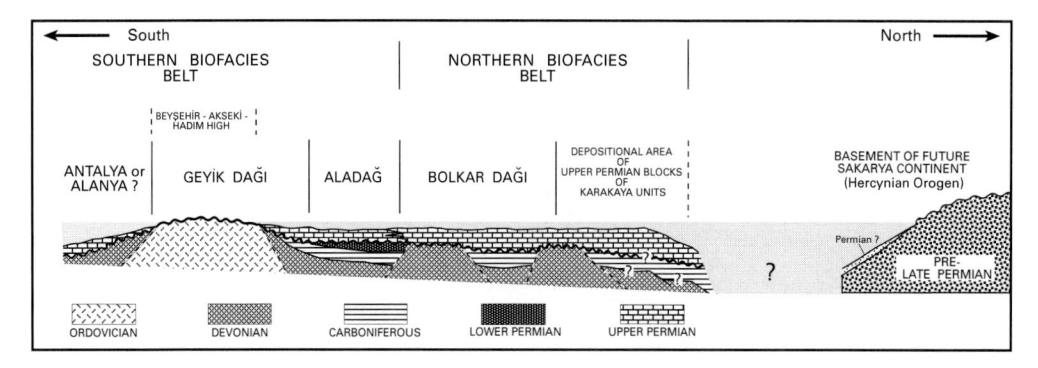

Fig. 8. Reconstruction of the Late Permian carbonate platform and its palaeogeographic position with respect to the basement of the future Sakarya Continent.

Reconstruction of the carbonate platform and conclusions on the Late Permian palaeogeography and tectonics of Turkey

The similarity of the biofacies characteristics of Upper Permian blocks in the Karakaya Orogen in north Turkey to those of Late Permian age in various tectonic units of the Taurides, combined with clear stratigraphic evidence proving the lateral continuity of the Southern and Northern Biofacies Belts in the Taurides, leads to reconstruction of one single carbonate platform model, comprising nearly all Upper Permian marine sedimentary rocks exposed in Turkey (Fig. 8). The lack of any transgressive Upper Permian deposits resting unconformably on the pre-Permian basement of northern Turkey [the basement of the future Sakarya Continent of Şengör & Yılmaz (1981)] is undoubtedly very significant. The presence of pelagic Permian blocks (Kozur & Kaya 1994; Okay & Mostler 1994) in the Karakaya units, in addition to this group's observations of blocks of Midian age displaying the character of a calciturbiditic slope deposit in the Bursa region, indicates that this platform was facing a trough or basin to the north. Such a basin must have been located between the carbonate platform in the south and the pre-Late Permian Sakarya basement rocks in the north. This conclusion would reject a Gondwana origin for the Sakarya Continent as claimed by Şengör & Yılmaz (1981), who defined it as a continental fragment of the Cimmerian Continent rifted from Gondwana during Permo-Triassic times and rotated anticlockwise northwards to close the Palaeotethys Ocean.

The marine deposition of the Late Permian carbonate platform is bounded to the south by a structural high that is here termed the Beyşehir–Akseki–Hadim High. This high formed during pre-Late Permian time, has been a linear emergent belt during most of the Late Palaeozoic (Monod 1977; Özgül 1997) and extended from the western Taurides through the Amonos Mountains into the southeast Anatolian Mardin High (Sungurlu 1974). Bordered to the north and south by incomplete Upper Palaeozoic successions of the Geyik Dağı (Özgül *et al.* 1973, 1991; Zaninetti *et al.* 1981) and Antalya or Alanya Units (Marcoux 1979; Lys & Marcoux 1978), respectively (Fig. 8), the Beyşehir–Akseki–Hadim High, made up mainly of Lower Palaeozoic (mostly Cambro-Ordovician) rocks, was not submerged during the remainder of the Palaeozoic, even during the vast Midian transgression in the Late Permian.

Successions belonging to the Southern Biofacies Belt are exposed in two main localities (Hazro and Hakkari) of the Arabian Platform (southeast Anatolia; Köylüoğlu & Altıner 1989). In both localities deposits were laid down bordering the Mardin High. The spatial distribution of the Southern Biofacies Belt from the Tauride belt to the Arabian Platform, confirmed by several faunal provinces (e.g. *Paradagmarita*), questions the validity of existing models (e.g. Stampfli *et al.* 1991; Dercourt *et al.* 1993) which generally favour separation to the north of the Arabian Platform and opening of a southerly Neotethys ocean during the Late Permian.

As shown by the Devonian–Permian stratigraphy of the Tauride units (Özgül 1976, 1984; Altıner 1984), the reconstructed Late Permian carbonate platform rests on a tectonically disturbed basement, most probably activated in Devonian and Carboniferous times and possibly extending into, or reactivated during, the Early Permian (Fig. 8). The presence of a non-depositional area represented by a structural high, probably reactivated during the Late Palaeozoic, and areas of incomplete Late Palaeozoic sedimentation (transgressive Upper Permian over either Devonian or Lower Carboniferous; Özgül *et al.* 1973, 1991; Monod 1977; Zaninetti *et al.* 1981; Altıner 1984; Demirtaşlı 1984; Özgül 1997) in the Geyik Dağı Unit, nearly complete sedimentation from Upper Devonian through Carboniferous to Asselo-Sakmarian (Lower Permian) in the Aladağ Unit (Monod 1977; Argyriadis 1978; Altıner 1981, 1984) and the outer platform Upper Permian successions over Devonian and Carboniferous units of the Bolkar Dağı Unit (Demirtaşlı *et al.* 1984; Özgül 1984, 1997) all suggest that this basement was probably rifted to give rise to differing stratigraphic sequences arranged in linear belts in southern Turkey. Relatively more complete successions, such as the Aladağ Unit, were laid down in fault-bounded, shallow-marine environments (Fig. 8). This basement configuration, preceding the Late Permian transgression, strongly implies the presence of a basin or trough to the north by Carboniferous time. If so, it might even suggest that some continental pieces rifted away from the Gondwana margin and drifted northwards to collide with the continuously growing orogenic belt in the north. In this model two important points emerge: the definition of the allochthonous terrane that drifted northwards (e.g. the Upper Permian of the Central Pontides, known as the Aktaş Unit, could be one of these drifted off fragments; Ustaömer & Robertson 1997) and the basin or trough left behind to the south of the orogen bounded by the Late Permian carbonate platform. The nature of the trough and related rock sequences connecting the Sakarya basement and the northernmost Taurides need to be determined by additional studies focusing on the Bolkar Dağı–Konya and Kütahya zones. These are still poorly defined because of the effects of metamorphism, the presence of a thick sedimentary cover and the lack of detailed stratigraphic and palaeontologic data.

The authors acknowledge the Scientific and Technical Research Council of Turkey (TÜBİTAK) for funding the project (ref. no. YBAG-0077/DPT) related to the subject of this paper.

References

AKYÜREK, B. & SOYSAL, Y. 1983. Basic geological features of the region south of the Biga peninsula (Savatepe-Kırkağaç–Bergama–Ayvalık). *Mineral Research and Exploration Institute of Turkey (MTA) Bulletin*, **95/96**, 1–13 [in Turkish with English abstract].

ALTINER, D. 1981. *Recherches stratigraphiques et micropaléontologiques dans le Taurus oriental au NW de Pınarbaşı (Turquie)*. Thèse, Université de Genève.

—— 1984. Upper Permian foraminiferal biostratigraphy in some localities of the Taurus Belt. *In*: TEKELİ, O. & GÖNCÜOĞLU, M. C. (eds) *Geology of the Taurus Belt*. Proceedings of an International Tauride Symposium, Mineral Research and Exploration Institute of Turkey (MTA) Publications, 255–268 [in Turkish with English abstract].

—— 1997. Origin, morphologic variation and evolution of Dagmaritin-type biseriamminid stock in the Late Permian. *In*: ROSS, C. A., ROSS, J. R. P. & BRENCKLE, P. L. (eds) *Late Palaeozoic Foraminifera; their biostratigraphy, evolution, and palaeoecology; and the Mid-Carboniferous boundary*. Cushman Foundation for Foraminiferal Research, Special Publication, **36**, 1–4.

—— & KOÇYİĞİT, A. 1993. Third Remark on the Geology of Karakaya Basin. An Anisian megablock in northern central Anatolia: micropalaeontologic, stratigraphic and tectonic implications for the rifting stage of Karakaya basin, Turkey. *Revue de Paléobiologie*, **12**, 1–17.

—— & ÖZKAN-ALTINER, S. 1998. *Baudiella stampflii*, n. gen., n. sp., and its position in the evolution of Late Permian ozawainellid fusulines. *Revue de Paléobiologie*, **17**, 163–175.

—— & ZANINETTI, L. 1977. *Kamurana brönnimanni*, n. gen., n. sp., un nouveau Foraminifère porcelané perforé du Permien supérieur du Taurus oriental, Turquie. *Notes du Laboratoire de Paléontologie de l'Université de Genève*, **1**, 1–6.

——, KOÇYİĞİT, A., FARINACCI, A., NICOSIA, U. & CONTI, M. A. 1991. Jurassic–Lower Cretaceous stratigraphy and palaeogeographic evolution of the southern part of northwestern Anatolia. *Geologica Romana*, **27**, 13–80.

ALTINLI, İ. E. 1975. Orta Sakarya Jeolojisi. *Proceedings of the 50th Anniversary of Turkish Republic Earth Sciences Congress*. Mineral Research and Exploration Institute of Turkey (MTA) Publications, 151–161 [in Turkish with English abstract].

ARGYRIADIS, I. 1978. *Le Permien alpino-méditerranean à la charnière entre l'Hercynien et l'Alpin*. Thèse, Université de Paris-Sud, Centre d'Orsay.

—— & LYS, M. 1977. La dynamique de la lithosphere au Permien supérieur et ses relations avec la biostratigraphie en mediterranée et au moyen-orient. *Communication au 6me Colloque sur la Géologie des Régions Egéenes*, Athènes, Septembre 1977, 1–20.

BAGHBANI, D. 1988. Shanita zone and its biostratigraphic significance in south and southwest Iran. *Revue de Paléobiologie, Special Volume*, **2**, 37.

BAUD, A., JENNY, C., PAPANIKOLAOU, D., SIDERIS, C. & STAMPFLI, G. 1991. New observations on Permian stratigraphy in Greece and geodynamic interpretations. *Proceedings of the 5th Congress of the Geological Society of Greece*. Geological Society of Greece, Special Publications, 187–206.

BÉRCZI-MAKK, A. 1992. Midian (Upper Permian) foraminifera from the large Mihalovits quarry at Nagyvisnyo (North Hungary). *Acta Geologica Hungarica*, **35**, 27–38.

BİNGÖL, E., AKYÜREK, B. & KORKMAZER, B. 1975. Geology of the Biga peninsula and some characteristics of the Karakaya blocky series. *Proceedings of the 50th Anniversary of Turkish Republic Earth Sciences Congress*. Mineral Research and Exploration Institute of Turkey (MTA) Publications, 70–77 [in Turkish with English abstract].

BRINKMANN, R. 1971. Jungpalaozoikum und alteres Mesozoikum in NW-Anatolien. *Mineral Research and Exploration Institute of Turkey (MTA) Bulletin*, **76**, 56–67.

—— 1976. *Geology of Turkey*. Elsevier.

BRÖNNIMANN, P., WHITTAKER, J. E. & ZANINETTI, L. 1978. Shanita, a new pillared miliolacean foraminifer from the Late Permian of Burma and Thailand. *Rivista Italiana Paleontologia*, **84**, 63–92.

CIARAPICA, G., CIRILLI, S., MARTINI, R. & ZANINETTI, L. 1986. Une microfaune d'age Permien remaniée dans le Trias moyen de l'Apennin méridional (Formation de Monte Facito, Lucanie occidentale); déscription de *Crescentia vertebralis*, n. gen., n. sp. *Revue de Paléobiologie*, **5**, 207–215.

COLLINS, A. S. & ROBERTSON, A. H. F. 1998. Process of Late Cretaceous to Late Miocene episodic thrust–sheet translation in the Lycian Taurides. *Journal of the Geological Society, London*, **155**, 749–772.

DE GRACIANSKY, P. C. 1972. *Recherches géologiques dans le Taurius lycien*. Thèse, Université de Paris-Sud, Centre d'Orsay.

DEMİRTAŞLI, E. 1984. Stratigraphy and tectonics of the area between Silifke and Anamur, Central Taurus Mountains. *In:* TEKELİ, O. & GÖNCÜOĞLU, M. C. (eds) *Geology of the Taurus Belt*. Proceedings of the International Tauride Symposium, Mineral Research and Exploration Institute of Turkey (MTA) Publications, 101–118 [in Turkish with English abstract].

——, TURHAN, N., BİLGİN, A. Z. & SELİM, M. 1984. Geology of the Bolkar Mountains. *In*: TEKELİ, O. & GÖNCÜOĞLU, M. C. (eds) *Geology of the Taurus Belt*. Proceedings of the International Tauride Symposium, Mineral Research and Exploration Institute and Geological Society of Turkey, 125–141.

DERCOURT, J., RICOU, L. E. & VRIELYNCK, B. 1993. *Atlas of Tethys Palaeoenvironmental Maps*. Gauthier-Villars.

ERK, S. 1942. *Etude Géologique de la Région Entre Gemlik et Bursa (Turquie)*. Mineral Research and Exploration Institute of Turkey (MTA), Special Publications, Series B, **9**, 1–295.

FLÜGEL, E., DI STEFANO, P. & SENOWBARI-DARYAN, B. 1991. Microfacies and depositional structure of allochthonous carbonate base-of-slope deposits: the Late Permian Pietra di Salomone Megablock, Sosio Valley (Western Sicily). *Facies*, **25**, 147–186.

GARGOURI, S. & VACHARD, D. 1988. Sur *Hemigordiopsis* et d'autres Foraminifères porcélanés du Murghabien du Tebega (Permien supérieur de Tunisie). *Revue de Paléobiologie Volume Special*, **2**, Benthos' **86**, 57–68.

GENÇ, Ş. C. & YILMAZ, Y. 1995. Evolution of the Triassic continental margin, northwest Anatolia. *Tectonophysics*, **243**, 193–207.

GLINTZBOECKEL, C. & RABATE, J. 1964. *Microfaunes et Microfacies du Permo-Carbonifère du Sud-Tunisien*. E. J. Brill.

GRANT, R. E., NESTELL, M. K., BAUD, A. & JENNY, C. 1991. Permian stratigraphy of Hydra Island, Greece. *Palaios*, **6**, 479–497.

HAAS, J., GOCZAN, F., OROVECZ-SCHAFFER, A., BARABAS-STUHL, A., MAJOROS, GY. & BERCZI-MAKK, A. 1986. Permian–Triassic boundary in Hungary. *Memoire della Società Geologica Italiana*, **34**, 221–241.

KAYA, O. 1991. Stratigraphy of the pre-Jurassic sedimentary rocks of the western parts of Turkey: type area study and tectonic considerations. *Newsletter on Stratigraphy*, **23**, 123–140.

KOÇYİĞİT, A. 1987. Tectonostratigraphy of the region of Hasanoğlan (Ankara): evolution of the Karakaya orogenic belt. *Hacettepe University Earth Sciences*, **14**, 269–294 [in Turkish with English abstract].

KÖYLÜOĞLU, M. & ALTINER, D. 1989. Micropaléontologie (Foraminifères) et biostratigraphie du Permien supérieur de la région d'Hakkari (SE Turquie). *Revue de Paléobiologie*, **8**, 467–503.

KOZUR, H. & KAYA, O. 1994. First evidence of pelagic Late Permian conodots from NW Turkey. *Neues Jahrbuch für Geologie und Palaeontologie Monatschefte*, **6**, 339–347.

KRUSHENSKY, R. D., AKÇAY, Y. & KARAEGE, E. 1980. Geology of the Karalar–Yeşiller area, Northwest Anatolia. *Geological Survey Bulletin*, **1461**, 1–72.

LEVEN, E. JA. 1992. Problems of Tethyan Permian stratigraphy. *International Geology Review*, **34**, 976–985.

—— & OKAY, A. İ. 1996. Foraminifera from the exotic Permo-Carboniferous limestone blocks in the Karakaya Complex, northwest Turkey. *Rivista Italiana Paleontologia e Stratigrafia*, **102**, 139–174.

LYS, M. & MARCOUX, J. 1978. Les niveaux du Permien supérieur des Nappes d'Antalya (Taurides occidentales, Turquie). *Comptes Rendus des Séances de l'Académie des Sciences, Paris*, **286**, 1417–1420.

MARCOUX, J. 1979. General features of Antalya Nappes and their significance in the palaeogeography of southern margin of Tethys. *Geological Society of Turkey Bulletin*, **22**, 1–5 [in Turkish with English abstract].

METCALFE, I. 1988. Origin and assembly of south-east Asian continental terranes. *In*: AUDLEY-CHARLES, M. G. & HALLAM, A. (eds) *Gondwana and Tethys*. Geological Society, London, Special Publications, **37**, 101–118.

MONOD, O. 1977. *Recherches géologiques dans le Taurus occidental au Sud de Beyşehir (Turquie)*. Thèse, Université de Paris-Sud, Centre d'Orsay.

MONTENAT, C., LAPPARENT, A. F., LYS, M., TERMIER, H., TERMIER, G. & VACHARD, D. 1976. La transgression Permienne et son substratum dans le Jebel Akhdar (Montagnes d'Oman, Peninsule Arabique). *Annales Société Géologique du Nord*, **96**, 239–258.

NAKAZAWA, K., ISHII, K.-I., KATO, M., OKIMURA, Y., NAKAMURA, K. & HARALAMBOUS, D. 1975. Upper Permian fossils from Island of Salamis, Greece. *Memoirs of the Faculty of Science, Kyoto University, Series of Geology and Mineralogy*, **41**, 21–44.

NERI, C. & PASINI, M. 1985. A 'mixed fauna' at the Permian–Triassic boundary – Tesero section, Western Dolomites (Italy). *Bollettino della Società Palaeontologica Italiana*, **23**, 113–117.

NESTELL, M. K. & PRONINA, G. P. 1997. The distribution and age of the genus *Hemigordiopsis*. *In*: ROSS, C. A., ROSS, J. R. P. & BRENCKLE, P. L. (eds) *Late Palaeozoic Foraminifera; their biostratigraphy, evolution, and palaeoecology; and the Mid-Carboniferous boundary*. Cushman Foundation for Foraminiferal Research, Special Publications, **36**, 105–110.

NOÉ, S. U. 1987. Facies and palaeogeography of the marine Upper Permian and of the Permian–Triassic Boundary in the Southern Alps (Bellerophon Formation, Tesero Horizon). *Facies*, **16**, 89–142.

OKAY, A. İ. & MOSTLER, H. 1994. Carboniferous and Permian radiolarite blocks in the Karakaya Complex in northwest Turkey. *Turkish Journal of Earth Sciences*, **3**, 23–28.

——, SİYAKO, M. & BÜRKAN, K. A. 1991. Geology and tectonic evolution of the Biga peninsula, northwestern Turkey. *Bulletin of the Technical University of İstanbul*, **44**, 91–256.

——, SATIR, M., MALUSKI, H., SİYAKO, M., MONIE, P., METZGER, R. & AKYÜZ, S. 1996. Palaeo- and Neo-Tethyan events in northwest Turkey: geological and geochronological constraints. *In*: YIN, A. & HARRISON, M. (eds) *Tectonics of Asia*. Cambridge University Press, 420–441.

OKLA, S. M. 1994. Fossil algae from Saudi Arabia revisited. *Rivista Italiana Palaeontologia e Stratigrafia*, **99**, 441–460.

ÖZGÜL, N. 1976. Some geological aspects of the Taurus orogenic belt–Turkey. *Geological Society of Turkey Bulletin*, **19**, 65–78 [in Turkish with English abstract].

—— 1984. Stratigraphy and tectonic evolution of the Central Taurides. *In*: TEKELİ, O. & GÖNCÜOĞLU, M. C. (eds) *Geology of the Taurus Belt*. Proceedings of the International Tauride Symposium, Mineral Research and Exploration Institute of Turkey (MTA) Publications, 77–90 [in Turkish with English abstract].

—— 1997. Stratigraphy of the tectono-stratigraphic units in the region Bozkır–Hadim–Taşkent (northern central Taurides). *Mineral Research and Exploration Institute of Turkey (MTA) Bulletin*, **119**, 113–174 [in Turkish with English abstract].

——, BÖLÜKBAŞI, S., ALKAN, H., ÖZTAŞ, Y. & KORUCU, M. 1991. Tectono-stratigraphic units of the Lake District, western Taurides. *In*: TURGUT, S. (ed.) *Proceedings of the Ozan Sungurlu Symposium*. Ozan Sungurlu Foundation for Science, Education and Aid, 213–237.

——, METİN, S., ERDOĞAN, B., GÖĞER, E., BİNGÖL, I. & BAYDAR, O. 1973. Cambrian–Tertiary rocks of the Tufanbeyli region, eastern Taurus, Turkey. *Geological Society of Turkey Bulletin*, **16**, 82–100 [in Turkish with English abstract].

PANTIC, S. 1969. Litostratigrafske i mikropalaeontološke karakteristike srednjeg i gornjeg perma zapadne Srbije. *Vesnik Zavoda Geološka za i geofizicka intrazivanja, Beograd*, **27**, 201–215.

PANZENELLI-FRATONI, R., LIMONGI, P., CIARAPICA, G., CIRILLI, S., MARTINI, R., SALVINI-BONNARD, G. & ZANINETTI, L. 1987. Les foraminifères du Permien supérieur remaniés dans le 'Complexe Terrigène' de la Formation Triasique du Monte Facito, Apennin méridional. *Revue de Paléobiologie*, **6**, 293–319.

PAPANIKOLAOU, D. & BAUD, A. 1982. Complexes à blocs et séries à caractère flysch au passage Permien–Trias en Attique (Grèce Orientale). *The 9th Reunion Annuelle des Sciences de la Terre*, Paris 1982, 492.

PICKETT, E. A. & ROBERTSON, A. H. F. 1996. Formation of the Late Palaeozoic–Early Mesozoic Karakaya Complex and related ophiolites in NW Turkey by Palaeotethyan subduction–accretion. *Journal of the Geological Society, London*, **153**, 995–1009.

REICHEL, M. 1946. Sur quelques foraminifères nouveaux du Permien Meditérranean. *Eclogae geologica Helvetiae*, **38**, 524–560.

RICOU, L. E., ARGYRIADIS, I. & MARCOUX, J. 1975. L'axe calcaire du Taurus, un alignement du fenêtres arabo-africaines sous les nappes radiolaritiques, ophiolitiques et métamorphiques. *Bulletin de la Société Géologique de France*, **7**, 1024–1044.

ROBERTSON, A. H. F. & DIXON, J. E. 1984. Introduction: aspects of the geological evolution of the Eastern Mediterranean. *In*: DIXON, J. E. & ROBERTSON, A. H. F. (eds) *The Geological Evolution of the Eastern Mediterranean*. Geological Society, London, Special Publications, **17**, 1–74.

——, CLIFT, P. D., DEGNAN, P. & JONES, G. 1991. Palaeogeographic and palaeotectonic evolution of the Eastern Mediterranean Neotethys. *Palaeogeography, Palaeoclimatology, Palaeoecology*, **87**, 289–344.

ŞENGÖR, A. M. C. 1984. *The Cimmeride Orogenic System and the Tectonics of Eurasia*. Geological Society of America, Special Papers, **195**.

—— & YILMAZ, Y. 1981. Tethyan evolution of Turkey: a plate tectonic approach. *Tectonophysics*, **75**, 181–241.

——, —— & SUNGURLU, O. 1984. Tectonics of the Mediterranean Cimmerides: nature and evolution of the western termination of Palaeotethys. *In*: DIXON, J. E. & ROBERTSON, A. H. F. (eds) *The Geological Evolution of the Eastern Mediterranean*. Geological Society, London, Special Publications, **17**, 77–112.

——, ALTINER, D., CİN, A., USTAÖMER, T. & HSÜ, K. J. 1988. Origin and assembly of the Tethyside orogenic collage at the expense of Gondwana Land. *In*: AUDLEY-CHARLES, M. G. & HALLAM, A. (eds) *Gondwana and Tethys*. Geological Society, London, Special Publications, **37**, 119–181.

SHENG, J. & HE, Y. 1983. Permian *Shanita–Hemigordius (Hemigordiopsis)* (foraminifera fauna in western Yunnan, China. *Acta Palaeontologica Sinica*, **22**, 55–60.

SKINNER, J. W. 1969. Permian Foraminifera form Turkey. *The University of Kansas Palaeontological Contributions*, **36**, 1–14.

—— & WILDE, G. L. 1966. Permian fusulinids from Sicily. *The University of Kansas Palaeontological Contributions*, **8**, 1–16.

—— & —— 1967. Permian Foraminifera from Tunisia. *The University of Kansas Palaeontological Contributions*, **30**, 1–22.

STAMPFLI, G. J., MARCOUX, J. & BAUD, A. 1991. Tethyan margins in space and time. *Palaeogeography, Palaeoclimatology, Palaeoecology*, **87**, 373–409.

SUNGURLU, O. 1974. VI. Bölge kuzeyinin jeolojisi ve petrol imkanları. *Proceedings of the 2nd Petroleum Congress and Exhibition of Turkey*. Turkish Associaton of Petroleum Geologists Publications, 85–107 [in Turkish with English abstract].

TEKELİ, O. 1981. Subduction complex of pre-Jurassic age, northern Anatolia, Turkey. *Geology*, **9**, 68–72.

USTAÖMER, T. & ROBERTSON, A. H. F. 1997. Tectonic–sedimentary evolution of the North Tethyan margin in the Central Pontides of Northern Turkey. *In*: ROBINSON, A. G. (ed.) *Regional and Petroleum Geology of the Black Sea and Surrounding Region*. AAPG Memoirs, **68**, 255–290.

VACHARD, D. & MICONNET, P. 1989. Une association à fusulinoïdes du Murghabien supérieur au Monte Facito (Apennin méridional, Italie). *Revue de Micropaléontologie*, **32**, 297–318.

—— & RAZGALLAH, S. 1993. Discussion sur l'age Murgabien ou Midien des séries Permiennes du Jebel Tebega (sud de la Tunisie). *Rivista Italiana Palaeontologia e Stratigrafia*, **99**, 327–356.

——, MARTINI, R. & ZANINETTI, L. 1995. Le Murgabien à fusulinoïdes des iles d'Hydra, Crète et Mytilène (Permien supérieur de Grèce). *Geobios*, **28**, 395–406.

——, ——, —— & ZAMBETAKIS-LEKKAS, A. 1993. Revision micropaléontologique (Foraminifères, Algues) du Permien inférieur (Sakmarien) et supérieur (Dorashamien) du Mont Beletsi (Attique, Grèce). *Bollettino della Società Palaeontologica Italiana*, **32**, 89–112.

WHITTAKER, J. E., ZANINETTI, L. & ALTINER, D. 1979. Further remarks on the micropalaeontology of the late Permian of eastern Burma. *Notes du Laboratoire de Paléontologie de l'Université de Genève*, **5**, 11–18.

YILMAZ, Y. 1981. Tectonic evolution of the southern margin of the Sakarya Continent. *İstanbul University Earth Sciences*, **1**, 33–52 [in Turkish with English abstract].

—— 1990. Allochthonous terranes in the Tethyan Middle East: Anatolia and the surrounding regions. *Philosophical Transactions of Royal Society, London, Series A*, **331**, 611–624.

ZANINETTI, L., ALTINER, D. & ÇATAL, E. 1981. Foraminifères et biostratigraphie dans le Permien supérieur du Taurus oriental, Turquie. *Notes du Laboratoire de Paléontologie de l'Université de Genève*, **7**, 1–37.

——, WHITTAKER, J. E. & ALTINER, D. 1979. The occurrence of *Shanita amosi* Brönnimann, Whittaker, and Zaninetti (Foraminiferida) in the Late Permian of the Tethyan region. *Notes du Laboratoire de Paléontologie de l'Université de Genève*, **5**, 1–9.

——, ALTINER, D., ÇATAL, E. & DECROUEZ, D. 1982. *Shanita brönnimanni*, n. sp., (Hemigordiopsidae, Foraminiferida), dans le Permien supérieur du Taurus oriental, Turquie: un exemple d'adaptation structurale à l'évolution regressive de la cavité loculaire chez les grands foraminifères porcélanés du Paléozoïque. *Revue de Paléobiologie*, **1**, 29–37.

Mesozoic–Tertiary tectonic–sedimentary evolution of a south Tethyan oceanic basin and its margins in southern Turkey

ALASTAIR H. F. ROBERTSON

Department of Geology and Geophysics, University of Edinburgh, Edinburgh EH9 3JW, UK (e-mail: Alastair.Robertson@glg.ed.ac.uk)

Abstract: This paper focuses on the Mesozoic–Tertiary tectonic evolution of southern Turkey and offshore areas of the easternmost Mediterranean. The area is discussed and interpreted utilizing three segments from west to east. In the far west, the Lycian Nappes represent emplaced remnants of mainly Mesozoic rift, passive margin and oceanic units that formed within a northerly strand of the Mesozoic (i.e. Neotethyan) ocean. Further east, the Hoyran–Beyşehir–Hadim Nappes, likewise encompass sedimentary and igneous units that formed within a northerly Neotethyan oceanic basin, although lithologies, structure and timing of emplacement differ from the Lycian Nappes. Further east (Adana region), ophiolites and ophiolitic mélange also formed in a northerly oceanic basin and were thrust southwards over the regionally extensive Tauride carbonate platform initially in latest Cretaceous time (e.g. Pozantı–Karsantı Ophiolite).

By contrast, further south the regionally important Antalya Complex records northerly areas of a separate, contrasting southerly Neotethyan oceanic basin. This comprised a mosaic of carbonate platforms and interconnecting seaways, similar to the Caribbean region today. In particular, an ocean strand separated Tauride carbonate platforms to the west (Bey Dağları) and east (e.g. Akseki Platform) within the Isparta Angle area. In the centre of southern coastal Turkey, the metamorphic Alanya Massif is interpreted as a Triassic rift basin bordered by two small platform units that was located along the northern margin of the southerly Neotethys which collapsed in latest Cretaceous and was finally emplaced in Early Tertiary time. Remnants of the southerly Neotethyan oceanic basin remain today in the non-emplaced continental margin of the Levant and North Africa, and neighbouring seafloor areas (e.g. Levant and Herodotus Basins).

In southern Turkey, emplaced Neotethyan units are unconformably overlain by a complex of mainly Miocene basins. These largely reflect the effects of southward directed crustal loading as convergence of Africa and Eurasia continued, although the basins were also influenced by an inferred more southerly subduction zone (near Cyprus). Further east, in southeastern Turkey, ophiolites, ophiolitic mélange and continental margin units were emplaced southwards onto the Arabian Margin, a promontory of North Africa in latest Cretaceous time. The south Neotethyan basin's north margin experienced northward subduction, accretion, arc volcanism and ophiolite emplacement in Late Cretaceous time. The intervening southerly Neotethyan oceanic basin remained partly open in the Early Tertiary, finally closing by diachronous collision in Eocene–Oligocene time, followed by further convergence and overthrusting in the Miocene. The Eocene later stages of convergence were marked by renewed arc volcanism and extensive subduction accretion (e.g. Maden Complex). In the west, subduction remained active in Late Oligocene–Early Miocene time giving rise to sedimentary mélanges (olistostromes) of the Misis–Andırın Mountains (Adana region) as an accretionary wedge. By the Miocene the subduction zone accommodating Africa–Eurasia convergence had been relocated to its present position south of Cyprus. Areas behind this subduction experienced crustal extension (e.g. Antalya and Adana–Cilicia Basins) from the Late Miocene onwards. After onset of westward 'tectonic escape' of the Turkish Plate in the Early Pliocene, southeastern Turkey was transected by the South Anatolian Transform Fault. Strike-slip was dissipated though the Kyrenia–Misis Lineament into Cyprus. Today, southeastern Turkey records a post-collisional setting, whereas areas to the west experience incipient collision of the African and Turkish Plates.

The hypothesis that the eastern Mediterranean (Fig. 1) originated as a small Neotethyan oceanic basin rests on three main lines of evidence. First, the present-day plate boundary of the African and Eurasian Plates runs through the eastern Mediterranean. This was long inferred (McKenzie 1978; Dewey & Şengör 1979) and more recently confirmed, mainly by seismic evidence (Kempler & Ben-Avraham 1987; Anastasakis & Kelling 1991). The location of the plate boundary south of Cyprus is now clarified by results obtained by drilling during Leg 160 (Emeis *et al.* 1996; Robertson *et al.* 1998). Secondly, seismic refraction studies, combined with

From: BOZKURT, E., WINCHESTER, J. A. & PIPER, J. D. A. (eds) *Tectonics and Magmatism in Turkey and the Surrounding Area.* Geological Society, London, Special Publications, **173**, 97–138. 1-86239-064-9/00/$15.00
© The Geological Society of London 2000.

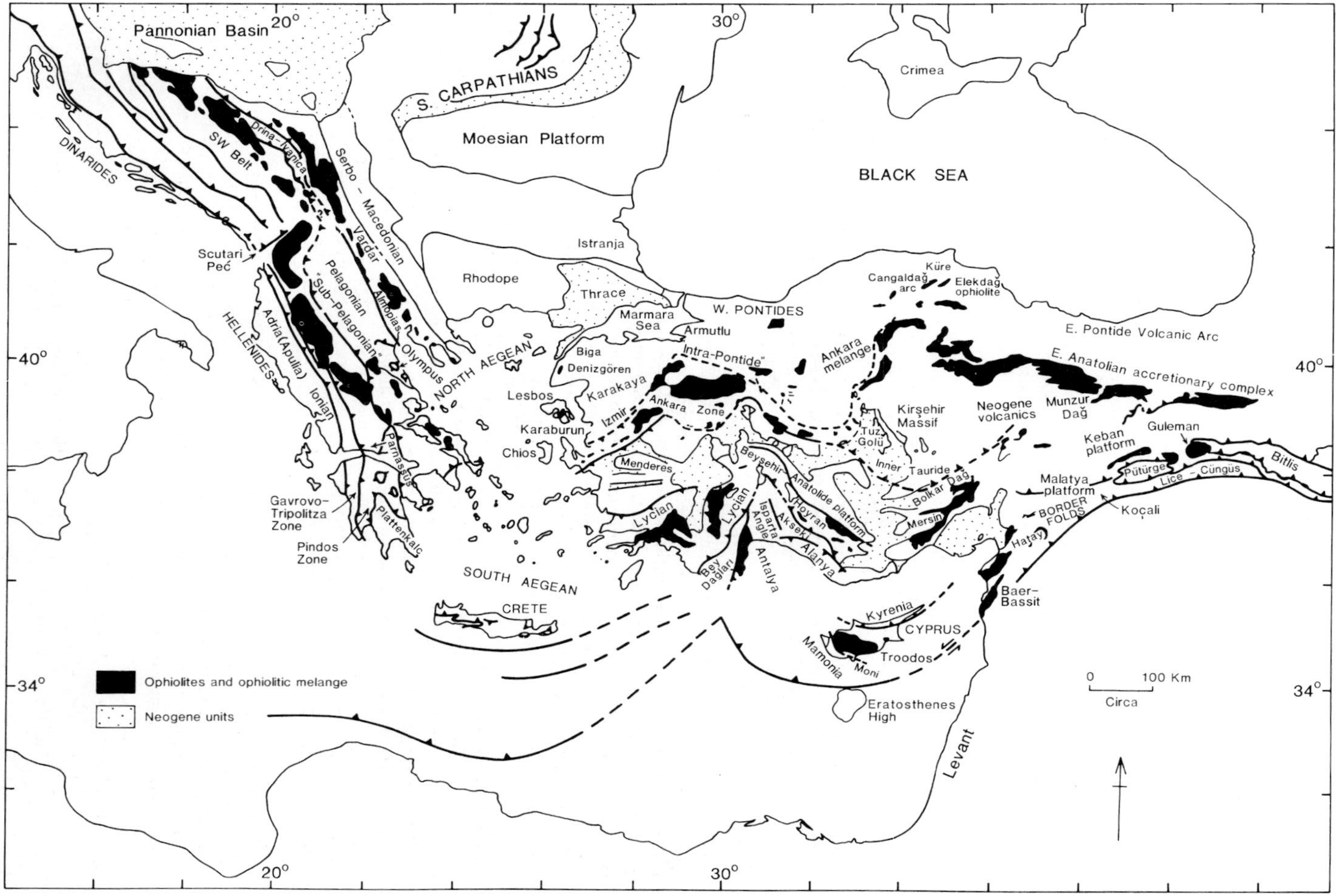

Fig. 1. Outline tectonic map of the easternmost Mediterranean showing the main tectonic features discussed in this paper. The discussion focuses in the area east of latitude 27°E and south of 40°N. More details of submarine features are shown in Figs 5 and 6.

regional gravity and magnetic anomaly patterns (Woodside 1977), strongly suggest that the eastern Mediterranean seafloor between North Africa and the Levant onshore is composed of Mesozoic oceanic crust (Makris *et al.* 1983; Ben-Avraham 1986; Ben-Avraham & Tibor 1994). The present-day Levant coast corresponds to a Mesozoic passive continental margin that passes oceanward into oceanic crust within the easternmost Mediterranean area (Garfunkel & Derin 1984). A continuum of geological processes exists within the easternmost Mediterranean area, beginning with rifting and continental break-up in the Permian–Triassic, followed by passive margin subsidence and then incipient continental collision of the African and Eurasian Plates along the Cyprus active margin. Thirdly, detailed field-based sedimentary and structural studies of northern Cyprus (Robertson & Woodcock 1979) and southwestern Turkey (Hayward & Robertson 1982; Woodcock & Robertson 1982; Robertson 1993) demonstrate that many of the Mesozoic ophiolites and related allochthonous units in these areas record emplaced remnants of a southerly Neotethyan ocean basin. Neotethys in the Mediterranean region was palaeogeographically varied and can be compared with the Caribbean or the southwestern Pacific regions in complexity. Indeed, some of the ophiolites and related units in the area considered were derived from a separate Neotethyan oceanic basin located to the north of an east–west belt of Mesozoic carbonate platform rocks that from the substratum of all of the allochthonous units in the region.

The aim here is to discuss the main geological information available for southern Turkey, extending from the easternmost Mediterranean (east of 28°E longitude) through southeastern Turkey (Fig. 1). Much of the information and interpretation presented here was previously published as chapter 54 (Synthesis section) of the Scientific Results of Ocean Drilling Program Leg 160, which included drilling on the Eratosthenes Seamount south of Cyprus (Robertson 1998). This contribution comprises a summary and synthesis of a wide area, including the easternmost Mediterranean Sea, Cyprus and the Levant region. In addition, a companion paper focusing on Cyprus, onshore and offshore, was recently published by the Geological Survey Department, Cyprus (Robertson 2000). Here, the focus is on the onshore geology and tectonics of southern Turkey that includes the deformed remnants of a southerly Mesozoic oceanic basin.

In general, the easternmost Mediterranean area includes important remnants of a southerly Neotethyan (mainly Mesozoic) oceanic basin which formed part of a larger Tethyan ocean (Le Pichon 1982). From Late Cretaceous to Holocene time, the eastern Mediterranean has been in a state of diachronous collision, whereas, areas to the east, in southeastern Turkey, are now in a post-collisional phase (Şengör & Yılmaz 1981; Dewey *et al.* 1986; Pearce *et al.* 1990), whereas areas to the west are still in an early collisional phase (south of Cyprus), or locally still in a pre-collisional phase (e.g. off southwest Cyprus) (Robertson & Grasso 1995). Thus, the present easternmost Mediterranean basin is a direct successor to a pre-existing Mesozoic Neotethyan oceanic basin.

Field evidence from southern Turkey shows that two separate Neotethyan oceanic basins developed in this region (southerly and northerly). The southerly oceanic basin system includes the present easternmost Mediterranean region. The northerly margins of this basin experienced deformation and ophiolite emplacement in the Late Cretaceous, whereas the southerly, North Africa–Levant Margin remained passive. By contrast, southern Turkey also includes far-travelled remnants of another, separate Neotethyan oceanic basin, represented by the Lycian Nappes in the west and the Beyşehir–Hoyran–Hadim Nappes further east. There is also evidence for southerly and northerly derived Neotethyan oceanic basins in southeastern Turkey, which will be briefly summarized.

This paper focuses on areas to the south of the drainage divide of the Taurus Mountains in southern Turkey. This region has seen much internationally based work over many years and is consequently the best documented and understood region of Turkey. A more complete database for the easternmost Mediterranean region as a whole, including Cyprus and the Levant, is published elsewhere (Robertson 1998).

In the Turkish literature, the Taurus Mountains are traditionally divided into western, central and eastern segments, based mainly on geography. The western segment is located west of the Isparta Angle, the central segment lies north of Cyprus, whereas the eastern segment is located east of the longitude of the Levant Margin. In this paper, for purposes of description and interpretation, the area is divided into three slightly different segments (Fig. 2). From west to east, these are: the *Western segment* from the Bozburun Peninsula in the west to near Anamur; the *Central segment* from

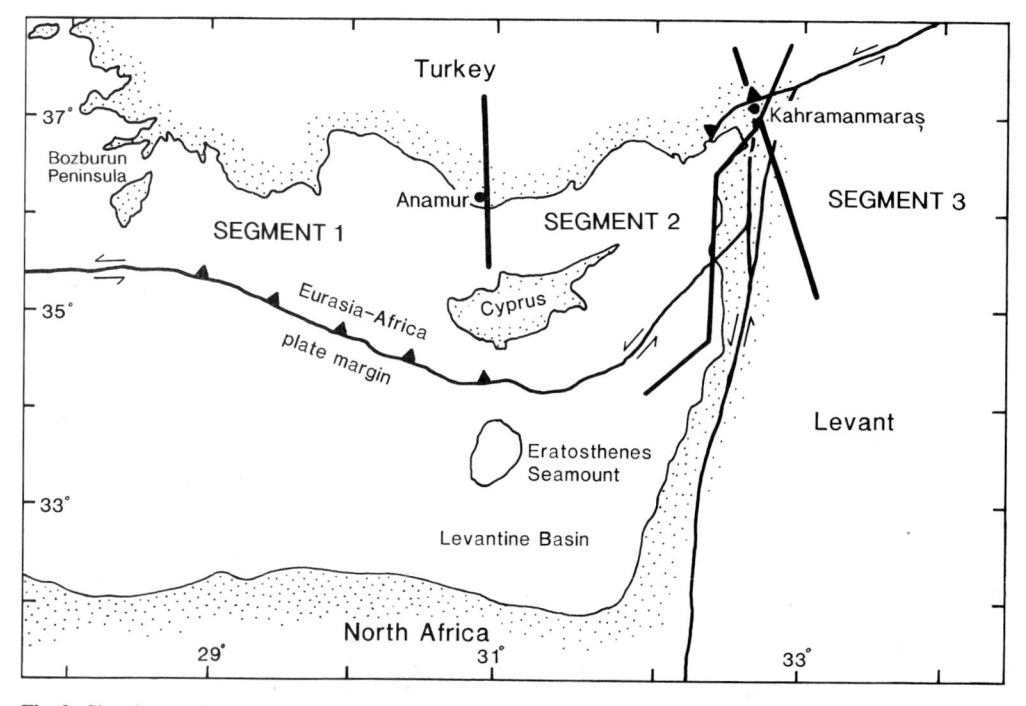

Fig. 2. Sketch map showing the three main segments discussed in this paper. Segment 1, the Western segment; segment 2, the Central segment; segment 3, the Eastern segment.

Anamur to near Kahraman Maraş; and the *Eastern segment* extending beyond this to the border with Iran. The main advantage of this informal classification is that it allows geologically similar units (e.g. Antalya region) to be treated together rather than splitting them based mainly on geography.

Before beginning the discussion of individual areas it should be noted that several substantially different interpretations exist for the southerly Neotethyan units in southern Turkey. The first is that all these units are substantially allochthonous and were thrust from a single Mesozoic ocean basin located far to the north in the Black Sea area ('internal hypothesis') (Ricou et al. 1984; Marcoux et al. 1989). If valid, the Neotethyan units of southern Turkey would have no connection with the tectonic evolution of the present-day eastern Mediterranean Sea area. A full discussion of why the 'internal hypothesis' is unlikely to be valid has been published elsewhere, especially focusing on the evidence of unbroken sedimentary successions overlying Mesozoic carbonate platforms in the west (Bey Dağları and Isparta areas) and the existence of structural evidence supporting an origin of continental margin and ophiolitic units within a southerly Neotethyan oceanic basin (Robertson & Woodcock 1980; Şengör & Yılmaz 1981; Robertson 1993; Robertson et al. 1996). The second alternative view accepts that the southern Turkish Neotethyan allochthonous units are southerly derived, but suggests that spreading in this basin did not take place until the Cretaceous, following successive rift phases in the Late Permian and Triassic (Dercourt et al. 1986; Dilek & Rowland 1993). A number of lines of evidence oppose this interpretation, including the presence of Triassic mid-ocean ridge basalt (MORB)-type lavas in the Antalya area, the existence of unbroken Triassic–Upper Cretaceous deep-water passive margin successions around the margins of the Tauride microcontinental units, without any identifiable Cretaceous break-up unconformity, and subsidence histories of the north Gondwana margin (e.g. the Levant), which do not indicate any major rift event after Permo-Triassic time (Garfunkel 1989). Indeed, recently, Stampfli et al. (2000) have proposed an even earlier, Late Permian time of initial spreading of a southerly Neotethys, although geological evidence for this is yet to be forthcoming. For the purposes of the following discussion it will be assumed that many (but not all) of the allochthonous units of southern Turkey represent part of a Neotethyan

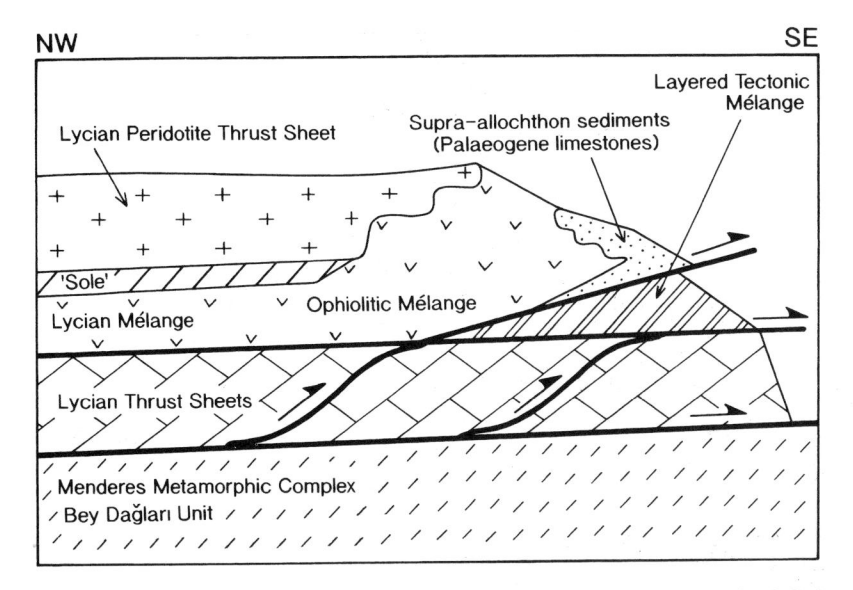

Fig. 3. Rock relations diagram for the Lycian Nappes; see text for explanation [after Collins & Robertson (1998)].

oceanic basin that opened in the Triassic followed by complete closure in Early Tertiary time.

The Western segment

The Western segment comprises the Lycian Nappes (northerly derived), the Antalya Complex (southerly derived), the metamorphic Alanya Massif and overlying Neogene sedimentary basins, both onshore and offshore.

The Lycian Nappes

In general, the Lycian Nappes are a composite unit dominated by ophiolites, accretionary prism-type volcanic–sedimentary units and sedimentary thrust sheets of Carboniferous–Permian to Late Cretaceous age (De Graciansky 1966; Poisson 1977, 1984; Okay 1990; Robertson et al. 1996; Collins & Robertson 1997, 1998, 1999).

As shown in Fig. 4, four major tectonostratigraphic units are present within the outcrop area of the Lycian Nappes as a whole from the structural base upwards: the Yavuz Thrust Sheet (lowest), the Karadağ Thrust Sheet, the Teke Dere Thrust Sheet and the Köyceğiz Thrust Sheet (highest), followed by mélanges (Lycian Mélange) and then an ophiolite (Lycian Peridotite). The coastal exposures are dominated by the higher parts of the

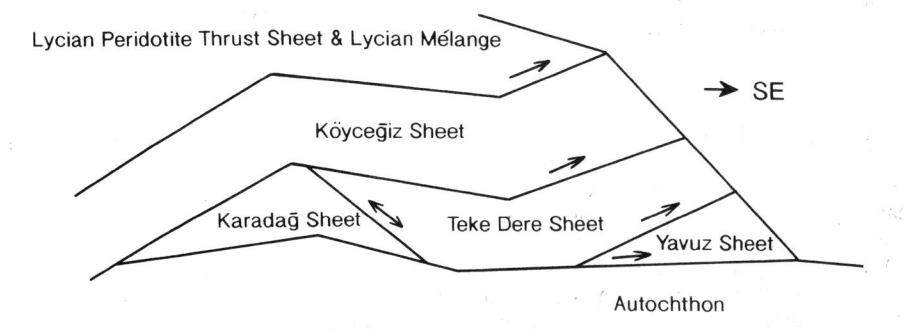

Fig. 4. Tectonostratigraphy of the Lycian Nappes, including the major thrust sheets making up the Lycian Allochthon. The Köyceğiz Thrust Sheet, the Lycian Mélange and the Lycian Peridotite are the main units exposed in southern coastal Turkey. After Collins & Robertson (1998, 2000).

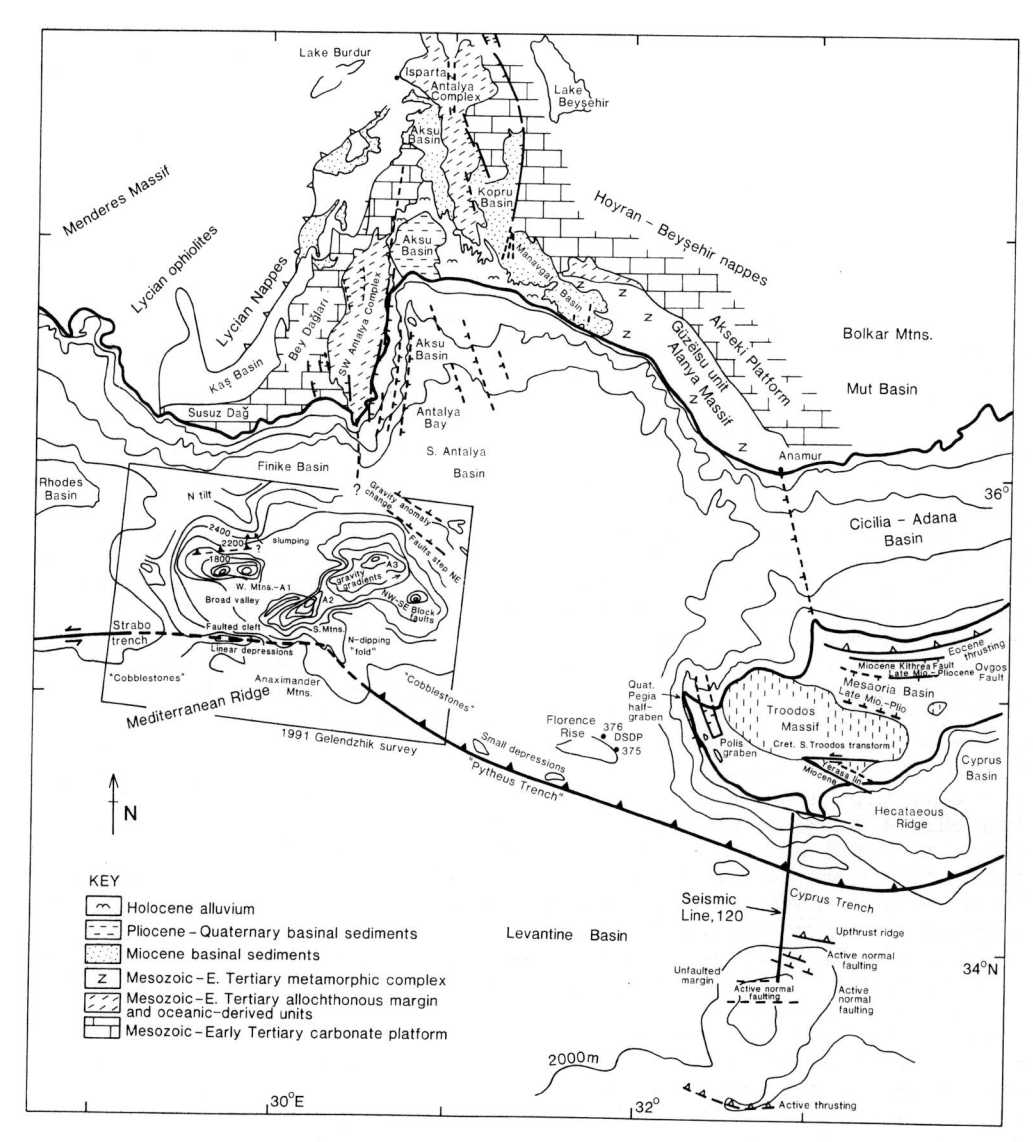

Fig. 5. Outline tectonic map of the easternmost Mediterranean area extending from southwestern Cyprus to southern Turkey. Data sources are specified in the text.

tectonostratigraphy, mainly the Köyceğiz Thrust Sheet, the Lycian Mélange and the Lycian Peridotite. The Köyceğiz Thrust Sheet begins with pillow basalts of chemically transitional type, interpreted as being formed along a Triassic rifted margin (northerly Neotethys), overlain by a Lower Jurassic shallow-water carbonate succession that then subsided to form a continental slope that survived until the Late Cretaceous. The Lycian Mélange, comprising both sediment-dominated and ophiolite-dominated units (Layered Tectonic Mélange and Ophiolitic Mélange, respectively), is interpreted as an accretionary prism related to subduction in latest Cretaceous (Campanian–Maastrichtian) time. The Lycian Peridotite is interpreted, based on geochemical evidence, as an Upper Cretaceous ophiolite formed by spreading above a subduction zone in the northerly branch of Neotethys (Collins & Robertson 1998). The crustal sequence of the Lycian ophiolite (sheeted dykes, lavas) is not preserved, presumably owing to erosion. However, the missing ophiolitic lithologies are present as inclusions within the underlying Lycian Mélange.

The Lycian Nappes are thrust over a Miocene, mainly terrigenous, turbiditic succession (Kaş Basin; Fig. 5; Hayward 1984). These Miocene sediments in turn overlie a Mesozoic carbonate platform, the Susuz Dağ, and its northward extension into the regional Bey Dağları–Menderes Unit, interpreted as a microcontinental unit (Collins & Robertson 1998). In addition, inland, the Lycian Nappes are transgressively overlain by Neogene sedimentary basins, contemporaneous with the later stages of emplacement of the Lycian Nappes (e.g. Tavas Basin), and postdating emplacement (e.g. Çameli Basin).

The Lycian Nappes are restored as a north facing Mesozoic rift and passive margin (De Graciansky 1966; Collins & Robertson 1998). Mainly deep-water sediments of Triassic–Late Cretaceous age, preserved as blocks within mélange units above the Lycian thrust sheets, are interpreted as deep-water sediments overlying Mesozoic Neotethyan oceanic crust. Northward subduction of the Neotethyan ocean basin began with accretion of oceanic-derived mélange and disrupted thrust sheets. Debris was shed into a continentward migrating flexural foredeep, initially located along the distal edge of the continental margin in the Campanian–Maastrichtian (represented by the highest stratigraphic levels of the Köyceğiz Thrust Sheet); the foredeep then propagated southwards over more proximal continental crust during Palaeocene time and, in turn, this was detached as the Teke Dere Thrust Sheet and the Karadağ Thrust Sheet (best exposed near the Aegean coast). The Lycian Allochthon was finally emplaced over the most proximal (southeasterly) foredeep (Kaş Basin) in Late Miocene time (Hayward 1984; Collins & Robertson 1997, 1998, 1999).

The Hoyran–Beyşehir–Hadim Nappes

To the east, the Lycian Nappes are replaced by a separate series of allochthonous units that extend from north of Lake Eğridir, through the Bolkar Mountains to near the Mediterranean coast between Anamur and Silifke. The lithology and timing of emplacement of these latter nappes differ markedly from that of the Lycian Nappes, discussed above. Based on regional mapping, much of the fundamental lithostratigraphy was established by Özgül (1984) and Monod (1977). These nappes regionally overlie a basement of Tauride platform units, locally the Mesozoic Akseki Platform in the west. Further east, the relatively autochthonous 'basement' includes successions of Cambrian–Permian platformal lithologies, unconformably overlain by Triassic and younger Mesozoic shelf units (Demirtaşlı 1984). Structurally above, the Beyşehir–Hoyran–Hadım Nappes comprise a number of contrasting tectonostratigraphic units of Late Palaeozoic–Mesozoic age. The structurally lowest thrust sheet in the west extends down as far as a succession of Permian platform carbonates. Structurally higher, Triassic–Upper Cretaceous units include volcanic rocks and volcaniclastic sediments, shallow-water to deeper water carbonates, radiolarian sediments, ophiolites, Upper Cretaceous ophiolitic mélange and Lower Tertiary flysch-type sediments related to emplacement (Monod 1977; Okay & Özgül 1984). Further east, smaller exposures near the coast are known as the Hadim Nappe (i.e. Aladağ Unit, Demirtaşlı et al. 1984). In contrast to the Lycian Nappes, there have been no extensive modern studies of these units, although such work is now beginning.

The Hoyran–Beyşehir–Hadim Nappes are restored to a position within a northerly, 'Inner Taurus ocean' (Görür et al. 1984), separate from the southerly Neotethyan oceanic basin (Şengör & Yılmaz 1981; Robertson & Dixon 1984; Şengör et al. 1984). The Hoyran–Beyşehir–Hadim Nappes are thus counterparts of the Lycian Nappes, which were also rooted in a northerly Neotethyan oceanic basin. However, the Hoyran–Beyşehir–Hadim Nappes differ in many ways lithologically and were finally emplaced in the Late Eocene, whereas the Lycian Nappes were only finally emplaced in Late Miocene time [see Özgül (1984) for a detailed discussion].

The Antalya Complex

The Antalya Complex is a regionally important allochthonous unit of mainly Mesozoic rocks exposed within the Isparta Angle area (Figs 6 and 7). This unit has close affinities with the Mamonia Complex of western Cyprus (Robertson & Woodcock 1980) and represents a critical part of the evidence of a southerly Neotethyan oceanic basin in the easternmost Mediterranean region. The Antalya Complex differs markedly in structure, lithology and timing of emplacement from the Lycian Nappes, described above. In particular, especially in the southwest of the area (southwest of Antalya city), high-angle structures dominate (Fig. 7b), in contrast to the regionally low-angle thrust sheets of the Lycian Nappes. This high-angle structure is interpreted

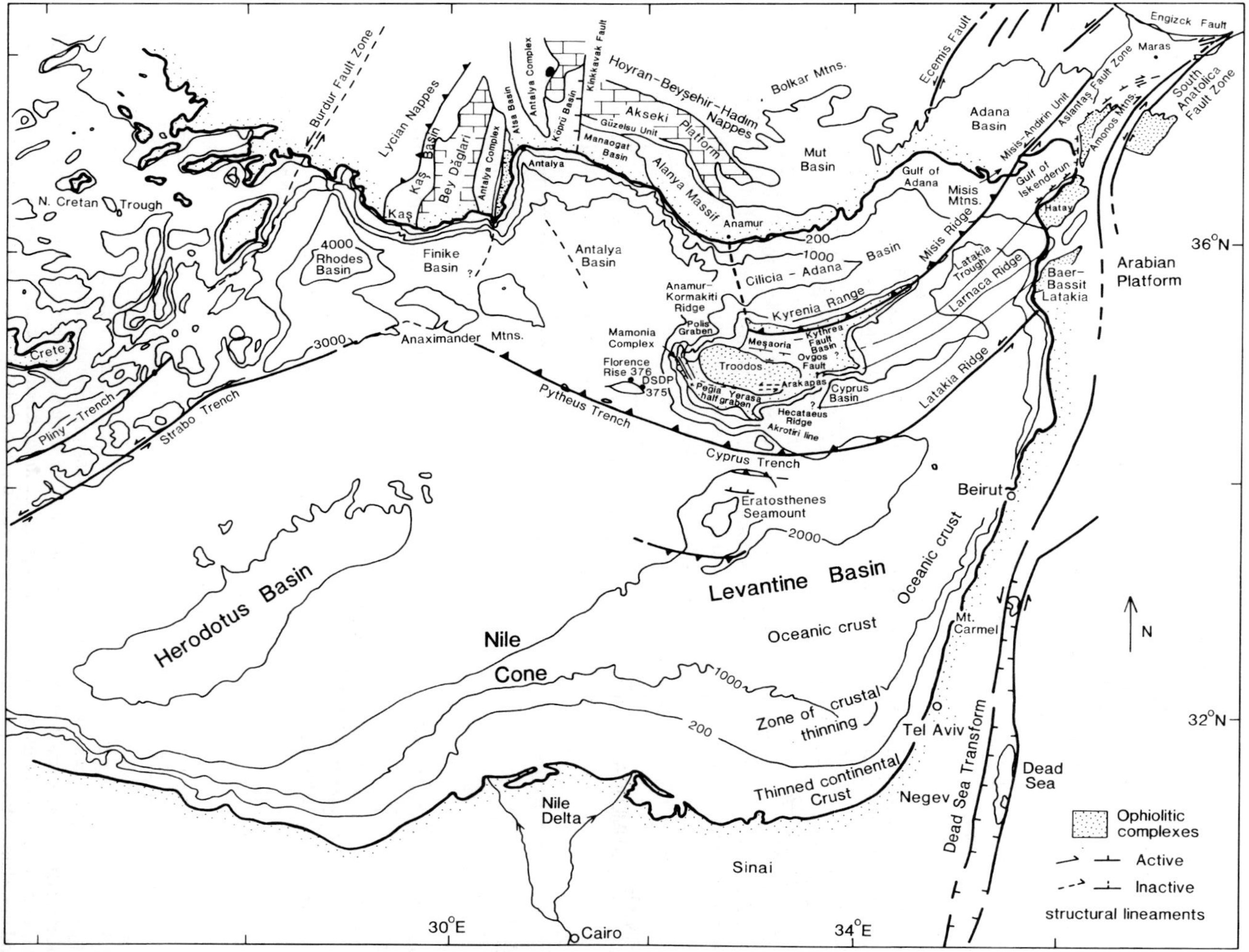

Fig. 6. Outline bathymetric and geologic map of part of the easternmost Mediterranean area. Detail of the area is shown in Fig. 5. More detailed bathymetric data for the Anaximander Mountain area are included [from Ivanov et al. (1992)].

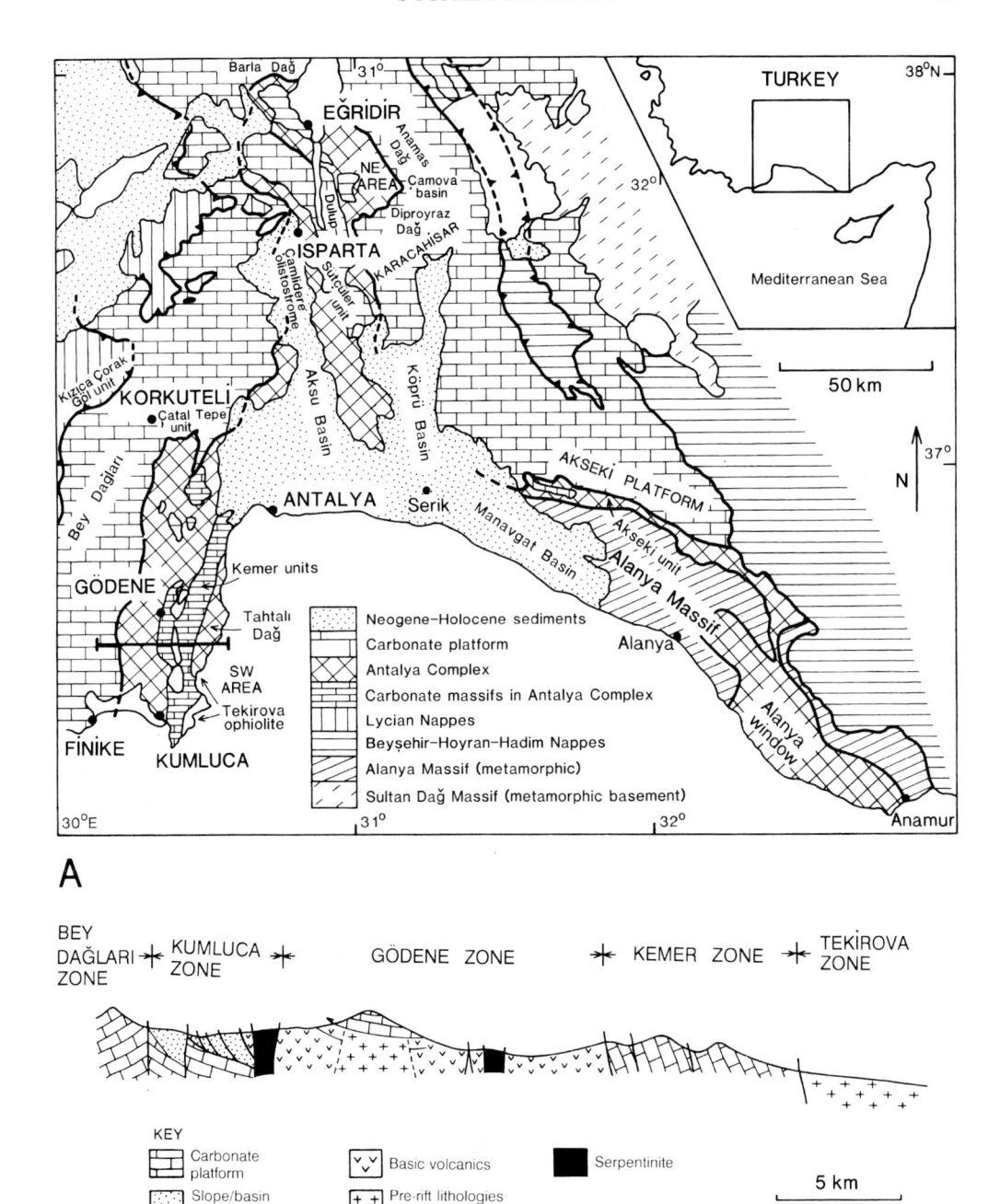

Fig. 7. Antalya Complex of southwestern Turkey. (**a**) Outline tectonic map of the Antalya area, showing particularly the large extent of relatively autochthonous Mesozoic carbonate platforms (e.g. Bey Dağları) and the Mesozoic Antalya Complex. Other units highlighted in the text are the Lycian Nappes, the Akseki Platform and the Alanya Massif [modified after Robertson & Woodcock (1984)]. Note: the internal structure of the Antalya Complex is not differentiated. (**b**) Cross-section of the Antalya Complex. This is interpreted as a transition from a carbonate platform in the west, across a continent–ocean transition zone, marked by ophiolitic rocks and carbonate build-ups on rifted blocks of Palaeozoic crust. The Antalya Complex (the southwestern area) was tectonically emplaced in the latest Cretaceous–Early Tertiary by a combination of westward thrusting and strike slip.

as evidence of a dominantly strike-slip mode of emplacement (Woodcock & Robertson 1982) and thus the earlier term of the Antalya Nappes (Delaune-Mayere et al. 1976) is not favoured. The Antalya Complex is exposed in segments around the periphery of the Isparta Angle. Exposures in the southwest (southwest of Antalya) constitute the type area, as all the main units are well exposed. High-angle structures dominate there and also along the southeastern outcrop of the Antalya Complex (near Serik). By contrast, exposures further north towards the apex of the Isparta Angle (e.g. Isparta–Eğridir area, not discussed here) are dominated by lower angle structures (Waldron 1984; Robertson 1993; Fig. 7).

In the west, the (relatively) autochthonous basement of the Antalya Complex, represented by the carbonate platform, is dominated by Jurassic–Lower Cretaceous shallow-water carbonates, overlain by Upper Cretaceous–Palaeogene pelagic carbonates (Poisson 1977). A transition to Upper Cretaceous deeper water carbonates is locally associated with sedimentary breccias (Robertson 1993), indicative of a possible control by coeval extensional faulting. Widespread shallow-water carbonate deposition ended in the Cenomanian, followed by onset of pelagic carbonate deposition in the Turonian, continuing until the Early Palaeocene when final emplacement of the allochthonous units (the Antalya Complex) took place (Poisson 1977, 1984; Robertson 1993). By contrast, synrift and pre-rift units of pre-Late Triassic age are exposed along the eastern limb of the Isparta Angle (Gutnic et al. 1979; Waldron 1984; e.g. Sütçüler Unit; Fig. 7).

The following main units are present in the classic southwestern segment of the Antalya Complex, from west to east (Robertson & Woodcock 1982; Fig. 7a and b). Firstly, the relatively autochthonous Mesozoic Bey Dağları carbonate Platform is overlain by Miocene foreland basin clastic sediments. Secondly, there is the Kumluca Zone, composed of thrust-imbricated deep-water passive margin sediments of Late Triassic–Late Cretaceous age. Thirdly, there is the Gödene Zone, composed of sheared ophiolitic rocks (commonly serpentinite), deep-sea sediments and large masses of shallow-water limestone, together ranging in age from Late Triassic to Late Cretaceous. Fourthly, further east, is the Kemer Zone, dominated by steeply dipping slices of Palaeozoic (Ordovician–Permian) sedimentary rocks overlain by Mesozoic shallow-water carbonates (Tahtalı Dağ; Fig. 7). Finally, there is the coastal Tekirova Zone (Fig. 7b), dominated by the deeper levels of an Upper Cretaceous ophiolite (Tekirova Ophiolite), which, however, lacks extrusives (Juteau 1970; Yılmaz 1984).

The Kumluca Zone is interpreted as the deformed deep-water, southeasterly passive margin of the Bey Dağları carbonate Platform to the west [the Menderes–Bey Dağları Unit of Collins & Robertson (1998)]. Lithologies include Upper Triassic turbiditic sandstones and pelagic *Halobia* limestones, Jurassic–Lower Cretaceous non-calcareous radiolarian sediments and silicified calciturbidites. The Gödene Zone is interpreted as relatively proximal oceanic crust, with both shallow- and deep-water sedimentary units. It includes Upper Triassic subalkaline pillow basalts up to 750 m thick (Juteau 1970). These lavas are intermediate in composition, between within-plate-type basalt and MORB-type basalt (Robertson & Waldron 1990). The extrusives are locally interbedded with, and overlain by, coarse siliciclastic turbidites and rare quartzose conglomerates of Late Triassic age, proving a near-continental margin origin (Robertson & Woodcock 1984). The associated shallow-water limestones are interpreted as carbonate build-ups constructed on rifted continental blocks located within a zone of proximal oceanic crust adjacent to a rifted margin. The Kemer Zone is viewed as one, or probably several, slivers of continental crust that were rifted from the larger Bey Dağları continental fragment in the Triassic. The pre-rift Ordovician–Carboniferous successions are interpreted as a part of the North African passive margin (Monod 1977). The Tekirova Zone is interpreted as oceanic crust and mantle of Late Cretaceous age formed within an Isparta Angle oceanic basin. It represents the most westerly preserved body of a zone of Upper Cretaceous ophiolites, including the Troodos Ophiolite, and those in southeastern Turkey and Syria (e.g. Hatay and Baer-Bassit). These ophiolites formed in a southerly Neotethyan oceanic basin, in contrast to the Lycian Nappes and the Hoyran–Beyşehir–Hadim Nappes (see below) that were derived from a Neotethyan oceanic basin to the north of the Tauride carbonate Platform (Şengör & Yılmaz 1981; Robertson & Dixon 1984). The exact tectonic affinities of the Tekirova Ophiolite are unclear, mainly because extrusive rocks (used for geochemical fingerprinting of tectonic setting) are not preserved. However, ductile shear fabrics within the plutonic rocks suggest formation in the vicinity of an oceanic transform fault (Reuber 1984).

The southwestern area of the Antalya Complex was initially deformed in the Late Cretaceous by a combination of wrench faulting and thrusting within the southerly Neotethyan ocean (Woodcock & Robertson 1982). The Antalya Complex was then emplaced generally westward over the relatively autochthonous Bey Dağları carbonate platform (part of the inferred Menderes–Bey Dağları microcontinent) in Late Palaeocene–Early Eocene time (Poisson 1977; Robertson & Woodcock 1982). Westward thrusting associated with foreland basin development carried the allochthonous units still further westwards in the Late Miocene, although evidence of this is seen only in the far southwest of the area (near Kumluca).

The southwest area of the Antalya Complex is restored as the southeasterly rifted passive margin of a large microcontinental unit within Neotethys (the Menderes–Bey Dağları microcontinent) that was finally emplaced westward onto a small Miocene foreland basin succession overlying shallow-water carbonates (Susuz Dağ) in the Late Miocene (Hayward & Robertson 1982). However, in the north (near Isparta; Fig. 7), the Antalya Complex is transgressed by Lower Miocene limestones and terrigenous turbidites, and is then overthrust by the leading edge of the Lycian Nappes, showing that the Antalya Complex and the Lycian Nappes were emplaced at different times (Poisson 1977; Gutnic et al. 1979; Fig. 7).

In addition, unmetamorphosed Mesozoic rocks of the Antalya Complex are locally exposed further east, adjacent to and beneath the metamorphic Alanya Massif. The Alanya Massif is a unique unit forming mountainous terrain north of the coastal town of Alanya (Fig. 7). A structurally coherent unit correlated with the Antalya Complex is exposed beneath the Alanya Massif in an important window (Fig. 7a). Successions of Palaeozoic age (e.g. Permian limestones) there pass upwards into Lower Triassic tuffaceous sediments and Middle–Upper Triassic radiolarian sediments (Ü. Ulu, pers. comm.), that relate to rifting of the southerly Neotethys and can be correlated with the main areas of the Antalya Complex further west.

Further north, unmetamorphosed units correlated with the Antalya Complex (Güzelsu Unit) emerge from beneath the Alanya Massif within a large east–west topographic depression, known as the Güzelsu Corridor (Monod 1977; Fig. 7a). The Güzelsu Unit includes highly deformed Mesozoic deep-sea sediments (e.g. radiolarites), minor volcanics, large masses of shallow-water limestone and ophiolitic fragments, especially sheared serpentinite. These Antalya Complex rocks were first deformed in the latest Cretaceous, but only finally thrust northward over a large carbonate platform, the Akseki Platform (Akseki Unit), in Palaeocene–Early Eocene time. This Akseki Platform and platform margin facies of Late Permian age (Altıner 1984; Altıner et al. 2000), overlain by Mesozoic carbonate platform successions, is interpreted as a Bahama-type carbonate platform located between Neotethyan oceanic basins to the north and south.

In summary, the Antalya Complex and related carbonate platform units (including the Güzelsu Unit in the east) record emplaced northerly parts of the southern Neotethyan oceanic basin. The Menderes–Bey Dağları continental fragment rifted from North Africa in the Late Permian–Early Triassic (Marcoux 1974, 1995), followed by genesis of oceanic crust in the easternmost Mediterranean area in Middle–Late Triassic time (Robertson & Woodcock 1984; Yılmaz 1984). The southerly oceanic basin opened in the Middle–Late Triassic to form a mosaic of carbonate platforms and basins (Fig. 8a), including the Bey Dağları Platform in the west and the Akseki Platform in the east. This was followed by genesis of additional oceanic crust in the Late Cretaceous, possibly in an above-subduction zone setting. Closure, probably associated with regional northward subduction, began in the Late Cretaceous (Maastrichtian) and persisted into Early Tertiary time, resulting in collision and amalgamation of platform units (Fig. 8b). Any remaining oceanic crust within the Isparta Angle area was eliminated by Early Palaeocene time, by collision of microcontinental units to the west and east, whereas the southerly Neotethys still remained open to the south.

The Alanya Massif

The metamorphic Alanya Massif is reported to be dominated by three thrust sheets, of which the lower and upper ones are composed of high-temperature–low-pressure (HT–LP) lithologies, whereas the middle unit comprises high-pressure–low-temperature (HP–LT) blueschists (Okay & Özgül 1984). The age of the blueschists is inferred to be pre-Maastrichtian, consistent with initial thrusting taking place in the Late Cretaceous. Later, the nappe stack was finally assembled by northward thrusting in Palaeocene–Eocene time. The high-pressure metamorphics of the Alanya Massif were exhumed prior to the Early Miocene, possibly related to a crustal extension event.

The Alanya Massif is provisionally restored as several platforms and basins, as shown in Fig. 8a. The latter is viewed as a small Neotethyan

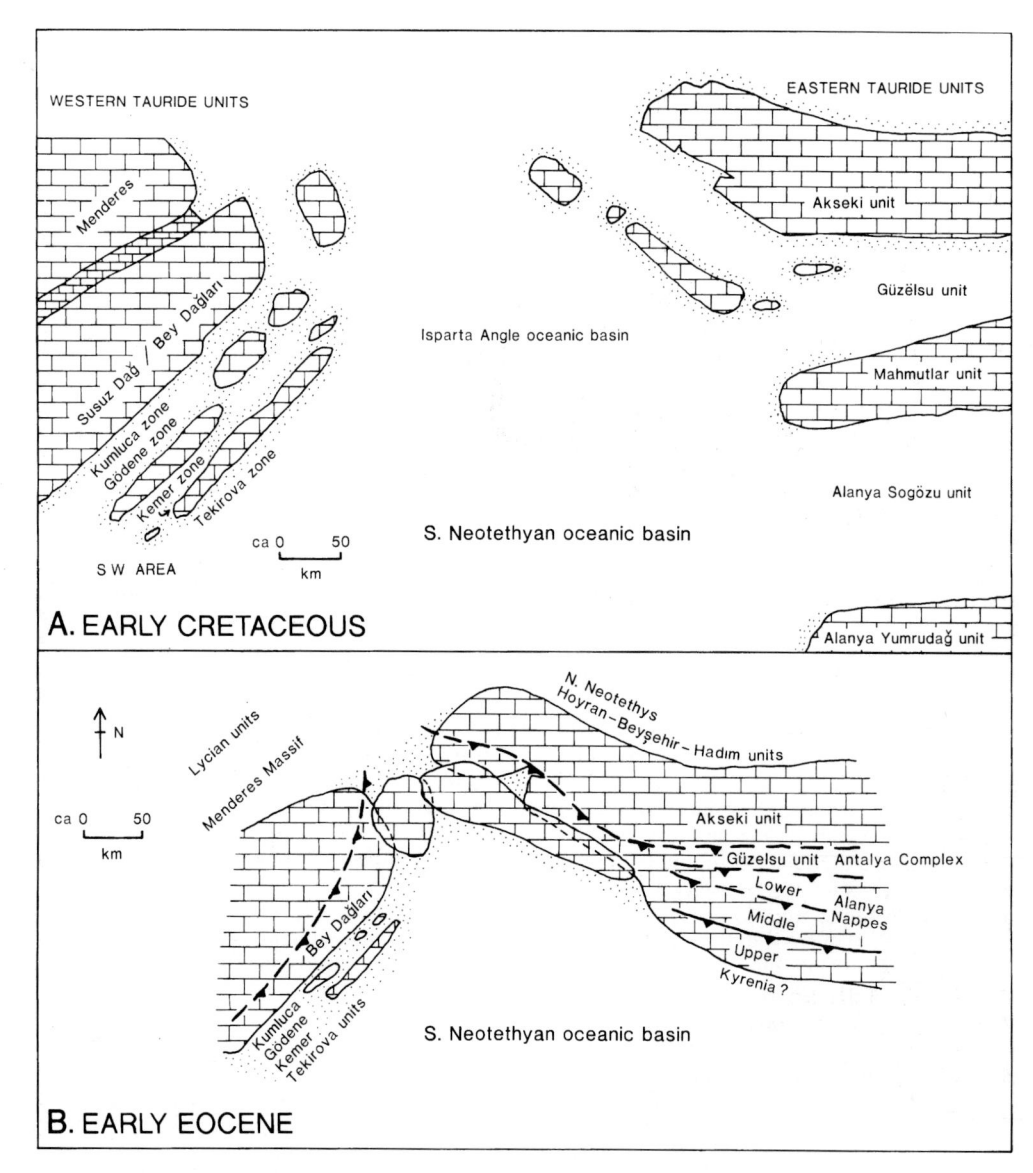

Fig. 8. Reconstruction of the palaeogeography and tectonic setting of the Antalya Complex, southwestern Turkey, in Early Cretaceous time. (**a**) Early Cretaceous. Large carbonate platforms were built on continental fragments rifted from Gondwana in the Triassic. In addition, a number of smaller satellite platforms existed. The reconstruction is based on unravelling the various tectonics units (Robertson 1993), but cannot be accurate in view of uncertainties, e.g. in the width of oceanic segments, the size and shape of platforms, etc. The diagram takes account of Neogene palaeomagnetic rotations, as explained in the text. Palaeolatitudes are those inferred by Dercourt et al. (1992) for the Taurus carbonate platform at these times; modified after Robertson (1993). (**b**) Early Eocene. Units on either side of the Isparta Angle oceanic basin have collided, resulting in thrusting both westwards onto the eastern margin and eastwards onto the westward margin. Neotethys still remained open to the south.

oceanic basin, or deep-rift basin (floored by volcanics), located between the Akseki carbonate Platform to the north and a carbonate platform that later became part of the Alanya Massif (Robertson et al. 1991). This restoration assumes that large-scale thrusting was essentially in-sequence, so that the thrust sheets can simply be pulled apart to reveal the pre-existing

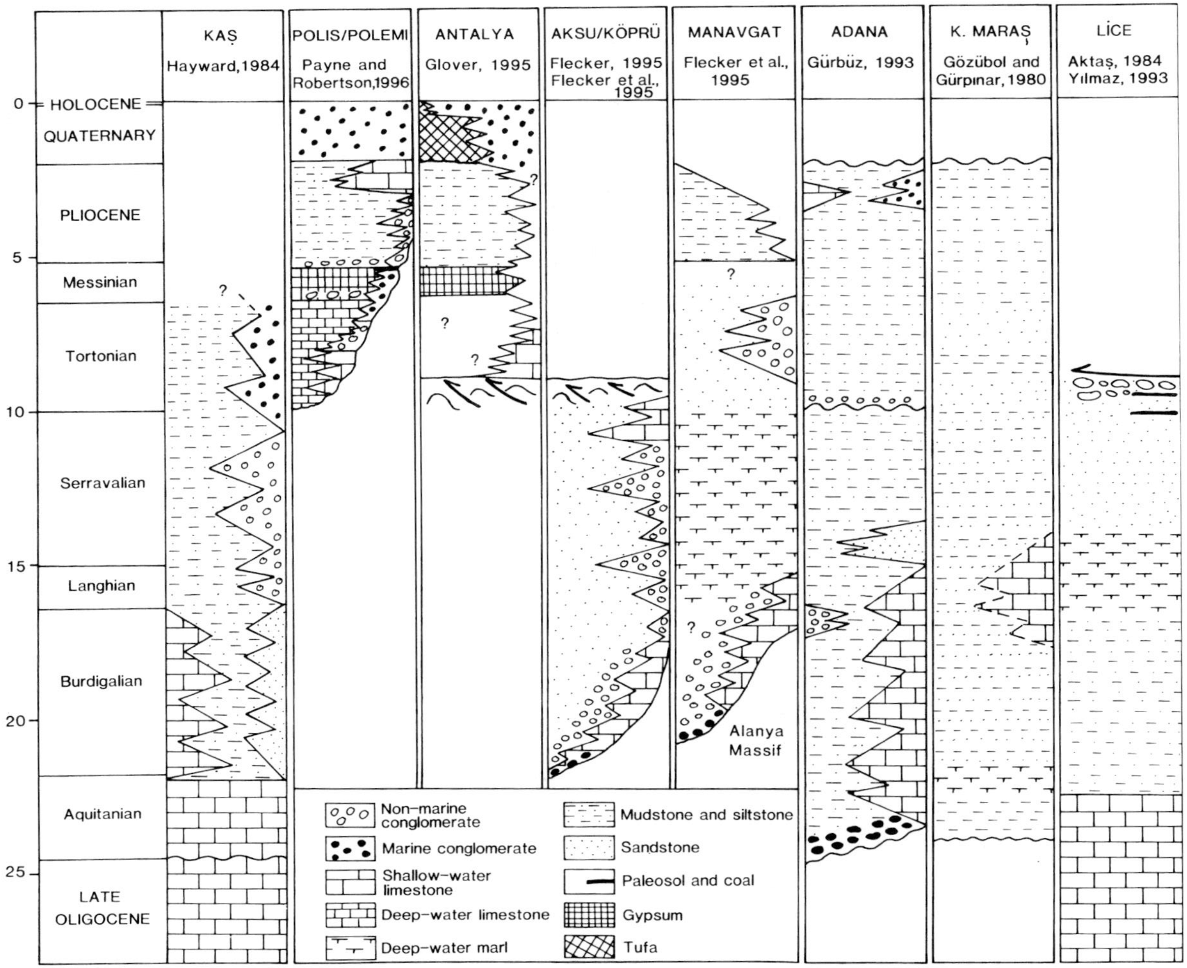

S TURKEY SYNTHESIS

Fig. 9. Summary of the onshore Miocene successions in southwestern, southern and southeastern Turkey. See text for explanation and data sources. Locations on maps in this paper are indicated.

palaeogeography. The HP–LT rocks of the Alanya Massif could thus represent another sutured small Neotethyan basin (rift or small oceanic) bordered by continental crust to the north and south (Fig. 8a). This basin would have closed in the Late Cretaceous, by northwest directed subduction, associated with initial deformation of the Antalya Complex. In the Late Eocene, the already assembled units of the Alanya Massif were thrust over the Antalya Complex and, in turn, finally over the Akseki carbonate platform to the north (Fig. 8b).

Miocene basins

Throughout the Western segment, from the front of the Lycian Nappes eastwards, the Mesozoic–Lower Tertiary Neotethyan units described above are unconformably overlain, or locally overthrust, by Miocene basinal units (Figs 7a and 9). Overthrusting is seen only in the west, where the frontal parts of the Lycian Nappes are thrust over Miocene turbidite successions (e.g. Kaş Basin), interpreted as a foreland basin (Hayward 1984). Locally, a klippen of the Lycian Nappes is seen overlying similar Lower Miocene (Burdigalian) turbidites near Isparta (Poisson 1977; Flecker 1995). Elsewhere, the Miocene basinal units unconformably overlie deformed Neotethyan units, i.e. the Tauride carbonate platforms, the Alanya Massif and the Antalya Complex. Within the Isparta Angle region, the Antalya Complex and adjacent relatively autochthonous Mesozoic 'basement' units (e.g. Bey Dağları) are unconformably overlain by two related north–south trending Miocene basins, the Aksu and Köprü Basins. Further southeast, the Alanya Massif is unconformably overlain by the Miocene Manavgat Basin (Fig. 7).

The Aksu Basin in the west begins with Lower Miocene transgressive limestones (Aquitanian), passing up into Burdigalian turbidites, exposed in the north, in the Isparta area (Poisson 1977; Hayward 1984; Flecker 1995; Fig. 9). Further south, the succession is dominated by Middle–Upper Miocene mudstones, turbidites and channelized conglomerates. These clastic sediments are interpreted as erosional debris from the front of the Lycian Nappes as they neared their final position (Flecker 1995; Flecker et al. 1995). The floor of the Miocene Aksu Basin was strongly faulted during sedimentation, possibly exploiting pre-existing lineaments within the underlying Mesozoic Antalya Complex (Flecker 1995). During the Late Miocene, the Lycian Nappes were thrust to their final position, deforming the Lower Miocene part of the foreland basin exposed in the north (near Isparta). Also in the Late Miocene, the easterly margin of the Aksu Basin was deformed by reverse faulting, thrusting and folding towards the west ([the 'Aksu phase' of Poisson (1977); Akbulut 1977; Frizon de Lamottte et al. 1995]. This deformation was generally coeval with counter-clockwise rotation of the western limb of the Isparta Angle and clockwise rotation of the eastern limb of the Isparta Angle (Kissel & Poisson 1986; Kissel et al. 1990; Morris & Robertson 1993). The Isparta Angle originated as a Mesozoic oceanic basin (with oceanic crust) separating several large carbonate platforms during the Mesozoic–Early Tertiary, as discussed above (Poisson 1984; Waldron 1984; Robertson 1993). The Isparta Angle was later tightened to form an 'oroclinal bend' during the Miocene, related to regional development of the southward curvature of the Aegean active margin (Kissel & Laj 1988). It was influenced by the final emplacement of the Lycian Nappes towards the southeast, and possibly also by westward movement of Anatolia caused by collision with the Arabian margin in southeastern Turkey.

Further east, the elongate Köprü Basin is divided into a western part, with coarse fan-delta conglomerates with local patch reefs, and an eastern part with thinly bedded sandy turbidites. Palaeocurrent data indicate sediment transport from north to south in the north of the basin, but southeastward flow in the southeast, towards the Manavgat Basin (described as part of the Central segment below) (Flecker et al. 1995). Recent Sr-isotope dating shows that the Köprü Basin was initiated in the Early Miocene (Early–Middle Burdigalian), slightly earlier than similar facies in the Manavgat Basin further east. The exact time of initiation of the Aksu Basin in the west is less well constrained.

The regional trend in the time of basin initiation and the subsequent facies are consistent with an origin as a foreland basin that migrated regionally southeastward with time related to the final stages of emplacement of the Lycian Nappes (Flecker et al. 1997). The Aksu–Köprü Basin System shows evidence of fault reactivation of old lines of crustal weakness in the Mesozoic 'basement', which helped divide the regional foredeep into two main semi-independent depocentres during sedimentation. The regional Aksu–Köprü Foreland Basin System was uplifted, faulted and eroded in Late Pliocene–Quaternary time to produce the present topography of Miocene sedimentary outcrops separated by Mesozoic limestone mountains.

Further east, the Alanya Massif is unconformably overlain by Miocene marine sediments of the Manavgat Basin (Akay et al. 1985; Flecker 1995; Flecker et al. 1995; Fig. 9). The succession in the Manavgat Basin begins with Lower Miocene shallow-water carbonates, including patch reefs, that developed along the southern margins of the Alanya Massif. During the Middle Miocene, deeper water turbidites and debris flows (with blocks derived from the Alanya Massif) accumulated, interbedded with bathyal hemipelagic carbonates. The Upper Miocene succession is dominated by relatively shallow-water muddy and siliciclastic sediments. Messinian evaporites are not exposed.

The Manavgat Basin reflects a phase of initial, Early Miocene, northward marine transgression over the metamorphic Alanya Massif, followed by abrupt subsidence that ushered in Middle Miocene deeper water deposition. Accommodation space was gradually filled and the basin progressively shallowed (Flecker et al. 1995). On a regional scale, the Manavgat Basin could be interpreted as: (1) an extensional basin dating from the Early Miocene (or earlier) behind a subduction trench to the south (Cyprus area; Fig. 6); (2) a flexural foreland basin related to generally southward thrusting of units in the Taurus Mountains to the north; or (3) part of a flexural foredeep related to southeastward thrusting of the Lycian Nappes (Flecker et al. 1997, 1998; Fig. 7). The timing of subsidence of the Manavgat Basin is compatible with an origin as essentially the distal, easterly part of a foreland basin that was related to the final stages of emplacement of the Lycian Nappes. An origin related to more southward thrusting from units directly to the north is unlikely as there is no evidence of (Miocene) thrusting of the Beyşehir–Hoyran–Hadim Nappes to the north after the Late Eocene (Monod 1977). An influence of a subduction zone to the south is also possible in view of possible evidence of northward subduction near Cyprus and in southern Turkey (i.e. Kyrenia–Misis Lineament in the Central segment discussed below).

Pliocene basins

The onshore Pliocene–Pleistocene successions of the Aksu Basin rests unconformably on the older sediments of the Miocene Aksu Basin (Akay et al. 1985; Fig. 9). The Pliocene record is dominated by a Messinian–Middle Pliocene extensional basin, infilled with shallow-marine to deltaic sediments up to several hundred metres thick (Glover & Robertson 1998a, b). Messinian evaporite is very locally present. The Pleistocene of the Aksu Basin (Fig. 9) is marked by extensive tufa deposits ('Antalya travertine', Bürger 1990) that precipitated from cool-water springs (Glover 1996). The Antalya tufa was built up on a thin (tens of metres) Upper Pliocene–Lower Pleistocene alluvial succession. Further east, the highest exposed levels of the Manavgat Basin comprise open-marine muddy sediments of Early Pliocene age (Glover & Robertson 1998a).

The Pliocene–Pleistocene Aksu Basin is interpreted in terms of an initial Late Miocene rift event (transtensional), followed by an Lower–Middle Pliocene sedimentary infill, then further rifting extension in the latest Pliocene–Early Pleistocene (Glover & Robertson 1998a, b) that was associated with uplift of the adjacent Taurus Mountains, and more generally linked to regional crustal extension and uplift of western Anatolia as a whole (Price & Scott 1994). Further east, the Lower Pliocene muddy sediments of the Manavgat Basin are interpreted as a shallow eastward extension of the Aksu Basin.

Offshore areas

The offshore area within the Western segment comprises, from west to east, the Rhodes Basin, the Anaximader Mountains and the Antalya Bay (Fig. 6).

In the far west, the seafloor is dominated by the Mediterranean Ridge (Fig. 5), now known to be a mud-dominated accretionary wedge overlying a northward (or northeastward) subduction zone in which Mesozoic oceanic crust has been consumed, probably since around Late Oligocene time (Kastens 1991; Kastens et al. 1992; Camerlenghi et al. 1995; Chaumillon & Mascle 1995; Robertson & Kopf 1998). In its central area, the Mediterranean Ridge is now in the process of collision with Cyrenaica, a large promontory of North Africa. However, further east in the Herodotus Basin area (Fig. 5), and west in the Ionian Sea, collision has not yet occurred. Small remnants of Mesozoic oceanic crust remain, overlain by very thick sediments. The Hellenic Trench, between the Mediterranean Ridge and Crete, was first postulated to be the present-day Africa–Eurasia Plate boundary (e.g. Hsü et al. 1978), but more recently has been reinterpreted as a flexural foredeep related to backthrusting of the Mediterranean Ridge accretionary complex (Camerlenghi et al. 1995). In the east, the Hellenic forearc zone includes the discrete Strabo and Pliny Trenches. It is

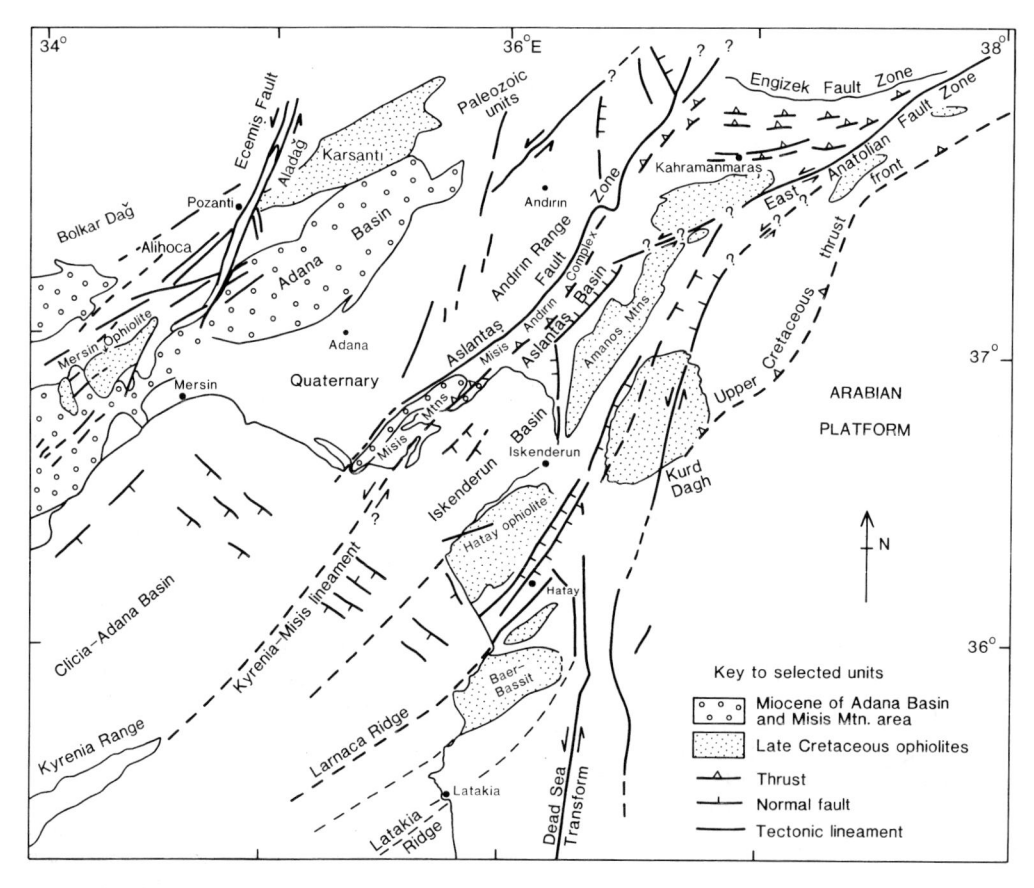

Fig. 10. Map of the main tectonic lineaments in the northeasternmost corner of the Mediterranean area. Onshore data from Karig & Kozlu (1990) and Kelling *et al.* (1987); offshore data from Aksu *et al.* (1992a).

widely believed that these features are dominated by left-lateral strike-slip (e.g. Le Pichon & Angelier 1979; Angelier *et al.* 1982). The Pliny Trench extends onshore as the important Burdur Fault Zone (Fig. 5), whereas the Strabo Trench extends eastward into the 'cleft' area south of the western Anaximander Mountains. The area to the west of the Anaximander Mountains connects with the Mediterranean Ridge accretionary complex via a series of fault lineaments dominated by left-lateral strike-slip. The southern boundary of the Anaximander Mountains (Fig. 6) is distinguished by an eastward extension of the 'cleft', marked by downfaulting that is especially marked on the northern side. The 'cleft lineament' then passes eastward into a broad fold structure that exhibits sediment deformation on side-scan sonar images (Ivanov *et al.* 1992; Woodside 1992). The Mediterranean Ridge to the south is marked by cobblestone topography (Fig. 6).

The Anaximander Mountains themselves comprise three main submarine highs, termed A-1 (Anaximander), A-2 (Anaximenes) and A-3 (Anaxogoras) from west to east (Woodside 1992). Recent results from dredging prove that the Anaximander Mountains represent an extension of southern Turkey in the Mesozoic and Early Tertiary (Woodside & Dumont 1997). The Anaximander Mountains are thus interpreted as a submarine extension of tectonic units exposed on land in southwestern Turkey, including lithologies derived from the seaward extension of the relatively autochthonous Susuz Dağ and the allochthonous Antalya Complex. The Anaximander Mountains exhibit a complex structure, including localized reverse faulting in the northwest, folding in the southeast and fault-related subsidence in the northeast. In general, this structure accommodates the intersection of a zone of left-lateral displacement (i.e. transpression) located within the Strabo Trench and an inferred northeastward dipping subduction zone beneath the Florence Rise (Fig. 6). The deep Finike and Rhodes Basins (Fig. 7) probably originated as rifted basins, similar to the Adana and Cilicia Basins further east (see below). These basins are now being actively deformed, in response to left-lateral transpressional deformation along the plate boundary to the south.

The deep structure of Antalya Bay, the offshore extension of the Isparta Angle (Fig. 6), remains poorly known. Seismic data indicate the presence of evaporites of presumed Messinian age and, thus, imply a pre-Messinian age of formation (Woodside 1977). Earthquake records suggest the existence of a detached oceanic slab beneath Antalya Bay (Rotstein & Kafka 1982; Jackson & McKenzie 1984, 1989), although the direction and timing of subduction remain unclear. Shallow-penetration seismic data indicate that the western margin of Antalya Bay is marked by a series of steep down-to-the east normal faults. The northeastern Antalya Bay area is characterized by more widely spaced faults that locally define a horst-and-graben structure (Glover & Robertson 1998a). By contrast, extensional faulting has produced a southwards widening rift basin (Antalya Basin) within the Antalya Bay area (Fig. 6). This basin is tentatively interpreted as the result of crustal extension behind a subduction zone linking the Florence Rise with the southern boundary of the Anaximander Seamount (Fig. 6).

The Central segment

The Central segment of the area under discussion includes the Bolkar Mountains (part of the Taurus Mountains) in the north, and associated ophiolites and ophiolitic mélange, the Miocene Mut and Adana Basins, in the east, and further south the Misis Mountains (westerly) and the Andırın Mountains (easterly). Offshore areas are the Adana–Cilicia Basin and the submerged Kyrenia–Misis Lineament connecting northern Cyprus and southern Turkey (Fig. 5).

The backbone of the Tauride Mountains in the Central segment is dissected by the prominent left-lateral Ecemiş Fault System that was active in Plio-Quaternary time and possibly earlier (Yetiş 1984; Fig. 10). The mountains to the west are the Bolkar Dağ and those to the east the Aladağ. Exposed, relatively autochthonous, units begin with a Palaeozoic pre-rift platformal unit, overlain by Mesozoic plaformal units, correlated with those of the Tauride carbonate platform further west (Özgül 1984). This platform unit is tectonically overlain by ophiolitic mélange and variably dismembered ophiolites. To the north of the Aladağ (in the east) is the large and relatively coherent Pozantı–Karsantı Ophiolite (Lytwyn & Casey 1995; Fig. 11). North of the Bolkar Dağ (further west), ophiolites are mainly dismembered as blocks within Upper Cretaceous ophiolitic mélange, although locally more coherent units are preserved (i.e. Alihoca Ophiolite; Dilek & Moores 1990). In addition, high-pressure blueschists occur locally along the contact between the ophiolite and the underlying Bolkar Dağ carbonate Platform (Dilek & Moores 1990).

To the south of the Bolkar Dağ a complete, relatively undeformed, Upper Cretaceous ophiolite (Mersin Ophiolite) is exposed (Fig. 10). Surprisingly, until recently this was largely ignored. The ophiolite is underlain by a metamorphic sole and, in turn, by polymict volcanic–sedimentary mélange that can be interpreted as accreted units, including deep-sea sediments and volcanics (Parlak et al. 1995). The Mersin Ophiolite has generally been correlated with allochthonous units further west (i.e. ophiolites of the Hoyran–Beyşehir–Hadim Nappes) that were finally thrust southward in the Late Eocene (Özgül 1984), possibly driven by the final stages of closure of a northerly Neotethyan oceanic basin ('Inner Taurus Ocean'; Şengör & Yılmaz 1981; Görür et al. 1984; Lytwyn & Casey 1995). However, Parlak et al. (1995) argued that the Mersin Ophiolite was emplaced northwards from a southerly Neotethyan oceanic basin in the Late Cretaceous, although this remains controversial and needs to be tested with more field evidence. If the Mersin Ophiolite was indeed emplaced from the north, this would have taken place in the latest Cretaceous rather than the Eocene, as the ophiolites and mélange along the north margin of the Bolkar Dağ are unconformably overlain by transgressive Palaeogene sediments, which are unconformably overlain by Neogene units (Demirtaşlı et al. 1984).

Miocene basins

To the south of the Taurus Mountains (Bolkar Dağ and Aladağ), the Miocene Adana Basin, the most intensively studied Neogene basin in southern Turkey, includes shallow-water carbonates, deep-water clastics, shallow-water clastics and fluvial sediments, as summarized in Fig. 9 [Görür 1977, 1992; Yalçın & Görür 1984; see Yetiş et al. (1995) for a recent review]. Seismic stratigraphic interpretation indicates the presence of three megasequences (Williams et al. 1995) within the Miocene Adana Basin. Oligocene sediments further north are also reported to exhibit evidence of compression that took

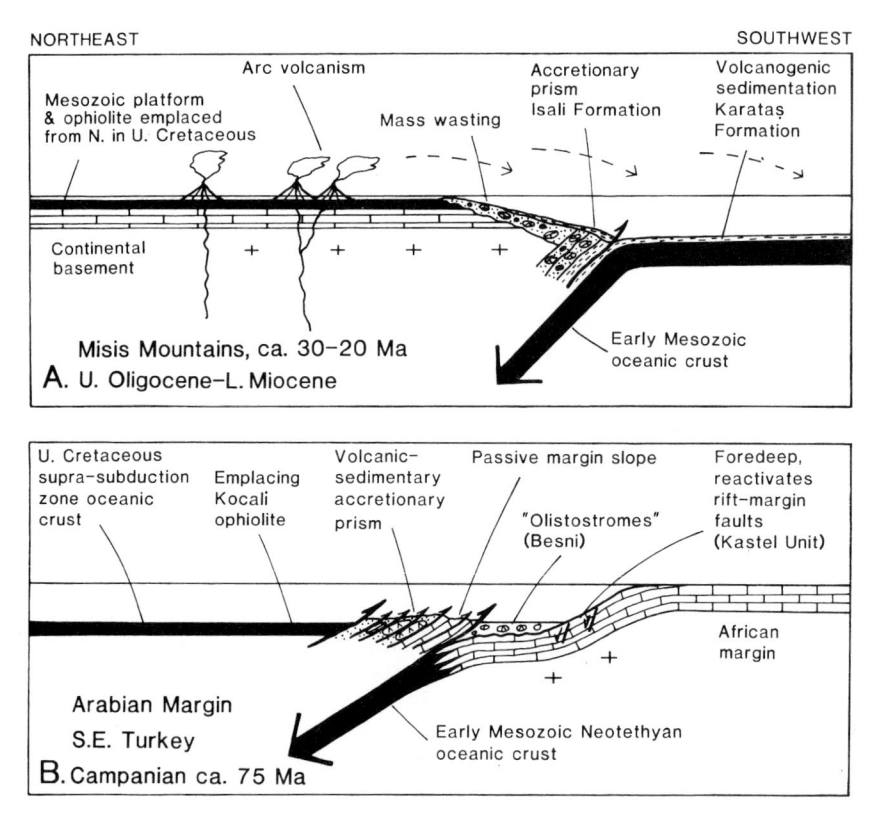

Fig. 11. Tectonic models for contrasting mélanges in the easternmost Mediterranean. (**a**) Upper Oligocene–Lower Miocene mélange in the Misis Mountains, southern Turkey (Misis–Andırın Complex), i.e. the northern margin of the southern Neotethys oceanic basin. (**b**) Upper Cretaceous mélange on the Arabian margin in southeastern Turkey, i.e. the southern margin of the southern Neotethys oceanic basin. See text for explanation.

place in the Late Oligocene. Williams *et al.* (1995) relate extensional faulting to renewed thrusting in the Taurus to the north. A problem, however, is that the sediments within the Adana Basin are everywhere mapped as unconformably overlying the basement (Mesozoic Tauride units). However, Williams *et al.* (1995) invoke the existence of concealed thrusts beneath the northern margin of the Adana Basin to cause flexural loading and faulting of the foreland. A possible alternative (or additional) control is that the faulting was related to regional extension that originated to the south, related to subduction (i.e. suprasubduction zone extension), rather than crustal loading driven from the north (see below).

North of the Adana Basin, Middle Eocene deformation was followed by erosion, then fault-controlled subsidence in the Oligocene, with initially terrestrial deposition in the Adana Basin (Fig. 9). Faulting continued into the Early Miocene, when Miocene reefs accumulated on rotated fault blocks, and shallow-marine to deeper marine clastics accumulated in intervening fault-controlled basins. A relative sea-level high in the mid-Miocene (Langhian) caused northward migration of reefs (Yalçın & Görür 1984). Mid-Miocene (Langhian–Serravalian) time was marked by deepening and deposition of turbidites, passing progressively upwards into shallow-marine, then continental, deposits. Palaeocurrents are initially towards the south and southeast, then later towards the southwest (Gökçen *et al.* 1988; Gürbüz 1993; Gürbüz & Kelling 1993).

Further west is the horseshoe-shaped Oligo-Miocene Mut Basin, bounded by the relatively subdued part of the Bolkar Dağ to the north and a rugged coastal area of mainly 'basement' rocks in the south (Gedik *et al.* 1979). The Mut Basin encompasses a very well-exposed succession beginning with Upper Oligocene–basal Miocene alluvial to lacustrine sediments unconformably overlying emplaced Neotethyan units. The succession was gently deformed in pre-Burdigalian time, followed by alluvial deposits, passing into lacustrine to shallow-marine mixed clastic–carbonate sediments. Thick reef limestones, including very well-exposed patch reefs developed during Langhian–Early Serravalian time during maximum transgression (Gedik *et al.* 1979). Any originally higher levels of the succession were later eroded.

The Misis–Andırın Complex

The Misis Mountains and parts of the Andırın Range further northeast are dominated by the Misis–Andırın Complex, for which very different names and origins have been proposed (Schiettecatte 1971; Kelling *et al.* 1987; Karig & Kozlu 1990).

The lower part of the Misis–Andırın Complex is well exposed in the Misis Mountains, where it is dominated by distal volcanogenic turbidites (Karataş Formation) of late Early Miocene–latest Middle Miocene age (Kelling *et al.* 1987; Gökçen *et al.* 1988). The tectonically overlying upper part of the Misis Complex is a mélange (İsalı Formation) containing numerous exotic blocks up to several hundred metres in size ['olistostromes' of Schiettecatte (1971)], embedded in calcareous claystone of locally Late Oligocene–earliest Miocene age. Blocks include ?Palaeozoic limestone, Mesozoic–Lower Tertiary limestones, ophiolitic rocks (serpentine, gabbros, pillow lava), volcaniclastic and radiolarian sediments (Schiettecatte 1971).

Abundant volcaniclastic sediments forming 'olistoliths' within the mélange include very disrupted units of massive volcaniclastics, thinner bedded volcaniclastics and more coherent successions of thin-bedded turbidites (Karataş Formation). Geochemical studies, involving stable trace element analysis, indicate that two types of volcanic rock are present: (1) evolved volcanics and siliceous tuffs of calc-alkaline composition, presumably erupted from a contemporaneous volcanic arc – fresh tuffaceous sediments are reported from in the upper part of the İsalı Formation and in the overlying Karataş Formation (Yetiş *et al.* 1995); and (2) basic volcanic rocks that appear on geochemical grounds to have been erupted above a subduction zone. Floyd *et al.* (1992) suggested that all these volcanic rocks formed in a Miocene volcanic arc and a related rifted back-arc basin. It is likely that these volcanics represent blocks within the mélange of possibly Late Cretaceous age, rather than coeval volcanics. The tuffaceous volcaniclastic sediments within deep-water turbidites might, however, have been derived from coeval arc volcanism.

The mélange was initially interpreted as an olistostrome related to southeastward sliding of heterogeneous blocks into a sedimentary basin (İsalı Formation). This was later emplaced over, or against, deep-water Miocene volcanogenic turbidites to the south (Karataş Formation). The setting of this deposition was the northern margin of the İskenderun Basin in Pliocene–Pleistocene time (Schiettecatte 1971). Kelling *et al.* (1987) regarded the mélange as olistostromes, shed into a compressional foredeep (related to southward thrusting) within an overall forearc or back-arc setting. The mélange was later deformed in a transpressional setting, while transtension prevailed further southwest. An alternative interpretation of the mélange is that it formed as a subduction–accretion complex (rather than simply a foreland basin), as shown in Fig. 11a.

A very different view emerged from a study of the Andırın Range, along-strike to the northeast (Karig & Kozlu 1990; Fig. 11). The Misis–Andırın Block exposes units including Upper Cretaceous carbonates, Eocene, mafic flows, agglomerates and volcaniclastics, bioclastic and argillaceous limestones, Eocene–Oligocene? hemipelagic silty marls, calcarenites calcilutites, and localized mass-flow units of Late Oligocene age. The assemblage thus appears to be more complete than that exposed in the Misis Mountains to the southwest (i.e. Aslantaş Formation).

The tectonic contact between the upper and lower units of the Misis–Andırın Range was mapped by Karig & Kozlu (1990) as a regionally important extensional fault (Aslantaş Fault Zone; Fig. 21) that separates mainly Mesozoic basement units of the Andırın Range to the north from the İskenderun Basin, and an extensional counterpart, the Aslantaş–İskenderun Basin, to the southwest (Fig. 21). The Misis–Andırın Complex is interpreted by these authors as a Late Oligocene–Middle Miocene half-graben rather than being related to a compressional foreland basin setting. Karig & Kozlu (1990) argue that final emplacement of regionally important allochthonous units in this area took place prior to the Miocene, followed by opening of a transtensional basin. The mélange is thus related to extensional detachment faulting in this model. More data are needed to test this hypothesis. However, an extensional origin is not in keeping with the evidence of regional convergence of the African–Eurasian Plates and related thrusting in southeastern Turkey and the Zagros Mountains during the Miocene.

After initiation of the East Anatolian Fault in the Early Pliocene, the northern boundary of the Aslantaş–İskenderun Basin switched to partitioned compression and strike-slip in Pliocene time, according to Karig & Kozlu (1990). Older thrust faults (e.g. in the Misis Mountains) were sealed by Upper Pliocene and Pleistocene sediments and volcanics, whereas fault patterns suggest a most recent switch to extension, or transtension. The timing of Late Miocene–Early Pliocene, inferred transpression onshore predates the initial uplift of the Kyrenia Range in Cyprus in Late Pliocene–Early Pleistocene (Robertson & Woodcock 1986).

The offshore Cilicia–Adana Basin

Between Turkey and northern Cyprus is the Cilicia–Adana Basin (Figs 5 and 10), characterized by east–west trending extensional faults, some of which appear to still be active (Aksu *et al.* 1992a, b). Available seismic reflection data show that the Cilicia–Adana Basin dates from pre-Messinian time. The basin is up to 3 km deep and contains thick Messinian evaporites, that are slightly deformed by growth faulting, overlain by Pliocene–Pleistocene muddy sediments. There is also evidence of localized salt diapirism (Aksu *et al.* 1992a, b). Estimated sediment thicknesses increase eastwards from c. 1 to 1.8 km in the middle part of the Cilicia–Adana Basin, to c. 3 km in the east, near Adana Bay, which experienced input from the Seyhan River. The lower sections of the Cilicia–Adana Basin are not imaged by available, relatively shallow, seismic data, but are assumed to date from the Late Oligocene–Early Miocene, the time of marine transgression as exposed in the Kyrenia Range (Baroz 1979; Robertson & Woodcock 1986). In the west, the Cilicia–Adana Basin is bounded by a fault lineament, the Anamur–Kormakiti Ridge, beyond which is the Antalya Basin (Fig. 5).

The offshore Kyrenia–Misis Lineament

The submerged Kyrenia–Misis Lineament is a major structural and topographic feature that runs from the Kyrenia Range under the sea and emerges as the relatively subdued Misis Mountains in southern Turkey (Evans *et al.* 1978; Aksu *et al.* 1992a, b; Fig. 10). Offshore, the Kyrenia–Misis Lineament forms a well-defined bathymetric ridge, the Misis Ridge, which remained mainly free of sediments, whereas basins were formed to the north and south along extensional (or transtensional) faults. Aksu *et al.* (1992a, b) interpreted the lineament as a narrow horst, bounded by strike-slip faults. Messinian evaporites pinch out against the Kyrenia–Misis Ridge, showing that it was already a positive feature in the Late Miocene. Kempler (1994) also interpreted the steep oblique faults bordering the Kyrenia–Misis bathymetric Ridge as reflecting strike-slip. The submerged Misis Ridge was seen as transtensional in origin, in contrast to the Kyrenia Range and Misis Mountains to the west and east, respectively, that were transpressional. Mapped Pliocene–Pleistocene extensional faults trend at right angles to the Misis–Kyrenia Ridge (Fig. 10). This extension is attributed to regional strike-slip, with individual listric faults soling out on Messinian evaporite (Aksu *et al.* 1992a, b).

Discussion: tectonic model

Exotic rocks, including Upper Cretaceous ophiolites, were regionally emplaced southwards from a more northerly Neotethyan ocean basin ('Inner Tauride Ocean') onto a Mesozoic carbonate platform in latest Cretaceous (Campanian–Maastrichtian) time, represented by the Bolkar Dağ and the Aladağ. Deformation recurred in the Eocene, related to final closure of the 'Inner Taurus Ocean' to the north. Continuing plate convergence was accommodated by crustal thickening in the north (possibly manifested in blind thrusting within basement), coupled with subduction of Neotethyan oceanic crust to the south, along a northward dipping subduction zone located in the vicinity of the Kyrenia Range of northern Cyprus. This subduction gave rise to Upper Eocene 'olistostromes' in the Kyrenia Range (Baroz 1979; Robertson & Woodcock 1986; Fig. 11b).

In response to continuing Africa–Eurasia convergence, northward dipping subduction was active in Late Oligocene–Early Miocene time, as evidenced in the Misis–Andırın Range. Deep-water volcanogenic sediments accumulated in a remnant Neotethyan oceanic basin to the south (Karataş Formation; Fig. 11a). The sedimentary mélange (İsalı Formation) is interpreted as a subduction trench deposit. Platform carbonates, ophiolitic and older arc rocks collapsed into the trench, as olistoliths. Sufficient subduction continued during the Miocene to fuel possible arc volcanism. The trench migrated southward to near its final position, south of Cyprus, by the start of the Miocene.

The Eastern segment: southeastern Turkey

The Eastern segment discussed here extends from near Kahraman Maraş in the west to near

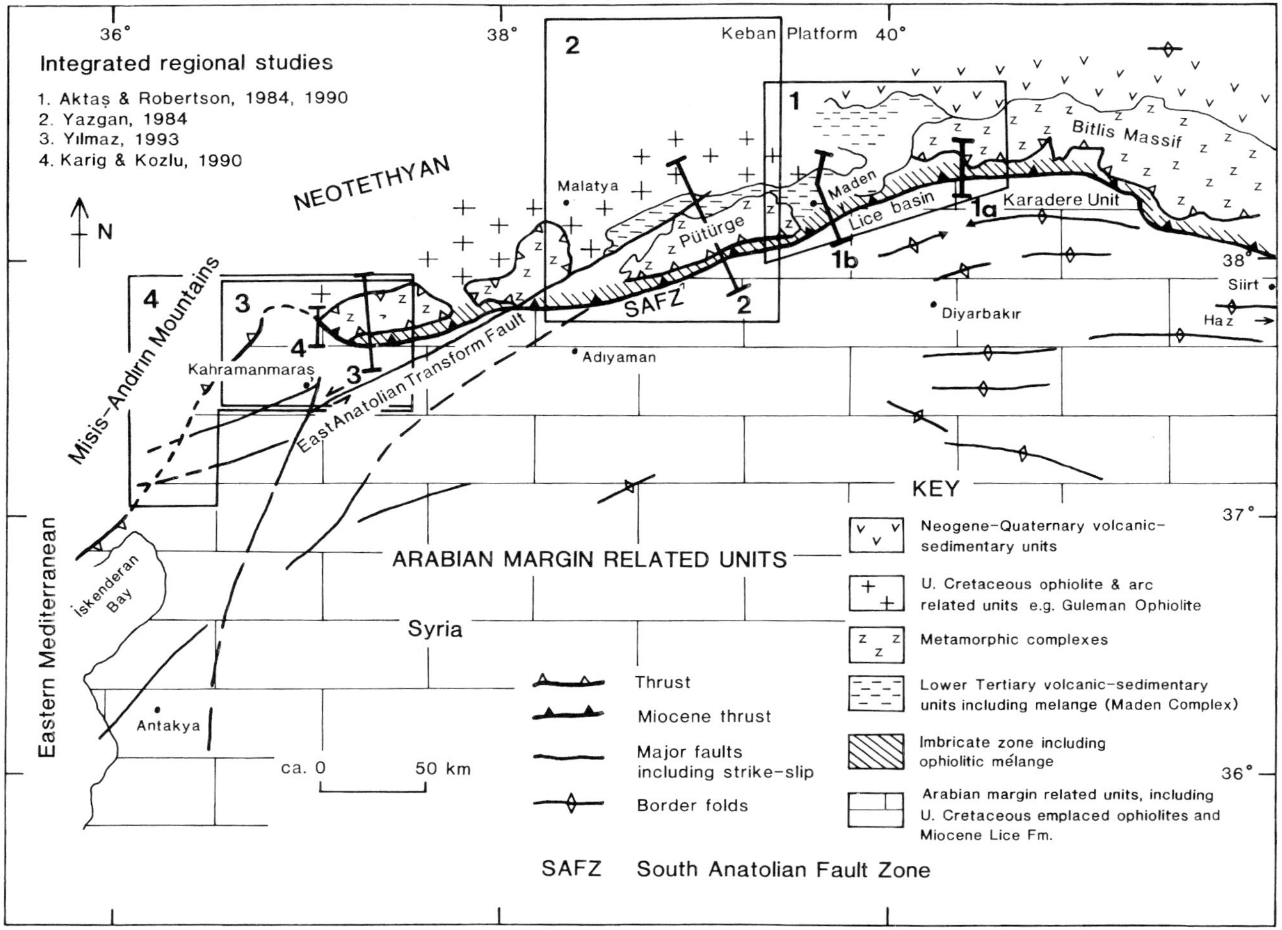

Fig. 12. Outline tectonic sketch map of the main units in southeast Turkey. Boxes show areas of cross-sections given in Fig. 13. Map based on 1:200 000 000 Geological Map of Turkey. See text for explanation and data sources.

Lake Van, c. 350 km away (Fig. 12). This area encompasses the Arabian foreland, including ophiolites (e.g. Hatay Ophiolite; Fig. 10) and mélange emplaced in the uppermost Cretaceous and Palaeogene–Neogene successions. It also includes large, structurally overlying, metamorphic units, the Pütürge and Bitlis Massifs, overlying ophiolites and magmatic arc-related units, and also structurally complex Palaeogene volcanic–sedimentary units (i.e. Maden Complex) sandwiched between the Arabian foreland and the overlying metamorphic massifs (Fig. 12).

This Eastern segment is divided into two fundamentally different units by a regionally extensive feature, termed the South Anatolian Suture Zone in the Turkish literature. The nongenetic term of the South Anatolian Fault Zone (SAFZ) is used here, as not all authors interpret this lineament simply as a suture zone. Units to the south form part of the Arabian continental margin, while those to the north of this fault zone include various allochthonous Neotethyan-related units. The SAFZ was generally interpreted as a major regional thrust of allochthonous Neotethyan units over the Arabian margin. However, Karig & Kozlu (1990) reinterpreted the westward continuation of this fault zone into the Misis–Andırın region (Fig. 12) as a strike-slip fault that was reactivated by late-stage, post-Miocene compression. A thrust-related origin fits the evidence, at least from southeastern Turkey. In addition, southeastern Turkey as a whole is transected by the Pliocene–Pleistocene East Anatolian Transform Fault. Displacement remains poorly constrained [< 25 km; see Şengör et al. (1985)], as precise correlations of units on either side have not been confirmed.

The Arabian foreland

Mesozoic time saw construction of a substantial carbonate platform along the southern margin of the Neotethyan ocean in a passive margin setting. Beneath are Palaeozoic shelf-type successions, inferred to have accumulated along the northern margin of Gondwana (as exposed in the Hazro Inlier, Fig. 12; Fourcade et al. 1991). During the latest Cretaceous, the passive margin gave way to a foredeep onto which ophiolites were emplaced from the north. The initial stages of southward obduction of the ophiolites resulted in subsidence of the Arabian Platform and incoming of Maastrichtian terrigenous sediments (Rigo di Righi & Cortesini 1964; Yılmaz 1993). This unit is tectonically overlain by sedimentary mélange (i.e. 'olistostromes'), then by tectonic mélange and finally by large ophiolite thrust slices. The Maastrichtian basin is interpreted as a foredeep related to ophiolite emplacement, as inferred for a similar unit (Muti Formation) in Oman (Robertson 1987). The tectonic mélange contains Upper Triassic–Upper Cretaceous units (Fourcade et al. 1991) and is interpreted as emplaced remnants of the former deep-water passive margin of the Arabian Plate. It is, for example, comparable with the para-autochthonous Sumeini slope carbonates in Oman (Watts & Garrison 1979). This mélange includes blocks of Permian and Triassic mafic volcanic rocks, limestone, radiolarite and serpentinite, set in a mainly pelitic matrix. Volcanic rocks and neritic limestones can be correlated with the Oman exotics, representing obducted oceanic seamounts (e.g. Searle & Graham 1982). The mélange is interpreted as a subduction–accretion complex related to northward subduction of Neotethyan oceanic crust in the Late Cretaceous, prior to its emplacement onto the Arabian Margin to the south (Fig. 11b). In southeastern Turkey, the overlying ophiolites (e.g. Koçalı Ophiolite) are smaller and less well exposed than their counterparts further west, notably the Hatay Ophiolite (Whitechurch et al. 1984).

Elsewhere within the Antalya region (Turkey) and northern Syria (Baer–Bassit region), discontinuous platform sedimentation continued into the Early Maastrichtian, followed by generally southward emplacement of allochthonous units. These comprise Upper Cretaceous ophiolites, of suprasubduction zone type, similar to the Troodos Ophiolite (Delaloye et al. 1980; Delaloye & Wagner 1984; Pişkin

---

Fig. 13. Simplified cross-sections across Neotethyan allochthonous units in southeast Turkey. (**1a**) From the Late Cretaceous arc, Guleman Ophiolite and their Lower Tertiary cover (Hazar Group), across a Palaeogene imbricate thrust wedge and mélange to the Arabian foreland below the basal thrust (Aktaş & Robertson 1984, 1990a, b). (**1b**) Through a slice of Middle Eocene volcanics and deep-sea sediments (further east), overthrust by the Bitlis Metamorphic Massif and underlain by the Arabian foreland (Aktaş & Robertson 1984, 1990a, b). (**2**) Across the entire thrust stack further west, including the Pütürge Metamorphic Massif (Yazgan 1984; Yazgan & Chessex 1991). (**3**) Section further west again, showing the inferred presence of a major Eocene volcanic arc unit near the base of the thrust stack (Yılmaz 1993). (**4**) In the far west, showing the notion of the Miocene basin on the Arabian foreland as a post-collisional extensional basin rather than a foreland basin (Karig & Kozlu 1990). See Fig.12 for location of section and the text for further explanation.

S TURKEY SYNTHESIS

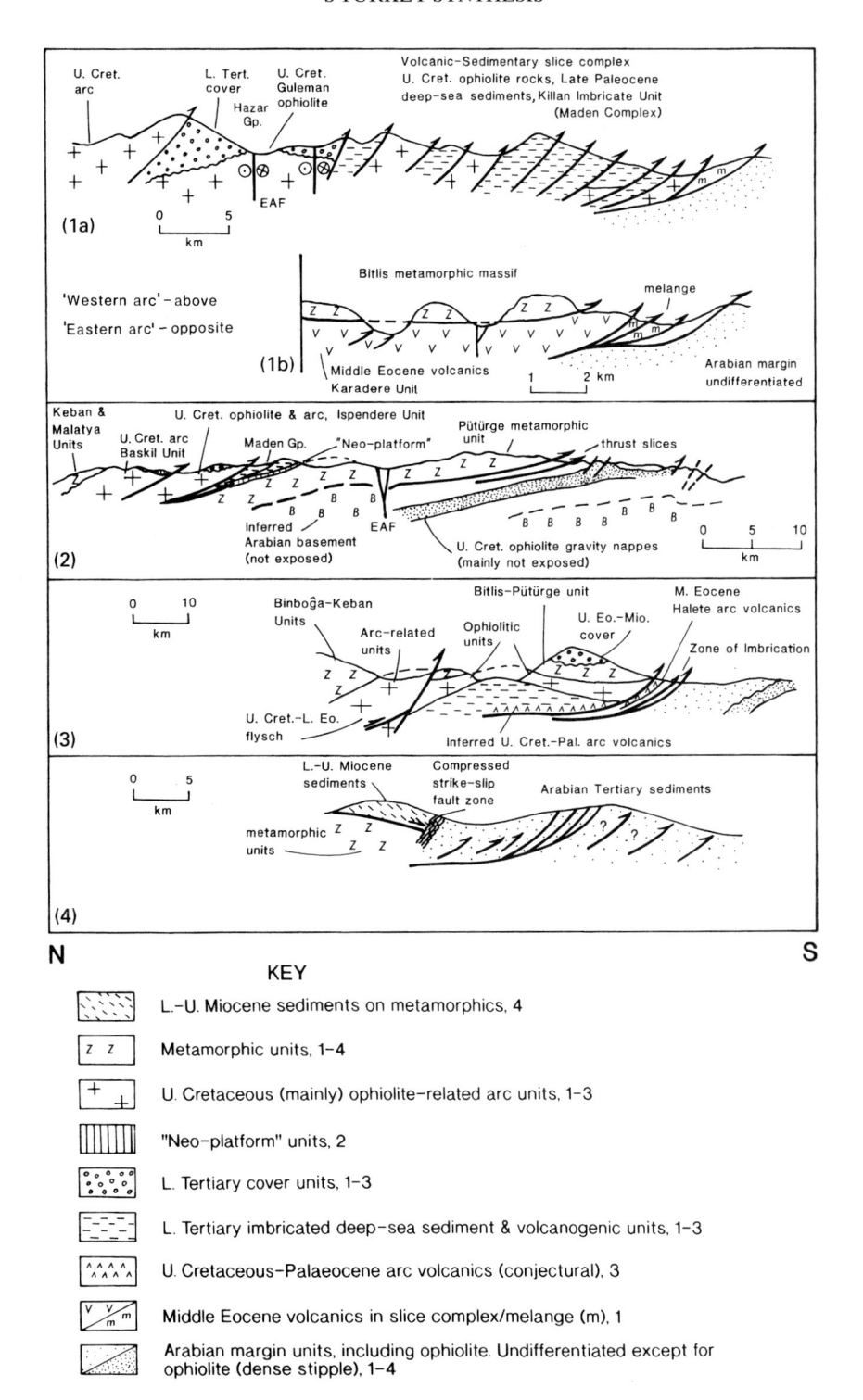

et al. 1986; Dilek *et al.* 1991; Dilek & Delaloye 1992). The emplaced ophiolite represents relatively intact oceanic lithosphere in Hatay (southern Turkey; Fig. 10), but the ophiolite was severely dismembered during emplacement in the Baer–Bassit and Kurd Dagh areas (in northern Syria). Recent geochemical studies suggest a suprasubduction zone origin for the Hatay Ophiolite and related structural studies point to an origin at a slow-spreading, rifted ocean ridge (Dilek & Thy 1998). The Baer–Bassit Ophiolite is tectonically underlain by deformed thrust sheets of deep-water sedimentary and volcanic rocks of Late Triassic–Cenomanian age that are interpreted as the emplaced northern margin of the Arabian Plate (Delaune-Mayere *et al.* 1977). In Hatay, the ophiolite is transgressed by Upper Cenomanian–Lower Maastrichtian fluvial clastics, then shallow-marine sediments, whereas in Baer–Bassit the oldest marine transgressive sediments are of Late Maastrichtian age (northern Syria) (Delaune-Mayere *et al.* 1977; Delaloye & Wagner 1984).

Following regional ophiolite emplacement, deposition resumed on a shelf in Early Tertiary time. The base of this succession unconformably overlies ophiolites that were emplaced in the Late Cretaceous. Calcareous sediments (c. 1000 m thick), of Oligocene–Early Miocene age, are interpreted to have accumulated on an unstable shelf prior to final suturing of Neotethys. Large areas of the northern margin of the Arabian Plate were overlain by the Lower Miocene terrigenous Lice Formation (Fig. 12), traditionally interpreted as a foredeep (Lice Basin) related to southward overthrusting of northerly derived Neotethyan units.

Northern margin units

Understanding the geological history of the structurally overlying allochthonous units to the north of the SAFZ is complicated by the different names used for often similar units in adjacent areas studied in the field by various workers. However, a relatively small number of major, grossly allochthonous, tectonostratigraphic units are present to the north of the SAFZ (Fig. 12).

Metamorphic complexes. The most obvious of these units are large metamorphic complexes, of which the Bitlis and Pütürge Massifs are the best known (Fig. 13). A polymetamorphic origin is generally accepted (Helvacı & Griffin 1984), including metamorphosed equivalents of the Upper Triassic volcanic–sedimentary units related to continental break-up (Perinçek 1979; Perinçek & Özkaya 1981). The metamorphics are intruded by granitic rocks in the Keban Platform (Fig. 12), dated at 85–76 Ma and Campanian–Maastrichtian, respectively], and assumed to document Late Cretaceous Andean-type magmatism. The Pütürge metamorphic Massif is depositionally overlain by Lower Tertiary–Middle Eocene shallow-water carbonates (Yazgan & Chessex 1991), passing up into deeper water volcanic–sedimentary units (Bingöl 1990; Hempton 1984).

One school of thought is that the metamorphic massifs represent the northernmost extension of the (unexposed) metamorphic basement of the Arabian Margin to the south (Yazgan 1984; Yazgan & Chessex 1991). However, most workers regard these metamorphic rocks as one, or several, continental fragments rifted from Gondwana in the Triassic, associated with opening of a southerly Neotethyan oceanic basin (Hall 1976; Şengör & Yılmaz 1981; Robertson & Dixon 1984; Yılmaz 1993).

Ophiolite and arc units. A second major unit north of the SAFZ is made up of remnants of mainly Upper Cretaceous ophiolitic- and arc-related units, known by different names in different areas (e.g. Yüksekova Arc, Baskil Arc, Guleman Ophiolite, İspendere Ophiolite). Exposures are most extensive in the area north of, and between, the Bitlis and Pütürge metamorphic Massifs (Fig. 12). One of the largest of these ophiolitic units is the Guleman Ophiolite, which, however, lacks an intact volcanic–sedimentary cover. Adjacent to this, and probably contiguous with it, is the large İspendere Ophiolite, which includes Upper Cretaceous arc-type intrusives (Yazgan 1984; Fig. 12). In addition, small slices of ophiolitic rocks are found structurally underlying the Bitlis and Pütürge metamorphic Massifs, and similar ophiolite slices are thrust-intercalated with a Lower Tertiary volcanic–sedimentary unit, known as the Maden Complex (Aktaş & Robertson 1984; Fig. 13.1b and 13.2; see below).

Volcanic–sedimentary units. A third major unit is the Maden Complex, or Maden Group. The Maden Complex is a composite tectonostratigraphic unit that was defined by Aktaş & Robertson (1984, 1990*a*, *b*), based on study of the type area (near Maden town; Fig. 12), as 'Upper Mesozoic ophiolitic rocks and Tertiary sediments, tectonic mélange and mafic volcanics . . .'. The term has been used for a variety of volcanic–sedimentary successions of mainly

Early Tertiary age (locally with Upper Cretaceous inclusions) in different areas and structural positions. Some of these are intact stratigraphic units, for which Maden Group is the preferred term, whereas others units are highly dismembered (mélange) for which the term Maden Complex is appropriate.

Along the northern margin of the Pütürge Metamorphic Massif, a deformed, but originally coherent, volcanic–sedimentary unit of Middle Eocene age is exposed (Maden Group; Hempton 1984; Fig. 13.1b and 13.2). Further east, relatively undeformed sediments of Maastrichtian–Middle Eocene age unconformably overlie the Upper Cretaceous Guleman Ophiolite and related units in the area between the Bitlis and Pütürge Metamorphic Massifs [Hazar Group of Aktaş & Robertson (1990a, b); Fig. 13.1A]. These latter sediments can be interpreted, in part, as proximal facies equivalents of Middle Eocene, mainly deeper water, facies described by Hempton (1984).

In the north, the northern margin of the Pütürge Metamorphic Massif is transgressed by shallow-water limestones (Maden Group; Fig. 13.2) that pass upwards into an intensely thrust-imbricated and sheared unit of deep-sea argillaceous and volcanogenic sediment (Hempton 1984). Further west, in the Kahraman Maraş area (Fig. 12), Yılmaz (1993) reported similar Eocene lithologies depositionally overlying Upper Cretaceous ophiolitic rocks in both southerly and northerly areas. (However, on his published cross-section the contact is shown as tectonic and the unit does not appear on the maps given, e.g. fig. 2a). Yılmaz (1993) used evidence that both the ophiolitic and metamorphic units are unconformably overlain by similar Middle Eocene sediments to argue that both units were structurally juxtaposed prior to the Middle Eocene.

Yılmaz (1993) also reported a small (c. 5 km wide) tectonic window of deep-sea, radiolarian and volcanogenic sediments of Middle Eocene age beneath a large thrust sheet of metamorphosed ophiolite in the westerly Kahraman Maraş area (Fig. 13.3). This relationship is crucial as it was taken to imply that the metamorphic units were emplaced over the Maden Group sediments after deposition in the Middle Eocene.

In the type area (i.e. near Maden town; Fig. 12), the Maden Complex is a c. 17 km thick imbricate slice complex, made up of thrust sheets and mélange composed of Upper Cretaceous ophiolitic rocks (basalts, intrusives and serpentinite), intercalated with deep-sea argillaceous, calcareous and volcanogenic sediments (Fig. 13.1a). The oldest recorded age of syntectonic sediments is earliest Late Palaeocene (Aktaş & Robertson 1990a, b). This unit is inferred to have formed originally further south than the coherent Eocene successions of the Maden Group mentioned above.

Eocene volcanic rocks. A fourth important unit is known structurally beneath the Bitlis Metamorphic Massif, both in the west (Karadere area) and in the east (Siirt area; Fig. 12). This unit is a substantial thrust sheet (up to 2 km thick), composed of subalkaline mafic volcanic rocks interbedded with terrigenous turbidites and depositionally overlain by pelagic carbonates of Middle Eocene age (Karadere Formation) (Aktaş & Robertson 1984; Fig. 13.1b). This unit was included within the Maden Complex by Aktaş & Robertson (1984), but is better recognized as a separate tectonic unit, termed the Karadere Unit, in view of its distinctive features.

The volcanic rocks of the Karadere Unit vary considerably within a single thrust sheet and include high Ti and Al levels, and evolved basalts (Aktaş & Robertson 1990a, b). Trace element geochemical data reveal a subduction zone imprint, but also imply the existence of an associated enriched source to explain the diverse lava types. A possible setting for eruption of these volcanic rocks is an above-subduction zone marginal basin, that included seamounts, similar to the Bransfield Strait in Antarctica (Weaver *et al.* 1977). However, such a setting might be discounted in view of the absence of coeval arc volcanic rocks or tuffaceous sediments within the Karadere Unit. Aktaş & Robertson (1990a) instead favoured formation in a transtensional pull-apart basin setting during the initial stages of collision of Arabian and Eurasian units. The subduction influence was seen as inherited from earlier Late Cretaceous subduction in the area. The pull-apart basin would have developed during oblique convergence of the African and Eurasian Plates, prior to complete collision. If such strike-slip took place, this would have implications for the entire northern Neotethyan margin at this time.

Disrupted thrust sheets and mélange. A final important unit is a zone of imbricate thrust sheets and mélange located at the lowest structural levels, above the Miocene Arabian margin (Lice Basin). Highly dismembered ophiolitic rocks and tectonic mélange predominate, with an estimated structural thickness of 25 km (Killan Unit). This mélange includes highly

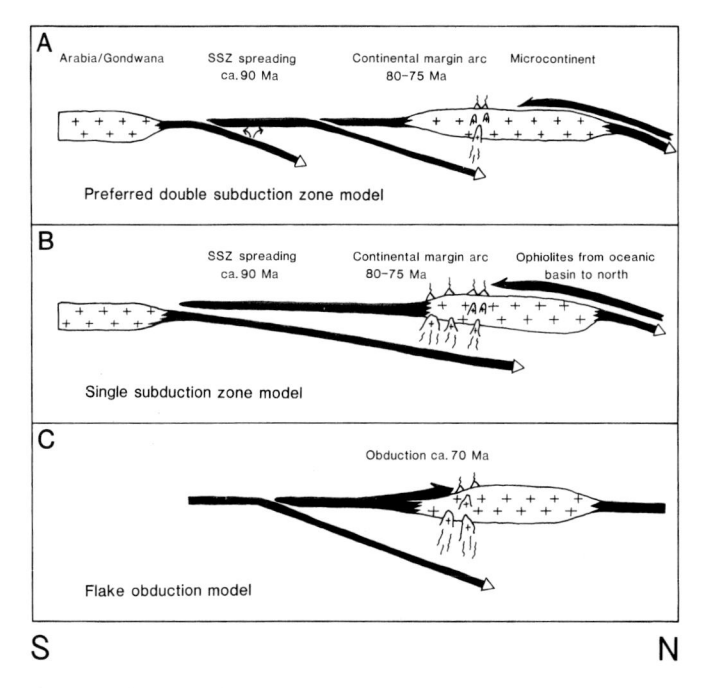

Fig. 14. Alternate plate tectonic models for genesis of Upper Cretaceous ophiolites. (**a**) Spreading above two subduction zones [i.e. an intra-oceanic one in the south and a continental margin one in the north (favoured model)]. (**b**) A single subduction zone. The main problems are that arc volcanism remained active after collision and little space remained for Tertiary subduction. (**c**) Emplacement of ophiolite from the south as an obducted flake. Not believed to be a viable mechanism.

dismembered equivalents of structurally overlying units within the type area of the Maden Complex (Aktaş & Robertson 1984). At two localities, ophiolitic basalt blocks contain radiolarians of Late Cretaceous age (Aktaş 1985; Fig. 13.1a and b).

Yılmaz (1993) reported an additional important unit of Eocene age at a similar structural level in the west (near Kahraman Maraş; i.e. the Halete Volcanics; Fig. 13.3). This unit forms a laterally continuous (> 20 km) thrust sheet, up to several kilometres thick, that is tectonically intercalated with chaotic Upper Eocene–Oligocene sediments below and an ophiolitic unit above, within an overall 'zone of imbrication'. The Halete Unit is composed of Middle–Upper Eocene andesite and associated pyroclastic rocks, with interbeds of reef limestone near the top of the succession (Yılmaz 1993). Eocene volcanics were also noted by Karig & Kozlu (1990) within their Misis–Andırın Unit. The overlying Miocene basin was then extensional in their view (Fig. 13.4).

Yılmaz (1993) interpreted the Halete Volcanic Unit as an Eocene volcanic arc related to the later stages of northward subduction of the Neotethyan ocean, *before* collision and final thrusting over the Arabian Margin along a South Anatolian basal thrust. This is important, as it requires that the southerly Neotethys remained sufficiently wide so as to fuel subduction and back-arc rifting in the Eocene. However, this hypothesis is difficult to evaluate as no geochemical data are given for the volcanic rocks. The Halete volcanic rocks could be equivalent to the Middle Eocene Karadere Unit volcanics further east, or possibly to younger (Palaeocene–Eocene) volcanic rocks in the Kyrenia Range of Cyprus (Baroz 1979; Robertson & Woodcock 1986). Geochemical evidence from both of these coeval units does not support a simple subduction-related origin.

Discussion: alternative tectonic models

The tectonostratigraphy of southeastern Turkey is now fairly well known, but controversies persist, particularly concerning the tectonic settings of formation and structural relationships of individual units. Alternative tectonic models are shown in Fig. 14.

The Eastern Bitlis Massif includes blueschists

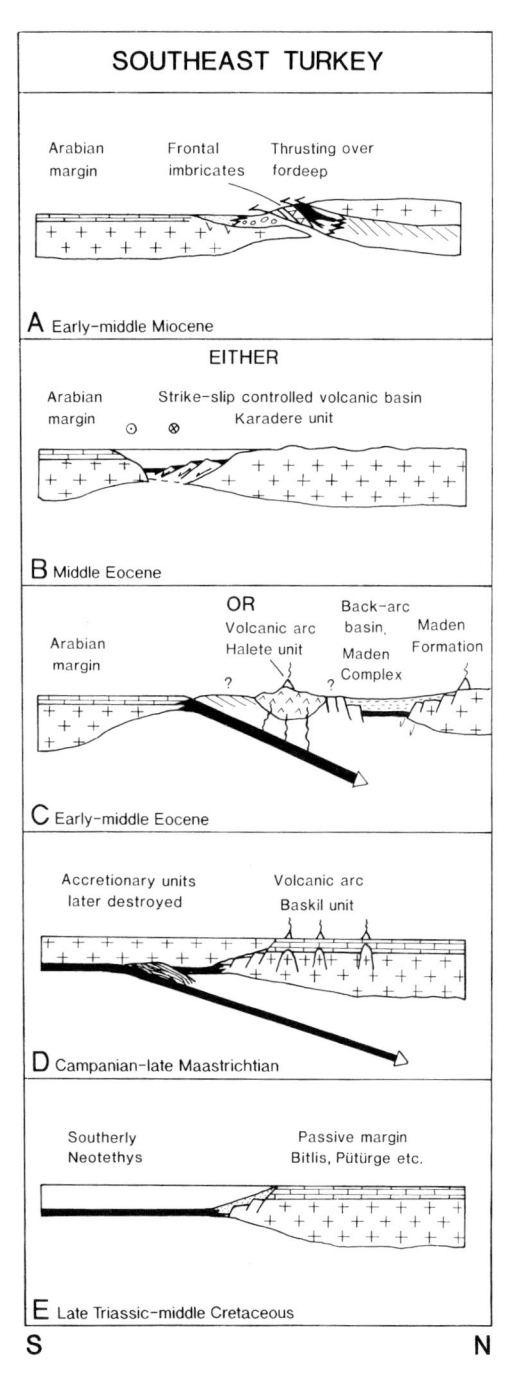

Fig. 15. Tectonic models for southeastern Turkey. This area formed parts of northern passive margin of Neotethys in the easternmost Mediterranean until Late Cretaceous northward subduction, which resulted in ophiolite genesis, accretion and emplacement. Convergence continued in the Early Tertiary. The Upper Palaeocene–Eocene Maden Complex (as defined here) formed either as a back-arc basin (Yılmaz 1993) (see H) or as a transtensional pull-apart basin (Aktaş & Robertson 1990a, b). The latter is preferred as it is likely that diachronous collision with Arabia was in progress by Early–Mid Eocene time and there is little field evidence of the 'Halete Volcanic Arc Unit'. The allocthon was emplaced over the Arabian Platform in Miocene time, marking the final collisional phase. See text for further explanation.

and is structurally underlain by an unmetamorphosed 'coloured mélange', including ophiolitic rocks and Upper Cretaceous radiolarites. Hall (1976) proposed that formation of blueschists was related to a northward dipping subduction zone located within a Neotethyan oceanic basin adjacent to the northern margin of Arabia. Based on extensive field mapping of the Pütürge and Bitlis Massifs further west, Perinçek (1979) identified evidence of Triassic metavolcanic rocks that could relate to rifting of the North African Margin. He also gave one of the first detailed interpretations of the Maden Complex, which he interpreted as an Early Tertiary back-arc basin, developed above a southward dipping subduction zone. This view was incorporated in Şengör & Yılmaz's (1981) early plate tectonic model. However, subsequent mapping compiled by Aktaş & Robertson (1984) and Yazgan (1984) showed conclusively that the structural grain was regionally northward dipping, and it was then generally accepted that any Late Cretaceous–Early Tertiary subduction in the area was northward dipping.

Accepting that any subduction was northward, the next question is whether ophiolitic rocks above the metamorphic massifs in the north (e.g. Guleman and İspendere Ophiolites) could be correlated with smaller slices beneath the Bitlis and Pütürge metamorphic Massifs (to the south), and interpreted as remnants of an originally much larger ophiolite thrust sheet that was emplaced southward over the Arabian Margin in the Late Cretaceous. A widespread early view was that only one Neotethyan ocean basin existed in eastern Turkey (Michard et al. 1984) and, in keeping with this, Yazgan (1984) correlated the northerly and southerly ophiolites, and suggested that they were all derived from north of the metamorphic massifs, which were then seen as regional extensions of the Arabian continental basement.

However, it is now clear that a southerly Neotethyan oceanic basin instead remained open in southeastern Turkey into the Tertiary. Ophiolites were incorporated into both the northern and southern margins of the Southern Neotethyan oceanic basin in the Late Cretaceous. Were the origin and emplacement of both of these northerly and southerly ophiolitic units genetically linked? Aktaş & Robertson (1984) favoured an open-ocean model, with no direct link between the genesis and emplacement of north and south margin ophiolites (Fig. 14a). The Lower Tertiary Maden Complex (Killan Unit of the Maden Complex; Fig. 13.1a) was interpreted as a subduction–accretion complex related to the later stages of closure of the southerly Neotethyan oceanic basin. Limited calc-alkaline volcanics of Eocene age to the north (Baskil Unit) suggested to them that significant subduction took place in the Early Tertiary. The Middle Eocene Karadere Volcanics were then seen as pull-apart basins developed in a pre-collisional setting, associated with oblique subduction.

The following model is proposed (Fig. 15). Ophiolites were emplaced onto the northern margin (i.e. the metamorphic units) of the Neotethyan oceanic basin in the Late Cretaceous. Ophiolites were also emplaced separately onto the Arabian margin, leaving a small Neotethyan oceanic basin surviving into Tertiary time (Fig. 15h). Northward subduction beneath the northern Neotethyan margin then resumed in the Late Palaeocene–Eocene (Killan Unit of the Maden Complex), backed by an unstable forearc basin (Hazar Group). A small extensional, volcanic-floored basin (Karadere Unit; Fig. 15g), opened briefly in the Middle Eocene, along the northerly active continental margin of the remaining southerly Neotethyan ocean basin. Collision of northerly units with the Arabian margin in eastern Turkey began in the Late Eocene. The Middle Eocene basin then collapsed and was overthrust by metamorphic units. The collision zone migrated westwards along the Arabian margin during the Oligocene–Miocene as remaining Neotethyan oceanic crust was consumed. The assembled stack was finally sutured with the Arabian margin (in Eastern Turkey) in the Late Miocene, preceded by formation of a Miocene flexural basin (Lice Formation; Fig. 15f).

Discussion: palaeotectonic evolution

When additional information from adjacent area, including Cyprus, the Levant and North Africa and intervening marine areas are taken in account, the plate tectonic evolution of the easternmost Mediterranean region can be summarized as in Figs 16–22 (see Robertson 1994, 1998). The reader is referred to earlier sections for specific data and literature that support interpretations as only references to previous syntheses are given below.

Late Permian–Late Triassic

Prior to Late Permian time, the easternmost Mediterranean area formed part of the northern margin of Gondwana. Remnants of this setting are preserved, for example, in the Antalya Complex. The first evidence of rifting is the development of regional highs that probably reflect rift-shoulder flexural uplift (e.g. Hazro in southeastern Turkey). Rifting began by the Late Permian, resulting in widespread marine transgression (e.g. Kantara Limestone in the Kyrenia Range, Cyprus). Further rifting took place in the Early Triassic, with the first evidence of widespread volcanism, recorded as tuffaceous sediments in the Antalya Complex (Fig. 16).

Final break-up took place in the Middle Triassic–Late Triassic time, marked by the incoming of deeper water radiolarian sediments, as exposed in the Antalya Complex, within the Alanya Massif window. Ocean-floor spreading was in progress during the Late Triassic (Carnian–Norian) when transitional to MORB-type volcanic rocks were erupted, as in the Antalya Complex. Associated terrigenous sediments show that the crust of the Antalya Complex formed near the rifted continental margin and cannot represent open-ocean Neotethyan lithosphere.

The Levant Margin represents either an orthogonally rifted or a transform passive margin segment. Oblique opening is depicted in Fig. 16. Crust north of the present latitude of 32°30′ is of MORB-type, while that to the south is probably thinned continental crust, or 'transitional' crust (e.g. including magmatic rocks intruded into stretched continental crust). The Eratosthenes Seamount is interpreted as a continental fragment, associated with mafic igneous rocks, that was rifted from the Levant continental margin in the Triassic (see Garfunkel 1998).

On a regional scale, large continental fragments were rifted from the north margin of Gondwana to form the northern margin of the (southerly) Neotethyan ocean basin. Carbonate platforms were established on these units from the Late Permian–Triassic onwards. Doubt persists as to which of these formed single microcontinental units as opposed to smaller fragments. However, in the west, the Bey Dağları carbonate Platform and the Menderes Massif are seen as a single palaeogeographic unit (Menderes–Bey Dağları microcontinent),

LATE TRIASSIC - LOWER JURASSIC

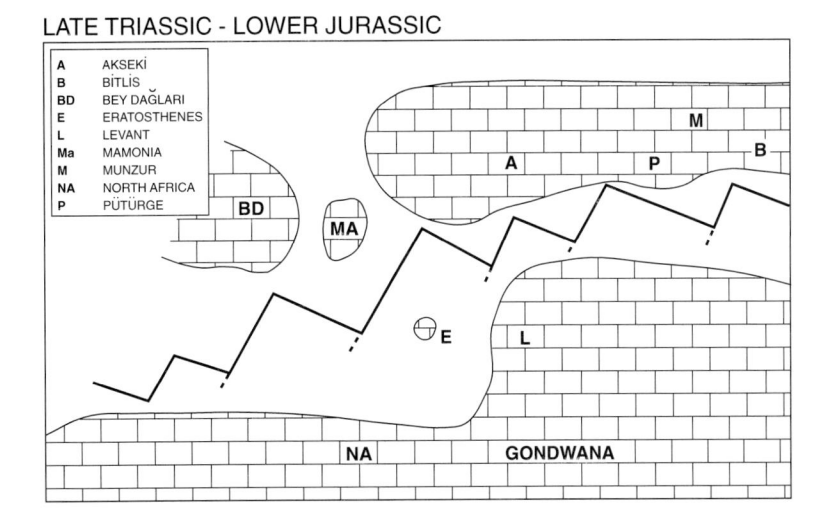

A	AKSEKİ
B	BİTLİS
BD	BEY DAĞLARI
E	ERATOSTHENES
L	LEVANT
Ma	MAMONIA
M	MUNZUR
NA	NORTH AFRICA
P	PÜTÜRGE

Fig. 16. Plate tectonic reconstruction for the Late Triassic–Early Jurassic. The Levantine oceanic basin mainly formed in the Late Triassic. The main uncertainty is over the identity of continental fragments in the east. The initial fit develops those proposed by Robertson & Woodcock (1980) and by Garfunkel & Derin (1984). The palaeolatitudes of North Africa in this and subsequent reconstructions are based on Dercourt *et al.* (1992). The palaeolatitude of the northern margin is poorly constrained.

separated from one, or several, fragments further east by a Neotethyan oceanic basin (represented by the Antalya Complex). These latter units now form the Taurus Mountains of southern Turkey (e.g. Akseki and Bolkar Dağ Units). Rifting and spreading to form more a northerly Neotethyan basin also took place in the Triassic, giving rise to the 'Northerly Neotethyan ocean' or 'Inner Tauride ocean', that is now represented by the Lycian Nappes and the Beyşehir–Hoyran–Hadim Nappes (Şengör & Yılmaz 1981; Robertson & Dixon 1984; Robertson *et al.* 1996).

Jurassic–Early Cretaceous

Both the northern and southerly margins of the southerly Neotethyan oceanic basin experienced passive margin subsidence during Jurassic–Early Cretaceous time. This is recorded, for example, by the Taurus carbonate platforms in the north. Deep-water passive margin successions accumulated around the margins of rifted continental fragments along the northern margin of the southerly Neotethys, and were later deformed and incorporated into allochthonous units, as seen in the Antalya Complex. No Jurassic oceanic crust is preserved. However, MORB-type volcanics of Early Cretaceous age are recognized in northeast and southwest areas of the Antalya Complex. Renewed ocean-floor spreading probably took place within the southerly Neotethyan ocean basin in Late Jurassic–Early Cretaceous time, but an oceanic basin already existed there since the Triassic. Continuous spreading from the Triassic to Late Cretaceous is, however, unlikely, as the southerly Neotethyan oceanic basin remained too small to be detected using palaeomagnetic methods (Morris 1996). Most of the oceanic crust was later subducted.

Late Cretaceous: Cenomanian–Turonian (Fig. 17)

From the late Early Cretaceous, the African and Eurasian Plate began to converge as the South Atlantic opened (Livermore & Smith 1984; Savostin *et al.* 1986). This convergence initiated a northward dipping subduction zone within the southerly Neotethyan oceanic basin. The cold, dense Triassic oceanic slab 'rolled back', allowing asthenosphere to well up and create suprasubduction zone ophiolites, *c.* 150 km wide, from west to east. These include the Tekirova and Hatay Ophiolites. From the Cenomanian onwards, the Tauride carbonate platforms to the north subsided and were overlain by pelagic carbonate. A possible explanation is that the platforms were subjected to regional crustal extension as the pre-existing oceanic plate began to 'roll back' to the south. Within the intra-oceanic subduction zone, deep-sea sediments and MORB-type volcanics were thrust

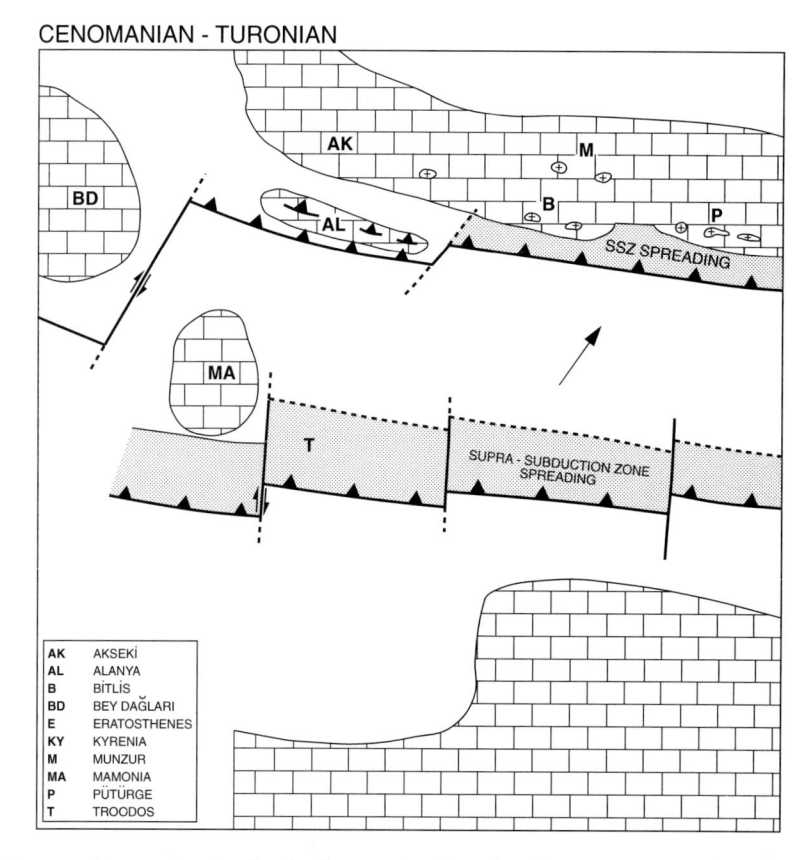

Fig. 17. Plate tectonic reconstruction for the Cenomanian–Turonian. The easternmost Mediterranean oceanic basin is closing through the activity of two northward dipping subduction zones, one intra-oceanic in the south and the other, a continental margin (i.e. Andean type), in the north. Above-subduction zone spreading occurs above both of these subduction zones. The Troodos Complex forms above the southerly subduction zone. Arc volcanism develops extensively in southeast Turkey, related to the northerly subduction zone.

beneath still hot Upper Cretaceous ophiolitic rocks to form dynamothermal metamorphic soles, as seen beneath the Hatay and Mersin Ophiolites.

Late Cretaceous: Turonian–Late Campanian

As northward intra-oceanic subduction continued, the trench neared the Arabian passive margin in the east, resulting in initial depositional hiatuses. Deep-water sediments, oceanic volcanics and carbonate build-ups were then accreted to the hanging wall of the trench to form subduction–accretion complexes, followed by underplating of deep-water passive margin and oceanic sediments and volcanics. The trench then collided with the Arabian passive margin, collapsing it to form a foredeep, and the entire assemblage was thrust southward to its final position, including the Koçali Ophiolite in southeastern Turkey. The ophiolites were briefly subaerially exposed, overlain by fluvial sediments, then transgressed by marine carbonates.

Late Campanian–Maastrichtian (Fig. 18)

The trench collided with the passive margin of Arabia in Late Campanian time (c. 70 Ma). However, the westward extension of the intra-oceanic subduction zone and its overriding suprasubduction zone ophiolite (i.e. Troodos Ophiolite) remained within a remnant of the Neotethyan oceanic basin to the west. As Africa–Eurasia convergence continued, this relict trench gradually pivoted counterclockwise, explaining the palaeorotation of a 'Troodos microplate' (Robertson 1990). Regional

LATE CAMPANIAN - MAASTRICHTIAN

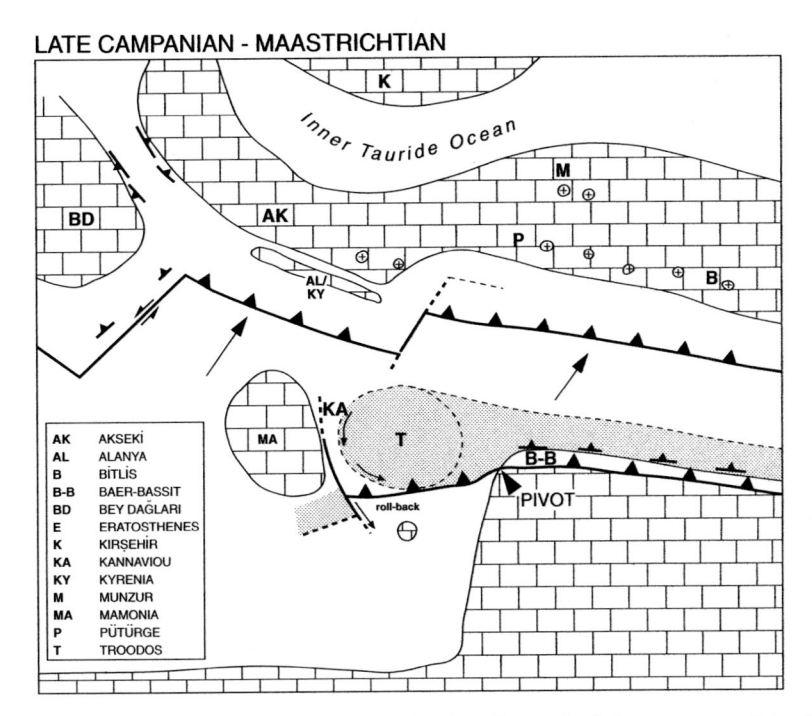

AK	AKSEKİ
AL	ALANYA
B	BİTLİS
B-B	BAER-BASSIT
BD	BEY DAĞLARI
E	ERATOSTHENES
K	KIRŞEHİR
KA	KANNAVIOU
KY	KYRENIA
M	MUNZUR
MA	MAMONIA
P	PÜTÜRGE
T	TROODOS

Fig. 18. Reconstruction for the Late Campanian–Maastrichtian. The southerly intra-oceanic subduction zone collides with the Arabian margin, emplacing ophiolites; whereas, the Troodos suprasubduction zone lithosphere remains in the ocean and progressively rotates counterclockwise as a microplate from Campanian to Early Eocene time. Northward subduction and arc volcanism continues along the more northerly active margin.

subduction was impeded by collision of the intra-oceanic subduction zone with the Arabian margin. Continued convergence of the African and Eurasia Plates generated an additional subduction zone along the northern margin of the Neotethyan ocean basin. Evidence for this northerly Andean-type arc is most widely developed in southeastern Turkey, as shown by extensive arc-type intrusive and extrusive activity (Baskil Arc Unit: 85–76 Ma.). The intrusions stitch both continental fragments (Keban, Bitlis and Pütürge Units) and ophiolitic crust (e.g. İspendere Ophiolite). This suggests that a strip of oceanic crust was isolated behind the subduction zone and preserved intact, whereas most of the Neotethyan oceanic crust was subducted.

During the Upper Cretaceous, the deep-water passive margin units adjacent to the rifted microcontinents units, as seen within the Taurus Mountains (e.g. Antalya Complex), were first deformed. In the case of the Antalya Complex, the cause was the collapse of small oceanic basins or broad rifts that separated a number of the carbonate platforms during the Mesozoic (e.g. Isparta Angle; Güzelsu Unit; a high-pressure unit of the Alanya Massif). A possible explanation is that this compression relates to southward emplacement of ophiolites and related units from the 'Inner Tauride ocean' onto the northern margin of the Taurus carbonate platforms (the Beyşehir–Hoyran–Hadim Nappes, which were rooted to the north). Further east, the Kyrenia–Misis–Andırın Lineament documents the history of the northern margin of the southern Neotethyan oceanic basin. In southeastern Turkey an Andean-type continental margin arc developed in Campanian–Maastrichtian time from 85–76 Ma.

Palaeocene–Early Eocene

Africa–Eurasia relative plate convergence may have paused during Early Tertiary time (70–48 Ma). There is no known record of subduction-related volcanism in southeastern Turkey from the Maastrichtian to the Early Eocene. However, Mesozoic oceanic crust still remained, both to the north, adjacent to the southern margin of Eurasia (Şengör & Yılmaz 1981; Robertson & Dixon 1984) and within the southern Neotethyan oceanic basin in the easternmost Mediterranean area. During the Eocene, convergence resumed and remaining

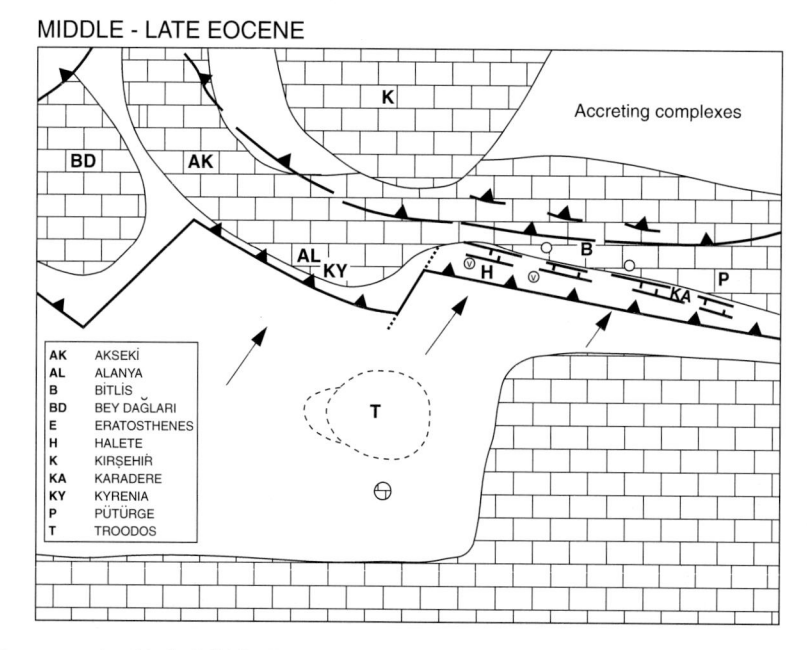

Fig. 19. Reconstruction for the Middle–Late Eocene. Subduction resumes along a partly amalgamated northern margin, together with limited arc volcanism. Middle Eocene volcanics in southeast Turkey either formed in a continental margin arc setting (i.e. Halete Volcanics) or in a transtensional pull-apart basin setting (i.e. Karadere Volcanics) or conceivably in both settings. The Isparta Angle Neotethyan oceanic basin is by then closed in the north but remains open in the south.

northerly Neotethyan oceanic crust (i.e. north of the Taurus carbonate platform) was subducted, resulting in diachronous collision. The deformation front then migrated progressively southward.

Late Palaeocene–Early Eocene time saw final closure of the Isparta Angle seaway, with expulsion of Mesozoic deep-sea and ophiolitic units, both eastward over the eastern limb of the Isparta Angle and westward over the eastern limb of the Isparta Angle. At this stage, the Isparta Angle was still a much more open feature than today.

Middle–Late Eocene (Fig. 19)

Northward subduction in southeastern Turkey resumed in latest Palaeocene time, giving rise to an accretionary prism (Killan Unit of the Maden Complex), together with limited Middle Eocene calc-alkaline volcanism further north (Baskil Unit). A deep-water volcanic-floored basin opened in the Middle Eocene (Karadere Unit) in an extensional, or strike-slip dominated, setting. Collision ensued from the Late Eocene onwards, tightening the suture zone, and the collision zone migrated westward toward the easternmost Mediterranean, where remnants of Neotethyan oceanic crust still remained. In the Kyrenia Range, northward subduction gave rise to an accretionary prism during Middle–Late Eocene time (Kalograi–Ardana 'Olistostromes'), including large blocks of Upper Permian shallow-water limestone (Kantara Limestone).

The Late Eocene was the time of final closure of the 'Inner Tauride ocean' to the north of the Tauride carbonate platforms (Görür *et al.* 1984; Robertson & Dixon 1984). As a result, the Beyşehir–Hoyran–Hadim Nappes were thrust southward over the Tauride carbonate platform to their most southerly locations. This southward thrusting deformed the northern margin of the remaining southern Neotethyan oceanic basin and triggered southward migration of the deformation front toward the remaining southern Neotethyan oceanic basin. In the west, the Beyşehir–Hoyran–Hadim Nappes were thrust southwestward, rethrusting the northeastern segment of the Antalya Complex. In response to regional north–south compression, the Antalya Complex (Güzelsu Unit) and the Alanya Massif were finally thrust northward over the neighbouring carbonate platform (Akseki Platform). The Kyrenia Range was thrust southward over a Middle

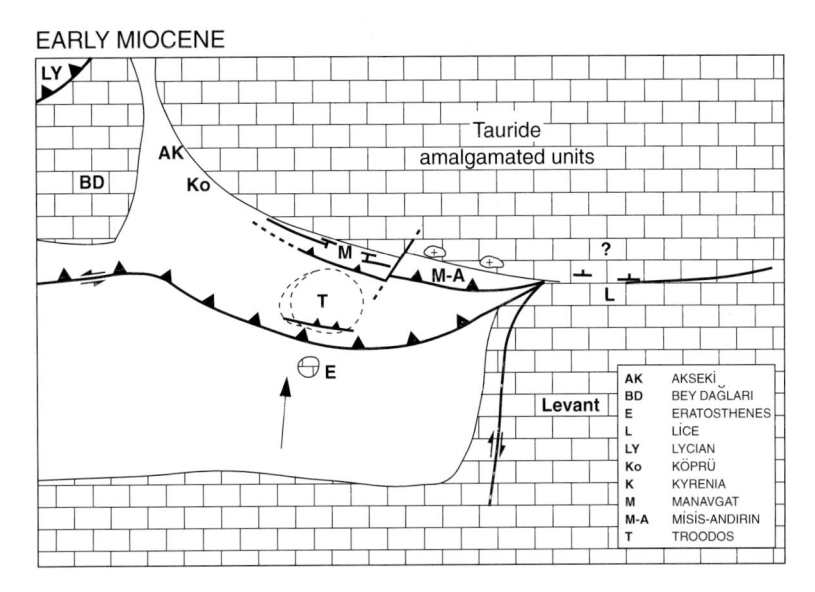

Fig. 20. Reconstruction for the Early Miocene. Final, localized subduction along northern margin, followed by a southward jump of the subduction zone to near its present relative location. Localized compression is related to establishment of a new subduction zone south of Cyprus. Thrusting occurs over the Arabian margin in the east with development of a flexural foredeep. The Lycian Nappes advance from the northwest, associated with inferred strike-slip and extensional basin formation, in the Isparta Angle area. Extension and subsidence take place along the southern margin of this northern collage.

Eocene inferred, trench–accretionary unit (Kalograi–Ardana Unit), as a stack of dismembered thrust sheets (Baroz 1979; Robertson & Woodcock 1986). In southeastern Turkey, the Middle Eocene deep-water volcanic-floored basin then collapsed and was overthrust by metamorphic massifs representing the continental margin to the north. In the northern Syria Platform, carbonate sedimentation experienced depositional hiatuses in the Middle Eocene, culminating in a major regional unconformity, followed by renewed deposition in the Early Miocene (Krasheninikov 1994). The Late Eocene was marked by a phase of inversion of the Syrian Arc in the Levant, suggesting that compression was translated from the Eurasian Plate to the Arabian Plate during that time, in response to progressive suturing of Neotethys. By contrast, the Troodos Ophiolite still lay to the south of the deformation front and deep-water pelagic carbonates (Lefkara Formation) continued to accumulate during this time interval (Robertson & Hudson 1974).

Oligocene

The Early Oligocene was a time of relative tectonic quiescence throughout the easternmost Mediterranean area, although faulting took place as Africa–Eurasia convergence continued. In the Late Oligocene, a marked change took place, manifested in the Kyrenia–Misis–Andırın Lineament. The onshore Adana Basin subsided, ushering in marine sedimentation influenced by block faulting. Sedimentary mélange of Late Oligocene–Early Miocene age (İsalı Unit) formed relatively further south, in the Misis–Andırın Complex, within an extensional forearc setting. The clasts were mainly derived from earlier emplaced units to the north (e.g. Mesozoic limestone, ophiolites etc.). Limited acidic, calc-alkaline, arc-related volcanism (represented by tuffs) perhaps took place at this time. Ophiolitic blocks in the mélange were reworked by gravitational processes from previously emplaced (?Late Cretaceous) units. The 'olistostrome' is interpreted as a trench–accretionary complex related to northward subduction of Neotethyan oceanic crust to the south.

Miocene (Figs 20 and 21)

Early Miocene time saw activity along the present subduction zone south of Cyprus, as part of the wider regional Africa–Eurasian Plate boundary, including the Aegean active margin (Aegean Arc). A possible explanation for southward migration of the subduction zone

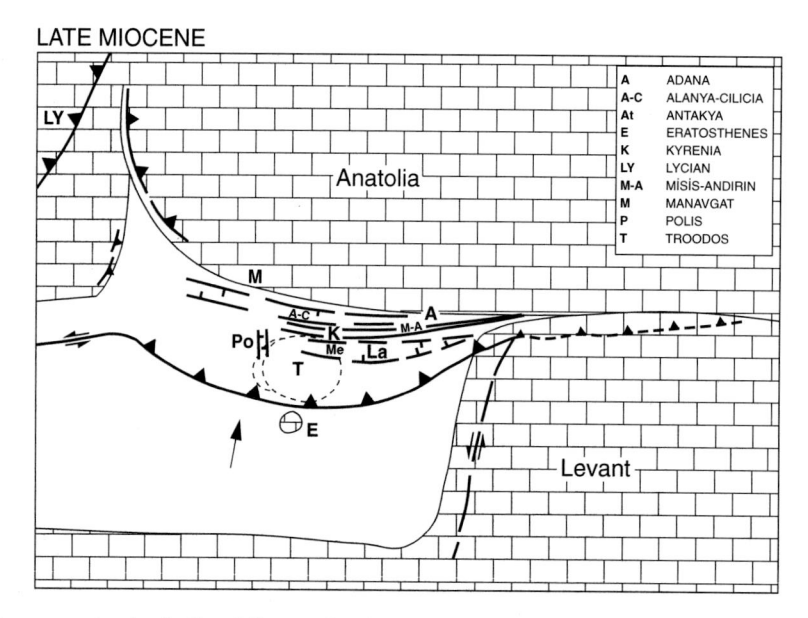

Fig. 21. Reconstruction for the Late Miocene. Southward thrust emplacement is completed in southeast Turkey. Strike-slip along the Dead Sea Transform is active; subduction continues in the Levantine Sea, with 'roll-back' giving rise to extensional basins in the overriding plate. During the Late Miocene, the Lycian Nappes are finally emplaced, associated with compression of the Isparta Angle area, and localized thrusting in the southwestern and eastern segments of the Antalya Complex.

is that remaining old (i.e. Early Mesozoic) Neotethyan crust was consumed, juxtaposing younger, more buoyant suprasubduction zone-type Upper Cretaceous oceanic crust with the active margin to the north. This was difficult to consume and the subduction zone then relocated itself within Lower Mesozoic oceanic crust further south.

During the Miocene, a division clearly existed between a pre-collisional setting in the easternmost Mediterranean (west of the longitude of the Levant Margin) and a collisional setting in southeastern Turkey. In the east, the Miocene Lice Basin originated as a foreland basin related to the final stages of southward thrusting of the overriding allochthonous eastern Taurus units. A foreland basin origin is less clear further west, and Karig & Kozlu's (1990) transtensional basin origin needs to be tested. However, it seems clear, on regional plate kinematic grounds (Livermore & Smith 1984; Savostin et al. 1986), that plate convergence was continuing during the Miocene and thus a convergent margin-type setting is probable.

Pliocene (Fig. 22)

Within the Aksu Basin, an extensional basin was infilled with a shallowing-upwards, shallow-marine to deltaic succession. Sedimentation in the Adana Basin was entirely continental by this time. Regional uplift took place in Late Pliocene–Early Pleistocene time, associated with further extensional faulting (Aksu Basin) and transtension (Adana Basin).

In southeastern Turkey, left-lateral strike-slip along the East Anatolian Fault (Hempton 1982) began in the latest Miocene (after suturing of the Arabian and Anatolian Plates) with a total displacement of < 25 km and was accompanied by more pronounced left-lateral motion along the North Anatolian Transform Fault (Barka & Hancock 1984; Dewey et al. 1986). The East Anatolian Fault Zone links eastward with the triple junction in the Kahraman Maraş area (Gözübol & Gürpınar 1980; Westaway & Arger 1996). Deformation in the Misis Mountains is mainly extensional in the Pliocene–Pleistocene. Westward escape of Anatolia during Pliocene–Pleistocene time was accompanied by a re-orientation of subduction from northward to more northeastward, consistent with the instantaneous Global Positioning System (GPS) data for southern Turkey (Oral et al. 1994; Toksöz et al. 1995).

Pleistocene–Holocene

Existing plate boundaries were active during this time. Northeastward subduction was active

LATE PLIOCENE - QUATERNARY

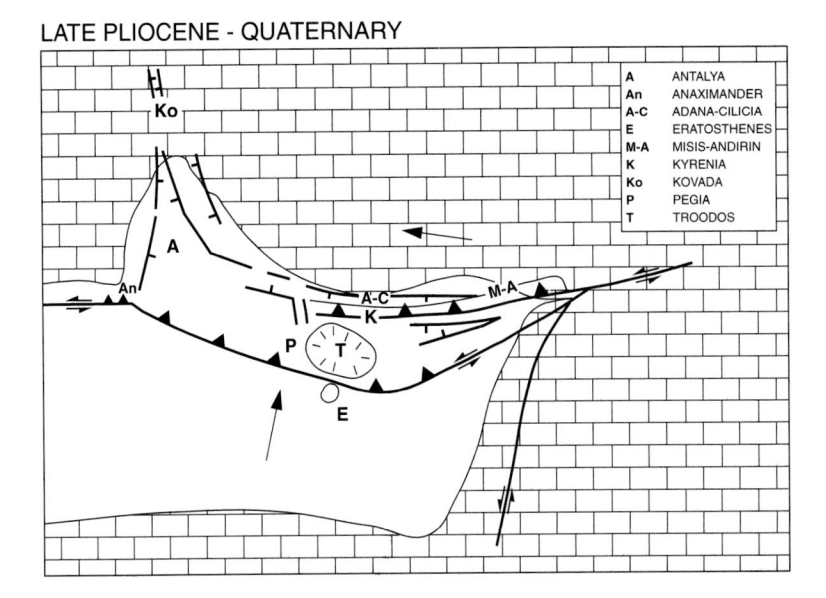

Fig. 22. Reconstruction for Late Pliocene–Pleistocene time. Following collision of Arabia with Eurasia, Anatolia underwent westward tectonic escape. The inferred direction of subduction in the Levantine Sea changes to the northeast. The Eratosthenes Seamount is being thrust beneath Cyprus. The Troodos Massif is strongly updomed in response to collision and related serpentinite diapirism. Extensional basinal formation continues in other areas of the overriding plate and locally onshore in the Aksu Basin (southwestern Turkey).

beneath southwest Cyprus and the southern Antalya Bay area. Crustal extension continued onshore in the Antalya Basin. The southern margin of the offshore Anaximander Mountains is seen as a left-lateral strike-slip zone that links with the Strabo Trench to the west. The active margin runs off south to the area where the Eratosthenes Seamount is in the process of active collision. The eastward extension of the Cyprus Trench is a zone of left-lateral strike-slip (Kempler & Garfunkel 1994), passing into a zone of distributed strike-slip, transtension, and transpression that links with the Levant Margin in area north of Latakia (Al-Rayami *et al.* 2000), extending to the Dead Sea Transform Fault, that in turn links northwards to the East Anatolian Fault Zone where left-lateral strike-slip continues. This simple northward linkage may, however, not be active today (Butler *et al.* 1997).

Conclusions

A review of the modern literature concerning the geology of southern Turkey indicates that it is dominated by evidence of the genesis and emplacement of a Neotethyan oceanic basin. This oceanic basin initially rifted in the Late Permian–Early Triassic, followed by opening of an oceanic basin in Middle–Late Triassic time. This basin was bordered by subsiding passive margins throughout Mesozoic time. The southerly ocean basin and its northerly margins were tectonically disrupted in latest Cretaceous time, but final ocean basin closure did not take place until Early Tertiary time. During the Mesozoic, the Taurus Mountains formed several microcontinents bordered by subsiding passive margins. Notably, in the west, the Isparta Angle existed as a small oceanic strand separating carbonate platforms to the west (Bey Dağları) and east (e.g. Central Taurus). In addition, a number of structurally higher allochthonous units, including Upper Cretaceous ophiolites (the Lycian Nappes and the Hoyran–Beyşehir–Hadim Nappes) were derived from a separate Neotethyan oceanic basin located to the north of the Taurus carbonate platforms.

The remnants of a southerly Neotethyan oceanic basin (e.g. Antalya Complex) have their counterparts elsewhere in the eastern Mediterranean region, including the Mamonia Complex of southwestern Cyprus, and the ophiolites and underlying mélange of the Baer–Bassit region in northern Syria (Al-Riyami *et al.* 2000). These units are discussed elsewhere (e.g. Robertson 1998). In general, the northerly margins of the southerly Neotethyan basin were tectonically active from the latest Cretaceous onwards, whereas the southern margins, as in the Levant and North Africa, remained as southerly passive margins until recent time.

The southerly Neotethyan oceanic basin extended eastwards through sutured units in southern Iran (Zagros) to connect with Oman, where important Late Cretaceous emplacement of ophiolitic and passive margin units also took place (Glennie *et al.* 1990). In addition, the remaining non-emplaced part of the southerly Neotethys in the Eastern Mediterranean links westwards beneath the Herodotus abyssal plain to the Ionian Sea, which preserves a further remnant of the Early Mesozoic Neotethys adjacent to North African.

Causes of the Triassic opening of a southerly Neotethyan oceanic basin are still under debate and include slab-pull related to northward subduction along the Eurasian margin (Robertson & Dixon 1984; Stampfli *et al.* 2000*a, b*), back-arc rifting above a southward subduction zone located within Palaeotethys to the north (Şengör & Yılmaz 1981), or activity of several possible plumes located along the northern Gondwana margin in Permo-Triassic time (Dixon & Robertson 1999).

References

AKAY, E., UYSAL, S., POISSON, A., CRAVETTE, J. & MÜLLER, C. 1985. Stratigraphy of the Antalya Neogene Basin. *Bulletin of the Geological Society of Turkey*, **28**, 105–119.

AKBULUT, A. 1977. *Etude géologique d'une partie du Taurus occidentale au sud d'Eğridir (Turkquie)*. Thèse, Université Paris-Sud, Orsay.

AKSU, A. E., CALON, T., PIPER, D. W. J., TURĞUT, S. & İZDAR, E. 1992*a*. Architecture of later orogenic Quaternary basins in northeastern Mediterranean Sea. *Tectonophysics*, **210**, 191–213.

——, ULUĞ, A., PIPER, D. W. J., KONUK, Y. T. & TURĞUT, S. 1992*b*. Quaternary sedimentary history of Adana, Cilicia and İskenderun Basins: northeastern Mediterranean Sea. *Marine Geology*, **104**, 55–71.

AKTAŞ, G. 1985. *The Maden Complex, S.E. Turkey: Sedimentation and volcanism along a Neotethyan active continental margin*. PhD Thesis, University of Edinburgh.

—— & ROBERTSON, A. H. F. 1984. The Maden Complex, SE Turkey: evolution of a Neotethyan continental margin. *In*: DIXON, J. E. & ROBERTSON, A. H. F. (eds) *The Geological Evolution of the Eastern Mediterranean*. Geological Society, London, Special Publications, **17**, 375–402.

—— & —— 1990*a*. Tectonic evolution of the Tethys suture zone in SE Turkey: evidence from the petrology and geochemistry of Late Cretaceous and Middle Eocene Extrusives. *In*: MOORES, E. M., PANAYIOTOU, A. & XENOPHONTOS, C. (eds) *Ophiolites – Oceanic Crustal Analogues*. Proceedings of the International Symposium, 'Troodos 1987', Geological Survey Department, Cyprus, 311–329.

—— & —— 1990*b*. Late Cretaceous–Early Tertiary fore-arc tectonics and sedimentation: Maden Complex, SE Turkey. *In*: SAVAŞÇIN, M. Y. & ERONAT, A. H. (eds) *Proceedings of the International Earth Sciences Congress on Aegean Regions*, İzmir, Turkey, **2**, 71–276.

AL-RIYAMI, K., ROBERTSON, A. H. F., XENOPHONTOS, C., DANELIAN, T. & DIXON, J. E. 2000. Mesozoic–Tertiary tectonic and sedimentary evolution of the Arabian continental margin in Baer–Bassit (northwest Syria). *In*: MALPAS, J., XENOPHONOTOS, C. & PANAYIDES, A. (eds) *Proceedings of the 3rd International Conference on the Geology of the Eastern Mediterranean*, Nicosia, Cyprus, September 23–26th, 1998. Geological Survey Department, Cyprus, in press.

ALTINER, D. 1984. Upper Permian foraminiferal biostratigraphy in some localities of the Taurus Belt. *In*: TEKELİ, O. & GÖNCÜOĞLU, M. C. (eds) *Geology of the Taurus Belt*. Proceedings of the International Tauride Symposium. Mineral Research and Exploration Institute of Turkey (MTA) Publications, 255–268.

——, ÖZKAN-ALTINER, S. & KOÇYİĞİT, A. 2000. Late Permian foraminiferal biofacies belts in Turkey: palaeogeographic and tectonic implications. *This volume*.

ANASTASAKIS, G. & KELLING, G. 1991. Tectonic connection of the Hellenic and Cyprus arcs and related geotectonic elements. *Marine Geology*, **97**, 261–277.

ANGELIER, J., LYBERIS, N., LE PICHON, X., BARRIER, E. & HUCHON, P. 1982. The tectonic development of the Hellenic arc and the sea of Crete, a synthesis. *Tectonophysics*, **86**, 159–196.

BARKA, A. A. & HANCOCK, P. L. 1984. Neotectonic deformation patterns in the convex-northwards arc of the North Anatolian fault zone. *In*: DIXON, J. E. & ROBERTSON, A. H. F. (eds) *The Geological Evolution of the Eastern Mediterranean*. Geological Society, London, Special Publications, **17**, 763–774.

BAROZ, F. 1979. *Etude géologique dans le Pentadaktylos et la Mesaoria (Chypre Septentrionale)*. Thèse, Université de Nancy, France.

BEN-AVRAHAM, Z. 1986. Multiple opening and closing of the Eastern Mediterranean and South China Basins. *Tectonics*, **8**, 51–362.

BİNGÖL, A. F. 1990. Petrological features of intrusive rocks of Yüksekova Complex in the Elazığ region (Eastern Taurus – Turkey). *Journal of Fırat University, Science and Technology*, **3**, 1–17.

BUTLER, R. W. H., SPENCER, S. & GRIFFITH, H. M. 1997. Transcurrent fault activity in Lebanon: implications for movement along the Dead Sea fault system. *Tectonics*, **9**, 1369–1386.

BÜRGER, D. 1990. The travertine complex of Antalya Southwest Turkey. *Zeitschrift für Geomorphologie. Neue Forschung, Supplementary Bulletin*, **77**, 25–46.

CAMERLENGHI, A., CITA, M. B., DELLA-VEDOVA, B., FUSI, N., MIRABILE, L. & PELLIS, G. 1995. Geophysical evidence of mud diapirism on the Mediterranean Ridge accretionary complex. *Marine Geophysical Research*, **17**, 115–141.

CHAUMILLON, E. & MASCLE, J. 1995. Variation laterale des fonts de déformation de la Ride Méditerranéenne (Méditerranée orientale). *Bulletin de la Société Géologique de France*, **166**, 463–478.

COLLINS, A. & ROBERTSON, A. H. F. 1997. Lycian Mélange, southwestern Turkey: an emplaced Late Cretaceous accretionary complex. *Geology*, **25**, 255–258.

—— & —— 1998. Processes of Late Cretaceous to Late Miocene episodic thrust-sheet translation in the Lycian Taurides, southwest Turkey. *Journal of the Geological Society, London*, **155**, 759–772.

—— & —— 1999. Evolution of the Lycian Allochthon, western Turkey as a north-facing Mesozoic rift and passive margin continental margin. *In*: BOZKURT, E. & ROWBOTHAM, G. (eds) *Advances in Turkish Geology, Part 1*: Tethyan evolution and fluvial-marine sedimentation. Geological Journal, in press.

DE GRACIANCKY, P. C. 1972. *Recherchers géologiques dans les Taurus Lycien*. DSc Thèse Université de Paris Sud.

DELALOYE, M. & WAGNER, J.-J. 1984. Ophiolites and volcanic activity near the western edge of the Arabian plate. *In*: DIXON, J. E. & ROBERTSON, A. H. F. (eds) *The Geological Evolution of the Eastern Mediterranean*. Geological Society, London, Special Publications, **17**, 225–233.

——, DE SOUZA, H. & WAGNER, J.-J. 1980. Isotopic ages of ophiolites from the Eastern Mediterranean. *In*: PANAYIOTOU, A. (ed.) *Ophiolites*. Proceedings of the International Symposium, Nicosia, Cyprus, 292–295.

DELAUNE-MAYERE, M. 1984. Evolution of a Mesozoic passive continental margin: Baer–Bassit (NW Syria). *In*: DIXON, J. E. & ROBERTSON, A. H. F. (eds) *The Geological Evolution of the Eastern Mediterranean*. Geological Society, London, Special Publications, **17**, 11–10.

——, MARCOUX, J., PAROT, J.-F. & POISSON, A. 1977. Model d'évolution Mesozoique de la palaéomarge Téthysienne au niveau des nappes radiolariques et ophiolitiques du Taurus Lycien, d'Antalya et du Baer–Bassit. *In*: BIJU-DUVAL, B. & MONTADERT, L. (eds) *International Symposium on the Structural History of the Mediterranean basins*, Split (Yugoslavia), October 1976, 79–94.

DEMIRTAŞLI, E., TURHAN, N., BILGIN, A. Z. & SELIM, M. 1984. Geology of the Bolkar Mountains. *In*: TEKELI, O. & GÖNCÜOĞLU, M. C. (eds) *Geology of the Taurus Belt*. Proceedings of the International Tauride Symposium. Mineral Research and Exploration Institute of Turkey (MTA) Publications, 101–118.

DERCOURT, J., RICOU, L. E. & VRIELYNCK, B (eds) 1992. *Atlas Tethys Palaeoenvironmental Maps*. Beicip-Franlab.

——, ZONENSHAIN, L. P., RICOU, L. E. ET AL. 1986. Geological evolution of the Tethys belt from the Atlantic to the Pamirs since the Lias. *Tectonophysics*, **123**, 241–315.

DEWEY, J. F. & ŞENGÖR, A. M. C. 1979. Aegean and surrounding regions: complex multi-late continuum tectonics in a convergent zone. *Geological Society of America Bulletin*, **90**, 89–92.

——, HEMPTON, M. R., KIDD, W. S. F., ŞAROĞLU, F. & ŞENGÖR, A. M. C. 1986. Shortening of continental lithosphere: the Neotectonics of Eastern Anatolia – a young collision zone. *In*: COWARD, M. P. & RIES, A. C. (eds) *Collision Tectonics*. Geological Society, London, Special Publications, **19**, 3–36.

DILEK, Y. & DELALOYE, M. 1992. Structure of the Kızıldağ ophiolite, a slow-spread Cretaceous ridge segment north of the Arabian promontory. *Geology*, **20**, 19–22.

—— & MOORES, E. M. 1990. Regional tectonics of the Eastern Mediterranean ophiolites. *In*: MALPAS, J., MOORES, E. M., PANAYIOTOU, A. & XENOPHONTOS, C. (eds) *Ophiolites – Oceanic Crustal Analogues*. Proceedings of the Symposium, 'Troodos 1987', Geological Survey Department, Cyprus, 295–309.

—— & ROWLAND, J. C. 1993. Evolution of a conjugate passive margin pair in the Mesozoic Antalya Complex, southern Turkey. *Tectonics*, **11**, 954–970.

—— & THY, P. 1998. Structure, petrology and seafloor spreading tectonics of the Kızıldağ Ophiolite, Turkey. *In*: MILLS, R.A. & HARRISON, K. (eds) *Modern Ocean Floor Processes and the Geological Record*. Geological Society, London, Special Publications, **148**, 43–69.

——, MOORES, E. M., DELALOYE, M. & KARSON, J. W. 1991. A magmatic extension and tectonic denudation in the Kızıldağ Ophiolite, South Turkey: implications for the evolution of Neotethyan oceanic crust. *In*: PETERS, T. J. (eds) *Ophiolite Genesis and the Evolution of Oceanic Lithosphere*. Kluwer, 458–500.

DIXON, J. E. & ROBERTSON, A. H. F. 1999. *Are Plumes Implicated in the Permo-Triassic Break-up of Neotethys*. European Union of Geosciences, Strasbourg.

EMEIS, K., ROBERTSON, A. H. F. & RICHTER, C. 1996. *Proceedings of the Ocean Drilling Program, Initial Reports*, **160**.

EVANS, G., MORGAN, P., EVANS, W. E., EVANS, T. R. & WOODSIDE, J. M. 1978. Faulting and halokinetics between Cyprus and Turkey. *Geology*, **6**, 392–396.

FLECKER, R. M. 1995. *Miocene basin evolution of the Isparta Angle, southern Turkey*. PhD Thesis, University of Edinburgh.

——, ROBERTSON, A. H. F., POISSON, A. & MÜLLER, C. 1995. Facies and tectonic significance of two contrasting Miocene basins in south coastal Turkey. *Terra Nova*, **7**, 213–220.

——, ELLAM, R. M., ROBERTSON, A. H. F., POISSON, A. & TURNER, J. 1997. Application of Sr isotope stratigraphy to the origin and evolution of the Isparta angle, southern Turkey. *Terra Nova*, **9**, (Abstract Supplement 1), 394.

——, ——, MÜLLER, C., POISSON, A., ROBERTSON, A. H. F. & TURNER, J. 1998. Application of Sr isotope stratigraphy and sedimentary analysis to the origin and evolution of the Neogene basins in the Isparta Angle, southern Turkey. *Tectonophysics*, **298**, 63–82.

FLOYD, P. A., KELLING, G., GÖKÇEN, S. L. & GÖKÇEN, N. 1992. Arc-related origin of volcaniclastic sequences in the Misis Complex, Southern Turkey. *Journal of Geology*, **100**, 221–230.

FOURCADE, E., DERCOURT, J., GÜNAY, Y. *ET AL*. 1991. Stratigraphie et paléogéographie de la marge septentrionale de la platforme arabe au Mesozoique (Turquie de Sud-East). *Bulletin de la Société Géologique de France*, **161**, 27–41.

FRIZON DE LAMOTTE, D., POISSON, A., AUBOURG, C. & TEMİZ, H. 1995. Chevauchements post Tortoniens vers L'ouest puis vers le sud au coeur de l' Angle d'Isparta (Taurus, Turquie). Consequences Géodynamique. *Bulletin de la Société Géologique de France*, **166**, 527–538.

GARFUNKEL, Z. 1989. Constraints on the origin and history of the Eastern Mediterranean basin. *Tectonophysics*, **298**, 5–35.

—— 1998. Constraints on the origin and history of the Eastern Mediterranean basin. *Tectonophysics*, **298**, 5–36.

—— & DERIN, B. 1984. Permian–early Mesozoic tectonism and continental margin formation on Israel and its implications for the history of the eastern Mediterranean. *In*: DIXON, J. E. & ROBERTSON, A. H. F. (eds) *The Geological Evolution of the Eastern Mediterranean*. Geological Society, London, Special Publications, **17**, 187–201.

GEDİK, A., BİRGİLİ, S., YILMAZ, H. & YOLDAŞ, R. 1979. Geology of the Mut–Ermenek–Silifke (Konya, Mersin) area and its petroleum possibilities. *Geological Society of Turkey Bulletin*, **22**, 7–26 [in Turkish with English abstract].

GLENNIE, K. W., HUGHES CLARKE, M. W., BOEF, M. W., PILAAR, M. G. H. & REINHARDT, B. 1990. Inter-relationship of the Makran–Oman Mountains belts of convergence. *In*: ROBERTSON, A. H. F., SEARLE, M. P. & RIES, A. C. (eds) *The Geology and Tectonics of the Oman Region*. Geological Society, London, Special Publications, **49**, 773–787.

GLOVER, C. 1996. *Plio-Quaternary sediments and Neotectonics of the Isparta Angle, southwest Turkey*. PhD Thesis, University Edinburgh.

—— & ROBERTSON, A. H. F. 1998a. Role of extensional processes and uplift in the Plio-Quaternary sedimentary and tectonic evolution of the Aksu Basin, southwest Turkey. *Journal of the Geological Society, London*, **155**, 365–368.

—— & —— 1998b. Neotectonic intersection of the Aegean and Cyprus arcs: extensional and strike-slip faulting in the Isparta Angle, southwest Turkey. *Tectonophysics*, **298**, 103–132.

GÖKÇEN, S. L., KELLING, G., GÖKÇEN, N. & FLOYD, P. A. 1988. Sedimentology of a Late Cenozoic collisional sequence: the Misis Complex, Adana, Southern Turkey. *Sedimentary Geology*, **59**, 205–235.

GÖRÜR, N. 1977. Depositional history of Miocene sediments on the northwest flank of the Adana Basin. *In*: İZDAR, E. & NAKOMAN, E. (eds) *Proceedings of the 6th Colloquium on Geology of the Aegean Region*. Piri Reis International Contribution Series, **2**, 185–208.

—— 1992. A tectonically controlled alluvial fan which developed into a marine fan-delta at a complex triple junction: Miocene Gildirli Formation of the Adana Basin, Turkey. *Sedimentary Geology*, **81**, 241–252.

——, OKTAY, F. Y. & ŞENGÖR, A. M. C. 1984. Palaeotectonic evolution of the Tuzgölü basin complex, Central Turkey: sedimentary record of a Neo-Tethyan closure. *In*: DIXON, J. E. & ROBERTSON, A. H. F. (eds) *The Geological Evolution of the Eastern Mediterranean*. Geological Society, London, Special Publications, **17**, 467–482.

GÖZÜBOL, A. M. & GÜRPINAR, O. 1980. The geology and the tectonic evolution of the north of Kahraman Maraş. *Proceedings of 5th Petroleum Congress and Exhibition of Turkey*, Turkish Association of Petroleum Geologists Publications, 21–29.

GUTNIC, M., MONOD, O., POISSON, A. & DUMONT, J.- F. 1979. Géologie des Taurus occidentales (Turquie). *Mémoire de la Société Géologique de France, Nouvelle série*, **58**, 1–112.

GÜRBÜZ, K. 1993. *Identification and evolution of Miocene submarine fans in the Adana Basin*. PhD Thesis, Keele University.

—— & KELLING, G. 1993. Provenence of Miocene submarine fans in the northern Adana basin, southern Turkey: a test of discriminant function analysis. *Geological Journal*, **28**, 277–293.

HALL, R. 1976. Ophiolite emplacement and the evolution of the Taurus suture zone, South-east Turkey. *Geological Society of America Bulletin*, **87**, 1078–1088.

HAYWARD, A. B. 1984. Miocene clastic sedimentation related to the emplacement of the Lycian Nappes and the Antalya Complex, southwest Turkey. *In*: DIXON, J. E. & ROBERTSON, A. H. F. (eds) *The Geological Evolution of the Eastern Mediterranean*. Geological Society, London, Special Publications, **17**, 287–300.

—— & ROBERTSON, A. H. F. 1982. Direction of ophiolite emplacement inferred from Cretaceous and Tertiary sediments of an adjacent autochthon, the Bey Dağları, S.W. Turkey. *Geological Society of America Bulletin*, **93**, 68–75.

HELVACI, C. & GRIFFIN, W. L. 1984. Rb–Sr geochronology of the Bitlis Massif, Avnik (Bingöl) area, SE Turkey. *In*: DIXON, J. E. & ROBERTSON, A. H. F. (eds) *The Geological Evolution of the Eastern Mediterranean*. Geological Society, London, Special Publications, **17**, 403–414.

HEMPTON, M. R. 1982. The North Anatolian fault and complexities of continental escape. *Journal of Structural Geology*, **4**, 502–504.

—— 1984. Results of detailed mapping near Lake Hazar (Eastern Taurus Mountains). *In*: TEKELİ, O. & GÖNCÜOĞLU, M. C. (eds) *Geology of the Taurus Belt*. Proceedings of International Tauride Symposium, Mineral Research and

Exploration Institute of Turkey (MTA) Publications, 223–228.

HSÜ, K. J., MONDADERT, L., BERNOULLI, D. ET AL. 1978. History of the Mediterranean salinity crisis. *Initial Reports of the Deep Sea Drilling Project.* US Government Printing Office (Washington, DC), **42A**, 1053–1078.

IVANOV, M. K., LIMONOV, M. K. & WOODSIDE, J. M. 1992. *Geological and geophysical investigations in the Mediterranean and Black Seas.* Initial results of the 'Training through Research' Cruise of R V Gelendzhik in the Eastern Mediterranean and the Black Sea (June–July, 1991), UNESCO Reports in Marine Science.

JACKSON, J. A. & MCKENZIE, D. M. P. 1984. Active tectonics of the Alpine–Himalayan belt between western Turkey and Pakistan. *Geophysical Journal of the Royal Astronomical Society*, **77**, 85–264.

—— & —— 1989. The relationship between plate motions and seismic moment tensors, and rates of active deformation in the Mediterranean and Middle East. *Geophysical Journal of the Royal Astronomical Society*, **93**, 45–73.

JUTEAU, T. 1970. Petrogenèse des ophiolites des Nappes d'Antalya (Taurus Lycien Oriental). Leur liaison avec une phase d'expansion oceanique active au Trias supérieur. *Mémoires de Sciences de la Terre*, **15**, 25–288.

KARIG, D. E. & KOZLU, H. 1990. Late Palaeogene evolution of the triple junction region near Maraş south-central Turkey. *Journal of the Geological Society, London*, **147**, 1023–1034.

KASTENS, K. 1991. Role of outward growth of the Mediterranean Ridge accretionary complex. *Tectonophysics*, **199**, 25–50.

——, BREEN, N. & CITA, M. B. 1992. Progressive deformation of an evaporite-bearing accretionary complex: SeaMARC 1, seabeam and piston-core observations from the Mediterranean Ridge. *Marine Geophysical Research*, **14**, 249–298.

KELLING, G., GÖKÇEN, S. L., FLOYD, P. A. & GÖKÇEN, N. 1987. Neogene tectonics and plate convergence in the Eastern Mediterranean: new data from southern Turkey. *Geology*, **15**, 425–429.

KEMPLER, D. 1994. *Tectonic patterns in the eastern Mediterranean.* PhD Thesis, Hebrew University of Jerusalem.

—— & BEN-AVRAHAM, Z. 1987. The tectonic evolution of the Cyprean Arc. *Annales Tectonicae*, **1**, 58–71.

—— & GARFUNKEL, Z. 1994. Structures and kinematics in the northeastern Mediterranean: a study of an irregular plate boundary. *Tectonophysics*, **234**, 19–32.

KISSEL, C. & LAJ, C. 1988. The Tertiary geodynamical evolution of the Aegean arc: a palaeomagnetic reconstruction. *Tectonophysics*, **146**, 183–201.

—— & POISSON, A. 1986. Etude paleomagnetique preliminaire des formations Neogene du bassin d'Antalya (Taurus occidentales–Turquie). *Compte Rendus de l'Académie Science de Paris*, **302**, Ser. 11 (10), 711–716.

——, AVERBUCH, O., FRIZON DE LAMOTTE, D., MONOD, O. & ALLERTON, S. 1990. First palaeomagnetic evidence of a post-Eocene clockwise rotation of the Western Taurus thrust belt, east of the Isparta re-entrant (Southwestern Turkey). *Earth and Planetary Science Letters*, **117**, 1–14.

KRASHENNINIKOV, V. A. 1994. Stratigraphy of the Maastrichtian and Cenozoic deposits of the coastal part of northwestern Syria (Neoautochthon of the Bassit ophiolite massif). *In*: HALL, J. K. (ed.) *Geological Structure of the Northeastern Mediterranean.* Cruise 5 of the Research Vessel 'Akademik Nokolaj Strakhov'. Historical Productions – Hall Ltd, Jerusalem, 265–276.

LE PICHON, X. 1982. Land-locked oceanic basins and continental collision: the Eastern Mediterranean as a case example. *In*: HSÜ, K. J. (ed.) *Mountain Building Processes.* Academic, 201–211.

—— & ANGELIER, J. 1979. The Hellenic arc and trench system: a key to the neotectonic evolution of the eastern Mediterranean area. *Tectonophysics*, **60**, 1–42.

LIVERMORE, R. A. & SMITH, A. G. 1984. Some boundary conditions for the evolution of the Mediterranean region. *In*: STANLEY, D. J. & WEZEL, F.-C. (eds) *Geological Evolution of the Mediterranean Basin.* Springer, 83–100.

LYTWYN, J. N. & CASEY, J. F. 1995. The geochemistry of postkinematic mafic dike swarms and subophiolitic metabasites, Pozantı–Karsantı ophiolite, Turkey: evidence for ridge subduction. *Geological Society of America Bulletin*, **107**, 830–850.

MCKENZIE, D. P. 1978. Active tectonism in the Alpine–Himalayan belt: the Aegean Sea and the surrounding regions (tectonics of the Aegean region). *Geophysical Journal of the Royal Astronomical Society*, **55**, 217–254.

MAKRIS, J., BEN-AVRAHAM, Z., BEHLE, A. ET AL. 1983. Seismic refraction profiles between Cyprus and Israel and their interpretation. *Geophysical Journal of the Royal Astronomical Society*, **75**, 575–591.

MARCOUX, J. 1974. Alpine type Triassic of the upper Antalya Nappe (western Taurus, Turkey). *In*: ZAPFE, H. (ed.) *Die Stratigraphie der Alpinmediterrannean Trias, Wien.* 145–146.

—— 1995. Initiation of the south-Neotethys margin in the Antalya Nappes (SW Turkey): Late Permian and Early Mid-Triassic rifting events, late Mid-Triassic oceanisation. *European Union of Geosciences, Abstracts*, 175.

——, RICOU, L.-E., BURG, J. P. & BRUNN, J. P. 1989. Shear-sense criteria in the Antalya thrust system (south-western Turkey): evidence for southward emplacement. *Tectonophysics*, **161**, 81–91.

MICHARD, A., WHITECHURCH, H., RICOU, L. E., MONTIGNY, R. & YAZGAN, E. 1984. Tauric subduction (Malatya–Elazığ provinces) and its bearing on tectonics of the Tethyan realm in Turkey. *In*: DIXON, J. E. & ROBERTSON, A. H. F. (eds) *The Geological Evolution of the Eastern Mediterranean.* Geological Society, London, Special Publications, **17**, 349–360.

MONOD, O. 1977. *Récherches géologique dans les*

Taurus occidental au sud de Beyşehir (Turquie). DSc Thèse Université de Paris-Sud, Orsay.

MORRIS, A. 1996. A review of palaeomagnetic research in the Troodos ophiolite, Cyprus. In: MORRIS, A. & TARLING, D. D. (eds) *Palaeomagnetism and Tectonics of the Mediterranean Region*. Geological Society, London, Special Publications, **105**, 311–324.

—— & ROBERTSON, A. H. F. 1993. Miocene remagnetisation of carbonate platform and Antalya complex units within the Isparta Angle, SW Turkey. *Tectonophysics*, **220**, 242–266.

OKAY, A. İ. 1990. The origins of the allochthons in the Lycian belt, southwestern Turkey. *Tectonophysics*, **177**, 367–379.

—— & ÖZGÜL, N. 1984. HP/LT metamorphism and the structure of the Alanya Massif. In: DIXON, J. E. & ROBERTSON, A. H. F. (eds) *The Geological Evolution of the Eastern Mediterranean*. Geological Society, London, Special Publications, **17**, 429–440.

ORAL, M. B., REILINGER, R., TOKSÖZ, M. N., KING, R.W., BARKA, A. A. & LENK, O. 1994. GPS measurements of crustal deformation in Turkey (1988–1992). Coherent rotation of the Anatolian Plate (Abstract). *American Geophysical Union*, **116**.

ÖZGÜL, N. 1984. Stratigraphy and tectonic evolution of the central Taurus. In: TEKELI, O. & GÖNCÜOĞLU, M. C. (eds) *Geology of the Taurus Belt*. Proceedings of the International Tauride Symposium. Mineral Research and Exploration Institute of Turkey (MTA) Publications, 77–90.

PARLAK, O., DELALOYE, M. & BİNGÖL, E. 1995. Origin of sub-ophiolitic metamorphic rocks beneath the Mersin ophiolite, Southern Turkey. *Ofioliti*, **20**, 97–110.

PARROT, J.-F. 1977. Assemblage ophiolitique de Baer–Bassit et termes éffusives du volcano-sedimentaire. *Travaux et Documents de L' ORSTROM*, **72**.

PEARCE, J. A., BENDER, J. F., DE LONG, S. E. ET AL. 1990. Genesis of post-collisional volcanism in Eastern Anatolia, Turkey. *Journal of Volcanological and Geothermal Research*, **44**, 189–229.

PERİNÇEK, D. 1979. *The Geology of Hazro–Korudağ–Cüngüş–Maden–Ergani–Hazar–Elazığ–Malatya Area*. Geological Society of Turkey, Special Publications [in Turkish with English abstract].

—— & ÖZKAYA, İ. 1981. Tectonic evolution of the northern margin of Arabian plate. *Hacettepe University Bulletin of Natural Sciences and Engineering*, **8**, 91–101 [in Turkish with English abstract].

PİŞKİN, O., DELALOYE, M. & WAGNER, J.-J. 1986. Guide to the Hatay geology. *Ofioliti*, **11**, 87–104.

POISSON, A. 1977. *Recherches géologiques dans les Taurus occidentales, Turquie*. Thèse, Université de Paris-Sud, Orsay.

—— 1984. The extension of the Ionian trough into southwestern Turkey. In: DIXON, J. E. & ROBERTSON, A. H. F. (eds) *Geological Evolution of the Eastern Mediterranean*. Geological Society, London, Special Publications, **17**, 241–250.

PRICE, S. P. & SCOTT, B. 1994. Fault-block rotations at the edge of a zone of continental extension: southwestern Turkey. *Journal of Structural Geology*, **16**, 381–392.

REUBER, I. 1984. Mylonitic ductile shear zones within tectonites and cumulates as evidence for an oceanic transform fault in the Antalya ophiolite, SW Turkey. In: DIXON, J. E. & ROBERTSON, A. H. F. (eds) *The Geological Evolution of the Eastern Mediterranean*. Geological Society, London, Special Publications, **17**, 319–334.

RICOU, L.-E., MARCOUX, J. & WHITECHURCH, H. 1984. The Mesozoic organisation of the Taurides: one or several oceanic basins. In: DIXON, J. E. & ROBERTSON, A. H. F. (eds) *The Geological Evolution of the Eastern Mediterranean*. Geological Society, London, Special Publications, **17**, 349–360.

RIGO DE RIGHI, M. & CORTESINI, A. 1964. Gravity tectonics in foothills structure belt of southeastern Turkey. *AAPG Bulletin*, **48**, 1911–1937.

ROBERTSON, A. H. F. 1986. The Hatay ophiolite (southern Turkey) in its Eastern Mediterranean tectonic context: a report on some aspects of the field excursion. *Ofioliti*, **11**, 105–119.

—— 1987. The transition from a passive margin to an Upper Cretaceous foreland basin related to ophiolite emplacement in the Oman Mountains. *Geological Society of America Bulletin*, **99**, 633–653.

—— 1990. Tectonic evolution of Cyprus. In: MALPAS, J., MOORES, E. M., PANAYIOTOU, A. & XENOPHONTOS, C. (eds) *Ophiolites – Oceanic Crustal Analogues*. Proceedings of the Symposium 'Troodos 1987', Geological Survey Department, Cyprus, 235–252.

—— 1993. Mesozoic–Tertiary sedimentary and tectonic evolution of Neotethyan carbonate platforms, margins and small ocean basins in the Antalya complex, S.W. Turkey. In: FROSTICK, L. E. & STEEL, R. (eds) *Special Publication of the International Association of Sedimentologists*, **20**, 415–465.

—— 1994. Role of the tectonic facies concept in orogenic analysis and its application to Tethys in the Eastern Mediterranean region. *Earth-Science Reviews*, **37**, 139–213.

—— 1998. Mesozoic–Tertiary tectonic evolution of the Easternmost Mediterranean area: integration of marine and land evidence. In: ROBERTSON, A. H. F., EMEIS, K.-C., RICHTER, C. & CAMERLENGHI, A. (eds) *Proceedings of the Ocean Drilling Program, Scientific Results*, **160**, 723–782.

—— 2000. Geological evolution of Cyprus: onshore and offshore evidence. In: MALPAS, J., XENOPHONOTOS, C. & PANAYIDES, A. (eds) *Proceedings of the 3rd International Conference on the Geology of the Eastern Mediterranean*, Nicosia, Cyprus – 1998. Geological Survey Department, Cyprus, in press.

—— & DIXON, J. E. 1984. Introduction: aspects of the geological evolution of the eastern Mediterranean. In: DIXON, J. E. & ROBERTSON, A. H. F. (eds) *The Geological Evolution of the Eastern*

Mediterranean. Geological Society, London, Special Publications, **17**, 1–74.

—— & GRASSO, M. 1995. Overview of the Late Tertiary tectonic and palaeo-environmental development of the Mediterranean region. *Terra Nova*, **7**, 114–127.

—— & HUDSON, J. E. 1974. Pelagic sediments in the Cretaceous and Tertiary history of the Troodos Massif, Cyprus. *In*: HSÜ, K. J. & JENKYNS, H. C. (eds) *Pelagic Sediments on Land and Under the Sea*. International Association of Sedimentologists, Special Publications, **1**, 403–436.

—— & KOPF, A. 1998. Tectonic setting and processes of mud volcanism on the Mediterranean Ridge accretionary complex: evidence from Leg 160. *In*: ROBERTSON, A. H. F., EMEIS, K.-C., RICHTER, C. & CAMERLENGHI, A. (eds) *Proceedings of the Ocean Drilling Program, Scientific Results*, **160**, 665–680.

—— & WALDRON, J. W. F. 1990. Geochemistry and tectonic setting of Late Triassic and Late Jurassic–Early Cretaceous basaltic extrusives from the Antalya Complex, south west Turkey. *In*: SAVAŞÇIN, M. Y. & ERONAT, A. H (eds) *Proceedings of International Earth Sciences Congress on Aegean Regions – 1990*, **2**, 279–299.

—— & WOODCOCK, N. H. 1979. The Mamonia Complex, south west Turkey: the evolution and emplacement of a Mesozoic continental margin. *Geological Society of America Bulletin*, **90**, 651–665.

—— & —— 1980. Tectonic setting of the Troodos massif in the east Mediterranean. *In*: PANAYIOTOU, A. (ed.) *Ophiolites. Proceedings of the International Symposium, Cyprus, 1979*. Cyprus Geological Survey Department, 36–49.

—— & —— 1982. Sedimentary history of the southwestern segment of the Mesozoic–Tertiary Antalya continental margin, south-western Turkey. *Eclogae Geologicae Helvetiae*, **75**, 517–562.

—— & —— 1984. The SW segment of the Antalya Complex, Turkey as a Mesozoic–Tertiary Tethyan continental margin. *In*: DIXON, J. E. & ROBERTSON, A. H. F. (eds) *The Geological Evolution of the Eastern Mediterranean*. Geological Society, London, Special Publications, **17**, 251–272.

—— & —— 1986. The geological evolution of the Kyrenia Range: a critical lineament in the Eastern Mediterranean. *In*: READING. H. G., WATTERSON, J. & WHITE, S. H. (eds) *Major Crustal Lineaments and their Influence on the Geological History of the Continental Lithosphere*. Philosophical Transactions the Royal Society, London, Series A, **317**, 141–171.

——, CLIFT P. D., DEGNAN, P. & JONES, G. 1991. Palaeogeographic and palaeotectonic evolution of the Eastern Mediterranean Neotethys. *Palaeogeography, Palaeoclimatology, Palaeoecology*, **87**, 289–344.

——, EMEIS, K.-C., RICHTER, C. & CAMERLENGHI, A. (eds) 1998. *Proceedings of the Ocean Drilling Program, Scientific Results*, **160**.

——, DIXON, J. E., BROWN, S., ET AL. 1996. Alternative models for the Late Palaeozoic–Early Tertiary development of Tethys in the eastern Mediterranean region. *In*: MORRIS, A. & TARLING, D. H. (eds) *Palaeomagnetism and Tectonics of the Mediterranean Region*. Geological Society, London, Special Publications, **105**, 239–263.

ROTSTEIN, Y & KAFKA, A. L. 1982. Seismotectonics of the southern boundary of Anatolia, Eastern Mediterranean region: subduction, collision, and arc jumping. *Journal of Geophysical Research*, **87**, 7694–7706.

SAVOSTIN, L. A., SIBUET, J. C., ZONENSHAIN, L. P., LE PICHON, X. & ROLET, J. 1986. Kinematic evolution of the Tethys belt, from the Atlantic to the Pamirs since the Triassic. *Tectonophysics*, **123**, 1–35.

SCHIETTECATTE, J. P. 1971. Geology of the Misis Mountains. *In*: CAMPBELL, M. (ed.) *Geology and History of Turkey, Tripoli, Libya*. Petroleum Exploration Society, 305–312.

SEARLE, M. P. & GRAHAM, G. M. 1982. 'Oman Exotics' – Oceanic carbonate build-ups associated with the early stages of continental rifting. *Geology*, **10**, 43–49.

ŞENGÖR, A. M. C. & YILMAZ, Y. 1981. Tethyan evolution of Turkey: a plate tectonic approach. *Tectonophysics*, **75**, 81–241.

——, GÖRÜR, N. & ŞAROĞLU, F. 1985. Strike-slip faulting and related basin formation in zones of tectonic escape. *In*: BIDDLE, K.T. & BLICK, N.C. (eds) *Strike-slip Deformation, Basin Formation and Sedimentation*. Society of Economic Palaeontologists and Mineralogists, Special Publications, **37**, 227–264.

——, YILMAZ, Y. & SUNGURLU, O. 1984. Tectonics of the Mediterranean Cimmerides: nature and evolution of the western termination of Palaeo-Tethys. *In*: DIXON, J. E. & ROBERTSON, A. H. F. (eds) *The Geological Evolution of the Eastern Mediterranean*. Geological Society, London, Special Publications, **17**, 77–112.

STAMPFLI, G. M. 2000. Tethyan oceans. *This volume*.

——, MOSAR, J., FAVRE, P., PILLEVUIT, A. & VANNAY, J.-C. 2000. Permo-Mesozoic evolution of the western Tethys realm. *In*: CAVAZZA, W., ZIEGLER, P. & ROBERTSON, A. H. F. (eds) *Peritethyan Rift/Wrench Basins and Passive Margins, IGCP 369*. Bulletin du Musée National d'Histoire Naturelle, Paris.

TOKSÖZ, M. N., REILINGER, R., KING, R., MCCLUSKY, B., SOUTER, B. & ORAL, B. 1995. 1994 GPS Measurements in Turkey and surrounding areas of the E. Mediterranean/Middle East (Abstract). *American Geophysical Fall Meeting*, 1995.

WALDRON, J. W. F. 1984. Evolution of carbonate platforms on a margin of the Neotethys ocean: Isparta angle, south-western Turkey. *Eclogae Geological Helvetiae*, **77**, 553–581.

WATTS, K. W. & GARRISON, R. E. 1979. Sumeini Group, Oman – evolution of a Mesozoic carbonate slope on a South Tethyan continental margin. *Sedimentary Geology*, **48**, 107–168.

WEAVER, S. D. SAUNDERS, A. D., PANKHURST, R. S. & TANER, J. 1977. A geochemical study of the Quaternary volcanoes of Bransfield Strait, from

South Shetland Islands, and of magmatism associated with the initial stages of backarc spreading. *Contributions to Mineralogy and Petrology*, **68**, 151–169.

WESTAWAY, R. & ARGER, J. 1996. The Gölbaşı basin, southeastern Turkey: a complex discontinuity in a major strike-slip fault zone. *Journal of the Geological Society, London*, **153**, 729–744.

WHITECHURCH, H., JUTEAU, T. & MONTIGNY, R. 1984. Role of the Eastern Mediterranean ophiolites (Turkey, Syria, Cyprus) in the history of Neo-Tethys. *In*: DIXON, J. E. & Robertson, A. H. F. (eds) *The Geological Evolution of the Eastern Mediterranean*. Geological Society, London, Special Publications, **17**, 301–318.

WILLIAMS, G. D., ÜNLÜGENÇ, U. C., KELLING, G. & DEMIRKOL, C. 1995. Tectonic controls on stratigraphical evolution of the Adana Basin, Turkey. *Journal of the Geological Society, London*, **152**, 873–882.

WOODCOCK, N. & ROBERTSON, A. H. F. 1982. Wrench and thrust tectonics along a Mesozoic–Cenozoic continental margin: Antalya Complex, S.W. Turkey. *Journal of the Geological Society, London*, **139**, 147–163.

—— & —— 1984. Structural variety of Tethyan ophiolite terrains. *In*: GASS, I. G., LIPPARD, S. J. & SHELTON, A. W. (eds) *Ophiolites and Oceanic Lithosphere*. Geological Society, London, Special Publications, **13**, 321–333.

WOODSIDE, J. M. 1977. Tectonic elements and crust of the Eastern Mediterranean Sea. *Marine Geophysical Research*, **3**, 317–354.

—— 1992. Area no. 2 (Anaximander Mountains). *Initial results of the 'Training through Research' cruise of R V Gelendzhik in the Eastern Mediterranean and the Black Sea (June–July, 1991)*. UNESCO, 116–170.

—— & DUMONT, J. F., 1997. The Anaximander Mountains are a southward rifted and foundered part of the southwestern Turkish Taurus. *Terra Nova*, **9**, 394.

YALÇIN, M. N. & GÖRÜR, N. 1984. Sedimentological evolution of the Adana basin. *In*: TEKELİ, O. & GÖNCÜOĞLU, M. C. (eds) *Geology of the Taurus Belt*. Proceedings of the International Tauride Symposium. Mineral Research and Exploration Institute of Turkey (MTA) Publications, 165–172.

YAZGAN, E. 1984. Geodynamic evolution of the Eastern Taurus region. *In*: TEKELİ, O. & GÖNCÜOĞLU, M. C. (eds) *Geology of the Taurus Belt*. Proceedings of the International Tauride Symposium. Mineral Research and Exploration Institute of Turkey (MTA) Publications, 199–208.

—— & CHESSEX, R. 1991. Geology and tectonic evolution of the Southeastern Taurus in the region of Malatya. *Turkish Association of Petroleum Geologists Bulletin*, **3**, 1–42.

YETİŞ, C. 1984. New observations on the age of the Ecemiş Fault. I. *In*: TEKELİ, O. & GÖNCÜOĞLU, M. C. (eds) *Geology of the Taurus Belt*. Proceedings of the International Tauride Symposium. Mineral Research and Exploration Institute of Turkey (MTA) Publications, 159–164.

——, KELLING, G., GÖKÇEN, S. L. & BAROZ, F. 1995. A revised stratigraphic frame for later Cenozoic sequences in the northeastern Mediterranean region. *Geologische Rundschau*, **84**, 794–812.

YILMAZ, P. O. 1984. Fossil and K–Ar data for the age if the Antalya Complex, southwest Turkey. *In*: DIXON, J. E. & ROBERTSON, A. H. F. (eds) *The Geological Evolution of the Eastern Mediterranean*. Geological Society, London, Special Publications, **17**, 335–348.

YILMAZ, Y. 1993. New evidence and model on the evolution of the southeast Anatolian orogen. *Geological Society of America Bulletin*, **105**, 251–271.

A geotraverse across northwestern Turkey: tectonic units of the Central Sakarya region and their tectonic evolution

M. CEMAL GÖNCÜOĞLU,[1] NECATİ TURHAN,[2] KAMİL ŞENTÜRK,[2] AHMET ÖZCAN,[2] ŞÜKRÜ UYSAL[2] & M. KENAN YALINIZ[3]

[1]*Middle East Technical University, Department of Geological Engineering, TR-06531 Ankara, Turkey (e-mail: mcgoncu@metu.edu.tr)*
[2]*Mineral Research and Exploration Institute of Turkey (MTA), Department of Geology, TR-06532 Ankara, Turkey*
[3]*Celal Bayar University, Department of Civil Engineering, Manisa, Turkey*

Abstract: In the Central Sakarya area of Turkey there are two main Alpine continental units, separated by a south verging ophiolitic complex which represents the root zone of the İzmir–Ankara Suture Belt.

The Central Sakarya Terrane in the north includes two 'Variscan' tectonic units in its basement. The Söğüt Metamorphic rocks represent a Variscan ensimatic arc complex and the Tepeköy Metamorphic rocks are characteristically a forearc–trench complex. The unconformably overlying Triassic Soğukkuyu Metamorphic rocks correspond to a part of the Karakaya Formation and are interpreted as a Triassic rift basin assemblage. These units are unconformably overlain by a transgressive sequence of Liassic–Late Cretaceous age that represents the northeastward deepening carbonate platform of the Sakarya Composite Terrane.

The middle tectonic unit (the Central Sakarya Ophiolitic Complex) comprises blocks and slices of dismembered ophiolites, blueschists and basic volcanic rocks with uppermost Jurassic–Lower Cretaceous radiolarite–limestone interlayers. Geochemical data from basalt blocks suggest mid-ocean ridge basalt (MORB)- and suprasubduction-type tectonic settings within the Neotethyan İzmir–Ankara Ocean.

The southern tectonic unit includes basal polyphase metamorphosed clastic rocks (Sömdiken Metamorphics), intruded by felsic and basic dykes and overlain by thick-bedded marbles. This assemblage is unconformably overlain by continental clastic rocks gradually giving way to thick-bedded recrystallized limestones, cherty limestones and pelagic limestones intercalated with radiolarites, and finally by a thick high pressure–low temperature (HP–LT) metamorphic synorogenic flysch sequence. This succession is identical to the passive continental margin sequences of the Tauride Platform. It is suggested that this passive margin was subducted during the Late Cretaceous in an intra-oceanic subduction zone and affected by HP–LT metamorphism. The emplacement of the allochthonous oceanic assemblages and the collision with the Central Sakarya Terrane was complete by the end of the Cretaceous.

Turkey is an east–west trending segment of the Alpine–Himalayan orogenic belt, and is located on the boundary between Gondwana in the south and Laurasia in the north. Within this belt, different continental and oceanic assemblages related to the opening and closure of the Palaeozoic and Mesozoic oceanic basins can be found. These oceanic basins are collectively named the Tethys Ocean.

One of the most critical problems on the Tethyan evolution of northwestern Turkey is the location of the oceanic suture root zones and the interrelations of Palaeo- and Neo-tethyan terranes. The Central Sakarya region (Fig. 1) is a key area where tectonic elements of both the Palaeo- and Neotethyan assemblages crop out.

It is frequently accepted that in this area three Alpine microplates occur (Şengör & Yılmaz 1981; Okay 1989; Göncüoğlu & Erendil 1990; Yılmaz 1990). The northernmost terrane is represented by the İstanbul Nappe of Şengör & Yılmaz (1981) or the İstanbul Zone of Okay (1989). It is separated from the Sakarya Microcontinent to the south (Şengör & Yılmaz 1981) by the Intra-Pontide Suture Zone. The major Neotethyan Vardar–İzmir–Ankara Suture Zone, on the other hand, separates the Sakarya Microcontinent from the Anatolides that represents the northern

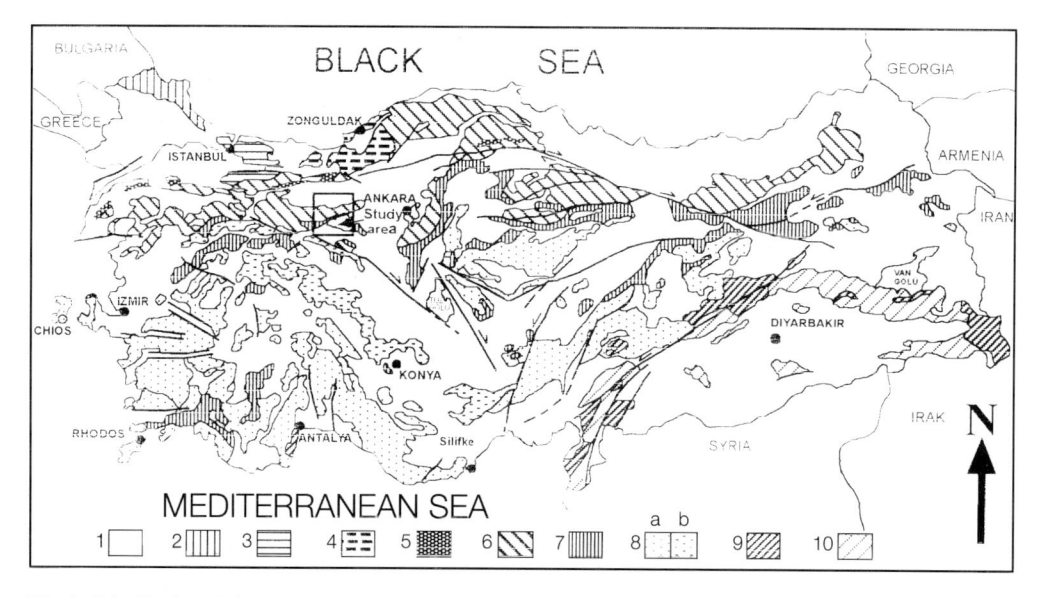

Fig. 1. Distribution of the main Alpine terranes in Turkey [after Göncüoğlu et al. (1996–1997)] and the location of the study area. 1, Tertiary cover; 2, Istranca Terrane; 3, İstanbul Terrane; 4, Zonguldak Terrane; 5, Intra-Pontide Suture Belt; 6, Sakarya Composite Terrane; 7, İzmir–Ankara–Erzincan Suture Belt; 8, Tauride–Anatolide Composite Terrane (a, undivided; b, Kütahya–Bolkardağ Belt); 9, Southeast Anatolian Suture Belt; 10, Southeast Anatolian–Arabian Plate.

margin of the Tauride–Anatolide Platform. Each of these Alpine units differ not only in their magmatic/metamorphic and depositional history but also in containing older units, and hence in their palaeogeographic positions during the Phanerozoic times.

Göncüoğlu et al. (1996–1997) recently reviewed the tectonic units of Turkey and described the Alpine fragments as 'terranes' or 'composite terranes' (*sensu* Howell 1989) with regard to their pre-Alpine history (Fig. 1).

According to this classification, the İstanbul Terrane in the north was probably located during the Palaeozoic at the passive margin of an isolated continental fragment within a Palaeozoic ocean to the north of Gondwana (Göncüoğlu 1997; Göncüoğlu & Kozur 1998). During the latest Mesozoic, it was attached to Laurasia and had an active margin setting to the north of the Neotethyan Intra-Pontide Ocean. The İstanbul Terrane overthrusts the Intra-Pontide Suture Zone (Göncüoğlu & Erendil 1990). The latter unit, located between the Sakarya Composite Terrane and the İstanbul Terrane, is characterized by ophiolites and a mélange complex with radiolarian cherts and pelagic limestones of Late Jurassic–Early Cretaceous age. It is thrust southward onto Upper Cretaceous flysch sequences of the Sakarya Composite Terrane. The Intra-Pontide Suture Zone probably extends westwards into the ophiolitic belt observed on Lesbos and the Eastern Rhodopian Ophiolites.

The Alpine 'Sakarya Composite Terrane', the Sakarya Zone of Okay et al. (1996), is generally accepted as an isolated carbonate platform in the Tethys during Middle–Late Mesozoic time. However, its pre-Alpine history is very complex and there is no consensus on the ages, extents and the geodynamic setting of more or less metamorphic tectonic fragments. These pre-Alpine elements mainly represent oceanic and continental fragments of Late Palaeozoic and/or Early Mesozoic age. The Central Sakarya 'Terrane' is one of these pre-Alpine units and will be one of the main topics of this paper.

The Vardar–İzmir–Ankara Suture Belt represents the northern branch of Neotethys. It is unequivocally accepted that huge allochthonous nappes of almost complete ophiolitic sequences and tectonic mélanges were generated during its closure and thrust southwards to the passive margin of the Tauride–Anatolide Platform. However, the location of the suture root zone in the Central Sakarya region has not been clearly identified (e.g. Monod et al. 1991), and will be evaluated in this paper. On the other hand, field and geochemical data from western Central Anatolia suggest that at least the İzmir–Ankara segment of the northern Neotethyan

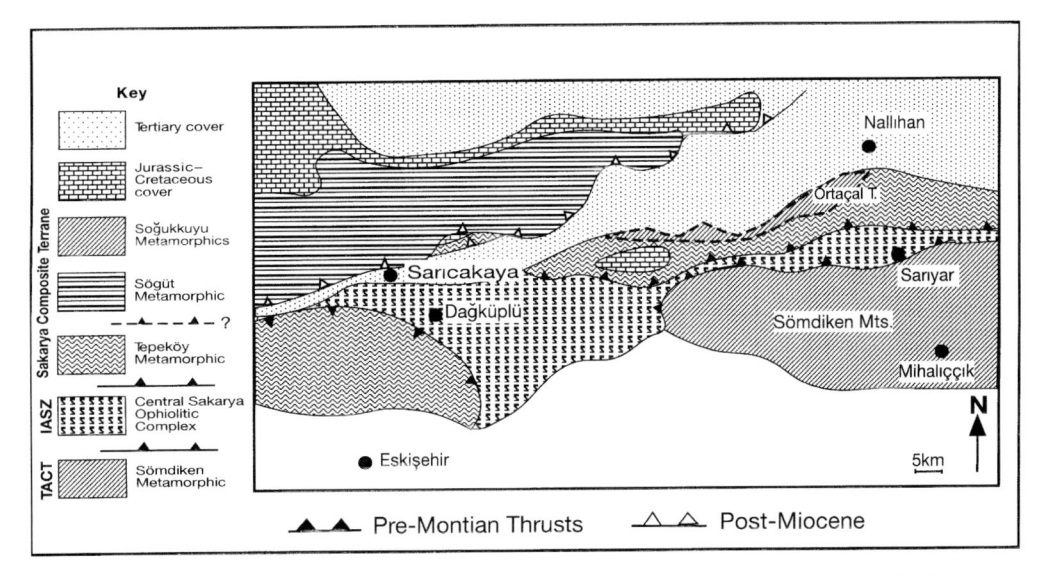

Fig. 2. Simplified structural map of the Central Sakarya area. TACT, Tauride Anatolide Composite Terrane; IASZ, İzmir–Ankara Suture Zone.

ocean was entirely consumed along a north dipping intra-oceanic subduction zone generating suprasubduction zone (SSZ)-type oceanic crust during early Late Cretaceous time (Göncüoğlu & Türeli 1993; Yalınız et al. 1996; Floyd et al. 1998). So, it is critically important to establish whether the intra-oceanic subduction had also played an important role during its closure in the western part of Anatolia as well.

The 'Tauride–Anatolide units' represent the Gondwana continental platform between the Neotethyan İzmir–Ankara–Erzincan Ocean to the north and the Southern Branch of Neotethys to the south. It comprises three units; from north to south these are: the Kütahya–Bolkardağ Belt, representing the tectonic slivers of the northern margin of the platform; the Menderes–Central Anatolian Unit, representing the metamorphic central part, and the Tauride Belt, consisting of a package of mainly non-metamorphic nappes. During the closure of the Vardar–İzmir–Ankara Ocean, ophiolites and slivers of slope-type rocks were thrust southwards (Lycian–Bozkır Nappes) onto the platform, whereas the rest of the platform margin was deeply subducted (Göncüoğlu & Türeli 1993; Okay & Kelley 1994), which resulted in high pressure–low temperature (HP–LT) metamorphism in northern and northwestern Anatolia. However, HP–LT metamorphism is not restricted to the Alpine event. Some authors have also suggested that part of the blueschists in northwestern Anatolia were formed during the closure of the 'Palaeotethys' (e.g. Şentürk & Karaköse 1981; Tekeli 1981; Okay et al. 1996), which was also checked by the present authors' detailed work in the Central Sakarya region.

In this study, the geology of the tectonic units along a geotraverse across the Sakarya Composite Terrane, the İzmir–Ankara Suture Zone and the northern part of the Tauride–Anatolide Composite Terrane in northwest Anatolia are described, and their Late Palaeozoic and Mesozoic geodynamic evolution interpreted. This study is mainly based on many years field work in the Armutlu Peninsula (Göncüoğlu et al. 1987; Göncüoğlu & Erendil 1990), the Central Sakarya region (Göncüoğlu et al. 1996) and the Kütahya–Bolkardağ Belt (Göncüoğlu et al. 1992).

Main tectonic units and their structural relations

The study area along the Sakarya Valley between Sarıcakaya and Nallıhan includes parts of two former continental plates: the Sakarya Composite Terrane to the north and the Kütahya–Bolkardağ Belt of the Tauride Platform to the south. They are separated by the ophiolites and mélange assemblages of the Tethyan İzmir–Ankara Suture Zone (Fig. 2). The Sakarya Composite Terrane is subdivided into the Tepeköy Metamorphics and the Söğüt Metamorphics. Both units were tectonically

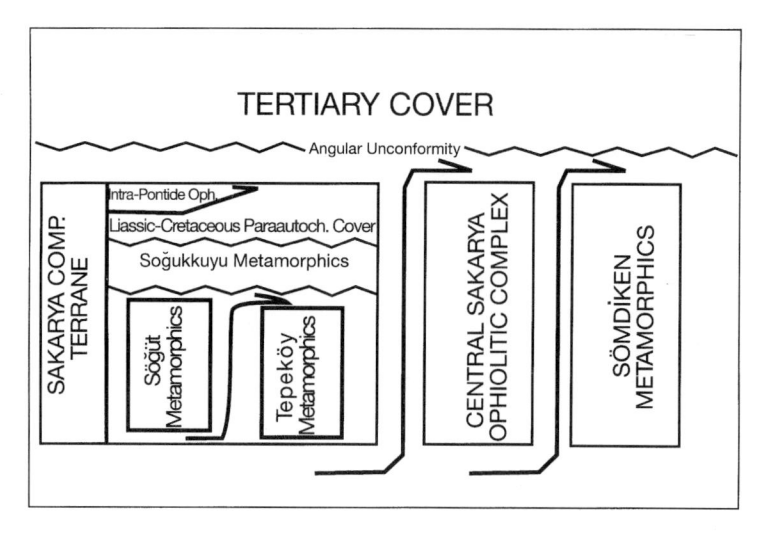

Fig. 3. Structural relations of the tectonic units in the Central Sakarya area.

superposed prior to Early Jurassic time (Fig. 3). The original contact between the Sakarya Composite Terrane and the Central Sakarya Ophiolitic Mélange is a south verging thrust fault that can be followed along-strike for > 75 km along the Central Sakarya Valley. The thrust plane is steep in the east (60–70° to the north), but in the west, the main body of the Tepeköy Metamorphics rests on the ophiolitic rocks with a thrust plane dipping north at c. 30°. The age of initial juxtaposition is probably post-Early Maastrichtian–pre-Middle Palaeocene, as molasse-type continental clastic rocks of Montian age unconformably cover both tectonic units. The original thrust planes must also have been reactivated during the Late Miocene compressional event, shown by north dipping inclined to overturned fold planes of the Lower–Middle Miocene Beypazarı Group and the thrusting of basement units on to the Palaeocene Kızılçay Group. The primary contact between the Central Sakarya Ophiolitic Mélange and Sömdiken Metamorphics of the Kütahya–Bolkardağ Belt (Tauride–Anatolide Platform) is again a south verging thrust, where ophiolitic rocks rest on a gently north dipping thrust plane overriding the metamorphic complex. In the eastern part of the study area, around Çalkaya Hill, the thrust plane is defined by a well-developed mylonitic zone.

Tectonostratigraphy

The structural relationship and generalized tectonostratigraphic features of the main tectonic units in the study area are summarized in Fig. 3.

The Central Sakarya Terrane

In the study area, the pre-Jurassic basement complex of the northern tectonic unit is represented by the Central Sakarya Terrane, a member of the Sakarya Composite Terrane. The latter is an east–west trending Alpine tectonic unit covering almost the whole of northern Anatolia (Fig. 1). It corresponds to a part of the Pontides of Ketin (1966) and the Rhodope–Pontide Fragment of Şengör et al. (1984). The Late Cretaceous closure of the Neotethyan oceanic basins and subsequent collisions of the microcontinents during latest Cretaceous–Palaeocene time has obscured the primary relationships of the pre-Alpine assemblages.

The Central Sakarya Terrane contains two metamorphic units termed the Söğüt Metamorphics and the Tepeköy Metamorphics, together with their cover, the Soğukkuyu Metamorphics (Fig. 3). The metamorphic units are exposed as east–west trending discontinuous tectonostratigraphic units along the Sakarya Valley.

The Söğüt Metamorphics. The Söğüt Metamorphics is the structurally higher unit in the basement of the Central Sakarya Terrane. It comprises paragneisses and granitic plutons. The gneisses comprise garnet amphibolites, sillimanite–garnet gneisses, biotite–amphibole

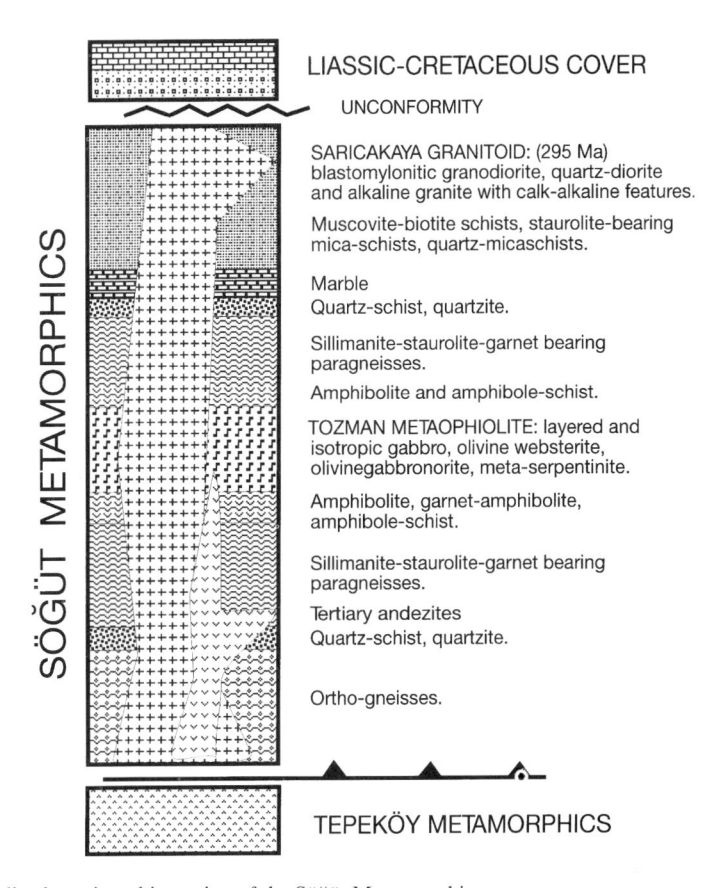

Fig. 4. Generalized stratigraphic section of the Söğüt Metamorphics.

gneisses, two-mica gneisses, sillimanite-bearing staurolite schists, and micaschists with very rare marble and quartzite interlayers (Fig. 4). Rapid alternations of different lithologies indicate that the protoliths of these rocks can be interpreted in terms of sedimentary and/or volcanogenic origin. Discontinuous, small and lens-shaped bodies of layered and cumulate metagabbros, associated with metaolivine pyroxenites (Tozman Metaophiolite; Göncüoğlu *et al.* 1997) and metaserpentinites, are generally concordant with the east–west trending foliation of the surrounding gneisses. These mafic–ultramafic lenses are rimmed by well-foliated amphibolites and high Mg-schists. The pre-Alpine metamorphism under medium–high amphibolite facies conditions of the Söğüt Metamorphics is suggested by the paragenesis: staurolite + almandine + biotite and biotite + almandine + sillimanite in the metapelites and the local presence of migmatites. Post-deformational thermal overprinting is represented by static andalusite porphyroblasts close to the contacts of the intruding calc-alkaline granitoids.

Numerous plutonic rocks of granitic–dioritic composition intrude the Söğüt Metamorphics (Yılmaz 1990). They form elliptical bodies with mylonitic–blastomylonitic textures, elongated east–west. Discordant stocks and dykes of granitoids cross-cut these rocks, and obviously postdate the main metamorphic and deformational event. Yılmaz (1979) recognized five different types of granitic rocks and indicated that they represent a magmatic arc. This suggestion is further supported by geochemical data (Kibici 1990). The granitoids yielded a zircon fission-track age of *c.* 295 Ma (Çoğulu *et al.* 1965). Development of cataclasis is related to late Alpine events.

The depositional age of the paragneisses in the Söğüt Metamorphics is suggested to be mid-Late Palaeozoic, based on regional correlations. Radiometric data from the intruding granitoids, and Lower–Upper Permian carbonates unconformably covering them in the Geyve area, clearly indicate that at least the main metamorphic event is pre-Permian in age (Göncüoğlu *et al.* 1987; Yılmaz 1990).

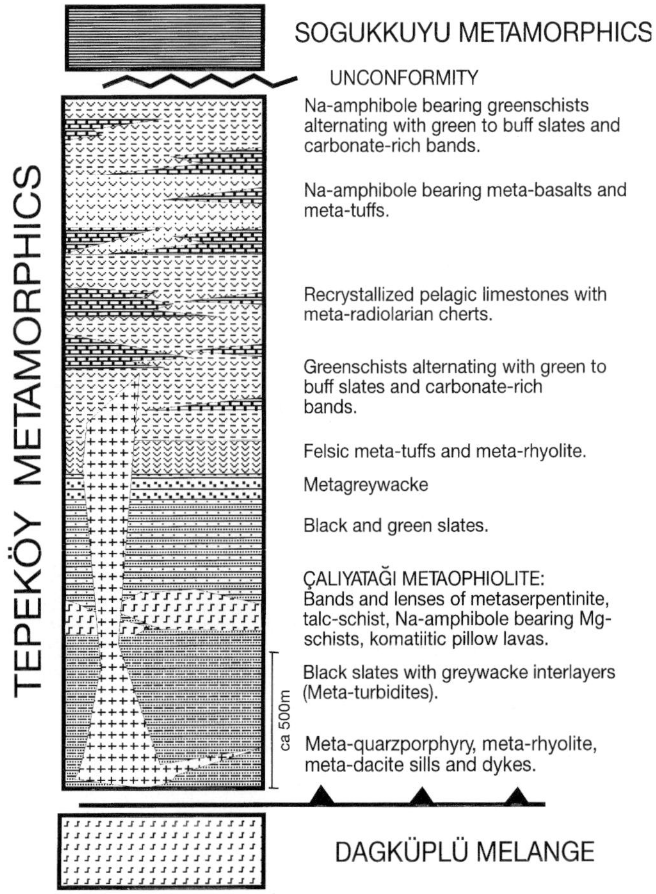
SOGUKKUYU METAMORPHICS
UNCONFORMITY
Na-amphibole bearing greenschists alternating with green to buff slates and carbonate-rich bands.
Na-amphibole bearing meta-basalts and meta-tuffs.
Recrystallized pelagic limestones with meta-radiolarian cherts.
Greenschists alternating with green to buff slates and carbonate-rich bands.
Felsic meta-tuffs and meta-rhyolite.
Metagreywacke
Black and green slates.
ÇALIYATAĞI METAOPHIOLITE:
Bands and lenses of metaserpentinite, talc-schist, Na-amphibole bearing Mg-schists, komatiitic pillow lavas.
Black slates with greywacke interlayers (Meta-turbidites).
Meta-quarzporphyry, meta-rhyolite, meta-dacite sills and dykes.
ca 500m
TEPEKÖY METAMORPHICS
DAGKÜPLÜ MELANGE

Fig. 5. Generalized stratigraphic section of the Tepeköy Metamorphics.

The variety of the metamorphic rock types, the presence of ophiolitic assemblages and the geochemical characteristics of the granitoids intruding them, strongly suggest an island-arc tectonic setting for the Söğüt Metamorphics.

The Tepeköy Metamorphics. This unit represents the lower tectonic slice of the Central Sakarya basement and comprises a belt > 100 km long extending along the Sakarya Valley (Fig. 2). Lithologically, it corresponds to the Sakarya Metabasites of Yılmaz (1979), the 'Greenschist–Marble Association' of Şentürk & Karaköse (1981). In northwestern Anatolia, a lithologically similar unit is known as the Nilüfer Unit (Okay *et al.* 1991).

The bulk of the Tepeköy Metamorphics is composed of metabasic rocks, metatuffs, metafelsic rocks, black phyllites, metagreywackes, metasandstones and recrystallized pelagic limestone with metaradiolarite interlayers (Fig. 5). The unit is > 3500 m thick and, in its lower part to the south-southwest of Nallıhan (where black and green phyllites and greywackes dominate), displays typical features of a 'sedimentary mélange' where depositional processes such as debris flows, gravity sliding and slumping, and tectonic deformation occur. Within these metaturbidites, pods and lenses of serpentinized mafic and ultramafic rocks (Çalıyatağı Metaophiolite; Göncüoğlu *et al.* 1996), and knockers of white, massive marbles are recognized to the northwest of the Sarıyar area. The upper part, along the Sakarya Valley, however, is represented by an alternation of metabasic rocks and carbonates. The carbonates (Eğriköyü Marbles; Göncüoğlu *et al.* 1996) are grey, medium- to thin-bedded and cut by diabase dykes and sills. Thin, dark red to black metachert interlayers are abundant in the upper part. The thickness of the recrystallized limestones locally reaches up to 150 m. The metabasic rocks make up > 75% of the upper part of the succession, including massive and pillowed lavas, volcanic breccia, and metatuffs interlayered with green to buff slates and carbonate-rich bands. The metafelsic rocks occur both as foliated intrusive rocks (quartz–porphyries and rhyolites–dacites) and volcaniclastic rocks.

The metamorphic mineral assemblage in the metagreywackes is actinolite (rimmed by Na-amphibole) + oligoclase + chlorite + epidote + white mica. The metabasalts and metadiabases contain relict phases of Ti-augite, olivine and brown hornblende, and metamorphic phases of actinolite + Na-amphibole + stilpnomelane + albite/oligoclase + epidote + garnet + chlorite. The Na-amphiboles occur both at the rim of actinolite, replacing primary hornblende, and along well-developed S_2 planes, where they are associated with stilpnomelane and chlorite. The basic metatuffs contain almost the same metamorphic mineral assemblage, but they are characterized by two sets of S planes, the younger one of which is represented by nematoblastic Na-amphibole. The metaultramafic blocks occur as lens-shaped bodies up to 200 m thick in the lower part of the unit and, in their core, contain relict primary mineral assemblages such as olivine, clinopyroxene, spinel and chromite. Towards the rim the metamorphic mineral assemblages – i.e. cummingtonite + anthophyllite + Mg-chlorite + gedrite + talc + chlorite + magnesite – dominate. The outermost rim is represented by Na-amphiboles replacing clino-amphiboles. Meta-olivine basalts, with relict porphyritic textures, occur within these meta-ultramafic blocks. These very peculiar glassy lavas display an earlier metamorphic mineral paragenesis of Mg-chlorite + cummingtonite (M_1) and a later one with Na-amphibole (M_2). Textural and mineralogical data strongly suggest an earlier regional upper greenschist facies metamorphism in the Tepeköy Metamorphics, followed by a HP–LT event producing the typical mineral paragenesis.

According to this group's preliminary geochemical data, the glassy lavas show typical geochemical features of boninites (TiO_2 < 0.5%; Zr < 15 ppm MgO $\cong$ 9%; Tokel, pers. comm.). The boninitic chemistry of the basic volcanic rocks, their association with radiolarian cherts and pelagic limestones, and the presence of felsic magmatic rocks intruding them all, suggest an intra-oceanic forearc setting for the Tepeköy Metamorphics. The appearance of mafic, ultramafic and pelagic limestone–chert blocks within a sedimentary complex, on the other hand, suggests an accretionary complex. Thus, it is assumed that the Tepeköy Metamorphics represents a forearc–trench complex. Tectonic units with similar rock assemblages are widespread in northwestern Anatolia. The HP–LT metamorphosed Nilüfer Unit of Okay et al. (1991, 1996) is interpreted as a subducted intra-oceanic arc to forearc sequence. Based on geochemical data, Pickett & Robertson (1996) interpreted the Nilüfer Unit, that consists of metamorphic (low to high greenschist facies) spilites and volcanogenic sediments, as a typical seamount sequence. In both units, continent-derived clastic material, and felsic volcanic and volcaniclastic rocks are noticeably absent. Disregarding the age constraints, the Tepeköy Metamorphics may be considered as a different metabasic unit due to the presence of a felsic volcanism.

No fossils have been reported yet in the Tepeköy Metamorphics, or similar units [e.g. the Nilüfer Unit of Okay et al. (1991)] in the western areas of the Sakarya Composite Terrane. Yılmaz (1981) proposed a Late Mesozoic formation age for the Tepeköy-type metabasic rocks just west of the study area. The Middle Triassic deposition age (Kaya & Mostler 1992), based on fossil findings, is probably not from Tepeköy type metabasites–metacarbonates but from the Soğukkuyu type sediments. In the study area, the primary contact between the Tepeköy Metamorphics and other units of the Central Sakarya Terrane is hard to recognize because of end Cretaceous and Early Miocene tectonics. However, southwest of Nallıhan (Ortaçal Tepe; Fig. 2), the Tepeköy Metamorphics are unconformably overlain by basal clastic rocks of the Soğukkuyu Metamorphics containing pebbles of the Tepeköy Metamorphics. In northwestern Anatolia, the contact between the Nilüfer Unit and the overlying Upper Triassic Hodul Unit has also been assumed to be a stratigraphic contact (Leven & Okay 1996). To the north of Central Sakarya, in the Geyve region, the basal clastic rocks of the Soğukkuyu Metamorphics rest on the Söğüt Metamorphics (Göncüoğlu et al. 1987) and their Permian cover. The same contact relations were reported by Genç (1992) and Genç & Yılmaz (1995) from west of Bilecik. On the other hand, the felsic magmatism in the Tepeköy Metamorphics is ascribed to the latest Carboniferous magmatic event observed in the Söğüt Metamorphics. This data indirectly suggest a pre-Permian formation and initial metamorphism ages for the Tepeköy and Söğüt Units. This suggestion does not exclude the presence of further metabasic rock units of

SOĞUKKUYU METAMORPHICS

LIASSIC-CRETACEOUS COVER

UNCONFORMITY

Recrystallized limestones with ?Upper Triassic algae.

Red to violet shales and cherts with greywacke interlayers (Meta-turbidites).

Basalts alternating with green to red slates and pelagic limestone bands with fragments of early Middle Triassic conodonts.

Metaclastics with blocks and pebbles of recrystallized limestone (Permian) and metabasics

ca 100m

Metaconglomerate

UNCONFORMITY

TEPEKÖY METAMORPHICS

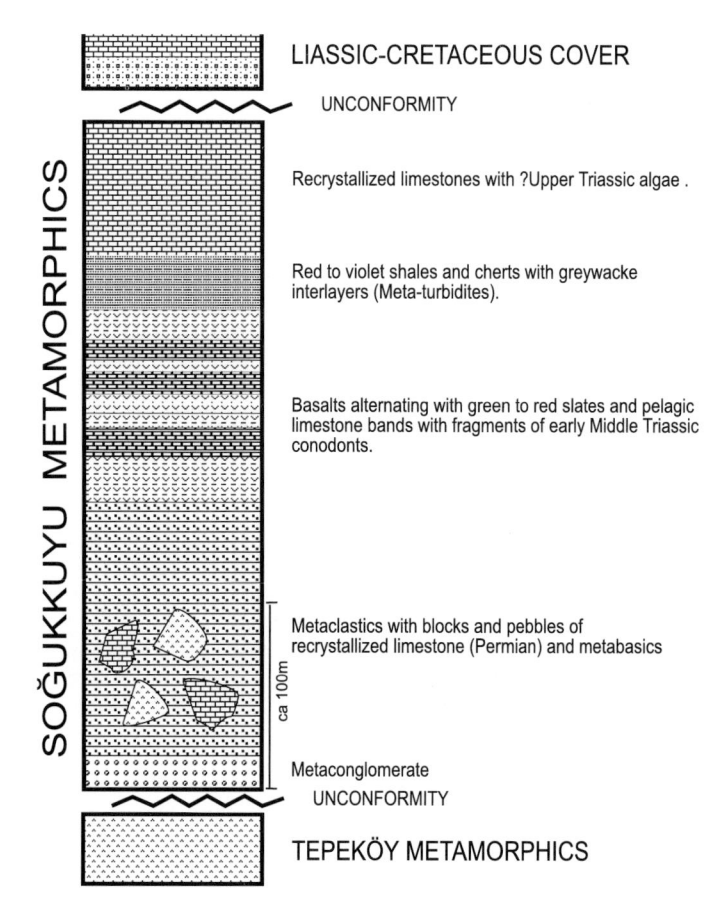

Fig. 6. Generalized stratigraphic section of the Soğukkuyu Metamorphics.

Triassic age within the Sakarya Composite Terrane, as will be discussed later within the framework of the geodynamic evolution of northwestern Anatolia.

The Soğukkuyu Metamorphics. This unit corresponds to part of the Karakaya Complex of Bingöl et al. (1975), and the Hodul and Çal Units of Okay et al. (1991).

In the Central Sakarya Area, the Soğukkuyu Metamorphics unconformably overlie both the Söğüt and Tepeköy Metamorphics (Fig. 6). Southwest of Nallıhan, the basal conglomerates include well-rounded pebbles of metabasalts, metacherts and recrystallized limestones. Southwest of Geyve, the basal clastic rocks of the Soğukkuyu Metamorphics consist of white arkosic sandstones with quartzite lenses (Göncüoğlu et al. 1987). The clasts are mainly derived from the underlying granitic rocks of the Söğüt Metamorphics. The basal clastic rocks are overlain by greywackes, debris flow conglomerates, and turbiditic siltstones–shales with knockers of metabasalts and recrystallized limestones of Permian age. The olistostromal unit, containing rare basaltic lava flows, is followed by a thick sequence where spilites, basaltic pillow lavas and red-violet mudstones alternate with pink-red pelagic limestones and pyroclastic rocks. Higher up, this volcano–sedimentary sequence grades vertically into thin-bedded pink-violet limestones, grey-white cherty limestones and, near the top, white, thick-bedded to massive recrystallized limestones with algae and gastropods (Ortaçal Tepe Limestone; Göncüoğlu et al. 1996). The pelagic limestones alternating with volcanic rocks contain poorly preserved conodont fragments of early Middle Triassic age (Keskin, pers. comm.), whereas the algae in the uppermost massive limestones resemble those in the Late Triassic (Kozur, pers. comm.).

The metamorphic mineral assemblage of basic volcanic rocks in the Soğukkuyu Metamorphics is chlorite + actinolite + epidote +

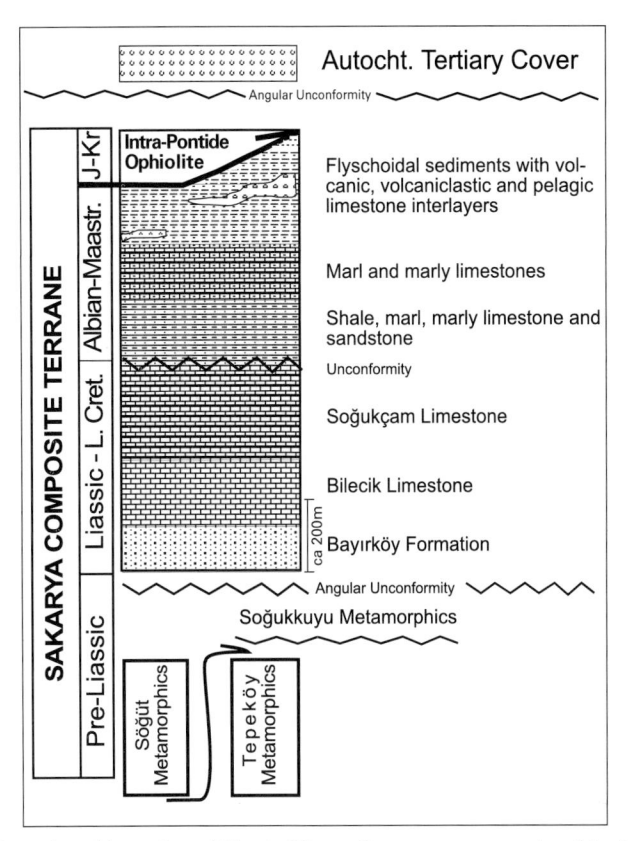

Fig. 7. Generalized stratigraphic section of Liassic–Upper Cretaceous cover units of the Central Sakarya Terrane.

albite, whereas the associated pelitic rocks contain only muscovite + chlorite. Only a single foliation and the absence of HP–LT paragenesis are characteristic features of the unit.

The rock units and their relations strongly suggest that Soğukkuyu Metamorphics were deposited in a rifted basin, which probably opened on the accreted Söğüt and Tepeköy Units and their Permian carbonate cover.

Jurassic–Cretaceous cover. The metamorphic basement units of the Central Sakarya Terrane are unconformably overlain by an unmetamorphosed sequence of Lower Jurassic and younger rocks (Fig. 7). This platform sequence has been mainly studied in detail in the Central Sakarya region (e.g. Altınlı 1975; Saner 1977; Altıner *et al.* 1991; Göncüoğlu *et al.* 1996).

The sequence starts with shallow-marine clastic rocks and carbonates of early Middle Liassic (Altıner *et al.* 1991; Göncüoğlu *et al.* 1996) age. They are followed by platform-type carbonates of Middle Jurassic–Early Cretaceous age that grade into slope-type deposits of Late Jurassic to Middle and Late Cretaceous age towards the northeast. During the Late Cretaceous (Maastrichtian), proximal turbidites with volcanic–volcaniclastic and calciturbiditic interlayers covered the northern part of the Central Sakarya Terrane. The upper part of this flysch succession, which represents a typical foreland sequence, is dominated by ophiolitic detritus. A slice of ophiolitic rocks (Intra-Pontide Ophiolites) overthrust the flysch rocks (Göncüoğlu & Erendil 1990; Yılmaz *et al.* 1995).

During the Middle–Late Mesozoic, the Central Sakarya Terrane as a whole represents a carbonate platform flanked by the Intra-Pontide Ocean in the north and the İzmir–Ankara Ocean in the south. Since the Middle Cretaceous, it was submerged and a foreland basin developed at its northern margin in front of the southward advancing ophiolitic nappes. The initial juxtaposition of the northern İstanbul–Zonguldak Terrane with the Central Sakarya Terrane was probably an oblique collision and

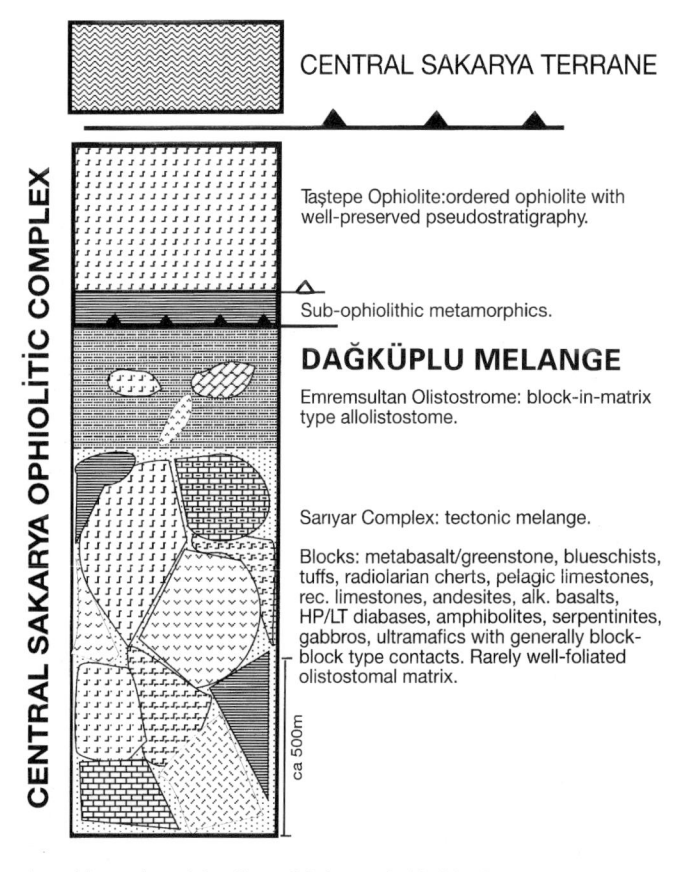

Fig. 8. Tectonostratigraphic section of the Central Sakarya Ophiolitic Complex.

must have taken place during the Early Senonian (Göncüoğlu & Erendil 1990).

The Central Sakarya Ophiolitic Complex. In the Central Sakarya region, the ophiolites and the mélange complex of the İzmir–Ankara Ocean are represented by an east–west trending tectonic sliver almost 100 km long. This ophiolitic complex is sandwiched between the Central Sakarya Terrane and the Tauride–Anatolide Composite Terrane, and represents the northernmost outcrops (and hence the root zone) of the İzmir–Ankara Suture in northwestern Anatolia (Göncüoğlu *et al.* 1997). The ophiolitic complex comprises an upper slice of more or less ordered ophiolite (Taştepe Ophiolite) with subophiolitic metamorphic rocks at its base and a lower disrupted slice (Dağküplü Mélange) (Fig. 8). The latter is further subdivided into mappable units (the Emremsultan Olistostrome and the Sarıyar Complex).

The Taştepe Ophiolite. This unit occurs as a nappe package almost 4 km thick which predominantly comprises slices of tectonites and mafic–ultramafic cumulates. The members of the dyke complex and lava sequence are only found as smaller slices between the main ultramafic body and the underlying mélange complex, or as huge blocks within the mélange. Discontinuous outcrops of metamorphic rocks, alternating garnet–amphibolites, and thin-bedded marble and metacherts, are exposed below the main ultramafic body. In the lowest part of the ophiolites, the dunites are relatively fresh. The partially serpentinized harzburgites contain pods and bands of chromite and display a distinct tectonite fabric defined by the alignment of the orthopyroxenes. The cumulates consist of dunite–clinopyroxenite/wehrlite–clinopyroxenite-gabbro bands. The layered gabbro, consisting of troctolites, two-pyroxene gabbros and gabbro-norites, in ascending order, is highly sheared. The upper part of the layered

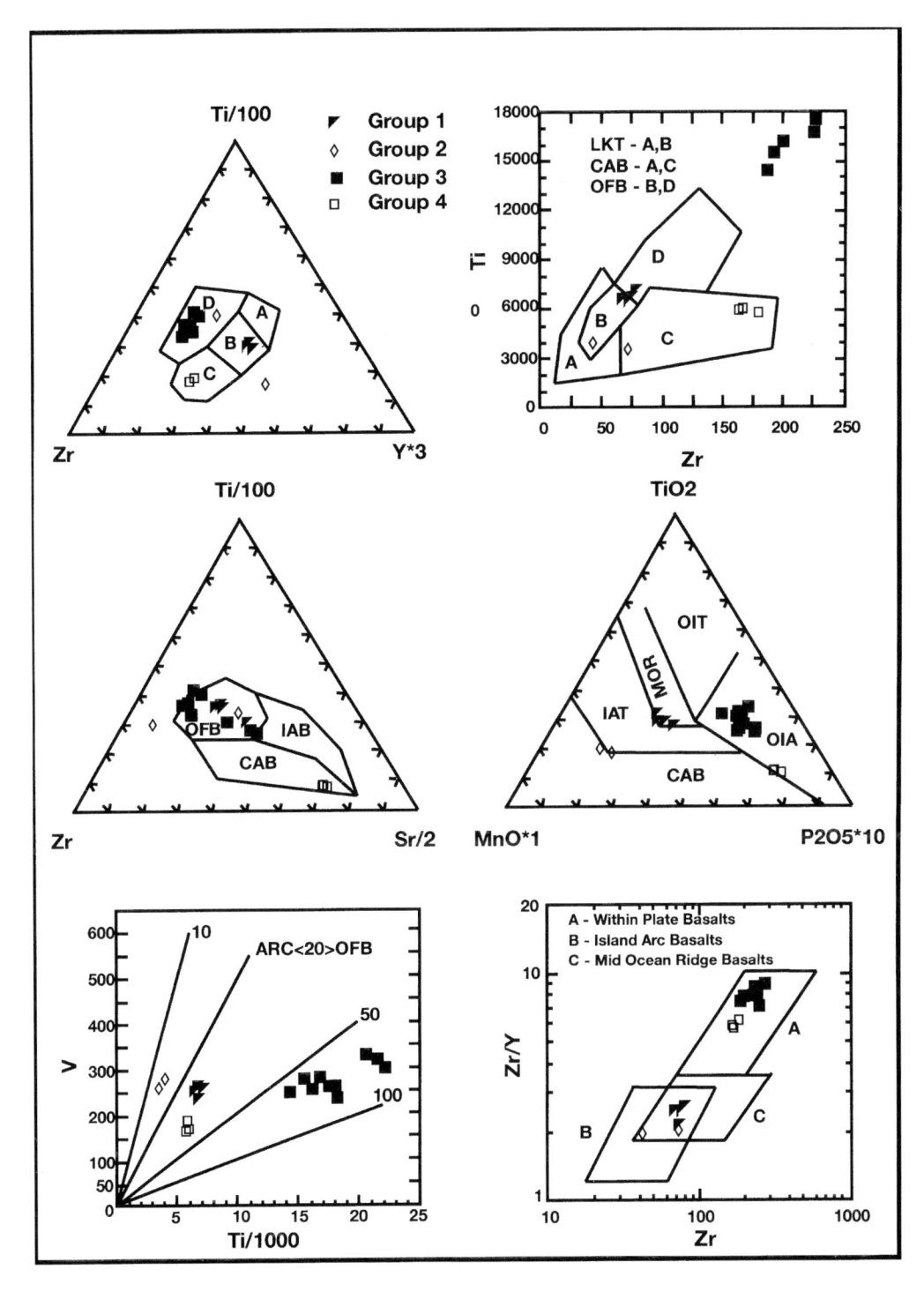

Fig. 9. Tectonic discrimination diagrams of the basic volcanic rocks of the Central Sakarya Ophiolite.

gabbro sequence contains low-grade metamorphic secondary phases such as uralite, chlorite, prehnite and pumpellyite (Asutay et al. 1989).

The Dağküplü Mélange. The Dağküplü Mélange is composed of the Sarıyar Complex and the Emremsultan Olistostrome. The former comprises blocks of spilitic metabasalts, glaucophane–lawsonite schists, radiolarian cherts, pelagic limestones, serpentinites and recrystallized neritic limestones of mainly Mesozoic age (Göncüoğlu et al. 1996). These lithologies make up c. 90% of the mélange blocks. Minor blocks are amphibolites, gabbros, pyroclastics and andesites–dacites (Fig. 8). The knockers are up to several kilometres across and, in general, display tectonic contacts. A well-foliated olistostromal matrix with south verging structural elements is only encountered southwest of Sarıyar. The Sarıyar Complex has a very complex imbricated internal structure with east–west trending shear zones and thrust faults, which is further complicated during subsequent compressional events masking its emplacement on to the Sömdiken Metamorphics and later events.

The neritic limestone blocks within the mélange are highly recrystallized, although their lithologies are very similar to the Middle Triassic–Jurassic carbonates of the Kütahya–Bolkardağ Belt of the Tauride–Anatolide carbonate Platform.

The dominant volcanic rock types are metabasalts. In different blocks of metabasic rocks, the metamorphic mineral assemblages range from typical parageneses of ocean-floor metamorphism to blueschist facies assemblages containing Na-pyroxene + Na-amphibole + lawsonite. The HP–LT parageneses are not restricted to the metabasalts, but were also observed in the metacherts and metatuffs. Na-amphibole nematoblasts were locally encountered in the highly sheared matrix.

Petrographical and geochemical investigations conducted on basic volcanic rocks within the mélange indicate the presence of four different compositional groups (Fig. 9) with distinctive magmatic affinities (Yalınız et al. 1998). The first group, represented by pillow lavas, displays geochemical affinity more akin to mid-ocean ridge basalts (MORB), with a flat pattern close to unity. The second group is made of HP–LT metamorphic pillow basalts, characterized by an island-arc tholeiite signature: strong depletion of high field strength elements (HFSE) and enrichment of large-ion lithophile elements (LILE) relative to MORB. The third group, the dominating block-type volcanic rocks, includes Ti-augite–phyric pillow basalts and breccias, and is characterized by signatures characteristic of within-plate alkaline magmatic series with enrichment of the more incompatible elements (e.g. Nb) relative to MORB. The fourth group, represented by olivine-poor, hornblende–clinopyroxene–phyric massive basalts, is characterized by a calc-alkaline signature, displaying a greater degree of enrichment in low field strength elements (LFSE) relative to MORB, depletion of Nb relative to LFSE and Ti relative to other LFSE and HFSE, and Ce and P enrichment relative to Zr, Ti and Y. The preliminary geochemical data therefore suggest the existence of a variety of magma types, ranging in composition from IAT to MORB to ocean island basalt (OIB) to CAB. The combination of these distinctive magma types suggests a subduction–accretion complex with blocks accreted from different oceanic settings, the most dominant one being an intra-oceanic suprasubduction zone environment.

The metabasalts alternate with radiolarian cherts that contain a rich radiolarian fauna. Spot samples from cherts associated with MORB-like basalts yielded *Dibolachras chandrika* Kocher, *Transhuum* sp., *Williriedellum* sp. aff. *W. carpathicum* Dumitrica, *Protunuma* sp. and *Stichocapsa* spp. *Amphipyndacidae indet.* of Late Bathonian–Early Tithonian age, and *Thanarla bruveri* TAN, *T. elegantissima* TAN and *T. gutta* JUD of latest Hauterivian–Early Aptian age (Tekin, pers. comm.). These age data suggest a Late Jurassic–Early Cretaceous age for the formation of MORB-type oceanic crust in the İzmir–Ankara oceanic branch of the Neotethys. The youngest ages obtained from the pelagic limestones within the mélange are Early Senonian (Asutay et al. 1989), indicating that the mélange formation, and hence the formation of the accretionary complex, lasted until the end of the Cretaceous.

The Emremsultan Olistostrome. The Emremsultan Olistostrome, a block-in-matrix-type allolistostrome, makes up the upper part of the Dağküplü Mélange and occurs mainly as a discrete tectonic sliver in the eastern part of the study area. It has a weakly deformed and greywacke-dominated matrix alternating with shales. The knockers are relatively small but are lithologically almost the same as in the Sarıyar Complex. The depositional features of the Emremsultan Olistostrome strongly suggest that it was formed in piggyback-type basins on the accretionary complex, represented by the Sarıyar Mélange.

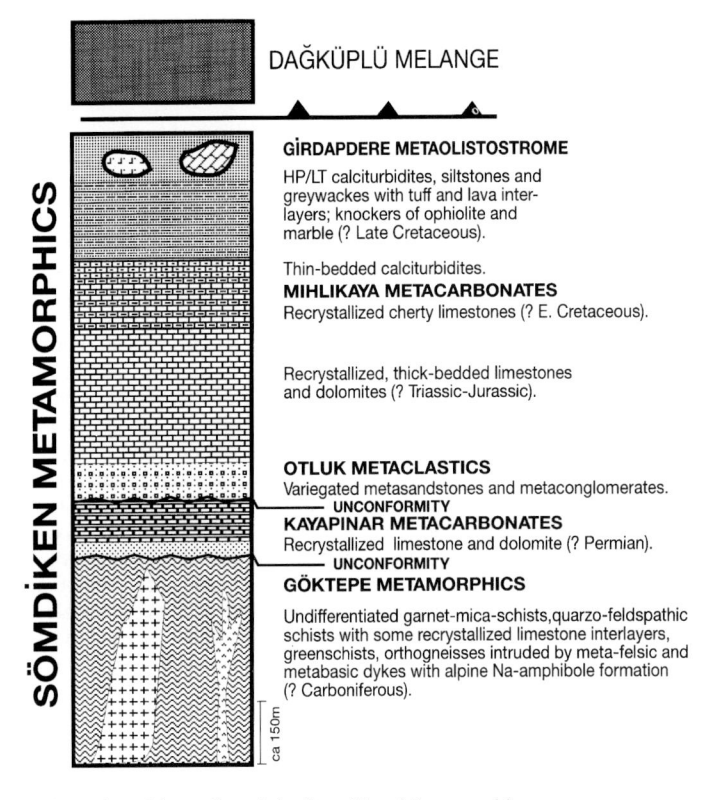

Fig. 10. Generalized stratigraphic section of the Sömdiken Metamorphics.

Sömdiken Metamorphics of the Tauride–Anatolide Platform

The Sömdiken Metamorphic rocks represent the lowermost structural unit in Central Sakarya. They are part of the Kütahya–Bolkardağ Belt that is characterized by slices of metamorphic units and represents the northern margin of the Gondwana Plate. Outcrops of the Sömdiken Metamorphics occur in a tectonic window southeast of the study area (Fig. 2). The generalized columnar section of this unit is given in Fig. 10.

The Göktepe Metamorphics. This unit occurs in the core of a south verging antiform in the Sömdiken Mountains (Fig. 2). It mainly comprises quarzofeldspathic gneisses in its lower part, overlain by quarzofeldspathic schists, pelitic micaschists with very rare carbonate and quartzite bands, and greenschists with a few lydite bands. The gneisses are blastomylonitic and display relict textures, indicating a granitic origin. The pelitic micaschists, > 2000 m thick, contain bands and lenses of greenschists. The whole sequence is cut by highly deformed basic and felsic dykes.

The Göktepe Metamorphic rocks have a polyphase metamorphic history. The micaschists are characterized by an earlier metamorphic assemblage comprising muscovite + biotite + chloritoid + garnet + chlorite, whereas the interlayered greenschists contain chlorite + epidote + actinolite + garnet. The later event, represented by a non-penetrative foliation in the micaschists, is characterized by the paragenesis muscovite + chlorite + albite, whereas the greenschists include Na-amphibole + phengite + stilpnomelane + albite. The basic dykes cutting the Göktepe Metamorphics are only affected by the latter HP–LT paragenesis.

The Göktepe Metamorphics are almost identical with the lower grade metamorphic İhsaniye Metamorphics of the Kütahya–Bolkardağ Belt in the Kütahya and Konya areas (Özcan et al. 1988; Göncüoğlu et al. 1992). In the latter localities, a low-grade metamorphic sequence with lydite-rich turbidites, felsic pyroclastics, olistostromes with fossiliferous Lower Silurian–Lower Carboniferous limestones and chert blocks are associated with metatrachyandesites, metadolerites and metagabbros, intruded by metagranites. Geochemical data on the trachyandesitic and doleritic rocks display a combination of within-plate and continental-arc settings, whereas metagabbros show MORB character (Kurt 1996). The sequence is unconformably overlain by Permian platform-type carbonates. Based on the lithologies of the fossil-bearing sedimentary rock associations, their depositional features and the geochemistry of the volcanic rocks, the İhsaniye Metamorphics were interpreted to be the product of a Carboniferous back-arc system (Özcan et al. 1988). The same tectonic setting is accepted for the Göktepe Metamorphics.

The Kayapınar Metacarbonates. A c. 150 m thick sequence of metacarbonates unconformably overlying the Göktepe Metamorphics is known as the Kayapınar Metacarbonates (Göncüoğlu et al. 1996). The lower part of the sequence is represented by 25 m thick quartzites with chloritoid-bearing micaschist and marble interlayers. The main body of the unit is made up of medium- to thick-bedded marbles with interlayers of calcareous schists.

Similar limestones to the Kayapınar Metacarbonates are well known in the Kütahya–Bolkardağ Belt (Eldeş Formation; Özcan et al. 1988) and in the Taurides (Özgül 1984), where they contain a rich fauna indicating a Middle–Late Permian depositional age.

The Otluk Metaclastites. The slightly metamorphosed epicontinental clastic rocks overlying the Göktepe and Kayapınar Formations, with angular unconformity, are named the Otluk Metaclastics. At the base there are red, violet and brownish massive conglomerates, with pebbles of orthogneiss, micaschists and marble, which pass upwards into an alternation of arkosic metasandstone, quartzite and metasiltstone. The metaclastites are c. 160 m thick and grade upwards into the Mıhlıkaya Metacarbonates. This very characteristic metaclastic unit occurs throughout the Kütahya–Bolkardağ Belt at the same stratigraphic level and has been dated in the Kütahya area as Scythian (Özcan et al. 1988). In this locality, the entire unit is interpreted as a transgressive sequence, starting with proximal alluvial fans and subsequently grading into meandering stream and coastal plain deposits, ending with intertidal sediments. Göncüoğlu et al. (1992) suggested that the deposition of these coarse clastic rocks was related to rapid uplift at the northern margin of the Tauride–Anatolide Unit, related to the initial rifting and subsequent opening of the İzmir–Ankara Branch of Neotethys during the Early Triassic.

The Mıhlıkaya Metacarbonates. This unit is represented by a c. 700 m thick carbonate sequence, which in the lower part contains medium- to thick-bedded, grey, white and black recrystallized limestones and dolomites. The upper part comprises an alternation of grey, beige and pink, thin-bedded recrystallized cherty limestones and cherts, that grade into a c. 50 m thick succession of pink and greenish grey cherts with thin-bedded reddish calc-schist and slate interlayers. This thick carbonate-dominated sequence grades upwards into Girdapere Metaolistostrome. The carbonates of the unit are recrystallized and only yielded ghost fossils in the study area. However, the Gökçeyayla Limestone in the Kütahya area has an identical stratigraphy and yields fossils indicating continuous carbonate deposition from Anisian to Late Jurassic time (Göncüoğlu et al. 1992). The cherty limestones and radiolarian cherts in the upper part of the unit, on the other hand, have been dated as Early Cretaceous–early Late Cretaceous. The same carbonate sequence is the most representative unit along the northern margin of the Tauride–Anatolide Platform. It is interpreted as a platform sequence on the north-facing margin of the Gondwanan Plate. The upper part of the sequence, represented by pelagic–hemipelagic condensed sediments, indicates foundering from platform to slope, and afterwards to basinal conditions.

The Girdapdere Metaolistostrome. The uppermost unit of the Sömdiken Metamorphics is represented by a c. 3000 m thick olistostromal sequence. The transitional lower part of the unit is characterized by a very thick succession of calciturbidites. The main body of the unit is made up of metamorphosed greywackes, calciturbidites, shales, siltstones, metatuffs, radiolarites, pelagic cherty limestones, turbiditic sandstones–conglomerates, and blocks of metabasalts, metaandesites–metadacites, recrystallized limestones and serpentinites. The Girdapdere Metaolistostrome is tectonically overlain by the Central Sakarya Ophiolitic Complex. It has undergone a HP–LT metamorphism, together with the rest of the Sömdiken Unit, with the development of three successive deformational phases. Textural data from the metatuff horizons suggest an earlier phase (S_1) with actinolite + epidote + chlorite formation, the second one (S_2) with syntectonic albite blasthesis and the last phase (S_3) with

Na-amphibole + albite + stilpnomelane + zoisite ± phengite ± lawsonite.

The unit is interpreted as a synorogenic metaclastic sequence formed on the northern margin of the Tauride–Anatolide Platform, in front of the advancing ophiolitic nappes derived from the closing İzmir–Ankara oceanic seaway. The depositional age of the olistostrome is only suggested by the presence of ?Lower Senonian ghost *Globotruncana* from the pelagic cherty limestone bands. However, in the Kütahya region, the lowest part of the synorogenic olistostromal sediments (Çoğurler Olistostrome; Özcan *et al.* 1988) yielded a rich microfauna indicating an early Late Maastrichtian depositional age. This may imply that the deposition of the flysch sediments became younger toward the south.

The presence of HP–LT assemblages implies that part of the continental margin has been deeply subducted, as previously suggested by Okay (1984) for the Tavşanlı area, west of the study area.

Post-tectonic cover

The oldest non-metamorphic sediments that unconformably overlie all the main tectonic units in the Central Sakarya area are represented by red continental conglomerates [Kızılçay Formation of Altınlı (1975)]. The lowest part of this unit contains boulder to block size clasts, derived from the Söğüt Metamorphics, the Dağküplü Mélange and the Sömdiken Metamorphics. The conglomerates are overlain by red to green mudstones, with dacitic to andesitic lavas and volcaniclastic rocks. Several plugs and dykes of dacitic composition occur within the sequence. Geochemical data of Kibici (1990) suggests a continental source and a post-collisional tectonic setting. The upper part of the unit is represented by fossiliferous marly limestones of Early Eocene age. For its lower part, Nebert *et al.* (1986) suggest a Palaeocene depositional age, which was confirmed by Göncüoğlu *et al.* (1996) who found fossils (*Laffitteina bibensis*, *L. mengaudi*, *Mississippina* sp., etc.) of Early Palaeocene age. This clastic-dominated unit is interpreted as a molasse-type depositional product.

These sediments are unconformably overlain by Middle Eocene–?Oligocene continental, shallow-marine sediments associated with andesitic volcanic rocks. The distribution of the sediments along east–west trending troughs, exhibiting very rapid lateral facies, changes strongly suggests that a transtensional system controlled the deposition.

The Miocene is characterized by continental sedimentation in two east–west trending basins, separated by a palaeohigh. The northern basin is characterized by the deposition of coarse clastics and shales, whereas the southern basin, around Beypazarı, is filled with conglomerates, marls and evaporites that contain lignite and trona deposits.

Tectonic evolution of the Central Sakarya area

The field, palaeontological and petrological data, outlined in previous chapters, indicate that the Central Sakarya area is composed of various tectonic units, which record four main orogenic events. The better known Alpine assemblage includes rocks from two oceanic seaways and three continental microplates. These are, from north to south: the İstanbul–Zonguldak Composite Terrane; the Intra-Pontide Suture; the Sakarya Composite Terrane; the İzmir–Ankara Suture; and the Tauride–Anatolide Composite Terrane. However, all of these tectonic units incorporate previous tectonic elements of Early Palaeozoic, Late Palaeozoic and Early Mesozoic events, respectively. The resulting picture is a very complex tectonic mosaic, involving not only continental microplates but also fore- and back-arc complexes, oceanic islands and subduction complexes. In the following section, the available data from northwestern Anatolia and the surrounding areas are interpreted to unravel this complicated history, starting with the Late Palaeozoic event. The northernmost tectonic unit considered in this scenario is the İstanbul Terrane. It differs from the İstanbul Nappe of Şengör *et al.* (1984) and the İstanbul unit of Okay (1989) in excluding the Zonguldak Terrane. The Early Palaeozoic history of the latter unit, involving a Cadomian event, has briefly been outlined by Göncüoğlu (1997) and Kozur *et al.* (1998). In this unit, the basement rocks are represented by an earliest Palaeozoic (c. 550 Ma; Ustaömer & Kipman 1997), intra-oceanic arc complex, unconformably overlain by Lower Ordovician–Upper Silurian platformal sediments. It was attached to the Moesian Platform during the Late Palaeozoic but has a quite different geological history (Göncüoğlu & Kozur 1998) to the İstanbul Terrane.

Late Palaeozoic (Variscan) events

It is commonly accepted that the İstanbul Terrane represents a south facing Palaeozoic

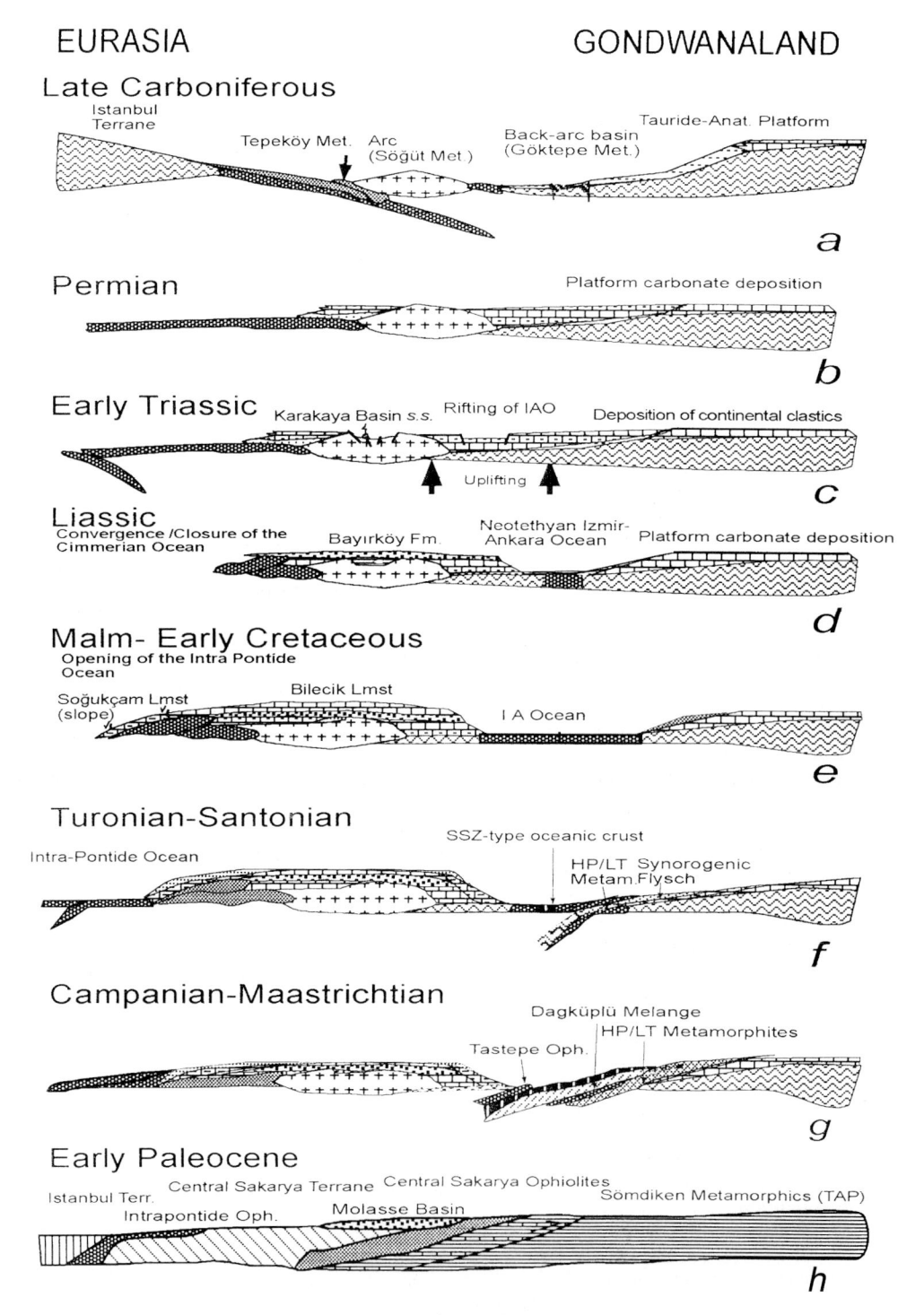

Fig. 11. Late Palaeozoic–Mesozoic evolution of the Central Sakarya region and its surroundings.

platform (Şengör et al. 1984), adjacent to a Palaeozoic oceanic branch (southern branch of the Rheic Ocean or Early Palaeozoic Tethys) that separated some Peri-Gondwanan microplates (e.g. Tauride–Anatolide, Balkan and Central Iran Terranes) from a mosaic of smaller ones, to which the İstanbul Terrane belonged (Göncüoğlu 1997). The Carboniferous rocks in this unit are synorogenic flysch deposits resting on the Devonian platform-slope carbonates, indicating a passive margin setting (Fig. 11a).

The intra-oceanic southward subduction within the Tethys during the Late Carboniferous is responsible for the arc- and forearc-type assemblages, represented by the Söğüt and Tepeköy Metamorphics of the Central Sakarya Terrane (Fig. 11a). Still further south, on the northern margin of the Tauride–Anatolide Carbonate Platform, a back-arc basin developed. The rock units of this basin occur at the base of the Sömdiken Unit (Göktepe Metamorphics) in the Central Sakarya region, as well as in most tectonic slices of the Kütahya–Bolkardağ Belt (e.g. the İhsaniye Metamorphics in the Kütahya area; Göncüoğlu et al. 1992; the 'turbidite-olistostrome unit' in the Karaburun Peninsula; Kozur 1997; and the Halıcı Formation at Konya; Özcan et al. 1988). Further south, on the Tauride–Anatolide Platform, carbonate deposition continued without any major change from the Devonian to the Permian, except for a rapid deepening and the influx of volcanic detritus during the lowest Carboniferous. Hence, during the Middle–Late Carboniferous, the Söğüt and Tepeköy arc–forearc units became attached to the northern margin of the Tauride–Anatolide Terrane. It was probably an accretion, or a gentle docking, rather than a forceful collision, evidenced by the weak deformation and metamorphism of the pre-Permian units (Kütahya–Bolkardağ Belt) of the Tauride–Anatolide Platform.

Autochthonous Lower Permian rocks, unknown in northwestern Anatolia, occur only on the Tauride Platform. However, the early Late Permian was a period of regional transgression on both the northern Tauride–Anatolide and Central Sakarya Units, implying that the Permian carbonate platform in the south also covered the northern units (Fig. 11b). It is not clear whether the İstanbul Terrane collided with the southern assemblages or whether the oceanic realm remained open during this period. However, the presence of latest Permian pelagic limestone blocks within the Triassic 'Karakaya Complex' (Kozur 1997) suggests the presence of a deep basin to the north of Central Sakarya Terrane at this time.

Early Mesozoic (Cimmerian) events

In the Central Sakarya area, Triassic deposition is represented by the Soğukkuyu Metamorphics, which unconformably overlie the pre-Triassic units and are interpreted as a rift sequence. At the initial phase of rifting, coarse clastic rocks and associated rift-related volcanic rocks (Genç & Yılmaz 1995) were formed (Fig. 11c). In the north, the closure of the Late Palaeozoic–Triassic ocean gave way to a very complex system involving subduction–accretionary complexes, ocean islands and/or intra-oceanic arcs, which were subsequently accreted to the Central Sakarya Terrane (Tekeli 1981; Okay et al. 1991; Genç & Yılmaz 1995; Pickett & Robertson 1996). The subduction polarity and the palaeogeographic positions of the microplates in northwest Anatolia are a matter of debate. Pickett & Robertson's (1996) model postulates that the Late Palaeozoic–Triassic ocean was located between the main trunk of Gondwana in the south and Gondwana derived continental fragments with Upper Permian carbonate platforms (e.g. the Sakarya Microcontinent) in the north. It closed by southward subduction, generating a subduction–accretion complex (the Karakaya Complex). The ophiolites [the Denizgören Ophiolite of Okay et al. (1991)] and the subduction–accretion assemblages were then emplaced northward, onto the Permian platform sequences [the Karadağ Unit of Okay et al. (1991)]. Recent work (Okay et al. 1996), however, has shown that the age of the intra-oceanic decoupling of the Denizgören Ophiolite is Early Cretaceous. Okay et al. (1996) also suggested that, during the Permian, the Sakarya Microcontinent was still attached to the Moesian Platform to the north and separated from the main body of Gondwana by the intervening Palaeotethyan oceanic basin. The formation of the Karakaya forearc–accretionary complex was attributed to southward subduction during the Triassic, followed by obduction of the accreted units onto the Sakarya Microcontinent prior to the Early Jurassic. In this study, the Late Palaeozoic–Triassic oceanic basin is located between the İstanbul Terrane and the Central Sakarya Terrane. The main evidence for a southerly location of the latter is its Late Palaeozoic evolution, indicating a continuity with the northern margin of the Tauride–Anatolide Unit prior to the Middle–Late Triassic opening of the Neotethyan branches. In the model presented here (Fig. 11c), the southward subduction model of Şengör et al. (1984) is adopted, and it is suggested that the Soğukkuyu Metamorphics were formed in rift-related marginal basins in the Central Sakarya Terrane. The presence of Nilüfer type oceanic assemblages, *sensu* Okay *et al.* (1991) with a latest Triassic HP–LT metamorphic event in northwestern Anatolia (phengite Ar–Ar ages of 192–214 Ma; Monod *et al.* 1996), does not conflict with the proposed model. Moreover, it may further support a deep intra-oceanic subduction to the north of the Central Sakarya Terrane, followed by accretion and southward backthrusting of the subduction–accretion assemblages towards the south prior to the Early Jurassic. The termination of this orogenic event is marked by deposition of Lower Jurassic epicontinental sediments unconformably covering the orogenic assemblages.

Early Triassic rifting in the Central Sakarya Terrane was accompanied by the formation of basin-and-range-type narrow continental basins on the northern Tauride–Anatolide Platform. Deposition of continental coarse clastics (Otluk Metaclastics) in the Sömdiken Unit, and similar formations further south, marks this event. These data do not support the model of Şengör & Yılmaz (1981), who postulated Jurassic rifting that gave way to the opening of the İzmir–Ankara Ocean. The rift basin north of the Sömdiken Unit, and hence at the northern margin of the Tauride–Anatolide Platform, must have evolved into an oceanic basin during the Middle–Late Triassic. The upper Middle–lower Upper Triassic carbonates in the Kütahya area are characteristically open-shelf to slope-type deposits which do not include rift-related volcanic rocks. However, rift-related and transitional MORB-type volcanic rocks, with basinal sediments of Carnian–Norian age, are found in the Lycian Nappes. It is unequivocally accepted that these nappes were derived from the northernmost margin of the Tauride–Anatolide Platform and emplaced during the closure of the İzmir–Ankara Ocean to the south. Thus, the original location of these nappes should be more internal than the Sömdiken and Kütahya Units.

Another important clue that during the Early Jurassic the Tauride–Anatolide Terrane was already separated from the Central Sakarya Terrane by the intervening İzmir–Ankara Oceanic basin is that the Early Mesozoic (Cimmerian) deformation is nowhere recorded on the Tauride–Anatolide Terrane. All the 'Cimmerian events' in the northwestern part of the Taurides are either based on structural misinterpretations or inaccurate age dating (e.g. Tavşanlı area: Akdeniz & Konak 1979). The 'Cimmerian orogenic events' of Monod & Akay (1984, fig. 2, locations 4–11) in the Taurides *s.s.*, on the other hand, are probably related to intraplatformal tectonic events.

In short, Triassic time designates the closure of the main oceanic branch to the north of the Sakarya Composite Terrane (Şengör *et al.* 1984; Ustaömer & Robertson 1993; Okay *et al.* 1991; Yılmaz *et al.* 1995), the opening and closure of the aborted 'Karakaya Rift Basin' on the Central Sakarya Terrane, and the opening of the Neotethyan İzmir–Ankara Branch between Central Sakarya and the Tauride–Anatolide Terrane by back-arc spreading.

Late Mesozoic (Alpine) events

From the Jurassic, the configuration of the Neotethyan plates is less ambiguous (Fig. 11d and e). One exception is the problem whether the Triassic ocean to the north of the Sakarya Composite Terrane was totally eliminated during the Cimmerian events or whether part of it remained open to develop into the Neotethyan Intra-Pontide Ocean. The former interpretation is supported here and it is proposed that the ocean reopened during the late Middle Jurassic, evidenced by the development of Upper Jurassic slope sequences in the northern Armutlu Carbonate Platform (Önder & Göncüoğlu 1989) and in the northeastern margin of the Central Sakarya Platform (Altı ner *et al.* 1991). In any case, during Jurassic–Early Cretaceous time, the Central Sakarya Terrane represents a carbonate platform limited by the Intra-Pontide Branch in the north and by the İzmir–Ankara Branch in the south. The MORB-type basaltic volcanic rocks associated with Upper Triassic–Lower Cretaceous radiolarian cherts (Göncüoğlu & Erendil 1990; Rojay *et al.* 1995; Bragin & Tekin 1996; the fossil data in this paper) indicate active spreading within these oceanic basins from the Late Triassic to the Early Cretaceous. The platform-type carbonate deposition on the northern margin of the Tauride–Anatolide Platform continued in the Jurassic with a slight change from open-shelf to open-slope conditions towards the end of the Jurassic and Early Cretaceous.

During the Early Cretaceous, the change in relative convergence between Gondwana and Eurasia to a more north–south orientation resulted in a convergence in the Intra-Pontide and İzmir–Ankara Branches of the Neotethys. The events relating to the closure of the Intra-Pontide Ocean have been evaluated by Göncüoğlu & Erendil (1990) and Yılmaz *et al.* (1995). Data from the northern part of the Central Sakarya area suggest that the regional subsidence was represented here by slope-type sediments of early Late Cretaceous age, followed by synorogenic flysch of Maastrichtian age containing ophiolitic detritus. The flysch sediments are overthrust by ophiolitic nappes derived from the Intra-Pontide Ocean. While it is generally accepted that the Intra-Pontide Ocean closed by northward subduction, the age of final collision of the Central Sakarya and İstanbul Terranes, however, is disputed. Based on field data, Göncüoğlu et al. (1987) suggested a Late Cretaceous age of collision, whereas Okay et al. (1994) preferred an Early Eocene collision.

During the Early Cretaceous convergent regime, within the İzmir–Ankara Branch and away from the passive margin of the Anatolide–Tauride Platform, northward intra-oceanic subduction was initiated (Fig. 11f) in Early–Middle Cretaceous times (Göncüoğlu & Türeli 1993). SSZ-type ophiolites were formed above this intra-oceanic subduction by the partial melting of the already depleted MORB-type İzmir–Ankara oceanic lithosphere during early Middle Turonian–Early Santonian times (Yalınız et al. 1996). The SSZ geochemical character of the basaltic rocks from the Dağküplü Mélange in the Central Sakarya area (Yalınız et al. 1998; this study), from the Kütahya region (Önen & Hall 1993) and from Central Anatolia (Yalınız et al. 1996), support this suggestion.

The subophiolitic amphibolites of the İzmir–Ankara Suture Zone from western Central Anatolia yielded mineral and isochron ages ranging from 101 to 90 Ma (Önen & Hall 1993; Harris et al. 1994), clearly indicating an early Late Cretaceous initial decoupling of oceanic crust. The upper level gabbros and dykes of SSZ-type oceanic crust in the Kütahya area, on the other hand, yielded isochron ages of c. 85 Ma (Önen & Hall 1993), which suggests that the formation of these ophiolites is penecontemporaneous with, or postdated, the deep intra-oceanic subduction and related HP–LT metamorphism. Due to their tectonic setting within the hanging wall, they probably escaped deep subduction (Fig. 11f), which would also explain the general absence of very HP parageneses in the SSZ-type ophiolites in both the study area and in Central Anatolia.

The blueschist-facies metamorphism related to this subduction has been the topic of copious studies in northwestern Anatolia (Yılmaz 1981; Okay et al. 1998 and refs cited therein). The HP–LT metamorphism recorded in the Dağküplü Mélange and the Sömdiken Group in the Central Sakarya area indicate, as in northwestern Anatolia, that not only the subduction–accretionary complex but also part of the passive continental margin was deeply subducted. In the Girdapdere Metaolistostrome, in this study area, the latest deformational phase is characterized by metamorphic conditions of c. 6 kbar and 200°C, which would correspond to a 20 km thick overburden of allochthonous material emplaced onto the passive margin of the Tauride–Anatolide platform.

The HP–LT metamorphism of the passive margin sequences in northwestern Anatolia [radiometric age data of Önen & Hall (1993) c. 90 ± 3 Ma is confirmed by Okay et al. (1998) who found Ar–Ar ages of c. 88 Ma] occurred in the Coniacian. However, the progressive southward younging of the metamorphosed synorogenic flysch sediments indicates that the emplacement of the oceanic material, subduction of the margin sequences, their metamorphism, exhumation and incorporation into foreland-type basins lasted until the Early Maastrichtian (Fig. 11g). The youngest flysch sediments unaffected by HP–LT metamorphism occur in the Kütahya area and were dated as early Late Maastrichtian–Early Palaeocene. In the Central Sakarya area, the molasse-type deposits of Middle Palaeocene (Montian) age unconformably cover all the main tectonic units (Fig. 11h). This indicates that the closure of the İzmir–Ankara oceanic basin, and the collision of the Central Sakarya Microcontinent and the Tauride–Anatolide Terrane occurred prior to the Middle Palaeocene.

The Middle Palaeocene–early Middle Eocene period in the Central Sakarya area is dominated by a tensional–transtensional regime, characterized by post-collisional magmatism and deposition of continental to shallow-marine sediments in fault-controlled basins. The Late Palaeogene in the study area is dominated by andesitic volcanism and deposition of alternating marine and terrestrial sediments. Renewed compression at the end of Miocene resulted in deformation of the Neogene basins and southward thrusting of the basement units.

Conclusions

The Late Palaeozoic–Mesozoic orogenic evolution (Fig. 11) of the Anatolian region and its surroundings can be summarized as a history of continuous convergence and divergence of microplates within the same main ocean, i.e. the Tethys. The durations, locations and names of the single branches (e.g. Prototethys, Palaeotethys, Karakaya Ocean, Neotethys, etc.) of this main ocean, as well as the nomenclature of the orogenic products, are still a matter of debate and beyond the scope of this study.

The Central Sakarya area is a key region in understanding the pre-Alpine and Alpine evolution of northwestern Anatolia. Along a north–south traverse in the Central Sakarya region, five different Alpine terranes were distinguished; from north to south these are: the İstanbul Terrane; the Intra-Pontide Suture Belt; the Sakarya Composite Terrane; the İzmir–Ankara Suture Belt; and the Tauride–Anatolide Composite Terrane. Among these, all of the continental units include amalgamated tectonic elements of Variscan and Cimmerian orogenic events.

The pre-Permian the Central Sakarya Terrane basement comprise two tectonic units: the Söğüt and Tepeköy Metamorphics. The former represents a Late Palaeozoic ensimatic arc complex whereas the latter is interpreted as a forearc–trench complex. These two basement units were juxtaposed during the 'Variscan' Orogeny due to the closure of a Palaeozoic oceanic branch by southward subduction. This subduction also produced a back-arc basin at the northern margin of the Peri-Gondwanan Tauride–Anatolide Platform. The Permian carbonates overlying this 'Variscan' orogenic assemblage characterize a period of platformal conditions prior to the Early Triassic rifting.

Triassic rifting is represented by a regional uplift and formation of the rift-basin assemblages (e.g. Soğukkuyu Metamorphics in Central Sakarya) or basin-and-range-type troughs in the northern margin of the Tauride–Anatolide Platform, filled with continental clastics. The rifting is related to the southward subduction of the 'Palaeotethyan' (*sensu* Şengör *et al.* 1984) oceanic crust. Starting from Middle–Late Triassic, one of the extensional basins on the northern Tauride–Anatolide Platform evolved in to the İzmir–Ankara Ocean and separated the Central Sakarya Terrane from the main body. In the north of the Central Sakarya Terrane, continuing subduction resulted in collision and amalgamation of oceanic assemblages to the northern margin of the Central Sakarya Terrane during Late Triassic and completed the 'Cimmerian' Orogeny.

Jurassic–Early Cretaceous time is represented in both microcontinents as a period of stable platform deposition. The only important event is Late Jurassic subsidence north of the Central Sakarya Terrane, probably related to the opening of the Intra-Pontide Ocean. In the Early Cretaceous, both branches of the northern Neotethys began to close by northward subduction.

Within the İzmir–Ankara oceanic seaway, northward intra-oceanic subduction was initiated in Early–Middle Cretaceous times, and SSZ-type ophiolites were formed in the upper plate during the early Late Cretaceous. First the subduction–accretionary complex, and later the passive margin of the Tauride–Anatolide Platform, were deeply subducted and affected by HP–LT metamorphism. Geological and geochronological data suggest that the subduction, blueschist metamorphism, exhumation and incorporation of ophiolite nappes onto foreland-type basins lasted until the Early Maastrichtian. The earliest post-tectonic molasse-type deposits in the Central Sakarya area are of Middle Palaeocene age, indicating that closure of the İzmir–Ankara oceanic basin, and collision of the Central Sakarya and the Tauride–Anatolide terranes predated the Middle Palaeocene.

Post-collisional compression and magmatism in the study area continued until the end of the Palaeogene, and the propagation of southward younging thrusts produced an immense crustal thickening which, in turn, gave rise to the metamorphism of the Menderes Massif.

This study is mainly based on our earlier fieldwork supported by Mineral Research and Exploration Institute of Turkey (MTA). The Scientific and Research Council of Turkey (TÜBİTAK) funded later excursions (project code no: YDABÇAG-85) to the study area. The authors thank K. Tekin, A. Işık and A. Keskin for their the palaeontological contributions, and H. S. Ling for providing the SEM images for the Radiolaria. Geochemical work on the ophiolitic rocks were supported by NATO Collaborative Research Grant 960549 and undertaken by P. A. Floyd, Department of Earth Sciences, Keele University. The authors are grateful to O. Monod and an anonymous reviewer whose suggestions have greatly improved the text.

References

Akdeniz, N. & Konak, N. 1979. *Simav, Emet, Tavşanlı, Kütahya dolaylarının jeolojisi.* Mineral Research and Exploration Institute of Turkey (MTA) Report, No. 6547 [in Turkish].

Altiner, D., Koçyiğit, A., Farinacci, A., Nicosia, U. & Conti, M. A. 1991. Jurassic–Early Cretaceous stratigraphy and palaeogeographic evolution of the southern part of North-Western Anatolia (Turkey). *Geologica Romana*, **27**, 13–80.

Altinli, E. 1975. Geology of the Central Sakarya. *Proceedings of the 50th Anniversary of the Turkish Republic Earth Science Congress*. Mineral Research and Exploration Institute of Turkey (MTA) Publications, 159–191 [in Turkish with English abstract].

Asutay, H. J., Küçüayman, A. & Gözler, Z. 1989. Dağküplü (Eskişehir kuzeyi) Karmaşığının stratigrafisi, yapısal konumu ve kümülatların petrografisi. *Mineral Research and Exploration Institute of Turkey (MTA) Bulletin*, **109**, 1–8 [in Turkish with English abstract].

BİNGÖL, E., AKYÜREK, B. & KORKMAZER, B. 1975. Biga yarımadasının jeolojisi ve Karakaya Formasyonunun bazı özellikleri. *Proceedings of the 50th Anniversary of the Turkish Republic Earth Science Congress*. Mineral Research and Exploration Institute of Turkey (MTA) Publications, 70–77 [in Turkish].

BRAGIN, N. Y. & TEKİN, K. 1996. Age of radiolarian–chert blocks from the Senonian ophiolitic Mélange (Ankara, Turkey). *Island Arc*, **5**, 114–122.

ÇOĞULLU, E., DELALOYE, M. & CHESSEX, R. 1965. Sur l'age de quelques roches plutoniques acides dans région d'Eskişehir, Turquie. *Archive Science de Genève*, **18**, 692–699.

FLOYD, P. A., YALINIZ, M. K. & GÖNCÜOĞLU, M. C. 1998. Geochemistry and petrogenesis of intrusive and extrusive ophiolitic plagiogranites, Central Anatolian Crystalline Complex, Turkey. *Lithos*, **42**, 225–241.

GENÇ, S. 1992. Geology of the Bursa Region. *In*: *A Geotraverse Across Tethyan Suture Zones in northwestern Anatolia*. Mineral Research and Exploration Institute of Turkey (MTA) Publications, 22–25.

—— & YILMAZ, Y. 1995. Evolution of the Triassic continental margin, northwest Anatolia. *Tectonophysics*, **243**, 193–207.

GÖNCÜOĞLU, M. C. 1997. Distribution of Lower Palaeozoic units in the Alpine Terranes of Turkey: palaeogeographic constraints. *In*: GÖNCÜOĞLU, M. C. & DERMAN, A. S. (eds) *Lower Palaeozoic Evolution in Northwest Gondwana*. Turkish Association of Petroleum Geologists, Special Publications, **3**, 13–24.

—— & ERENDİL, M. 1990. Pre-Late Cretaceous tectonic units of the Armutlu Peninsula. *Proceedings of the 8th Petroleum Congress and Exhibition of Turkey*. Turkish Association of Petroleum Geologists Publications, 161–168.

—— & KOZUR, H. 1998. Remarks on the pre-Variscan development in Turkey. *In*: LINNEMANN, U., HEUSE, T., FATKA, O., KRAFT, P., BROCKE, R. & ERDTMANN, B. D. (eds) *Prevariscan Terrane Analyses of 'Gondwanean Europa'*. Schriften des Staatlichen Museums, Mineralogie Geologie Dresden, **9**, 137–138.

—— & TÜRELİ, K. 1993. Orta Anadolu ofiyoliti plajiyogranitlerinin petrolojisi ve jeodinamik yorumu (Aksaray–Türkiye). *Turkish Journal of Earth Sciences*, **2**, 195–203 [in Turkish with English abstract].

——, DİRİK, K. & KOZLU, H. 1996–1997. Pre-Alpine and Alpine terranes in Turkey: explanatory notes to the terrane map of Turkey. *Annales Géologique de Pays Hellenique*, **37**, 515–536.

——, ÖZCAN, A., TURHAN, N. & IŞIK, A. 1992. Stratigraphy of the Kütahya Region. *In*: *A Geotraverse Across Suture Zones in northwestern Anatolia*. Mineral Research and Exploration Institute of Turkey (MTA) Publications, 3–8 [in Turkish with English abstract].

——, ERENDİL, M., TEKELİ, O., AKSAY, A., KUŞCU, İ. & ÜRGÜN, B. 1987. Geology of the Armutlu Peninsula. *In*: *Field Excursion along W-Anatolia*. Mineral Research and Exploration Institute of Turkey (MTA) Publications, 12–18.

——, TURHAN, N., ŞENTÜRK, K., UYSAL, Ş., ÖZCAN, A. & IŞIK, A. 1996. *Geological characteristics of the structural units in Central Sakarya between Nallihan and Saricakaya*. Mineral Research and Exploration Institute of Turkey (MTA) Report, No. 10 094 [in Turkish].

——, ——, ——, ——, —— & —— 1997. Rock units and geodynamic evolution of the Central Sakarya area and its correlation with the Serbo-Macedonian Terrane. *Proceedings of the 15th Carpato–Balcan Congress, Annales Géologique de Pays Hellenique*, **39**, 217–228.

HARRIS, N. B. W., KELLEY, S. P. & OKAY, A. İ. 1994. Post collision magmatism and tectonics in northwestern Turkey. *Contributions to Mineralogy and Petrology*, **117**, 241–252.

HOWELL, D. G. 1989. *Tectonics of Suspect Terranes – Mountain Building and Continental Growth*. Chapman & Hall.

KAYA, O. 1988. A possible Early Cretaceous thrust origin for the ancestral North Anatolian Fault. *METU Journal of Pure and Applied Sciences*, **21**, 105–126.

—— & MOSTLER, H. 1992. A Middle Triassic age for low-grade greenschist facies metamorphic sequence in Bergama (İzmir), western Turkey: the first palaeontological age assignment and structural–stratigraphic implications. *Newsletters on Stratigraphy*, **26**, 1–17.

KETİN, İ. 1966. Tectonic units of Anatolia. *Mineral Research and Exploration Institute of Turkey (MTA) Bulletin*, **66**, 23–34.

KIBICI, Y. 1990. Petrology and genetic interpretation of the Sarıcakaya (Eskişehir) volcanics. *Geological Society of Turkey Bulletin*, **33**, 69–78.

KOZUR, H. 1997. First discovery of *Muellerispharidae* (inc. sedis) and *Eoalbaillella* (radiolaria) in Turkey and the age of the siliciclastic sequence in Karaburun Peninsula. *Freiberger Forschungsheft*, **C466**, 33–59.

——, GÖNCÜOĞLU, M. C. & KOZLU, H. 1998. Caledonian and Hercynian history of Turkey. *The Third International Turkish Geology Symposium*, METU, Ankara, Abstracts, 110.

KURT, H. 1996. Geochemical characteristics of the meta-igneous rocks near Kadınhanı (Konya), Turkey. *Geosound*, **28**, 1–22.

LEVEN, E. JA. & OKAY, A. İ. 1996. Foraminifera from the exotic Permo-Carboniferous limestone blocks in the Karakaya Complex, northwestern Turkey. *Rivista Italiana di Palaeontologia e Stratigrafia*, **102**, 139–174.

MONOD, O. & AKAY, E. 1984. Evidence for a Late Triassic–Early Jurassic orogenic event in the Taurides. *In*: DIXON, J. E & ROBERTSON, A. H. F. (eds) *The Geological Evolution of the Eastern*

Mediterranean. Geological Society, London, Special Publications, **17**, 113–122.

——, ANDRIEUX, J., GAUTIER, Y. & KIENAST, J. R. 1991. Pontides–Taurides relationships in the region of Eskişehir (northwestern Turkey). *Bulletin of the İstanbul Technical University*, **44**, 257–278.

——, OKAY, A. İ., MALUSKI, H., MONIÉ, P. & AKKÖK, R. 1996. Schistes bleus du Trias supérieur en Turquie du NW: comment s'est fermée la Palaeo-tethys? *Proceedings of the 16th Réunion des Sciences de la Terre, Orléans, 10–12 Avril, 1996*. Société Géologique de France, edit Paris, 43.

NEBERT, K., BROSCH, F. J. & MÖRTH, W. 1986. Zur Geologie und plattentektonischen Entwicklung eines westlichen Teilabschnittes der Anatoliden–Pontiden–Sutur. *Jahrbuch der Geologisches Bundesamt*, **129**, 361–388.

OKAY, A. İ. 1984. Distribution and characteristics of the northwest Turkish blueschists. *In*: DIXON, J. E & ROBERTSON, A. H. F. (eds) *The Geological Evolution of the Eastern Mediterranean*. Geological Society, London, Special Publications, **17**, 455–466.

—— 1989. Tectonic units and sutures in the Pontides, northern Turkey. *In*: ŞENGÖR, A. M. C. (ed.) *Tectonic Evolution of the Tethyan Region*. Kluwer, 109–115.

—— & KELLEY, S. 1994. Jadeite and chloritoide schists from northwest Turkey: tectonic setting, petrology and geochronology. *Journal of Metamorphic Geology*, **12**, 455–466.

——, HARRIS, N. B. W. & KELLEY, S. P. 1998. Exhumation of blueschists along a Tethyan suture in northwest Turkey. *Tectonophysics*, **285**, 275–299.

——, ŞENGÖR, A. M. C. & GÖRÜR, N. 1994. Kinematic history of the opening of the Black Sea and its effect on the surrounding regions. *Geology*, **22**, 267–270.

——, SİYAKO, M. & BÜRKAN, K. A. 1991. Geology and tectonic evolution of the Biga Peninsula, northwestern Turkey. *Bulletin of the İstanbul Technical University*, **44**, 91–256.

——, SATIR, M., MALUSKI, H., SİYAKO, M., MONIÉ, P., METZGER, R. & AKYÜZ, R. 1996. Palaeo- and Neo-Tethyan events in northwest Turkey: geological and geochronological constraints. *In*: YIN, A. & HARRISON, M. (eds) *Tectonics of Asia*. Cambridge University Press, 420–441.

ÖNDER, F. & GÖNCÜOĞLU, M. C. 1989. Armutlu Yarımadasında (Batı Pontidler) Üst Triyas konodontları. *Mineral Research and Exploration Institute of Turkey (MTA) Bulletin*, **109**, 147–152 [in Turkish with English abstract].

ÖNEN, P. & HALL, R. 1993. Ophiolites and related metamorphic rocks from the Kütahya region, northwest Turkey. *Geological Journal*, **28**, 399–412.

ÖZCAN, A., GÖNCÜOĞLU, M. C., TURHAN, N., UYSAL, S. & ŞENTÜRK, K. 1988. Late Palaeozoic evolution of the Kütahya–Bolkardağ Belt. *METU Journal of Pure and Applied Sciences*, **21**, 211–220.

ÖZGÜL, N. 1984. Stratigraphy and tectonic evolution of the Central Taurides. *In*: TEKELİ, O. & GÖNCÜOĞLU, M. C. (eds) *Geology of the Taurus Belt*. Proceedings of International Tauride Symposim, Mineral Research and Exploration Institute of Turkey (MTA) Publications, 77–90 [in Turkish with English abstract].

PICKETT, E. & ROBERTSON, A. H. F. 1996. Formation of the Late Palaeozoic–Early Mesozoic Karakaya Complex and related ophiolites in northwestern Turkey by Palaeotethyan subduction–accretion. *Journal of Geological Society, London*, **153**, 995–1009.

ROJAY, B., YALINIZ, M. K. & ALTINER, D. 1995. Age and origin of some spilitic basalts from 'Ankara Mélange' and their tectonic implications to the evolution of northern branch of Neotethys, Central Anatolia. *International Earth Sciences Colloquium on the Aegean Region, Abstracts*, 82.

SANER, S. 1977. Geyve–Osmaneli–Gölpazarı–Taraklı Alanının Jeolojisi: Eski Çökelme Ortamları ve Çökelmenin Evrimi. PhD Thesis, İstanbul University [in Turkish with English abstract].

ŞENGÖR, A. M. C. & YILMAZ, Y. 1981. Tethyan evolution of Turkey: a plate tectonic approach. *Tectonophysics*, **75**, 181–241.

——, —— & SUNGURLU, O. 1984. Tectonics of the Mediterranean Cimmerides: nature and evolution of the western termination of Palaeo-Tethys. *In*: DIXON, J. E. & ROBERTSON, A. H. F. (eds) *The Geological Evolution of the Eastern Mediterranean*. Geological Society, London, Special Publications, **17**, 77–112.

ŞENTÜRK, K. & KARAKÖSE, C. 1981. Genesis and emplacement of the pre-Liassic ophiolites and blueschists of the Middle Sakarya region. *Geological Society of Turkey Bulletin*, **24**, 1–10.

TEKELİ, O. 1981. Subduction complex of pre-Jurassic age, northern Anatolia, Turkey. *Geology*, **9**, 68–72.

USTAÖMER, P. A. & KIPMAN, E. 1997. An example for a pre-Early Ordovician arc magmatism from northern Turkey: geochemical study of the Çaşurtepe Formation (Bolu – W Pontides). *Proceedings of the Symposium for the 20th Anniversary of Geology, Çukurova University Abstracts*, 84.

—— & ROBERTSON, A. H. F. 1993. A Late Palaeozoic–Early Mesozoic marginal basin along the active southern continental margin of Eurasia: evidence from the Central Pontides (Turkey) and adjacent regions. *Geological Journal*, **28**, 218–238.

YALINIZ, M. K., FLOYD, P. A. & GÖNCÜOĞLU, M. C. 1996. Supra-subduction zone ophiolites of Central Anatolia: geochemical evidence from the Sarıkaraman Ophiolite, Aksaray, Turkey. *Mineralogical Magazine*, **60**, 697–710.

——, GÖNCÜOĞLU, M. C. & FLOYD, P.A. 1998. Geochemistry and geodynamic setting of basic volcanics from the northernmost part of the İzmir–Ankara branch of Neotethys, Central Sakarya Region, Turkey. *Proceedings of the 3rd International Turkish Geology Symposium, METU-Ankara, Abstracts*, 174.

YILMAZ, Y. 1979. Söğüt–Bilecik bölgesinde polimetamorfizma ve bunların jeotektonik anlamı. *Geological Society of Turkey Bulletin*, **22**, 85–100 [in Turkish with English abstract].
—— 1981. Sakarya kıtası güney kenarının evrimi. *İstanbul University Earth Sciences*, **1/2**, 33–52 [in Turkish with English abstract].
—— 1990. Allochthonous terranes of Tethyan Middle East: Anatolia and the surrounding regions. *Philosophical Transactions of the Royal Society, London, Series A*, **331**, 611–624.
——, GENÇ, Ş. C., YİĞİTBAŞ, E., BOZCU, M. & YILMAZ, K. 1995. Geological evolution of a late Mesozoic continental margin of Northwestern Anatolia. *Tectonophysics*, **243**, 155–171.

Aspects of Jurassic radiolarite sedimentation in a ramp setting following the 'mid-Late Jurassic discontinuity', Barla Dağ area, Western Taurus, Turkey

ANNA FARINACCI,[1] ANDREA FIORENTINO[2] & VALERIO RIDOLFI[3]

[1]*Dipartimento di Scienze della Terra, Università La Sapienza, P.le Aldo Moro, 00185 Roma, Italy (e-mail: farinacci@axrma.uniroma1.it)*
[2]*Via Tacito 41, 00193 Roma, Italy*
[3]*Via Senofane 216, 00124 Roma, Italy*

Abstract: The Barla Dağ area of southwestern Turkey and its surroundings represent one of the most characteristic Tethyan regions in which the unique characteristics of the Jurassic radiolarite deposits permit detailed study of this enigmatic facies. Hitherto, radiolarites of Western Tethys have not been studied in sufficient detail to yield the information required for unequivocal interpretation of this siliceous sedimentary event. Moreover, few of the occurrences of Tethyan radiolarites during the Jurassic have been adequately explained by palaeoenvironmental causes deduced from facies analysis. In the Barla Dağ area, the main radiolarite episode began after the 'main gap' or mid-Late Jurassic discontinuity, a 25 Ma hiatus extending from the Early Bajocian to the Kimmeridgian. These radiolarites are interbedded with biocalcarenites characterized by shallow-water shells. They formed in a ramp environment subject to strong storm oscillatory movements and were deposited within, or just below, wave base.

Pre-existing platforms were converted into ramp settings by a widespread drowning episode, mainly following postulated regional warping that led to creation of the 'main gap'. Coincident with this event, the differentiation of rimmed platform lagoonal organisms and typical ramp inhabitants, such as *Tubiphytes*, took place. Furthermore, nearby platforms, unaffected by the extensional faulting (e.g. the Davras Dağ), were sites of carbonate accumulation receiving only a few radiolarians. On the other hand, displaced shallow-water organisms of the same age (typical of the restricted lagoons flanking the rimmed platforms such as pfenderinas, kurnubias and *Clypeina jurassica*) are absent from the sequences of calcarenites interbedded with radiolarian cherts. Replacement of deep basins by ramps is indicated by the changing depositional bathymetry of some radiolarites. It is tentatively attributed to the extension of shallow seas and narrowing of the oceanic realm between Eurasian and African Plates in Western Tethys.

Siliceous deposits are developed within the Mesozoic Bahamian-type platform carbonates of the Western Taurus (Farinacci & Köylüoğlu 1982). The Barla Dağ and its eastern surroundings (Fig. 1) represent one of the most characteristic areas in which the presence of Jurassic radiolarites, together with calcarenites, permit determination of the depositional bathymetry of some radiolarites. Here, as well as in similar Tethyan basins, radiolarite deposition represents the interaction of several events, each of which can be investigated. Until recently, such radiolarites have been widely interpreted in terms of deep bathymetry being the sole, or principal, environmental feature preventing biological carbonate deposition and permitting preservation of radiolarians on the seafloor.

In the Barla Dağ area, the Jurassic carbonate and siliceous deposits of cherty facies, developed during the main phases of the western Tethyan rifting, are extensively exposed. In this region, sedimentary environments and basin geometry have been strongly affected by rift tectonics and magmatic events. The interlinking of many physical features of these carbonate-chert sequences permits the construction of an environmental model in which tectonic effects predominate, even though the mechanism of rifting is not easily discerned because Tertiary compressional events have heavily overprinted the Jurassic structures.

This group's attempt at reconstructing the Jurassic basin setting of the Barla Dağ radiolarites relies exclusively upon field evidence obtained during this study, and from similar work elsewhere in the Western Tethys, and does not depend on existing models of ocean rifting mechanisms related to the Atlantic Ocean or to the Mediterranean Tethys.

From: BOZKURT, E., WINCHESTER, J. A. & PIPER, J. D. A. (eds) *Tectonics and Magmatism in Turkey and the Surrounding Area.* Geological Society, London, Special Publications, **173**, 163–170. 1-86239-064-9/00/$15.00
© The Geological Society of London 2000.

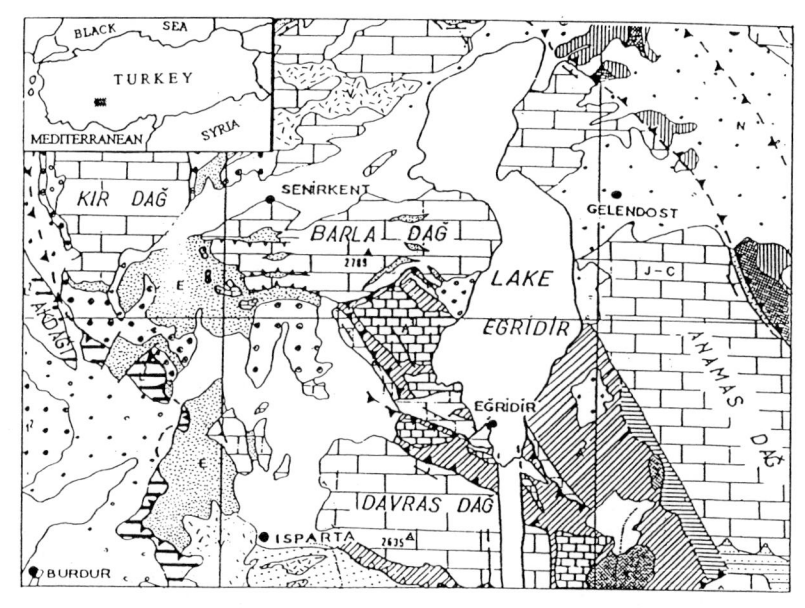

Fig 1. Location map of the Barla Dağ area. The large brick pattern represents Mesozoic shallow-marine limestones in autochthonous position, overlain by allochthonous units [after Brunn *et al.* (1971)].

Historical and theoretical framework

In the study of the Tethyan radiolarites, interpretation has frequently been confused with observation. Reconstructions of palaeoenvironments and basin morphology have commonly been based on widely held assumptions rather than direct palaeontological, sedimentological and geochemical evidence. However, recent progress in these disciplines has led to a substantial reappraisal of previously held views concerning the shape and morphology of many Jurassic radiolarite-bearing basins; e.g. the interpretation of water depth which was set in its historical frame by Haug (1900) has been briefly analysed by Garrison & Fischer (1969). The models proposed by these authors were mainly based on actualistic reconstructions and invoked depths below present levels of the carbonate compensation depth (CCD) as the sole or main cause of radiolarite sedimentation (Garrison & Fischer 1969 and refs cited therein). On the basis of this tenuous assumption, Garrison & Fischer (1969) invoked a wide Tethyan ocean, thousands of metres in depth, resulting from Jurassic rifting.

On the other hand, in a recent study of the Jarropa Radiolarite Formation in the Subbetic area of Southern Spain, which contains 'calciclastic strata with hummocky cross-stratification, indicating an outer carbonate ramp deposition', Molina *et al.* (1999) discuss different contemporary views on the subject. They summarize the opinions of workers in the Alps, Apennines, Carpathians, Dinarids, Hellenids, Rif, Tell, and elsewhere, who both defend and dispute the old palaeobathymetric scenarios. Separated by three decades of work, the two papers mentioned above (Garrison & Fischer 1969; Molina *et al.* 1999) are representative of the broad range of views concerning the sedimentation of radiolarites. The present study produces additional evidence and represents a further contribution to this problem. It augments previous studies carried out by the research team of Rome University, working in the Apennines (Farinacci *et al.* 1981; Farinacci 1988) and yields conclusions in broad agreement with the results of Molina *et al.* (1999).

The Barla Dağ radiolarites

Throughout the Barla Dağ area, radiolarites are extensively developed in the Jurassic–Cretaceous sedimentary succession and usually represent the thickest Jurassic sedimentary interval formed after drowning of the Yassiviran carbonate platform (composed of post-Triassic limestones and dolostones), with thicknesses ranging up to a few tens of metres (Fig. 2). Following the Late Pliensbachian drowning, appeared at the time of drowning. However, the top layers of the stratified nodular micritic limestones, which contain filaments and radiolarians, alternate oosparites with peloids, suggesting a further upward shallowing until the 'main gap', a hiatus which extended from Early Bajocian to Kimmeridgian times.

The Tınaz Tepe section typifies the Jurassic sedimentary evolutionary trend of the Barla Dağ and its surroundings following the Late Pliensbachian drowning. Differences among the various Barla Dağ sections can be discerned in the variable thicknesses of the biosequences and in their fossil assemblages. These differences reflect locally modified microenvironments within the Late Jurassic ramp setting, and especially the changes in their hydrodynamic conditions. In the sequences above the Bajocian–Kimmeridgian hiatus, the radiolarite interval comprises three well defined lithotypes: (1) radiolarian chert; (2) radiolarian calcareous mud; and (3) calcarenite. These three components result from the interaction of chemical, biochemical and mechanical events taking place in the environment, either simultaneously or alternating with a marked periodicity.

The siliceous episode

In the Jurassic, siliceous episodes marked by the abundance of radiolarians evidently start when the build-up of the carbonate-rimmed platform ceased in some areas. In the Barla Dağ, the siliceous interval can be considered to be a single sedimentary episode that began at the end of the Pliensbachian and reached its acme during the Kimmeridgian and Tithonian, where it is characterized by radiolarian chert.

Immediately after the initial appearance of radiolarians, at the time of the Late Pliensbachian drowning, calcareous thin-shelled posidoniids become widespread (forming the so-called 'packed filaments' of sedimentological terminology). The posidoniids include two thin-shelled species: *Bositra buchii* (Roemer) and *Lentilla humilis* Conti & Monari (Conti & Monari 1992). They reached their maximum development during the Aalenian, together with ammonites, nodosarids and gastropods, in addition, of course, to radiolarians. Thin-shelled bivalves can also be found associated only with radiolarians in some layers; these are present up to the appearance of calpionellids in the Early Cretaceous.

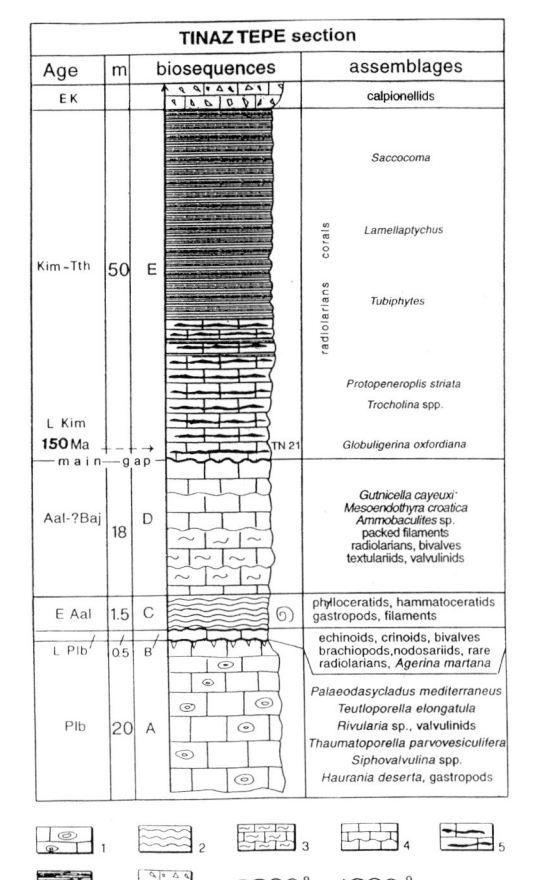

Fig 2. The Jurassic sedimentary episodes of the Barla Dağ sequence are well represented by the Tınaz Tepe section in which the depositional trend is clear. The section is subdivided into biosequences according to the recognized discontinuities. Note the relatively small thickness of sediments involved, changes in the fossil assemblages and the duration of the 'main gap'. Key to symbols: 1, algal limestones; 2, marly nodular limestones with ammonites; 3, packed filaments; 4, bedded nodular yellow limestones; 5, micritic limestones with yellow cherts; 6, radiolarian cherts interbedded with calcarenites; 7, breccia with small fragments; 8, discontinuity; 9, discontinuity with desiccation cracks. E, Early; L, Late. Modified from Farinacci et al. (1997).

several short depositional episodes, separated by hiatuses, occurred in the Early Aalenian and were characterized by a great abundance of crinoids. Similar short episodes in the Late Aalenian–Early Bajocian are marked by abundant posidoniid bivalves and a significant increase in radiolarians which had already

The base of the radiolarite and the 'main gap'

The well-characterized radiolarian bloom of the siliceous episode, represented in Tınaz by Biosequence E (Farinacci *et al.* 1997), is particularly evident immediately after the 'main gap', when sedimentation recommenced in the Kimmeridgian after *c.* 25 Ma of non-deposition. The radiolarites appear sharply above the top layers with packed filaments, containing *Mesoendothyra croatica* Gusic, of the underlying biosequence (the Tınaz Biosequence D, which ranges from Late Aalenian to Early Bajocian; Fig. 2). However, in the field this boundary is not conspicuous: the lithology changes from roughly nodular limestone with yellow clay to cherty limestone without clay. The first transgressive layers contain thin-shelled *Globuligerina oxfordiana* (Grigelis). Radiometric dating reveals that the 'main gap' between the two biosequences (D and E in Tınaz) represent *c.* 25 Ma (Farinacci *et al.* 1997).

In other areas of the Tethys, the 'main gap' is present but may be somewhat shorter or longer. It is known in the Central Apennines (Cecca *et al.* 1986, 1990), in the Ionian Zone of southwestern Albania (Cope *et al.* 1994; Dodona *et al.* 1994), and in the External Dinarids (Farinacci 1996). The main chemical event influencing radiolarite deposition is the silica supply in the sea water. This allows radiolarians to bloom, thus fixing SiO_2, and leads to the sedimentation of siliceous rocks. In the Barla Dağ, the radiolarian cherts are interbedded with calcarenites and with micrites, with the chert occurring as nodules and thin layers.

The interbedded tempestites

Storm deposits, represented by calcarenites, alternate with the radiolarian cherts. The cherty layers are laminated and increase in frequency toward the top of the Jurassic succession. The calcarenites consist of layers with abundant skeletal debris comprising a high-energy fauna of corals, thick-shelled bivalves, more or less broken cyanobacterial constructions, peloids (some of which are coated) and foraminifera with microgranular tests. The calcarenites also contain clasts of radiolarian micrites formed from the contemporaneous substrate and incorporated among the fragmentary shallow-water organisms. In the micritic layers, the mud is sometimes homogeneous, but may also be more granular in character and form peloids.

Frequent tempestites appear as repeated layers, the bottom surfaces of which display flame structures, but rarely have sharp interlayer boundaries, especially when deposition of these storm units has occurred above the storm wave-base level. Erosional basal surfaces are much more common in the lower part of the tempestite succession where the radiolarites are commonly intercalated with calcarenites. Corals are preserved as reefal fragments and the mixture of other organisms, together with the corals, suggests that reef bodies were laterally discontinuous and formed small patch reefs on local prominences within the ramp setting.

Tubiphytes is very common in the calcarenites interbedded with cherts and is the most widespread 'organism'. Although it is still an *incertae sedis* (supposedly a cyanobacterial construction), its palaeoenvironment is well known. It is common on the ramps or on the outer part of rimmed shelves, and is regarded as one of the more opportunistic organisms thriving after the storms and taking advantage of the available skeletal carbonate. The most frequent foraminifera are textularids, valvulinids and planispirally coiled forms. They are also present in the platform lagoons, where they are much bigger. Their reduced size in the ramp environment may be ascribed to the increased turbulence of that setting. *Protopeneroplis striata* Weynschenk and *Trocholina* spp. were also recovered. They are both typical of high-energy environments and are always found together on the ramps and on the outer marginal slope. Some calcarenites exhibit a different constitution: in addition to the shallow-water elements enumerated above, the remains of open-sea organisms, the so-called 'pelagic' elements (radiolarians, filaments, etc.), may contribute to the skeletal assemblage.

The succession at Tınaz Tepe continues upward to the top of the Jurassic sequence, with 'normal' open-sea Tethyan limestones containing successively packed filaments, then aptychi (among which *Lamellaptychus* is the most common) and then *Saccocoma* (Fig. 2). The benthonic lifestyle of *Saccocoma*, a stemless crinoid which lived on muddy substrates, has been demonstrated by Manni *et al.* (1997). It appears to have been 'an opportunist able to occupy a very selective and rapidly variable environment where other benthonic species cannot survive' (Manni *et al.* 1997, p. 131).

Morphological and other controls on siliceous sedimentation

All the evidence observed during our study points to localization of the Kimmeridgian radiolarite sedimentation in a carbonate ramp setting. The siliceous sediments are composed of fine radiolarian layers interbedded with calcarenites made up of fragmented skeletal elements of calcareous ramp organisms mixed with small clasts of radiolarian mud of comparable size, derived from the interbedded deposits. Fragmented remains in the calcarenites belong to organisms which normally did not live in the platform lagoons but were typical of ramp facies. Moreover, the gentle slope of the ramp favoured a clinoform bedding that has determined the sedimentary boudinage of the plastic chert and has even induced slumping.

In the shallow-water sectors of these ramps, the sedimentation of fragmented material, whether coming from outer margins of the platform or produced *in situ*, is normal. Ramps are particularly exposed to storm events and fragmentation might also be caused by induced oscillatory movements, ending with rapid deposition. Kimmeridgian times seem to be a critical period for Tethyan ramps due to a general increase in hydrodynamic energy. The presence of Upper Jurassic calcarenites is mentioned by Savu *et al.* (1995) and Dragastan (1997) in the Transylvanides, Farinacci (1996) in the Dinarids, Molina *et al.* (1997) in the Betic Cordilleras, and Ager (1974) in the Moroccan High Atlas. According to the scheme of Burchette & Wright (1992), who define the mid-ramp sediments as fine-grained, graded tempestites interbedded with laminated mudstone (Fig. 3), the Barla Dağ tempestites were deposited in a mid-ramp setting.

The processes which led to formation of the ramps started with the Late Pliensbachian drowning, ascribed to block faulting of the Taurus platform margin. Tectonics induced the formation of a tilted half-graben which controlled the type of facies and sedimentation, and strongly limited accumulation. On the higher parts of the half-graben, sedimentation was inhibited during the Toarcian and is almost absent from the drowned platform facies of the Barla Dağ, whereas the Aalenian is represented

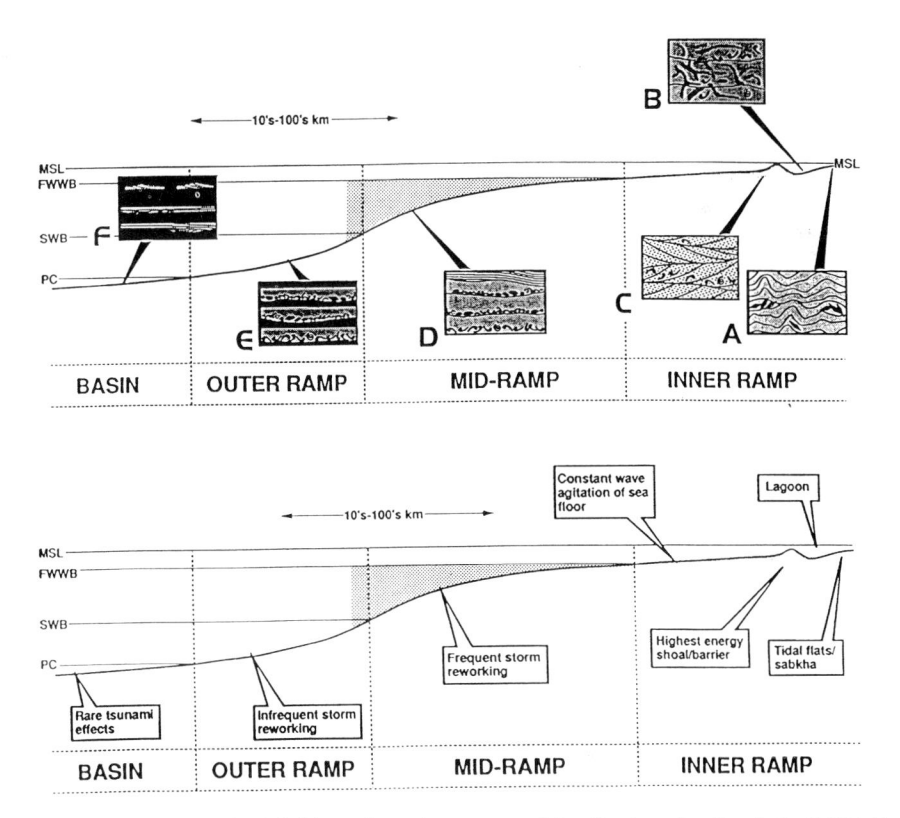

Fig 3. The main environmental subdivisions of a carbonate ramp [classification after Burchette & Wright (1992)]. All the evidence observed in Tınaz Biosequence E points to the localization of the Barla Dağ Kimmeridgian–Tithonian radiolarite sedimentation in a mid-ramp setting (D) between fair-weather wave base (FWWB) and storm wave base (SWB), in which sediment is frequently reworked by storms (shaded area).

by strongly oxidized and silty materials, probably related to subaerial exposure. This facies persist up to the Early Bajocian when sedimentation abruptly terminated.

After the 25 Ma hiatus of the 'main gap', the rifting process, although continuing in the mid–Late Jurassic by means of extensional faulting, resulted in important regional up-doming (Fig. 4). Subsidence almost ceased and the conditions for development of the ramps were created, both during the up-doming and subsequent deflation. At this point, further drowning took place and gave rise to rimmed shelves and ramps. Such an event is emphasized during the main phase of the rifting by a dramatic change in the marine chemistry. This favoured siliceous organisms which had already started to proliferate by the end of the Pliensbachian. In the Late Jurassic they out-competed other organisms on the ramps, but not on the adjacent rimmed shelves. Occasionally, radiolarians could even enter the restricted lagoons of the nearby platforms, as in the Davras Dağ area, and were carried there by storms, whereas lagoon organisms never reached the ramps. Over a period of time the storm-generated turbulence decreased and mud facies began to develop, supported by cyanobacterial activity, and benthonic stemless crinoids, such as *Saccocoma* sp., became widespread.

The correlation of SiO_2-radiolarians with magmatism has been noted by several authors (e.g. Conti & Marcucci 1986; Farinacci 1988; Pessagno *et al.* 1993) whilst investigating the origin of Jurassic radiolarites. The presence and preservation of radiolarians appear to record

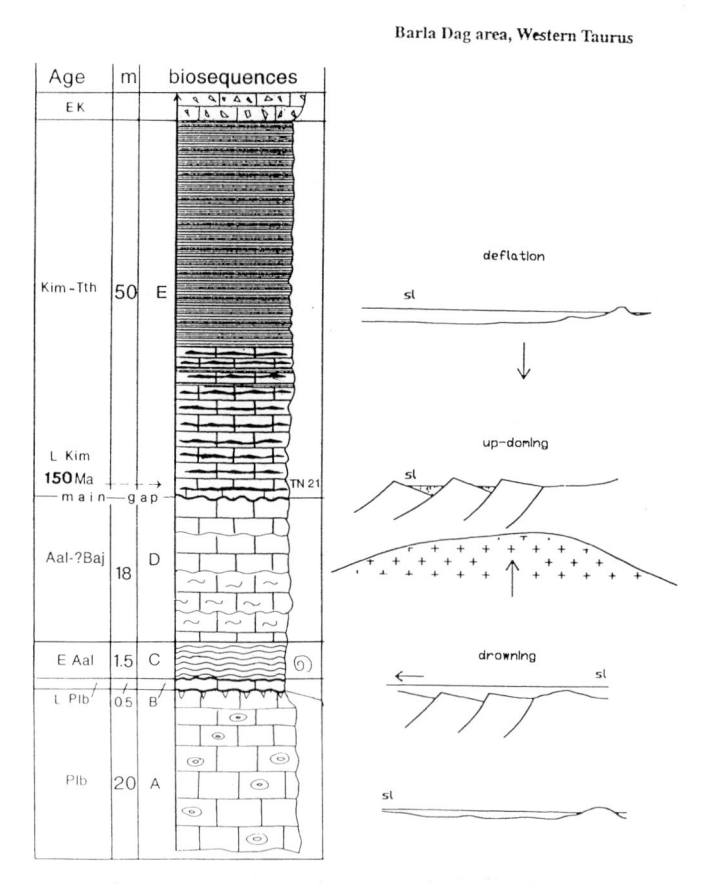

Fig 4. Tentative interpretation of the evolution of the Barla Dağ area from a Pliensbachian carbonate platform, through the drowning at the end of the Pliensbachian, the Toarcian gap corresponding to the emergent part of the graben, the Aalenian ammonite episode, the 'main gap' between the Early Bajocian and Kimmeridgian (interpreted to be due to thermal up-doming that brought the area near to the sea surface or to subaerial exposure), and finally to the deflation of the thermal dome inducing another drowning in which the ramp setting was created. The latter corresponds to the beginning of the chert–radiolarite–calcarenite sedimentation.

the effects on sea water of magmatism associated with the rifting.

The presence of calcareous mud in carbonate deposits, produced before the appearance of planktonic algae (coccoliths, nannoconids), is a problem which has been long debated. Detailed studies on modern cyanobacterial activity, aimed at understanding lime–mud sedimentation during Proterozoic, Early Palaeozoic and Jurassic times, induced Kazmierczak *et al.* (1996), based on observations in fine-grained, shallow-water Jurassic carbonates, to suggest a cyanobacterial origin for the production of micrite and peloidal limestone. In view of the similarity of interval 8 described by these authors (banded micritic/peloidal limestones with cherts) with the cherts and radiolarian mudstones observed in the studied sections, a similar origin for the associated lime–muds may be invoked in the present case.

Conclusions

Modern studies aimed at reconstructing depositional environments of radiolarites should focus mainly on the nature of associated sediments, the composition of palaeontological assemblages, inferred water chemistry and hydrodynamics, and the influence of magmatism. The 'main gap', from Bajocian to Kimmeridgian times, lasted *c*. 25 Ma in the Barla Dağ area. When sedimentation began again, it was characterized by alternations of radiolarian chert with interbedded calcarenites containing ramp organisms. The dominance of radiolarians may be attributed to several factors, including palaeobathymetric setting, current circulation, magmatism, up-welling, chemical equilibrium of sea water and nutrient supply.

The main results of the present study are: (1) the identification of extended sedimentary gaps due to subaerial exposure; and (2) the identification of the depositional environment of the radiolarites at a storm-influenced carbonate ramp setting. In conjunction with similar studies of other Tethyan basins, it is tentatively concluded that more extended shallow seas and a relatively narrow oceanic realm separated the Eurasian and African Plates in the Western Tethys.

We thank Sait Bölükbaşı for stimulating discussions in the 1995–97 joint field trips and the Turkish Petroleum Corporation (TPAO) direction for providing the logistical support for the fieldwork. We are deeply indebted to Gilbert Kelling for the detailed, critical and helpful review of the manuscript.

References

AGER, D. V. 1974. Storm deposits in the Jurassic of the Moroccan High Atlas. *Palaeogeography, Palaeoclimatology, Palaeoecology*, **15**, 83–93.

BRUNN, J. H., DUMONT, J. F., DE GRACIANSKY, P. CH. ET AL. 1971. Outline of the geology of the Western Taurides. *In*: CAMPBELL, A. S. (ed.) *Geology and History of Turkey*. The Petroleum Exploration Society of Libya, Tripoli, 225–255.

BURCHETTE, T. P. & WRIGHT, V. P. 1992. Carbonate ramp depositional systems. *Sedimentary Geology*, **79**, 3–57.

CECCA, F., CRESTA, S., PALLINI, G. & SANTANTONIO, M. 1986. Biostratigrafia ad ammoniti del Dogger-Malm di Colle Tordina Monti della Rossa, Appennino marchigiano. *Bollettino Servizio Geologico d'Italia*, **104**, 177–204.

——, ——, —— & —— 1990. Il Giurassico di Monte Nerone (Appennino marchigiano, Italia Centrale): biostratigrafia, litostratigrafia ed evoluzione palaeogeografica. *In*: PALLINI, G., CECCA, F., CRESTA, S. & SANTANTONIO, M. (eds) *Atti II Convegno Internazionale: Fossili, Evoluzione, Ambiente, Pergola*, **87**, 63–139.

CONTI, M. A. & MARCUCCI, M. 1986. The onset of radiolarian deposition in the ophiolite sequences of the Northern Apennines. *Marine Micropalaeontology*, **11**, 129–138.

—— & MONARI, S. 1992. Thin-shelled bivalves from the Jurassic Rosso Ammonitico and Calcari a Posidonia Formations of the Umbrian–Marchean Apennine (Central Italy). *Palaeopelagos*, **2**, 193–213.

COPE, J. C. W., DODONA, E., KANANI, J., NICOSIA, U. & TONIELLI, R. 1994. The sedimentary cover of the ophiolite of the inner Albanids and the age of Late Jurassic ocean floor spreading. *Palaeopelagos*, **4**, 3–12.

DODONA, E., FARINACCI, A., KANANI, J., NICOSIA, U. & TONIELLI, R. 1994. Mid Jurassic events in the Ionian Zone (SW Albania). *Palaeopelagos*, **4**, 73–85.

DRAGASTAN, O. 1997. Transylvanides – A Jurassic–Cretaceous palaeoenvironmental and depositional model. First Romanian National Symposium on Palaeontology. *Acta Palaeontologica Romaniae*, **1997**, 37–43.

FARINACCI, A. 1988. Radiolarites in a few Tethyan lacunose sequences and their relation to the Late Jurassic ophiolite event. *In*: ROCHA, R. B. & SOARES, A. F. (eds) *Proceedings of the 2nd International Symposium on Jurassic Stratigraphy*, Lisboa, 835–854.

—— 1996. Depositional discontinuity and biosequence concept. Examples from Mesozoic Western Tethys. *Palaeopelagos*, **6**, 211–227.

—— & KÖYLÜOĞLU, M. 1982. Evolution of the Jurassic–Cretaceous Taurus shelf (Southern Turkey). *Bolletino della Società Palaeontologica Italiana*, **21**, 267–276.

——, BÖLÜKBAŞI, S. & RIDOLFI, V. 1997. The Tethyan Jurassic 'main gap' in the Tınaz Tepe section of the Barla Dağ area, Western Taurus, Turkey. *Palaeopelagos*, **7**, 17–26.

——, MARIOTTI, N., NICOSIA, U., PALLINI, G. & SCHIAVINOTTO, F. 1981. Jurassic sediments in the Umbro–Marchean Apennines: an alternative model. *In*: FARINACCI, A. & ELMI, S. (eds) *Rosso Ammonitico Symposium*, Tecnoscienza, 334–398.

GARRISON, R. E. & FISCHER, A. G. 1969. Deep-water limestones and radiolarites of the Alpine Jurassic. *In*: FRIEDMAN, G. M. (ed,) *Depositional Environments in Carbonate Rocks*. Society of Economic Paleontologists and Mineralogists, Special Publications, **14**, 20–56.

HAUG, E. 1900. Les géosynclinaux et les aires continentales. Contribution à l'étude des regressions et des transgressions marines. *Bulletin de la Société Géologique de France*, **28**, 617–711.

KAZMIERCZAK, J., COLEMAN, M. L., GRUSZCZYNSKI, M. & KEMPE, S. 1996. Cyanobacterial key to the genesis of micrite and peloidal limestones in ancient seas. *Acta Palaeontologica Polonica*, **41**, 319–338.

MANNI, R., NICOSIA, U. & TAGLIACOZZO, L. 1997. *Saccocoma* as a normal benthonic stemless crinoid: an opportunistic reply within mud-dominated facies. *Palaeopelagos*, **7**, 121–132.

MOLINA, J. M., RUIZ-ORTIZ, P. A. & VERA, J. A. 1997. Calcareous tempestites in pelagic facies (Jurassic, Betic Cordilleras, Southern Spain). *Sedimentary Geology*, **109**, 95–109.

——, O'DOGHERTY, L., SANDOVAL, J. & VERA, J. A. 1999. Jurassic radiolarites in a Tethyan Continental Margin (Subbetic, Southern Spain): palaeobathymetric and biostratigraphic considerations. *Palaeogeography, Palaeoclimatology, Palaeoecology*, **150**, 309–330.

PESSAGNO, E. A. JR, BLOME, C. D., HULL, D. & SIX, W. M. 1993. Jurassic Radiolaria from the Josephine Ophiolite and overlying strata, Smith River subterrane (Klamath Mountains), northwestern California and southwestern Oregon. *Micropalaeontology*, **39**, 93–166.

SAVU, H., BOMBITA, G., UDRESCU, C. & GRABARI, G. 1995. Petrology, geochemistry and stratigraphic position of the Upper Jurassic volcanics on the bottom of the Transylvanian Basin. *Analele Universitatii Bucuresti, Geologie*, **44**, 17–26.

Structural correlation of the southern Transcaucasus (Georgia)–eastern Pontides (Turkey)

ALİ YILMAZ,[1] SHOTA ADAMIA,[2] ALEXANDER CHABUKIANI,[3] TAMARA CHKHOTUA,[3] KEMAL ERDOĞAN,[4] SEVİM TUZCU[4] & MUSTAFA KARABIYIKOĞLU[4]

[1]*Cumhuriyet University, Department of Environmental Engineering Department, TR-58140, Sivas, Turkey (e-mail: ayilmaz@cumhuriyet.edu.tr)*
[2]*Tbilisi State University, Department of Geology and Palaeontology, 1 Chavchavadze Avenue, 380 028, Tbilisi-Georgia*
[3]*Geological Institute of Academy of Sciences, 380 093, Tbilisi-Georgia*
[4]*Mineral Research and Exploration Institute (MTA), Department of Geological Research, TR-06520 Ankara, Turkey*

Abstract: The eastern Pontides (northeastern Turkey) and Transcaucasus (Georgia) belong to the same geological belt representing an active margin of the Eurasian continent. According to palaeotectonic–palaeogeographic reconstructions, based on regional geological, palaeomagnetic, palaeobiogeographical and petrological data, the eastern Pontides and the major part of the Transcaucasus, situated to the north of the North Anatolian–Lesser Caucasian ophiolitic suture, comprise island arc, forearc, back and interarc basins. The eastern Pontide segment of the belt consists of three structural units which, from north to south, are the northern, central and southern units. The northern unit, the southeastern Black Sea coast–Adjara–Trialeti Unit, represents a juvenile back arc basin formed during the Late Cretaceous (pre-Maastrichtian). This unit separates the southern and northern Transcaucasus zones. The central Artvin–Bolnisi Unit is also known as the northern part of the southern Transcaucasus and is characterized by Hercynian basement, unconformably overlying the Upper Carboniferous–Lower Permian molasse and Upper Jurassic–Cretaceous arc association. The southern unit is the imbricated Bayburt–Karabakh Unit and is known as the southern part of the southern Transcaucasus. This unit has a similar basement to the Artvin–Bolnisi Unit and also includes a chaotic assemblage; it unconformably overlies the Upper Jurassic–Cretaceous forearc association. The eastern Pontide system is interpreted as the product of interference between a spreading ridge and subduction zone during Late Jurassic–Cretaceous times. The North Anatolian–Lesser Caucasus Suture, comprising ophiolites, mélanges and an ensimatic arc association, separates the overlying system from the Anatolian–Iranian Platform in the south. Maastrichtian–Lower Eocene cover rocks in the region unconformably overlie all the other units. Middle Eocene rifting resulted in the formation of new basins, some of which closed during an Oligocene–Early Miocene regression. Others, such as the Black Sea and Caspian Basins, have survived to the present day as relict basins.

The Caucasus and Turkey have been divided into different tectonic units and investigated by many authors. It is generally accepted that the eastern Pontides of Ketin (1966; Fig. 1a) are equivalent to the southern Transcaucasus (International Geological Congress XXVII Session, Moscow 1984). This belt (Fig. 1) is bordered to the north by the Adjara–Trialeti Unit and to the south by the North Anatolian–Lesser Caucasus ophiolitic belt; the latter is a product of the final closure of the Neotethys Ocean (Şengör & Yılmaz 1981) during the Oligocene (Koçyiğit 1991).

Although complete correlative studies are rare, it is suggested that the Adjara–Trialeti Unit continues along the southeastern Black Sea coast in Turkey (A. Yılmaz 1989a; Adamia et al. 1995). After a joint geological compilation project in the border area, carried out between 1994 and 1996 (A. Yılmaz et al. 1996, 1997), it was concluded that the Adjara–Trialeti Unit can be traced along the Black Sea coast in Turkey as far west as the Sinop area. In the present study, tectonic division of the eastern Pontides and Transcaucasus is revised (Fig. 2) and characteristics of each unit, including the age, lithology and tectonic setting, are presented (Fig. 3). On the basis of this new division, the southeastern Black Sea coast–Adjara–Trialeti Unit represents the northern unit which is characterized

From: BOZKURT, E., WINCHESTER, J. A. & PIPER, J. D. A. (eds) *Tectonics and Magmatism in Turkey and the Surrounding Area.* Geological Society, London, Special Publications, 173, 171–182. 1-86239-064-9/00/$15.00
© The Geological Society of London 2000.

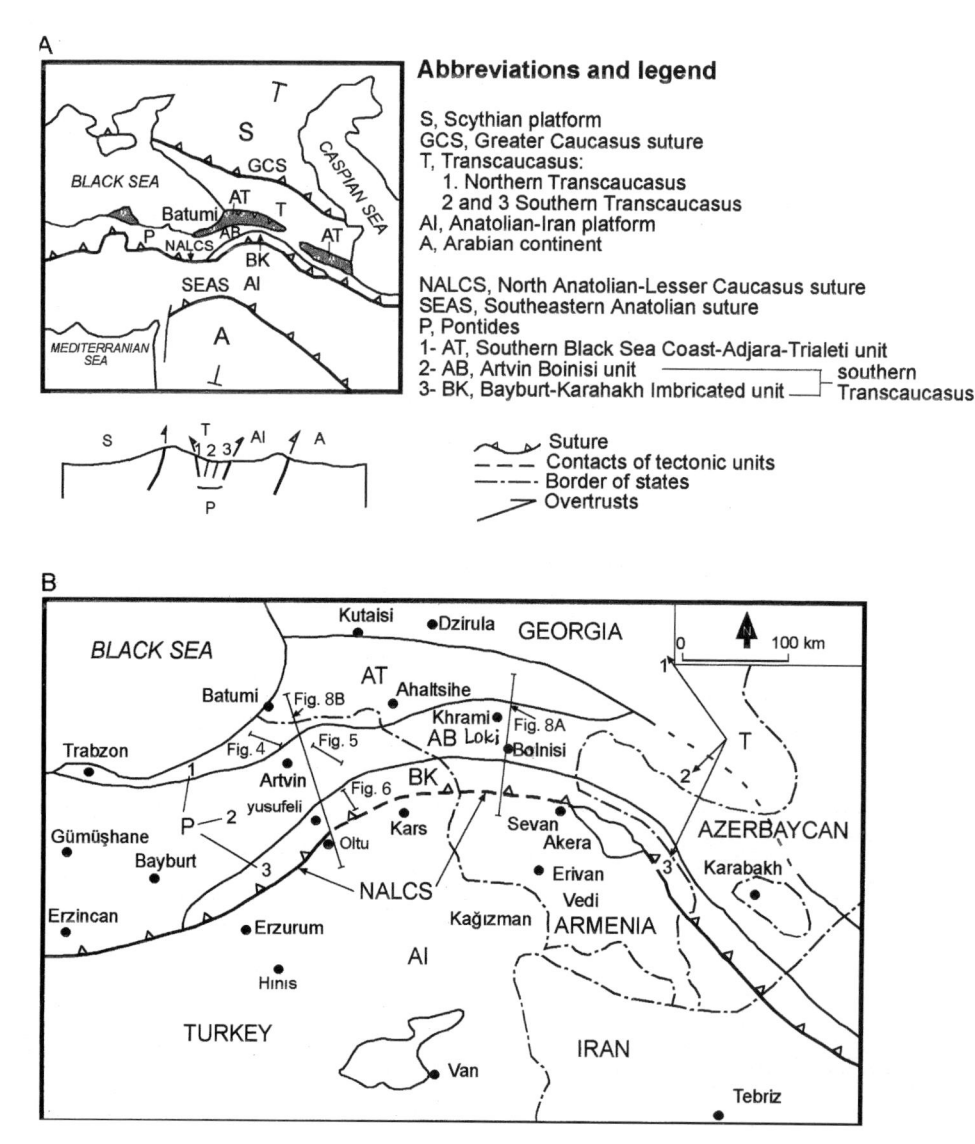

Fig. 1. Location-cross section (**a**) and sketch (**b**) maps of the study area [after A. Yılmaz et al. (1997) and new data].

by a juvenile back-arc association formed mainly during the Santonian–Campanian interval. The Artvin–Bolnisi Unit represents the central unit, which is characterized by an arc association formed mainly during the Liassic–Campanian interval. The Imbricated Bayburt–Karabakh Unit represents the southern unit and is characterized by a forearc association formed mainly during the Malm–Campanian interval. To the south the tectonic units are bordered by the North Anatolian–Lesser Caucasus Suture (Fig. 1). Maastrichtian–Tertiary sequences of the tectonic units can be correlated from north to south and mainly represent a terrigenous to continental unity. Although no adequate data are presented, this unity is also considered to be a post-collision sequence.

This paper deals with the palaeotectonic evolution of the border area and has been undertaken in conjunction with a joint project between Turkey and Georgia. It aims to: (1) introduce a new tectonic subdivision of the border area (Fig. 1); (2) describe Upper Cretaceous–Tertiary units and facies in detail; (3)

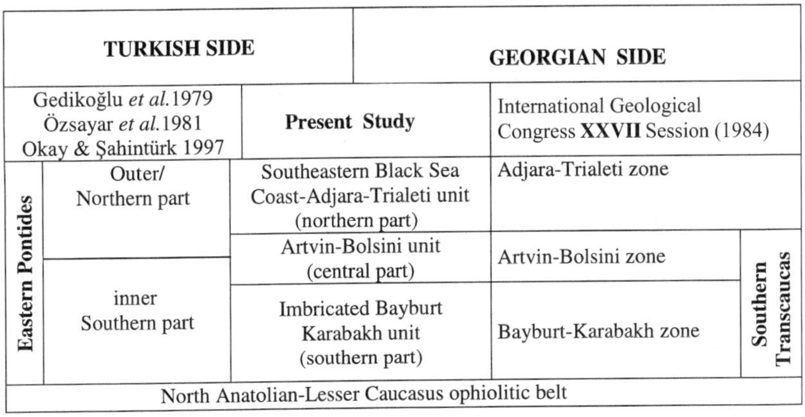

TURKISH SIDE		GEORGIAN SIDE
Gedikoğlu et al.1979 Özsayar et al.1981 Okay & Şahintürk 1997	**Present Study**	International Geological Congress **XXVII** Session (1984)
Eastern Pontides — Outer/Northern part	Southeastern Black Sea Coast-Adjara-Trialeti unit (northern part)	Adjara-Trialeti zone — Southern Transcaucas
Eastern Pontides — inner Southern part	Artvin-Bolsini unit (central part)	Artvin-Bolsini zone
	Imbricated Bayburt Karabakh unit (southern part)	Bayburt-Karabakh zone
North Anatolian-Lesser Caucasus ophiolitic belt		

Fig 2. Tectonic division of the study area.

evaluate the interference between a spreading ridge and subduction during closure of Neotethys; (4) present this group's views on the polarity of the eastern Pontide–southern Transcaucasus active continental margin; and (5) evaluate the geological evolution of the region.

This paper presents three measured sections studied during 1996–1997 (Figs 4–6), including lithological and faunal data from critical localities and a correlative table (Fig. 7) comparing generalized stratigraphic sections of each tectonic unit. Although there is no common formal lithostratigraphic nomenclature along the border area, some of the formations named by Adamia et al. (1995) and A. Yılmaz et al. (1997) have been used to facilitate correlations of the Turkish and Georgian sides in Figs 4–6. Samples have been collected at each location for

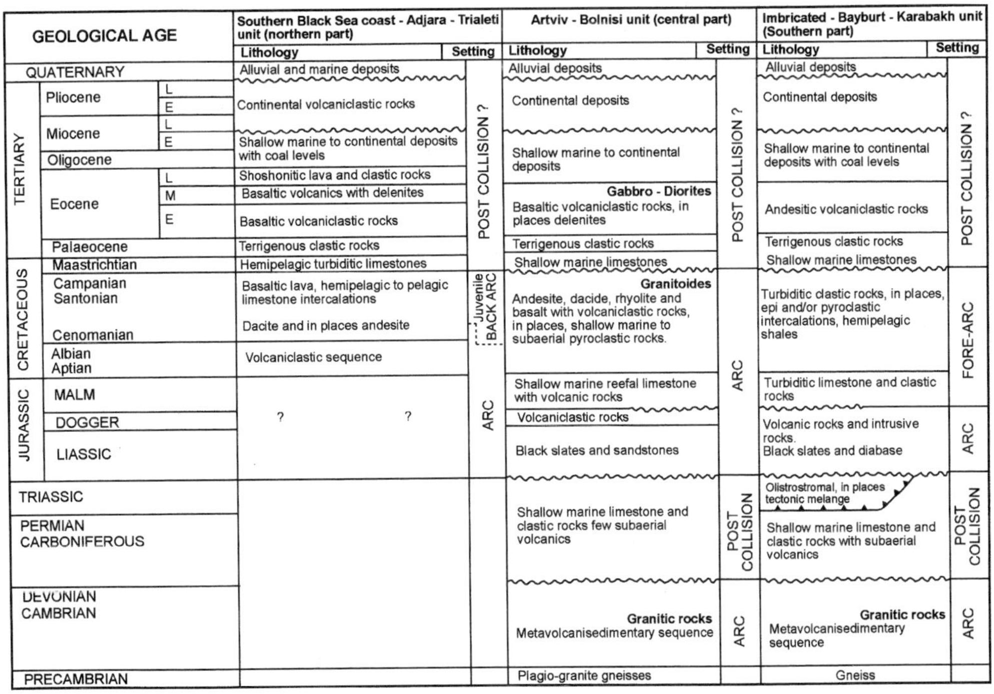

Fig 3. Correlation of tectonic units in the framework of age, lithology and tectonic setting. E, Early; M, Middle; L, Late.

stratigraphic analysis, dating and environmental interpretation. Approximate locations of the columnar sections are shown in Fig. 1.

The cross-sections (Fig. 8a and b) have been compiled from the Georgian and Turkish sides, respectively, based on new field observations and those of A. Yılmaz et al. (1997). Based on these sections, consecutive stages of the geological evolution of the region have been presented in terms of the Tethyan evolution.

The study area

The study area in the eastern Pontide and Transcaucasus belt represents the active margin of the Eurasian continent. Based on the tectonic division of the Caucasus, the Adjara–Trialeti Unit is a tectonic unit between the northern and southern Transcaucasus zones which extends along the southeastern Black Sea coast in Turkey (Fig. 1). The southern Transcaucasus zone comprises two tectonic units; the Artvin–Bolnisi Unit in the north and the Imbricated Bayburt–Karabakh Unit in the south. Hence, the study area comprises three tectonic subunits (Fig. 7). To the south, the Pontides–southern Transcaucasus zone is bordered by the North Anatolian–Lesser Caucasus ophiolitic belt.

The southeastern Black Sea coast– Adjara–Trialeti Unit

This unit represents the northern part of the eastern Pontides (Fig. 7) situated between northern Transcaucasus (the Georgian Dzirula Block) to the north and southern Transcaucasus (Artvin–Bolnisi Unit) to the south. It is a northeast–southwest trending structural unit and extends from the Lori River in the east to the southern Black Sea coast in the west (Fig. 1b). The northern and southern margins are defined by north and south facing overthrusts, respectively, delineated on the basis of dip angles of thrust planes (Fig. 8). Borehole data obtained in Georgia (Fig. 1b) show that Aptian–Cenomanian volcaniclastic rocks constitute the lowest level (Nadareishvili 1980, 1981), whereas the lowest level exposed in Turkey comprises acidic volcanic rocks conformably overlying an alternation of Santonian–Campanian basaltic lava and micritic limestone (A. Yılmaz et al. 1997). The alternation is named the Çağlayan Formation and the following pelagic forms (Fig. 4) have been determined from micritic limestones of the formation: *Globotruncana arca* (Cushman), *Globotruncana elevate* (Brotzen). Radiplaria are common in some levels of the limestone. Lithology and pelagic forms of the unit indicate a comparatively deeper environment than that of the arc association to the south. In addition, there is no level which is characterized by oceanic crust along the tectonic unit. Hence, it is suggested that this association was formed in a pelagic to hemipelagic environment which probably indicates a juvenile back-arc setting (Fig. 4). Maastrichtian clastic rocks and hemipelagic limestones, and Palaeocene–Lower Eocene turbiditic terrigenous clastic rocks, overlie the juvenile back-arc association conformably and, in places, pass gradationally upwards into Eocene volcanic rocks. The Maastrichtian sequence is named the Cankurtaran Formation (Fig. 4). It starts with a clastic level, including *Siderolites* sp., *Cuvillerina* sp. and algae, and, in turn, passes upwards to, turbiditic limestone and reddish micritic limestone. These include mainly *Globotruncanita stuartiformis* (Dalbiez), *G. stuarti* (de' Lapparent), *Globotruncana arca* (Cushman), *G. linneiana* (d'Orbigny), *Gansserina gansseri* (Bolli) and *Rosita contusa* (Cushman).

The Palaeocene–Lower Eocene sequence is named the Bakırköy Formation and is equivalent to the Borjomi Suite; it includes abundant nanno and foram fossils (Fig. 2). Dolerite dykes and sills are common in the Maastrichtian–Lower Eocene sequences but there are no volcanic interlayers in the Maastrichtian–Upper Eocene sequence (Fig. 4). Hence, it is concluded that the back-arc activity terminated during Maastrichtian–Early Eocene times. Middle Eocene volcanic rocks include, from bottom to top, an alternation of turbiditic rocks with basaltic volcaniclastic rocks up to 7 km in thickness and overlie the older sequences with a local unconformity mainly along the Black Sea coast on the Turkish side. Lordkipanidze et al. (1984) suggest that this sequence is a product of back-arc and/or interarc deposition, and is followed conformably by Upper Eocene shoshonitic volcanic rocks of a mature arc. However, similar volcanic rocks of the eastern Pontides with the same age and tectonic setting have been interpreted as the product of a post-collisional event (A. Yılmaz & Terzioğlu 1994; S. Yılmaz & Boztuğ 1996). Hence, the tectonic setting of the Eocene volcanic cycle is controversial and requires more study.

The Artvin–Bolnisi Unit

The Artvin–Bolnisi Unit is located between the southeastern Black Sea–Adjara–Trialeti Unit to the north and the Imbricated Bayburt–Karabakh Unit to the south. It represents the northern part of the southern Transcaucasus and the

GEOLOGIC AGE		LITHOLOGY	THICKNESS (m)	FORMATIONS AND DESCRIPTIONS	FAUNAL CONTENT	Interpretation
TERTIARY	EOCENE (M) (E)		>1000	Erenler and Kabaköy Formations, Likani and Peranga Suites Basaltic volcaniclastic sequence	Sphenolithus radions Deflandre Ericsonia cava (Hay and Mahler) E. ovalis Black E. subpertusa Hay and Mahler	
	PALAEOCENE		100	Bakırköy Formation–Barjomi Suite Turbiditic terrigenous clastic rocks Reddish to olive gray mudstone and claystone Diabasic dykes and sills	Prinsius bisulcus Hay and Mahler Fasciculithus sp. Discoaster sp. Morozorella cf. uncinata Bolli Planorotalites cf. ehrenbergi Bolli Acarinia sp.	ACTIVITY OF JUVENILE BACK ARC TERMINATED
CRETACEOUS (LATE)	MAASTRICHTIAN		~2500	Cankurtaran formations Reddish micritic limestone Turbiditic limestone Yellowish sandy limestone Diabasic dykes and sills	Globotruncanita stuartiformis (Daibiez) G. stuarti (de' Lapparent) Globotruncana arca (Cushman) G. linneiana (d' Orbigny) Gansserina gansseri (Bolli) Rosita contusa (Cushman)	
				Clastic rocks	Siderolites sp. Cuvilfferina sp. Rotaliiadae, Algae	
	PRE-MAAST-RICHTIAN		>1000	Çağlayan Formation: basaltic lavas and micritic limestone alternation	Globotruncana arca (Cushman) Globotruncanita elevata (Brotzen)	Juvenile back-arc

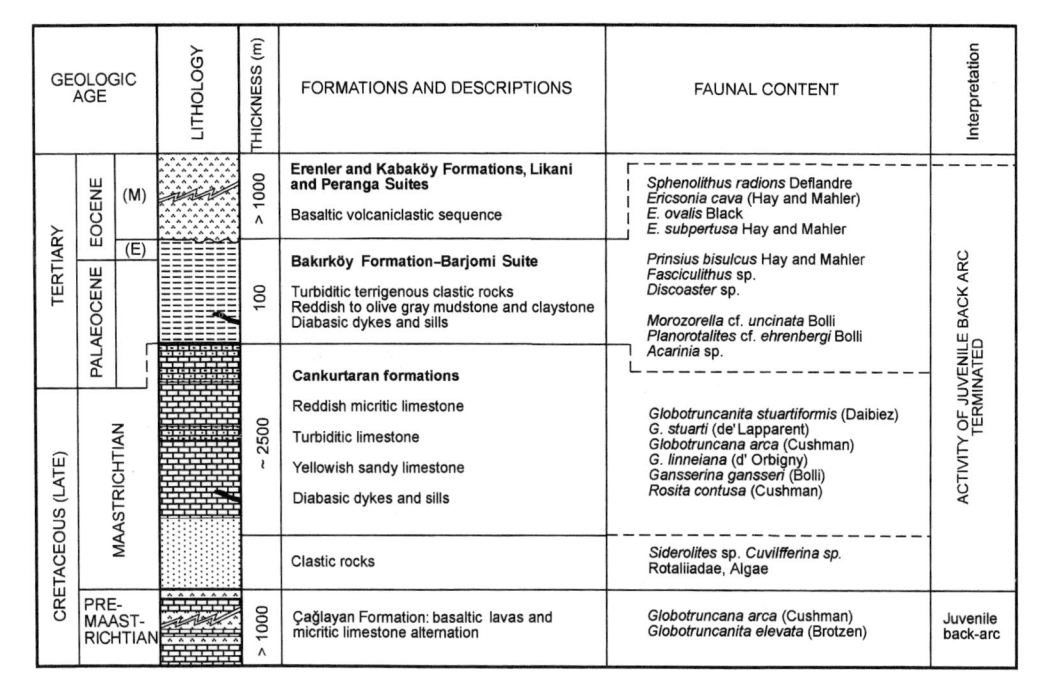

Fig. 4. Stratigraphic section in the Cankurtaran village area. (E, Early; M, Middle).

central part of the eastern Pontides (Figs 1 and 7). Its northern and southern margins are delineated by south facing overthrusts (Fig. 8).

The lowermost stratigraphic level is Pre-cambrian(?) and/or Lower Palaeozoic, with metamorphic rocks and Variscan granites cropping out around the Khrami and Artvin Massifs. These massifs are mainly built up of granite–gneisses and S-type plagiogranites (Belov et al. 1978; Adamia et al. 1983, 1995). A Carboniferous continental volcanic–sedimentary sequence overlies the older rocks unconformably and is followed upwards unconformably by Lower–Middle Jurassic volcaniclastic rocks, an Upper Jurassic–Lower Cretaceous shallow-marine limestone and volcaniclastic alternation, and an Upper Cretaceous calc-alkaline arc association (Gedikoğlu et al. 1979; Özsayar et al. 1981; Lordkipanidze et al. 1989). The latter association has been studied in detail along the Georgian side where it is divided into several suites and named the Varlık Group in the Artvin area (Fig. 5). It is made up of hemi-pelagic, shallow-marine to subaerial volcaniclastic rocks, andesite, dacite, rhyolite and basalt, and is intruded by granitoids. Coal levels and ignimbrite-type pyroclastic rocks can be seen mainly in the upper levels of the association. This sequence is followed by unconformably overlying Maastrichtian–Palaeocene shallow-marine limestones and turbiditic terrigenous clastic rocks that pass upwards into Lower Eocene clastic rocks (Fig. 5). The Maastrichtian sequence is named the Ziyarettepe Formation and is equivalent to the Tetritskaro Suite. It starts with cross-bedded conglomerate and shallow-marine to hemipelagic limestones, which include *Siderolites calcitrapoides* (Lamarck), *Orbitoides* sp., *Lepidorbitoides* sp. (and algae in the lower levels), *Globotruncana* gr. *linneiana* (d'Orbigny), *G. arca* (Cushman) and *G. stuarti* (de'Lapparent). The Palaeocene–Lower Eocene sequence is named the Kızılcık Formation and is made up of terrigenous clastic rocks, including abundant foraminifera (Fig. 5). The Maastrichtian–Lower Eocene sequence contains no volcanic rocks and it is concluded that arc activity had terminated by the Maastrichtian–Early Eocene. The Middle Eocene volcanic rocks show a similar succession to that in the Adjara–Trialeti Unit and appear to be 1.5–2 km or less in thickness. They overlie the older rocks unconformably and are themselves overlain conformably by Upper Eocene shallow-marine clastic rocks. The Eocene sequence is cut by the Karçal Intrusive Suite, which includes andesitic subvolcanics, diorite and gabbro. Similar intrusive rocks can be seen

GEOLOGIC AGE	LITHOLOGY	THICKNESS (m)	FORMATIONS AND DESCRIPTIONS	FAUNAL CONTENT	Interpretation
TERTIARY — EOCENE (L)(M)(E)		100	**Karçal Intrusive rocks:** subvolcanics diorite and gabbro (undivided) **Kabaköy and Taşköprü Formation– Peranga and Nagverevi Suites:** Basaltic to dacitic volcaniclastic sequence	*Planorotalites* cf. *pseudomenardii* Bolli *Morozovella* cf. *uncinata* Bolli *M.* cf. *pseudobulloides* (Plummer)	Debated Volcanic Cycle
TERTIARY — PALAEOCENE		50	**Kızılcık Formation:** terrigenous clastic rocks Parallel bedded marl and sandstone	*Globigerina* cf. *triloculinoides* (Plummer) *Acarinina* sp. Ostracoda	ACTIVITY OF THE ARC TERMINATED
CRETACEOUS (LATE) — MAASTRICHTIAN		75	**Ziyarettepe Formation–Tetritskaro Suite** Hemipelagic limestone Shallow-marine limestone	*Globotruncana* gr. *linneiana* (d'Orbigny) *G. arca* (Cushman) *G. stuarti* (de'Lapparent)	
		25	Conglomerate, cross-bedded and tabular cross-stratification	*Siderolites calcitrapoides* Lamarck *Orbitoides* sp. *Lepidorbitoides* sp. Algae	
PRE-MAASTRICHTIAN		100	**Varlık group–Shorsholeti Suite;** granitoides **Gasandami Suite** **Tandzia Suite** **Mashaveri Suite** **Didgverdi Suite**		Arc

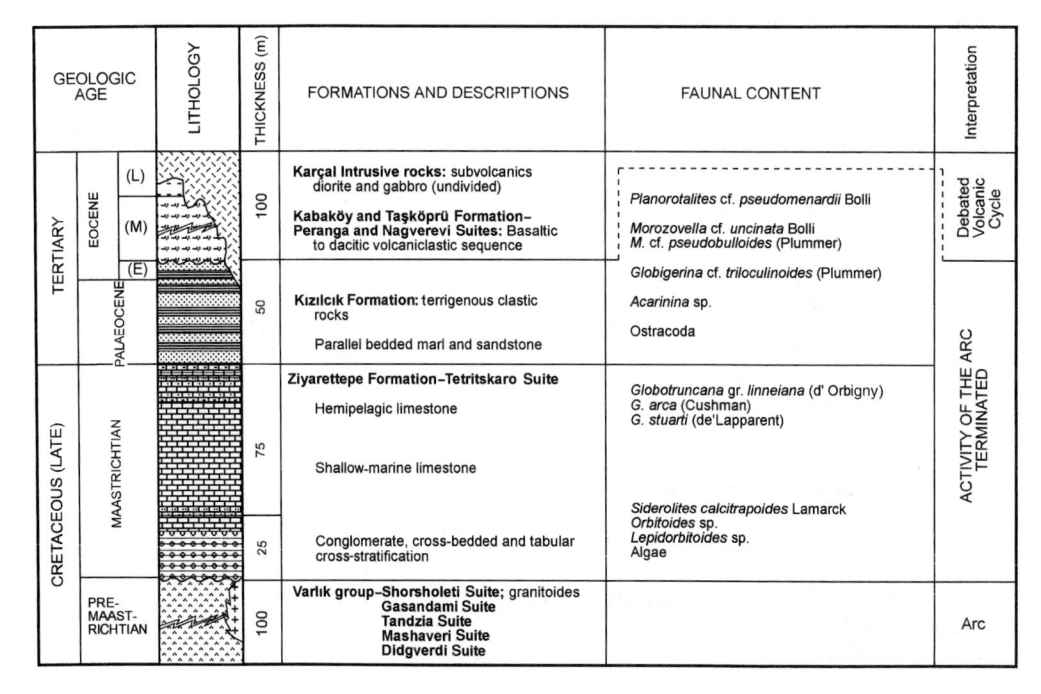

Fig. 5. Stratigraphic section in the Adagül–Karçal Dagh area (E, Early; M, Middle; L, Late).

along the southeastern Black Sea coast at the Adjara–Trialeti Unit and in the Imbricated Bayburt–Karabakh Unit as outcrops of variable size.

The Imbricated Bayburt–Karabakh Unit

This unit crops out between the Artvin–Bolnisi Unit to the north and the North Anatolian–Lesser Caucasus ophiolitic belt to the south. It represents the southern part of both the southern Transcaucasus and the eastern Pontides (Fig. 7). Its northern and, in places, southern margins are delineated by south facing overthrusts (Fig. 8).

Hercynian metamorphic rocks (Robinson et al. 1995) and associated granitic rocks (Y. Yılmaz 1976), and an unconformably overlying Upper Carboniferous–Lower Permian continental sequence (Akdeniz 1988), are also located along the southern margin of Laurasia; the latter contains distinct Euro-American fauna and flora assemblages (Okay & Leven 1996). In addition, it is suggested that there is a pre-Liassic chaotic association of oceanic products such as a pre-Liassic sheeted dyke complex in the Yusufeli area (Şengör et al. 1980; Y. Yılmaz et al. 1997a, b). However, there is no direct relationship between the dyke complex and lower or upper levels of the ophiolitic sequences in the Yusufeli area. In addition, the geochemical signature of the complex has not yet been determined. It is difficult to interpret the complex as the product of a relict ocean and it is concluded that there is no convincing data indicating Pre-Liassic ocean crust along the Imbricated Bayburt–Karabakh Unit.

Okay & Leven (1996) propose that the tectonic juxtaposition of these rock units occurred during the latest Triassic Cimmeride orogeny. Because Liassic volcaniclastic rocks overlie the older rocks unconformably and pass upwards into Upper Jurassic–Cretaceous forearc deposits with a local unconformity to the northeast of the Oltu area. Forearc deposits are represented by turbiditic limestones and clastics in the lower levels, turbiditic clastic rocks, and epi- and/or pyroclastic rock intercalations and hemipelagic shales in the upper levels; these are isoclinally folded and imbricated (Konak et al. 1995). Maastrichtian–Palaeocene turbiditic terrigenous clastics and limestone alternations overlie the forearc deposits unconformably; this sequence is named as the Atlılar Formation. It starts with a basal transgressive conglomerate and passes upwards into terrigenous clastic rocks and limestones with local turbiditic characteristics. In the lower levels of the sequence, corals and bivalves are common (Fig. 6). In the middle

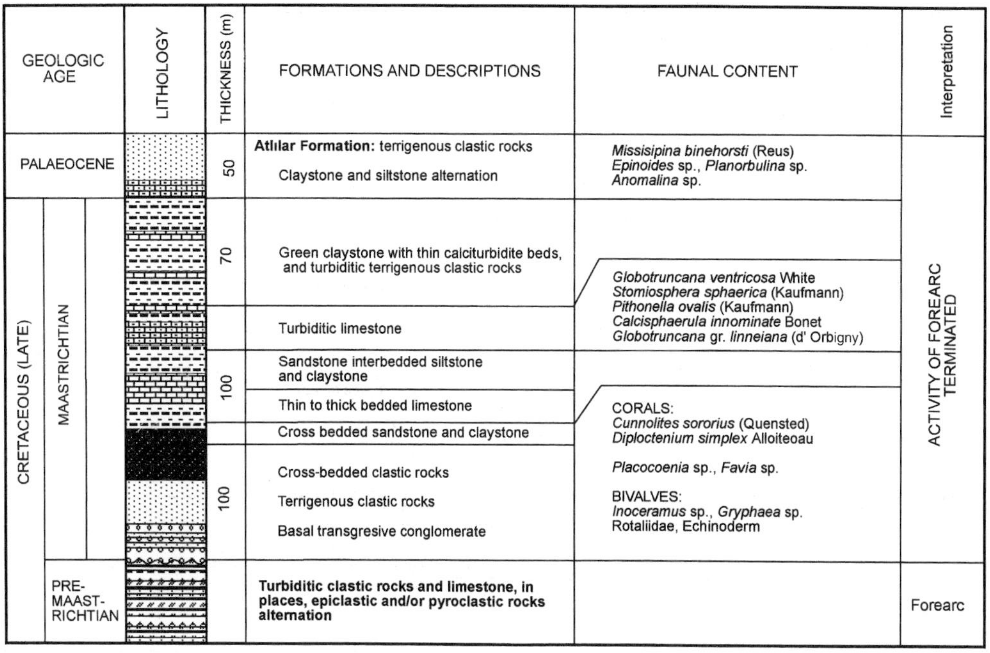

Fig. 6. Stratigraphic section in the Olur–Atlılar village area.

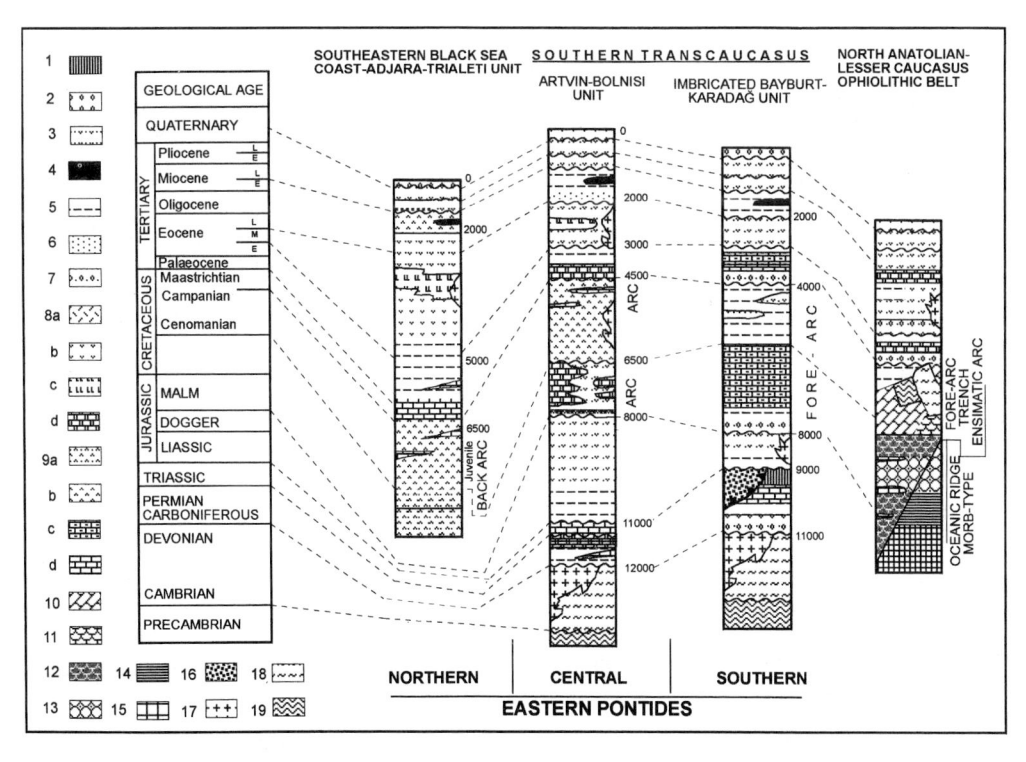

Fig. 7. Correlative table showing tectonic units of the study area. 1, Non-deposition; 2, alluvium and fluviatile deposits; 3, continental volcaniclastic rocks; 4, coal levels; 5, siltstone; 6, sandstone; 7, conglomerate; 8, shallow marine – a, volcaniclastic rocks; b, lavas; c, delenites; d, limestone; 9, deep marine – a, volcaniclastic rocks; b, lavas; c, turbiditic limestone; d, limestone; 10, ophiolitic mélange; 11, ensimatic arc lavas; 12, epiophiolitic lavas; 13, dyke complex; 14, gabbro; 15, serpentinite and peridotite; 16, pre-Liassic chaotic rocks; 17, intrusive rocks (undivided); 18, schist; 19, gneiss.

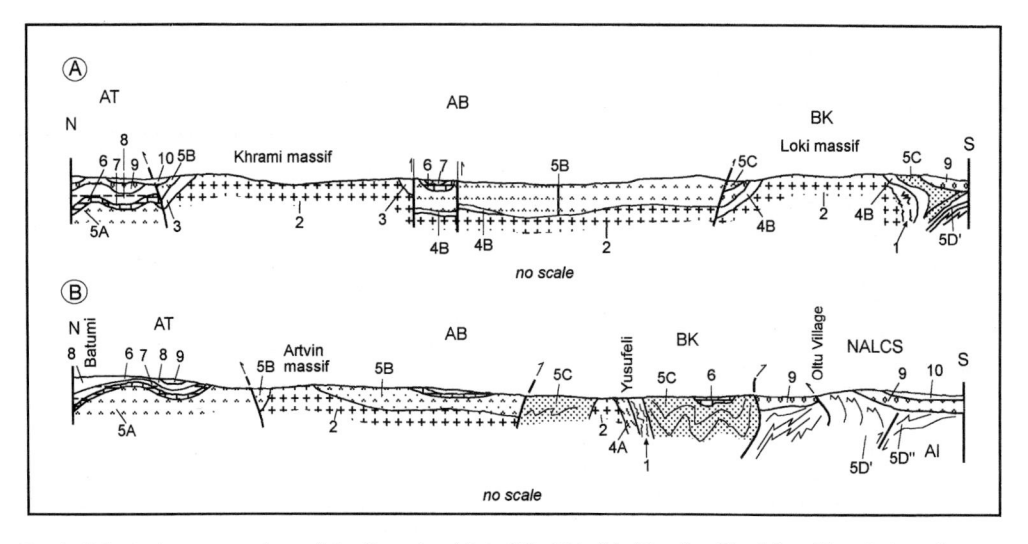

Fig. 8. Schematic cross-sections of the Georgian (**a**) and Turkish (**b**) sides. See Fig. 1 for abbreviations. 1, Precambrian–Cambrian metamorphic rocks; 2, Middle Palaeozoic granitoids; 3, Upper Carboniferous–Lower Permian molasse; 4A, Pre-Liassic dyke complex; 4B, Liassic–Dogger volcaniclastic and volcanic sequence; 5, Upper Jurassic–Cretaceous (Pre-Campanian) sequence – 5A, Juvenile back-arc volcanics; 5B, arc volcaniclastic rocks; 5C, forearc deposits; 5D, ophiolites, ophiolitic mélanges and ensimatic arc association – undivided; 5D', the Ankara–Erzincan/Sevan–Akera ophiolitic zone; 5D", the Northern Taurus–Erzurum–Kağızman–Vedi ophiolitic zone; 6, Maastrichtian–Palaeocene limestone; 7, Palaeocene–Lower Eocene clastic rocks; 8, Middle Eocene volcanic rocks and volcaniclastic rocks; 9, Upper Eocene–Lower Miocene clastic rocks; 10, Upper Miocene–Quaternary continental deposits.

levels *Globotruncana ventricosa* (White), *Stomiosphera sphaerica* (Kaufmann), *Pithonella ovalis* (Kaufmann), *Calcisphaerula innominate* Banet and *Globotruncana* gr *linneiana* (d'Orbigny) occur, and *Missisipina binehorsti* (Reus), *Epinoides* sp., *Planorbulina* sp. and *Anomolina* sp. are found in the upper levels. Finally, Middle Eocene volcaniclastics and Upper Eocene clastics overlie the older rocks unconformably (Fig. 7). The lithology and stratigraphic setting is similar to the Middle–Upper Eocene sequence of the Artvin–Bolnisi Unit.

The North Anatolian–Lesser Caucasus ophiolitic belt

This belt is a suture zone between the Pontian–southern Transcaucasus arc to the north and the Anatolian–Iranian Platform to the south. It is divided into two subzones, which resemble each other in lithology. The Ankara–Erzincan/Sevan–Akera ophiolitic zone is located to the north, whereas the Northern Taurus/Erzurum–Kağızman/Vedi ophiolitic zone is located to the south. Both zones are believed to be allochthonous (Fig. 8b) and continuous beneath the Cretaceous–Tertiary cover (Knipper 1980; Zakariadze *et al.* 1983).

Along this belt, ophiolites and ophiolitic mélanges of different ages crop out (Belov *et al.* 1978; Gasanov 1986; Tatar 1978; Koçyiğit 1990). The ophiolites comprise serpentinite, ultramafic rocks, layered gabbro, a sheeted dyke complex and volcano–sedimentary cover. Accretion and mixing of ophiolites of different ages occurred during Late Cretaceous tectonism. Jurassic–Lower Cretaceous volcanic rocks in the ophiolites are represented by mid-ocean ridge basalt (MORB)-type tholeiites, whereas Upper Cretaceous volcanic rocks in the ophiolites belong to the calc-alkaline island-arc basalt series (A. Yılmaz 1980, 1981; Buket 1982; Zakariadze *et al.* 1983). This arc may be regarded as ensimatic (Okay & Şahintürk 1997). Ophiolitic mélanges and olistostromes (Knipper 1980; Knipper *et al.* 1986), representing an accretionary prism (Koçyiğit 1991), formed during obduction of these ophiolites onto the southern Transcaucasus to the north and the Anatolian–Iranian Platform to the south. The formation and emplacement age of the mélange ranges from the Cenomanian to Early Coniacian in the Lesser Caucasus (Knipper & Khain 1980; Zakariadze *et al.* 1983) and is dated as pre-Late

Campanian in northeastern Anatolia (A. Yılmaz 1982). Therefore, the ophiolites in the region were all obducted before Late Coniacian–Campanian times.

Cover rocks

With the exception of the Adjara–Trialeti Unit, a Maastrichtian–Lower Eocene sequence overlies the older rock units unconformably and includes a great deal of terrigenous clastic rocks. It can be correlated from north to south along all tectonic units. No volcanic intercalations occur in this sequence. However, Middle Eocene volcanic rocks of the Adjara–Trialeti Unit have been interpreted as products of back-arc and/or interarc rifting (Lordkipanidze et al. 1984) with the Pontides as an arc (Tokel 1977). However, the A-type alkaline and M-type low-K tholeiitic characteristics have also been related to a post-collisional magmatic pulse in a tensional regime resulting from crustal thickening after collision (S. Yılmaz & Boztuğ 1996). Hence, the setting of the Middle Eocene volcanic rocks is controversial. In the study area, gabbro and diorite intrusions are common from north to south throughout all tectonic units and it can be concluded that Middle Eocene volcanic and intrusive rocks are products of a tensional event in a general sense. This may have lead to the formation of the eastern Black Sea (Okay & Şahintürk 1997). The evolution of these basins differs greatly from the western part of the Black Sea basin (Görür et al. 1993; Ustaömer & Robertson 1997). Late Eocene volcanic activity ceased gradually from south to north. Oligocene–Lower Miocene regressive shallow marine to continental deposits overlie the tectonic units with local unconformity. After this regression the formation of some basins, such as the Ahaltsihe and Oltu, was terminated while others, such as the Black Sea and Caspian Sea, have survived to the present day. Upper Miocene–Lower Pliocene and Upper Pliocene–Quaternary continental deposits overlie the older rock units unconformably.

Discussion and conclusion

Within the Pontide–Transcaucasus system there are important lateral and vertical facies/lithology differences. Previous hypotheses explain these differences inadequately in the frame of regional geodynamic evolution of the region. For instance, during the Late Cretaceous the northern segment the southeastern Black Sea coast–Adjara–Trialeti Unit represents a juvenile back-arc basin, whereas the western Pontides have been interpreted as a juvenile ocean and/or mature back-arc basin (Görür et al. 1993). This indicates that the western and eastern Pontides experienced a different geodynamic evolution. However, a reddish hemipelagic limestone unit is Maastrichtian in the Artvin–Cankurtaran area (Fig. 5) but Palaeocene in age in the Sinop area. This reddish limestone represents a characteristic horizon in the Pontides and indicates a transition from shallow-marine to pelagic environment, expressing the deepening of the basin. Within this framework, it is concluded that this level indicates a transgression occurring from east to west in the Artvin–Cankurtaran and Sinop areas during Maastrichtian–Palaeocene times.

The eastern Pontides have been interpreted as an arc (Peccerillo & Taylor 1975; Eğin & Hirst 1979; Gedikoğlu et al. 1979; Manetti et al. 1983) which was either north facing (Tokel 1972, 1977; Adamia et al. 1981, 1977; Knipper et al. 1986) or south facing (Bektaş 1984; Bektaş et al. 1984) between Palaeozoic and Tertiary times. Şengör & Yılmaz (1981) suggest that the Tethyan evolution of Turkey can be divided into two main phases – Palaeotethyan and Neotethyan – which partly overlap in time. In addition, they interpret the Pontide–Transcaucasus–Sanandaj–Sirjan (Podataksasi) zone as a north facing Palaeotethyan magmatic arc during Early Triassic–Late Cretaceous times (Şengör 1987). In spite of these differences, the Pontide Arc is, in general, considered to have been a typical arc mainly during the Late Cretaceous. This view is also acceptable for the western part of the eastern Pontides. In the eastern part of the eastern Pontides, in the Artvin area, Jurassic–Cretaceous volcanic rocks as a whole are also products of an arc-related system (Konak et al. 1995).

In the present study, tectonic division of the region is revised and, on the basis of this division, it is concluded that only the southern segment of the easternmost Pontides represents a forearc. The central segment represents the arc and northern segment represents a juvenile back-arc. On the basis of this reconstruction and the dipping angles of the thrust planes developed along the suture zone, the arc is inferred to have had a south facing setting during Late Jurassic–Cretaceous times. In the southern segment of the eastern Pontides, the Imbricated Bayburt–Karabakh Unit represents the forearc basin of an active continental margin during Late Jurassic–Cretaceous time, whereas in the western part of the eastern Pontides the Jurassic–Lower Cretaceous sequence was formed during a rifting event and indicates a passive continental margin (Görür et al. 1983; A. Yılmaz 1985). Thus, there is a progressive transition from east to west along the eastern Pontides. This may result from progressive interaction between a spreading ridge and a subduction zone, as suggested elsewhere by Dewey (1976); the consistent framework of this kinematic model includes a spreading ridge and a subduction zone together, as suggested by Cox & Hart (1986). Within this framework, the western part of the eastern Pontides was a passive continental margin during the Late Jurassic–Early Cretaceous, whilst during the Late Cretaceous (mainly pre-Maastrichtian) both western and eastern parts of the eastern Pontides became an active continental margin.

The Sevan Akera–Karadağ Suture of Şengör (1987) is a part of the North Anatolian (İzmir–Ankara–Erzincan) Lesser Caucasus Suture which separates the Anatolian–Iranian Platform from the eastern Pontide–Transcaucasus Arc. Ophiolites, mélanges and forearc deposits exposed in both sutures resemble each other in age and lithology. Due to the Tertiary cover, it is not possible to see outcrops of the Anatolian–Iranian Platform near the suture. However, the Akdağ metamorphic rocks of the Hınıs area and the Upper Palaeozoic–Lower Mesozoic carbonates of the Başkale (Van) and Nahcivan areas are components of the Anatolian–Iranian Platform. They represent the continental crust in East Anatolia and crop out as tectonic windows beneath obducted ophiolite and the ophiolitic mélange association (A. Yılmaz et al. 1989a,b, 1997).

On the basis of facies and data presented above, it is suggested that the easternmost Pontide and southern Transcaucasus represent, from north to south, juvenile back-arc, arc and forearc environments active mainly during the Late Jurassic–Cretaceous (pre-Maastrichtian) interval. This clearly indicates a northward subduction polarity in the region where lithologic and structural data show an east to west transition from active to passive continental margin. The absence of structural elements between the facies of active and passive margins suggests that interference of the spreading ridge and the subduction zone was a progressive event.

Based on structural correlation of the eastern Pontides (Turkey) and the southern Transcaucasus (Georgia), the following conclusions are drawn:

- In setting, the eastern Pontides and southern Transcaucasus belong to the same geological belt and represent the subduction zone of the Eurasian continent. They comprise, from north to south, juvenile back-arc, arc and forearc basins, formed mainly in the Late Jurassic–Campanian interval. In this context, the eastern Pontides are divided into three subzones, which, from north to south, are: the southeastern Black Sea coast–Adjara–Trialeti Unit, representing the juvenile back-arc; the Artvin–Bolnisi Unit, representing the arc; and the Imbricated Bayburt–Karabakh Unit, representing forearc environments. The region as a whole displays a clear northern polarity.
- On the basis of lateral facial distribution, a model comprising the interaction between a spreading ridge to the west and a subduction zone to the east of the region is preferable to other models.
- The activity of the system explained above ceased before the Maastrichtian and no evidence indicates volcanic activity in the Maastrichtian–Early Eocene interval. Debate continues on the setting of the Middle Eocene volcanism, which could have been erupted during a tensional period. This tensional event may be directly related to formation of the eastern Black Sea basin.

This study is a product of a joint project between the Turkish and Georgian workers. Present and previous directors of Mineral Research and Exploration Institute of Turkey (MTA), SDG and GIN fully supported the project throughout, for which we thank them. Thanks also to E. Bozkurt and O. Tatar for advice and critical reading of the manuscript. We also express our gratitude to the referees, B. A. Natal'in and A. H. F. Robertson, for their constructive reviews and useful suggestions.

References

ADAMIA, SH., BAYRAKTUTAN, S. & LORDKIPANIDZE, M. 1995. Structural correlation and Phanerozoic Evolution of the Caucasus–Eastern Pontides. *In*: ERLER, A., ERCAN, T., BİNGÖL, E. & ÖRÇEN, S. (eds) *Geology of the Black Sea Region, Proceedings of the International Symposium on the Geology of the Black Sea Region*. Mineral Research and Exploration Institute of Turkey (MTA) Publications, 69–75.
——, KEKELIA, M. & TSIMAKURIDZE, G. 1983. Pre-Variscan and Variscan granitoids of the Caucasus. *Newsletter*, **5**, 5–12.
——, LORDKIPANIDZE, M. B. & ZAKARIADZE, G. S. 1977. Evolution of an active continental margin as exemplified by the Alpine History of the Caucasus. *Tectonophysics*, **40**, 183–199.
——, CHKHOTUA, T., KEKELIA, M., LORDKIPANIDZE, M., SHAVISHVILI, I. & ZAKARIADZE, G. S. 1981. Tectonics of the Caucasus and adjoining regions:

implications for the evolution of the Tethys ocean. *Journal of Structural Geology*, **3**, 437–447.

AKDENİZ, N. 1998. The Demirözü Permo-Carboniferous and its importance in the regional structure. *Geological Society of Turkey Bulletin*, **31**, 71–80 [in Turkish with English abstract].

BEKTAŞ, O. 1984. Upper Cretaceous shoshonitic volcanism and its importance in the eastern Pontides. *Karadeniz Technical University, Earth Sciences Bulletin*, **3**, 53–62 [in Turkish with English abstract].

——, PELİN, S. & KORKMAZ, S. 1984. Mantle uprising and polygenetic ophiolites in the eastern Pontian (Turkey) back-arc basin. *Proceedings of the Ketin Symposium*. Geological Society of Turkey Publications, 175–188 [in Turkish].

BELOV, A. A., SOMIN, M. L. & ADAMIA, SH. 1978. Precambrian and Palaeozoic of the Caucasus (Brief Synthesis). *Jahrbuch für Geologie und Mineralogie*, **121**, 155–175.

——, —— & —— 1981. The main development stages and epochs of tectonic activity in the Mediterranean–Alpine folded area in Palaeozoic. *Newsletter*, **4**, 28–34.

BUKET, E. 1982. Petrochemical characteristics of the Erzincan–Refahiye ultramafic and mafic rocks and their correlations with other occurrences. *Hacettepe University Earth Sciences*, **9**, 43–65.

COX, A. & HART, R. B. 1986. *Plate Tectonics. How it Works*. Blackwell.

DEWEY, J. F. 1976. Ophiolite obduction. *Tectonophysics*, **31**, 93–120.

EĞİN, D. & HIRST, D. M. 1979. Tectonic and magmatic evolution of volcanic rocks from the northern Harşit River area, NE Turkey. *Proceedings of Geocome-1 (the 1st Geological Congress of the Middle East)*. Mineral Research and Exploration of Turkey (MTA) Publications, 56–93.

ELMAS, A. 1996. Geological evolution of Northeastern Anatolia. *International Geology Review*, **38**, 884–900.

GASANOV, T. A. 1986. Evolution of the Sevan–Akera Ophiolite Zone, Lesser Caucasus. *Geotectonics*, **20**, 147–156.

GEDİKOĞLU, A., PELİN, S. & ÖZSAYAR, T. 1979. The main lines of geotectonic development of the East Pontides in the Mesozoic Era. *Proceedings of Geocome-1 (the 1st Geological Congress of the Middle East)*. Mineral Research and Exploration of Turkey (MTA) Publications, 555–580.

GÖRÜR, N., ŞENGÖR, A. M. C., AKKÖK, R. & YILMAZ, Y. 1983. Sedimentological data regarding the opening of northern branch of the Neotethys. *Geological Society of Turkey Bulletin*, **26**, 11–19.

——, TÜYSÜZ, O., AKYOL, A., SAKINÇ, M., YİĞİTBAŞ, E. & AKKÖK, R. 1993. Cretaceous red pelagic carbonates of northern Turkey: their place in the opening history of the Black Sea. *Eclogae geologica Helvetica*, **86**, 819–838.

INTERNATIONAL GEOLOGICAL CONGRESS XXVII SESSION, MOSCOW. 1984. *Georgia SSR, Excursion Guidebook*. Publishing House 'Kholovneba'.

KETIN, İ. 1966. Tectonic units of Anatolia (Asia Minor). *Mineral Research and Exploration Institute of Turkey (MTA) Bulletin*, **66**, 23–34.

KNIPPER, A. L. 1980. The tectonic position of ophiolites of the Lesser Caucasus. *In*: PANAYIOTOU, A. (ed.) *Ophiolites*. Proceedings of the International Ophiolite Symposium, Cyprus–1979, 372–376.

—— & KHAIN, E. V. 1980. Structural position of ophiolites of the Caucasus. *Ofioliti*, **2**, 297–314.

——, RICOU, L. E. & DERCOURT, J. 1986. Ophiolites as indicators of the geodynamic evolution of the Tethyan Ocean. *Tectonophysics*, **123**, 213–240.

KOÇYİĞİT, A. 1990. Structural relations of three suture belts, west of Erzincan (NE Turkey), Karakaya, Intra-Tauride and Erzincan sutures. *Proceedings of the 8th Petroleum Congress and Exhibition of Turkey*. Turkish Association of Petroleum Geologists Publications, 152–160 [in Turkish with English abstract].

—— 1991. An example of an accretionary forearc basin from northern Central Anatolia and its implications for the history of subduction of Neotethys in Turkey. *Geological Society of America Bulletin*, **103**, 22–36.

KONAK, N., HAKYEMEZ, H. Y., BİLGİN, Z. R. & BİLGİÇ, T. 1995. *Oltu–Olur–Şenkaya Yöresi (Doğu Pontidler) Jeolojisi*. Mineral Research and Exploration Institute of Turkey (MTA) Report [in Turkish with English abstract].

LORDKIPANIDZE, M., MELIKSETIAN, B. & DJARBASHIAN, R. 1989. Mesozoic–Cenozoic magmatic evolution of the Pontiancrimean–Caucasus region. *Memoires de la Geologia France, Nouvella serie*, **154**, 103–124.

——, GUGUSHILLI, V. I., MIKADZE, G. A. & BATIASHVILI, I. 1984. Volcanism and postvolcanic processes on the rift and island arc stages of evolution of the Adjara–Trialetia. *Geoloqicke Prace Spravy*, **80**, 91–100.

MANETTI, P. Y., PECCERRILO, A., POLI, C. & CORSINI, F. 1983. Petrochemical constraints on models of Cretaceous–Eocene tectonic evolution of the Eastern Pontide Chain (Turkey). *Cretaceous Research*, **4**, 159–172.

NADAREISHVILI, G. S. 1980. Occurrence and evolution of Cretaceous volcanism. *In*: JANELIDZE, A. I. & TRALCHRELIDZE, A. A. (eds) *Volcanism and Formation of Mineral Resources in the Mobile Regions of the Earth*. Proceedings of the 5th International Volcanology Meeting Tbilisi, 127–145 [in Russian].

—— 1981. *Adjara–Trialeti Cretaceous Volcanism*. Academy of Sciences of Georgian Republic.

OKAY, A. İ. & LEVEN, E. JA. 1996. Stratigraphy and palaeontology of the Upper Palaeozoic sequence in the Pulur (Bayburt) region, Eastern Pontides. *Turkish Journal of Earth Sciences*, **5**, 145–155.

—— & ŞAHİNTÜRK, Ö. 1997. Geology of the Eastern Pontides. *In*: ROBINSON, A. G. (ed.) *Regional and Petroleum Geology of the Black Sea and Surrounding Region*. AAPG Memoirs, **68**, 291–311.

ÖZSAYAR, T., PELİN, S. & GEDİKOĞLU, A. 1981. Cretaceous of Eastern Pontides. *Karadeniz Technical University Bulletin of Earth Sciences*, **1**, 65–114 [in Turkish with English abstract].

—, —, —, EREN, A. A. & ÇAPKINOĞLU, S. 1980. Ardanuç (Artvin) yöresinin jeolojisi. *Proceedings of the TÜBİTAK VII Sciences Congress: Earth Sciences*, 189–205 [in Turkish with English abstract].

PECCERILLO, A. & TAYLOR, S. R. 1975. Geochemistry of Upper Cretaceous volcanic rocks from the Pontic chain, Northern Turkey. *Bulletin of Volcanology*, **39**, 557–569.

PEJATOVIC, S. 1971. Doğu Karadeniz Küçük Kafkasya bölgesindeki metalojenik zonlar ve bunların metalojenik özellikleri. *Mineral Research and Exploration Institute of Turkey (MTA) Bulletin*, **77**, 10–21 [in Turkish with English abstract].

ROBINSON, A. G., BANKS, C. J., RUTHERFORD, M. M. & HIRST, J. P. P. 1995. Stratigraphic and structural development of the Eastern Pontides, Turkey. *Journal of the Geological Society, London*, **152**, 861–872.

ŞENGÖR, A. M. C. 1987. An example of the importance of strike-slip tectonics in orogenic collages: Mesozoic evolution of Iran and surroundings. *Proceedings of the 7th Petroleum Congress and Exhibition of Turkey*. Turkish Association of Petroleum Geologists Publications, 50–64.

—— & YILMAZ, Y. 1981. Tethyan evolution of Turkey: a plate tectonic approach. *Tectonophysics*, **75**, 181–241.

—, — & KETİN, İ. 1980. Remnants of a Pre-Late Jurassic ocean in Northern Turkey: fragments of Permian–Triassic Palaeo-Tethys? *Geological Society of America Bulletin*, **91**, 599–609.

TATAR, Y. 1978. Tectonic study of the North Anatolian Fault zone between Erzincan and Refahiye region. *Hacettepe University Earth Sciences*, **4**, 201–236.

TOKEL, S. 1972. *Stratigraphical and volcanic history of the Gümüşhane region (NE Turkey)*. PhD Thesis, University College, London.

—— 1977. Eocene calc-alkaline andesites and geotectonism in the Eastern Black Sea region. *Geological Society of Turkey Bulletin*, **20**, 49–54 [in Turkish with English abstract].

—— 1987. Metallogeny of the Lesser Caucasus. *In*: JANKOVIÇ, S. (ed.) *Mineral Deposits of the Tethyan Eurasian Metallogenic Belt between the Alps and the Pamirs*. Belgrade, 119–127.

USTAÖMER, T. & ROBERTSON, A. H. F. 1997. Tectonic–sedimentary evolution of the North Tethyan margin in the Central Pontides of Northern Turkey. *In*: ROBINSON, A. G. (ed.) *Regional and Petroleum Geology of the Black Sea and Surrounding Region*. AAPG Memoirs, **68**, 255–290.

YILMAZ, A. 1980. *Origin and inner structure of the ophiolites and their relationships with other units along the region between Tokat and Sivas*. PhD Thesis, Ankara University [in Turkish with English abstract].

—— 1981. Inner structure and emplacement age of ophiolitic mélange of the area between Tokat and Sivas. *Geological Society of Turkey Bulletin*, **24**, 31–36 [in Turkish with English abstract].

—— 1982. Main geological characteristics and setting of the ophiolitic mélange of the area around Tokat (Dumanlı Dağı) and Sivas (Çeltek Dağı). *Mineral Research and Exploration Institute of Turkey (MTA) Bulletin*, **99/100**, 1–18 [in Turkish with English abstract].

—— 1985. Main geological characteristics and structural evolution of the region between the Kelkit Creek and the Munzur Mountains. *Geological Society of Turkey Bulletin*, **28/2**, 79–92 [in Turkish with English abstract].

—— 1989*a*. Tectonic zones of the Caucasus and their continuations in the northeastern part of Turkey: a correlation. *Mineral Research and Exploration Institute of Turkey (MTA) Bulletin*, **109**, 89–106 [in Turkish with English abstract].

——1989*b*. Thoughts on crustal structure of Eastern Anatolia. *28th International Geological Congress, Washington, USA, Abstracts*, **3**, 409.

—— & TERZİOĞLU, N. 1994. The geotectonic setting of Upper Cretaceous–Tertiary volcanism along the Eastern Pontian zone. *International Volcanological Congress, Abstracts*, 39.

—, AADAMIA, SH., ENGİN, T. & LAZARASHVILI, T. 1997. Geoscientific studies of the area along the Turkish–Georgian Border. Mineral Research and Exploration Institute of Turkey (MTA) Report, No. 10122 [in Turkish].

—, —, LORDKIPANIDZE, M. *ET AL.* 1996. A study of tectonic units of the area along the Turkish–Georgian border: *The 2nd International Symposium on the Petroleum Geology and Hydrocarbon Potential of the Black Sea area, Şile (İstanbul), Turkey, Abstracts*, 6.

YILMAZ, S. & BOZTUĞ, D. 1996. Space and time relations of three plutonic phases in the Eastern Pontides, Turkey. *International Geology Review*, **38**, 935–956.

YILMAZ, Y. 1976. Geology of the Gümüşhane granite (petrography). *İstanbul University Faculty of Science Bulletin Serie B*, **39**, 157–172.

—— 1980. *Transformation of a Spreading Ridge to a Subduction Zone. An example from Turkey*. Geological Society of Turkey, Special Publications.

—, TÜYSÜZ, O., YİĞİTBAŞ, E., GENÇ, Ş. C. & ŞENGÖR, A. M. C. 1997*a*. Geology and Tectonic Evolution of the Pontides. *In*: ROBINSON, A. G. (ed.) *Regional and Petroleum Geology of the Black Sea and Surrounding Region*. AAPG Memoir, **68**, 183–226.

—, SERDAR, H. S., GENÇ, Ş. C. *ET AL.* 1997*b*. The geology and evolution of the Tokat Massif, South-Central Pontides, Turkey. *International Geology Review*, **39**, 365–382.

ZAKARIADZE, G. S., KNIPPER, A. L., SOBOLEV, A. V., TSAMERIAN, O. P., DMITRIEV, L. V., VISHNEVSKAYA, V. S. & KOLESOV, G. M. 1983. The ophiolite volcanic series of the Lesser Caucasus. *Ofioliti*, **8**, 439–466.

Geochemical character and tectonic environment of Neotethyan ophiolitic fragments and metabasites in the Central Anatolian Crystalline Complex, Turkey

PETER A. FLOYD,[1] M. CEMAL GÖNCÜOĞLU,[2] JOHN A. WINCHESTER[1] & M. KENAN YALINIZ[3]

[1] *Department of Earth Sciences, University of Keele, Staffordshire, ST5 5BG, UK (e-mail: p.a.floyd@esci.keele.ac.uk)*
[2] *Department of Geological Engineering, Middle East Technical University, TR-06531 Ankara, Turkey*
[3] *Department of Civil Engineering, Celal Bayar University, Muradiye, Manisa, Turkey*

Abstract: The Central Anatolian Crystalline Complex (CACC) or Kırşehir Block is part of the metamorphosed leading edge of the Tauride–Anatolide Carbonate Platform. It contains oceanic remnants derived from the Neotethys Ocean (İzmir–Ankara–Erzincan branch) which separate it from the Sakarya microcontinent. Two tectonic units are distinguished: an amphibolite facies Mesozoic 'basement', dominated by platform marbles, over which is thrust a younger fragmented Upper Cretaceous ophiolite sequence. Three metabasite horizons were sampled to reconstruct the development of the oceanic components: (1) fragmented Upper Cretaceous (90–85 Ma) stratiform ophiolitic members comprising gabbros, sheeted dykes, basalt lavas and pelagic sediments thrust over all other units; (2) a tectonised admixture of basite, ultramafic and felsic blocks in an ophiolitic mélange (Upper Cretaceous matrix) thrust over the basement metamorphic rocks; and (3) amphibolites concordant with 'basement' marbles and minor pelagics of the largely (?)Triassic Kaleboynu Formation in the lower part of the carbonate platform.

Metabasalts and metagabbros from isolated fragments of the stratiform ophiolites form geochemically coherent groups and indicate the influence of a subduction component during their development. It is considered that the suprasubduction zone ophiolites record the association of a tholeiitic arc and an adjacent back-arc basin with more mid-ocean ridge basalt (MORB)-like compositions.

Metabasite blocks within the tectonised ophiolitic mélange slice are MORB like, together with minor ocean island basalt (OIB) and island arc basalts, and may be tectonically related to ophiolitic units within the accretionary wedge of the Ankara Mélange.

Concordant amphibolites of the Kaleboynu Formation are largely OIB types and reflect an early ensialic rifting stage of the Tauride–Anatolide Carbonate Platform. Small ocean basins also developed at this time, as recorded by the presence of MORB and associated pelagics.

The CACC block, together with parts of the Ankara Mélange, are considered to represent oceanic lithosphere (comprising both early spreading centre and latter subduction-influenced crust) and continental carbonate platform that were subsequently ejected from an accretionary–subduction complex on collision with the Sakarya microcontinent.

The Neotethyan area of the eastern Mediterranean records the development of a series of narrow oceanic seaways and microcontinent fragments that resulted from the fragmentation of the Gondwana margin during the Mesozoic (Robertson & Dixon 1984). Rifting of the northern margin of Gondwana, and the subsequent development and destruction of ocean floor, is documented by diverse tectonolithological associations and accompanying basaltic volcanism (Şengör & Yılmaz 1981; Robertson & Dixon 1984; Robertson *et al.* 1991). The Neotethys, initially generated during the Late Triassic, survived in small seaways up to Late Cretaceous time, when convergence of Africa and Eurasia largely closed this segment via northwards subduction under the Pontide active margin of Eurasia (Livermore & Smith 1984; Yılmaz *et al.* 1997).

The Turkish sector of Neotethys is characterized by two major east–west belts of ophiolitic fragments (Fig. 1, inset) that mark the presence of ancient suture zones and document the destruction of former ocean basins (Juteau

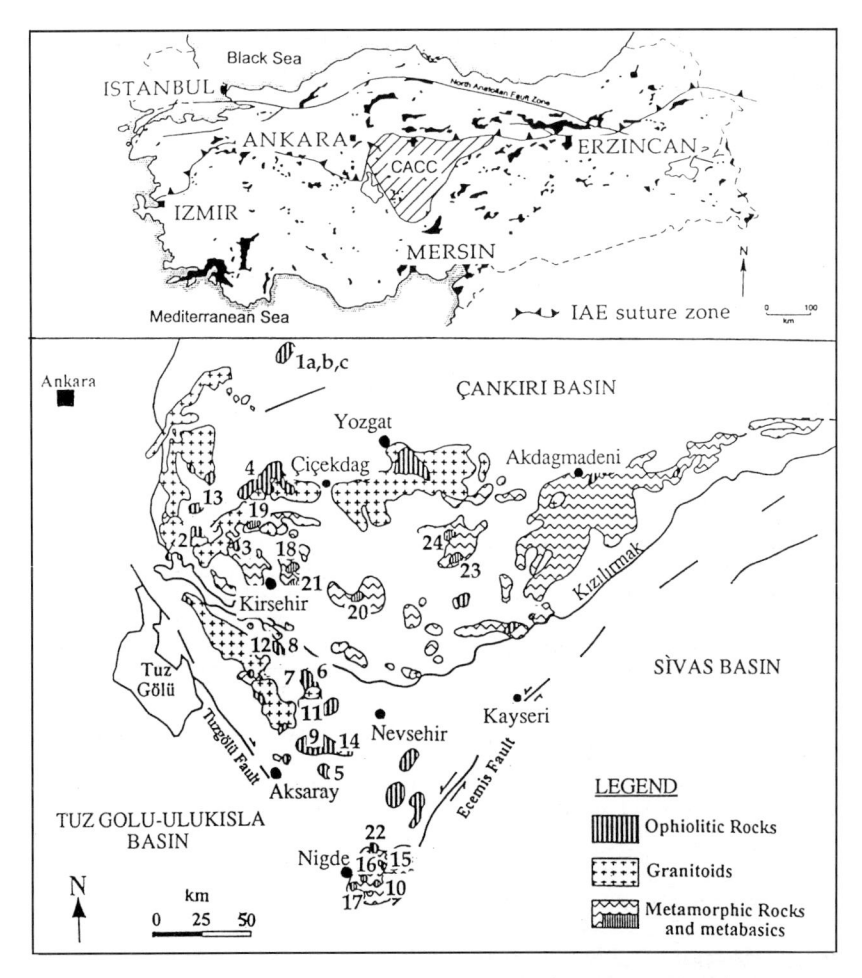

Fig. 1. Lithological sketch map of the CACC or Kırşehir Block (largely after surveying by Göncüoğlu et al. 1991; Yalınız & Göncüoğlu 1998) showing the location of the main SSZ ophiolitic fragments and the major metabasite bodies within the metamorphic basement. Inset shows Tethyan ophiolitic belts [after Juteau (1980)] and the location of the CACC in Turkey. The main sampling areas covered by this paper are shown by numbers and refer to the following localities: for Stratiform SSZ ophiolites – 4, Çiçekdağ Ophiolite; 6, Sarıkaraman Ophiolite: for isolated ophiolitic remnants – 1, Çankırı Basin (a, Ayvatlı; b, Alişeyhli; c, İneğazılı); 2, Kaman–Hirfanlı Dam; 3, Kurançalı; 5, Mamasin Dam; 7, Bozkır Dam; 8, Devedamı; 9, Alayhanı; 11, Yalıntaş; 12, Karataş; 13, Geyral; 14, Keskin; 15, Dokuzlar, 22, Aktaş Dam: for metamorphosed ophiolitic mélange – 10, Aşigediği; 18, Çimeli; 19, Köşker Yaylası; 20, Ayrıdağ: for the Kaleboynu Formation – 16, Söğütlütepe; 17, Göbettepe; 21, Kervansaray Dağı; 23, Çomakdağ; 24, Karaveli.

1980; Şengör & Yılmaz 1981). Each belt is considered to represent a separate Neotethyan ocean: (1) a northerly belt composed mainly of ophiolitic mélange (part of the 'Ankara Mélange') is representative of the İzmir–Ankara–Erzincan Ocean, located between the Sakarya microcontinent fragment and the leading edge of the Tauride–Anatolide Platform (TAP); and (2) a southern branch called either the Southern Neotethys Ocean (Şengör & Yılmaz 1981) or the Péri–Arabic Belt (Ricou 1971), which includes well-documented bodies such as Troodos, Mersin, Pozantı–Karsantı, and Hatay. The latter ocean strand separated the main body of the Gondwana continent from the rifted TAP to the north and was formed in a suprasubduction zone setting (Pearce et al. 1984) rather than at a major ocean-spreading centre.

Isolated, allochthonous metabasite bodies of possible ophiolitic affinity (termed the Central Anatolian Ophiolites; Göncüoğlu et al. 1991) are exposed in the triangular Central Anatolian

Crystalline Complex (CACC) or Kırşehir Block (Fig. 1), and are less well known than the Southern Neotethyan ophiolites. They are of particular interest to Neotethyan development in that they are considered to have been thrust southwards out of the İzmir–Ankara–Erzincan Ocean during closure (Şengör & Yılmaz 1981; Göncüoğlu et al. 1991), and preserve both spreading ridge and suprasubduction zone generated oceanic crust.

The Central Anatolian Crystalline Complex (CACC)

The CACC is situated north of the TAP proper and south of the İzmir–Ankara–Erzincan Suture (Fig. 1). A belt of glaucophanitic ophiolites situated around the southern margin of the CACC has been interpreted as an additional oceanic seaway (the Inner Tauride Belt; Görür et al. 1984) which once separated the CACC from the TAP. However, others considered that the CACC originally represented the northern passive margin of the TAP (Özgül 1976) and that the Inner Tauride ophiolites were also derived from the İzmir–Ankara–Erzincan Ocean to the north (Göncüoğlu 1986).

Apart from a cover of uppermost Cretaceous–Miocene volcanic and sedimentary rocks, the CACC (Fig. 1) comprises three fundamental lithological units (Göncüoğlu et al. 1991): (1) a mainly amphibolite facies metamorphic basement dominated by marble with subordinate pelitic to psammitic schists, gneisses and metabasites (the Central Anatolian Metamorphics); (2) fragmented ophiolitic remnants of pillow lavas, dykes and gabbros with minor ultramafic rocks and plagiogranites; and (3) two sets of granitoid rocks. Generalized stratigraphic relationships within the CACC are shown in Fig. 2 (not to scale).

The abundance of marbles with a carbonate-platform affinity in the basement suggests that it is the metamorphosed equivalent of the TAP. Due to recrystallization, precursor ages from fauna in the marbles are difficult to determine, but largely range in age from Triassic to Early Cretaceous, although some have suggested a Late Palaeozoic age (Silurian–Devonian) for the lower parts of the complex (e.g. Koçak & Leake 1994; Fig. 2). The regional metamorphism is generally considered to be pre-mid-Cretaceous (Göncüoğlu 1982). Two sets of acid plutonics are recognized: an early, but post-metamorphic, group of syncollisional (95 ± 11 Ma) and post-collisional granites (76–71 Ma; Rb–Sr whole rock) and a later group of cross-cutting syenites (71–70 Ma; Rb–Sr, K–Ar) (Ataman 1972; Göncüoğlu 1986; Göncüoğlu & Türeli 1993; Erler & Göncüoğlu 1996; Göncüoğlu et al. 1997). These are important markers relative to the formation age of the ophiolitic fragments in the CACC (see below) and demonstrate that, during collision, melting of continental crust both preceded suprasubduction-type ophiolite crust formation and postdated its subsequent obduction.

Metabasites and ophiolites

Pertinent to this overview are two tectonically separate magmatic groups: (1) metabasite bodies within the basement (Central Anatolian Metamorphics), as well as blocks from a separate ophiolitic mélange unit; and (2) massive stratiform ophiolites, together with various isolated remnants of suspected ophiolitic parentage.

Metabasites were sampled from the Kaleboynu Formation (that may, in part, be Triassic

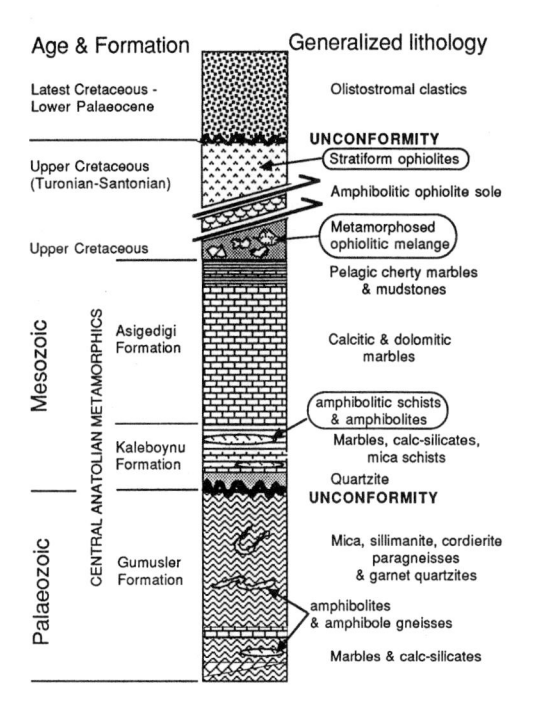

Fig. 2. Generalized stratigraphic column (not to scale) for the CACC (Göncüoğlu et al. 1991). The relative positions of the units sampled for the geochemical review are enclosed in envelopes: stratiform ophiolites, metamorphosed ophiolitic mélange and Kaleboynu Formation amphibolites.

in age) and the structurally higher ophiolitic mélange unit (matrix of possible Late Cretaceous age; Fig. 2). In the Kaleboynu Formation, the metabasites comprise multiple, relatively thin, concordant bodies often intimately associated with minor metapelites and highly siliceous sediments, representing pelagics and cherts, respectively. This is typical of the basement around Niğde (Fig. 1) which exhibits a high proportion of contorted marble with infolds of metabasites and metasediments. In contrast, various metabasite, meta-ultramafic and metafelsic rocks have been sampled from the metamorphosed ophiolitic mélange unit, where all three lithologies commonly occur as boudins, together with marble blocks derived from the Aşığediği Formation below, in a sheared metapelitic or calc-pelite matrix. This is a common occurrence in the northern part of the CACC around Kırşehir, where a relatively high proportion of metamagmatic lithologies are present in the tectonic mélange.

The other main set of basaltic rocks in the CACC are represented by variously fragmented ophiolites thrust/obducted over the basement (Fig. 2). These are low-grade greenschist facies, stratiform sequences of pillowed and massive metabasalts with sheeted dykes and associated gabbros (Yalınız et al. 1996). Plagiogranites developed throughout the plutonic and volcanic portions of the ophiolite are common (Göncüoğlu & Türeli 1993; Floyd et al. 1998). The largest ophiolites are the Sarıkaraman and Çiçekdağ Ophiolites (Fig. 1), which exhibit a typical suprasubduction zone (SSZ) chemistry (Yalınız et al. 1996; Yalınız & Göncüoğlu 1998). Scattered throughout the CACC are tectonically isolated remnants of pillow lavas or gabbros (both with dykes) that may also have an ophiolitic parentage. Based on the faunal content of pelagic sediments covering the volcanic portion of the massive stratiform ophiolites, formation ages are generally Middle Turonian–Early Santonian (90–85 Ma; Yalınız & Göncüoğlu 1998). These oceanic segments were short lived (5–10 Ma), as deep slicing and obduction over the metamorphic basement was rapidly followed by the intrusion of the late granitoids that cut the ophiolites at around c. 76–71 Ma (e.g. the Terlemez Monzogranite intrudes the Sarıkaraman Ophiolite; Floyd et al. 1998).

Objectives and comparisons

The main objectives of this paper are to: (1) review the chemical features of the metabasites within the CACC metamorphic basement and the fragmented stratiform ophiolites obducted over it; and (2) compare the basaltic compositions with reference units to determine their affinities and aid discrimination of their geotectonic environment. The chemical features of minor metafelsic and meta-ultramafic bodies in the basement will also be briefly discussed to decide whether they might be equivalent to ophiolitic plagiogranites and (mantle) cumulates, respectively.

In the light of current tectonic models, which suggest that the basement metabasites were derived from the İzmir–Ankara–Erzincan Ocean, comparisons will be made with metabasalts from the Ankara ophiolitic Mélange (Çapan & Floyd 1985; Floyd 1993). In a similar manner, those CACC obducted ophiolites with suprasubduction features (e.g. Sarıkaraman Ophiolite; Yalınız et al. 1996) will provide a chemical template for other isolated metabasic units with suspected ophiolitic affinities; e.g., pillow lavas in the Çankırı Basin and tectonically isolated gabbro bodies with associated dykes.

Determination of chemical affinity is relatively straightforward, although direct regional comparisons suffer from two problems: (1) that many of the carbonate sequences (marble) in the basement have not been dated and hence the age of associated volcanism is unknown; and (2) structural relationships are often complex and the assignment of metabasites to a particular basement formation has previously been based on apparently similar lithological associations. A broad comparison can be made with the carbonate platform of the TAP, but this features few volcanic rocks away from its northern edge. Thus, at best, this review, offers possible correlations and discriminations based on the data available and a model that emphasizes the chemical differences between the basement metabasites (possibly Late Triassic and/or broadly Jurassic–Early Cretaceous) and the later stratiform ophiolites, well documented as Late Cretaceous.

In summary, the following sampled units and groups of lithologies are chemically characterized and compared in this paper: (1) ophiolitic plagiogranites and basement metafelsic blocks (from the metamorphosed ophiolitic mélange); (2) basement meta-ultramafic rocks (from the Kaleboynu Formation and the metamorphosed ophiolitic mélange); (3) basement metabasites (from the Kaleboynu Formation and the metamorphosed ophiolitic mélange); and (4) stratiform ophiolites (the Sarıkaraman and Çiçekdağ Ophiolites), together with tectonically isolated remnants thought to have an ophiolitic parentage or affinity.

Petrographic summary of basic compositions

Stratiform ophiolites and tectonically isolated remnants

The Sarıkaraman and Çiçekdağ Ophiolites are the best representatives of fragmented stratiform bodies in the CACC; the former has already been described by Yalınız et al. (1996).

The volcanic portion of the Çiçekdağ Ophiolite has many features in common with the Sarıkaraman Ophiolite, exhibiting both massive and pillowed basalts cut by various feeder dykes, all of which have undergone low-grade greenschist or pumpellyite facies metamorphism. The Çiçekdağ sheeted dyke complex comprises aphyric and plagioclase–phyric metabasalts and dolerites with any minor interstitial glass replaced by chlorite. Carbonate, chlorite, quartz and minor epidote are common replacement minerals. The lavas may be aphyric, or exhibit a number of phenocryst assemblages, including olivine, plagioclase, plagioclase–clinopyroxene, all of which may be set in a variably quenched matrix of serrated plagioclase microlites, variolitic fans of crystallites and an opaque, originally glassy, matrix. Secondary assemblages are typically actinolite, chlorite or chlorite–smectite, carbonate, epidote and rarer pumpellyite. Interiors of pillow lavas and massive flows are coarser grained and generally holocrystalline, whereas the margins are glassy and often develop spherulites. Pillow breccias show similar textures and petrography. No systematic petrographic changes have yet been noted with stratigraphic height in the lava sequence, except that the upper basalts tend to be generally aphyric. Feeder dykes to the lavas are dominated by aphyric and plagioclase–phyric basalts and are commonly altered with variable proportions of epidote, actinolite, carbonate and chlorite.

The isolated outcrops of metagabbro and dykes seen at the Bozkır Dam Quarry and the Aktaş Dam and nearby Dokuzlar sites (locations 7, 22 and 15, respectively in Fig. 1) are also of assumed ophiolitic parentage. In each case, a variety of massive and tectonised metagabbros are exposed, cut by both basaltic and plagiogranite dykes. The gabbros have been variably amphibolitized and range from examples showing relict subophitic clinopyroxene and plagioclase surrounded by amphibole, to foliated hornblende–plagioclase amphibolites and massive granular-textured quartz-bearing amphibolites with rare pyroxene relicts. The basaltic dykes are generally plagioclase–phyric and variably altered.

Other tectonically isolated remnants include outcrops of pillow lavas, such as those which form a local basement to Tertiary basin-fill sediments in the Çankırı Basin (location 1 in Fig. 1). The pillow lavas, which are often highly altered by carbonate and chlorite, are represented by aphyric and plagioclase ± clinopyroxene–phyric metabasalts.

Basement metabasites

Although the metabasites are mainly amphibolites, some larger bodies retain relict textures and mineralogy indicating that they were originally gabbros or dolerites. While this is invariably true of the boudins found in the metamorphosed ophiolitic mélange unit, it is not seen in the concordant Kaleboynu Formation amphibolites. Throughout the basement, metamorphism was broadly amphibolite facies, although the variable colour of the hornblende (pale green, brownish green and blue-green) implies a range of conditions within this facies (e.g. Miyashiro 1973). The association of migmatitic sillimanite + garnet-bearing gneisses with amphibolites in the south of the CACC implies peak metamorphic conditions of 600–700°C at 4 kbar for the lowermost Gümüşler Formation (Koçak & Leake 1994).

Metabasite blocks from the Upper Cretaceous ophiolitic mélange (Fig. 2) range from coarse-grained amphibolites with strongly pleochroic hornblende (yellow to brownish green), granular plagioclase and minor quartz, to hornblende–quartz schists. Some coarser varieties of clear gabbroic parentage retain relict magmatic clinopyroxene and exhibit a granoblastic mosaic of sieve-textured clinopyroxene with plagioclase, quartz and abundant titanite. Progressive amphibolitization produced various hornblende schists, some with relict clinopyroxene, and abundant sphene which appears characteristic for this group of metabasites. Concordant Kaleboynu Formation metabasites from the Niğde area (Fig. 1) are dominated by fine-grained amphibolites or hornblende schists, and are characterized by variable secondary carbonate and epidote, probably resulting from penetration of these thin bodies by late circulating fluids from the adjacent marbles.

Analytical methods and alteration effects

Representative metabasaltic lavas were collected from the Çiçekdağ Ophiolite (Table 1) and tectonically isolated (presumed ophiolitic) gabbro bodies and pillow lava sequences (Table 2) for comparison with the Sarıkaraman Ophiolite [previously described by Yalınız et al. (1996)]. A suite of basement metabasites represented by foliated and massive amphibolites (Table 3), together with minor metafelsic and meta-ultramafic rocks, were collected from the Kaleboynu Formation and ophiolitic mélange. All were analysed for major and selected trace elements on an ARL 8420 X-ray fluorescence (XRF) spectrometer (Department of Earth Sciences, University of Keele), calibrated against both international and internal Keele standards of appropriate composition. Details of methods, accuracy and precision are given in Floyd & Castillo (1992).

The metabasite samples show varying degrees of low-grade mineralogical alteration (within the pumpellyite to amphibolite facies) and as such can be expected to have suffered selected element mobility, especially involving the large-ion lithophile (LIL) elements (e.g. Hart et al. 1974; Humphris & Thompson 1978; Thompson 1991). LIL element (e.g. K, Na, Rb, Ba, Sr) abundances are often highly variable, together with most major elements and ratios (e.g. FeO*/MgO), and are unreliable as indicators of petrogenetic relationships or tectonic discrimination. This is particularly true for the volcanic sequence of ophiolites where mineralogical and chemical alteration by submarine hydrothermal processes are well known (e.g. Gass & Smewing 1973; Pearce & Cann 1973; Spooner & Fyfe 1973; Smewing & Potts 1976). However, characteristic magmatic interelement relationships are often maintained by those elements that are considered relatively immobile during alteration, such as high field strength (HFS) elements and the rare earth elements (REE) (e.g. Pearce & Cann 1973; Smith & Smith 1976; Floyd & Winchester 1978). Under some circumstances, such as the extensive carbonatization of metabasites, the REE and HFS elements can also be mobilized and/or abundances diluted (e.g. Hynes 1980; Humphris 1984; Rice-Birchall & Floyd 1988), although this appears only to have seriously affected the meta-ultramafic rocks (see below).

Geochemistry of plagiogranites and basement metafelsics

Trondjemites and rhyolites (plagiogranite suite) of the Sarıkaraman Ophiolite associated with the gabbros and basalts, respectively, have compositions typical of other plagiogranites worldwide (Floyd et al. 1998). Both sets of plagiogranites from the stratiform Sarıkaraman and Çiçekdağ Ophiolites have typically low and uniform Nb values (Fig. 3). This feature distinguishes them from the late post-collisional granitoids (that cut the ophiolites) which display Zr/Nb ratios of c. 10. Various felsic or feldspar porphyry dykes that cut isolated pillow lava sequences and gabbro remnants also have similar chemical features to the plagiogranites. This suggests that they are also part of the plagiogranite suite and that their basic hosts represent tectonically isolated ophiolite remnants.

On the other hand, variably foliated metafelsic bodies within the basement (blocks from the metamorphosed ophiolitic mélange) have similar Zr/Nb ratios to the late granitoids, but tend to be less evolved (Fig. 3). Although these metafelsic rocks generally have low HFS element abundances they do not appear to belong to an earlier basement plagiogranite suite, but have features more akin to post-collisional, or possibly arc-related, 'granites' which commonly have Zr/Nb > 10 (Leat et al. 1986). It is suggested that the metafelsic rocks are not comagmatic with the associated basement metabasites but probably represent independent acid melts generated by crustal melting within an arc, or during continent collision, prior to incorporation in the ophiolitic mélange.

Geochemistry of meta-ultramafic bodies

Large boudins within segments of highly deformed ophiolitic mélange have many of the features typical of 'knockers' (Karig 1980). High MgO, Ni and Cr contents, together with a serpentine–talc–carbonate mineralogy, indicate an ultramafic parentage. However, most have suffered variable Ca and Sr metasomatism relative to fresh ultramafic compositions (Fig. 4), reflecting the mobility of these elements in the surrounding carbonate-rich mélange matrix during shearing. On the basis of their Ni and Cr contents (Floyd et al. 1993), the meta-ultramafic rocks comprise two separate groups, some of which have a clear ophiolitic affinity, while others represent cumulates associated with continental intrusive gabbro bodies. The low Ni and

Table 1. *Representative chemical analyses of basaltic lavas from the Sarıkaraman and Çiçekdağ Ophiolites*

	Sarıkaraman ophiolite basaltic lavas										Çiçekdağ ophiolite basaltic lavas								
Sample number	P-3	P-10	P-12	P-14	P-15	P-16	P-17	P-18	P-19	M-19	C-6	C-9	C-10A	C-10B	C-11	C-13	C-16	C-17	C-18	C-19
Major oxides (wt%) by XRF spectrometry
SiO_2	53.26	55.27	38.62	50.53	64.25	56.43	47.65	50.33	55.58	47.01	53.12	53.72	51.59	57.12	47.36	54.01	45.45	51.19	49.99	40.67
TiO_2	0.86	1.36	0.78	0.70	0.91	0.63	1.03	0.69	0.62	1.50	1.39	0.44	1.20	1.08	0.76	1.21	1.06	1.32	0.66	0.84
Al_2O_3	16.07	12.57	14.28	13.05	11.58	11.67	16.17	14.58	12.17	15.32	16.63	13.43	15.91	14.21	14.03	16.66	17.93	14.87	15.74	12.18
Fe_2O_3t	10.32	13.37	8.61	8.38	8.18	7.52	9.07	7.59	5.80	13.30	12.31	9.85	15.39	12.47	12.10	12.09	11.44	12.42	9.11	8.75
MnO	0.09	0.26	0.41	0.34	0.07	0.29	0.11	0.24	0.24	0.32	0.26	0.18	0.20	0.16	0.21	0.13	0.18	0.14	0.13	0.10
MgO	3.68	7.93	13.66	13.89	4.02	12.30	9.49	10.89	8.75	5.53	3.90	8.96	5.43	3.25	6.64	4.00	6.88	3.97	5.74	1.65
CaO	12.10	4.03	9.25	3.36	8.14	2.71	5.09	5.34	6.41	5.73	7.55	5.92	2.53	2.82	5.65	1.60	12.21	4.76	6.33	16.50
Na_2O	0.01	0.01	2.33	1.16	0.01	1.20	4.09	3.04	2.98	3.95	3.69	5.05	5.73	7.09	7.14	8.64	2.41	4.20	2.46	6.67
K_2O	0.01	0.01	0.01	0.01	0.01	0.01	0.01	0.01	0.01	0.01	0.43	0.09	0.17	0.07	0.08	0.20	0.37	0.04	0.18	0.09
P_2O_5	0.06	0.08	0.05	0.05	0.07	0.05	0.10	0.05	0.04	0.10	0.06	0.04	0.08	0.09	0.05	0.08	0.04	0.13	0.06	0.19
LOI	3.16	4.99	11.83	8.51	2.75	7.18	6.92	7.52	7.45	7.37	0.81	2.51	1.55	1.35	5.75	1.01	1.88	6.92	9.44	12.68
Total	99.62	99.88	99.83	99.98	99.99	99.99	99.73	100.28	100.05	100.14	100.15	100.19	99.78	99.71	99.77	99.63	99.85	99.96	99.84	100.32
Trace elements (ppm) by XRF spectrometry
Ba	12	7	5	14	9	7	11	20	5	8	73	36	25	76	47	182	53	17	55	24
Ce	10	1	3	3	10	9	1	2	10	8	3	1	10	3	1	6	3	5	2	2
Cl	50	54	46	86	59	54	58	107	72	44	226	1	5	5	3	5	4	2	3	1
Cr	116	49	487	460	116	452	204	311	359	24	21	345	24	9	254	38	127	27	110	242
Cu	14	18	9	10	10	3	7	8	16	32	89	162	161	170	74	50	68	45	109	37
Ga	17	15	10	12	16	10	15	12	9	17	20	10	13	18	15	13	19	20	12	8
La	1	1	1	1	2	1	1	1	1	3	3	3	1	1	2	2	1	4	1	1
Nb	2	3	1	1	2	1	1	1	1	3	2	1	2	2	2	2	4	2	1	2
Nd	19	8	13	11	11	7	8	14	5	11	5	7	15	1	7	11	9	16	3	6
Ni	35	18	186	164	46	187	76	115	124	14	7	55	17	18	70	12	54	16	50	85
Pb	10	6	6	6	8	7	4	6	6	7	75	9	9	4	15	6	30	2	5	12
Rb	3	5	4	3	4	3	3	3	3	4	17	2	2	5	3	6	6	3	8	2
S	65	72	70	76	70	82	75	80	85	88	210	1	2	19	10	9	3	23	12	48
Sr	325	103	76	34	192	27	78	99	75	31	190	92	66	134	108	124	737	41	34	106
V	327	442	219	262	258	217	260	245	164	528	301	281	576	648	433	385	356	451	332	251
Y	25	25	17	11	20	11	25	17	12	26	25	12	23	28	17	26	34	31	17	22
Zn	23	186	115	137	24	108	49	111	78	202	195	68	80	135	130	52	75	115	72	82
Zr	54	71	57	53	58	50	70	53	48	76	48	20	43	47	32	66	77	80	37	52

Table 2. *Representative chemical analyses of metamorphosed gabbros and basaltic pillow lavas from tectonically isolated ophiolitic fragments*

	Aktaş Dam gabbros							Aktaş Dam dykes					Çankırı Basin lavas					Çankırı Basin dykes		
Sample number	AK-1	AK-2	AK-3	AK-4	AK-5	AK-12	AK-13	AK-6	AK-7	AK-8	AY-5	AY-8	AY-9	AY-11	AY-18	AY-19	AY-20	AY-6	AY-10	AY-21

Major oxides (wt%) by XRF spectrometry

SiO_2	51.10	50.50	49.44	48.24	49.91	48.45	52.77	49.33	51.65	48.77	45.47	49.65	50.46	54.89	53.00	47.62	47.43	52.16	54.71	50.05
TiO_2	0.25	0.16	0.46	0.44	0.46	0.42	0.59	0.47	0.74	0.51	1.65	1.28	0.96	1.27	0.58	1.54	1.27	1.40	1.03	1.06
Al_2O_3	21.30	14.65	12.74	15.33	13.60	11.23	18.81	14.22	18.20	11.91	16.94	17.13	17.68	15.73	19.77	16.52	15.73	16.01	15.74	17.90
$Fe_2O_3^t$	5.76	5.92	9.76	9.89	9.43	8.99	7.36	10.40	12.18	17.19	11.86	9.60	10.30	4.54	13.65	12.88	13.07	7.46	9.22	8.75
MnO	0.13	0.11	0.18	0.17	0.15	0.15	0.14	0.18	0.18	0.23	0.21	0.25	0.16	0.15	0.11	0.22	0.20	0.13	0.28	0.22
MgO	5.10	11.60	13.89	9.95	12.22	15.64	4.75	11.28	5.28	11.95	6.39	5.42	4.18	4.38	1.69	4.37	5.29	3.40	3.50	5.82
CaO	12.08	14.82	10.84	13.68	10.62	12.00	10.55	11.72	9.70	12.15	1.96	4.78	6.16	4.31	4.49	7.55	5.44	4.01	5.17	4.70
Na_2O	2.11	0.78	1.30	0.94	1.53	1.09	3.53	1.76	3.67	1.44	5.74	5.70	6.14	5.99	4.69	4.10	5.32	6.94	6.66	5.38
K_2O	0.85	0.29	0.17	0.37	0.18	0.12	0.19	0.22	0.11	0.20	0.03	0.09	0.02	0.05	7.12	0.17	0.03	0.04	0.03	0.02
P_2O_5	0.01	0.01	0.03	0.02	0.03	0.03	0.43	0.04	0.08	0.01	0.12	0.15	0.12	0.16	0.25	0.14	0.16	0.17	0.16	0.14
LOI	1.01	1.14	0.97	1.25	1.54	1.37	0.78	1.25	0.42	0.53	4.63	3.28	4.14	2.70	4.19	5.98	5.73	2.67	5.47	5.46
Total	99.70	99.98	99.78	100.28	99.67	99.49	99.90	99.75	100.43	99.88	100.33	99.59	99.62	99.93	100.43	100.38	100.27	99.72	100.48	99.97

Trace elements (ppm) by XRF spectrometry

Ba	58	38	49	67	35	21	52	13	38	15	68	48	41	63	31	116	33	47	38	52
Ce	1	2	2	2	2	1	1	12	1	1	3	15	11	17	4	2	8	3	12	2
Cl	1	2	109	44	86	142	287	65	97	93	3	4	2	3	31	2	3	2	3	4
Cr	154	291	971	448	721	1443	167	647	35	699	1	19	38	20	42	13	62	1	59	123
Cu	12	22	29	40	45	41	36	23	48	22	18	95	15	31	79	54	20	20	14	40
Ga	17	10	13	10	12	11	18	13	17	14	17	17	14	14	20	17	14	18	14	15
La	1	2	2	2	1	2	5	2	3	1	2	2	3	2	1	3	1	1	4	2
Nb	3	3	1	1	2	2	3	2	3	3	5	5	4	5	4	3	5	6	6	5
Nd	11	6	2	2	3	2	6	18	1	4	6	15	11	17	6	8	10	11	17	4
Ni	30	107	266	55	207	241	33	191	15	172	11	18	24	13	22	8	33	9	27	45
Pb	20	16	6	10	9	8	3	11	7	11	2	2	5	5	9	6	7	5	5	3
Rb	31	10	7	16	5	8	8	5	2	3	3	3	2	1	1	3	1	1	1	1
S	1	2	2	1	41	3	4	3	2	1	2	27	3	6	6	17	1	2	2	2
Sr	163	79	165	92	84	56	215	94	187	64	66	149	138	126	345	170	72	77	88	104
V	119	116	248	303	244	229	116	247	257	344	420	356	246	308	295	420	351	368	219	252
Y	14	8	14	11	15	11	21	15	23	16	18	29	22	29	19	19	26	29	22	18
Zn	52	37	72	61	70	59	55	65	84	93	148	108	111	101	82	112	104	87	64	104
Zr	15	13	23	15	26	20	40	24	33	21	54	89	66	91	61	36	90	100	90	80

Table 3. Representative chemical analyses of metabasites (massive and foliated amphibolites) in different segments of the CACC metamorphic basement

	Metamorphosed ophiolitic mélange											Kaleboynu Formation								
Sample number	CM-9	CM-10	CM-11	CM-15	AD-1	AD-3	AD-5	ED-3	ED-4	ED-5	GT-8	GT-11	GT-14	GT-17	CO-2	CO-4	CO-5	KV-1	KV-2	KV-3
Major oxides (wt%) by XRF spectrometry																				
SiO$_2$	51.47	47.01	49.96	45.12	52.35	51.56	52.72	55.69	54.36	50.01	44.70	48.38	44.24	48.73	48.19	45.63	53.26	46.19	52.29	52.34
TiO$_2$	0.50	0.91	1.44	1.09	1.00	0.86	1.29	1.53	1.51	1.68	3.43	0.97	1.53	1.57	2.68	0.86	0.93	0.60	0.86	1.00
Al$_2$O$_3$	14.80	17.40	15.38	13.98	16.93	15.69	14.35	14.18	15.17	15.56	12.62	15.50	12.02	12.05	13.77	13.04	14.19	15.81	11.26	11.41
Fe$_2$O$_3$^{t}	8.65	8.90	12.21	13.08	9.65	9.25	9.80	12.31	12.59	13.39	15.32	8.05	13.27	12.85	16.26	6.65	6.97	12.05	7.31	6.94
MnO	0.17	0.17	0.18	0.21	0.17	0.17	0.20	0.16	0.18	0.20	0.16	0.19	0.27	0.23	0.20	0.11	0.11	0.20	0.21	0.18
MgO	8.82	7.03	6.32	9.84	4.14	5.67	6.09	4.12	4.34	5.67	10.17	6.68	10.05	8.37	6.00	4.19	3.67	7.43	5.60	6.21
CaO	9.96	12.31	9.20	11.14	9.20	9.82	9.29	7.78	8.37	8.68	9.00	11.48	12.67	11.11	9.10	21.06	16.55	13.62	17.49	17.06
Na$_2$O	4.21	3.98	4.47	2.23	4.99	4.68	3.84	3.47	2.93	3.23	2.73	3.90	1.85	2.72	3.17	1.55	1.30	1.85	2.32	2.18
K$_2$O	0.25	0.32	0.36	0.34	0.35	0.42	0.36	0.22	0.55	0.45	0.14	1.87	1.51	0.79	0.49	2.10	1.89	0.58	1.49	2.04
P$_2$O$_5$	0.08	1.64	0.08	0.01	0.27	0.24	0.24	0.14	0.11	0.14	0.41	0.13	0.23	0.29	0.28	0.15	0.15	0.10	0.16	0.20
LOI	0.93	0.58	0.64	2.84	1.35	1.50	1.55	0.28	0.39	0.73	1.59	3.15	2.74	1.29	0.23	4.68	1.51	0.94	1.51	1.03
Total	99.84	100.25	100.24	99.88	100.40	99.86	99.73	99.88	100.50	99.74	100.27	100.30	100.38	100.00	100.37	100.02	100.53	99.37	100.50	100.59
Trace elements (ppm) by XRF spectrometry																				
Ba	39	74	175	155	68	99	70	27	35	52	31	200	223	71	48	185	198	81	262	304
Ce	3	1	9	10	42	17	28	5	8	3	73	20	15	4	18	50	47	6	37	38
Cl	2	23	2	61	15	3	127	65	146	57	4	2	1	1	2	204	2	19	158	3
Cr	477	233	75	123	173	263	321	181	149	150	1129	357	513	486	248	193	240	184	324	319
Cu	62	20	194	17	112	93	70	38	53	52	33	68	54	63	29	22	15	123	24	66
Ga	14	17	16	13	18	17	14	18	19	20	22	15	17	16	19	18	19	15	16	13
La	2	1	1	1	15	14	11	2	4	4	26	4	5	7	6	23	33	3	14	26
Nb	2	3	4	6	9	9	21	5	17	6	38	8	21	23	7	22	29	10	16	24
Nd	8	12	13	16	28	12	24	11	21	10	40	17	14	7	15	28	22	35	25	22
Ni	112	61	34	80	35	56	106	31	21	31	244	104	305	276	64	60	65	9	121	129
Pb	7	22	18	16	10	8	10	5	8	6	12	1313	14	15	4	17	11	35	14	18
Rb	3	4	4	7	9	12	11	9	19	15	3	45	38	13	11	78	62	18	52	69
S	2	1	26	2	5	5	4	18	2	20	3	172	3	1	16	17	5	5	17	14
Sr	163	786	247	360	546	315	273	82	75	50	175	258	199	243	161	974	735	335	313	350
V	236	259	370	438	234	227	226	448	410	412	378	170	193	215	479	81	93	311	113	109
Y	15	25	30	31	25	21	29	35	37	41	27	20	19	20	58	27	29	14	27	27
Zn	58	75	80	83	84	65	89	80	91	106	122	172	121	107	135	94	109	78	112	90
Zr	24	41	85	85	105	92	119	88	93	108	204	77	89	89	173	148	179	23	118	147

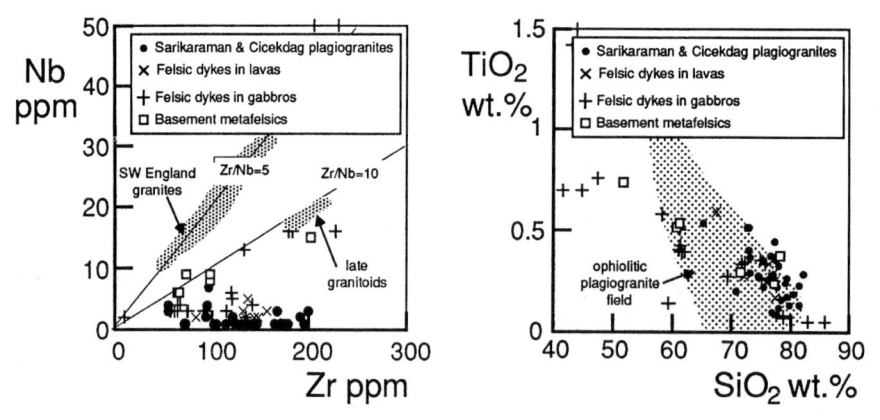

Fig. 3. Nb v. Zr and TiO$_2$ v. SiO$_2$ diagrams for various felsic rocks from the CACC: Sarıkaraman and Çiçekdağ ophiolitic plagiogranites [some data from Floyd et al. (1998)], metafelsics from the metamorphic basement, and late cross-cutting granitoids. Ophiolitic plagiogranites with very low and uniform Nb values are distinguishable from both basement metafelsics and the late granitoids. Ophiolitic plagiogranite field from Gerlach et al. (1981 and refs cited therein); data for southwest England granite field (high-level, post-collisional granites) from Exley et al. (1983).

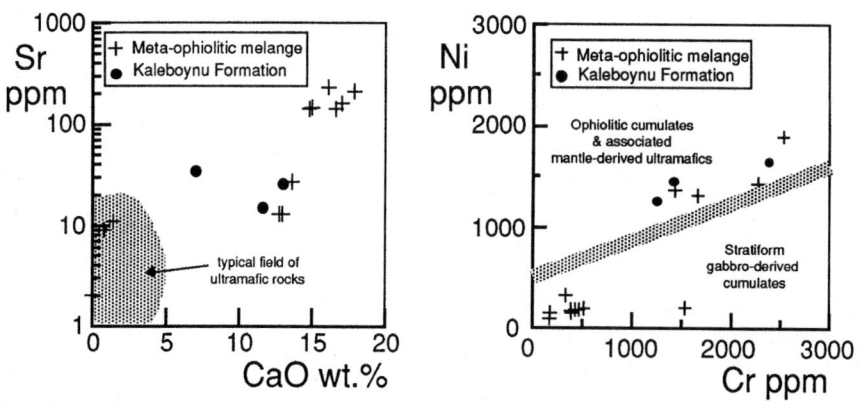

Fig. 4. Sr v. CaO and Ni v. Cr diagrams for meta-ultramafic bodies within the metamorphic basement complex of the CACC. Two chemical groups of meta-ultramafic rocks are probably present, some of which have subsequently been metasomatically enriched in Sr and Ca. Generalized ultramafic field from literature.

Cr values are not a consequence of metasomatic dilution as the Ca contents are variable throughout this group. The presence of ophiolitic ultramafics is to be expected as many of the associated metabasite blocks have mid-ocean ridge basalt (MORB) chemical characteristics (see below), both of which imply an oceanic regime. The low-Ni + Cr group, on the other hand, has chemical affinities with typical stratiform gabbros and might represent early gabbros intruded in a rifted continental setting prior to major ocean development.

Geochemistry of stratiform ophiolites

The Sarıkaraman Ophiolite has previously been shown to be a fragmented stratiform body with a volcanic section exhibiting typical suprasubduction zone (SSZ) characteristics similar to other Neotethyan ophiolites (Yalınız et al. 1996). The Çiçekdağ Ophiolite (location 4 in Fig. 1), of which the volcanic section also makes up the dominant exposed portion, has similar chemical features to the Sarıkaraman Ophiolite, with low, depleted incompatible element contents

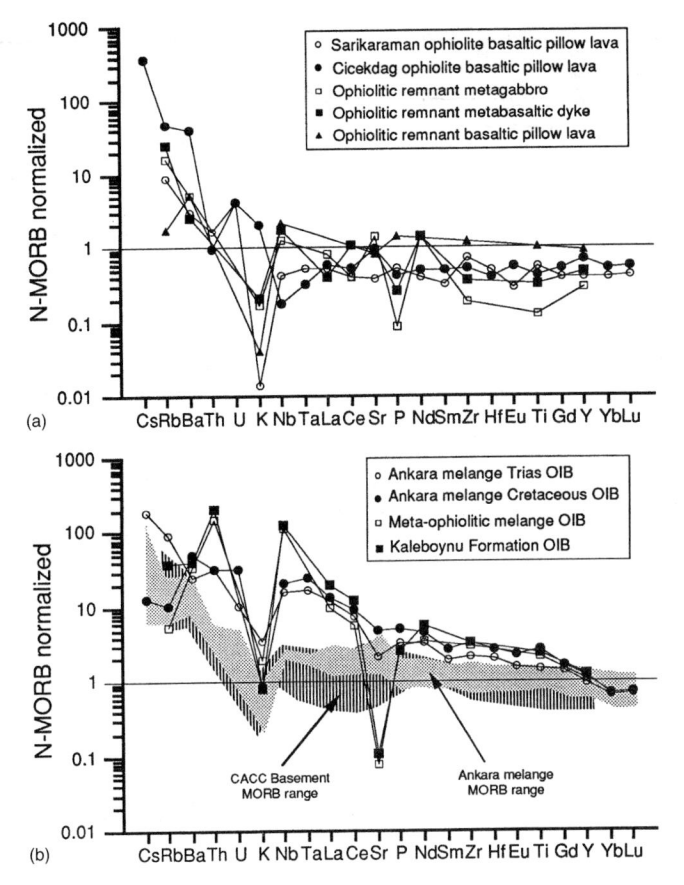

Fig. 5. (a) Representative multi-element patterns for Sarıkaraman and Çiçekdağ ophiolite pillow lavas (obducted stratiform ophiolites with suprasubduction zone features) compared with isolated remnants (gabbros with dykes, pillow lava sequences) of suspected ophiolitic affinity. (b) Normalized multi-element comparison between MORB and OIB in the CACC and Ankara Mélange. Sarıkaraman data from Yalınız *et al.* (1996), Ankara Mélange data from Floyd (1993). Normalization factors after Sun & McDonough (1989).

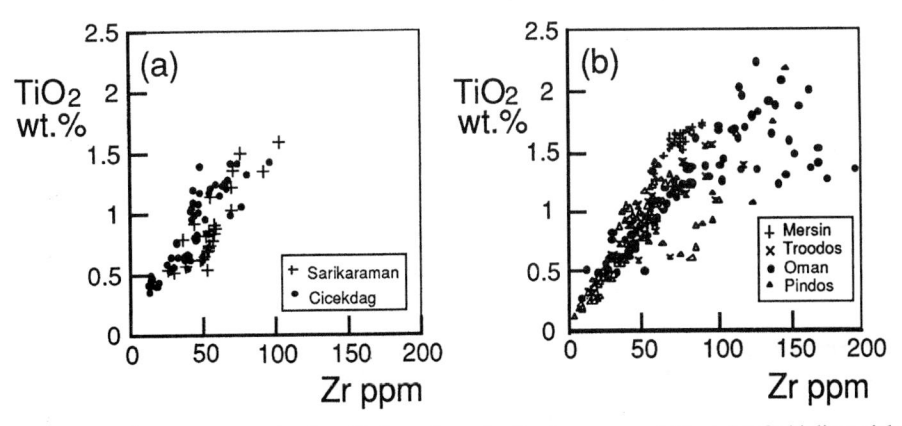

Fig. 6. TiO_2 v. Zr diagrams comparing basaltic lavas from the Sarıkaraman and Çiçekdağ Ophiolites of the CACC with other Neotethyan ophiolites with suprasubduction zone features. Data: Sarıkaraman (Yalınız *et al.* 1996); Mersin (Parlak 1996); Oman (Alabaster *et al.* 1982); Pindos (Valsami 1990); Troodos (Pearce 1975; Smewing & Potts 1976).

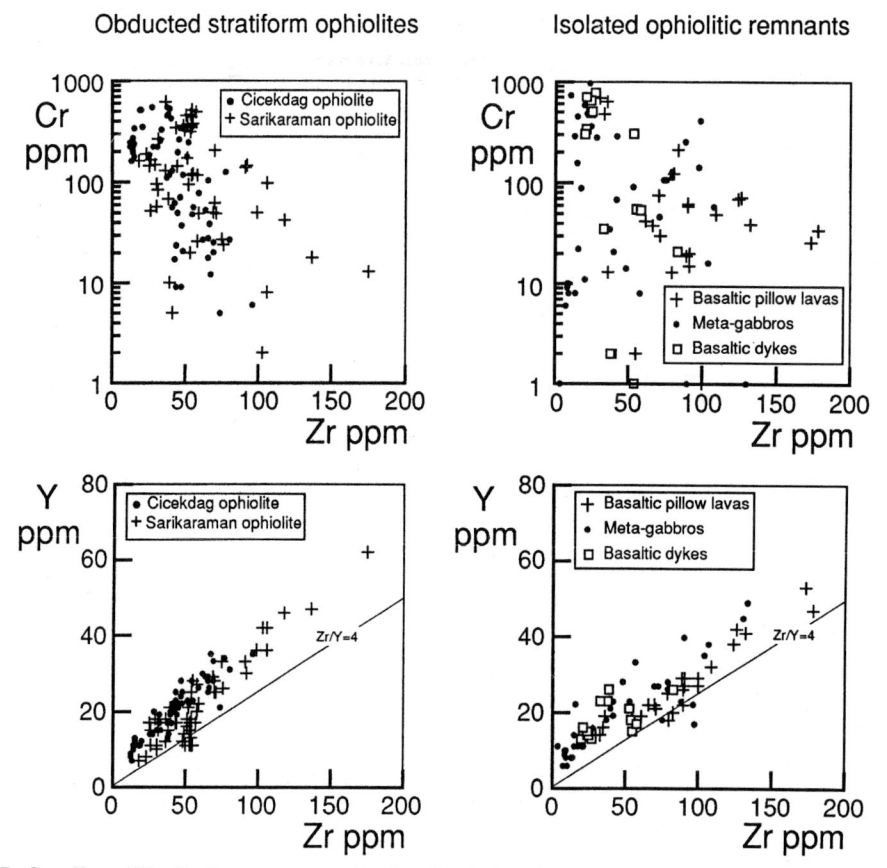

Fig. 7. Cr v. Zr and Y v. Zr diagrams comparing data for obducted stratiform ophiolites and tectonically isolated remnants represented by pillow lava sequences and gabbros with dykes, found within and adjacent to the CACC. The Çankırı Basin metabasalts define a similar trend to the pillow lavas of the ophiolites, but at a slightly higher ratio approaching Zr/Y = 4. Sarıkaraman data from Yalınız et al. (1996).

relative to MORB (Fig. 5) that reinforces their SSZ affinities. Both of these CACC stratiform ophiolites have features that can be directly compared with other ophiolites from the Neotethyan zone (e.g. Pindos, Mersin, Troodos, Oman). As seen in Fig. 6, they closely match the variation (in terms of TiO_2 and Zr) shown by the Pindos Ophiolite, but lack the degree of extensive chemical evolution shown by Mersin (highly evolved, but restricted in composition) or the range of Oman (to tholeiitic andesites).

Further evidence for the broad chemical overlap in composition between the two ophiolites is shown in Fig. 7, where the range in composition is largely a function of mafic ± plagioclase fractionation (decreasing Cr, with covariant and increasing Zr and Y). Although there is a dominant subduction-related signature similar to island-arc basalts (IAB) rather than MORB in terms of V v. TiO_2 distributions (Fig. 8), some metabasalts from both ophiolites spill over into the MORB field. On comparison with island-arc–back-arc basin pairs from the western Pacific (Woodhead et al. 1993), the more MORB-like components of the ophiolites appear to have back-arc characteristics with very low V/Ti ratios (Fig. 8). As many of these basalts are stratigraphically uppermost in the ophiolite lava sequence, the apparent change in chemical designation might suggest a transition to a back-arc environment.

Differences between the ophiolites are essentially restricted to the Çiçekdağ metabasalt lavas and dykes, showing a more limited range of Zr and Y contents (less fractionated), and lacking the special group of Sarıkaraman lavas characterized by uniform Zr values (50–60 ppm Zr) and variable Y contents (Fig. 7). However, as the formation ages and basalt chemistries are generally similar, they are considered to be

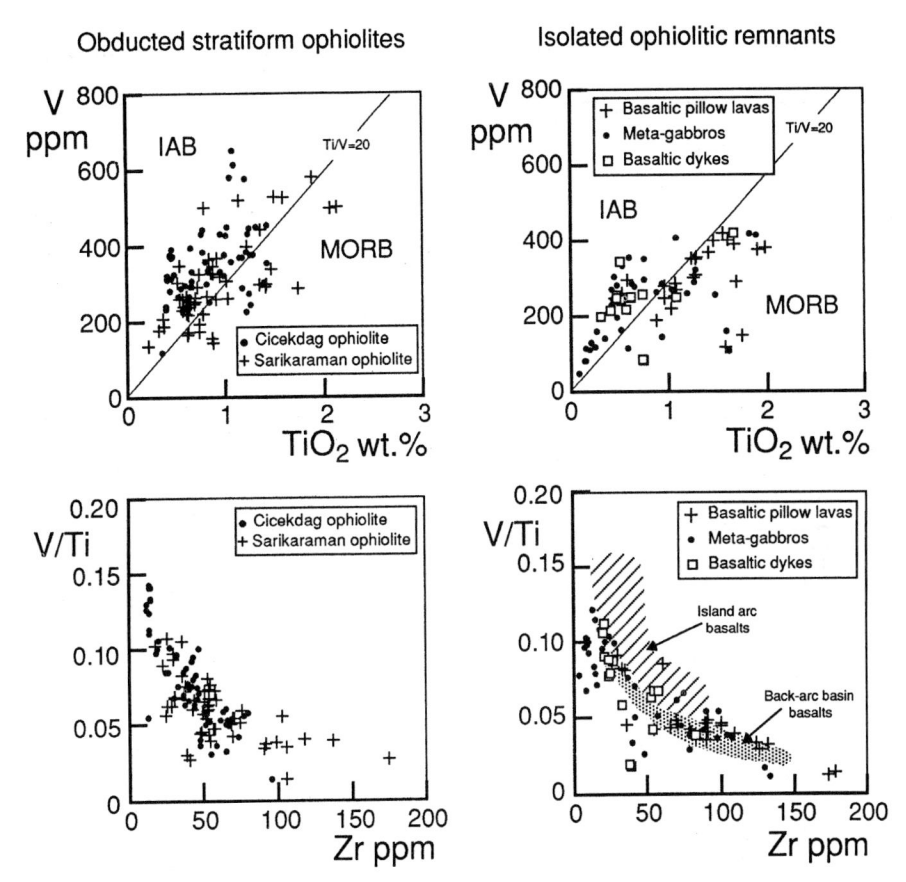

Fig. 8. V v. TiO$_2$ and V/Ti v. Zr diagrams comparing data for obducted stratiform ophiolites and tectonically isolated remnants represented by pillow lava sequences and gabbros with dykes, found within and adjacent to the CACC. The stratiform ophiolites show a dominant subduction-related signature (IAB field), although some late lavas and dykes (including the Çankırı metabasalts) have a more MORB-like affinity akin to back-arc basin basalts. Sarıkaraman data from Yalınız *et al.* (1996). IAB, island-arc basalts; MORB, mid-ocean ridge basalts; OIB, ocean island basalts; BABB, back-arc basin basalts. Ratios and fields in the V v. TiO$_2$ diagrams from Shervais (1982); IAB and BABB fields in the V/Ti v. Zr diagrams compiled from Woodhead *et al.* (1993).

different slices of the same obducted ophiolitic sheet.

Structurally isolated outcrops with suspected ophiolitic affinities have been grouped lithologically into pillow lavas, metagabbros and cross-cutting metabasaltic dykes, and plotted on companion diagrams for comparison with the stratiform ophiolites. As seen in Figs 5 and 7, the metagabbros and dykes have similar SSZ-type features to the massive ophiolites and may have originally been part of the same obducted sheet. However, basaltic pillow lavas and associated dykes from the Çankırı Basin generally form a different trend, with higher Zr/Y ratios (*c.* 4) and a MORB-like affinity. These chemical distinctions are emphasized in Fig. 8, where the pillow lavas predominantly have MORB-like compositions similar to those of back-arc basins, whereas the metagabbros and dykes are akin to IAB. These features alone are not sufficient to relate the Çankırı Basin lavas to the stratiform ophiolites, although they mirror the chemistry of the uppermost basalt lavas (with high Zr contents) of the Sarıkaraman Ophiolite (Fig. 8). However, evidence of an ophiolitic link is suggested by the composition of cross-cutting quartz–feldspar porphyry dykes (see above) which are similar to the Sarıkaraman and Çiçekdağ plagiogranites rather than late granitoids. It is therefore possible that the Çankırı Basin may have been floored by ophiolitic lavas of largely back-arc derivation. Support for this suggestion is provided by the association of the pillow lavas with bedded basic volcaniclastics, the latter of which are often a significant feature of the back-arc and forearc environments (Garcia 1978).

Geochemistry of basement metabasites

On the basis of current geotectonic models, the amphibolite facies metabasites are compared with the low-grade metabasalts of the Ankara Mélange which have been shown to be dominated by alkalic basalts of ocean island basalt (OIB) character, together with minor normal- and enriched-type MORBs (Figs 5 and 9) (Çapan & Floyd 1985; Floyd 1993). On the basis of stable Nb/Y ratios (Winchester & Floyd 1977), metabasites from the metamorphosed ophiolitic mélange are predominantly tholeiitic in character, whereas those from the Kaleboynu Formation are dominantly alkalic basalts. This latter group can be directly compared chemically with OIB from the Ankara Mélange with similar normalized patterns (Fig. 5) and Zr/Y of c. 8 (Fig. 9). However, although a chemical correspondence is indicated, no correlation is necessarily implied between the Karakaya Nappe of the Ankara Mélange (being a Sakarya unit) and the Kaleboynu Formation (a Tauride unit). In a similar fashion, the tholeiitic basalts largely have MORB-like features that mirror similar rocks in the Ankara Mélange with both depleted and enriched compositions (Figs 5 and 9).

One significant chemical feature to emerge from the basement metabasites is the presence of some clasts with an IAB chemistry exhibiting low TiO_2 and high V/Ti ratios (Fig. 10). The association of the three eruptive settings, represented by IAB–MORB–OIB, is discussed below.

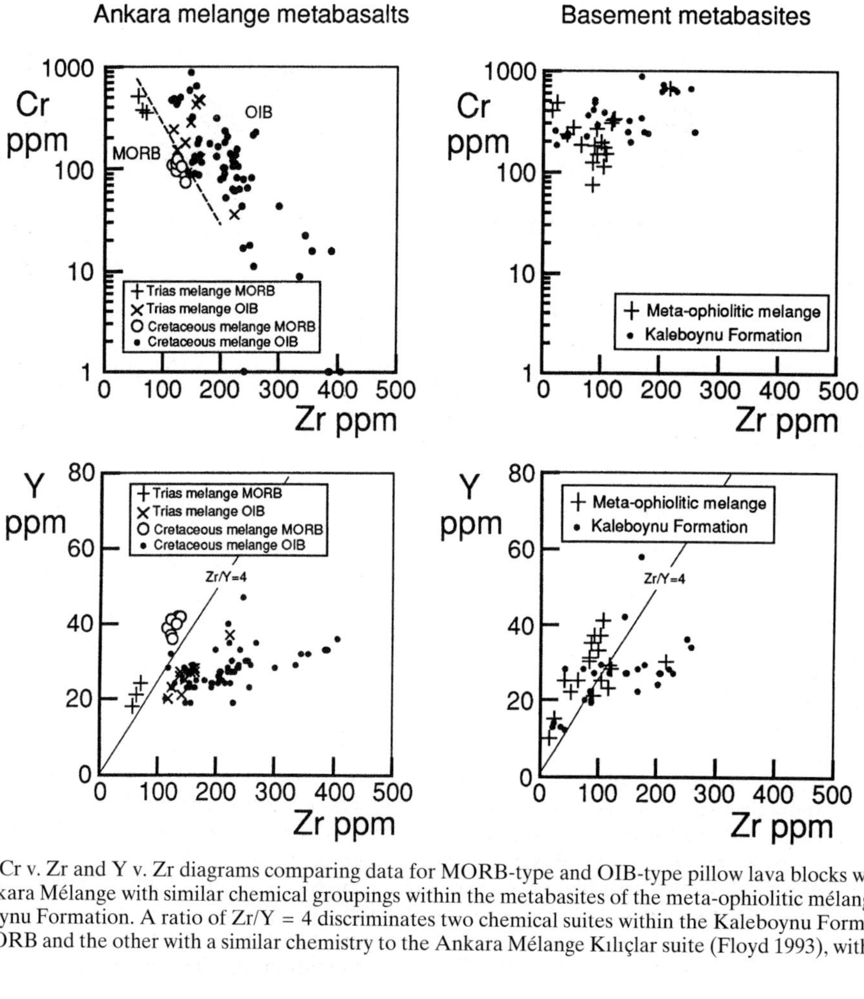

Fig. 9. Cr v. Zr and Y v. Zr diagrams comparing data for MORB-type and OIB-type pillow lava blocks within the Ankara Mélange with similar chemical groupings within the metabasites of the meta-ophiolitic mélange and Kaleboynu Formation. A ratio of Zr/Y = 4 discriminates two chemical suites within the Kaleboynu Formation: one MORB and the other with a similar chemistry to the Ankara Mélange Kılıçlar suite (Floyd 1993), with Zr/Y c. 8.

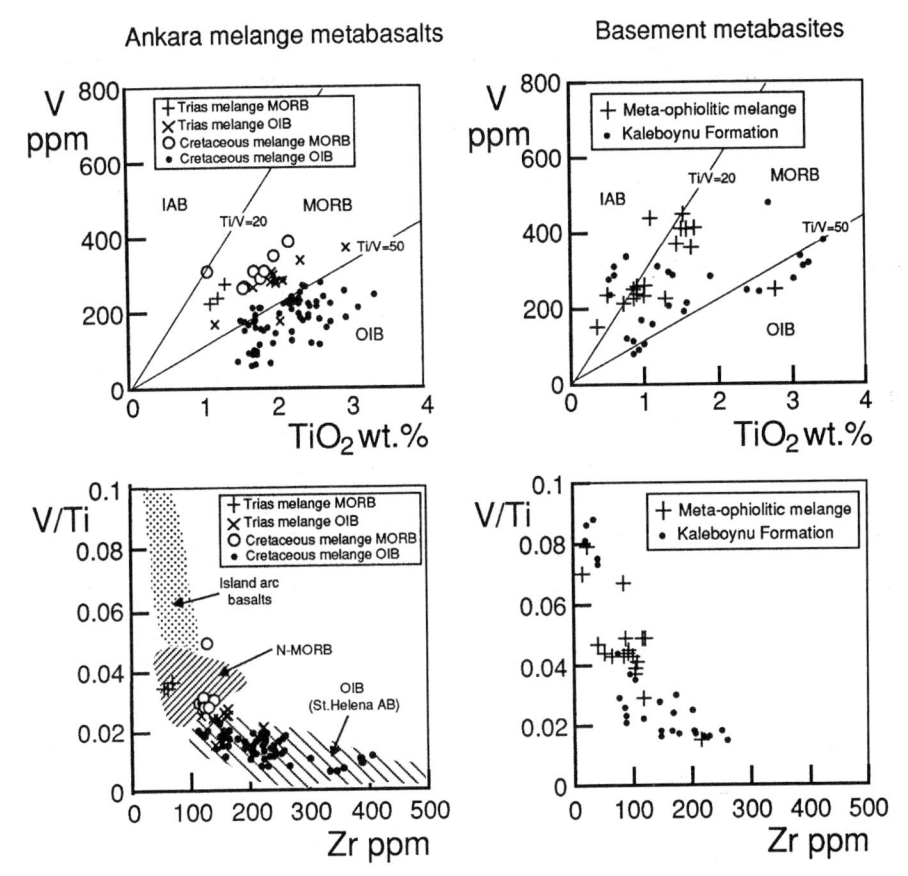

Fig. 10. V v. TiO$_2$ and V/Ti v. Zr diagrams comparing data for pillow lava blocks within the Ankara Mélange [data from Floyd (1993)] with similar chemical groupings within the metabasites of the meta-ophiolitic mélange and Kaleboynu Formation. Note the lack of subduction-related basaltic types relative to MORB and OIB in the Ankara Mélange. Fields in the V v. TiO$_2$ diagram from Shervais (1982); IAB from Woodhead *et al.* (1993); N-MORB from Floyd & Castillo (1992); representative OIB are alkali basalts (AB) from St Helena, from Chaffey *et al.* (1989).

Interpretation of geochemical features

This section summarizes the main geochemical characteristics of the stratiform ophiolites and basement metabasites, and interprets the magmatic associations in the light of their tectonic designations.

Stratiform ophiolites and tectonically isolated remnants

Main features. (1) The large obducted sheets of stratiform ophiolites (the Sarıkaraman and Çiçekdağ Ophiolites) have similar and overlapping basaltic compositions, with SSZ features similar to Neotethyan ophiolites of the eastern Mediterranean. (2) Both ophiolites have similar plagiogranite suites that differ chemically from late cross-cutting granitoids and syenitoids. (3) Isolated remnants of gabbroic bodies with basaltic and felsic dykes are chemically related to the main SSZ ophiolites. (4) Pillow lava sequences from the Çankırı Basin are chemically distinct from those in the stratiform ophiolites and display a more MORB-like affinity similar to back-arc basin basalts. These features are also shown by some lavas and dykes high in the ophiolite stratigraphic sequence.

The presence of a sheeted dyke complex indicates that the CACC stratiform ophiolites were generated in an extensional oceanic regime. Comparison with back-arc basin–island arc pairs suggests that the bulk of the SSZ ophiolite lavas have an island-arc character, whereas the (later) MORB-like lavas are more akin to back-arc basalts. As all the basaltic lavas are submarine, there is little evidence for the development of a major (subaerial) arc edifice, although the chemistry suggests the ophiolites represent a transition to, or an association with, a more back-arc-type environment just prior to obduction. The main evidence for the existence of a back-arc, however, is furnished by the separate Çankırı Basin pillow lavas which are also associated with volcaniclastic activity that might be expected in this setting. It is likely that the SSZ ophiolites and the Çankırı Basin basalts represent uncoupled or different obducted slices of oceanic crust from a broad arc-type setting.

Faunal evidence indicates that the SSZ-type ophiolites were generated during the Late Cretaceous (c. 90–85 Ma) and probably obducted soon afterwards, prior to late granitoid intrusion (c. 76–71 Ma). The main feature to emerge is that the SSZ-type ophiolites rapidly developed under the influence of a short-lived subduction zone and were subsequently obducted within 10 Ma or less.

Basement metamagmatic bodies

Main features. (1) Deformed metafelsic bodies in the ophiolitic mélange have enhanced HFS element abundances and high Zr/Nb ratios dissimilar to ophiolitic plagiogranites. (2) Boudinaged meta-ultramafic bodies are often Ca–Sr metasomatized by the adjacent carbonate-rich matrix. They appear to comprise two types with variable MgO + Ni + Cr contents, reflecting different origins: ultramafic rocks typically associated with ophiolite sequences and (possibly) cumulates related to continental stratiform gabbros. (3) Boudinaged ophiolitic mélange amphibolites ('knockers') are dominated by tholeiites with MORB-like features (both normal and slightly enriched types), although some minor OIB and IAB types are also present. (4) Kaleboynu Formation amphibolites, conformable with massive platform carbonates, are dominated by within-plate alkalic basalts of OIB type.

Although there is chemical correspondence between the basaltic components of the basement (ophiolitic mélange and Kaleboynu Formation) and units of the Ankara Mélange of broadly the same age, the main differences are in the proportion of MORB and OIB types, and the presence of IAB in the former. The features displayed by the ophiolitic mélange can be reconciled in a subduction–accretion setting where seamount structures, standing above the ocean floor, will be largely scraped off and accreted, whereas the ocean floor is more likely to be subducted. Thus, the dominant OIB in Cretaceous ophiolitic units of the Ankara Mélange represents the scraped off and accreted relics of alkalic seamounts (Floyd 1993), whereas the CACC ophiolitic mélange MORB and IAB represents initially subducted ocean floor, together with remnants of the adjacent arc, respectively. The close association of these three different basaltic components derived from the island arc, the ocean floor and ocean islands/seamounts is a feature of mélanges in forearc settings (Bloomer 1983; Johnson & Fryer 1990; Macpherson et al. 1990). Although this scenario could well reflect the association of metabasite clasts in the ophiolitic mélange, the origin of the OIB in the Kaleboynu Formation is unlikely to be seamount-derived as in the Ankara Mélange. The conformable relationship of many of the thin Kaleboynu Formation amphibolites with the surrounding marbles indicates that they were probably intrusive sheets and/or basic volcaniclastic accumulations in shallow rifted basins. Like the Ankara Mélange OIB, the Kaleboynu Formation OIB are alkalic basalts of within-plate character; only in the latter case does the geological situation suggest a different environmental setting – rifted carbonate platform relative to that of a seamount. That rifting had reached the stage of small basins floored with ocean crust in some cases is suggested by the presence of MORB compositions within the Kaleboynu Formation metabasites.

Overall, the chemistry of the different magmatic lithologies in the deformed crystalline basement (ophiolitic mélange and Kaleboynu Formation) suggests the admixture at different stages of what was originally major ocean, island-arc and rifted continent margin, environments. The oceanic setting is indicated by MORB-type compositions typical of spreading centres with scattered OIB-type seamounts dominated by alkalic basalts. This type of ocean floor is distinct from the SSZ characteristics of the later stratiform ophiolites. On subduction, the seamounts were fragmented and incorporated into the forearc accretionary prism. A few basaltic (as well as felsic) clasts within the ophiolitic mélange were derived from the overriding active island arc. The continental margin is represented by the metamorphosed TAP carbonate platform with associated within-plate alkalic basalts (Kaleboynu Formation OIB). The association of minor MORB in this predominantly continental setting suggests the development of small rifted submarine basins partly floored by ocean crust.

Tectonic model

On the basis of the above geochemical characterization, two features are clear: (1) the stratiform ophiolites with SSZ features are distinct from the metabasites in the underlying basement; and (2) both the ophiolitic mélange metabasite blocks and the concordant Kaleboynu Formation amphibolites have their corresponding chemical counterparts in different segments of the Ankara Mélange. The actual age of the basement metabasites is still a problem. Part of the Kaleboynu Formation is probably Triassic and similar in age to the Upper Karakaya Nappe ('metamorphic block mélange') of the Ankara Mélange (Koçyiğit 1991), although it is recognized that they belong to different crustal blocks – the Taurides and Sakarya unit, respectively. On the other hand, the basement ophiolitic mélange is probably equivalent to the Anatolian Nappe ('ophiolitic mélange') of Middle–Late Cretaceous age.

In many disrupted ophiolitic sequences, a metabasite block-bearing mélange tectonically underlies an ophiolite nappe with its attendant thrust-bound metamorphic sole (Parkinson 1996; Parlak 1996), and a genetic relation can often be demonstrated between the blocks and the ophiolite. A similar tectonic stacking sequence is indicated here for the CACC (Fig. 2), although the N-MORB-dominated ophiolitic mélange blocks appear to be chemically unrelated to the SSZ-type ophiolites. In the model below, however, a relationship between the two magmatic groups is based on the premise that the MORB-type blocks are representative of an original (and older) oceanic crust on which the later SSZ-type ophiolites developed.

The following tectonic model (illustrated in Fig. 11) is based on the geochemical interpretation of the data above and attempts to integrate the CACC into a broad model for the development of the Neotethys Ocean (e.g. Robertson et al. 1991; Yılmaz et al. 1997). A number of stages are envisaged:

- The northern margin of Gondwanaland rifted in the Late Triassic with the eventual development of small ocean basins. The association of volcanic rocks and carbonates of Triassic age indicate that the marginal platforms were being subjected to an extensional regime at this time. The concordant OIB-type metabasites (mainly alkali basalts) of the

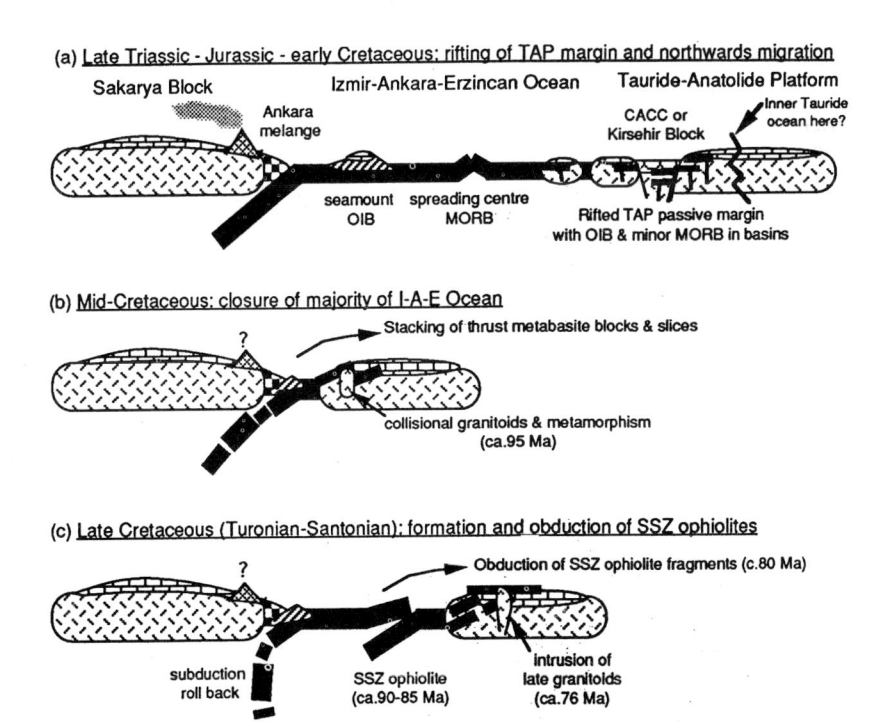

Fig. 11. Cartoon illustrating the closure of the İzmir–Ankara–Erzincan (I–A–E) Ocean by the collision of the CACC with the Sakarya microcontinental block during the Late Cretaceous. The original eruptive settings of various ophiolitic bodies (SSZ) and metabasites (MORB, OIB) now found in the CACC are exhibited.

Kaleboynu Formation are evidence for within-plate rifting, although the presence of MORB indicates that continental crust attenuation had been sufficient to generate some true ocean floor during the Triassic.

- By the Early Jurassic and throughout the Cretaceous, as the Neotethys Ocean gradually widened, a progression of microcontinental slices (Sakarya, Kırşehir–CACC), that had been rifted off Gondwanaland, migrated northwards towards Eurasia (Fig. 11a). According to Özgül (1976), the CACC remained attached to the carbonate TAP, although Görür et al. (1984) think that these two blocks were separated by another seaway, the Inner Tauride Ocean (Fig. 11a). The CACC block lagged behind the Sakarya Block and became separated from it by the İzmir–Ankara–Erzincan oceanic strand. Most of this ocean was subducted under the southern active margin of the Sakarya microcontinent, so that evidence for its existence is now largely found in tectonic mélanges. Taken together, metabasaltic blocks in segments of the Ankara Mélange and the basement ophiolitic mélange record the existence of MORB, OIB and minor IAB compositions, typical of forearc accretionary wedges. MORB compositions represent the subducting ocean crust, whereas the alkali basalts were derived either from volcanic seamounts developed on the ocean floor and/or earlier rift-related ensialic environments. The huge rafts of platform carbonate seen in part of the Ankara Mélange suggest that much of the alkalic basalt could be associated with the original Trias rifting, although OIB pillow lavas intimately associated with Cretaceous pelagic limestones indicate a true seamount construct (Floyd 1993). The minor IAB and subduction-related metafelsic blocks were probably derived by tectonic erosion of the subduction zone hanging wall or intersection of deep-arc faults.
- As the CACC block approached the Sakarya subduction zone, the intervening ocean crust was largely subducted, whereas the seamount constructs and rafted carbonate platforms were mainly accreted into the trench mélange. By the mid-Cretaceous, the leading promontories of the CACC carbonate platform collided with the Sakarya microcontinent. Deeply subducted material, metamorphosed to amphibolite facies, was then buoyantly disgorged and obducted southwards over the CACC margin (Fig. 11b). The higher grade material (amphibolite facies) is now represented by the Central Anatolian Metamorphics of the CACC, whereas the lower grades (pumpellyite and greenschist facies) are found in the Ankara Mélange. Southward stacking of the contents of the subduction zone took place before the emplacement of post-collisional granitoids at c. 95 Ma (Fig. 11b).
- Due to the irregular margin of the advancing CACC microcontinent after initial collision, small segments of oceanic crust remained that had not yet been subducted. It was in these remnants that the Late Cretaceous (90–85 Ma) SSZ-type ophiolites would develop (Fig. 11c). It is speculated that, on initial collision, subduction zone roll-back occurred, thereby inducing extension in the remaining ocean crust and the development of a new subduction zone (Fig. 11c). Another possibility is that asymmetric collapse of any remaining spreading ridge in the oceanic segment engendered a subduction zone (e.g. Clift & Dixon 1998). Either way, an incipient arc developed with a limited degree of back-arc spreading – environments which are now represented by the SSZ ophiolites and the Çankırı Basin pillow lavas, respectively. This phase lasted between 5 and 10 Ma.
- Further compression produced slicing and imbrication of the SSZ oceanic lithosphere and the eventual obduction of ophiolitic fragments over the CACC metamorphic basement (Fig. 11c). The whole sequence was then intruded by late granitoids at c. 76 Ma.

In conclusion, the CACC and the Ankara Mélange are considered to represent variably tectonized and subducted oceanic lithosphere and continental carbonate platform that were subsequently ejected from an accretionary–subduction complex on collision with the Sakarya microcontinent.

Funding for this collaborative project was provided by NATO (grant CRG 960549) and the Scientific and Research Council of Turkey (TÜBİTAK) (code no: YDABÇAG-85) grant to whom thanks are due. Analytical data production was greatly helped by the expertise of D. W. Emley and M. Aikin at the Department of Earth Sciences, University of Keele, UK. We acknowledge the helpful and expert comments of P. Clift and M. Delaloye.

References

ALABASTER, T., PEARCE, J. A. & MALPAS, J. 1982. The volcanic stratigraphy and petrogenesis of the Oman ophiolite complex. *Contributions to Mineralogy and Petrology*, **81**, 168–183.

ATAMAN, G. 1972. Ankara'nın güneydoğusundaki granitik–granodiyoritik kütlelerden Cefalık Dağın radyometrik yaşı hakkında ön çalışma. *Hacettepe University Bulletin of Natural Sciences and Engineering*, **2**, 44–49 [in Turkish with English abstract].

BLOOMER, S. H. 1983. Distribution and origin of igneous rocks from the landward slopes of the Mariana Trench: implications for its structure and evolution. *Journal of Geophysical Research*, **88**, 7411–7428.

ÇAPAN, Ü. & FLOYD, P. A. 1985. Geochemical and petrographic features of metabasalts within units of the Ankara Mélange, Turkey. *Ofioliti*, **10**, 3–18.

CHAFFREY, D. J., CLIFF, R. A. & WILSON, B. M. 1989. Characterisation of the St. Helena magma source. In: SAUNDERS, A. D. & NORRY, M. J. (eds) *Magmatism in Ocean Basins*. Geological Society, London, Special Publications, **42**, 257–276.

CLIFT, P. D. & DIXON, J. E. 1998. Jurassic ridge collapse, subduction initiation and ophiolite obduction in the southern Greek Tethys. *Eclogae Geologicae Helvetiae*, **91**, 123–138.

ERLER, A. & GÖNCÜOĞLU, M. C. 1996. Geologic and tectonic setting of the Yozgat batholith, northern Central Anatolian Crystalline Complex, Turkey. *International Geology Review*, **38**, 714–726.

EXLEY, C. S., STONE, M. & FLOYD, P. A. 1983. Composition and petrogenesis of the Cornubian granite batholith and post-orogenic volcanic rocks in southwest England. In: HANCOCK, P. L. (ed.) *The Variscan Foldbelt in the British Isles*. Adam Hilger, 153–177.

FLOYD, P. A. 1993. Geochemical discrimination and petrogenesis of alkalic basalt sequences in part of the Ankara Mélange, central Turkey. *Journal of the Geological Society, London*, **150**, 541–550.

—— & CASTILLO, P. R. 1992. Geochemistry and petrogenesis of Jurassic ocean crust basalts, ODP Leg 129, Site 801. In: LARSON, R., LANCELOT, R. ET AL (eds) *Proceedings of ODP, Scientific Results*, **129**, 361–388. College Station, Texas.

—— & WINCHESTER, J. A. 1978. Identification and discrimination of altered and metamorphosed volcanic rocks using immobile elements. *Chemical Geology*, **21**, 291–306.

——, EXLEY, C. S. & STYLES, M. T. 1993. *Igneous Rocks of South-west England*, Geological Conservation Review Series, **5**. Chapman & Hall, London.

——, YALINIZ, M. K. & GÖNCÜOĞLU, M. C. 1998. Geochemistry and petrogenesis of intrusive and extrusive ophiolitic plagiogranites, Central Anatolian Crystalline Complex, Turkey. *Lithos*, **42**, 225–241.

GARCIA, M. O. 1978. Criteria for the identification of ancient volcanic arcs. *Earth Science Reviews*, **14**, 147–165.

GASS, I. G. & SMEWING, J. D. 1973. Intrusion, extrusion and metamorphism at constructive margins: evidence from the Troodos massif, Cyprus. *Nature*, **242**, 26–29.

GERLACH, D. C., LEEMAN, W. P. & LALLEMENT, A. H. G. 1981. Petrology and geochemistry in the Canyon Mountain ophiolite, Oregon. *Contributions to Mineralogy and Petrology*, **77**, 82–92.

GÖNCÜOĞLU, M. C. 1982. Zircon U/Pb ages from the Niğde Massif. *Geological Society of Turkey Bulletin*, **25**, 61–66.

—— 1986. Geochronological data from the southern part (Niğde area) of the Central Anatolian massif. *Mineral Research and Exploration Institute of Turkey (MTA) Bulletin*, **105/106**, 111–124.

—— & TÜRELI, K. 1993. Petrology and geodynamic interpretation of plagiogranites from Central Anatolian ophiolites. *Turkish Journal of Earth Sciences*, **2**, 195–203.

——, KÖKSAL, S. & FLOYD, P. A. 1997. Post-collisional A-type magmatism in the Central Anatolian Crystalline Complex: petrology of the İdiş Dağı Instruives (Avanos, Turkey). *Turkish Journal of Earth Sciences*, **6**, 65–76.

——, TOPRAK, V., ERLER, A. & KUŞCU, İ. 1991. *Orta Anadolu Masifi Batı Kesiminin Jeolojisi, Bölüm I: Güney Kesim*. Turkish Petroleum Corporation (TPAO) Report, No. 2909 [in Turkish].

GÖRÜR, N., OKTAY, F. Y., SEYMEN, İ. & ŞENGÖR, A. M. C. 1984. Palaeotectonic evolution of Tuz Gölü Basin complex, central Turkey. In: DIXON, J. E. & ROBERTSON, A. H. F. (eds) *The Geological Evolution of the Eastern Mediterranean*. Geological Society, London, Special Publications, **17**, 81–96.

HART, S. R., ERLANK, A. J. & KABLE, E. J. D. 1974. Sea floor basalt alteration: some chemical and Sr isotopic effects. *Contributions to Mineralogy and Petrology*, **44**, 219–230.

HUMPRIS, S. E. 1984. The mobility of the rare earth elements in the crust. In: HENDERSON, P. (ed.) *Rare Earth Element Geochemistry*. Elsevier, Amsterdam, 317–342.

—— & THOMPSON, G. 1978. Trace element mobility during hydrothermal alteration of oceanic basalts. *Geochimica et Cosmochimica Acta*, **42**, 127–136.

HYNES, A. 1980. Carbonitization and mobility of Ti, Y, and Zr in Ascot Formation metabasalts, S. E. Quebec. *Contributions to Mineralogy and Petrology*, **75**, 79–87.

JOHNSON, L. E. & FRYER, P. 1990. The first evidence for MORB-like lavas from the outer Mariana forearc: geochemistry, petrography and tectonic implications. *Earth and Planetary Science Letters*, **100**, 304–316.

JUTEAU, T. 1980. Ophiolites of Turkey. *Ofioliti*, **2**, 199–237.

KARIG, D. E. 1980. Material transport in accretionary prisms and the 'knocker' problem. *Journal of Geology*, **88**, 27–39.

KOÇAK, K. & LEAKE, B. E. 1994. The petrology of the Ortaköy district and its ophiolite at the western edge of the Middle Anatolian Massif, Turkey. *Journal of African Earth Sciences*, **18**, 163–174.

KOÇYIĞIT, A. 1991. An example of an accretionary forearc basin from northern Central Anatolia and its implications for the history of subduction of Neo-Tethys in Turkey. *Geological Society of America Bulletin*, **103**, 22–36.

LEAT, P. T., JACKSON, S. E., THORPE, R. S. & STILLMAN, C. J. 1986. Geochemistry of bimodal basalt–subalkaline/peralkaline rhyolite provinces within the southern British Caledonides. *Journal of the Geological Society, London,* **143**, 259–273.

LIVERMORE, R. A. & SMITH, A. G. 1984. Some boundary conditions for the evolution of the Mediterranean region. *In*: STANLEY, D. J. & WEZEL, F. C. (eds) *Geological Evolution of the Mediterranean Basin.* Springer, 83–110.

MACPHERSON, G. J., PHIPPS, S. P. & GROSSMAN, J. N. 1990. Diverse sources for igneous blocks in Franciscan mélanges, California Coast Ranges. *Journal of Geology,* **98**, 845–862.

MIYASHIRO, A. 1973. *Metamorphism and Metamorphic Belts.* Allen & Unwin.

ÖZGÜL, N. 1976. Toroslarin bazı temel jeoloji özellikkeri. *Geological Society of Turkey Bulletin,* **19**, 65–78 [in Turkish with English abstract].

PARKINSON, C. D. 1996. The origin and significance of metamorphosed tectonic blocks in mélanges: evidence from Sulawesi, Indonesia. *Terra Nova,* **8**, 312–323.

PARLAK, O. 1996. Geochemistry and geochronology of the Mersin ophiolite within the eastern Mediterranean tectonic frame (southern Turkey). *Terre et Environnement,* **6**, 1–242.

PEARCE, J. A. 1975. Basalt geochemistry used to investigate past tectonic environments on Cyprus. *Tectonophysics,* **25**, 41–68.

—— & CANN, J. R. 1973. Tectonic setting of basaltic volcanic rocks determined using trace element analyses. *Earth and Planetary Science Letters,* **19**, 290–300.

——, LIPPARD, S. J. & ROBERTS, S. 1984. Characteristics and tectonic significance of suprasubduction zone ophiolites. *In*: KOKELAAR, B. P. & HOWELLS, M. F. (eds) *Marginal Basin Geology.* Geological Society, London, Special Publications, **16**, 77–94.

RICE-BIRCHALL, B. & FLOYD, P. A. 1988. Geochemical and source characteristics of the Tintagel Volcanic Formation. *Proceedings of the Ussher Society,* **7**, 52–55.

RICOU, L. E. 1971. Le croissant ophiolitique périarabe: une ceinture de nappes mises en place au Crétacé supérieur. *Review of Geographical, Physical and Geological Dynamics,* **13**, 327–349.

ROBERTSON, A. H. F. & DIXON, J. E. 1984. Introduction: aspects of the geological evolution of the eastern Mediterranean. *In*: DIXON, J. E. & ROBERTSON, A. H. F. (eds) *The Geological Evolution of the Eastern Mediterranean.* Geological Society, London, Special Publications, **17**, 1–74.

——, CLIFT, P. D., DEGNAN, P. J. & JONES, G. 1991. Palaeogeographic and palaeotectonic evolution of the Eastern Mediterranean Neotethys. *Palaeogeography, Palaeoclimatology, Palaeoecology,* **87**, 289–343.

ŞENGÖR, A. M. C. & YILMAZ, Y. 1981. Tethyan evolution of Turkey: a plate tectonic approach. *Tectonophysics,* **75**, 181–241.

SHERVAIS, J. W. 1982. Ti-V plots and the petrogenesis of modern and ophiolitic lavas. *Earth and Planetary Science Letters,* **59**, 101–118.

SMEWING, J. D. & POTTS, P. J. 1976. Rare-earth abundances in basalts and metabasalts from the Troodos Massif, Cyprus. *Contributions to Mineralogy and Petrology,* **57**, 245–258.

SMITH, R. E. & SMITH, S. E. 1976. Comments on the use of Ti, Zr, Y, Sr, K, P and Nb in classification of basaltic magmas. *Earth and Planetary Science Letters,* **32**, 114–120.

SPOONER, E. T. C. & FYFE, W. S. 1973. Sub-seafloor metamorphism, heat and mass transfer. *Contributions to Mineralogy and Petrology,* **42**, 287–304.

SUN, S. S. & McDONOUGH, W. F. 1989. Chemical and isotopic systematics of oceanic basalts: implications for mantle composition and processes. *In*: SAUNDERS, A. D. & NORRY, M. J. (eds) *Magmatism in Ocean Basins.* Geological Society, London, Special Publications, **42**, 313–345.

THOMPSON, G. 1991. Metamorphic and hydrothermal processes: basalt–seawater interactions. *In*: FLOYD, P. A. (ed.) *Oceanic Basalts.* Blackie, 148–173.

VALSAMI, E. 1990. *Geochemistry and petrology of hydrothermal discharge zones in the Pindos and Othris ophiolites, Greece.* PhD thesis, University of Newcastle.

WINCHESTER, J. A. & FLOYD, P. A. 1977. Geochemical discrimination of different magma series and their differentiation products using immobile elements. *Chemical Geology,* **20**, 325–343.

WOODHEAD, J., EGGINGS, S. & GAMPLE, J. 1993. High field strength and transition element systematics in island arc and back-arc basin basalts: evidence for multiphase melt extraction and a depleted mantle wedge. *Earth and Planetary Science Letters,* **114**, 491–504.

YALINIZ, M. K. & GÖNCÜOĞLU, M. C. 1998. General geological characteristics and distribution of the Central Anatolian Ophiolites. *Hacettepe University Earth Sciences,* **20**, 1–12.

——, FLOYD, P. A. & GÖNCÜOĞLU, M. C. 1996. Suprasubduction zone ophiolites of Central Anatolia: geochemical evidence from the Sarıkaraman ophiolite, Aksaray, Turkey. *Mineralogical Magazine,* **60**, 697–710.

YILMAZ, Y., TÜYSÜZ, O., YIĞITBAŞ, E., GENÇ, Ş. C. & ŞENGÖR, A. M. C. 1997. Geology and tectonic evolution of the Pontides. *In*: ROBINSON, A. G. (ed.) *Regional and Petroleum Geology of the Black Sea and Surrounding Region.* AAPG Memoir, **68**, 183–226.

Geochemistry of volcanic rocks from the Çiçekdağ Ophiolite, Central Anatolia, Turkey, and their inferred tectonic setting within the northern branch of the Neotethyan Ocean

KENAN M. YALINIZ,[1] PETER A. FLOYD[2] & M. CEMAL GÖNCÜOĞLU[3]

[1]*Department of Civil Engineering, Celal Bayar University, Muradiye, Manisa, Turkey*
(e-mail: mukenan@anet.net.tr)
[2]*Department of Earth Sciences, University of Keele, Staffordshire, ST5 5BG, UK*
[3]*Department of Geological Engineering, Middle East Technical University, TR-06531 Ankara, Turkey*

Abstract: The Central Anatolian Ophiolites (CAO) comprise a number of little studied Upper Cretaceous ophiolitic bodies that originally represented part of the northern branch of the Neotethyan ocean. The Çiçekdağ Ophiolite (CO) is an dismembered example of this ophiolite group that still retains a partially preserved magmatic pseudostratigraphy. The following units (bottom to top) can be recognized: (1) layered gabbro; (2) isotropic gabbro; (3) plagiogranite; (4) dolerite dyke complex; (5) basaltic volcanic sequence; and (6) a Turonian–Santonian epi-ophiolitic sedimentary cover. The magmatic rock units (gabbro, dolerite and basalt) form part of a dominant comagmatic series of differentiated tholeiites, together with a minor group of primitive unfractionated basalts. The basaltic volcanics mainly consist of pillow lavas with a subordinate amount of massive lavas and rare basaltic breccias. Petrographic data from the least altered pillow lavas indicate that they were originally olivine-poor, plagioclase–clinopyroxene phyric tholeiites. Immobile trace element data from the basalt lavas and dolerite dykes show a strong subduction-related chemical signature. Relative to N-mid-ocean ridge basalt the Çiçekdağ basaltic rocks (allowing for the effects of alteration) have typical suprasubduction zone features with similarities to the Izu-Bonin Arc, i.e. enriched in most large-ion lithophile elements, depleted in high field strength elements and exhibiting depleted light rare earth element patterns. The geochemical characteristics are similar to other eastern Mediterranean Neotethyan SSZ-type ophiolites and suggest that the CO oceanic crust was generated by partial melting of already depleted oceanic lithosphere within the northern branch of the Neotethyan ocean. The Çiçekdağ body, along with the other fragmented CAO, is thus representative of the Late Cretaceous development of new oceanic lithosphere within an older oceanic realm.

Turkey occupies a critical segment in the Alpine–Himalayan orogenic system where remnants of both the Palaeotethyan and Neotethyan ocean basins crop out along broadly linear tectonic belts. Turkey is divided into three main east–west trending belts of separate ocean basins (Şengör & Yılmaz 1981): (1) a northern belt representing remnants of an Intra-Pontide ocean; (2) a median belt representing allochthonous units derived from the Vardar–İzmir–Ankara-Erzincan (VIAE) ocean [although Görür et al. (1984) also consider a separate derivation for the Inner Tauride Belt]; and (3) a southern belt, variably called the Peri-Arabic Belt or southern Neotethyan ocean (Şengör & Yılmaz 1981) (Fig. 1, inset a).

Most of the ophiolites occurring in these belts have been studied to varying degrees, although the isolated outcrops of allochthonous ophiolitic rocks (termed the Central Anatolian Ophiolites; CAO) found in the Central Anatolian Crystalline Complex (CACC) have received little attention until now. The CACC represents the northern passive margin of the Mesozoic Tauride–Anatolide Platform that faced the VIAE ocean (Göncüoğlu 1986; Göncüoğlu et al. 1991) (Fig. 1). The CAO are significant in that they were initially generated within the VIAE ocean and emplaced southwards onto the CACC (Yalınız & Göncüoğlu 1998). The majority of these ophiolites are tectonically dismembered, although stratigraphic reconstruction reveals that they originally represented a complete ophiolitic sequence (Yalınız & Göncüoğlu 1998) (Fig. 1, inset b).

In terms of the pseudostratigraphic relationship of the magmatic units and their chemical designation, the CAO exhibit a suprasubduction zone (SSZ) character within the VIAE segment of the Neotethys (Yalınız 1996; Yalınız &

From: BOZKURT, E., WINCHESTER, J. A. & PIPER, J. D. A. (eds) *Tectonics and Magmatism in Turkey and the Surrounding Area.* Geological Society, London, Special Publications, **173**, 203–218. 1-86239-064-9/00/$15.00
© The Geological Society of London 2000.

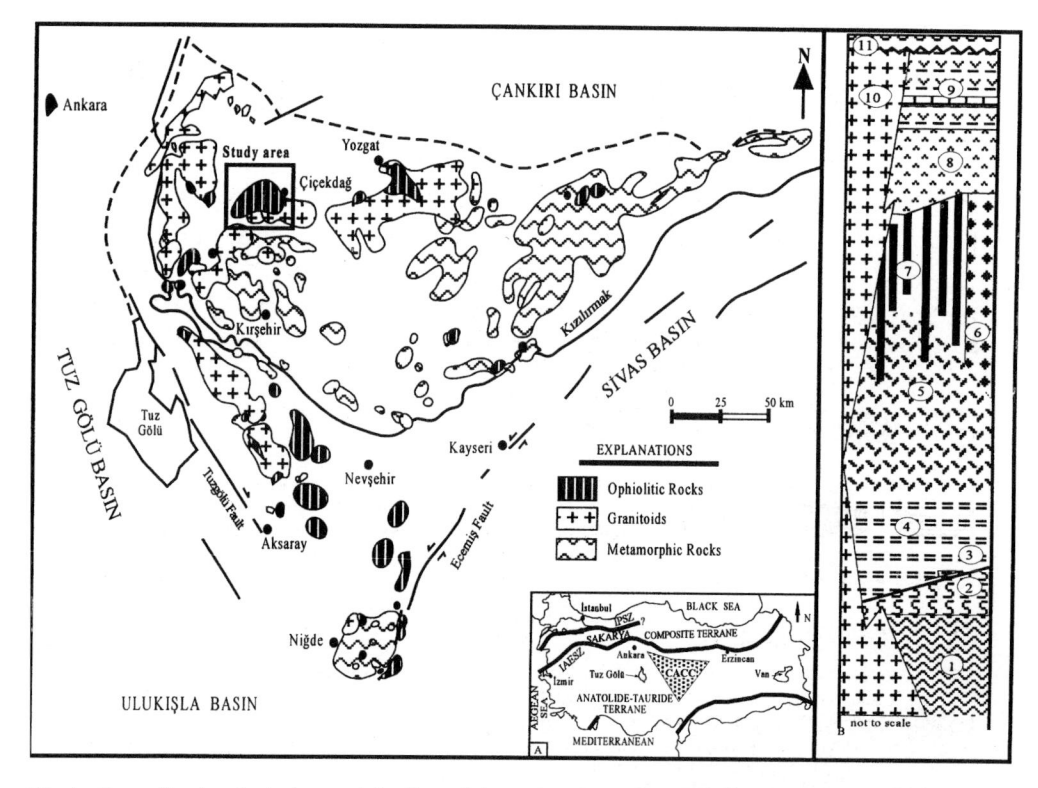

Fig. 1. Generalized geological map of the Central Anatolian Crystalline Complex and the location of the Çiçekdağ Ophiolite. (**a**) Neotethyan suture map of Turkey and the location of the Central Anatolian Crystalline Complex (CACC); IPSC, Intra-Pontide Suture Zone; IAESZ, İzmir–Ankara–Erzincan Suture Zone; SEASZ, Southeast Anatolian Suture Zone. (**b**) Simplified columnar section of the Çiçekdağ Ophiolite: 1, Central Anatolian Metamorphics; 2, metamorphic ophiolite-bearing olistostrome; 3, thrust boundary; 4, layered gabbro; 5, isotropic gabbro; 6, plagiogranites; 7, dyke complex; 8, massive and pillow lavas; 9, epiophiolitic sediments and lavas; 10, granitoids; 11, Lower Tertiary cover sediments.

Göncüoğlu 1998; Yalınız et al. 1996; Floyd et al. 1998a).

Objectives

The Çiçekdağ Ophiolite (CO) is one of the most complete and best exposed of the CAO bodies. Although it has not been studied in detail, little geochemical data being available, its general character invites comparison with other better known CAO, such as the Sarıkaraman Ophiolite (SO). New and extensive geochemical data on the least altered volcanic rocks of the CO are presented here. The purpose of this study is to describe the compositional features of the volcanic rocks of the CO as a basis for determining their original tectonic setting and a comparison with other better known SSZ-type Central Anatolian ophiolites. A further aim is to discuss the genesis and the palaeotectonic setting of the CO within the VIAE ocean to provide a fuller picture for the geodynamic evolution of the northern branch of the Neotethys.

Distribution of rock types and stratigraphy

The CO tectonically overlies the CACC basement along a moderately steep northward dipping thrust just to the south of the Çiçekdağ Massif. Previous field studies in the Çiçekdağ area were mainly carried out by Ketin (1959), Erdoğan et al. (1996) and Yılmaz-Şahin & Boztuğ (1997). Typical ophiolitic sequences are exposed around Akçakent, in the core of a roughly north–south trending antiform in the Çiçekdağ region (Fig. 2). Well-preserved and unfaulted successions are rare, although the section along the Akçakent–Kilimli road exhibits much of the original structure.

The CO pseudostratigraphy consists (from top to bottom) of mainly pink pelagic cherty

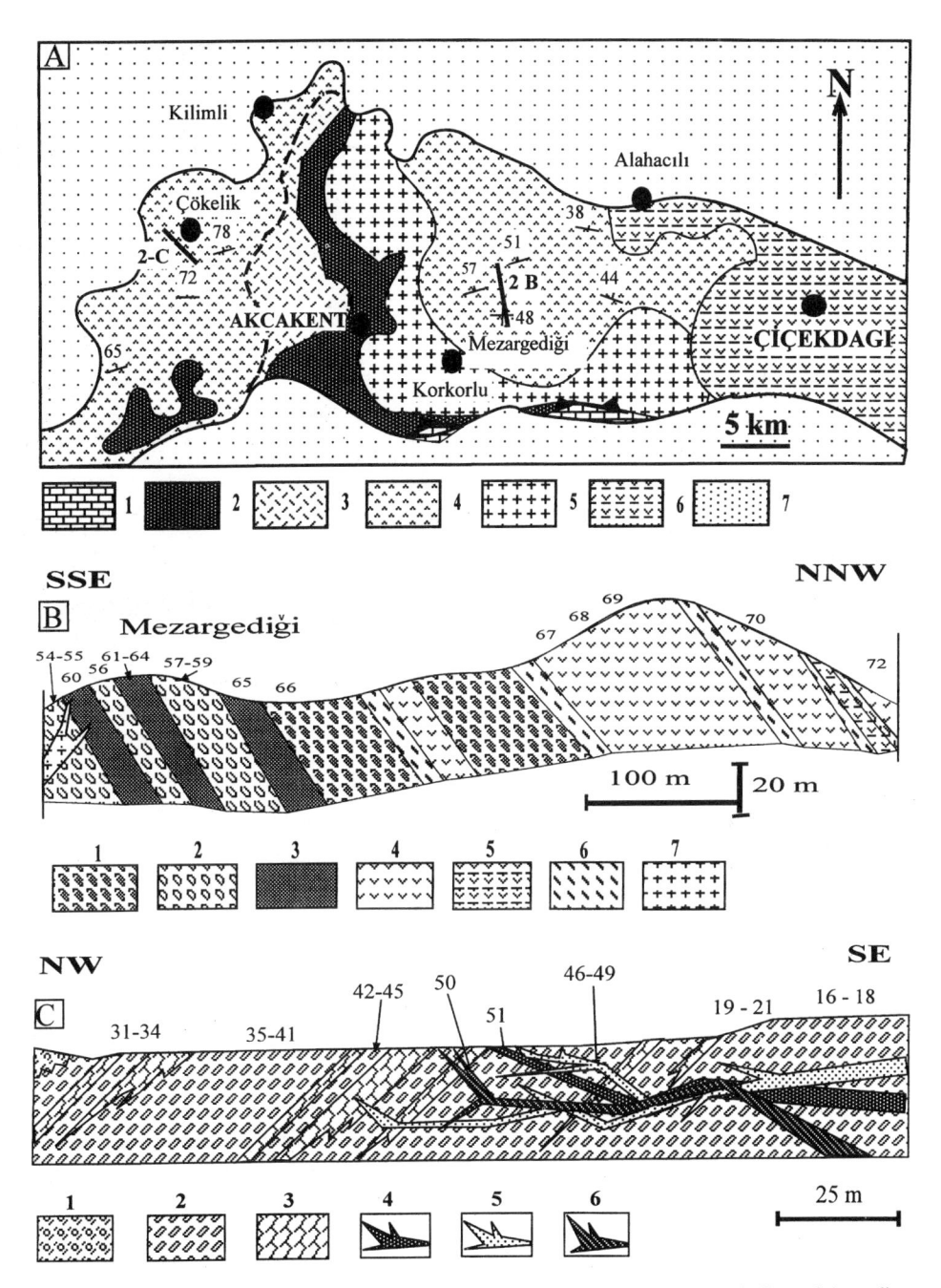

Fig. 2. (a) Simplified geological map of the Çiçekdağ Ophiolite and surrounding area. 1, Central Anatolian Metamorphics; 2, layered and isotropic gabbros; 3, dyke complex; 4, lavas and epi-ophiolitic sediments; 5, granitoids; 6, Eocene cover units; 7, Neogene continental clastics; 8, locations of geological cross-sections (S-1, Çökelik section; S-2, Mezargediği section). (b) Generalized geological cross-section from the Mezargediği Pass; 1, spherulitic pillow lava; 2, pillow lava; 3, pillow breccia; 4, massive lava; 5, lavas interlayered with sediments; 6, pelagic sediments; 7, feldspathoidal granitoids; 8, sample locations (numbers correspond to samples listed in Table 1). (c) Generalized geological cross-section from Çökelik village; 1, Tertiary cover; 2, pillow lavas and volcaniclastics; 3, pelagic sediments; 4, Group III dykes; 5, Group II dykes; 6, Group I dykes; 7, sample locations (numbers correspond to samples listed in Table 1).

limestone and pillow intercalations of Turonian–Santonian age (Erdoğan et al. 1996), pillowed and massive basalts with basaltic breccias, sheeted dyke complex, isotropic to layered gabbros and rare associated plagiogranites (Fig. 1, inset b). The lavas, which exhibit typical pillow structures, are best developed at the top of the volcanic sequence where volcanoclastic intercalations are also present. Massive basaltic flows, ranging in thickness from c. 1 to 3 m, are also present. The volcanic–volcanosedimentary successions are known as the Çökelik Volcanics (Erdoğan et al. 1996), which outcrop on the flanks of the antiformal structure. The areas immediately to the northeast of Korkorlu village and southwest of Çökelik village on the western flank, and the Mezargediği area on the eastern flank of the antiformal structure provide well-exposed sections through the basaltic extrusives. Due to the rarity of volcanic rock exposures in the vicinity of Çiçekdağ, these locations are particularly important and were chosen for detailed sampling (Fig. 2a and b). The sheeted dyke complex is poorly exposed relative to the gabbros but locally contains plagiogranite dykes. The gabbroic sequence is mainly composed of isotropic gabbros and subordinate layered gabbros. The ultramafic part of the ophiolite sequence is not observed in the Çiçekdağ area.

The ophiolitic succession is intruded by early monzogranitic, and later syenitic, intrusions. To the east and northeast of Korkorlu village, gabbroic and basaltic rocks are found as roof pendants within the granitoids. Dykes of feldspathoidal syenites are the last products of the Late Cretaceous collision-type felsic magmatism and cross-cut the volcanic carapace. The earliest fossiliferous cover sediments in the Çiçekdağ region are Lutetian in age (Ketin 1959).

Petrography

The basaltic rocks of the CO are weakly to strongly phyric, with plagioclase and subordinate clinopyroxene as phenocryst phases set in a variably quenched matrix of serrated plagioclase microlites, variolitic fans of crystallites and an opaque matrix. In the Çökelik area, basalts intercalated with the sediments higher in the ophiolite sequence are distinguished from the lower flows by the presence of sparse 'olivine' phenocrysts, which are usually pseudomorphosed by orange-red calcite and hematite in these pervasively altered rocks.

Basaltic samples are all hydrothermally altered to varying degrees. In many sections, the effects of alteration are so intense that primary mineral phases and magmatic textures are strongly overprinted by secondary minerals. As well as the matrix, secondary minerals are also abundant in vesicles, veins and pillow interstices, and include albite, chlorite, epidote, quartz, calcite, actinolite and iron oxides. Fresh clinopyroxene phenocrysts are the only relict igneous phase, although they occasionally show marginal alteration to actinolite. Actinolitic amphibole crystals are also identified as needles in the groundmass. Throughout the Çiçekdağ basaltic lavas, plagioclase laths and phenocrysts are mostly pseudomorphosed by albite. Chlorite commonly occurs in the groundmass of these lavas and also as massive, radiating fibrous aggregates infilling vesicles and veinlets. Epidote is found in vesicles at the margins of lava flows, along the walls of fractures and in veinlets. Where it occurs as a vesicle infill, it forms fan-shaped crystal aggregates often reaching 2–3 cm in diameter, commonly surrounded by subhedral quartz crystals which line the vesicles. Quartz is generally deposited together with epidote and forms a vesicle infill. Calcite is much less abundant than chlorite, epidote and quartz in the Çiçekdağ lavas, but similarly occurs in rock spaces rather than replacing previous minerals. Pillow breccias show similar petrography and textures.

The dykes display an equigranular texture, are medium to fine grained, and are characterized by ophitic, subophitic and intergranular textures. They are mostly aphyric to sparsely plagioclase–phyric dolerites. Although the primary igneous mineralogy of the dykes consists of clinopyroxene, calcic plagioclase and opaque minerals, most of these are replaced by a greenschist facies assemblage made up of actinolite, chlorite, epidote, plagioclase, quartz, sphene and secondary Fe–Ti oxides and hydroxides. Rare core relicts of clinopyroxene are also locally identified at the base of the sheeted dyke complex. Most dykes retain their primary igneous textures but there are some that are more severely altered to equigranular textured quartz- and epidote-rich epidosites. These are often green coloured compared to the more normal grey dykes that are usually characterized by dominantly albite–actinolite–chlorite assemblages.

Sampling and analytical methods

Sixty-three samples were collected from basaltic lavas and dykes, many selected from the Mezargediği and Çökelik traverses (see location and relative sampling positions in Fig. 2). Mafic samples were taken from the crystalline

Table 1. *Representative chemical analyses of basaltic lavas and dykes from the Çiçekdağ Ophiolite*

Sample	C-6	C-9A	C-10A	C-10B	C-11	C-66	C-76	C-77	C-78	C-79	C-31	C-32	C-33	C-34	C-35	C-36	C-37	C-39	C-40	C-41	C-42
Rock type*	p.lava	p.lava	p.lava	p.lava	p.lava	p.lava	p.lava	p.lava	p.lava	p.lava	p.lava	p.lava	p.lava	p.lava	p.lava	p.lava	p.lava	p.lava	p.lava	p.lava	p.lava
Major oxides (wt%)																					
SiO_2	53.12	53.72	51.59	57.12	47.36	48.64	52.41	52.88	49.37	50.52	50.59	51.06	44.43	45.34	49.41	45.27	56.45	41.75	43.38	51.78	52.10
TiO_2	1.39	0.44	1.20	1.08	0.76	0.67	0.96	0.81	0.59	0.65	0.44	0.42	0.41	0.41	0.44	0.46	0.36	0.47	0.46	1.18	1.25
Al_2O_3	16.63	13.43	15.91	14.21	14.03	14.04	15.44	15.45	15.43	15.22	14.60	14.07	12.73	13.12	14.41	13.69	13.11	14.30	15.27	15.40	15.84
Fe_2O_3t	12.31	9.85	15.39	12.47	12.10	10.34	9.77	9.49	10.90	10.00	10.34	10.03	8.56	9.20	10.54	9.26	7.77	10.30	11.02	9.92	10.34
MnO	0.26	0.18	0.20	0.16	0.21	0.20	0.23	0.12	0.18	0.15	0.14	0.16	0.18	0.19	0.13	0.16	0.09	0.18	0.17	0.17	0.17
MgO	3.90	8.96	5.43	3.25	6.64	10.21	5.90	4.83	6.48	7.57	7.84	7.60	4.67	4.93	8.37	3.80	1.24	9.29	10.19	5.72	4.69
CaO	7.55	5.92	2.53	2.82	5.65	11.09	7.69	9.84	9.18	8.53	5.48	9.11	12.37	10.93	5.07	11.01	7.53	9.18	6.47	4.32	6.87
Na_2O	3.69	5.05	5.73	7.09	7.14	1.75	4.92	4.17	3.57	3.09	3.29	3.32	4.95	5.00	3.71	5.99	6.99	3.32	3.45	5.27	3.60
K_2O	0.43	0.09	0.17	0.07	0.08	0.32	0.05	0.04	0.99	1.38	0.22	0.20	0.06	0.07	0.19	0.17	0.14	0.11	0.09	0.27	0.19
P_2O_5	0.06	0.04	0.08	0.09	0.05	0.06	0.10	0.10	0.06	0.06	0.03	0.03	0.04	0.03	0.03	0.04	0.06	0.03	0.03	0.09	0.15
LOI	0.81	2.51	1.55	1.35	5.75	2.65	2.41	2.18	3.43	2.88	7.35	3.71	11.90	10.97	7.83	10.26	6.39	10.94	9.31	5.45	4.51
Total	100.15	100.19	99.78	99.71	99.77	99.97	99.88	99.91	100.18	100.05	100.32	99.71	100.30	100.19	100.13	100.11	100.13	99.87	99.84	99.57	99.71
Trace elements (ppm)																					
Ba	73	36	25	76	47	37	25	7	156	87	22	11	15	41	20	20	19	2	27	18	50
Ce(XRF)	3	1	10	3	1	2	2	3	1	1	3	2	1	3	2	1	8	1	2	1	12
Cl	226	1	5	5	3	28	3	3	2	2	4	2	2	1	1	3	2	2	2	1	1
Cr	21	345	24	9	254	531	179	194	182	330	272	220	220	244	245	177	160	240	334	119	77
Cu	89	162	161	170	74	147	112	13	78	94	27	23	86	38	637	52	29	57	85	16	25
Ga	20	10	13	18	15	13	15	15	11	13	13	13	11	10	13	10	8	12	15	15	19
La(XRF)	3	3	1	1	2	2	2	2	2	1	3	1	1	2	2	1	1	1	3	2	3
Nb	2	1	2	2	2	2	2	2	2	0.5	1	2	1	2	2	2	1	1	1	2	2
Nd(XRF)	5	7	15	1	7	4	6	7	5	1	2	1	7	2	13	1	16	6	13	7	17
Ni	7	55	17	18	70	165	47	47	66	95	53	59	47	52	60	32	52	62	60	28	27
Pb	75	9	9	4	15	8	6	10	10	6	2	4	5	4	2	3	3	3	1	1	4
Rb	17	2	2	5	3	8	1	1	21	24	5	5	3	2	4	3	4	2	2	4	3
S	210	1	2	19	10	47	1	3	2	2	22	14	29	26	32	63	35	26	27	3	1
Sr	190	92	66	134	108	72	188	151	203	164	135	160	72	76	110	101	49	154	160	145	176
V	301	281	576	648	433	292	330	298	296	293	371	312	319	311	377	367	119	373	389	286	272
Y	25	12	23	28	17	20	23	19	14	15	11	11	9	8	11	13	7	12	11	22	26
Zn	195	68	80	135	130	92	83	59	64	109	71	70	71	71	74	64	67	73	80	82	86
Zr	48	20	43	47	32	39	51	44	25	33	14	14	12	12	14	15	13	15	15	48	59

Table 1. Continued

Sample	C-43	C-44A	C-44B	C-45	C-54	C-55	C-56	C-57	C-58	C-59	C-67	C-68	C-69	C-70	C-72	C-13	C-16	C-17	C-18	C-19	C-21
Rock type*	p.lava	p.lava	p.lava	p.lava	m.lava	m.lava	m.lava	m.lava	m.lava	m.lava	m.lava	m.lava	m.lava	m.lava	m.lava	breccia	breccia	breccia	breccia	breccia	breccia

Major oxides (wt%)

SiO_2	51.21	52.38	55.64	51.98	49.52	53.38	54.30	54.85	52.32	53.04	53.95	54.56	50.05	50.85	51.01	54.01	45.45	51.19	49.99	40.67	48.04
TiO_2	0.46	0.40	0.99	1.26	1.09	0.40	0.65	1.05	0.77	0.83	0.62	0.55	0.80	0.78	0.53	1.21	1.06	1.32	0.66	0.84	0.56
Al_2O_3	14.50	12.04	13.31	16.90	13.65	11.64	14.14	14.47	14.97	14.86	14.79	14.79	15.80	13.81	13.05	16.66	17.93	14.87	15.74	12.18	16.75
Fe_2O_3t	10.09	8.94	13.50	10.71	15.29	8.94	9.32	12.69	9.99	11.70	10.08	9.00	11.86	14.60	10.40	12.09	11.44	12.42	9.11	8.75	8.73
MnO	0.20	0.20	0.15	0.17	0.21	0.20	0.15	0.16	0.14	0.24	0.16	0.16	0.25	0.20	0.23	0.13	0.18	0.14	0.13	0.10	0.15
MgO	9.28	10.76	5.28	5.65	6.00	10.76	3.77	3.65	4.66	5.98	6.93	5.99	7.74	6.57	10.22	2.00	6.88	3.97	5.74	1.65	8.04
CaO	7.46	9.58	3.35	5.50	4.21	8.69	6.91	4.59	4.69	5.70	5.27	7.49	5.48	7.18	9.07	1.60	12.21	4.76	6.33	16.50	10.56
Na_2O	2.53	2.17	2.53	3.65	5.20	2.06	6.43	6.53	6.19	5.55	4.99	4.22	4.59	3.27	2.60	8.64	2.41	4.20	2.46	6.67	3.41
K_2O	0.08	0.49	0.18	0.36	0.07	0.69	0.13	0.08	0.12	0.23	0.53	1.50	0.15	0.15	0.69	0.20	0.37	0.04	0.18	0.09	0.12
P_2O_5	0.03	0.02	0.10	0.14	0.07	0.02	0.08	0.08	0.07	0.08	0.05	0.05	0.08	0.07	0.05	0.08	0.04	0.13	0.06	0.19	0.05
LOI	4.18	3.06	4.57	3.28	4.12	3.06	3.98	1.58	4.90	1.80	2.30	1.57	2.53	2.36	2.31	1.01	1.88	6.92	9.44	12.68	3.62
Total	100.02	100.04	99.60	99.60	99.43	99.84	99.86	99.73	99.82	100.01	99.67	99.88	99.33	99.84	100.16	97.63	99.85	99.96	99.84	100.32	100.03

Trace elements (ppm)

Ba	29	111	52	91	34	98	46	30	39	125	98	262	19	39	116	182	53	17	55	24	42
Ce(XRF)	2	2	7	2	7	1	13	3	2	9	4	7	3	5	4	6	3	5	2	2	1
Cl	1	1	2	1	2	2	1	1	2	201	26	60	28	59	24	5	4	2	3	1	1
Cr	197	519	20	103	9	515	219	17	265	71	128	118	336	259	537	38	127	27	110	242	223
Cu	172	74	112	46	70	75	249	81	256	104	82	143	84	53	14	50	68	45	109	37	61
Ga	14	11	15	22	18	10	14	18	12	14	13	13	16	15	11	13	19	20	12	8	15
La(XRF)	2	2	1	1	2	3	1	1	2	2	1	2	2	1	1	2	1	4	1	1	2
Nb	1	1	3	2	2	2	1	2	0.41	2	2	0.41	2	2	2	1	4	2	2	2	2
Nd(XRF)	6	9	15	13	14	6	22	12	9	12	8	13	8	13	8	11	9	16	3	6	6
Ni	53	120	16	28	21	124	47	17	63	44	46	43	95	83	187	12	54	16	50	85	92
Pb	3	5	3	4	3	7	58	6	25	9	7	8	7	7	11	6	30	2	5	12	3
Rb	2	18	4	5	2	18	3	2	4	7	10	27	3	3	15	6	6	3	8	2	2
S	2	6	13	3	2	3	5	2	17	2	3	26	62	169	2	9	3	23	12	48	20
Sr	148	131	70	169	101	130	115	113	106	138	76	84	28	65	91	124	737	41	34	106	161
V	312	241	322	245	613	232	395	580	443	334	244	286	357	384	266	385	356	451	332	251	228
Y	11	11	28	28	25	11	20	22	19	22	21	19	22	21	14	26	34	31	17	22	15
Zn	73	66	44	88	132	67	79	76	95	106	87	72	100	88	113	52	75	115	72	82	63
Zr	14	19	69	65	43	18	28	42	32	46	41	39	46	45	27	66	77	80	37	52	30

Table 1. Continued

Sample	C-60	C-61	C-62	C-63	C-64	C-65	C-46	C-47	C-48	C-49	C-50	C-51	C-23	C-24	C-26	C-27	C-28	C-29	C-30
Rock type*	breccia	breccia	breccia	breccia	breccia	breccia	dyke	dyke	dyke	dyke	dyke	dyke	sh.dyke	sh.dyke	sh.dyke	sh.dyke	sh.dyke	sh.dyke	sh.dyke

Major oxides (wt%)

SiO_2	0.82	58.27	54.06	57.82	59.06	60.11	48.68	47.45	43.37	45.54	49.66	53.25	59.50	48.06	52.46	51.49	52.74	51.67	46.85
TiO_2	1.42	1.24	1.26	1.28	1.15	1.23	0.64	0.62	0.63	0.66	0.49	1.21	1.43	1.02	0.99	1.01	1.42	0.96	1.17
Al_2O_3	16.84	14.49	15.98	15.63	14.41	15.68	14.53	14.94	15.14	15.09	14.12	14.33	14.82	14.64	15.78	15.86	15.07	15.30	15.57
Fe_2O_3t	3.22	12.38	13.05	12.65	11.59	11.55	10.12	10.10	9.84	9.78	9.45	9.02	10.11	10.70	11.78	11.90	12.96	11.22	11.49
MnO	0.27	0.13	0.17	0.05	0.15	0.10	0.21	0.23	0.26	0.19	0.15	0.14	0.33	0.34	0.32	0.32	0.30	0.33	0.34
MgO	4.88	2.08	2.89	0.56	2.33	1.02	8.96	8.84	8.78	8.65	8.04	3.14	3.43	5.78	6.24	7.43	6.04	7.02	6.54
CaO	2.21	2.48	2.54	1.82	2.58	1.50	5.00	5.32	7.64	6.89	8.36	6.34	1.45	6.70	3.19	1.48	2.57	3.49	6.46
Na_2O	6.86	7.90	7.58	9.35	7.73	7.62	3.28	3.56	3.83	3.64	2.99	7.37	6.14	4.96	5.45	5.33	4.96	4.57	5.19
K_2O	0.16	0.12	0.30	0.08	0.20	0.18	0.04	0.04	0.05	0.11	0.08	0.05	0.01	0.04	0.03	0.09	0.05	0.06	0.01
P_2O_5	0.05	0.12	0.10	0.15	0.10	0.09	0.05	0.05	0.05	0.06	0.03	0.10	0.32	0.07	0.08	0.14	0.13	0.07	0.09
LOI	2.40	0.85	1.48	0.31	0.76	0.43	8.45	8.63	10.08	9.57	6.28	5.18	2.61	7.60	3.64	3.99	3.62	5.13	6.45
Total	99.13	100.06	99.41	99.70	100.06	99.51	99.96	99.78	99.67	100.18	99.65	100.13	100.15	99.91	99.96	99.04	99.86	99.82	100.16

Trace elements (ppm)

Ba	113	56	91	64	156	150	14	19	15	17	9	10	25	33	14	29	30	31	16
Ce(XRF)	5	6	5	5	2	2	1	3	2	2	9	5	15	2	3	7	9	18	11
Cl	33	18	338	154	2	5		3		1	3	52	18	8	5	2	93	80	1
Cr	5	18	28	12	27	52	543	453	467	423	201	56	6	57	50	37	25	62	48
Cu	39	14	25	21	28	69	133	188	328	68	194	9	200	21	48	16	52	48	16
Ga	22	14	16	12	13	11	14	13	12	14	13	12	17	15	16	14	17	16	17
La(XRF)	3	2	2	1	1	1	2	3	1	1	1	4	1	1	2	1	2	3	1
Nb	2	3	2	5	2	2	2	1	1	2	1	2	0.68	1	2	1	2	2	2
Nd(XRF)	8	6	9	14	3	10	4	11	7	8	2	6	18	4	9	13	16	22	16
Ni	18	13	18	14	18	13	172	147	145	145	59	21	4	26	23	23	17	23	37
Pb	4	5	48	5	14	9	1	1	1	1	2	2	4	8	1	1	2	1	5
Rb	5	3	12	2	6	5	1	2	1	2	1	2	0.3	2	1	2	2	2	2
S	2	9	9	8	39	21	16	14	16	14	21	5	66	141	100	990	319	149	22
Sr	77	92	88	75	143	127	127	135	143	120	148	74	35	79	78	68	69	74	64
V	354	447	376	377	369	430	266	254	237	306	324	224	118	449	375	426	455	430	367
Y	21	29	25	35	30	29	13	13	14	17	10	23	35	22	22	24	33	20	28
Zn	115	54	90	25	54	50	72	73	75	76	75	55	228	171	199	236	394	191	268
Zr	74	65	65	67	62	64	37	37	38	40	14	55	96	41	44	47	69	42	54

* p.lava, Pillow lava; m.lava, massive lava; breccia, pillow lava breccia; dyke, individual dykes cross-cutting lavas; sh.dyke, sheeted dyke complex.

Table 2. Additional chemical data on Çiçekdağ basaltic rocks

Sample Rock type	C-58 Massive lava	C-68 Massive lava	C-79 Pillow lava	C-23 Sheeted dyke
Incompatible elements (ppm)				
Os	3.08	2.67	0.97	0.13
Hf	0.73	0.83	0.73	1.08
Sc	38.77	38.33	46.67	42.40
Ta	0.03	0.04	0.04	0.05
Th	0.06	0.11	0.07	0.07
U	0.11	0.20	0.12	0.15
Rare earth elements (ppm)				
La	1.59	1.47	1.18	1.43
Ce	3.63	3.94	3.39	4.30
Pr	0.65	0.67	0.57	0.76
Nd	3.64	3.73	3.17	4.39
Sm	1.27	1.34	1.13	1.63
Eu	0.50	0.56	0.45	0.60
Gd	1.83	1.97	1.64	2.42
Tb	0.34	0.37	0.31	0.45
Dy	2.22	2.41	2.05	2.96
Ho	0.49	0.54	0.45	0.65
Er	1.46	1.60	1.32	1.87
Tm	0.25	0.26	0.22	0.32
Yb	1.50	1.61	1.35	1.92
Lu	0.24	0.26	0.22	0.31

interiors of pillows, pillow breccias, massive lavas and dykes, and selected in order to minimize the effect of vesiculation, alteration and weathering. All samples were analysed for major and selected trace elements, and a subset of four samples for rare earth element (REE), and Hf, Ta, Th and U (Tables 1 and 2). Analysis for major oxides and most trace elements were determined on an ARL 8420 X-ray fluorescence spectrometer (Department of Earth Sciences, University of Keele, UK) calibrated against both international and internal Keele standards of appropriate composition, whereas the REE etc. were determined by instrumental neutron activation analysis (Activation Laboratories Ltd, Canada). Details of methods, accuracy and precision are given in Floyd & Castillo (1992).

Chemical discrimination

As outlined above, all the basaltic lavas and dykes in the CO have undergone greenschist facies alteration and metamorphism. Thus, for chemical discrimination of the volcanics, only elements that are immobile during alteration should be used. There is general agreement that the high field strength elements (HFSE: P, Ti, Y, Zr, Nb, Hf, Ta; Saunders *et al.* 1980), Th (Wood *et al.* 1979), the transition metals (Sc, V, Cr, Ni) and REE are essentially immobile during all but the most severe seafloor or hydrothermal alteration (e.g. Pearce 1975; Shervais 1982). SiO_2 and most large-ion lithophile elements (LILE: Na, K, Cs, Rb, Ba, Sr) are almost certainly mobile under these alteration conditions (Smith & Smith 1976; Coish 1977; Humphris & Thompson 1978*a, b*). Thus, chemical discrimination mainly relies on the HFSE and REE, which tend to be the least mobile in aqueous fluids.

Geochemistry

General features

On the basis of their Nb/Y ratios (Winchester & Floyd 1977), all the mafic samples are tholeiitic (Nb/Y = 0.04–0.20) with generally low HFSE contents (Table 1). Metabasaltic lavas from the Mezargediği and Çökelik traverses have overlapping compositions in terms of incompatible element abundances, similar Zr/Y and Zr/Ti ratios, and depleted REE patterns. These features suggest that they represent a similar lava sequence, possibly related to a single chemical group, but now separated by tectonic dismemberment. However, direct chemical correlation is not possible between the two lava sequences throughout, as two characteristics found in the Çökelik traverse are missing in the Mezargediği

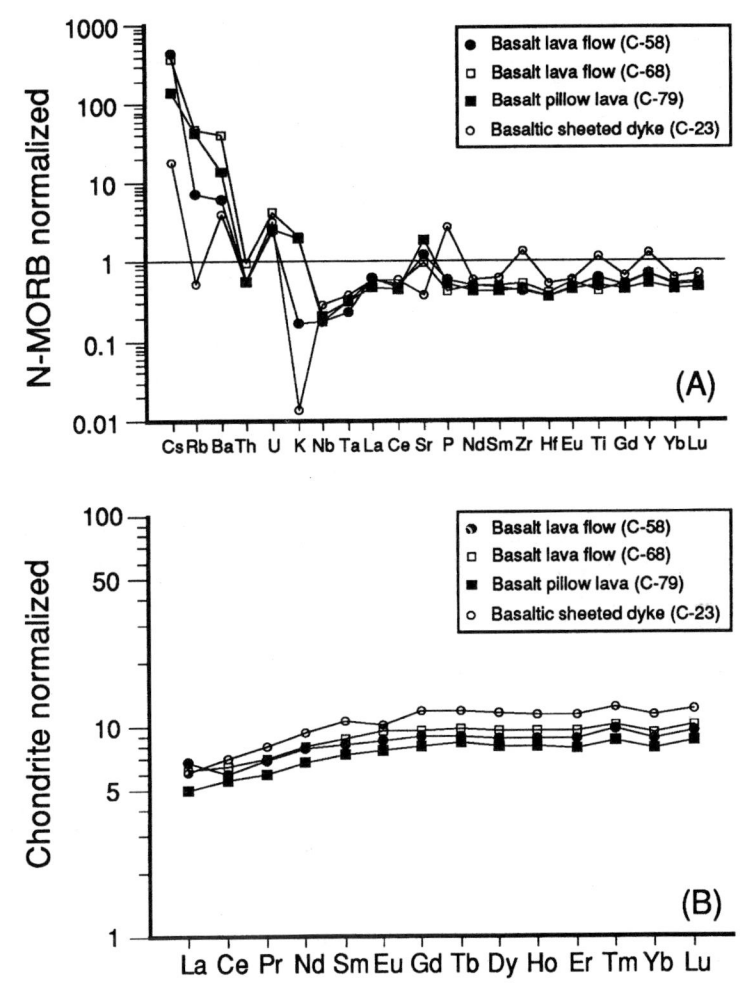

Fig. 3. (a) N-MORB normalized multi-element diagram for Çiçekdağ basalts showing HFSE element depletion and LILE enrichment. (b) Chondrite-normalized REE patterns of Çiçekdağ basalts showing light REE depletion. Normalization factors from Sun & McDonough (1989).

traverse. In particular, the former sequence contains numerous feeder dykes towards the base and a set of primitive lavas with low Zr and Y contents that are missing from the Mezargediği sequence. On the grounds that the lower volcanic portions of an ophiolite should contain a greater density of feeder dykes than the upper, it is suggested that the Çökelik sequence records a generally lower volcanic section than the Mezargediği sequence.

Normalized patterns for incompatible elements and REE (Fig. 3) illustrate a number of features typical of the CO. All basaltic samples show broadly parallel and depleted light REE patterns with (La/Yb)N of c. 0.6, suggesting that the samples plotted are probably related by simple fractionation. Relative to N-mid-ocean ridge basalt (MORB), the normalized incompatible element patterns show the depleted HFSE and enriched LILE features of SSZ ophiolites, although the later variable abundances (apart from Th) are largely governed by alteration effects.

Internal variation

Samples from the lavas and dykes of the Mezargediği and Çökelik traverses, and isolated locations elsewhere, show internal variation that is largely governed by fractional crystallization. Using Zr as a fractionation index, increasing FeO^*/MgO (Fig. 4a) and decreasing Cr (Fig.

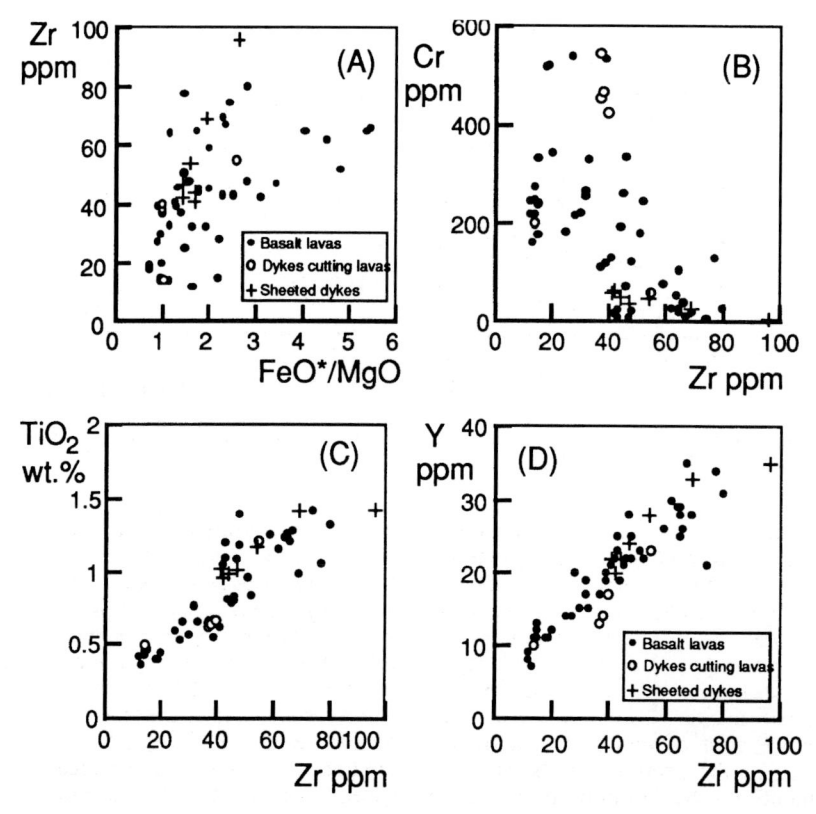

Fig. 4. Chemical diagrams illustrating internal variation for Çiçekdağ lavas and dykes. (**a**) and (**b**), Effects of mafic fractionation; (**c**) and (**d**), linear covariance suggests a single comagmatic suite for the basalts.

4b) implies mafic fractionation, as indicated by the observed relict phenocryst phases. Some of the scatter here is the consequence of alteration, whereas the covariance of immobile elements (Fig. 4c and d) indicate that all the samples probably represent a single, fractionated, comagmatic chemical group. The very low Zr (< 20 ppm) and Y (c. 10 ppm) set of samples, however, might be a different chemical group from the rest, especially in view of their position in the Çökelik traverse magmatic stratigraphy (see below).

Chemostratigraphy

Systematic sampling along the Mezargediği and Çökelik traverses allows some monitoring of chemical variability with height in the volcanic portion of the CO. Due to tectonic dismemberment, stratigraphic height is only relative, being measured from the arbitrary base of the traverse in both cases. As seen in Fig. 5, the basal part of each traverse is highly variable in terms of the elements plotted, whereas the uppermost

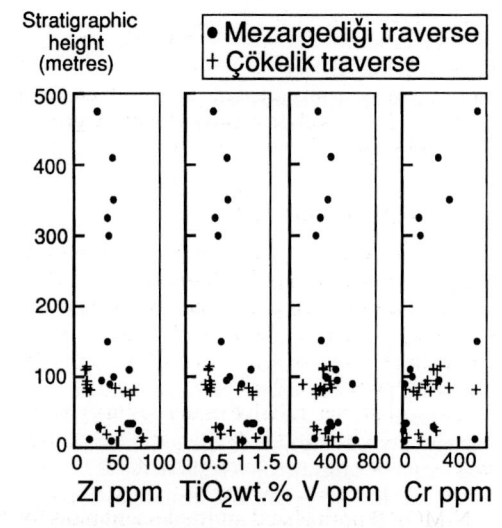

Fig. 5. Chemical variation with height in the Çiçekdağ lavas. Height is relative and arbitrary starting at the base of the traverse in both the Mezargediği and Çökelik sequences.

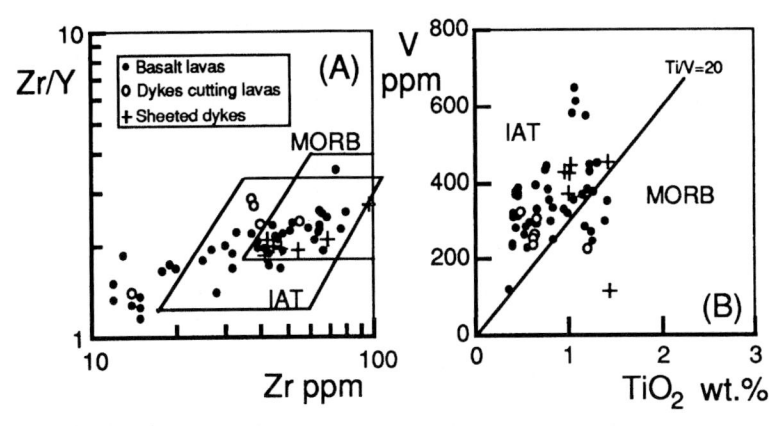

Fig. 6. Discrimination of tectonic setting using chemical parameters. (**a**) Zr/Y v. Zr [after Pearce & Norry (1979)]; (**b**) V v. TiO$_2$ [after Shervais (1982)].

portion is more 'steady state', although this effect may be more apparent than real (fewer samples were collected here). The chemical stratigraphy is interpreted here in terms of the periodic tapping of a well-fractionated magma chamber which, in the light of the similarity in composition of the two traverses, might be the same ponded magma body. However, this simple picture must be modified to allow for the presence of the primitive lavas with low HFSE contents at c. 80–110 m from the base in the Çökelik traverse. It is suggested that these lavas represent an influx of pristine, largely unfractionated, melt derived from another source which were not involved with the main fractionated magma chamber.

In general, the similarity of basaltic compositions indicate that open-system processes were limited throughout the section available and only one (or possibility two) chemical groups were involved.

Tectonic discrimination

A number of plots can be used to discriminate the provenance of the CO basalt samples (Fig. 6). Despite some compositional overlap with the island arc tholeiites (IAT) and MORB designated fields, most samples clearly fall within the former, especially in the TiO$_2$ v. V diagram. It is also apparent that, relative to the majority of the samples, the very low Zr–Y Çökelik samples plot outside the IAT field (Fig. 6), again emphasizing their different nature and origin.

N-MORB normalized multi-element plots for the CO basalts (Fig. 3) show the depletion of HFSE (Nb, Ta, Zr, Hf, Ti, P; generally < 1) together with the relative LILE (K, Ba, Sr, Rb, Th) enrichment features characteristic of subduction-related arc magmas (e.g. Wood *et al.* 1979; Wood 1980; Pearce 1982; Pearce & Parkinson 1993; Pearce *et al.* 1984). Within the generally mobile LILE group, Th is a relatively stable and reliable indicator, whose enrichment relative to other incompatible elements (especially Nb–Ta) is taken to represent the subduction zone component (e.g. Wood *et al.* 1979; Pearce 1983). This feature is attributed to the modification of the mantle source region of the Çiçekdağ basalts by a 'subduction component', i.e. metasomatism by aqueous fluids derived from the underlying subducting slab (e.g. Pearce 1982; Pearce *et al.* 1984; Hawkesworth *et al.* 1977, 1993; Saunders & Tarney 1984; McCulloch & Gamble 1991). In general, the CO has all the features typical of suprasubduction oceanic crust (Pearce *et al.* 1984).

Normalized REE patterns (Fig. 3) for basalts do not generally make good discriminants due to similar patterns being generated in different environments. For example, both MORB and back-arc basin basalts (BABB) are characterized by patterns that display slight light REE depletion relative to the heavy REE and are generally enriched 10–20 times relative to chondrite. Distinguishing between basalts formed in these two eruptive settings is therefore not possible solely on the basis of REE patterns. REE patterns of most IAT also display a large degree of overlap with MORB-type patterns, although the former can exhibit varying degrees of light REE enrichment and depletion (e.g. White & Patchett 1984; Brouxel *et al.* 1989). Since normal IAT may exhibit variable light REE patterns from extremely depleted (e.g. Mariana forearc; Hickey & Frey 1981; Crawford *et al.* 1986) to substantially flat (e.g. Tonga–Kermadec; Ewart *et al.* 1977: Yap Trench;

shown by the CO can best be matched with arc basalts similar to those found in immature island-arc sequences such as the Izu-Bonin Arc.

Comparison with the Sarıkaraman and other Neotethyan ophiolites

The CO and the Sarıkaraman Ophiolite (SO) are representative members of the CAO (Göncüoğlu et al. 1991). Both are characterized by the lack of ultramafic rocks in direct contact with the rest of the ophiolitic slab. The lowest section of the CO is composed of layered gabbros, whereas isotropic gabbros are found at the base of the SO (Yalınız et al. 1996). The SO is remarkable in displaying a high portion of the plagiogranite suite (Floyd et al. 1998a) and a late set of isolated dykes (Yalınız et al. 1996). Neither of these features are as well developed in the CO, nor exposed to the same degree. Both ophiolites are overlain by a sequence of Lower Turonian–Campanian pelagic sediments and are also intruded by Upper Cretaceous–Lower Palaeocene granitoids (Göncüoğlu et al. 1991; Yalınız et al. 1996).

It has previously been shown that the volcanic section of the SO exhibits typical SSZ characteristics similar to other Neotethyan ophiolites (Yalınız et al. 1996), including the CO. The CO closely matches the variation shown by the SO but lacks the extensive degree of chemical evolution (Fig. 7); the lavas sampled are less chemically evolved. A significant difference is also exhibited by the latter cross-cutting dolerite dykes of the SO which has a MORB-like affinity (Yalınız et al. 1996). In general, minor differences are apparent in their stratigraphy, although overall the basalts have a similar chemistry and formation age.

The CAO are also part of a long chain of Tethyan ophiolites that extend from the western Mediterranean to the Far East. Petrological and geochemical studies have shown that there are important differences along the Tethyan ophiolitic belt. Several workers have divided the Tethyan ophiolites into two groups: (1) Jurassic ophiolites in the western and central area (Alps, Appennines, Carpathians, Dinarides, Hellenides) which display MORB affinities; and (2) Cretaceous ophiolites at the eastern Mediterranean end (Troodos, Semail, Baer-Bassit, Hatay) which exhibit SSZ features (e.g. Pearce et al. 1984). Since the chemical signature of the Cretaceous Çiçekdağ lavas also indicate that they were generated in a SSZ, this ophiolite can be fitted into the regional tectonic framework, displaying a similar tectonic setting to other late

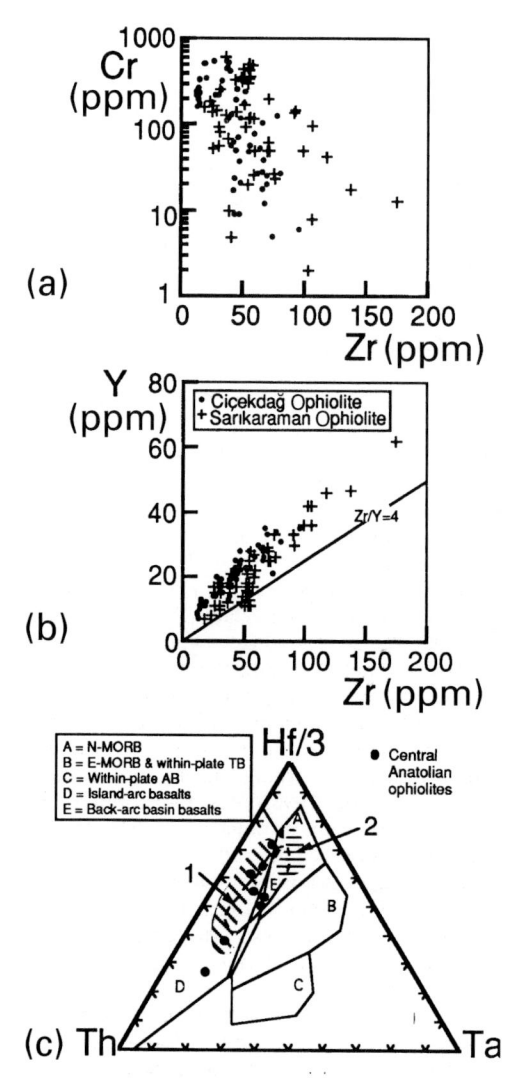

Fig. 7. (a) and (b) Chemical comparisons of Çiçekdağ and Sarıkaraman ophiolitic basalts; (c) Comparison of Çiçekdağ and Sarıkaraman basalts with SSZ-type eastern Mediterranean ophiolites (area 1) and MORB-type western Mediterranean ophiolites (area 2). Comparison data selected from the Tethyan ophiolite literature (e.g. Pearce et al. 1984). Discrimination fields after Wood (1980).

Beccaluval & Serri 1988) to slightly enriched (e.g. Aleutians and Sunda; Kay 1977), it follows that the components added to IAT mantle sources generally carry minor light REE in addition to K, Rb, Ba, Sr, Th, U, Cs and H_2O. However, strongly REE-depleted tholeiites are generally interpreted as a product of immature island arcs (Crawford et al. 1986; Brouxel et al. 1989). It is suggested that the REE patterns

Cretaceous Neotethyan ophiolites in the eastern Mediterranean area (Fig. 7).

Genesis and emplacement of SSZ-type Central Anatolian ophiolites

The geodynamic evolution of the İzmir–Ankara branch of Neotethys has been the topic of copious studies (e.g. Robertson & Dixon 1984; Tüysüz et al. 1995; Okay et al. 1998). All of these studies are based mainly on very general approaches and broadly consider the geological–geochemical aspects. Any model explaining the tectonic evolution of the İzmir–Ankara branch of Neotethys must be consistent with the following local and regional constraints. Firstly, the stratigraphic succession in the Central Anatolian Metamorphic Complex is almost identical with that of the less metamorphosed Palaeozoic–Mesozoic platform margin sequences of the Taurides–Anatolides (Göncüoğlu et al. 1991). It differs from the later only by a complex high-temperature–low-pressure (HT–LP) metamorphic overprint and the presence of collision-type magmatism. Secondly, the upper part of the metamorphic succession is represented by an ophiolite-bearing olistostromal complex, representing synorogenic flysch deposition on the platform margin (Göncüoğlu 1981). The ophiolitic knockers within this meta-olistostrome display mainly MORB characteristics with minor ocean island basalt (OIB) and IAT (Floyd et al. 1998b). Thirdly, the metamorphic rocks are overthrust by dismembered ophiolites with SSZ-type geochemical characteristics akin to other eastern Mediterranean ophiolites (Yalınız et al. 1996; Floyd et al. 1998a, b). Structural data clearly indicate an emplacement direction towards the south and southwest. The formation age is early Middle Turonian–Early Santonian, based on palaeontological data from the associated epi-ophiolitic sedimentary rocks (Yalınız et al. 1997). The markedly high volume of epiclastic volcanogenic sediments in this succession suggests the involvement of an island arc. Finally, the accretionary wedge complex (the Ankara Mélange) to the north of the CACC, which was formed within the converging İzmir–Ankara branch of Neotethys, contains slivers of ophiolitic rocks dominated by MORB, OIB, and minor IAT compositions, characteristic of forearc accretionary wedge complexes (Floyd 1993; Floyd et al. 1998b). Associated sediments yield Late Triassic–Early Cretaceous fossils.

These data suggest: (1) the oceanic crust of the İzmir–Ankara branch of Neotethys has been partly consumed along a north dipping intra-oceanic subduction zone (Göncüoğlu et al. 1991); and (2) the SSZ-type ophiolitic bodies, observed both as allochthonous thrust sheets on the Taurides–Anatolides and as blocks in the mélange complexes, were generated within the MORB-type old hanging-wall block of the İzmir–Ankara oceanic crust and emplaced southwards onto the Tauride–Anatolide passive margin.

However, oceanic crust formation in a SSZ setting can involve quite different tectonic settings, such as active island arcs, back-arc and forearc basins. Thus, for a better understanding of the evolution of the Neotethys it is critical to determine the specific environment of lithosphere generation. The following model (Fig. 8) considers not only the geochemical and structural data, but also the geological relations to adjacent terranes.

- The rapid convergence between Eurasia and Africa, due to the opening of the south Atlantic in the Early Cretaceous, resulted in a compressional regime in the Tethyan realm. This compression caused the formation of an intra-oceanic subduction within the İzmir–Ankara branch of the Neotethys, which led to formation of a juvenile ensimatic arc (Göncüoğlu & Türeli 1993; Tüysüz et al. 1995) on MORB-type oceanic crust during the Cenomanian (Fig. 8a).
- The old MORB-type oceanic crust, together with trapped OIB-type volcanics and accretion prism material, was obducted onto the Tauride–Anatolide passive margin and generated the ophiolite-bearing olistostromal sediments that overlie the slope sediments of the CACC. The emplacement of ophiolitic nappes resulted in crustal thickening and early stage syncollisional granitoids (Erler & Göncüoğlu 1996). The trench rollback generated through subduction of the forearc basin resulted in the extension of the overlying forearc region and generation of SSZ-type oceanic crust by the partial melting of already depleted MORB-type lithosphere (Fig. 8b).
- Continuing convergence during the Campanian caused the complete subduction of the old oceanic slab and final collision of the trench with the buoyant margin of the CACC. This led to a switch from extension to compression in the forearc region. The new and hot SSZ-type oceanic crust is prevented from subducting and will obduct onto the CACC (Fig. 8c).
- Complete obduction of SSZ-type oceanic crust is followed by post-collisional uplift at

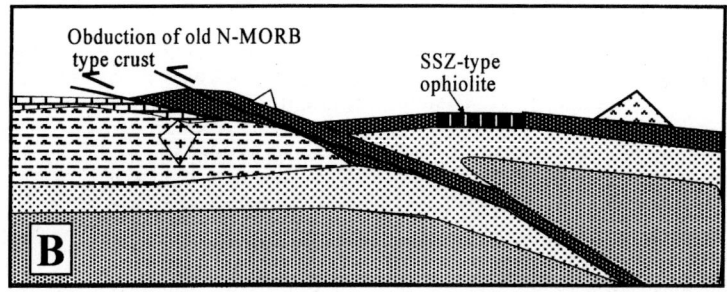

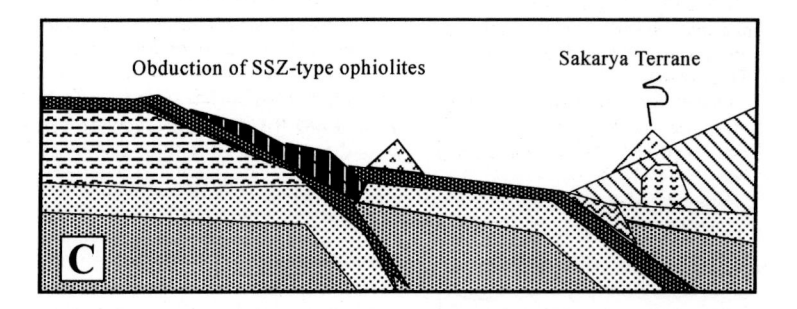

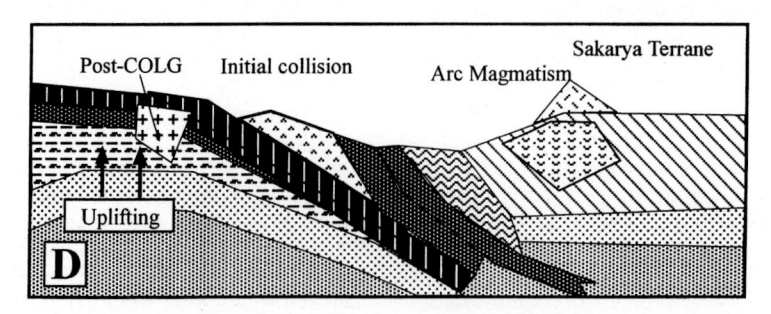

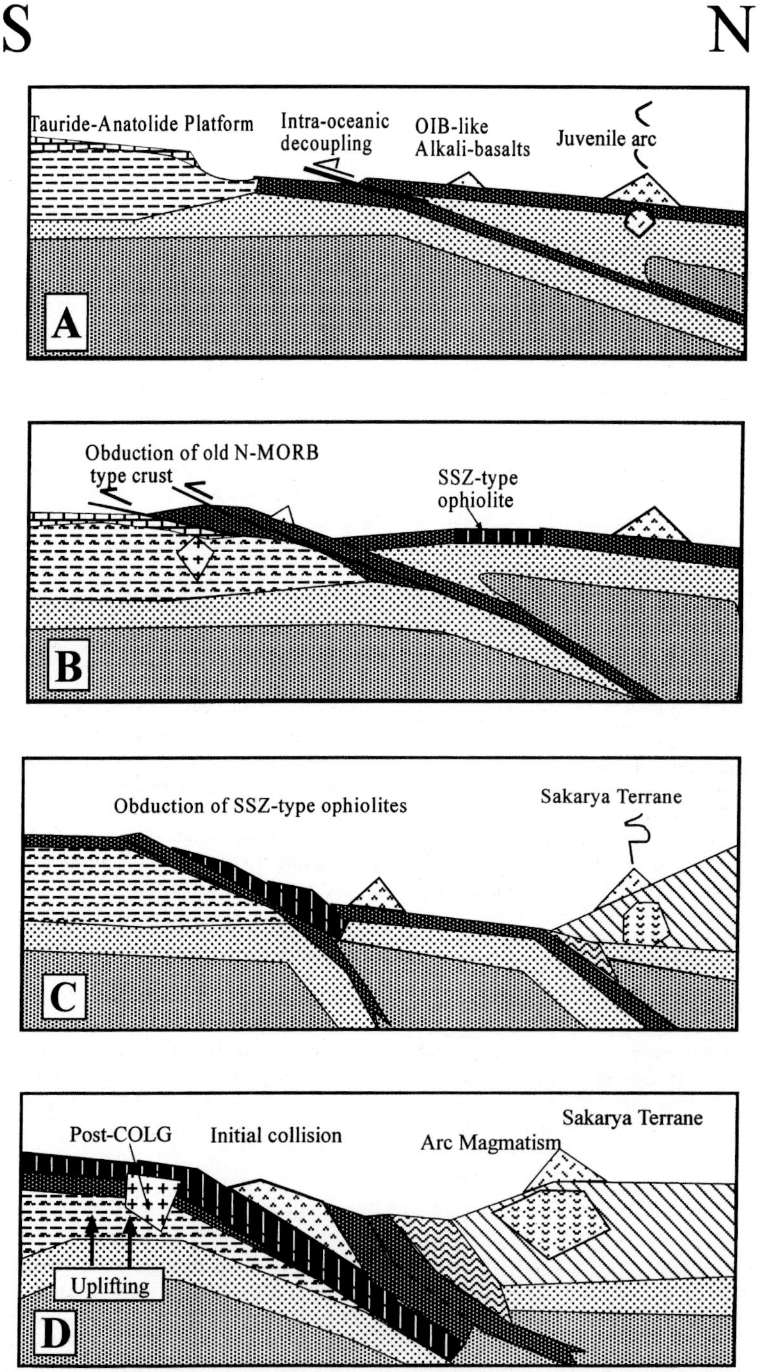

Fig. 8. Schematic representation of the tectonic evolution of the Central Anatolian ophiolites during the Late Mesozoic; (**a**) > 90 Ma: intra-oceanic decoupling within the İzmir–Ankara branch of Neotethys and the formation of an ensimatic arc; (**b**) 90–80 Ma: emplacement of old MORB-type oceanic crust onto the Tauride–Anatolide passive margin resulting in deformation and formation of SSZ-type Central Anatolian ophiolites in an arc-related setting; (**c**) 80–70 Ma: emplacement of SSZ-type ophiolites onto the CACC; (**d**) 70–65 Ma: initial collision of the arc, post-collisional extension and intrusion of post-collisional granites.

the CACC margin with Late Cretaceous post-collisional intrusions that cut both the CACC and the obducted SSZ-type ophiolites.

In conclusion, based on the geological and geochemical data, it is suggested that the fragmented Central Anatolian ophiolites, with their distinct SSZ geochemical features, were generated above an intra-oceanic subduction zone within the converging İzmir–Ankara branch of the Neotethys and emplaced southwards onto the passive margin of the CACC.

Funding for this collaborative project was provided by NATO (grant CRG 960549) and the Scientific and Research Council of Turkey (TÜBİTAK; project no. YDABÇAG-85) which are gratefully acknowledged. Analytical data production was helped by D. W. Emley and M. Aikin, Department of Earth Sciences, University of Keele, UK.

References

BECCALUVAL, L. & SERRI, G. 1988. Boninitic and low-Ti subduction-related lavas from intraocenic arc–backarc system and low-Ti ophiolites: a reappraisal of their petrogenesis and original tectonic setting. *Tectonophysics*, **146**, 291–315.

BROUXEL, M., LECUYER, C. & LAPIERRE, H. 1989. Diversity of Magma types in a Lower Palaeozoic island arc/marginal basin system (Eastern Klamath Mountains, California, U.S.A.). *Chemical Geology*, **77**, 251–264.

COISH, R. A. 1977. Ocean floor metamorphism in the Betts Cove ophiolite, Newfoundland. *Contributions to Mineralogy and Petrology*, **60**, 255–270.

CRAWFORD, A. J., BECCALUVA, L., SERRI, G. & DOSTAL, J. 1986. Petrology, geochemistry and tectonic implications of volcanics dredged from the intersection of the Yap and Mariana trenches. *Earth and Planetary Science Letters*, **80**, 265–280.

ERDOĞAN, B., AKAY, E. & UĞUR, M. S. 1996. Geology of the Yozgat Region and evolution of the collisional Çankırı Basin. *International Geology Review*, **38**, 788–806.

ERLER, A. & GÖNCÜOĞLU, M. C. 1996. Geologic and tectonic setting of the Yozgat Batholith, Northern Central Anatolian Crystalline Complex, Turkey. *International Geology Review*, **38**, 714–726.

EWART, A., BROTHERS, R. N. & MATEEN, A. 1977. An outline of the geology and geochemistry, and the possible petrogenetic evolution of the volcanic rocks of the Tonga–Kermadec–New Zealand island arc. *Journal of Volcanology and Geothermal Research*, **2**, 205–250.

FLOYD, P. A. 1993. Geochemical discrimination and petrogenesis of alkali basalt sequences in part of the Ankara Mélange, central Turkey. *Journal of the Geological Society, London*, **150**, 541–550.

—— & CASTILLO, P. R. 1992. Geochemistry and petrogenesis of Jurassic ocean crust basalts, ODP Leg 129, Site 801. *In*: LARSON, R. &

LAUNCELOT, Y. *ET AL.* (eds) *Proceedings of ODP, Scientific Results*, **129**, 361–388.

——, YALINIZ, M. K. & GÖNCÜOĞLU, M. C. 1998a. Geochemistry and petrogenesis of intrusive and extrusive ophiolitic plagiogranites, Central Anatolian Crystalline Complex, Turkey. *Lithos*, **42**, 225–241.

——, WINCHESTER, J. A., GÖNCÜOĞLU, M. C., YALINIZ, M. K. & PARLAK, O. 1998b. Geochemical overview and relations of metabasites from the Central Anatolian Crystalline Complex. *Proceedings of the Third International Turkish Geology Symposium*, METU–Ankara, Abstracts, 171.

GÖNCÜOĞLU, M. C. 1981. Origin of the viridine–gneiss in the Niğde Massif. *Geological Society of Turkey Bulletin*, **24**, 45–51.

—— 1986. Geochronological data from the southern part (Niğde Area) of the Central Anatolian Massif. *Mineral Research and Exploration Institute of Turkey (MTA) Bulletin*, **105/106**, 83–96.

—— & TÜRELİ, K. 1993. Petrology and geodynamic setting of plagiogranites from Central Anatolian Ophiolites (Aksaray, Turkey). *Turkish Journal of Earth Sciences*, **2**, 195–203.

——, TOPRAK, V., ERLER, A. & KUŞCU, İ. 1991. Geology of western Central Anatolian Massif, Part I, southern areas. Turkish Petroleum Corporation (TPAO), Report No. 2909 [in Turkish].

GÖRÜR, N., OKTAY, F. Y., SEYMEN, İ. & ŞENGÖR, A. M. C. 1984. Palaeotectonic evolution of Tuz Gölü Basin complex, central Turkey. *In*: DIXON, J. E. & ROBERTSON, A. H. F. (eds) *The Geological Evolution of the Eastern Mediterranean*. Geological Society, London, Special Publications, **17**, 81–96.

HAWKESWORTH, C. J., GALLAGHER, K., HERGT, J. M. & MCDERMOTT, F. 1993. Mantle and slab contributions in arc magma. *Annual Reviews of Earth and Planetary Sciences*, **1**, 175–204.

——, O'NIONS, R. K., PANKHURST, R. J., HAMPTON, P. J. & EVENSEN, N. M. 1977. A geochemical study of island-arc and back-arc tholeiites from the Scotia Sea. *Earth and Planetary Science Letters*, **36**, 253–267.

HICKEY, R. L. & FREY, F. A. 1981. Rare-earth element geochemistry of Mariana fore-arc volcanics: Deep-Sea Drilling Project Site 458 and Hole 459B. *Initial Reports of the Deep-Sea Drilling Project*, **60**, 735–741.

HUMPHRIS, S. E. & THOMPSON, G. 1978a. Hydrothermal alteration of oceanic basalts by seawater. *Geochimica et Cosmochimica Acta*, **42**, 107–125.

—— & —— 1978b. Trace element mobility during hydrothermal alteration of oceanic basalts. *Geochimica et Cosmochimica Acta*, **42**, 127–136.

KAY, R. W. 1977. Geochemical constraints on the origin of Aleutian magmas. *In*: TALWANI, M. & PITMAN, W. C. (eds) *Island Arcs, Deep Sea Trenches and Back-Arc Basins*. American Geophysical Union, Washington, DC, 229–242.

KETIN, İ. 1959. Uber Alter und Art der kristallinen Gesteine und Erzlagerstatten in Zentral-Anatolien. *Berg und Hüttenmaennschen Monatshefte*, **104**, 163–169.

MCCULLIOCH, M. T. & GAMBLE, J. A. 1991.

Geochemical and geodynamical constraints on subduction zone magmatism. *Earth and Planetary Science Letters*, **102**, 358–374.

OKAY, A. İ., HARRIS, N. B. W. & KELLEY, S. P. 1998. Exhumation of blueschists along a Tethyan suture in northwest Turkey. *Tectonophysics*, **285**, 275–299.

PEARCE, J. A. 1975. Basalt geochemistry used to investigate past tectonic environments on Cyprus. *Tectonophysics*, **25**, 41–67.

—— 1982. Trace element characteristics of lavas from destructive plate boundaries. *In*: THORPE, R. S. (ed.) *Andesites: Orogenic Andesites and Related Rocks*. John Wiley, Chichester, 525–548.

—— 1983. Role of the subcontinental margins. *In*: HAWKESWORTH, C. J. & NORRY, M. J. (eds) *Continental Basalts and Mantle Xenoliths*. Shiva Publishing, Nantwich, 230–249.

—— & CANN, J. R. 1973. Tectonic setting of basic volcanic rocks determined using trace element analyses. *Earth and Planetary Science Letters*, **19**, 290–300.

—— & NORRY, M. J. 1979. Petrogenetic implications of Ti, Zr, Y, and Nb variations in volcanic rocks. *Contributions to Mineralogy and Petrology*, **69**, 33–47.

—— & PARKINSON, I. J. 1993. Trace element models for mantle melting: applications to volcanic arc petrogenesis. *In*: PRICHARD, H. M., ALABASTER, T., HARRIS, N. B. W. & NEARY, C. R. (eds) *Magmatic Processes and Plate Tectonics*. Geological Society, London, Special Publications, **76**, 373–403.

——, LIPPARD, S. J. & ROBERTS, S. 1984. Characteristics and tectonic significance of suprasubduction zone ophiolites. *In*: KOKELAAR, B. P. & HOWELLS, M. F. (eds) *Marginal Basin Geology*. Geological Society, London, Special Publications, **16**, 77–94.

ROBERTSON, A. H. F. & DIXON, J. E. 1984. Introduction: aspects of the geological evolution of the eastern Mediterranean. *In*: DIXON, J. E & ROBERTSON, A. H. F. (eds) *The Geological Evolution of the Eastern Mediterranean*. Geological Society, London, Special Publications, **17**, 1–74.

SAUNDERS, A. D. & TARNEY, J. 1984. Geochemical characteristics of basaltic volcanism within back-arc basin. *In*: KOKELAAR, B. P. & HOWELLS, M. F. (eds) *Marginal Basin Geology*. Geological Society, London, Special Publications, **16**, 59–76.

——, MARSH, N. G. & WOOD, D. A. 1980. Ophiolites as ocean crust or marginal basin crust: a geochemical approach. *In*: PANAYIOTOU, A. (ed.) *Ophiolites*. Proceedings of an International Symposium – 1979, Cyprus. Cyprus Geological Survey Department, 193–204.

ŞENGÖR, A. M. C. & YILMAZ, Y. 1981. Tethyan evolution of Turkey: a plate tectonic approach. *Tectonophysics*, **75**, 181–241.

SHERVAIS, J. W. 1982. Ti–V plots and the petrogenesis of modern and ophiolitic lavas. *Earth and Planetary Science Letters*, **59**, 101–118.

SMITH, R. E. & SMITH, S. E. 1976. Comments on the use of Ti, Zr, Y, Sr, K, P and Nb in classification of basaltic magmas. *Earth and Planetary Science Letters*, **32**, 114–120.

SUN, S. S. & MCDONOUGH, W. F. 1989. Chemical and isotopic systematics of oceanic basalts: implications for mantle composition and processes. *In*: SAUNDERS, A. D. & NORRY, M. J. (eds) *Magmatism in Ocean Basins*. Geological Society, London, Special Publications, **42**, 313–345.

TÜYSÜZ, O., DELLALOĞLU, A. A. & TERZİOĞLU, N. 1995. A magmatic belt within the Neo-Tethyan suture zone and its role in the tectonic evolution of northern Turkey. *Tectonophysics*, **243**, 173–191.

WHITE, W. M. & PATCHETT, P. 1984. Hf–Nd–Sr and incompatible element abundances in island arcs: implications for magma origins and crust–mantle evolution. *Earth and Planetary Science Letters*, **67**, 167–185.

WINCHESTER, J. A. & FLOYD, P. A. 1977. Geochemical discrimination of different magma series and their differentiation products using immobile elements. *Chemical Geology*, **20**, 325–343.

WOOD, D. A. 1980. The application of a Th–Hf–Ta diagram to problems of tectonomagmatic classification and to establishing the nature of crustal contamination of basaltic lavas of the British Tertiary volcanic province. *Earth and Planetary Science Letters*, **50**, 11–30.

——, JORON, J. L. & TREUIL, M. 1979. A re-appraisal of the use of trace elements to classify and discriminate between magma series erupted in different tectonic settings. *Earth and Planetary Science Letters*, **45**, 326–336.

YALINIZ, M. K. 1996. *Petrology of the Sarıkaraman Ophiolite (Aksaray–Turkey)*. PhD Thesis, Middle East Technical University.

—— & GÖNCÜOĞLU, M. C. 1998. General geological characteristics and distribution of the Central Anatolian Ophiolites. *Hacettepe University Earth Sciences*, **20**, 1–12.

——, FLOYD, P. A. & GÖNCÜOĞLU, M. C. 1996. Suprasubduction zone ophiolites of central Anatolia: geochemical evidence from the Sarıkaraman ophiolite, Aksaray, Turkey. *Mineralogical Magazine*, **60**, 697–710.

——, PARLAK, O., ÖZKAN-ALTINER, S. & GÖNCÜOĞLU, M. C. 1997. Formation and emplacement ages of supra-subduction-type ophiolites in Central Anatolia; Sarıkaraman Ophiolites, Central Turkey. *Proceedings of the Çukurova University, 20th Annual of Geological Education*, Adana, Abstracts, 61–62.

YILMAZ-ŞAHİN, S. & BOZTUĞ, D. 1997. Petrography and whole rock chemistry of the gabbroic, monzogranitic and syenitic rocks from the Çiçekdağ region, N Kırşehir, Central Anatolia, Turkey. *In*: BOZTUĞ, D., YILMAZ-ŞAHİN, S., OTLU, N. & TATAR, S. (eds) *Alkaline Magmatism, Theoretical Considerations and a Field Excursion in Central Anatolia. Proceedings.* TÜBİTAK-BAYG/NATO-D Program-Cumhuriyet University Publication, 29–42.

Suprasubduction zone origin of the Pozantı–Karsantı Ophiolite (southern Turkey) deduced from whole-rock and mineral chemistry of the gabbroic cumulates

OSMAN PARLAK,[1,2] VOLKER HÖCK[2] & MICHEL DELALOYE[3]

[1]Çukurova University, Department of Geological Engineering, TR-01330 Adana, Turkey (e-mail: parlak@mail.cu.edu.tr)
[2]University of Salzburg, Department of Geology and Palaeontology, A-5020 Salzburg, Austria
[3] University of Geneva, Department of Mineralogy, CH-1211 Geneva, Switzerland

Abstract: The Pozantı-Karsantı Ophiolite Complex is situated in the eastern Tauride Belt and represents a remnant of the Mesozoic Neotethyan Ocean. It consists of three distinct nappes: (1) an ophiolitic mélange; (2) a metamorphic sole; and (3) ophiolitic rocks. The oceanic lithosphere section of the Pozantı–Karsantı Ophiolite comprises mantle tectonites, ultramafic–mafic cumulates, isotropic gabbros, sheeted dykes and basaltic volcanic rocks. These units are cut by isolated microgabbro–diabase dykes at all structural levels. New results are presented on the whole-rock and mineral chemistry of the gabbroic cumulates. Well-layered, low-Ti gabbroic cumulates, showing adcumulate to mesocumulate textures, are represented exclusively by gabbronorites. The mineral chemistry of gabbronorites from the Pozantı–Karsantı Ophiolite indicates that these cumulate rocks have been produced by the low-pressure crystal fractionation of basaltic liquid. Magnesium numbers (Mg-numbers) of clinopyroxene, orthopyroxene and amphibole range from 89 to 73, 80–66 and 80–72, respectively. Plagioclase compositions range from An_{94} to An_{84}. The coexistence of calcic plagioclase, magnesian clinopyroxene and orthopyroxene indicates that the cumulate gabbronorites from the Pozantı–Karsantı Ophiolite were formed in an arc environment. The covariation of Al_2O_3 and Mg-numbers of both clinopyroxene and orthopyroxene show features typical of low-pressure igneous intrusions such as the Skaergaard and Tonsina Complexes, but differ from the high-pressure ultramafic cumulates found in the same arc. The cumulate gabbronorites probably represent shallower levels in the arc which were subsequently juxtaposed against deeper level ultramafic cumulates either during accretion or later faulting.

Mesozoic ophiolites in southern Turkey are located either along the continent–continent collision zone (Bitlis–Zagros Suture Zone) or on both sides of the Tauride calcareous axis (Ricou 1971; Juteau 1980; Şengör & Yılmaz 1981; Dilek & Moores 1990) (Fig. 1a). These ophiolites were subsequently obducted onto the passive Gondwanaland continental margin as a result of closure of the Neotethyan Ocean during the Late Cretaceous. The Tauride ophiolites, resting in tectonic contact with underlying platform carbonates, are generally associated with an ophiolitic mélange and metamorphic sole (Juteau 1980; Dilek & Moores 1990; Parlak & Delaloye 1999).

The ophiolites in the western Alps differ from the eastern Mediterranean equivalents in terms of their age and geodynamic environment of formation. The Jurassic ophiolites in the western Alps display mid-ocean ridge basalt (MORB)-like composition (Koller & Höck 1990), whereas the Upper Cretaceous ophiolitic bodies of Troodos in Cyprus, Baer-Bassit in Syria, and Hatay, Mersin, Pozantı–Karsantı and Sarıkaraman in Turkey, suggest a suprasubduction zone tectonic setting (Miyashiro 1973; Pearce et al. 1984; Hebert & Laurent 1990; Robertson 1994; Lytwyn & Casey 1993, 1995; Parlak et al. 1996; Yalınız et al. 1996; Parlak & Höck 1998). On the other hand, the Jurassic ophiolites along the Hellenic–Dinaric Belt exhibit both MORB and island arc tholeiite (IAT) character (Smith 1993).

Well-preserved ultramafic–mafic cumulates in the eastern Mediterranean ophiolites, namely Oman (Pallister & Hopson 1980, 1981; Browning 1984), Troodos (Thy 1987a, b, 1990; Hebert & Laurent 1990) in Cyprus, and Hatay (Pişkin et al. 1990), Mersin (Parlak et al. 1996, 1997), Antalya (Juteau & Whitechurch 1980) and Pozantı–Karsantı (Parlak & Höck 1998) in Turkey, show evidence of high-pressure crystal

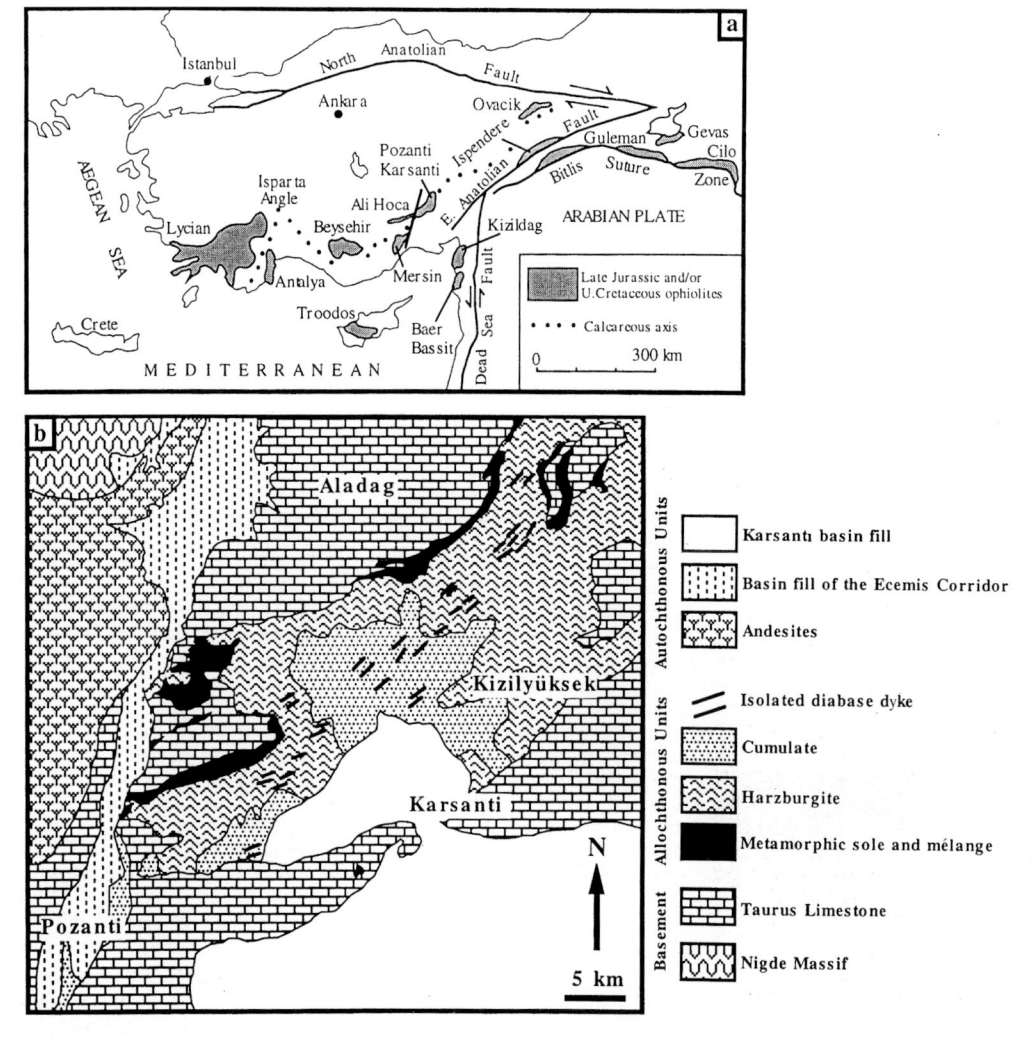

Fig. 1. (a) Distribution of ophiolite bodies in southern Turkey [modified from Dilek & Moores (1990)]. (b) Geological map of the Pozantı–Karsantı Ophiolite (Bingöl 1978).

fractionation beneath an island-arc tectonic setting. Knowledge of igneous processes occurring in an island-arc setting is very important in understanding lower and upper crustal growth (DeBari & Coleman 1986, 1989). Xenoliths of gabbroic cumulates have been reported from well-known volcanic arcs, namely the Aleutian islands (Conrad & Kay 1984), New Guinea (Gust & Johnson 1981), the Philippines (Newhall 1979), Lesser Antilles (Arculus & Wills 1980), Japan (Aoki & Kuno 1971), Central America (Walker 1984), Indonesia (Morrice *et al.* 1983) and the Marianas (Stern 1979; Meijer & Reagan 1981). Beard (1986) documented xenoliths of gabbroic cumulates from plutonic rocks in eroded arcs and classified them into three groups: (1) olivine–gabbro; (2) gabbro-norite; and (3) amphibole–gabbro.

The scope of this study is to present whole-rock and mineral chemistry of the gabbronorites from the cumulates of the Pozantı–Karsantı Ophiolite and evaluate their geodynamic environment of formation within the eastern Mediterranean tectonic frame.

Geology of the Pozantı–Karsantı Ophiolite

The Pozantı–Karsantı Ophiolite, exposed in the eastern Tauride Belt (Aladağ region) in southern Turkey, is one of the largest Late

Cretaceous oceanic lithospheric remnants (Juteau 1980; Dilek & Moores 1990; Polat & Casey 1995) (Fig. 1b). It covers an area of $c.$ 1300 km^2 (80 km in length and 25 km in width) between the left-lateral Ecemiş Strike-slip Fault to the west and the left-lateral East Anatolian Strike-slip Fault to the east (Bingöl 1978; Çakır 1978; Çataklı 1983; Tekeli et al. 1983; Polat & Casey 1995) (Fig. 1b).

An imbricated stack of thrust sheets (upper and lower) resting upon the eastern Taurus Autochthon is exposed in the Aladağ region (Blumenthal 1947; Polat & Casey 1995). The upper thrust sheets are characterized by ophiolite-related rock assemblages, namely ophiolitic mélange, dynamothermal metamorphic sole and oceanic lithospheric section (Fig. 2); whereas the lower thrust sheets consist of platform-type carbonates, ranging in age from Late Devonian to Early Cretaceous (Tekeli et al. 1983). The Upper Campanian–Maastrichtian unmetamorphosed ophiolitic mélange is composed of a variety of igneous, metamorphic and sedimentary blocks structurally dispersed in a serpentinitic to pelitic matrix (Tekeli et al. 1983; Polat & Casey 1995). The mid-Upper Cretaceous dynamothermal metamorphic sole displays a typical inverted metamorphic sequence, grading from amphibolite facies directly beneath the highly sheared harzburgitic tectonite to lower greenschist facies near the mélange contact (Thuizat et al. 1978; Lytwyn & Casey 1995). The Pozantı–Karsantı Ophiolite is composed of harzburgitic to dunitic tectonites, ultramafic and mafic cumulates, isotropic gabbro, sheeted dykes and pillow lavas (Bingöl 1978; Çakır 1978; Çataklı 1983). Swarms of microgabbro and diabase dykes cut the ophiolitic units at different structural levels as well as the metamorphic sole. The dykes are observed not to cut the underlying mélange and Tauride platform rocks, indicating that some dyke emplacement postdated formation of the ophiolite and the metamorphic sole, but predated its final obduction onto the Tauride Platform (Çakır et al. 1978; Thuizat et al. 1981; Lytwyn & Casey 1995).

Petrography

The gabbroic cumulates in the studied section of the Pozantı–Karsantı Ophiolite are composed almost exclusively of gabbronorite (two-pyroxene gabbro). The gabbronorite exhibits adcumulate–mesocumulate textures and shows magmatic accumulation features such as igneous lamination, crystal size grading and rhythmic layering. Cumulus orthopyroxene (En$_{65-79}$Fs$_{19-34}$Wo$_{1-3}$), with grain sizes of between 0.4 and 2.6 mm, forms 15–30% of the gabbroic rocks as euhedral–subhedral crystals and is roughly equal in abundance to clinopyroxene. Clinopyroxene (En$_{40-48}$ Fs$_{6-16}$ Wo$_{41-47}$), with a grain size of between 0.4 and 2 mm, is generally present as subhedral–anhedral crystals and forms 20–30% of the gabbronorite. Plagioclase typically forms 40–60% of the gabbronorite and is present as unzoned cumulus or intercumulus grains. Minor amounts (commonly < 5%) of edenitic amphibole with high magnesium numbers, (Mg-number; 0.70) are present in the gabbronorites. Olivine is virtually absent in the studied section whereas olivine-rich gabbroic cumulates are documented in the western part of the massif (Çataklı 1983).

Isotropic gabbro in the study area is represented by plagioclase (40–55%), clinopyroxene

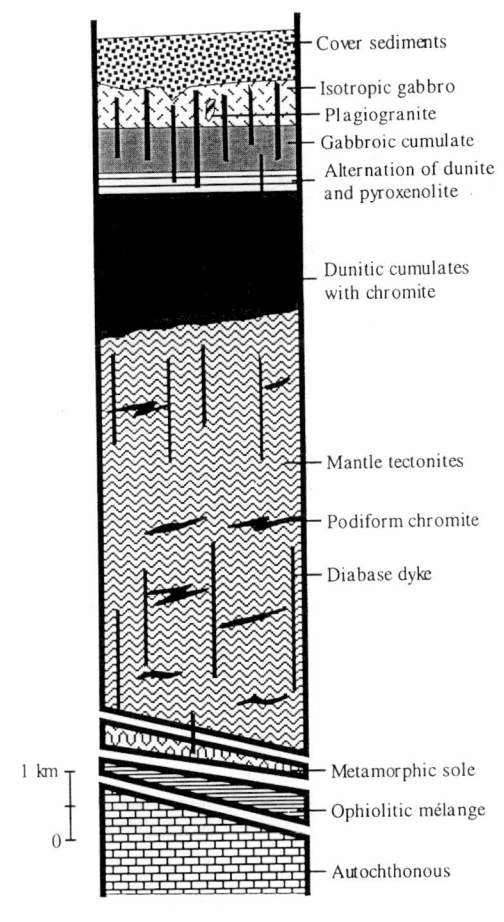

Fig. 2. Synthetic log of the Pozantı–Karsantı Ophiolite (Bingöl 1978).

(25-35%) and amphibole (10-15%), and exhibits a doleritic-subophitic texture. The primary mineral assemblages have been modified by hydrothermal alteration processes characterized by secondary albite, hornblende, epidote and chlorite.

Geochemistry

Analytical method

A total of 20 samples from the cumulate gabbronorites and five samples from the isotropic gabbros were analysed for major and trace element contents. Rare earth element (REE) analyses were carried out on three plutonic rocks by the inductively coupled plasma-atomic emission spectroscopy (ICP-AES) technique (Voldet 1993). Major and trace element analyses were carried out at the University of Geneva. Major elements were determined by X-ray fluorescence (XRF) spectroscopy on glass beads fused from ignited powders to which $Li_2B_4O_7$ was added (1:5), in a gold-platinum crucible, at 1150°C. Trace elements were analysed on pressed powder pellets by the same method. A total of 20 representative polished sections were used for electron microprobe analysis on a JEOL JXA-8600 in the Geology and Palaeontology Department at Salzburg University. The analytical conditions for the elements were 13 s (10 s for peak and 3 s for background) counting interval, a beam current of -20 nA and on acceleration voltage of -15 kV.

Whole rock

Major and trace element concentrations of the cumulate gabbros and the isotropic gabbros are given in Table 1. The cumulate gabbronorites are represented by low TiO_2 (0.09-1.32%), K_2O (0-0.02%), P_2O_5 (0.01-0.03%), large-ion lithophile (LIL) elements such as Ba (13-33 ppm), Rb (10-11 ppm), other incompatible elements, i.e. Zr (4-8 ppm), Y (6-16 ppm), Nb (2-5 ppm), and REE. In contrast, the concentrations of Al_2O_3 (9.4-20.8%), CaO (11.8-15.1%) and MgO (6.2-18.0%) are high (Table 1). The low concentrations of the incompatible trace elements are attributed to the high proportion of cumulate minerals and low amounts of intercumulus-trapped liquid in the magma chamber of the Pozantı-Karsantı Ophiolite. Parlak et al. (1996) reported similar observations for the Mersin Ophiolite in southern Turkey. The isotropic gabbros are characterized by higher concentrations of major and minor elements, such as TiO_2 (0.55-1.36%), K_2O (0.09-0.37%), P_2O_5 (0.09-0.13%), and incompatible trace elements, compared to the cumulate gabbros (Table 1). Low Nb/Y ratios in the isotropic gabbros (0.04-0.1) and in the cumulate gabbronorites (0.1-0.5) are characteristic of subalkaline (tholeiitic) basalts (Winchester & Floyd 1977).

Major element compositions of the cumulate gabbros and isotropic gabbros are plotted on the AFM diagram of Beard (1986). The gabbronorites plot in the arc-related mafic cumulate field, whereas the isotropic gabbros plot in the arc-related non-cumulate field (Fig. 3). This implies that both the cumulate and non-cumulate gabbroic rocks formed in a suprasubduction zone tectonic setting.

REE concentrations of two isotropic gabbros and one cumulate gabbronorite are given in Table 2. They exhibit slightly light rare earth element (LREE) depleted to flat patterns (La/Yb_n = 0.8-1.1) and overall REE abundances of between 5X and 14X chondritic (Fig. 4), suggesting slightly differentiated comagmatic tholeiitic suites. This reveals that the plutonic rocks in the Pozantı-Karsantı Ophiolite are compositionally similar to tholeiitic basalts and basaltic andesites found in modern island-arc environments.

Mineral chemistry

Clinopyroxene. Representative clinopyroxene compositions from cumulate gabbronorites are given in Table 3. The clinopyroxenes show limited compositional variations (En_{40-48} $Wo_{41-47}Fs_{6-16}$) and plot in the diopside field of the pyroxene ternary diagram (Fig. 5). On the basis of clinopyroxene compositions, the gabbronorites from the Pozantı-Karsantı Ophiolite are consistent with the gabbroic rocks formed in an island-arc tectonic setting (Fig. 5). Mg-numbers of the clinopyroxenes reach as high as 89 at the stratigraphic base and diminish to 75 at the upper levels.

The Cr_2O_3 in clinopyroxene is plotted v. Mg-number in Fig. 6a; the cumulate gabbronorites exhibit a trend of decreasing Cr_2O_3 with decreasing Mg-number. The clinopyroxenes plot in the field of low-pressure Cpx derived from 1 atm experimental studies of N-type MORB (Elthon 1987). The Cr_2O_3 content and Mg-number of clinopyroxenes from the ultramafic cumulates are characterized by higher values, consistent with the ultramafic cumulates that formed in a high-pressure environment

Table 1. Major and trace element analyses of the cumulate and isotropic gabbro from the Pozantı–Karsantı Ophiolite (southern Turkey)

Sample	H-30	H-31	H-37	Y-1	Y-2	Y-4	Y-7	Y-9	Y-10	Y-11	Y-12	Y-13	Y-14	Y-15	Y-16	Y-17	Y-18	Y-19	Y-20	Y-21	H-33	H-36	Y-3	Y-6	Y-8
									Cumulate gabbro														Isotropic gabbro		
SiO₂	50.03	50.78	41.57	47.96	50.39	47.75	47.06	47.07	44.96	46.90	48.58	50.30	50.04	49.17	48.70	46.46	43.38	50.42	47.47	47.40	50.77	50.73	49.98	54.56	51.99
TiO₂	0.22	0.19	1.32	0.30	0.16	0.11	0.12	0.10	0.61	0.17	0.29	0.21	0.19	0.18	0.13	0.09	0.67	0.16	0.09	0.17	1.36	1.17	1.17	0.55	0.94
Al₂O₃	11.06	10.32	17.80	11.78	9.90	17.74	17.86	20.83	18.12	19.35	12.04	7.68	9.40	11.49	16.75	19.04	18.62	6.17	19.11	17.34	15.65	15.14	15.18	14.52	15.23
FeO*	10.84	10.18	17.48	11.63	9.51	7.68	6.60	6.23	13.30	7.91	12.83	10.08	9.95	8.62	8.10	6.45	14.05	8.66	6.69	8.45	12.23	10.56	11.33	8.30	10.09
MnO	0.20	0.19	0.28	0.20	0.19	0.15	0.13	0.12	0.18	0.15	0.23	0.20	0.19	0.17	0.14	0.12	0.15	0.17	0.13	0.16	0.20	0.19	0.18	0.14	0.16
MgO	14.22	15.43	6.15	13.89	16.22	12.23	10.53	8.51	8.71	8.90	13.61	16.04	16.39	14.56	12.11	9.20	7.81	18.00	10.21	10.00	4.83	6.84	6.40	5.50	5.98
CaO	13.04	12.35	12.47	12.53	13.00	11.75	13.91	14.70	13.25	14.32	12.00	14.01	12.64	14.21	13.34	13.75	13.70	15.10	14.19	14.12	8.09	10.07	10.32	9.10	9.69
Na₂O	0.60	0.65	1.11	0.42	0.33	0.66	0.71	0.64	0.64	0.59	0.39	0.25	0.34	0.37	0.45	0.67	0.58	0.22	0.45	0.60	4.19	2.93	2.44	2.87	2.47
K₂O	0.00	0.00	0.02	0.00	0.00	0.01	0.00	0.01	0.00	0.01	0.00	0.01	0.00	0.00	0.00	0.01	0.00	0.00	0.01	0.01	0.37	0.15	0.18	0.09	0.18
P₂O₅	0.02	0.02	0.03	0.01	0.01	0.01	0.01	0.01	0.01	0.01	0.01	0.01	0.01	0.02	0.01	0.01	0.01	0.02	0.01	0.01	0.12	0.11	0.09	0.13	0.09
Cr₂O₃	0.20	0.19	0.01	0.09	0.20	0.11	0.11	0.02	0.02	0.05	0.08	0.22	0.22	0.15	0.07	0.01	0.02	0.30	0.01	0.01	0.01	0.02	0.01	0.02	0.01
NiO	0.02	0.03	0.00	0.02	0.03	0.02	0.03	0.01	0.00	0.01	0.02	0.02	0.03	0.02	0.02	0.01	0.01	0.05	0.01	0.01	0.00	0.01	0.01	0.00	0.01
LOI	0.00	0.41	1.33	1.26	0.80	1.53	2.44	0.87	0.56	0.94	0.31	1.27	0.85	0.62	0.68	4.48	0.89	1.01	1.09	1.16	2.27	2.10	2.90	3.62	3.35
Sum	100.45	100.73	99.56	100.09	100.73	99.71	99.50	99.13	100.27	99.31	100.39	100.30	100.25	99.57	100.49	100.29	99.90	100.26	99.47	99.45	100.08	100.02	100.20	99.39	100.21
Nb	2	2	1	1	3	1	1	1	1	1	1	5	4	1	1	1	1	6	1	1	3	2	1	3	1
Zr	8	8	6	6	6	5	6	5	6	6	7	7	7	6	6	4	6	8	4	6	84	76	58	84	57
Y	15	16	10	8	9	7	8	6	8	8	9	10	10	9	7	7	6	11	6	8	31	27	24	33	24
Sr	51	57	117	44	33	77	98	77	75	73	47	19	35	40	56	64	71	16	62	66	241	129	119	119	102
U	4	4	5	6	9	7	2	6	11	11	4	8	8	8	8	5	9	8	10	11	12	10	9	2	6
Rb	10	10	4	11	11	10	10	11	11	11	10	11	11	11	2	10	11	11	2	6	12	7	5	9	3
Th	4	2	5	6	7	2	2	2	2	2	4	4	4	2	2	2	2	2	2	2	2	2	2	2	12
Pb	24	24	18	27	27	17	13	13	22	17	27	29	29	27	20	9	21	6	10	19	6	10	14	10	12
Ga	15	15	19	15	16	16	16	16	16	16	15	16	15	15	15	16	17	34	15	16	20	12	14	18	18
Zn	97	94	134	59	59	56	41	42	74	46	76	59	64	55	61	49	74	60	55	56	103	18	18	73	79
Ni	155	205	11	133	229	99	177	72	13	59	148	175	186	164	108	50	57	437	59	53	14	95	81	26	30
Co	54	59	53	67	59	50	40	33	65	44	70	61	59	52	48	43	59	65	45	43	43	49	39	30	38
Cr	1458	1289	39	575	1377	617	723	134	44	301	459	1442	1491	982	445	64	106	1911	57	114	39	42	43	105	40
V	209	199	555	407	197	113	114	90	437	168	293	250	190	186	176	119	785	184	132	161	403	104	81	131	309
Ce	3	3	3	3	3	3	3	3	5	3	3	3	3	4	3	3	3	3	3	3	7	319	350	4	3
Nd	4	8	4	4	4	8	9	7	10	9	4	8	3	4	5	4	4	3	4	9	12	12	10	4	3
Ba	19	4	13	16	9	4	9	4	9	9	33	26	4	9	4	4	19	9	9	4	83	13	35	8	11
La	4	4	4	4	4	4	4	4	4	4	4	4	4	4	4	4	4	5	4	11	4	4	4	6	4
Hf	8	9	4	6	13	9	6	5	5	6	7	11	12	9	8	4	3	12	5	6	7	8	7	8	7
Sc	47	49	61	53	49	31	23	20	30	21	57	52	55	38	29	23	30	46	23	31	47	43	39	39	40
Mg-number	56.744	60.249	26.01	54.429	63.03	61.42	61.45	57.73	39.584	52.92	51.471	61.413	62.213	62.8	59.902	58.971	35.7	67.508	60.4	54.2	28.309	39.314	36.09	39.9	37.222

* Total Fe is expressed as FeO.

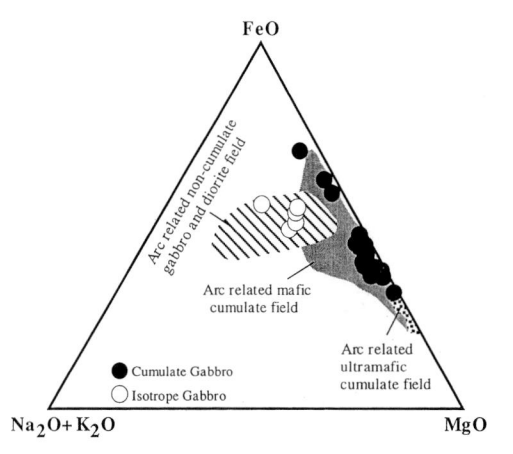

Fig. 3. AFM compositions of cumulate gabbronorites (filled circles) and isotropic gabbro (open circles) in the Pozantı–Karsantı Ophiolite. Fields of cumulate and non-cumulate rocks are from Beard (1986).

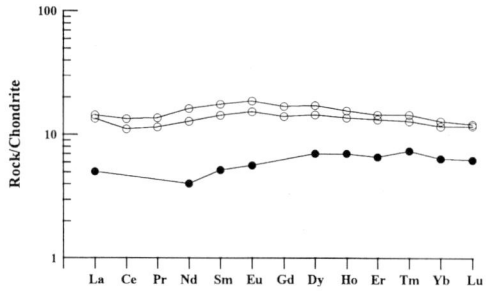

Fig. 4. REE patterns of the cumulate gabbronorite (filled circles) and isotropic gabbros (open circles) from the Pozantı–Karsantı Ophiolite. Normalizing values are from Sun & McDonough (1989).

such as Bay of Islands (Elthon et al. 1982) and Mersin (Parlak et al. 1996).

Elthon (1987) suggested that clinopyroxenes, formed at a moderate or higher pressure, also have high Al_2O_3 (> 3%) and TiO_2 (> 1%) contents. The Al_2O_3 and TiO_2 contents of the clinopyroxenes from the cumulate gabbronorites in the Pozantı–Karsantı Ophiolite are plotted in Fig. 6b. The data from the gabbronorites are represented by low TiO_2 (< 0.4%) and Al_2O_3 (< 2.5%), and plot within the field of low-pressure clinopyroxenes from the Semail Ophiolite. The Al_2O_3 content of clinopyroxenes in the gabbronorites is very stable and does not vary with increasing Mg-number. In Fig. 7b, the variation of Al_2O_3 v. Mg-number is perpendicular to the trend of a typical high-pressure field defined by Medaris (1972), but is parallel to the trend of typical low-pressure cumulates such as the Semail Ophiolite, the Skaergaard Intrusion and high-level gabbros from the Tonsina Complex (DeBari & Coleman 1989). All the evidence based on the clinopyroxene mineral chemistry suggest that the cumulate gabbronorites from the Pozantı–Karsantı Ophiolite formed under low-pressure conditions representative of shallow levels of an island arc.

Orthopyroxene. Representative orthopyroxene analyses from gabbronorites are given in Table 4. Orthopyroxene compositions are $En_{79-65}Wo_{3-1}Fs_{34-20}$. Mg-numbers vary between 80 and 66. Orthopyroxenes are major cumulus phases in the cumulate gabbronorites of the Pozantı–Karsantı Ophiolite. The Cr_2O_3 content of orthopyroxenes is negligible (< 0.2%) in gabbronorites, whereas in ultramafic cumulates the Cr_2O_3 content is higher (0.3%–0.5%), related to magmatic differentiation in the parent liquid. They are hypersthene in composition and are represented by high FeO (15–21%) and low Al_2O_3 (< 2%). The compositions of the orthopyroxenes are consistent with the compositions of gabbroic rocks that formed in an island arc environment (Fig. 5).

The Al_2O_3 content of orthopyroxenes are plotted v. Mg-numbers in Fig. 7a. As observed for the clinopyroxenes, the Al_2O_3 contents of the orthopyroxenes are very restricted in compositional range with continuous differentiation and almost parallel to the trend of well-known plutonic intrusions (Skaergaard and high-level gabbros from the Tonsina Complex),

Table 2. *REE results of cumulate and isotropic gabbros from the Pozantı–Karsantı Ophiolite (southern Turkey)*

Samples	Y-3	H-36	H-31
La	3.20	3.40	1.20
Ce	6.80	8.20	n.d.
Pr	1.10	1.30	n.d.
Nd	6.00	7.60	1.90
Sm	2.20	2.70	0.80
Eu	0.89	1.09	0.33
Gd	2.90	3.50	n.d.
Dy	3.70	4.40	1.80
Ho	0.78	0.89	0.40
Er	2.2	2.40	1.10
Tm	0.33	0.37	0.19
Yb	2.00	2.20	1.10
Lu	0.30	0.31	0.16

Y-3 and H-36 are isotropic gabbros; H-31 is cumulate gabbronorite.
n.d., Not determined.

Table 3. *Representative clinopyroxene analysis from the cumulate gabbronorites*

Sample	Y1-1c	Y2-1c	Y4-1c	Y7-1c	Y9-1c	10-1c	11-1c	12-1c	13-1c	14-1c	15-1c	16-1c	17-1c	18-1c	19-1c	20-1c	H31-1c	H31-2c
SiO_2	52.20	52.40	52.70	52.70	52.10	52.50	51.50	51.90	50.60	52.00	52.30	51.50	50.90	51.20	51.90	51.80	52.20	52.20
Al_2O_3	2.44	2.27	2.51	2.31	1.94	2.03	2.02	2.24	2.36	2.52	2.49	2.20	2.11	2.18	2.41	2.13	2.35	2.43
TiO_2	0.33	0.24	0.33	0.32	0.32	0.38	0.36	0.38	0.23	0.32	0.31	0.32	0.27	0.32	0.22	0.24	0.36	0.33
FeO*	8.50	6.50	7.60	7.00	7.30	7.60	8.50	8.60	7.40	7.50	6.70	7.20	8.20	8.80	5.60	7.30	7.30	7.30
MnO	15.10	15.10	15.80	15.10	14.80	15.10	14.40	14.40	15.00	15.70	15.30	15.40	15.70	14.60	16.20	15.30	14.70	14.70
MgO	15.10	15.10	15.80	15.10	14.80	15.10	14.40	14.40	15.00	15.70	15.30	15.40	15.70	14.60	16.20	15.30	14.70	14.70
Cr_2O_3	0.12	0.31	0.27	0.23	0.08	0.07	0.00	0.00	0.27	0.28	0.24	0.16	0.00	0.00	0.40	0.00	0.33	0.34
CaO	21.70	22.80	21.10	22.40	22.70	22.40	22.10	22.30	22.70	21.90	23.00	22.40	21.50	21.40	22.30	22.50	22.10	22.00
Na_2O	0.17	0.19	0.21	0.21	0.27	0.27	0.23	0.29	0.21	0.23	0.26	0.23	0.21	0.33	0.16	0.22	0.34	0.35
K_2O	0.00	0.00	0.00	0.00	0.00	0.00	0.00	0.00	0.00	0.00	0.00	0.00	0.00	0.00	0.00	0.00	0.00	0.00
Sum	100.86	99.98	100.72	100.42	99.68	100.56	99.34	100.36	98.96	100.65	100.80	99.55	99.12	99.15	99.36	99.69	99.84	99.85
Si	1.93	1.93	1.93	1.94	1.93	1.93	1.92	1.92	1.89	1.90	1.91	1.91	1.89	1.91	1.92	1.92	1.93	1.93
Al IV	0.07	0.07	0.07	0.06	0.07	0.07	0.08	0.08	0.11	0.10	0.09	0.09	0.11	0.09	0.08	0.08	0.07	0.07
Al VI	0.03	0.03	0.04	0.04	0.02	0.02	0.01	0.02	-0.01	0.01	0.02	0.01	-0.01	0.01	0.02	0.01	0.04	0.04
Ti	0.01	0.01	0.01	0.01	0.01	0.01	0.01	0.01	0.01	0.01	0.01	0.01	0.01	0.01	0.01	0.01	0.01	0.01
Fe^{3+}	0.03	0.02	0.02	0.01	0.05	0.05	0.06	0.06	0.11	0.07	0.06	0.08	0.12	0.08	0.05	0.08	0.03	0.02
Fe^{2+}	0.23	0.18	0.21	0.20	0.18	0.19	0.21	0.20	0.12	0.16	0.14	0.14	0.14	0.19	0.12	0.15	0.20	0.20
Mn	0.01	0.01	0.01	0.00	0.01	0.01	0.01	0.01	0.01	0.01	0.01	0.00	0.01	0.01	0.01	0.01	0.01	0.01
Mg	0.83	0.83	0.86	0.83	0.82	0.83	0.80	0.79	0.83	0.86	0.83	0.85	0.87	0.81	0.89	0.84	0.81	0.81
Cr	0.00	0.01	0.01	0.01	0.00	0.00	0.00	0.00	0.01	0.01	0.01	0.00	0.00	0.00	0.01	0.00	0.01	0.01
Ca	0.85	0.90	0.83	0.88	0.90	0.88	0.88	0.88	0.91	0.86	0.90	0.89	0.86	0.86	0.88	0.89	0.88	0.87
Na	0.01	0.01	0.01	0.01	0.02	0.02	0.02	0.02	0.02	0.02	0.02	0.02	0.02	0.02	0.01	0.02	0.02	0.03
K	0.00	0.00	0.00	0.00	0.00	0.00	0.00	0.00	0.00	0.00	0.00	0.00	0.00	0.00	0.00	0.00	0.00	0.00
Ens	42.58	42.87	44.71	42.89	41.92	42.44	10.93	40.69	42.17	43.91	42.86	43.24	43.76	41.64	45.68	42.88	42.28	42.34
Fors	13.45	10.62	12.38	11.39	11.87	12.32	13.92	14.06	11.97	12.08	10.84	11.56	13.18	14.50	9.13	11.80	12.04	12.12
Wol	43.97	46.51	42.91	45.72	46.21	45.24	45.14	45.28	45.86	44.01	46.30	45.20	43.06	43.86	45.19	45.32	45.68	45.54
Mg-number	76.43	80.55	78.75	79.36	78.33	77.98	75.13	74.91	78.33	78.87	80.28	79.22	77.34	74.73	83.76	78.89	78.21	78.21

Number of ions on the basis of 6(O).
*Total Fe is expressed as FeO.
c indicates core.

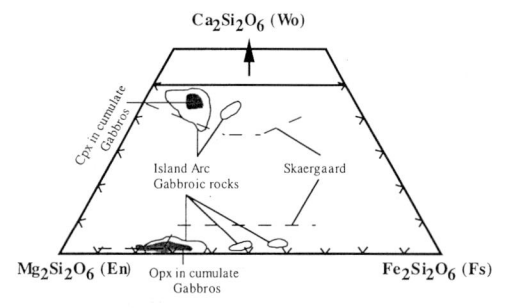

Fig. 5. Pyroxene ternary diagram showing clinopyroxene and orthopyroxene compositions from the gabbronorites. Fields of island-arc gabbroic rocks and Skaergaard trends are from Burns (1985).

interpreted as low-pressure igneous rocks (Fig. 7a). This implies that the gabbronorites were formed at shallow levels, or under low pressures, in an island arc.

Plagioclase. Representative plagioclase analyses from cumulate gabbronorites are given in Table 5. Plagioclase is not observed in the basal ultramafic cumulates and appears as a major cumulus phase in the gabbronorites (Parlak & Höck 1998). The plagioclase compositions are highly calcic, ranging from An_{94} to An_{84}, and do not show variations in individual grains from core to rim (Table 5). Highly calcic plagioclase, and its limited compositional range in the gabbronorites from the Pozantı–Karsantı Ophiolite, is not consistent with gabbros formed at mid-oceanic ridges (Burns 1985; Hebert 1982), whereas they are comparable with gabbroic rocks forming in a suprasubduction zone tectonic environment. Similar observations have been published from cumulate gabbros of different ophiolite complexes (Mersin Ophiolite – Parlak et al. 1996, 1997; Troodos Ophiolite – Hebert & Laurent 1990) and arc-related igneous rocks (Arculus & Wills 1980; Dupuy et al. 1982; Beard 1986; DeBari & Coleman 1986; Fujimaki 1986; DeBari et al. 1987).

Amphibole. Representative amphibole analyses are given in Table 6. Amphiboles are present as post-cumulus phases in the cumulate gabbronorites of the Pozantı–Karsantı Ophiolite. Amphiboles, usually forming 2–3% of the rock volume, are characterized by edenite based on the Leake classification (Leake 1978). Edenite and edenitic hornblende are reported from the Peninsular Range Batholith (Smith et al. 1983). Mg-numbers of the amphiboles ranges from 80 to 72; amphiboles in the gabbronorites are characterized by low Ti (0.05–0.2) contents. Green & Ringwood (1968) suggested that formation of low Ti amphiboles requires equilibrium with Ti-depleted magma which contains high amounts of water. Hydrous magmas are related to arc regions where amphiboles are commonly formed (Jakes & White 1972; Arculus & Wills 1980; Foden 1983; Hebert & Laurent 1990), whereas the water content of ocean floor basalt magmas is very low (0.3%) (Moore 1970).

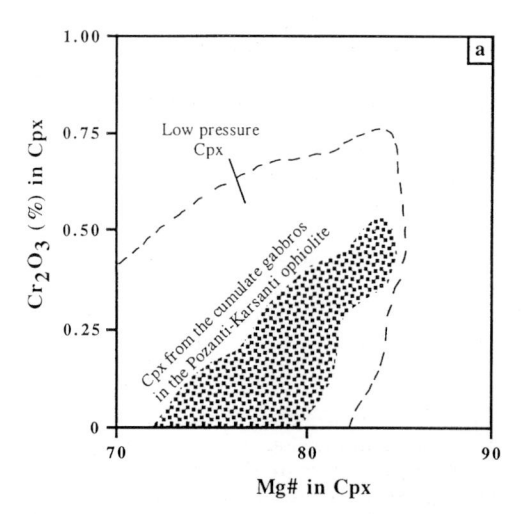

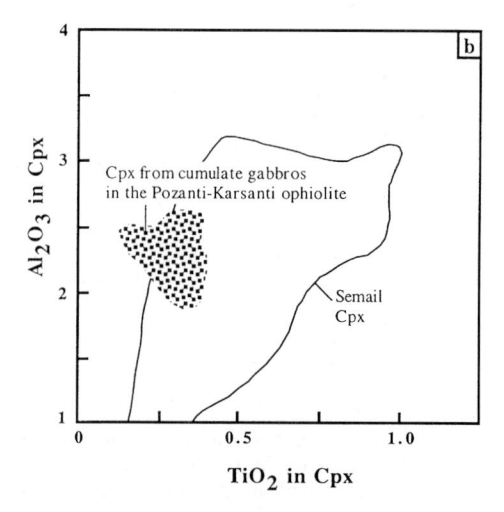

Fig. 6. (a) Cr_2O_3 v. Mg-number in clinopyroxenes of the gabbroic cumulate. Low-pressure Cpx field from 1 atm experimental studies of N-MORB (Elthon 1987). (b) Al_2O_3 v. TiO_2 contents in clinopyroxenes. The field of the Semail Ophiolite is from Pallister & Hopson (1981).

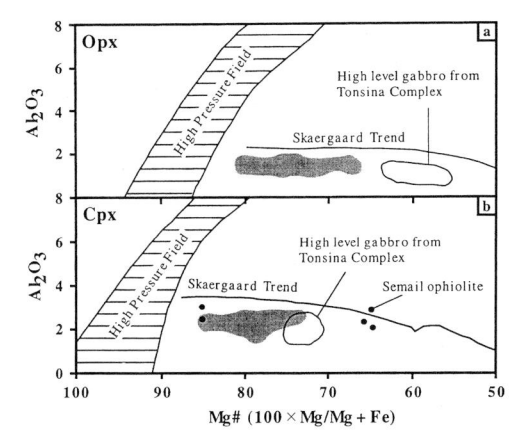

Fig. 7. Plot of Al_2O_3 v. Mg-number in the clinopyroxenes and orthopyroxenes from the gabbronorites. The field of high-pressure peridotite after Medaris (1972). The trend of gabbroic cumulates is parallel to the trend of low-pressure cumulates such as Skaergaard, the Semail Ophiolite and the Tonsina Complex [from DeBari & Coleman (1989)].

Petrogenesis of the Pozantı–Karsantı Ophiolite

Magmas erupted in destructive plate boundaries are commonly considered to be derived from sources initially more depleted than the sources of MORB and subsequently selectively enriched by components derived from subducted crust (Knittel & Oles 1994). Magnesian olivine (Fo_{92-94}) and Cr-rich chromite (Cr-number > 0.65) in primitive island-arc basalts and low abundances of high field strength (HFS) elements are additional evidence for initially depleted sources (Woodhead et al. 1993).

On the basis of both petrographic and mineralogical studies, the plutonic rocks in the Pozantı–Karsantı Ophiolite evolved by fractional crystallization and differentiation. Progressive decrease of Cr_2O_3 and TiO_2 in ortho- and clinopyroxenes (Fig. 6), with decreasing Mg-number, are indicative of these processes. Low Al contents in ortho- and clinopyroxenes suggest relatively high SiO_2 activity in the parental and derivative liquids, indicating that the magma had a tholeiitic or calc-alkaline affinity (Le Bas 1962; Hebert & Laurent 1990). Crystallization of calcic plagioclase (An_{94-84}) and a limited range of clinopyroxene compositions are similar to the restricted compositional variations observed in orogenic andesite suites within an island arc (Hebert & Laurent 1990). The presence of amphibole and An-rich plagioclase in the plutonic rocks of the Pozantı–Karsantı Ophiolite indicates hydrous conditions at the time of their magmatic differentiation. According to the experimental work of Spulber-Dixon & Rutherford (1983), amphibole is stable in basaltic magmas under a total pressure > 2 kbar (3–6 km). Conrad et al. (1983) stated that temperatures of amphibole stability range from 1050 to 700°C.

In order to constrain pressure–temperature (= depth) conditions of the ophiolitic magma chamber where accumulation was taking place during the generation of the Pozantı–Karsantı Ophiolite, several calculation methods were used on single and coexisting mineral pairs from the gabbroic cumulates. A Cpx–Opx geothermometer, based on the calculations of Wood & Banno (1973), yielded 950°C. An Opx geothermometer (Brey & Köhler 1990) yielded temperatures of 805–1033°C. Application of a Cpx–Pl geobarometer (Ellis 1980) to the cumulate gabbronorites yielded pressures of 2.3–3.6 kbar, suggesting a depth of 3–6 km for the crystallization of cumulate gabbronorites at the time of oceanic crust generation.

All the evidence presented so far suggests that the primary magma which created the Pozantı–Karsantı Ophiolite was compositionally similar to that observed in modern island-arc tholeiitic sequences.

Tectonic implications

Ophiolites in the eastern Mediterranean region and the Middle East, including Oman, Baer Bassit in Syria, the Zagros ophiolites in Iran, and Kızıldağ (Hatay), Mersin, Sarıkaraman and Pozantı–Karsantı in Turkey, originated in a suprasubduction zone tectonic environment, based on the geochemical studies of volcanic and layered plutonic rock units (Pearce et al. 1984; Lytwyn & Casey 1993; Robertson 1994; Parlak et al. 1996, 1997; Yalınız et al. 1996).

Lytwyn & Casey (1995) and Polat et al. (1996) concluded that the Pozantı–Karsantı Ophiolite probably formed along a mid-ocean ridge north of the Tauride Platform and constituted part of the forearc mantle wedge as already formed oceanic crust during the intra-oceanic subduction zone, and was finally obducted onto the Tauride Platform during Late Cretaceous or Early Palaeocene time. They also pointed out that the doleritic and gabbroic dyke swarms intruding the Pozantı–Karsantı Ophiolite and metamorphic sole were produced during intra-oceanic subduction and thus are compositionally similar to island-arc basalts and basaltic andesites. Whole-rock and mineral chemistry of

Table 4. *Representative orthopyroxene analysis from the cumulate gabbronorites*

Sample	Y1-1c	Y2-1c	Y4-1c	Y7-1c	Y9-2c	10-1c	11-1c	12-1c	13-1c	14-1c	15-1c	16-1c	17-1c	18-2c	Y19-1c	20-1c	H31-1c	H31-2c	H31-3c
SiO_2	54.20	54.40	54.60	54.40	53.90	53.60	53.10	52.10	53.90	53.70	53.40	53.60	53.50	52.80	53.80	53.60	53.70	53.80	53.90
Al_2O_3	1.25	1.59	1.58	1.35	1.16	1.13	1.13	1.29	1.42	1.63	1.52	1.47	1.44	1.21	1.56	1.31	1.29	1.42	1.32
TiO_2	0.15	0.10	0.13	0.10	0.12	0.16	0.13	0.14	0.13	0.14	0.15	0.16	0.15	0.20	0.09	0.11	0.15	0.21	0.11
FeO*	17.90	15.90	15.90	16.50	18.40	18.60	21.30	20.30	17.50	16.70	16.70	17.50	18.00	20.20	13.00	17.40	17.70	17.40	17.70
MnO	0.42	0.37	0.38	0.33	0.47	0.41	0.55	0.44	0.37	0.40	0.39	0.39	0.40	0.46	0.26	0.37	0.37	0.38	0.39
MgO	25.60	26.70	27.10	26.80	25.60	25.90	23.80	24.20	26.50	27.20	27.30	26.70	26.20	24.30	29.50	26.50	25.60	25.70	25.70
Cr_2O_3	0.00	0.25	0.15	0.15	0.00	0.00	0.00	0.00	0.15	0.19	0.19	0.14	0.00	0.00	0.22	0.00	0.21	0.21	0.23
CaO	0.63	0.74	0.76	0.60	0.88	0.69	0.87	1.12	0.58	0.74	0.70	0.87	1.17	0.76	0.88	0.99	0.85	0.95	0.75
Na_2O	0.00	0.00	0.00	0.00	0.00	0.00	0.00	0.00	0.00	0.00	0.00	0.00	0.00	0.00	0.00	0.00	0.00	0.00	0.00
K_2O	0.00	0.00	0.00	0.00	0.00	0.00	0.00	0.00	0.00	0.03	0.00	0.00	0.00	0.00	0.00	0.00	0.00	0.00	0.00
Sum	100.15	100.05	100.60	100.23	100.53	100.49	100.88	99.59	100.55	100.73	100.35	100.83	100.86	99.93	99.31	100.28	99.87	100.07	100.10
Si	1.97	1.96	1.96	1.96	1.96	1.94	1.94	1.92	1.94	1.93	1.92	1.93	1.93	1.94	1.92	1.94	1.96	1.96	1.96
Al (IV)	0.03	0.04	0.04	0.04	0.04	0.06	0.06	0.08	0.06	0.07	0.08	0.07	0.07	0.06	0.08	0.06	0.04	0.04	0.04
Al (VI)	0.03	0.03	0.03	0.02	0.01	-0.01	-0.01	-0.02	0.01	-0.01	-0.01	-0.01	-0.01	0.00	-0.01	-0.01	0.01	0.02	0.02
Ti	0.00	0.00	0.00	0.00	0.00	0.00	0.00	0.00	0.00	0.00	0.00	0.00	0.00	0.01	0.00	0.00	0.00	0.01	0.00
Fe^{3+}	0.00	0.00	0.00	0.01	0.03	0.06	0.05	0.09	0.04	0.07	0.08	0.07	0.08	0.05	0.08	0.06	0.01	0.01	0.01
Fe^{2+}	0.55	0.49	0.47	0.49	0.53	0.51	0.60	0.54	0.49	0.43	0.42	0.45	0.47	0.57	0.31	0.46	0.53	0.52	0.53
Mn	0.01	0.01	0.01	0.01	0.01	0.01	0.02	0.01	0.01	0.01	0.01	0.01	0.01	0.01	0.01	0.01	0.01	0.01	0.01
Mg	1.39	1.44	1.45	1.44	1.38	1.40	1.30	1.33	1.43	1.45	1.46	1.43	1.41	1.33	1.57	1.43	1.39	1.39	1.39
Cr	0.00	0.01	0.00	0.00	0.00	0.00	0.00	0.00	0.00	0.01	0.01	0.00	0.00	0.00	0.01	0.00	0.01	0.01	0.01
Ca	0.02	0.03	0.03	0.02	0.03	0.03	0.03	0.04	0.02	0.03	0.03	0.03	0.05	0.03	0.03	0.04	0.03	0.04	0.03
Na	0.00	0.00	0.00	0.00	0.00	0.00	0.00	0.00	0.00	0.00	0.00	0.00	0.00	0.00	0.00	0.00	0.00	0.00	0.00
K	0.00	0.00	0.00	0.00	0.00	0.00	0.00	0.00	0.00	0.00	0.00	0.00	0.00	0.00	0.00	0.00	0.00	0.00	0.00
Ens	70.46	73.43	73.68	73.08	69.53	69.88	64.88	66.05	71.73	72.87	73.01	71.46	70.12	66.69	78.52	71.27	70.43	70.69	70.63
Fors	28.29	25.11	24.83	25.75	28.75	28.78	33.42	31.76	27.14	25.70	25.64	26.86	27.63	31.81	19.80	26.81	27.89	27.44	27.89
Wol	1.25	1.46	1.48	1.18	1.72	1.34	1.70	2.20	1.13	1.42	1.35	1.67	2.25	1.50	1.68	1.91	1.68	1.88	1.48
Mg-number	71.63	74.58	75.24	74.33	71.27	71.29	66.58	68.00	72.97	74.38	74.45	73.12	72.18	68.20	80.18	73.08	72.06	72.48	72.13

Number of ions on the basis of 6 (O).
* Total Fe is expressed as FeO.
c indicates core.

Table 5. Representative plagioclase analysis from the cumulate gabbronorite

Sample	Y1-1c	Y2-1c	Y4-1c	Y7-1c	Y7-5c	Y9-1c	10-1c	11-1c	12-1c	13-1c	14-1c	15-1c	16-1c	17-1c	18-1c	20-1c	31-3c	31-4c
SiO_2	45.10	45.20	45.00	45.20	45.40	44.90	45.20	45.20	44.90	44.10	43.80	44.30	44.70	44.50	45.50	44.40	46.90	46.30
Al_2O_3	35.20	35.00	35.30	35.20	35.20	34.50	34.40	34.70	34.50	34.60	34.40	35.00	35.00	35.00	34.60	34.90	33.60	34.00
FeO	0.34	0.33	0.28	0.35	0.33	0.37	0.31	0.39	0.42	0.35	0.31	0.37	0.49	0.46	0.37	0.42	0.28	0.29
CaO	18.60	18.40	18.80	18.60	18.50	18.50	18.40	18.50	18.60	19.10	18.90	18.50	18.20	18.50	18.20	18.90	16.90	17.30
Na_2O	0.83	1.00	0.78	0.94	0.99	1.09	1.10	1.05	1.09	0.72	0.84	0.90	0.94	0.82	1.11	0.82	1.83	1.52
K_2O	0.00	0.00	0.00	0.00	0.00	0.00	0.04	0.00	0.00	0.00	0.00	0.00	0.00	0.00	0.00	0.00	0.00	0.00
Sum	100.07	99.93	100.16	100.29	100.42	99.36	99.45	99.84	99.51	98.87	98.25	99.07	99.33	99.28	99.78	99.44	99.51	99.41
Si	2.08	2.09	2.07	2.08	2.08	2.08	2.09	2.09	2.08	2.06	2.06	2.06	2.08	2.07	2.10	2.06	2.16	2.14
Al(IV)	0.92	0.91	0.93	0.92	0.92	0.92	0.91	0.91	0.92	0.94	0.94	0.94	0.92	0.93	0.90	0.94	0.84	0.86
Al(VI)	0.99	0.99	0.99	0.99	0.99	0.97	0.97	0.98	0.96	0.97	0.96	0.98	0.99	0.99	0.99	0.97	0.99	1.00
Fe^{2+}	0.01	0.01	0.01	0.01	0.01	0.01	0.01	0.02	0.02	0.01	0.01	0.01	0.02	0.02	0.01	0.02	0.01	0.01
Ca	0.92	0.91	0.93	0.92	0.91	0.92	0.91	0.92	0.92	0.96	0.95	0.92	0.91	0.92	0.90	0.94	0.84	0.86
Na	0.07	0.09	0.07	0.08	0.09	0.10	0.10	0.09	0.10	0.07	0.08	0.08	0.08	0.07	0.10	0.07	0.16	0.14
K	0.00	0.00	0.00	0.00	0.00	0.00	0.00	0.00	0.00	0.00	0.00	0.00	0.00	0.00	0.00	0.00	0.00	0.00
Or	0.00	0.00	0.00	0.00	0.00	0.00	0.23	0.00	0.00	0.00	0.00	0.00	0.00	0.00	0.00	0.00	0.00	0.00
Ab	7.47	8.95	6.98	8.38	8.83	9.63	9.74	9.31	9.59	6.39	7.44	8.09	8.55	7.43	9.94	7.28	16.38	13.72
An	92.53	91.05	93.02	91.62	91.17	90.37	90.03	90.69	90.41	93.61	92.56	91.91	91.45	92.57	90.06	92.72	83.62	86.28

Number of ions on the basis of 16 (O).
Total Fe is expressed as FeO.
c indicates core.

Table 6. *Representative amphibole analysis from the cumulate gabbronorites*

Sample	11-1c	11-3c	11-4c	11-6c	Y9-1c	Y9-2c	Y9-3c	Y9-4c	Y9-5c	13-1c	13-2c	13-3c	13-4c	Y1-1c	Y2-2c	Y4-1c	Y4-2c	Y4-5c	Y4-6c	Y7-1c
SiO$_2$	50.50	50.00	48.50	53.40	48.70	49.40	50.80	50.40	48.80	48.90	49.40	47.80	49.50	49.60	49.40	48.80	47.80	50.60	51.00	48.80
Al$_2$O$_3$	5.99	6.30	7.53	3.63	8.00	6.75	6.14	6.22	6.86	7.09	6.53	7.05	6.57	7.27	7.80	8.60	8.90	6.79	6.39	8.10
TiO$_2$	1.05	1.07	0.75	0.43	0.73	1.01	1.02	0.93	0.98	0.81	0.71	0.87	0.74	1.02	0.76	0.90	0.98	0.71	0.68	1.02
FeO	11.10	10.50	11.20	9.00	10.60	11.10	10.50	10.40	11.30	9.30	9.40	9.30	9.10	9.80	8.70	9.20	9.20	8.10	8.00	9.20
Cr$_2$O$_3$	0.00	0.00	0.00	0.00	0.08	0.00	0.00	0.08	0.08	0.20	0.20	0.39	0.26	0.12	0.45	0.29	0.33	0.29	0.35	0.34
MnO	0.19	0.15	0.23	0.15	0.15	0.12	0.16	0.13	0.13	0.09	0.13	0.12	0.11	0.12	0.12	0.09	0.10	0.10	0.10	0.10
MgO	16.80	16.70	16.60	18.40	16.50	16.50	17.30	16.90	16.50	17.20	17.40	17.00	17.20	16.90	17.40	16.90	16.80	17.80	18.30	16.60
CaO	11.40	11.90	11.20	12.30	12.00	11.90	11.80	12.10	11.70	12.00	11.80	12.10	12.10	11.80	11.80	12.00	12.10	12.00	12.00	11.90
Na$_2$O	0.80	0.73	1.21	0.37	1.02	0.82	0.70	0.72	0.89	0.84	0.72	0.91	0.83	0.91	0.97	1.22	1.19	0.95	0.93	1.03
K$_2$O	0.05	0.10	0.17	0.07	0.22	0.08	0.06	0.06	0.09	0.08	0.06	0.07	0.06	0.04	0.00	0.04	0.00	0.00	0.03	0.10
Sum	97.88	97.45	97.39	97.75	98.00	97.68	98.48	97.94	97.33	96.51	96.35	95.61	96.47	97.58	97.40	98.04	97.40	97.34	97.78	97.19
Si	7.23	7.18	7.01	7.56	6.99	7.11	7.21	7.20	7.06	7.07	7.15	7.00	7.15	7.10	7.05	6.95	6.87	7.19	7.22	7.01
Al(IV)	0.77	0.82	0.99	0.44	1.01	0.89	0.79	0.80	0.94	0.93	0.85	1.00	0.85	0.90	0.95	1.05	1.13	0.81	0.78	0.99
Al(VI)	0.24	0.25	0.30	0.16	0.34	0.25	0.24	0.25	0.23	0.28	0.26	0.22	0.27	0.32	0.36	0.39	0.37	0.33	0.28	0.38
Ti	0.11	0.12	0.08	0.05	0.08	0.11	0.11	0.10	0.11	0.09	0.08	0.10	0.08	0.11	0.08	0.10	0.11	0.08	0.07	0.11
Fe^{2+}	1.33	1.26	1.35	1.06	1.27	1.34	1.25	1.24	1.37	1.12	1.14	1.14	1.10	1.17	1.04	1.10	1.10	0.96	0.95	1.10
Cr	0.00	0.00	0.00	0.00	0.01	0.00	0.00	0.01	0.01	0.02	0.02	0.05	0.03	0.01	0.05	0.03	0.04	0.03	0.04	0.04
Mn	0.02	0.02	0.03	0.02	0.02	0.01	0.02	0.02	0.02	0.01	0.02	0.01	0.01	0.01	0.01	0.01	0.01	0.01	0.01	0.01
Mg	3.59	3.58	3.58	3.88	3.53	3.54	3.66	3.60	3.56	3.71	3.75	3.71	3.70	3.60	3.70	3.59	3.60	3.77	3.86	3.55
Ca	1.75	1.83	1.74	1.86	1.84	1.83	1.79	1.85	1.81	1.86	1.83	1.90	1.87	1.81	1.80	1.83	1.86	1.83	1.82	1.83
Na	0.22	0.20	0.34	0.10	0.28	0.23	0.19	0.20	0.25	0.24	0.20	0.26	0.23	0.25	0.27	0.34	0.33	0.26	0.26	0.29
K	0.01	0.02	0.03	0.01	0.04	0.01	0.01	0.01	0.02	0.01	0.01	0.01	0.01	0.01	0.00	0.01	0.00	0.00	0.01	0.02
Mg-number	0.73	0.74	0.73	0.78	0.74	0.73	0.75	0.74	0.72	0.77	0.77	0.77	0.77	0.75	0.78	0.77	0.77	0.80	0.80	0.76

Number of ions on the basis of 23 (O).
Total Fe is expressed as FeO.
c indicates core.

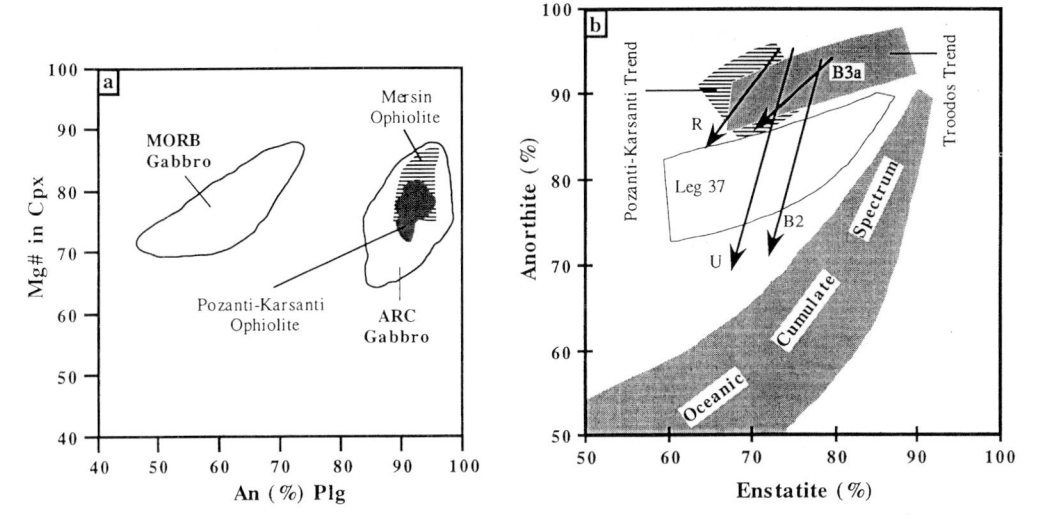

Fig. 8. (**a**) Composition of coexisting plagioclase (An mol%) and clinopyroxene (Mg-number) in the gabbronorite of the Pozantı–Karsantı Ophiolite. Fields of MORB and arc gabbro are from Burns (1985). The field of the Mersin Ophiolite is from Parlak *et al.* (1996). (**b**) Anorthite content in plagioclase (mol%) v. enstatite content in orthopyroxene (mol%) for the Pozantı–Karsantı Ophiolite. The Troodos Trend is from Hebert & Laurent (1990). R, Rindjami Volcano (Foden 1983); B3a, B2, Boisa Volcano (Gust & Johnson 1981); U, Usa Volcano (Fujimaki 1986).

cumulate gabbronorites from the Pozantı–Karsantı Ophiolite indicate an arc rather than mid-ocean ridge affinity. Figure 8a shows the covariation of plagioclase and clinopyroxene from both the Pozantı–Karsantı and the Mersin Ophiolites. Gabbroic rocks which originated in an arc environment differ from those formed in a mid-ocean ridge environment in terms of their high anorthite content of plagioclase (Burns 1985; Hebert & Laurent 1990; Parlak *et al.* 1996). The gabbroic cumulates of the Pozantı–Karsantı Ophiolite clearly plot in the arc gabbro field defined by Burns (1985). As indicated in the petrography section, olivine does not occur in the studied gabbroic section. Figure 8b presents the covariation of plagioclase and enstatite from the Pozantı–Karsantı and Troodos Ophiolites, and known island arcs. This diagram clearly indicates that covariation of coexisting minerals evolved similarly to the trends of the Troodos Ophiolite and island arcs, and differs from the oceanic cumulate spectrum. All the evidence presented so far suggests that the Pozantı–Karsantı Ophiolite originated during intra-oceanic subduction and is compositionally similar to the plutonic rocks of the Mersin Ophiolite (Parlak *et al.* 1996) in southern Turkey and the Troodos Ophiolite (Hebert & Laurent 1990) in Cyprus.

To summarize, in the north of the Anatolide–Tauride Platform, the Pozantı–Karsantı Ophiolite represents an important element for regional reconstruction. It originated at the beginning of Late Cretaceous time in a suprasubduction zone tectonic setting related to the north dipping subduction of the northern branch of Neotethyan Ocean. This intra-oceanic subduction led to the formation of a metamorphic sole followed by dykes intruding both the metamorphic sole and the oceanic crust. This sole contains remnants of ocean island rocks and oceanic crust carried into the subduction zone (Lytwyn & Casey 1995). The isolated diabase/dolerite dykes exhibit the chemical signature of island-arc basalts (Lytwyn & Casey 1995). The Pozantı–Karsantı Ophiolite continued to accrete mélange after the dyke emplacement and was finally obducted over the Tauride Platform during Late Cretaceous or Early Palaeocene time (Lytwyn & Casey 1995; Polat & Casey 1995).

Conclusions

- The major and trace element (including REE) geochemistry, as well as mineral chemistry, of the plutonic rocks suggests that the primary magma generating the Pozantı–Karsantı Ophiolite is compositionally similar to those observed in modern island-arc tholeiitic sequences.
- The mineral chemistry of the basal ultramafic cumulates (dunite, wehrlite, websterite)

show evidence of high-pressure crystallization (c. 10 kbar), typical for the base of an island-arc system, whereas the mineral chemistry of the cumulate gabbronorites exhibits lower pressure crystallization (2.3–3.6 kbar), owing to the presence of amphibole. These gabbronorites probably represent shallower levels in the same arc which were juxtaposed against the basal ultramafic cumulates either during accretion or later faulting.
- The results of this work suggest that the Pozantı–Karsantı Ophiolite had a similar tectonic setting to other Late Cretaceous Neotethyan ophiolites, namely the Sarıkaraman (Yalınız et al. 1996) and Mersin (Parlak et al. 1996, 1997) ophiolites in Turkey and the Troodos ophiolite in Cyprus (Hebert & Laurent 1990), in the eastern Mediterranean region, rather than along a mid-oceanic ridge during the opening of Neotethyan ocean as suggested by Lytwyn & Casey (1995).

OP gratefully acknowledges the financial support of the Austrian Science Foundation (Lise Meitner Fellowship, Project No. M444-GEO). The authors would like to thank Fabio Capponi for performing major and trace element analyses. Dan Topa is thanked for his guidance during the microprobe analysis at Salzburg University. P. A. Floyd, M. C. Göncüoğlu and E. Bozkurt are thanked for their constructive and very helpful comments.

References

AOKI, K. & KUNO, H. 1971. Gabbro–quartz diorite inclusions from Izu–Hakone region, Japan. *Bulletin Volcanologique*, **36**, 164–173.

ARCULUS, R. J. & WILLS, K. J. A. 1980. The petrology of plutonic blocks and inclusions from the Lesser Antilles island arc. *Journal of Petrology*, **21**, 743–799.

BEARD, J. S. 1986. Characteristic mineralogy of arc-related cumulate gabbros: implications for the tectonic setting of gabbroic plutons and for andesite genesis. *Geology*, **14**, 848–851.

BİNGÖL, A. F. 1978. *Pétrologie du massif ophiolitique de Pozantı–Karsantı (Taurus cilicien, Turquie): Étude de la partie orientale*. Thèse 3e cycle, Université Strasbourg.

BLUMENTHAL, M. M. 1947. *Belemedik Palaeozoyik Pençeresi ve Bunun Mesozoyik Kalker Çevresi (Klikya Torosları)*. Mineral Research and Exploration Institute of Turkey (MTA) Publications, Serie D, **3** [in Turkish with English abstract].

BREY, G. P. & KÖHLER, T. 1990. Geothermobarometry in four-phase lherzolites. II. New thermobarometers, and practical assessment of existing thermobarometers. *Journal of Petrology*, **31**, 1353–1378.

BROWNING, P. 1984. Cryptic variation within the cumulate sequence of the Oman ophiolite: magma chamber depth and petrological implications. *In*: GASS, I. G., LIPPARD, S. J. & SHELTON, A.W. (eds) *Ophiolites and Oceanic Lithosphere*. Geological Society, London, Special Publications, **13**, 71–82.

BURNS, L. E. 1985. The Border Ranges ultramafic and mafic complex, south central Alaska: cumulate fractionates of island arc volcanics. *Canadian Journal of Earth Sciences*, **22**, 1020–1038.

ÇAKIR, U. 1978. *Pétrologie du massif ophiolitique de Pozantı–Karsantı (Taurus Cilicien, Turquie): Étude de la partie centrale*. Thèse doctorat, Université Louis Pasteur.

——, JUTEAU, T. & WHITECHURCH, H. 1978. Nouvelles preuves de l'ecaillage intra-oceanique precoce des ophiolites tethsiennes: les roches metamorphiques infra-peridotitiques du massif de Pozantı–Karsantı (Turquie). *Bulletin de la Société Géologique de France*, **7**, 61–70.

ÇATAKLI, A. S. 1983. *Assemblage ophiolitique et roches associées de la partie occidentale du massif de Pozantı–Karsantı (Taurus cilicien, Turquie)*. Thèse d'Etat, Université de Nancy I.

CONRAD, W. K. & KAY, R. W. 1984. Ultramafic and mafic inclusions from Adak island: crystallisation history and implications for the nature of primary magmas and crustal evolution in the Aleutian arc. *Journal of Petrology*, **25**, 88–125.

——, KAY, S. M. & KAY, R. W. 1983. Magma mixing in Aleutian arc: evidence from cognate inclusions and composite xenoliths. *Journal of Volcanology and Geothermal Research*, **18**, 279–295.

DEBARI, S. M. & COLEMAN, R. G. 1986. Petrologic aspects of gabbros from the Tosnia complex, Chugach mountains, Alaska: evidence for deep magma chambers under an island arc. *Geological Society of America Abstracts and Programs*, **18**, 99.

—— & —— 1989. Examination of the deep levels of an island arc: evidence from the Tonsina ultramafic–mafic assemblage, Tonsina, Alaska. *Journal Geophysical Research*, **94**, 4373–4391.

——, KAY, S. M. & KAY, R. W. 1987. Ultramafic xenoliths from Adagdak volcano, Adak, Aleutian islands, Alaska: deformed igneous cumulates from the Moho of an island arc. *Journal of Geology*, **95**, 329–341.

DİLEK, Y. & MOORES, E. M. 1990. Regional tectonics of the eastern Mediterranean ophiolites. *In*: MALPAS, J., MOORES, E., PANAYIOTOU, A. & XENOPHONTOS, C. (eds) *Ophiolites – Oceanic Crustal Analogues*. Proceedings of Troodos Ophiolite Symposium, 1987, 295–309.

DUPUY, C., DOSTAL, J., MARCELOT, G., BOUGAULT, H., JORON, J. L. & TREUIL, M. 1982. Geochemistry of basalts from central and southern New Hebrides arcs: implication for their source rock composition. *Earth and Planetary Science Letters*, **60**, 207–225.

ELLIS, D. J. 1980. Osumilite–sapphirine–quartz granulites from Enderby Land, Antarctica: *P-T* conditions of metamorphism, implications for garnet–cordierite equilibria and the evolution of the deep crust. *Contributions to Mineralogy and Petrology*, **74**, 201–210.

ELTHON, D. 1987. Petrology of gabbroic rocks from the Mid-Cayman rise spreading centre. *Journal of Geophysical Research*, **92**, 658–682.

——, CASEY, J. F. & KOMOR, S. 1982. Mineral chemistry of ultramafic cumulates from the North Arm Mountain massif of the Bay of Islands ophiolite: evidence of high pressure crystal fractionation of oceanic basalts. *Journal of Geophysical Research*, **87**, 8717–8734.

FODEN, J. D. 1983. The petrology of the calc-alkaline lavas of Rindjani volcano, east Sunda arc: model for island arc. *Journal of Petrology*, **24**, 98–130.

FUJIMAKI, H. 1986. Fractional crystallisation of the basaltic suite of Usa volcano, southwest Hokkaido, Japan, and its relationships with the associated felsic suite. *Lithos*, **19**, 129–140.

GREEN, T. H. & RINGWOOD, A. E. 1968. Genesis of the calc-alkaline igneous rock suite. *Contributions to Mineralogy and Petrology*, **18**, 105–162.

GUST, D. A. & JOHNSON, R. W. 1981. Amphibole bearing cumulates from Boisa island, Papua New Guinea: evaluation of the role of fractional crystallisation in an andesitic volcano. *Journal of Geology*, **89**, 219–232.

HEBERT, R. 1982. Petrography and mineralogy of oceanic peridotites and gabbros: some comparisons with ophiolite examples. *Ofioliti*, **7**, 299–324.

—— & LAURENT, R. 1990. Mineral chemistry of the plutonic section of the Troodos ophiolite: new constraints for genesis of arc-related ophiolites. *In*: MALPAS, J., MOORES, E., PANAYIOTOU, A. & XENOPHONTOS, C. (eds) *Ophiolites–Oceanic Crustal Analogues*. Proceedings of Troodos Ophiolite Symposium, 1987, 149–163.

JAKES, P. & WHITE, A. J. R. 1972. Hornblendes from calc-alkaline volcanic rocks of island arcs and continental margins. *American Mineralogist*, **57**, 887–902.

JUTEAU, T. 1980. Ophiolites of Turkey. *Ofioliti*, **2**, 199–235.

—— & WHITECHURCH, H. 1980. The magmatic cumulates of Antalya (Turkey): evidence of multiple intrusions in an ophiolitic magma chamber. *In*: PANAYIOTOU, A. (ed.) *Ophiolites*. Proceedings of International Ophiolite Symposium, Cyprus, 1979, 377–391.

KNITTEL, U. & OLES, D. 1994. Basaltic volcanism associated with extensional tectonics in the Taiwan–Luzon island arc: evidence for non-depleted sources and subduction zone environment. *In*: SMELLIE, J. L. (ed.) *Volcanism Associated with Extension at Consuming Plate Margins*. Geological Society, London, Special Publications, **81**, 77–93.

KOLLER, F. & HÖCK, V. 1990. Mesozoic ophiolites in the eastern Alps. *In*: MALPAS, J., MOORES, E., PANAYIOTOU, A. & XENOPHONTOS, C. (eds) *Ophiolites–Oceanic Crustal Analogues*. Proceedings of Troodos Ophiolite Symposium, 1987, 253–263.

LEAKE, B. E. 1978. Nomenclature of amphiboles. *Canadian Mineralogist*, **16**, 501–520.

LE BAS, M. J. 1962. The role of aluminium in igneous clinopyroxenes with relation to their parentage. *American Journal of Science*, **260**, 267–288.

LYTWYN, J. N. & CASEY, J. F. 1993. The geochemistry and petrogenesis of volcanics and sheeted dikes from the Hatay (Kızıldağ) ophiolite, southern Turkey: possible formation with the Troodos ophiolite, Cyprus, along fore-arc spreading centres. *Tectonophysics*, **223**, 237–272.

—— & —— 1995. The geochemistry of postkinematic mafic dike swarms and subophiolitic metabasites, Pozantı–Karsantı ophiolite, Turkey: evidence for ridge subduction. *Geological Society of America Bulletin*, **107**, 830–850.

MEDARIS, L. G. 1972. High-pressure peridotites in south western Oregon. *Geological Society of America Bulletin*, **83**, 41–58.

MEIJER, A. & REAGAN, M. 1981. Petrology and geochemistry of the island of Sarigan in the Marianas arc: calc-alkaline volcanism in an oceanic setting. *Contributions to Mineralogy and Petrology*, **77**, 337–354.

MIYASHIRO, A. 1973. The Troodos ophiolitic complex was probably formed in an island arc. *Earth and Planetary Science Letters*, **19**, 218–224.

MOORE, J. C. 1970. Water content of basalt erupted on the ocean floor. *Contributions to Mineralogy and Petrology*, **28**, 272–279.

MORRICE, M. G., JEZEK, P. A., GILL, J. B., WHITFORD, D. J. & MONOAIFA, M. 1983. An introduction to the Sangihe arc: volcanism accompanying arc–arc collision in the Molucca Sea, Indonesia. *Journal of Volcanology and Geothermal Research*, **19**, 135–165.

NEWHALL, C. G. 1979. Temporal variations in the lavas of Mayon Volcano, Philippines. *Journal of Volcanology and Geothermal Research*, **6**, 61–83.

PALLISTER, J. S. & HOPSON, C. A. 1980. Semail ophiolite magma chamber: I. Evidence from gabbro phase variation, internal structure and layering. *In*: PANAYIOTOU, A. (ed.) *Ophiolites*. Proceedings of International Ophiolite Symposium, Cyprus, **1979**, 402–404.

—— & —— 1981. Semail ophiolite plutonic suite: field relation, phase variation, cryptic variation and layering, and a model of a spreading ridge magma chamber. *Journal of Geophysical Research*, **86**, 2593–2644.

PARLAK, O. & DELALOYE, M. 1999. Precise $^{40}Ar/^{39}Ar$ ages from the metamorphic sole of the Mersin ophiolite (Southern Turkey). *Tectonophysics*, **301**, 145–158.

—— & HÖCK, V. 1998. Suprasubduction zone origin of a Neotethyan ophiolite: petrographical and geochemical evidences from the Pozantı–Karsantı ophiolite, southern Turkey. *Proceedings of the Third International Turkish Geology Symposium*, METU–Ankara, Abstracts, 173.

——, DELALOYE, M. & BİNGÖL, E. 1996. Mineral chemistry of ultramafic and mafic cumulates as an indicator of the arc-related origin of the Mersin ophiolite (southern Turkey). *Geologische Rundschau*, **85**, 647–661.

——, —— & —— 1997. Phase and cryptic variation through the ultramafic–mafic cumulates in the

Mersin ophiolite (Southern Turkey). *Ofioliti*, **22**, 81–92.

PEARCE, J. A., LIPPARD, S. J. & ROBERTS, S. 1984. Characteristics and tectonic significance of supra-subduction zone ophiolites. *In*: KOKELAAR, B. P. & HOWELLS, M. F. (eds) *Marginal Basin Geology*. Geological Society, London, Special Publications, **16**, 77–89.

PİŞKİN, Ö., DELALOYE, M., MORITZ, R. & WAGNER, J. J. 1990. Geochemistry and geothermometry of the Hatay complex Turkey: implication for genesis of the ophiolite sequence. *In*: MALPAS, J., Moores, E., PANAYIOTOU, A. & XENOPHONTOS, C. (eds) *Ophiolites–Oceanic Crustal Analogues*. Proceedings of Troodos Ophiolite Symposium, 1987, 329–337.

POLAT, A. & CASEY, J. F. 1995. A structural record of the emplacement of the Pozantı–Karsantı ophiolite onto the Menderes–Taurus block in the late Cretaceous, eastern Taurides, Turkey. *Journal of Structural Geology*, **17**, 1673–1688.

——, —— & KERRICH, R. 1996. Geochemical characteristics of accreted material beneath the Pozantı-Karsantı ophiolite, Turkey: intra-oceanic detachment, assembly and obduction. *Tectonophysics*, **263**, 249–276.

RICOU, L. E. 1971. Le croissant ophiolitique périarabe, une ceinture de nappes mises en place au Crétacé supérieur. *Revue de Géologie Dynamique et de Géographie Physique*, **13**, 327–349.

ROBERTSON, A. H. F. 1994. Role of the tectonic facies concept in orogenic analysis and its application to Tethys in the eastern Mediterranean region. *Earth Science Review*, **37**, 139–213.

ŞENGÖR, A. M. C. & YILMAZ, Y. 1981. Tethyan evolution of Turkey: Plate tectonic approach. *Tectonophysics*, **75**, 181–241.

SMITH, A. G. 1993. Tectonic significance of the Hellenic–Dinaric ophiolites. *In*: Pichard, H. M., Alabaster, T., Harris, N. B. W. & Neary, C. R. (EDS) *Magmatic Processes and Plate Tectonics*. Geological Society, London, Special Publications, **76**, 213–243.

SMITH, T. E., HUANG, C. H., WALLAWENDER, M. J., CHEUNG, P. & WHEELER, C. 1983. The gabbroic rocks of the Peninsular Ranges Batholiths, southern California: cumulate rocks associated with calc-alkalic basalts and andesites. *Journal of Volcanology and Geothermal Research*, **18**, 249–278.

SPULBER DIXON, S. & RUTHERFORD, M. J. 1983. The origin of rhyolite and plagiogranite in oceanic crust: an experimental study. *Journal of Petrology*, **24**, 1–25.

STERN, R. J. 1979. On the origin of andesite in the northern Mariana island arc: implications from Agrigan. *Contributions to Mineralogy and Petrology*, **68**, 207–219.

SUN, S. S. & MCDONOUGH, W. F. 1989. Chemical and isotopic systematics of oceanic basalts: implications for mantle composition and processes. *In*: SAUNDERS, A. D. & NORRY, M. J. (eds) *Magmatism in the Ocean Basins*. Geological Society, London, Special Publications, **42**, 313–347.

TEKELİ, O., AKSAY, A., ÜRGÜN, B. M. & IŞIK, A. 1983. Geology of the Aladağ Mountains. *In*: TEKELİ, O. & GÖNCÜOĞLU, M. C. (eds) *Geology of the Taurus Belt*. Proceedings of International Tauride Symposium. Mineral Research and Exploration Institute of Turkey (MTA) Publications, 143–158.

THUIZAT, R., MONTIGNY, R., ÇAKIR, U. & JUTEAU, T. 1978. K–Ar investigations on two Turkish ophiolites. *In*: ZARTMAN, R. E. (ed.) *Short Papers of the 4th International Conference on Geochronology, Cosmochronology, Isotope Geology*. Geological Survey of America Open File Report, **78–701**, 430–432.

——, WHITECHURCH, H., MONTIGNY, R. & JUTEAU, T. 1981. K–Ar dating of some infra-ophiolitic metamorphic soles from the eastern Mediterranean: new evidence for oceanic thrusting before obduction. *Earth and Planetary Science Letters*, **52**, 302–310.

THY, P. 1987*a*. Petrogenetic implications of mineral crystallisation trends of Troodos cumulates, Cyprus. *Geological Magazine*, **124**, 1–11.

—— 1987*b*. Magmas and magma chamber evolution, Troodos ophiolite, Cyprus. *Geology*, **15**, 316–319.

—— 1990. Cryptic variation of a cumulate sequence from the plutonic complex of the Troodos ophiolite. *In*: MALPAS, J., Moores, E., PANAYIOTOU, A. & XENOPHONTOS, C. (eds) *Ophiolites–Oceanic Crustal Analogues*. Proceedings of Troodos Ophiolite Symposium, **1987**, 165–172.

VOLDET, P. 1993. From neutron activation to inductively coupled plasma-atomic emission spectrometry in the determination of rare-earth elements in rocks. *Trends in Analytical Chemistry*, **12**, 339–344.

WALKER, J. A. 1984. Volcanic rocks from the Nejapa and Granada cinder cone alignments, Nicaragua, Central America. *Journal of Petrology*, **25**, 299–342.

WINCHESTER, J. A. & FLOYD, P. A. 1977. Geochemical discrimination of different magma series and their differentiation products using immobile elements. *Chemical Geology*, **20**, 325–343.

WOOD, B. J. & BANNO, S. 1973. Garnet–orthopyroxene and orthopyroxene–clinopyroxene relationships in simple and complex systems. *Contributions to Mineralogy and Petrology*, **42**, 109–124.

WOODHEAD, J., EGGINS, S. & GAMBLE, J. 1993. High field strength and transition element systematics in island arc and back arc basin basalts: evidence for multi-phase melt extraction and depleted mantle wedge. *Earth and Planetary Science Letters*, **114**, 491–504.

YALINIZ, M. K., FLOYD, P. A. & GÖNCÜOĞLU, M. C. 1996. Supra-subduction zone ophiolites of Central Anatolia: geochemical evidence from the Sarıkaraman ophiolite, Aksaray, Turkey. *Mineralogical Magazine*, **60**, 697–710.

Early stages of evolution of the Black Sea

V. G. KAZMIN,[1] A. A. SCHREIDER[1] & A. A. BULYCHEV[2]

[1] Nakhimovsky, Prospect 36, Institute of Oceanology, Russian Academy of Science,
Moscow 117851, Russia (e-mail: vkazmin@geo.sio.rssi.ru)
[2] M.V. Lomonosov Moscow State University, Moscow, Russia

Abstract: Comparison of two sets of structural and thickness maps of the Black Sea Basin produced by Russian and Italian workers revealed important differences in the interpretation of thickness and structure of the lower sedimentary unit, referred to in both works as 'Palaeocene–Eocene'. The map based on the Italian data shows two depocentres with > 5 km of sediment in the westernmost part of the Western Black Sea Basin (WBSB), while in the rest of the WBSB and in the Eastern Black Sea Basin (EBSB) the thickness is 2–3 km. Analysis of the land and submarine geology suggests that depocentres correspond to two segments of an early back-arc rift system formed in the late Early Cretaceous. The third segment of this system is represented by the Karkinit Graben on the northern shelf of the Black Sea. Submarine studies reveal that the graben originated behind an Early Cretaceous volcanic arc situated on the present day continental slope and rise.

Most of the WBSB and the EBSB opened in the Eocene. For the EBSB this age is supported by data on its landward extension – the Adjaro–Trialet Basin. The EBSB could not open due to anticlockwise rotation of the Shatsky Rise because there was no corresponding subduction or shortening in the Greater Caucasus Basin. An alternative hypothesis is that of simultaneous opening of the EBSB and the WBSB as a result of southward drift of the Pontides and clockwise rotation of the Andrusov Rise.

Sedimentary fill of the Black Sea Basin has been studied by seismic methods since the 1950s. At the end of the 1980s the great volume of accumulated data was analysed and processed by two groups of workers in Russia and Italy. The Russian group, led by D. A. Tugolesov, published a set of structural and thickness maps for the Cenozoic sequence of the basin in 1985 and 1989 (Tugolesov et al. 1985; Tugolesov 1989) (isopachs and depths were presented in metres). Italian work was originally based on a joint Italian–Russian project and a collection of maps was compiled and published in Italy (Finetti et al. 1988). In this publication a two-way time travel (TWTT) scale was used for maps of thicknesses and structures.

The different methods of presentation made comparison between the two sets of maps extremely difficult. At the same time, it was clear that important differences existed in interpretation of the distribution and structure of sedimentary sequences. To facilitate comparison, the maps of the Italian authors were recalculated by converting the TWTT to a depth scale, using interval velocities adopted by Finetti and co-authors (Finetti et. al. 1988, plates 6–10). (Transformation and computer drawing of maps was carried out at the Moscow State University by a group of geophysicists led by V. P. Melikhov and A. A. Bulychev.)

The next step was to establish correlation between seismostratigraphic schemes used in each analysis. At first sight the differences in the stratigraphic volume of the mapped units prevented any meaningful correlation (Fig. 1). Only the lower unit termed 'Palaeocene–Eocene', was interpreted identically by Russian and Italian authors. In the Eastern Black Sea Basin (EBSB) and at the margins of the Western Black Sea Basin (WBSB), its lower boundary coincides with the top of the pre-Cenozoic (Cretaceous?) sequence and in the central part of the WBSB – presumably with the top of the basaltic crust (Neprochnov et al. 1974). The top of the unit is marked by a strong reflector at the contact between the Palaeocene–Eocene carbonates and Maikopian clays. The stratigraphic volumes of other units do not coincide. However, close inspection of publications (Finetti et al. 1988; Tugolesov 1993) clearly shows that in both schemes the same reflectors were mapped as unit boundaries, although the ages ascribed to them were different. The reason for discrepancies was very simple: Russian interpretators used drillholes and outcrops on the northern and northwestern shelves and on the Caucasus coast of the Black Sea for seismic profile calibration, while the stratigraphic scheme applied by the Italian geophysicists tied the profiles to drillholes on the Bulgarian or Turkish shelves and also used local stratigraphy. For example, the top of Maikopian was identified in the same

From: BOZKURT, E., WINCHESTER, J. A. & PIPER, J. D. A. (eds) *Tectonics and Magmatism in Turkey and the Surrounding Area.* Geological Society, London, Special Publications, **173**, 235–249. 1-86239-064-9/00/$15.00
© The Geological Society of London 2000.

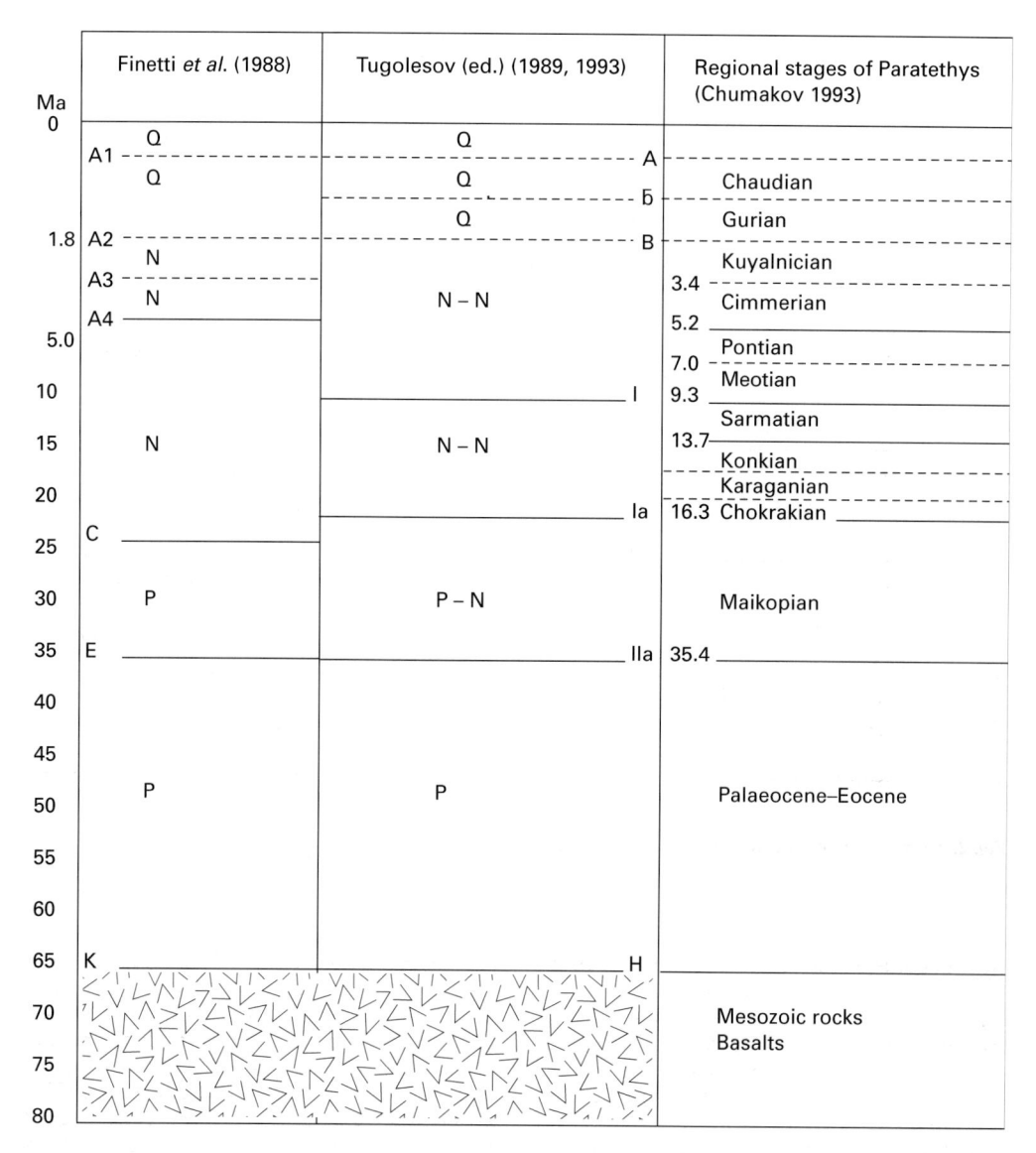

Ma	Finetti et al. (1988)	Tugolesov (ed.) (1989, 1993)	Regional stages of Paratethys (Chumakov 1993)
0	A1 Q	Q	A
	Q	Q	Б Chaudian
1.8	A2	Q	B Gurian
	N		Kuyalnician
	A3	N – N	3.4
	N		Cimmerian
	A4		5.2
5.0			Pontian
			7.0
10			Meotian 9.3
		I	Sarmatian
15	N	N – N	13.7
			Konkian
20			Karaganian
		Ia	16.3 Chokrakian
25	C		
30	P	P – N	Maikopian
35	E		IIa 35.4
40			
45			
50	P	P	Palaeocene–Eocene
55			
60			
65	K		H
70			Mesozoic rocks
75			Basalts
80			

Fig. 1. Main reflectors and Cenozoic seismic complexes.

way in both works, but in the Russian version it was dated as the top of Early Miocene, whereas in the Italian one as the top of the Oligocene. The situation with other seismostratigraphic boundaries is the same. Their correlation is presented in Fig. 2, with the Russian version of stratigraphy accepted as being supported by better biostratigraphic evidence.

In this paper only the data on the lower (Palaeocene–Eocene) part of the basin sequence are examined. In the compared publications the differences concerning this part are most striking and have the most important implications.

The thickness and structure of the Palaeocene–Eocene sediments and their implications for the early stages of the Black Sea opening

The deep basin of the Black Sea consists of two sub-basins: the WBSB and the EBSB (Fig. 3). Until the Middle Miocene they were separated

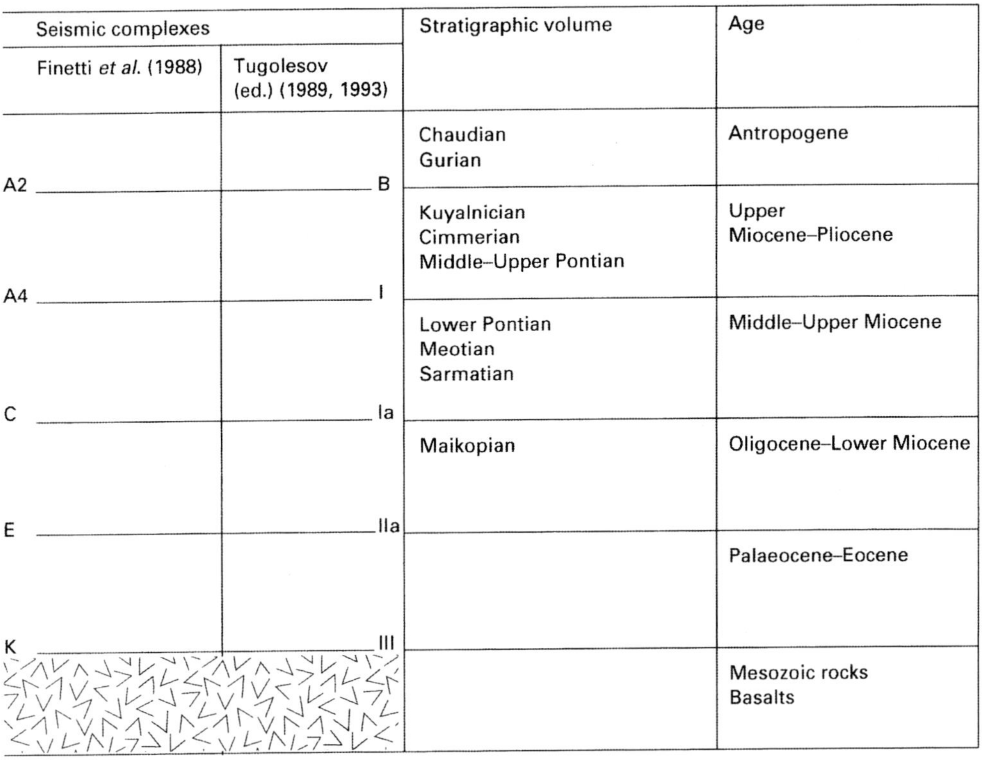

Seismic complexes		Stratigraphic volume	Age
Finetti et al. (1988)	Tugolesov (ed.) (1989, 1993)		
A2 ————	———— B	Chaudian Gurian	Antropogene
A4 ————	———— I	Kuyalnician Cimmerian Middle–Upper Pontian	Upper Miocene–Pliocene
C ————	———— Ia	Lower Pontian Meotian Sarmatian	Middle–Upper Miocene
E ————	———— IIa	Maikopian	Oligocene–Lower Miocene
			Palaeocene–Eocene
K ————	———— III		Mesozoic rocks Basalts

Fig. 2. Correlation of reflectors and Cenozoic seismic complexes.

by a massive elongated crustal block including the Andrusov and Arkhangelsky Rises. In many publications the block is described as the Central Black Sea Rise (CBR). Since Middle Miocene times, sediments overlapped the rise and the sub-basins have merged into a single deepwater basin.

According to Tugolesov and co-authors (Tugolesov et al. 1985; Tugolesov 1989, 1993), the Palaeocene–Eocene sedimentation pattern in the two sub-basins was different (Fig. 4). In the EBSB the sediments, 2–3 km thick, reveal a rather complicated structure with deeps and rises of a roughly northeast–southwest trend. A greater thickness (up to 6 km) is characteristic only for the southeastern corner of the EBSB, which forms an extension of the Adjaro–Trialet Basin of the Transcaucasus (Adamia et al. 1974; Kopp & Shcherba 1998).

The maximum thickness of the Palaeocene–Eocene sediments in the WBSB is 5–6 km; this is characteristic of the central and eastern portion of the sub-basin. In its western part the thickness decreases from 5 to 1 km. The greatest thickness, up to 6 km, occurs in a narrow zone in the extreme southwest, roughly corresponding to the Lower Kamchian foredeep. Another small area of 6 km thick sediments is in the eastern corner of the WBSB between the CBR and the Pontides. The much greater thickness of Palaeocene–Eocene sediments in the WBSB compared with the EBSB could be interpreted as an indication of significant differences in the history of two sub-basins.

A very different picture appears from the map based on Finetti's data (Fig. 5). As in Tugolesov's map, the configuration of isopachs is more complex in the EBSB. However, in most of the WBSB and the EBSB the thickness of Palaeocene–Eocene sediments is similar and varies between 2 and 3 km. Only in the western part of the WBSB are there two very distinctive depocentres with a maximum thickness of the Palaeocene–Eocene up to 5–6 km and, perhaps, even more. The depocentres are elongated northeast–southwest and are situated at the

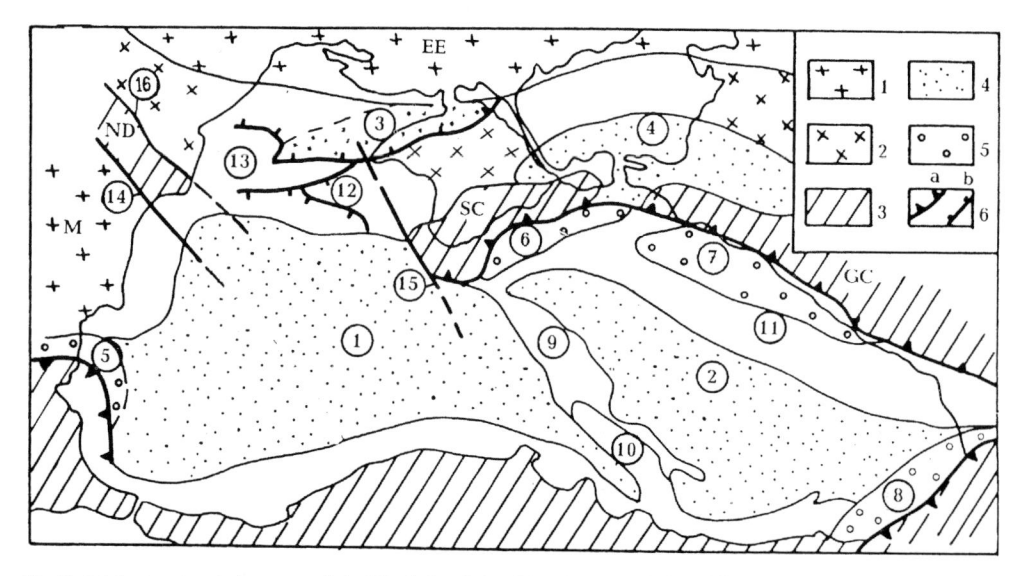

Fig. 3. Main structural elements of the Black Sea [after Tugolesov *et al.* (1985)]. 1, Precambrian platform; 2, Palaeozoic (Scythian) platform; 3, fold belts; 4, major sedimentary basins; 5, foredeeps; 6, faults – a, nappes; b, normal faults; thick line, transverse fault. EE, East European Platform; GC, Greater Caucasus; M, Moesian Platform; ND, North Dobrogea. Encircled numbers: 1, Western Black Sea; 2, Eastern Black Sea; 3, Karkinit Graben; 4, Indolo-Kuban Basin. Foredeeps: 5, Kamchian; 6, Sorokin; 7, Tuapse; 8, Gurian and Rioni. Uplifted blocks: 9, Andrusov; 10, Arkhanglevsky; 11, Shatsky; 12, Kalamit; 13, Kilian. Transverse faults: 14, Pechenega–Camena; 15, West Crimea; 16, St George Fault.

base of the Black Sea western continental slope. The northern depocentre is partly overlapped by sediments of the Danube Delta, which form the prograding shelf. However, its position at the base of the palaeoslope is clearly seen in the reconstructed map (Tugolesov *et al.* 1985; Fig. 6).

The depocentres are separated by a saddle lying along-strike of a major continental fault zone – the Pechenega–Camena Fault. This zone forms the northeastern limit of the Moesian Plate, separating it from the early Cimmerian North Dobrogea Fold Belt. Along the supposed extension of the Pechenega–Camena Fault, the depocentres are sinistrally disposed. The steep northern slope of the northern depocentre is also along-strike of one of the most important faults in the adjacent portion of the continent –

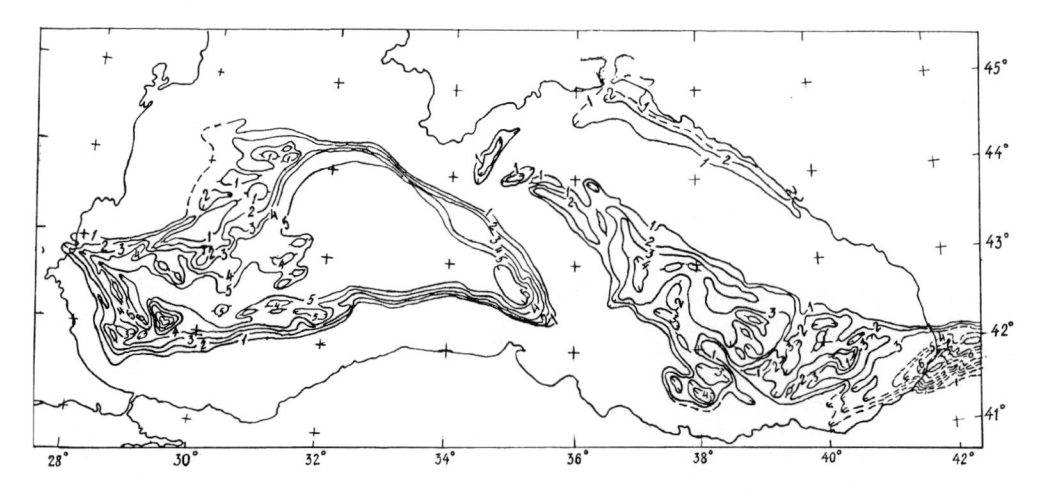

Fig. 4. Palaeocene–Eocene thickness map from Tugolesov (1989). Isopachs are at 1 km intervals.

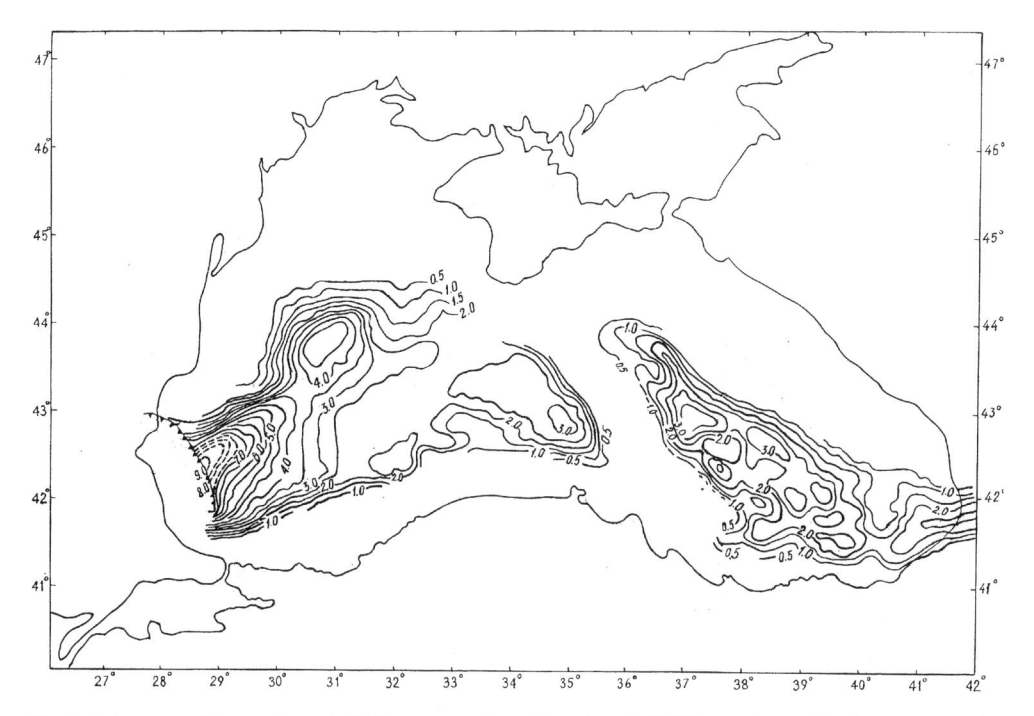

Fig. 5. Palaeocene–Eocene deposit thickness maps from Finetti et al. (1988) transformed by the present authors into the metric thickness scale. Isopachs are drawn with 0.5 km intervals.

the St George Fault. This fault bounds the North Dobrogea Orogen from the northeast and, as well as the Pechenega–Camena Fault, is traced across the shelf and up to the western continental slope of the Black Sea (Finetti *et al.* 1988). It seems, that in the structure of the depocentres, both faults acted as transverse or transcurrent features.

The maps provide important information about the relief of the basement top. In Tugolesov's map (Tugolesov 1989) the basement (seismic horizon K) in most of the WBSB is at 14 km or more depth, the deepest area coinciding with the flat central-eastern part of the sub-basin (Fig. 7). In the area of the depocentres, described above, the basement shallows to *c.* 9–10 km; by contrast, the EBSB displays a more complex basement relief with several highs and lows up to 11–12 km deep.

The map based on Finetti's data (Fig. 8) shows quite a different pattern. A depression of 15 km, i.e. the deepest part in the whole Black Sea, corresponds to the southwestern depocentre of the Palaeocene–Eocene sediments. This depression is separated from the rest of the WBSB by a vaguely defined barrier or ridge with a northwesterly trend in line with the Pechenega–Camena Fault. A less deep (up to 14 km), but still pronounced, basement depression corresponds to the northeastern depocentre. In the northeast this depression is limited by a northwesterly trending scarp, which sharply displaces the continental slope of the WBSB. In other words, both depocentres coincide with deep depressions in the pre-Cenozoic basement.

Speculating on the tectonic nature of the depressions and the corresponding depocentres, one can consider the possibility that they represent the oldest part of the Black Sea. Opening of the WBSB began in the Barremian–Aptian (Okay *et al.* 1994; Robinson *et al.* 1996), or in the Albian (Kazmin 1997; Schreider *et al.* 1997), as a result of back-arc rifting in the rear of the Pontide volcanic arc. If the depressions are early rifts they may contain not only the Palaeocene–Eocene sediments but also older (Lower and Upper Cretaceous) syn- and post-rift deposits; thus explaining the great thickness of the sedimentary pile in the depocentres. The shape of the depocentres, their relative displacement and apparent control by transcurrent faults favours a rift origin.

Important data in support of this concept

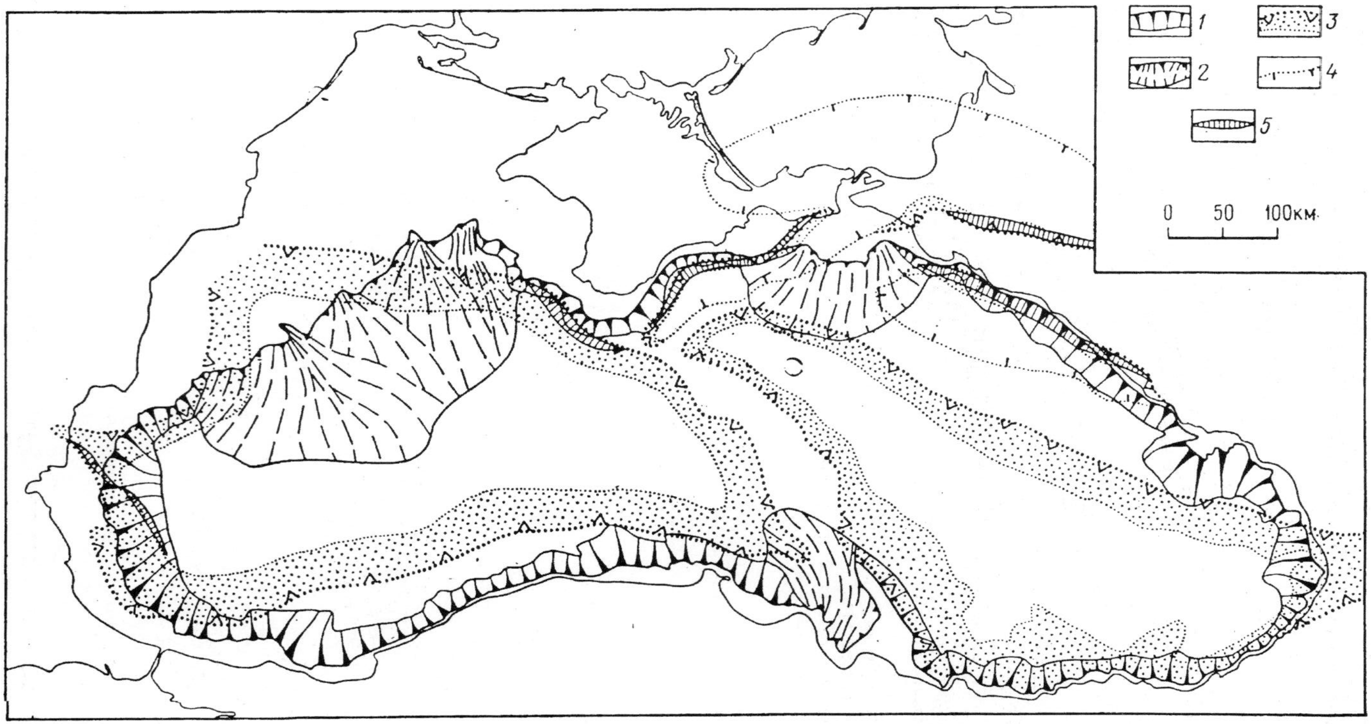

Fig. 6. Continental palaeoslopes of the Black Sea Basin [after Tugolesov *et al.* (1985)]. 1, Present-day continental slope; 2, major present-day delta; 3, ancient (Cretaceous–Palaeogene) continental slope; 4, outer limits of the Oligocene–Quaternary foredeeps; 5, concentration of faults on the slope.

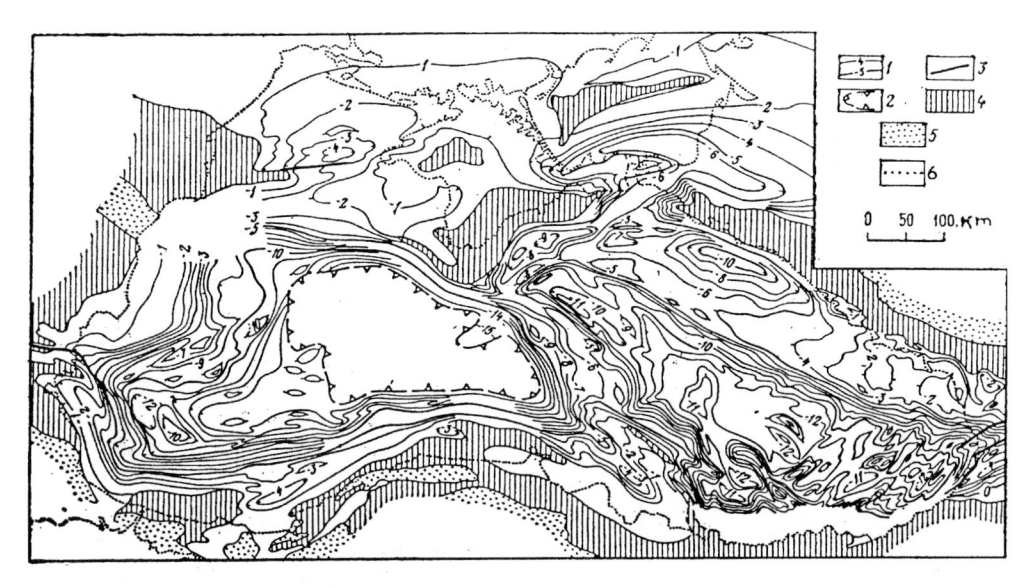

Fig. 7. Depth to the top of the pre-Cenozoic (in places pre-Late Cretaceous?) basement [after Tugolesov et al. (1985)]. 1, Isolines in km; 2, areas where basement was not resolved; 3, faults; 4, outcrops of Mesozoic rocks; 5, outcrops of pre-Mesozoic rocks; 6, present coastline.

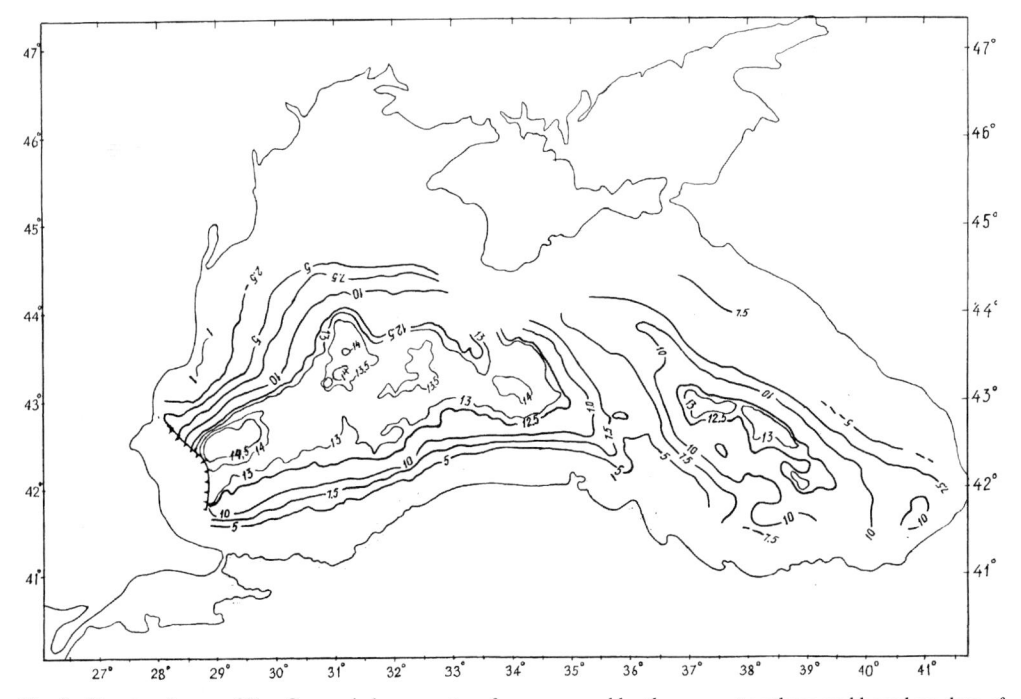

Fig. 8. Structural map of Pre-Cenozoic basement surface prepared by the present authors and based on data of Finetti et al. (1988). Isolines of the surface depth drawn with 2.5 km interval (and additional with 0.5 km interval). Tooth-line, front of folded Balkanides.

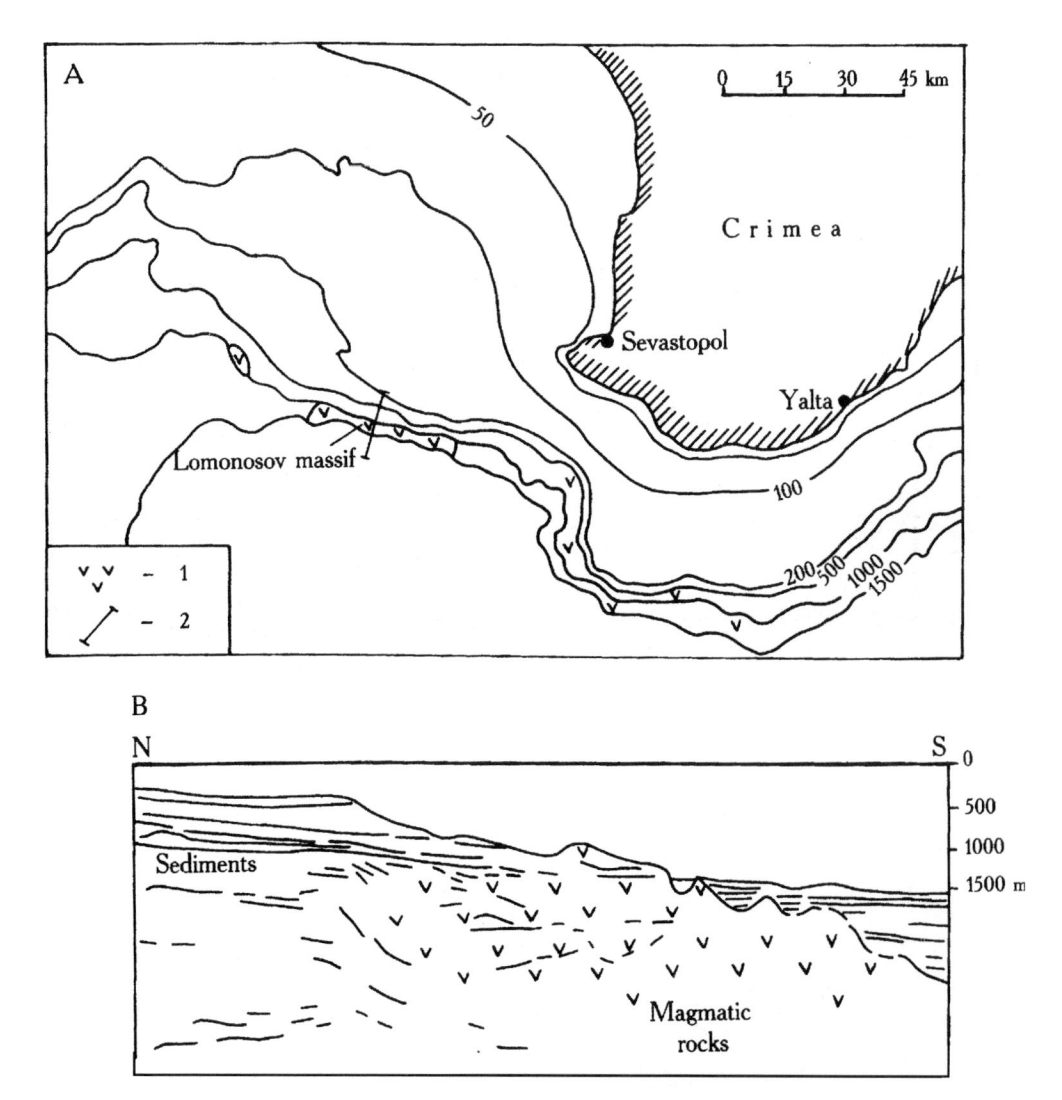

Fig. 9. (a) Position of the Lomonosov Massif (arc). 1, magmatic rocks; 2, position of cross-section. (b) Cross-section, based on seismic line [after Shnyukov et al. (1997)].

comes from the northern shelf and continental slope of the Black Sea, where the main tectonic feature is a system of grabens extending from Karkinit Bay and Tarkhankut Peninsula of Crimea, westwards for > 200 km. The system consists of at least two half-grabens with major faults on their southern sides (Bogaetz et al. 1986). A major phase of subsidence was apparent in Albian time when a sharp transition from thin Aptian shallow-water clastics to very thick (up to 2000 m) clastic, volcanic and volcaniclastic rocks of Albian age occurred. Volcanic rocks (mainly tuffs) belong to the basalt–andesite–dacite association and are usually referred to as 'calc-alkaline' (Bogaetz et al. 1986). Wide development of calc-alkaline material indicates the proximity of a volcanic arc, presumably south of the Karkinit Graben system (Zonenshain & Le Pichon 1986).

Indeed, the existence of this arc was proved by submarine geological studies. Early traverses by scientists from the Institute of Oceanology (Moscow), in the manned submersible *Argus*, discovered outcrops of andesitic–basaltic rocks on the continental slope of western Crimea (Zhigunov 1986). Even more important data were derived by Ukrainian scientists during submarine studies of the continental slope of the Black Sea west of Crimea (Fig. 9). Here extensive outcrops of basalts, andesito-basalts, dacites and rhyolites of the Lomonosov Massif were discovered at the base of the slope in the course of submarine traverses. Lavas and tuffs associated with intrusive rocks, including plagiogranites, tonalites, quartz diorites, diorites and gabbros (Shnyukov *et al.* 1997); the outcrops were traced for several tens of kilometres.

Geochemistry and petrology show that these rocks belong to boninitic and calc-alkaline series. The relationships between them were not established; however, the island-arc nature of the magmatic complexes is clear.

K–Ar age determinations of 36 samples yielded ages ranging from 140 to 70 Ma, with two peaks in the histogram corresponding to 130–110 and 90–70 Ma (Fig. 10). The most reliable determinations, based on quartz diorites, yielded ages from 103 to 126 Ma (Shnyukov *et al.* 1997), proving that the arc was of Cretaceous, most probably Aptian–Albian, age. However, a possibility remains that volcanism continued until the end of the Cretaceous up to *c.* 70 Ma (Maastrichtian).

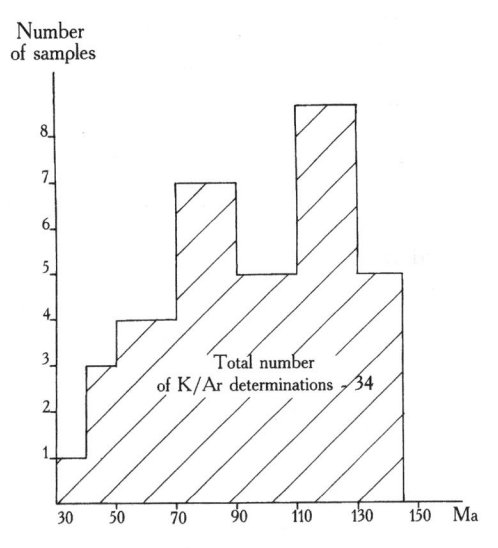

Fig. 10. Histogram of K–Ar ages obtained from magmatic rocks [after Shnyukov *et al.* (1997)].

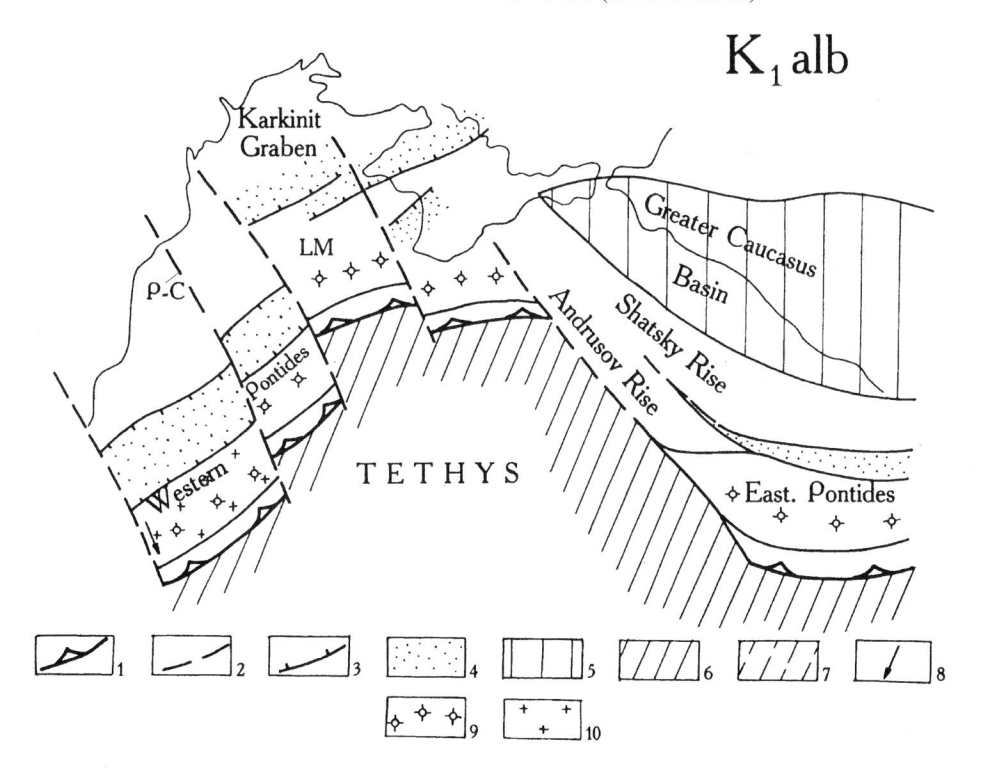

Fig. 11. Palaeotectonic reconstruction for Albian time. 1, subduction zone; 2, transverse fault; 3, normal fault; 4, Early Cretaceous grabens; 5, Early Mesozoic back-arc basin; 6, ocean; 7, Eocene back-arc basin; 8, block movement direction relative to Eurasia; 9, volcanic arc; 10, fragment of Moesian Platform; LM, Lomonosov Fault; P–C, Pechenega–Camena Fault.

Relative to this arc, the Karkinit grabens occupied an apparent back-arc position. Following initial rifting in the Albian (or Aptian?), the Karkinit Basin went through a long period of post-rift subsidence in the Late Cretaceous and Palaeocene, accompanied by accumulation of c. 4000 m of marine sediments (Bogaetz et al. 1986). In this respect it resembles the two depocentres in the west of the WBSB, where Palaeocene–Eocene and Upper Cretaceous(?) sediments attain a comparable thickness. The present authors believe, therefore, that the Karkinit grabens constituted the third segment of the early rift system of the WBSB, a suggestion strongly supported by the disposition of the described structures (Fig. 11).

The continental slope of the Lomonosov Massif has a step-like morphology, suggesting the presence of normal faults (Shnyukov et al. 1977). These faults are also interpreted from multichannel seismic lines, crossing the Odessa Shelf and continental slope (Finetti et al. 1988). The large Lomonosov Fault was detected at the base of the slope (Shnyukov et al. 1997), which was apparently formed in the course of rifting. This rifting split the volcanic arc and led to the opening of the greater portion of the WBSB, which is interpreted here as an interarc basin. The time of opening was definitely post-Albian and possibly, if the K–Ar determinations are taken at face value, latest Cretaceous or post-Cretaceous.

The Early Cretaceous history of the EBSB is not clear. According to Görür (1988), opening of the EBSB and the WBSB was simultaneous, starting in the Aptian–Albian but mainly occurring in the Cenomanian. However, in contrast to the Western-Central Pontides, the Eastern Pontides lack shallow-marine clastic rocks of Aptian–Albian age, interpreted as synrift deposits. In the Cenomanian–Campanian, the Eastern Pontides were overthrust from the south by an ophiolite nappe and display no evidence of extension and/or rifting (Okay & Şahintürk 1997).

Some indication of Albian(?) opening of the EBSB comes from Transcaucasus. In a north–south cross-section from the Gagra–Java zone (landward continuation of the Shatsky Rise), towards Adjaro–Trialetia, a transition from thin shallow-marine carbonates to thick clastic and volcaniclatic rocks can be seen (Tugolesov et al. 1985). These were most probably deposited in a back-arc basin behind the Eastern Pontides Arc (Fig. 12). This basin apparently constituted the landward extension of the EBSB, inverted later in the course of the Arabia–Eurasia collision. The time of opening of the remaining (and greater) portion of the EBSB is uncertain. In seismic lines interpreted by Finetti et al. (1988), tilted blocks of Mesozoic rocks onlapped by the Palaeocene–Eocene sediments can be observed. Systems of tilted blocks are interpreted on both margins of the EBSB (the Andrusov and Shatsky Rises) from where they subside below thick sediments of its axial zone. Consequently, the age of rifting was defined by the Italian authors as the end of Mesozoic to the beginning of Cenozoic.

Robinson et al. (1996) dated the major rifting in the EBSB as Early Palaeocene. They did not present any direct evidence for this age but emphasized that Eocene sediments onlap the Upper Cretaceous rocks on the Shatsky Rise and in the Pontides. Important evidence for the age of opening comes from land observations. According to Adamia et al. (1974), the EBSB continued eastwards to the Adjaro–Trialet Basin, which originated in the Eocene and was filled with a thick succession of volcanic and volcaniclastic rocks. The basin occupied a back-arc position relatively to the East Pontides volcanic arc and perhaps evolved with some limited spreading in the Middle Eocene (Adamia et al. 1974). Thus, the probable age of the EBSB is Early(?)–Middle Eocene.

If this is the case, the opening immediately followed a major compressional event in the Pontides in the Palaeocene–Early Eocene. According to Okay & Şahintürk (1987), at this time the Eastern Pontides collided with Taurides. The major compressional event, caused by this collision, was followed by regional extension and basaltic volcanism in the Middle Eocene coinciding with the opening of the EBSB. The situation with back-arc extension immediately following collision of an arc with a terrane is very common in the Tethyan belt and elsewhere (Kazmin 1999).

Discussion

In the early models of the Black Sea opening a simple solution was usually accepted (Zonenshain & Le Pichon 1986; Finetti et al. 1988; Kazmin 1997). Both the WBSB and EBSB were considered to have opened simultaneously in the Late Cretaceous, or earlier, due to southward drift of the Pontides Arc and clockwise rotation of the Andrusov Rise. This interpretation was revised by Okay et al. (1994), who suggested that in Albian–Cenomanian time a block, called the 'İstanbul Zone', rifted from the northern passive margin of the Black Sea by back-arc extension. In the east, this block was bounded by the West Crimea Fault. In

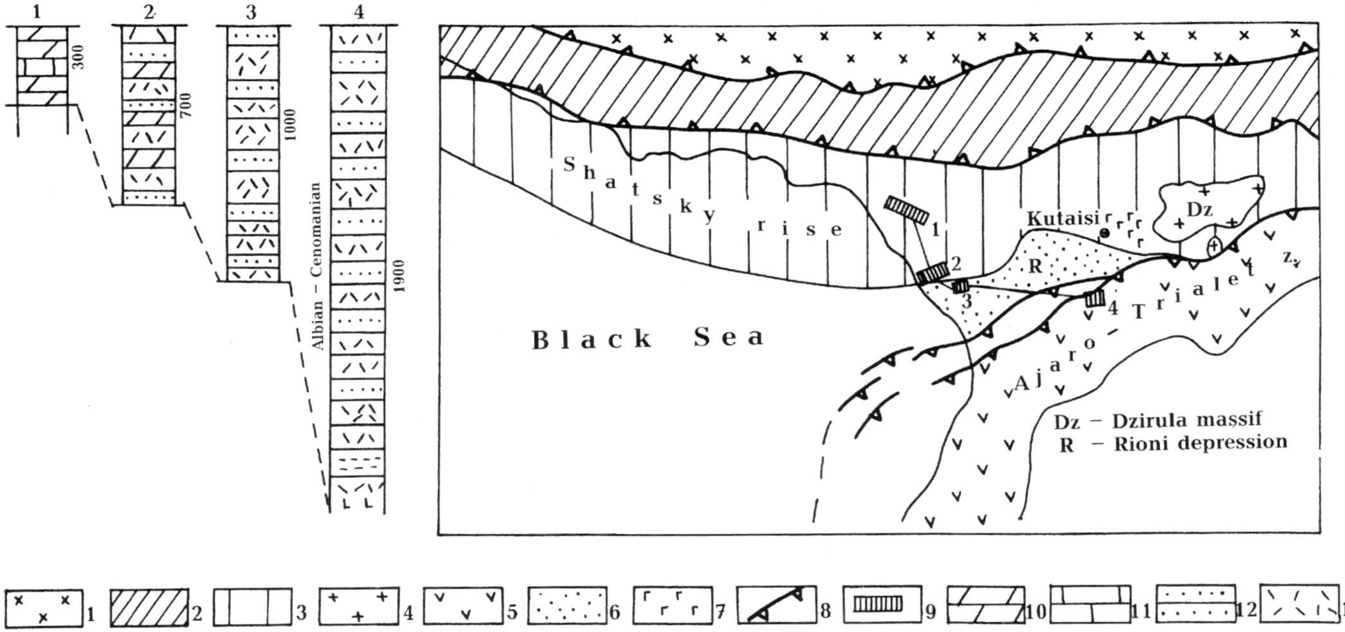

Fig. 12. Albian–Cenomanian sections of the Shatsky Rise and the Rhioni Depression [after Tugolesov *et al.* (1985)]. 1, Overthrust complexes of the Scythian Platform; 2, complexes of the Greater Caucasus marginal Basin; 3, Shatsky Rise; 4, Dzirula Massif; 5, Adjaro–Trialet volcanic zone; 6, foredeep; 7, alkaline basalts; 8, overthrust; 9, prospecting areas; 10, marl; 11, limestone; 12, sandstone; 13, volcanic rocks.

Santonian–Campanian time, the İstanbul Zone drifted southwards, leaving behind it the open WBSB. Simultaneously, the Mid Black Sea Ridge (the Andrusov Rise) and the Shatsky Rise were separated by a rift, i.e. the opening of the EBSB. According to this scheme, the Shatsky Rise rotated anticlockwise, implying subduction and/or shortening in the Greater Caucasus Basin.

The weak point of the model, from the kinematic point of view, is the mechanism of the EBSB opening. There is no evidence of Senonian subduction or compressional deformation in the Greater Caucasus. Moreover, there is no indication of what kind of tectonic regime caused anticlockwise rotation of the Shatsky Rise.

In another interpretation, suggested by Robinson et al. (1995, 1996), the kinematic model is a simplified version of the solution proposed by Okay et al. (1994). Robinson and co-authors open the WBSB by back-arc rifting of the Pontides from the northern Black Sea margin in the Mid-Cretaceous (Late Barremian–Albian). Using data on syn- and post-rift sequences in the rift structures of Western Pontides, they suggest that rifting in the WBSB was followed by spreading in Cenomanian time. However, in another interpretation the beginning of spreading is dated as middle–late Senonian (Tüysüz 1998). According to Robinson et al. (1996), the EBSB is much younger than that and opened in the Late Palaeocene and Early–Middle Eocene, again due to anticlockwise rotation of the Shatsky Rise. The last point meets the same objection as given above, i.e. there is no evidence of subduction or of any significant compressional deformation in the Greater Caucasus Basin at that time (Kopp & Shcherba 1998). As will be discussed below, the diachronous opening of the WBSB and the EBSB, implied by the model, is also doubtful. However, the Eocene age of the EBSB opening finds support in both marine and land geology. The Cenozoic sediments of the EBSB (and also the WBSB) were not affected by compressional deformation older than Late Eocene (Finetti et al. 1988; Kazmin 1997). The Late Eocene (pre-Maikopian) compression caused overthrusting of the Crimea, Pontides and Caucasus onto the EBSB and the WBSB, formation of the Oligocene–Miocene foredeeps (Tuapse, Gurian, Sorokin) and thrust faulting on the northwestern shelf of the Black Sea (Robinson et al. 1996). If the Cainozoic sedimentary pile of the EBSB included Palaeocene or older rocks, some evidence of the strong Palaeocene compression, known in the Pontides (Okay & Şahintürk 1997), should be seen in the basin. This, in combination with data from the Adjaro–Trialetia (Adamia et al. 1974) firmly points to an Eocene age for the EBSB.

According to interpretation in Figs 5 and 11, the width of the early post-rift basins in the WBSB did not exceed 150 km; the Early Cretaceous rifts at their base were probably narrower still. It is not likely that the transition to spreading occurred in these narrow grabens, which were probably continental rifts of the same type as the Karkinit Graben.

Data on the Lomonosov Massif show that opening of the greater part of the WBSB resulted from splitting of the Cretaceous volcanic arc – the time of this event is uncertain. It could have immediately followed the opening of the Karkinit back-arc basin in post-Albian time. On the other hand, volcanic activity in the Lomonosov Massif possibly continued until the end of Cretaceous times (Fig. 10) and so the splitting could be post-Cretaceous.

Of great importance in this respect are data on the thickness of the oldest sedimentary sequence in the central-eastern part of the WBSB. Assuming similar rates of terrigenous supply and subsidence, these data point to similar histories of WBSB and EBSB opening. In other words, both basins were actively opening in the Eocene.

In the western Pontides there is evidence of strong compression at the end of the Late Cretaceous (Yiğitbaş et al. 1998), possibly related to the Kırşehir–Pontides collision. After that sedimentary basins were opened under a transtensional regime and closed at the end of Early Eocene time. A period of extension followed in Middle Eocene time and it seems likely that this corresponds to opening of the central-eastern WBSB.

As mentioned above, the Eocene opening of the EBSB can hardly be explained by anticlockwise rotation of the Shatsky Rise. A simple alternative model suggests clockwise rotation of the Andrusov Rise, implying simultaneous opening of both the EBSB and the WBSB, associated with southward drift of the Pontides (Fig. 13). This model seems attractive because it has analogues in present-day rift structures. Two rift depressions are often separated by a diagonal crustal block or hinge and open simultaneously (e.g. the Danakil Block in the southern Red Sea and the Academician Ridge in Baikal). The driving force in the suggested model is also common: it is oceanward motion of the Pontides Arc induced by roll-back or other similar processes.

Finally, one more point should be discussed.

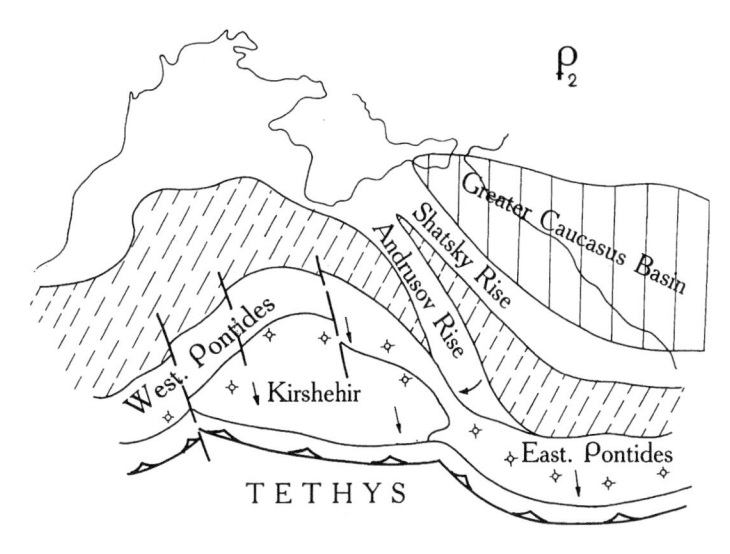

Fig. 13. Palaeotectonic reconstruction for Eocene time. Symbols as in Fig. 11.

A model was developed recently to describe syn- and post-rift stratigraphies and subsidence history of the WBSB and the EBSB (Robinson et al. 1995; Spadini et al. in press). This model is based on the assumption of basic differences in the evolution of the WBSB and the EBSB. For the WBSB, Mid-Cretaceous rifting and Cenomanian spreading are assumed, while for the EBSB the age of opening is Late Palaeocene–Eocene. Another assumption is that the thickness of sediments in the WBSB is much greater than in the EBSB (13 and 11 km, respectively). To explain these variations in sedimentation rates, thicknesses and facies, the authors accept the age and thickness of rifted lithosphere as basic parameters of the model (Robinson et al. 1995, table 1). They suggested that the WBSB opened due to rifting of thick (200 km), cold and mechanically strong lithosphere of the Moesian Platform, while in the case of the EBSB the rifted lithosphere was only 80 km thick, hot and weak.

It is easy to demonstrate that only a small portion of the WBSB originated due to rifting of the Moesian Platform. The latter is bound to the northeast by the Pechenega–Camena Fault (Figs 3 and 11), hence only the southwestern segment of the rift system shown in Fig. 11 is likely to have 'Moesian' basement. The remaining (and much greater) part of the WBSB originated by rifting of the much younger ('epihercynian') basement of the Scythian Platform. Similar basement is known in the Eastern Pontides (Şengör et al. 1984; Okay & Şahintürk 1997) and in the landward extension of the Shatsky Rise (the Dzirula Massif of Georgia). There is no doubt that the EBSB also formed on the 'epihercynian' basement and in this respect did not differ from the WBSB. In both basins this pre-rift basement was modified by a long succession of Mesozoic–Cainozoic rifting and collision events, and suprasubductional magmatism. So, it is unlikely that any part of the rifted lithosphere in the Black Sea region, except the westernmost segment of the WBSB, was thick, cold and mechanically strong. Since lithospheric properties are crucial for the discussed model, it can hardly be accepted without serious reconsideration.

Conclusion

The thickness map of the lower seismostratigraphic unit in the Black Sea Basin and the basement surface topography map (Finetti et al. 1988) were converted to a depth (metric) scale and compared with analogous maps of Tugolesov and co-authors (Tugolesov et al. 1985; Tugolesov 1989). Important differences were recognized between the two sets of maps.

In maps based on Finetti et al.'s (1988) data, two deep sub-basins were revealed in the western part of the WBSB and were interpreted as post-rift depocentres which evolved above Mid-Cretaceous rifts. A third segment of the early rift system is represented by the Karkinit grabens.

The rifts opened in Barremian–Albian times behind the Pontide volcanic arc, remnants of which were recently discovered on the northern continental slope of the WBSB. The central-eastern part of the WBSB opened as an interarc basin after splitting of the Cretaceous volcanic arc. The EBSB opened in Eocene times, mainly in the Middle Eocene, following a compressional phase in the Pontides (Okay & Şahintürk 1997). This opening, and the simultaneous opening of the central-eastern part of the WBSB, resulted from the southward drift of the Pontides, accompanied by clockwise rotation of the Andrusov Rise, a diagonal barrier connecting the two rift depressions.

The authors are very grateful for both reviewers of the paper for criticism and editing of the text. They are also greatly indebted to Erdin Bozkurt for his help in their work.

References

ADAMIA, SH., GAMKRELIDZE, I. P., ZAKARIADZE, G. S. & LORDKIPANIDZE, M. B. 1974. Adjharo–Trialet trough and the Black Sea formation problem. *Geotectonik*, **1**, 78–94 [in Russian].

BELOUSOV, V. V. & VOLVOVSKY, B. S. (eds) 1992. *Structure and Evolution of the Black Sea Earth Crust*. Nauka, Moscow [in Russian].

BOGAETZ, A. T., BONDARCHUK, G. K., LESKIN, I. V., NOVOSILETSKY, R. M., PAVLYUK, PALYI, A. M. ET AL. 1986. *Geology of the Ukrainian Shelf*. Kiev, Naukova Dumka [in Russian].

CHUMAKOV, I. S. 1993. The Paratethys Late Cenozoic Radiometric scale. *Priroda*, **12**, 68–75 [in Russian].

DERCOURT, J., ZONENSHAIN, L. P. RICOU, L.-E., KAZMIN, V. G., LE PICHON, X ET AL. 1986. Geological evolution of the Tethys belt from the Atlantic to the Pamir since the Lias. *Tectonophysics*, **123**, 241–315.

FINETTI, I., BRICCHI, G., DEL BEN, A., PIPAN, M. & XUAN, Z. 1988. Geophysical study of the Black Sea. *Bollettino di Geofisika Teorica ed Applicata*, **30**, 197–324.

GÖRÜR, N. 1988. Timing of opening of the Black Sea basin. *Tectonophysics*, **147**, 247–262.

HARLAND, W. B., ARMSTRONG, R. L., COX, A.V., CRAIG, L. E., SMITH, A. G. & SMITH, D. G. 1989. *A Geologic Time Scale*. Cambridge University Press.

KAZMIN, V. G. 1987. *The East African Rifts – Continental Break-up and the Origin of the Ocean*. Nauka, Moscow [in Russian].

—— 1990. Early Mesozoic reconstructions of the Black Sea–Caucasus region. *In*: RAXÚS, M. ET AL. (eds) *Evolution of the Northern Margin of Tethys*. Mémoires de la Société Geologique de France, Paris, Nouvelle Serie **154** (III, pt. 1), 147–158.

—— 1997. *Mesozoic to Cenozoic history of the back-arc basins in the Black Sea – Caucasus region*. Cambridge Arctic Shelf Program (CASP), Report, N656, Cambridge.

—— 1999. Mobility of subduction zones and subduction belts. *Doklady Rossiyskoi Akademii Nauk*, **366**, 525–529 [in Russian].

KOPP, M. L. & SHERBA, I. G. 1998. Caucasian basin in the Paleogene. *Geotektonika*, **2**, 29–50 [in Russian].

NEPROCHNOV, YU. P. 1960. Deep seismic structure of the crust under the Black Sea. *Bulletin Moskovskogo Obshchestva Ispytatelei Prirody (MOIP), Otdel Geologicheskiy*, **35**, 30–35 [in Russian].

——, NEPROCHNOVA, A. F. & MIRLIN, E. E. 1974. Deep structure of the Black Sea basin. *In*: DEGENS, E. T. & ROSS, D. A. (eds) *The Black Sea Geology, Chemistry and Biology*. American Association of Petroleum Geologists (AAPG) Publications, Tulsa, 35–49.

OKAY, A. İ. & ŞAHINTÜRK, O. 1997. Geology of the Eastern Pontides. *In*: ROBINSON, A. G. (ed.) *Regional and Petroleum Geology of the Black Sea and Surrounding Region*. American Association of Petroleum Geologists (AAPG) Memoir, **68**, 291–311.

——, ŞENGÖR, A. M. C. & GÖRÜR, N. 1994. Kinematic history of the opening of the Black Sea and its effect on the surrounding regions. *Geology*, **22**, 267–270.

ROBINSON, A. G., RUDAT, J., BANKS, C. & WILES, R. 1996. Petroleum geology of the Black Sea. *Marine and Petroleum Geology*, **13**, 195–233.

——, SPADINI, G., CLOETINGH, S. & RUDAT, J. 1995. Stratigraphic evolution of the Black Sea: inferences from basin modelling. *Marine and Petroleum Geology*, **12**, 821–835.

SANDULESCU, M. 1995. *Dobrogea within the Carpathian Foreland. IGCP Project No. 369. Field Guidebook: Central and North Dobrogea*. Geological Institute of Romania, 1–4.

SCHREIDER, A. A., KAZMIN, V. G. & LYGIN, V. S. 1997. Magnetic anomalies and the Black Sea basin age problem. *Geotektonika*, **1**, 59–70 [in Russian].

ŞENGÖR, A. M. C., YILMAZ, Y. & SUNGURLU, O. 1984. Tectonics of the Mediterranean Cimmerids: nature and evolution of the western termination of Paleo-Tethys. *In*: DIXON, J. A. & ROBERTSON, A. H. F. (eds) *The Geological Evolution of the Eastern Mediterranean*. Geological Society, London, Special Publications, **14**, 77–112.

SHCHERBA, I. G. 1994. The Black Sea Paleogene basin. *Bulletin Moskovskogo Obshchestva Ispytalelei Prirody (MOIP), Otdel Geologicheskiy*, **69**, 71–80 [in Russian].

SHNYUKOV, E. F., SHCHERBAKOV, I. B. & SHNYUKOVA, E. E. 1997. *Palaeo-island Arc in the Northern Black Sea*. Ukrainskaya Akademiya Nauk, Kiev [in Russian].

SPADINI, G., ROBINSON, A. & CLOETHING, S. Western versus Eastern Black Sea tectonic evolution: Pre-rift lithspheric controls on basin formation. *Marine and Petroleum Geology*, in press.

TUGOLESOV, D. A. (ed.) 1989. *Album of Structural and Thickness Maps of Black Sea Cenozoic Sediments*. Glavnoe Upravlenie Geodezii I Kartgrafii (GUGK), Moscow [in Russian].

—— (ed.) 1993. *Explanatory Notes to the Album of Structural and Thickness Maps of the Black Sea*

Cenozoic Deposits. Gosudarstvennoe Predpriyatie Nauchno-Issledovatelskiy i Proektny Institut (GP NIPI) Oceangeofizika, Gelendjik [in Russian].

——, GORSHKOV, A. S. & MEISNER, L. B. 1985. *Tectonics of the Black Sea Mesozoic Deposits*. Nedra, Moscow [in Russian].

TÜYSÜZ, O. 1998. Cretaceous–Eocene tectonic evolution of western Pontide sedimentary basins. *The Third International Turkish Geology Symposium*, METU – Ankara, Abstracts, 249.

YIĞITBAŞ, E., ELMAS, A. & YILMAZ, Y. 1998. The geological evolution of the western Pontides. *The Third International Turkish Geology Symposium*, METU – Ankara, Abstracts, 250.

ZHIGUNOV, A. S. 1986. Mesozoic deposits of the Alushta sector of the Crimea continental slope. *Okeanologia*, **326**, 655–666 [in Russian].

ZONENSHAIN, L. P. & LE PICHON, X. 1986. Deep basins of the Black Sea and Caspian Sea as remnants of Mesozoic back-arc basins. *Tectonophysics*, **123**, 181–211.

Neogene Paratethyan succession in Turkey and its implications for the palaeogeography of the Eastern Paratethys

NACİ GÖRÜR,[1] NAMIK ÇAĞATAY,[2] MEHMET SAKINÇ,[1] REMZİ AKKÖK,[1] ANDREY TCHAPALYGA[3] & BORIS NATALİN[1]

[1]*İstanbul Teknik Üniversitesi, Maden Fakültesi, Jeoloji Mühendisliği Bölümü, Ayazağa, TR-80626, İstanbul, Turkey (e-mail: cagatay@itu.edu.tr)*
[2]*İstanbul Üniversitesi, Deniz Bilimleri ve İşletmeciliği Enstitüsü, Vefa, TR-34470, İstanbul, Turkey*
[3]*Institute of Geography, Russian Academy of Sciences, 10917 Moscow, Russia*

Abstract: The Neogene marginal succession of the Eastern Paratethys (EP) crops out along the southern Black Sea coast and in the Marmara region of Turkey, and provides important clues to the tectono–sedimentary and palaeoceanographic conditions.

In the Tarkhanian stage, the southern margin of the EP basin was largely a carbonate platform covered by warm, marine waters. From the end of the Tarkhanian to the Early Chokrakian there was an overall emergence throughout the basin, which is indicated by an influx of siliciclastic sediments. The fossil assemblage indicates that normal marine conditions persisted during most of this period, except for a salinity reduction towards the end due to an eustatic isolation of the basin, which in turn led to anoxic bottom water conditions. The Late Chokrakian isolation became even more severe during the Karaganian as indicated by the endemic fossil assemblage indicating brackish–marine conditions. Carbonate platform conditions prevailed in the northern Pontides during this time. In the Early Konkian, the basin was reconnected briefly with the world ocean by a transgression from the Indo-Pacific Ocean. In the Late Konkian there was a return to brackish–marine conditions.

Lower Sarmatian sediments are absent in the southern margin of the EP, but elsewhere in the basin this stage is characterized by a widespread marine transgression. In the Middle–Late Sarmatian, the EP basin was partially isolated with freshening and anoxic bottom-water conditions. During this time there was a brief marine transgression from the Mediterranean into the Marmara region, but it did not reach the Paratethyan basin. The Pontian is characterized by an extensive transgression from the EP that inundated the Marmara and northeastern Aegean regions. The connection with the Marmara Basin was cut off during the Kimmerian and re-established during the Late Akchagylian, when the EP basin was inundated by the Mediterranean waters via the Sea of Marmara as a result of increased North Anatolian Fault activity and a short-term global sea level rise.

The Neotethys Ocean, which is identical with Suess's (1888, 1893) classical Tethys, was obliterated mainly during the Late Cretaceous–Tertiary by the collision of the dispersed pieces of Gondwanaland with Eurasia (Şengör 1984, 1987; Şengör et al. 1988). In the Neogene, this obliteration resulted in the formation of several basins of various sizes, including Paratethys and the Mediterranean Sea (Buchbinder & Gvirtzman 1976; Nevesskaya et al. 1985; Rögl et al. 1978).

The Paratethys occupied a vast area between the Rhone Valley in the west and the Aral Sea in the east (Fig. 1). It was generally a land-locked basin, although recurring connections to the Mediterranean Sea and the Indo-Pacific Ocean were established throughout its history. During isolation from these marine realms, its palaeoceanographic conditions were drastically changed and distinct facies with endemic fossils accumulated. As a result, a concept of regional stages for this intracontinental marine realm was established (Nevesskaja et al. 1985; Muratov & Nevesskaja 1986). Owing to the development of different ecosystems along its length, the east–west trending Paratethys was subdivided into three parts, namely the Western, Central and Eastern Paratethys (EP) (Senes 1959; Rögl & Steininger 1984). The Western Paratethys extended from the Rhone Valley to Bavaria. The Central Paratethys occupied the Alpine–Carpathian foredeep, and the intermontane basins between Austria and Ukraine. Finally, the EP covered the Ponto–Caspian or Euxinian regions, including the areas of the Black, the Caspian and the Aral Seas. A large number of studies have been carried out on the Paratethys in general and the EP in particular

From: BOZKURT, E., WINCHESTER, J. A. & PIPER, J. D. A. (eds) *Tectonics and Magmatism in Turkey and the Surrounding Area.* Geological Society, London, Special Publications, **173**, 251–269. 1-86239-064-9/00/$15.00
© The Geological Society of London 2000.

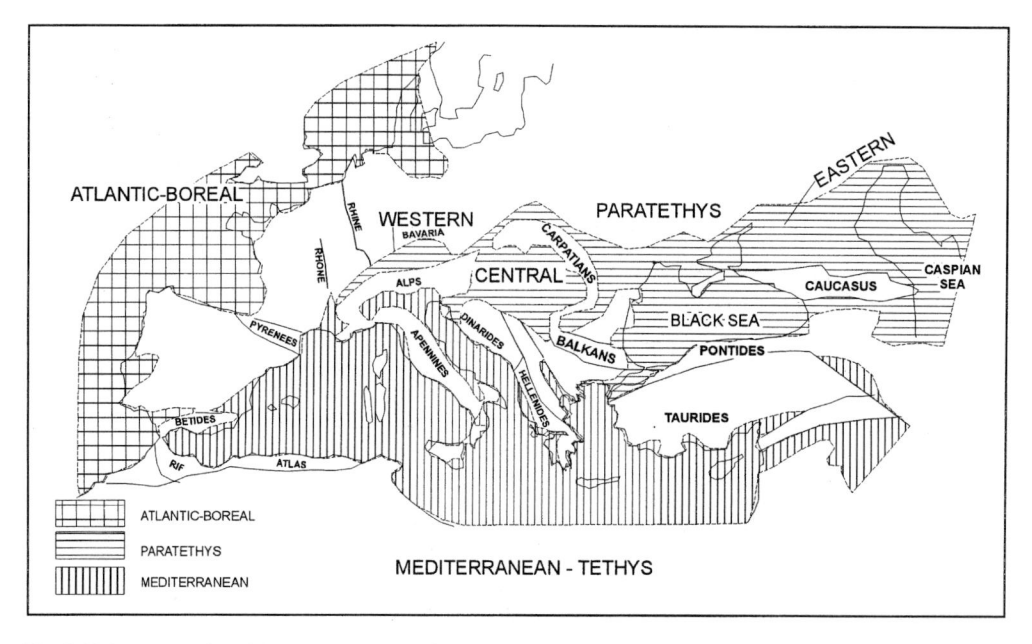

Fig. 1. Paleogeography of Paratethyan basins [after Rögl & Steininger (1983)].

(e.g. Andrussov 1918; Senes 1959, 1971; Gillet 1961; Jiricek 1975; Steininger *et al.* 1975; Dumitrica 1978; Khrushchov & Petrichenko 1979; Paramonova *et al.* 1979; Semenenko 1979; Baldi 1980, 1982; Cita 1981; Vavra 1981; Popov & Voronina 1983; Rögl & Steininger 1983, 1984; Nevesskaja *et al.* 1985; Muratov & Nevesskaja 1986; Jones & Simmons 1996, 1997). However, this topic has not attracted much interest in Turkey and therefore the Paratethyan geology of the country is relatively unknown (Chaput & Gillet 1938; Yalçınlar 1958; Özsayar 1977; Görür *et al.* 1998).

Neogene sediments representing the southern margin of the EP are well exposed in Turkey along the Black Sea coast and in the Sea of Marmara region (Fig. 2). They range in age from Tarkhanian to Akchagylian (Middle Miocene–Pliocene), with each stage being represented by distinct facies and diagnostic fauna. Their detailed examination provides important clues to both tectonosedimentary and palaeoceanographic conditions that prevailed during the Neogene on the southern margin of the EP in Turkey. The aim here is both to unravel these conditions and, by combining this group's and published data from the northern Black Sea region, reach more general conclusions on the palaeogeographic evolution of the EP as a whole.

Although in this paper mainly original data from the Pontides of northern Turkey are discussed, this group's palaeogeographic synthesis data from all around the Paratethys are also included. All these data are presented using seven palaeogeographic maps for the Tarkhanian–Cimmerian time interval (Figs 4–10). For the northern part of the EP region the main sources of data are from Muratov & Nevesskaya (1986) and Vinogradov (1967).

The Eastern Paratethyan succession along the Black Sea coast of Turkey

This succession crops out in the Sinop Peninsula, Trabzon, Akçaabat, Pazar and İğneada (Fig. 2). It is well exposed and most complete in the Sinop Peninsula, where it ranges in age from Tarkhanian to Sarmatian (Middle–Late Miocene). The succession is divided into eight formations and, for the first time, reliable data on the presence of the Konkian (Middle Miocene) and Cimmerian (Pliocene) stages in this region are reported (Fig. 3).

The Akliman Formation

This formation crops out in several sea cliffs in Akliman, 10 km west of Sinop, with a thickness of >15 m. It rests on Upper Cretaceous volcanic rocks above an angular unconformity and is itself disconformably overlain by the Kurtkuyusu Formation of Chokrakian age (Middle

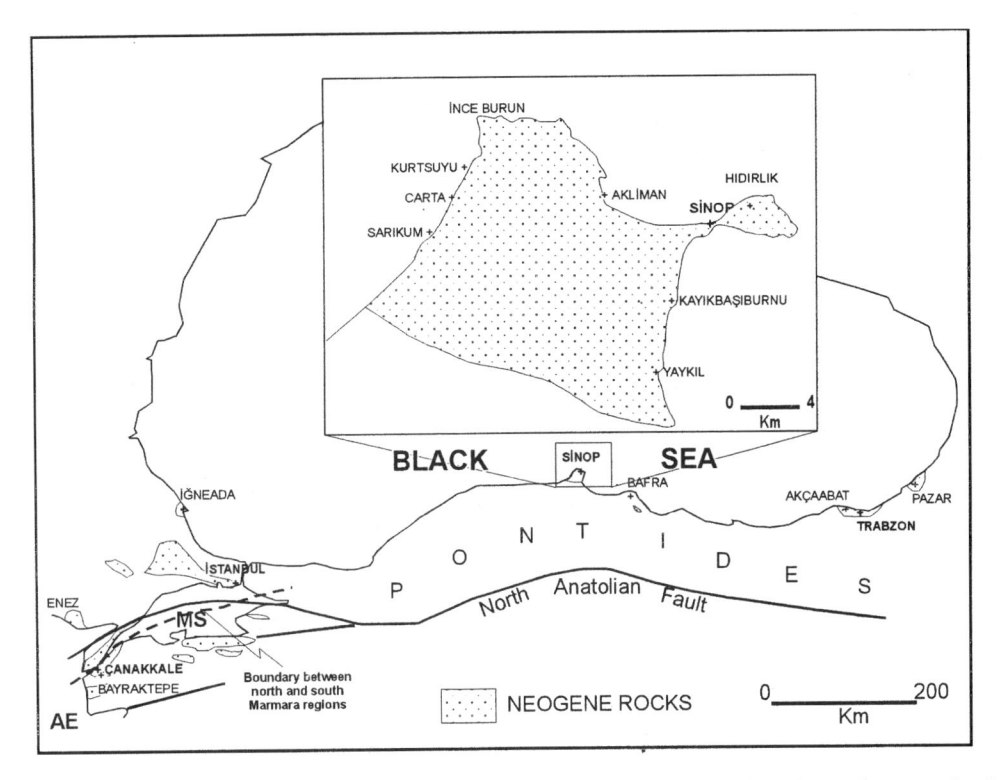

Fig. 2. Map showing the location of stratigraphic sections and observation points along the southern margin of the Eastern Paratethys (AE, Aegean Sea; MS, Sea of Marmara).

Miocene). The formation consists mainly of yellow, medium- to thickly bedded bioclastic and oolitic limestones that become marly upwards in the section. The bioclasts are represented by bivalves, echinoderms and small benthic foraminifera. The bivalves are particularly common in the limestones and include the following species: *Crassostrea gryphoides, Ostrea digitalina, Chlamys opercularis domgeri, Chlmys digitalina domgeri* and *Gibbula bajarunasi*. The foraminifera are mostly found in the marly beds and belong mainly to amphisteginid, rotaliid and globigerinid types. Based on these fossils, a Middle Tarkhanian (Terskian) age is attributed to the Akliman Formation. This fossil assemblage also leaves no doubt that this formation was deposited in a carbonate-shelf environment with normal salinity.

The Kurtkuyusu Formation

This formation is well exposed in Kurtkuyusu, on the western side of the Sinop Peninsula, where it is *c.* 36 m thick and disconformably overlies the Akliman Formation. At the base are dark coloured mudstones, containing silicified wood and plant fragments, and a small number of blocks and pebbles from the Akliman Formation. The mudstones are in part organic rich, pyritized, and contain thin fossil horizons with abundant bivalves, ostracods, worm tubes, bryzoans and miliolids. In the lower part of the formation, they alternate with thin, yellow to grey, friable bioclastic and oolitic limestones, which become dominant higher in the section. The bivalves, which were collected from both the mudstones and the limestones, contain: *Nuculana fragilis, Gibbula chokrakensis, Nassarius limatus, Bella costulata, Bittium digitatum, Bittium agibelicum, Spiratella tarchanensis, Abra scythica, Alvenius nitidus, Ditrupa* sp. and *Cardium* sp., and indicate a Chokrakian–(?)Karaganian age (Middle Miocene). A Chokrakian age is assigned to the formation because the overlying sediments contain Lower Karaganian fossils.

As indicated by the presence of both the silicified wood fragments and the limestone blocks derived from the underlying formation, the boundary between the Kurtkuyusu and

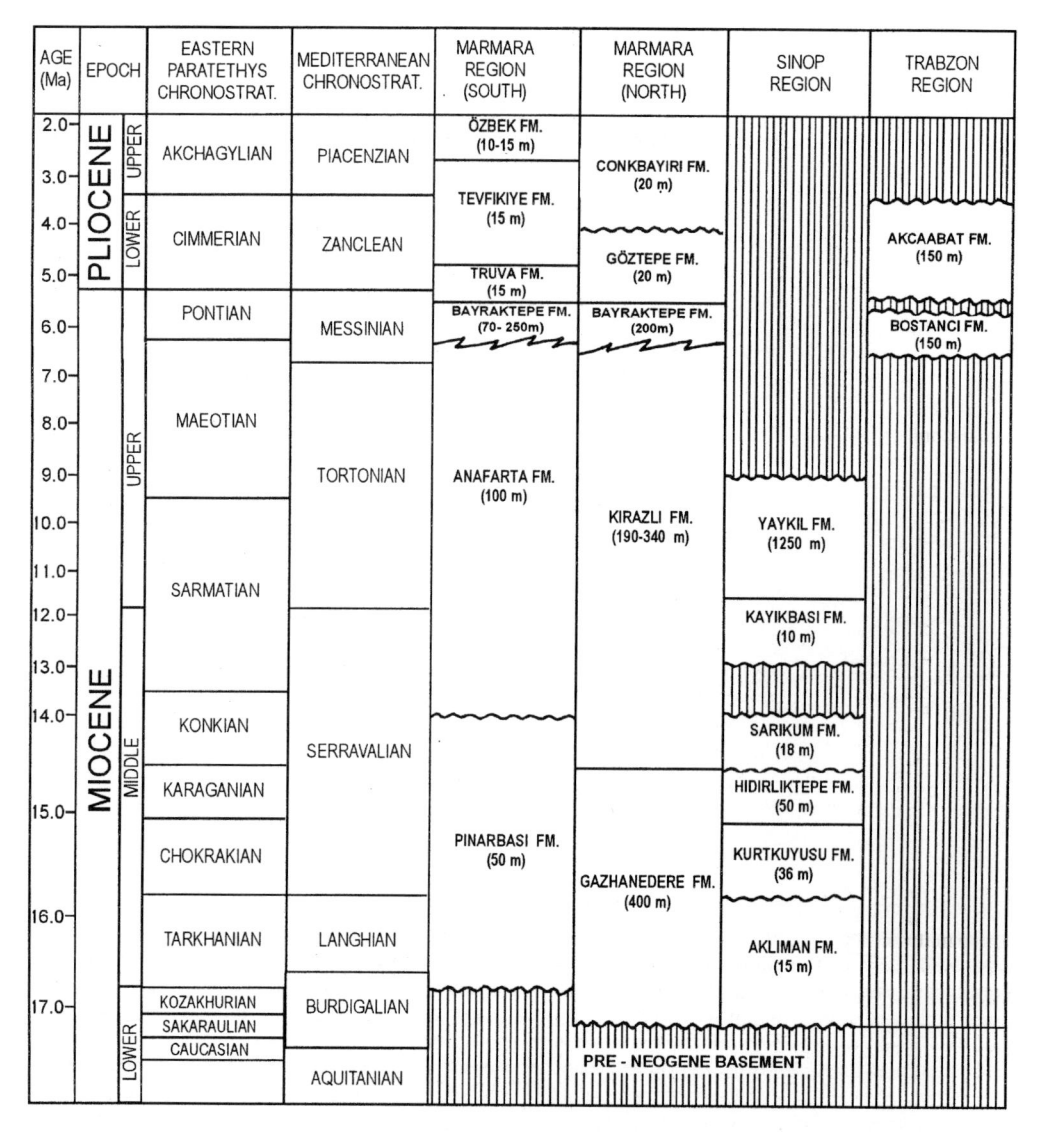

AGE (Ma)	EPOCH	EASTERN PARATETHYS CHRONOSTRAT.	MEDITERRANEAN CHRONOSTRAT.	MARMARA REGION (SOUTH)	MARMARA REGION (NORTH)	SINOP REGION	TRABZON REGION
2.0	PLIOCENE UPPER	AKCHAGYLIAN	PIACENZIAN	ÖZBEK FM. (10-15 m)	CONKBAYIRI FM. (20 m)		
3.0							
4.0	PLIOCENE LOWER	CIMMERIAN	ZANCLEAN	TEVFIKIYE FM. (15 m)	GÖZTEPE FM. (20 m)		AKCAABAT FM. (150 m)
5.0				TRUVA FM. (15 m)			
6.0		PONTIAN	MESSINIAN	BAYRAKTEPE FM. (70-250m)	BAYRAKTEPE FM. (200m)		BOSTANCI FM. (150 m)
7.0	MIOCENE UPPER	MAEOTIAN	TORTONIAN	ANAFARTA FM. (100 m)	KIRAZLI FM. (190-340 m)	YAYKIL FM. (1250 m)	
8.0							
9.0							
10.0							
11.0		SARMATIAN					
12.0						KAYIKBASI FM. (10 m)	
13.0	MIOCENE MIDDLE						
14.0		KONKIAN	SERRAVALIAN			SARIKUM FM. (18 m)	
15.0		KARAGANIAN				HIDIRLIKTEPE FM. (50 m)	
		CHOKRAKIAN		PINARBASI FM. (50 m)	GAZHANEDERE FM. (400 m)	KURTKUYUSU FM. (36 m)	
16.0		TARKHANIAN	LANGHIAN			AKLIMAN FM. (15 m)	
17.0	MIOCENE LOWER	KOZAKHURIAN SAKARAULIAN CAUCASIAN	BURDIGALIAN AQUITANIAN		PRE-NEOGENE BASEMENT		

Fig. 3. Generalized stratigraphic sections for the Neogene in southern Black Sea and Marmara regions.

Akliman Formations is a disconformity. The lithology of the Kurtkuyusu Formation represents a shallowing-upwards succession. Its upper parts were deposited under reduced salinity conditions.

The Hıdırlık Tepe Formation

This formation crops out in the Sinop Peninsula and İğneada. It is c. 50 m thick and well exposed at Hıdırlık Tepe in Sinop, where it consists of yellow to beige, medium- to thinly bedded, and locally cross-bedded, bioclastic and oolitic limestones. The formation rests conformably on the Kutkuyusu Formation. Allochems of these grain-supported carbonates include bioclasts, oolites, peloids, grapestones, lithoclasts and intraclasts. The bioclasts predominate, together with the oolites, over the other grains and are represented mostly by bivalves, benthonic and encrusting-type foraminifera, bryozoa and algae. In İnceburun (Sinop) the formation is 15 m thick and is composed mainly of fossiliferous sandstones and limestones (Özsayar 1977).

The following bivalves are found in these sediments in the Hıdırlık Tepe and Akliman

Formations, yielding an Early Karaganian (Arkhashenian) age: *Spaniodontella gentilis*, *S. (Davitaschvilia) intermedia*, *S. umbonata* and *Sevanella andrussovi* (Özsayar 1977). These molluscan species are endemic to the Paratethys and lived during the complete isolation of this sea from the world oceans (Nevesskaja *et al.* 1985).

The Sarıkum Formation

This formation crops out in the Sarıkum area on the western coast of the Sinop Peninsula and in Akçaabat, west of Trabzon. It disconformably overlies the Hıdırlık Tepe Formation and is, in turn, disconformably overlain by the Kayıkbaşı Burnu Formation. At its type locality in Sarıkum, it is *c.* 15 m thick and composed mainly of grey and laminated to thinly bedded mudstones and siltstones, which contain abundant well-rounded quartz grains and faecal pellets with some fossils, such as bivalves, echinoderms and ostracods. The bivalves are represented by *Acanthocardia andrussovi* and *Abra alba*, indicating a Konkian age for these sediments. In Akçaabat, the formation is lithologically similar but contains pelagic foraminifera (Özsayar 1977). The character of fossils in the Sarıkum Formation in general indicates that it was deposited in a marine environment with normal salinity.

The Kayıkbaşı Burnu Formation

This formation crops out at Kayıkbaşı Burnu on the eastern coast of the Sinop Peninsula. It is >10 m thick, although its exact thickness is not known because the base is not exposed. At the base of the exposed part of the formation it consists of conglomerates, passing upwards into a yellow and pebbly limestone package. The conglomerates are moderately sorted and clast supported. Most of the clasts are small and rounded quartz pebbles. The limestone package is represented by bivalves- and foraminifera-bearing limestones in the lower part, and by pebbly and oolitic limestones in the upper part. The bivalves consist mostly of *Mactra fabreana*, *Cardium fittoni*, *Mytilaster volhynicus*, *Barbotella intermedia* and *Calliostema hommairei*. The foraminifera are characterized by various species of Elphididae, including *Elphidium reginum*, *E. flexuoseum grilli*, *E. Koberi* and *E. aceleqtum minoriformis*. This fauna yields a Middle Sarmatian age for these sediments (Bessarabian, Middle–Late Miocene). The basal conglomerate and the Middle Sarmatian age indicates that this formation disconformably overlies the Sarıkum Formation (Fig. 3). The nature of both the bivalves and the foraminifera indicates that the Kayıkbaşı Burnu Formation was laid down in a brackish–marine environment.

The Yaykıl Formation

This formation is well exposed in coastal cliffs 12 km south of Sinop. It comprises > 1300 m thick bluish grey mudstones with sandy, silty, marly and carbonaceous intervals. In Yaykıl village, it dips at *c.* 60° to the south-southeast and is overlain with angular unconformity by Quaternary terrace deposits. In Kayıkbaşı Burnu, the Yaykıl Formation conformably overlies the Kayıkbaşı Burnu Formation. Thin bedding, lamination and burrowing are common in the Yaykıl mudstones. These fine-grained sediments contain no diagnostic fossils other than some species of Elphidiidae (mostly *Elphidium reginum*). Therefore, a Late Sarmatian age was attributed to the Yaykıl Formation mainly on the basis of the age of the underlying Kayıkbaşı Burnu Formation.

The Bostancı Formation

This formation crops out in Trabzon and Bafra. In its type locality in the Bostancı district of Trabzon, it is *c.* 120 m thick and consists of laminated to thinly bedded marl, clay (mostly smectite) and silt, locally with slump structures. It overlies Eocene volcanic rocks with angular unconformity and is itself unconformably overlain by Quaternary terrace deposits of Uzunlarian age (Pleistocene). It contains a fauna consisting of *Eupatorina littoralis*, *Paradacna abichi*, *Congeria* sp., *Dreissena* sp., *Valenciennius* sp. and *Prosodacna*, indicating an Early Pontian age (Novorussian, Late Miocene).

East of Bafra, the Bostancı Formation is composed mainly of 10–20 m thick, beige marls with a rich ostracod fauna of *Caspiolla* cf. *acuminata*, *C.* cf. *acronasuta*, *Loxoconcha* cf. *graciella*, *L.* cf. *subgranifera*, *Pontoniella* cf. *acuminata*, *Typhlocyprela* sp., *Hastacondana* sp., *Reticulocondana* cf. *reticulata*, *Euxinocythere* div. sp., *Euxinocythere* aff. *multicristata* and *Candona* sp. (Özsayar 1977). This ostracod fauna also supports a Pontian age for the Bostancı Formation. The lithology and slump structures suggest a possible slope environment for the deposition of this formation.

The Akçaabat Formation

This formation crops out in the Osmanbaba district of Akçaabat where it unconformably rests on Cretaceous volcanic rocks and is itself unconformably overlain by Quaternary deposits. The formation consists of 150 m thick, very coarse clastics, including mainly boulder and pebble conglomerates with some gravel and sand intercalations. The upper part of the formation is fossiliferous and contains *Dreissena inequivalvis* and *Viviparus* cf. *bifarcinatus*. This bivalve fauna gives a Kimmerian age (Early Pliocene). The facies changes indicate a transition from an alluvial fan in the lower part to a near-shore coastal environment in the upper part of the formation.

The Eastern Paratethyan succession in the Marmara region

In the Early–Middle Miocene, the Sea of Marmara region was characterized by a fluvio–lacustrine environment, where variegated conglomerate, sandstone, mudstone and fanglomerates were deposited (the Gazhanedere and Pınarbaşı Formations) (Fig. 3) (Görür et al. 1997; Sümengen et al. 1987; Siyako et al. 1989). The fluvio-lacustrine conditions continued during the deposition of the lower part of the siliciclastic Anafarta and Kirazlı Formations. In the early Late Miocene (Late Serravalian–Early Tortonian), a marine transgression took place along the northern coast of the future Sea of Marmara, depositing *Ostrea*-bearing calcareous sandstone, with shale interbeds (the Çınarlı Member of the Kirazlı Formation) (Görür et al. 1997); this was the first marine transgression from the Aegean Sea. It followed the shear zone of the incipient North Anatolian Fault (Şengör et al. 1985) but did not reach the EP basin. This conflicts with earlier studies that infer connections of the EP basin with the Mediterranean via the Marmara region during the Chokrakian, Karaganian and Konkian (e.g. Muratov & Nevesskaya 1986).

Following this first marine transgression, the Marmara region became a transitional waterway between the Paratethys in the north and the Mediterranean in the west (Fig. 2). The Paratethyan succession in this region is characterized by both marine and quasi-marine facies, in places alternating with one another. In the southern Marmara region, the succession consists of the Pontian (latest Miocene) Bayraktepe Formation and the Akchagylian–Chaudian (Pliocene–Pleistocene) Özbek Formation (Fig. 3), separated by the Truva and Tevfikiye Formations (latest Miocene–early Late Pliocene). The latter two formations consist of fluvio–lacustrine clastic sediments and algal limestones that indicate disconnection with both the Paratethyan and the Mediterranean realms. In the northern Marmara region, the only Paratethyan formation is the Bayraktepe Formation, which is there overlain by the Mediterranean *Ostrea* bank facies of the Göztepe Formation (Çağatay et al. 1998), in turn overlain by alluvial fan deposits of the Conkbayırı Formation (Yaltırak 1995).

The Bayraktepe Formation

The Bayraktepe Formation crops out extensively in circum-Sea of Marmara regions (Chaput & Gillet 1938; Sümengen et al. 1987; Siyako et al. 1989; Görür et al. 1997; Çağatay et al. 1998) (Fig. 2). Around the Dardanelles Strait, it conformably overlies the Anafarta and Kirazlı Formations (Fig. 3). The formation consists mainly of up to 250 m thick oolitic and bioclastic limestones with basal clastic rocks. The bioclastic limestones are rich in *Mactra*, gastropods and ostracods. In Enez, these limestones contain abundant *Dreissena* sp., *Paradacna abichi* and *Cardium edulis*. The molluscan fauna of the Pontian Bayraktepe Formation is endemic to Paratethys and indicates deposition in a shallow, brackish–marine environment.

The Özbek Formation

The Özbek Formation crops out south of the Dardanelles Strait and is intercepted in offshore wells in the Sea of Marmara (Marathon Petroleum 1975; Erol & Çetin 1995; Toker & Şengüler 1995; Görür et al. 1997) (Fig. 2). It consists of fossiliferous conglomerate and sandstone at the base and limestone at the top. The fauna in the clastic rocks is predominantly brackish–marine gastropods (e.g. *Bittium reticulatum*) of Parathethyan affinity, whereas the fauna in the limestone includes both brackish and normal marine (Mediterranean) types, such as bivalves (e.g. *Veneripus aurea* and *Ostrea edilus*), foraminifera and nannofossils (*Braarudosphaera bigelowi, Coccolithus pelagicus, C. pliopelagicus, Helicosphaera kaptneri, Gephyrocapsa caribbeanica, Pseudoemiliana lacunosa* and *Discoaster brouweri*) (Toker & Şengüler 1995). On the basis of this faunal content, the Özbek Formation is of Akchagylian age. The depositional conditions were initially shallow and brackish–marine but later became marine.

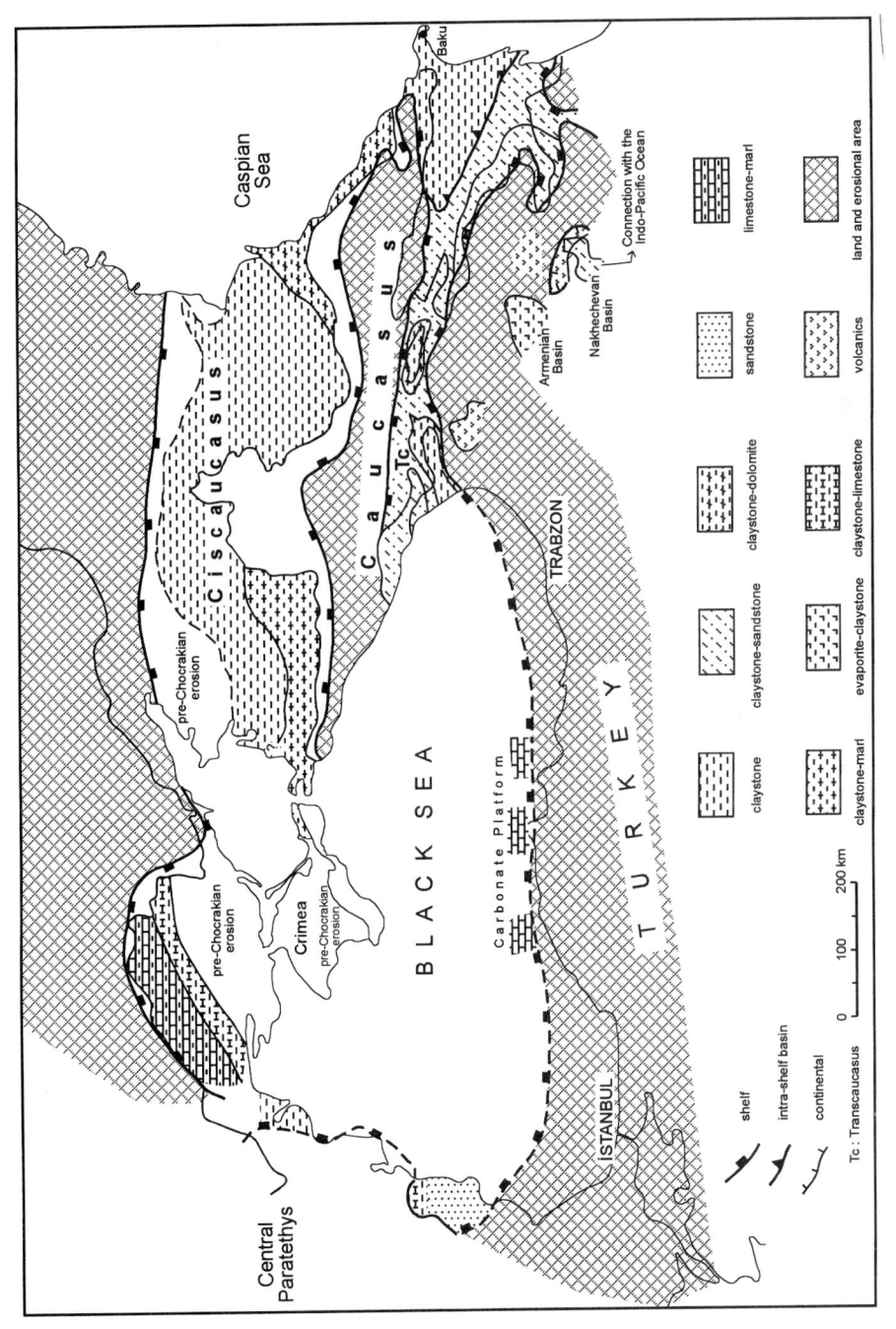

Fig. 4. Palaeogeographic map of the Eastern Paratethys and adjacent regions during the Tarkhanian (16.5–15.5 Ma BP).

Palaeogeographic evolution

The Tarkhanian (16.5–15.5 Ma BP)

The Tarkhanian fauna of the EP indicates that during this time this basin was largely marine with warm waters (Fig. 4). A connection to the world's oceans probably existed through the Western Paratethys, as the Bitlis–Zagros Ocean was being obliterated during Tarkhanian time (Şengör 1979; Şengör & Yılmaz 1981; Dewey et al. 1986; Görür 1988). Various sedimentary facies accumulated in the EP. In the northern Black Sea region, including the Crimea, Ciscaucasus and the Transcaucasus, large amounts of mixed deposits, comprising claystones, marls and limestones, were deposited. However, two contrasting facies occurred in the Armenian and the Nakhechevan Basins of the Transcaucasus. In the former, the Tarkhanian (and actually the whole Neogene) was represented by volcanic and volcaniclastic rocks [Palaeontological Institute (RAS) 1996], and in the latter by calcareous sandstones (Fig. 4). At this time, the Nakhechevan Basin was a northern embayment of the Indo-Pacific Ocean.

In the northern Black Sea region, the Tarkhanian sediments accumulated on the Maykopian facies with a transitional contact, indicating continuation of marine sedimentation from the Early Oligocene. The angular unconformity at the base of the equivalent deposits in the Pontides (the Akliman Formation) implies that the depression did not extend to this region in the south. During the Maykopian, the Pontides were elevated and eroded. Their northernmost part was transgressed by the sea in Middle Tarkhanian time and became a carbonate platform with bioclastic and oolitic limestones (the Akliman Formation). The absence of detrital sediments in the lower part of the carbonate succession indicates that the northern Pontides were beyond their reach during Early Tarkhanian time. Such sediments began entering the area at the end of this time, as indicated by the marly facies in the upper part of the carbonate succession.

The Chokrakian (15.5–13.8 Ma BP)

At the end of the Tarkhanian, there was probably an overall emergence throughout the EP region, indicated by a decrease in salinity, the regional disconformity between the Tarkhanian and the overlying Chokrakian deposits, and, in places (e.g. Ciscaucasus), by erosion of the Tarkhanian sediments. In the Transcaucasus region, however, the Tarkhanan–Chokrakian succession is continuous.

The presence of Tarkhanian limestone clasts in the Chokrakian sediments of the northern Pontides (the Kurtkuyusu Formation) implies that erosion was also associated with emergence on the southern margin of the EP. The reason for this emergence was probably tectonic because, following the final closure of the Neotethys Ocean in the south, the tectonic regime in both the Pontides and the Transcaucasus switched from extensional to compressional during the Miocene (Şengör & Yılmaz 1981; Banks et al. 1997; Okay & Şahintürk 1997; Yılmaz et al. 1997; Görür et al. 1998). Thrust faulting and folding in the outer part of this belt began in the Early Miocene and reached a peak during the Late Miocene. However, emergence may also result in part from a significant eustatic contribution, since the global sea-level curve also shows a worldwide drop in sea level at the end of the Tarkhanian (Haq et al. 1988).

During the Chokrakian, the isolation of the EP continued and the palaeogeographic picture of the EP was similar to that of the Tarkhanian. However, thicker accumulation of fine clastic sediments persisted and extended over a larger area (Fig. 5). The carbonate platform in the northern Pontides was also converted into a clastic shelf. In all of these areas, deposition took place mainly under quasi-marine conditions with a gradual decrease of salinity, as inferred from the fossil content of the Chokrakian sediments. Benthonic foraminifera, such as *Elphidium*, *Nonion* and *Quinqueloculina*, in the upper part of the Chokrakian succession indicate a significant salinity reduction, particularly towards the end of this time interval (Nevesskaya et al. 1985; Jones & Simmons 1997). Isolation from the open oceans led, in part, to the development of H_2S-rich anoxic bottom waters, as indicated by the dark colour, and organic-rich and authigenic mineral (i.e. pyrite and jarosite) contents of the Chokrakian sediments. It also enhanced the endemic faunal evolution throughout the EP (Goncharova 1989). The stratigraphic signature of this isolation appears to be in a good agreement with the global sea-level chart (Haq et al. 1988), thus indicating a eustatic contribution to this event.

The Karaganian (13.8–12.5 Ma BP)

In contrast with the dominant carbonate deposition in the northern Pontides (the Hıdırlık Tepe Formation), most of the EP

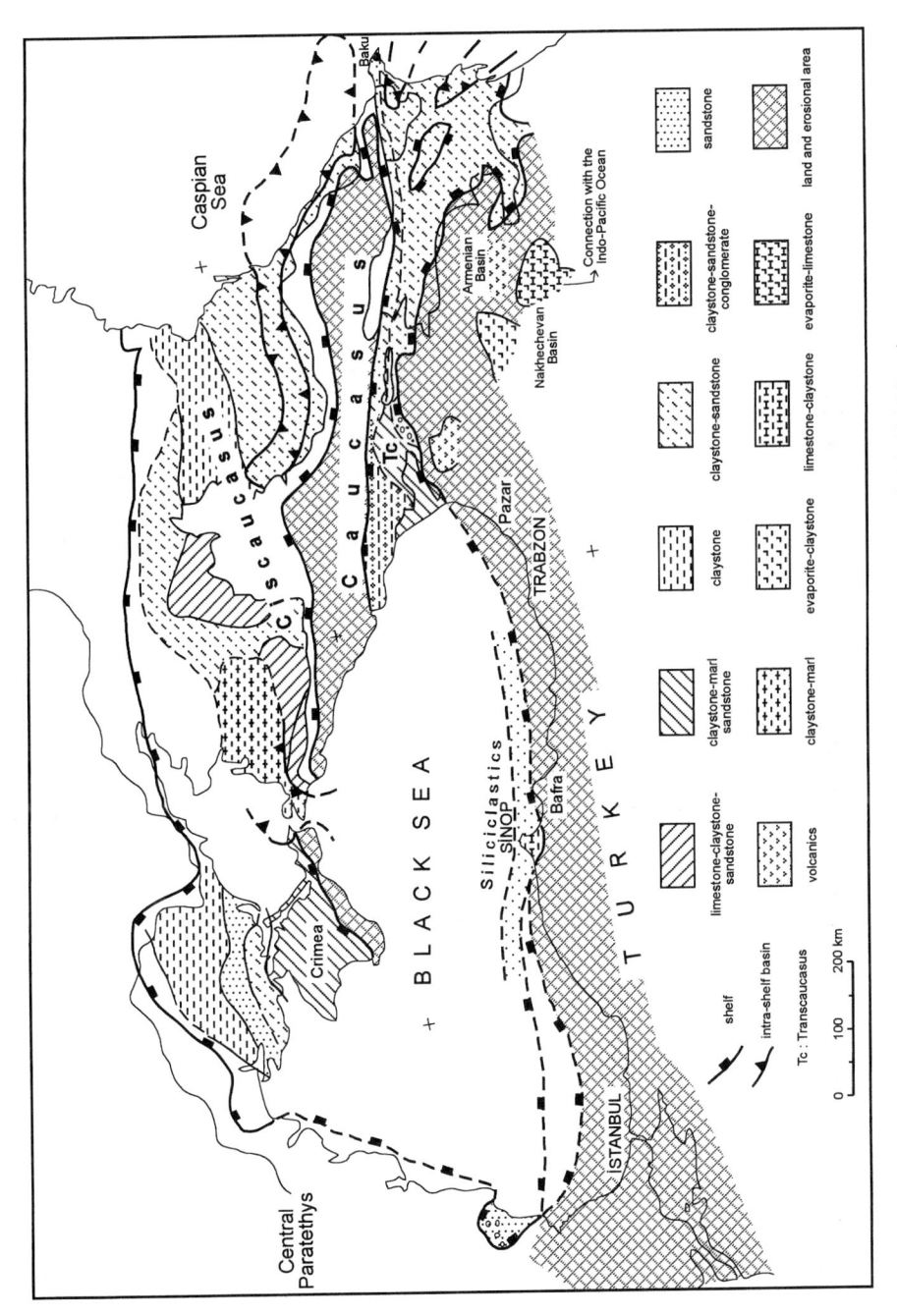

Fig. 5. Palaeogeographic map of the Eastern Paratethys and adjacent regions during the Chokrakian (15.5–13.8 Ma BP).

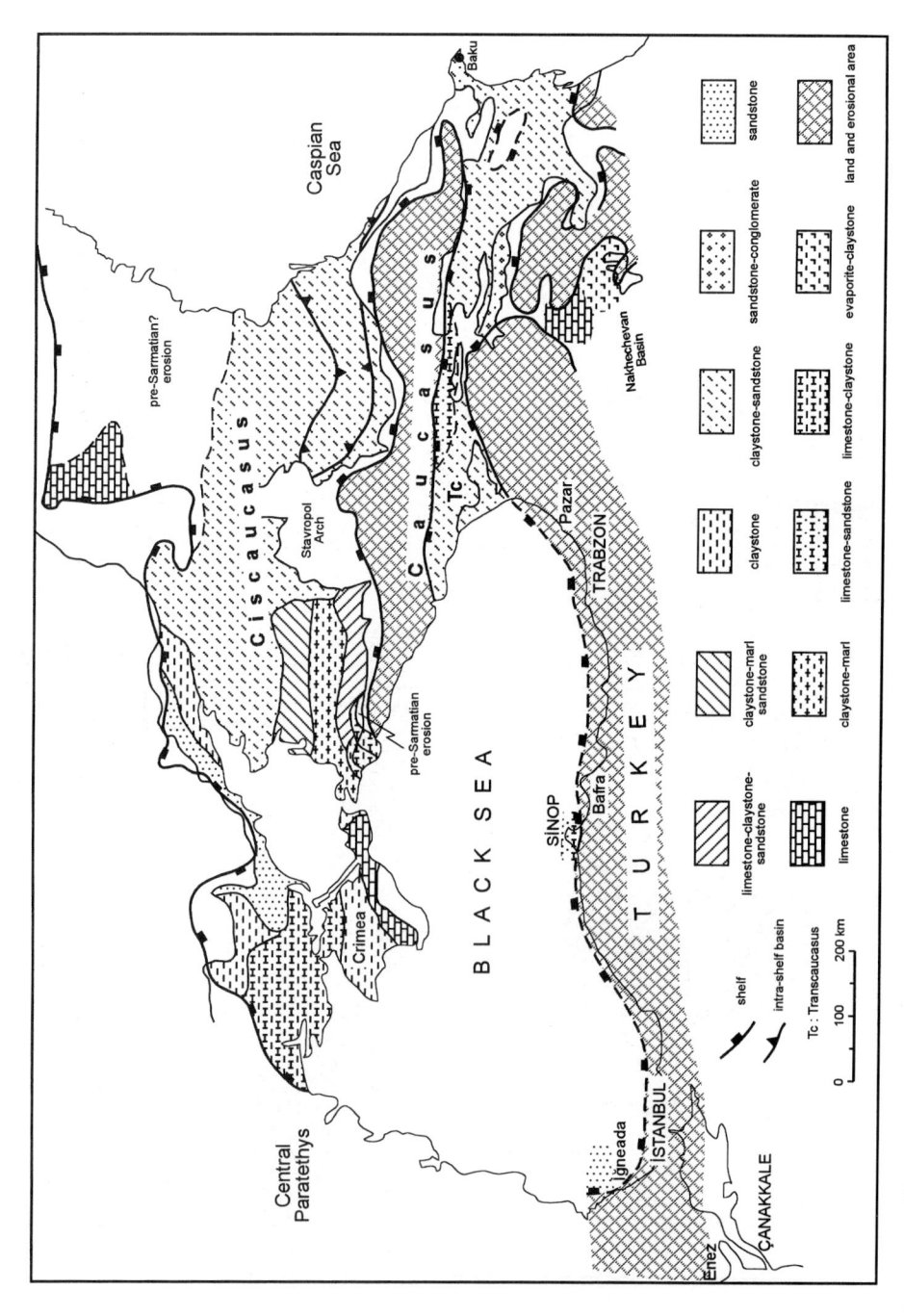

Fig. 6. Palaeogeographic map of the Eastern Paratethys and adjacent regions during the Karaganian (13.8–12.5 Ma BP).

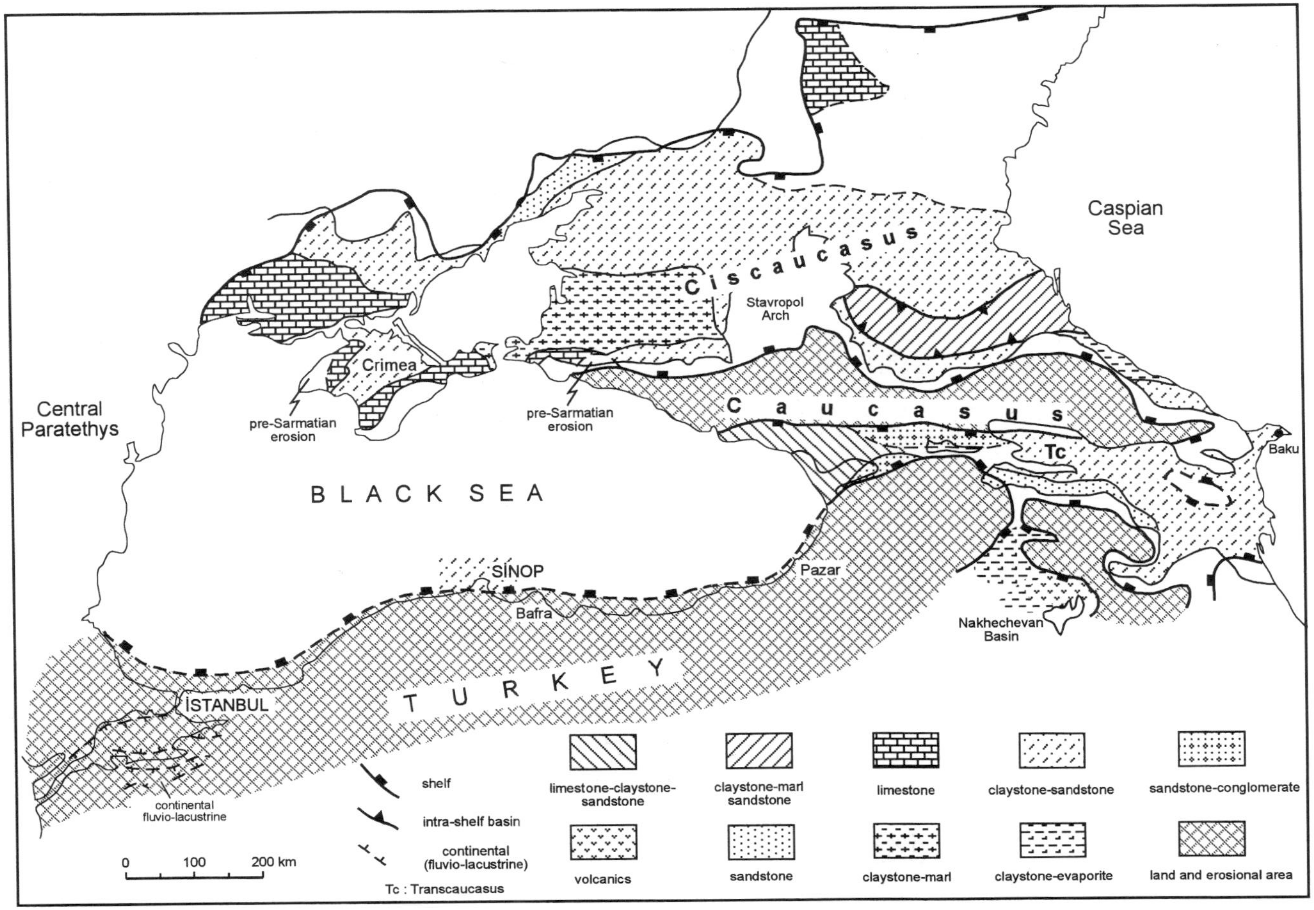

Fig. 7. Palaeogeographic map of the Eastern Paratethys and adjacent regions during the Konkian (12.5–10.5 Ma BP).

remained during the Karaganian as a site of mainly fine-grained siliciclastic sedimentation with little carbonate deposition (Fig. 6). The fossil assemblage of these sediments indicates that sedimentation took place under brackish–marine conditions. The Chokrakian isolation of the EP from the neighbouring basins became more effective during this time, except perhaps in the Late Karaganian when a brief connection with the world ocean was established (Muratov & Nevesskaya 1986). A similar restriction also developed in the Central Paratethys where thick evaporites were deposited in the northern part of the basin during the Middle Badenian, including the Karaganian (Rögl et al. 1978; Jinaridze 1979; Khrushchov & Petrichenko 1979; Rögl & Steininger 1984). Evaporite deposition was otherwise not common in the EP and seems to have been restricted since the Tarkhanian mostly to the Nakhechevan Basin of the Transcaucasus. This basin established its first connection with the EP during this time.

The Konkian (12.5–10.5 Ma BP)

Although the Konkian sediments of the Northern Pontides mostly contain normal marine fauna, the equivalent deposits in most parts of the EP include both marine and brackish marine fossils; the latter occur in the upper part of the Konkian sections. This distribution of the fauna indicates that communication with the world's oceans was re-established at the beginning of this time, with the inflow of saline waters into the EP. A coeval transgression also took place in the neighbouring Central Paratethys. The transgression was most probably from the Indo-Pacific Ocean because, by this time, the Bitlis–Zagros Ocean had already closed and the Sea of Marmara was a site of continental deposition (Buchbinder & Gvirtzman 1976; Rögl et al. 1978; Görür et al. 1995, 1997) (Fig. 7). The Early Konkian transgression flooded the previous brackish–marine areas and deposited, in most places, dark coloured claystones and marls with sandstones and limestones. The equivalent rocks in the Armenian and Nakhechevan Basins were characterized by volcanic rocks and brown claystones, with minor amounts of sandstones, siltstones and salts. The normal marine conditions were short lived because, towards the end of the Konkian, brackish–marine conditions prevailed throughout the EP, as indicated by the brackish–marine fauna of the Upper Konkian sediments.

The Sarmatian (10.5–8.2 Ma BP)

The faunal content and the stratigraphy of the Sarmatian sections from various regions of the EP indicate that, during this time, notable changes in both the depositional environment and the palaeogeographical framework took place. In the Early Sarmatian, relatively deeper, normal (or near-normal) marine conditions were widespread (Fig. 8). Under these conditions coarsening-upward, regressive siliciclastic successions developed. Towards the end of the Sarmatian, the salinity of the EP was greatly reduced (Papp 1963) and most of its basins were filled with lagoonal or fluvio–lacustrine sediments. At this time, in the Black Sea, as a part of the EP, mainly black clays were deposited (Hsü 1978). It was probably cut off from coarse clastic sources and devoid of biogenic sediments. Similar conditions also existed in the northern Pontides (the Yaykıl Formation). The great thickness of the Yaykıl Formation can be explained by subsidence caused by the Neogene thrusting, which is well established by seismic profiling on the southern Black Sea margin (Finetti et al. 1988). The Black Sea became isolated from the Caspian Sea for the first time towards the end of the Sarmatian (Tchapalyga 1985; Jones & Simmons 1997).

The marine to quasi-marine character of the Early Sarmatian fauna indicates that, after the Late Konkian isolation, a well-marked and far-reaching transgression took place in the EP region. Following this transgression, the huge intracontinental Paratethyan sea was progressively freshened during the Middle–Late Sarmatian, with a rapidly changing endemic faunal development. At the same time, a brief incursion of Mediterranean waters into the future Marmara Basin occurred along the broad shear zone of the incipient North Anatolian Fault, but it did not reach the EP basin (Görür et al. 1997) (Fig. 8). Therefore, at this time, the EP basin was isolated from any marine influence and began breaking down into small local basins.

The Pontian (6.9–5.5 Ma BP)

Brackish–marine conditions prevailed over the EP during the Pontian and resulted in the accumulation of fine-grained clastic sediments, marls and limestones [Palaeontological Institute (RAS) 1996] (Fig. 9). The existence of these sediments around the Dardanelles Strait and north of the Saros Gulf (the Bayraktepe Formation) indicates that, during the Pontian the Paratethys was linked with the incipient Sea of Marmara and the northeastern Aegean region

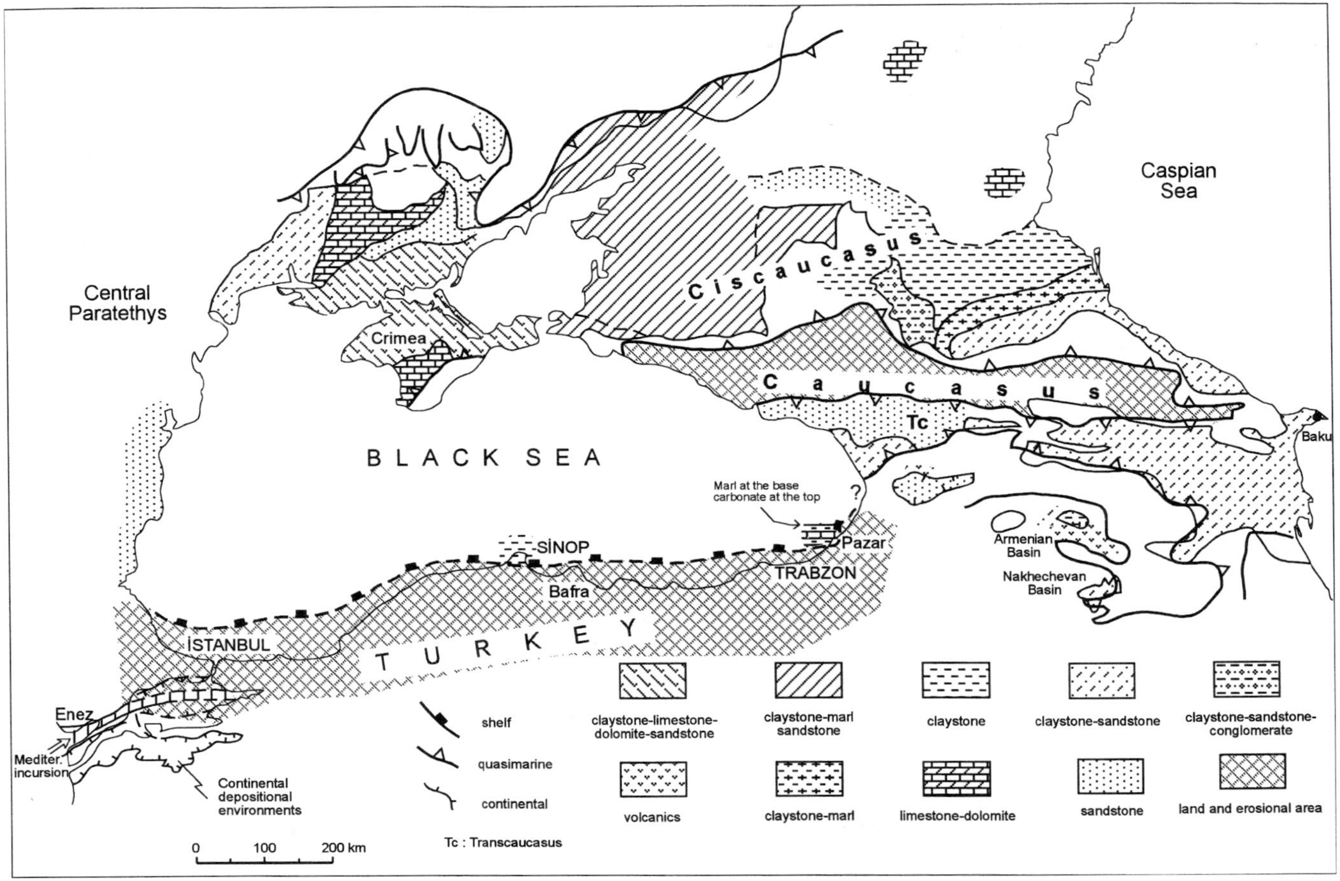

Fig. 8. Palaeogeographic map of the Eastern Paratethys and adjacent regions during the early Middle–Late Sarmatian (10.5–9.2 Ma BP).

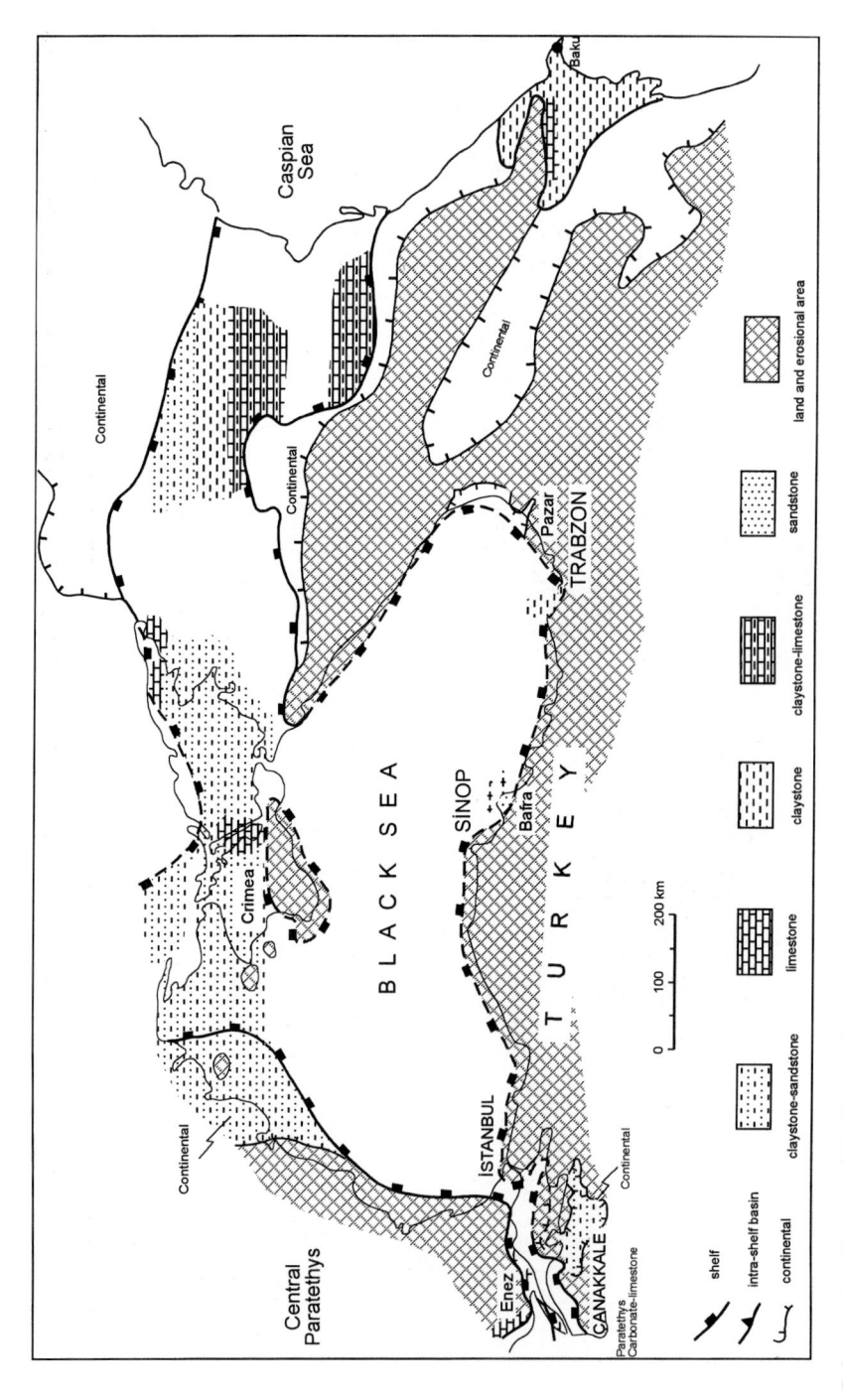

Fig. 9. Palaeogeographic map of the Eastern Paratethys and adjacent regions during the Pontian (6.9–5.5 Ma BP).

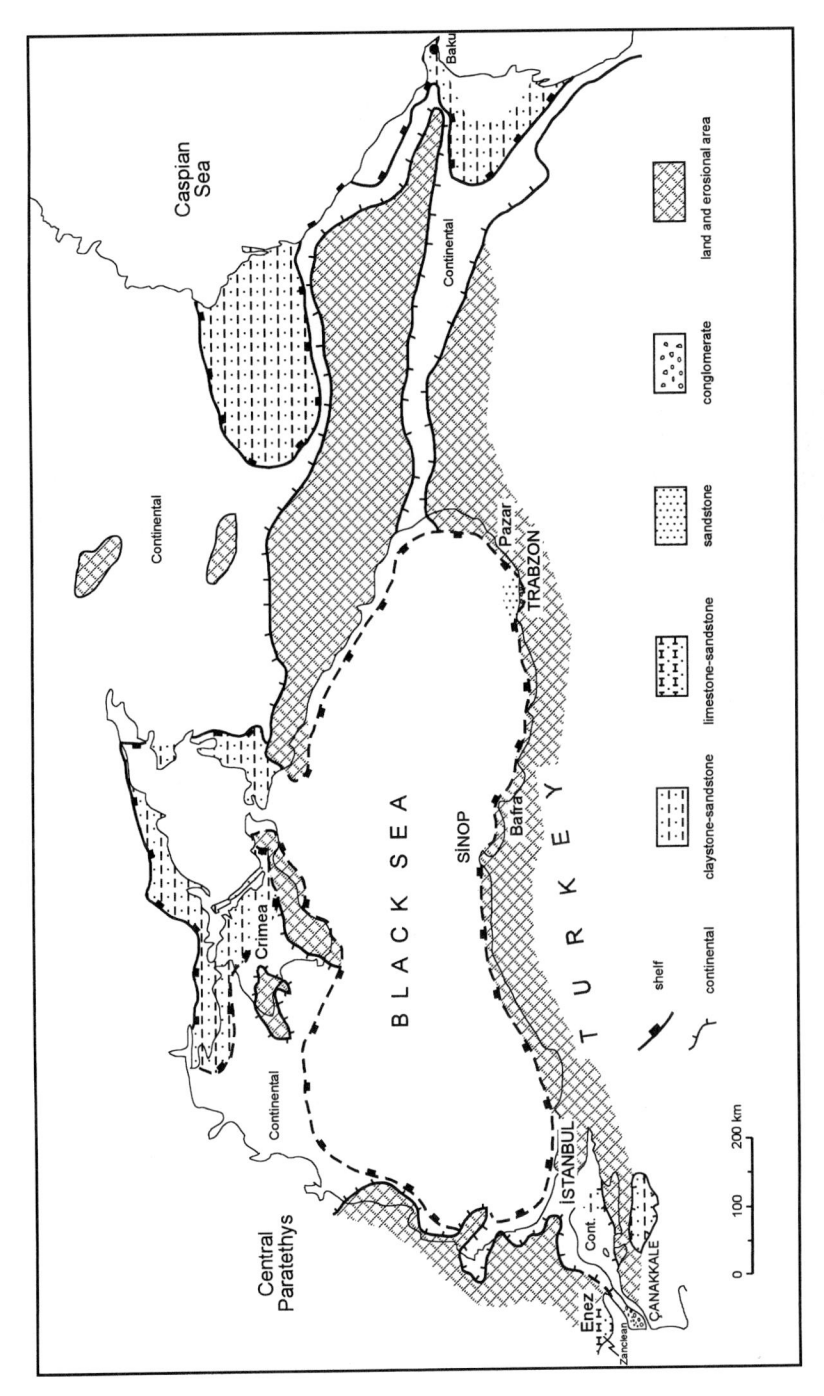

Fig. 10. Palaeogeographic map of the Eastern Paratethys and adjacent regions during the Cimmerian (5.5–3.2 Ma BP).

(Görür et al. 1997; Çağatay et al. 1998). This link was established when the Mediterranean was experiencing a salinity crisis in the Messinian (Hsü 1972, 1974, 1978; Ryan & Cita 1978).

The Cimmerian (5.5–3.2 Ma BP)

During the Cimmerian, brackish–marine and continental conditions prevailed over the EP basin (Fig. 10) [Degens et al. 1986; Palaeontological Institute (RAS) 1996]. The faunal content of the Cimmerian sediments indicates lower salinity than that in the Pontian. The nature of the Cimmerian sediments in the northern Pontides indicates that, at this time, the Pontides served as the main source of coarse detrital sediments along the southern coast of the Black Sea. This region and the Marmara region were perhaps being elevated during this period. The previously marine areas during Pontian time in this region became continental areas with fluvio–lacustrine sedimentation (Fig. 10). However, a marine transgression occurred in the northeastern Aegean region during the Kimmerian, depositing the *Ostrea* bank facies of the Göztepe Formation. The transgression, related to the Pliocene reconnection of the Mediterranean with the Atlantic Ocean after the Messinian desiccation (Hsü & Bernoulli 1978), did not reach either the Sea of Marmara or the EP basin (Fig. 10).

The Akchagylian (Kuyalnikian) (3.2–1.8 Ma BP)

Sediments of this age are found only in the southern Marmara region (the Özbek Formation). Their age equivalents in the northern Black Sea region consist of brown ironstone, sand, silt and clay deposits. They contain a rich bivalve fauna, including *Lymnocardium limanicium*, *Dreissena theodori* and *Unio* in the lower part of the sections, and *Avimactra subcaspia* and *Cerastoderma dombra* in the upper part of the sections [Palaeontological Institute (RAS) 1996]. The presence of zonal coccolith species (e.g. *Discoaster brouweri*) in these beds (Toker & Şengüler 1995) allows their correlation with the Özbek Formation in the Marmara region.

Both the stratigraphy and the faunal assemblage of the Akchagylian sediments indicate that brackish–marine conditions prevailed during the Early Akchagylian, followed by marine conditions and then a return to brackish–marine conditions during the Late Akchagylian (Nevesskaya et al. 1985). Most of the Akchagylian bivalves are of Mediterranean origin, indicating the connection with this sea through the Sea of Marmara, as shown by the faunal content of the Özbek Formation. The Sea of Marmara became a waterway between the EP and the Mediterranean, being first inundated by the Paratethys and then by the Mediterranean. These connections with the adjacent seas resulted from the increased activity of the North Anatolian Fault and a rise in the short-term global sea level (Haq et al. 1987; Görür et al. 1997).

Conclusions

The Neogene marginal succession of the EP, cropping out along the Black Sea coast and in the Marmara region of Turkey, provides important evidence about the tectonosedimentary and palaeoceanographic history of the basin. Despite the differences in the lithofacies between the southern and northern marginal successions, the timing of marine transgressions and isolations from the world ocean, as well as the tectonic events, can be correlated across the EP basin.

In the Tarkhanian, the southern margin of the EP basin was largely a carbonate platform with normal marine waters. During the Late Tarkhanian–Early Chokrakian there was an overall emergence of the basin. However, normal marine conditions prevailed during most of this period, except for a salinity reduction and stagnation induced by a eustatic isolation towards the end. This isolation became even more severe during the Karaganian, when carbonate sedimentation prevailed in the northern Pontides. The basin was briefly inundated by marine waters from the Indo-Pacific Ocean in the Early Konkian, with a return to brackish–marine conditions at the end of this stage.

After a widespread transgression in the Early Sarmatian, the EP basin was partially isolated in the Middle–Late Sarmatian, with freshening water and anoxic bottom-water conditions. The EP was connected with the Marmara and northeastern Aegean regions during the Pontian, as a result of an extensive transgression from the Paratethys itself. The passage of the Mediterranean waters to the EP basin via the Marmara region did not take place until the Late Akchagylian. These results disagree with the earlier claims that the Marmara region was a gateway between the Mediterranean and the Parathethys during most of the Middle Miocene.

This research is a part of the National Marine Geology and Geophysics Programme (NG, Coordinator) supported by TÜBİTAK (The Scientific and Research Council of Turkey) (YDABÇAG 591/G project). Thanks are extended to the Turkish Academy of Sciences (TÜBA) for additional financial support to NG.

References

ANDROSOV, N. I. 1918. Vzaimootnoshenie of the Euksinskogo i Kaspiysyskogo bassieinov Neogenovuyu epokhu [Relationships of the Euxinian and Caspian Basins in the Neogene epoch]. *Union of Soviet Socialist Republics Academy of Sciences*, **6**, (12, 8) 749–760

BALDI, T. 1980. The Early History of the Paratethys–Földt Közl. *Bulletin of the Hungarian Geological Society*, **110**, 456–472.

—— 1982. Mid-Tertiary tectonic and paleogeographic evolution of the Carpathian–East Alpine–Pannonian System. Discussion. *Paleontology*, **28**, 79–155.

BANKS, S., ROBINSON, A. G. & WILLIAMS, M. P. 1997. Structure and regional tectonics of the Achara–Trialet Fold Belt and the adjacent Rioni and Kartli foreland Basins, Republic of Georgia. *In*: ROBINSON, A. G. (ed.) *Regional and Petroleum Geology of the Black Sea and Surrounding Region*. AAPG Memoir, **68**, 331–346.

BUCHBINDER, B. & GVIRTZMAN, G. 1976. The breakup of the Tethys ocean into the Mediterranean, the Red Sea and the Mesopotamian Basin during the Miocene. A succession of fault movements and desiccation events. *The 1st Congress Pacific Neogene Stratigraphy, Abstracts*, Tokyo, 32–35.

ÇAĞATAY, M. N., GÖRÜR, N., ALPAR, B. *ET AL*. 1998. Geological evolution of the Gulf of Saros, NE Aegean Sea. *Geo-Marine Letters*, **18**, 1–9.

CHAPUT, E. & GILLET, S. 1938. Les faunes de Mollusques des terrains a Hipparion gracile de Küçükçekmece près d'İstanbul. *Bulletin de la Société Géologique de France*, **5**(VIII), 5–6.

CITA, M. B. 1981. Distribution of the evaporites in the Neogene of the Mediterranean, the Paratethys and the Middle East. Background and motivation. *Annales Geology Pays Helleniques*, **4**, 219–231.

DEGENS, E. T., WONG, H. K. & WIESNER, M. G. 1986. The Black Sea region: sedimentary facies, tectonicis and potential. *Mittlungen der Geologische–Paläontologische Institute der Universität Hamburg*, **60**, 127–147.

DEWEY, J. F., HEMPTON, M. R., KIDD, W. S. F., ŞAROĞLU, F. & ŞENGÖR, A. M. C. 1986. Shortening of continental lithosphere. The neotectonics of Eastern Anatolia – a young collision zone. *In*: COWARD, M. P. & RIES, A. C. (eds) *Collision Tectonics*. Geological Society, London, Special Publications, **19**, 3–36.

DUMITRICA, P. 1978. Badenian radiolaria from the central Paratethys. *In*: PAPP, A., CICHA, I., SENES, Y. & STEININGER, F. (eds) *M₄Badenien*. Serie Chronostratigraphy, Neostratotypen, Bratislava (SAV), **6**, 231–261.

EROL, O. & ÇETİN, O. 1995. Marmara Denizi'nin Geç Miyosen–Holosen'deki Evrimi. *In*: MERIÇ, E. (ed.) *İzmit Körfezi Kuvaterner İstifi*. Deniz Harp, Tuzla, Istabul, 313–342 [in Turkish with English abstract].

FINETTI, I., BRICHI, G., DEL BEN, A., PIPAN, M. & XUAN, Z., 1988. Geophysical study of the Black Sea. *Bollettino di Geofisica Teorica Applicata*, **30**, 117–118 and 197–234.

GILLET, S. 1961. Essai de palaéogeographie du Néogene et du Quaternarie inférieur d'Europe orientale. *Revue de Géographie Physique et Géologie Dynamique*, **2**, 218–250.

GONCHAROVA, I. A. 1989. Bivalve molluscs of the Tarkhanian and Tschokrakian basins. *Proceedings of the Paleontological Institute*.

GÖRÜR, N. 1988. Timing of opening of the Black Sea Basin. *Tectonophysics*, **147**, 247–262.

——, SAKINÇ, M., BARKA, A., AKKÖK, R. & ERSOY, Ş. 1995. Miocene to Pliocene palaeogeographic evolution of Turkey and its surroundings. *Journal of Human Evolution*, **28**, 309–324.

——, AKKÖK, R., ÇAĞATAY, M. N., SAKINÇ, M., NATAL'IN, B. A. & TCHAPALYGA, A. 1998. Miocene palaeogeographic evolution of the Eastern Paratethys. *The Third International Turkish Geology Symposium, METU – Ankara, Abstracts*, 208.

——, ÇAĞATAY, M. N., SAKINÇ, M., SÜMENGEN, M., ŞENTÜRK, K., YALTIRAK, C. & TCHAPALYGA, A. 1997. Origin of the Sea of Marmara as deduced from the Neogene to Quaternary palaeogeographic evolution of its frame. *International Geological Review*, **39**, 342–352.

HAQ, B. U., HARDENBOL, J. & VAIL, P. 1987. Chronology of the fluctuating sea level since the Triassic. *Science*, **225**, 1156–1167.

——, —— & —— 1988. *Mesozoic and Cenozoic Chronostratigraphy and Cycles of Sea-level Change*. Society of Economic Paleontologists and Mineralogists, Special Publications, **42**, 71–108.

HSÜ, K. J. 1972. Origin of saline giants: a critical review after the discovery of the Mediterranean evaporite. *Earth Science Reviews*, **8**, 371–396.

—— 1974. Mélanges and their distinction from olistostromes. *In*: DOTT, R. H. J. & SLTAVER, S. (eds) *Modern and Ancient Geosynclinal Sedimentation*. Society of Economic Paleontologists and Mineralogists, Special Publications, **19**, 321–333.

—— 1978. When the Black Sea was drained. *Scientific America*, **238**, 52–63.

—— & BERNEOULLI, D. 1978. Genesis of the Tethys and the Mediterranean. *Initial Reports of the Deep-Sea Drilling Project*, **42**, 943–950.

JINARIDZE, N. M. 1979. Peculiarities of time-spatial differentiation for halogen deposits of Carpathian region and problem of deposits genesis. *Annales Geology Pays Helleniques*, **2**, 569–583.

JIRICEK, R. 1975. *Biozonen der Zentralen Paratethys*. NAFTA, Gbely.

JONES, R. W. & SIMMONS, M. D. 1996. A review of the stratigraphy of EP (Oligocene–Holocene). *Bulletin of the British Museum (Natural History). Geology series*, **52**(1), 25–49.

—— & —— 1997. A review of the stratigraphy of EP (Oligocene–Holocene) with particular emphasis on the Black Sea. *In*: ROBINSON, A. G. (ed.) *Regional and Petroleum Geology of the Black Sea and Surrounding Region*. AAPG Memoir, **68**, 39–52.

KHRUSHCHOV, D. P. & PETRICHENKO, O. I. 1979. Evaporite formations of Central Paratethys and conditions of their sedimentation. *Annales Geology Pays Helleniques*, **2**, 595–612.

MAROTHON PETROLEUM 1975. *Marmara No. I. Final Well Report*. Marathon Petroleum Ltd, Ankara.

MURATOV, M. B. & NEVESSKAJA, L. A. (eds) 1986. *Neogenovaya Sistema (The Neogene)*. Nedra, Moscow, **1**, 414–443.

NEVESSKAYA, L. A., TIKHOMIROV, V. V. & FEDEROV, P. V. 1985. Stratigraphy of Neogene–Quaternary deposits of the euxinian area. *The VII Congress: Regional Commission on the Mediterranean Region, Neogene Stratigraphy*, Budapest, 413–415.

OKAY, A. İ & ŞAHİNTÜRK, Ö. 1997. Geology of the Eastern Pontides. *In*: ROBINSON, A. G. (ed.) *Regional and Petroleum Geology of the Black Sea and Surrounding Region*. AAPG Memoir, **68**, 291–312.

ÖZSAYAR, T. 1977. *A Study on Neogene Formations and their Molluscan Fauna Along the Black Sea Coast*. Karadeniz Technical University Publications, **79** [in Turkish].

PALAENTOLOGICAL INSTITUTE (RUSSIAN ACADEMY OF SCIENCE) 1996. *Neogene Stratigraphy and Palaentology of the Taman and Kerch Peninsulas*. Excursion guidebook, IGCP Project No. 329.

PAPP, A. 1963. Das verhalten neogener Molluskenfaunen bei verschiedenen Salzgehalten. *Fortschrifte der Geologie Rheinland Westfelien*, **10**, 35–48.

PARAMONOVA, N. P., ANANOVA, E. N., ANDREEVA GRIGROVIC, A. S. *ET AL*. 1979. Paleontological characteristics of the Sarmatian sensu lato and Meotian of the Ponto–Caspian area and possibilities of correlation of the Sarmatian sensu stricto and Pannonian of Central Paratethys. *Annales Geology Pays Helleniques*, **1979**, 961–971.

POPOV, S. V. & VORONINA, A. A. 1983. Kozachurian stage of development of Eastern Paratethys. *Union of Soviet Socialist Republics Academy of Sciences, Geology*, **1**, 58–67.

RÖGL, F. & STEININGER, F. F. 1983. Vom zerfall der Tethys zu Mediterran und Paratethys. Die Neogene Paleogeographie und Palinspastik des zirkum–Mediterranean Raumes. *Annalen Naturhistorie Museum Wien*, **85A**, 135–163.

—— & —— 1984. Neogene Paratethys, Mediterranean and Indo-Pacific Seaways. *In*: BRANCHLEY, P. J. (ed.) *Fossils and Climate*. Wiley, 171–200.

——, —— & MÜLLER, C. 1978. Middle Miocene salinity crisis and palaeogeography of the paratethys (Middle and Eastern Europe*) Initial Reports of the Deep-Sea Drilling Project*, **42**, 985–990.

RYAN, W. B. F. & CITA, W. B. 1978. The nature and distribution of Messinian erosional surfaces. Indicators several kilometers deep in the Mediterranean in the Miocene. *Marine Geology*, **27**, 193–230.

SEMENENKO, V. N. 1979. Correlation of the Mio-Pliocene of the EP and Tethys. *Annales Geology Pays Helleniques*, **3**, 1101–1111.

SENES, J. 1959. Unsere Kenntnisse Über die Palaeogeographie der Zentral Paratethys. *Geologicke Práce*, **55**, 83–108.

—— 1971. Korrelation des Miozans der Zentralen Paratethys (Stand 1970*). Geologica Carpatica*, **22**, 3–9.

ŞENGÖR, A. M. C. 1979. The North Anatolian Transform Fault, its age, offset and tectonic significance. *Journal of the Geological Society, London*, **136**, 269–282.

—— 1984. *The Cimmeride Orogenic System and the Tectonics of Eurasia*. Geological Society of America, Special Paper, **195**.

—— 1987. Tectonics of the Tethysides: Orogenic collage Development in a collisional setting. *Earth and Planetary Sciences*, **15**, 213–244.

—— & YILMAZ, Y. 1981. Tethyan evolution of Turkey: a plate tectonic approach. *Tectonophysics*, **75**, 181–241.

——, GÖRÜR, N. & ŞAROĞLU, F. 1985. Strike-slip faulting and related basin formation in zones of tectonic escape: Turkey is a case study. *In*: BIDDLE, K. T. & CHRISTIE-BLICK, N. (eds) *Strike-slipe Deformation, Basin Formation and Sedimentation*. Society of Economic Palaeontologists and Mineralogists, Special Publications, **37**, 227–264.

——, ALTINER, D., CİN, A., USTAÖMER, T & HSÜ, K. J. 1988. Origin and assembly of the Tethyside orogenic collage at the expense of Gondwana Land. *In*: AUDLEY-CHARLES, M. G. & HALLAM, A. (eds) *Gondwana and Tethys*. Geological Society, London, Special Publications, **37**, 119–181.

SİYAKO, M., BÜRKAN, K. A. & OKAY, A. İ. 1989. Biga ve Gelibolu Yarımadalarının Tersiyer Jeolojisi ve Hidrokarbon Olanakları. *Turkish Association of Petroleum Geologists Bulletin*, **1**, 183–199 [in Turkish with English abstract].

STEININGER, F. F., RÖGL, F. & NEVESSKAJA, L. A. 1985. Sediment distribution maps for selected time intervals through the Neogene. *In*: *Neogene of the Mediterranean Tethys and Paratethys: Stratigraphic Correlation Tables and Sediment Distribution Maps*. Vienna, 91–102.

SUESS, E. 1888. *Das Antliz der Erde, Zweiter Band*. F. Tempsky, Leipzig.

—— 1893. Are great ocean depths permanent? *Natural Science*, **2**, 180–187.

SÜMENGEN, M., TERLEMEZ, İ., ŞENTÜRK, K. *ET AL*. 1987. *Gelibolu Yarımadası ve Güneybatı Trakya Havzasının Stratigrafisi, Sedimantolojisi ve Tektoniği*. Mineral Research and Exploration Institute of Turkey (MTA) Report, No. 8128 [in Turkish].

TCHAPALYGA, A. L. 1985. Climatic and eustatic flactuations in the Paratethys basin history. Abstracts of the VII Congress. *In*: *Regional Commission on Mediterrranean Region Neogene Stratigraphy*.

Geological Institute of Romania, Budapest, 34–36.

TOKER, V. & ŞENGÜLER, İ. 1995. İzmit Körfezi (Hersek Burnu–Kababurun) Kuvaterner İstifinin Nannoplankton florası. *In*: MERIÇ, E. (ed.) *İzmit Körfezi Kuvaterner İstifi*. Deniz Harp, Tuzla, Istabul, 163–173 [in Turkish with English abstract].

VAVRA, N. 1981. Bryozoa from the Eggenburgian (Early Miocene, Central Paratethys) of Austria. *In*: LARWOOD, G. P. & NIELSEN, C. (eds) *Recent and Fossil Bryozoa*. Olsen and Olsen, 273–280.

VINAGRADOV, A. P. 1967. *Atlas of the Lithological–Palaeontological Maps of the USSR*. Paleogene, Neogene and Quaternary IV. Ministry of Geology of the USSR, Moscow.

YALÇINLAR, Y. 1958. Samsun Bölgesinin Neojen ve Kuvaterner Kıyı Depoları. *İstanbul University, Institute of Geography Bulletin*, **5**, 11–21 [in Turkish with English abstract].

YALTIRAK, C. 1995. Gelibolu Yarımadası'nda Pliyo-Kuvaterner sedimantasyonunu denetleyen tektonik mekanizma. *Proceedings of the Nezihi Canıtez Symposium*, **10**, 103–106 [in Turkish with English abstract].

YILMAZ, Y., TÜYSÜZ, O., YIĞITBAŞ, E., GENÇ, C. & ŞENGÖR, A. M. C. 1997. Geology and tectonics evolution of the Pontides. *In*: ROBINSON, A. G. (ed.) *Regional and Petroleum Geology of the Black Sea and Surrounding Region*. AAPG Memoir, **68**, 183–226.

Tectonosedimentary evolution of the Miocene Manavgat Basin, western Taurides, Turkey

MUSTAFA KARABIYIKOĞLU,[1] ATTİLA ÇİNER,[2] OLIVIER MONOD,[3] MAX DEYNOUX,[4] SEVİM TUZCU[1] & SEFER ÖRÇEN[5]

[1]*Mineral Research and Exploration Institute (MTA), Department of Geological Research, TR-06520 Ankara, Turkey (e-mail: karabyk@mta.gov.tr)*
[2]*Hacettepe University, Department of Geological Engineering, Beytepe, TR-06532 Ankara, Turkey*
[3]*Lab. Géologie Structurale UMR 6530, CNRS-Université d'Orléans, BP 6759, 45067-Orléans Cedex 2, France*
[4]*Ecole et Observatoire des Sciences de la Terre, UMR 7517, CNRS-Université Louis Pasteur, 1 rue Blessig, 67084 Strasbourg Cedex, France*
[5]*Kocaeli University, Department of Geological Engineering, Kocaeli, Turkey*

Abstract: The Manavgat Basin is a northwest–southeast oriented basin that developed on the eastern side of the Isparta Angle, south of the Late Eocene thrust belt of the western Taurides. The Miocene fill of the basin lies unconformably on an imbricated basement, comprising a Mesozoic para-authocthonous carbonate platform overthrust by the Antalya Nappes and Alanya Massif metamorphics. The sedimentary fill is represented by clastic-dominated deposits consisting of, in ascending order, a conglomeratic wedge, reefal shelf carbonates, limy mudstones, and calciturbidites with subordinate breccias and conglomerates.

Process-oriented facies analysis of the basin fill indicates a variety of depositional environments ranging from fluvial/alluvial fan and fan-delta complexes through reefal carbonate shelf and forereef slope to slope fan and basin floor. Fluvial/alluvial fan and fan-delta deposits are Burdigalian–Early Langhian in age and represent the initial conglomeratic valley-fill sedimentation during a relative sea-level rise balanced by important sediment supply from relief in the north-northeast hinterland. The continuous relative sea-level rise and a decreasing rate of sediment supply allowed the deposition of transgressive reefal shelf carbonates of Langhian age. Tectonic activity demonstrated by synsedimentary faults resulted in block faulting of the narrow carbonate shelf and foundering of the basin. The rest of the sedimentation consists of the fill of newly created accommodation space. The overall coarsening-upward succession consists of Upper Langhian–Serravallian limy mudstones–calciturbidites and debris flows, overlain by Tortonian coarse-grained fan-delta deposits. The gravity induced character of most of this progradational wedge implies a progressive uplift of the hinterland.

Situated between the Late Miocene Hellenic–Lycian Arc and the Late Eocene Taurus Arc, the Isparta Angle is a conspicuous syntaxis within the Alpine Chain in southern Turkey (map insert in Fig. 1). It has resulted from a complex tectonic history related to the emplacement of different tectonic units (e.g. Lycian Nappes, the Beyşehir–Hoyran Nappes, the Antalya Nappes and the Alanya Massif) from Late Cretaceous to Late Miocene (Bizon *et al.* 1974; Monod 1977; Gutnic *et al.* 1979; Poisson *et al.* 1983; Marcoux 1987; Kissel *et al.* 1993; Şenel *et al.* 1993; Frizon *et al.* 1995; Barka *et al.* 1997). It is presently occupied in part by the Miocene and Pliocene sediments of the Antalya Gulf which opens southward to the Mediterranean Sea. As shown below, the Miocene sedimentation in the Isparta Angle was strongly dependent upon pre-existing structures, and its content and deformation reflect the latest stages of the orogenic development of the chain. In this paper a broad outline of the sedimentary sequences of the eastern part of the Antalya Gulf (Manavgat Basin) is presented, illustrating the tectonosedimentary relationships of the area with some regional implications.

Structural and stratigraphic setting

The Manavgat Basin is located on the eastern flank of the 'Isparta Angle' (Blumenthal 1963). As shown by the map and cross sections (Figs 1

From: BOZKURT, E., WINCHESTER, J. A. & PIPER, J. D. A. (eds) *Tectonics and Magmatism in Turkey and the Surrounding Area.* Geological Society, London, Special Publications, **173**, 271–294. 1-86239-064-9/00/$15.00
© The Geological Society of London 2000.

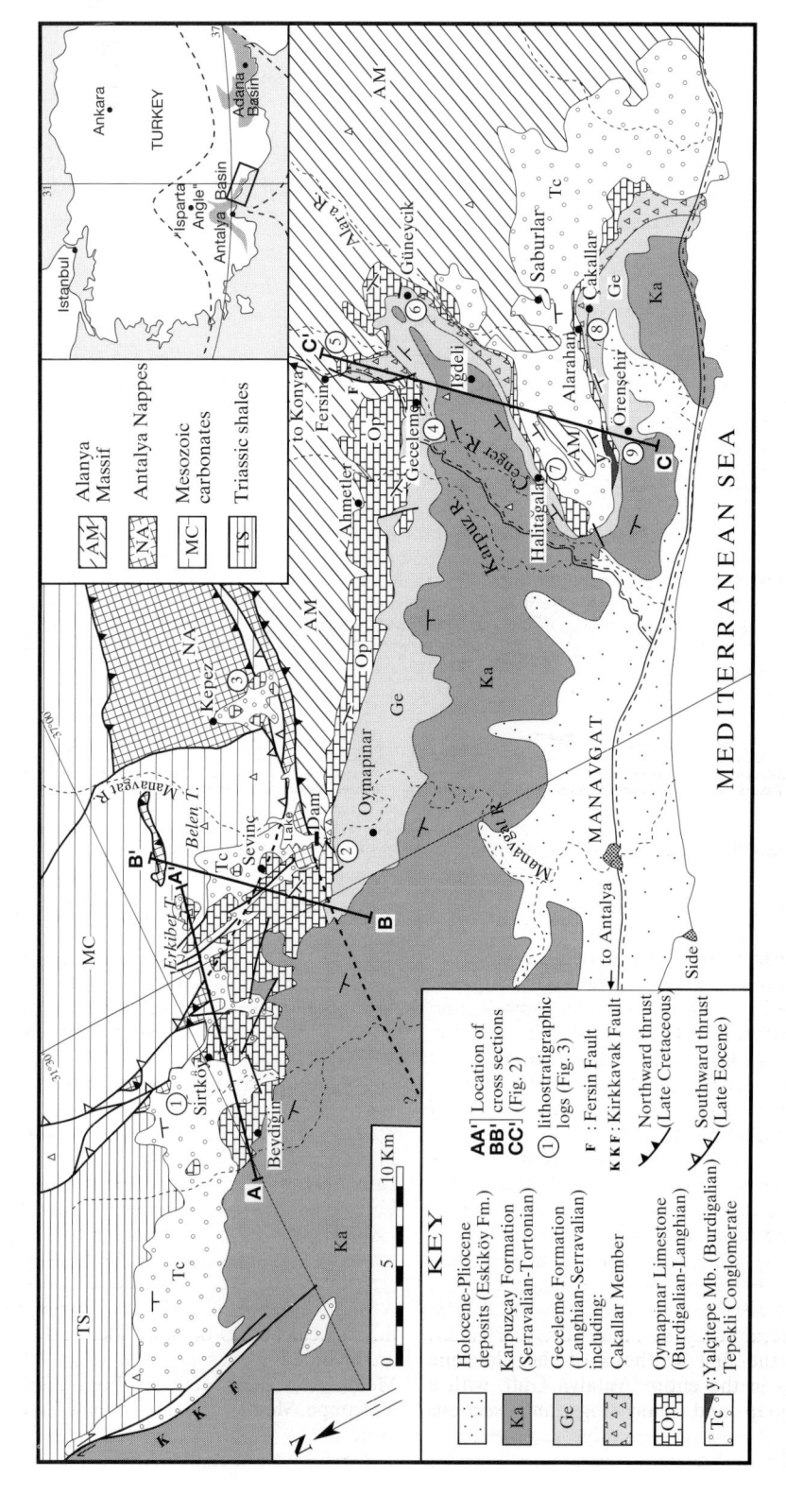

Fig. 1. Geological map of the Manavgat Basin, modified after Akay & Uysal (1985).

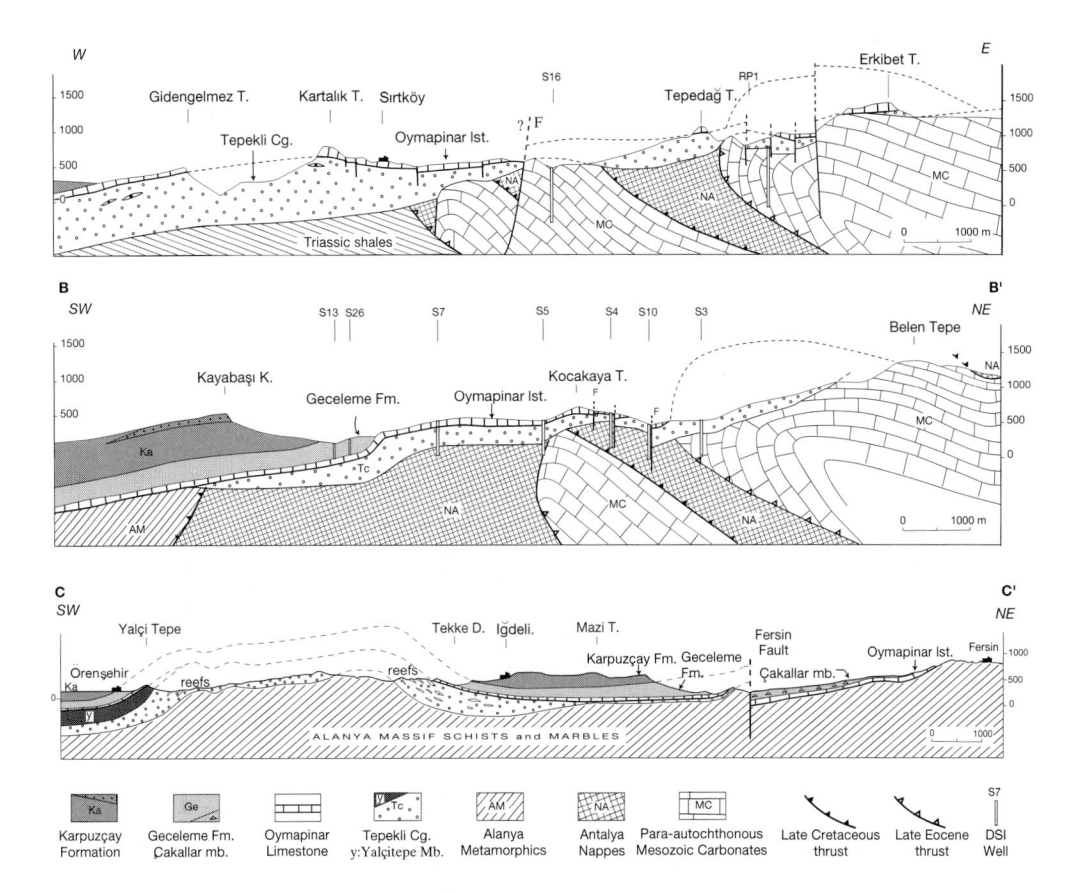

Fig. 2. Three representative cross-sections from the Manavgat Basin. Location in Fig. 1.

and 2, respectively), the Miocene deposits of the Manavgat Basin unconformably lie on a Mesozoic carbonate platform overthrust by the Antalya Nappes units and the Alanya Massif metamorphics. This tectonic imbrication occured during the Late Eocene. In contrast, the overlying Miocene formations are relatively undeformed, although the highest Miocene deposits have been uplifted to an altitude of 1401 m (Erkibet Tepe; Figs 1 and 2).

The basic knowledge of the geology of the Manavgat Basin was established by Blumenthal (1951), who mapped the area and defined four of the main formations with a rough Miocene chronostratigraphy. Subsequently, many refinements to the dating of the formations have been made (Bizon *et al.* 1974; Erk *et al.* 1995) and a general study by Akay & Uysal (1985) produced the first synthesis of the Neogene formations in the entire Antalya Gulf, with a detailed map and many logs and sections (Akay *et al.* 1985). Recently, a general interpretation of Miocene sedimentary features in the Antalya Basin was presented by Flecker (1995) and Flecker *et al.* (1995, 1998). On tectonic grounds, Frizon *et al.* (1995) have pointed out two major compressive events postdating the Tortonian which are responsible for the main present-day physiographic features of the Antalya area.

Several formation names have been used for the four lithological units that have been recognized in the Manavgat Basin [see review in Flecker (1995)]. Except for the lowermost conglomerates, the nomenclature used by Akay & Uysal (1985) in their synthetic report is followed here. From base to top these are: the Tepekli Conglomerate, the Oymapınar Limestone, the Geceleme Formation and the Karpuzçay Formation. Additional names refer to lithological bodies of more local extent which are included here as members (e.g. the Çakallar Member in the Geceleme Formation or the Yalçıtepe Member in the Tepekli Conglomerate).

Three cross-sections across the Manavgat

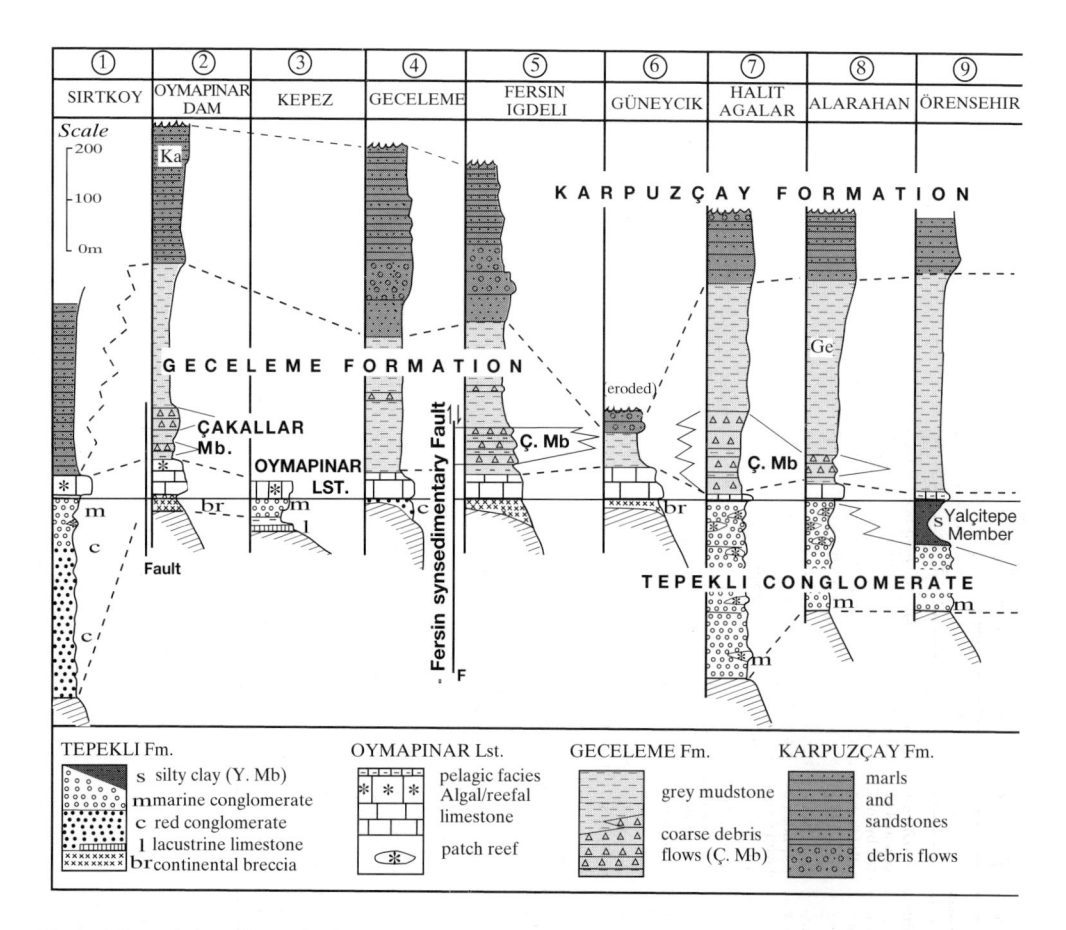

Fig. 3. Schematic logs illustrating lateral and vertical relationships between the formations and the inferred depositional environments. Log numbers refer to their location in Fig. 1.

Basin (Fig. 2) illustrate the present-day geometry of the Miocene deposits. Additionally, nine schematic logs, from northwest to southwest, are presented in Fig. 3 to illustrate the thickness and facies changes. Beneath the Oymapınar Limestone, which is chosen here as a datum line, there is a spectacular variation in thickness of the Tepekli Conglomerate (Fig. 2). Wells drilled by DSI (the State Water Works) for the dam and outcrop constraints indicate that these conglomerates vary from 0 (west of Erkibet Tepe) to 280 m (S7) and over 600 m west of Sırtköy. The Oymapınar Limestone, although present in all sections, also varies considerably, from thick reef buildups in the northwest (the Sırtköy–Oymapınar area, logs 1 and 2 in Fig. 3), to thin pelagic facies in the southeast (Örenşehir, log 9 in Fig. 3). Above, the Geceleme mudstones are thickest in the southeast (logs 7–9 in Fig. 3), but cannot be defined in the northwest (log 1 in Fig. 3) where turbidite and debris flow deposits characteristic of the Karpuzçay Formation prevail instead. Locally, breccias of the Çakallar Member are intercalated in the lower half of the Geceleme Formation (logs 2, 5, 7 and 8 in Fig. 3).

Several normal faults of limited importance offset the Miocene formations. In most cases these do not appear to be related to pre-existing faults. In particular, none of the large thrusts of Late Eocene age that are responsible for the imbrication of the Antalya Nappes were reactivated during the Miocene or subsequently, as shown in sections A and B (Fig. 2). In the southeastern part of the basin, between Örenşehir and Halitağalar (Fig. 1), a broad anticlinal fold is present (section CC' in Fig. 2). Its east–west orientation is roughly parallel to the tectonic boundary of the Alanya Massif to the north. The age of this anticlinal folding may be

Table 1. *Facies description and interpretation of depositional environments of the Tepekli Conglomerate and Oymapınar Limestone*

Facies	Description	Interpretation
Facies 1 *Microcodium* limestone breccia and pebbly limestone	Highly fractured, metalimestone of Alanya Massif (AM) with *Microcodium*-bearing lime mudstone infills and well-bedded, monogenetic metalimestone breccia and subrounded pebbly limestone containing borings. The matrix of limestones is represented by fossiliferous lime wackestone–packstone comprising benthic foraminifers, bioclasts, peloids and fine clastics	*Microcodium*-bearing limestone represents deposition under continental conditions. Limestone breccia indicates gravity induced mass transport including rock avalanches and slides on steep subaerial slopes. Pebbly limestones with borings represent deposition in a near-shore environment resulting from reworking of breach gravels
Facies 2 Matrix-supported disorganized conglomerate	Massive to thickly bedded, poorly sorted, boulder conglomerates with muddy matrix. Well-rounded coarse pebbles and abundant boulders (1 m) with large clasts floating within the bed or protruding into the overlying bed. Conglomerates are disorganized with no apparent clast fabric and sedimentary structures. Sharp flat bases and tops with occasional small-scale scours	Gravity induced deposition en masse. Matrix and large floating clasts indicate rapid deposition from high-viscosity flows with matrix strength. Cohesive debris flows representing sedimentation in a proximal alluvial fan and/or high gradient braided stream environment
Facies 3 Clast-supported organized conglomerate	Poorly to moderately well-sorted, massive to crudely bedded, well-rounded pebble–cobble conglomerate. Laterally extensive, thick, tabular to lenticular beds or thick channel fills. Flat to irregular lower and upper surfaces. Represented by two sub-facies: (1) unfossiliferous intergranular matrix; (2) locally fossiliferous (marine bivalves and oysters) calcareous sandy matrix	Subaerial and subaqueous processes resulting from hyperconcentrated flows and/or stream flows. Deposition in a high-gradient braided stream and fan-delta front environment
Facies 4 Sandstone–siltstone interbedded mudstone	Red massive to flat-bedded mustone with thin lenticular fine-grained sandstone–siltstone interbeds (up to 20 cm in thickness) characterized by flat to scoured bases and flat to slightly convex tops with rare asymmetrical ripples	Flood flows forming amalgamated beds or overbank (flood plain) and crevasse splay deposits
Facies 5 Peloid mollusc–echinoid wackestone–packstone	Thick, parallel bedded very fine-grained and moderately well-sorted, blue-grey peloidal wackestone–packstone. Contains carbonized (lignite?) fragments, some subangular to subrounded intraclasts and bioclasts. Flat upper and lower bedding surfaces. Well-preserved fragments of gastropods, bivalves and echinoids. Local coquinoid beds up to 20 cm thick	Deposition in a low-energy restricted shelf lagoon with open-marine influence. Carbonized grains may suggest transportation from a nearby marshy (peat) environment
Facies 6 Algal wackestone–packstone	Massive to thickly parallel-bedded, limestone comprising algae (mainly *Lithothamnium* and *Lithophyllum*), benthic foraminifers, bivalves, gastropods, echinoids, bryzoa and coral fragments. Disarticulated to finely fragmented and subrounded bioclasts, rhodoliths (up to 4–5 cm) and oncoliths (up to 1 cm)	Intertidal to shallow-marine environment. Oncoliths and rhodoliths are indicative of limited depth and turbulent environments
Facies 7 Benthic foraminiferal wackestone–packstone	Thick, parallel-bedded and poorly to moderately sorted, coarse wackestone–packstone containing relatively rich and diversified benthic foraminifers and macrofauna with some peloids and rare intraclasts. The bioclasts are characteristically disarticulated, fragmented or abraded. They include skeletal remains of coralline algae, *Halimeda*, foraminifers, bivalves, gastropods, echinoids, annelid tubes, bryozoa and corals. Also contains small (*Miliolid, Borelis, Rotalid, Textularia, Elphidium* and *Gypsina*) and large (*Heterostegina, Operculina, Acervulina, Miogypsina, Amphistegina, Archaias, Peneroplid* and *Victoriellid*) bethic foraminifers	Low to moderate energy, normal salinity shallow shelf close to wave base. Small benthic foraminifera-rich wackestone–packstone represent shelf lagoon with restricted water circulation. Large benthic foraminifera bearing wackestone–packstone indicate deeper water on an open shelf
Facies 8 Benthic–planktic foraminiferal wackestone–packstone–grainstone	Poorly to well-sorted, fine- to medium-grained limestone comprising mainly *Heterostegina, Operculina*, with rare *Amphistegina, Textularia, Rotalid, Globigerinid*. Parallel-bedded, low-angle inclined accretionary beds and locally developed *Callianasa* burrows	Relatively high-energy beach and open-shelf environment with a normal salinity
Facies 9 Massive coral-algal boundstone	Small, laterally delimited, massive mound-like limestone bodies with flat bases and convex-up tops. Rich and diverse *in situ* hermatypic coral colonies with minor solitary corals in places	Coral-algal patch reef developed in fan-delta front on a normal shallow-marine shelf. The local presence of solitary corals may represent coral reef development in relatively deeper water settings
Facies 10 Reefal rudstone–floatstone and debris	Fine- to coarse-grained clast-supported (rudstone) or matrix-supported (floatstone) coral fragments and very coarse reefal debris. Massive or thin to thick inclined beds with flat to variably scoured bases and flat tops	Very local redeposition of coral-algal sediments eroded from the reefs. Rudstones mainly represent deposition on reef flats and reef talus, whereas floatstones represent deposition in base of forereef slopes

Table 2. *Facies description and interpretation of depositional environments of the Geceleme Formation, the Çakallar Member and the Karpuzçay Formation*

Class – Group – Facies	Description	Interpretation
A: Conglomerates, muddy conglomerates and conglomeratic mudstones *A1: Disorganized* A1.1: Disorganized conglomerates A1.2: Muddy conglomerates A1.3: Conglomeratic (blocky) mudstones	2–6 m thick, tabular to lenticular beds with scoured bases. Very poorly sorted, very angular to subangular limestone clasts (3–10 cm) of Oymapınar Limestone (OP) and Alanya Massif (AM), and well-rounded clasts of Mesozoic carbonate. Clast to matrix supported. Some blocks > 1 m may be present in Facies A1.3	Non-cohesive to cohesive (muddy) debris flows in fan-delta and/or slope apron
A2: Organized A2.1: Stratified conglomerates A2.2: Inversely graded conglomerates A2.3: Normally graded conglomerates A2.7: Normally graded pebbly sandstones	1–4 m thick, tabular to lenticular beds with basal erosion. Clast-supported, poorly sorted, very well-rounded pebbles (1–10 cm) and often 60 cm to 3 m blocks projecting into the overlying muddy and/or calcarenitic beds. Normal and inverse grading	High-density turbidity currents in a slope fan
B: Calcarenites *B2: Organized* B2.1: Parallel-stratified calcarenites	5–20 cm thick, amalgamated, tabular to wavy calcarenite beds. No mudstones. Few centimetre long groove casts	High-concentration turbidity currents in a fan-delta and/or submarine fan
C: Calcarenite–mudstone couplets *C2: Organized* C2.1: Very thick/thick bedded C2.2: Medium bedded C2.3 Thin-bedded calcarenite–mudstone couplets	Well- to poorly sorted, fine- to medium-grained calcarenite–mudstone couplets with changing bed thicknesses. Laterally very continuous beds and variable mudstone ratios. Bouma divisions are common mostly in Facies C2.3	High-concentration turbidity current deposits (C2.1) and relatively dilute current deposits (C2.3). Deposition in a submarine fan lobe
D: Siltstones, siltstone–mudstone couplete *D1: Disorganized* D1.1: Structureless siltstones	Thick- to very thick-bedded, parallel-sided structureless siltstones with sharp bases and tops. High carbonate content	High (D1.1) to low (D2.1) concentration silt-dominated turbidity currents in a submarine fan lobe to basin plain
D2: Organized D2.1: Graded stratified siltstones	Thin- to medium-bedded siltstones alternating with massive to laminated mudstones. Mudstone/siltstone ratio is 1:1	
E: Mudstones *E2: Organized* E2.1: Graded mudstones	Up to 2 m thick, laterally continuous, organic-rich mudstone. Sharp upper and lower contacts with well-developed bioturbation	Low-concentration turbidity currents in a submarine fan lobe or basin plain
F: Chaotic deposits *F1: Extra-clasts* F1.1: Extra-clast limestone breccia F1.2: Isolated reefal blocks	Chaotic assemblage of angular to subangular, clast-supported reefal limestone (OL) and basement-derived clasts (AM) (5–40 cm) and blocks (3–8 m) Subangular to subrounded blocks (up to several metres) embedded in fine-grained, parallel-stratified sediments	Submarine rock falls/avalanches or slides in forereef slope talus/apron Rock falls or slides in lower slope to basin plain
F2: Contorted/disturbed strata F2.1: Coherent folded and contorted strata F2.2: Brecciated strata	Coherently folded and contorted calcarenite–siltstone–mudstone layers of a few centimetres to tens of centimetres thick; found together with Class C deposits Chaotic mixture of brecciated and balled strata. Crude parallel stratification. Overlying and underlying sediments are mostly Class C deposits	Slumps or slump-generated debris flows in a fan-delta Gravity induced sliding with internal deformation in fan-delta

constrained between the age of the youngest beds affected (Messinian) and the oldest, unaffected sediments, the overlying Pliocene Eskiköy Conglomerate (Akay & Uysal 1985). The CC′ section also cuts through the Fersin normal fault which is interpreted here as a synsedimentary feature responsible of the polymictic breccias (the Çakallar Member) found along that fault (see below).

The present authors are well aware that, in the northwest area (see Fig. 1), the transition from the Manavgat Sub-basin to the Köprüçay Sub-basin is not satisfactorily understood. According to Dreyer (pers. comm.), the disapearance of the Oymapınar Limestone under the onlapping Karpuzçay Formation reflects an erosional input of detrital material into the Manavgat Basin. Flecker (1995) also considers a western origin for the onlapping Karpuzçay Formation. This problem requires additional work along the Kırkkavak Fault (KKF in Fig. 1) and in the Köprüçay Sub-basin.

Facies, depositional processes and environments

The Miocene fill of the Manavgat Basin is represented by a thick succession of terrestrial to marine clastics (the Tepekli Conglomerate), reefal shelf carbonates (the Oymapınar Limestone) followed by deeper marine clastics (the Geceleme and Karpuzçay Formations). The complete and detailed description of the facies is beyond the scope of this paper, however, an outline of the facies types is presented in Tables 1 and 2.

Within the Tepekli Conglomerate and the shallow-marine Oymapınar Limestone a total of ten facies, representing both subaerial and subaquaous depositional processes, have been recognized (Table 1; Fig. 4). They are named descriptively following the schemes developed by Miall (1978) for clastics, and Dunham (1962) and Wilson (1975) for carbonates.

Several facies classifications exist for deep-water clastic sediments (Mutti & Ricchi 1972, 1978; Mutti 1977). For the Çakallar Member, and the Geceleme and Karpuzçay Formations, the hierarchical classification by Pickering et al. (1986, 1989) is used. In this scheme, facies classes (A–G) are composed of two or more facies groups (A1, A2, B1, etc.), which are further subdivided into constituent facies (A1.1, A1.2, A1.3, etc.). In the deep-marine clastics of the Manavgat Basin a total of six classes, nine groups and 18 facies were recognized (Table 2; Fig. 5).

The Tepekli Conglomerate

The Tepekli Conglomerate is a pebble-dominated clastic formation exposed in two separate areas along the northwest and southeast margins of the basin. The spatial distribution, overall geometry and facies changes within the conglomerate bodies reflect two distinct depositional settings: (1) fluvial (braided stream) to alluvial fan; (2) southward prograding fan-deltas.

In the northwestern part of the basin, west of Sırtköy and in the Sevinç area (Fig. 1), the Tepekli Conglomerate is highly variable in thickness over short distances (0–600 m). This variation appears to be directly related to the lithology of the underlying formations; the Alanya Massif metalimestones and the Mesozoic carbonates support little or no conglomerates; the maximum conglomerate thickness is recorded where they overlie the Triassic shales or sandstones of the Antalya Nappes. This suggests that deep morphologic depressions were carved into the softer formations and subsequently filled by the Tepekli Conglomerate, as already proposed by Flecker (1995).

In the northwest, the dominant facies consists of a poorly to moderately sorted pebble–cobble conglomerate with well-rounded clasts and poorly developed polymodal granular matrix (Facies 3 in Table 1). It occurs as laterally extensive amalgamated tabular units, a few metres thick, or as thick channel fills intercalated with poorly developed red mudstone with sandstone–siltstone interbeds and nodular calcrete horizons (Facies 4). A crude horizontal lamination due to pebble alignment and imbrication is locally visible in the conglomerates (Fig. 4b). Boulder conglomerates with a sandy muddy matrix (Facies 2) are also present in places, generally near the base of the formation. Such a succession suggests high-energy braided streams with partly preserved overbank deposits where soil horizons developed. The clasts are polymict, comprising Mesozoic limestones from the nearby platform carbonates, but also volcanic clasts originating from distant source areas such as the Huğlu Volcanics from the Beyşehir–Hoyran Nappes, 50 km to the northeast. The facies characteristics, together with the presence of well-rounded, far-travelled clasts suggest an alluvial fan environment. Upsection and southwards (e.g. 2.5 km north of Beydiğin; Fig. 1), large patch reefs were found embedded within winnowed clast-supported conglomerates. This vertical evolution of the Tepekli Conglomerate reflects the passage from alluvial fan to fan-delta deposits, the latter being widely exposed in the

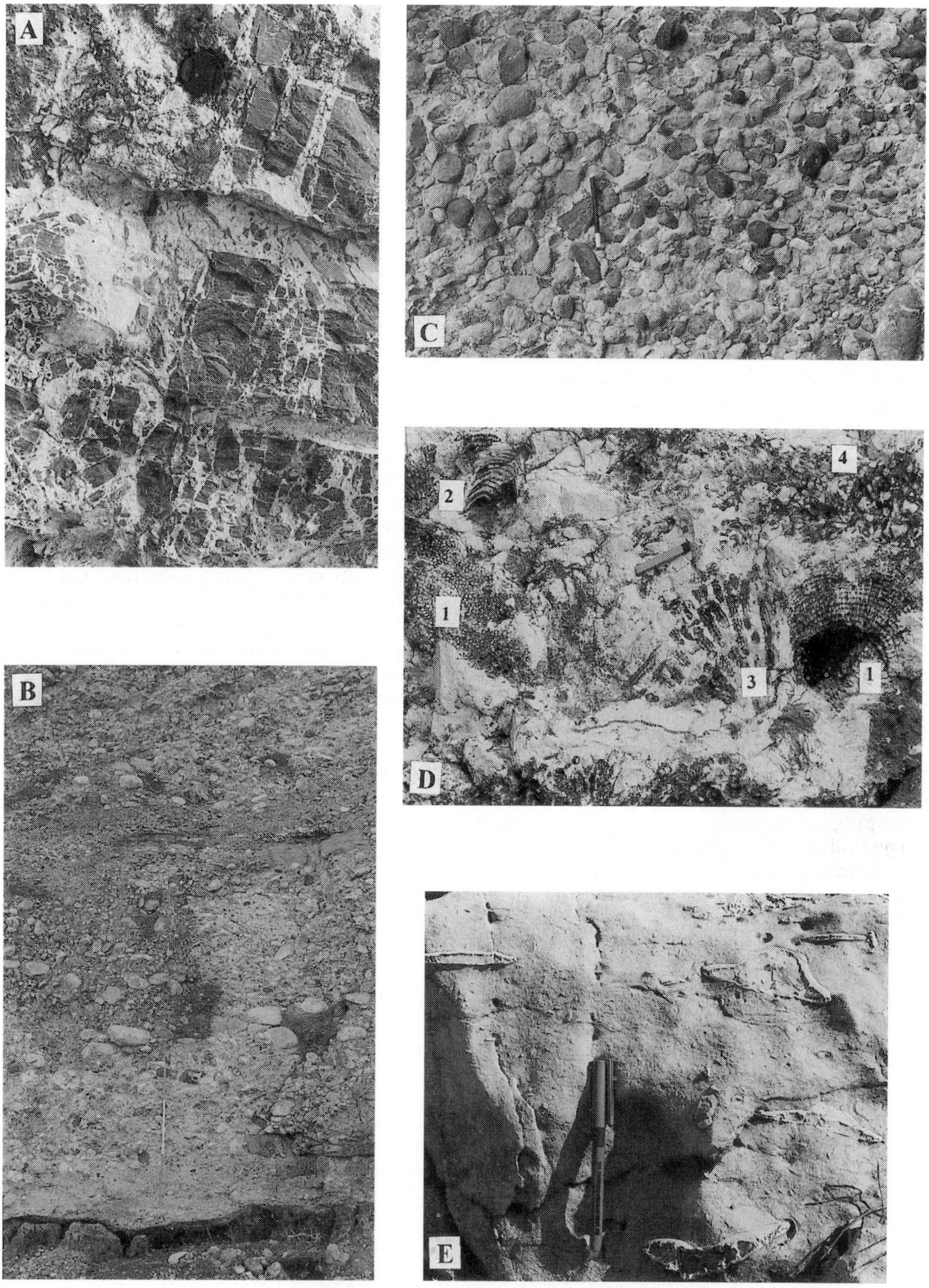
A
B
C
D
1
2
3
4
1
E

southeastern part of the basin (the Alarahan–Halitağalar area).

The section exposed below Alarahan Castle (Fig. 6c) displays thickly bedded, low-angle clinoforms of clast-supported polymict conglomerates (Facies 3) and subordinate matrix-supported conglomerate (Facies 2), with coral–algal patch reefs and foraminiferal wackestone/packstone interbeds (Facies 7). The section is c. 180 m thick and represents the upper part of the Tepekli Conglomerate. The basal part is exposed near Saburlar village (Fig. 1), resting on the marbles of the Alanya Massif, the surface of which is karstified, bored by *Lithophaga* and has oyster shell fragments cemented on it. In the Alarahan section, the conglomerates are moderately to well sorted and polymict, with dominant subrounded to well-rounded metamorphic and non-metamorphic limestone and sandstone pebbles, and rare occurrences of red chert clasts (Fig. 4c), forming a suitable hard substratum for the reef growth. The section terminates in low-angle, cross-bedded foraminiferal packstone–grainstones (Facies 8), c. 2–3 m thick, representing a possible beach environment truncated by the overlying transgressive Oymapınar Limestone. Throughout the section, five coral patch reefs, up to 5–6 m thick and 2–30 m in lateral extent, are present. The initiation of reef growth appears to have developed preferentially on the matrix-free, flat surfaces of the clast-supported conglomerate beds (Facies 3), suggesting reef development on a delta front subjected to winnowing.

The Halitağalar section (Fig. 6b) represents a lateral development of the Alarahan fan-delta. It is composed of coarse conglomerates interbedded with thick bioclastic limestones and isolated patches of coral reefs. The conglomerates directly overlie the Alanya Massif. This section differs markedly from the Alarahan section, as the clasts are almost exclusively derived from the underlying Alanya Massif metalimestones. The interbedded bioclastic limestones are characterized by algal, foraminiferal wackestone–packstone facies (Facies 6 and 7). The faunal content is dominated by larger benthic foraminifera such as *Miogypsina*, *Heterostegina*, *Operculina*, smaller benthics (*Textularia*) and planktics (*Globigerinids*), together with molluscan bioclasts, echinoid spines, coral and bryozoa fragments. These composite assemblages suggest a fan-delta progradation into a relatively deeper water outer-shelf environment. In both sections the patch reefs are mainly characterized by rich and diverse coral assemblages composed of massive domal and thin branching coral frameworks, reflecting a relatively shallow, moderate-energy, normal-marine environment. The coral framework is characteristically composed of densely packed, *in situ* coral assemblages dominated by *Tarbellastraea*, *Heliastraea*, *Porites* and *Stylophora* (Fig. 4d), with some large massive coral colonies reaching up to 60 cm in size. Some broken and overturned colonies are observed within the framework, which is bound by encrusting coralline algae (*Lithothamnium*, *Lithophyllum*). Intrareef sediments filling the space between the frame builders are mainly of bioclastic wackestone–packstones containing rhodoliths and coral fragments. The bioclasts include green algae (*Halimeda*), benthic foraminifera, bryozoa, bivalves, gastropods, echinoid plates and spines. The reef bodies are flat-based domal forms exhibiting smooth to highly irregular convex-up upper surfaces, without distinct coral zonation. A Burdigalian age is suggested in the Alarahan section by *Miogypsina*-bearing assemblages intercalated within the highest conglomeratic beds.

On both flanks of the Halitağalar-Örenşehir Anticline, the Tepekli Conglomerates were deposited in a marine environment, as shown by numerous reef intercalations (Fig. 3, logs 7 and 9). On the southern flank (Yalçı Tepe;

Fig. 4. Outcrop photographs representative of some facies forming the Tepekli and Oymapınar Formations. (**a**) Intensely fractured and brecciated metacarbonate of the Alanya Massif (Facies 1; Table 1) exposed on the western bank of the Oymapınar Dam site. Note the *Microcodium*-bearing lime mudstone infill between the fractures, indicating a terrestrial environment. Lens cap for scale. (**b**) Two clast-supported fining-upwards units within a channel fill of the Tepekli Conglomerate (Facies 3) in the Sevinç area. Each unit starts with an erosive horizon of coarse-grained sandstone. Note the clast imbrication. Scale bar is 60 cm long. (**c**) Bimodal clast-supported conglomerate with coarse-grained sandy matrix, representing wave-reworked fan-delta front deposits of the Tepekli Conglomerate (Facies 3) in the Alarahan section. Note the well-sorted and rounded aspect of the metalimestone pebbles from the Alanya Massif. Pencil 15 cm long. (**d**) Reef core facies (Facies 9) from the Oymapınar Limestone on the western bank of the Oymapınar Dam site. Note densely packed, high diversity coral reef framework, mainly consisting of massive and domal growth forms (1, *Heliastraea* sp.; 2, *Porites* sp.; 3, *Caulastraea* sp.; 4, *Stylophora* sp.). Scale is 6 cm long. (**e**) Echinoid-rich bioclastic wackestone–packstone (Facies 5) of the east bank of the Oymapınar Dam site, representing a lagoonal environment with an open-marine influence. Pencil 15 cm long.

A
B
C
Olistolith
Reefal Limestone Block
D
E
Reefal limestone block
F

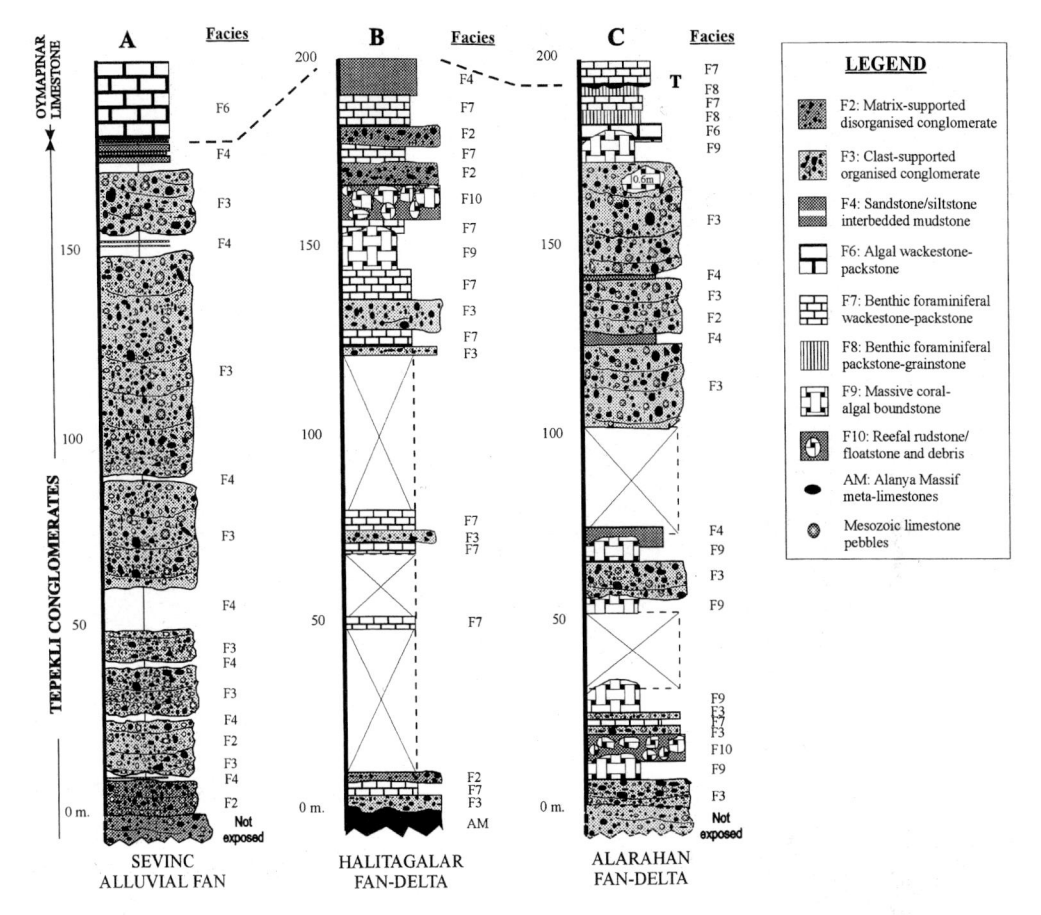

Fig. 6. Generalized logs of the Alarahan and Halitağalar fan-deltas, and the Sevinç alluvial fan.

Fig. 2), the conglomerates grade upwards into marine silty clays (S in Fig. 3, log 9), 100 m thick, with a few fine sandstone beds, containing sparse mollusc shells. This new member (the 'Yalçıtepe Member') provided the opportunity for dating precisely the levels immediately underlying the Oymapınar Limestone, the age of which was uncertain (cf. Flecker et al. 1998). The microfauna (det. A. Hakyemez, MTA) contains the following assemblage: *Globigerinoides altiaperturus* Bolli, *G. trilobus sacculifer* (Brady), *G. trilobus trilobus* (Reuss), *Globorotalia continuosa* Blow, *G. acrostoma* Wezel, *Globigerina falconensis* Blow, *G.* cf. *woodi woodi* Jenkins, *Catapsydrax dissimilis* Cushman and Bermudez: which can confidently be assigned to the N6 zone of Blow's scale, nearly equivalent to NN3 (Moullade, pers. comm.).

Fig. 5. Outcrop photographs representative of some facies from the Geceleme and Karpuzçay Formations along the Manavgat–Konya road. (**a**) Calcarenite–mudstone couplets (Group C2; Table 2) overlain by inversely graded conglomerates in the Karpuzçay Formation (Facies A2.2). Person for scale. (**b**) Normally graded conglomerates with gutter cast in the Karpuzçay Formation (Facies A2.3). Hammer for scale. (**c**) Extraclast limestone breccia (Facies F1.1) within the Çakallar Member. The olistolith (dotted line) is 8 m in length. (**d**) Isolated reefal block (Facies F1.2) in thin-bedded calcarenites and mudstones (Facies C2.3) of the Geceleme Formation. Person for scale. (**e**) Brecciated strata, including reefal limestone block, overlain by calcarenite–mudstone couplets (Facies C2.2) of the Karpuzçay Formation. Person for scale. (**f**) Coherently folded and contorted strata within the calciturbidite and mudstone deposits of the Geceleme Formation. Note the shallow-channel form made up of normally graded calcarenites. Person for scale.

Fig. 7. Probable configuration of fan-deltas during the deposition of the Tepekli Conglomerate.

This age (19–17.5 Ma, Early–Middle Burdigalian) is consistent with the Sr isotope chronology performed by Flecker *et al.* (1998) in the nearby Alarahan section on benthic and planktic foraminifera from the Oymapınar Limestone and the overlying mudstones.

Where the Tepekli Conglomerates are absent, as in the central area of the map presented in Fig. 1, coarse and poorly stratified breccias, up to 60 m thick, are found instead. The material of the breccias is derived from the underlying black Alanya Massif metalimestones and the fragments are embedded in a pale red to red calcareous matrix (Fig. 4a). The angular shape of the fragments and the size of the largest blocks [up to several metres, as can be observed south of Oymapınar Dam (Fig. 9)] demonstrate their proximal origin. In the matrix, the occasional presence of *Microcodium* indicates that deposition took place in subaerial conditions within the water-table oscillations. Thus, these breccias are considered to represent continental scree deposited at the foot of steep slopes surrounding the Lower Miocene Alanya Massif.

A tentative representation of the Manavgat Basin during the deposition of Tepekli Conglomerate is shown in Fig. 7. Two fan-deltas are depicted, separated by an elevated area (the 'Alanya High') attested by the locally preserved continental scree deposits extending from Oymapınar Dam to south of Fersin. In the northwest, the alluvial-fan recorded in the Sırtköy–Sevinç area extends southwards into a (probably) large fan-delta, with a narrow branch coming from the area of Kepez village. In the southeast, the Alarahan fan-delta was mainly fed from the east and locally from the north (Halitağalar), according to clast composition, current direction and facies distribution, in good agreement with Flecker (1995) and Flecker *et al.* (1998).

The Oymapınar Limestone

The Oymapınar Limestone is exposed in a northwest–southeast trending narrow belt (Fig. 1) which onlaps the Tepekli Conglomerate and the Alanya Massif northwards (Fig. 2). It is of variable thickness (20–150 m; Fig. 3) and is represented by an overall upwards shallow-marine carbonate shelf succession (Figs 3 and 8).

At the southeastern end of the basin (the Alarahan section; Fig. 6) the Oymapınar Limestone is *c.* 20–30 m thick and overlies the Alarahan fan-delta deposits abruptly with a flat truncation surface (T in Fig. 6c). The section is made up of coarse-grained large benthic foraminiferal wackestone–packstone (Facies 7) containing abundant *Operculina*. The 6–8 m thick lower part is heavily bioturbated, while upwards, the wackestone–packstone becomes well bedded and comprises small isolated coral reefs (Facies 9). The coral reefs are almost entirely composed of rich and highly diversified hermatypic corals with minor solitary forms. The overall sequence suggests carbonate deposition in a relatively deep-water outer-shelf environment, promoting only limited

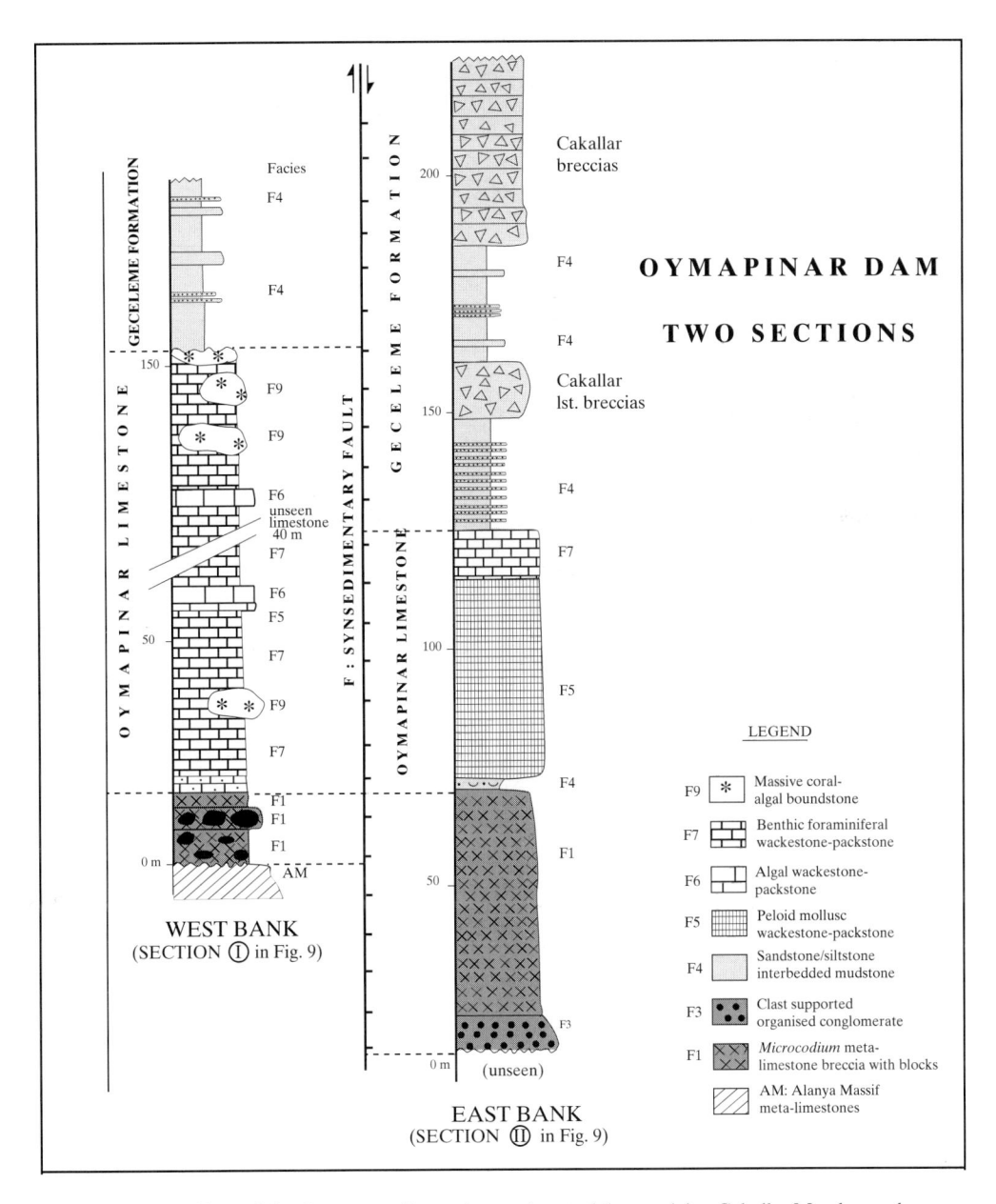

Fig. 8. Generalized logs of the Oymapınar Formation, and part of the overlying Çakallar Member and Geceleme Formation in the Oymapınar Dam area.

development of coral reefs. The truncation at the base of the heavily bioturbated basal unit marks the onset of a transgressive period in the area.

Further to the northwest, along the road to Ahmetler village (Fig. 1), the Oymapınar Limestone directly overlies the metacarbonates of the Alanya Massif. There, the Oymapınar Limestone is dominated by a succession of algal and large benthic foraminiferal wackestone–packstone facies associations (Facies 7), with small coral reef patches (Facies 9). The detail of its internal geometry, bedding configuration and facies characteristics reveal the presence of a shelf-margin algal-mound complex with well-developed basin- and shelfward-dipping beds

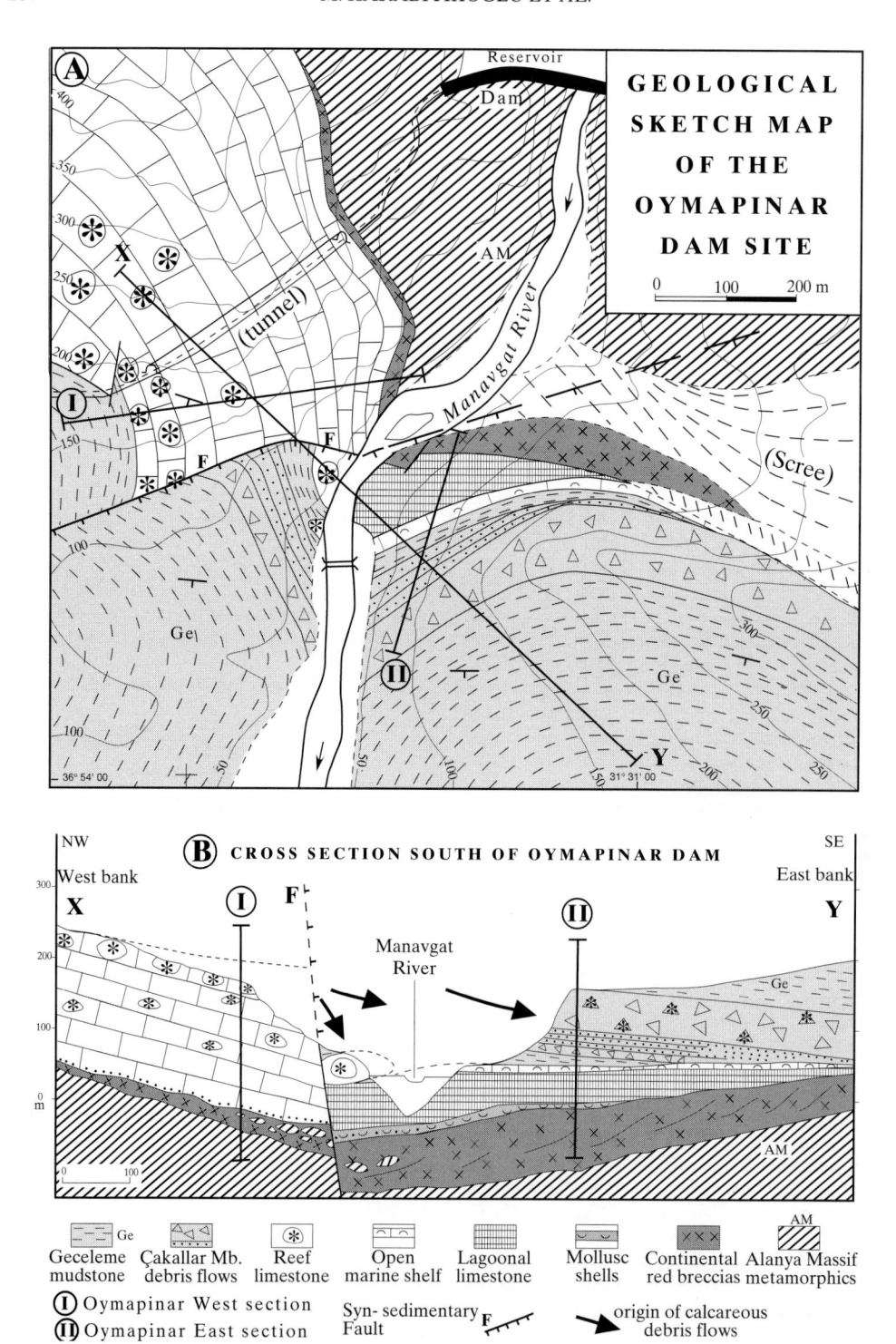

Fig. 9. (a) Oymapınar Dam site interpretative map and (b) X–Y cross-section.

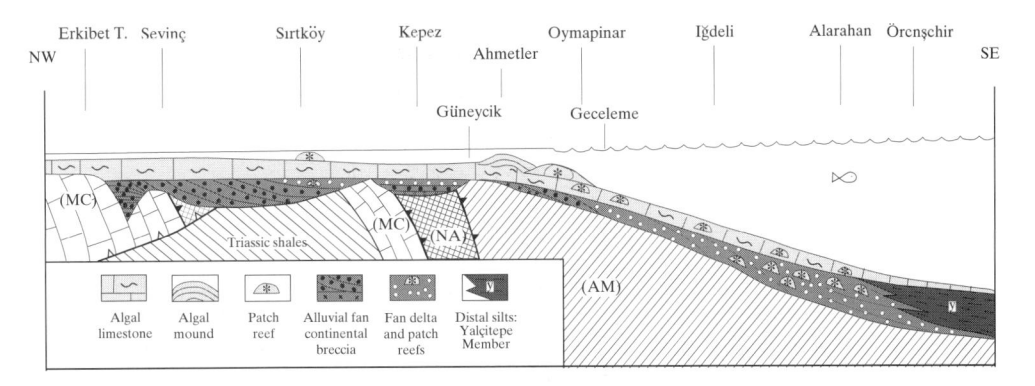

Fig. 10. Interpretative view of the facies distribution of the Tepekli Conglomerate and the Oymapınar Limestone from the narrow shelf to talus slope.

with horizontally stratified intermound beds. The core of the algal mound is characterized by relatively thick, flat-topped and steep-sided massive, rhodolithic algal limestone that is mainly made up of poorly sorted yellow foraminiferal algal wackestone–packstone (Facies 7). It contains well-preserved, but randomly scattered, bodies of echinoderms, thick-walled ostrea shells, disarticulated or fragmented bivalves and a large number of large rhodolites reaching up to 4–5 cm in size.

At the Oymapınar Dam site, the eponymous limestones are represented by two contrasting facies developments, separated by a synsedimentary fault (see below):

- On the western bank (section I; Figs 8 and 9), the succession shows a gradual transition from pebbly bioclastic packstones (15 m thick) of a high-energy near-shore environment to a thick succession (120 m) of well-bedded foraminiferal wackestone–packstone (Facies 7) with coral–algal patch reefs, reflecting a normal salinity carbonate shelf. At the top, the presence of globigerinids, and rare solitary corals in the reef, indicates a relatively deeper water environment immediately preceding the overlying Geceleme Formation mudstones.
- In contrast, along the eastern bank of the Manavgat River (section II; Figs 8 and 9), the marine succession commences with a 7 m thick gastropod-rich bioturbated mudstone (Facies 4) representing a siliciclastic-dominated lagoonal deposit. This is abruptly overlain by a 25 m thick peloid mollusc blue-grey wackestone–packstone (Facies 5) with silt-size quartz and carbonized fragments, suggesting a restricted shelf lagoon. The rest of the section is represented by 35 m of thick similar carbonates showing a deepening upwards trend according to the rare occurrence of benthic and planktonic foraminifera. This succession is abruptly overlain by mudstones of the Geceleme Formation and reefal shelf-derived debris flows of the Çakallar Member.

This contrasting development in the Oymapınar Limestones across the river and the Çakallar Member will be discussed below (fault related; Fig. 9).

Figure 10 shows an interpretative view of the facies distribution of the Oymapınar Limestone and the Tepekli Conglomerate along a synthetic 2D section. In the northwest, the Tepekli Conglomerate is represented as a subaerial alluvial fan filling deeply incised valleys, as seen in the Sevinç area. Southwards, the conglomerates progressively pass into a marine environment (fan-delta) indicated by patch reefs (Halitağalar and Alarahan sections). Silty clays (S) appear in the southeasternmost section (Örenşehir) and are interpreted as a deeper and distal facies of the Tepekli Conglomerate (Yalçıtepe Member). Above, the distribution of patch reefs and algal mounds in the Oymapınar Limestone suggests a shallow-marine carbonate platform, deepening and thinning to the south and southeast (Örenşehir), with an inner shelf algal facies and rarer patch reefs in the north (Sevinç).

The Geceleme Formation

The Geceleme Formation is mostly composed of hemipelagic mudstones (Class E; Table 2) with intercalated calciturbidites and siltstone beds (Classes B–D), and occasional chaotic deposits (Class F). It is exposed in the central and eastern parts of the Manavgat Basin (Fig. 1)

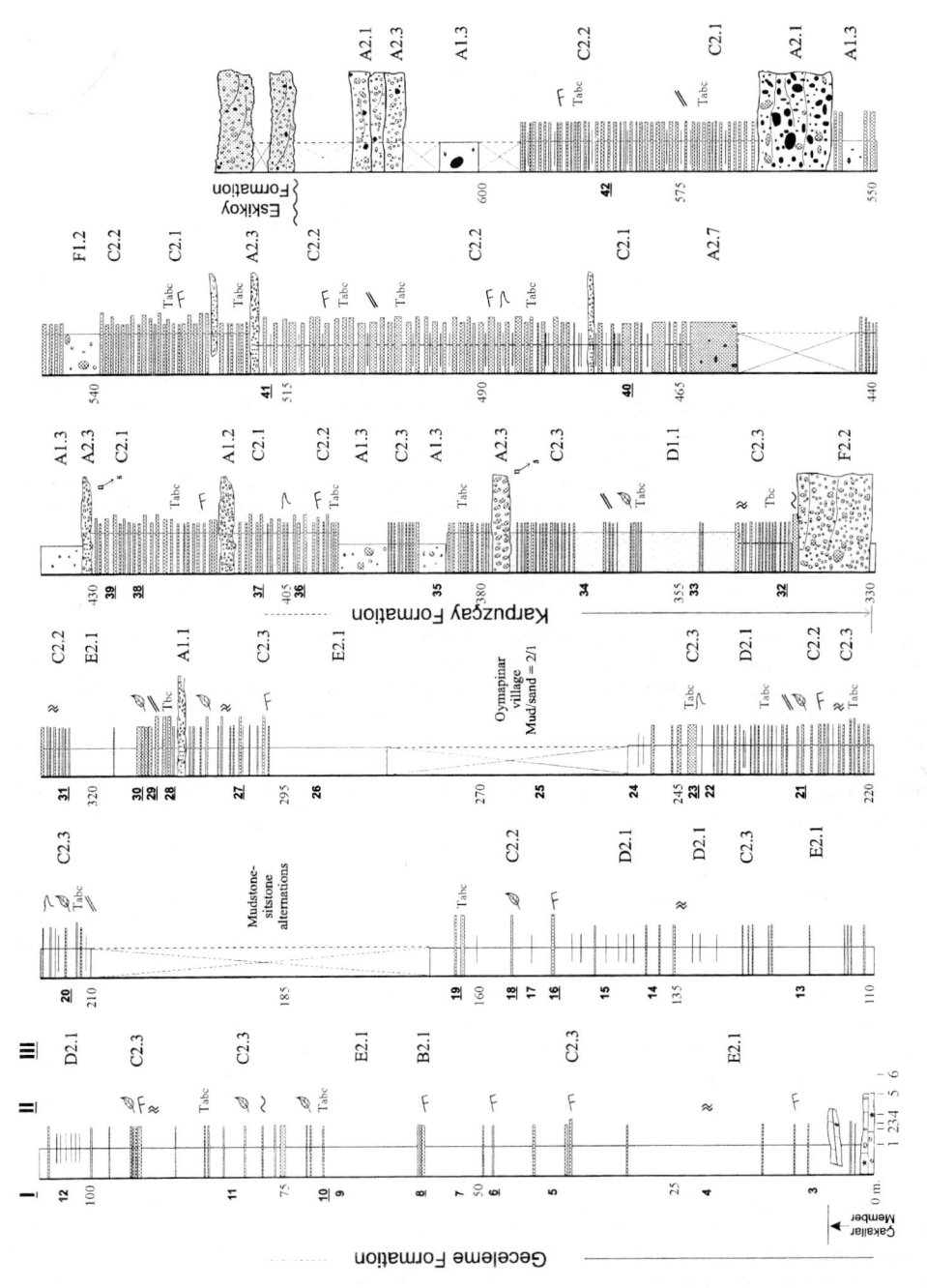

Fig. 11. Measured section of the Geceleme and Karpuzçay Formations along the Manavgat River road. See Fig. 13 for legend.

where it conformably overlies the Oymapınar Limestone. It passes progressively upwards into the coarser Karpuzçay Formation. Conglomerates and breccias that are locally present in the lower portion of the Geceleme Formation form the Çakallar Member (see below).

The measured section (320 m) along the Manavgat River, on the road to the Oymapınar Dam, clearly shows the mudstone-dominated character of the formation (Fig. 11). Planktonic foraminifera are abundant and belong to the *Orbulina universa* and *Globigerina nepentes* biozones, indicative of the Early and Late Serravallian, respectively. The Geceleme mudstones are black to dark grey in colour and locally show well-developed bioturbation. The calcarenite intercalations are fine to medium grained, thin to medium bedded, and mostly tabular and laterally very continuous (Facies C2.3 and C2.2). They frequently contain brown coloured plant debris in their lower part and exhibit current ripples on their upper surfaces. Bouma Ta, Tb and Tc divisions are common, especially in medium-bedded calciturbidites. In places, normally graded coarse-grained calcarenites form lenticular intercalations, suggesting shallow-channel structures (Fig. 5f). Thin- to medium-bedded siltstones (Facies D2.1) are also present. Intervals with higher concentrations of calciturbidites, with 1/1 calcarenite–mudstone ratios, form 15–30 m thick conspicuous levels upsection (Fig. 11; 210–245 and 300–318 m). This slight upwards coarsening marks a gradual transition to the Karpuzçay Formation, the base of which is taken at the first significant conglomeratic horizon (Fig. 11; 330 m).

Another well-exposed section of the Geceleme Formation (290 m) is located 20 km further east, along the Manavgat–Konya road (Fig. 12), immediately south of Geceleme village (cf. Flecker 1995). As in the preceding section, the dominant lithology is hemipelagic mudstone with thin- to medium-bedded calciturbidite intercalations. Concentrations of calciturbidites also occur in the upper part of the formation. This section differs from the preceding one by containing thick siltstone beds (Facies D1.1), conglomeratic horizons (Facies A1.1), folded and contorted horizons (Facies F1.2; Fig. 5f) and large-scale isolated reef–limestone blocks (Facies F2.2) within the well-bedded mudstone–fcalcarenite deposits (Fig. 5d) in the lower half of the section. This association denotes an unstable margin.

The overall hemipelagic character of the sedimentation, with occasional slumps and rock falls/slides (Fig. 5d–f), as well as the local occurrence of the Çakallar Member (see below), suggest deposition in a fault-bounded base-of-slope to basin-floor setting. The amalgamation of calcarenite beds in the upper part of the formation suggest a shallowing trend and may have formed in fine-grained and/or distal submarine fans preceding the high-density turbidity currents and debris flows which characterize the overlying Karpuzçay Formation.

The Çakallar Member

Although the Çakallar 'Formation' was defined by Akay & Uysal (1985) near Çakallar village (Fig. 1), these locally occuring deposits are here considered as a member within the lower part of the Geceleme Formation, in contrast to Flecker (1995) who identified it only at the boundary between the Oymapınar Limestone and the Geceleme Formation. It is beautifully exposed along the main road from Manavgat to Konya, 2 km south of the village of Fersin. Along this section (Fig. 3, log 5, and Fig. 5c) the Çakallar Member is *c.* 110 m thick and directly overlies the Oymapınar Limestone. It is represented by an alternation of sharp flat-based or channelled disorganized conglomerates (Group A1) and extraclast and intraclast limestone breccias (Facies F1.1), embedded in calcarenite–mudstone intercalations (Class C). Two thick polymict breccia horizons at the base and in the middle of the Çakallar exposure (15 and 30 m, respectively; Fig. 13) exhibit a chaotic assemblage of angular to subangular, clast-supported reef fragments and dark metalimestone clasts, with enormous blocks reaching 8 m in length (Fig. 5c). These chaotic deposits thin out within a few hundred metres southwards into the Geceleme mudstones.

A conspicuous fault scarp, which can be traced over 2.5 km south of Fersin, offsets the Oymapınar Limestone > 400 m (Fig. 14) and constitutes the western boundary of the Çakallar Member outcrop. South of Çenger Stream, the continuation of this fault loses displacement within the Geceleme mudstones; the overlying Karpuzçay Formation is not affected (see Fig. 1). Further south, on both sides of the Örenşehir-Halitağalar Anticline, the Çakallar Member thins and disappears westwards (Fig. 15). The synsedimentary character of the Fersin Fault, the limited extent of the Çakallar Member and the northward increase of block size, strongly suggest that the Çakallar breccias resulted from proximal redeposition of blocks fallen from the upthrown Oymapınar Shelf and Alanya basement at the beginning of the deposition of the Geceleme Formation mudstones.

In a similar way, the Oymapınar Dam area

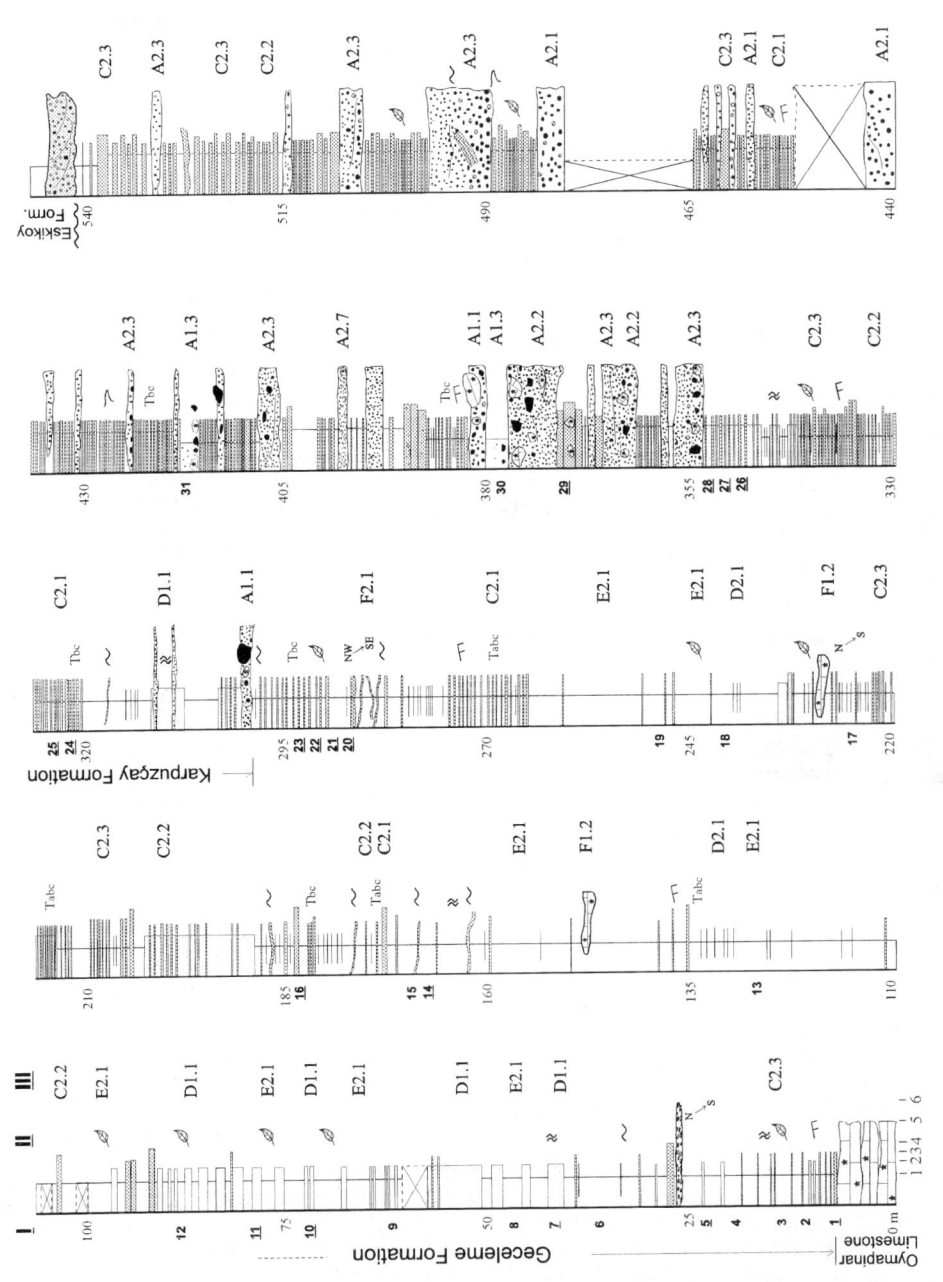

Fig. 12. Measured section of the Geceleme and Karpuzçay Formations along the Manavgat–Konya road. See Fig. 13 for legend.

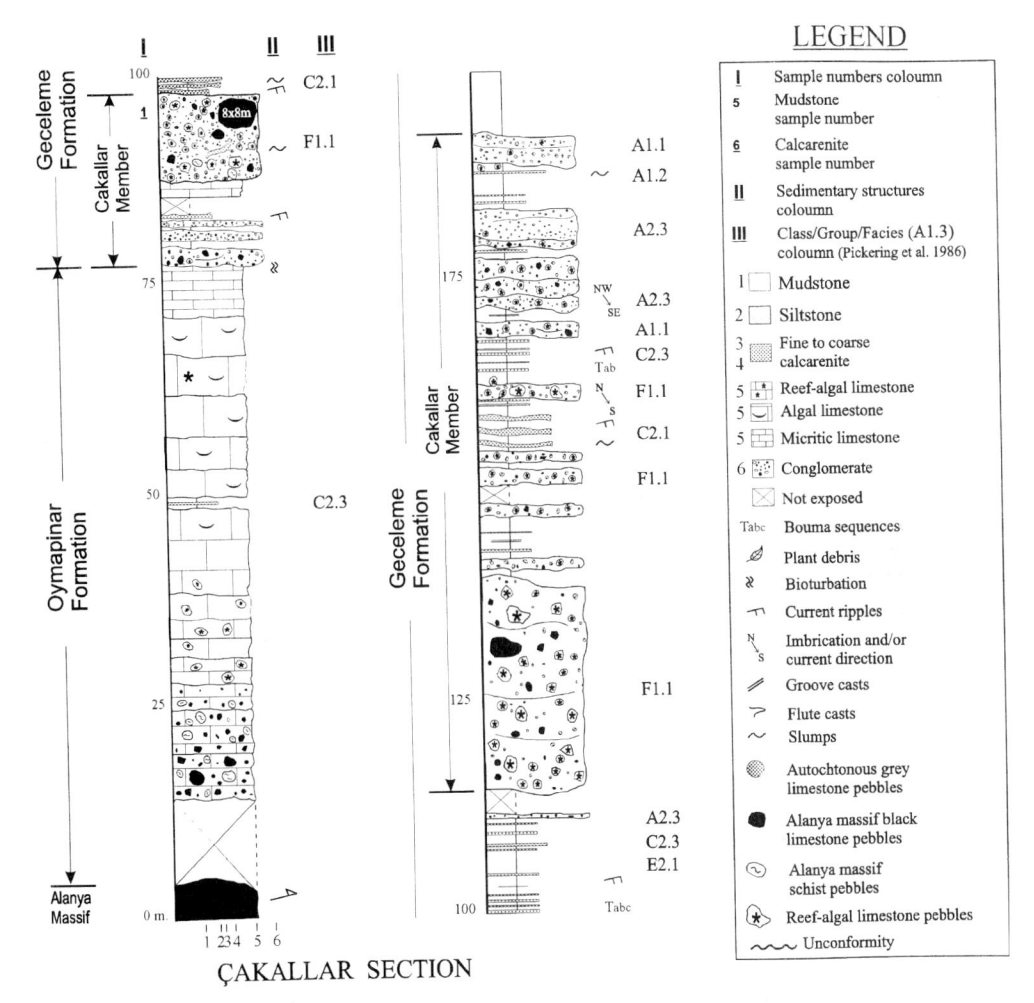

Fig. 13. Measured section of the Çakallar Member along the Manavgat–Konya main road, 2 km south of Fersin village.

(Fig. 9) also yields a conspicuous, but limited, outcrop of Çakallar breccias, 20 m thick, containing Oymapınar reefal blocks several metres long floating in the Geceleme mudstones. Surprisingly, whereas the Çakallar deposits are similar on both sides of Manavgat Gorge, striking lithological differences appear in the underlying Oymapınar Limestones. As seen above, shelf carbonates with abundant patch reefs are well developed on the west bank (c. 170 m thick; section I; Fig. 8), whereas on the opposite bank (section II), the limestones exhibit only 60 m thick lagoonal facies without reefs. Careful examination of the site disclosed a southwest–northeast oriented normal fault (F in Fig. 9) which separates the successions. This fault is interpreted here as a synsedimentary feature which was active when the Geceleme Basin formed, and was shedding coral blocks and shelfal limestone fragments into the Geceleme Formation mudstones. Accordingly, west of that fault, the Çakallar breccias are totally absent within the Geceleme mudstones, as shown in Fig. 9. In detail, a more complex picture of the fault mechanics may be suspected since the presently upthrown side of the fault (west) contains the reefal facies, whereas the downthrow side (east) is attributed to protected, lagoonal facies.

The Karpuzçay Formation

The Karpuzçay Formation outcrops as a large continuous belt from west to east (Fig. 1). It

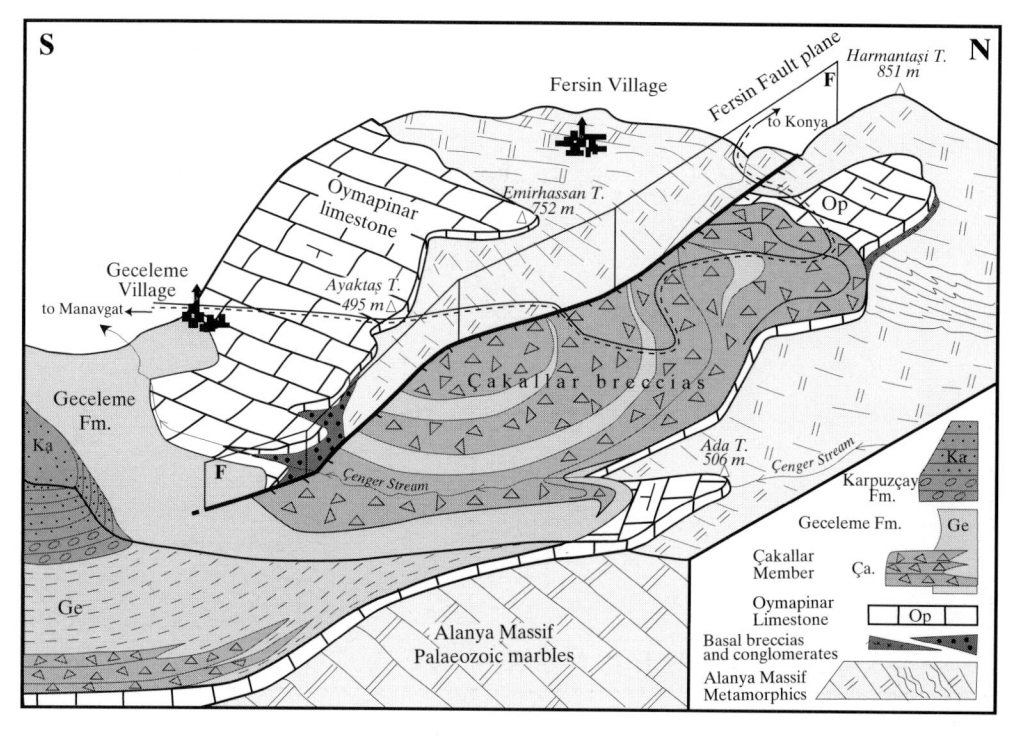

Fig. 14. Present outcrop view of the synsedimentary Fersin Fault east of Geceleme village.

represents the final stage of the Manavgat Basin fill during the Tortonian–Messinian (*Globorotalia acostaensis acostaensis* biozone), and higher. It is unconformably overlain by fluvial deposits of the Eskiköy Formation (?Early Pliocene).

As shown in the Manavgat River section (Fig. 11) and Konya road section (Fig. 12), the Karpuzçay Formation comprises thicknesses > 300 and 240 m, respectively. It consists of interlaminated calciturbidites and mudstones or siltstones (Group C2), commonly interrupted by a few to several metres thick erosional intercalations of clast- to matrix-supported conglomeratic horizons (Class A; Fig. 5a), and occasional chaotic deposits (Class F) up to several metres thick. Compared to the Geceleme Formation, the mudstones are much less dominant with respect to the calcarenites. Plant debris, groove casts and Bouma sequences are still common in the calcarenite–mudstone couplets. The clast-supported conglomeratic horizons (Facies A2.3) form either tens of centimetres to metres thick amalgamated beds with irregular erosional soles, or large-scale channellized structures pinching out laterally into overbank coarse-grained calciturbidites. The clasts, which range from granule to boulder grade, are randomly orientated and commonly display fining-upward trends. The matrix-supported conglomerates, or blocky mudstones (Facies A1.3), form normally or inversely graded (Fig. 5a and b), up to few metres thick, sheet-like bodies with slightly erosive bases. Whatever the type of conglomerate, the clasts are subrounded to well rounded, except for the largest ones (metre scale) which come from the Alanya Massif. In the Manavgat River section, pebbles are mostly composed of Mesozoic carbonates, which are mixed with clasts from the Alanya Massif only in the highest part of the section. In contrast, in the Manavgat–Konya road section most of the clasts are derived from the Oymapınar Limestones or the Alanya Massif metalimestones.

The Karpuzçay Formation does not represent direct continuation of the fault-controlled sedimentation illustrated locally by the Çakallar breccias, but reflects a more long-ranging phase of tectonic activity which caused uplift of the hinterland and the shelfal area north of the basin. The coarser facies can be interpreted in terms of a variety of mass-flow processes, including high-density turbidity currents, slumps and debris flows in an upper slope setting.

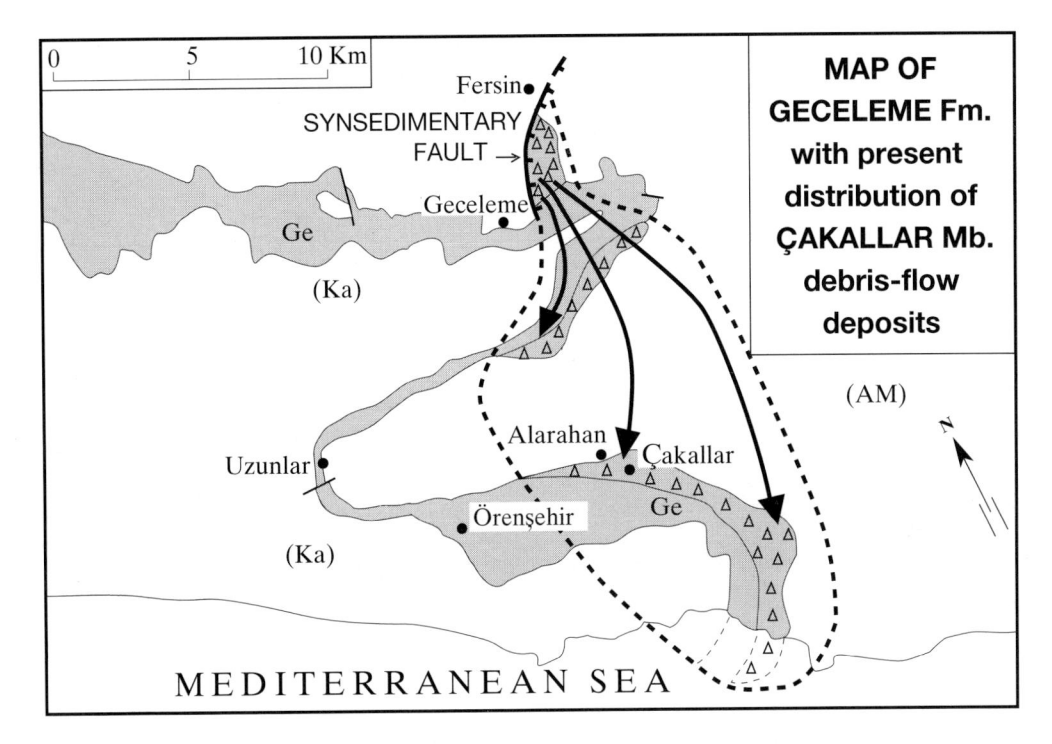

EVOLUTION OF MANAVGAT BASIN, WESTERN TAURIDES 291

Fig. 15. Distribution of debris flow deposits during Çakallar time in the Fersin–Çakallar area.

On the elongated İğdeli–Güneycik Plateau (Fig. 1), the Geceleme mudstones are unconformably overlain by a 20–100 m thick conglomerate. This conglomerate is made up of 1–10 m thick sheet-like beds with sharp erosional bases and fining-upwards trends. Well-rounded clasts, up to 30 cm in size, pass progressively upwards into coarse calcirudite. Imbricated clasts are frequent in the coarsest part of the beds, which also comprise rip-up clasts of calcarenite. Clasts belong to the Alanya Massif and, quite a large proportion, the Oymapınar Formation, including patch reef-derived debris. An overall fining-upwards trend is also visible and coarse-grained calcarenites with well-defined horizontal or low-angle lamination form most of the higher located beds near İğdeli village. Although this conglomerate may resemble those of the Tepekli Conglomerate, it never shows evidence of subaerial overbank deposits (e.g. red mudstone or nodular calcrete horizons). Clast imbrications indicate southwestward directed palaeocurrents and the whole conglomeratic unit forms a huge northeast–southwest oriented erosive channel-like structure, while the underlying Geceleme facies thins out towards the northeast (c. 50 m near Güneycik; Fig. 3). It is suggested here that the İğdeli–Güneycik conglomerate corresponds to the last locally preserved fan-delta that filled up the Miocene Manavgat Basin. It might represent the proximal equivalents of the above described slope facies and could thus be included in the Karpuzçay Formation. A similar notion regarding fan-delta deposition in the Karpuzçay Formation was suggested by Flecker (1995, her 'Taşkesiği Group').

Discussion and conclusion

The initial coarse deposits (the Tepekli Conglomerate, Burdigalian in part) are irregularly distributed, and fill a pre-existing topography that marks an important period of relative sea-level fall and subsequent erosion. Two major fan-like systems have been identified: the Sevinç alluvial fan and fan-delta in the northwest, and the Alarahan fan-delta in the southeast (Fig. 7). Between these two drainage systems, a mountainous country is implied by the discontinuous deposits of red monomict breccias of continental origin (scree), implying steep slopes in the immediate vicinity. According to the clast provenance, the Sevinç and Alarahan fan-deltas were the output of two major drainage systems tapping into source areas in the newly created mountainous area in the northern and northeastern parts of the western Taurus. The fan-delta complexes prograded into a shallow shelfal area in which the distal silts of the Yalçıtepe Member accumulated.

A sharp rise of relative sea level and a decreasing rate of sediment supply, due to the progressive denudation of relief, resulted in the deposition of the transgressive Oymapınar Limestone (Late Burdigalian–Langhian) which onlaps the fan-delta deposits and the substratum, with a gentle southward deepening trend documented by the distribution of the reefs (Fig. 10).

After this tectonically quiescent episode, a sudden deepening is documented by the onset of deposition of the Geceleme mudstones with pelagic faunas overlying the patch reefs of the Oymapınar shelf carbonates. The Çakallar breccias and debris flows that locally appear in the lower part of the Geceleme Formation demonstrate the tectonic origin of this sudden deepening. Synsedimentary faulting can be precisely documented by the presence of a large amount of debris from the Oymapınar shelf carbonates, by the distribution of the breccias (Fig. 15) and by the identification of the fault planes themselves (Fig. 14), all implying fragmentation and sinking of the Oymapınar carbonate shelf. The foundering of this basin, however, has no positive counterpart inland, as shown by the persistent predominance of fine mudstone facies throughout the Geceleme Formation.

The succeeding sedimentation comprised the filling of the newly created accommodation space with an overall coarsening-upwards succession from the Geceleme to Karpuzçay Formations. The gravity induced character of most of the Karpuzçay Formation, the sharp passage from high-density currents and debris flows to a turbulent coarse-grained fan-delta, suggests that the sedimentation was largely controlled by progressive uplift of the hinterland, from the Tortonian onwards.

A differential uplift may be inferred from clast composition of the debris flows, which implies a strong erosion of the Alanya Massif and its Miocene carbonate cover in the east (Alarahan area). In contrast, Mesozoic limestone clasts predominate in the west of the basin, reflecting a larger uplift in this part of the Taurus chain. Although discontinuous, the presence of debris flow deposits throughout the Karpuzçay Formation implies repeated influx of coarse material from nearby sources, and suggests a persistent elevation inland until the end of Miocene [Messinian; cf. Flecker et al. (1995)]. The resulting uplift may be compared to the Mut region, 150 km further east, where horizontal marine Miocene strata are presently found at an elevation of 2000 m. The filling of the Manavgat Basin ended with the Messinian (Bizon et al. 1974), and a north–south compressional phase subsequently produced large open folds in the Miocene deposits before the deposition of the undeformed Pliocene fluvial conglomerates (the Eskiköy Formation).

Flecker et al. (1998) (Fig. 8) consider that the load of the Lycian Nappes arriving in the western part of the Antalya Gulf may have induced flexural effects in the lithosphere, and these effects should have influenced the Neogene sedimentation in the three sub-basins situated in front of the nappes (Aksu, Köprüçay and Manavgat), as suggested by contrasting drainage patterns in the first two basins, although their orientation is oblique relative to the thrust front. The influence of the advancing Lycian Nappes is not disputable in the Miocene basin situated along the thrust front (Poisson 1977; Hayward 1984), and it may be considered in the Aksu and Köprüçay Basins, 30–60 km away. However, the case of the Manavgat Sub-basin is more questionable, owing to its orthogonal orientation and its distance from the front of the Lycian Nappes, presently > 100 km away.

In contrast, the lithologies and stratigraphic ages of the main formations in the Manavgat Basin may easily be connected with the northern part of the Adana Basin (Fig. 1), where a very similar evolution has been reported (cf. Görür 1992; Naz et al. 1992, 1998; Williams et al. 1993; Gürbüz 1999). Three distinct episodes are especially noticeable there, corresponding to fan-delta deposition in the Early Miocene, followed by Burdigalian–Serravalian reefal limestone and subsequent thick turbiditic filling. Extensional tectonics during the Early–Middle Miocene is also emphasized by normal faulting, and an important uplift at the end of Miocene is documented by the present altitude (2000 m) of Middle Miocene debris flows north of Adana (Derman 1998). In that way, the Manavgat Basin may be seen as the westernmost part of the Adana Basin.

This paper is the result of a joint project, TÜBİTAK–CNRS No. 5785. The authors express their gratitude to E. Akay, R. Flecker, M. Şenel, S. Erk, A. Hakyemez, Ch. Perrin, S. Altuğ and M. Moullade for providing most interesting documents, and helpful discussions. The French Embassy in Ankara and the MTA headquarters in Konya are acknowledged for ongoing support. As reviewers, R. Flecker and T. Dreyer made numerous criticisms and helpful suggestions to improve the manuscript.

References

AKAY, E. & UYSAL, S. 1985. *Orta Toroslarιn Batιsιndaki (Antalya) Neojen Çökellerinin Stratigrafisi, Sedimantolojisi ve Yapιsal Jeolojisi.* Mineral Research and Exploration Institute of Turkey (MTA) Report, No. 2147 [in Turkish].

——, ——, POISSON, A., CRAVATTE, J. & MÜLLER, C. 1985. Stratigraphy of the Antalya Neogene Basin. *Geological Society of Turkey Bulletin*, **28**, 105–119.

BARKA, A. A., REILINGER, R., ŞAROĞLU, F. & ŞENGÖR, A. M. C. 1997. The Isparta Angle: its importance in the neotectonics of the Eastern Mediterranean region. *In*: PIŞKIN, O., ERGÜN, M., SAVAŞÇIN, N. Y. & TARCAN, G. (eds) *Proceedings of the International Earth Science Colloquium on the Aegean*, 9–14 October 1995, İzmir, **1**, 1–18.

BIZON, G., BIJU-DUVAL, B., LETOUZEY, J., MONOD, O., POISSON, A., ÖZER, B. & ÖZTÜMER, E. 1974. Nouvelles précisions stratigraphiques concernant les bassin tertiaires du sud de la Turquie (Antalya, Mut, Adana). *Revue de l'Institut Français du Pétrole, Paris*, **29**, 305–320.

BLUMENTHAL, M. 1951. *Recherches géologiques dans le Taurus occidental dans l'arrière pays d'Alanya.* Mineral Research and Exploration Institute of Turkey (MTA) Publications, Series D, **5**.

—— 1963. Le système structural du système sud-Anatolien. *In*: '*Livre à la mémoire du Pof. P. Fallot*'. Memorie Hors Série Société Géologique de France, 611–662.

DERMAN, S. 1998. Characteristics of downlapping beds along a maximum flooding surface in a Miocene sequence, B. Koraş village, Karaman, Turkey. *Proceedings of the 12th International Petroleum Congress and Exhibition of Turkey*, 217–228.

DUNHAM, R. J. 1962. Classification of carbonate rocks according to depositional texture. *In*: HAM, W. E. (ed.) *Classification of Carbonate Rocks.* AAPG Memoirs, 108–121.

ERK, S., AKGA, N. & ERTUĞ, K. 1995. Manavgat Miyosen havzası'nın biyostratigrafisi. *Karadeniz Technical University, 30th Anniversary of Geological Engineering Symposium*, Abstracts, 23.

FLECKER, R. 1995. *Miocene basin evolution of the Isparta Angle, southern Turkey.* PhD Thesis, Edinburgh University.

——, ROBERTSON, A. H. F., POISSON, A. & MÜLLER, C. 1995. Facies and tectonic significance of two contrasting Miocene basins in south coastal Turkey. *Terra Nova*, **7**, 221–232.

——, ELLAM, R. M., MÜLLER, C., POISSON, A., ROBERTSON, A. H. F. & TURNER, J. 1998. Application of Sr isotope stratigraphy and sedimentary analysis to the origin and evolution of the Neogene basins in the Isparta Angle, southern Turkey. *Tectonophysics*, **298**, 83–101.

FRIZON DE LAMOTTE, D., POISSON, A., AUBOURG, C. & TEMIZ, H. 1995. Chevauchements post-Tortonien vers l'Ouest puis vers le Sud au coeur de l'Angle d'Isparta (Taurus, Turquie). Conséquences géodynamiques. *Bulletin de la Société Géologique de France*, **166**, 59–67.

GÖRÜR, N. 1992. A tectonically controlled alluvial fan which developed into a marine fan-delta at a complex triple junction: Miocene Gildirli Fm. at the Adana Basin, Turkey. *Sedimentary Geology*, **81**, 243–252.

GUTNIC, M., MONOD, O., POISSON, A. & DUMONT, J.-F. 1979. Géologie des Taurides occidentales (Turquie). *Mémories de la Société Géologique de France*, 137.

GÜRBÜZ, K. 1999. Regional implications of structural and eustatic controls in the evolution of submarine fans: an example from the Miocene Adana Basin, southern Turkey. *Geological Magazine*, **136**, 311–319.

HAYWARD, A. B. 1984. Miocene clastic sedimentation related to the emplacement of the Lycian Nappes and the Antalya Complex, SW Turkey. *In*: DIXON, J. E. & ROBERTSON, A. H. F. (eds) *The Geological Evolution of the Eastern Mediterranean.* Geological Society, London, Special Publications, **17**, 287–300.

KISSEL, C., AVERBUCH, O., FRIZON DE LAMOTTE, D., MONOD, O. & ALLERTON, S. 1993. First palaeomagnetic evidence for a post-Eocene clockwise rotation of the Western Taurides thrust belt east of the Isparta reentrant (SW Turkey). *Earth and Planetary Science Letters*, **117**, 1–14.

MARCOUX, J. 1987. *Histoire et topologie de la Neotethys.* Thèse, University of Paris.

MIALL, A. D. 1978. Lithofacies types and vertical profiles models in braided river deposits. A summary. *In*: MIALL, A. D. (ed.) *Fluvial Sedimentology.* Canadian Society of Petroleum Geologists Memoirs, **5**, 597–604.

MONOD, O. 1977. *Recherches géologiques dans le Taurus occidental au sud de Beyşehir (Turquie).* Thèse,University of Paris Sud-Orsay.

MUTTI, E. 1977. Distinctive thin-bedded turbidite facies and related depositional environments in the Eocene Hecho Group (south central Pyrenees, Spain). *Sedimentology*, **24**, 107–131.

—— & RICCI LUCCHI, F. 1972. Le torbiditi dell'Apennino settentrionale: introduzione all'analisi di facies. *Geological Society of Italy Memoirs*, **11**, 161–199.

—— & —— 1978. Turbidites of the northern Apennines: introduction to facies analyses. *International Geology Review*, **20**, 125–166.

NAZ, H., KOZLU, H. & DEMIRKOL, C. 1998. Common Neogene sedimentary architectures in the Adana and Manavgat Basin successions. *Proceedings of the 12th International Petroleum Congress and Exhibition of Turkey*, Abstracts, 214.

——, ALKAN, H., ERK, S., AKGA, N., ERTUĞ, K. & DEMIR, E. 1992. *Manavgat Baseni (XVI. Bölge) Miyosen İstifinin Lito-biyostratigrafisi, Fasiyes ve Dizilim Analizi ve Hidrokarbon Potansiyelinin Değerlendirilmesi.* Turkish Petroleum Corporation (TPAO) Report, No. 1734 [in Turkish].

PICKERING, K. T., HISCOTT, R. N. & HEIN, F. J. 1989. *Deep Marine Environments: Clastic Sedimentation and Tectonics.* Unwin Hyman.

——, STOW, D. A. V., WATSON, M. & HISCOTT, R. N. 1986. Deep-water facies, processes and models: a review and classification scheme for modern and ancient sediments. *Earth Science Review*, **22**, 75–174.

POISSON, A. 1977. *Recherches géologiques dans les Taurides occidentales (Turquie)*. Thèse, University of Paris Sud-Orsay.

——, AKAY, E., CRAVATTE, J., MÜLLER, C. & UYSAL, S. 1983. Données nouvelles sur la chronologie de mise en place des nappes d'Antalya (Taurides occidentales, Turquie). *Comptes Rendus des Séances de l'Académie des Sciences Paris*, **296**, 923–925.

ŞENEL, M., DALKILIÇ, H., GEDİK, S. ET AL. 1993. *Eğridir–Yenişarbademli–Gebiz ve Geris-Köprülü (Isparta–Antalya) Arasında Kalan Alanlarin Jeolojisi*. Mineral Research and Exploration Institute of Turkey (MTA)–Turkish Petroleum Corporation (TPAO) Report, No. 3132 [in Turkish].

WILLIAMS, G. D., ÜNLÜGENÇ, U. C., KELLING, G. & DEMİRKOL, C. 1993. Structural controls on stratigraphic evolution of the Adana Basin, Turkey. *Journal of the Geological Society, London*, **152**, 873–882.

WILSON, J. L. 1975. *Carbonate Facies in Geologic History*. Springer.

Palaeostress inversion in a multiphase deformed area: kinematic and structural evolution of the Çankırı Basin (central Turkey), Part 1 – northern area

NURETDİN KAYMAKÇI,[1] STANLEY H. WHITE[2] & PAUL M. VAN DIJK[1]

[1]*ITC, Hengelosestr 99, PO Box 6, 7500 AA Enschede, The Netherlands*
(email: nuri@itc.nl)
[2]*Utrecht University, Earth Sciences Faculty, Budapestlaan 4, 3508 TA Utrecht, The Netherlands*

Abstract: The kinematic and structural evolution of the major structures affecting the Çankırı Basin, central Turkey, has been deduced from a palaeostress inversion study. Four palaeostress tensor configurations indicative of four-phase structural evolution have been constructed from the fault slip data collected from the Çankırı Basin. The first two phases indicate the dominant role of thrusting and folding, and are attributed to the collision between the Pontides and the Taurides, the proposed interface of which is straddled by the Çankırı Basin. Phase 1 occurred in the pre-Late Palaeocene and Phase 2 in the Late Palaeocene–pre-Burdigalian. The third phase is dominated by extensional deformation in the Middle Miocene. The latest phase has been active since then and is characterized by regional transcurrent tectonics.

Palaeostress analysis is the estimation of the principal stress orientations using fault slip data obtained by field measurements of the orientations of populations of fault planes together with slip data. Slip directions are generally inferred from the orientations of frictional grooves or fibrous lineations, termed slickensides (Fleuty 1974). However, they can also be deduced from the focal mechanism of earthquakes (Angelier 1984; Gephart & Forsyth 1984; Carey-Gailhardis & Mercier 1987) and from the orientations of mechanical twins in calcite (Lacombe *et al.* 1990, 1992).

A number of methods have been developed for the palaeostress inversion and the separation of stress tensors in multiphase deformation situations, following the initial graphical and numerical methods of Arthaud (1969) and Carey & Bruner (1974), respectively. Graphical methods have been further developed by Alexandrowski (1985) (modified M-plane method) and Krantz (1988) (odd-axis method). However, they are only applicable if special conditions are fulfilled. For instance, the faults to be analysed using the M-plane method should have developed under uniaxial stress conditions in which two of the principal stress magnitudes are equal and are manifest, in plan view, in a radial or concentric pattern of faults. The odd-axis method is applicable in triaxial strain conditions where two pairs of conjugate fault sets develop and display orthorhombic symmetry. On the other hand, numerical methods are more robust and have been more widely used (e.g. see Angelier 1979, 1984, 1994; Etchecopar *et al.* 1981; Angelier *et al.* 1982; Armijo *et al.* 1982; Gephart & Forsayth 1984; Michael 1984; Carey-Gailhardis & Mercier 1987; Reches 1987; Hardcastle 1989; Gephart 1990; Marret & Almandinger 1990; Fleischman & Nemcok 1991; Will & Powell 1991; Yin & Ranalli 1993; Nieto-Samaniego & Alaniz-Alvarez 1997).

All numerical methods are based on the Wallace (1951)–Bott (1959) assumption that slip occurs parallel to the maximum resolved shear stress and is also presumed to be parallel to the slickenline direction. A further assumption is that a given tectonic event is characterized by one regional homogeneous stress field. This implies that the slip direction on a fault plane is determined by a single-stress deviator and that all faults which slipped during one tectonic event moved independently but in a way consistent with this single-stress deviator (Will & Powell 1991). After determining the stress tensor with respect to operative fault planes, it is transformed to a regional coordinate system using standard computational transformations (Means 1976; Angelier 1994).

The assumptions upon which the numerical methods are based are an oversimplification of the situation encountered in the field. Inhomogeneity and anisotropic material properties, fault interactions (especially in strike-slip settings), the presence of rotational deformation, and the non-coaxial stress and strains

From: BOZKURT, E., WINCHESTER, J. A. & PIPER, J. D. A. (eds) *Tectonics and Magmatism in Turkey and the Surrounding Area.* Geological Society, London, Special Publications, **173**, 295–323. 1-86239-064-9/00/$15.00
© The Geological Society of London 2000.

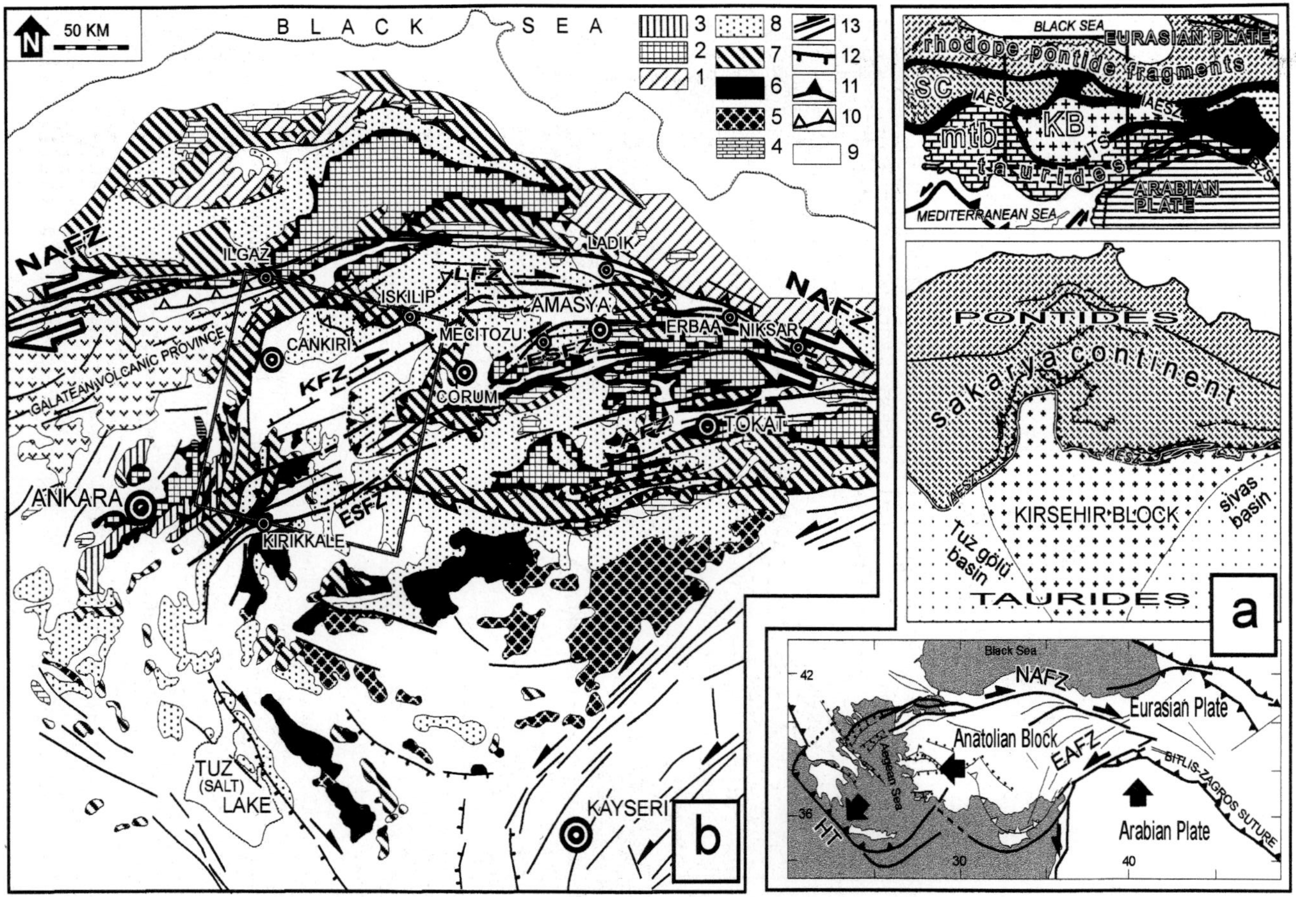

Fig. 1. (a) Inset map showing the geological outline of the eastern Mediterranean area [modified after Şengör et al. (1984)]. BSZ, Bitlis–Zagros Suture; IAESZ, İzmir–Ankara–Erzincan Suture Zone; ITS, Intra-Tauride Suture; KB, Kırşehir Block; MTB, Taurus–Menderes Block; SC, Sakarya continent [after Şengör et al. (1984)]. (b) Tectonostratigraphic map of central Turkey. The box shows the location of the study area. AFZ, Almus Fault Zone (Bozkurt & Koçyiğit 1995, 1996); ESFZ, Ezinepazarı–Sungurlu Fault Zone; KFZ, Kızılırmak Fault Zone (Kaymakçı & Koçyiğit 1995); LFZ, Laçin Fault Zone [after Özçelik (1994)]; NAFZ, North Anatolian Fault Zone. 1, Metamorphic basement of the Pontides; 2, pre-Jurassic metamorphic basement of the Sakarya continent; 3, Triassic Karakaya Complex; 4, Jurassic–Cretaceous platform carbonates and clastic rocks; 5, pre-Upper Cretaceous metamorphic rocks of the Kırşehir Block; 6, granites; 7, ophiolitic units; 8, Lower Tertiary clastic rocks and volcanic rocks; 9, cover units; 10, reverse faults; 11, thrust faults; 12, normal faults; 13, active strike-slip faults. (c) Geological map of the Çankırı Basin; 1–4 are the locations of the subareas.

EVOLUTION OF THE ÇANKIRI BASIN

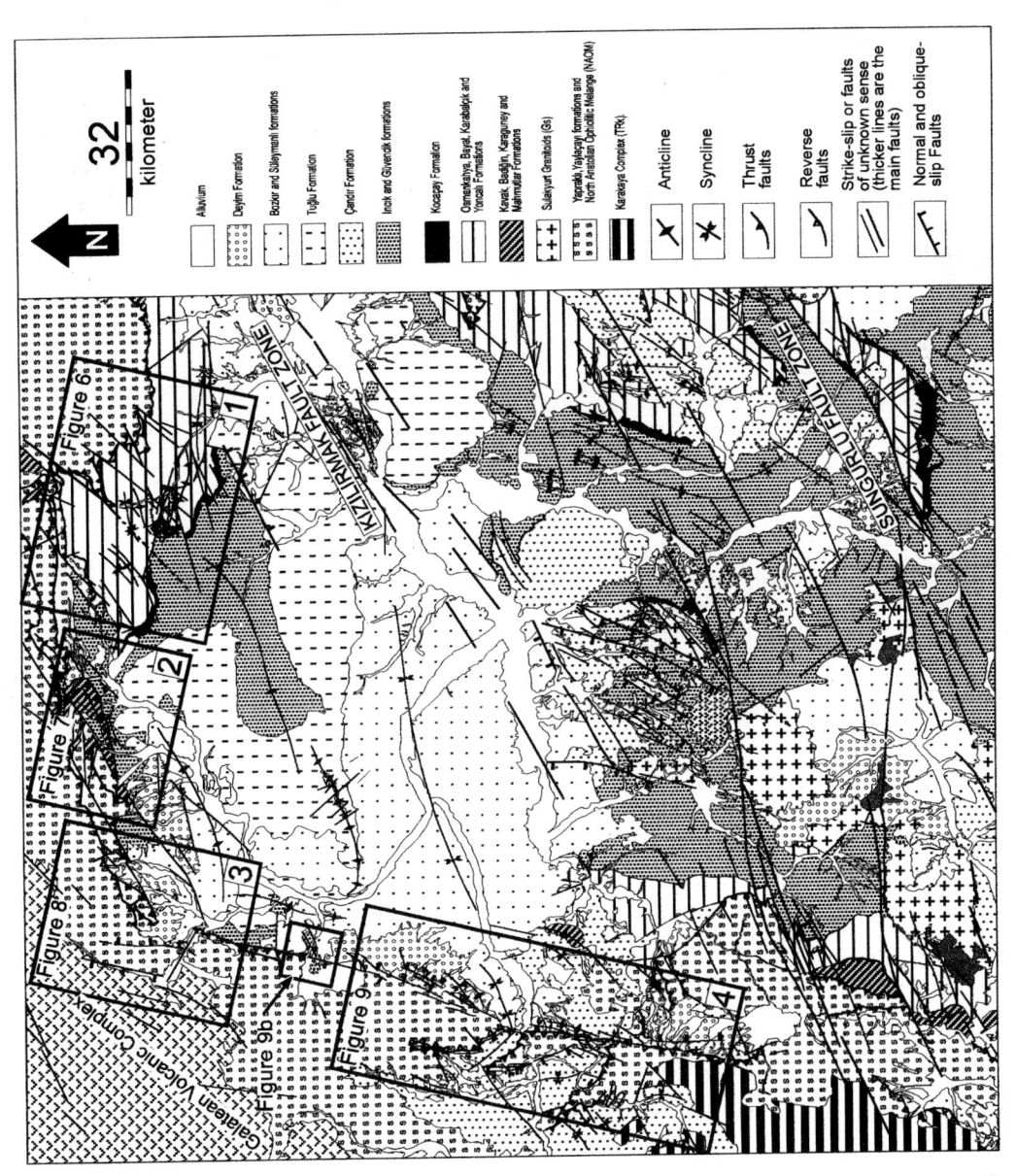

[monoclinic or triclinic symmetry of Twiss & Unruh (1998)] can either cause local variations in the stress field or a very high deviation between the maximum resolved shear stress and the slip direction (Pollard et al. 1993). This decreases the reliability of the stress-inversion procedure and makes the identification of different deformation phases more difficult, but is helped if constrained by stratigraphic controls and overprinting and cross-cutting relationships (Nemcok & Lisle 1995; Hardcastle 1989; Angelier 1994).

Fault reactivation is another source of difficulty as not all inversion procedures can cope with it. For example, Angelier's (1979, 1984) method is best suited for reactivated systems, but it has limitations because faults with pure dip-slip and strike-slip components yield unreliable results because the intermediate stress will be perpendicular to the slip direction and it will have no effect on the inversion procedure [see Angelier (1994) for the details].

Although the basic assumptions underlying stress inversion procedures have been criticized (e.g. Pollard et al. 1993; Twiss & Unruh 1998), empirical observations and theoretical analyses (e.g. Duphin et al. 1993) show that the shear stress vectors and the slip vectors on a single isolated fault plane vary little in orientation from those predicted, i.e. the average slip remains parallel to the average shear stress, thus agreeing with the Wallace–Bott assumption (Angelier 1994). Stress inversion techniques have been applied to fault slip data from a variety of tectonic settings and have produced results that are consistent and interpretable (Pollard et al. 1993).

The aim of this study is to use the palaeostress inversion procedure to delineate the kinematic evolution of the structures within the northern part of the Çankırı Basin with respect to its structural and tectonic evolution.

Background

The Çankırı Basin (Fig. 1) is thought to be located in a zone where the Sakarya continent attached to the Pontides and the Kırşehir Block of the Taurides collided and sutured along the İzmir–Ankara–Erzincan Suture Zone (İAESZ), which demarcates the former position of the northern branch of Neotethys (Şengör & Yılmaz 1981). The timing of collision is under debate. Okay (1984) argued that it occurred at the end of the Late Cretaceous. Şengör & Yılmaz (1981), when reviewing the plate tectonic evolution of Turkey from the Precambrian to the present, proposed a Late Palaeocene–Middle Eocene age for the timing of the collision. Görür et al. (1984) proposed a Middle Eocene origin based on their study of the Tuzgölü (Salt Lake) Basin which is located in the western margin of the Kırşehir Block and is thought to have a similar stratigraphical and evolutionary history to the Çankırı Basin (Fig. 1). It is possible that these different dates reflect a diachronous collision which may be due to irregularities at the promontories of colliding blocks and oblique collision (Dewey 1977)

Besides being affected by collisionary process, the Çankırı Basin was subjected to further deformation in post-middle Miocene being a part of the Anatolian wedge caught between the expulsive transcurrent motions on the North and East Anatolian Faults. This has resulted a number of northwards, convex, dextral strike-slip faults which bifurcate from the North Anatolian Fault Zone (NAFZ) (Barka & Hancock 1984; Şengör et al. 1985; Kaymakçı & Koçyiğit 1995). The Kızılırmak and Sungurlu Fault Zones are the two major splays of the NAFZ, which partly controlled the Late Miocene evolution of the Çankırı Basin (Fig. 1c).

Fig. 2. Generalized tectonostratigraphic column of the units exposed in and around the Çankırı Basin. MN zones in the age column are obtained from Hans de Bruijn (pers. comm.). 1, North Anatolian Ophiolitic Mélange (NAOM) (ophiolitic mélange); 2, Yaylaçayı Formation (distal fore-arc sequence); 3, Yapraklı Formation (proximal forearc facies); 4, Sulakyurt Granites of the Kırşehir Block that intruded in pre-Palaeocene time; 5, Kavak Formation (red clastics and carbonates); 6, Badiğin Formation (neritic limestones); 7, Karagüney Formation (clastics derived mainly from the Kırşehir Block) 8, Mahmatlar Formation (clastic rocks derived from the Sulakyurt Granite); 9, Dizilitaşlar and Hacıhalil Formations (mainly turbiditic clastic rocks and intercalated limestones); 10, Yoncalı Formation (Eocene flysch); 11, Karabalçık Formation (distributary channel conglomerates and sandstones with coal seams); 12, Bayat Formation (Eocene volcanic rocks and volcaniclastic rocks); 13, Osmankahya Formation (mixed-environment clastics and red beds); 14, Kocaçay Formation (Middle Eocene nummulitic limestone covering both basin infill and the Sulakyurt Granites), 15, İncik Formation (Upper Eocene–Middle Oligocene continental red clastic rocks); 16, Güvendik Formation (Middle Oligocene evaporites); 17, Kılçak Formation; 18, Altıntaş Formation (fluvial red clastics exposed only in the Hancılı Basin); 19, Hancılı Formation (lacustrine deposits exposed only in the Hancılı Basin); 20, Çandır Formation; 21, Tuğlu Formation (Lower–Upper Miocene evaporites and lacustrine shale–marl); 22; Süleymanlı Formation (fluvio-lacustrine red clastic rocks); 23, Bozkır Formation (evaporites); 24, Deyim Formation (fluvial clastic rocks); 25, alluvium.

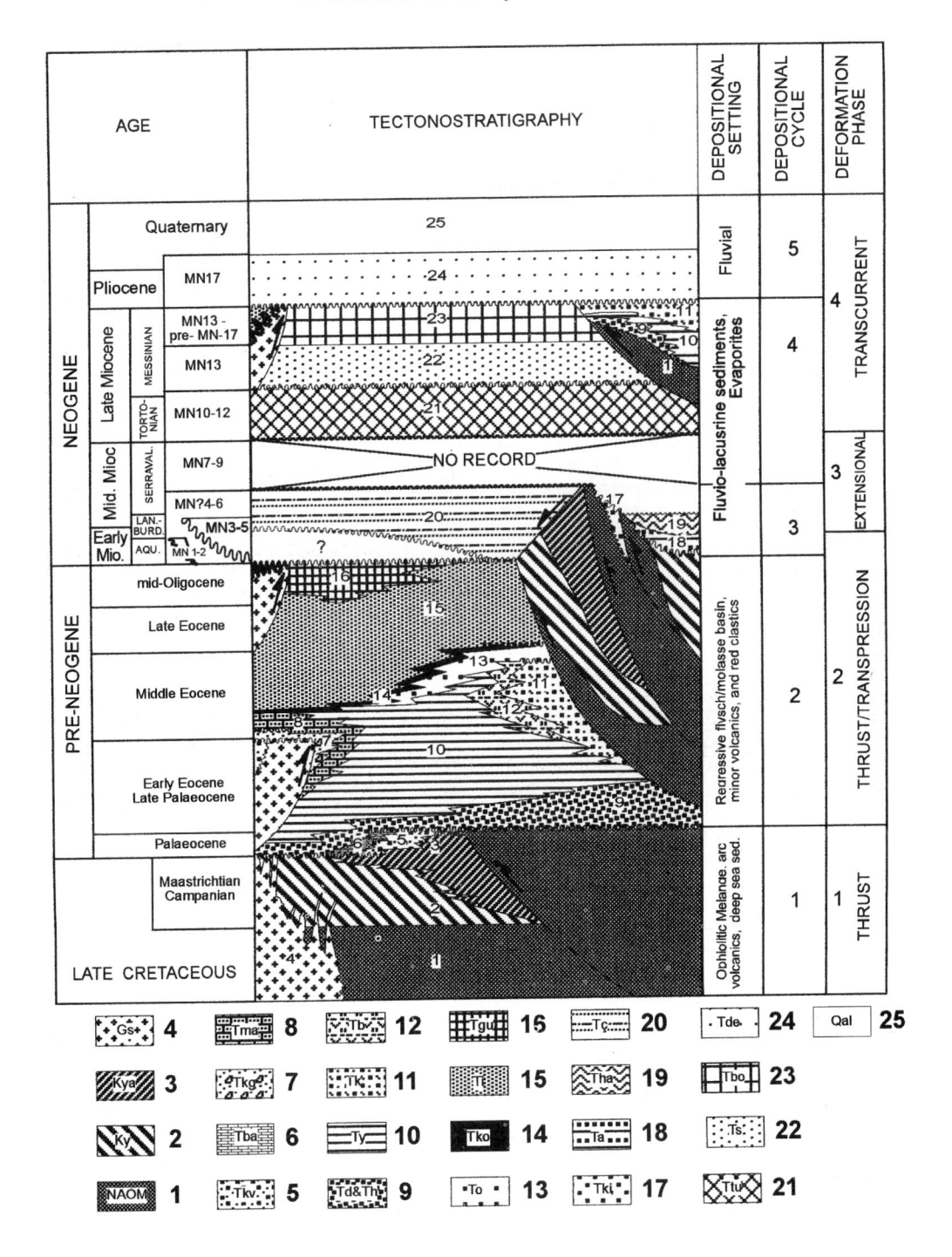

EVOLUTION OF THE ÇANKIRI BASIN

Geological setting

The Çankırı Basin has an Ω-shape (Fig. 1b), with the main outcrops lying in the west, north and east. In the south it is delineated by the granitoids of the Kırşehir Block of the Taurides (Fig. 1). The rim of the Çankırı Basin is marked by an ophiolitic mélange, the North Anatolian Ophiolitic Mélange (NAOM) [terminology after Rojay (1993, 1995)], which is thought to underlie the sedimentary infill of the Çankırı Basin. The basement of the mélange is thought to be the Kırşehir Block in the southern part but is unknown in the north, i.e. the İAESZ may lie below the Çankırı Basin rather than skirting around its northern margin as indicated in Fig. 1.

The fill of the Çankırı Basin is > 4 km thick and accumulated in five different cycles of sedimentation (Fig. 2). The oldest cycle comprises Upper Cretaceous volcaniclastic rocks and regressive shallow-marine units, and Palaeocene mixed-environment red clastics and carbonates (Özçelik 1994). The subsequent cycles have been partly studied by Dellaloğlu *et al.* (1992) and their scheme is followed in this study. The second cycle is a Late Palaeocene–mid-Oligocene regressive flysch to molasse sequence overlain by a widespread, thin (< 100 m) nummulitic limestone of Middle Eocene age which passes up into a very thick (up to 2000 m) Upper Eocene–mid-Oligocene continental red clastic sequence intercalated by mid-Oligocene evaporites. The third cycle is represented by fluvio–lacustrine clastic rocks deposited in the Early–Middle Miocene. The fourth cycle is represented by deposits laid down under Late Miocene fluvio–lacustrine conditions and are frequently alternating with evaporites. The Plio-Quaternary alluvial fan deposits and recent alluvium locally overlie all of these units (Fig. 2). Names of the formal units are mostly adopted after Dellaloğlu *et al.* (1992).

The main structures shaping the current geometry of the Çankırı Basin (Fig. 1c) are the thrust faults defining its western and northern rims. The eastern margin is defined by a belt of north-northeast striking folds. In the south, the basin fill onlaps on to the Kırşehir Block. Other major structures affecting the Çankırı Basin are the dextral Kızılırmak Fault Zone, oriented southwest–northeast in the central part of the basin, and the Sungurlu Fault Zone in the southeast. Both are regarded as splays of the NAFZ. The south-central area of the basin is dominated by a number of curvilinear faults oriented approximately northeast—southwest (Fig. 1c).

Methodology

Data collection

The relative ordering of fault motions and related deformation was established from overprinting and cross-cutting relations. The age constraints were applied through careful documentation and analysis of the fault structures in each of the above stratigraphical horizons. To avoid problems due to relative block and fault plane interactions [see Pollard *et al.* (1993) and Twiss & Unruh (1998)] sampling sites were as small as possible and structurally homogeneous (Hancock 1985). In addition, the displacement should also be as small as possible (a few centimetres) so that it should not accommodate significant strain (Hardcastle 1989), therefore the principal strain and stress axes should remain parallel.

Most Lower Tertiary infill of the Çankırı Basin is only exposed in the three belts forming the western, northern and eastern rims of the basin (Fig. 1). The central parts of the basin are covered mainly by evaporites that are very susceptible to gravity induced ductile deformation and were not included in the analysis. As a result, the study was limited to the northern and western margins which form a convex arcuate belt (see Fig. 1). Four subareas were selected for detailed study. Two of them lie in the northern margin of the basin. In subarea 1 (1 in Fig. 1c) the infill of the basin is well exposed and the boundary is affected by dextral northeast trending transcurrent faults (Fig. 1c). The second subarea (2 in Fig. 1c) is dominated by anastomosing east-northeast trending thrust faults. In addition, Upper Cretaceous–Palaeocene units are better exposed in this area than in any other part of the basin. The third subarea lies where the rim of the Çankırı Basin turns from an overall east–west trend to a north–south one (3 in Fig. 1c). In this subarea, the basin units are post-Middle Eocene in age. The fourth subarea (4 in Fig. 1c) covers the western margin of the basin and the Neogene Hancılı Basin. In this area, mainly Upper Cretaceous units and Miocene–Recent units are exposed; the Lower Tertiary units are missing or not exposed.

In order to have structurally homogeneous data (Hancock 1985), the size of a sampling site has been restricted to < 50 m diameter. Areas larger than this were subdivided into subsites and analysed independently. More than 600 slickenline data from 72 sites have been collected. For each fault measured in the field, the following features were noted: (1) the attitude of the plane; (2) the stratigraphic units which were displaced; (3) whenever possible, the relative order of movement; (4) the amount of offset; (5) type of slickenline; (6) type of shear sense indicators; (7) evidence of ductility (i.e. breccia v. mylonite); and (8) degree of planarity. Each was given a confidence value of 1–4 (excellent to poor) [as explained in Hardcastle (1989)]. If no movement sense could be deduced, the fault was not used in the analyses, which applied to c. 20% of the data.

The stress-inversion procedure and separation of movement phases

It was found that most of the measured faults had undergone reactivation, as seen from overprinting kinematic indicators which were subsequently used to order the different phases of movement (Fig. 3). The maximum number of slickenline overprinting and/or overgrowth patterns observed in any fault plane was three, which was encountered in 10% of the faults, c. 25% had two overprinting sets.

The relative age of each movement phase was determined independently for each fault from a given subarea, carefully correlated with data from other subareas to form a regional subset and then processed (see Fig. 4) by Angelier's method (1989) using his computational procedures. During the analyses, the data were carefully examined and data from each site were correlated with those from other sites in a subarea such that the slip data with the same order of occurrence could be grouped together for preliminary stress-inversion processing (see flowchart in Fig. 4 for the steps followed in the stress-inversion procedure). The computational procedures used were: direct inversion (INVD); right dihedra (P- & T-dihedra); and iterative methods (R4DT, R4DS, R2DT, R2DS).

After preliminary processing, faults giving spurious results were re-examined. If the results still remained spurious they were separated from the data set and treated separately. After removal of spurious data, the stress tensor was recomputed and reanalysed using the software developed by Hardcastle & Hills (1991) for the automated separation of stress tensors associated with the different deformation phases as indicated from the field observations. Concordant data were then taken, so indicating that the computed stress tensor is most likely to be correct. The initial spurious data were recomputed by the Hardcastle & Hills (1991) method and if again spurious they were deleted; if concordant they were included and reanalysed.

In order to determine the mean stress tensor

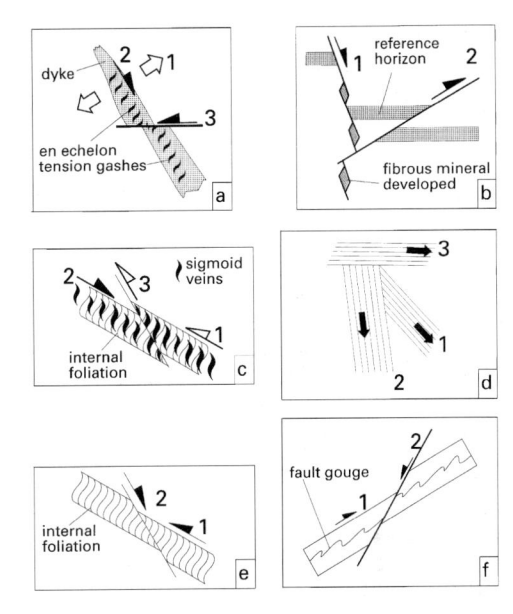

Fig. 3. Schematic illustrations of the criteria used to date relative occurrences of slip data. Numbers 1–3 indicate the sequences of deformation, oldest to youngest, respectively. (**a**) A dyke indicating extension (1) (large opposite arrows), then tension veins indicating dextral movement (2), all displaced sinistrally (3). (**b**) Hybrid joints with fibrous mineral development having normal separation (1) are displaced by reverse faults (2). (**c**) Internal foliation within the shear zones with reverse separation (1) sigmoidal veins indicating that an opposite sense of movement (2) developed, they are then displaced by a reverse fault with opposite movement to the veins (3). (**d**) Cross-cutting slickensides. (**e**) Shear zone with internal foliation indicating reverse separation (1) displaced by a fault with an opposite sense of movement (2). (**f**) Folded fault gouge with reverse separation displaced by a fault with an opposite sense of movement. The numbers are relative to each other and do not necessarily correspond to the order of the regional deformation phases.

configuration for a given movement phase in each subarea, all the data from each site and each phase were grouped and the above procedure was repeated. After the mean stress orientations were determined, all of the raw data was reprocessed using Hardcastle & Hills' (1991) approach and the results were compared with those obtained using the direct-inversion method. As the minimum number of slickenline data required in the direct-inversion method is four (Angelier 1979), the data from sites containing fewer than this number were used only in the construction of the mean stress tensor for the subarea within which it was located. After the stress tensors for each set at each site were

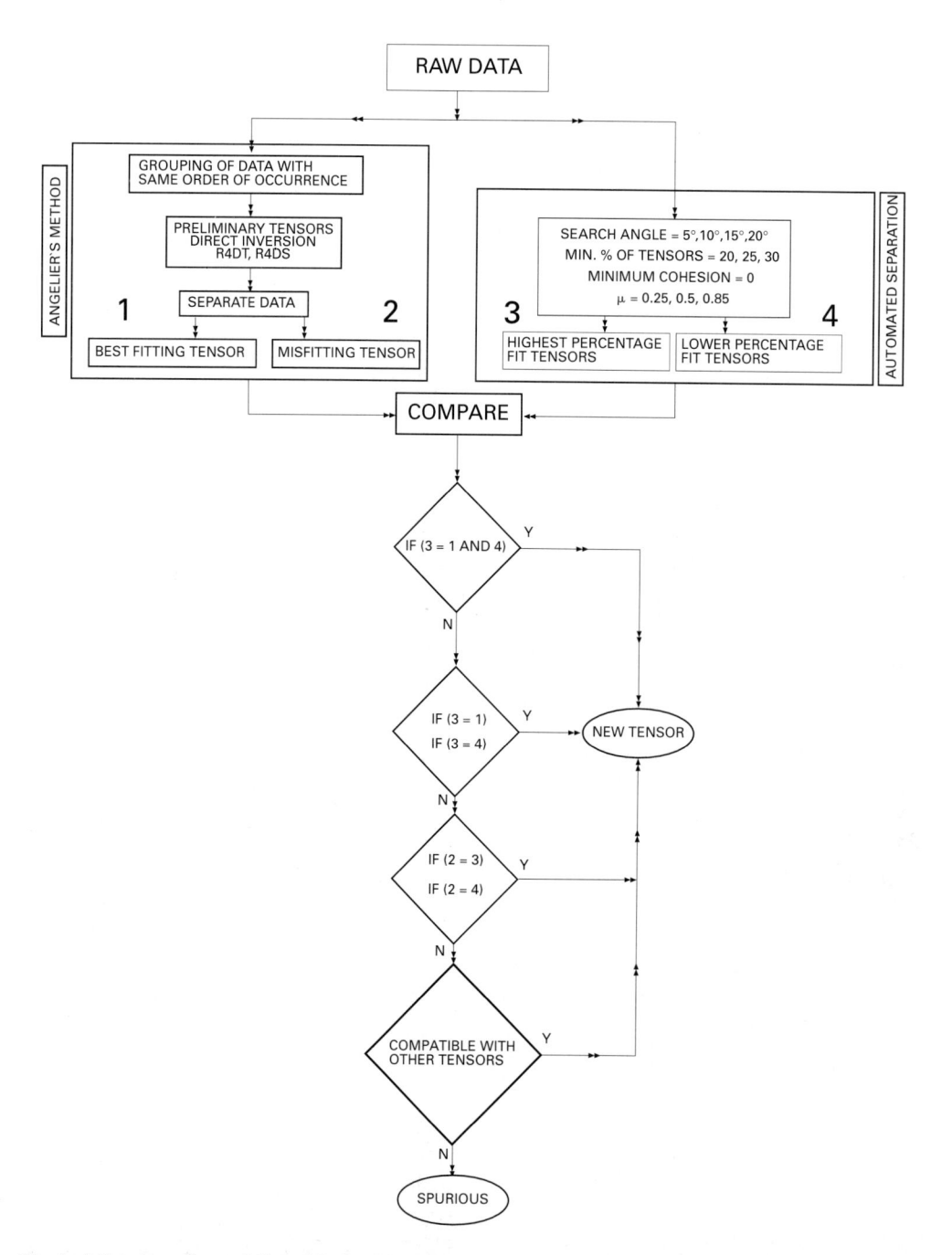

Fig. 4. A flowchart of steps followed during the analysis of the data. The data were analysed independently using two different methods which were then compared. After determination of tensors and separation of the data into best-fitting and misfitting faults, the results were compared with the automated method. If best-fitting and misfitting tensors of Angelier's method fell within the highest percentage fit tensors of the automated method, then each subset was ascribed two separate acceptable tensors. If not, and if they fell within the lower percentage fit tensor, then further analysis was carried out until reasonable and acceptable tensors were obtained (i.e. lowest misfit angles of < 15° obtained for all groups. Only faults with > 45° are accepted as spurious (< 2% of the whole data). This process is repeated for each class of structures (number of relative chronology) for each phase and for determining mean regional (subarea-based) stress tensor configurations.

Table 1. *Field characteristics of sites in sub area 1*

Site[†]	Shear	Vein	OP	S.line	CC	Unit[†]	No. move.
60	–	–	2	SP	2	K	2
61	–	–	2	SP	2	K	2
62	–	+	3	SP, Ca	4	K	4
63	+	+	3	SP, Ca	4	K	4
64	+	+	3	SP, Ca	4	K	4
65	–	–	2	–	2	Tb	2
66	+	–	–	–	–	K	1
77	+	+	2	Ca	3	Ty	3
78	+	–	2	–	–	To	2
79	–	+	2	–	2	Tko	2
80	–	+	2	Ca	2	To	2
81	–	+	2	Ca	2	Ti	2
82	–		2	–	2	To	2
115	–	+	2	Ca	2	Ti	2
116	–	+	2	Ca	2	Ti	2
117	–	+	2	Ca	2	Ti	2
118	–	+	–	–	–	Ti	1
119	–	+	2	Ca	2	Ti	3

OP, Number of overprinting slickenline sets; s.line, fibrous slickensides associated with the sampled faults (SP, serpentine; Ca, calcite); CC, number of cross-cutting relationships either with veins, shear zones or other faults; no. move., number of movement sets encountered in each site (the numbers do not necessarily correspond to the order of the regional deformation phases). *+, exists; –, not observed.[†] K, Upper Cretaceous units (NAOM; Ky; Kya); Ty, Yoncalı Formation; To, Osmankahya Formation; Tko, Kocaçay Formation; Ti, İncik Formation.

determined, they were correlated with other sites. By combining the stratigraphic information and relative order of different sets, the stress tensors were arranged into ordered deformation phases.

In both the direct-inversion and the Hardcastle & Hills' (1991) methods, 15° was chosen as the maximum angular deviation acceptable for the computation of a given stress tensor. Faults with greater angular deviations were considered as spurious and deleted.

The tensors computed for a subarea should be more reliable than site-based tensors. The reasons for this are obvious but include deviations due to a particular site being located at the termination of a fault or in an area where two or more structures interact. Both will cause a deviation of the local stress tensor from the regional tensor (Pollard *et al.* 1993) and will tend to cancel out in the regional compilation (Angelier 1994).

Results for subarea 1

The location of subarea 1 is shown in Fig. 1c. The main structures in this area are the east-northeast trending thrust faults (TF1) along which the Upper Cretaceous NAOM was thrust over the Yaylaçayı Formation which, in turn, was thrust onto the Late Palaeocene–Middle Eocene Yoncalı and Karabalçık Formations (Fig. 5a). Relatively, the oldest slickensides which record the first movement on TF1 include pitches with dominant thrust fault character (> 45°) with a dextral lateral component. This thrust belt is displaced dextrally by a number of later north-east–south-west oriented strike-slip faults that cut through both the hanging-wall and footwall blocks. Further to the southwest, the Yoncalı Formation was thrust over the İncik Formation and the Middle Eocene units – the Karabalçık, Bayat, Osmankahya and Kocaçay Formations – along fault TF3 (Fig. 5a). A number of synsedimentary unconformities were observed in the İncik Formation during the field studies (especially near sites 81–82 and 115 in Fig. 5a), implying that TF2 (Fig. 5a) operated during the deposition of the İncik Formation in Late Eocene–pre-mid-Oligocene times. The main fault planes of TF2 and TF3 are dominated by overprinting slickensides. Relatively, the oldest of these slickensides have pitches ranging between 15 and 35° with a sinistral sense of movement. Likewise, the younger slickensides have pitches dipping at < 20° with a dextral sense of movement. These relations indicate that TF2 and TF3 were developed as transpressional sinistral strike-slip faults and later reactivated into dextral strike-slip faults.

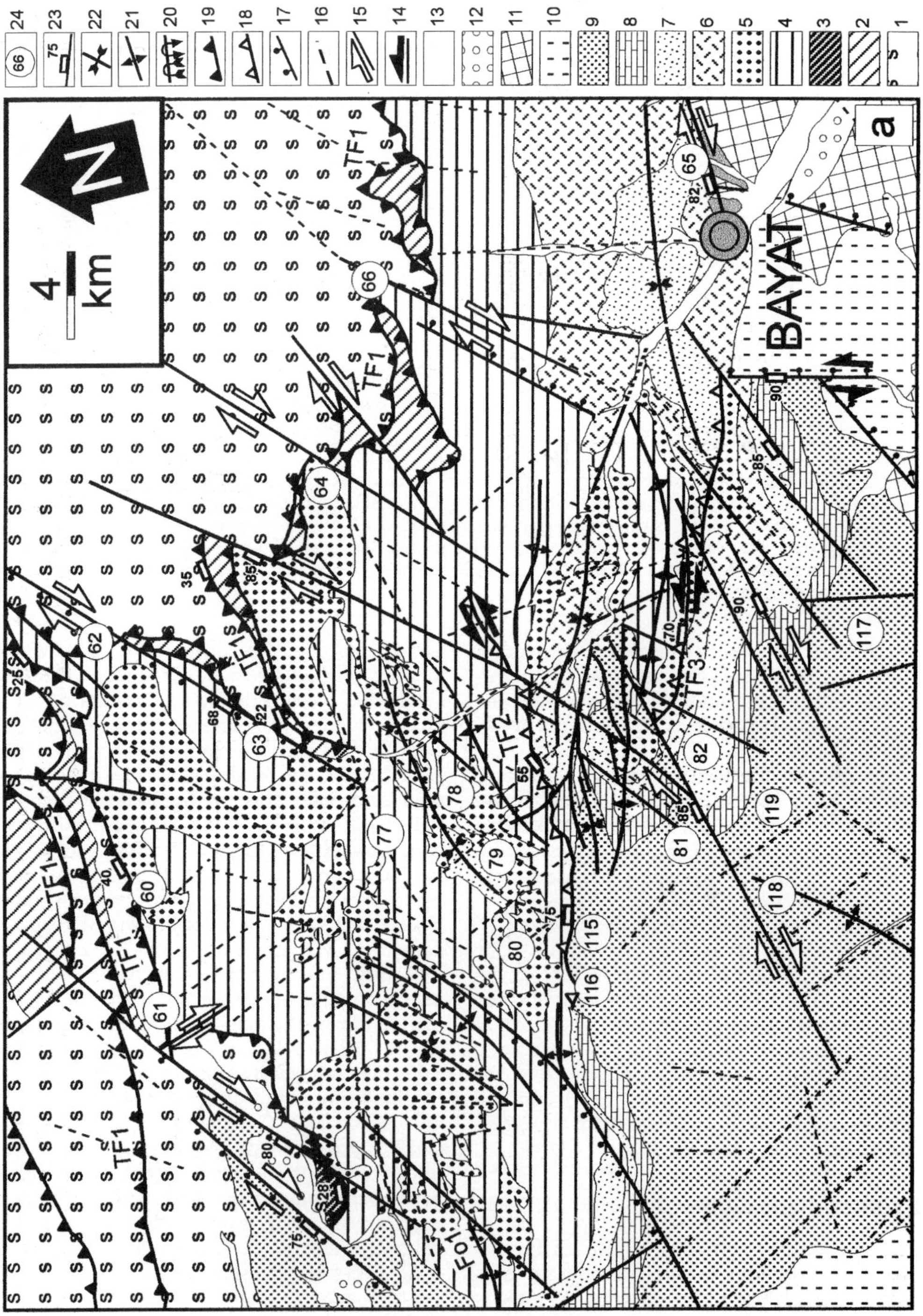

a
BAYAT
4 km
N
TF1
TF2
TF3

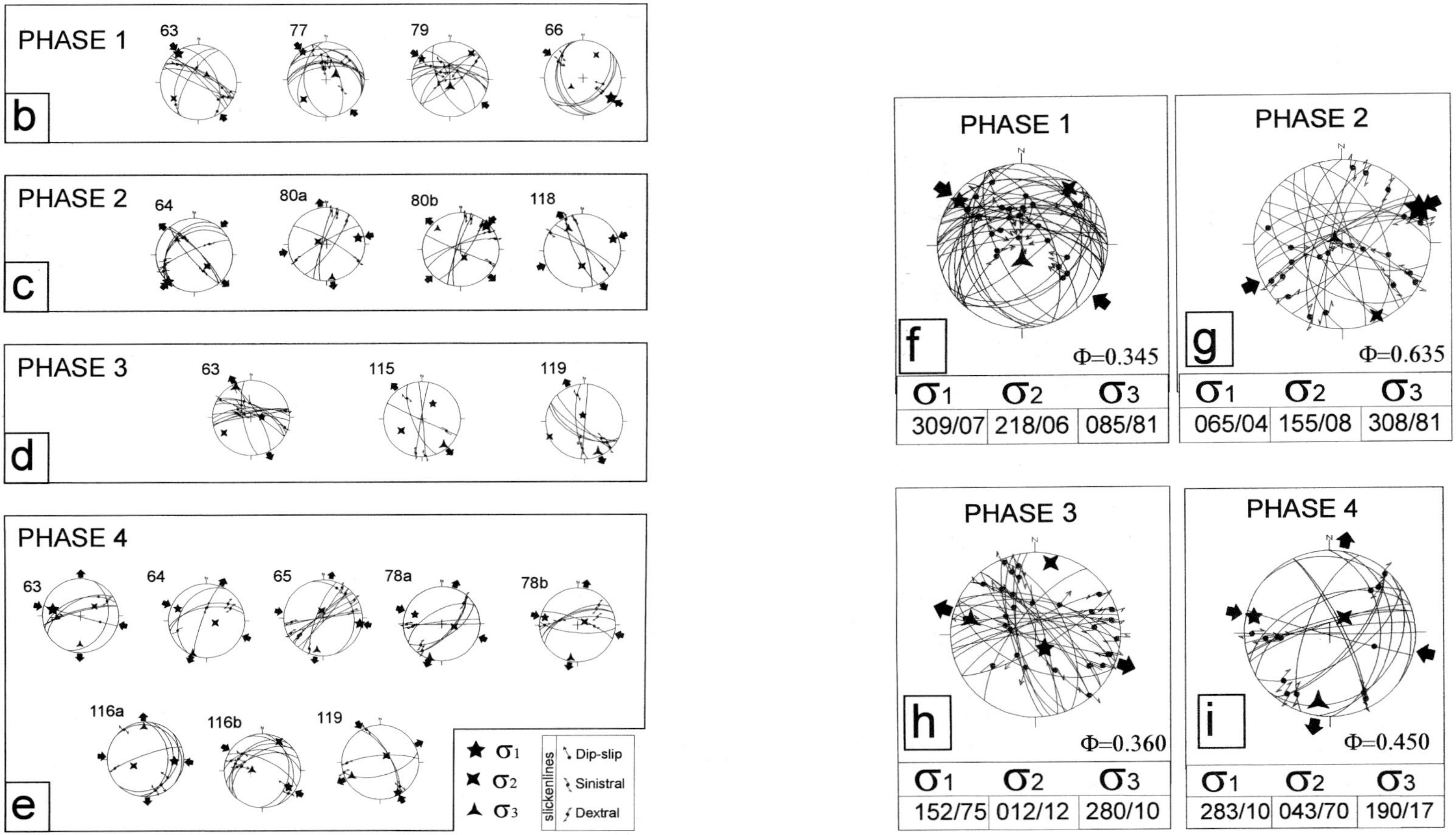

Fig. 5. (a) Geological map and sample locations in subarea 1. 1, NAOM; 2, Yaylaçayı and Yaprakli Formations; 3, Dizilitaşlar Formation; 4, Yoncalı Formation; 5, Karabalçık Formation; 6, Bayat Formation; 7, Osmankahya Formation; 8, Kocaçay Formation; 9, İncik Formation; 10, Tuğlu Formation; 11, Süleymanlı Formation; 12, Deyim Formation; 13, alluvium; 14, sinistral strike-slip faults; 15, dextral strike-slip faults; 16, photo-lineaments; 17, normal faults; 18, reverse faults; 19, thrust faults; 20, overturned folds; 21, anticline; 22, syncline; 23, dips of faults where they are best observed in the field; 24, sample site locations. Plots of fault planes: (b)–(e) fault planes, slickenlines and stress orientations for each site; (f)–(i) whole data in particular phases and subareas (lower hemisphere equal-area projections).

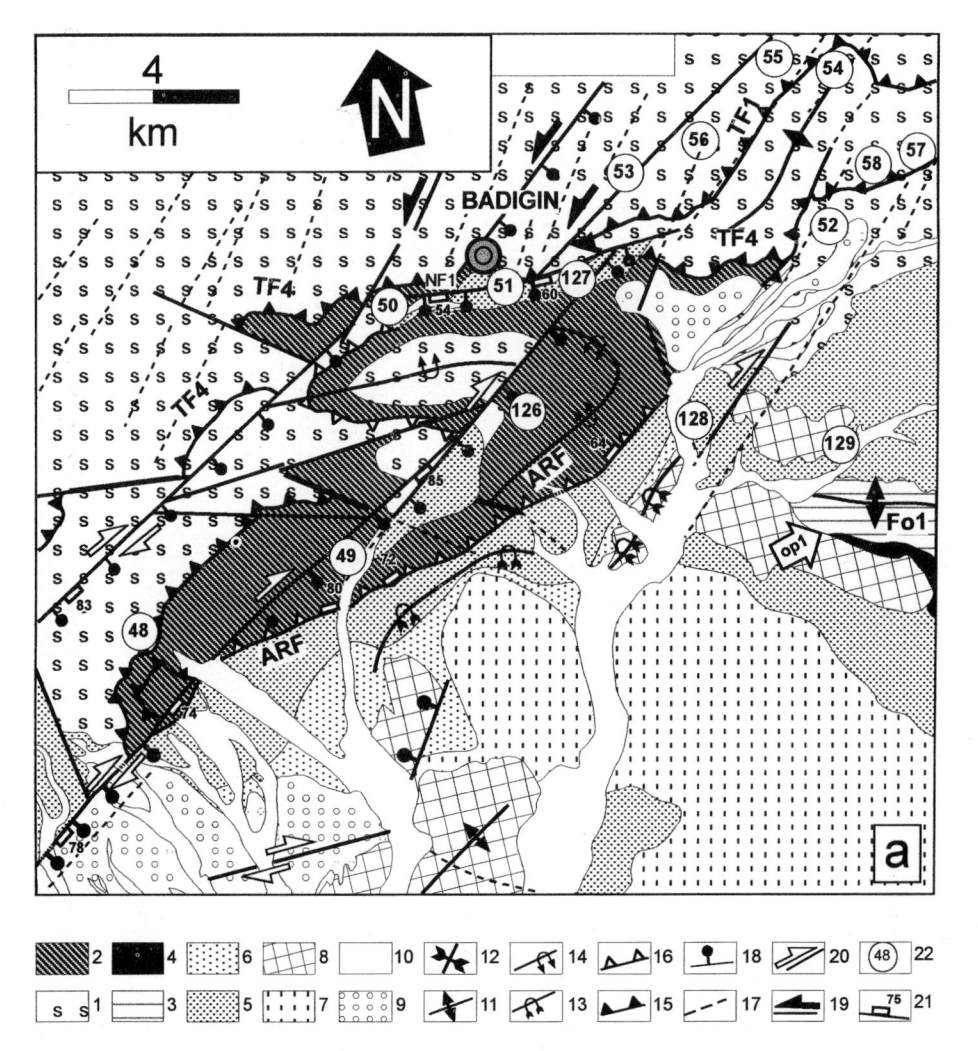

Fig. 6. (a) Geological map and sample locations in subarea 2. 1, NAOM; 2, Yapraklı, Kavak and Bağdiğin Formations; 3, İncik Formation; 4, Çandır Formation; 5, Tuğlu Formation; 6, Süleymanlı Formation; 7, Deyim Formation; 8, alluvium; 9, anticline; 10, syncline; 11, overturned syncline; 12, overturned anticline; 13, thrust faults; 14, reverse faults; 15, photo-lineaments; 16, normal faults; 17, sinistral strike-slip faults; 18, dextral strike-slip faults; 19, dips of faults where they are best observed in the field; 20, sample site locations. Plots of fault planes: (b)–(e) fault planes, slickenlines and stress orientations for each site (f)–(i) whole data in a particular phase and subarea (lower hemisphere equal-area projections).

EVOLUTION OF THE ÇANKIRI BASIN

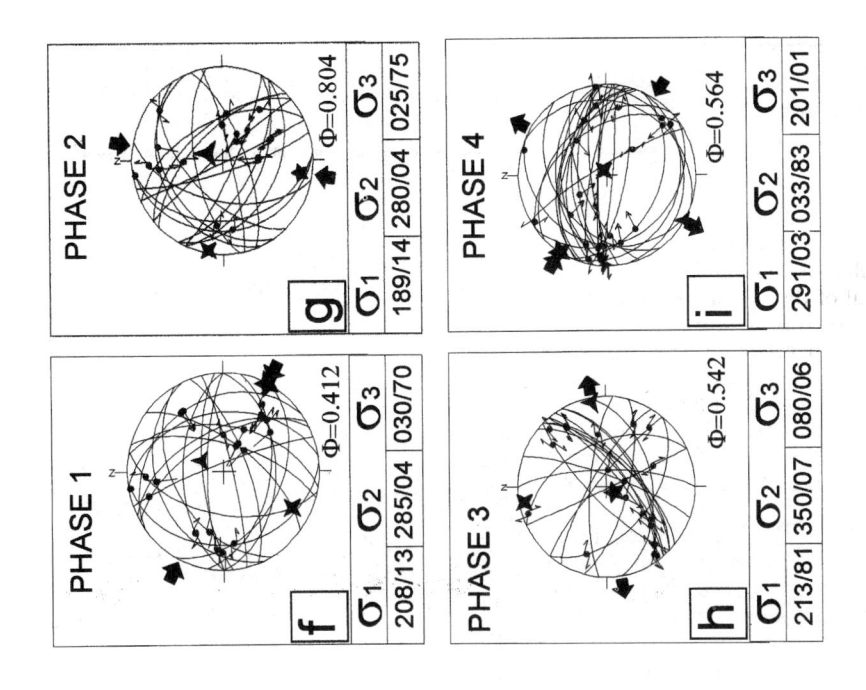

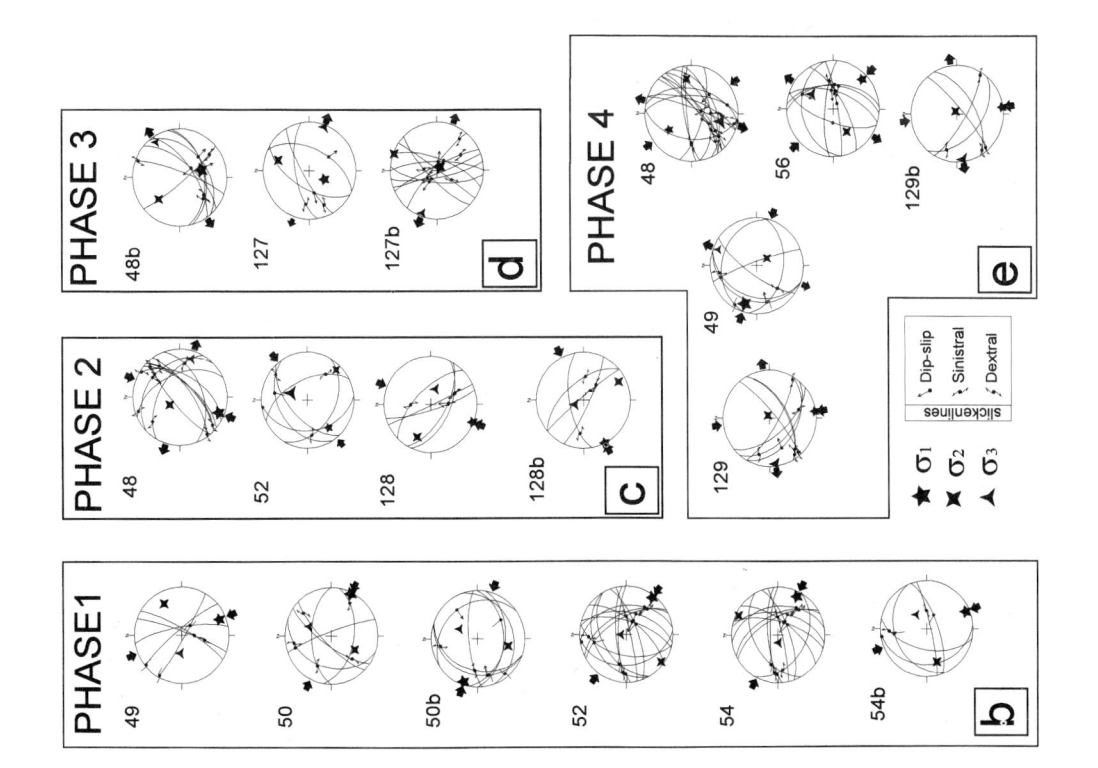

Palaeostress inversion

Eight sites were selected (see Table 1) and, from the field analyses, four phases of fault activity were recognized using the criteria outlined in Fig. 3. In some of the sites (62–64) the sampled faults have three sets of overprinting slickenlines. These faults, in turn, cross-cut other structures such as shear zones, én-echelon veins and other sets of faults (as indicated in Fig. 3). A combination of all of these relationships led to the identification of four sets of fault movements. Based on their relative timing, the sets of faults having similar movements were directly assigned to the deformation phases arranged from older to younger (Nemcok & Lisle 1995). These sites were used as a reference in the analysis of other sites that have fewer sets of fault movements.

From the 18 sites sampled, 12 sites have sufficient slip data for the construction of site-based palaeostress tensors (Fig. 5b–e). From these sites, 19 palaeostress configurations have been constructed. The data from the remaining eight sites were combined in constructing the mean stress tensor for the whole subarea (see Fig. 5f–i).

Phase 1. Only four sites had sufficient data for construction of a site-based stress tensor (Fig. 5b). The average orientations of the principal stresses and the stress ratio for the subarea are: $\sigma_1 = 309°N/07°$, $\sigma_2 = 218°N/06°$, $\sigma_3 = 085°N/81°$; and $\Phi = 0.345$ (Fig. 5f). The major stress direction is north-northwest–south-southeast and the minor stress direction (σ_3) is subvertical, indicating thrust tectonics in this phase. The orientation of σ_1 is approximately perpendicular to the main northeast–southwest striking segments of TF1 and oblique to the other segments (Fig. 5a).

Phase 2. The second phase was recognized in four sites (Fig. 5c), which also give comparable results. One site, 64b, gives a slight deviation; this site is very close to the intersection of TF1 and a northeast–southwest trending oblique-slip fault. The deviation may be due to interaction of these faults [as explained by Pollard *et al.* (1993)]. In the subarea, both σ_1 and σ_2 are subhorizontal, and σ_3 is subvertical in all sites and the subarea-based tensor indicates thrusting during this deformation phase. The orientation of subarea-based principal stresses and the stress ratio are: $\sigma_1 = 065°N/04°$, $\sigma_2 = 155°N/08°$, $\sigma_3 = 308°N/81°$; and $\Phi = 0.635$ (Fig. 5g). The orientation of σ_1 is almost perpendicular to the northwest–southeast striking segments of TF1 and TF3, and the northwest striking folds east of the TF2 fault.

Phase 3. Three sites had sufficient slip data for the construction of site-based stress tensors. The orientations of the principal stresses are relatively consistent in each site (Fig. 5d). The horizontal components of σ_3 are oriented north-northeast–southwest. The orientation of subarea-based average stress tensors and the stress ratio are: $\sigma_1 = 152°N/75°$, $\sigma_2 = 012°N/12°$, $\sigma_3 = 280°N/10°$; and $\Phi = 0.360$ (Fig. 5h), indicating extensional deformation in this phase.

Phase 4. This phase is recognized in seven sites (Fig. 5e). Except for site 119, all other sites yielded compatible results. The horizontal component of σ_1 is oriented approximately west-northwest–east-southeast, which is also parallel to the subarea-based stress tensor. The orientations of subarea-based principal stresses and the stress ratio are: $\sigma_1 = 283°N/10°$, $\sigma_2 = 043°N/70°$, $\sigma_3 = 190°N/17°$; and $\Phi = 0.450$ (Fig. 5i), indicating strike-slip deformation in this phase.

Results for subarea 2

The location of this subarea is shown in Fig. 1c. This area is also dominated by northeast trending thrust faults. Upper Cretaceous units are thrust over the Palaeocene units by TF1 near Badiğin and TF4 to the east (Fig. 6a). Relatively, the oldest slickenlines observed on TF4 indicate a dominant thrust fault character with dextral lateral component. The Palaeocene units (Kavak and Badiğin Formations; see Fig. 2) are thrust over the Upper Eocene–Middle Oligocene İncik Formation along the Ayseki Reverse Fault (ARF). In the central parts of the subarea, the Palaeocene units, the İncik Formation and the Upper Miocene units are folded and overturned parallel to the ARF (Figs 1c and 6a).

In the southeastern part of the study area, an east–west trending fold (Fo1), observed within Eocene–Oligocene units (Ty and Ti), is unconformably overlain by the relatively undisturbed Upper Miocene units (op1 in Fig. 6a).

Near Badiğin village, TF1 and TF4 are displaced by an east–west trending normal fault, NF1, which is, in turn, displaced by northeast–southwest trending faults with an apparent sinistral offset and normal component. These faults also displace Upper Miocene units (Süleymanlı and Bozkır Formations) and the Plio-Quaternary Deyim Formation (Fig. 6a), which proves their post-Late Miocene activity.

Table 2. *Field characteristics of sites in subarea 2.*

Site	Shear	Vein	OP	S.line	CC	Unit*	No. move.
48	+	+	2	SP, Ca	4	K	4
49	−	+	2	SP, Ca	2	K	3
50	+	+	2	SP	4	K	4
51	+	−	3	SP, Tl	3	K	4
52	+	+	2	SP	2	K	3
53	−	+	−	Ca	2	K	2
54	−	−	3	Ca	3	K	3
55	−	−	2	Ca	2	K	2
56	−	−	−	−	−	K	1
57	−	+	−	Ca	2	K	2
58	−	−	−	−	−	K	1
126	−	−	−	−	−	Ti + Tkv	1
127	−	+	2	Ca	2	Ti + Tkv	2
128	−	−	2	Ca	2	Ti	2
129	−	+	2	Ca	2	Ti	2

* Tkv, Kavak Formation. Other abbreviations as in Table 1.

Palaeostress inversion

Four deformation phases were recognised in this subarea. The same criteria were used for ordering the deformation phases as described in Fig. 4. The details of sites located in this subarea are given in Table 2 and results are illustrated in Fig. 6b–e.

Phase 1. Six sites had sufficient data to construct site-based stress tensors for this subarea. Although the angular discrepancy between the orientations of σ_1 constructed for each site and averaged for the subarea is *c.* 45°, there is a great discrepancy between the orientations of σ_2 and σ_3. The orientations of the σ_1 vary from north-northwest–south-southeast to west-northwest–east-southeast and are subhorizontal in each site. σ_2 and σ_3 are oblique, although σ_3 is more vertical than σ_2 (Fig. 6b). Orientation of the stresses and the stress ratio for the subarea are: $\sigma_1 = 208°\text{N}/13°$, $\sigma_2 = 285°\text{N}/04$, $\sigma_3 = 030°\text{N}/70°$; and $\Phi = 0.412$ (Fig. 6f). Having σ_3 subvertical and the other stresses subhorizontal indicates thrusting in this phase.

Phase 2. Only four sites had sufficient data for construction of the site-based stress tensors. The orientation of site- and subarea-based tensors is relatively compatible. Orientation of σ_1 ranges from north-northwest–south-southeast to north-east–southwest (Fig. 6c). The mean subarea-based principal stress orientations and the stress ratio are: $\sigma_1 = 189°\text{N}/14°$, $\sigma_2 = 280°\text{N}/04°$, $\sigma_3 = 025°\text{N}/75°$; and $\Phi = 0.804$ and indicates compressive deformation. The orientation of σ_1 is perpendicular to Fo1 (Fig. 6g).

Phase 3. Only three sites had sufficient slip data for construction of the site-based tensors (Fig. 6d). In sites 127 and 48, σ_2 is oblique, implying local transtension which is not observed in the subarea-based stress tensor. Only site 48 is compatible with the subarea-based tensor; others deviate from it. This relation may indicate local stress perturbations, e.g. site 127 is very close to the normal fault NF1. The orientation of the mean subarea-based principal stresses and the stress ratio are: $\sigma_1 = 213°\text{N}/81°$, $\sigma_2 = 350°\text{N}/07°$, $\sigma_3 = 080°\text{N}/06°$; and $\Phi = 0.542$ (Fig. 6h). Having σ_1 subvertical and other stresses subhorizontal indicates extensional deformation in this phase.

Phase 4. Five sites had sufficient slip data for the construction of site-based stress tensors (Fig. 6e). In almost all the sites, σ_2 is subvertical and σ_1 ranges from northwest–southeast to north-northwest–south-southeast; this relation indicates strike-slip deformation. The orientation of the subarea-based mean stresses and the stress ratio are: $\sigma_1 = 291°\text{N}/03°$, $\sigma_2 = 033°\text{N}/83°$, $\sigma_3 = 201°\text{N}01°$; and $\Phi = 0.564$ (Fig. 6i). Having σ_2 vertical and the other stresses horizontal indicates regional strike-slip deformation in this phase. Most of the thrust and reverse faults and folds trending northeast–southwest are almost perpendicular to σ_1. In addition, the folds within the Upper Miocene units (near site 128) are perpendicular to σ_1.

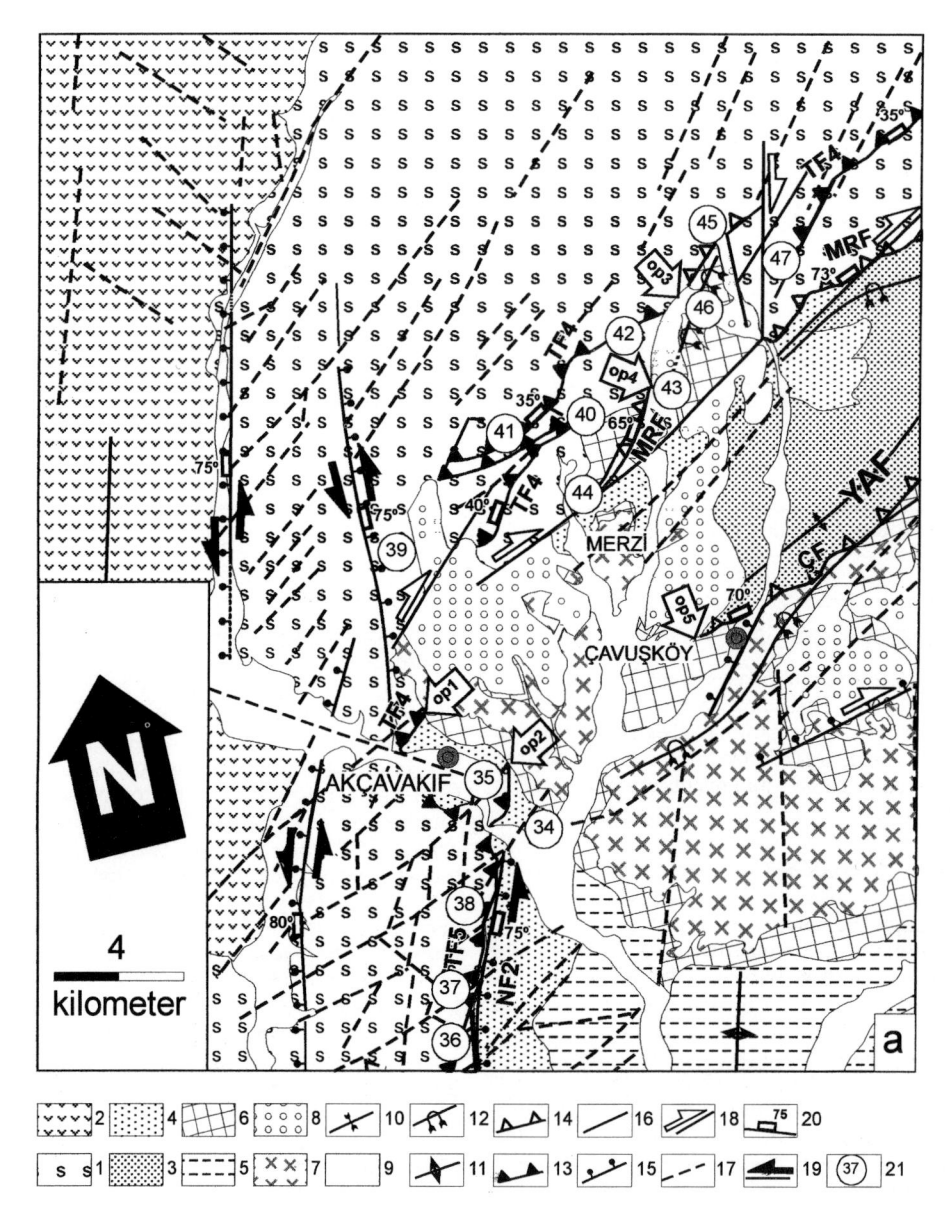

Fig. 7. (a) Geological map and sample location of subarea 3. 1, NAOM; 2, Galatean Volcanic Province [see Tankut *et al.* (1995) and Toprak *et al.* (1996)]; 3, İncik Formation; 4, Çandır Formation; 5, Tuğlu Formation; 6, Süleymanlı Formation; 7, Bozkır Formation; 8, Deyim Formation; 9, alluvium; 10, syncline; 11, anticline; 12, overturned syncline; 13, thrust faults; 14, reverse faults; 15, normal faults; 16, strike-slip faults or faults with an unknown sense of movement; 17, photo-lineaments; 18, dextral strike-slip faults; 19, sinistral strike-slip faults; 20, dips of faults where they are best observed in the field; 21, sample site locations. Plots of fault planes: (**b**)–(**e**) fault planes, slickenlines and stress orientations for each site; (**f**)–(**i**) whole data in a particular phase and subarea (lower hemisphere equal area projections).

EVOLUTION OF THE ÇANKIRI BASIN 311

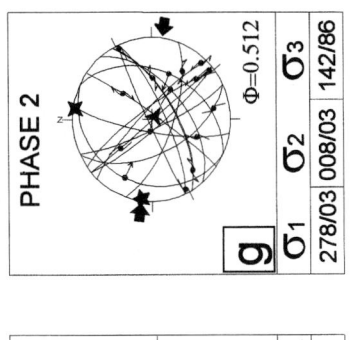

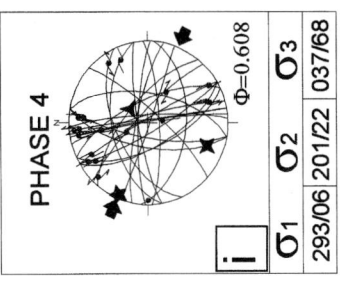

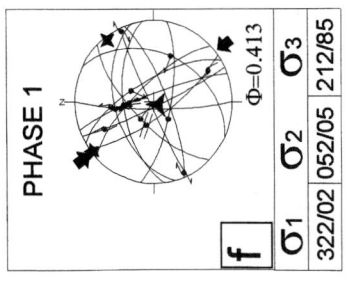

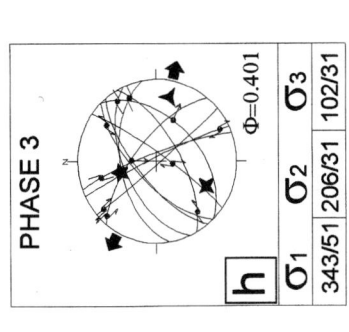

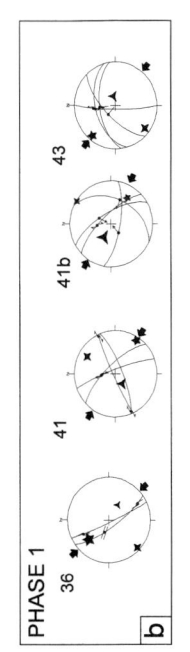

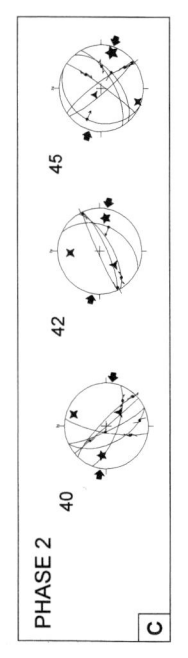

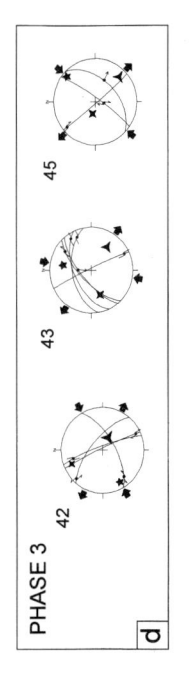

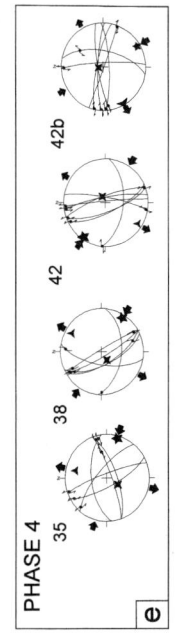

Table 3. *Field characteristics of sites in subarea 3*

Site	Shear	Vein	OP	S.line	CC	Unit*	No. move.
34	–	+	2	Ca	2	Ti	3
35	+	+	2	SP, Ca	2	K	2
36	–	–	2	SP	2	K	2
38	–	–	2	–	2	K	2
39	+	–	–	SP	2	K	2
40	+	–	2	SP	2	K	2
41	–	–	2	SP	2	K	2
42	+	+	2	Ca	2	K	3
43	+	+	3	SP, Ca	3	K	4
44	–	–	–	Ca	–	K	1
45	+	+	–	SP, Ca	2	K	2
46	–	–	–	–	–	K, Tç	1
47	+	+	2	SP	4	K, Tç	4

* Tç, Çandır Formation. Other abbreviations as in Table 1.

Results for subarea 3

This area is located at the northwest corner of the Çankırı Basin (3 in Fig. 1) where the NAOM and the thrust faults bounding the western margin of the basin change their strike from north-northeast–south-southwest to northeast–southwest. The area in which the sharpest change occurs is hidden below the Plio-Quaternary units (the Deyim Formation) (Figs 1 and 7a). The thrust faults along which the ophiolites and Upper Cretaceous units thrust over the Middle Miocene Çandır Formation are covered by the Upper Miocene Süleymanlı and Bozkır Formations (op1 and op2 in Fig. 7a). This relation indicates that thrust activity along TF4 and TF5 took place after the Middle Miocene and prior to the Late Miocene. Along the Merzi Reverse Fault (MRF), the NAOM is thrust over the Upper Miocene Süleymanlı Formation (op3 in Fig. 7a) and the thrust contact is covered by the Plio-Quaternary Deyim Formation (op3 and 4 in Fig. 7a). TF5 is displaced by NF2, indicating that two distinct tectonic regimes gave rise to the development of these structures. The first one is thrusting that resulted in the development of TF5 and the second one is an extensional one which gave rise to NF2, which is a normal fault with a strike-slip component (Fig. 7a).

In the central-eastern part of the subarea, near Çavuşköy, the post-Middle Eocene İncik Formation was thrust over the Upper Miocene–Pliocene(?) Bozkır Formation along the Çavuşköy Reverse Fault (ÇF) and the reverse fault contact is covered by the Plio-Quaternary Deyim Formation (op5 in Fig. 7a), indicating post-Late Miocene–Pliocene(?) formation of the reverse fault. The Deyim Formation, in turn, is displaced by northeast–southwest trending dextral strike-slip faults (Fig. 7a).

Palaeostress inversion

In this subarea, like the previous sites, four phases of fault movement are observed. These movements are assigned to deformation phases and they are ordered according to their occurrence as explained previously (see Fig. 3). The details of each site are given in Table 3 and the results are presented in Fig. 7b–e.

Phase 1. Only four of the 14 sites had sufficient data for the construction of site-based stress tensors (Fig. 7b). The orientations of the principal stresses are consistent on a site basis and they are also consistent with the constructed regional stress tensor (Fig. 8b). The orientation of the principal stresses and the stress ratio are: $\sigma_1 = 322°N/02°$, $\sigma_2 = 052°N/05°$, $\sigma_3 = 212°N/85°$; and $\Phi = 0.413$ (Fig. 7f) and indicate a compressive deformation. The orientation of σ_1 is almost perpendicular to the trace of TF4 and oblique to the trace of TF5 and NF2.

Phase 2. Only three sites had sufficient slip data for the construction of site-based stress tensors (Fig. 7c). The orientation of principal stresses are consistent with each other on a site basis and, when averaged for all subareas, the orientation of the principal stresses and the stress ratio are: $\sigma_1 = 278°N03°$, $\sigma_2 = 008°N/03°$, $\sigma_3 = 142°N/86°$; and $\Phi = 0.512$ (Fig. 7g) and indicate a compressive deformation. σ_1 is almost perpendicular to TF5 and NF2, and oblique to MRF, TF4 and ÇF.

Phase 3. Only three sites had sufficient data for the construction of the site-based stress tensors, although the horizontal component of σ_3 is consistent for individual sites and with the constructed mean subarea stress tensor (Fig. 7d). Orientation of regional stresses and the stress ratio are: $\sigma_1 = 343°N/51°$, $\sigma_2 = 206°N/31°$, $\sigma_3 = 102°N/31°$; and $\Phi = 0.401$ (Fig. 7h). None of the principal stresses are oriented vertically or horizontally and there are stress permutations (Angelier 1994) between the sites. This relation may indicate the state of so-called 'tri-axial strain conditions' (Reches 1978*a, b*). The orientation of the horizontal component of σ_3 is almost perpendicular to NF2 and other major normal faults with a sinistral strike-slip component (Fig. 7a).

Phase 4. Four sites had sufficient slip data for the construction of site-based stress tensors (Fig. 7e). The orientations of site-based tensors are relatively consistent with each other. In each site, σ_2 is subvertical and the horizontal component of σ_1 ranges from west-northwest–east-southeast to northwest–southeast. The orientations of subarea-based mean stresses and the stress ratio are: $\sigma_1 = 293°N/06°$, $\sigma_2 = 201°N/22°$, $\sigma_3 = 037°N/68°$; and $\Phi = 0.608$ (Fig. 7i) and indicate a compressive deformation. The horizontal component of σ_1 is almost perpendicular to TF5, NF2 and MRF, and oblique to TF4 and ÇF.

Results for subarea 4

This subarea includes the western margin of the Çankırı Basin and extends into the adjacent Hancılı Basin, which is separated from the Çankırı Basin by TF5 and NF2 (Figs 1c and 8a).

Along TF5, the Upper Cretaceous units are thrust over the Middle Miocene Çandır Formation and the fault contacts are covered by Upper Miocene units in the north outside subarea 4 (op1–3 in Fig. 8b). Along TF6, Upper Cretaceous rocks are thrust over the Middle Miocene Hancılı Formation. Along TF7 and TF8, the Upper Cretaceous rocks are thrust over the Lower–Middle Miocene Aslantaş and Hancılı Formations; these units are locally overturned along TF8. TF7 is covered by the Plio-Quaternary Deyim Formation, indicating pre-Plio-Quaternary activity of the fault. Along TF9, the Lower–Middle Miocene Aslantaş and Hancılı Formations are thrust over Upper Cretaceous rocks. TF10 is a set of thrust faults developed within the Upper Cretaceous and Palaeocene rocks. It is covered by the Plio-Quaternary Deyim Formation, indicating pre-Plio-Quaternary activity of this fault set.

TF5 is displaced by a number of approximately north-northeast–south-southwest striking oblique-slip faults with normal components (e.g. NF3) that strike almost parallel to TF5. TF6–TF9 are displaced by a number of approximately northeast–southwest striking strike-slip faults, some of which have a normal component of movement (Fig. 8a).

In the Çankırı Basin, the folds are oriented in two directions. The folds developed in the Çandır Formation are oriented north–south to north-northeast–south-southwest and the ones developed in the Upper Miocene units are oriented northeast–southwest. This relation indicates two phases of folding. The earlier folding postdates the deposition of the Middle Miocene Çandır Formation and predates the Upper Miocene units, and the latter postdates the deposition of Upper Miocene rocks.

The folds in the Hancılı Basin are oriented in two different directions. One set is oriented northwest–southeast, parallel to the thrust faults TF6–TF9. This relation indicates thrust-related folding of the units in the Hancılı Basin after deposition of Hancılı Formation in the Early Middle Miocene. The second set is oriented northeast–southwest, parallel to the folds in the Çankırı Basin affecting the Upper Miocene units (Süleymanlı and Bozkır Formations).

Palaeostress inversion

Twenty-one sites were selected for the construction of site-based stress tensors. As in the previous subareas, four phases of fault movements were again recognized. These movements are assigned into deformation phases and are ordered according to their occurrence as explained previously (see Fig. 4). The details of each site are given in Table 4 and the results are presented in Fig. 8c–f.

Phase 1. Seven sites had sufficient slip data for the construction of site-based stress tensors. Except for site 136 (Fig. 8c), the orientation of the principal stresses are consistent on a site basis and in all other sites; σ_3 is subvertical, σ_1 is oriented in a northwest–north-northwest to southeast–south-southeast direction and σ_2 is subhorizontal. The orientation of the mean regional principal stresses and the stress ratio are: $\sigma_1 = 165°N/01°$, $\sigma_2 = 255°N/16°$, $\sigma_3 = 071°N/74°$; and $\Phi = 0.372$ (Fig. 8g), indicating that thrusting occurred during this phase.

Phase 2. Only four sites had sufficient slip data for the construction of site-based stress tensors

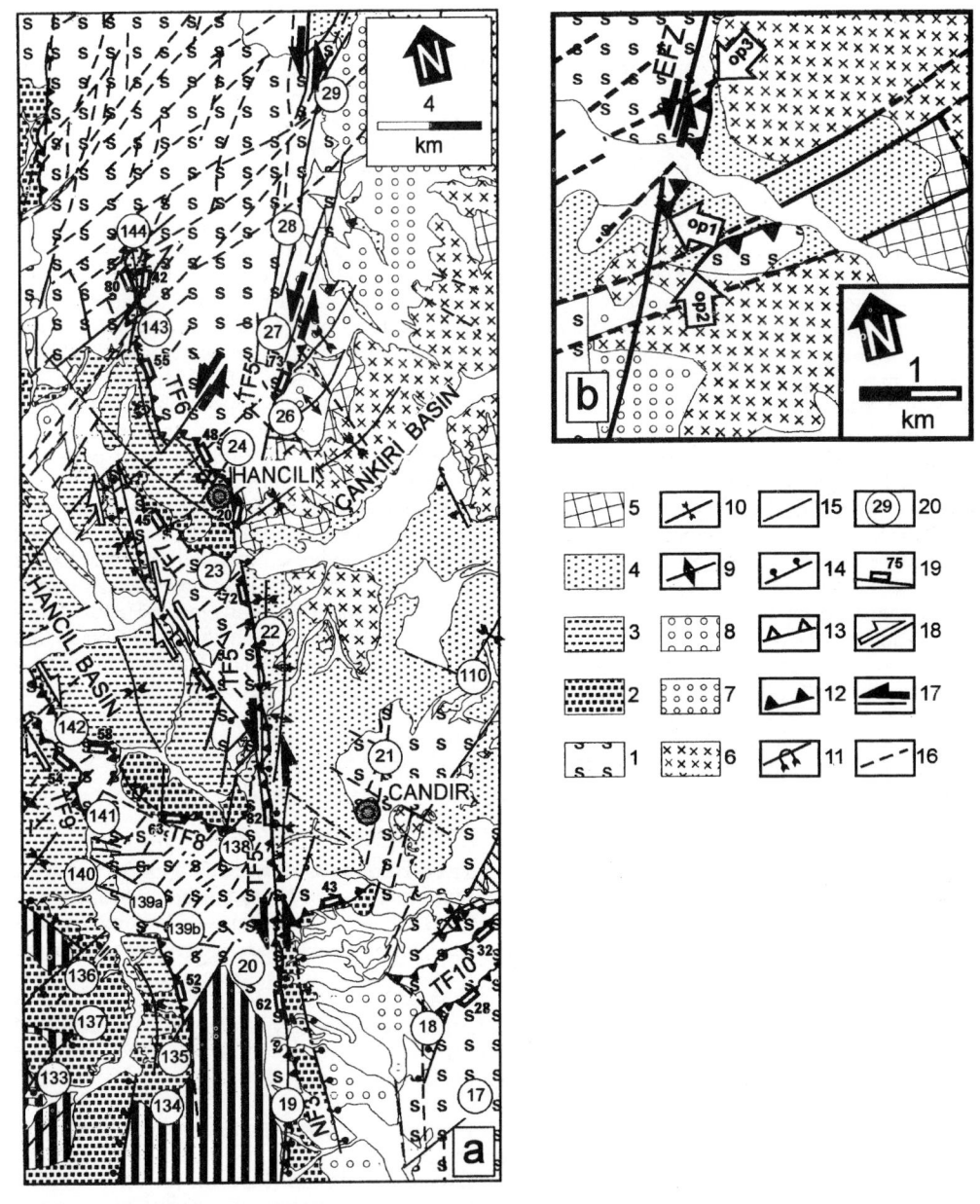

Fig. 8. (**a**) Geological map and sample location of subarea 4. 1, NAOM; 2, Aslantaş and Kılçak Formations; 3, Hancılı Formation; 4, Çandır Formation; 5, Süleymanlı Formation; 6, Bozkır Formation; 7, Deyim Formation; 8, alluvium; 9, anticline; 10, syncline; 11, overturned syncline; 12, thrust faults; 13, reverse faults; 14, normal faults; 15, strike-slip faults or faults with an unknown sense of movement; 16, photo-lineaments; 17, sinistral strike-slip faults; 18, dextral strike-slip faults; 19, dips of faults where they are best observed in the field; 20, sample site locations. (**b**) Map showing the relationship between thrusting of the NAOM onto the Çandır Formation and covering of the fault contact by the Bozkır Formation (see Fig. 1c for the location of the map). Plots of fault planes: (**c**)–(**f**) fault planes, slickenlines and stress orientations for each site; (**g**)–(**j**) whole data in a particular phase and subarea (lower hemisphere equal-area projections).

EVOLUTION OF THE ÇANKIRI BASIN 315

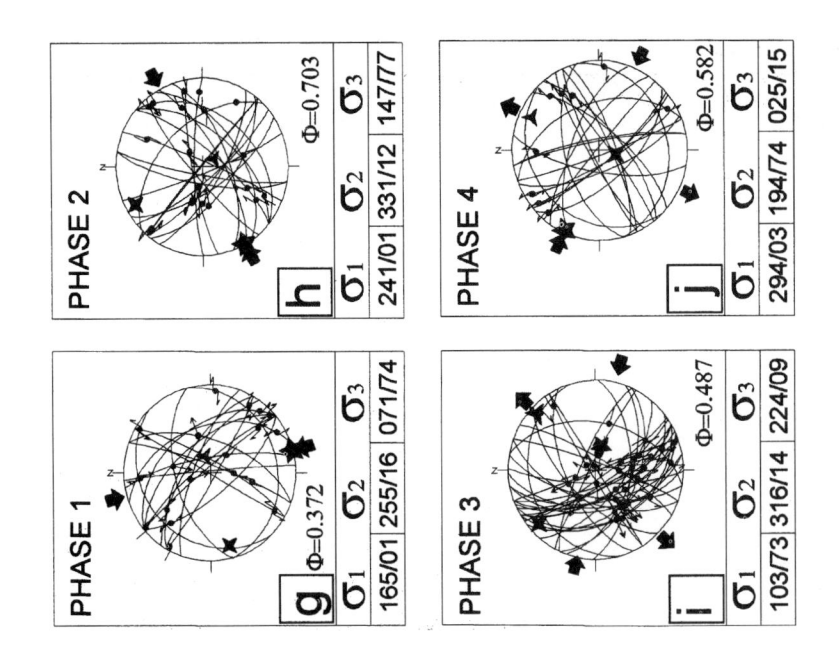

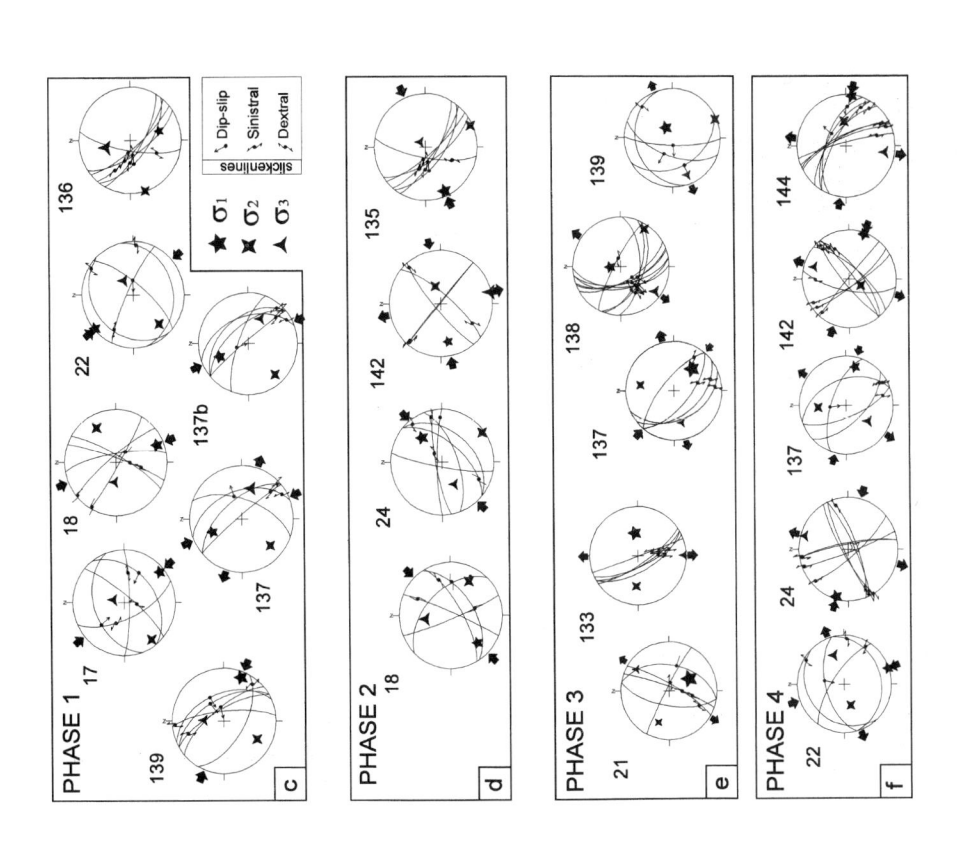

Table 4. Field characteristics of sites in subarea 4.

Site	Shear	Vein	OP	S.line	CC	Unit*	No. move.
17	−	−	2	−	2	Ti	2
18	+	+	−	Ca	2	K	3
19	+	+	−	SP, Ca	3	K	3
20	−	+	−	−	−	K	2
21	+	−	2	−	2	K	3
22	+	+	2	Ca	2	K	3
23	−	−	−	−	−	K	1
24	+	+	2	−	3	K	4
25	−	−	−	−	−	−	−
26	+	−	−	−	2	K	2
27	−	−	−	−	−	K	1
28	−	−	−	−	−	K	1
29	−	−	−	−	−	K	1
133	−	−	−	−	−	Ta	1
134	−	+	−	Ca	−	Tha	1
135	−	+	−	Ca	2	Tha	2
136	−	+	−	Ca	2	Tha	2
137	+	−	2	−	2	TRK, Ta	3
138	+	+	2	−	2	K, Tha	3
139a	+	+	2	−	2	K	3
139b	+	+	2	SP, Ca	2	K, Ta	3
140	−	−	−	−	−	Ta	1
141	+	−	−	−	2	Tha	2
142	+	−	−	−	2	Tha	2
143	−	−	−	−	−	Tha	1
144	−	−	2	Ca	2	Tha	2

* Ta, Aslantaş Formation; Tha, Hancılı formation. Other abbreviations as in Table 1.

(Fig. 8d). The orientation of σ_2 and σ_3 is variable in each site, whilst σ_1 is relatively consistent and oriented east-northeast–west-southwest to northeast–southwest. The orientation of the averaged subarea-based stresses and the stress ratio are: $\sigma_1 = 241°N/41°$, $\sigma_2 = 331°N/12°$, $\sigma_3 = 147°N/77°$; and $\Phi = 0.703$ (Fig. 8h), indicating thrusting during this phase.

Phase 3. Six sites had sufficient slip data for the construction of site-based stress tensors (Fig. 8e). In all the sites, σ_1 is subvertical and other stresses are subhorizontal. The horizontal component of σ_3 is relatively consistent in each site, but other stresses are variable in orientation. This may be because the magnitudes of the σ_2 and σ_3 are very close to each other, resulting in stress permutations (Angelier 1994). In addition, the orientation of the horizontal component of σ_3 is approximately perpendicular to the thrust faults (TF6–TF9) in the Hancılı Basin whence most of the data came. The orientation of the mean regional principal stresses and the stress ratio are: $\sigma_1 = 130°N/73°$, $\sigma_2 = 316°N/14°$, $\sigma_3 = 224°N/09°$; and $\Phi = 0.487$ (Fig. 8i). Having σ_1 subvertical and other stresses subhorizontal indicates extensional deformation in this phase.

Phase 4. Five sites had sufficient slip data for the construction of site-based stress tensors (Fig. 8f). The orientation of the principal stresses are variable, whilst σ_1 is relatively consistent in each site except for site 22. It is oriented west-northwest–east-southeast, while the horizontal component of σ_3 is oriented north-northeast–south-southwest. The orientation of mean subarea-based stresses and the stress ratio are: $\sigma_1 = 294°N/03°$, $\sigma_2 = 194°N/74°$, $\sigma_3 = 025°N/25$; and $\Phi = 0.582$ (Fig. 8j), indicating strike-slip deformation in this phase.

Discussion

The results obtained are summarized in Fig. 9, in which both the site and subarea stress tensors are presented. They are important to the structuring of the Çankırı Basin subarea stress tensors and the discussion will concentrate on them.

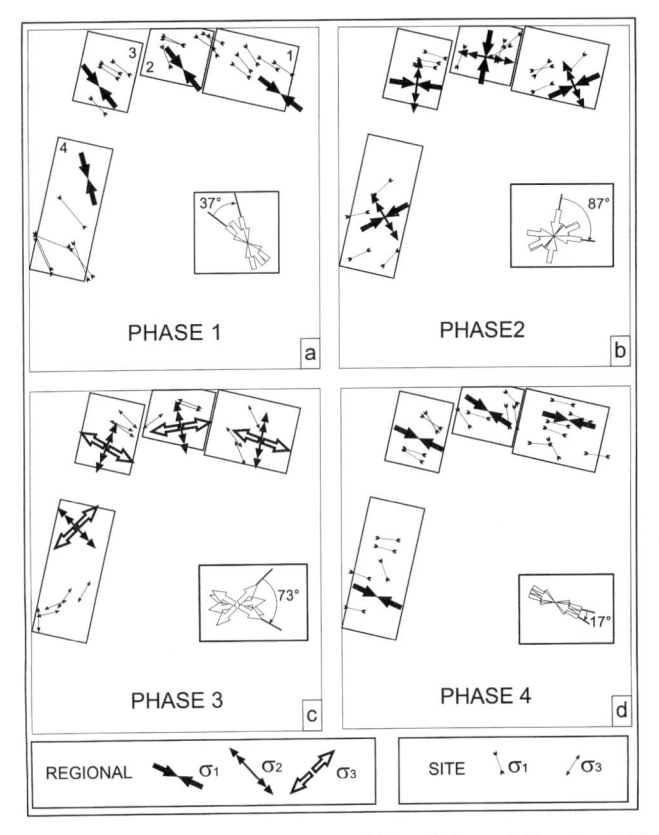

Fig. 9. Plots of horizontal components of σ_1 [(**a**), (**b**) and (**d**)], σ_2 [(**b**) and (**c**)] and σ_3 (**c**) in different phases and subareas (1–4).

Phase 1

As stated above, Phase 1 is characterized by compressional deformation in which the orientation of σ_1 is subhorizontal and that of σ_3 is subvertical. When the results are considered it can be seen that σ_1 has a consistent northwest trend in the northern and northwestern margins but a north-northwest trend in the western margin; this gives an overall discrepancy of 37° (Fig. 9a). The orientation of σ_1 in the north (subareas 1–3) is almost perpendicular to the main thrust faults and oblique to the ones in the western margin of the basin (Fig. 10). The youngest units affected in this phase are pre-Late Palaeocene; therefore, this deformational phase operated until pre-Late Palaeocene time.

Phase 2

This phase is also characterized by compressional deformation in which σ_1 is subhorizontal and σ_3 is subvertical, although some sites do indicate local strike-slip deformation in which σ_1 and σ_3 are both subhorizontal and σ_2 is subvertical. The averaged trend of σ_1 is very variable, changing from northeast–southwest in subareas 1 and 2 to almost north–south in subarea 2, to east–west in subarea 3 (Fig. 10b); the angular discrepancy between the subareas is 87° (Fig. 9b). The fault slip data ascribed to this phase are observed only in the pre-Burdigalian units, therefore, a Middle Eocene–pre-Burdigalian age is ascribed to this phase.

Phase 3

This phase is characterized by extensional deformation in which σ_1 is subvertical and σ_3 is subhorizontal. The orientation of σ_3 changes from west-northwest in subareas 1 and 3, to east-northeast in subarea 2 and northeast trend in subarea 4 (Fig. 9c); this gives an overall angular discrepancy of 73° (Fig. 9c). The orientation of the horizontal component of σ_3 in each subarea is oblique to the faults, which were previously supposed to be active in this deformation phase (Fig. 10c). The fault slip data

Fig. 10. Plot of horizontal components of σ_1 for phases 1, 2 and 4 (converging large arrows) and σ_3 for Phase 3 (diverging large arrows), and for the structures proposed to have developed in a corresponding phase [(**a**)–(**c**) and (**e**)]. (**d**) Length-weighted rose diagram prepared from the faults proposed to have developed in deformation Phase 3 and an idealized stereographic projection of largest populations (note that they display two sets of conjugate faults). (**f**) Length-weighted rose diagram prepared from the faults proposed to have developed in deformation Phase 4.

ascribed to this deformation phase are obtained mainly from Lower to Middle Miocene units in subarea 4. In other subareas they overprint the slickenlines that were previously ascribed to older phases and overprinted by the latest phase. Therefore, an Early–Middle Miocene age is assigned to this phase. The length-weighted rose diagram (Fig. 10d), prepared from the normal faults supposedly developed in this deformation phase, indicates two sets of conjugate pairs of normal faults. Considering the oblique nature of the principal stresses discussed above, it is proposed that these faults are developed in so-called 'tri-axial strain conditions', which assumes that σ_1 is subvertical and other stresses are oblique to the horizontal plane (Reches 1978a, b; Krantz 1988).

Phase 4

This phase is characterized by strike-slip deformation in which σ_2 is subvertical and σ_1 and σ_3 are subhorizontal. The orientations of σ_1 trends are relatively consistent in the subareas and trend west-northwest with 17° of overall discrepancy (Fig. 9d). In Fig. 10e the structures

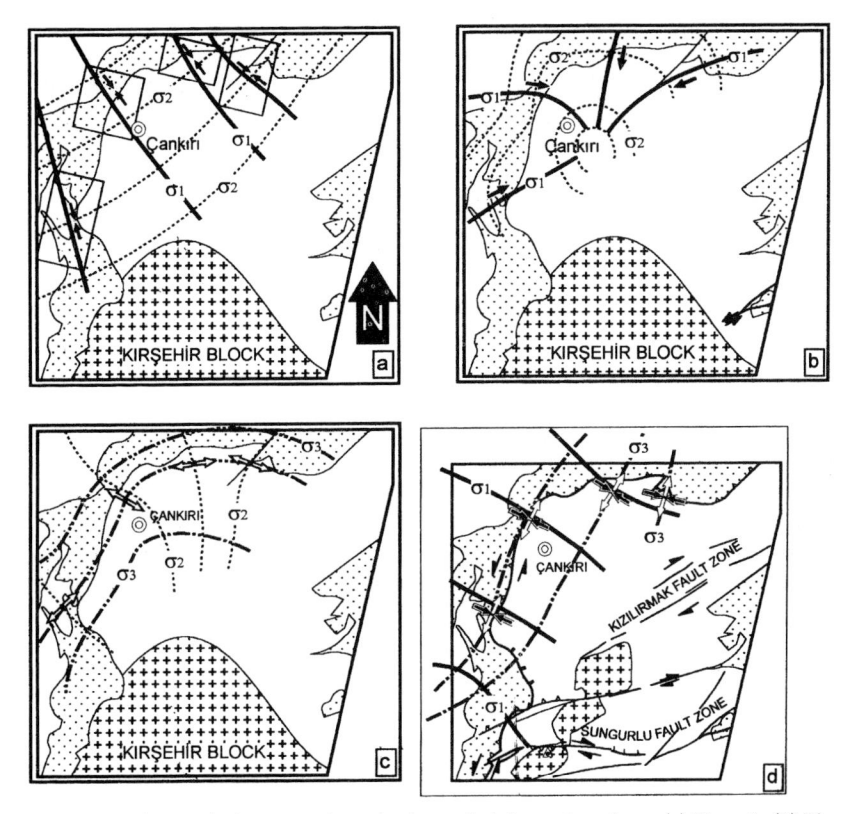

Fig. 11. Interpreted and smoothed stress trajectories for each deformation phase. (**a**) Phase 1; (**b**) Phase 2; (**c**) Phase 3; (**d**) Phase 4.

that may have developed in this phase are illustrated. It is obvious that most of these structures were reactivated in this deformation phase along inherited planes of weaknesses, although the variation of σ_1 between subareas is almost negligible. This relationship indicates that pre-existing planes of weakness do not play a major role in the stress-inversion procedure. The slip data attributed to this phase include the latest overprinting slickensides and the data collected from the faults that affected the Upper Miocene and younger units; therefore, a post-Middle Miocene age is assigned to this phase.

Stress trajectories and the models

Using the subarea-based principal stresses, smoothed stress trajectories of each phase are plotted (Fig. 11). The stress trajectories in Phase 1 display a mesh-like pattern in which the σ_1 trajectories are oriented northwest–southeast while σ_2 trajectories are curvilinear and convex southeastwards (Fig. 11a). In Phase 2 they display radial σ_1 and concentric σ_3 patterns (Fig. 11b). In deformation Phase 3, the σ_2 trajectories display radial pattern while σ_2 are concentric around the rim of the Çankırı Basin and exposed parts of the Kırşehir Block (Fig. 11c). The concentric pattern of σ_3 trajectories in deformation Phase 3 indicates uniaxial extension (Carey & Bruner 1974) which is characteristic for areas of regional doming (Means 1976) and multidirectional extension (Arlegui-Crespo & Simon-Gomez 1998). In deformation Phase 4, σ_1 and σ_3 trajectories display a mesh like pattern oblique to the western and northern rim, and to the Kızılırmak and Sungurlu Fault Zones (Fig. 11d).

The subduction of Neotethys took place northwards under the Pontides along a roughly east–west trending trench (Şengör & Yılmaz 1981; Görür et al. 1984; Koçyiğit et al. 1988; Koçyiğit 1991; Dellaloğlu et al. 1992) in Late Cretaceous–Early Tertiary time, i.e. during deformation Phase 1. Considering the east–west oriented zone of convergence, the orientation of σ_1 (in Phase 1) is therefore oblique to the

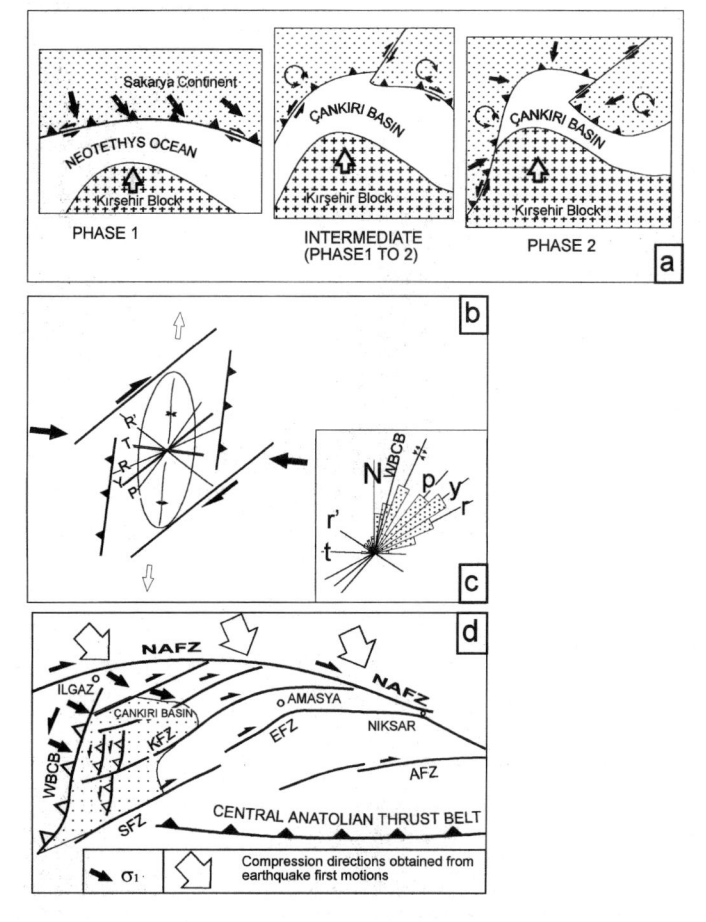

Fig. 12. (**a**) Cartoons illustrating the possible development of the Çankırı Basin through deformation Phases 1 and 2. Note block rotations and the response of principal stress orientations as the Kırşehir Block drives northwards. Black arrows are σ_1; white arrows are the relative movement sense of the Kırşehir Block (Pontides assumed fixed). (**b**) Riedel pattern of deformation [after Biddle & Christie-Blick (1985)] and a plot of horizontal components of σ_1 (convergent large arrows) proposed to explain the structures developed in deformation Phase 4. (**c**) Length-weighted rose diagram prepared from the structures proposed to have developed in this phase and corresponding Riedel shears – r, primary synthetics shear; r″, antithetic shear; p, secondary synthetic shear; y, principal displacement zone; t, extensional structures. (**d**) Schematic illustration of the structures in north-central Turkey plotted to explain the structural development of the Çankırı Basin in deformation Phase 4 – AFZ, Almus Fault Zone; EFZ, Ezinepazarı Fault Zone; KFZ, Kızılırmak Fault Zone; NAFZ, North Anatolian Fault Zone; WBCB, western boundary fault of the Çankırı Basin [modified after Şengör et al. (1985)].

direction of convergence (Fig. 12a). This relationship may indicate that subduction had a dextral strike-slip component in this part of the Tethys Ocean (Fig. 12a), where the Sakarya continent and the Kırşehir Block eventually collided and amalgamated. Palaeomagnetic studies undertaken separately indicate that the western part of the Çankırı Basin rotated c. 30° anticlockwise, while the eastern margin rotated c. 50° clockwise, which resulted in the Ω-shape of the basin in Eocene–Oligocene times.

This relationship may also be the reason for the radial σ_1 pattern and concentric σ_2 pattern developed with σ_3 subvertical in deformation Phase 2 (Fig. 11b).

In the Early Miocene, the collision and further convergence of the Sakarya continent and the Kırşehir Block was completed. Subsequently, the compressional regime was replaced by an extensional regime (in deformation Phase 3), which may be due to gravitational collapse (Dewey 1988). This gave rise to the formation of multidirectional normal faulting and the deposition of the Aslantaş, Hancılı and Çandır Formations within graben complexes. Extensional deformation, driven by gravitational collapse, has already been postulated for western Turkey and the Aegean area (Seyitoğlu *et al.* 1992; Bozkurt & Park 1994, 1997; see Lips 1998 and Walcott 1998 for Aegean references). Therefore, it can be proposed that Early–Middle Miocene extension in western Anatolia extended as far east as the Çankırı Basin area in central Turkey.

Having σ_2 subvertical and σ_1 oriented northwest–southeast during Phase 4, indicates that the Sungurlu (SFZ) and Kızılırmak Fault Zones (KFZ) have been the two major strike-slip faults which deformed the Çankırı Basin and displaced its rims dextrally. The length-weighted rose diagrams, prepared from these structures, indicate the dominance of northeasterly trends which display a Riedel deformation pattern (Figs 10f, 12b and c), which commonly develops in regions of regional transcurrent deformation and along strike-slip fault zones (Biddle & Christie-Blick 1985). During this phase, the western margin of the basin reactivated as a sinistral strike-slip fault zone as the conjugate of the KFZ and the SFZ. As the western margin was dominated by a pre-existing thrust fault belt, it was reactivated into a zone of sinistral transpression (Fig. 12d). In addition, the orientations of the constructed principal stresses (Fig. 12d) are parallel to the compressive and tensile axes obtained from recent earthquakes along the NAFZ (Jackson & McKenzie 1984; Dewey *et al.* 1986). This relationship is consistent with the results given here.

Conclusions

- Four deformation phases have been recognized and their palaeostress configurations are constructed.
- The first phase is characterized by northwest–southeast oriented σ_1 and subvertical σ_3, indicating compressional deformation characterized by thrusting.
- The second phase is characterized by radial σ_1 and concentric σ_2 patterns with subvertical σ_3, indicative of thrusting.
- Concentric σ_3 and subvertical σ_1 are indicative of extensional deformation in the third deformation phase.
- The final phase is characterized by a northwest–southeast oriented σ_1 pattern with very little variation of σ_1 orientations between the subareas.

- The structures, which were active in each deformation phase, are plotted. The structures, which were active in the latest deformation phase, display a Riedel deformation pattern that is consistent with the strike-slip deformation that has been in operation since the Late Miocene in the area.

We thank J. Angelier and K. Hardcastle for supplying their software; Y. Özçelik, Y. Öztaş, H. de Bruijn, A. van der Meulen, E. Ünay and G. Saraç for their help in the field and the Neogene age determinations. C. Duermeijer is thanked for palaeomagnetics. We appreciate the constructive criticism of K. Hardcastle, who twice reviewed early drafts of this paper. NK thanks A. Zanchi, who introduced Angelier's software. This work was conducted under the programme of the Vening Meinesz Research School of Geodynamics (VMSG). NK also received a grant from Kocaeli University, İzmit–Turkey, which is gratefully acknowledged. We thank R. J. Lisle and A. Koçyiğit for their constructive reviews of the manuscript.

References

ALEXANDROWSKI, P. 1985. Graphical determination of principal stress directions for slickenside lineation populations: an attempt to modify Arthoud's method. *Journal of Structural Geology*, **7**, 73–82.

ANGELIER, J. 1979. Determination of the mean principal directions of stress for a given fault population. *Tectonophysics*, **56**, T17–T26.

—— 1984. Tectonic analysis of fault slip data sets. *Journal of Geophysical Research*, **89**, 5835–5848.

—— 1989. From orientation to magnitudes in palaeostress determination using fault slip data. *Journal of Structural Geology*, **11**, 37–50.

—— 1994. Fault slip analysis and palaeostress reconstruction. *In*: Hancock, N. L. (ed.) *Continental Deformation*. Pergamon Press, 53–100.

——, TARANTOLA, A., VALETTE, B. & MANOUSSIS, S. 1982. Inversion of field data in fault tectonics to obtain the regional stress – 1. Single phase fault populations: a new method of computing the stress tensor. *Geophysical Journal of Royal Astronomical Society*, **69**, 607–621.

ARLEGUI-CRESPO, L. E. & SIMON-GOMEZ, J. L. 1998. Reliability of palaeostress analysis from fault striations in near multidirectional extension stress fields. Examples from the Ebro Basin, Spain. *Journal of Structural Geology*, **20**, 827–840.

ARMIJO, R., CAREY, E. & CISTERNAS, A. 1982. The inverse problem in microtectonics and the separation of tectonic phases. *Tectonophysics*, **82**, 145–169.

ARTHAUD, F. 1969. Méthode de détermination graphique des directions de reccourcissement d'allongement et intermédiare d'une population de failles. *Bulletin de la Société Géologique de France* **11**, 739–757.

BARKA, A. A. & HANCOCK, P. L. 1984. Neotectonic deformation patterns in the convex-northwards arc of the North Anatolian Fault Zone. *In*: DIXON, J. E. & ROBERTSON, A. H. F. (eds) *The Geological Evolution of the Eastern Mediterranean*. Geological Society, London, Special Publications, **17**, 763–774.

BIDDLE, K. T. & CHRISTIE-BLICK, N. 1985. *Strike-slip Faulting and Basin Formation*. Society of Economic Palaeontologists and Mineralogists, Special Publications, **37**.

BOTT, M. P. H. 1959. The mechanics of oblique-slip faulting. *Geological Magazine*, **96**, 109–117.

BOZKURT, E. & KOÇYİĞİT, A. 1995. Almus Fault Zone: its age, total offset and relation to the North Anatolian Fault Zone. *Turkish Journal of Earth Sciences*, **4**, 93–104.

—— & —— 1996. The Kazova Basin: an active negative flower structure on the Almus Fault Zone, a splay fault system of the North Anatolian Fault Zone, Turkey. *Tectonophysics*, **265**, 239–254.

—— & PARK, R. G. 1994. Southern Menderes Massif: an incipient metamorphic core complex in western Anatolia, Turkey. *Journal of the Geological Society, London*, **151**, 213–216.

—— & —— 1997. Evolution of a mid-Tertiary extensional shear zone in the southern Menderes Massif, western Turkey. *Bulletin de la Société Géologique de France*, **168**, 3–14.

CAREY, E. & BRUNER, B. 1974. Analyse théorique et numérique d'une modèle méchanique élémentaire appliqué a l'étude d'une population de failles. *Comptes Rendus de l'Académie des Sciences de Paris*, **279**, 891–894.

CAREY-GAILHARDIS, E. & MERCIER, J. L. 1987. A numerical method for determining the state of stress using focal mechanisms of earthquake populations: application to Tibetan teleseisms and microseismicity of southern Peru. *Earth and Planetary Science Letters*, **82**, 165–179.

DELLALOĞLU, A. A., TÜYSÜZ, O., KAYA, İ. H. & HARPUT, B. 1992. *Kalecik (Ankara)–Eldivan–Yapraklı (Çankırı)–İskilip (Çorum) ve Devrez Çayı Arasındaki Alanın Jeolojisi ve Petrol Olanakları*. Turkish Petroleum Corporation (TPAO) Report, No. 3194.

DEWEY, J. F. 1977. Suture zone complexities: a review. *Tectonophysics*, **40**, 53–67.

—— 1988. Extensional collapse of orogens. *Tectonics*, **7**, 1123–1139.

——, HEMPTON, M. R., KIDD, W. S. F., ŞAROĞLU, F. & ŞENGÖR, A. M. C. 1986. Shortening of continental lithosphere: the neotectonics of Eastern Anatolia – a young collisional zone. *In*: COWARD, M. P. & RIES, A. C. (eds) *Collision Tectonics*. Geological Society, London, Special Publications, **19**, 3–36.

DUPHIN, J. M, SASSI, W. & ANGELIER, J. 1993. Homogeneous stress hypothesis and actual fault slip: a distinct element analysis. *Journal of Structural Geology*, **15**, 1033–1043.

ETCHECOPAR, A., VISSEUR, G. & DAIGNIERES, M. 1981. An inverse problem in microtectonics for the determination of stress tensors from fault striation analysis. *Journal of Structural Geology*, **3**, 51–65.

FLEISCHMAN, K. H. & NEMCOK, M. 1991. Palaeostress inversion of fault-slip data using the shear stress solution of Means (1989). *Tectonophysics*, **196**, 195–202.

FLEUTY, M. J. 1974. Slickensides and slickenlines. *Geological Magazine*, **112**, 319–322.

GEPHART, J. W. 1990. Stress and direction of slip on fault planes. *Tectonophysics*, **8**, 845–858.

—— & FORSYTH, D. W. 1984. An improved method for determining the regional stress tensor using earthquake focal mechanism data: application to the San Fernando earthquake sequence. *Journal of Geophysical Research*, **89**, 9305–9320.

GÖRÜR, N., OKTAY, F. Y., SEYMEN, İ. & ŞENGÖR, A. M. C. 1984. Palaeotectonic evolution of the Tuzgölü Basin complex, central Turkey: sedimentary record of a Neotethyan closure. *In*: DIXON, J. E. & ROBERTSON, A. H. F. (eds) *The Geological Evolution of the Eastern Mediterranean*. Geological Society, London, Special Publications, **17**, 81–96.

HANCOCK, P. L. 1985. Brittle microtectonics: principles and practice. *Journal of Structural Geology*, **7**, 437–457.

HARDCASTLE, K. C. 1989. Possible palaeostress tensor configurations derived from fault-slip data in eastern Vermont and eastern New Hampshire. *Tectonics*, **8**, 265–284.

—— & HILLS, L. S. 1991. BRUTE3 & SELECT: QuickBasic 4 programs for determination of stress tensor configurations and separation of heterogeneous populations of fault-slip data. *Computer and Geoscience*, **17**, 23–43.

JACKSON, J. A. & MCKENZIE, D. P. 1984. Active tectonics of the Alpine–Himalayan Belt between western Turkey and Pakistan. *Geophysical Journal of the Royal Astronomical Society*, **77**, 185–264.

KAYMAKÇI, N. & KOÇYİĞİT, A. 1995. Mechanism and basin generation in the splay fault zone of the North Anatolian Fault Zone. *EUG-8 Abstracts*.

KOÇYİĞİT, A. 1991. An example of an accretionary forearc basin from north Central Anatolia and its implications for the history of subduction of Neotethys in Turkey. *Geological Society of America Bulletin*, **103**, 22–36.

——, ÖZKAN, S. & ROJAY, B. 1988. Examples from the fore-arc basin remnants at the active margin of northern Neotethys; development and emplacement age of the Anatolian Nappe, Turkey. *METU Journal of Pure and Applied Sciences*, **21**, 183–120.

KRANTZ, R.W. 1988. Multiple fault sets and three-dimensional strain: theory and application. *Journal of Structural Geology*, **10**, 225–237.

LACOMBE, O., ANGELIER, J. & LAURENT, PH. 1992. Determining palaeostress orientations from faults and calcite twins: a case study near the Sainte–Victoire Range (southern France). *Tectonophysics*, **201**, 141–156.

——, ——, ——, BERGERAT, F. & TOURNERET, C. 1990. Joint analysis of calcite twins and fault slips as a key for deciphering polyphase tectonics: Burgundy as a case study. *Tectonophysics*, **202**, 83–93.

LIPS, A. L. W. 1998. *Temporal constraints on the kinematics of the destabilisation of an orogen: syn- to post-orogenic extensional collapse of the northern Aegean region.* PhD Thesis, Utrecht University (166).

MARRET, R. & ALMANDINGER, R. W. 1990. Kinematic analysis of fault slip data. *Journal of Structural Geology*, **12**, 973–986.

MEANS, W. D. 1976. *Stress and Strain*. Springer.

MICHAEL, A. J. 1984. Determination of stress from slip data: faults and folds. *Journal of Geophysical Research*, **89**, 11 517–11 526.

NEMCOK, M. & LISLE, R. J. 1995. A stress inversion procedure for polyphase fault/slip data sets. *Journal of Structural Geology*, **17**, 1445–1453.

NIETO-SAMANIEGO, A. F. & ALANIZ-ALVAREZ, S. A. 1997. Origin and tectonic interpretation of multiple fault patterns. *Tectonophysics*, **270**, 197–206.

OKAY, A. İ. 1984. Distribution and characteristics of the northwest Turkish blueschists. *In:* DIXON, J. E. & ROBERTSON, A. H. F. (eds) *The Geological Evolution of the Eastern Mediterranean*. Geological Society, London, Special Publications, **17**, 429–440.

ÖZÇELIK, Y. 1994. *Tectono-stratigraphy of the Laçin area (Çorum – Turkey)*. MSc Thesis, Middle East Technical University.

PASQUARE, G., POLI, S., VEZZOLI, L. & ZANCHI, A. 1988. Continental arc volcanism and tectonic setting in Central Anatolia, Turkey. *Tectonophysics*, **146**, 217–230.

POLLARD, D. D., SALTZER, S. D. & RUBIN, A. M. 1993. Stress inversion methods: are they based on faulty assumptions? *Journal of Structural Geology*, **15**, 1045–1054.

RECHES, Z. 1978a. Analysis of faulting in three-dimensional strain field. *Tectonophysics*, **47**, 109–129.

—— 1978b. Faulting of rocks in three-dimensional strain field – II; theoretical analysis. *Tectonophysiscs*, **95**, 133–156.

—— 1987. Determination of the tectonic stress tensor from slip along faults that obey the Coloumb yield creterion. *Tectonics*, **6**, 849–861.

ROJAY, B. 1993. *Tectonostratigraphy and neotectonic characteristics of the southern margin of Merzifon–Suluova Basin (Central Pontides, Amasya)*. PhD Thesis, Middle East Technical University.

—— 1995. Post-Triassic evolution of Central Pontides: evidence from Amasya region, Northern Anatolia. *Geologica Romana*, **31**, 329–350.

ŞENGÖR, A. M. C. & YILMAZ, Y. 1981. Tethyan evolution of Turkey: a plate tectonic approach. *Tectonophysics*, **75**, 181–241.

——, ŞAROĞLU, F. & GÖRÜR, N. 1995. Strike-slip deformation and related basin formation in zones of tectonic escape: Turkey as a case study. *In:* BIDDLE, K.T. & BLICK, N.C. (eds) *Strike-slip Deformation, Basin Formation and Sedimentation*. Society of Economic Palaeontologists and Mineralogists, Special Publications, **37**, 227–264.

——, YILMAZ, Y. & SUNGURLU, O. 1984. Tectonics of the Mediterranean Cimmerides: nature and evolution of the western termination of Palaeo-Tethys. *In:* DIXON, J. E. & ROBERTSON, A. H. F. (eds) *The Geological Evolution of the Eastern Mediterranean*. Geological Society, London, Special Publications, **17**, 77–112.

SEYİTOĞLU, G., SCOT, B. & RUNDLE, C. 1992. Timing of Cenozoic extensional tectonics in west Turkey. *Journal of the Geological Society, London*, **149**, 533–538.

TANKUT, A., SATIR, M., GÜLEÇ, N. & TOPRAK, V. 1995. *Galatya Volkaniklerinin Petrojenezi*. TÜBİTAK Project No. YBAG-0059 [in Turkish with English abstract].

TOPRAK, V., SAVAŞÇIN, Y., GÜLEÇ, N. & TANKUT, A. 1996. Structure of the Galatean Volcanic Province, Turkey. *International Geology Review*, **38**, 747–758.

TWISS, R. J. & UNRUH, J. R. 1998. Analysis of fault slip inversion; do they constrain stress or strain rate? *Journal of Geophysical Research*, **103**, 12 205–12 222.

WALCOTT, C. R. 1998. *The Alpine evolution of the Thessaly (NW Greece) and Late Tertiary kinematics*. PhD Thesis, Utrecht University (162).

WALLACE, R. E. 1951. Geometry of shearing stress and relation to faulting. *Journal of Geology*, **69**, 118–130.

WILL, T. M. & POWELL, R. 1991. A robust approach to the calculation of palaeostress fields from fault plane data. *Journal of Structural Geology*, **13**, 813–821.

YIN, Z. M. & RANALLI, G. 1993. Determination of tectonic stress field from fault slip data, toward a probabilistic model. *Journal of Geophysical Research*, **98**, 12 165–12 176.

Cenozoic extension in Bulgaria and northern Greece: the northern part of the Aegean extensional regime

CLARK B. BURCHFIEL,[1] RADOSLAV NAKOV,[2] TZANKO TZANKOV[2] & LEIGH H. ROYDEN[1]

[1]*Department of Earth, Atmospheric, and Planetary Sciences, Massachusetts Institute of Technology, Cambridge, MA 02139, USA (e-mail: bcburch@mit.edu)*
[2]*Institute of Geology, Bulgarian Academy of Sciences, Acad. Boncev Str., B. 24, 1113 Sofia, Bulgaria*

Abstract: The well-known Cenozoic Aegean extensional regime, initiated at *c.* 25 Ma, thinned the crust so that most of it now lies submerged. North of the western continuation of the North Anatolian Fault the Aegean extensional regime is present in central and southern Bulgaria, northern Greece, Former Yugoslavian Republic (FYR) of Macedonia and eastern Albania. Here the system is exposed on land and offers an opportunity to reconstruct the extensional evolution of the system. The southern Balkan peninsula forms the northern part of the Aegean extensional system; deformation is not as great as in the Aegean, but reconstruction of this part of the extensional regime will provide important constraints on its dynamics.

Following a period of arc-normal extension associated with Late Eocene–Late Oligocene magmatism, major lithospheric extension appears have begun between 26 and 21 Ma in northern Greece, involving east-northeast–west-southwest extension east of Mount Olympos, on the Island of Thasos and near Kavala. This period of extension may have been accompanied by a short period of coeval compression north of the arc during Early Miocene time or perhaps a little earlier in the Thrace Basin of northwestern Turkey. Northeast–southwest directed Middle–Late Miocene extension appears to have developed obliquely to the older magmatic arc and migrated northward into southwestern Bulgaria in the Sandanski Graben (and perhaps also into the Mesta and Padesh Grabens) by 16.3–13.6 Ma, and in the Blagoevgrad and Djerman Grabens by *c.* 9 Ma. Extension in southwestern Bulgaria was reorganized by *c.* 5 Ma and in northern Greece extension on the Strymon Valley detachment fault ended by *c.* 3.5 Ma, but extension continued on new fault systems. From limited structural and stratigraphic data, it is speculated here that related extension may have also occurred during this time in FYR Macedonia and eastern Albania. This northeast–southwest extension is interpreted to be related to trench roll-back along the northern part of the subduction boundary in the western Hellenides.

North-south extension along east–west striking faults in central Bulgaria began only after extension was well underway in northern Greece and the Sandanski Graben of southwestern Bulgaria. Within the Sofia Graben, the Sub-Balkan grabens, and grabens to their east, north–south extension began at *c.* 9 Ma, and may have begun about the same time in the Plovdiv, Zagore and Tundja Grabens of the northern Thracian Basin: north–south extension has continued to the present in these grabens. The cause of the north–south extension is unclear and may be related to trench roll-back along the central part of the subduction zone in the Hellenides, or more local causes of clockwise and counterclockwise rotation of the western Hellenides and western Turkey, respectively.

By Late Pliocene time a major erosion surface, the sub-Quaternary surface, was developed over a large area of central Bulgaria creating a major unconformity that marks the beginning of Quaternary deposition in the basinal areas. Many large and small graben-bounding faults in west-central Bulgaria displace this erosion surface and demonstrate the widespread extent of Quaternary north–south extension. North–south extension extended westward, with probably decreasing magnitude, across the older northwest trending graben of southwest Bulgaria (the Simitli and Djerman Grabens) and into eastern FYR Macedonia.

During latest Pliocene(?) and Quaternary time, northern Greece developed a complex pattern of northeast–southwest extension associated with northeast to east–west striking right-lateral faults forming transfer faults between more local extensional areas. This system of faults overprinted the older northwest trending extensional faults, such as the Strymon Detachment, and may be related to the propagation of the right-lateral North Anatolian Fault into the north Aegean Sea and formation of parallel faults to its north. These two different tectonic regimes extend into FYR Macedonia, where a third regime of

From: BOZKURT, E., WINCHESTER, J. A. & PIPER, J. D. A. (eds) *Tectonics and Magmatism in Turkey and the Surrounding Area.* Geological Society, London, Special Publications, 173, 325–352. 1-86239-064-9/00/$15.00
© The Geological Society of London 2000.

east–west extension in western FYR Macedonia and eastern Albania is present, and where extension may represent the continuation of the east–west extensional regime initiated in Middle–Late Miocene time.

Active deformation determined from seismicity and Global Positioning System studies suggest northern Greece, and perhaps southwest Bulgaria and FYR Macedonia, is dominated by north–south extension. This pattern of deformation must have developed as recently as perhaps Late Quaternary time.

Except for mountains near the Adriatic Sea all of the mountainous topography in the southern Balkan region may be the result of Miocene–Recent extension.

The Alpine orogenic system passes through the southern Balkan region in two branches, the Hellenides to the southwest (e.g. Aubouin et al. 1963; Jacobshagen 1986; Papanikalaou 1993) and the Carpathian–Balkan chain to the northeast (e.g. Burchfiel 1980; Gocev 1986, 1991; Boyanov et al. 1989) and continues eastward into the complex Tauride and Pontide chains in Turkey (e.g. Şengör & Yılmaz 1981; Yılmaz 1997). Both branches were deformed by shortening, strike-slip and local extensional deformation from Late Jurassic to Recent time, driven by complex convergence between Africa and Eurasia. Following collision and closing of oceanic regions within the inner part of the orogens, shortening proceeded toward the margins of these orogenic belts, particularly toward the southern and western margins (e.g. Aubouin et al. 1963; Jacobshagen et al. 1978; Jacobshagen 1986; Papanikalaou 1993). Coeval extension began at c. 25 Ma within the Aegean region, including western Turkey and the southern

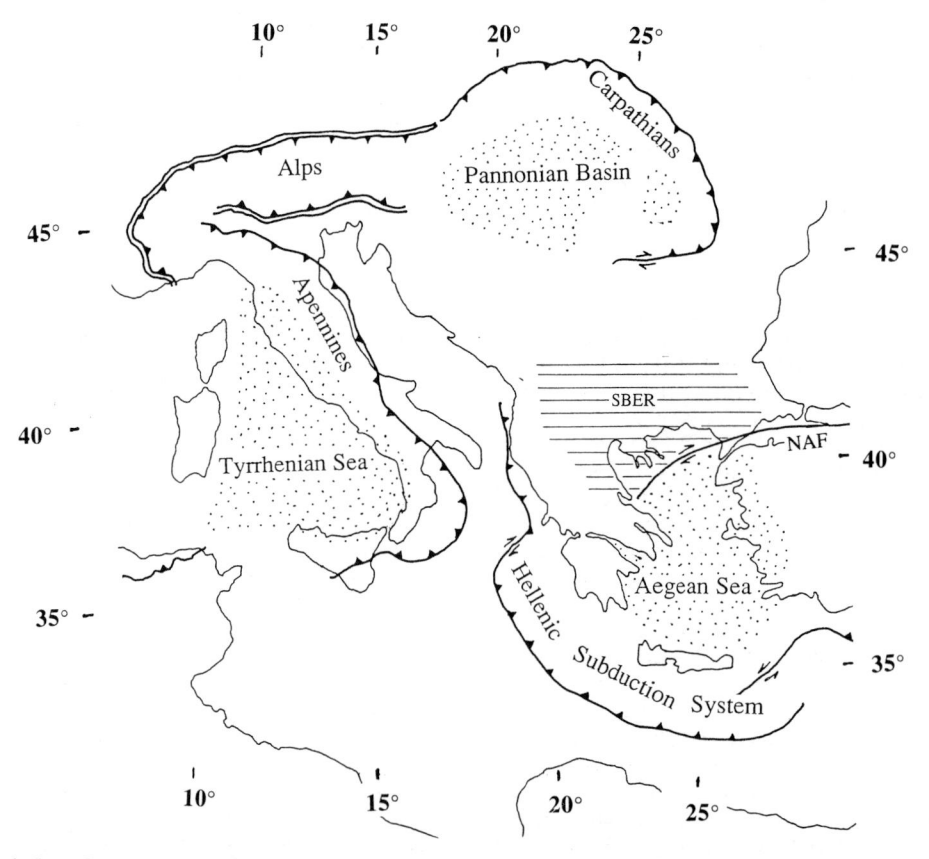

Fig. 1. Late Cenozoic tectonic features of the eastern Mediterranean region showing location of suprasubduction extensional areas (dotted areas) that formed coeval with adjacent areas of subduction and shortening (single barbed lines). The location of the southern Balkan extensional region (SBER) discussed here is shown by horizontal lines. Areas of shortening without coeval extension are bounded by doubled barbed lines.

Balkan region (e.g. Lister et al. 1984; Buick 1991; Avigad 1993; Gautier et al. 1993; Schermer 1993; Vandenberg & Lister 1996; Avigad et al. 1997; Jolivet et al. 1998; Wawrzenitz & Krohe 1998). Extension, superposed diachronously on the convergent structures of the orogen during Middle and Late Cenozoic time, is probably polygenetic, driven by trench roll-back, gravitational instabilities in an evolving lithosphere and extrusion from post-suturing convergence in the Middle East (McKenzie 1978; Dewey & Şengör 1979; Angelier et al. 1982; Royden 1993; Jolivet et al. 1998). The extensional system follows, and cuts across, crustal anisotropies developed during convergence.

The Aegean extensional regime is of interest not only because young extensional structures are spectacularly exposed in northern Greece (Dinter & Royden 1993; Kaufman 1995) and Bulgaria (Tzankov et al. 1996), but also because the Aegean system represents an active example of extension in the hanging wall of an adjacent subduction zone (Royden 1993). Other Mediterranean examples include the Tyrrhenian–Apennine/Calabrian system and the Pannonian–Carpathian system (Fig. 1). The many similarities between these basin–thrust belt pairs strongly suggest that the space–time relationships between extension and convergence are not incidental, but rather reflect similar dynamic processes operating in each system. Yet despite the systematic spatial and temporal relationships that can be broadly documented on a regional scale, the dynamic and detailed kinematic processes remain poorly understood.

The extensional history of the Aegean region shows superposed events, often with different extensional directions and apparent causes [e.g. see Vandenberg & Lister (1996); and Jolivet et al. [1998]]. Unravelling the complex kinematic history of extension, strike-slip and rotational deformation within the Aegean is necessary to understand fully the dynamic regime responsible for its formation. However, within the Aegean this will be difficult because extension has thinned the crust (Sachpazi et al. 1997; Tsokas & Hansen 1997) so that most of it now lies submerged beneath the Aegean Sea. North of the western continuation of the North Anatolian Fault, in the southern Balkan region, the Aegean extensional regime is exposed on land in central and southern Bulgaria, northern Greece, the Former Yugoslavian Republic (FYR) of Macedonia and eastern Albania (Figs 1 and 2), and offers an opportunity to reconstruct the extensional evolution of this part of the system in some detail. Although the southern Balkan Peninsula forms the northern part of the Aegean regime, the magnitude of extension in this region may not as great as that further south. Nevertheless, reconstruction of this part of the extensional regime will provide important constraints on its dynamics.

To date, this group's studies of the southern Balkan extensional regime have focused on the youngest part of the extensional history in Bulgaria and northern Greece, from Middle Miocene to Recent time, and current studies of Late Eocene–Middle Miocene extension are at present incomplete.

Bulgaria and northern Greece

The continuation of the Alpine orogenic belt through Bulgaria has long been recognized and its nature hotly debated (e.g. Burchfiel 1980; Şengör 1984; Gocev 1986; Boyanov et al. 1989). Until recently, less attention has been paid to the widespread Cenozoic extensional deformation that has overprinted much of the earlier compressional tectonics. Indeed, many structural and morphological features of Bulgaria and the southern Balkan region, commonly ascribed to earlier compressional events, now appear to be the result of young extensional deformation. Some of these features are spectacularly exposed along the southern flank of the Stara Planina (Tzankov et al. 1996) and along the western margin of the Rhodope Massif in southwestern Bulgaria (Fig. 2; Zagorchev 1992, 1995), allowing an excellent opportunity to study the extensional processes and their relationship to regional tectonic events within Bulgaria and in the surrounding regions of Greece, FYR Macedonia, Albania and Turkey.

In a modern plate tectonic setting, Turkey, Greece, the Aegean Sea, and parts of Bulgaria, FRY Macedonia, Albania and southern Yugoslavia may be considered dynamically as a single tectonic unit that will be referred to here as the Aegean extensional province (Fig. 1). To the east, the region of central Turkey lying between the North and East Anatolian Faults is actively moving westward relative to Eurasia as a coherent unit (Reilinger et al. 1997; McClusky et al. 2000), but whether this has been the case throughout Late Cenozoic time remains unclear. In western Turkey this westward to southwestward moving region is affected by extensional tectonism that extends into the Aegean Sea and mainland Greece. From examination of regional geologic and topographic maps, extension occurs as far north as central Bulgaria, FYR Macedonia and eastern Albania

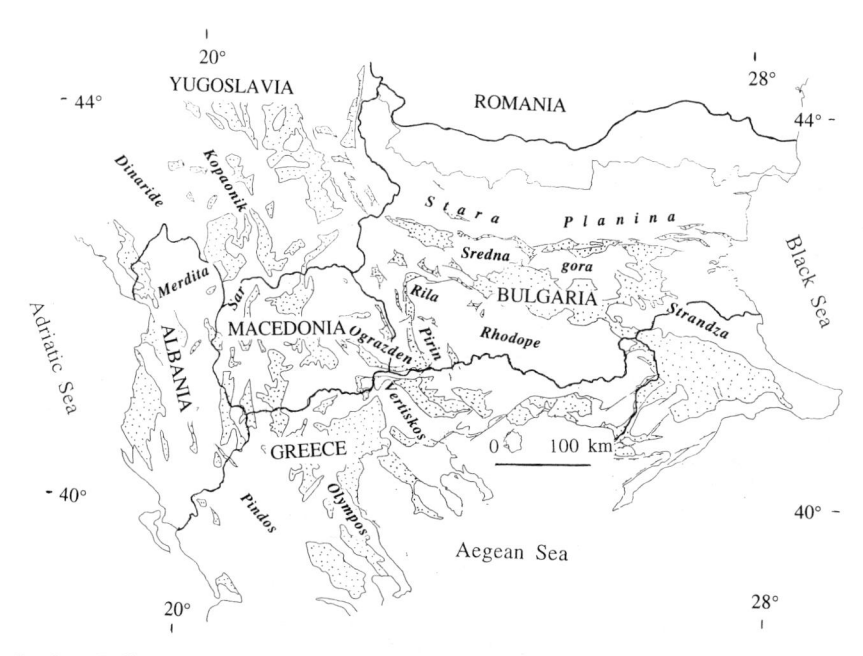

Fig. 2. Southern Balkan extensional region showing country boundaries, basin locations (stippled) and some mountain ranges mentioned in the text.

(Figs 1 and 2). To the west and south the extensional region is bounded by the Hellenic subduction zone, along which the lithosphere of the Adriatic and eastern Mediterranean is subducted northeastward beneath Albania, Greece, Crete and southern Turkey (Fig. 1). With the exception of a narrow compressional belt adjacent to the subduction zone, the entire region appears to have experienced extension during at least the last 20–25 Ma.

In most studies, the North Anatolian Fault, its westward extension through the northern Aegean Sea and projected extension across northwestern Greece to the Hellenic Trench are considered to be the northern boundary of the Middle East–Balkan tectonic system (Kahle et al. 1995; Fig. 1). In his original work on this tectonic system, McKenzie (1972, 1978) clearly recognized that the northern boundary of the system lay in the southern Balkan Peninsula, but because of lack of data on active and Late Cenozoic tectonics and difficulties of working in the Balkan region, he was not able to document the nature of deformation within Bulgaria and the southern Balkans. Many workers have omitted the northern part of the system from their tectonic syntheses (e.g. Kahle et al. 1995).

Late Eocene–Late Miocene deformation

Although this study has focused on Late Miocene and younger extensional tectonism, deformation from Late Eocene to Late Miocene time includes both extension and shortening. This older deformation is an important part of the evolutionary sequence that must be considered when evaluating Late Cenozoic extension in Bulgaria, northern Greece and surrounding areas.

During most of Late Mesozoic and Early Cenozoic time, Bulgaria and northern Greece lay within a largely convergent regime dominated by crustal shortening (except for a major extensional event of Middle–Late Cretaceous age that formed the east–west trending Sredna Gora Volcanic Rift Zone through central Bulgaria, extending eastward into the Black Sea; Boyanov et al. 1989). Several shortening events are recorded in latest Cretaceous–Late Eocene time (Boyanov et al. 1989).

During Late Eocene, Oligocene and perhaps even into the earliest part of Early Miocene time, a west–northwestern trending magmatic belt developed in southern Bulgaria, northern Greece, northwesternmost Turkey and central FYR Macedonia. This belt is characterized by

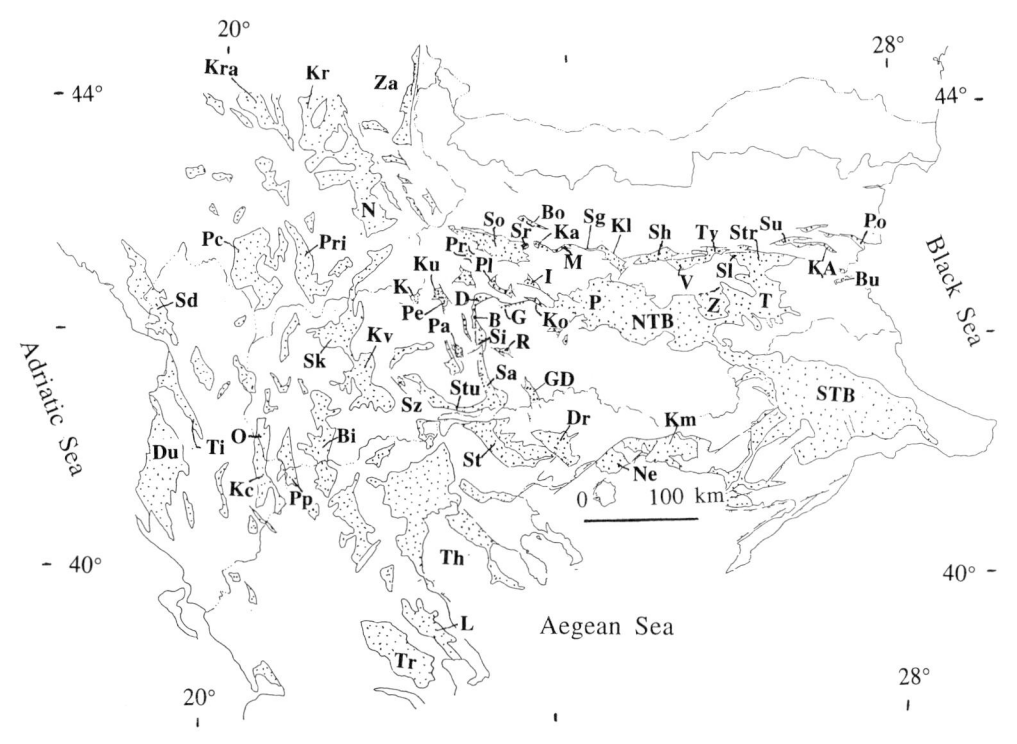

Fig. 3. Neogene basins of the southern Balkan extensional regime. A few Upper Palaeogene basins are included. B, Blagoevgrad; Bi, Bitola; Bo, Botevgrad; Bu, Burgas; D, Djerman; Dr, Drama; Du, Dures; G, Govedartsi; GD, Gotse Delchev; I, Ihtiman; K, Kamenitsa; Ka, Kamartsi; KA, Karnobat–Aitos; Kc, Korce; Ko, Kostenets; Kl, Karlıovo; Km, Komotnin; Kr, Kraguleva; Kra, Kraljevo; Ku, Kustendil; Kv, Kavadarci; L, Larisa; M, Mirkovo; NTB, northern Thracian Basin; N, Nis; Ne, Nestos; O, Ohrid; P, Plovdiv; Pa, Padesh; Pc, Pec; Pe, Persink–Bobov dol; Pl, Palakaria; Po, Prosenik; Pp, Prespansko; Ps, Prespansko; Pr, Pernik; Pri, Pristina; R, Razlog; Sa, Sandanski; Sd, Skadarsko; Sg, Srednagorie; Sh, Sheinovo; Si, Simitli; Sk, Skopje; Sl, Sliven; So, Sofia; Sr, Sarantsi; STB, southern Thracian Basin; St, Strymon; Str, Straldja; Stu, Strumeshnitsa; Su, Sungurlare; Sz, Strumitza; T, Tundja; Te, Tetovo; Th, Thermaikos; Ti, Tirana; Tr, Trikola; Tv, Tvarditsa; V, Vetren; Z, Zagore; Za, Zajecar.

subaereal intermediate to acid volcanic flows, ignimbrites, shallow-level intrusive bodies and thick sequences, locally 1–2 km thick, of dominantly coarse clastic sedimentary rocks (Figs 3 and 4; e.g. Harkovska *et al.* 1989). The northern boundary of this magmatic belt lies along, and locally just north of, the northern Rhodopian Mountains; a small Oligocene subvolcanic stock lies just south of the Sofia Basin. These volcanic rocks are interbedded with thick conglomerate and sandstone units which locally contain large blocks of debris flow and landslide origin [e.g. Ivanov *et al.* (1977), although they ascribed the emplacement of the large sheet-like olistostromes to gravity thrusting]. These rocks continue westward into FYR Macedonia (Boev *et al.* 1995) and eastward into the southern Thrace Basin of northern Turkey, where they are interbedded with marine turbidites and are as old as Middle Eocene (Doust & Arıkan 1974; Perinçek 1991; Turgut *et al.* 1991).

This complex magmatic belt is probably partly related to the final northward subduction along the Vardar–İzmir–Ankara Zone and to melting of thickened continental crust north of the suture (e.g. Harkovska 1984). Faulting and the coarse-grained nature of the sediments associated with the igneous activity suggest that extension occurred during and following magmatism, but it is not clear whether extension was partly coeval with shortening along the east–west trending fold-thrust belt in northern Bulgaria or was entirely post-shortening. (This group's studies on faulting associated with the Palaeogene magmatism have not yet been completed and a more detailed analysis will be given in the future.) Harkovska (1984) showed that the main magmatic conduits trend parallel to the magmatic arc and suggested a

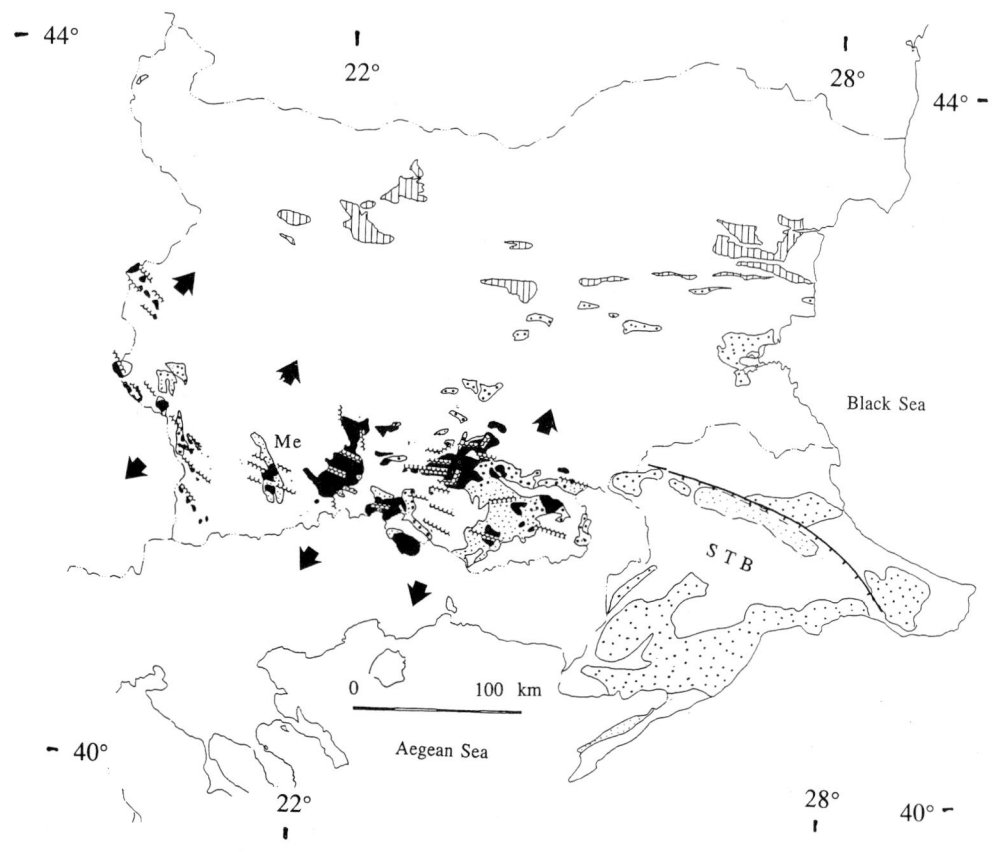

Fig. 4. Upper Eocene–Upper Oligocene rocks in Bulgaria and northwestern Turkey. Volcanic (black) and volcaniclastic sediments (fine dots) record the presence of a magmatic arc trending northwest–southeast through southern Bulgaria into northwest Turkey. (The magmatic rocks continue northwest into FYR Macedonia and south into northern Greece but are not shown.) Scalloped lines show the orientation of conduits for the volcanic rocks and are interpreted to have formed normal to regional extension, as shown by the large arrows. Data in Bulgaria are largely taken from Havkovska (1984). Large dots are non-volcanic sedimentary rocks which are continental with marine intercalation within Bulgaria, and marine within the southern Thracian Basin (STB) in northwestern Turkey. Volcaniclastic sediments (fine dots) in the southern Thracian Basin crop out along its southern boundary and are known in the subsurface from its northeastern part (outlined by dotted line). The northeastern margin of the southern Thracian Basin is marked by a large south dipping normal fault (heavy ticked line) active during sedimentation. Sediments of the same age in northern Bulgaria shown by vertical lines are within and north of the foreland fold thrust belt of the Forebalkan and are not discussed here. Regional convergence was within the Hellenides at this time, along the thrust faults that involve the Pindos Zone.

north-northeast to northeast extension direction, perpendicular to the trend of the magmatic arc, a conclusion supported here. Largely non-volcanic sediments were deposited in the northern Thrace Basin north of the magmatic belt, although ash beds are present in Oligocene sediments as far north as the northeastern part of Stara Planina near Varna (Alexiev 1959; Figs 2 and 4). Strata along the northern edge of the magmatic arc, and further north, are continental and marine, suggesting that the magmatic arc probably had a moderately low elevation.

At the end of Oligocene time, volcanism and associated sedimentation largely ceased along the volcanic belt in southern Bulgaria, although sedimentation continued locally to the north into Early Miocene time in the southern Thrace Basin, into earliest Middle Miocene in parts of the northern Thrace Basin and into late Early Miocene time [based on dating of sediments by spores and pollen by Cernjavska (1977)] in the Pernik and Persink–Bobov Dol Basins of west-central Bulgaria (Figs 3 and 5). Prior to Late Miocene time, the last shortening event took

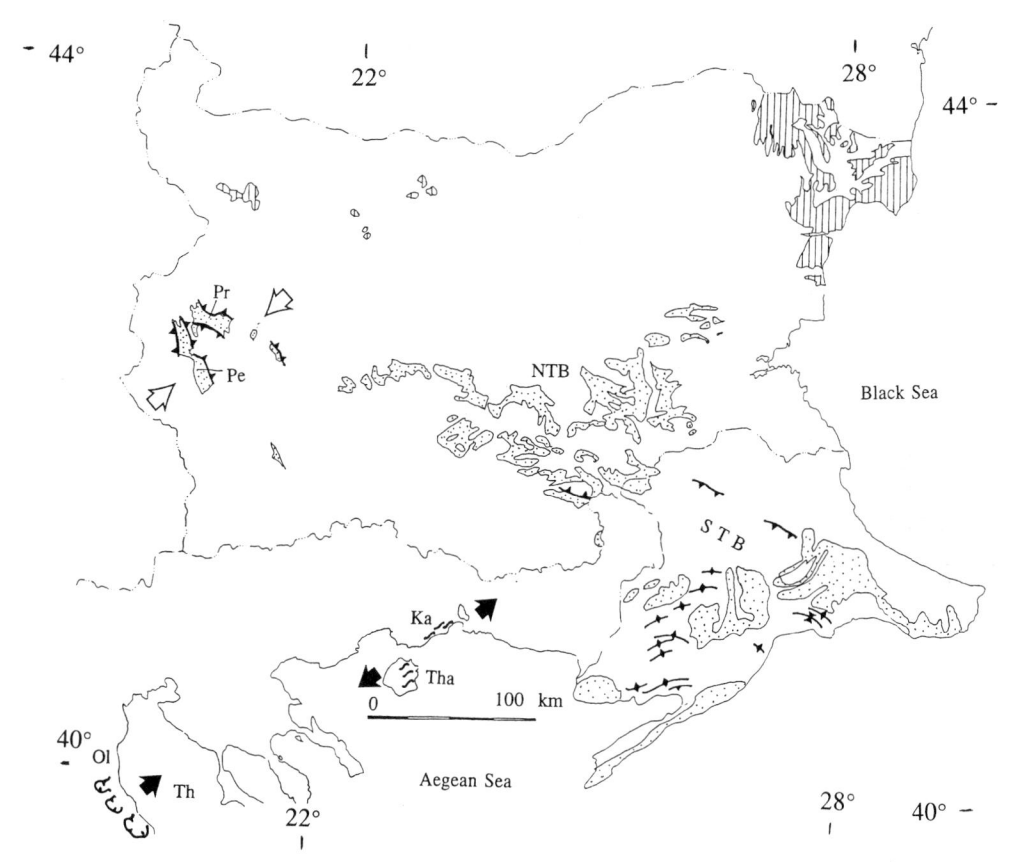

Fig. 5. Sedimentary rocks and deformational events of latest Oligocene and Early Miocene (locally ?lowermost Middle Miocene) age. Outcrops of largely continental sedimentary rocks in central Bulgaria, and continental and marine sedimentary rocks in the south Thracian Basin in northwestern Turkey are stippled. Post-tectonic continental and marine sedimentary rocks above the Forebalkan foreland fold and thrust belt, and its foreland, are shown by vertical lines (not discussed here). Folds and thrust faults that deform Lower Miocene rocks, and are overlain by Upper Miocene rocks, in central Bulgaria are shown by heavy barbed lines. Folds and thrust faults in the southern Thracian Basin (STB) of northwestern Turkey are largely in the subsurface and taken from Schindler (1997); these structures may be slightly older than those in Bulgaria. Northeast vergent extension in the Mount Olympos area (Ol) and northeast–southwest lineated ductile mylonitic rocks on Thasos (Tha) and near Kavala (Ka), with no clear vergence, are shown by heavy ticked lines, and sigmoidal lines respectively. NTB, north Thracian Basin; Pr, Pernik Basin; Pe, Persink–Bobov dol Basin; Th, Thermaikos Basin. The direction of inferred shortening is shown by open arrows and the directions of extension are shown by black arrows. Regional convergence at this time was within the Hellenides to the west along the thrust faults at the base of the Gavrovo tectonic unit.

place within western and central Bulgaria [constrained by folded and thrusted lowest Lower Miocene rocks unconformably overlain by Upper Miocene rocks, and regarded as Early Miocene folding by Zagorchev (1992); Fig. 5]. Strong folding and thrust faulting occurred in the Pernik and Persink–Bobov Dol Basins, and locally weak deformation occurred within the northern Thrace Basin (Fig. 5). Similar folding and faulting, but perhaps slightly older, occurred in the southern Thrace Basin in Turkey (Perinçek 1991; Schindler 1997). Due to the uncertainty in timing, it is unclear if the Early Miocene folding occurred contemporaneously with the oldest extension in the region or occurred during a short interval when extension ceased. As outlined below, this folding may have been coeval with the earliest extension further to the south in northern Greece.

There is a major break in sedimentation across much of Bulgaria in Middle and early Late Miocene time. This marks a time of important erosion, a probable reduction in topographic relief and an important change in the pattern of deformation. These studies have largely focused on the tectonic evolution of Bulgaria and adjacent regions following this important unconformity, although some discussion of older deformation and sedimentation will be made where it is important.

Miocene–Recent evolution of the extensional system in Bulgaria and northern Greece

Between latest Oligocene and the beginning of Late Miocene time there is a transition from older intra-arc extension to extension that is apparently unrelated to magmatism in Bulgaria and northern Greece. Extension began in northern Greece on the Island of Thasos in the latest Oligocene (Wawrzenitz & Krohe 1998), in Early Miocene time in the Kavala area (Dinter et al. 1995) and perhaps in the Mount Olympos area (Schermer et al. 1990; Schermer 1993). During Late Miocene–Recent time, extension developed with complex temporal changes in the strikes of faults and direction of extension in northern Greece, central and southern Bulgaria, and probably FYR Macedonia.

Earliest extensional events

Recent work on Thasos has demonstrated that top-to-the-southwest extension began there along a major detachment system not later than 26–23 Ma and appears to have continued from 21 Ma to at least 13 Ma (Fig. 5; Wawrzenitz & Krohe 1998). This detachment system continues northward as the Strymon Valley detachment system of Dinter & Royden (1993), which was active until $c.$ 3.5 Ma. Although this detachment system appears to be related to granitic plutons in the Kavala and Vrondu areas, it is suggested here that this detachment system begins a transition from intra-arc to non-arc extension because: (1) the northwest trend of the faults is generally oblique to the arc trend in this area; (2) the younger displacement on the detachment occurs without coeval magmatism; and 3) the detachment system propagates northward into Bulgaria, where it also trends oblique to the Eocene–Oligocene magmatic arc.

In the Kavala area, directly north of Thasos, Dinter & Royden (1993) and Dinter et al. (1995) have shown that northeast–southwest directed, ductile mid-crustal extension may have begun at $c.$ 21 Ma (Fig. 5). These ductilely deformed rocks were subsequently unroofed by the northwest trending, top-to-the-southwest, Strymon Valley detachment, which shows a transition from ductile to brittle behaviour between $c.$ 16 to 3.5 Ma. The earliest mid-crustal ductile extension occurs in the $c.$ 21 Ma Symvolon Pluton.

At Mount Olympos, Schermer et al. (1990) document extension with top-to-the-northeast shear beginning between 23 and 16 Ma (Fig. 5). This generally northwest-trending extensional fault system formed without associated magmatism, and parallels the extensional fault system of Thasos and the Strymon Valley. As suggested by Dinter (1998), it may be part of the same period of latest Oligocene–Early Miocene extension.

These areas document that the beginning of extension is at least as old as, shortening described above and may predate, and appears to have been active through, the Early or Middle Miocene. Although not firmly constrained, it is likely that shortening and extension were coeval, but spatially separated, with shortening occurring further north.

Late Miocene–Pliocene extensional events

In northern Greece, extension continued into Late Miocene or Pliocene time and is expressed by several Late Miocene northwest trending basins (Kojumdgieva 1987; Figs 6 and 7). In Bulgaria these basins are dated as Badenian–Maeotian ($c.$ 16–5 Ma) in age, and are nearly identical in age, structure and stratigraphy to basins along the Strymon Valley detachment system in Greece. For example, the structure and stratigraphy of the Bulgarian Sandanski Basin, initiated in Middle Miocene time, is summarized below to illustrate the type of information that can be obtained from these basins and to show their relevance in understanding regional tectonic events.

Extension in the Sandanski and related basins. Sediments in the Sandanski Basin (Kojumdgieva et al. 1982; Nedjalkov et al. 1986; Zagorchev 1992) begin with marl, siltstone, sandstone and fine-grained conglomerate, containing several well-dated horizons of brackish-water clay of Badenian–Sarmatian age ($c.$ 13.6–16.3 Ma; Fig. 6). The next overlying unit, > 1 km thick, consists mainly of sandstone and siltstone with several thick beds of conglomerate. A prominent layer of limestone breccia emplaced in Pontian time occurs at the top of this unit, followed by a coarse

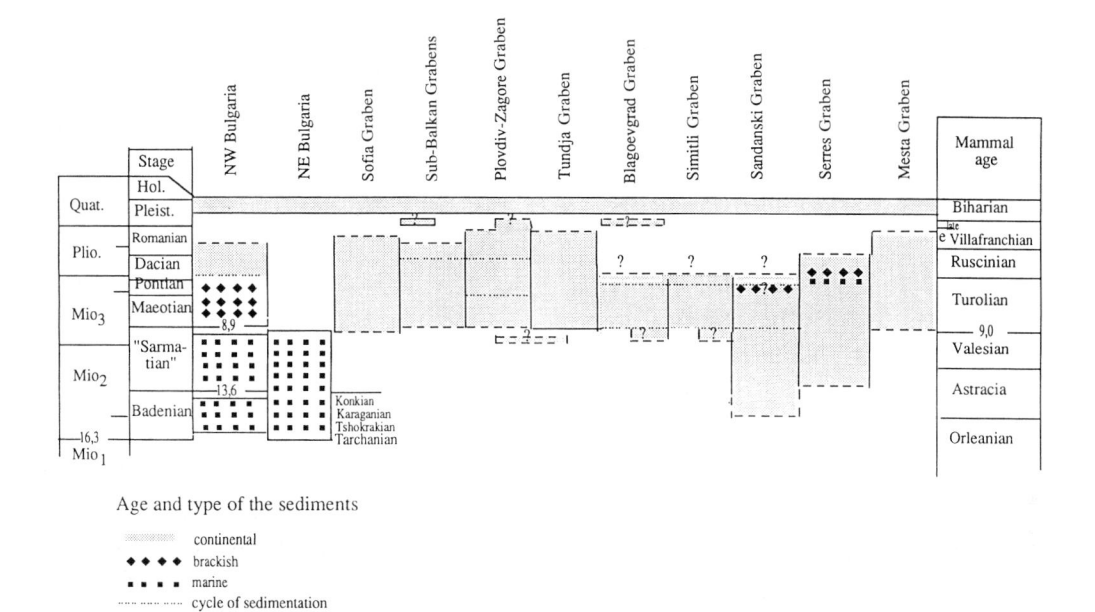

Fig. 6. Generalized time intervals and environments of deposition for some of the major Late Cenozoic graben systems in Bulgaria.

conglomerate unit with a sandstone matrix and containing large (up to 1 m) well-rounded, matrix-supported clasts of granite. Limestones similar to those within the breccia at the top of the second unit are not present directly east of the Sandanski Graben and probably have a source many kilometres to the east. This suggests that the west dipping normal faults that bound the east side of the Sandanski Basin have many kilometres of Middle–Late Miocene displacement. The youngest sediments in the basin are Early Dacian (i.e. Early Pliocene, c. 5 Ma; Fig. 6). Sediment thicknesses vary greatly, with the thickest and coarsest strata occurring along the east side of the basin. The maximum thickness is probably > 1700 m.

The Sandanski Basin is an east tilted half-graben. Sediments dip east 5–20° and are bounded on the east side by a fault of large displacement and on the west side by a fault of small displacement. In its southern part, the basin is bound by a gently (c. 15–20°) west dipping normal fault that juxtaposes Middle Miocene sediments in the hanging wall against the predominantly crystalline basement of the Rhodope Massif in the footwall. Slickensides along this fault surface, and to the south in Greece, indicate an east-northeast–west-southwest direction of extension, with the hanging wall moving down to the southwest. The northern part of the basin is bound on the east side by a steeper west dipping normal fault. This may be the northern continuation of the Melnik Fault that lies west of the basin-bounding, low-angle fault to the south; the Melnik Fault may have displaced the original low-angle fault along the northern part of the basin.

Strata similar to those found in the upper part of the Sandanski Basin occur as the basal units of the Simitli, Blagoevgrad and Djerman Basins to the north (Figs 3 and 8). They consist of conglomerate, sandstone and siltstone, with the coarsest grained rocks present along the eastern or southeastern sides of the basins. The strata range in age from Meotian to Early Dacian (perhaps c. 9–5 Ma). The eastern side of the Blagoevgrad Basin is marked by a prominent southwest dipping fault that is related to basin development, but the Simitli and Djerman Basins have been modified by younger east-northeast–west-southwest striking faults. The formation of these grabens indicates that extension in Bulgaria began in the Sandanski Graben and later migrated northward to form the Simitli, Blagoevgrad and Djerman Grabens.

The Sandanski Basin was the northern continuation of the Seres Basin (Fig. 7) in pre-Pliocene time and contains sedimentary rocks similar to the basinal sediments underlying the Strymon Valley in northern Greece (Dinter &

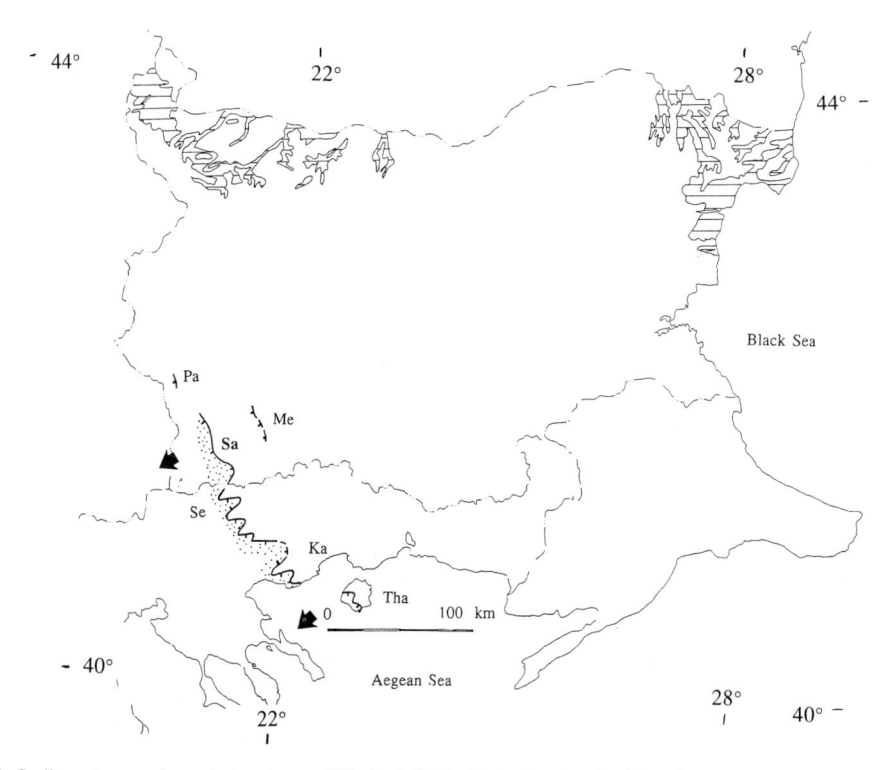

Fig. 7. Sedimentary rocks and structures of Early–Middle Badenian age (c. 15 Ma). The major structure that developed at this time was the southwest vergent Strymon Valley Detachment Fault, shown by the heavy ticked line. The detachment extends from southwestern Bulgaria southward through northern Greece to near Kavala (Ka) and reappears on the Island of Thasos (Tha). The supradetachment sediments were deposited in the Sandanski Basin (Sa) in Bulgaria and the Serres Basin (Se) and other areas in northern Greece, and are shown by dotted areas. This pattern of deposition and faulting formed from c. 16 to 10 Ma. Other areas where normal faulting of this age is suspected, but not proven, are in the Mesta (Me) and Padesh (Pa) Grabens. Sediments of the eastern Paratethys realm are shown by horizontal lines in northern Bulgaria and are not discussed here. Black arrows indicate the directions of extension. Regional shortening was within the Hellenides to the west within the Gavrovo and Ionian Zones.

Royden 1993; Kaufman 1995; Dinter 1998). Pliocene–Quaternary uplift has created an intervening, east–west trending mountain range along the Bulgarian–Greek border (Kojumdgieva *et al.* 1982; Nedjalkov *et al.* 1988). Late Miocene–Pliocene (c. 16–3 Ma) strata of the Seres Basin and along the Strymon Valley lie above the major west dipping, low-angle Strymon Valley detachment. The detachment continues into the northern Aegean where it is exposed on Thasos (Dinter *et al.* 1995; Wawrzenitz & Krohe 1998; Fig. 7). These data suggest that extension began first in northern Greece at 22–26 Ma, migrated northwards into the Sandanski Graben at c. 16 Ma, and into the Simitli, Blagoevgrad and Djerman Grabens at c. 9–5 Ma. Displacement on the graben-bounding faults appears to decrease from south to north, with c. 80 km of apparent displacement near the Aegean coast and c. 40 km near Mount Vrondou (see Dinter & Royden 1993; Kaufman 1995). Extension is unquantified but appears to be less in the Sandanski Graben and ends north of the Djerman Graben. The decrease of extension northward indicates an important rotational component to the extension with a pole of rotation not far north of the Djerman Graben, close to the area that was shortened in Early or Middle Miocene time (see above).

Other extensional structures in Bulgaria and northern Greece may be related to this period of extension. The northeast dipping normal faults at Mount Olympos in northern Greece (Fig. 5; Schermer 1993) were probably active contemporaneously with at least the older part of the Strymon Valley detachment fault, and younger faults east of Mount Olympos continued the activity to the present day. The oldest normal faults of the Olympos region were important in the early opening of the Thermaikos Basin. Geological maps show only northeast dipping thrust faults continuing northward from Mount Olympos into southern FYR Macedonia along the west side of the Vardar Zone, but Schermer's (1993) work suggests that these faults need to be re-examined. Some of them may be east dipping normal faults, so that important extension of Early Miocene–Pliocene age may be present in southern FYR Macedonia.

Within the Mesta Graben, located east of the Sandanski Graben, Upper Eocene–Oligocene volcanic and sedimentary rocks dip east very steeply and in places are vertical (Fig. 7). They are unconformably overlain by the gently folded Pontian–Dacian strata (c. 7–5 Ma) of the Gotse Delchev and Razlog Grabens. In the Padesh Graben, located northwest of the Sandanski Graben, similar Upper Eocene–Oligocene strata dip up to 50°E (Fig. 7). The steep dip of these Palaeogene strata suggest that they have been rotated in the hanging walls of gently west dipping normal faults, but the age of the faulting is unclear. The possibility that this faulting was related to Late Miocene–Pliocene extension in the Sandanski and Strymon Valley areas remains to be examined.

Deposition and active extension between c. 16 and c. 9 Ma were largely restricted to the Sandanski Graben (and its continuation into northern Greece), but at c. 9 Ma(?) sedimentation became widespread within southwest and central Bulgaria (Figs 6 and 8).

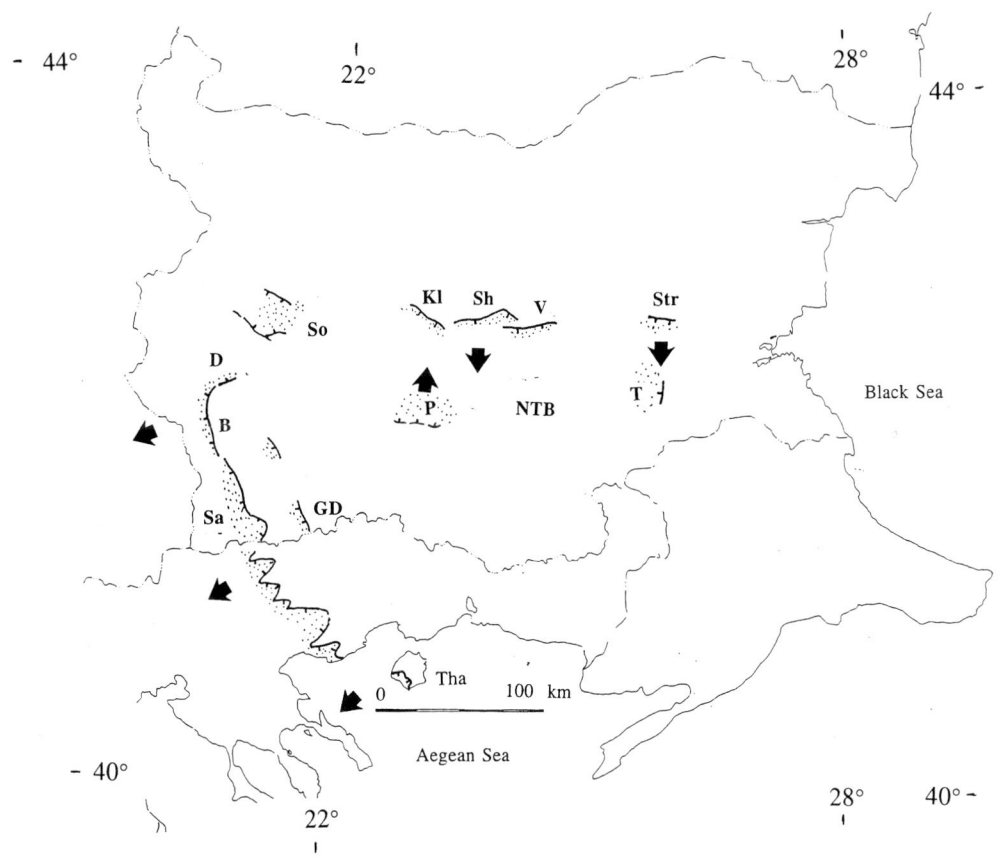

Fig. 8. Sedimentary rocks and structures of Late Sarmatian–Early Meotian age (c. 11–8 Ma). During this transitional time period, faulting continued in northern Greece on the Strymon Valley Detachment Fault from Thasos (Tha) through the northwest trending Sandanski (Sa) Graben and extended northward through the Blagoevgrad (B) and Djerman (D) Grabens. GD, Gotse Delchev Graben. East–west trending grabens began to form in the Sub-Balkan graben system [Vetren (V), Sheinovo (Sh) and Karlovo (KL) Graben], the eastern part of the Sofia (So) Graben, the Straldja (Str) Graben, and the Plovdiv (P) and Tundja (T) Grabens of the north Thracian Basin (NTB). The directions of extension are shown by black arrows. Regional convergence was occurring within the Ionian and Apulian Zones in the western Hellenides at this time.

Sofia Basin. The strata in the Sofia Basin form a single continental succession that ranges in age from Maeotian to Early Romanian (*c.* 9–4 Ma; Kamenov & Kojumdgieva 1983), but the basal unit is poorly dated and could be as old as Late Sarmatian (*c.* 10–11 Ma; Figs 6 and 8). The oldest strata are here regarded as Maeotian, and that they are related to faulting and basin subsidence because they form a basal conglomerate sequence that grades upward into overlying strata also dated as Maeotian. The present basin shape is the result of latest Pliocene–Recent faulting, and Upper Neogene basin sediments are unconformably overlain by Quaternary and Holocene strata. Because pre-Quaternary strata are largely restricted to the subsurface, and information on their thickness and sedimentary characteristics are mainly from borehole and seismic data, reconstruction of the Late Miocene–Pliocene tectonic history of the basin has been difficult. The basin has steep margins on the eastern and southwestern sides, a less steep margin on the northeast side and a gentle southeast-dipping slope on the northwestern side.

The oldest undated sediments are present only in the eastern part of the basin. They consist of locally deposited coarse conglomerate, suggesting that faulting was coeval with the initiation of basin subsidence. However, the trend of faults associated with these strata are unknown. There is weak evidence that north-northeast trending faults, with a strike-slip component, may have played a role in formation of a small, north–south extending, pull-apart basin at this time. Younger alluvial, fluvial and lacustrine strata of Meotian–Romanian age vary in thickness as the locus of subsidence shifted during deposition. The greatest thickness of Neogene strata is > 1300 m and occurs in the east-central part of the basin. The fine-grained character of the sediments, and the fact that local remnants of similar strata are found south of the basin, indicate that faulting along the present margins of the basin was not intense during latest Miocene and Pliocene times.

Strata in the eastern part of the Sofia Basin extend eastward into the Sarantsi Graben, where it can be shown that graben formation did not begin until Quaternary time. Thus, at least some of the sediments in the Sofia Basin were deposited over a broader region than the present graben and their present distribution is partly a result of Quaternary faulting. However, the strata that were deposited outside the modern Sofia Graben are much thinner than those in the graben, indicating that subsidence was greatest in the area of the graben. Miocene–Pliocene strata within the Sofia Basin, and the remnants of similar strata to the south, are unconformably overlain by latest Pliocene–earliest Quaternary (Villafranchian) conglomerate (Kamenov 1965; Popov 1968). This unconformity is widespread in Bulgaria and appears to separate sediments deposited during a time of weak(?) to moderate extension during Meotian–Romanian time from those deposited during intense normal faulting and topographic development during Quaternary time.

The Sub-Balkan graben system

The Sub-Balkan graben system consists of nine east–west trending grabens located along the southern boundary of the Stara Planina Mountains of central Bulgaria and forms the northern boundary to the Aegean extensional system (Figs 8–10). These grabens have been discussed in detail elsewhere (Tzankov *et al.* 1996) and only a brief summary of their important features is presented here.

Extension began in the central part of the Sub-Balkan system in the Karlıovo (Angelova *et al.* 1991*a*, *b*), Sheinovo and Vetren Grabens, where the oldest known strata are dated as Meotian–Pontian (Figs 6 and 8), and no older than *c.* 9 Ma. The basal strata in the grabens young to the west and east, with Dacian–Romanian and then Upper Pleistocene strata forming the basin sediments within the graben system. Graben fill consists of non-marine strata with conglomerate, sandstone, siltstone and lacustrine strata. These sediments dip north into south dipping basin margin faults. Older strata dip more steeply than younger strata and coarser grained rocks lie along the north side of the basins, demonstrating that sedimentation and faulting were contemporaneous. The graben fill is *c.* 600 m thick in the central Sheinovo Graben and becomes thinner (< 100 m) towards the east and west. In all grabens, Pleistocene strata lie disconformably above Pliocene rocks.

The Sub-Balkan grabens are complexly faulted (Tzankov *et al.* 1996) with faults on both flanks, but the main faults occur along the northern side. Master graben-forming faults are south dipping, low-angle normal faults (dipping *c.* 30° or less) which have been cut by steeper (60–80°) young to active normal faults at the base of the Stara Planina. A south dipping Eocene thrust fault lies below the basin-bounding, low-angle normal fault, perhaps influencing the position and dip of the normal faults. Thus, this is one area where pre-existing crustal

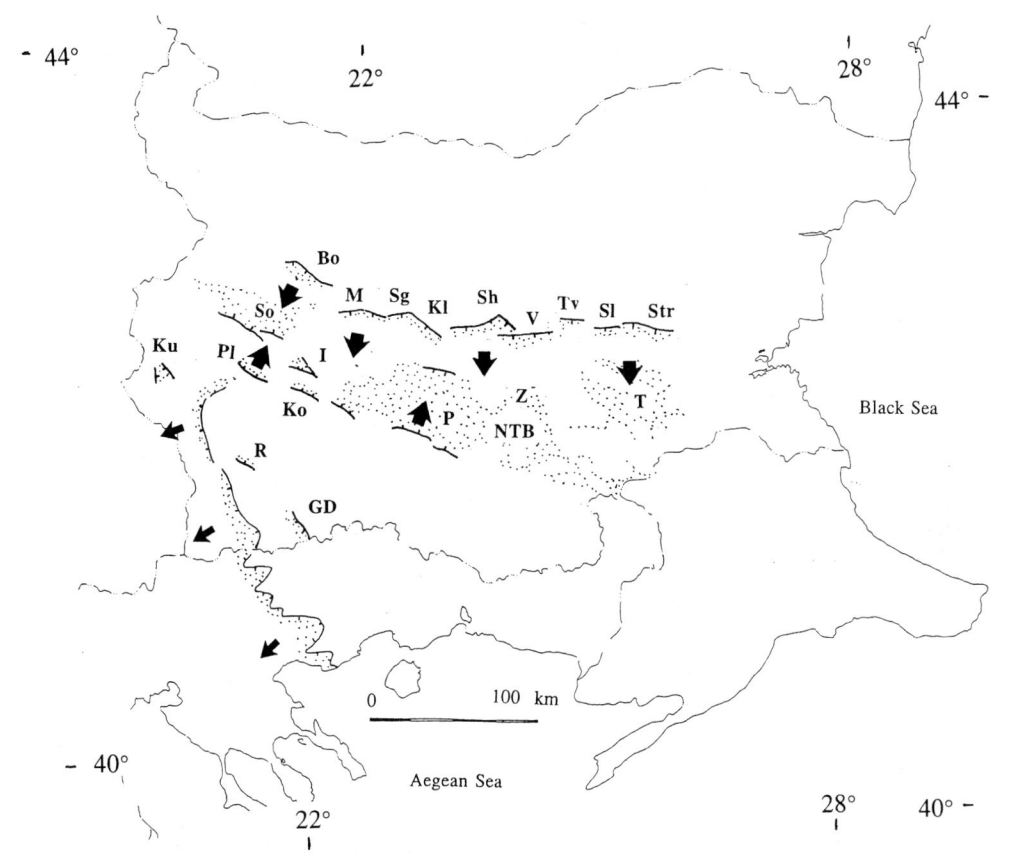

Fig. 9. Sedimentary rocks and structures of Pontian–Dacian (c. 7–3.5 Ma). Sedimentation and faulting expanded in central Bulgaria in the north Thracian Basin (NTB) – with local increased subsidence in the Plovdiv (P), Zagore (Z) and Tundja (T) areas – the Sub-Balkan graben system – KL, Karlovo; M, Mirkovo; Sg, Srednagorie; Sh, Sheinovo; Sl, Sliven; Str, Straldja; Tv, Tvarditsa; V, Vetren – the Botevgrad (Bo), Ihtiman (I), Kostenets (Ko), Kustendil (Ku), Palakaria (Pl) and Sofia (So) Grabens, and in southwest Bulgaria – GD, Gotse Delchev; R, Razlog. The northwest trending grabens in southwestern Bulgaria ended their activity in the north at c. 5 Ma and in northern Greece by c. 3.5 Ma. Heavy ticked lines are normal faults and black arrows indicate the directions of extension. Coeval shortening occurred in the Apulian Zone within western Greece.

anisotropy may have played a role in the localization of the extensional faults.

East of the Sliven Graben is the Straldja Graben, which Tzankov et al. (1996) did not include in the Sub-Balkan graben system (Fig. 8). Strata in the Straldja Graben are claystone, siltstone, sandstone and conglomerate, similar in character to rocks dated as Meotian–Pontian (9–5 Ma), Dacian, Romanian and Early Pleistocene in the Vetren Graben, and similar to rocks in the Tundja Graben located to the south within the northern Thracian Basin. The strata of the Straldja Graben dip north and are thicker to the north (c. 300 m), indicating that sedimentation and faulting were contemporaneous. Three other north dipping en echelon half-grabens – the Sungurlare, Karnobat-Itos and Prosenik Grabens (Fig. 3) – are present east of the Straldja Graben. Strata in these grabens are generally c. 100 m thick and poorly known. Based on the similarity of their sequences to other well-dated rocks, they are inferred to be latest Pliocene and Quaternary [data from Kanchev discussed in Yovchev 1971]. The age of these grabens is mainly Quaternary but faulting may have begun in latest Pliocene time.

In summary, during Meotian–Late Pliocene time, the northern boundary of the Aegean extensional system in central Bulgaria is marked by a series of east–west trending grabens whose initiation was no older than 9 Ma (and perhaps no older than c. 6.5 Ma). Not all these grabens formed at the same time and some (Sarantzi, Kamartsi, Mirkovo, Sliven,

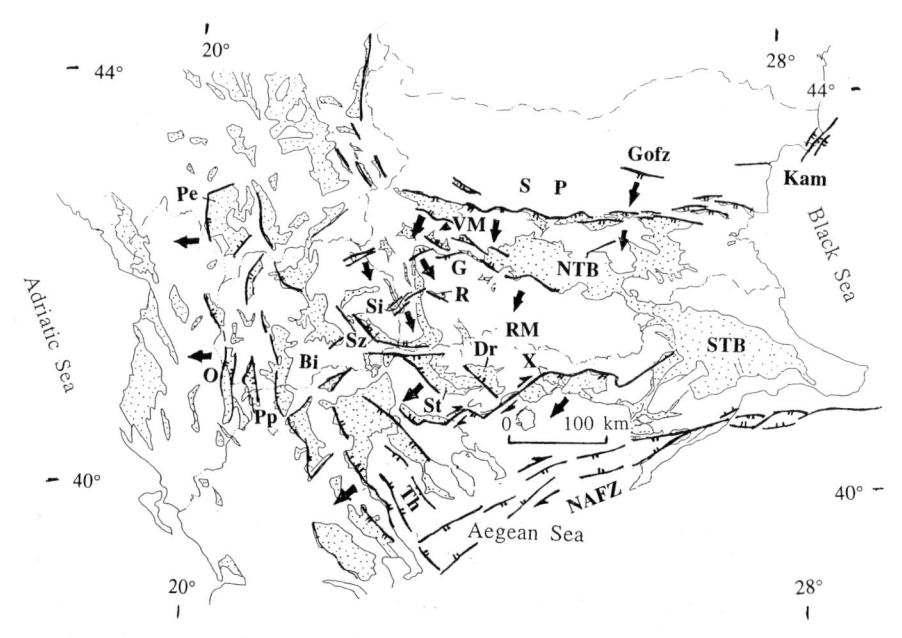

Fig. 10. Sedimentary rocks and structures of Pleistocene age. The area has expanded to include FYR Macedonia and Albania based on the topography of basins with Pleistocene deposition. Bi, Bitola; Dr, Drama; G, Govedartsi; GOfz, Gorna Orijahovista fault zone; Kam, Kamchia River; NATF, North Anatolian Fault; O, Ohrid; Pe, Pec; Pp, Prespansko; R, Razlog; RM, Rhodope Mountains; SP, Stara Planina Mountains; Si, Simitli; St, Strymon; Sz, Stumitza; Th, Thermaikos; VM, Vitosha Mountain; X, Xanthi–Komotini.

Sungurlare, Karnobat-Aitos and Prosenik Grabens) are only as old as latest Pliocene–earliest Pleistocene. Some of the grabens (Sungurlare, Sheinovo, Vetren, Tvarditsa and Straldja Grabens) contain more steeply dipping Eocene and Oligocene strata that are unconformable below Meotian–Pliocene sediments. These Palaeogene rocks represent remnants of more extensive sedimentary rocks whose present distribution is controlled by extensional faulting and erosion prior to Late Neogene graben formation.

The Northern Thracian Basin. The northern Thracian Basin covers a large area of east-central Bulgaria. It is irregularly and unconformably underlain by an older sequence of rocks ranging from Middle Eocene to Early Miocene in age. Below the southern part of the basin these strata locally contain abundant volcanic rocks and form the northern volcaniclastic edge of the Palaeogene magmatic arc; northward these strata are dominated by sedimentary rocks containing less volcanic material.

Middle Eocene–Lower Miocene strata are overlain unconformably or disconformably by strata of Middle(?) or Late Miocene to Recent age. Deposition of this younger succession was coeval with Late Neogene extension. Although the present shape of the northern Thacian Basin is roughly east–west, the depocentres for Upper Miocene and Pliocene strata are more equidimensional and shift through time. Three depocentres can be identified: an older one near Plovdiv, and two younger ones near Zagore and Tundja (Figs 3 and 8). Shallow areas separating these depocentres are either covered by thin basin sediments or unconformably overlain by Quaternary strata. The northern Thracian Basin is underlain by a complex Neogene graben structure which was first described by Bonchev & Bakalov (1928), but the age of the individual faults is difficult to determine. Quaternary and active faults are morphologically expressed, but identification of Late Neogene faults is more difficult because the evidence is mainly subsurface. Geophysical studies and data from wells delineate a complex pattern of subsurface faults that cut both the Palaeogene and Neogene sedimentary rocks.

There are two major successions of Cenozoic strata in the northern Thracian Basin, separated by an unconformity. The oldest succession, described above, is preserved in different parts of the basin and ranges from Eocene to earliest Middle Miocene in age. The younger succession is Meotian–Pontian to earliest Pleistocene, although there are rare remnants of strata that lie unconformably between the two successions. These remnant strata consist of sandstone, siltstone and rare gravel conglomerate in the Plovdiv depocentre (Dragomanov et al. 1988), and conglomerate in the Zagore depocentre (Nenov & Dragomanov 1987). They have not yielded fossils but lie between dated earliest Middle Miocene and Maeotian rocks, and thus must be of late Middle–early Late Miocene age, i.e. 'Sarmatian' (13.6–9.0 Ma).

The youngest succession of strata in the northern Thracian Basin is assigned to the Ahmatovo Formation of Maeotian–Pontian and (?)earliest Pleistocene age (Kojumdgieva & Dragomanov 1979; Dragomanov et al. 1981, 1984). The Ahmatovo Formation consists mainly of sandstone and conglomerate. The oldest part of the formation has a depocentre near Plovdiv and the youngest part of the formation has a depocentre near Zagore (Figs 8 and 9). These strata were deposited in alluvial, fluvial and locally marsh environments; rivers flowed to the east in the southern part of the basin but to the west in the northern part (Dragomanov et al. 1981).

The Ahmatovo Formation is thickest, c. 280–300 m, near Plovdiv (Dragomanov et al. 1981) and c. 350 m thick near Haskovo (Dragomanov et al. 1984), but is generally < 100 m thick. The shapes of the depocentres are not well defined but appear to be equidimensional rather than elongate. The sedimentation realm of the northern Thracian Basin expanded aerially with time, so that younger strata in the Ahmatovo Formation are exposed beyond the margins of the thicker basal deposits and rest unconformably on metamorphic rocks of the Rhodope and Sredna Gora Mountains to the south and north, respectively, and on Cretaceous and older rocks to the east. Younger strata of the Ahmatovo Formation appear to have been connected through the Zagore Basin to the eastern basins of the Sub-Balkan grabens. These younger strata define the greatest extent of Maeotian–Late Pliocene sedimentation in the northern Thracian Basin.

Structures related to Maeotian–Romanian deposition are difficult to identify. Subsurface data from the Tundja Basin (eastern part of the northern Thracian Basin; Fig. 8) suggest that the eastern boundary of thick sediments lies along a north–south trending fault with thin basinal sediments to the east. Data from drill holes suggest that the Elhovo Formation [described by Kojumdgieva et al. (1984)], equivalent to the Ahmatovo Formation, which consists of alluvial, fluvial and locally lacustrine claystone, sandstone and rare conglomerate with local lignite, diatomite and limestone, thickens eastward toward a north–south trending feature that can be interpreted as a synsedimentary fault. The formation, 150–200 m thick but locally reaching 300 m, is Maeotian–Pliocene in age (9–5 Ma). However, recently Angelova et al. (1991a, b) have suggested that the uppermost limestone (their Prustenic Formation) is Eopleistocene (upper Villafranchian). Displacement on faults appears to have been essentially vertical with little tilting of strata.

Although subsurface information suggests that the basin sediments are complexly faulted, it is difficult to determine the age of the faults. Along the northern and western margins of the basin the uppermost strata of the Ahmatovo Formation unconformably onlap basement rocks. Along the gentle slopes that dip into the basin there are morphological steps that step down to the basin, but it is not clear if these steps are bounded by faults and, if they are, whether the faults are Late Neogene or Quaternary in age. Along the southern margin of the basin, the uppermost part of the Ahmatovo Formation onlaps basement on the northern slopes of the Rhodope Mountains and is cut by basin-bounding faults of Quaternary age. Whether some of these faults were active during Late Neogene time is difficult to prove, but because only the uppermost parts of the Ahmatovo Formation onlap the basement south of these faults, and because the sedimentary sections are thicker and more complete to the north, some of these east–west trending faults were probably synsedimentary. The equidimensional shape of the basins suggests that north–south trending faults may have also played a role in basin subsidence.

The connection of the Thracian Basin with the Sub-Balkan grabens – where synsedimentary, east–west trending faults are proven – suggest that subsidence of the northern Thracian Basin may have occurred during north–south extension during Late Neogene time. Because the sedimentary sequence of the central Sub-Balkan grabens is not similar to rocks in northwestern part of the Thracian Basin, it is likely that the east–west trending Sredna Gora already uplifted and separated the two depositional realms. It is important to note that the eastern margin of the Thracian Basin lies just east of the present Tundja River and that it did not extend to the Black Sea. In fact, it appears that the Late Neogene, and even Quaternary, tectonics of Bulgaria are little affected by events in the Black Sea, except along the coastal parts of Bulgaria.

Most of the sediments in the northern Thracian Basin are fine grained and deposited in fluvial, lacustrine and marsh environments; locally the sequence contains coal and diatomite. These sediments suggest that subsidence was slow, that there was little development of high relief and that the rate of faulting that controlled basin formation was slow. Only along its northern border, where the Thracian Basin merges with the eastern part of the Sub-Balkan graben system, does deposition of coarse-grained sediments suggest faulting at a higher rate and development of high relief. The magnitude of extension within the northern Thracian Basin during Maeotian–Romanian time was probably small because sediment thicknesses of this age are less than $c. < 400$ m, and in many places $< c.$ 200 m. This contrasts with the Quaternary sedimentation which indicates that more active tectonism was associated with Quaternary basin subsidence. In latest Pliocene–earliest Pleistocene time, the older strata were displaced and locally eroded, so that they crop out on ridges separating the modern rivers. Everywhere an erosional unconformity separates Quaternary rocks from the underlying Ahmatovo (and Elhovo) Formation.

The formation of the grabens in central Bulgaria, the northern Thracian Basin, the Sub-Balkan grabens and the en echelon grabens to their east are partly contemporaneous with the northwest trending grabens of southwest Bulgaria, being coeval with the youngest part of the Sandanski Graben, and about the same age as its northern extension in the Simitli, Blagoevgrad and Djerman Grabens. The younger development of the central Bulgarian grabens is also coeval with formation of the Gotse Delchev and Razlog Grabens (Figs 8 and 9). The youngest faulting and basin formation in these central Bulgarian grabens outlasts the development of the northwest trending grabens of southwest Bulgaria which appears to end at $c.$ 5 Ma.

Basins between the northern Thracian and Sofia Basins. Between the Sofia and Thracian Basins are several small but important basins that generally trend east–west (Figs 8 and 9). These basins formed in Quaternary time, but also contain sediments of Late Neogene age that are unconformable below Villafranchian strata. In the Palakaria Graben (Fig. 9) there are $c.$ 500 m of Late Miocene, Pontian–Dacian to (?)Romanian strata consisting of conglomerate, sandstone and claystone with a middle unit containing coal and diatomite (Antimova & Kojumdgieva 1991). Similar strata are present outside the graben and are cut by the Pliocene–Quaternary graben-forming faults (Bojadjiev & Sapoundjieva 1960). The Ihtiman Graben (Fig. 9) contains $c.$ 200 m of probable Upper Neogene sediments unconformable below Quaternary strata, but little is known of the sequence or its age. The Kostenets Graben (Fig. 9) contains $c.$ 400–500 m of coal-bearing strata. Here, Antimova & Kojumdjieva (1991) described three lithostratigraphic units of Pliocene age that may correlate with similar strata in the Palakaria Graben. All of these grabens were faulted during and after Neogene deposition, subjected to erosion and faulted again during deposition of Quaternary strata. Although the distribution of the pre-Quaternary strata remains uncertain, their lithologic similarities suggest that they were deposited in an originally widespread, larger sedimentary realm, and that this thin sedimentary sequence has been preserved by Quaternary faulting and erosion.

The sub-Quaternary erosion surface

Throughout much of central Bulgaria, important erosion surfaces are preserved over large areas and help to bracket tectonic events. The youngest, and apparently most widespread, surface occurs around the Palakaria, Ihtiman and Kostenets Grabens in the region of the western Sredna Gora. A similar surface is present in many areas of Bulgaria, such as in the western and central Stara Planina, west of the Sofia Graben, the northern and eastern Rhodope Mountains, in Sakar, and in the low hills west of the Trundja River. In all of these areas the surface is locally overlain by a Villafranchian fluvial conglomerate. The surface displays low relief and in many places is cut on Maeotian–Pliocene sedimentary rocks. It is marked by an unconformity or disconformity between Upper Pliocene and Lower Quaternary strata. This surface will be referred to here as the sub-Quaternary surface. The surface completed its development in Late Pliocene or earliest Quaternary time. In some areas, where it is cut on pre-Miocene rocks, its formation may have begun with Miocene erosion.

Although the history of formation of the sub-Quaternary surface is unclear, the surface can be used to decipher the age of faulting in many areas. For example, in the Kostenets Basin (Fig. 9), Upper Miocene–Pliocene rocks are tilted up to 30°S along the north dipping fault that forms the northern border of the Rhodope Mountains.

In this area the sub-Quaternary erosion surface can followed from the metamorphic rocks of the Sredna Gora southward across the tilted Pliocene strata of the basin. This indicates that the fault along the south side of the Kostentz Basin was active in Late Miocene–Pliocene time; the presence of the surface high in the Rhodopian Mountains indicates significant Quaternary displacement at the margin of the mountains. This sub-Quaternary surface, and the associated unconformity, are present throughout the northern Thracian Basin where Upper Quaternary strata overlie Palaeogene and Neogene strata of different ages, and clearly indicates that faulting took place in latest Pliocene time.

Quaternary–Recent basin and tectonic development in Bulgaria

Quaternary–Recent faulting was more intense than during Late Neogene time and much of the present topography of Bulgaria is an expression of Quaternary extensional tectonism (Tzankov et al. 1996). Morphologically, the sub-Quaternary erosion surface forms broad valleys and gently rounded, low-relief hills and low mountains, suggesting that Late Neogene extension was generally of insufficient magnitude to create high relief or to severely disrupt the sub-Quaternary surface (except locally). In contrast, this surface was strongly disrupted by Quaternary faulting, producing relief of > 1.0–1.5 km in many places. It is because of the greater magnitude of Quaternary faulting that the Late Neogene evolution of Bulgaria has been difficult to determine.

Important Quaternary faults occur along the northern and southern boundaries of the northern Thracian Basin, the Sub-Balkan graben system, the Sofia Basin, the Strumitza, Simitli, Razlog, Djerman and Kustendil Grabens of southwest Bulgaria, and the Drama, Strymon and other basins of northern Greece, and bound numerous small basins north and south of the Sofia Basin (such as the Botevgrad, Palakaria, Govedartsi, Ihtiman and Kostenets Grabens; Figs 3 and 10). The general trend of these basins is east–west, but the faults form a complex and intersecting pattern of east–west, west–northwest and northwest trends. Faults in the Razlog and Gotse Delchev Grabens (Fig. 9) trend more northwesterly than faults in the older Mesta Graben. In the southwestern part of Bulgaria, east–west or west–southwest trending faults bound the Govedartsi, Djerman and Kamenitsa Grabens, and similar trending faults along the Simitli and Stumeshnitsa Grabens, cut older northwest-trending grabens along the Struma River (cf. Figs 8 and 10). The westernmost of these faults continue into eastern FYR Macedonia. Along the border with Greece, east–west trending faults separate the southern end of the Sandanski Graben from the Siderokastro Basin of northern Greece.

At least one, and possibly three, east–west trending faults occur north of the Stara Planina. In north-central Bulgaria, seismic activity occurs along the Gorna Orijahovista Fault, but is difficult to interpret (Figs 10 and 11). Van Eck & Stoyanov (1996) suggest that the focal mechanisms indicate north-northwest–south-southeast extension. This fault is marked topographically by an east–west trending valley and disrupts north flowing rivers. There is no evidence that the faults north of the Stara Planina have a Late Neogene history; it is suggested here that they are very young and represent the northernmost migration of normal faulting into the Moesian crust. Further east, landscape morphology along the river valley near and west of Varna and along the Kamchia River suggest the presence of east–west trending faults (Fig. 11). In the offshore Black Sea, east of the Kamchia River mouth, the east–west trending Kamchia Depression, marked by the Bliznatizi flexure Fault Zone, contains a thick (300–400 m) section of Neogene sedimentary rocks (Dachev et al. 1988). These faults appear to have been active in Late Neogene time.

Many Quaternary faults show evidence for active displacement and are marked by fault scarps, triangular facets, steep topographic gradients and active alluvial fan development (Fig. 11). This includes many of the faults along the Sub-Balkan grabens, along the southwest side of the Sofia Basin, some faults along the southern side of the northern Thracian Basin, along the Strumeshnitsa, Simitli and Razlog Grabens, and along the southwestern side of the Djerman Graben. Not all basins with Quaternary sediments appear to be bound by active faults, although many of the faults may have a subtle morphological expression and a slow slip rate, and thus do not show clear evidence for active faulting.

In west-central Bulgaria, the sub-Quaternary erosion surface is displaced by many Quaternary faults, which are well developed around the Sofia Graben where the surface is present north and south of the graben at an elevation of c. 1000–1200 m, while the floor of the Sofia Graben is at c. 500 m above sea level (asl). The Iskar River cuts into the sub-Quaternary surface and forms a deep canyon south of the graben. The river has a knick point at Iskar Lake, to the

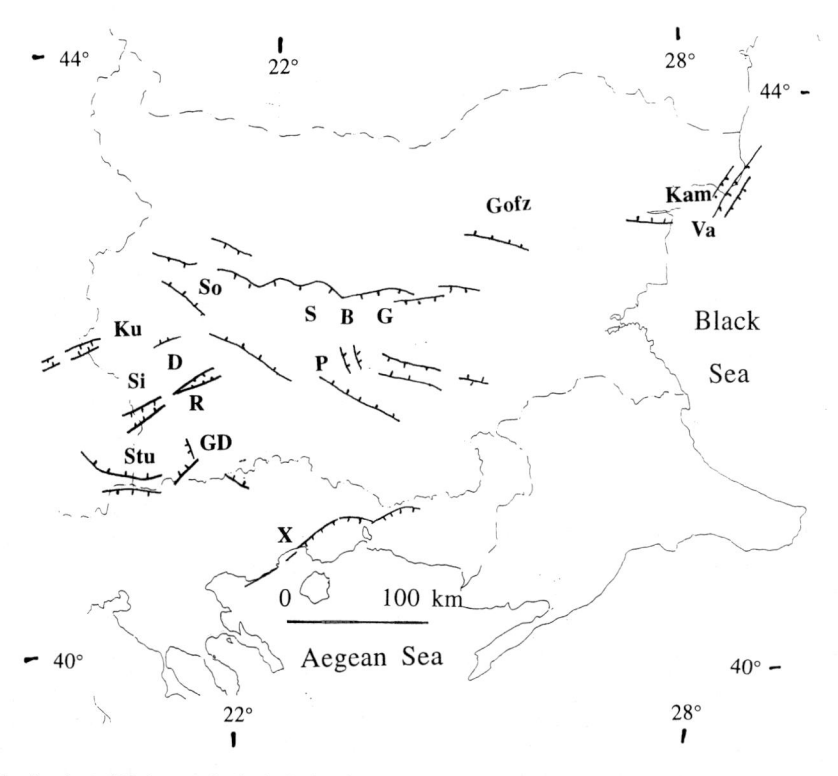

Fig. 11. Distribution of Holocene faults in Bulgaria and adjacent areas. D, Djerman; GD, Gotse Delchev; GOfz, Gorna Orijahovista fault zone; Kam, Kamchia River; Ku, Kustendil; P, Plovdiv; R, Razlog; SBG, Sub-Balkan graben system; Si, Simitli; So, Sofia; Stu, Strumeshnitsa; Va, Varna; X, Xanthi–Komotini.

south of which the river and its tributaries flow on the sub-Quaternary surface. The Iskar River flows north across the Sofia Graben and cuts a gorge through the western Stara Planina which is capped by the sub-Quaternary surface to reach the Danube River at an elevation of c. 100 m asl. These relations indicate that the Iskar River is antecedent to the Sofia Graben and that the relative uplift of the mountains on either side is the result of Quaternary faulting. Because the sub-Quaternary surface is present around the western, southern and eastern flanks of the 2290 m high Vitosha Mountain south of Sofia (Fig. 10), this peak appears to have been a peak rising above the sub-Quaternary erosion surface.

South of Vitosha Mountain, the sub-Quaternary erosion surface slopes gently south to the steep northern slope of the Rhodope Mountains. Locally, the surface dips gently to the northeast and forms a broad valley. These gentle slopes are interrupted by the Djerman, Palakaria, Govedardtsi, Kostenets and Ihtiman Grabens, indicating that these grabens (Figs 3 and 10) were formed after the development of the sub-Quaternary surface and that their bounding faults are of Quaternary age. It also suggests that some, or perhaps most, of the relative vertical movement of the northern Rhodope Mountains is of Quaternary age. Remnants of an erosion surface, or surfaces, are present on the Rhodope Mountains, but the age of these surfaces has been difficult to determine because of extensive erosion.

A similar erosion surface, of perhaps the same age, can be traced from south of Vitosha Mountain westward into the headwaters of the Struma River, where the river has been incised several hundred metres into the surface. This surface defines a broad valley that is cut on bedrock and locally, west of the Rila Mountains, on the Pliocene Djerman Formation. This surface cuts across the eastern boundary fault of the Blagoevgrad Graben and can be traced high onto the western slope of the Rila Mountains. The eastern boundary fault of the Blagoevgrad Graben does not displace the erosion surface, indicating that the graben-bounding fault did not have Quaternary movement. Much of the present topographic relief along the northern part of the Struma River is due to downcutting of the river into this surface.

Further south, where the Struma River cuts through the Sandanski Graben, is a series of terraces along the east side of the river valley (Zagorchev 1992). The higher terraces are cut on bedrock of the Pirin Mountains and the lower terraces are cut on sediments of the Sandanski Graben and are not displaced by its graben-bounding faults. The highest terrace occurs at an elevation of 1900–2200 m asl. Zagorchev (1992) assigned the terrace to probable Late Miocene age and showed that there has been c. 3 km of vertical separation along the eastern side of the Sandanski Graben during its formation. This also suggests that development of the sub-Quaternary erosion surface may have begun in Late Miocene time.

South of Plovdiv, at least three surfaces are present on the northern slope of the Rhodope Mountains, at elevations of 500, 700 and > 1000 m asl. The lower surface is cut on coarse-grained deposits of the marginal facies of the Pliocene Ahmatovo Formation and is overlain by Villafranchian conglomerate. This is clearly the sub-Quaternary surface, elevated by Quaternary faulting, and it is cut by the small east–west trending Biaga Graben. There is no information concerning the age of the two higher surfaces. The presence of these surfaces, and those of the Rila Mountains, suggest there may be erosion surfaces of more than one age. More detailed study is necessary to be sure which of the surfaces at different elevations may be the same surface displaced by faulting.

Data from the Quaternary faults suggest general north–south extension in Bulgaria. The east–west trending faults show only evidence for dip-slip movement and there is little or no field data to support strike-slip displacement on most of the faults. Some faults of northeastern and northwestern strike show evidence for strike-slip and others can be interpreted to have a strike-slip component. For example, the northwestern striking fault along the southwestern side of the Razlog Graben shows evidence for right-lateral displacement and the northeastern striking fault along the Simitli Graben shows evidence for left-lateral strike-slip (Fig. 10). The Palakaria Graben ends abruptly at both ends against north-northeastern striking faults, and the eastern fault displaces Pliocene sediments and thus has Quaternary movement. Both faults should have a strike-slip component as they are nearly perpendicular to the north–south extension that formed the graben. Strike-slip movement on these faults is consistent with the fact they have little surface expression because of a lack of a vertical component. The fault that trends southwest from Kustendil into FYR Macedonia (Fig. 11) may be interpreted to have left-lateral displacement because it connects two extensional basins in the appropriate positions to be pull-apart structures. A similar relationship is present along the fault that bounds the southern margin of the Strumeshnitsa Graben, where it trends west into the northwest tending Strumitsa Graben in southeastern FYR Macedonia (Fig. 10); here the fault is interpreted to have a right-lateral component. These strike-slip faults are mainly in southwestern Bulgaria and most of them cut obliquely across the older late Neogene northwest trending grabens. This suggests an important tectonic change in this part of Bulgaria during Quaternary time; similar changes in tectonic regime occur in northern Greece at about the same time (see below), marking an important reorganization of extensional features within the northern Aegean extensional regime.

Studies of earthquake focal mechanisms also indicate that active tectonism is dominated by north–south extension, with rare indications for strike-slip (Van Eck & Stoyanov 1996). The general pattern of active faulting is similar to that based on geological data for Quaternary time (cf. Figs 10 and 11). Zones of seismic activity trend east–west along the southern margin of the northern Thracian Basin and in several short zones north of the Stara Planina (along the Gorna Orijahovista Fault, near Varna, and along the Kamchia River). Although there is a region of complex seismicity in southwestern Bulgaria, near the Simitli, Razlog and Gotse Delchev Grabens, there is a surprising lack of seismicity along the Sub-Balkan graben system of central Bulgaria, where numerous young scarps indicate active faulting.

Quaternary tectonic activity in northern Greece: possible extension into FYR Macedonia

In northern Greece, north–south extension is present but is associated with other directions of extension and strike-slip displacements. East–west trending faults bound the southern flank of the Strumeshnitsa Basin and the northern end of the Strymon Basin (Figs 10 and 11). The northern of the two faults trends west into the northwest trending extensional Strumitsa Basin in southeastern FYR Macedonia, a geometry which suggests that the Strumitza Basin is a pull-apart basin and that the east–west trending fault may have a component of right-lateral strike-slip.

Dinter & Royden (1993) have shown that the youngest strata involved in the west dipping Strymon Valley detachment system are c. 3.5 Ma old. Since that time, a new set of northwest trending and northeast dipping normal faults has displaced the detachment and supradetachment sediments, and formed the boundary faults for the modern Strymon and Drama Basins (Figs 3 and 10). Vertical displacement on these faults reaches at least 3.5 km (Dinter 1998).

The Drama and Strymon Basins are part of a series of extensional basins bounded by Quaternary northwest trending faults in northern Greece and the northern Aegean Sea (Thermaikos, and the parallel Kassandreias, Siggitikos and Strymonikos Basins to the east; Fig. 10). Associated with these basins are east-northeast trending faults, at least two of which can be shown to have right-lateral displacements (Dinter 1998). Further east, a similar pattern of Pliocene–Quaternary faulting was

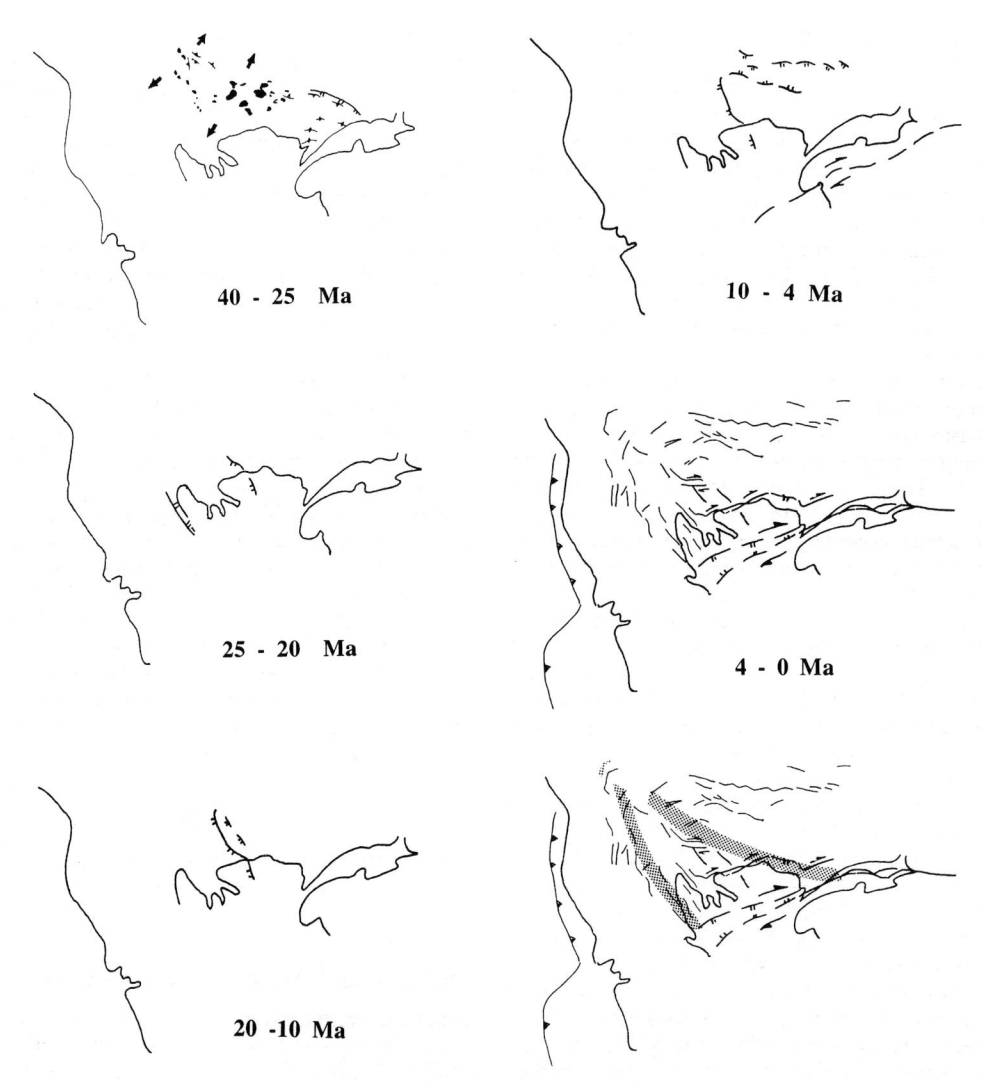

Fig. 12. Summary of the evolution of extensional faulting in the southern Balkan extensional regime discussed in this paper. The bottom right-hand figure shows the three areas of different structural regions active during Pliocene–Quaternary time.

suggested by Koukouvelas & Doutsos (1990) for the northeast trending Xanthi–Komotini Fault Zone. Here, northeast trending faults have both dip-slip and strike-slip components, and have segments that are connected by northwest trending normal faults or terminate basins bounded by northwest trending normal faults. All of these northeast trending faults parallel the western strands of the North Anatolian Fault within the Aegean Sea.

Seismic activity along the North Anatolian Fault shows dominantly right-lateral displacement (Jackson & McKenzie 1988). In northern Turkey, recent Global Positioning System (GPS) measurements demonstrate 22 ± 3 mm a^{-1} right slip across the east-northeast trending North Anatolian Fault (Reilinger et al. 1997; Straub et al. 1997). Within the northern Aegean are a series of subsea basins that lie within and parallel to the North Anatolian Fault, which indicate there is an important component of extension normal to the fault zone. Inferred here is that Pliocene–Quaternary faults in northern Greece, which parallel the North Anatolian Fault, may have both normal and dextral components.

It is suggested that dip-slip, southwest–northeast extension on northwest–southeast tending normal faults and right-slip on northeast tending faults has led to the development of the present basin morphology in northern Greece (beginning probably in latest Pliocene–Quaternary time; Dinter & Royden 1993). This intersecting pattern of faults and basins forms an irregular tectonic boundary near the Bulgarian–Greek border, of which the Xanthi–Komontini Fault and the faults along the northern and southern sides of the Belasitsa (Kerkini) Mountains are the most prominent examples (Fig. 12). (A similar pattern, but less well developed, may extend through the Mesta and Kustendil Valleys.) This fault pattern appears to have formed in latest Pliocene–Quaternary time and separated the north–south extensional regime of Bulgaria from the east-northeast extensional regime of northern Greece.

Thus, during latest Pliocene and Quaternary time Bulgaria and northern Greece appear to have been affected by somewhat different extensional regimes, involving general north–south extension in Bulgaria and both north–south and northeast extension in northern Greece. A complex zone of strike-slip and extensional faults in southwest Bulgaria may form a diffuse boundary between the two regimes. Although reconnaissance observations are from FYR Macedonia only, it appears that the Pliocene–Quaternary extensional regime of northern Greece is present in southeastern FYR Macedonia (along the Strumitza Graben). It is not yet known if older northwest trending extensional faults (similar to the Strymon, Struma and Olympos Fault Systems) are present in central FYR Macedonia, or if they were overprinted by the Pliocene–Quaternary regime of normal and strike-slip faults of northern Greece.

It is speculated that a third extensional regime is present in southwestern FYR Macedonia and eastern Albania. Here, north–south trending grabens, such as the Ohrid and Prespanski Grabens, are suggestive of east–west extension (Figs 3 and 10). The morphological expression of these grabens suggests that they are active, and were active during Quaternary time. Seismic activity in this area also supports east–west extension (McKenzie 1972; Jackson & McKenzie 1988). It is the authors' opinion that this zone of east–west extension may have been present since the initial stages of extension in Middle–Late Miocene time, and has not undergone tectonic reorganization in Late Pliocene–Quaternary time. The transition between these three regimes in central FYR Macedonia remains unresolved and Fig. 12 suggests diffuse boundaries between them.

Evidence from seismicity and GPS studies in northern Greece suggest that Holocene deformation is different from the Pliocene–Quaternary deformation. Focal mechanisms for earthquakes in northern Greece indicate a dominance of north–south extension with only rare evidence for right-lateral strike-slip faulting in the region north of the North Anatolian Fault (Jackson & McKenzie 1988). Recent GPS studies by McClusky et al. (2000) show slow southward velocities, relative to stations near the Greek–Bulgarian border, for stations in the southern part of northern mainland Greece. These data suggest that active deformation in northern Greece is mainly north–south extension. The Pliocene–Quaternary pattern of deformation probably began c. 3.5 Ma ago (Dinter & Royden 1993), thus the active pattern of deformation is younger and must have developed in Late Quaternary time.

Summary of Miocene–Recent tectonism

From the analysis presented above, the following summary of events in Bulgaria and northern Greece, including speculation about events in FYR Macedonia and eastern Albania, is suggested (Fig. 12). Following the Eocene–Oligocene volcanism and related arc-normal extension, a new extensional system appears have begun between 26 and 21 Ma in northern Greece, involving east-northeast–west-southwest extension east of Mount Olympos, on the Island of Thasos and near Kavala (Fig. 12). Extension migrated northward into southwestern Bulgaria with extension in the Sandanski Graben (and perhaps also into the Mesta and Padesh Grabens) by 16.3–13.6 Ma, and in the Blagoevgrad and Djerman Grabens by c. 9 Ma. Extension in southwest Bulgaria was reorganized by c. 5 Ma and in northern Greece by c. 3.5 Ma. It is speculated that related extension may also have occurred during this time in FYR Macedonia and eastern Albania.

North–south extension along east–west trending faults in central Bulgaria began only after extension was well underway in northern Greece and the Sandanski Graben of southwest Bulgaria. Within the Sofia Graben, the Sub-Balkan grabens, and grabens to their east, north–south extension began at c. 9 Ma, and may have begun about the same time in the Plovdiv, Zagore and Tundja Grabens of the northern Thracian Basin. Associated with extension are north-northeast and north–south trending faults that may have had a strike-slip component, but have only local topographic expression. North–south extension has continued to the present in these grabens.

By Late Pliocene time a major erosion surface, the sub-Quaternary surface, was developed over a large area of central Bulgaria, creating a major unconformity that marks the beginning of Quaternary deposition in the basinal areas. Many large and small graben-bounding faults in west-central Bulgaria displace this erosion surface and demonstrate the widespread extent of Quaternary north–south extension. North–south extension extended westwards, with decreasing magnitude, across the older northwest trending graben of southwest Bulgaria (the Simitli and Djerman Grabens) and into eastern FYR Macedonia, and this region of southwestern Bulgaria became affected by a change in tectonic regime best developed in northern Greece.

During latest Pliocene(?) and Quaternary time, northern Greece developed a complex pattern of northeast extension associated with northeast to east–west striking right-lateral faults forming transfer faults between more local extensional areas. This system of faults overprints the older northwest trending extensional faults such as the Strymon Detachment. These two different tectonic regimes extend into FYR Macedonia where a third regime of east–west extension in western FYR Macedonia and eastern Albania is present, and where extension may represent the continuation of the east–west extensional regime initiated in Middle–Late Miocene time.

Active deformation in northern Greece, and perhaps southwest Bulgaria and adjacent FYR Macedonia, is dominated by north–south extension. This pattern of deformation may have developed as recently as Late Quaternary time.

Tectonic interpretation and discussion

The oldest Cenozoic extension considered here began in Late Eocene–Early Oligocene time and occurred within a magmatic arc that can be traced from northwestern Turkey, through southern Bulgaria, northern Greece and FYR Macedonia. The extension direction was approximately normal to the arc. At least part of this extension occurred during intracontinental convergence after final closure of the Vardar–İzmir–Ankara Zone by northward subduction in early Middle Eocene time in Turkey (Yılmaz *et al.* 1997). It is difficult to determine if convergence continued throughout this period of extension but the evidence suggests that it did. Convergent deformation occurred from Late Eocene to Late Oligocene in east central Bulgaria (Boyanov *et al.* 1989), in the Mount Olympos area (Schermer 1990) and along the Nestos Thrust (Dinter 1998) of northern Greece. Convergent deformation of Late Oligocene–Early Miocene deformation took place in the Thrace Basin of northwestern Turkey and Early–Middle Miocene convergent deformation occurred in central Bulgaria (see above). Thus, the period from Late Eocene to Early (or even Middle) Miocene is interpreted as a period of intra-arc extension contemporaneous with convergence in Greece, FYR Macedonia, Bulgaria and Turkey, perhaps induced by crustal weakening due to magmatic and radiogenic heating of thickened crust. It is also possible that reduction in intracontinental convergence, such as might have been related to a reduction in convergent plate velocity, also facilitated the shallow lithospheric extension, but this remains untested.

In Early or Middle Miocene to early Late Miocene time, intra-arc extension appears to have been transformed into a more regional extension. This transitional period is similar in age in Bulgaria and northern Greece, and similar to the oldest known extension within the greater Aegean region at c. 25–20 Ma. Extensional fabrics within rocks on Thasos Island, and within the Symvolon Pluton in the adjacent onshore, may be related to continued intra-arc extension or may belong to the second period of extension with local magmatic activity caused by extension-related pressure reduction (Jones et al. 1992). Other coeval areas of extension are amagmatic, such as at Mount Olympos, suggesting that the transition to a second period of extension had begun, at least in some places. This second period of extension begins with local development of associated magmatic rocks but continued extension in northern Greece and southwestern Bulgaria (Struma Valley and related half-grabens) occurs in a non-magmatic environment. This second phase of extension occurs along northwest trending extensional structures that trend obliquely, but at a low angle, to the older magmatic arc; thus, it is suggested that these features relate to a new tectonic regime. This system of extension is interpreted to continue to the present in western Greece and Albania, where it is logically related to trench roll-back (see below); thus, the second period of extension is interpreted to be related to roll-back of the Hellenic Trench.

Following this transition period of Oligocene, to perhaps Early Miocene, time, tectonism in the Aegean region and its continuation into the southern Balkans became dominated by extension related to roll-back of the Hellenic Trench, as expressed by the southward migration of the Hellenic Volcanic Arc (Jolivet et al. 1998). By Early–Middle Miocene time, east-north-east–west-southwest directed extension was expressed by movement on major low-angle normal faults, such as at Mount Olympos and along the Strymon Valley detachment in northern Greece (Fig. 13; note: all directions in this section are given in present-day coordinates, but significant clockwise rotation has occurred during Neogene time – see below). The Strymon Detachment appears to have propagated northward into Bulgaria at c. 16 Ma. Other major faults, such as at Mount Olympos, may also have propagated northward into central FYR Macedonia. Northeast–southwest to east–west extension migrated further north along the Struma Valley and by c. 9 Ma (beginning of Maeotian time) reached the Blagoevgrad and Djerman Grabens, which continued their activity to c. 5 Ma.

At c. 9 Ma, north–south extension began in central Bulgaria along the Sofia, Sub-Balkan and northern Thracian Basins. Extension apparently began slowly and continued from Late Pliocene to Recent more quickly. The cause of this extension is difficult to determine. It is suggested here that it may represent the northernmost effects of the north–south extension that occurred along the western coast of Turkey beginning in Tortonian time [c. 10 Ma; see Yılmaz (1997)], and may be related to the south to south–southwest movement of the Aegean region. The ultimate source for this motion is best ascribed to trench roll-back along the southern part of the Hellenic subduction system. The problem with this interpretation is the localization of this extension in central Bulgaria, because it is not obvious that north–south extension is continuous from the well-developed grabens in western Turkey to central Bulgaria. Some workers (e.g. Schindler 1997) have suggested that the southern strand of the North Anatolian Fault became active in Tortonian time. If so, such movements could have begun the current counterclockwise motion of northwest Turkey relative to Eurasia which, coupled with the clockwise rotation in northwestern Greece and Albania (see below), suggests southward motion within the region between, i.e. central Bulgaria. Schindler (1997) reports the magnitude of this early motion on the southern branch of the North Anatolian Fault to be small, consistent with low rates of coeval extension in central Bulgaria. However, most other reports on the geology of northwest Turkey have not documented a pre-Pliocene history for the North Anatolian Fault, although Barka (1997) suggests that there could have been faulting in the Marmara region during Late Miocene time.

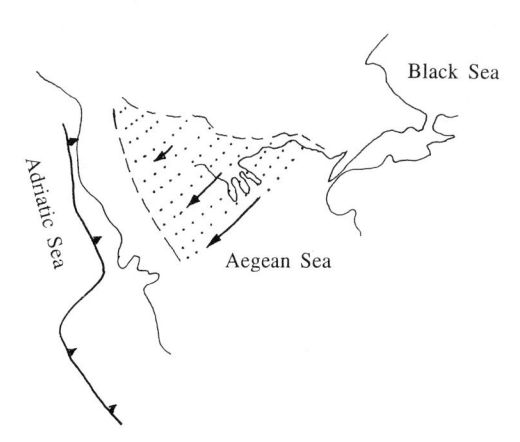

Fig. 13. Clockwise rotation of the western part of the southern Balkan extensional regime based on the palaeomagnetic data of Kissel & Laj (1988). Based on this analysis, most rotation must be accommodated by northwest trending extensional faults in southwestern Bulgaria, the low-angle normal fault at Mount Olympos, and additional unrecognized, large-magnitude normal faults in FYR Macedonia, eastern Albania, and northwestern Greece. The region of post-Eocene extension is shown by dashed lines.

North–south extension in central Bulgaria appears to extend westward, with decreasing magnitude, through Quaternary time, cutting across older northwest trending grabens (e.g. the Simitli, Kustendil and, possibly, the Strumeshnitsa Grabens). The area affected by north–south extension expanded and extension rates increased, at least in central Bulgaria, during Quaternary time; this is interpreted as being related to the development of the North Anatolian Fault (see below).

In Late Pliocene time (c. 3–4 Ma) the northeast to east–west extensional regime in southwestern Bulgaria and northern Greece was affected by another tectonic regime, which is expressed by continued northeast to north–south extension, but is distinguished by the development of segmented northeast to east–west striking right-lateral strike-slip faults which functioned as transfer faults between areas of extension. At about the same time, right-lateral displacement on the east–west trending North Anatolian Fault developed in western Turkey (Barka 1997) and probably extended into the northern Aegean Sea (Dinter & Royden 1993). GPS results indicate a 22 ± 3 mm a^{-1} rate on the North Anatolian Fault (Straub et al. 1977; Reilinger et al. 1997; McClusky et al. 2000); if this has been constant through time, its total offset of c. 80 km, still somewhat uncertain, indicates that the fault had its inception at c. 4 Ma. Beginning east of the Sea of Marmara, the North Anatolian Fault begins to branch westward and several strands of the fault can be traced across the northern Aegean Sea (McKenzie 1972, 1978). The northeast trending faults in northern Greece are interpreted here to be a northern expression of regional dextral shearing, which is mainly accommodated along the North Anatolian Fault. These northeast trending faults formed an interference pattern with the northwest trending extensional faults in northern Greece. These northeast to east–west trending right-slip faults affect an area that extends northward into FYR Macedonia. There is a diffuse northwest trending boundary between the region affected by the right-slip faults and the region of continued north–south extension in central Bulgaria (Fig. 12). The right-slip faults in this region are part of larger area undergoing counterclockwise rotation relative to Europe south of the North Anatolian Fault (Reilinger et al. 1997; McClusky et al. 2000). Western Greece and eastern Albania have rotated clockwise in Late Neogene time (see below) and it is suggested here that the continued north–south extension in Bulgaria is a result of continued southward movement of southern Bulgaria between two regions undergoing opposite senses of rotation. The propagation of the right shear along the North Anatolian Fault system into the southern Balkan region during the last 4 Ma is interpreted to be the cause of the increase in intensity of the north–south extension in central Bulgaria. Most of the northeast trending faults in the southern Balkans also show a component of extension so that both these faults, and the northwest trending extensional faults, bound mountains and adjacent valleys.

General east–west extension in eastern Albania, western FYR Macedonia and northern Greece is related to continued trench roll-back along the northern part of the Hellenic subduction system. In this part of the subduction zone, convergence continues along the east coast of the Adriatic Sea, where thicker crust has entered the subduction zone (which is probably in the process of being terminated). Thus, during the last 4 Ma, with coeval north–south extension in central Bulgaria, coupled strike-slip and northeast–southwest extension in southwest Bulgaria, northern Greece and central FYR Macedonia, and east–west extension in western FYR Macedonia and eastern Albania, there have been three areas of different styles of extensional tectonism related to different causes in the southern Balkan region bounded by diffuse boundaries (Fig. 12).

Studies of focal mechanisms and GPS studies in northern Greece, north of the North Anatolian Fault, suggest dominantly active north–south extension, different from the Pliocene–Quaternary deformation. These data suggest there has been a recent, perhaps Late Quaternary, change in deformation within northern Greece, perhaps further north into southwestern Bulgarian and FYR Macedonia.

The geometry of deformation within the southern Balkan and northern Aegean region discussed above requires important rotations of the crust about a vertical axis. Recent faulting in western FYR Macedonia and eastern Albania indicate active east–west extension, continued from Early Neogene time. Thrusting has been continuous throughout Neogene time along the northern part of the Hellenic Trench, as shown by the development of the Albanian fold and thrust belt, and magmatic activity in central FYR Macedonia. Thus the Neogene–Recent extension in western FYR Macedonia and eastern Albania is interpreted to be the result of trench roll-back, with western Albania moving westward relative to Bulgaria during middle Miocene–Recent time. The magnitude of northeast–southwest extension in northern

Greece and southwestern Bulgaria, beginning in Early–Middle Miocene time, decreases northward as expressed by the decrease in depth of basins, the decrease in separation of tectonic units and the increase in crustal thickness. This suggests that western Albania rotated clockwise relative to regions to the east at the same time as northwestern Anatolia (and the eastern Aegean crust) rotated counterclockwise.

Palaeomagnetic studies during the past decade are consistent with such rotations. Recent work in the western Dinarides, and the Albanian fold and thrust belt (Kissel et al. 1995; Marton 1987), have shown that, relative to Africa, the western fold and thrust belt of the Hellenides, from the Peloponnesus to the Tirana Basin, has rotated 45° clockwise in post-Eocene time and c. 25° since the Early Pliocene. Palaeomagnetic data from the southern Dinarides do not show post-Eocene rotation, so that rotation ends near the Scutari–Pec Line. Kondopoulou & Westphal (1986) showed that the Eocene–Oligocene intrusive rocks from Chalkidiki in northern Greece have rotated clockwise by c. 25°. In contrast, some studies (e.g. Dolapchieva 1994) show that most of Bulgaria has rotated counterclockwise from Neogene time by c. 10°, but with local rotations of up to 39°, and with only small and local clockwise rotation in southeastern Bulgaria. These palaeomagnetic data support the relative rotations suggested by the tectonic interpretation of the evolution of Middle–Late Miocene to Recent faulting from the southern Balkan region presented here.

At present there is insufficient palaeomagnetic and tectonic data to quantify the amount of total extension and rotation from this region, but preliminary estimates appear reasonable. During the extension in northern Greece, relative westward movement along the Strymon Valley detachment was at least 80 km (Dinter & Royden 1993; Dinter 1998). The magnitude of relative eastward movement along the western side of the Gulf of Thermaikos, near Mount Olympos, is unknown, but extension has cut out all the Upper Cenozoic nappe succession (Schermer 1993) and may have been tens of kilometres. Thus, an estimate of 120–150 km may not be unreasonable. The distance from the palaeomagnetic pole of rotation near the Scutari–Pec Line in northern Albania to northern Greece, where this extension occurs, is c. 200 km, yielding a clockwise rotation of at least 30–35° between central Bulgaria and the western Hellenides. However, it is difficult to separate local and regional rotations. For example, studies by one of the authors (TT) on southeastern Bulgaria indicates that the north–south Neogene–Recent extension has involved relative rotations of small crustal blocks both clockwise and counterclockwise, and Pavlides et al. (1989) have shown the complexity of palaeomagnetic rotations from small blocks in the Serbo–Macedonian Massif of northern Greece.

These studies of the Neogene–Recent evolution in the southern Balkan region are still preliminary, but are beginning to place important constraints on possible geodynamic models for the region. Any dynamic model for this region will have to explain the evolution of deformation in the southern Balkan region which involves a complex interplay between the Hellenic Trench, extrusion of Anatolia along the North Anatolian Fault and the development of north–south extension in central Bulgaria, and large-scale crustal rotation about a vertical axis. It appears clear that there are different processes that cause extension in the southern Balkan extensional province within an overall extensional regime.

This work has been part of a cooperative study between the Institute of Geology and the Institute of Geodesy, Bulgarian Academy of Sciences, Sofia, and the Massachusetts Institute of Technology, Cambridge MA, USA. It has been supported by the Bulgarian Academy of Sciences and by the National Science Foundation, grant EAR 9628225, USA. The authors thank A. M. C. Şengör and D. Dinter for very thorough reviews.

References

ALEXIEV, B. 1959. Pyroclastic sedimentary rocks of Oligocene age from the district of Varna. *Geological Institute of the Bulgarian Academy of Sciences Bulletin*, **VII**, 99–117.

ANGELIER, J., LYBERIS, N., LE PICHON, X., BARRIER, E. & HUCHON, P. l982. The tectonic development of the Hellenic arc and the Sea of Crete: a synthesis. *Tectonophysics*, **86**, 159–196.

ANGELOVA, D., POPOV, N. & MIKOV, E. 1991a. Stratigraphy of the Quaternary sediments in the Tunja depression. *Reviews of the Bulgarian Geological Society*, **LII**, 99–105.

——, RUSEVA, M. & TZANKOV, T. 1991b. On the structure and evolution of the Karlovo Graben. *Geotectonics, Tectonophysics and Geodynamics*, **23**, 26–46.

ANTIMOVA, TZ. & KOJUMDGIEVA, E. 1991. The Neogene of the Palakaria basin (lithostratigraphic subdivision and geological evolution). *Palaeontology, Stratigraphy and Lithology*, **29**, 45–59 (in Bulgarian with abstract in English).

AUBOUIN, J., BRUNN, J. H., CELET, P., DERCOURT, J., GODFRIAUX. I. & MERCIER, J. 1963. 'Esquisse ce la geologie de la Grece'. *In*: *Livre a la Mémoire du*

Professeur Paul Fallot. Mémoire Hors-serie, Geological Society of France, **II**, 583–610.

AVIGAD, D. 1993. Tectonic juxtaposition of blueschists and greenschists in Sifnos Island (Aegean Sea) – implications for the structure of the Cycladic blueschist belt. *Journal of Structural Geology*, **15**, 1459–1469.

——, GARFUNKEL, Z., JOLIVET, L. & AZANON, J. M. 1997. Back arc extension and denudation of the Mediterranean eclogites. *Tectonics*, **16**, 924–941.

BARKA, A. 1997. Neotectonics of the Marmara region: *In*: SCHINDLER, C. & PFISTER, M. (eds) *Active Tectonics of Northwestern Anatolia – The MARMARA Poly-Project; A Multidisciplinary Approach by Space-Geodesy, Geology, Hydrogeology, Geothermics and Seismology*. vdf Hochschulerlag AG an der ETH Zurich, 55–87.

BOEV, B., STOYANOV, R. & SRAFIMOVSKI, T. 1995. Tertiary magmatism in the south-western parts of the Carpatho–Balkanides with a particular reference on magmatism in the area of the Republic of Macedonia. *Geologica Macedonia*, **9**, 23–38.

BOJADJIEV, S. T. & SPOUNDJIEVA, V. 1960. The Pliocene in the Palakaria, Samokov area. *Reviews of the Bulgarian Geological Society*, **XXI**, 86–92.

BONCHEV, ST. & BAKALOV, P. 1928. The earthquakes in South Bulgaria on the 14 and 18 of April 1928. *Review of bulgarian Geological Society*, **1**, 50–63 (in Bulgarian and French with abstract in English).

BOYANOV, I., DOBOVSKI, C. H., GOCHEV, P., HARKOVSKA, A., KOSTADINOV, V., TZANKOV. T. Z. & ZAGORCHEV, I. 1989. A new view of the Alpine tectonic evolution of Bulgaria. *Geologica Rhodopica*, **1**, 107–121.

BURCHFIEL, B. C. 1980. The eastern Alpine orogen of the Mediterranean: an example of collision tectonics. *Tectonophysics*, **63**, 31–610.

BUICK, I. S. 1991. The Late Alpine evolution of an extensional shear zone, Naxos, Greece. *Journal of the Geological Society, London*, **148**, 93–103.

CERNJAVSKKA, S. 1977. Palynolofical studies on Palaeogene deposits in South Bulgaria. *Geologica Balcanica*, **7**, 3–26.

DACHEV, C., STANEV, V. & BOKOV, P. 1988. Structure of the Bulgarian Black Sea area. *Bulletin Geofisica Teorica ed Applicata*, **XXX**, 79–107.

DEWEY, J. F. & ŞENGÖR, A. M. C. 1979. Aegean and surrounding regions: complex multiplate and continuum tectonics in a convergent zone. *Geological Society of America Bulletin*, **90**, 84–92.

DINTER, D. A. 1998. Late Cenozoic extension of the Alpine collisional orogen, northeastern Greece: origin of the north Aegean basin. *Geological Society of America Bulletin*, **110**, 1208–1230.

—— & ROYDEN, L. H. 1993. Late Cenozoic extension in northeastern Greece: Strymon Valley detachment and Rhodope metamorphic core complex. *Geology*, **21**, 45–48.

——, MACFARLANE, A. M., HAMES, W., ISACHSEN, C., BOWRING, S. & ROYDEN, L. H. 1995. U–Pb and $^{40}Ar/^{39}Ar$ geochronology of the Symvolon granodiorite: implications for the thermal and structural evolution of the Rhodope metamorphic core complex, northeastern Greece. *Tectonics*, **14**, 886–908.

DOLAPCHIEVA, P. 1994. Review and taxonomy of data and results of the Palaeomagnetic investigations in Bulgaria performed by Senior Researcher Peter Nozharov and his team. *Bulgarian Geophysical Journal*, **XX**, 40–50.

DOUST, H. & ARIKAN, Y. 1974. The geology of the Thrace Basin. *In*: OKAY, H. & DILEKÖZ, E. (eds) *Proceedings of the 2nd Petroleum Congress and Exhibition of Turkey*. Turkish Association of Petroleum Geologists Publications, 227–248.

DRAGOMANOV, L., NOVGORODCEV, A., TONCHEV. T. & GARELOVA, M. 1988. Middle Miocene sedimentary rocks in the area of Bjala Reka and Filevo, Plovdiv district. *Reviews of the Bulgarian Geological Society*, **XLIX**, 99–103.

——, ANGELOV, G., KOJUMDGIEVA, E., NIKOLOV, K. & KOMOROGOVA, I. 1984. The Neogene in Haskovo district. *Palaeontology, Stratigraphy and Lithology*, **20**, 71–75.

——, KAZARINOV, V., KOJUMDGIEVA, E., NIKOLOV, I., ENCHEV, E. & CHRISTOV, C. H. 1981. Paleogeography of the Neogene in Plovdiv and Pazardzik districts. *Palaeontology, Stratigraphy and Lithology*, **14**, 65–75.

GAUTIER, P., BRUN, J. P. & JOLIVET, L. 1993. Ductile crust exhumation and extensional detachments in the central Aegean (Cyclades and Evvia Islands). *Tectonics*, **12**, 1180–1194.

GOCEV, P. M. 1986. Modèle pour une nouvelle synthèse tectonique de la bulgarie. *Dari de Seama, Part 5, Tectoica si Geolgoie Regionala*, **70–71**, 97–107.

—— 1991. The Alpine orogen in the Balkans – a polyphase collisional structure. *Geotectonics, Tectonophysics and Geodynamics*, **22**, 3–44.

HAVKOVSKA, A. 1984. Tertiary magmotectonic zones in southwest Bulgaria: in magmatism of the molasse-forming epoch and its relation to endogenous mineralisation. *In*: VOZAR, J. (ed.) *Magmatism of the Molasse-forming Epoch and its Relation to Endogenous Mineralisation*. Geologicky ustav Dioyza Stura, Bratislava, 9–34.

——, YANEV, Y. & MARCHEV, P. 1989. General features of the Palaeogene orogenic magmatism in Bulgaria. *Geologica Balcanica*, **19**, 37–72.

IVANOV, Z., MOSKOVSKI, ST. & SIRAKOV, N. 1977. Olistostromes in the Palaeogene of the Hvoina basin. *Annuaire de l'Université de Sofia, Faculté de géologie et géographie*, **70**, 17–52.

JACKSON, J. & MCKENZIE, D. P. 1988. The relationship between plate motions and seismic tremors, and the rates of active deformation in the Mediterranean and Middle East. *Geophysical Journal of the Royal Astronomical Society*, **93**, 45–73.

JACOBSHAGEN, V. 1986. *Geologie von Greichenland*. Gebruder Borntraeger.

——, DÜRR, S., KOCKEL, R., KOPP, K.-O., KOWALCZYK, G., BERCKHEMER, H. & Buttner, D. 1978. Structure and geodynamic evolution of the Aegean region: *In*: CLOSS, H., ROEDER, D. & SCHMIDT, K. (eds) *Alps, Apennines, Hellenides*. Stuttgart

Inter-Union Commission on Geodynamics, Scientific Report No. 38, E., Schweizerbart'sche Verlagsbuchhandlung, 537–564.

JANKOVIC, S., SERAFIMOVSKI, T., JELENKOVIC, R. & CIFLIGANEC, V. 1997. Metallogeny of the Vardar zone and Serbo-Macedonian Mass. *In*: BOEV, B. & SERAFIMOVSKI, T. (eds) *Proceedings, Magmatism, Metamorphism and Metallogeny of the Vardar Zone and Serbo-Macedonian Massif.* Stip-Dojoran, 29–63.

JOLIVET, L., FACCENNA, C., GOFFÉ, B. *ET AL.* 1998. Midcrustal shear zones in post orogenic extension: example from the northern Tyrrhenian Sea. *Journal of Geophysical Research*, **103**, 12 123–12 160.

JONES, C. E., TARNEY, J., BAKER, J. H. & GEROUKI, G. 1992. Tertiary granitoids of Rhodope, northern Greece: magmatism related to extensional collapse of the Hellenic orogen. *Tectonophysics*, **210**, 295–314.

KAHLE, H.-G., MÜLLER, M. V., GEIGER, A. *ET AL.* 1995. The strain field in northwestern Greece and the Ionian Islands: results inferred from GPS measurements. *Tectonophysics*, **249**, 41–52.

KAMENOV, B. 1965. The boundary Pliocene–Pleistocene in the Sofia kettle (valley). *Reviews of the Bulgarian Geological Society*, **XXVI**, 112–114.

—— & KOJUMDGIEVA, E. 1983. Stratigraphy of the Neogene in the Sofia basin. *Palaeontology, Stratigraphy and Lithology*, **18**, 69–85.

KAUFMAN, P. S. 1995. *Extensional tectonic history of the Rhodope Metamorphic Core Complex, Greece and geophysical modelling of the Halloran Hills, Califonia.* PhD Thesis, Cambridge Massachusetts Institute of Technology.

——, KONDOPOUILOU D., LAJ, C. & PAPADOPOULOS, P. 1986. New Palaeomagnetic data from Oligocene formations of northern Aegean. *Geophysical Research Letters*, **13**, 1039–1042.

KISSEL, C. & LAJ, C. 1988. The Tertiary geodynamical evolution of the Aegean arc: a palaeomagnetic reconstruction. *Tectonophysics*, **146**, 183–201.

——, SPERANZA, F. & MILICEVIC, V. 1995. Palaeomagnetism of external southern and central Dinarides and northern Albanides: implications for the Cenozoic activity of the Scutari–Pec transverse zone. *Journal of Geophysical Research*, **100**, 14 999–15 007.

KOJUMDGIEVA, E. 1987. Evolution geodynamique du bassin Egéen pendant le Miocene superieur et ses relations a la Paratethys orientale. *Geologica Balcanica*, **17**, 3–14.

——, NIKOLOV, I., NEDJALKOV, P. & BUSEV, B. 1982. Stratigraphy of the Neogene in the Sandanski Graben. *Geologica Balcanica*, **12**, 69–81.

KOJUMDGIEVA, I. & DRAGOMANOV, L. 1979. Lithostratigraphy of the Oligocene and Neogene sediments in Plovdiv and Pazardzik districts. *Palaeontology, Stratigraphy and Lithology*, **11**, 49–61.

——, STOYKOV, S. T. & MARKOVA, S. T. 1984. Lithostratigraphy of the Neogene sediments in the Tundza (Elhovo–Jambol) basin. *Reviews of the Bulgarian Geological Society*, **XLV**, 287–295.

KONDOPOULOU, D. & WESTPHAL, M. 1986. Palaeomagnetism of the Tertiary intrusives from Chalkidiki (northern Greece). *Journal of Geophysics*, **59**, 62–66.

KOTOPOULI, C. N. & PE-PIPER, G. 1988. Geochemical characteristics of felsic intrusive rocks within the Hellenic Rhodope: a comparative study and petrogenetic implications. *Neues Jahrbuch für Mineralogie Abhandlung*, **161**, 141–169.

KOUKOUVELAS, I. & DOUTSOS, T. 1990. Tectonic stages along a traverse cross cutting the Rhodopian zone (Greece). *Geologische Rundschau*, **79**, 753–776.

—— & PE-PIPER, G. 1991. The Oligocene Xanahi pluton, northern Greece: a granodiorite emplaced during regional extension. *Journal of the Geological Society, London*, **148**, 749–756.

LE PICHON, X. & ANGELIER, J. 1981. The Aegean Sea. *Philosophical Transactions of the Royal Society, London, Series A*, **300**, 357–372.

LISTER, G. S., BANGA, G. & FEENSTRA, A. 1984. Metamorphic core complexes of Cordilleran type in the Cyclades, Aegean Sea, Greece. *Geology*, **12**, 221–225.

MCCLUSKY, S., BALASSANIAN, S., BARKA, A. *ET AL.* 2000. GPS constraints on plate motions and deformations in the eastern Mediterranean: implications for plate dynamics. *Journal of Geophysical Research*, in press.

MCKENZIE, D. P. 1972. Active tectonics of the Mediterranean region. *Geophysical Journal of the Royal Astronomical Society*, **30**, 109–185.

—— 1978. Active tectonics of the Alpine–Himalayan belt: the Aegean sea and surrounding regions (tectonics of the Aegean region). *Geophysical Journal of the Royal Astronomical Society*, **55**, 217–254.

MARTON, E. 1987. Palaeomagnetism and tectonics in the Mediterranean region. *Journal of Geodynamics*, **7**, 33–57.

NEDJALKOV, P., TCHEREMISIN, N., KOJUMDGIEVA, E., TZATZEV, B. & BUSEV, A. 1986. Facies and palaeogeographic features of Neogene deposits in the Sandanski graben. *Geologica Balcanica*, **16**, 69–80.

——, KOJUMDGIEVA, E. & BOZKOV, I. 1988. Sedimentation cycles in the Neogene grabens along the Struma Valley. *Geologica Balcanica*, **18**, 61–66.

NENOV, T. & DRAGOMANOV, L. 1987. Iron concretions in the Neogene of the Zagora depression. *Reviews of the Bulgarian Geological Society*, **XLVIII**, 93–97.

PAPANIKOLAOU, D. 1993. Geotectonic evolution of the Aegean. *Bulletin of the Geological Society of Greece*, **XXVIII**, 33–48.

PAPAZACHOS, B. C., KIRATZI, A. A., HATZIDIMITRIOU, P. M. & ROCCA, A. C. 1984. Seismic faults in the Aegean area. *Tectonophysics*, **106**, 71–85.

PAVLIDES, S. B., KONDOPOULOU, D. P., KILIAS, A. A. & WESTPHAL, M. 1988. Complex rotational deformations in the Serbo-Macedonian massif (north Greece): structural and Palaeomagnetic evidence. *Tectonophysics*, **145**, 329–335.

PERİNÇEK, D. 1991. Possible strand of the North

Anatolian Fault in the Thrace Basin, Turkey – an interpretation. *AAPG Bulletin*, **75**, 241–257.

POPOV, N. 1968. Pleistocene. *In*: TZANKOV, V. & SPASOV, C. H. R. (eds) *Stratigraphy of Bulgaria*. Tehnika, 382–386.

REILINGER, R. E., MCCLUSKY, S. C., ORAL, M. B. ET AL. 1997. Global positioning system measurements of present-day crustal movements in the Arabia–Africa–Eurasia plate collision zone. *Journal of Geophysical Research*, **102**, 9983–9999.

ROYDEN, L. H. 1993. The tectonic expression of slab pull at continental convergent boundaries. *Tectonics*, **12**, 303–325.

SACHPAZI, M., HIRN, A., NERCESSINA, A., AVEDIK, R., MCBRIDGE J., LOUCOYANNAKIS, M., NICOLICH, R. & THE STREAMERS-PROFILES GROUP. 1997. A first coincident normal-incidence and wide-angle approach to studying the extending Aegean crust. *Tectonophysics*, **270**, 301–312.

SCHERMER, E. R. 1990. Mechanisms of blueschist creation and preservation in an A-type subduction zone, Mount Olympos region, Greece. *Geology*, **18**, 1130–1133.

—— 1993. Geometry and kinematics of continental basement deformation during the Alpine orogeny, Mount Olympos region, Greece. *Journal of Structural Geology*, **15**, 571–591.

——, LUX, D. R. & BURCHFIEL, B. C. 1990. Temperature–time history of subducted continental crust, Mount Olympos region, Greece. *Tectonics*, **9**, 1165–1195.

SCHINDLER, C. 1997. Geology of Northwestern Turkey: results of the MARMARA Poly-Project. *In*. SCHINDLER, C. & PFISTER, M. (eds) *Active Tectonics of Northwestern Anatolia – The MARMARA Poly-Project; a Multidisciplinary Approach by Space-Geodesy, Geology, Hydrogeology, Geothermics and Seismology*. vdf Hochschulerlag AG an der ETH Zurich, 329–373.

ŞENGÖR, A. M. C. 1984. *The Cimmeride Orogenic System and the Tectonics of Eurasia*. Geological Society of America Special Papers, **195**.

—— & YILMAZ, Y. 1981. Tethyan evolution of Turkey: a plate tectonic approach. *Tectonophysics*, **75**, 181–241.

STRAUB, C., KAHLE, H.-G. & SCHINDLER, C. 1997. GPS and geologic estimates for the tectonic activity in the Marmara Sea region, northwestern Anatolia. *Journal of Geophysical Research*, **102**, 27 587–27 601.

THOMPSON, S. N., STOCKHERT, B. & BRIX, M. R. 1998. Thermochronology of the high-pressure metamorphic rocks of Crete, Greece: implications for the speed of tectonic processes. *Geology*, **26**, 259–262.

TSOKAS, G. N. & HANSEN, R. O. 1997. Study of the crustal thickness and the subducting lithosphere in Greece from gravity data. *Journal of Geophysical Research*, **102**, 20 585–20 597.

TURGUT, S., TÜRKASLAN, M. & PERINÇEK, D. 1991. Evolution of the Thrace sedimentary basin and its hydrocarbon prospectivity. *In*: SPENCE, A. M. (ed.) *Generation, Accumulation and Production of Europe's Hydrocarbons*. European Association of Petroleum Geoscience, Special Publications, **1**, 415–437.

TZANKOV, T. Z., ANGELOVA, D., NAKOV. R., BURCHFIEL, B. C. & ROYDEN, L. H. 1996. The Sub-Balkan graben system of central Bulgaria. *Basin Research*, **8**, 125–142.

VANDENBERG, L. C. & LISTER, G. S. 1996. Structural analysis of basement tectonites from the Aegean metamorphic core complex of Ios, Cyclades, Greece. *Journal of Structural Geology*, **18**, 1437–1454.

VAN ECK, R. & STOYANOV, T. 1996. Seismotectonics and seismic hazard modelling for Southern Bulgaria. *Tectonophysics*, **262**, 77–100.

WAWRZENITZ, N. & KROHE, A. 1998. Exhumation and doming of the Thasos metamorphic core complex (S Rhodope, Greece): structural and gechronological constraints. *Tectonophysics*, **285**, 310–332.

YILMAZ, Y. 1997. Geology of Western Anatolia. *In*: SCHINDLER, C. & PFISTER, M. (eds) *Active Tectonics of Northwestern Anatolia–The MARMARA Poly-Project; a Multidisciplinary Approach by Space-Geodesy, Geology, Hydrogeology, Geothermics and Seismology*. vdf Hochschulerlag AG an der ETH Zurich, 31–53.

——, TÜYSÜZ, E., YİĞİTBAŞ, S., GENÇ, Ş. C. & ŞENGÖR, A. M. C. 1997. Geology and tectonic evolution of the Pontides. *In*: ROBINSON, A. G. (ed.) *Regional and Petroleum Geology of the Black Sea and Surrounding Region*. AAPG Memoirs, **68**, 183–226.

YOVCHEV, Y. (ed.) 1971. *Tectonic Structure of Bulgaria*. Technika.

ZAGORCHEV, I. 1992. Neotectonics of the central parts of the Balkan Peninsula: basic features and concepts. *Geologicshe Rundschau*, **81**, 635–654.

—— 1995. *Geological Guidebook, Geological Institute, Bulgarian Academy of Sciences*. Academic Publishing House, Sofia.

When did the western Anatolian grabens begin to develop?

YÜCEL YILMAZ,[1] Ş. CAN GENÇ,[1] FEVZİ GÜRER,[2] MUSTAFA BOZCU,[3] KAMİL YILMAZ,[3] ZEKİYE KARACIK,[1] ŞAFAK ALTUNKAYNAK[1] & ALİ ELMAS[4]

[1] *İstanbul Technical University, Mining Faculty, Department of Geological Engineering, Maslak, TR-80626 İstanbul, Turkey (e-mail: yilmazy@itu.edu.tr)*
[2] *Kocaeli University, Engineering Faculty, Department of Geological Engineering, Kocaeli, Turkey*
[3] *Süleyman Demirel University, Faculty of Engineering, Department of Geological Engineering, Isparta, Turkey*
[4] *İstanbul University, Engineering Faculty, Department of Geological Engineering, Avcılar, TR-34850 İstanbul, Turkey*

Abstract: To solve a long-lasting controversy on the timing and mechanism of generation of the western Anatolian graben system, new data have been collected from a mapping project in western Anatolia, which reveal that initially north–south trending graben basins were formed under an east–west extensional regime during Early Miocene times. The extensional openings associated with approximately north–south trending oblique slip faults provided access for calc-alkaline, hybrid magmas to reach the surface. A north–south extensional regime began during Late Miocene time. During this period a major breakaway fault was formed. Part of the lower plate was uplifted and cropped out later in the Bozdağ Horst, and above the upper plate approximately north–south trending cross-grabens were developed. Along these fault systems, alkaline basalt lavas were extruded. The north–south extension was interrupted at the end of Late Miocene or Early Pliocene times, as evidenced by a regional horizontal erosional surface which developed across Neogene rocks, including Upper Miocene–Lower Pliocene strata. This erosion nearly obliterated the previously formed topographic irregularities, including the Bozdağ elevation. Later, the erosional surface was disrupted and the structures which controlled development of the Lower–Upper Miocene rocks were cut by approximately east–west trending normal faults formed by rejuvenated north–south extension. This has led to development of the present-day east–west trending grabens during Plio-Quaternary time.

Figure 1 is a summary map of the geology of western Anatolia and shows that the region is characterized by a number of approximately east–west trending, subparallel, normal fault zones bordering a set of grabens and intervening horst blocks. Seismic activity is intense and has been recorded by a network of instruments roughly encircling the active faults. Motions on the faults confirm that extension is in a north–south direction. Western Anatolia and the Aegean regions have long been known to represent a broad zone of extension (Phillipson 1910–1915) stretching from Bulgaria in the north to the Hellenic arc in the south (McKenzie 1972).

There are about ten approximately east–west orientated grabens in western Anatolia. The best-developed grabens are Büyük Menderes, Gediz, Edremit, Gökova and Bergama. They are c. 100–150 km long and 5–15 km wide. In each graben, one margin is characterized by steeper topography, associated with surface breaks. On the footwall margins of the grabens, planar faults are readily observed.

The data available on the timing and histories of development for the east–west grabens are conflicting. There are ongoing debates on two major problems associated with the geology of the graben regions and the related structures: (1) when did the grabens begin to develop?; 2) what event triggered their initiation?

Two different views have been proposed for the timing of the development of the east–west grabens. According to one view, the grabens began to develop during Late Oligocene–Early Miocene times and have progressively enlarged since then (Seyitoğlu & Scott 1991, 1994, 1996). The other view proposes that the grabens are relatively young and no older than Late Miocene (Şengör *et al.* 1985; Görür *et al.* 1995). Resolution of this question requires a fuller understanding of the geology of the region, as

From: BOZKURT, E., WINCHESTER, J. A. & PIPER, J. D. A. (eds) *Tectonics and Magmatism in Turkey and the Surrounding Area.* Geological Society, London, Special Publications, **173**, 353–384. 1-86239-064-9/00/$15.00
© The Geological Society of London 2000.

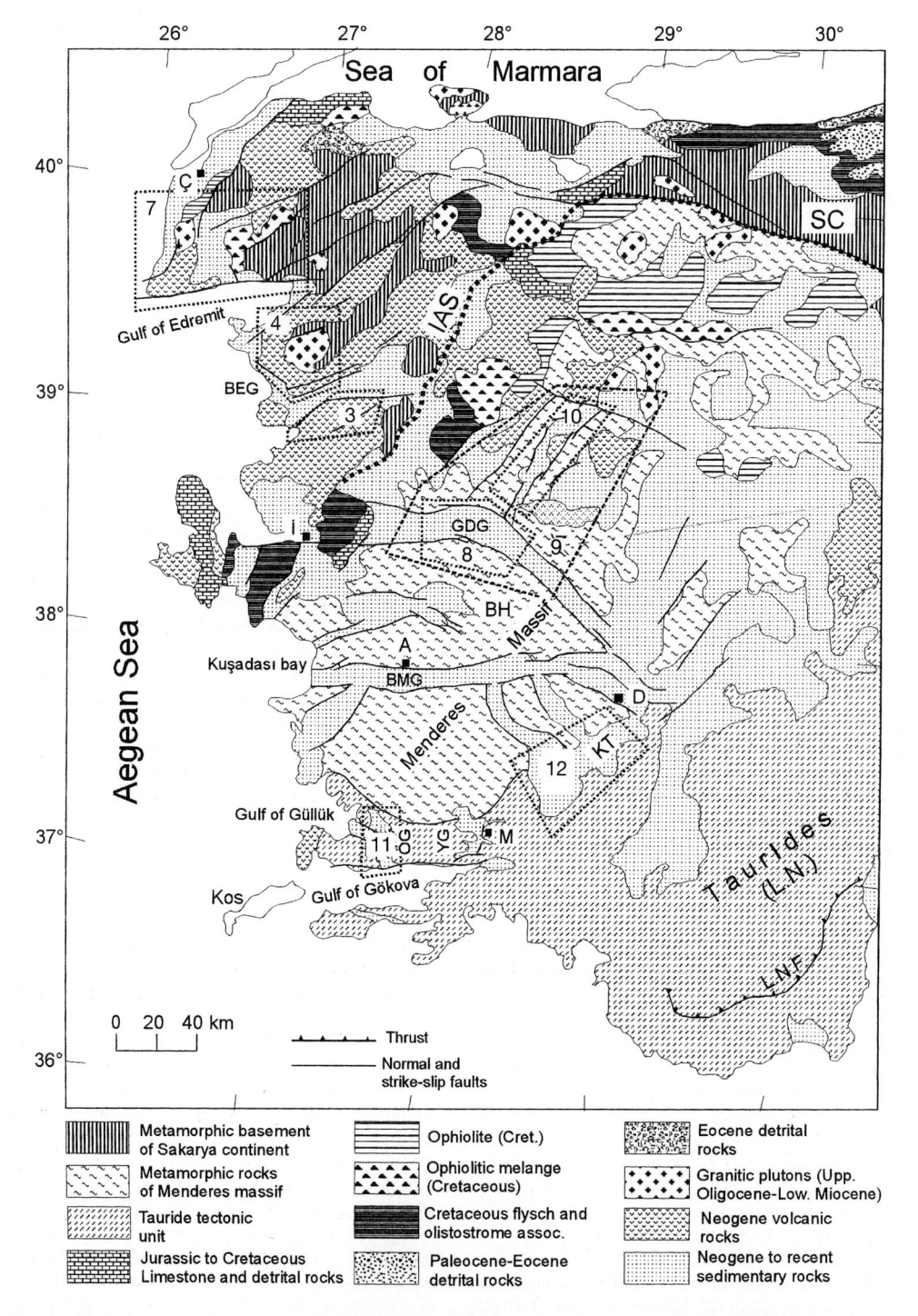

Fig. 1. Geological map of western Anatolia. Numbers indicate the locations of the maps in Figs 2–4 and 7–12. BEG, Bergama Graben; GDG, Gediz Graben; BMG, Büyük Menderes Graben; KT, Kale–Tavas Basin; IAS, İzmir–Ankara Ophiolite Suture; SC, Sakarya Continent; LN, Lycian Nappes; LNF, Lycian Nappe front; BH, Bozdağ Horst. A, Ç, D, I and M are the cities of Aydın, Çanakkale, Denizli, İzmir and Muğla, respectively.

well as the detailed geology of the graben areas.

To solve the questions noted above the following approaches may be adopted: (1) obtain stratigraphic data directly from the graben floor to date the initiation of the grabens – this is often difficult due to the presence of a young and thick alluvial cover; (2) collect data to establish the timing of the compressional tectonic phase prior to the extensional regime; and (3) use geophysical and structural data from the active physical processes to determine the length of time which has elapsed since the beginning of the graben development.

In this study all three approaches have been utilized. Since detailed information on the tectonostratigraphy of the magmatic and sedimentary associations of the grabens and surrounding region are needed to constrain models for the evolution of the region, a field-based project was undertaken in western Anatolia which has lasted for 7 years. During the study work was begun in the northern areas and the study area then enlarged towards the south to cover the entire rifted region. More than 60 sheets of detailed geological maps, at the scale of 1/25 000, have been produced. The areas that will be outlined in this paper are selected from four major east–west trending grabens, two of which are onshore and two of which are offshore. From north to south these are: the Edremit, the Bergama (Pergamon), the Gediz and the Gökova Grabens (Fig. 1). In this paper, new data is presented and in the light of these data development of the grabens is discussed. The four regions will be treated separately below. Geology of the Bergama Graben region will be documented first, because this region provides the most complete sedimentary and magmatic succession, and the more varied structural features.

The Bergama Graben region

The Bergama Graben (Figs 1 & 2) is one of the major east–west grabens of western Turkey. It is $c.$ 60 km long and 5 km wide: the Bakırçay River flows west along the valley floor. The topography is asymmetric, being steeper along the northern side of the valley where Mount Kozak rises steeply to over 800 m from the graben floor, which is 50–80 m above sea level. At the southern side, the topography is more subdued, toward the Yuntdağ Mountain (Fig. 2). The graben is an approximately east–west striking structural low but, in detail, is a V-shaped depression trending north-northeast–south-southwest between Bergama and Göçbeyli, and northwest–southeast between Altınova and Dikili (Fig. 2). The width of the graben increases from its centre to the east and to the west, where the graben merges with the north–south trending Soma Graben and the northeast–southwest trending Altınova Depression (Fig. 2).

The faults along both sides of the graben, around the Bergama area (Fig. 2), indicate mainly normal faulting. Occasionally, a small component of strike-slip motion, sinistral to the west and dextral to the east of the Bergama, is observed. This fault pattern is compatible with a north–south extensional regime. The major faults along the northern side of the graben dip steeply ($> 70°$) to the south. Along the south margin they are antithetic with respect to the main fault system, north dipping and subparallel to east–west striking normal faults (Figs 2 and 4). They cause abrupt steps in the topography. The sediments, which have been deposited on the downthrown blocks, are slightly back-tilted. The main active faults place young alluvium against the older Lower–Middle Miocene succession. Near the town of Bergama, the dip-slip on the normal fault is estimated to be > 500 m because the Lower Miocene succession, which is $c.$ 500 m thick, is missing on the upthrown northern block near the fault; alluvium here is faulted against the metamorphic basement.

The geology of the sectors north and south of the Bergama Graben are outlined below.

The northern sector

A simplified geology map of the northern sector is illustrated in Fig. 3. Northeast–southwest trending oblique faults dominate this region (Fig. 2) and more than 50 fault planes have been measured. They strike approximately N25–40°E, and the slickenlines on the fault planes plunge 50°SE to N30–50°E. Faults divide the region into subparallel, northeast–southwest elongated horsts and grabens, such as the Ayvalık–Burhaniye (Altınova Depression) and Örenli–Eğiller Grabens, and the intervening Kozak Horst (Fig. 2). The horst and grabens are cut abruptly by faults bounding the Bergama–Dikili Graben (Fig. 2).

The Kozak Horst is the most prominent morphological feature of the region and Mount Kozak rises to over 1000 m within a distance of 30 km from the gulfs of Edremit and Dikili (Fig. 2). In the horst, metamorphic, plutonic, volcanic and the Neogene sedimentary rocks crop out (Fig. 3). Below the cover rocks (Fig. 5a) is a metamorphic association known as the

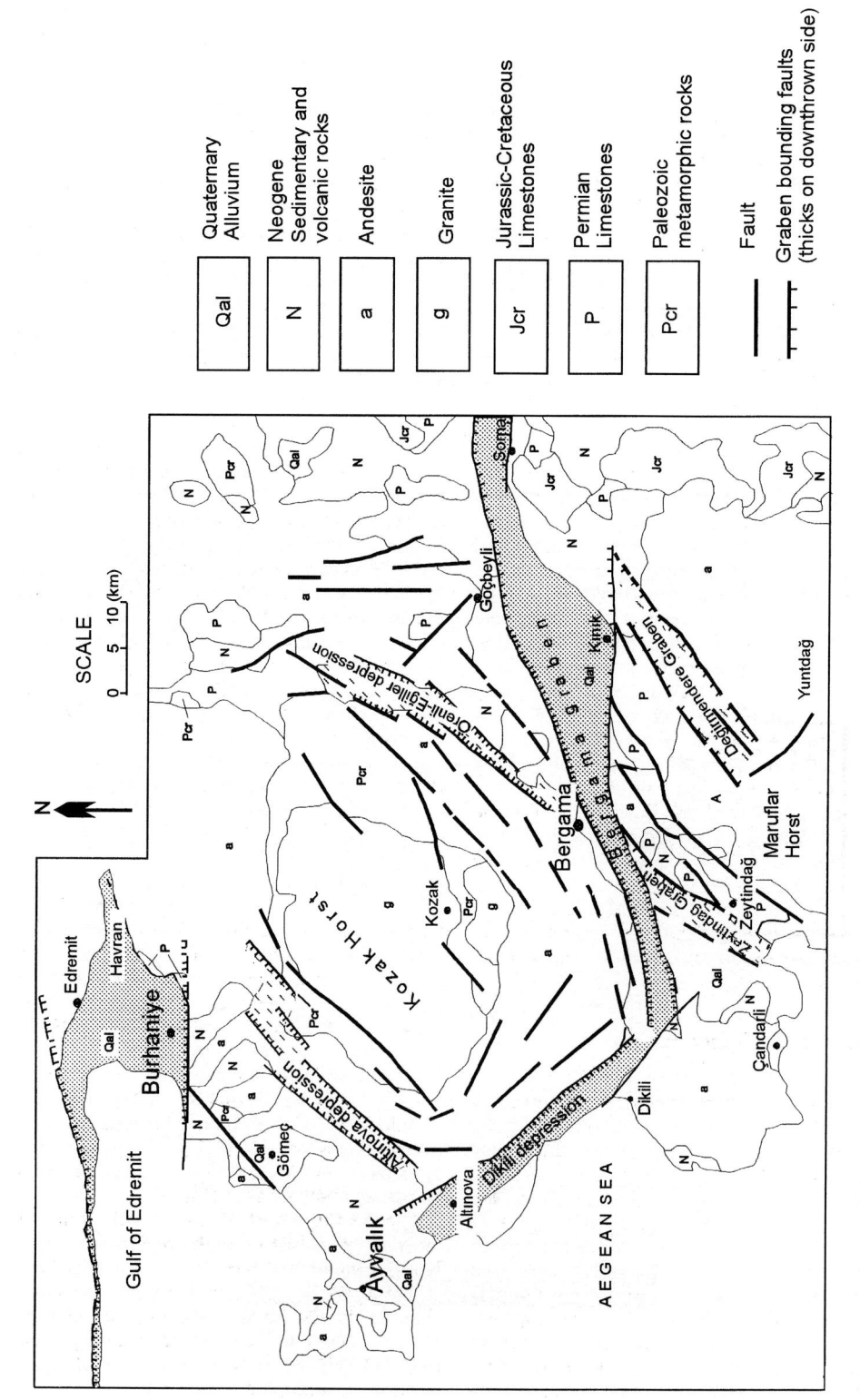

Fig. 2. Tectonic map of the Bergama Graben and surrounding regions.

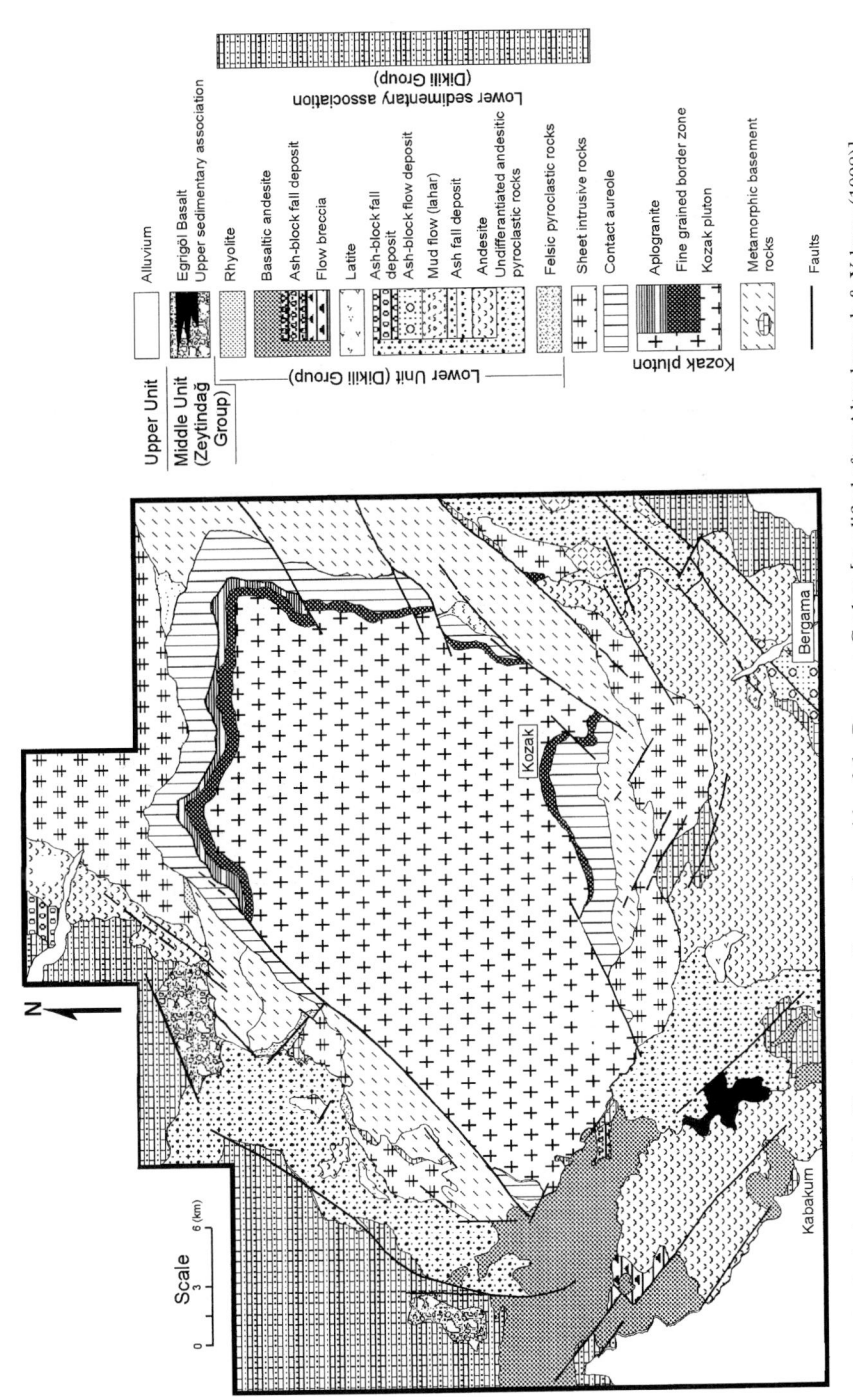

Fig. 3. Geological map of the Kozak region on the northern side of the Bergama Graben [modified after Altunkaynak & Yılmaz (1998)].

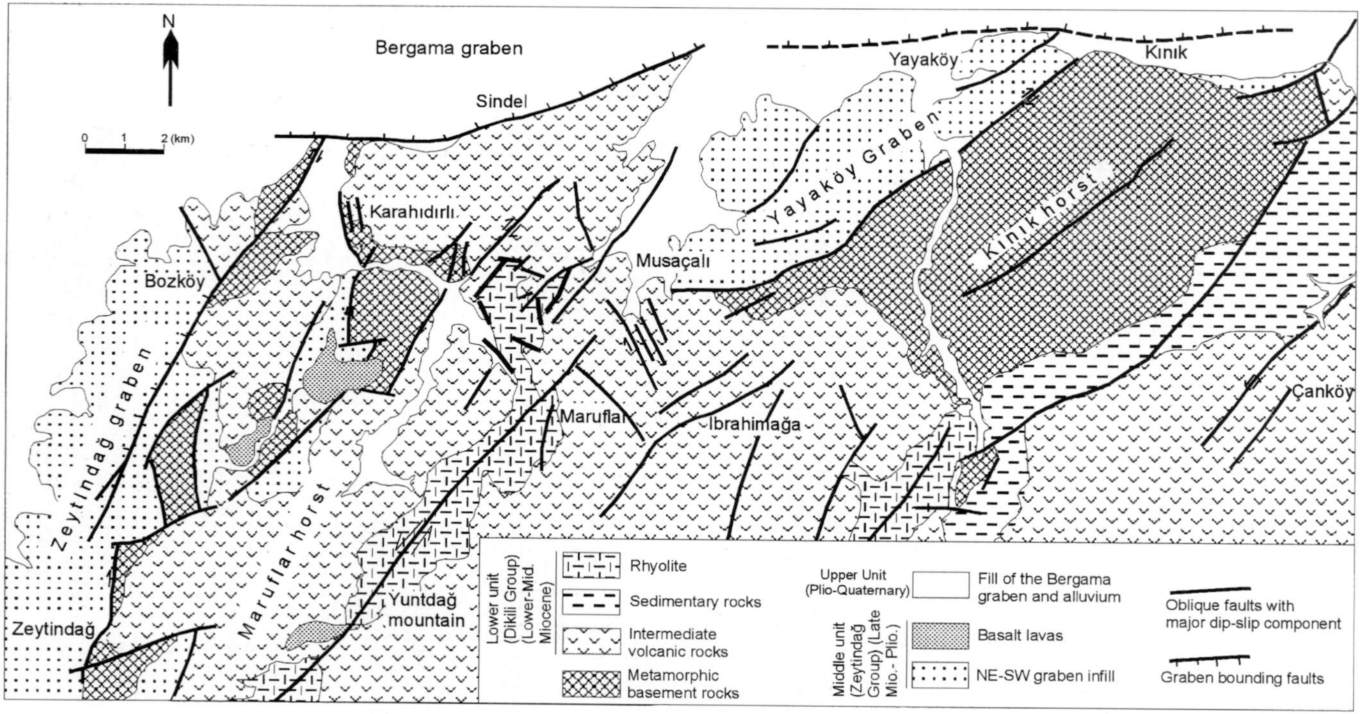

Fig. 4. Geological map of the Zeytindağ–Maruflar region on the southern side of the Bergama Graben.

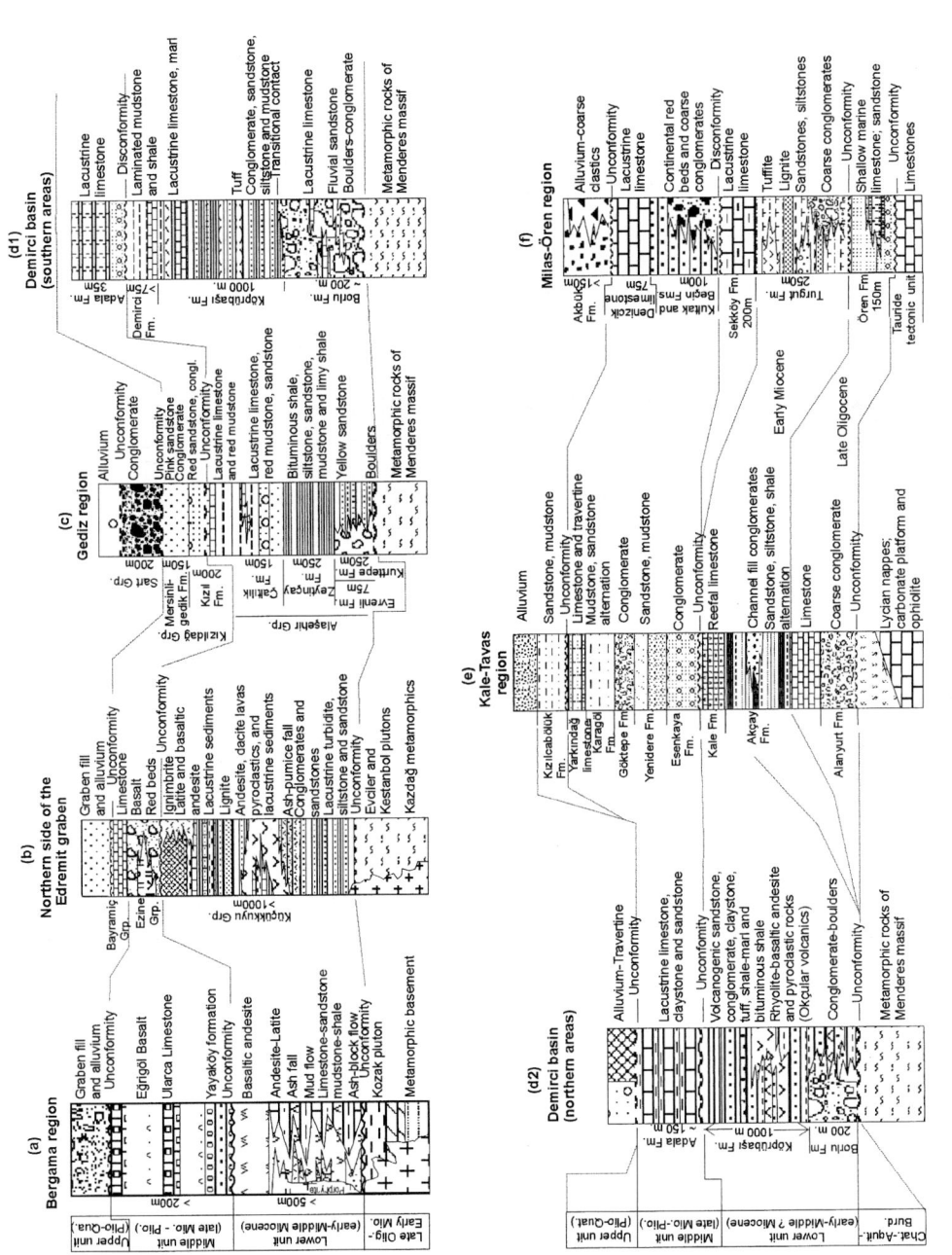

Fig. 5. Representative stratigraphic sections from different graben regions of western Anatolia.

Karakaya Formation (Bingöl et al. 1973), metamorphosed during latest Triassic times (Akyürek & Soysal 1983; Kaya & Mostler 1992; Genç & Yılmaz 1995; Altunkaynak 1996).

In the central part of the Kozak Horst, a granitic pluton (the Kozak Granite) was emplaced into the metamorphic rocks during latest Oligocene–Early Miocene times (20–23 Ma: Ataman 1974; Bingöl et al. 1982). The pluton consists mainly of coarse-grained, porphyritic diorites and is surrounded by a number of hypabyssal intrusives in the form of dykes and cone sheets varying in width from 5 to 100 m (Fig. 3). The sheets are also represented by porphyritic microdiorites and microgranodiorites, and are petrographically similar to the plutonic rocks. The granite and the surrounding geochemically related hypabyssal and volcanic rocks formed in a collapsed caldera environment (Altunkaynak & Yılmaz 1998).

The Neogene and younger rocks are made up of three lithostratigraphic units, separated from one another by unconformities (Fig. 5a). These are, from base to top, the Dikili Group, the Zeytindağ Group and an upper unit consisting of the sediment fill of the Bergama Graben. Of these three rock units only the lower one is exposed in the Kozak Horst. The middle unit is present only within the surrounding northeast–southwest grabens. The Dikili Group is c. 750 m thick and rests unconformably on the metamorphic rocks. It comprises a volcano–sedimentary assemblage of Early–Middle Miocene age. A direct stratigraphic contact of the Dikili Group with the plutonic rocks is only rarely exposed, indicating that the pluton was unroofed during deposition of the Dikili Group. The succession is detailed in Fig. 5a and only a brief summary is presented here. At the bottom of the Dikili Group are dark siltstones, purple mudstones, and white, finely laminated and bituminous shales. The shales are predominant. These rocks are interpreted as low energy lacustrine deposits (Akyürek & Soysal 1983; Altunkaynak 1996). The Dikili Group is widespread in western Anatolia and not restricted to the northeast–southwest or east–west horst–grabens of the Bergama region. In the Kozak area it is exposed from the top of the horsts into the surrounding graben depressions.

Volcanic rocks, spatially and temporally associated with the sedimentary rocks of the Dikili Group, consist predominantly of andesite, latite, dacite lavas and their pyroclastic equivalents (Figs 3 and 5a). The volcanic centres are aligned in a northeast–southwest direction (Fig. 3). The lavas, lahar breccias and pyroclastic fall-out deposits are arranged in a decreasing order away from the volcanic axes. Isotopic dates on the andesite and dacite lavas range from 19 to 15 Ma (Ercan et al. 1984b, 1985; Altunkaynak & Yılmaz 1997 and refs cited therein). The palaeontological data obtained from the intercalated sedimentary rocks yield Early–Middle Miocene ages (Akyürek & Soysal 1983; Ercan et al. 1984b).

The Kozak Horst is separated from surrounding grabens by a set of en echelon, oblique-slip and transtensional faults, displaying left-lateral strike-slip and dip-slip components. The faults strike mainly in northeast–southwest, north-northwest–south-southeast and north–south directions (Figs 2 and 4). The northeast–southwest striking faults are predominant and they dip steeply with an average angle of 70°.

The stratigraphic sequence in the northeast–southwest grabens surrounds the Kozak Horst and forms the Zeytindağ Group illustrated in Fig. 5a. Lying on top of the Dikili Group, or resting directly on the metamorphic basement, this sequence begins with the commonly internally chaotic coarse clastic Yayaköy Formation. The strata are fault-scree, slope waste, debris flow and lateral fan deposits that occur along the steeply dipping faults, and were sourced from fault-elevated blocks. Away from the faults towards the graben axes they grade into sandstones, siltstones, mudstones, marls and white lacustrine limestones of the Ularca Formation. The Zeytindağ Group is commonly undeformed, except near the fault zones where the units are back-tilted towards the horst blocks at an average angle of 15°.

Andesitic volcanic activity waned before development of the northeast–southwest trending grabens because their fill is devoid of intermediate volcanic rocks. Only few and scattered basaltic lava flows (the Eğrigöl Basalts) formed during this phase. These are interbedded with lacustrine limestones of Late Miocene age (Nebert 1978; Akyürek & Soysal 1983; Ercan et al. 1984b). The lavas were fed from fissures, associated mainly with the northeast–southwest graben-bounding faults and are dated by the Rb/Sr method at 9–6 Ma (Borsi et al. 1972; Ercan et al. 1985).

The upper group is represented by the infill of the Bergama Graben (Figs 3 and 4). The present graben-floor sediments are related to ongoing active processes controlled by the graben-bounding faults, and are fan deposits formed along the northern graben margin and fluvial deposits. The fan deposits are sourced from high hills bordering the east–west graben. The complete graben fill of the Bergama Graben is nowhere exposed, but is calculated from gravity data (The Turkish Petroleum Co., unpublished) to be thicker than 500 m. The precise age of the graben fill is not known, however, it may be inferred from stratigraphic data to be younger than Late Miocene–Early Pliocene because the graben-bounding faults cut and elevate rock units on both north and south horsts, which contain strata as young as Late Miocene–Early Pliocene age. No volcanic rocks were extruded during development of the Bergama Graben basin.

The southern sector

The geology of the southern sector can also be divided into northeast–southwest trending horsts and grabens (Fig. 4). These structures, which are cut abruptly by the active graben-bounding faults of the Bergama Graben, appear to be the southerly continuation of horsts and grabens of the same trend observed in the northern sector (Fig. 2). Among the structural zones of the southern sector, the Maruflu Horst, located at the centre of the region, is the most prominent morphological feature. This horst is bordered by the Zeytindağ and Yayaköy Grabens to the west and the east, respectively.

The Maruflar Horst has an average elevation of 300 m. Toward the Zeytindağ Graben, in the northwest, the topography steps down along en echelon oblique faults striking northeast–southwest with normal slip as well as a dextral strike-slip component (Fig. 4). The Maruflar Horst is geologically similar to the Kozak Horst, but elevation is lower and the cover rocks are more intact. In the horst, rhyolitic domes and associated ignimbrites are present and distributed along short curvilinear faults which delimit the ponded ignimbrites. The distribution pattern, and the time and space association of the volcanic rocks, suggest a link to a shallow-level granitic pluton in a caldera environment, similar to the Kozak Pluton. The rock association of the southern sector, as illustrated on the geology map and the stratigraphic column (Figs 4 and 5a) is the same as for the northern sector.

The data from the Bergama region may be interpreted as follows. The three major rock groups of the region are separated by unconformities and were deposited in three successive superimposed basins of different orientation. Consecutive stages of basin development are described in Fig. 6. The lower unit is deformed with approximately east–west trending open folds. Rare reverse faults verging commonly north are also recognized. The middle and upper units are commonly undeformed.

The Edremit Graben region

The Edremit Graben is one of the largest east–west trending, offshore grabens of western Anatolia (Fig. 1), and is 80 km long and 5–20 km wide. The northern margin of the graben is bounded by the linear mountain front of the Kazdağ Mountain (Fig. 7), rising steeply from sea level to over 1000 m. The southern margin has a more subdued topography where the coastline displays bays and inlets. This morphology suggests that the northern side has been recently elevated by the graben-bounding faults, an interpretation supported by the presence of raised beaches $c.$ 100 m asl. Along the southern margin of the graben, north dipping ($> 50°$), listric normal faults are closely developed ($c.$ 1 km apart). The unconsolidated sediments deposited on the footwall blocks are back-tilted by $c.$ 15°. Slickenlines on well-polished fault planes have been measured from 25 individual faults and reveal major normal slip with an average N60°E slip vector, plunging by $c.$ 50°. Faults on the southern side are antithetic to the main northern margin fault system.

The Edremit Graben area is close to the area of influence of the North Anatolian Transform Fault Zone (NAFZ), which is presently the most active major structure of Anatolia. This dextral fault, which splits into several strands in northwest Anatolia, has branches that trend toward the Edremit Graben (Fig. 1). Possibly related to motions on the NAFZ, the faults situated to the north of the Edremit Graben display both strike-slip and dip-slip components. A petroleum exploration well drilled in the Edremit Gulf area, has revealed strike-slip displacement in the offshore areas, as evidenced by offsets of the Neogene units, from the land areas into the gulf (Turkish Petroleum Corporation, unpublished). Seismic, gravity and drilling data have shown > 2700 m of Neogene and younger sediments in the gulf. Of these, $c.$ 700 m are weakly consolidated and unconsolidated clastic rocks, deposited only within the present graben floor. A geology map of the northern side of the Edremit Graben (Fig. 7) shows that the region is also divided into northeast–southwest trending horsts and intervening grabens, similarly to the Bergama Graben region. These are the Kazdağ and Ezine Horsts and the Etili and Gülpınar Grabens.

The most prominent topographic feature of the region is the Kazdağ Mountain (Horst). Along this horst, basement metamorphic rocks of Palaeozoic–Triassic age (Bingöl et al. 1973) are exposed. The basement rocks include high-grade gneisses, schists, migmatites,

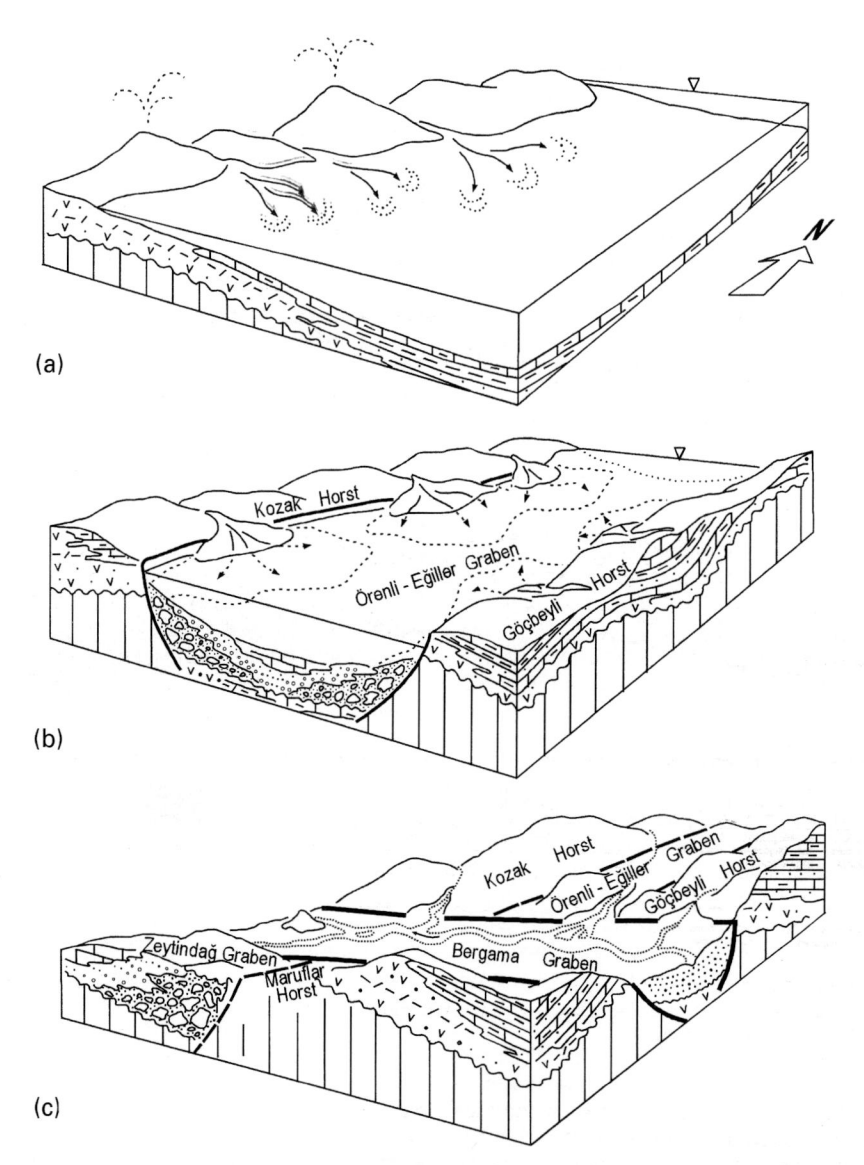

Fig. 6. Block diagrams showing subsequent stages of the geological evolution of the Bergama Graben region. (**a**) During Early Miocene time the region was covered by a vast lake basin(s) in which low-energy lacustrine sediments, such as micritic limestones and shales, were commonly deposited. The volcanoes, trending northeast–southwest, delimited the lake basin. Intermediate lavas, pyroclastic flows and fall outs were extruded from the volcanoes. (**b**) The andesitic volcanic activity waned at the end of the Middle Miocene. A new set of northeast–southwest trending oblique faults, along the similar trend of the volcanoes, were formed under a transtensional regime during the Late Miocene. This created horsts and narrow grabens. Along the graben margins, coarse clastics were deposited and sourced from the fault-elevated blocks. Away from the graben margins the conglomerates grade into fine clastics and then into lacustrine limestones, deposited along the centres of the grabens. (**c**) During latest Miocene–Early Pliocene time the horst–graben morphology was considerably reduced as a result of severe erosion. Consequently, a low-relief plateau surface was established. This was followed by the development of a set of approximately east–west trending normal faults formed under the north–south extension during Plio-Pleistocene time. These faults cut, and clearly succeed, the northeast–southwest horsts and grabens.

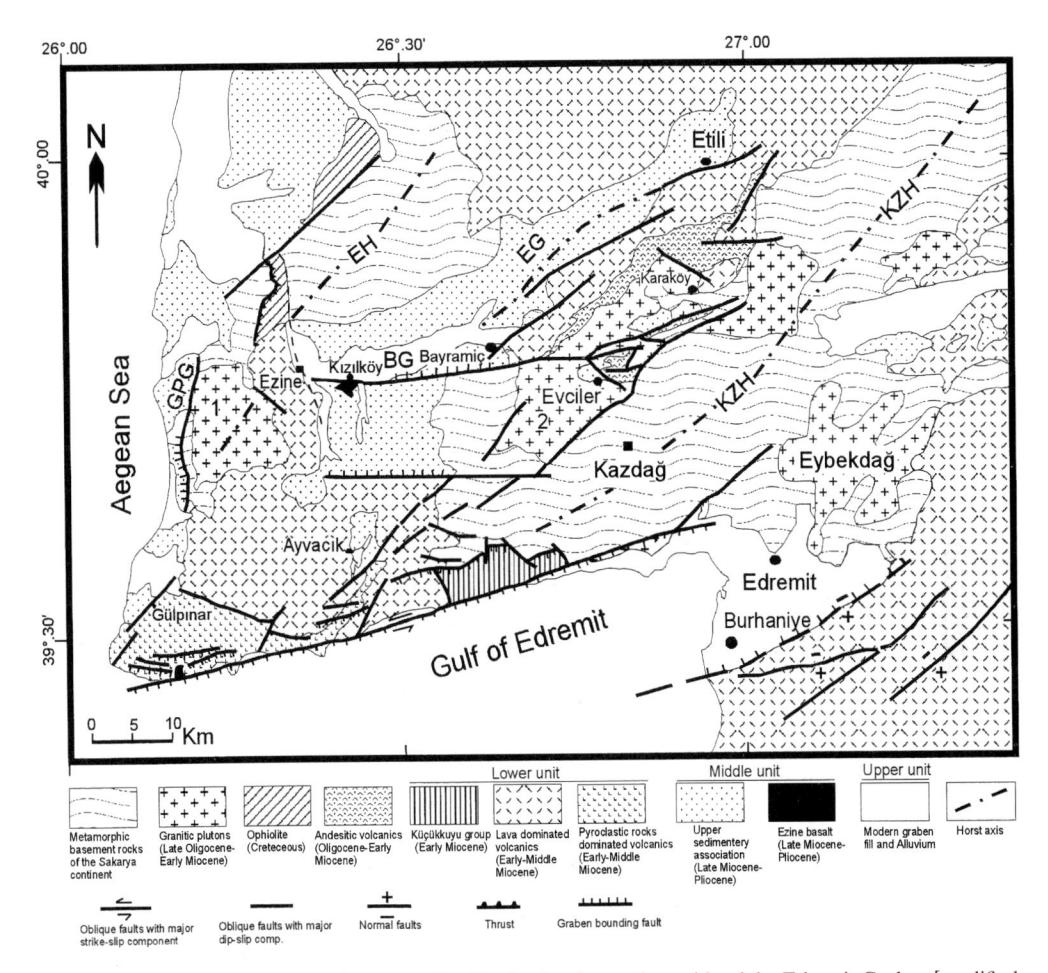

Fig. 7. Geological map of the southwestern Biga Peninsula, the northern side of the Edremit Graben [modified after Karacık & Yılmaz (1998)]. KZH, Kazdağ Horst; EH, Ezine Horst; EG, Etili Graben; GPG, Gülpınar Graben; BG, Bayramiç Graben.

metagabbros, and low-grade phyllites, marbles and recrystallized limestones. Plutonic rocks were emplaced into the metamorphic rocks during Late Oligocene–Early Miocene times, i.e. the Karaköy–Evciler and Kestanbolu Plutons, isotopically dated at 25 ± 0.2 (Birkle & Satır 1995) and 28 ± 0.88 Ma (Fytikas et al. 1976), respectively. The plutons are elliptical magmatic bodies with northeast–southwest long axes, and composed mainly of granodiorites and diorites enveloped by fine-textured hypabyssal and volcanic rocks of similar compositions (Karacık 1995; Karacık & Yılmaz 1998).

In the northern sector, three major lithostratigraphic units, separated by unconformities, may be distinguished (Figs 5b and 7); a lower unit (the Küçükkuyu Group), a middle unit (the Ezine Group) and an upper unit (the Bayramiç Group). Of the three rock units, only the Küçükkuyu Group is exposed on the horst blocks. The Küçükkuyu Group is composed of a thick (> 300 m) volcano–sedimentary association. At the base of the sequence is a thick bituminous white shale unit (> 250 m); thin sandstone beds alternate with shales in the middle of the succession. This turbiditic sandstone–shale alternation is interpreted as a flysch deposited in a lacustrine environment during the Early Miocene (İnci 1984; Siyako et al. 1989). Towards the top of the flysch, pyroclastic beds representing early products of volcanic activity are intercalated with the sediments. During this period the northeast–southwest trending en echelon, oblique faults with dextral strike-slip (N45–60°E) components, were formed. The evidence for this is: (1) red coloured, internally chaotic, coarse clastic rocks deposited on top of flysch, or resting directly on elevated basement metamorphic rocks – these were formed as debris flow deposits and fluvial pebble conglomerates; (2) red beds deposited next to the fault planes, and apparently confined to the vicinity of the northeast–southwest trending, steeply (> 60°) northwest dipping faults.

The coarse clastic rocks grade laterally and vertically into a lacustrine white marl–shale succession. The sediments have coal seams and beds containing the Eskihisar sporomorph association that yielded an Early Miocene age (Benda 1971). Intermediate lavas are intercalated with the sediments; they are latite and andesite lava flows, flow breccias and lahar breccias, and accumulated along the northeast–southwest oriented faults. The lavas are dated by the K/Ar method at 19–17 Ma (Borsi et al. 1972; Ercan et al. 1985).

The Ezine Group is present in the Etili Basin which formed as a half-graben. There is no distinct fault system along its northern side, where the topography is subdued. The southern edge of the graben is bounded by northeast–southwest trending en echelon oblique-slip faults. These faults are commonly steep, with an average altitude of N65–75°E, 70–85°NW.

The Ezine Group rests unconformably on the lower unit. At its base it consists of fault scree deposits and fluvial pebble–cobble conglomerates. These internally chaotic rocks were apparently derived from the elevated fault blocks and deposited along the basin margin. Some of these strata may be of fan-delta origin. The coarse clastic rocks pass laterally and vertically into fine-grained sandstones and siltstones, which grade rapidly into white limestones interpreted as lake deposits (Siyako et al. 1989). The limestones are the dominant lithology in the Ezine Group. At the top of the sequence, red mudstones and siltstones alternate with, and gradually replace, the limestones. There are a few scattered subaerial, vesicular basalt lava flows (the Ezine Basalts) within the group (Figs 5b and 7) which represent fissure eruptions fed from vents along the faults running parallel to the basin margin. No age diagnostic fossils were found to date the sedimentary rocks of the Ezine Group, however, basalts alternating with the sediments were dated by K/Ar and Ar/Ar methods at 6–4 Ma (Borsi et al. 1972; Ercan et al. 1985; authors' unpublished data).

The southern boundary of the Etili Graben is cut and bounded by a set of east–west trending normal faults that are younger than Late Miocene–Early Pliocene. These faults reduced the size of the rift forming a new east–west trending small graben, known as the Bayramiç Graben, and modified its orientation from northeast–southwest to east–west (Fig. 7). The east–west trending faults are steep, north dipping and segmented over distances < 10 km. Elevation of the Kazdağ Mountain to the present structural position occurred possibly during this stage, because the influx of coarse clastics, sourced from the Kazdağ High, were deposited as unconsolidated materials (the Bayramiç Group) into the Bayramiç Graben basin along the northern margin of the Kazdağ Horst. No age-diagnostic fossils have been found in the present graben fill, therefore, ages of the present graben fill can only be estimated from stratigraphic data, but graben-bounding faults cut and displace rocks as young as Early Pliocene; hence, the graben fill is clearly younger than this.

The present east–west trending Bayramiç Graben appears to be cogenetic with the Edremit Graben, because both are parallel and have been formed by the young east–west trending faults (Fig. 7). The southern sector of the Edremit Graben region corresponds to the northwestern slope of the Kozak Horst – geological maps and stratigraphic sections of this region are shown in Figs 3 and 5a.

The Gediz Graben region

This region is located in the central part of western Anatolia where one of the major east–west trending grabens, the Gediz Graben, is located (Fig. 1). The Gediz Graben is c. 140 km long and 10–15 km wide, and forms an arc-shaped structural pattern; the graben trends approximately east–west and west-northwest–east-southeast from the Ahmetli area to the west and to the east, respectively (Fig. 1). The graben faults are described in considerable detail by Patton (1992), Cohen et al. (1995) and Yılmaz et al. (1999), and only major geological features will be outlined here. The graben is asymmetrical, with the southern margin steeper and seismically more active. Along the southern margin rises the steep northern flank of Bozdağ Mountain, bounded by a major fault zone consisting of a number of steeply (> 70°) north dipping normal faults, one of which moved during a major earthquake near Alaşehir in 1969 (Eyidoğan & Jackson 1985). The major fault, although observed along the whole length of the graben, is segmented on a short-length scale (Fig. 8); each segment ≤ 10 km. Drainage on the northern slope of Bozdağ Mountain is mainly through fault-parallel linear valleys and

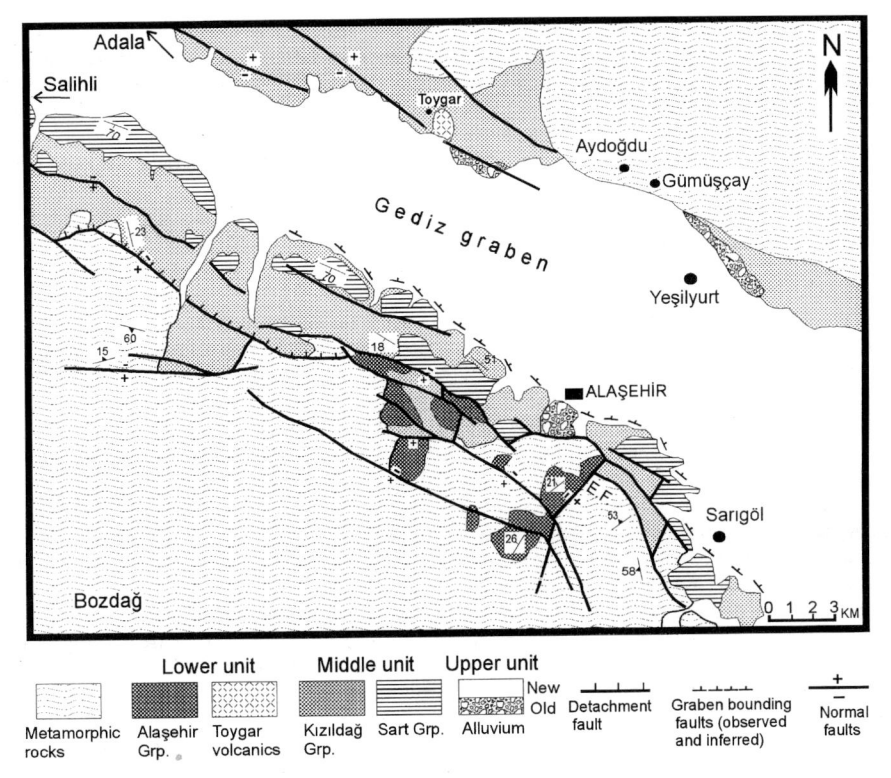

Fig. 8. Simplified geological map of the Gediz Graben [modified after İztan & Yazman (1990) and Yılmaz *et al.* (1997b)]. EF, Evrenli Fault.

subsequent stream valleys. Headward erosion by the streams has not yet reached the flat-lying plateau at the top of Bozdağ Mountain.

The northern edge of the Gediz Graben is also fault bounded but topography along this margin is subdued. The faults are morphologically less marked and seismically inactive. In Bozdağ Mountain metamorphic rocks of the Menderes Massif, crop out as uplifted basement against the Neogene units (Fig. 8) and consist of schists, gneisses, migmatites and phyllites. Neogene, and younger, sedimentary rocks overlie the metamorphic rocks of the Menderes Massif; they are different in the northern and southern sectors around the Gediz Graben and are outlined separately below.

The southern sector

Cover rocks of the Menderes Massif consist of three distinct lithostratigraphic units – a lower unit (the Alaşehir Group), a middle unit (the Kızıldağ Group) and an upper unit (the Sart Group); these may also be subdivided into formations and members (Fig. 5c). The collective thickness of the succession is > 1500 m.

Geology of the southern sector is described briefly below. Detailed descriptions of the lithostratigraphical units, the lithofacies and structures are given in İztan & Yazman (1990), Seyitoğlu & Scott (1992), Cohen *et al.* (1995) and Yılmaz *et al.* (1997b, 1999).

At the base of the Alaşehir Group are polygenic, 5–100 m thick, cobble–pebble conglomerates of the Evrenli Formation, derived from underlying schists and gneisses. This coarse (up to 60 cm) clastic unit is poorly sorted and subrounded to subangular, suggesting proximity to the source. To the east, the coarse clastic rocks are terminated by the major north-northeast–south-southwest striking Evrenli Fault (Fig. 8). The conglomerates are interpreted as fault scree and fan deposits (Cohen *et al.* 1995). Although weak, the palaeocurrent directions, obtained from clast imbrication, trend mostly north-northeast and north–south, and are subparallel to the Evrenli Fault. The lithologies of the Alaşehir Group collectively display a fining-upward profile. This is observed clearly along the east–west sections, where rapid lateral transitions are observed from coarse conglomerate to sandstone, siltstone and then to shale of the Zeytinçay Formation. The sandstones contain alternating lignite beds (c. 15 cm) and the shales are bituminous. The fine detrital units are interpreted as lacustrine facies (İztan & Yazman 1990; Seyitoğlu & Scott 1992, 1996; Cohen et al. 1995). The lithological ordering indicates a rapid transition from a high-energy fluvial depositional environment to a low-energy lacustrine environment.

Toward the top of the Evrenli Group, shales are replaced gradually by sandstones, fine-grained channelled conglomerates and rare limestone lenses (Çaltılık Formation), interpreted as the late-stage deposition in a lake basin. These rocks are overlaid by red mudstones and sandstones, suggesting that the lake basin was gradually filled, and rise above the lake level.

Above the Evrenli Group, the Kızıldağ Group begins with coarse clastic rocks, which rest in many places directly on the Menderes metamorphic rocks. The Kızıldağ Group may be divided into two parts. The lower part is represented dominantly by red, massive, coarse-grained (5–15 cm) poorly sorted conglomerates (Kızıl Formation). In the upper part of the sequence, grain size diminishes, a faint sorting is recognized and the colour turns gradually to yellow (Mersinligedik Formation). The lower part (> 250 m thick) is interpreted to be a fan deposit with associated river channel fills (İztan & Yazman 1990). The upper part (400 m thick) is regarded as sediments formed in a linear fluvial system (Cohen et al. 1995). The palaeocurrent direction, measured from clast imbrication, is either to the west or northwest (Cohen et al. 1995).

The Kızıldağ Group, which formed following deposition of sediments in a low-energy lacustrine environment, marks the beginning of major tectonic activity. The thick, coarse clastic rocks in the lower part of the succession were apparently formed in a high-energy depositional environment. The energy of the environment decreased gradually with time, as evidenced by the fining-upward profile recognized in the succession. The tectonic activity may be associated with development of the major normal fault along which the Menderes Massif was elevated; this is evidenced by the constant supply of coarse clastic materials from the metamorphic massif into the structurally low-lying basin in front of it. The fault is a major breakaway (detachment) fault and its present dip is < 25° to the north. Above the fault plane red clastics of the Kızıldağ Group are steeply back-tilted (> 45°) to the south. The fault scarp is exposed discontinuously between Karadut and Allah-diyen villages, where erosion has partly removed the red clastic rocks from the hanging wall block. The polished fault scarp is barely eroded by the modern streams, suggesting that the fault plane has been exposed recently. Under the present fill of the Gediz Graben, a subhorizontal fault plane has been imaged on seismic data (Eyidoğan & Jackson 1985; Turkish Petroleum Corporation, unpublished seismic profile along the Gediz Graben, 1999). This may be interpreted as part of the detachment fault which has been cut and displaced on downthrown blocks by steeply dipping, younger, normal faults of the Gediz Graben.

To the south, the detachment fault is traced to the top of Bozdağ Mountain, where a flat-lying erosional surface, developed above the Menderes metamorphic rocks and the cover sediments, including the Upper Miocene red clastics, is recognized. This erosion surface represents a denudation phase that obliterated pre-existing topography. For this reason, the erosional surface may be used as a stratigraphic marker to distinguish earlier and later geological events. The rise of the Menderes Massif to its present structural position and topographic elevation is apparently younger than this erosional phase, because the erosional surface and the gently dipping detachment fault have been faulted and displaced by a younger set of steeply dipping (> 70°) faults (Fig. 8). Remnants of this erosion surface and detachment are presently observed in the downthrown blocks. These faults initiated development of the present Gediz Graben system and controlled the subsequent sediment deposition.

The upper unit (Sart Group) is confined to Gediz Graben Depression. This group consists of thick (> 500 m), semi-lithified, buff-coloured, coarse (> 3 cm), poorly bedded, alluvial conglomerates with minor sandstones and clays. They rest unconformably on either the metamorphic rocks or on the Kızıldağ Group (Fig. 8). The Sart Group borders the graben floor along its southern margin, where it is uplifted by the graben-bounding faults, and forms steep linear hills (Fig. 8). The conglomerates are back-tilted to the south, at varying angles (10–45°) depending on their proximity to the faults. The back-tilting is due to rotation on the north dipping normal faults. The clastic content of the Sart Group may be interpreted as large fan and fluvial sediments sourced from high hills bordering the graben. The clast imbrication indicates an east to west palaeocurrent direction (Cohen et al. 1995).

According to seismic data of the Turkish Petroleum Corporation (unpublished), the thickness of the sediment fill in the Gediz Graben changes abruptly from c. 2.5 to 3 km to < 1 km along the graben axis, across approximately north–south trending faults which are hidden under the unconsolidated sediments. The faults appear to define grabens and intervening horst blocks trapped within the Gediz Graben. On the horst blocks, only the middle and upper units are present; the lower unit is confined to the grabens. This observation is in close agreement with the present authors' field observations from the southern sector, where the approximately north–south striking Evrenli Fault delimits the Evrenli Group and controlled its deposition.

There is controversy about the ages of the three rock groups discussed above. According to Seyitoğlu & Scott (1992, 1996) the three rock groups form a continuous succession, and the lower and middle units are Early Miocene in age, based on sporomorph associations from a lignite specimen from the top of the Kızıldağ Group. Later studies in the region (e.g. Cohen et al. 1995) have used this age evidence extensively. This Early Miocene age for the middle rock unit is in apparent conflict with the age data obtained from the same rock unit by other studies. For example, İztan & Yazman (1990) and Ediger et al. (1996) considered the red clastic rocks of the Kızıldağ Group to be Late Miocene in age by assigning Kızılhisar pollen assemblages from the mudstone layers to the Late Miocene. Emre (1996) collected a gastropod fauna from the same rock group and assigned a Late Miocene–Early Pliocene age. Recent studies (Ediger, Batı & Sarıca, pers. comm.) orientated towards resolving the age controversy have gathered new fossil data supporting the Late Miocene–Pliocene age. The stratigraphical data from this study, as displayed in Fig. 5c, also favour the Late Miocene age assignment for the reasons given below. (1) There is a time gap between development of the three rock units, as represented by the unconformities (Fig. 5c). Each one of these rock units was formed under an entirely different tectonic environment. For example, shales of the Evrenli Group were deposited in a low-energy, lacustrine environment. The overlying Kızıldağ Group, as represented by the coarse red conglomerates mostly of fluvial origin, formed in a tectonically active environment. (2) The lignite specimen, from which the Lower Miocene Eskihisar sporomorph association was identified, was collected from the sandstone–conglomerate unit, lying at the top of the Kızıldağ Group. These rocks were deposited in a highly oxidizing, high-energy, fluvial environment unfavourable for coal formation. The lignite fragment is small (20 × 5 cm) and trapped within the sandstone–conglomerate beds with its long axis lying obliquely to the strike of the host rock. It therefore seems likely that the lignite represents a transported fragment. (3) The red clastic rocks pass laterally into lacustrine limestones of the Adala Formation in the northern side of the graben, which yield Upper Miocene–Lower Pliocene fossils. Ünay et al. (1995) list fossils of Pliocene–Pleistocene age from the bottom of the Sart Group, indicating an age range from Pliocene to the present.

In the light of these data the following geological evolution may be envisaged for the southern sector of the Gediz Graben. (1) The Alaşehir Group was deposited within an approximately north–south trending graben basin which formed prior to the development of the east–west graben system. This unit is Early–Middle Miocene in age according to the sporomorph association. (2) The Kızıldağ Group was possibly formed in response to rapid uplift of the Menderes Massif, along a major breakaway fault. The red clastics were shed from the Menderes structural high and deposited within structural lows surrounding the massif (Fig. 8). This began during the Late Miocene and possibly continued during Early Pliocene time (İztan & Yazman 1990). (3) The initial uplift of the Menderes Massif and the consequent sediment deposition was followed by a major erosional phase. During this phase, topographic irregularities were considerably denuded. (4) The erosional surface has been fragmented by east–west trending steep faults, which also cut older structures, and produced the present Gediz Graben. This began during Pliocene–Pleistocene times and has continued to the present. The Sart Group was deposited during this phase as the infill of the Gediz Graben.

The northern sector

A wide variety of rock groups crop out separately along the northern edge of the graben. Previous studies (e.g. İztan & Yazman 1990; Seyitoğlu et al. 1992, 1996; Cohen et al. 1995) concentrating on this narrow zone failed to correlate these rocks and consequently links between groups on the northern and southern sides of the Gediz Graben have not been established. Along the northern edge of the Gediz Graben four major rock units may be distinguished. From west to east these are: the Adala Limestone; the Toygar Volcanics; the Aydoğdu Formation; the Yeşilyurt Formation. The Adala Limestone is mainly a white

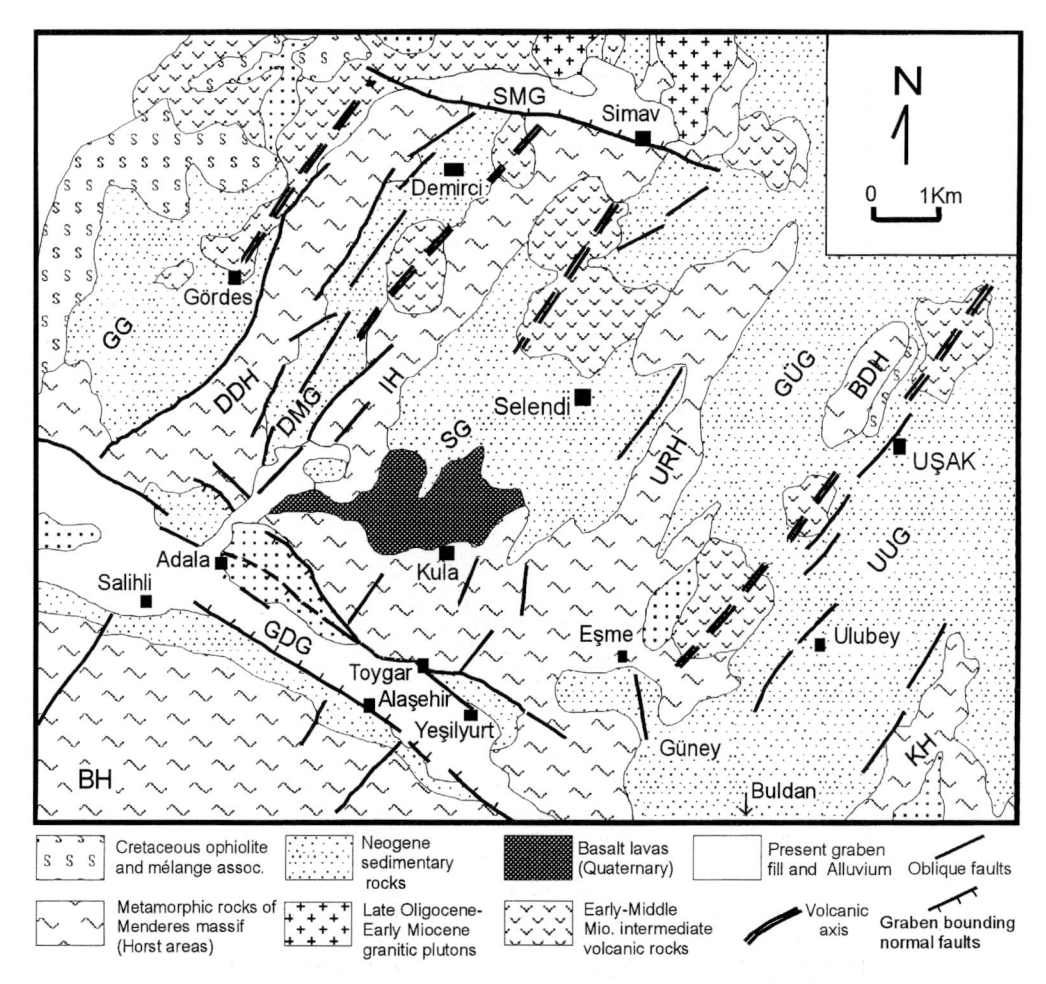

Fig. 9. Simplified geological map of the region lying between the Simav and Gediz Grabens. SMG, Simav Graben; GDG, Gediz Graben; GG, Gördes Graben; DMG, Demirci Graben; SG, Selendi Graben; GÜG, Güre Graben; UUG, Uşak–Ulubey Graben; DDH, Dilekdağ-Demircidağ Horst; IH, İcikler Horst; URH, Umurbaba–Rahmanlar Horst; BDH, Beydağ Horst; KH, Karadağ Horst; BH, Bozdağ Horst.

limestone which crops out in the Adala area (Fig. 8) and unconformably overlies high-grade metamorphic rocks of the Menderes Massif. The Adala Limestone is represented by thick- (> 100 cm) to medium-bedded micritic lacustrine limestones and is overlain unconformably by a poorly lithified, sandstone–conglomerate unit.

The Toygar Volcanics are represented by a dacite dome 150–200 m in diameter which crops out near Toygar village and is dated by the K/Ar method at 15 Ma (Yazman &. Ercan, pers. comm.; Seyitoğlu, pers. comm.). The volcanic rocks are petrochemically similar to the widespread Lower–Middle Miocene volcanic associations of northwest Anatolia.

The Aydoğdu Formation is a thick (> 200 m) clastic unit that consists mainly of poorly bedded cobble–pebble conglomerates, which are commonly subrounded to subangular, poorly sorted and poorly lithified. To the north, the Aydoğdu Formation is bounded by steeply dipping, east–west trending, normal faults extending along the graben margins, on which are sited hot springs and associated travertine deposits. Away from the faults the coarse clastics pass laterally into sandstones and are interpreted as debris flow deposits and fan conglomerates. The succession is repeated, and becomes younger, towards the north, suggesting that faulting migrated to the north with time. The Aydoğdu Formation is inferred to be

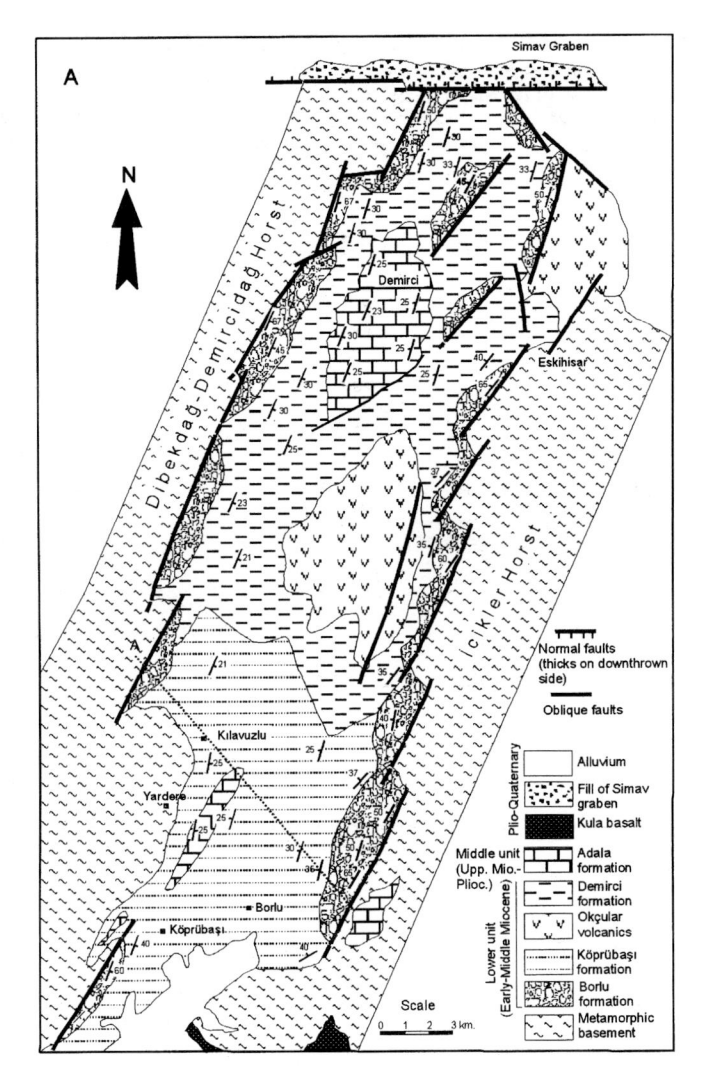

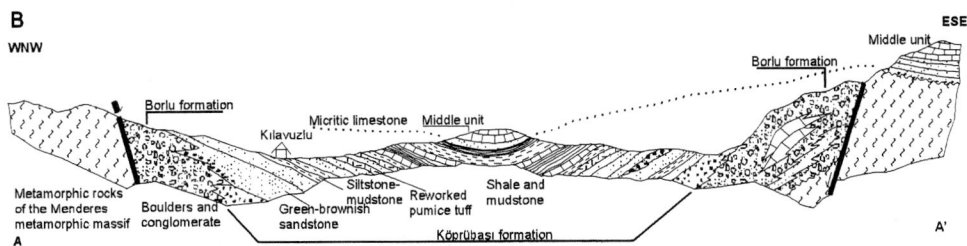

Fig. 10. (a) Simplified geological map of the Demirci Graben and the surrounding areas. A–A′, Direction of the accompanying cross-section in (b). (b) Geological cross-section across the A–A′ direction.

Plio-Quaternary in age by Ercan *et al.* (1978) and Yusufoğlu (1996).

The Yeşilyurt Formation is a red mudstone–siltstone alternation that crops out extensively to the north of Yeşilyurt (Fig. 8). Between Yeşilyurt and Uşak (Fig. 9) it contains a number of channelled conglomerate lenses. Around Yeşilkonak village, near Eşme, red clastic rocks pass upwards into, or alternate with, white limestones. The limestones are 20–100 m thick and display similar lithofacies characteristics to the Adala Limestone.

Further away from the graben margin, the northern region is divided into approximately north-northeast–south-southwest trending horsts and grabens (Fig. 9). From west to east, the major grabens are Gördes, Demirci, Selendi and Uşak–Ulubey. The grabens and the intervening horsts share many similar lithological and structural features and, therefore, only the geology of one horst–graben pair is summarized here – the Demirci Graben and the adjacent İcikler Horst.

The geology map of the Demirci Graben, and the accompanying geological section across the graben, are illustrated in Figs 10a and b. The graben is separated from the adjacent horsts by a series of steep, discontinuous en echelon oblique faults. None of the fault segments is longer than 7–8 km. The majority of the faults display clear right-lateral strike-slip and subordinate normal-slip displacements. The fault planes are steeper than 70° and the slickenlines plunge < 30°.

The generalized stratigraphic sections of the Demirci Graben and the İcikler Horst are presented in Fig. 5d1. At the base of the graben fill is a cobble (up to 1 m in diameter) and pebble conglomerate unit (50–300 m thick) with subrounded to subangular clasts, derived from underlying high-grade schists and gneisses, belonging to the Menderes Massif. The strata are internally chaotic, poorly sorted clastic rocks – the Borlu Formation – and are interpreted as debris flows and alluvial fan deposits formed in association with elevations of the adjacent fault blocks. The coarse clastics are bounded by the north-northeast–south-southwest trending faults. Away from the fault zone the clast size decreases rapidly; the cobblestones pass laterally into a well-bedded sandstone–siltstone alternation – the Köprübaşı Formation (< 250 m). At the top of the sequence these are replaced by marls and shales of the Demirci Formation. A few limestone lenses also occur at the top of the succession.

In the southern part of the Demirci Graben, the fine-grained detrital rocks alternate with green coloured, ash-fall tuff horizons (Fig. 10b). The pyroclastic beds increase in thickness and abundance towards the north, where the two volcanic centres are located along the approximately northeast striking faults. Around the volcanic centres, latite and dacite lavas, flow breccias and lahar breccias of the Okçular Volcanics dominate. These were extruded during Early–Middle Miocene time when most of petrochemically similar intermediate volcanic rocks of northwestern Turkey were generated (Ercan et al. 1985; Yılmaz 1989, 1990, 1997; Seyitoğlu et al. 1997). Isotopic dates on volcanic rocks from this and neighbouring areas range from 18 to 14 Ma.

In addition to the northeast–southwest striking graben-bounding faults, there is also a subordinate set of northwest–southeast trending sinistral strike-slip faults in this region (Fig. 10a). The two sets of faults may be interpreted as a conjugated pair formed under north–south compression, which affected the region during the Early–Middle Miocene period. These two sets of faults were cut and displaced by the east–west striking faults during development of the Gediz and Simav Grabens, located to the south and north, respectively (Figs 9 and 10a). The region bounded by the Gediz and Simav Grabens forms a giant east–west elongated horst block. This horst was gently north tilted during development of the east–west trending grabens, as shown by the following evidence: (1) the post-Late Miocene–Early Pliocene low-relief erosional surface, which is observed extensively on the plateaux, developed above the horst and has been tilted northeast by 10–15°; (2) the metamorphic basement rocks underlying the Neogene cover have been uplifted to higher elevations in the south than in the north; consequently, the cover rocks have been mostly removed by erosion in the south and the metamorphic rocks crop out more extensively along the southern part of the horst (Fig. 9).

The volcano–sedimentary association of Early–Middle Miocene age is overlain disconformably by the Adala Limestone [the Demirci Formation of İnci (1984)] and, in places, rests directly on the metamorphic rocks (Fig. 10b). The Adala Limestone begins locally with conglomerates and sandstones sourced from the adjacent fault-elevated structural high (Yılmaz et al. 1997b). A further indication of the disconformity is the presence of an erosional surface, recognized between the Adala Limestone and the underlying units. According to palynological data, the Adala Limestone is Late Miocene–Early Pliocene in age (İnci 1984).

In the light of the data documented above, correlations among the units of the northern and southern sectors may be summarized as follows: (1) the Toygar Volcanics have the same petrochemical characteristics and formed during the same time period as the volcanic rocks of the Demirci Graben; (2) in the Demirci Graben the lavas alternate with sedimentary rocks which closely resemble the Alaşehir

AGE OF WESTERN ANATOLIAN GRABENS

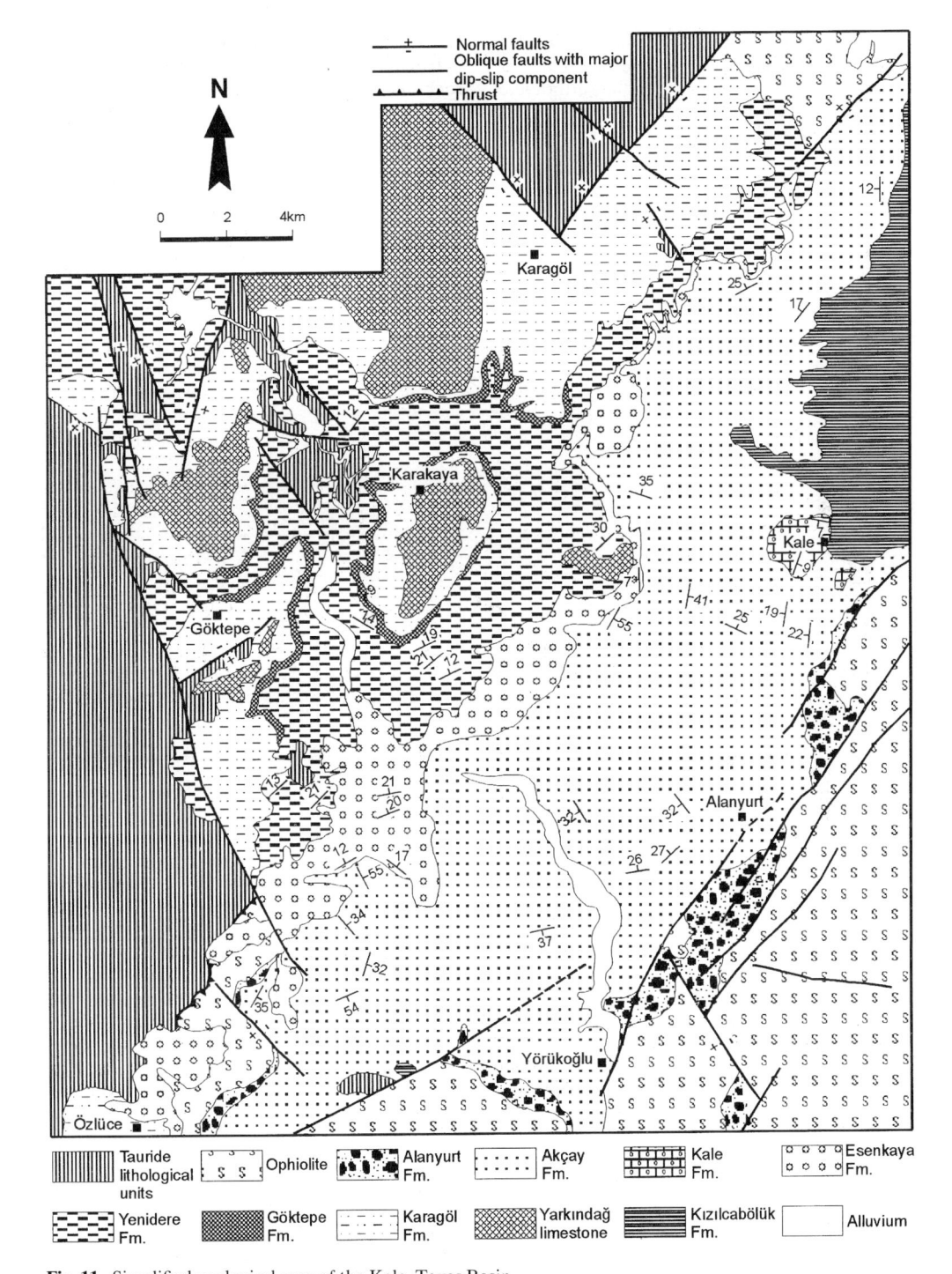

Fig. 11. Simplified geological map of the Kale–Tavas Basin.

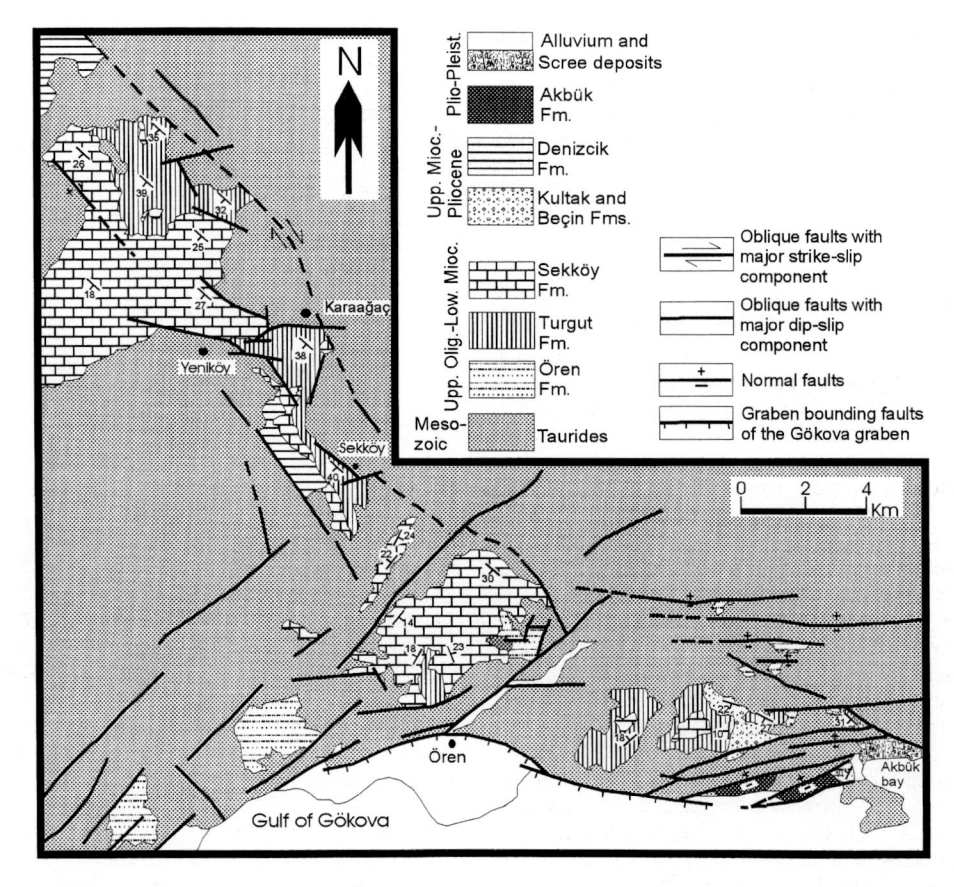

Fig. 12. Simplified geological map of the Ören Graben.

Group of the southern sector in overall lithological and structural characteristics. Both sedimentary units are Early–Middle Miocene in age.

The coeval Upper Miocene–Pliocene red clastic rock units of the southern and northern sectors – the Kızıldağ Group and the Yeşilyurt Formation, respectively – may be regarded as laterally equivalent successions because they grade into one another in the Sarıgöl–Buldan area east of the Bozdağ Horst. The Kızıldağ Group has coarser clastic material than the Yeşilyurt Formation. The difference in grain size appears to be due to the proximity to the source, the Bozdağ Horst. Yeşilyurt Formation, in turn, grades into the Adala Limestone in the northern regions, i.e. Güre, near Eşme (Fig. 9). Among these rock units, only the Aydoğdu Formation, with restricted outcrop along the northern edge of the graben, is genetically connected to the Gediz Graben. The Aydoğdu Formation and the Sart conglomerates of the southern graben margin may thus be regarded as laterally equivalent deposits formed within the Gediz Graben basin.

The Gökova Graben region

The geology of the Gökova Graben and surrounding regions (Fig. 1) has previously been described in considerable detail (De Graciansky 1972; Atalay 1980; Ersoy 1991; Görür et al. 1995). Here, critical additional information obtained during our field studies is presented.

In the Gökova region, basins of various ages and orientations have been distinguished (Figs 11–13). The oldest basin, the Kale–Tavas molasse Basin of Şengör & Yılmaz (1981), is Late Oligocene–Early Miocene in age and orientated east-northeast–west-southwest. The youngest basin is the modern Gökova Graben. Between the development of these approximately east-northeast–west-southwest or east–west basins, roughly north–south trending basins formed during the Early Miocene and

Late Miocene times. Locations of the basins are shown in the simplified geological maps of the region (Figs 1 and 13); the stratigraphic columns are displayed in Fig. 5e and f. Major geological characteristics of the basins are briefly described below.

The Kale–Tavas Basin

The infill of the Kale–Tavas Basin is observed discontinuously from the east of Denizli City in the east to the north of Gökova Graben in the west (Fig. 1). The Kale–Tavas Basin units rest on the Lycian Nappes with angular unconformity (Fig. 11). No stratigraphic contacts of the basin units are observed with the Menderes Massif. The sequence may be summarized as follows (Fig. 5e). The lowermost unit is a red, thick, massive to poorly bedded and poorly sorted coarse conglomerate – the Alanyurt Formation. It is composed predominantly of ophiolitic material derived from the underlying ophiolite, which forms the uppermost tectonic slice in the Lycian Nappe pile. The red continental clastics are devoid of fossils. They were accumulated to the north of an approximately northeast–southwest trending fault zone (Figs 11 and 13a) as debris flow and fluvial deposits sourced from the fault-induced structural highs. The clast imbrication indicates an approximately southeast to northwest palaeocurrent direction. The grain size decreases toward the north, where the red clastics pass laterally and vertically into grey conglomerates, which in turn give way to grey and well-sorted sandstones with some limestone lenses. In places they pass laterally into grey shales containing lignite beds. These are lagoonal and shallow-marine clastics which contain gastropods, bivalves and benthic foraminifers of Late Oligocene–Early Miocene age (Koçyiğit 1984; Hakyemez 1989; Akgün & Sözbilir 2000).

The sandstones are followed upwards by a flysch-like sequence, the Akçay Formation, composed mainly of alternating sandstones and marls, which contain frequent lenses of coarse conglomerates formed as fluvial channel fills. Upward in the section, the high energy of the environment of deposition decreases, as evidenced by the fining-upward succession, and eventually the clastics are gradually replaced by limestones of the Kale Formation. The fauna identified from the limestones yields Aquitanian–Burdigalian ages (Hakyemez 1989; Akgün & Sözbilir 2000). The younger part of the succession consists of fluvial and lacustrine deposits of Late Miocene–Pliocene age and rests unconformably on the marine units. This begins with a thick conglomerate unit – the Esenkaya Formation – which passes laterally and vertically into red mudstone–sandstone units comprising the Yenidere and Karagöl Formations; these contain coarse conglomerate lenses – the Göktepe Formation. The top of the succession is a white lacustrine limestone – the Yarkındağ Limestone – which covers the entire region.

The Ören and Yatağan Basins

The Ören and Yatağan Basins are approximately north-northwest–south-southeast trending subparallel basins (Figs 1, 12 and 13) which display identical strata of Early Miocene age. Commonly, the basin units rest unconformably on slightly metamorphosed successions of the western Taurides, composed dominantly of Mesozoic platform carbonates and underlying older phyllitic rocks. In places, the lowermost clastic rocks rest unconformably on fossiliferous marine limestones of the Ören Formation, which possibly formed within the Kale–Tavas Basin prior to development of the north-northwest–south-southeast trending basins. Formation of these basins was apparently controlled by an oblique fault system displaying major dip-slip and subordinate dextral strike-slip components. The Lower Miocene units are disrupted by the east–west trending normal faults around the Gulf of Gökova, which opened the Gökova Graben (Fig. 12).

The Ören and Yatağan Basin infills consist mainly of two rock units (Fig. 5f). The lower unit is composed predominantly of clastic rocks, beginning with coarse conglomerates. This unit is a massive to poorly bedded and poorly sorted grey conglomerate with well-rounded clasts of the underlying recrystallized limestones and phyllites. These are of debris flow and fluvial origin. Upward in the succession, the coarse conglomerates are replaced by sandstones. The upper unit is a shale–marl dominated, fine-clastic association assigned to the Turgut Formation; it has a number of lignite beds. The top of the sequence is a white, marl and limestone unit – the Sekköy Formation (Atalay 1980).

The north-northwest–south-southeast trending basin fills are commonly represented by sediments, deposited in a fluvial and lacustrine environment (Görür et al. 1995). However, towards the south where the east-northeast–west-southwest trending, partly coeval, marine Kale–Tavas Basin lies, a number of marine incursions into the north–south troughs occurred intermittently, as evidenced by marine bivalve- and gastropod-bearing sandstone layers identified within the lacustrine succession. The marine beds wedge out towards the north.

The white lacustrine succession is replaced upwards by brown to red continental mudstones and conglomerates of the Kultak Formation. In places, the red continental clastics rest directly on the Lycian Nappe units. The red beds yield a rich mammal fauna with an age span from Middle Astracian to Turolian in the Muğla region (Atalay 1980). The coarse clastic unit is locally > 500 m thick. It is composed of fluvial and debris flow deposits, laid down within fault-bounded, approximately north–south trending, basins. Fine clastic rocks replace the conglomerates both laterally and vertically. The coarse and fine clastic rocks were previously named the Yatağan Formation (Atalay 1980). The top of the sequence is a white lacustrine limestone – the Denizcik Formation – which is not confined to the limits of the fault-controlled depressions where the lower clastic rocks accumulated. It covers a vast region as a capping limestone, extending from the gulf areas in the west to the Denizli region in the east (Figs 11 and 13b). This unit is the equivalent of the Yarkındağ Limestone of the Kale–Tavas region. This indicates that the north–south graben depressions were filled and lost their topographic expression before deposition of the limestones.

The Gökova Graben

This is the southernmost graben of western Anatolia (Figs 1 and 12). It is c. 150 km long and enlarges westward from c. 5 km to > 30 km. A major part of the graben is offshore, forming the Gulf of Gökova. The northern margin is bounded by a linear mountain front, which rises steeply to > 1000 m. The southern margin is topographically less steep and marked by many bays and small offshore islands. East–west trending listric normal faults, together with a set of N60–80°E trending oblique faults, characterize the northern margin. The east–west striking faults commonly cut and offset the oblique faults. Earthquakes were registered along both of the fault systems (Ambrasseys 1988; Jackson & McKenzie 1988; Taymaz et al. 1991). Due to the development of the two sets of faults making an acute angle with one another, the shoreline is not straight but zigzags, due to breaks in the normal faults at c. 5–10 km intervals. The oblique fault system forms deep gorges inland containing the major drainage of the interior region. The alluvial fans enlarging seaward have developed at intersections of the two fault systems, i.e. the Ören alluvial Fan (Fig. 12). The fault scarp is observed clearly from the Ören area to the eastern end of the graben; to the west of Ören it is less well marked.

The graben-bounding faults of the Gökova Graben have apparently controlled deposition of a thick post-Miocene–Lower Pliocene(?) sedimentary sequence comprising the Akbük Formation. The east–west faults cut the north-northeast–south-southwest trending faults and truncate the various rock groups, including the Upper Miocene–Lower Pliocene(?) strata. The sediments, which have been deposited along the margin of the Gökova Graben, consist of coarse clastics, formed as scree deposits, unconsolidated slope debris and lateral fan deposits. Their source is undoubtedly the uplifted horst block lying in the immediate vicinity. The graben fill is rotated gently (5–10°) northwards due to rotation of the major east–west trending listric faults. The age of the unconsolidated sediments may be inferred to be post-Late Miocene–Early Pliocene(?) from the stratigraphic evidence because debris from rocks of this age sourced from the adjacent horst blocks have been incorporated into the present graben fill.

In conclusion, structures and sediments of the northern Gökova region suggest different episodes of basin development. During the first episode (in Early Miocene time), interconnected north-northwest–south-southeast and east-northeast–west-southwest trending basins were formed. The basins were filled and lost their morphological expression at the end of Middle Miocene time. A new group of basins were developed along the rejuvenated north-northwest–south-southeast trending faults during Late Miocene times and possibly survived into the Early Pliocene. The modern Gökova Graben developed later, along east–west trending normal faults which cut and truncate the older units and their associated structures.

Discussion and conclusions

Considering the geology of the graben areas outlined above, the following major stages may be distinguished in the geological evolution of western Anatolia: (1) a pre-graben stage; (2) an east–west extensional stage; (3) an earlier stage of north–south extension; (4) an interrupting stage of north–south extension; and (5) a later stage of north–south extension.

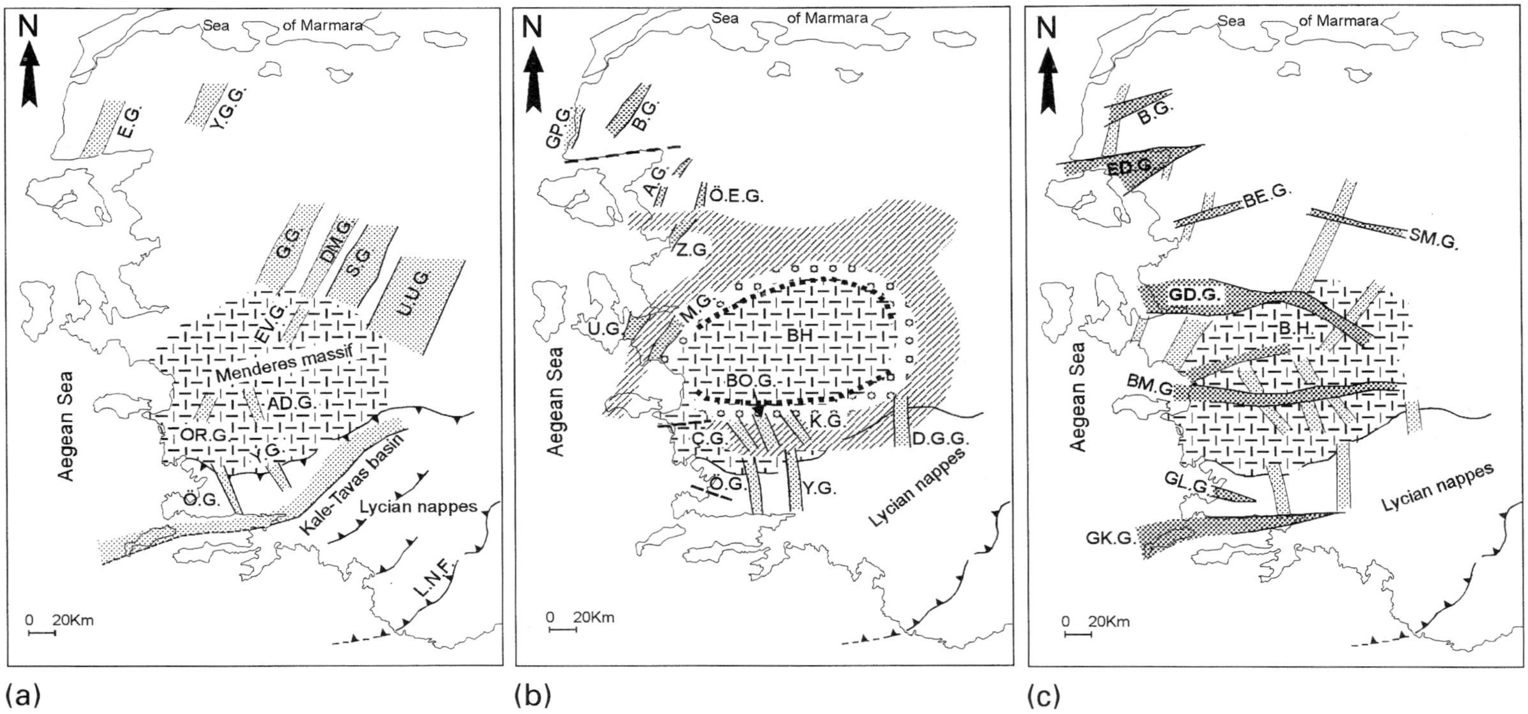

Fig. 13. Palaeogeological maps depicting the tectonic evolution of western Anatolian grabens from Early Miocene to present. The present shape of the Menderes Massif is displayed in the maps for reference purposes. (a) Early Miocene: EG, Etili Graben; YGG, Yenice–Gönen Depression; GG, Gördes Graben; DMG, Demirci Graben; SG, Selendi Graben; UUG, Uşak–Ulubey Graben; ORG, Ortaklar Graben; ADG, Aydın-Dalama Graben; ÖG, Ören Graben; YG, Yatağan Graben. (b) Late Miocene: GPG, Gülpınar Graben; BG, Bayramiç Graben; AG, Altınova Depression; ÖEG, Örenli–Eğiller Graben; ZG, Zeytindağ Graben; UG, Urla Graben; MG, Mustafa Kemalpaşa Graben; ÇG, Çine Graben; BOG, Bozdoğan Graben; KG, Karacasu Graben; DGG, Denizli Güney Graben. (c) Plio-Pleistocene: BG, Bayramiç Graben; EDG, Edremit Graben; BEG, Bergama Graben; SMG, Simav Graben; GDG, Gediz Graben; BMG, Büyük Menderes Graben; GLG, Güllük Graben; GKG, Gökova Graben; BH, Buldan Horst. O, Clastic materials, derived from the Bozdağ Horst into the surrounding low lands. The obliquely ruled area represents the lake basin(s) surrounding the Bozdağ Horst.

The pre-graben stage

This stage corresponds to the Late Cretaceous–pre-Miocene time period. In western Anatolia the following events have been recorded during this period. (1) Collision between the continental fragments of the Pontides and the Taurides along the İzmir–Ankara Suture (Fig. 1) during the Late Cretaceous–pre-Middle Eocene interval (Şengör & Yılmaz 1981; Yılmaz et al. 1997a). (2) Following collision, convergence continued and the region underwent compression. Consequently, the thrusting propagated northwards and southwards across the Pontides and the Taurides, respectively. Thrusting in the north continued until the end of Late Eocene–Oligocene times (Yılmaz et al. 1997a; Yılmaz & Polat 1998). In the south, within the western Taurides and along the frontal thrust zone of the Lycian Nappes, it continued until Late Miocene time (De Graciansky et al. 1967; Şengör 1982; Hayward 1984; Şengör et al. 1985). The Lycian Nappes form a nappe package, consisting of slices of the western Taurus metamorphic basement and the overlying Mesozoic platform carbonates together with the dismembered ophiolites that form the uppermost nappe. The nappe package is known to have travelled southwards during Early Miocene time and was finally emplaced onto the Lower Miocene basin fill of the Antalya Basin before Late Miocene time.

The amount of crustal shortening within western Anatolia is difficult to ascertain because: (1) the amount of shortening is not known precisely across the middle, deep crustal, ductilely deformed metamorphic rocks; (2) to balance and restore cross-sections across the orogen is difficult because of lateral movements during, and after, collision. However Şengör et al. (1985) and Şengör (1993) estimated a minimum shortening of 200 km from the Thrace region in the north to the Mediterranean in the south.

No sediments were deposited in central-western Anatolia during the Late Eocene–Late Oligocene interval, suggesting that the region was a subaerial landmass and topographically high, possibly as a consequence of the continuing convergence. Since the continental crust is presently c. 30–32 km thick (Ezen 1991), it is assumed to have been > 50 km thick before the onset of north–south extension (Le Pichon & Angellier 1981; Şengör 1982, 1993; Jackson & McKenzie 1988).

Western Anatolia underwent high-temperature, medium–high-pressure metamorphism during Eocene times (50–36 Ma) (Şengör et al. 1984; Satır & Friedrischen 1986; Hetzel & Reischmann 1996), and widespread upper mantle and crustal melting (45–20 Ma) is shown by the mantle-derived, hybrid-intrusive granites (Bingöl et al. 1982; Yılmaz 1989; Harris et al. 1994). The sillimanite–kyanite schists and gneisses (Bozkurt & Park 1994) of the Menderes Massif formed at pressures of 5–8 kbar and depths of 15–25 km. The geochronological and structural constraints suggest that crustal thickening and timing of the peak metamorphism of the Menderes Massif were partly synchronous.

The fluvial and lacustrine sediments, which were formed extensively during Early Miocene times, rest unconformably on high-grade metamorphic rocks of the Menderes Massif. Their outcrops are observed from the northern (the Demirci–Simav region) to the southern (the Çine–Yatağan region) edges of the massif (Fig. 1), and suggest that the main uplift of the Menderes Massif, resulting in exhumation of middle and lower crustal rocks, occurred before Early Miocene time. Approximately 20 km of material was removed from the massif, and this apparently happened before it was elevated once again under the north–south extension during Late Miocene time. Bozkurt & Park (1994) demonstrated a detachment fault of possible Oligocene age from the southern border of the massif. However, this age is widely debated (e.g. Hetzel & Reischman 1996), mainly because it is not substantiated by radiometric age data. The problem of how the Menderes Massif was exhumed during Eocene–Oligocene(?) time remains unsolved. This may have occurred along the low-angle thrusts, back thrusts and the associated normal faults recognized extensively in the massif.

The east–west extension stage

There are about 20 major, approximately north–south trending, grabens in western Anatolia (Fig. 1, 13a and b) between the Sea of Marmara in the north and the Gulf of Gökova in the south. In detail, their trends display small diversity between north-northeast and north-northwest: the grabens do not appear to extend east of longitude 29°E (Fig. 1). Some of these grabens were formed during Early Miocene time and others during Late Miocene time (Figs 13a and b).

The southernmost representative of the north–south Grabens are the Ören and Yatağan Grabens, which merge into the approximately east-northeast-west-southwest trending Kale–Tavas Basin (Figs 11 and 13a). Within this

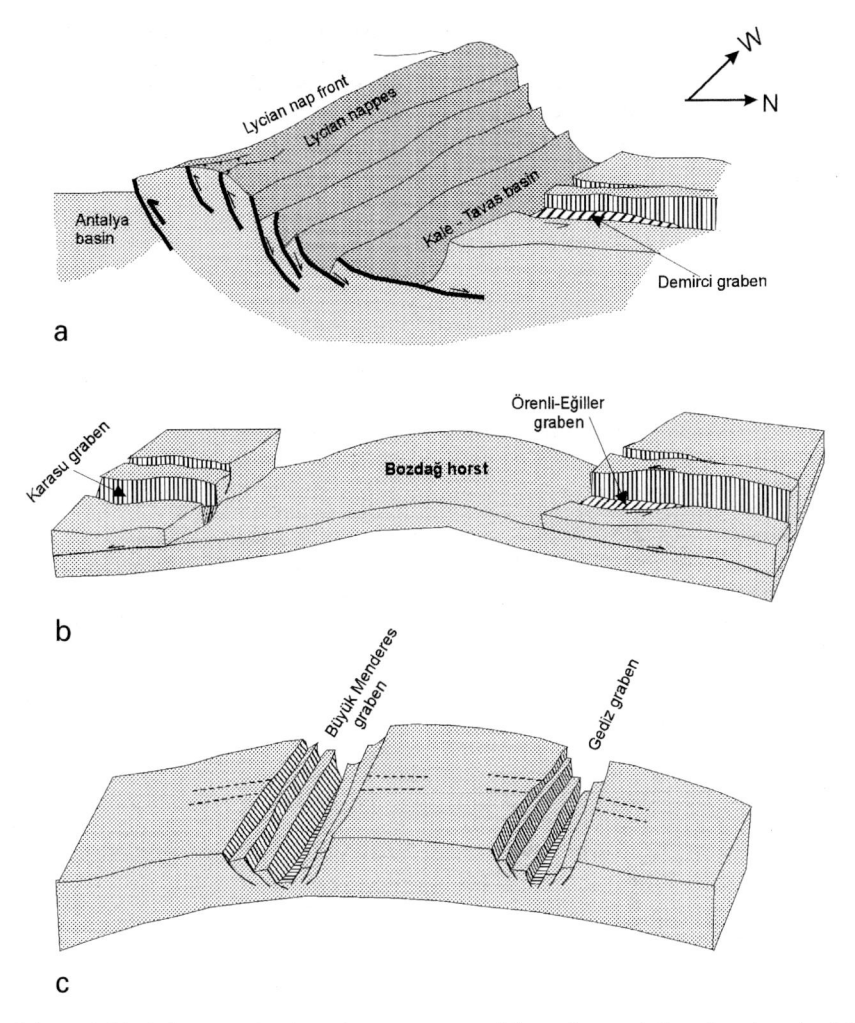

Fig. 14. Schematic block diagrams showing subsequent stages of the graben evolution of western Anatolia from the Early Miocene to the present. (a)–(c) represent the southern, central and entire western Anatolia regions, respectively. (**a**) During the Late Oligocene the Kale–Tavas Basin was formed above the southerly transporting Lycian Nappes. In association with this movement, the northwest–southeast trending (i.e. the Ören and Yatağan) grabens were developed (see Fig. 13a). Further north, approximately north–south trending grabens (see Fig. 13a) began to develop under east–west extension, possibly associated with ongoing north–south compression. The north–south basins of the northern regions may thus be regarded as Tibetan-type grabens. (**b**) During the Late Miocene, north–south extension began. The Bozdağ Horst was elevated and a major breakaway fault was formed. Above the detachment surface, approximately north–south trending grabens began to form as cross-grabens (see Fig. 13b). (**c**) After the Late Miocene–Early Pliocene(?), the east–west trending modern Graben basins began to form under rejuvenated north–south extension after a brief period of erosion. The east–west grabens have truncated older structures and associated rock units, including Upper Miocene–Lower Pliocene(?) strata.

basin, sediment deposition began during the Late Oligocene (Chattian; Hakyemez 1989; Akgün & Sözbilir 2000), indicating that this graben began to form slightly earlier than the north–south grabens. The Kale–Tavas Basin extends westwards toward Crete and may extend even further west to the Greek mainland (Papanikolau 1984).

Mostly ophiolite-derived, coarse clastic materials of the Kale–Tavas Basin were supplied from a structural high formed from the Lycian Nappes located to the south of the basin. The southern margin of the basin with the structural high is characterized by east-northeast–west-southwest trending normal faults (Fig. 13a). In this respect, the Kale–Tavas Basin, which was situated above the contemporaneously southerly transported Lycian Nappe package, may be regarded as a piggyback basin (Fig. 14a). The previous view (Şengör & Yılmaz 1981), which regards it as a molasse basin with respect to the Menderes Dome, located to the north, is refuted by the following field data: (1) no metamorphic debris from the Menderes Massif was transported into the Kale–Tavas Basin, thus, there is no evidence to suggest that the Menderes Massif was topographically higher during this period; (2) there is no stratigraphic contact of the Kale–Tavas Basin sediments with the Menderes Massif. The data thus suggest that this part of the Menderes Massif remained buried under the Lycian Nappes during this period.

The north–south trending grabens are commonly bounded by oblique slip faults, i.e. strike-slip faults with dip-slip components. They form a conjugated pair, which possibly developed in a north–south compressional stress field. This view is supported further by gentle east–west trending folds recognized extensively in the Lower Miocene successions. There are also locally developed reverse faults verging north and south.

The earlier stage of the north–south extension

The north–south extension began during Late Miocene time. During this period the east–west trending Bozdağ Horst, located in the middle of the Menderes Massif (Figs 1 and 12) was elevated. Low-angle (10–15°) detachment faults are recognized along both the southern and northern flanks of the horst. High-grade metamorphic rocks and pre-tectonic granites of Miocene age (19 Ma; Hetzel et al. 1995) along the footwalls of the detachment fault are juxtaposed against unmetamorphosed Upper Miocene continental red beds (The Kızıldağ Group) in the hanging walls. The detachment faults appear to have been active mainly during the Late Miocene. The Upper Miocene fluvial lateral fan deposits derived from the horst were transported into the surrounding low topography as coarse clastic materials. Away from the horst, the coarse clastics pass into the lacustrine limestones, suggesting that the horst was surrounded by possibly interconnected lake basins (Fig. 13b). Further away from the horst, to the south and to the north, approximately north–south trending cross-grabens began to develop on the upper plates of the detachment faults (Fig. 14b). These were long and narrow troughs bounded by the oblique-slip faults which controlled development of Upper Miocene sediment deposition. Some of these faults appear to be spatially associated with the Early Miocene basin-bounding fault system, as exemplified by the Ören Graben. This suggests that these faults represent reactivated structures, as described by Şengör et al. (1984), which had already formed during the Early Miocene period (Bozkurt 1998). The initial stage of development of the east–west trending normal faults also occurred during this period. This may be documented from the northern (the Edremit Graben: Karacık & Yılmaz 1998), the central (the Söke Basin faults: Yılmaz et al. 1997b) and the southern (the Güllük Fault: Ercan et al. 1984a) regions. In addition to the stratigraphical data, the evidence for the Late Miocene age of this faulting is the basalt lavas dated at 9–6 Ma extruded from these fault zones.

The interruption stage of the north–south extension

Continuation of the north–south extension appears to have been interrupted at the end of Late Miocene–Early Pliocene(?) time when a region wide, low-relief erosional surface formed on rocks up to, and including, Upper Miocene strata. Limestones and the red clastic rocks of Late Miocene age, flat or tilted, were eroded and lie below this surface. Remnants of the erosional surface are presently observed on the plateau areas almost continuously from the Kazdağ Mountain in the north to the Bozdoğan–Karacasu plateaux in the south of the Menderes Massif (Fig. 1).

The later stage of the north–south extension

Following development of the erosional surface, north–south extension was rejuvenated and the east–west grabens began to form (Fig. 14c). As a result, the erosional surface, as well as the older structures and the Miocene strata, was cut by faults bounding the east–west trending grabens. The continuity of structures and strata are abruptly interrupted and/or offset across the east–west grabens (Figs 2–4, 6, 7, 9, 10a and 11). The low-relief erosional surfaces on the plateaux make a sharp edge and a steep angle with the slopes of newly developed graben valleys. Headward erosion from the graben floors to the top of the plateaux has not yet been fully developed; the stream valleys across the graben walls are presently in an incipient stage. On the plateaux the streams still display meandering profiles, despite the presence of the nearby deep valleys. These morphological data collectively indicate that the east–west graben valleys are very young features and that the time elapsed since their formation has been quite short.

The geology of the graben regions outlined above, when evaluated collectively, leads to the evolutionary history described below.

(1) Thick volcano–sedimentary associations were formed in western Anatolia during the Early–Middle Miocene period. They share many common geological features, which may be summarized as follows: (a) a number of approximately north–south trending, fault-bounded basins were formed in an east–west extensional regime; (b) magmas were emplaced along the tensional openings associated with these fault zones; (c) initial deposits in the north–south grabens were coarse clastics. Later shale-dominated successions were deposited extensively in a low-energy lacustrine environment. These rocks contain extensive lignite beds having a common pollen assemblage known as the Eskihisar sporomorph association (Benda 1971; Lutting & Steffens 1976).

The coeval sedimentary rocks of similar lithofacial characteristics crop out extensively and suggest that possibly interconnected lake basin(s) invaded the entire west of Anatolia, from the Çanakkale region in the north to the Aydın-Muğla regions in the south (Fig. 1) (Benda 1971; Benda *et al.* 1974; Lutting & Steffens 1976; Seyitoğlu & Scott 1991). Contrary to earlier claims (Seyitoğlu & Scott 1991, 1996; Cohen *et al.* 1995), the evidence is conclusive and therefore the Lower–Middle Miocene lacustrine units have no genetic relationship to the east–west graben basins. The Lower Miocene sediments rest nonconformably on the high-grade metamorphic rocks of the Menderes Massif and suggest that the Menderes Dome was not yet elevated to its present position during this period.

According to some previous workers (e.g. Şengör *et al.* 1984; Seyitoğlu & Scott 1994), the north–south and east–west grabens, such as the Demirci and Gediz Grabens (Fig. 9), were formed contemporaneously under ongoing north–south extension during the Early Miocene. They suggested that motions along the east–west trending major detachment faults triggered development of north–south trending cross-faults on the hanging walls as accommodation structures. The field data presented above oppose this view, particularly for the north–south trending Demirci, Gördes and Selendi Graben basins, for the following reasons. (a) The east–west grabens cut and displace the north–south grabens. Consequently, the north–south grabens remained as hanging grabens on the footwall blocks. This indicates that the north–south grabens were continuous across the east–west graben areas before development of the latter (Fig. 13c). Clear evidence for their continuity is displayed by seismic data, obtained from the file of the Turkish Petroleum Corporation along and across the east–west grabens. This reveals that within the east–west grabens are remnants of the trapped north–south trending grabens. (b) The post-Oligocene stratigraphic sequence, as shown above, is discontinuous in western Anatolia and marked unconformities are commonly recognized, particularly between the Middle and Upper Miocene and the Upper Miocene–Lower Pliocene(?) and younger units. These intervals correspond to region-wide denudational phases.

(2) In western Anatolia, two magmatic episodes are readily distinguished (Borsi *et al.* 1972; Fytikas *et al.* 1984; Yılmaz 1989, 1990, 1997). An intermediate to felsic association was first formed during Oligocene–Early Miocene times, during which granitic plutons intruded into shallow crustal levels such as in the Kozak and Ezine–Kestanbol areas. They were surrounded by temporally, spatially and genetically associated hypabyssal and volcanic rocks (Yılmaz 1989, 1990; Altunkaynak & Yılmaz 1998; Genç 1998; Karacık & Yılmaz 1998). The second magmatic phase occurred during the Late Miocene–Pliocene period, during which alkaline basalt lavas were sporadically erupted (Fytikas *et al.* 1984; Ercan *et al.* 1985; Yılmaz 1989, 1990, 1997; Seyitoğlu *et al.* 1997).

The magmatic associations of the first phase display common petrochemical characteristics, being high-K, calc-alkaline and hybrid. Their compositions reveal crystallization from mantle-derived magmas contaminated by up to 50% crustal materials (Borsi *et al.* 1972; Yılmaz 1989, 1990, 1997; Güleç 1991). The high amount of the crustal components appears to be related to an over-thickened crust, through which the magmas passed during Oligocene–Early Miocene time. This magmatic event may thus be regarded as a Tibetan or East Anatolian type, and late–post-collisional with respect to Tethyan convergence (Şengör & Kidd 1979; Yılmaz 1989, 1990; Pearce *et al.* 1990; Genç 1998).

The Late Miocene–Pliocene basalts display petrochemical features which are similar to the lavas formed within a continental-rift environment (Fytikas et al. 1984; Ercan et al. 1985; Yılmaz 1990, 1997). These data are in a close agreement with the tectonic environment in western Anatolia during this period. The first magmatic phase died out c. 14 Ma ago (Yılmaz 1989, 1997). According to the radiometric ages obtained from the various volcanic rocks from the entire western Anatolian region, there appears to be a brief non-volcanic interval, corresponding to a period between 14 and 10 Ma (Yılmaz 1990, 1997). This period coincides approximately with the transition from north–south compression to north–south extension. This event may be evaluated as late orogenic collapse following excessive crustal thickening, which is widely recognized in orogenic belts (Dewey 1988; Burchfiel & Royden 1991). An extensional regime often follows development of the orogenic belts and the extensional driving forces are known to be related to its pre-extensional history. In many cases, the area of extension invades the area of shortening with time (cf. Burchfiel & Royden 1991 and refs cited therein). During this period major breakaway faults began to form around the Bozdağ Horst of the Menderes Massif (Figs 13b and 14b). Above the detachment faults, the hanging wall blocks underwent large extensional strain as they moved away from the horst (Fig. 14b). The depositional response of the elevation of Bozdağ is the Kızıldağ Group in the Alaşehir–Salihli region (Fig. 9). The red, coarse clastics were deposited along the periphery of Bozdağ, as exemplified in a number of areas, including Yatağan-Muğla in the south [the Yatağan lithofacies of the Becker-Platen (1970)], Mustafa Kemalpaşa near İzmir in the west, Ortaklar–Aydın (Cohen et al. 1995) in the south and near Buharkent–Buldan towns of Denizli in the east (Figs 1 and 13b).

The triggering of the north–south extension in the Aegean region has been variously attributed to different mechanisms, as discussed at length by Şengör (1993). The views may be classified under three main groups: (1) roll-back of the subducted slab beneath the Hellenic Trench (Le Pichon & Angellier 1979; McKenzie and Yılmaz, 1991); (2) westward escape of the Anatolian Plate toward the Aegean region (McKenzie 1972; Dewey & Şengör 1979); (3) gravitational collapse of the over-thickened continental crust (Dewey 1988; Seyitoğlu & Scott 1991).

Timing of the roll-back of the slab is not yet clearly known. Seyitoğlu & Scott (1991) argue that the thermal collapse of the Menderes Dome is the single cause of the north–south extension. They also claim that the north–south extension began during the Late Oligocene and has continued uninterruptedly for the last c. 25–30 Ma. The collapse mechanism has undoubtedly played role in accelerating the extension, however, the Late Oligocene–Early Miocene timing of inception of east–west extension is not supported by the geological data documented above. Tectonic escape of Anatolia, which began during Late Miocene–Pliocene time (Şengör 1979; Şengör & Kidd 1979; Şaroğlu & Yılmaz 1987), appears to be more closely synchronous with the north–south extension. Therefore, it may be responsible for the beginning of the north–south extension.

Determination of the time of initiation of the modern graben basins can be estimated from the present strain rate and the total amount of extension. Using a variety of techniques, the present rate of extension has been calculated to be of the order of c. 2.5–6 cm^{-1} over a distance of c. 800 km between Bulgaria and the Mediterranean (Le Pichon & Angellier 1979; Jackson & McKenzie 1988; Ekström & England 1989; Main & Burton 1989; Sellers & Cross 1989; Westaway 1994). The ß factor of extension has been calculated from various sets of data, including: (1) topographic data, employing the Airy isostatic balance; (2) gravity data (Makris & Stobbe 1984; Meissner et al. 1987); and (3) seismic data (Makris & Stobbe 1984; Mindevalli & Mitchell 1989), obtained particularly from the wide grabens of western Anatolia (i.e. the Büyük Menderes and Gediz Grabens). This ranges from 1.2 to 1.6 in the land areas to 2 in the Aegean Sea. Using these data, the time period extrapolated for the amount of extension is < 5 Ma. This length of time is in close agreement with the post-Late Miocene–Early Pliocene age for the period of development of the east–west grabens, as revealed by this study's field data. This conclusion is supported further by the palaeomagnetic data, obtained from the northern Aegean region, indicating that the main rotations on opposite sides of the major east–west grabens occurred after the Late Miocene (Kissel & Laj 1988; Orbay et al. 1998). The palaeomagnetic studies on the Oligocene–Lower Miocene volcanic rocks indicate that clockwise rotations up of to 30° in the northern Aegean and northwestern Anatolia occurred before opening of the east–west grabens (Kissel & Laj 1988; Orbay et al. 1998). This rotation restores the Lower Miocene basins to an approximate north–south trend (Fig. 13a).

This paper is a part of the National Marine Geology and Geophysics project supported by the Turkish Scientific and Technical Research Council (TÜBİTAK). We are indebted to the Turkish Petroleum Corporation (TPAO) and the Turkish Academy of Sciences (TÜBA), which funded and supported different phases of this seven year long, field-based project. We would particularly like to thank our colleagues, Mr M. Yazman, H. S. Serdar, Ö. Balkaş and K. Saka, for accompanying us on some field trips, and for fruitful discussions. The TPAO is acknowledged for permission to use some of their seismic and other geological data. We are grateful to C. B. Burchfiel and A. Koçyiğit who critically read and reviewed the paper. Their numerous comments have greatly improved the text. We also thank the guest editors of this special volume, in particular E. Bozkurt for his great enthusiasm and his efforts in the preparation of this paper.

References

AKYÜREK, B. & SOYSAL, Y. 1983. Biga yarımadası güneyinin (Savaştepe–Kırkağaç–Bergama–Ayvalık) temel jeoloji özellikleri. *Mineral Research and Exploration Institute of Turkey (MTA) Bulletin*, **95/96**, 1–12 [in Turkish with English abstract].

AKGÜN, F. & SÖZBİLİR, H. (2000). A palynostratigraphic approach to the SW Anatolian molasse basin: Kale–Tavas molasse and Denizli molasse. *Geodinamica Acta*, in press.

ALTUNKAYNAK, Ş. 1996. *Bergama–Ayvalık Dolayında Genç Volkanizma Plütonizma İlişkilerinin Jeolojik ve Petrolojik Araştırılması*. PhD Thesis, İstanbul Technical University [in Turkish with English abstract].

—— YILMAZ, Y. 1998. The Mount Kozak magmatic complex, western Anatolia. *Journal of Volcanology and Geothermal Research*, **85**, 211–231.

AMBRASEYS, N. N. 1988. Engineering seismology, earthquake. *Engineering Structural Dynamic*, **17**, 1–105.

ATALAY, Z. 1980. Muğla–Yatağan ve yakın dolayı karasal Neojeninin stratigrafi araştırması. *Geological Society of Turkey Bulletin*, **23**, 93–99 [in Turkish with English abstract].

ATAMAN, G. 1974. Revue géochronologique, des massifs plutoniques et métamorphiques de l'Anatolie. *Hacettepe University Bulletin of Natural Sciences and Engineering*, **3**, 518–523.

BECKER-PLATEN, J. D. 1970. Lithostratigraphisce Untersuchungen im Känozoikum Südwest – Anatoliens (Türkei). *Beihefte zum geologischen Jahrbuch*, **97**, 244.

BENDA, L. 1971. Principles of the palynologic subdivision of the Turkish Neogene. *Newsletter on Stratigraphy*, **1**, 23–26.

—— INNOCENTI, F., MAZZUOLI, R., RADICATI, F. & STEDDENS, P. 1974. Stratigraphic and radiometric data of the Neogene in northwest Turkey. *Zeitschrift der deutschen geologischen Gesellschaft*, **125**, 183–193.

BİNGÖL, E., AKYÜREK, B. & KORKMAZER, B. 1973. Biga yarımadasının jeolojisi ve Karakaya formasyonunun bazı özellikleri. *Proceedings of the 50th Anniversary of Turkish Republic Earth Sciences Congress*. Mineral Research and Exploration Institute of Turkey Publications, 70–75 [in Turkish with English abstract].

——, DELALOYE, M. & ATAMAN, G. 1982. Granitic intrusion in western Anatolia: a contribution to the geodynamic study of this area. *Eclogea Geologisch Helvetica*, **75**, 437–446.

BIRKLE, P. & SATIR, M. 1995. Dating, geochemistry and geodynamic significance of the Tertiary magmatism of the Biga Peninsula, northwest Turkey. *In:* ERLER, A., ERCAN, T., BİNGÖL, E. & ÖRÇEN, S. (eds) *Geology of the Black Sea Region*. Proceedings of International Symposium on the Geology of the Black Sea Region, Mineral Research and Exploration Institute of Turkey (MTA) Publications, 171–180.

BORSI, S., FERRARA, G., INNOCENTI, F. & MAZZUOLI, R. 1972. Geochronology and petrology of recent volcanics in the Eastern Aegean Sea. *Bulletin of Volcanology*, **36**, 473–496.

BOZKURT, E. 1998. Origin of N-trending basins in western Turkey: 'replacement' vs 'revolutionary' structures. *Third International Turkish Geology Symposium, METU-Ankara*, Abstracts, 197.

—— & PARK, R. G. 1994. Southern Menderes Massif: an incipient metamorphic core complex in western Anatolia, Turkey. *Journal of the Geological Society, London*, **151**, 213–216.

BURCHFIEL, B. C. & ROYDEN, L. H. 1991. Antler orogeny: a Mediterranean-type orogeny. *Geology*, **19**, 66–69.

COHEN, H. A., DART, C. J., AKYÜZ, H. S. & BARKA, A. A. 1995. Syn-rift sedimentation and structural development of the Gediz and Büyük Menderes grabens, western Turkey. *Journal of the Geological Society, London*, **152**, 629–638.

DE GRACIANSKY, P. C. 1972. *Recherches géologiques dans le Taurus Lycien occidental*. Thèse Doctorat d'Etat, Université de Paris-Sud Orsay.

——, LEMOINE, M., LYS, M. & SIGAL, J. 1967. Une coupe stratigraphique dans le Paléozoique Supérieur et le Mésozoique á l'extremité occidentale de la chaine sud-anatolienne. *Mineral Research and Exploration Institute of Turkey (MTA) Bulletin*, **65**, 10–33.

DEWEY, J. F. 1988. Extensional collapse of orogens. *Tectonics*, **7**, 1123–1139.

—— & ŞENGÖR, A. M. C. 1979. Aegean and surrounding regions: complex multiplate and continuum tectonics in a convergent zone. *Geological Society of America Bulletin*, **190**, 84–92.

EDİGER, V. Ş., BATI, Z. & YAZMAN, M. 1996. Palaeopalynology of possible hydrocarbon source rocks of the Alaşehir–Turgutlu area in the Gediz graben (western Anatolia). *Turkish Association of Petroleum Geologists Bulletin*, **9**, 11–23.

EKSTRÖM, G. A. & ENGLAND, P. C. 1989. Seismic strain rates in regions of distributed continental deformation. *Journal of Geophysical Research*, **94**, 231–257.

EMRE, T. 1996. Gediz grabeninin jeolojisi ve tektoniği.

Turkish Journal of Earth Sciences, **5**, 171–185 [in Turkish with English abstract].

ERCAN, T. GÜNAY E. & TÜRKECAN, A. 1984a. Bodrum Yarımadası'nın jeolojisi. *Mineral Research and Exploration Institute of Turkey (MTA) Bulletin,* **97/98**, 1–12 [in Turkish with English abstract].

——, DİNÇEL, A., METİN, S., TÜRKECAN, A. & GÜNAY, E. 1978. Uşak yöresindeki Neojen havzalarının jeolojisi. *Geological Society of Turkey Bulletin,* **21**, 97–106 [in Turkish with English abstract].

——, SATIR, M., KREUZER, H., ET AL. 1985. Batı Anadolu Senozoyik volkanitlerine ait yeni kimyasal, izotopik ve radyometrik verilerin yorumu. *Geological Society of Turkey Bulletin,* **28**, 121–136 [in Turkish with English abstract].

——, TÜRKECAN, A., AKYÜREK, B. ET AL. 1984b. Dikili–Bergama–Çandarlı (Batı Anadolu) yöresinin jeolojisi ve magmatik kayaçların petrolojisi. *Geological Engineering,* **20**, 47–60 [in Turkish with English abstract].

ERSOY, Ş. 1991. Datça (Muğla) yarımadasının stratigrafisi ve tektoniği. *Geological Society of Turkey Bulletin,* **34**, 1–14 [in Turkish with English abstract].

EYİDOĞAN, H. & JACKSON, J. A. 1985. A seismological study of normal faulting in the Demirci, Alaşehir and Gediz earthquakes of 1969–70 in western Turkey: implications for the nature and geometry of deformation in the continental crust. *Geophysical Journal of the Royal Astronomical Society,* **81**, 569–607.

EZEN, Ü. 1991. Crustal structure of western Turkey from Rayleigh wave dispersion. *Bulletin of International Instate of Seismology and Earthquake Engineering,* **25**, 1–21.

FYTIKAS, M., GIULIANO, O., INNOCENTI, F., MARINELLI, G. & MAZZUOLI, R. 1976. Geochronological data on recent magmatism of the Aegean sea. *Tectonophysics,* **31**, 29–34.

——, INNOCENTI, F., MANETTI, P., MAZZUOLI, R., PECCERILLO, A. & VILLARI, L. 1984. Tertiary to Quaternary evolution of volcanism in the Aegean region. *In:* DIXON, J. E. & ROBERTSON, A. H. F. (eds) *The Geological Evolution of the Eastern Mediterranean.* The Geological Society, London, Special Publications, **17**, 687–700.

GENÇ, Ş. C. 1998. Evolution of the Bayramiç magmatic complex, northwestern Anatolia. *Journal of Volcanology and Geothermal Research,* **85**, 233–249.

—— & YILMAZ, Y. 1995. Evolution of the Triassic continental margin, Northwest Anatolia. *Tectonophysics,* **243**, 193–207.

GÖRÜR, N., ŞENGÖR, A. M. C., SAKINÇ, M. ET AL. 1995. Rift formation in the Gökova region, southwest Anatolia: implications for the opening of the Aegean Sea. *Geological Magazine,* **132**, 637–650.

GÜLEÇ, N. 1991. Crust–mantle interaction in western Turkey: implications from Sr and Nd isotope geochemistry of Tertiary and Quaternary volcanics. *Geological Magazine,* **128**, 417–435.

HAKYEMEZ, Y. H. 1989. Geology and stratigraphy of the Cenozoic sedimentary rocks in the Kale–Kurbalık area. Denizli–southwestern Turkey. *Mineral Research and Exploration, Institute of Turkey (MTA) Bulletin,* **109**, 1–14.

HARRIS, N. B. W., KELLEY, S. & OKAY, A. İ. 1994. Post collision magmatism and tectonics in northwest Anatolia. *Contribution to Mineralogy and Petrology,* **117**, 241–252.

HAYWARD, A. B. 1984. Miocene clastic sedimentation related to the emplacement of the Lycian Nappes and the Antalya complex, SW Turkey. *In:* DIXON, J. E. & ROBERTSON, A. H. F. (eds) *The Geological Evolution of the Eastern Mediterranean.* The Geological Society, London, Special Publications, **17**, 287–300.

HETZEL, R. & REISCHMANN, T. 1996. Intrusion age of Pan-African augen gneisses in the southern Menderes Massif and the age of cooling after Alpine ductile extensional deformation. *Geological Magazine,* **133**, 565–572.

——, RING, U., AKAL, C. & TROESCH, M. 1995. Miocene north-northeast-directed extensional unroofing in the Menderes Massif, southwestern Turkey. *Journal of the Geological Society, London,* **152**, 639–654.

İNCİ, U. 1984. Demirci ve Burhaniye bitümlü şeyllerinin stratigrafisi ve organik özellikleri. *Geological Society of Turkey Bulletin,* **5**, 27–40 [in Turkish with English abstract].

İZTAN, H. & YAZMAN, M. 1990. Geology and hydrocarbon potential of the Alaşehir (Manisa) area, western Turkey. *Proceedings of an International Earth Sciences Congress on Aegean Regions,* İzmir, 327–338.

JACKSON, J. A. & MCKENZIE, D. 1988. Rates of active deformation in the Aegean Sea and surrounding areas. *Basin Research,* **1**, 121–128.

KARACIK, Z. 1995. *Ezine-Ayvacık (Çanakkale) Dolayında Genç Volkanizma Plütonizma İlişkileri.* PhD Thesis, İstanbul Technical University [in Turkish with English abstract].

—— & YILMAZ, Y. 1998. Geology of the ignimbrites and the associated volcano–plutonic complex of the Ezine area, northwestern Anatolia. *Journal of Volcanology and Geothermal Research,* **85**, 251–264.

KAYA, O. & MOSTLER, H. 1992. A Middle Triassic age for low grade greenschist facies metamorphic sequence in Bergama (İzmir), western Turkey: the first palaeontological age assignment and structural–stratigraphic implications. *Newsletter on Stratigraphy,* **26**, 1–17.

KISSEL, C. & LAJ, C. 1988. *Palaeomagnetic Rotations and Continental Deformation.* Kluwer.

KOÇYİĞİT, A. 1984. Tectono-stratigraphic characteristics of Hoyran Lake region (Isparta Bend). *In:* TEKELİ, O. & GÖNCÜOĞLU, M. C. (eds) *Geology of the Taurus Belt.* Proceedings of International Tauride Symposium, Mineral Research and Exploration Institute of Turkey (MTA) Publications, 53–67.

LE PICHON, X. & ANGELLIER, J. 1979. The Hellenic arc and trench system: a key to the neotectonic evolution of the eastern Mediterranean area. *Tectonophysics,* **60**, 1–42.

—— & —— 1981. The Aegean Sea. *Philosophical*

Transactions of the Royal Society, London, Series A, **300**, 357–372.

LUTTING, G. & STEFFENS, P. 1976. Explanatory notes for the palaeogeographic atlas of Turkey from the Oligocene to the Pleistocene. *Bundesanstalt für Geowissenschaften und Rohstoffer, Hannower.*

MCKENZIE, D. 1972. Active tectonics of the Mediterranean regions. *Geophysical Journal of the Royal Astronomical Society*, **30**, 109–185.

—— & YILMAZ, Y. 1991. Deformation and volcanism in western Turkey and the Aegean. *Bulletin of the İstanbul Technical University*, **44**, 345–373.

MAIN, I. G. & BURTON, P. W. 1989. Seismotectonics and the earthquake frequency–magnitude distribution in the Aegean area. *Geophysical Journal of the Royal Astronomical Society*, **98**, 575–586.

MAKRIS, J. & STOBBE, C. 1984. Physical properties and state of the crust and upper mantle of the eastern Mediterranean Sea deduced from geophysical data. *Marine Geology*, **55**, 347–363.

MEISSNER, R., WEVER, T. H. & FLÜH, E. R. 1987. The Moho in Europe: implications for crustal development. *Annales Geophysicae*, **513**, 357–364.

MINDEVALLI, O. Y. & MITCHEL, B. J. 1989. Crustal structure and possible anisotrophy in Turkey from seismic wave dispersion. *Geophysical Journal International*, **98**, 93–106.

NEBERT, K. 1978. Linyit içeren Soma Neojen bölgesi, Batı Anadolu. *Mineral Research and Exploration Institute of Turkey (MTA) Bulletin*, **90**, 20–69 [in Turkish with English abstract].

ORBAY, N., SANVER, M., TAPIRDAMAZ, C., ÖZÇEP, F., İŞSEVEN, T. & HİSARLI, M. 1998. Güney Trakya ve kuzey Biga yarımadasının palaeomagnetizması. *İstanbul University Earth Sciences*, **11**, 1–21 [in Turkish with English abstract].

PAPANIKOLAU, D. J. 1984. Three metamorphic belts of the Hellenides: a review and a kinematic interpretation. *In:* DIXON, J. E. & ROBERTSON, A. H. F. (eds) *The Geological Evolution of the Eastern Mediterranean.* The Geological Society, London, Special Publications, **17**, 501–562.

PATTON, S. 1992. Active normal faulting, drainage patterns and sedimentation in southwestern Turkey. *Journal of the Geological Society, London*, **149**, 1031–1044.

PEARCE, J. A., BENDER, J. F., DE LONG, S. E. ET AL. 1990. Genesis of collision volcanism in eastern Anatolia, Turkey. *Journal of Volcanology and Geothermal Research*, **44**, 189–229.

PHILLIPSON, A. 1910–1915. *Reisen und Forschungen im westlichen Kleinasien.* Ergänzungshefte 167, 172, 177, 180, 183 der Petermanns Mitteilungen, Gotha, Justus Perthes.

ŞAROĞLU, F. & YILMAZ, Y. 1987. Geological evolution and basin models during neotectonic episode in the eastern Anatolia. *Mineral Research and Exploration Institute of Turkey (MTA) Bulletin*, **107**, 74–94.

SATIR, M. & FRIEDRISCHSEN, H. 1986. The origin and evolution of the Menderes Massif, western Turkey: rubidium/strontium and oxygen isotope study. *Geologische Rundschau*, **75**, 703–714.

SELLERS, P. C. & CROSS, P. A. 1989. 1986 and 1987 Wegener–Medlas baselines determined using the pseudo-short arc technique. *Proceedings of the International Conference WEGENER-MEDLAS Project*, Scvheviningen.

ŞENGÖR, A. M. C. 1979. The North Anatolian Transform Fault: its age, offset and tectonic significance. *Journal of the Geological Society, London*, **136**, 269–282.

—— 1982. Egenin neotektoniğini yöneten etkenler. *In:* EROL, O. & OYGÜR, V. (eds) *Batı Anadolunun Genç Tektoniği ve Volkanizması.* Geological Society of Turkey Publications, 59–71 [in Turkish with English abstract].

—— 1993. Some current problems on the tectonic evolution of the Mediterranean during the Cainozoic. *In:* BOSCHI, E., MANTOVANI, E. & MORELLI, A. (eds) *Recent Evolution and Seismicity of the Mediterranean Region.* Kluwer, Dordrecht, 1–51.

—— & KIDD, W. S. F. 1979. Post-collisional tectonics of Turkish–Iranian plateau and a comparison with Tibet. *Tectonophysics*, **55**, 361–376.

—— & YILMAZ, Y. 1981. Tethyan evolution of Turkey: a plate tectonic approach. *Tectonophysics*, **75**, 181–241.

——, GÖRÜR, N. & ŞAROĞLU, F. 1985. Strike-slip faulting and related basin formation in zones of tectonic escape: Turkey as a case study. *In:* BIDDLE, K. T. & BLICK, N. C. (eds) *Strike-slip Deformation, Basin Formation and Sedimentation.* Society of Economic Palaeontologists and Mineralogists, Special Publications, **37**, 227–264.

——, SATIR, M. & AKKÖK, R. 1984. Timing of tectonic events in the Menderes Massif, western Turkey: implications for tectonic evolution and evidence for Pan-African basement in Turkey. *Tectonics*, **3**, 693–707.

SEYİTOĞLU, G. & SCOTT, B. C. 1991. Late Cenozoic crustal extension and basin formation in West Turkey. *Geological Magazine*, **128**, 155–166.

—— & —— 1992. Late Cenozoic volcanic evolution of the northeastern Aegean region. *Journal of Volcanology and Geothermal Research*, **54**, 157–176.

—— & —— 1994. Late Cenozoic basin development in West Turkey: Gördes basin: tectonics and sedimentation. *Geological Magazine*, **131**, 631–637.

—— & —— 1996. Age of the Alaşehir graben (West Turkey) and its tectonic implications. *Geological Journal*, **31**, 1–11.

——, —— & RUNDLE, C. C. 1992. Timing of Cenozoic extensional tectonics in west Turkey. *Journal of the Geological Society, London*, **149**, 533–538.

——, ANDERSON, D., NOWELL, G. & SCOTT, B. 1997. The evolution from Miocene Potassic to Quaternary sodic magmatism in Western Turkey: implications for enrichment processes in the litospheric mantle. *Journal of Volcanology and Geothermal Research*, **76**, 127–147.

SİYAKO, M., BÜRKAN, K. A. & OKAY, A. İ. 1989. Biga ve Gelibolu Yarımadalarının Tersiyer Jeolojisi ve Hidrokarbon olanakları. *Turkish Association of Petroleum Geologists Bulletin*, **1**, 183–199 [in Turkish with English abstract].

TAYMAZ, T., JACKSON, J. & MCKENZIE, D. 1991. Active

tectonics of the north and central Aegean Sea. *Geophysical Journal International*, **106**, 433–490.

ÜNAY, E., GÖKTAŞ, F., HAKYEMEZ, H. Y., AVŞAR, M. & ŞAN, Ö. 1995. Büyük Menderes grabeni'nin kuzey kenarındaki çökellerin Arvicolidae (Rodentia, Mammalia) faunasına dayalı olarak yaşlandırılması. *Geological Society of Turkey Bulletin*, **38**, 75–80 [in Turkish with English abstract].

WESTAWAY, R. 1994. Present-day kinematics of the Middle East Mediterranean. *Journal of Geophysical Research*, **99**, 12 071–12 090.

YILMAZ, Y. 1989. An approach to the origin of young volcanic rocks of western Turkey. *In:* ŞENGÖR, A. M. C. (ed.) *Tectonic Evolution of the Tethyan Region*. Kluwer, 159–189.

—— 1990. Comparison of young volcanic associations of western and eastern Anatolia under compressional regime; a review. *Journal of Volcanology and Geothermal Research*, **44**, 69–87.

—— 1997. Geology of western Anatolia. Active tectonics of northwestern Anatolia. *The Marmara Poly-Project, A Multidisciplinary Approach by Space-geodesy, Geology, Hydrogeology, Geothermics and Seismology*. vdf Hochschulverlag AG an der ETH Zurich, 31–53.

—— & POLAT, A. 1998. Geology and evolution of the Thrace volcanics, Turkey. *Acta Volcanologica*, **10**, 293–303.

——, TÜYSÜZ, O., YİĞİTBAŞ, E., GENÇ, Ş. C. & ŞENGÖR, A. M. C. 1997*a*. Evolution of the Pontides. *In:* ROBINSON, A. G. (ed.) *Regional and Petroleum Geology of the Black Sea and Surrounding Region*. AAPG Memoir, 183–226.

——, GENÇ, Ş. C., GÜRER, Ö. F., BOZCU, M., GÖRÜR, N. & AKKÖK, R. 1997*b*. Batı Anadolu Neojen istiflerinin karşılaştırılması. *Ulusal Deniz Araştırmaları Projesi*. Tübitak, Ankara, 1–32.

——, ——, ——, ET AL. 1999. Ege denizi ve Ege bölgesinin jeolojisi ve evrimi. *In:* GÖRÜR, N. (ed.) *Türkiye Denizleri Devlet Planlama Teşkilati, TÜBİTAK Publications*, Ankara. 211–337 [in Turkish with English abstract].

YUSUFOĞLU, H. 1996. Northern margin of the Gediz graben: age and evolution, West Turkey. *Turkish Journal of Earth Sciences*, **5**, 11–23.

Timing of extension on the Büyük Menderes Graben, western Turkey, and its tectonic implications

ERDİN BOZKURT

Middle East Technical University, Department of Geological Engineering, Tectonic Research Unit, TR-06531 Ankara, Turkey (e-mail: erdin@metu.edu.tr)

Abstract: The Büyük Menderes Graben is one of the most prominent structures of western Anatolia (Turkey) and borders the Aegean. New structural and stratigraphic evidence demonstrates that the (?)Miocene fluvio–lacustrine, coal-bearing red clastic sediments exposed along the northern margin of the graben are northward back-tilted, locally folded and overlain unconformably by horizontal terraced Pliocene–Pleistocene sediments. Also, there is no evidence that these red clastics at the base of the Neogene sequence were deposited during neotectonic extension. It is suggested here that these sediments cannot be regarded as passive neotectonic graben-fill deposits.

This new evidence further indicates that the age of the modern Büyük Menderes Graben is Pliocene, younger than previously considered (Early–Middle Miocene) and that initiation of north–south neotectonic extensional tectonics in the graben, and thus in western Anatolia, is unlikely to have resulted from orogenic collapse. The Pliocene estimate of the start of extension is in close agreement with the start of slip on the North Anatolian Fault Zone. The north–south extensional tectonics, and associated east–west faulting and basin formation, commenced during the Pliocene due to the effect of westward tectonic escape of the Anatolian block along the North and East Anatolian Faults. New mammal evidence also constrains the start of slip on the younger faults which bound the present-day graben floor to *c.* 1 Ma.

The Büyük Menderes Graben has experienced a two-stage extension. An initial extension (latest Oligocene–Early Miocene) along initially moderately, steeply dipping normal faults was superseded by movement on steeper normal faults during the (?)Pliocene. The two phases of deformation appear to reflect significant changes in the tectonic setting of western Anatolia and are attributed to orogenic collapse followed by tectonic escape.

Western Anatolia (Turkey) is a region presently dominated by approximately north–south directed continental extension. It is part of a zone of distributed extensional deformation affecting a large area (the Aegean extensional province) that includes the Aegean Sea, Greece, Macedonia, Bulgaria and Albania, and is bound by the Hellenic Trench in the south (Fig. 1). Regional Global Positioning System (GPS) data show that the central Aegean is currently moving southwestwards, relative to Eurasia, at a rate of *c.* 30–40 mm a^{-1} (Le Pichon *et al.* 1995; Barka & Reilinger 1997; Reilinger *et al.* 1997 and refs cited therein), whilst Anatolia, which is undergoing counterclockwise rotation, is escaping westwards from eastern Anatolia at a rate of *c.* 30 mm a^{-1} and is being expelled onto the African oceanic Plate along the Hellenic Trench. This all results from the collision of the Eurasian and Arabian Plates (Barka & Reilinger 1997; Reilinger *et al.* 1997 and refs cited therein).

In western Anatolia, east–west and west-northwest–east-southeast grabens (e.g. the Gökova, Büyük Menderes, Gediz, Bakırçay, Simav and Kütahya Grabens) and their related active normal faults are the most prominent neotectonic features (McKenzie 1978; Dewey & Şengör 1979; Şengör *et al.* 1985; Şengör 1987; Jackson & McKenzie 1988; Seyitoğlu & Scott 1991, 1992, 1996; Emre & Sözbilir 1995; Görür *et al.* 1995; Emre 1996; Koçyiğit *et al.* 1999) (Fig. 2). The activity of these structures is shown by the numerous historical earthquakes which have occurred along the faults (e.g. Ambraseys 1988; Ambraseys & Jackson 1998 and refs cited therein). In addition to these structures, north–south basins (e.g. the Gördes, Demirci, Selendi and Uşak-Güre basins), characterized by the widespread occurrence of Neogene sediments, are also important features (Fig. 2).

Apart from the high-angle, active graben-bounding normal faults and their role in the neotectonics of the region, evidence is available that initially moderately steeper, but presently low-angle, inactive normal faults also played an important role in exhuming the metamorphic rocks of the Menderes Massif, in controlling

From: BOZKURT, E., WINCHESTER, J. A. & PIPER, J. D. A. (eds) *Tectonics and Magmatism in Turkey and the Surrounding Area.* Geological Society, London, Special Publications, **173**, 385–403. 1-86239-064-9/00/$15.00
© The Geological Society of London 2000.

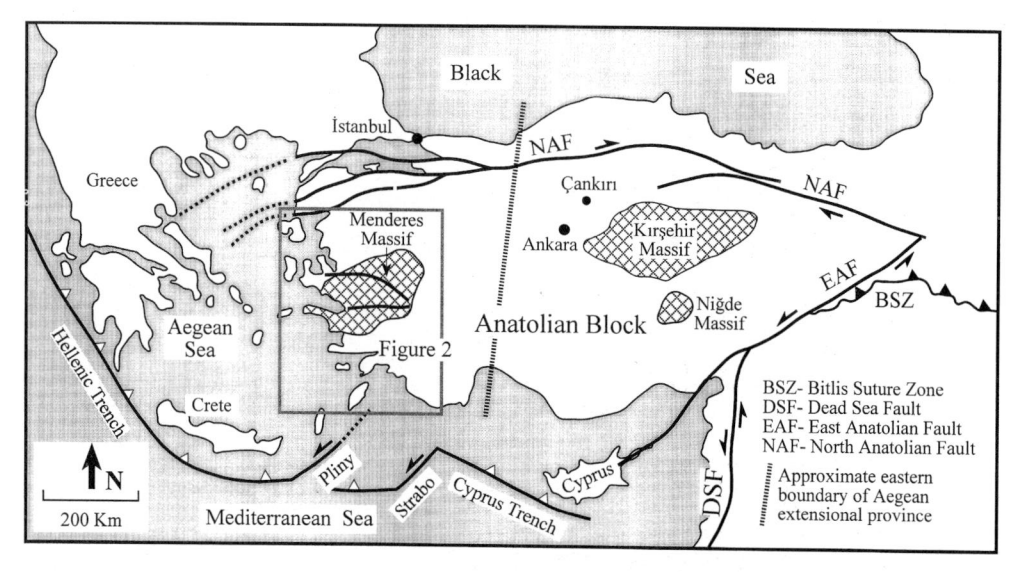

Fig. 1. Simplified tectonic map of the eastern Mediterranean region showing major tectonic elements [simplified from Barka & Reilinger (1997)].

sedimentation in the hanging-wall basins and in the consequent extension during latest Oligocene–Early Miocene phase of orogenic collapse (Bozkurt & Park 1994, 1997; Emre & Sözbilir 1995; Hetzel et al. 1995, 1998; Emre 1996).

The origin and age of crustal extension in the Aegean have been subjects of controversy for many years. Extension in this region has been explained by three different models: (1) the *tectonic escape model* – the westward escape of the Anatolian block along its boundary structures, the dextral North and sinistral East Anatolian Faults, since the Late Serravalian (12 Ma) following collision of the Arabian and Eurasian Plates across the Bitlis Suture Zone (Dewey & Şengör 1979; Şengör 1979, 1987; Şengör et al. 1985; Görür et al. 1995); (2) the *back-arc spreading model* – back-arc extension caused by the south-southwestward migration of the Hellenic Trench System [the mechanism of subduction roll-back; see McKenzie (1978), Le Pichon & Angelier (1979) and Meulenkamp et al. (1988)]. However, there is no common agreement among scientists on the inception date for the subduction roll-back process and proposed ages range between 60 and 5 Ma (McKenzie 1978; Le Pichon & Angelier 1979, 1981; Kissel & Laj 1988; Meulenkamp et al. 1988); (3) the *orogenic collapse model* – localized extension induced by late orogenic gravitational collapse of overthickened crust following the latest Palaeocene collision across Neotethys along the İzmir–Ankara–Erzincan Suture Zone during the Late Oligocene–Early Miocene (Seyitoğlu & Scott 1991, 1992).

More recently, Koçyiğit et al. (1999) proposed an *'episodic, two-stage graben model'*, with an intervening phase of short-term compression for the evolution of the Gediz Graben: a Miocene–Early Pliocene first stage occurred as a consequence of orogenic collapse and a second phase of north–south extension originated from westward escape of the Anatolian block, triggered by the commencement of seafloor spreading along the Red Sea during the Early Pliocene. They consider that the intervening short-term compressional episode resulted from a change in the kinematics of the Eurasian and African Plates.

The Büyük Menderes Graben is bounded by one of the principal active normal fault zones in western Turkey. The main aspect of this paper is to propose that neotectonic extension in the Büyük Menderes Graben began in the Pliocene or later, rather than in the Early–Middle Miocene as others have claimed in recent literature (Seyitoğlu & Scott 1991, 1992). The previous age proposal was based on the age of sediments and volcanic rocks at (or near) the base of the sedimentary sequence. It is argued instead that these sediments, exhumed with respect to the present-day graben floor, have nothing to do with neotectonic extension prevailing in the region. The purpose of this paper, based on

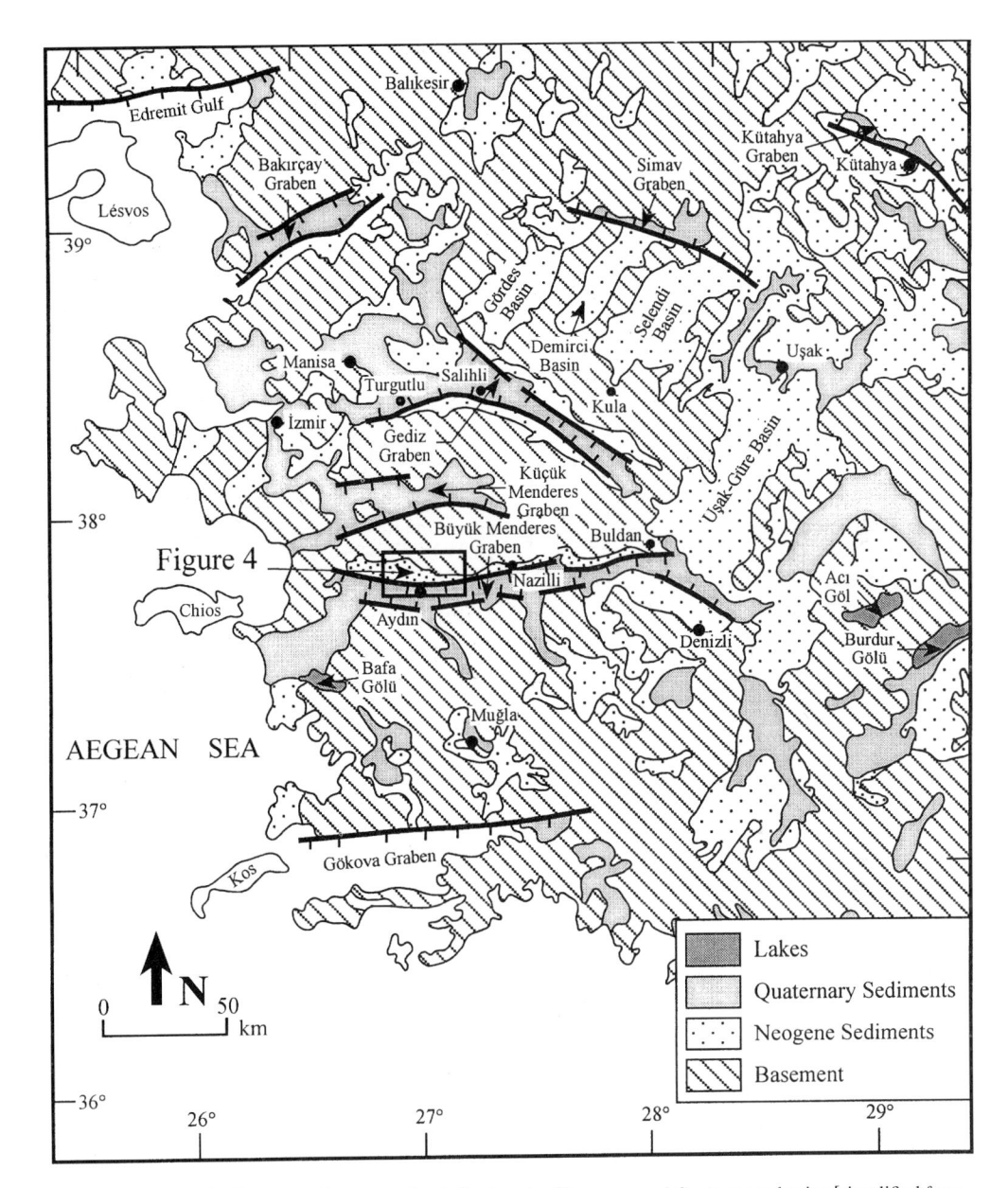

Fig. 2. Outline geological map of western Anatolia showing Neogene and Quaternary basins [simplified from Bingöl (1989)]. Note that the (?)Miocene and Pliocene sediments are not differentiated due to lack of data.

mapping, field observations and the reassessment of available literature, is therefore to present new structural and stratigraphic information from the area around Aydın (Figs 2 and 3) that bears influence on the age of the Büyük Menderes Graben, and to discuss its implications for the age and cause of neotectonic extension in western Anatolia.

Büyük Menderes Graben

Established knowledge

The Büyük Menderes Graben, one of the major east–west grabens in western Anatolia, is a structure *c.* 125 km long and 8–12 km wide. The plain in the interior of the graben consists of an

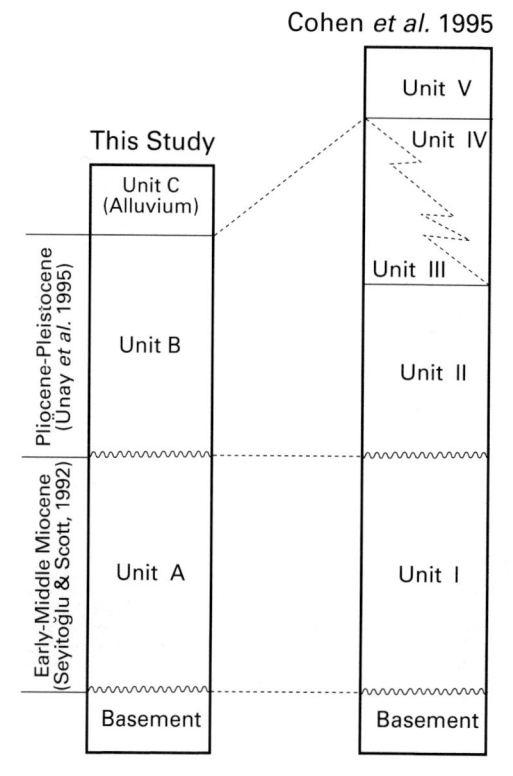

Fig. 3. Stratigraphy of the Büyük Menderes Graben around Aydın and its correlation with that of Cohen *et al.* (1995).

axial fluvial depocentre bounded to the north by a segmented, moderately steep, south dipping active normal fault. Some parts of this fault have slipped in recent times, recorded by instrumental records and historical earthquakes [e.g. the 1899 Nazilli–Denizli Earthquake, the 1956 Söke-Balat Earthquake and the 1965 Denizli Earthquake; see Ambraseys (1988), Westaway (1993) and Ambraseys & Jackson (1998 and refs cited therein). In other parts, the fluvial depocentre is bounded to the south by less important antithetic normal faults. For most of its length, the uplifting footwall of this active normal fault on the north side of the graben floor contains a narrow (c. 5–10 km) former depocentre which is now eroding. This depocentre is bounded to the north, at the southern edge of the outcrop of Menderes Massif metamorphic rocks, by a straight mountain front controlled by another segmented south dipping normal fault and its depositional substrate accumulated in the hanging wall when this fault was active. This contact is known to be a low-angle (at present) normal fault, as in places it is possible to measure its slip sense (e.g. Westaway 1990*a, b*; this study). However, there is no evidence (e.g. from seismicity) that it is active at present. It has long been suggested that neotectonic extension began on this more northerly fault zone (e.g. Jackson & McKenzie 1988; Seyitoğlu & Scott 1991, 1992)

In addition to the axial fluvial sedimentation, many small lateral rivers cut through the uplifted basin on the northern flank of the Büyük Menderes Graben. These have caused erosion of the Menderes Massif and uplifted Neogene basin, and deposition of alluvial fans on the valley floor where they are interbedded with the axial fluvial sediments. As the same pattern is evident within much of the sequence of eroding sediments of the uplifted western Anatolian Neogene basins (e.g. Roberts 1988; Paton 1992; Cohen *et al.* 1995), it is assumed that the same sedimentary and geomorphological environment existed at the time when these latter basins were infilled.

Sedimentary sequence

Fluvio–lacustrine sediments in and around the Büyük Menderes Graben are best exposed in a 2–5 km wide zone along its northern margin. These sediments are exhumed along the footwall of the south facing active normal faults with respect to the present-day graben floor. Three main lithological associations, based on their distinct structure, have been mapped in the Aydın area: (1) northwards tilted sediments (unit A); (2) almost flat-lying, terraced sediments (unit B); and (3) marginal alluvial fans and present-day graben-floor sediments (unit C; see Figs 3 and 4). Each unit contains vertical and lateral variations and displays various relationships of interfingering and intergradations.

Unit A. This unit consists mainly of northwards tilted continental clastic sediments located between the metamorphic rocks of the Menderes Massif in the north and the present-day graben-bounding faults in the south (Fig. 4). The basal lithology is a reddish, coarse-grained, well-cemented, poorly sorted, polygenetic conglomerate composed of clasts derived from the underlying metamorphics and minor but widespread interbedded lignites. Above the conglomerates, the unit is composed of siltstone, mudstone and shale alternations, together with conglomerates and pebbly sandstones. Lateral and vertical transitions from one lithology to another are very common throughout this sequence, which is also characterized by numerous scour-and-fill structures filled with channel

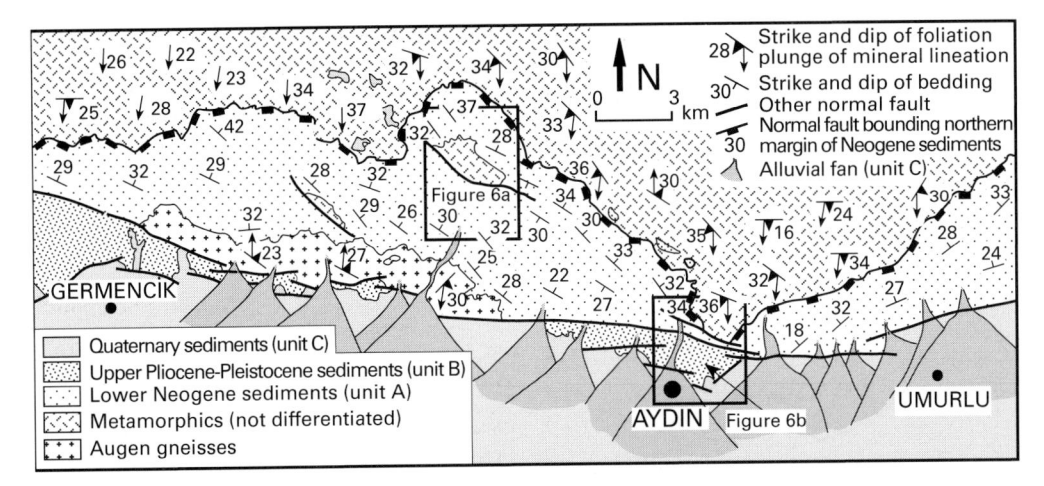

Fig. 4. Simplified geological map of the northern margin of the Büyük Menderes Graben in the area between Germencik and Umurlu.

conglomerates. This unit comprises a broadly coarsening-upwards sequence with a total thickness of c. 2 km (Cohen et al. 1995). Laminar to trough-like cross-bedding, pebble imbrication, graded bedding and normal-type growth faults are commonly observed synsedimentary structures in this unit. More details are given in Cohen et al. (1995).

Unit B. This unit comprises approximately horizontal, massive, cobble to boulder conglomerates with alternations of sandstone, siltstone, mudstone and claystone which crop out to the south of the tilted sediments of unit A. Unit B is bound by approximately east–west trending, high-angle normal faults along the contacts, both with the deformed sediments of unit A to the north and the younger basin-fill sediments (unit C) to the south (Fig. 4).

Unit C. These sediments, with the present-day configuration of the Büyük Menderes Graben, are juxtaposed with unit B sediments along high-angle graben-bounding normal faults. They are composed mainly of marginal alluvial fan and graben-floor sediments. The northern margin of the Büyük Menderes Graben is marked by many steep, well-developed, alluvial fans of diverse size, aligned in a narrow zone (Fig. 4). The source of the alluvial fan sediments is the metamorphic basement and exhumed unit A and B sediments. The alluvial fans grade into fine-grained basin-floor sediments along the Büyük Menderes River. In places, the alluvial fans coalesce and degrade and result in a fault-parallel alluvial fan apron (Fig. 4). The coarse-grained nature of the marginal sediments and the steepness of the alluvial fans indicate rapid uplift of the source mountains, accompanied by erosion and rapid sedimentation, attesting to the activity of these graben-bounding faults.

Structure

Three types of major structures occur along the northern margin of the Büyük Menderes graben: (1) an inactive, presently low-angle, normal fault; (2) west-northwest–east-southeast to northwest–southeast folds within the unit A sediments; and (3) approximately east–west high-angle, graben-bounding normal faults.

Table 1. *Measurements of slickensides and slickenlines on the presently low-angle normal fault*

Location	Dip direction (°N)	Dip amount (°)	Rake (°)	Sense
1	192	22	86	Normal
2	192	30	85	Normal
3	220	36	72	Normal
4	200	28	87	Normal
5	202	32	82	Normal
6	190	26	80	Normal
7	192	28	76	Normal
8	194	29	88	Normal
9	192	33	75	Normal
10	195	30	78	Normal
11	204	32	76	Normal
12	206	34	85	Normal
13	194	29	86	Normal

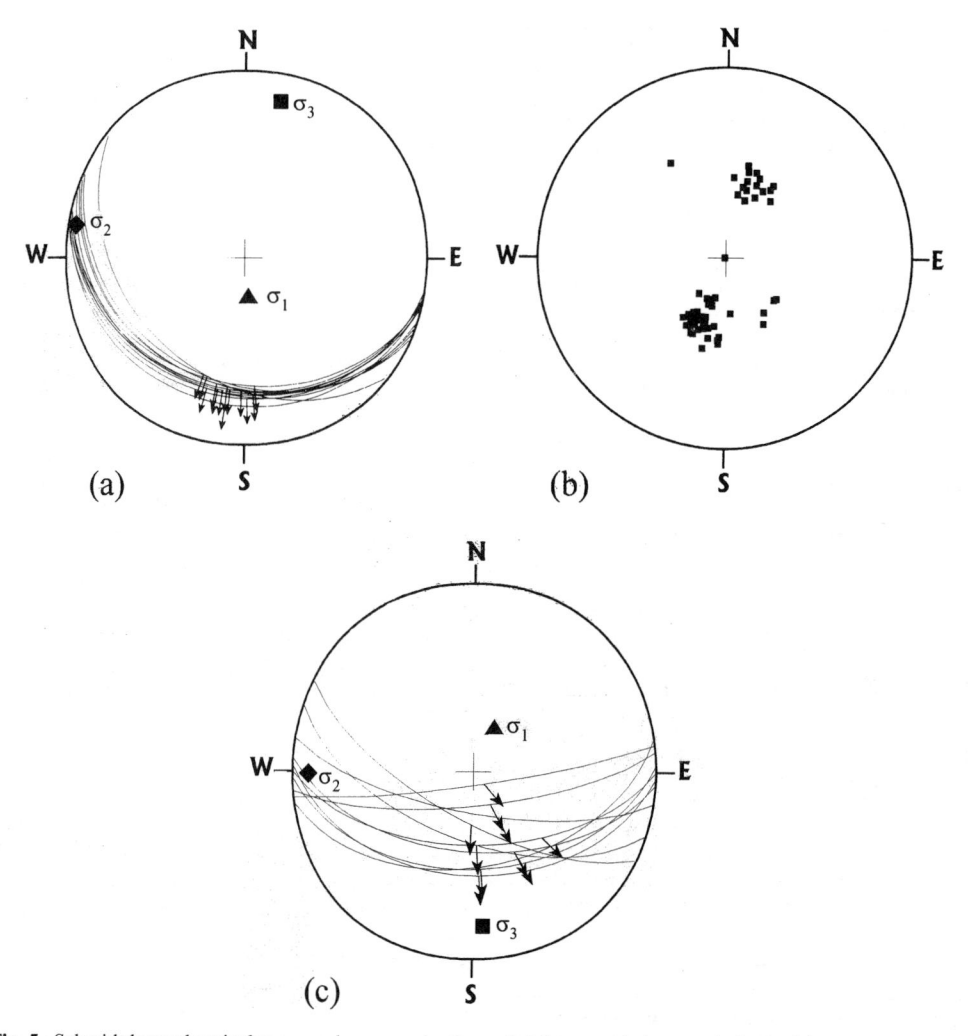

Fig. 5. Schmidt lower hemisphere equal-area projections of: (**a**) presently low-angle fault; (**b**), poles to bedding planes in Lower–Middle Miocene sediments; (**c**) graben-bounding, high-angle active normal faults. Great circles in (a) and (c) are fault surfaces, the arrows are striations (see Tables 1 and 2 for details).

The northern boundary of the unit A sediments is a major south facing, low-angle (22–34°; see Table 1; Fig. 5a), inactive, normal fault that separates them in the hanging wall from ductilely deformed metasediments and in the footwall from metagranite of the Menderes Massif to the north (Fig. 4). This fault has been cut by steeper graben-bounding active normal faults (as discussed below) and the metamorphics have been progressively uplifted, mylonitized and exhumed in the footwall. The deformed unit A sediments may thus be regarded as being deposited in a basin that was situated on the upper plate of this fault. Another, but circumstantial, piece of evidence of contemporaneous sedimentation and faulting is that the unit A sediments dip to the north, suggesting rotation of both the fault and the strata during the evolution of this fault. The calculated σ_1 trend, from stratum on this presently low-angle fault plane, is 163° and plunges steeply at 71°, whereas σ_2 and σ_2 axes plunge at 5 and 15°, respectively (Fig. 5a). These estimates of stress field orientations and others elsewhere in this study are calculated from observed slip vector orientations using the computer program of Caputo (1989).

Dips of beds within the unit A sediments vary throughout the basin. The available data (Fig. 6a) are interpreted as evidence of folding with

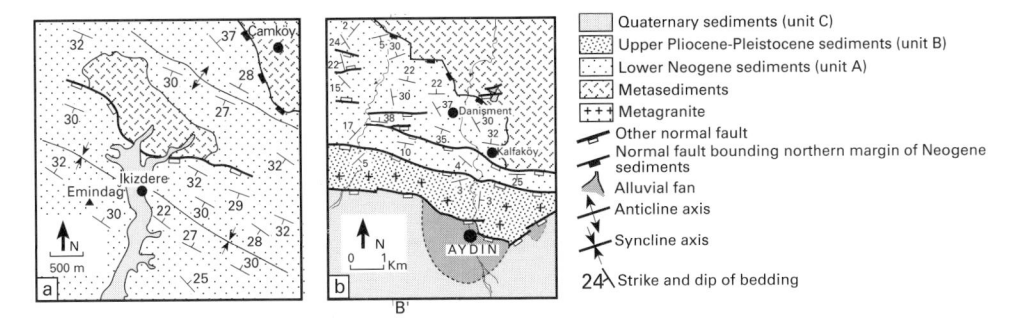

Fig. 6. Geological maps of (**a**) İkizdere and (**b**) Aydın areas [simplified and interpreted from Cohen *et al.* (1995)] showing the folds in unit A sediments and their boundary relationships with the Upper Pliocene–Pleistocene fluvial sediments (unit B).

Table 2. *Measurements of slickensides and slickenlines on the high-angle graben-bounding faults*

Location	Dip direction (°N)	Dip amount (°)	Rake (°)	Sense
1	208N	69	80	Normal
2	200N	60	78	Normal
3	180N	58	60	Normal
4	190N	72	76	Normal
5	172N	84	86	Normal
6	170N	48	77	Normal
7	174N	75	84	Normal
8	182N	44	86	Normal
9	178N	47	88	Normal
10	184N	55	70	Normal

west-northwest–east-southeast axes. This folding can be seen directly in the field at the locations covered in Fig. 6a, where dips of beds change direction systematically over scales of typically several hundred metres. This folding thus occurs on a much larger scale than the minor folding noted by Cohen *et al.* (1995) and causes lateral variations in dip on a scale of a few metres. These folds are open structures with vertical to inclined axial planes and gently plunging axes that run parallel to the graben-bounding normal faults (Figs 4, 5b and 6a). The structures are observed to fold the bedding planes of the unit A sediments. The dip of beds averages 30° but in areas close to the inactive normal fault this may increase to 35–40° (Fig. 5b).

Although the unit A sediments are northward tilted and locally folded, the unit B strata in the basin show a different evolution. They are deformed only by graben-bounding normal faults, which are the most conspicuous features of the northern margin of the Büyük Menderes Graben. These faults, which dip southwards (Fig. 5c; Table 2), form the boundary between the deformed unit A sediments and the approximately horizontal unit B and younger basin fill (unit C; see Figs 4 and 6). These faults dip southwards at angles of 44–84° (Table 2) and show normal faulting with minor components of left-lateral slip (Fig. 5c). Computed results of slip data measurements on these fault planes define an approximately vertical σ_1 trending 25° and plunging steeply at 75°, and σ_2 and σ_3 dipping gently at 10 and 15°, respectively (Fig. 5c). The unit A sediments have been exhumed along the footwall of these active structures and provide the source of both the terraced unit B and younger basin-fill sediments (unit C). As these faults control rapid changes in the morphology and the drainage pattern, and are marked by triangular facets, fault scarps and active and extensive development of steep alluvial fans, they may have a neotectonic origin. The activity of these structures is indicated by the recent earthquakes that have occurred along them (see Established knowledge).

Interpretation

Sedimentary unit A

The relatively steeply north dipping fluvial and fluvial fan sediments with red weathering, which are situated at the base of the young sedimentary sequence in the uplifted basin north of the Büyük Menderes Graben near Aydın, are called unit A. This unit can be correlated with unit I of Cohen *et al.* (1995; Fig. 3). Cohen *et al.* (1995) tentatively accepted these sediments as equivalent to the lignite-bearing sediments, also with red weathering, from near Nazilli (Fig. 2). Seyitoğlu & Scott (1992) assigned to these red

Table 3. *The results of mammal sites dated by Ünay et al. (1995) and Ünay & De Brujin (1998)*

Location	Name	Situation	Mammal age
1	Söke	BM valley floor	(a) Early Biharian (Early Pleistocene)
			(b) Toringian (Middle–Late Pleistocene)
2	Ortaklar	Uplifted basin outside BMFZ	Late Villanian (Late Pliocene)
3		BM valley floor	Late Pliocene–Early Pleistocene
4	Germencik	BM valley floor	Late Biharian–Early Toringian (Early–Middle Pleistocene)
5		BM valley floor	Late Pliocene–Pleistocene
6	Kurttepe	Uplifted basin N of BMFZ	Late Villanian–Early Biharian (Late Pliocene–Early Pleistocene)
7	Bozköy	Uplifted basin N of BMFZ	Late Villanian (Late Pliocene)
8	Nazilli–Şevketin Dağ	Uplifted basin N of BMFZ	Late Villanian–Early Biharian (Late Pliocene–Early Pleistocene)

sediments the Eskihisar sporomorph assemblage which is dated Early–Middle Miocene (20–14 Ma; Benda & Meulenkamp 1979). Because these are the oldest Neogene sediments in the Nazilli area, Seyitoğlu & Scott (1992) inferred that they mark the start of neotectonic extension in the Büyük Menderes Graben. Similarly, Lower–Middle Miocene coal-bearing sediments were reported from different parts of the Büyük Menderes Graben to the east of the present study area in some earlier studies (Karamanderesi 1972; Emre & Sözbilir 1995). There are, of course, red Neogene sediments in a lot of other places in western Turkey, including sites outside extensional basins (e.g. Becker-Platen 1971; Sickenberg & Tobien 1971; Kaya 1981; Gökçen 1982; Steininger & Rögl 1984). However, it is not obvious that all of these sediments are the same age. Most of the correlations in the previous works were based on red weathering. This suggests only that at some time since the youngest of these sediments were deposited the climate was subtropical for a while, and thus oxidized whatever sediments happened to be already exposed. It is known from the literature that climates favourable to hematite genesis in western Turkey persisted until the early part of the Late Miocene, or possibly even later (e.g. Steininger & Rögl 1984; Robertson et al. 1991). Thus, there is no convincing evidence that the red fluvial sediments near Aydın have the same age as the red lignite-bearing sediments near Nazilli. The argument related to Neogene climate change places a lower bound to the age of the sediments at Aydın.

More recently, a mammal site from the previously mapped Lower–Middle Miocene sediments (Seyitoğlu & Scott 1991) in the Nazilli area (Şevketin Dağ: Table 3; Fig. 7, location 8) yielded Late Pliocene–Early Pleistocene ages (Late Villanian–Early Biharian) from these (approximately horizontal) sediments (Ünay et al. 1995; Ünay & De Brujin 1998). This means that the Miocene ages for the sediments quoted by Seyitoğlu & Scott (1992) are no longer tenable and revision of their stratigraphy and its interpretation are urgently required. Moreover, another mammal site (Bozköy: Table 3; Fig. 7, location 7) from the northward tilted red clastics, designated as unit A sediments in the present study, yielded a Late Pliocene (Late Villanian) age (Ünay et al. 1995; Ünay & De Brujin 1998). However, Ünay et al. (1995) and Ünay & De Brujin (1998) reported neither the stratigraphy at these locations nor the positions of the dated sites in the succession. Pending the necessary revision of the Seyitoğlu & Scott (1991) stratigraphy and taking account of the other points already mentioned, the age of the unit A sediments will be quoted in this paper as Early Neogene (?Early–Middle Miocene or even as young as ?Pliocene).

As noted by Cohen et al. (1995), the presence of abundant small-scale normal faults are another characteristic feature in the unit A sediments. The cut-off angle with the bedding suggests that these structures formed as high-angle faults (70° to the vertical) but rotated to lower angles during northward back-tilting of the unit A sediments. My own observations strongly suggest a syndepositional origin for these structures, confirming the conclusion of Cohen et al. (1995). Evidence for syndepositional fault activity includes: (1) abrupt and rapid termination of fault displacements upwards in the stratigraphy; (2) thickness variations in the lithologies across the faults where

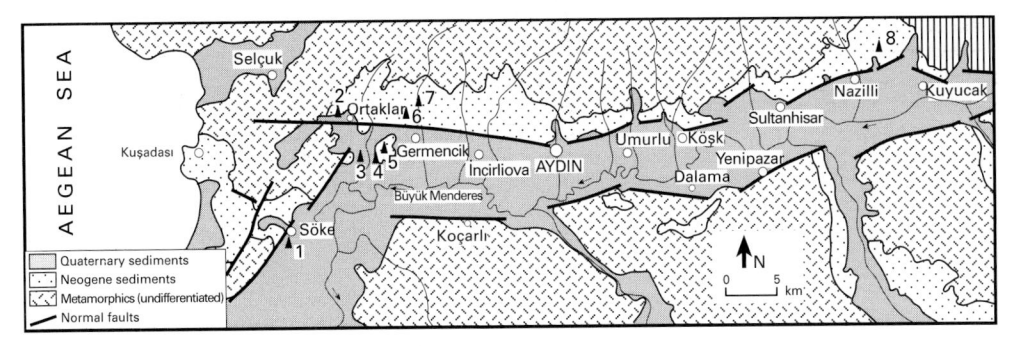

Fig. 7. Simplified map of the Büyük Menderes Graben showing mammal sites dated by Ünay et al. (1995) and Ünay & De Brujin (1998). See Table 3 for details.

the sediments (usually, relatively coarse grained) are thicker in their hanging walls than in their footwalls; and (3) wedging of these hanging-wall sediments, which thin away from the fault.

Sedimentary unit B

This unit can be correlated with units II–IV of Cohen et al. (1995) (Fig. 3). It is not weathered red and is thus younger than the time of any climate which allowed that style of weathering to happen. Evidently, an axial river existed when it was deposited, so it presumably post-dates the start of neotectonic extension within the Büyük Menderes Graben. On the other hand, it is not back-tilted nor does the bedding diverge as would be expected if it thickened towards a normal fault. This can be interpreted in two ways: (1) unit B was deposited over a sloping palaeoland surface and so it did not thicken towards any active fault and does not indicate the palaeohorizontal; (2) there was an initial phase of extension on the more northerly fault bordering the Menderes Massif metamorphics which tilted unit A, followed by a pause during which unit B was deposited; finally, extension resumed. The second interpretation is preferred here.

The clear structural difference between the unit B sediments and the older Neogene fluvial sediments of unit A (Cohen et al. 1995; this study) suggest a structural discontinuity between them. This, in turn, indicates either a substantial time gap while the older unit was eroded (see Discussion) or a sudden erosional event, in which case the most probable cause is a reduction in the base level of the river which drained this area at the time. One possibility is that this change relates to the start of cyclic drawdown in global sea level $c.$ 2.5–2 Ma, caused by the first development of northern hemisphere ice sheets. Another possibility is that it reflects the Messinian drawdown in the level of the Mediterranean at the end of the Miocene (e.g. Robertson et al. 1991 and refs cited therein). Many of the existing palaeogeographic maps show a land bridge in the way, such that this region either drained internally or northwards into the Paratethys [i.e. Black Sea; see e.g. Robertson et al. (1991 and refs cited therein)]. If so, changes in its base level could have responded to climate-induced changes in the level of the Black Sea. The angular unconformity between Miocene and Pliocene, and/or younger sediments, has long been known and was previously reported from the area to the east of Aydın (Karamanderesi 1972).

Until recently, no diagnostic fossil evidence existed from the post-Miocene sediments of the Büyük Menderes Graben. Ünay et al. (1995) and Ünay & De Brujin (1998) presented such evidence from eight sites within the graben and assigned, on the basis of mammal faunas, a Late Pliocene–Pleistocene age to these fluvial sediments. The dated sites, except for location 7, lie outside the study area, but their results are summarized in Table 3 and the locations are given in Fig. 7. The important note to be added here is that the dated samples (except for location 7) are from horizontal sediments, which are designated as unit B in this study. The obvious interpretation is that sedimentation within the present fluvial depocentre has been continuous since at least the Late Pleistocene, but sedimentation in what is now the uplifted basin north of this modern depocentre had ceased by the late Early Pleistocene or Middle Pleistocene. In other words, this fossil evidence supports the view that the present set of normal faults bounding the modern depocentre became active around the end of the Early Pleistocene, i.e. $c.$ 1 Ma.

This interpretation is consistent with that of

Jones & Westaway (1991) who made the first tentative estimate of c.1 Ma for the timing of transition to the modern set of faults. Subsequently, similar timings have been proposed for other Aegean normal faults, notably in the Gulf of Corinth where timing is constrained by well-dated marine sediments (Westaway 1996). Westaway (1994a, b) first suggested that this timing may be the same throughout the region. Later, Westaway (1996, 1998) proposed a possible physical mechanism.

Tilting and folding of Unit A

Unit A is tilted northwards which is thought to be the result of back-tilting beside a set of south dipping normal faults, as suggested by Cohen et al. (1995). This further means that extension was occurring on a fault system which approximates to the fault that bounds the oldest Neogene sediments (in its hanging wall, with the Menderes Massif metamorphics in its footwall) sometime after and/or during deposition of unit A but before unit B was deposited. This fault was later cut and locked up when slip began on the now active fault zone along the edge of the fluvial depocentre. This change occurred for some reason connected with the observation that slip on the initial fault zone back-tilted it and thus changed its orientation so that slip could no longer be maintained in the regional stress field (e.g. Jackson & McKenzie 1988). Similar abandoned young depocentres are evident in the footwalls of other active normal fault zones in western Turkey and central Greece (e.g. Roberts & Jackson 1991; Westaway 1998), suggesting that a systematic effect affected fault systems throughout this region. However, in some localities in the Büyük Menderes Graben, such as around Aydın, there are three generations of faults instead (Fig. 6b), indicating a more complex pattern.

The fluvial sediments of unit A are also observed to be folded (Figs 4 and 6). One particular area has been chosen in which to study these structures (Fig. 6a) and it is described here for the first time in the literature. It is a particularly good locality to study because access is relatively easy.

The age of this folding is uncertain. Because of this uncertainty in timing it is also not clear whether the folding occurred synchronously with extension during the deposition of unit A sediments or during a short time interval following the deposition of unit A. There are, of course, plenty of possible mechanisms for folding during extension due to: (1) differential compaction; (2) draping; (3) fault 'drag'; and (4) lateral variations in tilt caused by individual fault segments dying out along-strike. However, the scale of the folding and the lack of any clear relationship to the normal faults do not indicate a synsedimentary cause (see below).

Before going further, it is important to decide whether folds are local, i.e. unique to the Büyük Menderes Graben, or regional. Similar structures within the Neogene sediments have long been known. There are reports from many Aegean islands, particularly those located close to the coast of western Turkey (e.g. Kos, Samos, Chios, Paros, Naxos, Mykonos, Anafi and Milos) (Angelier 1976, 1978; Angelier & Tsoflias 1976; Mercier 1976, 1979, 1981; Mercier et al. 1976, 1979; Jackson et al. 1982; Boronkay & Doutsos 1994) and many of the western Anatolian grabens. In most of these studies it has been emphasized that extension in the Aegean was interrupted, at least in some places, by one or more shorter periods of compression involving folding and/or thrusting. In contrast, Jackson et al. (1982) propose that these shortening structures can be satisfactorily explained by uplift in the footwall blocks of normal faults and do not require regional compression. They also suggested that the compressional episodes are probably not regional in extent and may not be truly compressional in origin, but are more likely to be a consequence of considerable rotation due to internal deformation of blocks bounded by major normal faults. However, more recent studies, particularly that of Boronkay & Doutsos (1994), report evidence from the central Aegean region which suggests that crustal shortening occurred during the Miocene and that the resulting transpressive structures controlled the evolution of sedimentary basins.

Moreover, personal field observations near the eastern end of the Büyük Menderes Graben to the south of Buldan (Fig. 2), and in other east–west grabens, and the integration of available literature (e.g. Nebert 1960, 1978; Ercan et al. 1978; Dumont et al. 1979; Boray et al. 1985; Yalçın et al. 1985; İnci 1991; Yağmurlu 1991; Koçyiğit et al. 1995, 1999; Bozkuş 1996; Seyitoğlu 1997; Yılmaz 1997; Altunkaynak & Yılmaz 1998; Koçyiğit & Bozkurt 1998; Yılmaz et al. 2000) confirm that the Miocene deposits in many of the western Anatolian basins (regardless of their size and orientation) are deformed and folded, strongly suggesting a regional event.

This information favours the second possibility that folding occurred during a time interval after deposition of unit A ceased but before the deposition of unit B sediments began. This event is constrained between the age of deformed Lower Neogene sediments (unit A) and unconformable Pliocene–Pleistocene sediments (unit B). Given the available information (already discussed), on the timing of unit A deposition, this folding event can be dated sometime after the Middle Miocene but before the Late Pliocene. It is noteworthy that this time interval corresponds to a major break in sedimentation and magmatism, and a regional folding event, across many of the western Anatolian basins (see Discussion).

Nevertheless, whatever the cause of folding in unit A sediments, the important point is that unit B sediments do not bear any sign of deformation. The obvious interpretation is that the Lower Neogene sediments, exhumed on the shoulders of present-day Büyük Menderes Graben, have nothing to do with the age of initiation of neotectonic extension in the graben as was previously thought by Seyitoğlu & Scott (1992). Instead, it is the younger Pliocene–Pleistocene undeformed sediments which are coeval with formation of modern Büyük Menderes Graben.

This interpretation thus indicates two distinct phases of extension. The first phase of extension involved slip on the presently low-angle normal fault bounding the northern edge of the Lower Neogene sediments (Fig. 4). This extension appears to have accompanied the deposition of unit A sediments. It was followed by an interval during which unit A sediments were folded. Later still, extension resumed and led to the modern geometry of the Büyük Menderes Graben.

Initial dip of the low-angle normal fault

As already mentioned, the northern boundary of unit A sediments is a major normal fault, with a present-day dip of 22–34° (Fig. 5a). The present-day dips can be related to the dips of the steepest dipping Lower Neogene sediments, which reach 30–35°. The most appropriate way to restore such dips is [following Westaway & Kusznir (1993a)] to assume that the rocks deformed during extension by distributed vertical shear. If α and β are initial and present-day dips of the fault, respectively, and δ is the present-day dip of the oldest hanging-wall sediments, then:

$$\tan \alpha = \tan \beta + \tan \delta \quad (1)$$

With β = 22—34° and δ = 30–35°, the initial dip of the fault plane (α) can be calculated as 44–54°, i.e. 49 ± 5°. Dips of this order are common for many faults in the Aegean region (e.g. Westaway 1993; Westaway & Kusznir 1993a, b) and are explained by conventional theory. Thus, this particular boundary fault is a normal fault with an expected initial dip which has been back-tilted as a result of substantial extension. This agrees very closely with the estimates by Cohen et al. (1995) for the Büyük Menderes Graben and those given by Westaway (1993) and Westaway & Kusznir (1993a, b) for the initial dip of Denizli Normal Fault (Fig. 2).

Age of the Büyük Menderes Graben

The above observations demonstrate that unit A sediments, since the earliest Neogene ones along the northern margin of the Büyük Menderes Graben are back-tilted northwards and folded, cannot correspond to the early graben fill as previously suggested by Seyitoğlu & Scott (1992). Moreover, the clear angular difference between the tilted beds of Lower Neogene and horizontal Pliocene–Pleistocene sediments implies the presence of a major regional unconformity. However, they were previously considered to form a single continuous megasequence (Seyitoğlu & Scott 1992). The early sediments of the neotectonic graben must therefore correspond to the horizontal terrace sediments exhumed in the footwall of the graben-bounding normal faults (unit B), thus indicating a Pliocene or younger age for the initiation and formation of the present-day graben.

The important erosional surface that developed on the Lower Neogene sedimentary rocks (unit A) is not unique to the Büyük Menderes Graben but also occurs in many of the east–west trending grabens in western Turkey, such as the Gediz Graben (Yağmurlu 1987; Cohen et al. 1995; Koçyiğit et al. 1999), the Gökova Graben (Görür et al. 1995) and the Kütahya Graben (Koçyiğit & Bozkurt 1998). In all of these cases, this surface is overlain unconformably by Pliocene fluvial conglomerates. There are, of course, many places in Turkey where sediments such as those at the base of the Büyük Menderes sequence are found in sag basins or ovas (plains). In principle, it seems reasonable to consider the possibility that, in the Büyük Menderes Graben, the situation may be one where young normal faults have cut an older sag basin and the early sedimentation was previously misinterpreted.

Another way of attempting to estimate the age of Büyük Menderes Graben is to divide the extension across it (i.e. the sum of heaves of the graben-bounding normal faults) by the extension rate. The total heave (= horizontal slip) across the graben-bounding normal faults are measured, as accurately as possible, on a structural cross-section based on fig. 11b of Cohen

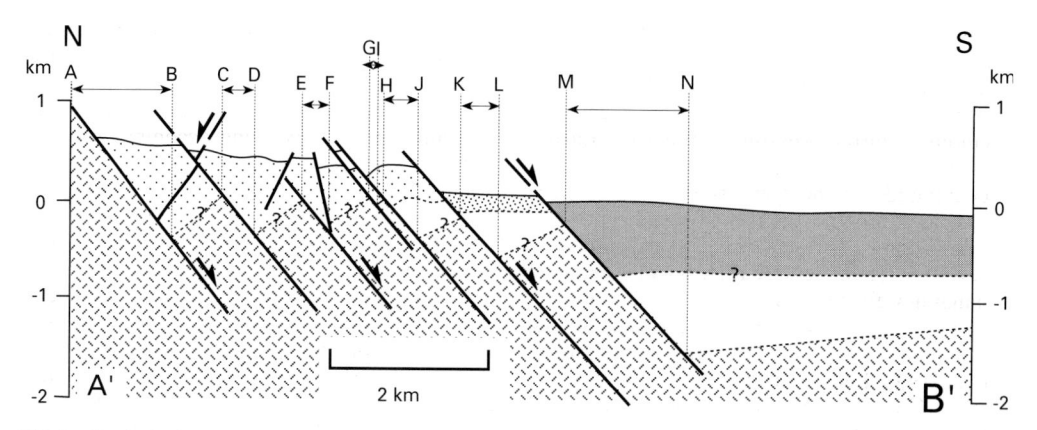

Fig. 8. Geological cross-section of the northern margin of the Büyük Menderes Graben (see Fig. 6b for location) based on fig. 11b of Cohen *et al.* (1995). This cross-section indicates a total of *c.* 5 km of extension. Assuming a uniform extension rate, the age of the fault zone is (*c.* 5 km/1 mm a^{-1}) 5 Ma. One could partition this with a possible *c.* 3 km of extension on the first set of faults, during the *c.* 5–2 Ma interval, then *c.* 1 km on the second fault set during the *c.* 2–1 Ma interval, then another *c.* 1 km on the present set since *c.* 1 Ma. A–B, 1.32 km; C–D, 0.4 km; E–F, 0.35 km; G–H, 0.3 km; I–J, 0.45 km; K–L, 0.5 km; M–N, 1.62 km.

et al. (1995) (Fig. 8). Westaway (1994*a, b*) argued that a reasonable present-day extension rate is *c.* 1 mm a^{-1}. This cross-section indicates a total of *c.* 5 km of extension. Assuming a uniform extension rate, the age of the fault zone is c. 5 km/1 mm a^{-1}, or 5 Ma (see Fig. 8 caption).

It was pointed out earlier that the deposition of unit A and other Lower–Middle Miocene sediments exposed along the northern margin of the Büyük Menderes Graben may have accompanied an early phase of extension along the presently low-angle normal fault at the northern margin of the depocentre. Furthermore, it is not clear how much of the tilting of these sediments, and the slip, occurred during such an earlier phase and how much occurred later. The same normal fault surface, active in the first phase of extension, may have been reactivated during the early part of the second phase of extension. This is quite logical since it is already known that new structures commonly follow pre-existing planes of weakness.

The Pliocene initiation age for the Büyük Menderes Graben is in close agreement with those suggested for the Gediz Graben (Early Pliocene: Koçyiğit *et al.* 1999), for the Gökova Graben (latest Miocene–Pliocene: Kurt *et al.* 1999) and for the whole of western Turkey (Pliocene: Yılmaz *et al.* 2000). Furthermore, Burchfiel *et al.* (2000) confirm that the initiation of east–west trending grabens (that mark the northern boundary of Aegean graben system) in central Bulgaria is no older than 9 Ma (perhaps no older than *c.* 6.5 Ma).

Discussion

Seyitoğlu & Scott (1991, 1992) proposed that the initiation of Büyük Menderes Graben, and therefore the neotectonic north–south extensional tectonics in western Anatolia, occurred in the Early Miocene. They thus suggested that this extension involved the spreading and thinning of crust which had previously thickened as a result of the Late Palaeogene continental collision following closure of the Neotethys Ocean. A Pliocene inception age for the graben, proposed here, clearly contradicts these previous conclusions. It is worth mentioning that Seyitoğlu & Scott (1992) provided no evidence that the lignite-bearing red clastics were deposited during extension. The only basis for their model is the assumption that Neogene sediments record the initiation age of the graben [following previous contentions by Şengör & Yılmaz (1981), Şengör *et al.* (1985) and Şengör (1987)] and the reassessment of the sediment age using the newly proposed age span of the Eskihisar sporomorph association (Benda & Meulenkamp 1979) contained within the lignite layers.

In contrast, other lines of evidence suggest that the age of the present-day extension prevailing in western Anatolia is Pliocene and is therefore unlikely to be the consequence of orogenic collapse.

- Evidence from the Niğde Massif in central Anatolia (Whitney & Dilek 1997, 1998)

suggests that Late Oligocene–Early Miocene extensional collapse and core-complex formation is not unique to western Anatolia but is widespread and affects larger areas, including central Anatolia. Similarly, Early–Middle Miocene extension has also been postulated for the Çankırı Basin in central Turkey (Kaymakçı et al. 2000).
- The presence of Early Miocene normal faults and associated sedimentary basins in their hanging wall is not limited to Turkey. Similar zones are recognized in the Cycladic Massif (e.g. Lister et al. 1984; Urai et al. 1990; Faure et al. 1991; Lee & Lister 1992; Gautier et al. 1993; Gautier & Brun 1994; Vandenberg & Lister 1996) and in the Rhodope Massif (e.g. Dinter & Royden 1993; Tzankov et al. 1996; Dinter 1998; Burchfiel et al. 2000). Therefore, it can be proposed that extensional deformation driven by gravitational collapse was widespread, affecting larger areas including the central Aegean, western Turkey and as far east as central Anatolia. In contrast, north–south neotectonic extension in Turkey is limited to western Anatolia (Fig. 1).
- Using recent GPS measurements, Barka & Reilinger (1997) and Reilinger et al. (1997) further demonstrated that central Anatolia [the Ova Province of Şengör et al. (1985)], previously affected by orogenic collapse-accommodated extension, is now undergoing an approximately north–south or north-northeast—south-southwest shortening and anticlockwise rotation due to slip along the dextral North Anatolian Fault.
- Although Miocene continental sediments are widespread in western Anatolia, within both the north–south and east–west grabens around and within the Menderes Massif, the Late Miocene–Pliocene sediments are confined to the east–west grabens (e.g. Paton 1992; Görür et al. 1995; Yılmaz 1997; Yusufoğlu 1998; Koçyiğit et al. 1999; Yılmaz et al. 2000), supporting the view that the east–west grabens developed during the Pliocene.
- It has been concluded, based on palaeomagnetic data and extrapolation of modern strain rates, that most of the total extension in the Aegean has occurred since the Early Pliocene (Jackson & McKenzie 1988; Kissel & Laj 1988).
- Although the Miocene sediments in western Turkey are usually deformed (as mentioned above), the Pliocene and younger sediments show no sign of such a style of deformation.
- Lastly, movements dated to 6–7 Ma (from ^{40}Ar–^{39}Ar laser probe experiments on white mica; Lips 1998) have been documented along the southern margin of the Gediz Graben. This implies that a phase of Early Miocene extension (Hetzel et al. 1995) was followed by latest Miocene extension at this locality, both slips being accommodated on the same fault (as is proposed in this study for the Büyük Menderes Graben).

This evidence suggests that the neotectonic phase of north–south extension in western Anatolia commenced during the latest Miocene or Pliocene. This age is in agreement with the inception of dextral movement along the North Anatolian Fault Zone (sometime in the Pliocene; Tokay 1973; Barka 1984, 1997; Barka & Kadinsky-Cade 1988; Koçyiğit & Rojay 1988; Toprak 1988; Dirik 1991; Rojay 1993; Tatar 1993; Westaway 1994a; Bozkurt & Koçyiğit 1995, 1996; Westaway & Arger 1996, 1998; Ünay et al. 1998; Yürür et al. 1998). This phase of extension can thus be attributed to a 'tectonic escape' mechanism. It is either a direct consequence of slip on the North Anatolian Fault Zone or an indirect consequence caused by another change – e.g. a change in the geometry of subduction along the Hellenic Trench – caused by slip on the North Anatolian Fault Zone.

It is thus suggested here that western Anatolia may be an example of a region which has experienced two modes of extension: a 'core-complex mode' and a 'wide-rift mode' (cf. Buck 1991). The first phase may be related to the latest Oligocene–Early Miocene gravitational collapse of the orogenically thickened crust and core-complex formation, following the Palaeogene collision across the Neotethyan ocean (Seyitoğlu & Scott 1991, 1992). This event is considered to have commenced c. 18–20 Ma [for details see Seyitoğlu & Scott (1992), Seyitoğlu et al. (1992) and Hetzel et al. (1995)]. It has been suggested that, during this collapse, the Menderes Massif was exhumed in the lower plate of presently low-angle normal fault(s) (Bozkurt & Park 1994, 1997; Hetzel et al. 1995, 1998). Deposition of the oldest sediments, which now crop out along the Büyük Menderes Valley, may have occurred in the hanging wall of such a fault during the 14–20 Ma time interval (Seyitoğlu & Scott 1991, 1992; Emre & Sözbilir 1995; Emre 1996; Seyitoğlu 1997). Collapse and related extension ceased by the mid-Late Miocene (c. 12 Ma) according to ^{40}Ar–^{39}Ar biotite cooling ages from the synextensional Salihli and Turgutlu Granodiorites emplaced along the footwall of the north facing, presently low-angle, normal fault at the southern margin of the Gediz Graben (Hetzel et al. 1995). This timing is also consistent with the age of the Miocene sediments (14–20 Ma; Seyitoğlu & Scott 1992; Seyitoğlu 1997).

A comparison of the present-day graben-fill sediments (Pliocene–Pleistocene) and radiometric constraints from the extensional shear zone mylonites (c. 12–20 Ma: Hetzel et al. 1995), and the hanging-wall basin fill (c. 14–20 Ma; Seyitoğlu & Scott 1991, 1992; Seyitoğlu 1997), suggest a time gap lasting c. 6–8 Ma, separating the development of major normal faults and the associated metamorphic complexes from the development of the horst-and-graben systems currently observed in western Anatolia. The latter interval may reflect a change in style of extension or correspond to a time of regional erosion that created the major disconformity between unit A and B sediments, and the folding of Miocene sediments, in the Büyük Menderes Graben and in other Miocene basins in western Turkey. This time interval also corresponds to a major break in magmatism in western Turkey (c. 14–10 Ma; Yılmaz et al. 2000) between the calc-alkaline, high-K Oligocene–Early Miocene first stage which died out at c. 14 Ma (Yılmaz 1989, 1990, 1997; Altunkaynak & Yılmaz 1999); and also to the second phase of Late Miocene–Pliocene rift-related basaltic volcanism. [Readers are referred to Yılmaz et al. (2000) for a detailed discussion.]

Following the initiation of strike-slip movement along the dextral North and sinistral East Anatolian Faults, the Anatolian block began to move westwards. The effect of westward tectonic escape resulted in a north–south upper crustal extension on active east–west normal faults during the Pliocene. The start of neotectonic extension is also marked by a change in the nature of the volcanism from dominantly calc-alkaline in the Middle Miocene to alkaline, including the Pliocene Kula basalts (7.5 ± 0.22–0.00025 Ma K–Ar ages; Ercan et al. 1985).

The above discussion further demonstrates that the east–west and north–south basins in western Turkey have not developed coevally, as was thought, for instance, by Şengör (1987) and Seyitoğlu (1997). In contrast, the north–south basins are cut by later east–west trending high-angle normal faults (see Yılmaz et al. 2000). This interpretation is further supported by the fact that the north–south grabens are elevated along the footwalls of the east–west normal faults. Thus, the episodic two-stage extension model suggested for the Gediz Graben by Koçyiğit et al. (1999) also seems to be supported in adjacent regions. Firstly, it explains why extension in western Anatolia and the central Aegean have occurred during and since the Pliocene (McKenzie 1978; Dewey & Şengör 1979; Koçyiğit et al. 1999). Secondly, it also accounts for the differences between the calculated initial dips of the low-angle normal faults controlling the core-complex mode of extension (c. 30° for the Gediz Graben; Hetzel et al. 1995: 49 ± 5° for the Büyük Menderes Graben; this study) and the high-angle graben-bounding faults which are active at present (up to 82°; Hancock & Barka 1987; Westaway 1990a, b, 1993; Jones & Westaway 1991; Paton 1992; Cohen et al. 1995; this study).

Conclusions

- There is no evidence that the red clastic sediments at the base of the Neogene sequence were deposited during neotectonic extension.
- The Büyük Menderes Graben exhibits evidence for two stages of extension – an initial extension on moderately steeply dipping normal faults was superseded by later steeper normal faults. This can support the following interpretation: during the first stage (latest Oligocene–Early Miocene), the Lower Neogene fluvial clastics were deposited in the basin that sits on the hanging wall of the normal fault(s), and the metamorphic rocks of the Menderes Massif were deformed, mylonitized and progressively exhumed in the footwall. The main graben-bounding normal faulting (Pliocene?) that overprints this early phase cuts and offsets the Miocene units and the normal faults.
- The fault that has controlled the early phase of extension may have been reactivated during the second phase.
- Mammal evidence constrains the start of slip on these younger faults to c. 1 Ma, as has been believed from other evidence.
- The Pliocene estimate for the start of extension is in close agreement with the start of slip on the North Anatolian Fault Zone, which further suggests that the extension can be explained as a geometrical consequence of this strike-slip movement. This further implies that the whole crust in western Turkey and the Aegean has since undergone extension due to the effect of extrusion processes along the North and East Anatolian Faults (tectonic escape).
- The two phases of deformation appear to reflect significant changes in the tectonic setting of western Anatolia, which can be attributed to orogenic collapse followed by tectonic escape.

The work reported here was supported by the Scientific and Technical Research Council of Turkey (TÜBİTAK); project YDABÇAG-221/A) and the National Marine Geology and Geophysics project (coordinated by N. Görür). Thanks are due to G. Kelling, A. Koçyiğit and B. F. Rojay for fruitful discussions. S. Mittwede helped with the English. L. Royden is thanked for her comments on the manuscript. The author is also greatly indebted to R. Westaway, whose suggestions have greatly improved the manuscript.

References

ALTUNKAYNAK, İ. & YILMAZ, Y. 1998. The Mount Kozak magmatic complex, western Anatolia. *Journal of Volcanology and Geothermal Research*, **85**, 211–231.

—— & —— 1999. The Kozak Pluton and its emplacement. In: BOZKURT, E. & ROWBOTHAM, G. (eds) *Advances in Turkish Geology: Regional Geology and Tectonic Evolution: Part I*. Geological Journal, **34**.

AMBRASEYS, N. N. 1988. Engineering seismology. *Earthquake Engineering, Structure and Dynamics*, **17**, 1–105.

—— & JACKSON, J. A. 1998. Faulting associated with historical and recent earthquakes in the Eastern Mediterranean region. *Geophysical Journal International*, **133**, 390–406.

ANGELIER, J. 1976. Sur l'aternance mio-plio-quaternaire de mouvements extensif et compressifs en Egee orientale: l'ile de Samos (Grèce). *Comptes Rendus de l'Académie des Sciences de Paris*, **283**, 463–466.

—— 1978. Tectonic evolution of the Hellenic arc since the late Miocene. *Tectonophysics*, **49**, 22–36.

—— & TSOFLIAS, P. 1976. Sur les mouvements mio-plio-quaternaires et la seismicité historique dans l'ile de Chios (Grèce). *Comptes Rendus de l'Académie des Sciences de Paris*, **283**, 1389–1391.

BARKA, A. A. 1984. Kuzey Anadolu Fay Zonundaki bazı Neojen-Kuvaterner havzaların jeolojisi ve tektonik evrimi. In: *Proceedings of the Ketin Symposium*. Geological Society of Turkey Publications, 209–227 [in Turkish with English abstract].

—— 1997. Neotectonics of the Marmara region. In: SCHINDLER, C & PFISTER, M. (eds) *Active Tectonics of Northwestern Anatolia – The MARMARA Poly Project; A Multidisciplinary Approach by Space Geodesy, Geology, Hydrogeology, Geothermics and Seismology*. Vdf. Hochschulerl, an der ETH Zurich, 55–87.

—— & KADINSKY-CADE, C. 1988. Strike-slip fault geometry in Turkey and its effect on earthquake activity. *Tectonics*, **7**, 663–684.

—— & REILINGER, R. 1997. Active tectonics of the Eastern Mediterranean region: deduced from GPS, neotectonic and seismicity data. *Annali Di Geofisica*, **XL**, 587–610.

BECKER-PLATEN, J. D. 1971. Stratigraphic division of the Neogene and oldest Pleistocene in southwest Anatolia. *Newsletters on Stratigraphy*, **1**, 19–22.

BENDA, L. & MEULENKAMP, J. E. 1979. Biostratigraphic correlations in the Eastern Mediterranean Neogene. 5. Calibration of sporomorph associations, marine microfossils and mammal zones, marine and continental stages and the radiometric scale. *Annales Géologique Des Pays Helleniques*, (hors er.) **1**, 61–70.

BİNGÖL, E. 1989. *Geological Map of Turkey at 1:2 000 000 Scale*. Mineral Research and Exploration Institute of Turkey (MTA) Publications.

BORAY, A., ŞAROĞLU, F. & EMRE, Ö. 1985. Isparta büklümünün kuzey kesiminde Doğu-Batı daralma için bazı veriler. *Geological Engineering*, **23**, 9–20 [in Turkish with English abstract].

BORONKAY, K. & DOUTSOS, T. 1994. Transpression and transtension within different structural levels in central Aegean region. *Journal of Structural Geology*, **16**, 1555–1573.

BOZKURT, E. & KOÇYİĞİT, A. 1995. Almus Fault zone: its age, total offset and relation to the North Anatolian Fault Zone. *Turkish Journal of Earth Sciences*, **4**, 93–104.

—— & —— 1996. Kazova basin: an active negative flower structure on the Almus Fault Zone, a splay fault system of the North Anatolian Fault Zone, Turkey. *Tectonophysics*, **265**, 239–254.

—— & PARK, R. G. 1994. Southern Menderes Massif: an incipient metamorphic core complex in western Anatolia, Turkey. *Journal of the Geological Society, London*, **151**, 213–216.

—— & —— 1997. Evolution of a mid-Tertiary extensional shear zone in the Southern Menderes Massif, Western Turkey. *Bulletin de la Société Géologique de France*, **168**, 3–14.

BOZKUŞ, C. 1996. Kavacık (Dursunbey-Balıkesir) Neojen grabeninin stratigrafisi ve tektoniği. *Turkish Journal of Earth Sciences*, **5**, 161–170 [in Turkish with English abstract].

BUCK, R. 1991. Modes of continental lithospheric extension. *Journal of Geophysical Research*, **96**, 20 161–20 178.

BURCHFIEL, C. B., NAKOV, R., TZANKOV, T. & ROYDEN, L.H. 2000. Cenozoic extension in Bulgaria and Northern Greece: the northern part of the Aegean extensional regime. *This volume*.

CAPUTO, R. 1989. *Fault: A Programme for Structural Analysis*. University of Florence.

COHEN, H. A., DART, C. J., AKYÜZ, H. S. & BARKA, A. A. 1995. Syn-rift sedimentation and structural development of Gediz and Büyük Menderes Grabens, western Turkey. *Journal of the Geological Society, London*, **152**, 629–638.

DEWEY, J. F. & ŞENGÖR, A. M. C. 1979. Aegean and surrounding regions: complex multiplate and continuum tectonics in a convergent zone. *Geological Society of America Bulletin*, **90**, 84–92.

DINTER, D. A. 1998. Late Cenozoic extension of the Alpine collisional orogen, northeastern Greece: origin of the north Aegean basin. *Geological Society of America Bulletin*, **110**, 1208–1230.

—— & ROYDEN, L.H. 1993. Late Cenozoic extension in northern Greece: Strymon Valley detachment system and Rhodope metamorphic core complex. *Geology*, **21**, 25–49.

Dirik, K. 1991. *Tectonostratigraphy of the Vezirköprü Area (Samsun–Turkey)*. Phd Thesis, Middle East Technical University.

Dumont, J. F., Uysal, Ş., Şimşek, Ş., Karamanderesi, İ. H. & Letouzcy, F. 1979. Güneybatı Anadolu'daki grabenlerin oluşumu. *Mineral Research and Exploration Institute of Turkey (MTA) Bulletin*, **92**, 7–17.

Emre, T. 1996. Gediz grabeninin jeolojisi ve tektoniği. *Turkish Journal of Earth Sciences*, **5**, 171–185 [in Turkish with English abstract].

—— & Sözbilir, H. 1995. Field evidence for metamorphic core complex, detachment faulting and accommodation faults in the Gediz and Büyük Menderes grabens, western Anatolia. *In*: Pişkin, Ö., Ergün, M., Savaşçın, M. Y. & Tarcan, G. (eds) *Proceedings of the International Earth Science Colloquium on the Aegean region*, 9–14 October 1995, İzmir–Güllük, Turkey, **1**, 73–93.

Ercan, T., Dinçel, A., Metin, S., Türkecan, A. & Günay, E. 1978. Uşak yöresindeki havzaların jeolojisi. *Geological Society of Turkey Bulletin*, **21**, 97–106 [in Turkish with English abstract].

——, Satır, M., Kreuzer, H. *et al.* 1985. Batı Anadolu Senozoyik volkanitlerine ait yeni kimyasal, izotopik ve radyometrik verilerin yorumu. *Geological Society of Turkey Bulletin*, **28**, 121–136 [in Turkish with English abstract].

Faure, M., Bonneau, M. & Pons, J. 1991. Ductile deformation and syntectonic granite emplacement during the late Miocene extension of the Aegean (Greece). *Société Géologique de France Bulletin*, **162**, 3–11.

Gautier, P. & Brun, J.-P. 1994. Ductile crust exhumation and extensional detachments in the central Aegean (Cyclades and Evvia islands). *Geodinamica Acta (Paris)*, **7**, 57–85.

——, Brun, J. P. & Jolivet, L. 1993. Structure and kinematics of upper Cenozoic extensional detachment on Naxos and Paros (Cyclades islands, Greece). *Tectonics*, **12**, 1180–1194.

Gökçen, N. 1982. The ostracoda biostratigraphy of the Denizli-Muğla Neogene sequences. *Hacettepe University Earth Sciences*, **9**, 111–131.

Görür, N., Şengör, A. M. C., Sakinç, M. *et al.* 1995. Rift formation in the Gökova region, southwest Anatolia: implications for the opening of the Aegean Sea. *Geological Magazine*, **132**, 637–650.

Hancock, P. L. & Barka, A. A. 1987. Kinematic indicators on active normal faults in western Turkey. *Journal of Structural Geology*, **9**, 573–584.

Hetzel, R., Ring, U., Akal, C. & Troesch, M. 1995. Miocene NNE- directed extensional unroofing in the Menderes Massif, southwestern Turkey. *Journal of the Geological Society, London*, **152**, 639–654.

——, Romer, R. L., Candan, O. & Passchier, C. W. 1998. Geology of the Bozdağ area, Central Menderes Massif, SW Turkey: Pan-African basement and Alpine deformation. *Geologische Rundschau*, **87**, 394–406.

İnci, U. 1991. Torbalı (İzmir) kuzeyindeki Miyosen tortul istifinin fasiyes ve çökelme ortamları. *Mineral Research and Exploration Institute of Turkey (MTA) Bulletin*, **112**, 13–26 [in Turkish with English abstract].

Jackson, J. A. & McKenzie, D. P. 1988. Rates of active deformation in the Aegean Sea and surrounding regions. *Basin Research*, **1**, 121–128.

——, King, G. & Vita-Finzi, C. 1992. The neotectonics of the Aegean: an alternative view. *Earth and Planetary Science Letters*, **61**, 303–318.

Jones, M. & Westaway, R. 1991. Microseismicity and structures of the Germencik area, west Turkey. *Geophysical Journal International*, **106**, 293–300.

Karamanderesi, İ. H. 1972. *Aydın Nazilli–Çubukbağ Arası Jeotermal Olanakları Hakkında Jeolojik Rapor*. Mineral Research and Exploration Institute of Turkey (MTA) Report, No. 5224 [in Turkish].

Kaya, O. 1981. Miocene reference sections for the coastal parts of west Anatolia. *Newsletters on Stratigraphy*, **10**, 164–191.

Kaymakçı, N., White, S. H. & van Dijk, P. M. (2000). Palaeostress inversion in a multiphase deformed area: kinematic and structural evolution of the Çankırı Basin (central Turkey), Part 1. *This volume*.

Kissel, C. & Laj, C. 1988. Tertiary geodynamical evolution of the Aegean arc: a palaeomagnetic reconstruction. *Tectonophysics*, **146**, 183–201.

Koçyiğit, A. & Bozkurt, E. 1998. Kütahya-Tavşanlı çöküntü alanının neotektonik özellikleri. The Scientific and Research Council of Turkey (TÜBİTAK) Project No. YDABÇAG-126 [in Turkish with English abstract].

—— & Rojay, F. B. 1988. Geological setting, origin, type and age of the Merzifon–Suluova basin, N Turkey. *Symposium for the 20th Anniversary of Earth Sciences at Hacettepe University*, October 25–27, Beytepe–Ankara, Abstracts, 42.

——, Yusufoğlu, H. & Bozkurt, E. 1999. Evidence from the Gediz graben for episodic two-stage extension in western Turkey. *Journal of the Geological Society, London*, **156**, 605–616.

——, Türkmenoğlu, A., Beyhan, A., Kaymakçı, N. & Akyol, E. 1995. Post-collisional tectonics of Eskişehir–Ankara–Çankırı segment of the İzmir–Ankara–Erzincan Suture Zone (İAESZ): Ankara orogenic phase. *Turkish Association of Petroleum Geologists Bulletin*, **6**, 69–86 [in Turkish with English abstract].

Kurt, H., Demirbağ, E. & Kuçu, İ. 1999. Investigation of the submarine active tectonism in the Gulf of Gökova, southwest Anatolia–southeast Aegean Sea, by multi-channel seismic reflection data. *Tectonophysics*, **305**, 477–496.

Le Pichon, X. & Angelier, J. 1979. The Hellenic arc and trench system: a key to the neotectonic evolution of the Eastern Mediterranean area. *Tectonophysics*, **60**, 1–42.

—— & —— 1981. The Aegean Sea. *Philosophical Transactions of the Royal Society of London, Series A*, **300**, 357–372.

——, Chamot-Rooke, C., Lallemant, S., Noomen, R.

& VEIS, G. 1995. Geodetic determination of the kinematics of Central Greece with respect to Europe: implications for Eastern Mediterranean tectonics. *Journal of Geophysical Research*, **100**, 12 675–12 690.

LEE, J. & LISTER, G. S. 1992. Late Miocene ductile extension and detachment faulting, Mykonos, Greece. *Geology*, **20**, 607–610.

LIPS, A. L. W. 1998. *Temporal constraints on the kinematics of the destabilitization of an orogen; syn- to post-orogenic extensional collapse of the Northern Aegean region*. Geologica Ultraiectina, Mededelingen van de Faculteit Aardwetenschappen, Universiteit Utrecht, No. 166.

LISTER, G.S., BANGA, G. & FEENSTA, A. 1984. Metamorphic core complexes of Cordilleran type in the Cyclades, Aegean Sea, Greece. *Geology*, **12**, 221–225.

MCKENZIE, D. P. 1978. Active tectonics of the Alpine–Himalayan belt: the Aegean sea and surrounding regions (tectonics of Aegean region). *Royal Astronomical Society Geophysical Journal*, **55**, 217–254.

MERCIER, J.-L. 1976. La néotectonique–ses méthodes et ses buts. Un exemple: l'arc agéen (Méditerranée orientale). *Revue de Géographie Physique et de Géologie Dynamique*, **18**, 323–346.

—— 1979. Signification néotectonique de l'arc agéen. Une revue des idées. *Revue de Géographie Physique et de Géologie Dynamique*, **21**, 5–16.

—— 1981. Extensional–compressional tectonics associated with the Aegean Arc: comparison with the Andean Cordillera of S. Peru–N. Bolivia. *Philosophical Transactions of the Royal Society of London, Series A*, **300**, 337–355.

——, CAREY, E., PHILIP, H. & SOREL, D. 1976. La néotectonique plio-quaternaire de l'arc agéen externe et de la Mer agéen et ses relations avec seismicité. *Bulletin de la Société Géologique de France*, **18**, 159–176.

——, DELIBASSIS, N., GAUTHIER, A. *ET AL.* 1979. Le néotectonique de l'arc agéen. *Revue de Géographie Physique et de Géologie Dynamique*, **21**, 67–92.

MEULENKAMP, J. E., WORTEL, W. J. R., VAN WAMEL, W. A., SPAKMAN, W. & HOOGERDUYN STRATING, E. 1988. On the Hellenic subduction zone and geodynamic evolution of Crete in the late middle Miocene. *Tectonophysics*, **146**, 203–215.

NEBERT, K. 1960. Tavşanlı'nın batı ve kuzeyindeki linyit ihtiva eden Neojen sahasının mukayeseli stratigrafisi ve tektoniği. *Mineral Research and Exploration Institute of Turkey (MTA) Bulletin*, **54**, 7–35 [in Turkish with English abstract].

—— 1978. Linyit içeren Soma Neojen bölgesi, Batı Anadolu. *Mineral Research and Exploration Institute of Turkey (MTA) Bulletin*, **90**, 20–69 [in Turkish with English abstract].

PATON, S. 1992. Active normal faulting, drainage patterns and sedimentation in southwestern Turkey. *Journal of the Geological Society, London*, **149**, 1031–1044.

REILINGER, R. E., MCCLUSKY, S. C., ORAL, M. B. *ET AL.* 1997. Global Positioning System measurements of present-day crustal movements in the Arabia–Africa–Eurasia plate collision zone. *Journal of Geophysical Research*, **102**, 9983–9999.

ROBERTS, S. C. 1998. *Active normal faulting in Central Greece and Western Turkey*. PhD Thesis, University of Cambridge.

—— & JACKSON, J. A. 1991. Active normal faulting in central Greece: an overview. *In*: ROBERTS, A. M., YIELDING, G. & FREEMAN, B. (eds) *The Geometry of Normal Faults*. Geological Society, London, Special Publications, **56**, 125–142.

ROBERTSON, A. H. F., CLIFT, P. D., DEGNAN, P. & JONES, G. 1991. Palaeogeographic and palaeotectonic evolution of the eastern Mediterranean region. *Palaeogeography, Palaeoclimatology, Palaeoecology*, **87**, 289–344.

ROJAY, F. B. 1993. *Tectonostratigraphy and Neotectonic characteristics of the southern margin of Merzifon–Suluova Basin (central Pontides, Amasya)*. PhD Thesis, Middle East Technical University.

ŞENGÖR, A. M. C. 1979. The North Anatolian transform fault: its age, offset and tectonic significance. *Journal of the Geological Society, London*, **136**, 269–282.

—— 1987. Cross-faults and differential stretching of hanging-walls in regions of low-angle normal faulting: example from Western Turkey. *In*: COWARD, M. P., DEWEY, J. F. & HANCOCK, P. L. (eds) *Continental Extensional Tectonics*. Geological Society, London, Special Publications, **28**, 575–589.

—— & YILMAZ, Y. 1981. Tethyan evolution of Turkey: a plate tectonic approach. *Tectonophysics*, **75**, 181–241.

——, GÖRÜR, N. & ŞAROĞLU, F. 1985. Strike-slip deformation, basin formation, and sedimentation. *In*: BIDDLE, K. T. & CHRISTIE-BLICK, N. (eds) *Strike-slip Faulting and Basin Formation*. Society of Economic Paleontologists and Mineralogists, Special Publications, **37**, 227–264.

SEYİTOĞLU, G. 1997. Late Cenozoic tectono-sedimentary development of the Selendi and Uşak–Güre basins: a contribution to the discussion on the development of east–west and north trending basins in western Turkey. *Geological Magazine*, **134**, 163–175.

—— & SCOTT, B. C. 1991. Late Cenozoic extension and basin formation in west Turkey. *Geological Magazine*, **128**, 155–166.

—— & —— 1992. The age of Büyük Menderes Graben (west Turkey) and its tectonic implications. *Geological Magazine*, **129**, 239–242.

—— & —— 1996. The age of the Alaşehir graben (west Turkey) and its tectonic implications. *Geological Journal*, **31**, 1–11.

——, —— & RUNDLE, C. C. 1992. Timing of Cenozoic extensional tectonics in west Turkey. *Journal of Geological Society, London*, **149**, 533–538.

SICKENBERG, O. & TOBIEN, H. 1971. New Neogene and Lower Quaternary vertabrate faunas in Turkey. *Newsletters on Stratigraphy*, **1**, 51–61.

STEININGER, F. F. & RÖGL, F. 1984. Palaeogeography and palinspastic reconstruction of the Neogene of the Mediterranean and Paratethys. *In*: Dixon, J. E. & Robertson, A. H. F. (eds) *The Geological Evolution of the Eastern Mediterranean*. Geological Society, London, Special Publications, **14**, 659–668.

Tatar, O. 1993. *Neotectonic structures in the east central part of the North Anatolian Fault Zone, Turkey*. PhD Thesis, Keele University.

Tokay, M. 1973. Kuzey Anadolu Fay Zonu'nun Gerede ile Ilgaz arasındaki kısımında jeolojik gözlemler. *Proceedings of the Symposium on the North Anatolian Fault and Earthquake Belt*. Mineral Research and Exploration Institute of Turkey (MTA) Publications.

Toprak, V. 1988. Neotectonic characteristics of the North Anatolian Fault Zone between Koyulhisar and Suşehri (NE Turkey). *METU Journal of Pure and Applied Sciences*, 155–168.

Tzankov, T. Z., Angelova, D., Nakov, R., Burchfiel, B. C. & Royden, L. H. 1996. The Sub-Balkan graben system of central Bulgaria. *Basin Research*, **8**, 125–142.

Urai, J. L., Schuiling, R. D. & Jansen, J. B. H. 1990. Alpine deformation on Naxos (Greece). *In*: Knipe, R. J. & Rutter, E. H. (eds) *Deformation Mechanisms, Rheology and Tectonics*. Geological Society, London, Special Publications, **54**, 509–522.

Ünay, E. & De Bruijn, H. 1998. Plio-Pleistocene rodents and lagomorphs from Anatolia. *Mededelingen Nederlands Instituut voor Toegepaste Geowetenschappen TNO*, **60**, 431–466.

——, Emre, Ö., Erkal, F. & Keçer, M. 1998. The age on the basis of Arvicolidae (Rodentia, Mammalia) of the Adapazarı pull-apart basin in the western part of the north Anatolian Fault Zone (Turkey): the preliminary results. *Third International Turkish Geology Symposium*, METU – Ankara, Abstracts, 219.

——, Göktaş, F., Hakyemez, H. Y., Avşar, M. & Şan, Ö. 1995. Büyük Menderes Grabeni'nin kuzey kenarındaki çökellerin Arvicolidae (Rodentia, Mammalia) faunasına dayalı olarak yaşlandırılması. *Geological Society of Turkey Bulletin*, **38**, 75–80 [in Turkish with English abstract].

Vandenberg, L. C & Lister, G. S. 1996. Structural analysis of basement tectonites from the Aegean metamorphic core complex of Ios, Cyclades, Greece. *Journal of Structural Geology*, **18**, 1437–1454.

Westaway, R. 1990a. Block rotation in western Turkey. 1. Observational evidence. *Journal of Geophysical Research*, **95**, 19 857–19 884.

—— 1990b. Block rotation in western Turkey. 2. Theoretical models. *Journal of Geophysical Research*, **95**, 19 885–19 901.

—— 1993. Neogene evolution of the Denizli region of western Turkey. *Journal of Structural Geology*, **15**, 37–53.

—— 1994a. Present-day kinematics of the Middle East and Eastern Mediterranean. *Journal of Geophysical Research*, **99**, 12 071–12 090.

—— 1994b. Evidence for dynamic coupling of surface processes with isostatic compensation in the lower crust during active extension of western Turkey. *Journal of Geophysical Research*, **99**, 20 203–20 223.

—— 1996. Quaternary elevation change in the Gulf of Corinth of central Greece. *Philosophical Transactions of Royal Society, London, Series A*, **354**, 1125–1164.

—— 1998. Dependence of active normal fault dips on lower-crustal flow regimes. *Journal of Geological Society, London*, **155**, 233–253.

—— & Arger, J. 1996. The Gölbaşı basin, southeastern Turkey: a complex discontinuity in a major strike-slip fault zone. *Journal of Geological Society, London*, **153**, 729–743.

—— & —— 1998. Kinematics of the Malatya–Ovacık Fault Zone. *The Third International Turkish Geology Symposium*, METU-Ankara, Abstracts, 197.

—— & Kusznir, N. J. 1993a. Fault and bed 'rotation' during continental extension: block rotation or vertical shear? *Journal of Structural Geology*, **15**, 753–770.

—— & —— 1993b. Correction to "Fault and bed 'rotation' during continental extension: block rotation or vertical shear?". *Journal of Structural Geology*, **15**, 1391.

Whitney, D. L. & Dilek, Y. 1997. Core complex development in central Anatolia, Turkey. *Geology*, **25**, 1023–1026.

—— & —— 1998. Metamorphism during Alpine crustal thickening and extension in central Anatolia, Turkey: Niğde metamorphic core complex. *Journal of Petrology*, **39**, 1385–1403.

Yağmurlu, F. 1987. Salihli güneyinde üste doğru kabalaşan Neojen yaşlı alüvyon yelpaze çökelleri ve Gediz grabeninin tektono-sedimanter gelişimi. *Geological Society of Turkey Bulletin*, **30**, 33–40 [in Turkish with English abstract].

—— 1991. Yalvaç-Yarıkkaya Neojen havzasının tektonosedimanter özellikleri ve yapısal evrimi. *Mineral Research and Exploration Institute of Turkey (MTA) Bulletin*, **112**, 1–12 [in Turkish with English abstract].

Yalçın, H., Semelin, B. & Gündoğdu, N. 1985. Geological investigation of Emet lacustrine basin of Neogene age (south of Hisarcık). *Hacettepe University Earth Sciences*, **12**, 39–52.

Yilmaz, Y. 1989. An approach to the origin of young volcanic rocks of western Turkey. *In*: Şengör, A. M. C. (ed.) *Tectonic Evolution of the Tethyan Region*. Kluwer, 159–189.

—— 1990. Comparisons of the young volcanic associations of the west and the east Anatolia under the compressional regime: a review. *Journal of Volcanology and Geothermal Research*, **44**, 69–87.

—— 1997. Geology of Western Anatolia. *In*: Schindler, C. & Pfister, M. (eds) *Active Tectonics of Northwestern Anatolia – The MARMARA Poly Project; A Multidisciplinary Approach by Space Geodesy, Geology, Hydrogeology, Geothermics and Seismology*. Vdf. Hochschulerl, an der ETH Zurich, 31–53.

——, Genç, Ş. C., Gürer, F. *et al.* 2000. When did the

western Anatolian grabens begin to develop? *This volume.*

YUSUFOĞLU, H. 1998. *Palaeo- and Neo-tectonic characteristics of the Gediz and Küçük Menderes Grabens in West Turkey*. PhD Thesis, Middle East Technical University.

YÜRÜR, T., KÖSE, O., BUKET, E., DEMİRBAĞ, H. & GÜVEN, A. R. 1998. Recent tectonics and volcanism in the eastern vicinity of Karlıova junction zone, eastern Turkey. *Third International Turkish Geology Symposium*, METU – Ankara, Abstracts, 96.

Episodic graben formation and extensional neotectonic regime in west Central Anatolia and the Isparta Angle: a case study in the Akşehir–Afyon Graben, Turkey

ALİ KOÇYİĞİT,[1] ENGİN ÜNAY[2] & GERÇEK SARAÇ[3]

[1] *Middle East Technical University, Department of Geological Engineering, Tectonic Research Unit, TR-06531 Ankara, Turkey (e-mail: akoc@metu.edu.tr)*
[2] *Cumhuriyet University, Department of Anthropology, TR-58140 Sivas, Turkey*
[3] *Mineral Research and Exploration Institute of Turkey (MTA), Geology Department, TR-06520 Ankara, Turkey*

Abstract: Central and Western Anatolia form a continental back-arc region related to the Hellenic–Cyprus convergent plate boundary of the Anatolian and African Plates. The Akşehir–Afyon Graben (AAG), the easternmost extension of the west Anatolian horst–graben system, is located at the junction of Central Anatolia and eastern limb of the Isparta Angle. The AAG is 4–20 km wide and 90 km long. It trends west-northwest–east-southeast and is an actively growing rift containing two sedimentary infills of continental fluvio-lacustrine origin bounded on both sides by oblique-slip normal faults. The older infill is folded, thrust faulted and early Late Miocene in age. The younger infill, which is nearly horizontally bedded, is Plio-Quaternary in age and rests on the older infill with angular unconformity. The deformation of the older infill and the angular unconformity indicate a Late Miocene phase of compression, which separates two extensional periods. The second phase of extension has lasted since the Pliocene and is part of the current extensional neotectonic regime of both west Central Anatolia and the Isparta Angle, despite being previously reported as a compressional neotectonic regime.

The graben-bounding Akşehir Fault Zone (AFZ) and the Karagöztepe Fault Zone display well-preserved fault surfaces and slickenlines. Although stereographic plots of the fault slip data show that the graben-bounding structures are oblique-slip normal faults, the AFZ has also been described as a single reverse fault. Both the field and seismic data, particularly the 1921 Argıthanı–Akşehir and 1946 Ilgın–Argıthanı earthquakes, indicate that the AAG is an active neotectonic structure. However, it can also be interpreted to lie in a seismic gap when its rate of seismicity is compared with that of the Gediz–Simav Graben forming its west-northwestern extension.

There is still no common agreed explanation for the neotectonic regime in Turkey. Four principal explanations have been proposed: (1) tectonic escape (Dewey & Şengör 1979; Dewey *et al.* 1986); (2) back-arc spreading (Le Pichon & Angelier 1979; Meulenkamp *et al.* 1988); (3) orogenic collapse (Dewey 1988; Seyitoğlu & Scott 1992); and (4) a two-stage graben model incorporating episodic graben formation (Koçyiğit *et al.* 1999). In the tectonic escape model, the west-southwestward tectonic escape of the Anatolian Platelet has produced an extensional neotectonic regime leading to development of the horst–graben system in west Turkey just after collision of Arabia and Eurasia in Late Serravalian time (*c.* 12 Ma). In this model the onset age of the current extensional neotectonic regime is late Middle Miocene. In the back-arc spreading model, south-southwestward migration of the Hellenic Trench System is assumed to have created an extensional neotectonic regime which formed the horst–graben system in the Aegean and west Turkey. Estimates of the onset age of Hellenic subduction range from 5 to 60 Ma (McKenzie 1978; Le Pichon & Angelier 1979; Kissel & Laj 1988; Meulenkamp *et al.* 1988). In this case, the onset age of the extensional neotectonic regime is cryptic and could range from 60 to 5 Ma. In the orogenic collapse model, the origin of the extensional neotectonic regime is related to spreading and stretching of overthickened crust. The onset age and formation of the horst–graben system in west Anatolia is then predicted to be Late Oligocene–Early Miocene following cessation of the Palaeocene collision, and the shortening and overthickening of the Aegean–Anatolian crust along the İzmir–Ankara–Erzincan Neotethyan Suture. According to this model, graben formation is a continuous event lasting without interruption from the Late Oligocene–Early

From: BOZKURT, E., WINCHESTER, J. A. & PIPER, J. D. A. (eds) *Tectonics and Magmatism in Turkey and the Surrounding Area.* Geological Society, London, Special Publications, 173, 405–421. 1-86239-064-9/00/$15.00
© The Geological Society of London 2000.

Miocene when the current extensional tectonic regime began.

According to the two-stage graben model, a combination of back-arc spreading, tectonic escape, orogenic collapse and roll-back processes is responsible for formation of the graben–horst systems. Indeed, the idea of a two-stage graben can explain the temporal evolution of the graben–horst systems. The graben–horst systems in both West and Central Anatolia developed in two (Early–Middle Miocene and Plio-Quaternary) extensional phases, interrupted by an intervening Late Miocene and/or Early Pliocene compressional phase. The onset age of the current extensional neotectonic regime in West and Central Anatolia is then interpreted as early Middle Pliocene.

In the context of the neotectonics of Turkey, Central Anatolia is a geologically critical area for two main reasons. Firstly, the western half of Central Anatolia is dominated by a series of graben and horst structures bounded by active oblique-slip normal faults. In contrast, its eastern half is characterized by a strike-slip tectonic regime and related structures, such as sinistral to dextral active strike-slip faults and basins (Koçyiğit 1984; Şaroğlu et al. 1987; Pasquare et al. 1988; Koçyiğit & Beyhan 1998). Thus, Central Anatolia forms a broad transitional tectonic zone between the extensional neotectonic regime of Aegean–West Anatolia and the prominent strike-slip tectonic regime of East Anatolia (Koçyiğit 1985; Dewey et al. 1986; Y. Yılmaz et al. 1987). Secondly, Central Anatolia is the continental back-arc of the north dipping Hellenic–Cyprus subduction zone. Most of the geological structures of Central Anatolia and the Taurides, including the Isparta Angle, have been sourced from tectonic and magmatic events related to this active convergent plate boundary (Glover & Robertson 1998). For this reason, data obtained from Central Anatolia and the Isparta Angle contribute to wider solutions of problems related to development of the East Mediterranean.

The current tectonic regime of Central Anatolia was previously termed the 'Ova Regime' by Şengör (1980). He defined the term 'Ova' as a basin bounded by more than two oblique faults. The geological mapping and case studies carried out here, at a number of locations in Central Anatolia, indicate that the depressions, or 'Ova' of Şengör (1980), in the western half of Central Anatolia are grabens bounded by oblique-slip normal faults but are strike-slip basins in the eastern half. Hence, in this paper the term 'Central Anatolian neotectonic regime' is preferred in place of the term 'Ova Regime'.

The current study area is the Aşehir–Afyon Graben (AAG) at the southwestern corner of Central Anatolia and is sited on the northeastern outer limb of the Isparta Angle (Fig. 1). Some basic geological problems of both the Isparta Angle and Central Anatolia still under discussion include: (1) the interpretation of most of previous workers (Boray et al. 1985; Şaroğlu et al. 1987; Barka et al. 1995) that basin fill of the Akşehir–Afyon Depression is Late Miocene–Pliocene in age and that its southern margin-bounding structure is a reverse fault; (2) whilst the current tectonic regime at the eastern limb of the Isparta Angle is indeed compressional, this group's recent field data indicate an extensional neotectonic regime with oblique-slip normal faulting and a Plio-Quaternary basin fill for both the AAG and the Isparta Angle. The first goal of this paper is to test above-mentioned models of extension. The second goal is to bring reliable observation-based solutions to problems of the Isparta Angle and Central Anatolia using recent data from a case study carried out in the AAG, and thereby to contribute to a solution of some regional problems related to the tectonic evolution of the East Mediterranean.

Tectonic setting

Major neotectonic structures of central Turkey and adjacent areas include the sinistral Dead Sea Fault Zone, the sinistral East Anatolian Transform Fault Zone and the dextral North Anatolian Transform Fault Zone. Since Pliocene times, the Anatolian Platelet has been moving west-southwest along the latter two structures, towards the easily subductable oceanic lithosphere of the African Plate southwest of the Hellenic to Cyprus Arc (Fig. 1). In addition to the major structures, there are two groups of second-order intracontinental faults cutting across Central Anatolia. These comprise northeast–southwest and northwest–southeast striking faults (Koçyiğit 1984; Pasquare et al. 1988; Koçyiğit & Beyhan 1998). The first group of faults splays off mostly from the North Anatolian Transform Fault and displays both sinistral and dextral strike-slip movement with considerable normal to reverse-slip components due to variations in the general trend of the faults – well-defined examples of this group of faults are illustrated in Fig. 1. The second group of faults is confined entirely to Central Anatolia and comprises mostly oblique-slip normal faults with considerable the Akşehir–Afyon Graben bordered by the Emirdağ and Sultandağ Horsts (AAG, EH and SH, respectively, in Fig. 1). Other major geological elements of Central Anatolia are the Kırşehir Block, the Cretaceous–Early Tertiary İzmir–Ankara Suture Zone, the Campanian–Lower Quaternary Galatean Volcanic Complex (Koçyiğit 1991; Keller *et al.* 1992), the Middle Miocene–Quaternary Cappadocian Volcanic Plateau and its isolated composite volcanoes (Pasquare *et al.* 1988), and the Lower–Middle Miocene Bolvadin, Erenlerdağ (Innocenti *et al.* 1975) and Pliocene Isparta Volcanic Complexes (Fig. 1). The Akşehir–Afyon Graben, with its boundary faults and basin fill, are the main the topics of this paper.

The Akşehir–Afyon Graben (AAG)

The AAG is 4–20 km wide, 90 km long and trends west-northwest–east-southeast. It is an actively growing rift, as indicated by the 26th September 1921 Argıthanı and the 21st February 1946 Ilgın–Argıthanı Earthquakes (Eyidoğan *et al.* 1991). The rift is asymmetric and contains the Eber and Akşehir Lakes (Fig. 1).

Graben fill

The graben fill consists of two major sequences: (1) a deformed (folded to thrust faulted) fluvio-lacustrine sequence of Early–early Late Miocene age, which is over 300 m thick and divided into the Köstere and Gölyaka Formations; (2) an undeformed fluvio–lacustrine sequence of Plio-Quaternary age, which is 670 m thick and divided into the Doğancık, Gözpınarı, Taşköprü and Dursunlu Formations. These two sequences are separated from each other by an angular unconformity (Fig. 2).

The Köstere Formation. This starts with a basal conglomerate on an erosional surface of pre-Jurassic metamorphic rocks and continues upward with an alternation of various lithofacies. At the top, it is first thrust by the same basement and then overlain unconformably by red continental clastic rocks of the Pliocene Doğancık Formation (Fig. 2). The basal conglomerates are red-yellow in colour, thick bedded to massive, unsorted and polygenetic in composition. Pebbles are rounded to sub-rounded and consist mostly of quartz, marble and schist. The total thickness ranges from a few metres up to 30 m. These basal conglomerates are succeeded by numerous rock packages of dissimilar facies and thicknesses (up to 22 m).

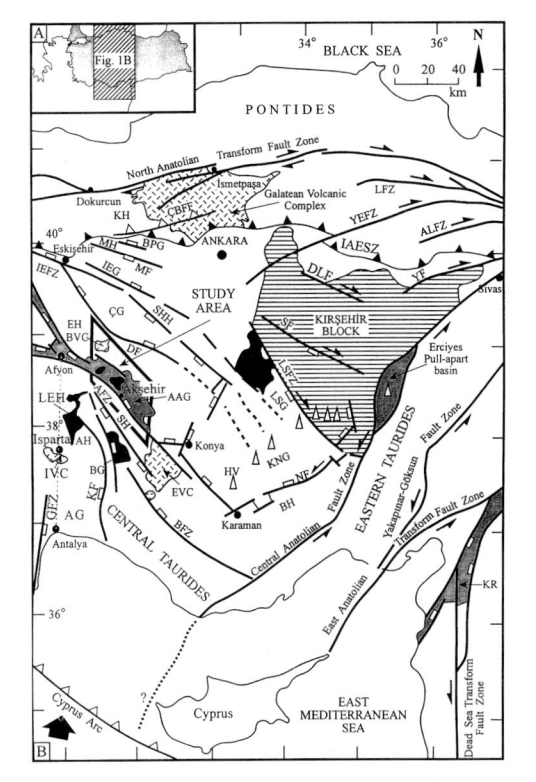

Fig. 1. (**A**) Location of central Turkey. (**B**) Simplified neotectonic map of central Turkey. AH, Anamas Horst; BH, Bolkardağ Horst; DE, Emirdağ Horst; KH, Köroğlu Horst; MH, Mihalıccık Horst; SH, Sultandağ Horst; SHH, Sivrihisar Horst; AG, Antalya Graben; AAG, Akşehir–Afyon Graben; BG, Beyşehir Graben; BPG, Beypazarı Graben; ÇG, Çifteler Graben; İEG, İnegöl–Eskişehir Graben; KNG, Konya–Niğde Graben; LSG, Lake Salt Graben; KR, Karasu Rift; ALFZ, Almus Fault Zone; BFZ, Beyşehir Fault Zone; ÇBFZ, Çayırhan–Beypazarı forced Fold Zone; DF, Davulga Fault; DLF, Delice Fault; GFZ, Göynük Fault Zone; HV, historical volcanoes; IEFZ, İnönü–Eskişehir Fault Zone; KF, Kırkkavak Fault; LSFZ, Lake Salt Fault Zone; LFZ, Laçin Fault Zone; MF, Mihalıccık Fault; NF, Niğde Fault; SF, Salanda Fault; YEFZ, Yağmurlu–Ezinepazarı Fault Zone; YF, Yıldızeli Fault; BVC, Bolvadin Volcanic Complex; EVC, Erenlerdağ Volcanic Complex; IVC, Isparta Volcanic Complex; LEH, Lake Eğirdir–Hoyran; IAESZ, İzmir–Ankara–Eskişehir Suture Zone. Note: the east of Afyon–Antalya Line is the eastern limb of the Isparta Angle.

dextral strike-slip components – well-defined examples of this second group are also seen in Fig. 1. Owing to displacements along this second group of faults, the western half of Central Anatolia has been divided into a number of horsts and grabens, one of which is

AGE	UNIT	THICK. (m)	LITHOLOGY	DESCRIPTION	TECTONIC PERIOD
HOL.		100		D. Silt, clay, lime (marsh-lake deposits) C. Delta deposits B. Recent fan deposits (proximal part) A. Red palaeosoil with potter's field —— Local Disconformity ——	NEOTECTONIC PERIOD
PLEISTOCENE	A–B. Taşköprü Formation C. Dursunlu Formation	320		C. Alternation of claystone, marl, limy mudstone with coal seam in the nature of peat (lacustrine facies) B. Graded- to cross-bedded conglomerate, sandstone, siltstone alternation with abundant *Dreissensia* sp. (delta-beach facies) A. Loose conglomerates (proximal part of fan facies) —— Local Disconformity ——	
PLIOCENE	A. Doğancık formation B. Gözpınarı Formation	c. 250		B. Alternation of mudstone, marl, claystone and limestone (lacustrine facies) A. Unsorted, polygenetic conglomerates with sandstone lenses; cross-bedded, graded bedded conglomerates, sandstones and siltstones (fan-flood plain-delta facies) —— Angular Unconformity ——	
EARLY–EARLY LATE MIOCENE	A. Köstere Formation B. Gölyaka Formation	> 300		C. Pre-Jurassic metamorphic rocks ← Yakasinek reverse fault B. Alternation of limestone, marl, mudstone, shale, siltstone with coal seam (lacustrine facies) A. Alternation of sandstone, siltstone, conglomerate, porous algal limestone, channel conglomerate and mudstone (marginal facies) BC. Basal conglomerates —— Nonconformity ——	PALAEOTECTONIC PERIOD
PRE-MIOCENE				Metamorphic rocks (mostly marble, calc-schist, quartzite)	

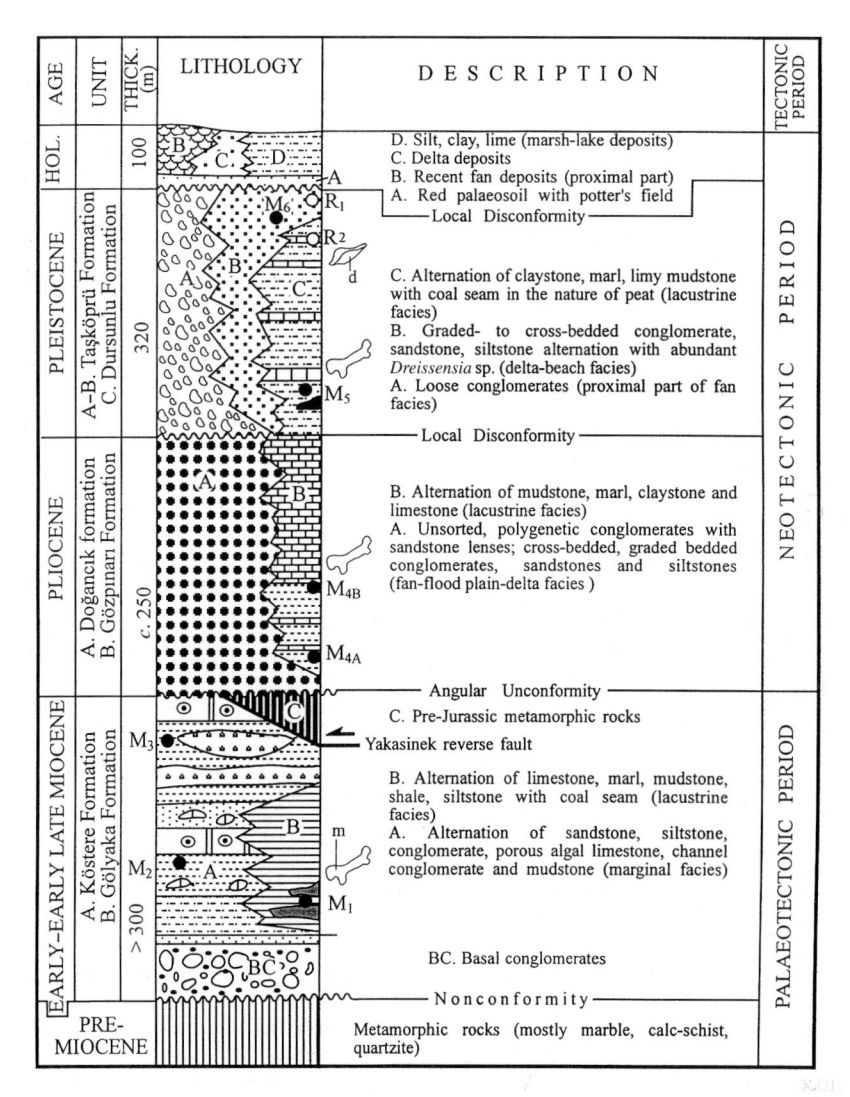

Fig. 2. Simplified tectonostratigraphic column of the graben fill. M1–M6 are horizons of mammalian fossils; R1 and R2 are sites of radiometric age datings.

The latter are yellow-pinkish sandstone–siltstone, blue-black organic-rich shale–marl, limy marl, siltstone–mudstone with abundant angular limestone clasts, massive sandstone, sandy nodular limestone, limy sandstone, yellow-white porous and highly fractured algal limestone, and thin–medium-bedded micritic limestone. The Köstere Formation also includes 2 m thick and 150–200 m long lensoidal channel conglomerate intercalations. It displays well-developed syndepositional features, such as cross-bedding, pebble imbrication, broken formation (olistostrome) and normal growth faults with displacements up to 2 m (Fig. 3). These primary features and lithofacies of the Köstere Formation indicate a tectonically active fan to flood plain type of depositional setting at the margin of a fluvio–lacustrine depositional system.

At different horizons of the Köstere Formation, abundant macro- and micromammalian fossils have been identified. These are: *Byzantinia bayraktepensis*, *Byzantinia* cf. *ozansoyi*, *Cricetulodon* sp., *Pliospalax* cf. *canakkalesis*, *Spermophillinus* cf. *bredai*, *Schizogalerix* sp., *Myocricetodon* sp. and *Myomimus* sp. at the M2 stratigraphic horizon (Köstere and Kırca outcrops), and *Byzantinia* sp., *Cricetulodon* sp.,

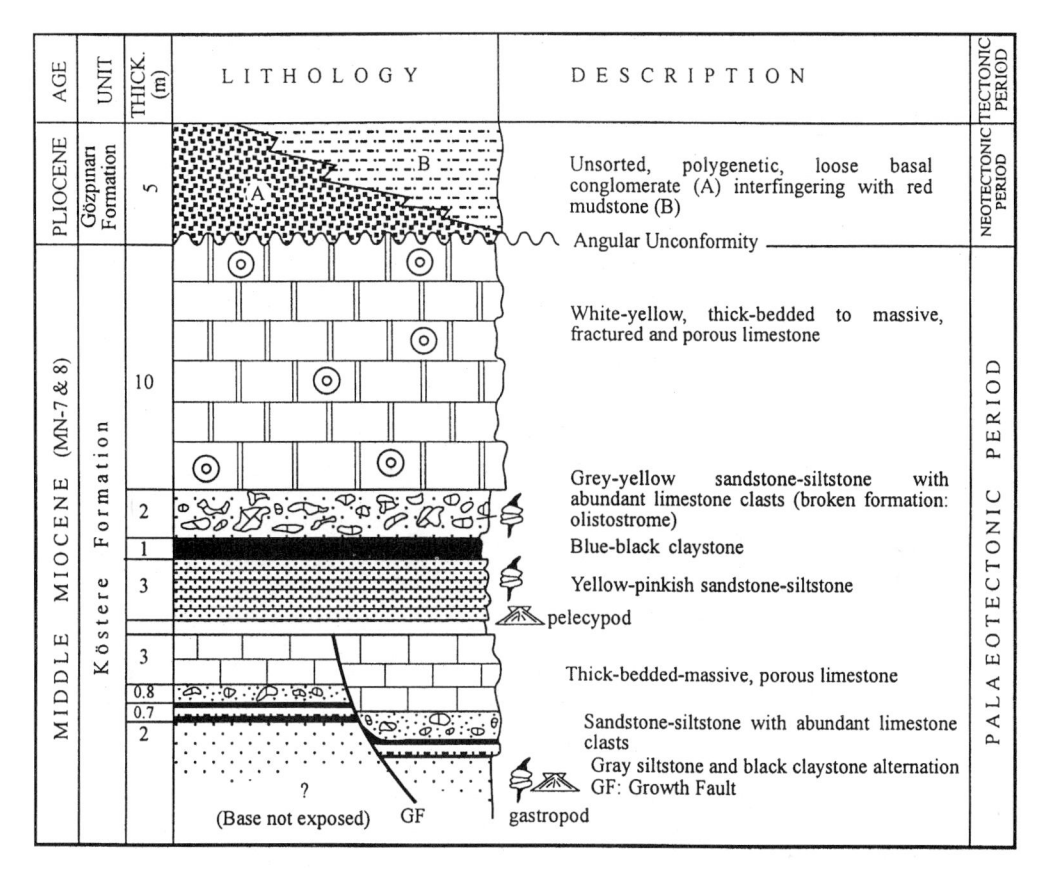

Fig. 3. Stratigraphical column showing various lithofacies, some syndepositional features (growth faults and broken formations: olistostromes) and erosional top contact of the Köstere Formation (Köstere Outcrop).

Progonomys sp., *Ramys* cf. *multicrestatus*, *Schizogaleri* cf. *sinapensis* and Soricidae at the M3 stratigraphical horizon (Dığrak–Aşağıçiğil outcrops). Middle Miocene (MN 7 and 8) and early Late Miocene (MN 9–12) ages are assigned to the middle and upper parts of the Köstere Formation, respectively, on the basis of this fauna (Fig. 2). Formerly, the Köstere Formation, particularly at Kırca, was dated as Late Miocene–Pliocene by Boray *et al.* (1985) in the absence of fossil evidence.

The Gölyaka Formation. The Gölyaka Formation is a lateral and lacustrine equivalent of lower and middle parts of the Köstere Formation (Fig. 2). It is a grey-green-blue claystone resting on siltstones and sandstones of the Köstere Formation; the claystone is succeeded by an alternation of massive blue marl, medium–thick-bedded limestone, organic-rich black claystone–shale, pinkish siltstone–mudstone and biotite-rich tuffite. A red palaeosoil with abundant caliche nodules occurs at the top.

The Gölyaka Formation also contains a few coal seams up to 10 m in thickness at its lower levels. The total thickness is *c.* 80 m. Immediately above and below the coal seams (M1 stratigraphical horizon of the Gölyaka outcrop), a rich macro- and micromammalian fossil assemblage occurs, consisting of *Galerix sartji*, *Paleosciurus* sp., *Cricetulodon* sp., *Democricetodon* sp., *Mirabella anatolica*, *Eumyarion carbonicus*, *Bransatoglis complicatus*, *Glis transversus*, *Glirus* aff. *ekremi*, *Vasseuromys duplex*, *Gliridinus haramiensis*, *Tapirus* sp. and *Dorcatherium* sp. Based on this fossil assemblage, an Early Miocene (MN 2) age was assigned to lower levels of the Gölyaka Formation. Thus, it is concluded that the first sequence of the graben fill was deposited in a fluvio–lacustrine system during an Early–early Late Miocene extensional regime. Later, this sequence was deformed by folding and thrusting during a short-lived compressional regime, which interrupted the normal faulting and terminated formation of the AAG.

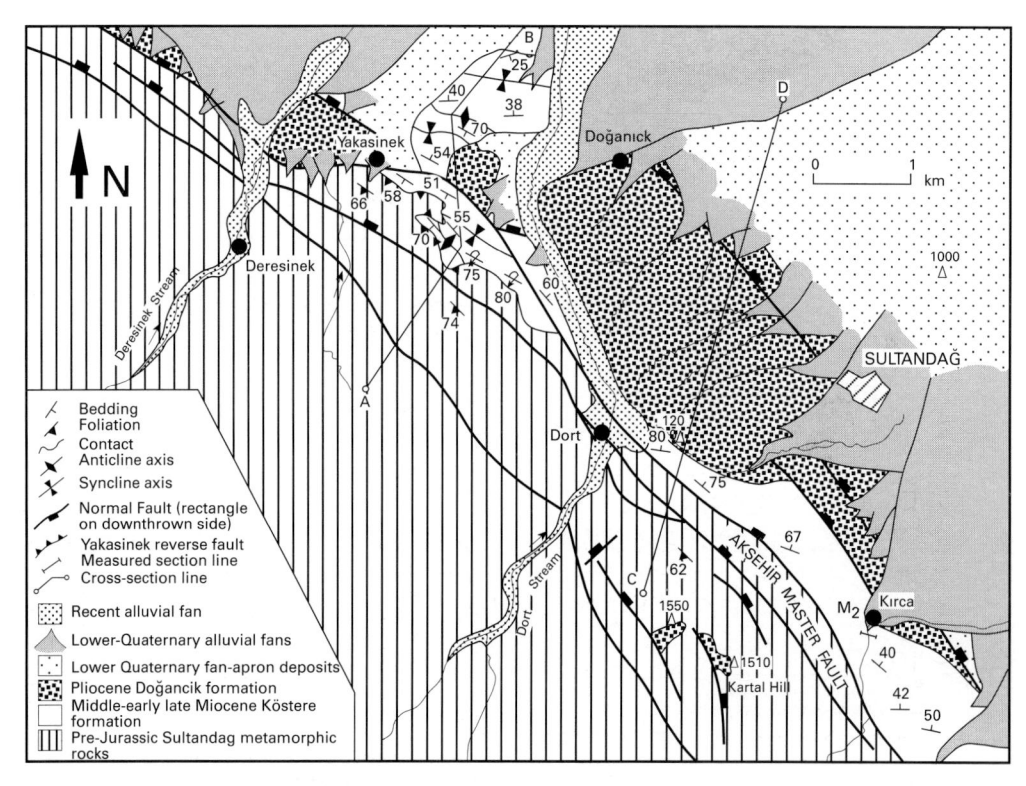

Fig. 4. Geological map of the Sultandağ area.

The Doğancık Formation. The Doğancık Formation represents the marginal part of a new fluvio–lacustrine depositional system, and occurs in variable sized and discontinuous outcrops at both fault-bounded margins of the AAG. It has been faulted, uplifted and tilted as fault terraces lying at various elevations up to 570 m above the present-day elevation (980 m) of the AAG floor (Fig. 4).

The type locality is Doğancık village, located at the southwestern margin of the AAG. At this locality it consists of fan conglomerates deposited by debris flow and braided streams. The conglomerates are yellow to red, thick bedded (up to 3 m) to massive, unsorted and polygenetic. They consist of subrounded to angular cobbles to boulders up to 2 m in diameter, set in a sandy matrix bound firmly by iron-rich calcite and quartz cement. Pebbles in the conglomerate are marble, phyllite, schist, quartzite and quartz, together with algal limestone derived from the Miocene Köstere Formation. Although the conglomerates are generally structureless, they locally display poorly developed graded bedding, cross-bedding and pebble imbrication. The measured total thickness of the Doğancık Formation is 250 m.

No precise age data for the Doğancık Formation could be obtained. However, the following field evidence implies an Early–Middle Pliocene age for the lower part: (1) it unconformably overlies the Köstere Formation of Middle–early Late Miocene age; (2) it is nearly flat lying; (3) it is disconformably overlain by the Pleistocene Taşköprü Formation; (4) the Doğancık Formation is the coarse-grained marginal equivalent, and interfingers with fine-grained depocentral facies of the Gözpınarı Formation, and the lower half of the Gözpınarı Formation contains mammalian fossils of early Middle Pliocene age; and (5) it has been a source area for Lower Quaternary fans with a well-developed drainage pattern (Figs 2 and 4).

The Gözpınarı Formation. This forms the lateral and lacustrine equivalent of the Doğancık Formation (Fig. 2). Basal, thin (up to 5 m) red clastic rocks, resting on the erosional surface of the Köstere Formation, are overlain by an alternation of siltstone, mudstone, yellow-green-blue marl, black shales, thin- to thick-bedded lacustrine limestones. At the top, it shows a transitional or locally erosional contact with the overlying Pleistocene Dursunlu

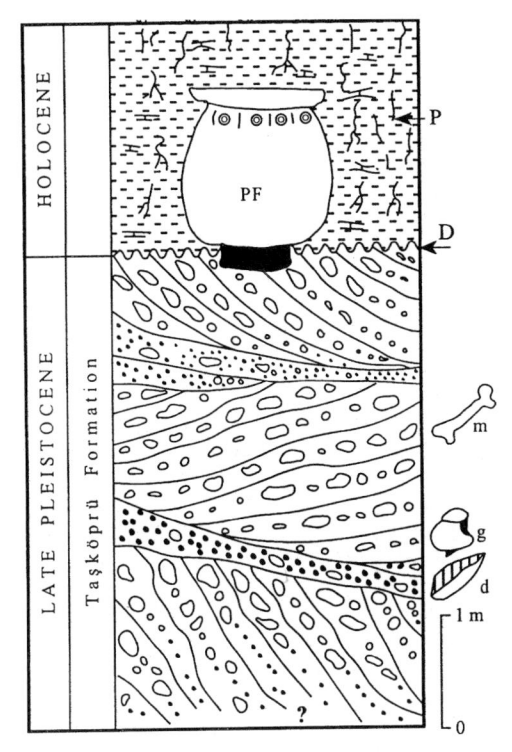

Fig. 5. Measured stratigraphical column showing cross-bedded internal structure. d, *Dreissensia*; g, gastropod; and m, mammalian fossil contents of the delta deposits and their erosional top contact relationship with a key horizon of palaeosoil, including potter's field (PF) (250 m west of Taşköprü village); D, disconformity.

Numerous ancient alluvial fans of different sizes occur along both the southern and northern margins of the AAG. They are aligned parallel to the margin-bounding faults and range in size from a few kilometres to 20 km in length. Alluvial fans consist of partly lithified, unsorted and polygenetic boulder to pebble conglomerates in the proximal parts, where clasts range from a few centimetres to 1.5 m in diameter, set in a sandy matrix bound by an iron-rich calcite and quartz cement, and coarse-grained sandstone to siltstone in the distal parts. These ancient alluvial fans are deeply eroded and locally overlain by newly formed fans, implying a recent motion along the margin-bounding faults.

The southern marginal areas of the AAG are covered by a broad and thick blanket of fan-apron deposits resulting from the coalescence of ancient alluvial fans. These deposits are dominated by loose conglomerates comprising boulders of varied sizes and pebbles set in a sandy–clayey matrix. Both the fan and fan-apron conglomerates grade into fine-grained flood plain deposits consisting mainly of an alternation of fine-grained sandstone, siltstone and mudstone, with lensoidal channel conglomerates in some places. Maximum thicknesses of the fan and flood plain deposits are 319 and 280 m, respectively, based on data from boreholes drilled through the fans and flood plains (Çuhadar 1977).

Delta deposits occur as uplifted and terraced conglomerates within a 1–2 km wide belt paralleling the present-day shorelines of the Akşehir and Eber Lakes, and located 4–5 km away from them. These ancient delta deposits consist of seven terraces located at different elevations between 7 and 42 m above the present-day level (958 m) of the lakes (Atalay 1975). They consist mostly of medium–thick-bedded (up to 2.3 m) polygenetic conglomerates interbedded with 10–30 cm thick yellow sandstone horizons and 5–10 cm thick bands containing *Dreissensia* sp. and gastropods. They show well-developed, large-scale planar to trough cross-bedding, graded bedding, top-set beds and pebble imbrications (Fig. 5). Pebbles in the conglomerates are well rounded and range in size from 0.5 to 10 cm. They are cemented together by an iron-rich yellow-red calcite matrix. The observed thickness of the delta deposits is 42 m. The upper levels of delta deposits contain abundant dreissensia and rare mammalian fossil, such as *Dreissensia polymorpha*, *Dreissensia* sp., *Dreissensia buldurensis* and *Bos* sp. (M6 in Fig. 2), yielding a Late Pleistocene age for the upper parts of the Taşköprü Formation. This is

Formation. It measures c. 320 m in total thickness. Fine-grained clastic lithologies of the Gözpınarı Formation are full of gastropod, pelecypod and micromammalian fossils, including, in its lower and middle stratigraphic horizons (M4A and M4B, respectively, in Fig. 2) Arvicolidae and *Mimomy*s sp. Based on these fossils, an Early–Middle Pliocene age is assigned to this formation.

The Taşköprü Formation. Marginal parts of a new fluvio–lacustrine depositional system, developed since the Pliocene, are distinguished as the Taşköprü Formation. In this last regime the margins were broad but lakes were small and isolated. One of them, Dursunlu Lake, has been uplifted and dried out by the present day, whereas the other (Akşehir and Eber) lakes, have persisted. The Taşköprü Formation consists of four major lithofacies, namely fan conglomerates, fan-apron deposits, flood plain deposits and delta deposits.

confirmed by ^{14}C ages of 12 725–14 125 a BP obtained from the same deposits [R2 in Fig. 2; see Kazancı *et al.* (1997)].

The Dursunlu Formation. The Dursunlu Formation is well exposed in the eastern part of the AAG and covers an area of 30 km^2. It is a lateral lacustrine equivalent of the Taşköprü Formation (Fig. 2) and comprises alternations of lithofacies packages. The latter include fine-grained conglomerate, sandstone, green-blue and bioturbated claystone to siltstone, mudstone, thin-bedded to laminated carbonates and yellow-green marl, including peat horizons up to 5 m in thickness. The fine-grained horizons and peats are full of gastropods and mammalian fossils (M5 in Fig. 2) including: *Lepus* sp., *Ochotona* sp., *Mimomys savini*, *Lagurus arankae*, *Microtus (Allophaiomys) nutiensis*, *Ellobius* sp., *Apodemus* sp., *Micromys* sp., *Allactaga euphratica*, *Cricetulus migratorius*, *Mesocricetus auratus*, *Spalax leucodon*, *Spermophilus* sp., *Trogontherium cuvieri*, *Carnivora* sp., *Mammuthus trogentherii*, *Equus* sp., *Hippopotamus* sp., *Cricetulus* sp. and *Bos primigenius*. Based on this fossil assemblage, an Early Pleistocene age is assigned to lower levels of the Dursunlu Formation (Güleç *et al.* 1997). In addition, the topmost carbonate horizon, which is overlain conformably by upper levels of delta deposits of the Taşköprü Formation, has a ^{14}C age of 23 245–31 785 a BP [R1 in Fig. 2; see Kazancı *et al.* (1997)]. Thus, the Dursunlu and Taşköprü Formations must be Pleistocene in age.

The two formations are overlain disconformably by a 0.3–1.5 m thick, red-brown palaeosoil horizon including carbonate patches (caliche) and a potter's field, a graveyard inherited from an ancient civilization living at *c.* 4000–6000 BC (Figs 2 and 5). This palaeosoil is a widespread and a key horizon indicating climatic change and a short-term erosional period separating Pleistocene and Holocene sedimentation (Fig. 2). The Holocene units consist of actively growing and unconsolidated fan to flood plain sediments at, or near to, the fault-bounded margins, and fine-grained, organic-rich silt, clay and lime along the graben floor close to Akşehir and Eber Lakes (Fig. 2). Their total thickness is *c.* 100 m based on data obtained from exploration boreholes (Çuhadar 1977).

Graben structure

Structures shaping the AAG and playing key roles in its episodic evolutionary history fall into three categories: (1) folds; (2) reverse faults; and (3) step-like normal faults.

Folds. Most previous studies (Atalay 1975; Demirkol *et al.* 1977; Öğdüm *et al.* 1991) report that the Miocene–Pliocene infill of the AAG is only tilted. However, this group's field geological mapping shows that Miocene units comprising the Köstere and Gölyaka Formations are folded and thrusted. Folds occur as a series of anticlines and synclines with parallel to subparallel axes; these are curvilinear in pattern and range in length from a few kilometres to 15 km. In general, they trend in two directions, north-northwest–south-southwest and west-northwest–east-southwest. These folds are well exposed in Yakasinek village (Fig. 4), where they are cut and displaced by the Akşehir Master Fault, and buried by nearly horizontal Plio-Quaternary graben fill (Fig. 6). The general trend of these folds indicates an approximately northeast–southwest directed compression attributable to the post-Tortonian Aksu Phase of Poisson & Akay (1981).

The Yakasinek Reverse Fault. Faults occur at the southern margin of the AAG. One is map scaled and well exposed in Yakasinek village, and is therefore termed the Yakasinek Reverse Fault. It is *c.* 2 km in length, displays a curved to curvilinear fault trace and its nature changes from an overturned contact to a high-angle thrust fault dipping south-southwest at 60–80°. At outcrop it occurs as a *c.* 20 m wide cataclastic zone made up of a series of parallel to subparallel and closely spaced reverse shear planes, along which both the lowermost Upper Miocene continental sequence and pre-Jurassic metamorphic rocks are intensely folded, sheared and crushed. Within this shear zone, pre-Jurassic metamorphic rocks are thrust from south-southwest to north-northeast over lowermost Upper Miocene clastic rocks. In addition, in Yakasinek village, the fault is cut and downthrown by the Akşehir Master Fault, and buried beneath a thick Plio-Quaternary graben infill (Figs 4 and 6). This indicates that faulting is younger than early Late Miocene but older than Pliocene.

In both the Isparta Angle and west Central Anatolia, most contacts between the Miocene and older rocks are faulted. Some faults display more than one set of fault plane-related kinematic indicators, such as slickensides, slickenlines, mineral fibres and fault steps. They indicate normal, thrust and strike-slip faulting in turn. The contact between the Miocene and older rocks varies from normal to nearly vertical, overturned and reverse faulted, as indicated by the Yakasinek Fault in the study area (Fig. 4). In addition, Miocene sequences are full of

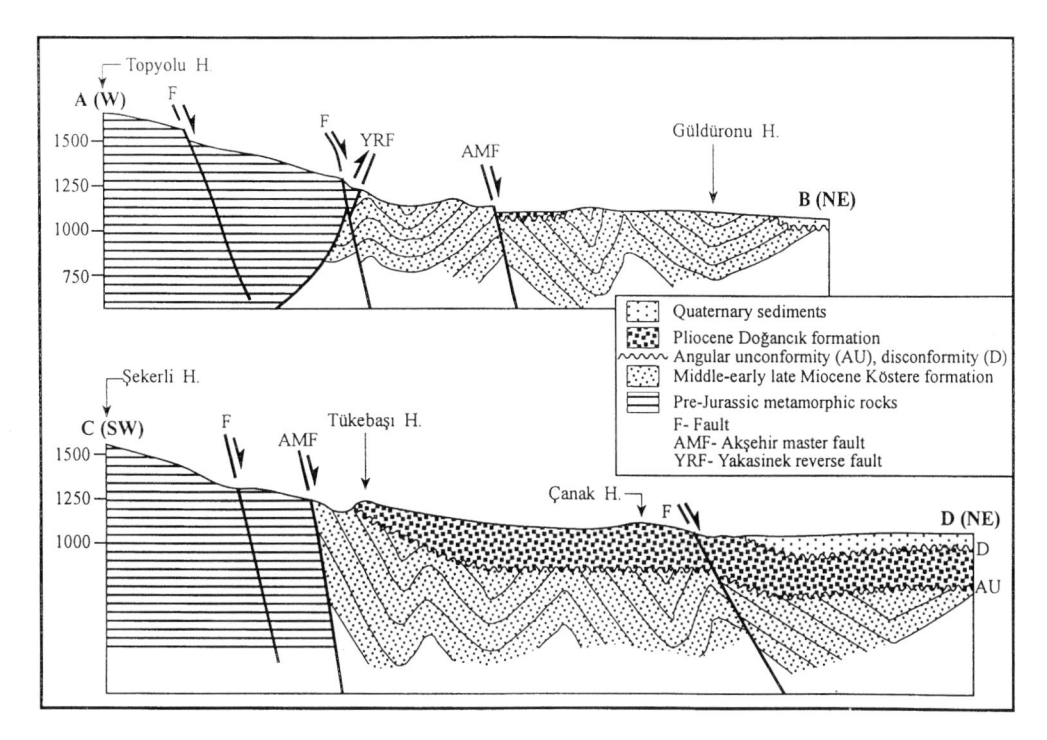

Fig. 6. Geological cross-sections showing Late Miocene compressional structures (reverse fault, folds) and younger normal faults cutting across them in the Yakasinek–Doğancık–Kırca area (see Fig. 4 for locations of cross-sections).

normal growth faults, olistostromes and 1–20 m thick intercalations of fault breccias. The latter interfinger with fine-grained facies of the Miocene sequence and wedge out in the direction away from the faulted contact. These field data can be attributed to a phase of extension and normal fault-controlled sedimentation in Miocene times. The normal faults might have reactivated as reverse faults because of the inversion in the earlier extensional regime after Miocene sedimentation. Thus, this compressional tectonic regime resulted in the Yakasinek Fault and terminated the first phase of extension.

Some previous workers (e.g. Boray et al. 1985; Yağmurlu 1991) have reported that the graben fill is folded, thrust faulted and Late Miocene–Pliocene in age. They also attribute compressional structures to a neotectonic regime. However, this group's detailed field mapping and stratigraphical–palaeontological studies clearly reveal that there are two different basin fills in the AAG, with the oldest deformed in Early–early Late Miocene times. This implies that the extensional evolutionary history of the AAG was interrupted by a compressional phase.

Step-like normal faults. Normal faults of varying sizes occur at both margins of the seismically active AAG (Fig. 7). They are well exposed as short (> 1–10 km) and long (10 km) fault segments cutting older basement (pre-Jurassic metamorphic rocks), the Lower–lowermost Upper Miocene Köstere and Gölyaka Formations, and their folded and thrust structures. They also juxtapose these two formations with either older basement or Plio-Quaternary graben fill. They display a graben-facing step-like pattern dominated by first-order major and second- to third-order synthetic to antithetic normal faults. The Akşehir and Karagöztepe Fault Zones bounding the southern and northern margins of the AAG, respectively, played a key role in the development of the AAG (Fig. 7).

Most previous studies of this area (Boray et al. 1985: Öğdüm et al. 1991; Şaroğlu et al. 1987; Barka et al. 1995) have reported that the Akşehir Fault Zone, at the southern margin of the AAG, is a reverse fault. However, this group's mapping and kinematic analysis clearly shows it to be an oblique-slip normal fault zone, which is termed here the Akşehir Fault Zone

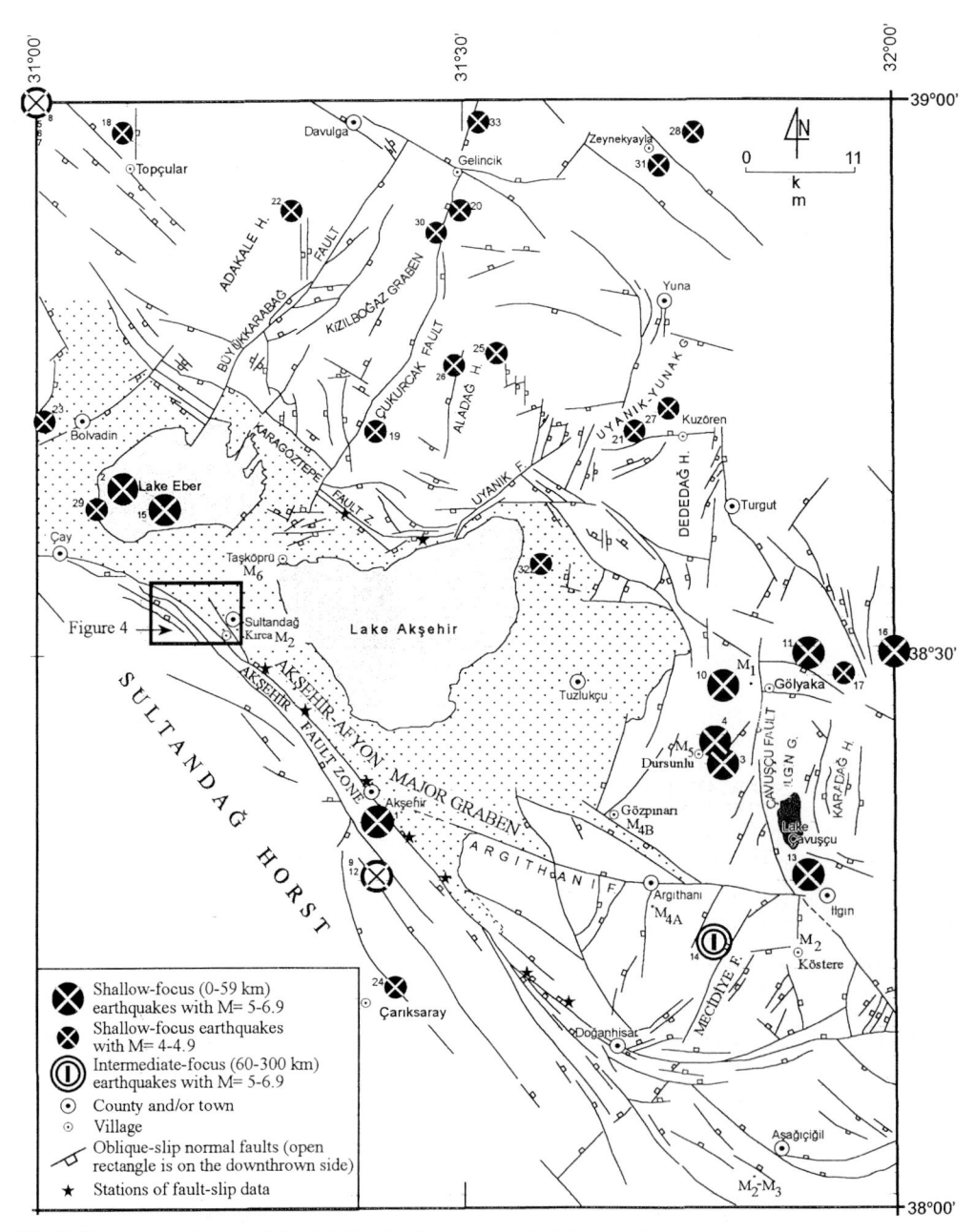

Fig. 7. Seismotectonic map of the AAG and adjacent areas. M1–M6 are sites of mammalian fossils.

(AFZ). It comprises a 2–7 km wide, 120 km long, graben-facing step-like normal fault zone, and the strike changes from 285°W in the west to 320°W in the central part and *c.* 270°E in the east to produce an S-shaped outcrop pattern. The fault zone includes numerous second- to third-order, closely spaced, synthetic normal faults ranging from 2 to 50 km in length which display well-developed fault scarps, triangular facets, and well-preserved fault slickensides and slickenlines (Table 1). Their stereographic plots show that the southern margin-bounding structure of the AAG is an oblique-slip normal fault zone dipping on average 59°NE, with a minor

Table 1. *Measurement of slickensides and slickenlines on the Akşehir Master Fault*

	Fault plane (strike and dip)	Rake	Type of movement
1	N60°W, 44°NE	85°SE	Normal
2	N65°W, 70°NE	79°SE	Normal
3	N17°W, 61°NE	45°NW	Normal
4	N21°W, 72°NE	63°NW	Normal
5	N27°W, 65°NE	46°NW	Normal
6	N16°W, 48°NE	84°NW	Normal
7	N80°W, 29°NE	73°SE	Normal
8	N79°W, 46°NE	74°SE	Normal
9	N69°W, 56°NE	57°NW	Normal
10	N45°W, 58°NE	45°NW	Normal
11	N43°W, 42°NE	72°NW	Normal
12	N43°W, 47°NE	68°NW	Normal
13	N58°W, 30°NE	90°	Normal
14	N77°W, 30°NE	73°SE	Normal
15	N74°W, 34°NE	78°SE	Normal
16	N44°W, 31°NE	73°NW	Normal
17	N59°W, 32°NE	85°NW	Normal
18	N47°W, 42°NE	87°NW	Normal
19	N55°W, 43°NE	77°NW	Normal
20	N48°W, 82°NE	85°NW	Normal
21	N65°W, 64°NE	85°NW	Normal
22	N73°W, 78°NE	87°NW	Normal
23	N48°W, 50°NE	85°NW	Normal
24	N46°W, 38°NE	88°NW	Normal
25	N59°W, 74°NE	49°NW	Normal
26	N62°W, 70°NE	43°NW	Normal
27	N41°W, 68°NE	53°NW	Normal
28	N42°W, 74°NE	55°NW	Normal
29	N79°W, 47°NE	58°NW	Normal
30	N42°W, 40°NE	73°SE	Normal
31	N35°W, 42°NE	61°SE	Normal
32	N58°W, 40°NE	83°SE	Normal
33	N29°W, 52°NE	72°SE	Normal
34	N32°W, 49°NE	85°SE	Normal
35	N42°W, 72°NE	84°NW	Normal
36	N42°W, 65°NE	69°NW	Normal
37	N49°W, 77°NE	82°NW	Normal
38	N56°W, 85°NE	89°NW	Normal
39	N65°W, 84°NE	90°	Normal
40	N55°W, 77°NE	83°NW	Normal
41	N39°W, 76°NE	81°NW	Normal
42	N52°W, 67°NE	85°NW	Normal
43	N46°W, 82°NE	72°NW	Normal
44	N55°W, 78°NE	80°NW	Normal
45	N30°W, 74°NE	73°SE	Normal
46	N40°W, 40°NE	20°NW	Normal
47	N55°W, 48°NE	60°NW	Normal
48	N62°W, 85°NE	74°NW	Normal
49	N78°W, 70°NE	57°SE	Normal
50	N77°W, 83°NE	57°SE	Normal
51	N59°W, 68°NE	62°SE	Normal
52	N75°W, 82°NE	55°SE	Normal

amount of dextral and/or sinistral strike-slip component of motion (Fig. 8a).

The Pliocene Doğancık Formation is cut, terraced and elevated by up to 570 m above the present-day elevation (980 m) of the graben floor (Fig. 4). A borehole drilled through a Lower Quaternary alluvial fan accumulated in the downthrown block of the Akşehir Master

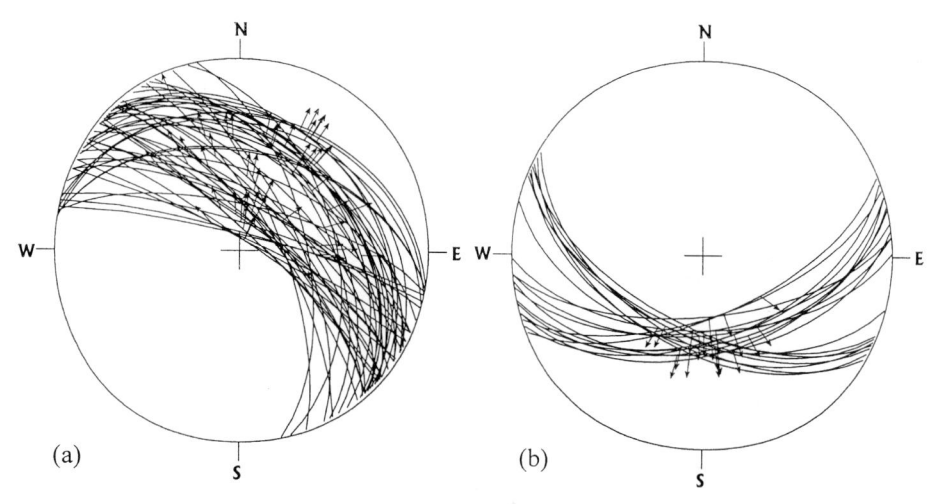

Fig. 8. (a) Stereographic plots of the Akşehir Master Fault slip data on a Schmidt lower hemisphere net. (b) Stereographic plots of the Karagöztepe Master Fault slip data on the Schmidt lower hemisphere net.

Fault encountered the same formation at a depth of 320 m below the present-day elevation of the AAG. These data indicate that total throw on the AFZ has been c. 870 m (550 + 320 = 870 m) since at least the Late Pliocene. This value more or less equals the total amount of downcutting of several consequent streams, which flow into the graben and cut deeply (up to 750–1100 m) into their beds along the southern fault-bounded margin (Öğdüm et al. 1991). Thus, it may be concluded that the average rate of motion along the AFZ is c. 0.3 mm a^{-1}.

The northern margin of the AAG is more complicated than the southern margin because it is shaped by three sets of oblique-slip normal faults, first detected in this study. These comprise northwest, north–south and northeast striking fault sets, and consist of short (1–5 km) to long (up to 40 km), closely spaced, synthetic to antithetic normal fault segments. Major faults shaping the northern margin are the Karagöztepe, Büyükkarabağ, Çukurcak, Uyanık and Çavuşcu Faults. Owing to the latter three fault sets, the northern margin has been segmented into several second-order horsts and grabens, namely the Adakale, Aladağ, Dededağ and Karadağ Horsts, with intervening Kızılboğaz, Uyanık–Yunak and Ilgın Subgrabens (Fig. 7).

The main structure playing a key role in the development history of the northern margin of the AAG is the Karagöztepe Fault Zone, a 1–10 km wide, 90 km long, west-northwest and east-northeast striking discontinuous fault zone. It comprises numerous closely spaced, short (1 km) to long (up to 25 km), synthetic to antithetic fault segments which are cut and displaced in both dextral and sinistral directions up to 5 km by northeast striking fault sets (Fig. 7). The Karagöztepe Fault Zone cuts across older basement and a continental sedimentary sequence of Early–early Late Miocene age and juxtaposes them with the Plio-Quaternary graben fill. Faults of the Karagöztepe Fault Zone display well-exposed fault slickensides with a relief of 25 m and well-preserved slickenlines (Table 2). Their stereographic plots show that the northern margin-bounding structure of the AAG is an oblique-slip normal fault zone striking east-northeast and west-northwest, and dipping south-southeast and south-southwest at average angles of 56–61° (Fig. 8b).

As in the case of southern margin of the AAG, the Pliocene Doğancık Formation and the Lower Quaternary delta deposits of the Taşköprü Formation are cut, terraced and tilted by up to 60°, and uplifted by up to 200 m with respect to their present-day position in the downthrown block of the Karagöztepe Master Fault along which recent graben floor sediments are tectonically juxtaposed with older rocks. The total throw of the northern margin-bounding faults of the AAG since the Early Quaternary is therefore 200 m. This yields a rate of subsidence of c. 0.2 mm a^{-1}, which compares with the rate of 0.3 mm a^{-1} for the southern margin. Hence, the AAG has experienced an asymmetrical evolutionary history during the extensional neotectonic period, the second phase of extension.

The current activity of graben-bounding structures is indicated by: (1) numerous faulted,

Table 2. *Measurements of slickensides and slickenlines on the Karagöztepe Master Fault*

	Fault plane (strike and dip)	Rake	Type of movement
1	N75°E, 50°SE	70°SW	Normal
2	N89°E, 58°SE	90°	Normal
3	N84°E, 64°SE	90°	Normal
4	N80°E, 58°SE	90°	Normal
5	N82°E, 48°SE	90°	Normal
6	N84°E, 56°SE	90°	Normal
7	N70°E, 64°SE	90°	Normal
8	N68°E, 54°SE	90°	Normal
9	N72°E, 48°SE	90°	Normal
10	N66°E, 64°SE	75°NE	Normal
11	N72°E, 54°SE	78°NE	Normal
12	N76°E, 52°SE	80°NE	Normal
13	N60°W, 50°SW	75°SE	Normal
14	N65°W, 56°SW	70°SE	Normal
15	N63°W, 60°SW	80°SE	Normal
16	N58°W, 54°SW	90°	Normal
17	N66°W, 58°SW	78°SE	Normal
18	N74°W, 52°SW	74°SE	Normal
19	N62°W, 56°SW	90°	Normal
20	N64°W, 51°SW	82°SE	Normal

terraced and fault-parallel ancient to recent alluvial fans, with apices adjacent to the graben-bounding faults; (2) faulting and terracing of the Plio-Quaternary graben fill; (3) shifting of stream courses; (4) deep erosion by flowing consequent streams; (5) tectonic juxtaposition of Plio-Quaternary graben fill with older rocks; and (6) cold to hot water springs with actively growing travertines.

Seismicity

In the period of 1901–1986, 33 seismic events with magnitude of 4–6.8 took place in the AAG and adjacent areas (Table 3). Their epicentral distribution shows that both the northwest striking graben-bounding faults and some of the northeast striking second-order faults are active (Fig. 7). In addition, all these seismic events are shallow-focus earthquakes, as indicated by the cross-section showing the distribution of earthquake foci between northern latitudes of 34–39° (Harsch et al. 1981). Two seismic events are the 26th September 1921 Argıthanı–Akşehir and the 21st February 1946 Ilgın–Argıthanı earthquakes, with magnitudes of 5.4 and 5.5, respectively, which took place in the eastern part of the AAG as indicated by numbers 4 and 14, respectively, in Table 3. During these earthquakes, 12 people died, nine were injured, 285 houses and some parts of railway track were heavily damaged, and 153 houses including the railway station collapsed entirely in Argıthanı–Ilgın–Doğanhisar counties and their villages. Unfortunately, there is no fault plane solution of these two earthquakes and other seismic events due to the lack of accurate seismic records.

The west-northwesternmost part of the AAG is the Gediz–Simav Graben where the Gediz segment was ruptured during the 28th March 1970 Gediz Earthquake with a magnitude of 7.1 (Ambraseys & Tchalenko 1972). Despite this large seismic event, during the past 29 years seismicity of the AAG has been very low compared with the Gediz–Simav Graben, which implies that the AAG is the site of a seismic gap.

Discussion and conclusions

This paper documents how the Akşehir–Afyon Graben (AAG) has developed episodically rather than continuously. In general, the AAG contains two major infilling successions, namely: (1) a continental sedimentary sequence of Early–early Late Miocene age (Köstere and Gölyaka Formations); and (2) a Plio-Quaternary continental sedimentary sequence (Doğancık, Gözpınarı, Taşköprü and Dursunlu Formations, and recent graben floor sediments). The older infill is dominated by a sequence over 300 m thick, including red debris flow conglomerates, long and lensoidal channel conglomerates, coal seams (up to 10 m in thickness) and the normal growth faults. It accumulated in both the marginal and axial parts of a fluvio–lacustrine depositional system

Table 3. List of earthquakes which occurred in the Akşehir-Afyon Graben and adjacent areas in the period of 1901–1986*

No	Longitude (X)	Latitude (Y)	dd–mm–yy	Depth (m)	Ms	Mb	Ml	Relative magnitude	Relative depth
1	31.4	38.35	?–4–1901	0	5.1			medium	shallow
2	31.1	38.65	1914	0	5.7			medium	shallow
3	31.8	38.4	13–4–1921	30	5.2	0	0	medium	shallow
4[†]	31.79	38.42	26–9–1921	10	5.4	0	0	medium	shallow
5	31	39	14–09–1925	0	4.9			small	shallow
6	31	39	20–09–1925	0	4.9			small	shallow
7	31	39	20–12–1926	0	4.9			small	shallow
8	31	39	07–02–1927	15	5.2			medium	shallow
9	31.4	38.3	11–01–1931	0	4.9			small	shallow
10	31.8	38.47	12–1–1931	20	5	0	0	medium	shallow
11	31.9	38.5	12–1–1931	30	5	0	0	medium	shallow
12	31.4	38.3	28–01–1931	0	6.8			medium	shallow
13	31.9	38.3	09–04–1931	0	6			medium	shallow
14[‡]	31.79	38.24	21–2–1946	60	5.5	0	0	medium	medium
15	31.15	38.63	16–7–1946	40	5.1	0	0	medium	shallow
16	32	38.5	01–02–1953	0	5			medium	shallow
17	31.94	38.48	22–6–1956	40	4.6	0	0	small	shallow
18	31.1	38.97	3–11–1966	9	4.6	0	0	small	shallow
19	31.4	38.7	28–3–1970	0	0	0	4.8	small	shallow
20	31.5	38.9	28–3–1970	0	0	0	4.4	small	shallow
21	31.7	38.7	18–4–1970	0	0	4	0	small	shallow
22	31.3	38.9	6–5–1970	0	0	0	4.1	small	shallow
23	31.01	38.71	28–10–1975	23	0	4.4	0	small	shallow
24	31.42	38.2	23–7–1976	0	0	4.7	0	small	shallow
25	31.54	38.77	21–4–1977	0	0	4	0	small	shallow
26	31.49	38.76	26–4–1978	0	0	4.1	0	small	shallow
27	31.74	38.72	27–4–1978	0	0	4.1	0	small	shallow
28	31.77	38.97	26–5–1980	29	0	4	0	small	shallow
29	31.07	38.63	14–3–1982	0	0	4.3	4	small	shallow
30	31.47	38.88	27–4–1984	0	0	0	4	small	shallow
31	31.73	38.94	23–6–1984	0	0	0	4	small	shallow
32	31.59	38.58	17–1–1986	0	0	4.7	4.1	small	shallow
33	31.52	38.98	26–2–1986	10	0	4.5	0	small	shallow

* These data were taken from Seismological division, Department of Earthquake Research, General Directorate of Disaster Affairs, Ankara Turkey, and Gençoğlu et al. (1990).
† 26 September 1921 Argıthanı–Akşehir Earthquake. ‡ 21 February 1946 Ilgın–Argıthanı Earthquake.

established on relatively low-lying parts of an uneven erosional surface inherited from a long-term erosional period (Late Eocene–Oligocene). This stage postdates the final collision, uplift and overthickening of crust along the İzmir–Ankara–Erzincan Suture Zone (İAESZ) through Central and western Anatolia. The younger infill is characterized by a 670 m thick sequence consisting mainly of fan and fan-delta conglomerates interfingering with a coal-bearing fine-grained lacustrine facies. The older infill is folded and thrust faulted, and overlain by nearly horizontal younger infill with angular unconformity. This is not a local situation confined only to the AAG, but is also observed in numerous other basins and young grabens, such as the Karaman, Uşak, Senirkent–Hoyran, Refahiye, Lake districts, İnönü–İnegöl, in the Ankara region, the Yarıkkaya, Soma, Gökova, Kavacık, Gediz and Büyük Menderes Grabens in both Central and Western Anatolia (Koçyiğit 1976, 1983, 1996; Ercan et al. 1978; Yağmurlu 1991; Gemici et al. 1991; Görür et al. 1995; Koçyiğit & Kaymakçı 1995; Koçyiğit et al. 1995; Bozkuş 1996; Yusufoğlu 1996). This regional compressional deformation pattern (folds and thrust faults) and the angular unconformity, which separates older and younger infills, clearly reveal a Late Miocene phase of compression reflecting the well-known post-Tortonian Aksu phase of compression which prevailed through the eastern limb of the Isparta Angle where the AAG is now located. The Late Miocene phase of compression

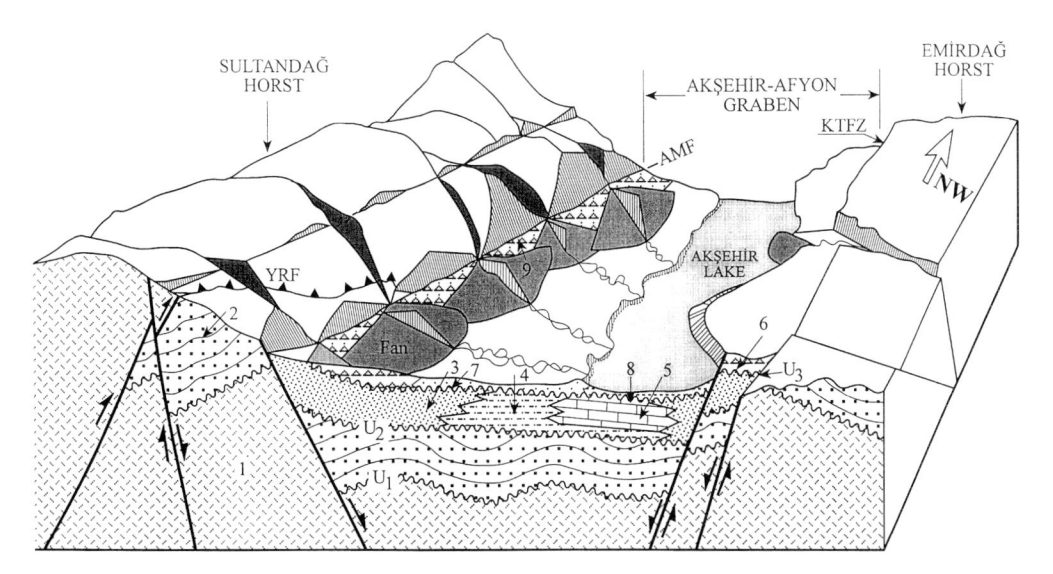

Fig. 9. Block diagram depicting a tentative episodic evolutionary history of the Akşehir–Afyon Graben. 1, Pre-Jurassic metamorphic rocks; 2, older infill (Köstere and Gölyaka Formations of Early–early Late Miocene age); 3 and 4, Doğancık and Taşköprü Formations of Plio-Quaternary age; 5, Gözpınarı and Dursunlu Formations of Plio-Quaternary age; 6, uplifted and terraced delta deposits of Late Pleistocene age; 7, ancient fan-apron deposits of Late Pleistocene age; 8, silt, mud and lime; 9, slope scree deposits; AMF, Akşehir Master Fault; KTFZ, Karagöztepe Fault Zone; YRF, Yakasinek Reverse Fault; U1 and U2, angular unconformities; D, disconformity: 3–9 form younger infill (neotectonic fill).

separates the two extensional periods and implies that the AAG has an episodic evolutionary history (Koçyiğit 1996).

The older infill (Köstere and Gölyaka Formations) was deposited in a fluvio–lacustrine depositional system which emerged after a long-term erosional period under the control of first phase of extension related to orogenic collapse along the İAESZ (Dewey 1988; Seyitoğlu & Scott 1996). Later, it was replaced by a short phase of east-northeast–west-southwest directed compression resulting from a probable variation in the kinematics of the Eurasian and African Plates in Late Miocene time. As explained by Dewey et al. (1986), final collision involving the entire demise of the oceanic crust, and emergence of the Bitlis Suture Zone between the Arabian–African Plates and the Eurasian Plate, took place in Late Serravalian (c. 12 Ma) times. This terminal collision and suturing were followed by a transitional period with a very low rate of intracontinental convergence, uplift and thickening of the crust until the Late Miocene. Thus, the Late Miocene interval corresponds to a short-term phase of compression in the eastern limb of the Isparta Angle, and was responsible for deformation of the older infill into a series of anticlines and synclines; it also produced local high-angle faulting (the Yakasinek Reverse Fault).

In Early Pliocene times, sea-floor spreading began along the central line of the Red Sea (Hempton 1987) and the wedge-shaped Anatolian Platelet and its boundary faults, namely the dextral North Anatolian and the sinistral East Anatolian Transform Faults, were initiated. The consequent west-southwestward escape of Anatolia may have led to an increase in the relative motion with respect to the African Plate along the Hellenic–Cyprus convergent plate boundary. It has also been proposed that variations in relative motion of plates may create roll-back processes (Froitzheim et al. 1997) and such processes may have triggered initiation of the second phase of extension in the Pliocene which is still acting as a neotectonic regime through Central and West Anatolia, as well as in the AAG, even though it has been claimed that the eastern limb of the Isparta Angle, including the AAG, is under the control of a compressional tectonic regime (Boray et al. 1985; Barka et al. 1995).

The younger infill (the Doğancık, Gözpınarı, Taşköprü and Dursunlu Formations, and the recent graben-floor sediment) was deposited on the erosional surface of early formed and deformed older infill under the control of this second phase of extension and related oblique-slip normal faulting. These events and their products are depicted in Fig. 9, where, during the second phase of extension (the neotectonic period), oblique-slip normal faulting and sedimentation have still been active, as indicated by both the seismicity and field evidence.

We are indebted to the president of Middle East Technical University for financial support of field studies. We also thank Erdin Bozkurt, who reorganized and improved the illustrations.

References

AMBRASEYS, N. N. & TCHALENKO, J. S. 1972. Seismotectonic aspects of the Gediz, Turkey, earthquake of March, 1972. *Geophysical Journal of the Royal Astronomical Society*, **30**, 229–252.

ATALAY, A. 1975. Akşehir, Eber ve Karamık gölleri havzalarının Kuvaterner depoları ve Jeomorfolojisi. *Proceedings of the 50th Anniversary of the Turkish Republic Earth Sciences Congress*. Mineral Research and Exploration Institute of Turkey (MTA) Publications, 365–385 [in Turkish with English abstract].

BARKA, A., REILINGER, R., ŞAROĞLU, F. & ŞENGÖR, A. M. C. 1995. The Isparta Angle: its importance in the neotectonics of the Eastern Mediterranean Region. *Proceedings of the International Earth Sciences Colloquium on the Aegean Region*, **1**, 3–18

BORAY, A. ŞAROĞLU, F. & EMRE, Ö. 1985. Isparta Büklümünün kuzey kesiminde Doğu–Batı daralma için bazı veriler. *Geological Engineering*, **23**, 9–20 [in Turkish with English abstract].

BOZKUŞ, C. 1996. Kavacık (Dursunbey–Balıkesir) Neojen grabeninin stratigrafisi ve tektoniği. *Turkish Journal of Earth Sciences*, **5**, 161–170 [in Turkish with English abstract].

ÇUHADIR G. 1977. *Akarçay Havzası Hidrojeolojik Etüd Raporu*. State Water Works (DSİ) Report [unpublished, in Turkish].

DEMİRKOL, C., SİPAHİ, H. & ÇİÇEK, S. 1977. *Sultandağının stratigrafisi ve jeolojik evrimi*. Mineral Research and Exploration Institute of Turkey (MTA) Report, No. 6305 [in Turkish].

DEWEY, J. F. 1988. Extensional collapse of orogens. *Tectonics*, **7**, 1123–1139.

—— & ŞENGÖR, A. M. C. 1979. Aegean and surrounding regions: complex multiple and continuum tectonics in a convergent zone. *Geological Society of America Bulletin*, **90**, 84–92.

——, HEMPTON, M. R., KIDD, W. S., ŞAROĞLU, F. & ŞENGÖR, A. M. C. 1986. Shortening of continental lithosphere: the neotectonics of eastern Anatolia – a young collision zone. *In*: COWARD, M. P. & RIES, A. C. (eds) *Collisional Tectonics*. Geological Society, London, Special Publications, **19**, 2–36.

ERCAN, T., DİNÇEL, A., METİN, S., TÜRKECAN, A. & GÜNAY, E. 1978. Uşak yöresindeki Neojen havzalarının jeolojisi. *Geological Society of Turkey Bulletin*, **21**, 97–106 [in Turkish with English abstract].

EYİDOĞAN, H., GÜÇLÜ, U., UTKU, Z. & DEĞİRMENCİ 1991. *Türkiye Büyük Depremleri Makro Sismik Rehberi (1900–1988)*. İstanbul Technical University, Faculty of Mining, Department of Geophysical Engineering Publications [in Turkish with English abstract].

FROITZHEIM, N., CONTI, P. & VAN DAALEN, M. 1997. Late Cretaceous, synorogenic, low-angle normal faulting along the Schlinig fault (Switzerland, Italy, Austria) and its significance for the tectonics of the Eastern Alps. *Tectonophysics*, **280**, 267–293.

GEMİCİ, Y., AKYOL, E., AKGÜN, F. & SEÇMEN, Ö. 1991. Soma kömür havzası fosil makro ve mikroflorası. *Mineral Research and Exploration Institute of Turkey (MTA) Bulletin*, **112**, 161–178 [in Turkish with English abstract].

GENÇOĞLU, S., İNAN, E. & GÜLER, H. 1990. *Türkiye'nin Deprem Tehlikesi*. Chamber of Turkish Geophysical Engineers Publications.

GÖRÜR, N., ŞENGÖR, A. M. C., SAKINÇ, M. *ET AL*. 1995. Rift formation in the Gökova region, southwest Anatolia: implications for the opening of the Aegean Sea. *Geological Magazine*, **132**, 637–650.

GLOVER, C. & ROBERTSON, A. 1998. Neotectonic intersection of the Aegean and Cyprus Tectonic arcs: extensional and strike-slip faulting in the Isparta Angle, SW Turkey. *Tectonophysics*, **298**, 103–132.

GÜLEÇ, E., CLARK, D., CURTIS, G. *ET AL*. 1997. The Early Pleistocene lacustrine deposits of Dursunlu. Preliminary Results. *An International Symposium on the Late Quaternary in the Eastern Mediterranean: Programme and Abstracts*. Mineral Research and Exploration Institute of Turkey (MTA) Publications, **19**.

HARSCH, W., KUPFER, T., RAST, B. & SAGASSER, R. 1981. Seismotectonic considerations on the nature of the Turkish–African plate boundary. *Geologische Rundschau*, **70**, 368–385.

HEMPTON, M. R. 1987. Constraints on Arabian plate motion and extensional history of the Red Sea. *Tectonics*, **6**, 687–705.

INNOCENTI, F., MAZZUOLI, R., PASQUARE, G., RADICATIDIVROZOLO, F. & VILLARI, L. 1975. The Neogene calc-alkaline volcanism of Central Anatolia: geochronological data on Kayseri–Niğde area. *Geological Magazine*, **112**, 349–360.

KAZANCI, N., NEMEC, W., İLERİ, Ö., KARADENİZLİ, L. & CHRISTIAN BRISEID, H. 1997. Palaeoclimatic significance of the Late Pleistocene deposits of Akşehir Lake, west Central Anatolia. *An International Symposium on the Late Quaternary in the Eastern Mediterranean: Programme and Abstracts*. Mineral Research and Exploration Institute of Turkey (MTA) Publications, 58–59.

KELLER, D., JUNG, F. J., ECKHARD, T. & KREUZER, H. 1992. Radiometric ages and chemical characterisation of the Galatean andesite massif, Pontus, Turkey. *Acta Vulcanologica, Marinelli volume*, **2**, 267–276.

KISEEL, C. & LAJ, C. 1988. The Tertiary geodynamical evolution of the Aegean arc: a palaeomagnetic reconstruction. *Tectonophysics*, **146**, 183–201.

KOÇYİĞİT, A. 1976. Karaman–Ermenek (Konya)

bölgesinde ofiyolitli melanj ve diğer oluşuklar. *Geological Society of Turkey Bulletin,* **19**, 103–116 [in Turkish with English abstract].

—— 1983. Hoyran gölü (Isparta Büklümü) dolayının tektoniği. *Geological Society of Turkey Bulletin,* **26**, 1–10 [in Turkish with English abstract].

—— 1984. Güneybatı Türkiye ve yakın dolayında levhaiçi yeni Tektonik gelişim. *Geological Society of Turkey Bulletin,* **27**, 1–16 [in Turkish with English abstract].

—— 1985. Muratbağı–Balabantaş (Horasan) arasında Çobandede Fay kuşağının jeotektonik özellikleri ve Horasan–Narman depremi yüzey kırıkları. *Cumhuriyet University, Faculty of Engineering, Earth Sciences,* **2**, 17–33 [in Turkish with English abstract].

—— 1991. An example of an accretionary forearc basin from northern Central Anatolia and its implications for the history of Neo-Tethys in Turkey. *Geological Society of America Bulletin,* **103**, 22–36.

—— 1996. Superimposed basins and their relations to the recent strike-slip fault zones: a case study of the Refahiye superimposed basin adjacent to the North Anatolian Transform Fault, northeastern Turkey. *International Geology Review,* **38**, 701–713.

—— & BEYHAN, B. 1998. A new intracontinental transcurrent structure: the Central Anatolian Fault Zone, Turkey. *Tectonophysics,* **284**, 317–336.

—— & KAYMAKÇI, N. 1995. İnönü–İnegöl superimposed basins and initiation age of the extensional neotectonic regime in west Turkey. *International Earth Sciences Colloquium on the Aegean Region,* Abstracts, 33.

——, YUSUFOĞLU, H. & BOZKURT, E. 1999. Evidence from the Gediz graben for episodic two-stage extension in western Turkey. *Journal of the Geological Society, London,* **156**, 605–616.

——, TÜRKMENOĞLU, A., BEYHAN, A., KAYMAKÇI, N. & AKYOL, E. 1995. Post-collisional tectonics of Eskişehir–Ankara–Çankırı segment of the İzmir–Ankara–Erzincan Suture Zone (İAESZ): Ankara orogenic phase. *Turkish Association of Petroleum Geologists Bulletin,* **6**, 69–86.

LE PICHON, X. & ANGELIER, J. 1979. The Hellenic arc and trench system: a key to the neotectonic evolution of the eastern Mediterranean area. *Tectonophysics,* **60**, 1–42.

MCKENZIE, D. 1978. Active tectonics of the Alpine–Himalayan belt: the Aegean Sea and surrounding regions (tectonics of Alpine region). *Geophysical Journal of the Royal Astronomical Society,* **55**, 217–245.

MEULENKAMP, J. E., WORTEL, W. J. R., VAN WAMEL, W. A., SPASKMAN, W. & HOOGERDUYN STRATING, E. 1988. On the Hellenic subduction zone and geodynamic evolution of Crete in the late Middle Miocene. *Tectonophysics,* **146**, 203–215.

ÖĞDÜM, F., KOZAN, T., BIRCAN, A., BOZBAY, E. & TÜFEKÇİ, K. 1991. *Sultandağları ile Çevresindeki Havzaların Jeomorfolojisi ve Genç Tektoniği.* Mineral Research and Exploration Institute of Turkey (MTA) Report, No. 9123 [in Turkish].

PASQUARE, G., POLI, S., VEZZOLI, L. & ZANCHI, A. 1988. Continental arc volcanism and tectonic setting in Central Anatolia. *Tectonophysics,* **146**, 217–230.

POISSON, A. & AKAY, E. 1981. Modalités de la transgression Miocene dans le basin de Gökbük–Çatallar (Tauris Occidentales – Turquie). *General Directorate of Petroleum Works,* **25**, 217–222.

ŞAROĞLU, F., EMRE, Ö. & BORAY, A. 1987. *Türkiye'nin Diri Fayları ve Depremselliği.* Mineral Research and Exploration Institute of Turkey (MTA) Report, No. 8174 [in Turkish].

ŞENGÖR, A. M. C. 1980. Türkiye'nin Neotektoniğinin Esasları. *Geological Society of Turkey Conference Series,* **2** [in Turkish with English abstract].

SEYİTOĞLU, G. & SCOTT, B. C. 1992. The age of Büyük Menderes graben (west Turkey) and its tectonic implications. *Geological Magazine,* **129**, 239–242.

—— & —— 1996. The age of the Alaşehir graben (west Turkey) and its tectonic implications. *Geological Journal,* **31**, 1–11.

YAĞMURLU, F. 1991. Yalvaç–Yarıkkaya Neojen havzasının tektonosedimanter özellikleri ve yapısal evrimi. *Mineral Research and Exploration Institute of Turkey (MTA) Bulletin,* **112**, 1–12 [in Turkish with English abstract].

YILMAZ, Y., ŞAROĞLU, F. & GÜNER, Y. 1987. Initiation of the neomagmatism in East Anatolia. *Tectonophysics,* **134**, 177–199.

YUSUGOĞLU, H. 1996. Northern margin of the Gediz graben, age and evolution, west Turkey. *Turkish Journal of Earth Sciences,* **5**, 11–23.

Palaeomagnetic study of the Erciyes sector of the Ecemiş Fault Zone: neotectonic deformation in the southeastern part of the Anatolian Block

ORHAN TATAR,[1] JOHN D. A. PIPER[2] & HALİL GÜRSOY[1]

[1]*Department of Geological Engineering, Cumhuriyet University, TR-58140 Sivas, Turkey (e-mail: tatar@cumhuriyet.edu.tr)*
[2]*Geomagnetism Laboratory, Department of Earth Sciences, University of Liverpool, Liverpool L69 7ZE, UK*

Abstract: In the Turkish sector of the Afro-Eurasian collision zone, continuing northward motion of the Arabian promontory is extruding the Anatolian region to the west. Although the East Anatolian Fault Zone (EAFZ) is recognized as the southern boundary of this tectonic escape, distributed deformation is occurring across a broad zone extending for at least 300 km to the northwest, which includes a number of major dextral and sinistral fault lineaments. The longest of these lineaments is the Ecemiş Fault Zone, although the long-term significance of deformation, and specifically strike-slip, across this zone is disputed. This zone is also the locus of major volcanic activity in the Kayseri region. The present paper reports a palaeomagnetic study of young (1–2 Ma) lava flows which has aimed to identify recent block rotations resulting from regional deformation in this region. Rock magnetism shows the lavas to be dominated by low-Ti magnetite assemblages of primary cooling related origin. Although grain properties are predominantly multidomain, significant fractions of single domains are always present and responsible for a stable thermoremanence of normal and reversed polarity. Whilst group mean directions show that the blocks in this sector of Anatolia show the typical counterclockwise rotation resulting from tectonic escape, in this case by $c.$ 10° during the last 1 Ma, larger differential rotations in both senses are identified across the Sultansazlığı Depression. Comparable differential rotations recorded by the 2.8 ± 0.2 Ma İncesu Ignimbrite have resulted from the pull-apart in this sector of the Ecemiş Zone which has accommodated emplacement of the Erciyes volcanic centre since the termination of ignimbrite activity. Within the wedge-shaped terrane confined between the EAFZ in the south and the North Anatolian Fault Zone in the north, the degree of counterclockwise rotation during tectonic escape within the last 2–3 Ma has diminished from $c.$ 25° in the east to $c.$ 10° towards the southwest. This corresponds to a transition from highly strained to a less-strained lithosphere as the width of the semi-plastic Anatolian terranes confined between the Arabian–Eurasian pincer broadens out to the west.

Asia Minor has formed by successive docking of allochthonous terranes as the Palaeotethys and Neotethys Oceans have closed since Early Mesozoic times. This (palaeotectonic) phase of orogenic deformation has been progressively replaced since Late Miocene times by continuing (neotectonic) deformation as the Arabian sector of the African Plate has impinged into the Pontide Orogen along the Bitlis Suture Zone (BSZ; Fig. 1). Hence, progressive deformation of the collage of accreted terranes has been driven by the northward motion of Arabia at a rate of $c.$ 2.5 cm a^{-1} (Barka & Reilinger 1997) relative to the African Plate ($c.$ 1.0 cm a^{-1}). Differential motion is taken up mainly along the Dead Sea Fault Zone (DSFZ in Fig. 1) and is responsible for westward extrusion of the crust in central Turkey. This escape tectonics is accommodated primarily by dextral motion along the North Anatolian Fault Zone (NAFZ; Fig. 1), a major intracontinental transform defining the tectonic boundary between the Eurasian Plate to the north and the Anatolian terranes to the south.

Geological data indicate that the relative motion on the NAFZ increases to the west (Şengör et al. 1985; Andrieux et al. 1995), and the Global Positioning System (GPS) (Oral et al. 1995; Barka & Reilinger 1997) gives a range of 1.6–2.7 cm a^{-1} (Kiratzi 1993). Consequent internal deformation of the Anatolian Block is complex: in part, it is probably accommodated by block rotations along side splays such as the Kırıkkale–Erbaa and Almus Fault Zones (KEF and AF, respectively; Fig. 1). In addition, continuing impingement of Arabia into Anatolia

From: BOZKURT, E., WINCHESTER, J. A. & PIPER, J. D. A. (eds) *Tectonics and Magmatism in Turkey and the Surrounding Area*. Geological Society, London, Special Publications, **173**, 423–440. 1-86239-064-9/00/$15.00
© The Geological Society of London 2000.

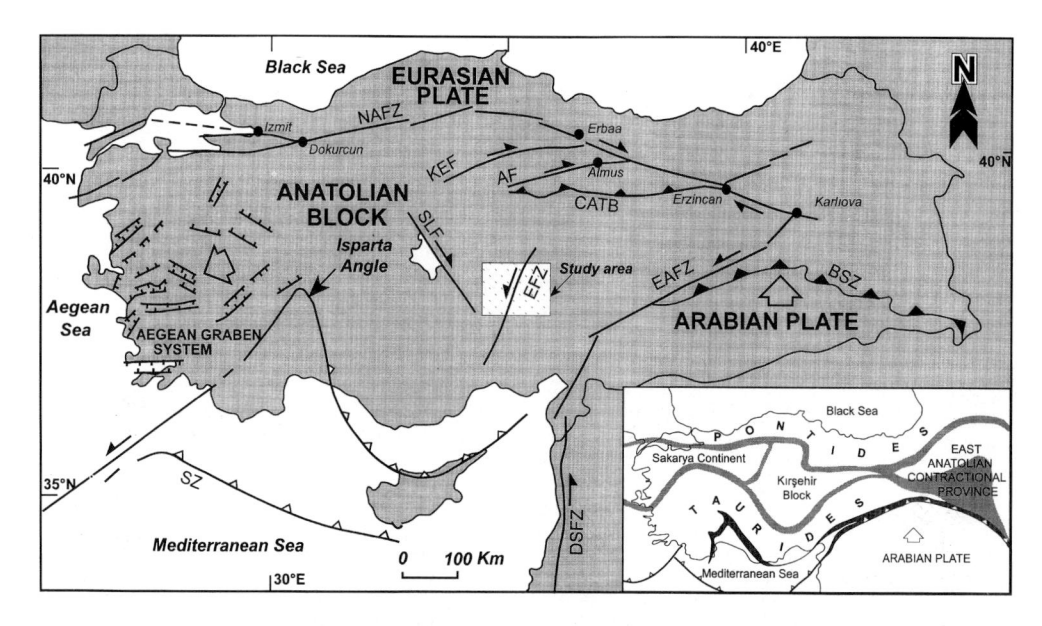

Fig. 1. The tectonic divisions and distribution of major lineaments in Turkey and adjoining regions. The large open arrows show relative motions of the plates and the smaller half-arrows are directions of movement on major strike-slip faults. SLF, Salt Lake Fault Zone; KEF, Kırıkkale–Erbaa Fault Zone; AF, Almus Fault Zone; EFZ, Ecemiş Fault Zone; SZ, East Mediterranean subduction zone; CATB, Central Anatolian Thrust Belt; DSFZ, Dead Sea Fault Zone. The political boundary of Turkey is also shown, together with boundaries of areas illustrated in subsequent figures. The inset figure shows the major terrane divisions of Turkey juxtaposed by the collisional events preceding the neotectonic phase of deformation.

and Eurasia has produced complex deformation within the wedge-shaped eastern sector sited between the NAFZ and the East Anatolian Fault Zone (EAFZ; Fig. 1); this is accompanied by sinistral strike slip on the latter fault at a slower rate ($c.$ 0.9 cm a^{-1}) than along the NAFZ.

Although the EAFZ is recognized as the southern boundary of the westward escaping Anatolian region (Barka & Reilinger 1997), a zone of distributed deformation extends for at least 300 km to the northwest and includes several major fault lineaments with both dextral and sinistral senses of motion (Tchalenko 1977; Ambraseys & Finkel 1987). The longest of these lineaments, and the most distant from the EAFZ, is the EFZ (Fig. 1), which has been a zone of sporadic seismic activity in Recent times (Ambraseys 1989; Jackson 1994), although its longer term significance is unclear. The structural features of the fault zone have been studied by Koçyiğit & Beyhan (1998). They refer to the northeast trending intracontinental structure, that cuts across the Anatolian Plateau between Düzyayla in the northeast and Anamur county in the southwest, as the Central Anatolian Fault Zone (CAFZ).

In the short term (10–10^2 a), crustal deformation in regions such as Turkey is resolved by GPS measurements and ground surveying. In the medium term (10^3–10^5 a) it is resolved by geomorphic study of features such as offset stream courses or changing elevations of erosional and depositional surfaces. In the longer term (> 10^5 a) deformation must be mainly resolved by palaeomagnetic studies which identify divergences of magnetization directions from expected geomagnetic field directions. Such divergences are the signature of block rotation. Combined geological and palaeomagnetic information may then aid in resolving the size and integrity of the fault blocks.

To apply this method, studies need to focus on rocks that act as good long-term recorders of the ancient geomagnetic field. Lava flows are amongst the most suitable targets because they usually record the field at the time of post-emplacement cooling as a strong thermal remanent magnetization (TRM); fortunately, such rocks are widely distributed within Anatolia. In this study, lavas which have erupted during the last 1–2 Ma have been used to identify block rotations in the vicinity of the EFZ, which corresponds to the locus of volcanic activity

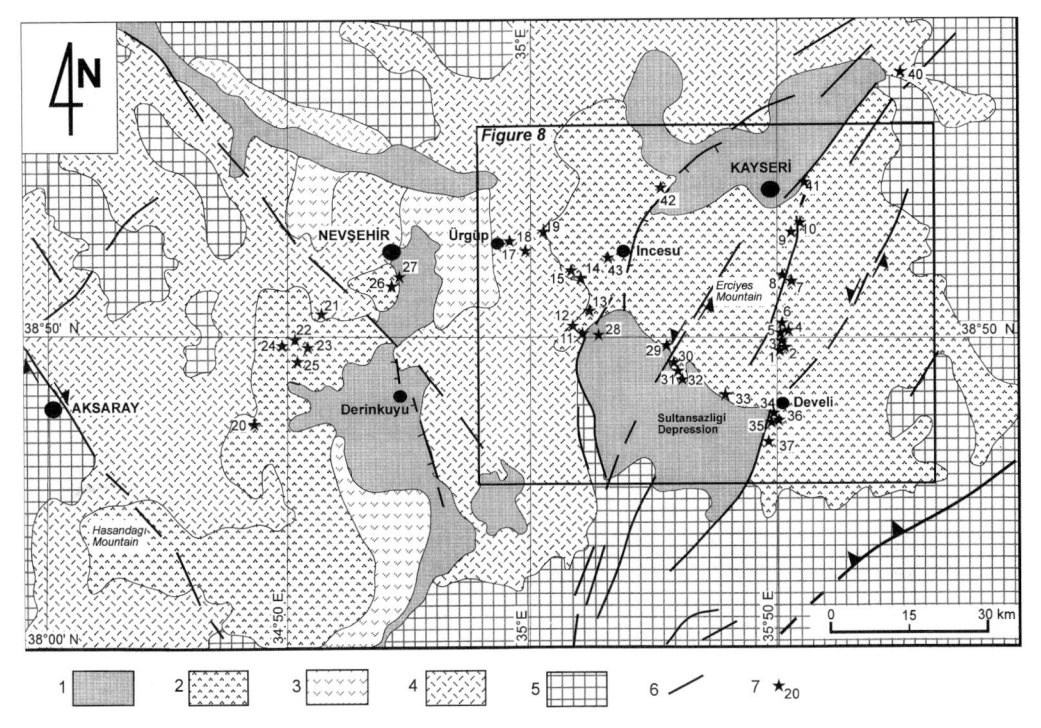

Fig. 2. Outline geological map of Kayseri region showing the Central Anatolian Volcanic Province (CAVP) and the adjoining sector of the Ecemiş Fault Zone. The palaeomagnetic sampling sites are also shown. 1, Quaternary alluvium; 2, Plio-Quaternary subareal volcanic rocks; 3, Neogene subareal volcanic rocks; 4, Neogene tuffs and pyroclastic rocks with some lava; 5, Pre-Neogene basement; 6, faults; 7, sampling sites.

dominated by the Erciyes Strato Volcanic Complex near Kayseri (Fig. 2).

Geological framework

The Anatolide and Tauride Belts (Fig. 1) formed as a consequence of continent–continent collision along the BSZ. Subsequent suturing has produced three tectonic provinces, classified by Şengör (1980) as the Aegean Graben System, the Central Anatolian 'Ova' (= plain) Province and the East Anatolian Contractional Province. Neogene volcanic activity in Anatolia has occurred in four main regions, geographically located in eastern, central and western Anatolia, and in the Galatean Volcanic Province bordering the NAFZ along the northern margin of the block. The Central Anatolian Volcanic Province (CAVP) is located within the Ova Province. It comprises some 20 eruptive centres extending for 300 km along a northeast–southwest trend. Activity appears to be concentrated at the intersections of major fault zones such as the (dextral) Salt Lake and (sinistral) Ecemiş Faults (Fig. 1; Pasquare *et al.* 1988; Toprak & Göncüoğlu 1993). Volcanic and volcaniclastic successions have been erupted onto Neogene sedimentary basins (Salt Lake, Ulukışla and Sivas) or older metamorphic complexes (Kırşehir, Niğde).

Volcanic activity in the CAVP falls into three major cycles of calc-alkaline magmatism (Pasquare *et al.* 1988) comprising: (1) an initial phase of basaltic and andesitic eruption between 13.5 and 8.5 Ma; (2) an ignimbrite-producing phase between Middle Miocene and Early Pliocene times (*c.* 11 and 2 Ma); and (3) a continuing cycle characterized by growth of major stratovolcanoes (Erciyes, Hasandağı) and numerous small monogenetic volcanoes erupting lavas and tuffs predominantly of basaltic composition. The present study is concerned with products of the latter phase and largely with lavas erupted from the Erciyes Stratovolcano Complex (ESC), the largest regional example of the stratovolcanoes of phase (3).

The ESC rises 3917 m above sea level. It was

initiated by basaltic fissure eruptions coeval with faulting and segmentation of the Cappadocian Volcanic Province and initial sinking of the Erciyes Depression. The stratovolcano is emplaced onto the Cappadocian Ignimbrite succession and is therefore entirely younger than Late Pliocene in age. Subsequent eruptive episodes have included basaltic and andesitic lavas, radial dyke emplacement and formation of dacite–rhyodacite lava domes, and emplacement of various hyaloclastite and pumiceous deposits. The latter activity has continued into Prehistoric times (15 500 ± 2500 a; Innocenti et al. 1975) and the stratovolcano is flanked by many morphologically young lava flows. The geochemical composition of the volcanic rocks is predominantly calc-alkaline (Batum 1978).

The Erciyes Depression has resulted from a releasing type of sinistral double bending on the CAFZ since Early Pliocene times (Koçyiğit & Beyhan 1998). It is now bounded by marginal fault systems recording accumulative downthrows > 1000 m; the eastern margin, bounds fault systems which branch into two subsystems (Dundarlı–Erciyes and Develi) crossing the pull-apart basin (Fig. 2).

Field and laboratory methods

Lava flows are typically high-quality recorders of the magnetic field at the time of cooling. Since their tectonic orientation can usually be constrained by flow structures or interbedded strata, they are also effective recorders of subsequent tectonic rotation. Their main limitation is that rapid cooling will usually result in a near-instantaneous record of palaeosecular variation. It is therefore important to draw tectonic conclusions from groups of sites yielding a mean approximating to a time-averaged palaeomagnetic direction. A second qualification which can limit the value of some individual results, and increase scatter of inclinations in the overall sample, is uncertainty in the tilt adjustment: in the present case, lavas have probably flowed down primary slopes and they may also have consolidated since initial cooling; when orientation information is based on structures within the flow, the precise amount of the tilt adjustment therefore remains uncertain.

The lavas were cored in the field, at sites summarized in Table 1 and shown in Fig. 2, using a motorized drill and oriented by sun and magnetic compasses. Six or seven separate cores were distributed across several metres of lava flow outcrop. Core collection was accompanied by assessment of the local geological setting to evaluate orientation of the sampled units; the sampled lava flows of this study have tectonic tilts which are too low to be accurately discernible and palaeomagnetic directions are therefore treated in situ.

In the laboratory, the field cores were sliced into 2.3 cm long cylinders for routine palaeomagnetic measurements. Except for a few units that exhibited instability to cleaning treatment, the specimens had magnetic structures dominated by discretely defined characteristic remanent magnetizations (ChRM). These were typically contaminated by only minor amounts of viscous remnant magnetization (VRM). No significant difference was observed between ChRM defined by alternating field (a.f.) and thermal cleaning, and approximately equal numbers of cores were subjected to each cleaning technique.

All specimens were progressively demagnetized in steps of 50 or 100°C and 5 or 10 milliTesla (mT) until components were subtracted, or directional behaviour ceased to be stable. Remanence directions were calculated by principal-component analysis and common-site populations grouped to yield the site means listed in Table 2. Lowest blocking temperature/coercivity components are laboratory or drilling-related acquisitions, or are in the present field direction; these are removed in the first one or two steps of treatment and are interpreted as VRM. Higher and/or distributed demagnetization spectra are regarded as ChRM and included in Table 3.

Rock magnetism

The most important ferromagnetic mineral in basaltic and andesitic lavas is usually a titanomagnetite with a composition of about TM50 near the middle of the magnetite–ulvöspinel solid solution series and with a low Curie point of c. 250°C. However, during cooling, such grains typically undergo subsolidus exsolution to produce ilmenite lamellae hosted by Ti-poor magnetite. These lamallae may then be oxidized to produce pseudobrookite and hematite if the lava is permeated by fluids of high oxygen fugacity during deuteric stages of cooling. A general consequence of this alteration is to subdivide the effective magnetic grain sizes and enhance the magnetic stability, although the recorded palaeomagnetic field direction will be substantially the same as at the time of eruption. Stages of petrologic alteration are subdivided into seven classes (Haggerty 1976), ranging from unoxidized to completely oxidized.

Table 1. *Summary of palaeomagnetic sampling sites in the Erciyes volcanic district and eastern Cappadocia, central Turkey.*

Site no.		Age	N	M_0
Lavas – eastern flank of the Erciyes Volcanic Complex				
1	Basalt, Develi	Pliocene–Quaternary	6	130–139
2	Basalt, Develi	Pliocene–Quaternary	6	104–188
3	Basalt, Develi	Pliocene–Quaternary	6	38–558
4	Basalt, Develi	Pliocene–Quaternary	6	33–520
5	Basalt, Develi	Pliocene–Quaternary	6	20–289
6	Basalt, Develi	Pliocene–Quaternary	6	51–372
7	Basalt, Kayseri	Pliocene–Quaternary	6	52–149
8	Basalt, Kayseri	Pliocene–Quaternary	6	24–131
9	Basalt, Kayseri	Pliocene–Quaternary	6	23–132
10	Basalt, Kayseri	Pliocene–Quaternary	6	62–80
Lavas – SW Kayseri, western flank of the Erciyes Complex				
11	Basalt, İncesu	Pliocene–Quaternary	8	9–14
12	Basalt, İncesu	Pliocene–Quaternary	6	68–220
13	Basalt, İncesu	Pliocene–Quaternary	6	8–47
14	Basalt, İncesu	Pliocene–Quaternary	6	30–89
15	Andesitic tuff	Pliocene–Quaternary	6	14–25
16	Basalt, İncesu	Pliocene–Quaternary	6	0.3–9
17	Basalt, İncesu	Pliocene–Quaternary	6	24–136
18	Basalt, İncesu	Pliocene–Quaternary	6	95–124
19	Andesitic tuff	Pliocene–Quaternary	6	3–64
Lavas – central Cappadocia				
20	Basalt, Acıgöl	Quaternary	7	6–54
21	Scoria, Boğazköy	Quaternary	7	232–941
22	Basalt, Acıgöl	Quaternary	7	378–2994
23	Basalt, Acıgöl	Quaternary	7	7–21
24	Basalt, Acıgöl	Quaternary	7	182–333
25	Basalt, Kurugöl	Quaternary	7	119–292
26	Basalt, Güvercinlik	Quaternary	7	29–65
27	Basalt, Göre	Quaternary	7	4–26
Lavas – Sultansazlığı Depression, South Erciyes Complex				
28	Basalt, İncesu	Quaternary	7	445–339
29	Basalt, İncesu	Quaternary	7	30–166
30	Basalt, Yeşilhisar	Quaternary	7	93–124
31	Basalt, Çayırözü	Quaternary	7	129–2492
32	Basalt, Çayırözü	Quaternary	7	125–2890
33	Basalt, Soysaldı	Quaternary	7	73–327
34	Basalt, Develi	Quaternary	7	2–52
35	Basalt, Develi	Quaternary	7	208–428
36	Basalt, Develi	Quaternary	7	88–683
37	Basalt, Develi	Quaternary	7	33–108
İncesu Ignimbrite				
38	Mudflow, Bünyan	Pliocene, 2.8 Ma	9	39–74
39	Ignimbrite, Bünyan	Pliocene, 2.8 Ma	12	192–258
40	Ignimbrite, Sarımsaklı	Pliocene, 2.8 Ma	8	90–137
41	Ignimbrite, Kayseri	Pliocene, 2.8 Ma	7	24–45
42	Ignimbrite, Saraycık	Pliocene, 2.8 Ma	8	37–44
43	Ignimbrite, İncesu	Pliocene, 2.8 Ma	10	7–14

Ages of the lavas are estimated from aspect and setting. N is the number of separately oriented cores. Intensities of the Natural Remanent Magnetization (NRM), M_0, are $\times 10^{-4}$ A m^2 kg^{-1}.

Table 2. *Summary of thermomagnetic and hysteresis properties of lavas in the Erciyes–eastern Cappadocia region, central Turkey*

Site no.	T_{c1}	T_{c2}	RM	M_S	M_{RS}	H_C	M_{RS}/M_S	X_{MD}
1	–	570	0.94	96.54	13.05	14.50	0.14	0.75
2	–	570	0.82	106.19	12.73	12.00	0.12	0.79
3	540	585	0.80	69.53	7.86	10.50	0.11	0.81
4	555	585	–	95.39	11.38	13.50	0.12	0.79
5	550	–	0.90	89.79	13.09	13.00	0.15	0.73
6	–	585	0.88	191.75	29.04	18.50	0.15	0.79
7	555	–	1.00	90.75	15.74	19.00	0.17	0.77
8	–	580	1.01	106.08	12.20	10.50	0.12	0.79
9	560	590	0.90	114.66	12.04	12.00	0.11	0.81
10	419	494	0.92	176.29	9.15	4.50	0.05	0.93
11	–	570	0.68	204.22	27.49	13.50	0.13	0.77
12	320	580	1.34	H	14.99	13.50	–	*
13	–	585	0.69	103.24	11.97	12.00	0.12	0.79
14	565	565	0.84	254.97	21.45	10.00	0.08	0.88
15	570	570	0.69	128.95	9.27	10.00	0.07	0.90
16	–	–	–	H	18.93	23.50	0.07	0.90
17	380	560	1.55	72.13	14.55	10.00	0.20	0.63
19	230	(525)	2.55	140.01	18.48	8.00	0.13	0.77
20	180	440	0.72	200.59	15.02	3.50	0.07	0.89
21	–	595	0.64	258.18	14.13	4.00	0.05	0.94
22	–	575	0.91	239.89	51.31	15.00	0.21	0.44
23	–	570	0.80	467.66	20.11	6.00	0.04	0.96
24	–	550	0.95	430.92	30.87	4.50	0.07	0.90
25	–	580	1.00	364.40	55.54	8.50	0.15	0.73
26	–	570	0.90	134.25	14.83	10.00	0.11	0.81
27	–	560	1.00	169.34	28.76	21.00	0.17	0.69
28	–	570	0.90	142.74	39.46	30.50	0.28	0.46
29	–	580	0.79	87.98	10.80	19.00	0.12	0.79
30	–	570	0.85	125.10	14.77	15.00	0.12	0.79
31	–	570	0.84	341.13	93.40	32.00	0.27	0.48
32	275	570	0.81	125.61	20.02	9.50	0.16	0.71
33	–	545	0.99	69.27	11.56	17.00	0.17	0.69
34	405	580	0.44	107.27	17.83	20.00	0.17	0.69
35	–	550	0.90	324.42	55.62	20.00	0.17	0.69
36	–	570	0.80	198.28	15.52	9.00	0.08	0.88
37	–	540	1.00	62.23	6.59	7.00	0.10	0.85

T_{C1} and T_{C2} are successive Curie temperatures; temperatures in brackets are observed in the cooling cycle only. RM is the ratio at 100°C of the magnetization during cooling to the magnetization during heating. Units of M_S and M_{RS} are $\times 10^{-2}$ A m^2 kg^{-1}, and H_C is in mT. * A constriction in the hysteresis curve due to a mixture of hard and soft magnetic phases. The values of M_S and M_{RS}/M_S are not given if the sample failed to saturate in a 1 T field.

Strong-field thermomagnetic analysis uses the Curie temperature to indicate the nature of the titanomagnetite; it may also provide information about the degree of alteration that occurs during heating. In this study, saturation magnetization was measured from room temperature to 700°C using a computer-controlled horizontal Curie balance. The Curie temperature was estimated from the thermomagnetic curve using the method of Grommé *et al.* (1969); the change in saturation magnetization at 100°C, following heating and cooling (the ratio RM in Table 2), is a measure of alteration that occurs during heating.

The majority of thermomagnetic curves in these young lavas illustrate a smooth fall to define the Curie point of a Ti-poor titanomagnetite, usually just below the Curie point of pure magnetite (*c.* 590°C; Fig. 3). In some specimens, from lavas on the eastern flank of the Erciyes Complex, two distinct magnetite phases are discernible; one is a Ti-bearing phase with a Curie point of *c.* 540°C and the other a pure magnetite. A possible signature of low-temperature maghemitization resulting in the formation of metastable non-stoichiometric titanomaghemite (cation-deficient titanomagnetite) is a weak inflection in a few examples, defining which is the most common feature of this collection (Fig. 3, Table 2), is probably due to incipient oxidation of magnetite to hematite by the heating. Uncommon RM values of > 1 are usually present when a low-temperature phase is discernible and a ferromagnetic mineral, possibly maghemite, is converted to magnetite.

Hysteresis data yield information about mineralogy and domain states: titanomagnetites and titanomaghemites saturate in fields of 300 mT or less, whereas hematites require much larger fields to achieve saturation. The following hysteresis parameters were determined using a Molspin vibrating sample magnetometer (VSM): saturation magnetization, M_S, saturation remanence, M_{RS}; and coercive force, H_C. With a maximum field of 1 T it is not possible to saturate hematite and M_S cannot be determined when this mineral is present in significant amounts. However, only two of these young lava flows failed to saturate, thus confirming the dominance of titanomagnetite (Table 2). Parameters M_S and M_{RS} are dependent on the concentration and type of the magnetic minerals present; M_S is independent of grain size but M_{RS} is smaller for multidomain (MD) than for single domain (SD) minerals. Hence the ratio of M_{RS}/M_S is a useful indicator of domain states: if $M_{RS}/M_S < 0.1$, the sample is dominated by MD grains; if $M_{RS}/M_S > 0.1$, small but significant fractions of SD grains are indicated within a dominant MD assemblage. Provided that these mixed domain sizes are present within magnetite only (indicated by M_{RS}/M_S values of 0.02–0.5), the equation

$$X_{MD} = [0.5 - (M_{RS}/M_S)]/0.48$$

yields an estimate of the fraction of MD grains present (Thomas 1992).

In these lavas M_{RS}/M_S is always $\leqslant 0.5$, confirming that titanomagnetite is the dominant ferromagnetic mineral. The grains are predominantly in a MD state, although 10–20% fractions of SD grains are present, as indicated by the ratio X_{MD} (Table 2), and are presumably the main carriers of stable remanence.

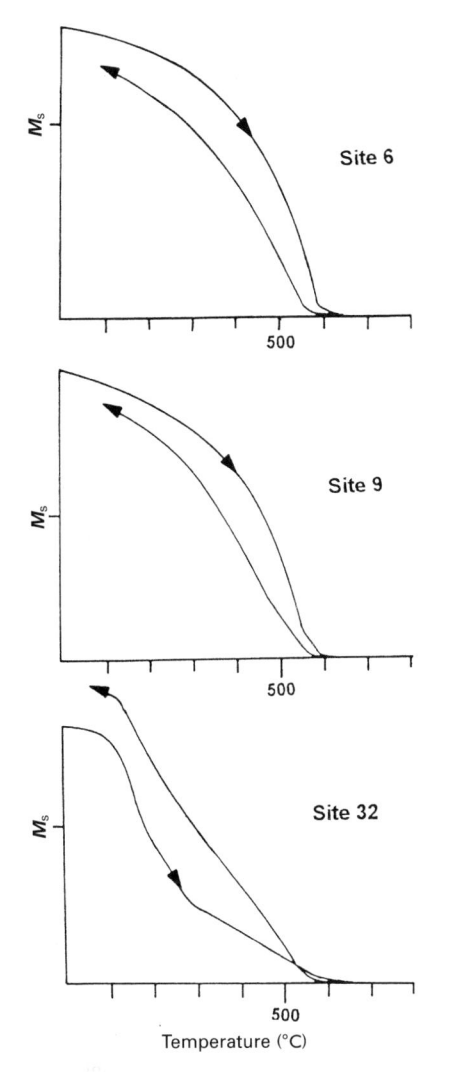

Fig. 3. Examples of thermomagnetic curves [saturation magnetization (Ms) v. temperature (in °C)] from young lavas of this study; the scales on the magnetisation axes are equivalent to 500 mT. Site 6 is an example of a single magnetite Curie point whereas site 9 show a distinct Curie point and site 32 has low Curie point phase (probably maghemite) which converts to magnetite on heating.

Curie points of between c. 180 and 350°C. This is observed in only five specimens and bears no apparent relationship to magnetic stability (Table 3). Hematite is discernible in none of the thermomagnetic curves, although it can escape detection because it has a saturation magnetization only 2% of that of magnetite. The reduction of saturation magnetization (M_s) with heating, to yield RM values of c. 0.7–0.9,

Palaeomagnetic results

Examples of a.f. and thermal demagnetization behaviours are shown in Figs 5 and 6. Component trajectories are usually dominated by one convergent ChRM subtracted at, or just above, the Curie point of pure magnetite. These components are of both polarities but tend to have declinations significantly different from the present day field direction. In the

Table 3. *Palaeomagnetic results from lavas of the Erciyes sector of the Ecemiş Fault Zone, central Turkey*

Site no.	N/n	R	K	α_{95}	D	I
Lavas – eastern flank of the Erciyes Complex						
1	6/6	5.93	7.52	7.3	355.44	59.9
2	6/6	5.99	492.5	3.0	4.7	58.6
3	6/6	5.99	418.2	3.3	172.5	−55.3
4	6/6	5.98	271.4	4.1	171.0	−51.4
5	6/6	5.92	60.7	8.7	332.1	16.2*
6	6/4	3.97	102.8	9.1	339.5	15.8*
7	6/6	5.83	29.2	12.6	359.6	28.6
8	6/5	4.93	56.2	11.1	48.8	35.7*
9	6/6	5.99	495.4	3.0	330.5	54.5
10	6/6	5.99	462.2	3.1	356.1	55.5
Lavas – SW Kayseri, western flank of the Erciyes Complex						
11	8/8	7.91	7.73	6.3	151.3	−33.8
12	6/6	6.00	1168.9	2.0	193.5	−58.4
13	6/6	5.89	43.7	10.2	192.7	−54.2
14	6/6	5.97	196.4	4.8	308.2	41.7
15	6/5	5.00	1420.8	2.0	175.1	−42.2
16	6/2	2.00	582.2	–	5.2	42.7
17	6/6	5.89	43.9	10.2	338.2	−0.4*
18	6/4	3.98	161.9	7.2	339.2	44.3
19	6/6	5.99	494.4	3.0	197.4	−62.1
Lavas – central Cappadocia						
20	7/6	5.84	32.2	12.0	174.8	−63.9
21	7/4	3.80	19.1	21.6	109.7	−57.9*
22	7/5	4.65	11.3	23.8	203.6	−54.2
23	7/3	3.00	1212.2	3.5	213.9	−30.9
24	7/7	6.94	97.7	6.1	359.1	62.1
25	7/5	4.90	41.6	12.0	20.5	53.3
26	7/5	4.89	34.8	13.2	155.9	−43.0
27	7/3	2.97	79.3	13.9	96.1	23.9*
(ii)	7/3	2.95	37.9	20.3	94.8	−31.9*
Lavas – Sultansazlığı Depression, South Erciyes Complex						
28	7/7	6.96	170.6	4.6	188.1	−40.4
29	7/7	6.97	182.5	4.5	17.8	55.6
30	7/6	5.98	316.2	3.8	278.9	66.2*
31	7/4	3.93	43.5	14.1	218.7	−42.1
32	7/7	6.86	42.4	9.4	29.7	45.1
33	7/7	6.98	338.7	3.3	185.7	−74.8
34	7/4	3.96	77.4	10.5	324.2	54.4
(ii)	7/6	5.96	132.4	5.8	158.9	−60.9
35	7/7	6.94	97.9	6.1	171.9	−54.4
36	7/4	3.98	145.6	7.6	324.5	54.0
37	7/7	6.93	83.2	6.7	180.0	54.0
Group mean results						
Eastern flank of the Erciyes Complex						
7 lavas		6.86	42.5	9.4	353.1	52.5
Western flank of the Erciyes Complex						
8 lavas		7.62	18.6	13.2	350.1	49.7
Central Cappadocia						
6 lavas		5.73	18.5	16.0	9.7	53.3
Sultansazlığı Depression						
Sites 28–33		4.83	24.0	16.0	22.2	52.2
Sites 34–37		4.94	70.6	9.2	339.9	56.4

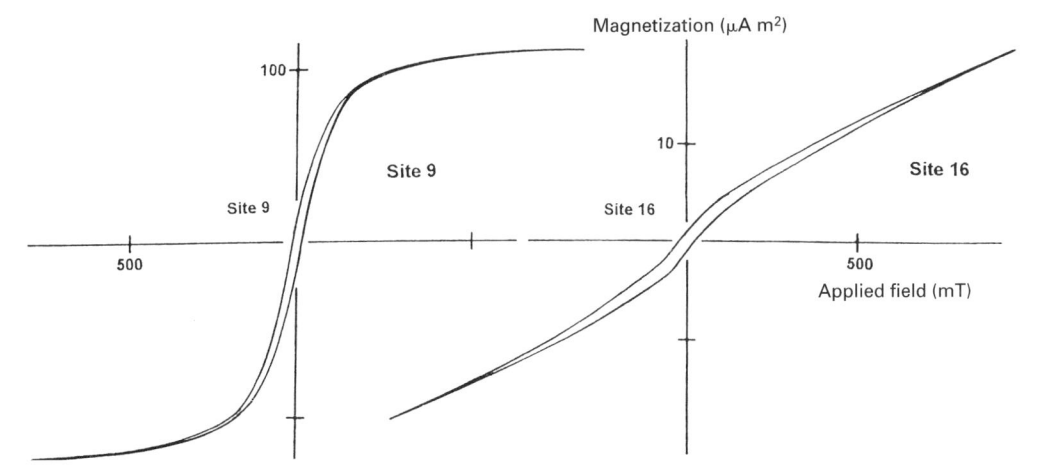

Fig. 4. Examples of saturated and unsaturated hysteresis curves from lavas of this study.

context of this observation and the rock magnetic data, they are therefore interpreted as thermal remanent magnetization (TRM) acquired during initial cooling, which record instantaneous stages of Late Pliocene–Recent secular variation and may also have been subsequently rotated.

Plio-Quaternary lava flows from the eastern flank of the Erciyes Volcano (sites 1–10) were a.f. demagnetized and show a VRM removed within the first one or two steps to subtract single convergent components of both normal and reversed polarity (cf. specimens 2-2 and 4-6; Fig. 5). Declinations are within 30° of the Recent field axis (Table 3). Sites 5 and 6 have anomalously shallow inclinations in relatively low coercivity components; magnetic properties are not unusual at these sites (Table 2) and they seem to have either recorded intermediate directions during a polarity transition or have been subjected to deformation during collapse. Excluding these anomalous sites and a single, possibly large, clockwise rotated direction at site 8, sites in this sector combine to yield a mean of $D/I = 353/53°$ ($\alpha_{95} = 9°$; Table 3) with reversed sites inverted to a common polarity. (D, mean declination; I, mean inclination; α_{95}, radius of the cone of 95% confidence about the mean direction.)

Lavas on the western side of the Erciyes Volcano are also dominated by single components following the removal of a soft component in initial steps. The latter are usually random, although some examples are evidently VRM in the present-day field, such as specimen 19-5 (Fig. 5), where this component is removed to recover a reversed polarity ChRM. Polarities in this sector of the volcano are predominantly reversed and therefore at least > 0.78 Ma in age. Declinations are also more variable and show both clockwise (specimen 13-3; Fig. 5) and counterclockwise (specimen 11-5; Fig. 5) rotation. Site 16 showed complex directional behaviour with only two specimens grouping near the present field direction and was therefore rejected from further calculation. The very weak intensities here (Table 1) are attributed to a dominant paramagnetism rather than a transitional field record (Table 2; Fig. 4). Excluding site 17, with anomalously shallow inclination, the group mean of these lavas is $D/I = 350/50°$ ($\alpha_{95} = 13°$); the declination would not be significantly changed by inclusion of site 17.

The lavas from the latest eruptive phase in Central Cappadocia postdate the ignimbrite-producing episode (Pasquare *et al.* 1988) and are hence younger than 1.1 ± 0.1 Ma (Innocenti *et al.* 1975; Mues-Schumacher &

D and I are the mean declination and inclination derived from N specimens from site populations given in Table 1; in a few cases N includes two cylinders cut from the same field core. R is the magnitude of the resultant vector derived from the n component directions from a site population of N and k in the Fisher precision parameter $[= (n-1)/(n-R)]$; α_{95} is the radius of the cone of 95% confidence about the mean direction in degrees. Where more than one component is recognized at a site they are listed in order of lower to higher blocking temperature/coercivity.* Site mean directions excluded from group mean calculations.

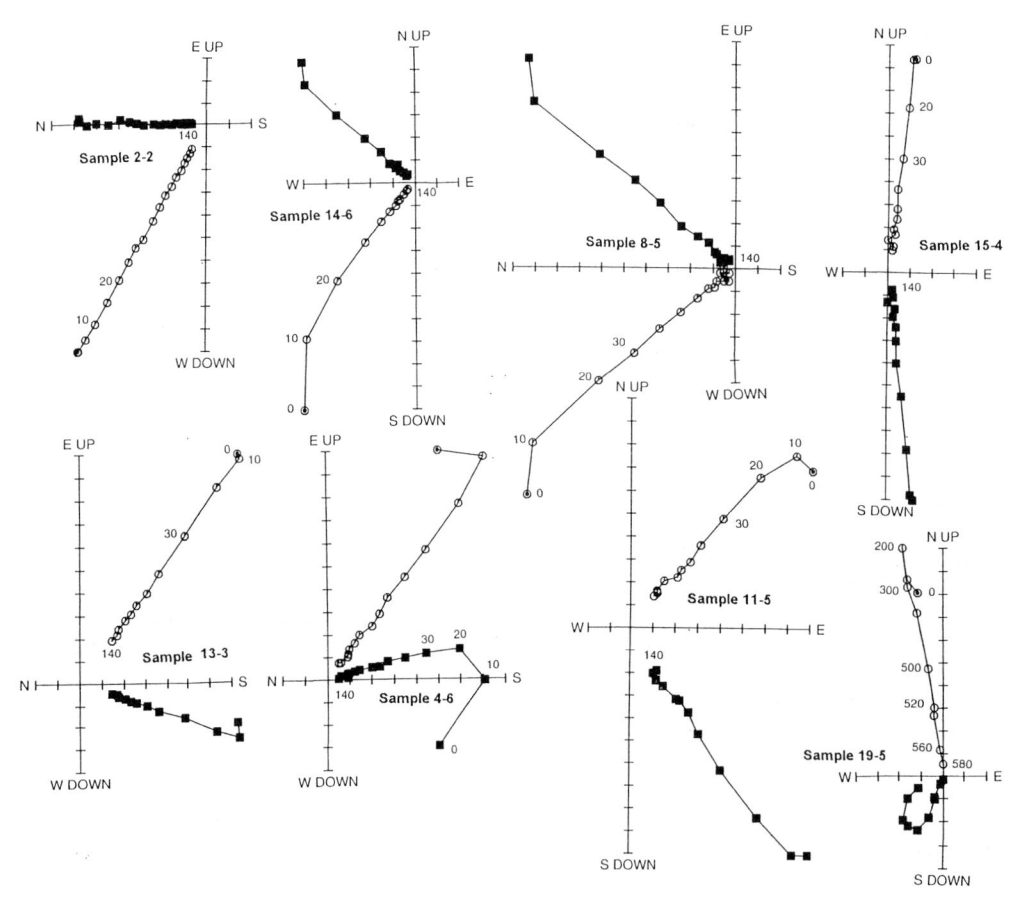

Fig. 5. Alternating field (a.f.) and thermal demagnetization (sample 19-5) results from lavas in the eastern and western sectors of the Erciyes Volcano. The demagnetization behaviours are shown *in situ* as orthogonal projections of the magnetization vector onto the horizontal (closed squares) and vertical (open circles) planes. The demagnetization steps are listed in temperatures (°C) or peak a.f. (mT).

Schumacher 1996). Four lavas are reversed and were therefore erupted during the latter part of the Matuyama Chron (specimens 22-4 and 26-4; Fig. 6). Site 27 has anomalous easterly declinations with inclinations falling into two groups: the intensities are very weak for Quaternary basalt (Table 1) and this lava appears to have been magnetized during a polarity transition. The reverse-magnetized sites have considerable scatter but if site 21 (where declination is anomalous and possibly indicative of large local counterclockwise rotation) is excluded, a coherent mean of $D/I = 10/-53°$ ($\alpha_{95} = 16°$) is derived (Table 3) for this region.

Sites 28–37 comprise a west to east traverse across the Sultansazlığı depression in the southern part of the Erciyes pull-apart basin (Koçyiğit & Beyhan 1998). In the first five lavas, declinations in both normal- and reverse-magnetized flows on the west side of the Dündarlı–Erciyes Subfault are rotated clockwise from the present field direction (specimen 28-2; Fig. 6), with the exception of site 30 which has an anomalous westerly declination. Directions in the four lava flows at the eastern part of the profile are, however, rotated counterclockwise; this sense of rotation includes both low and high blocking temperature components of opposite polarity at site 34. Thus, it appears that opposite block rotations are defined by these two sectors (Fig. 8), summarized by contrasting mean directions of $D/I = 10/52°$ ($\alpha_{95} = 16°$) and $D/I = 340/56°$ ($\alpha_{95} = 9°$) in Table 3.

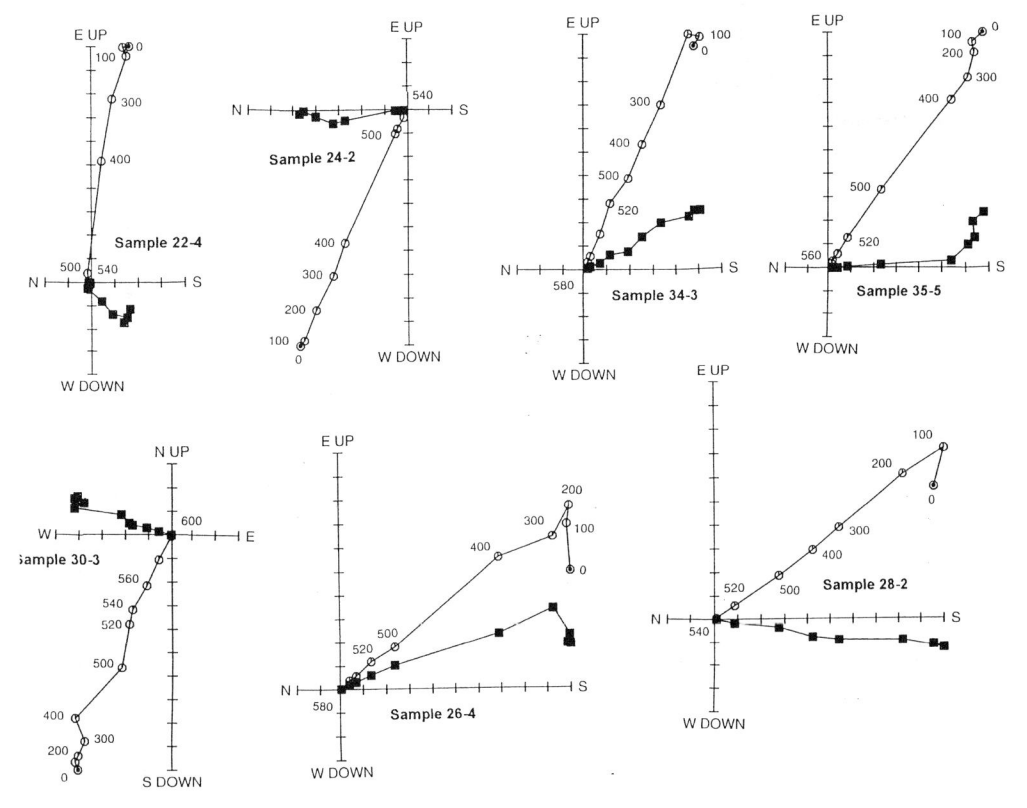

Fig. 6. Examples of thermal demagnetization of young lavas from Cappadocia and the Sultansazlığı Depression. Directions are *in situ* and symbols are as for Fig. 5.

Regional interpretation

For tectonic interpretation, the group mean palaeomagnetic directions may be compared with the present-day average dipole field in this region. Derived from the mean latitude of 38.4°N, this has a direction of $D/I = 0/+57.5°$ (normal) and $D/I = 180/-57.5°$ (reversed); the regional mean inclinations are within 15° of this field direction (Table 3). The bias towards shallower inclinations is presumed to be an expression of the far-sided dipole effect very widely recognized in Cenozoic lava studies (Wilson 1971). The tectonic rotation, R', is determined by comparison of the mean directions with the predicted palaeofield axis. Normally, both observed and reference directions have confidence limits (ΔD and DD_{ref}, respectively). Beck (1980) proposed a confidence limit on R' defined by $\Delta R' = \sqrt{(\Delta D^2 + \Delta D_{ref})}$. Demarest (1983) concluded that this value overestimates the errors and showed that, provided α_{95} is small and preferably < 10°, a standard correction factor is applicable. For $n \geq 6$ the correction factor lies between 0.78 and 0.80; $\Delta R'$ can then be derived from the equation $\Delta R' = 0.8\sqrt{(\Delta D^2 + \Delta D_{ref}^2)}$. Only if $R' > \Delta R'$ can a rotation be regarded as significant.

Five estimates of regional rotations are derived from the present study. These are: eastern Erciyes, $-7 \pm 8°$; western Erciyes, $-10 \pm 10°$; central Cappadocia, $+10 \pm 13°$; western Sultansazlığı, $+22 \pm 13°$; eastern Sultansazlığı, $-20 \pm 7°$. Thus, the young volcanic regions around the Erciyes Volcano seem to have rotated counterclockwise during the last 10^5–10^6 a, although the amount of rotation is close to the error limits of the data. It is poorly constrained for western Erciyes where the mean includes three clockwise rotations of reversed polarity (Table 3). There is an indication that the sense of movement by block rotation changes to the west and south, as shown by results from central Cappadocia (Gürsoy *et al.* 1998) and the western Sultansazlığı Depression. The large (c. 1000 m) subsidence associated with the Sultansazlığı Depression is evidently linked to significant regional differential block

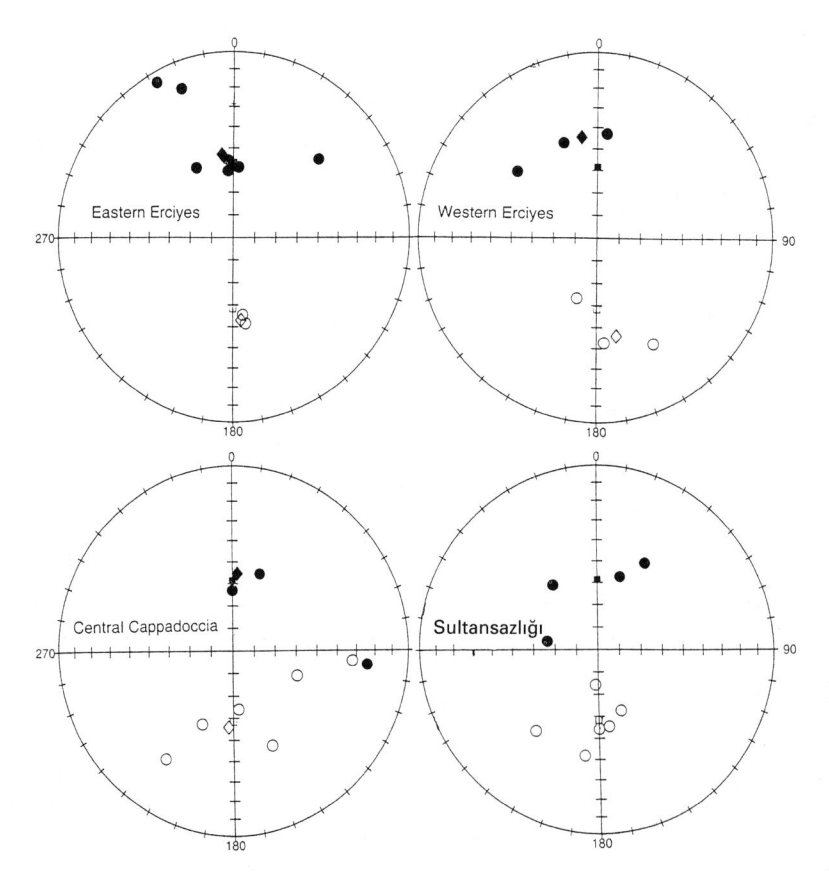

Fig. 7. Summary of site mean directions for the four regions investigated in this study. Open symbols are upper hemisphere projections and closed symbols are lower hemisphere projections; the squares are the direction of the mean present-day dipole field and the diamonds are group mean directions (see Table 3).

rotation, as shown by the palaeomagnetic results from young (and in some cases largely uneroded) lava flows.

To clarify possible differential rotation across this sector of the fault zone, the palaeomagnetism of the İncesu Ignimbrite has been studied. This is one of the youngest members of the Cappadocian ignimbrite province, dated at 2.8 ± 0.2 Ma; it is also one of the most voluminous, extending across the EFZ for at least 85 km east of Kayseri (Mues-Schumacher & Schumacher 1996). It comprises a distinctive densely welded, brown to pink unit that now underlies the Erciyes volcanic edifice.

The magnetization in this unit is dominated by a highly stable single component (Fig. 9) residing mostly in magnetite, although an unblocking spectra continuing to 630°C in some samples suggests that hematite also contributes. The components are of typical normal polarity (attributable to the Gauss normal polarity chron) and yield high within-site precisions (Table 4). Rock magnetic and magnetic fabric data comprise part of a larger study of the Cappadocian ignimbrite province and will be reported elsewhere. Alpha-95 values of as little as 2–3° make this ignimbrite unit a high fidelity recorder of block rotations after 2.8 Ma. Declinations show a consistent regional variation across the pull-apart basin (Fig. 10): the palaeofield declinations illustrate the typical counterclockwise rotations identified in the Sivas Basin (Gürsoy *et al.* 1997) on the north and east side of the Erciyes Complex, but change to *c.* 5° (clockwise to the east and south east of Erciyes. This accords with the clockwise rotations observed in the younger lavas further south in the Sultansazlığı Depression (Fig. 10). Continuing further to the west, counterclockwise rotation is again recognized when the west margin bounding

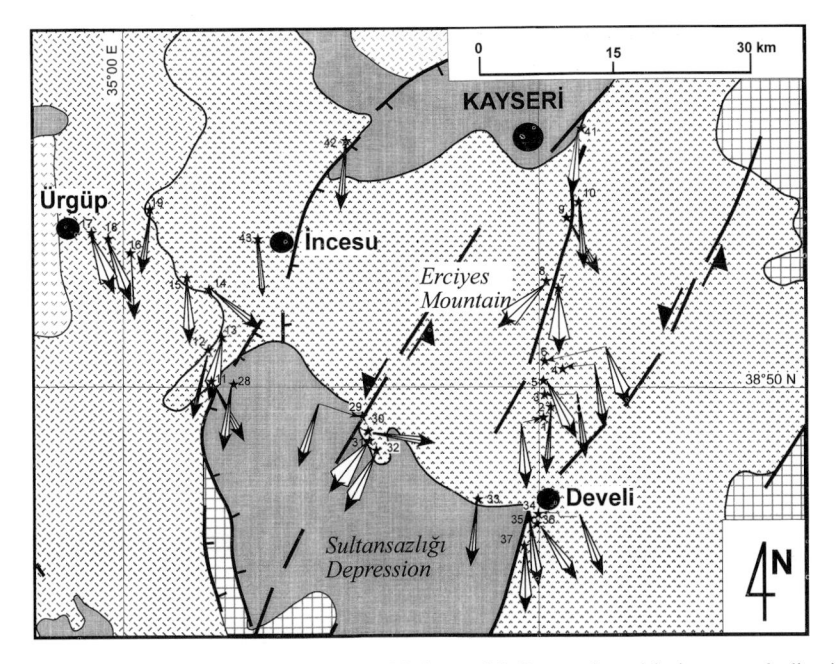

Fig. 8. Geological map of the Erciyes Mountain and Sultansazlığı Depression with site mean declinations of magnetizations; the error limits are 95% confidence limits. Keys are as for Fig. 2.

fault system is crossed (site 43); comparable counterclockwise rotation is present in all of the Cappadocian ignimbrites to the west (11.2–1.1 Ma; unpublished data). Thus, the İncesu Ignimbrite clearly records differential rotation of up to 10° since this eruptive event occurred 2.8 Ma ago. The deformation is most obviously explained in terms of pull-apart in this sector of the fault zone, with local extension accommodating emplacement of the Erciyes volcanic centre since the termination of ignimbrite magmatism c. 1.1 Ma ago.

A summary of Cenozoic palaeomagnetic results from the central Anatolian region which may be used for mapping regional variations of tectonic rotation is given in Table 5; the data are plotted in Fig. 11. These data are all based on regional or formation studies which may be assumed to have averaged the effects of secular variation and therefore record palaeomagnetic directions. Additional data based on isolated sites are given in Platzman et al. (1998) and are in general agreement with the pattern of regional directions shown in Fig. 11.

A dominance of regional, but variable, counterclockwise rotation is evident from these data. Rock units older than Middle Miocene in age were emplaced during the palaeotectonic history and their cumulative rotations could therefore be a composite of deformation during both the palaeotectonic and neotectonic regimes. Nevertheless, rotations recognized in Eocene units are comparable over a large area of central Anatolia (Tatar et al. 1996; Fig. 8)

Table 4. *Palaeomagnetic results from sites in the İncesu Ignimbrite (2.8 Ma) in the Erciyes region*

Site no.	N	R	K	α_{95}	D	I
38	6	6.00	1114	2.0	352.7	49.3
39	7	6.99	442	2.9	351.5	50.3
40	10	9.95	196	3.5	359.8	47.4
41	9	8.92	96	5.3	5.5	41.9
42	7	6.99	487	2.7	5.2	46.3
43	9	8.98	429	2.5	354.7	52.0

Symbols are as for Table 3.

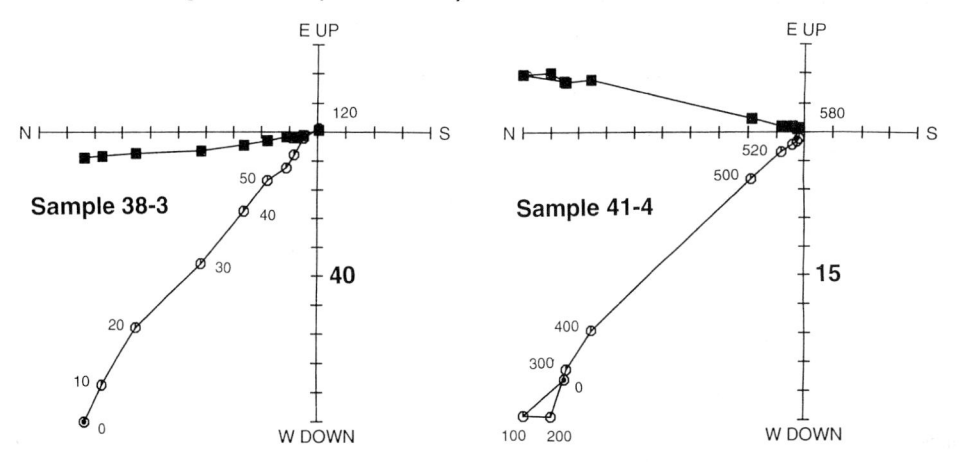

Fig. 9. Typical orthogonal projections of samples from sites in the İncesu Ignimbrite illustrating clockwise and counterclockwise rotation (see Table 4). Symbols are as for Fig. 5.

and north of the NAFZ (Sarıbudak 1989; Piper et al. 1996, 1997). They prove to be similar to rotations recognized in neotectonic rock units (Fig. 11), an observation which implies that major differential rotation did not occur during continental collision associated with the emplacement of Pontide and Tauride orogenic belts.

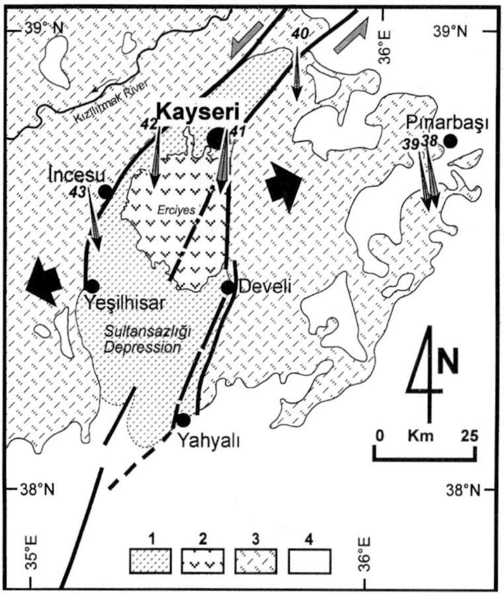

The palaeomagnetic study of lava flows, which are demonstrably late in the volcanic–tectonic history, is important because it can help to constrain the timing of rotations associated with regional deformation. In the Sivas Basin, bordering the Kayseri Fault Zone 400 km to the northeast (Fig. 11), mean rotations resolved from Miocene, Pliocene and Quaternary volcanic rocks of −34, −25 and −28°, respectively, show that the bulk of regional counterclockwise rotation associated with lateral extrusion of Anatolia has been concentrated within the last part of the neotectonic history (Gürsoy et al. 1997). The implication is that post-collisional deformation was initially accommodated by crustal thickening during Miocene and Pliocene times, whilst regional rotations recording lateral escape of the Anatolian Block commenced in a major way when this thickening could no longer be sustained.

The lava flows grouped as 'Quaternary' in the Sivas Basin study are of young aspect but not usually interleaved with datable strata. This c. 25° of anticlockwise rotation is therefore constrained only to the last 2–3 Ma. The present study indicates that the magnitude of counterclockwise rotation diminishes to 10° or less in the west and south of Anatolia [see also Gürsoy et al. 1998], although it has all occurred during the last 1–2 Ma. This decrease in the amount of rotation is clearly linked to an increase in width, and decrease in strain, across the zone of semiplastic lithosphere comprising the Anatolian

Fig. 10. Outline geological map of the Erciyes sector of the Ecemiş Fault Zone showing magnetic declinations and 95% arcs at sites in the 2.8 Ma İncesu Ignimbrite. 1, Quaternary basin fill deposits; 2, Plio-Quaternary subareal volcanic rocks; 3, Neogene tuffs and pyroclastic rocks with some lava; 4, Pre-Neogene basement.

Table 5. *Summary of palaeomagnetic directions of Cenozoic age derived from studies in central Turkey*

District/fault block	Age*	Location° E	Location° N	N	Palaeomagnetic direction D	Palaeomagnetic direction I	α_{95}	Reference
Results from rock units emplaced during the palaeotectonic regime								
• *North of the NAFZ*								
Mesudiye (N)	E	37.7	40.6	7	165.9	−50.4	6.7	Baydemır 1990
Mesudiye (R)	E	37.7	40.6	8	185.6	−45.1	15.1	Baydemır 1990
Niksar	E	37.0	40.7	10	152.4	−42.5	11.3	Tatar *et al.* 1995
Erbaa	E	36.3	40.8	7	194.6	−48.8	15.3	Piper *et al.* 1996
Mesudiye	E	37.7	40.5	7	166.8	−49.9	9.6	Orbay & Bayburdı 1979
Kusuri	E	35.7	41.5	5	161.2	−48.4	12.9	Sarıbudak 1989
Kastamonu	E	33.7	41.2	3	164.2	−31.7	10.6	Piper *et al.* 1996
• *South of the NAFZ*								
İmranlı	E	38.4	39.8	10	146.0	−34.2	6.3	Baydemır 1990
Almus	E	36.9	40.4	8	144.1	−47.5	7.6	Tatar *et al.* 1995
Kalehisar	E	34.5	40.2	5	160.4	−51.0	5.2	Piper *et al.* 1996
Akdağmadeni	E–M	35.8	39.7	3	133.9	−43.5	5.2	Tatar *et al.* 1996
Yozgat	E	34.7	39.8	5	158.4	−46.5	18.6	Tatar *et al.* 1996
Bayat	E	34.2	40.7	7	172.1	−47.1	7.6	Piper *et al.* 1997
Galatean Region	M	31.8	40.6	15	196.8	−56.5	6.9	Gürsoy *et al.* 1998
Results from rock units emplaced during the neotectonic regime								
CAT[†]	M–Q	36.5	39.8	5	169.7	−47.6	19.9	Gürsoy *et al.* 1997
Yıldızeli	M–Q	36.9	39.8	12	150.9	−50.8	11.9	Gürsoy *et al.* 1997
Şarkışla	M–Q	37.0	39.5	18	147.9	−55.0	10.9	Gürsoy *et al.* 1997
Kangal	M–Q	37.0	39.0	6	129.6	−47.9	15.6	Gürsoy *et al.* 1997
Gürün	P–Q	37.2	38.8	7	157.0	−54.1	14.0	Gürsoy *et al.* 1997
Gemerek (R)	M	36.0	39.2	35	111.9	−60.8	9.6	Krijgsman *et al.* 1996
Gemerek (N)	M	36.0	39.2	41	307.9	32.2	6.7	Krijgsman *et al.* 1996
İnkonak (R)	M	37.0	39.4	30	129.8	−37.3	7.7	Krijgsman *et al.* 1996
İnkonak (N)	M	37.0	39.4	30	307.1	49.6	10.0	Krijgsman *et al.* 1996
Yeniköy (R)	M	36.4	39.1	49	126.3	−33.3	8.8	Krijgsman *et al.* 1996
Yeniköy (N)	M	36.4	39.1	32	329.6	47.7	7.2	Krijgsman *et al.* 1996
Haramiköy (R)	M	31.8	38.5	37	190.1	−40.9	5.3	Krijgsman *et al.* 1996
Haramiköy (N)	M	31.8	39.5	45	2.4	37.2	4.3	Krijgsman *et al.* 1996
Karaman	P	33.2	37.2	13	174.3	−51.8	8.6	Gürsoy *et al.* 1998
Karapınar	Q	33.6	37.5	5	156.9	−42.2	14.9	Gürsoy *et al.* 1998
Karacadağ	P	33.7	37.7	13	177.5	−57.6	7.0	Gürsoy *et al.* 1998
Hasandağı	M	34.2	38.0	7	170.1	−57.6	10.9	Gürsoy *et al.* 1998
Ignimbrites	M–P	34.8	38.6	15	169.3	−47.1	4.4	Unpublished data
E Erciyes	Q	35.6	38.6	7	173.1	−52.5	9.4	This paper
W Erciyes	Q	35.0	38.5	8	170.1	−49.7	13.2	This paper
Cappadocia	Q	34.7	38.3	6	189.7	−53.2	16.0	This paper
Sultansazlığı 1	Q	35.2	38.5	5	202.2	−52.2	16.0	This paper
Sultansazlığı 2	Q	35.5	38.4	5	159.9	−56.4	9.2	This paper

The directions given are the reversed palaeofield directions and the location listed is the centre of the study area. N is the number of separate units included in the calculated mean and α_{95} is the radius of the cone of 95% confidence about the mean direction. * Ages are: C, Upper Cretaceous; E, Eocene; M, Miocene; P, Pliocene; Q, Quaternary. [†] CAT, Central Anatolian Thrust Belt. +, Mean calculated from five sites in Late Cretaceous–Eocene volcanic rocks.

composite terranes as it broadens out to the west between the rigid pincer blocks of Arabia and Eurasia (Fig. 11).

Following Cummings (1976), the fault patterns developed within such a system are related to the pattern developed within a wedge-shaped body (Anatolia) in a Prandtl cell compressed between two oblique rigid forelands (Arabia and Eurasia) by Gürsoy *et al.* (1997) and Barka & Reilinger (1997). The correspondence is not a direct one because the Prandtl cell predicts that the fault systems in Anatolia should be concave

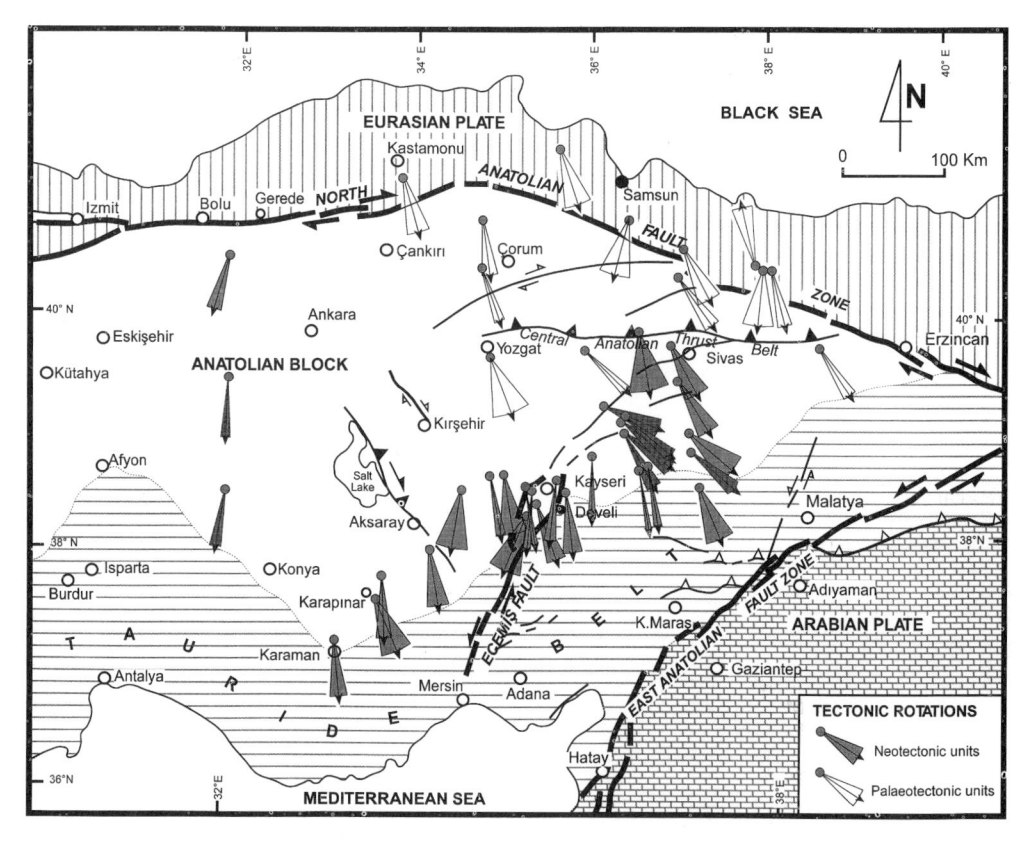

Fig. 11. Group mean palaeomagnetic directions with 95% confidence limits from Late Cretaceous and Cenozoic rock units in central Turkey based on the compilation of Table 4. All vectors are shown as directions of the reversed polarity field (with the exception of the third result from the Reşadiye area for clarity). Note that results from rock units emplaced in the palaeotectonic regime may include rotations imparted during continental collision, whereas rotations of rock units emplaced during the neotectonic regime are assigned to this latter regime only.

to the northwest, whereas in practice they are dominantly convex (Figs 1 and 11). This extension of the Prandtl cell to a lithosphere scale could be complicated by the following points: (1) the compressive/strike slip regime in Anatolia merges into an extensional regime in western Turkey where the wedge is unconstrained; the predictions of the Prandtl cell more closely correspond on the eastern side of the Arabian Promontory (Barka & Reilinger 1997, fig. 12c) where the crust is constrained as in the Prandtl cell; (2) fracture systems may be progressively rotated and/or change their shape as continuing escape modifies the extruded terrane; the curvatures of the Prandtl fractures may then be expected to change progressively during ensuing deformation; (3) the Eurasian Plate incorporates the Black Sea Basin which developed as a back-arc basin behind the subducting Intra-Pontide Ocean in mid-Cretaceous times (Görür et al. 1994) immediately to the north of Turkey; Eurasia north of the NAFZ may not, therefore, be acting as a truly rigid foreland.

The importance of temporal change in the regional stress system with continuing deformation is suggested by the results from Cappadocia. Here, the ignimbrites which erupted between c. 11 and 2 Ma ago have a mean declination of $D = 350°$ (unpublished data); this is recorded in the youngest units and implies that counterclockwise rotation of c. 10° in this southern sector of Anatolia was concentrated within the last 1–2 Ma. The accompanying sinistral double bending at the Erciyes sector of the EFZ has produced pull-apart during this same interval and permitted emplacement of the stratovolcanic complex. The pull-apart is recognized by clockwise rotation of young lavas in the Sultansazlığı

Depression and the İncesu Ignimbrite east of the western boundary fault system.

Whilst counterclockwise rotation is a key feature of continuing motions deduced from the GPS in the Anatolian region (Oral et al. 1995; Barka & Reilinger 1997), the palaeomagnetic results imply that the average short-term rotations of $1.2°$ Ma^{-1} deduced from GPS data are exceeded by an order of magnitude on a regional scale by fault blocks in the c. 10–100 km size range.

This study has been facilitated by a link between Cumhuriyet University of Sivas and the Geomagnetism Laboratory of the University of Liverpool supported by the British Council, the Scientific and Technical Research Council of Turkey (TÜBİTAK), project number 198Y008 and NATO Scientific Affairs Division (CRG.972054). We are very grateful to Rolf Schumacher and Ulricke Mues-Schumacher for identifying sampling sites in the İncesu Ignimbrite for palaeomagnetic study, and to Tony Morris and Arno Patzelt for reviewing of the manuscript. We also thank E. Bozkurt for his great enthusiasm and effort invested in the preparation of this Special Volume and this paper.

References

AMBRASEYS, N. N. 1989. Temporary seismic quiescence, S. E. Turkey. *Geophysical Journal International*, **96**, 411–431.

—— & FINKEL, C. F. 1987. Seismicity of the northeast Mediterranean region during the early 20th century. *Annales Geophysicae*, **5B**, 701–726.

ANDRIEUX, J., ÖVER, S., POISSON, A. & BELLIER, O. 1995. The North Anatolian Fault Zone: distributed Neogene deformation in its northwest convex part. *Tectonophysics*, **243**, 135–154.

BARKA, A. A. & REILINGER, R. 1997. Active tectonics of the Eastern Mediterranean region: deduced from GPS, neotectonic and seismicity data. *Annali Di Geofisica*, **40**, 587–610.

BATUM, İ. 1978. Nevşehir güneybatısındaki Göllüdağ ve Acıgöl yöresi volkanitlerinin jeoloji ve petrografisi. *Hacettepe University Earth Sciences*, **4**, 50–69 [in Turkish with English abstract].

BECK, M. E. 1980. Palaeomagnetic record of plate margin processes along the western edge of North America. *Journal of Geophysical Research*, **87**, 7115–7131.

BAYDEMİR, N. 1990. Palaeomagnetism of the Eocene volcanic rocks in the eastern Black Sea region. *İstanbul University, Faculty of Engineering Journal of Earth Sciences*, **7**, 167–176 [in Turkish].

CUMMINGS, D. 1976. Theory of plasticity applied to faulting, Mojave Desert, Southern California. *Geological Society of America Bulletin*, **87**, 720–724.

DEMAREST, H. H. 1983. Error analysis for the determination of tectonic rotation from palaeomagnetic data. *Journal of Geophysical Research*, **88**, 4321–4328.

GÖRÜR, N., TÜYSÜZ, A., AYKOL, M., SAKINÇ, M., YIĞITBAŞ, E., AKKÖK, R. & YILMAZ, Y. 1994. Cretaceous red pelagic carbonates of northern Turkey: their place in the opening history of the Black Sea. *Eclogae Geologica Helvetica*, **86**, 819–838.

GROMMÉ, S., WRIGHT, T. L. & PECK, D. L. 1969. Magnetic properties and oxidation of iron-titanium oxide minerals in Alea and Makaopuhi lava lakes, Hawaii. *Journal of Geophysical Research*, **74**, 5277–5293.

GÜRSOY, H., PIPER, J. D. A., TATAR, O. & TEMİZ, H. 1997. A palaeomagnetic study of the Sivas Basin, Central Turkey: crustal deformation during lateral extrusion of the Anatolian Block. *Tectonophysics*, **271**, 89–105.

——, ——, —— & MESCI, L. 1998. Palaeomagnetic study of the Karaman and Karapınar volcanic complexes, central Turkey: Neotectonic rotation in the south-central sector of the Anatolian Block. *Tectonophysics*, **299**, 191–211.

HAGGERTY, S. E. 1976 Oxidation of opaque mineral oxides in basalts. *In:* RUMBLE, D. (ed.) *Oxide Minerals.* Mineralogical Society of America, Short Course Notes, **3**, 1–98.

INNOCENTI, F., MAZZUOLI, R., PASQUARE, G., REDICATI DE BROZOLO, F. & VILLARI, L. 1975. The Neogene calc-alkaline volcanism of central Anatolia: geochronological data from the Kayseri–Niğde area. *Geological Magazine*, **112**, 349–360.

JACKSON, J. 1994. Active tectonics of the Aegean region. *Annual Review of Earth and Planetary Science*, **22**, 239–271.

KIRATZI, A. A. 1993. A study on the active crustal deformation of the North and East Anatolian Fault Zones. *Tectonophysics*, **225**, 191–203.

KOÇYİĞİT, A. & BEYHAN, A. 1998. A new intracontinental transcurrent structure: the Central Anatolian Fault Zone, Turkey. *Tectonophysics*, **284**, 317–336.

KRIJGSMAN, W., DUERMEIJER, C. E., LANGEREIS, C. G., BRUIJN, H., SARAÇ, G. & ANDRIESSEN, P. A. M. 1996. Magnetic polarity stratigraphy of late Oligocene to middle Miocene mammal-bearing continental deposits in central Anatolia (Turkey). *Newsletter on Stratigraphy*, **34**, 13–29.

MUES-SCHUMACHER, U. & SCHUMACHER, R. 1996. Problems of stratigraphic correlation and new K-Ar data for ignimbrites from Cappadocia, Central Turkey. *International Geology Review*, **38**, 737–746.

ORAL, M. B., REILINGER, R. E., TOKSÖZ, M. N., KING, R. W., BARKA, A. A., KINIK, İ. & LENK, O. 1995. Global positioning system offers evidence of plate motions in eastern Mediterranean. *EOS, Transactions of the American Geophysical Union*, **76**, 9–11.

ORBAY, N. & BAYBURDİ, A. 1979. Palaeomagnetism of dykes and tuffs from the Mesudiye region and rotation of Turkey. *Geophysical Journal of the Royal Astronomical Society*, **59**, 437–444.

PASQUARE, G., POLI, S., VENZOLLI, L. & ZANCHI, A.

1988. Continental arc volcanism and tectonic setting in Central Anatolia. *Tectonophysics*, **146**, 217–230.

PIPER, J. D. A., TATAR, O. & GÜRSOY, H. 1997. The deformational behaviour of continental lithosphere deduced from block rotations across the North Anatolian Fault Zone in Turkey. *Earth and Planetary Science Letters*, **150**, 191–203.

——, MOORE, J., TATAR, O., GÜRSOY, H. & PARK, R. G. 1996. Palaeomagnetic study of crustal deformation across an intracontinental transform: the North Anatolian Fault Zone in Northern Turkey. *In*: MORRIS, A. & TARLING, D. H. (eds) *Palaeomagnetism of the Mediterranean Regions*. Geological Society, London, Special Publications, **105**, 299–310.

PLATZMAN, E. S., TAPIRDAMAZ, C. & SANVER, M. 1998. Neogene anticlockwise rotation of Anatolia (Turkey): preliminary palaeomagnetic and geochronological results. *Tectonophysics*, **299**, 175–189.

SARIBUDAK, M. 1989. New results and a palaeomagnetic overview of the Pontides in northern Turkey. *Geophysical Journal International*, **90**, 521–531.

ŞENGÖR, A. M. C. 1980. *Türkiye'nin Neotektoniğinin Esasları*. Geological Society of Turkey Conference Series, **2** [in Turkish with English abstract].

——, GÖRÜR, N. & ŞAROĞLU, F. 1985. Strike-slip faulting and related basin formation in zones of tectonic escape: Turkey as a case study. *In*: BIDDLE, K. T. & BLICK, N. C. (eds) *Strike-slip Deformation, Basin Formation and Sedimentation*. Society of Economic Palaeontologists and Mineralogists, Special Publications, **37**, 227–264.

TATAR, O., PIPER, J. D. A., GÜRSOY, H. & TEMİZ, H. 1996. Regional significance of Neotectonic counterclockwise rotation in Central Turkey. *International Geology Review*, **38**, 692–700.

——, ——, PARK, R. G. & GÜRSOY, H. 1995. Palaeomagnetic study of block rotations in the Niksar overlap region of the North Anatolian Fault Zone, Central Turkey. *Tectonophysics*, **244**, 251–266.

TCHALENKO, J. S. 1977. A reconnaissance of the seismicity and tectonics at the northern border of the Arabian Plate (Lake Van Region). *Revue de Géographie Physique et Géologie Dynamique*, **19**, 189–208.

THOMAS, D. N. 1992. *Rock magnetic and palaeomagnetic investigations of the Gardar Lava Succession, South Greenland*. PhD thesis, University of Liverpool.

TOPRAK, V. & GÖNCÜOĞLU, M. C. 1993. Tectonic control on the development of the Neogene–Quaternary Central Anatolian Volcanic Province, Turkey. *Geological Journal*, **28**, 357–369.

WILSON, R. L. 1971. Dipole offset – the time average palaeomagnetic field over the past 25 million years. *Geophysical Journal of the Royal Astronomical Society*, **22**, 481–504.

S–I–A-type intrusive associations: geodynamic significance of synchronism between metamorphism and magmatism in Central Anatolia, Turkey

DURMUŞ BOZTUĞ

Department of Geological Engineering, Cumhuriyet University, TR-58140 Sivas, Turkey (e-mail: boztug@cumhuriyet.edu.tr)

Abstract: The Central Anatolian crystalline region, comprising metasediments, ophiolitic slabs and numerous intrusives, has been called the Kırşehir Massif, the Kırşehir Block or Central Anatolian Crystalline Complex. Intrusive associations in Central Anatolia are summarized as: (1) a syncollisional, S (or C_{ST})-type, peraluminous and two-mica leucogranitic association; (2) a post-collisional, I (or H_{LO})-type, metaluminous, high-K calc-alkaline, typically K-feldspar megacrystic monzonitic association; and (3) an A-type, post-collisional and within-plate alkaline association comprising a high-K and silica-oversaturated alkaline, K-feldspar megacrystic monzonitic and syenitic subgroup, and a silica-undersaturated alkaline feldspathoid–sodalite syenite porphyry subgroup.

Radiometric data suggest that metamorphism and magmatism in Central Anatolia were synchronous during the Late Cretaceous. This implies that metamorphism may have been generated by inverted metamorphism induced by the Anatolide–Pontide collision and was linked to a decreasing metamorphic grade from the north (i.e. from the main suture zone) towards the south (i.e. Taurides). Collision-related intrusive rocks within this crystalline body reflect differences in geological setting, mineralogical–chemical composition and associated ore deposition. Published data considering the distribution and timing of metamorphism and magmatism in the Central Anatolia suggest that both are related to the Anatolide–Pontide collision along the İzmir–Ankara–Erzincan Suture Zone in the Late Cretaceous. Various magmatic episodes affected along the passive margin of the Anatolides, including a syncollisional peraluminous episode, a post-collisional calc-alkaline hybrid and a post-collisional within-plate alkaline episode.

The juxtaposition of two plates by collision at the end of an orogenic cycle takes several tens of millions of years [c. 30–50 Ma according to Bonin (1990)]. This time span is particularly characterized by a distinctive geological record that embraces magmatism, metamorphism and sedimentation (Coward & Ries 1986). Turkey constitutes an important part of the Alpine–Himalayan collision system. It is widely recognized that the Neotethyan oceanic realm opened in this region during the Triassic and closed in the Late Cretaceous (Şengör & Yılmaz 1981; Poisson 1986). This has been the important post-Palaeozoic convergent system responsible for the evolution of Turkey. The Late Cretaceous subduction and the subsequent collision of the Neotethyan oceanic crust with the Rhodope–Pontide fragment resulted in the formation of the İzmir–Ankara–Erzincan Suture Zone, the eastern Pontide arc magmatism, and crustal thickening and metamorphism during Late Campanian–Maastrichtian times (Şengör & Yılmaz 1981). Post-collisional structures include the Sivas Basin (Cater *et al.* 1991; Yılmaz 1994) and other Central Anatolian basins (Görür *et al.* 1984); collision-related magmatism includes granitoides (Akıman *et al.* 1993; Boztuğ *et al.* 1994; Göncüoğlu & Türeli 1994; Erler & Bayhan 1995; Erler & Göncüoğlu 1996; İlbeyli & Pearce 1997). Recent studies (Alpaslan & Boztuğ 1997; Ekici & Boztuğ 1997; Boztuğ 1998; Tatar & Boztuğ 1998) show that the collision-related granitoids can also be subdivided into some associations on the basis of temporal and spatial distribution related to their geological–geodynamic setting, and their mineralogical–chemical characteristics. These include S-type (Chappell & White 1974, 1992; White & Chappell 1988) or C_{ST}-type (crustal-shearing, thrusting type; Barbarin 1990), a syncollisional (Harris *et al.* 1986) peraluminous leucogranitic association; an I-type (Chappell & White 1974, 1992; Chappell & Stephens 1988) or H_{LO}-type (hybrid-late orogenic type; Barbarin 1990), a post-collisional (Harris *et al.* 1986) high-K calc-alkaline monzonitic association; and an A-type (Collins *et al.* 1982; Whalen *et al.* 1987), post-collisional, within-plate alkaline association. A concomitant Late Cretaceous regional metamorphic event is proposed by various authors from the global geological setting (Şengör & Yılmaz 1981) and K–Ar radiometric

From: BOZKURT, E., WINCHESTER, J. A. & PIPER, J. D. A. (eds) *Tectonics and Magmatism in Turkey and the Surrounding Area.* Geological Society, London, Special Publications, **173**, 441–458. 1-86239-064-9/00/$15.00 © The Geological Society of London 2000.

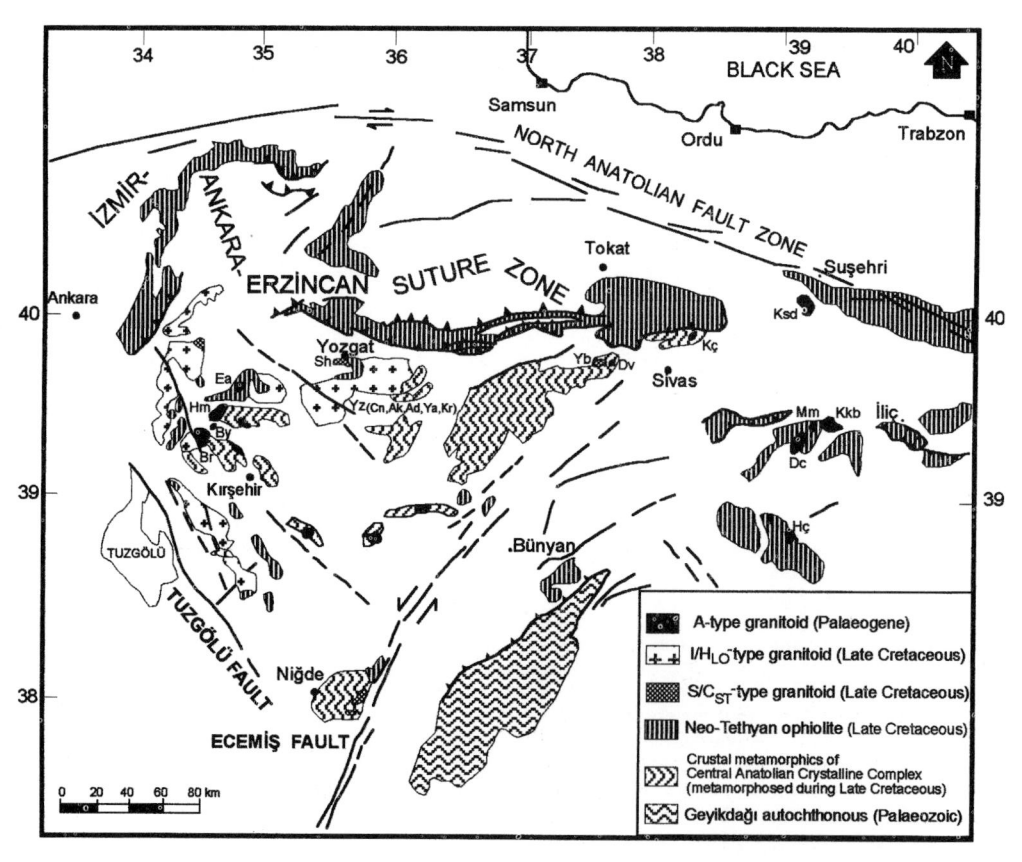

Fig. 1. Simplified distribution of plutonic and metamorphic rocks in Central Anatolia, Turkey [modified after Bingöl (1989)]. The abbreviations for plutons (from west to east) are: Hm, Hamit; By, Bayındır; Br, Baranadağ; Ea, Eğrialan; Sh, Sarıhacılı; Yz (Cn, Ak, Ad, Ya, Kr), composite Yozgat Batholith consisting of Cankılı Monzodiorite, Akçakoyunlu quartz Monzodiorite, Adatepe quartz Monzonite, Yassıağıl Monzogranite, Karakaya Monzogranite); Yb, Yücebaca; Dv, Davulalan; Kç, Karaçayır; Ksd, Kösedağ; Dc, Dumluca; Mm, Murmana; Kkb, Karakeban; Hç, Hasançelebi.

data (Erkan & Ataman 1981; Göncüoğlu 1986; Alpaslan et al. 1996) from different parts of Central Anatolia.

This paper deals essentially with the main mineralogical and whole-rock major and trace element geochemical characteristics of the S–I–A-type magmatic rock associations and the geodynamic significance of the Late Cretaceous synchronization of these collisional-related granitoids and metamorphism in Central Anatolia. All the geochemical analyses data have been obtained at the Mineralogical–Petrographical and Geochemical Research Laboratories (MIPJAL) of the Department of Geological Engineering of Cumhuriyet University in Sivas, Turkey, with a Rigaku E-WDS-3270 X-ray fluorescence spectrometer. Both the major and trace element concentrations have been analysed with pressed pellet pastilles produced by a hydraulic press sample preparation apparatus, working with a pressure of 12 tonnes, following mixture of 10 g of rock powder and 10 drops of water, including 10% dissolved polyvinylpyrolidon [poly (1-vinyl-2-pyrolidon)] binding material. Calibration was conducted using USGS and CRPG rock standards (Govindaraju 1989).

General overview to metamorphism and collision-related intrusives in Central Anatolia

Many studies have been carried out on the metamorphism in Central Anatolia (Fig. 1) by different authors. Among the most important

Table 1. *The geochronology data of intrusive and metamorphic rocks in Central Anatolia, Turkey*

Studied rock unit	Method	Age (Ma)	Reference
Intrusive rocks			
Baranadağ Q Monzonite	Total Pb in zircon mineral	54	Ayan (1963)
Cefalıkdağ Pluton	Rb–Sr whole-rock biotite isochrone	71	Ataman (1972)
Kösedağ Batholith	Rb–Sr whole-rock isochrone	42 ± 4	Kalkancı (1974)
Üçkapılı Granitoid	Rb–Sr whole-rock isochrone	95 ± 11	Göncüoğlu (1986)
	Rb–Sr whole-rock biotite isochrone	77.8 ± 1.2	Göncüoğlu (1986)
	K–Ar biotite	74.9 ± 1.2	Göncüoğlu (1986)
	K–Ar biotite	76.2 ± 1.2	Göncüoğlu (1986)
	K–Ar muscovite	78.5 ± 1.2	Göncüoğlu (1986)
Bayındır neph. Syenite	Rb–Sr whole-rock isochrone	70.7 ± 1.1	Gündoğdu *et al.* (1988)
Murmana Pluton	Rb–Sr whole-rock isochrone	110 ± 5	Zeck & Ünlü (1987)
Ağaçören Granitoid	Rb–Sr whole-rock isochrone	108 ± 3	Güleç (1994)
Metamorphic rocks			
Kalkanlıdağ–Kırşehir region	K–Ar mineral (biotite, amphibole)	69–74	Erkan & Ataman (1981)
Niğde Massif	K–Ar mineral (biotite)	76–77	Göncüoğlu (1986)
Yıldızeli–Sivas region	K–Ar mineral (biotite, muscovite)	68–77	Alpaslan *et al.* (1996)

studies are: Erkan (1976, 1977, 1978, 1981); Seymen (1984); Tolluoğlu (1987); and Tolluoğlu & Erkan (1989). These studies conclude that the protolith comprised Palaeozoic sedimentary units, sometimes intercalated with basic igneous rocks before metamorphism. Metamorphic units have been identified, in stratigraphic order, as the Kalkanlıdağ Formation, the Kargasekmez Quartzite Member, the Naldökendağ Formation and the Bozçaldağ Formation; these are considered to be part of the Kırşehir Block (KB) (Görür *et al.* 1984; Poisson 1986) and the Central Anatolian Crystalline Complex (CACC) (Göncüoğlu *et al.* 1991). Most of these studies identify a decrease in metamorphic grade from north to south in Central Anatolia. They also suggest two different metamorphic episodes: the first is a progressive regional dynamothermal deformation stage, ranging from greenschist to amphibolite facies, and the second is a retrograde cataclastic process related to uplift.

Widespread magmatic intrusive associations are emplaced into the metamorphic rocks of the CACC, the pre-Maastrichtian Central Anatolian Ophiolite (Yalınız *et al.* 1996) and other units of Cretaceous–Early Tertiary age. They are unconformably covered by Eocene or younger sedimentary units in Central Anatolia (Fig. 1). Hence, these plutons are considered to have been emplaced in a time interval between the Late Cretaceous (Maastrichtian) and the Eocene.

Radiometric age determinations on some of these plutons and crustal metamorphic rocks in Central Anatolia, by various authors, are summarized in Table 1. On the basis of existing literature, the Central Anatolian intrusive rocks may be subdivided into pre-orogenic to synorogenic to post-orogenic suites, comprising: (1) some gabbroic rocks and oceanic plagiogranites (Göncüoğlu & Türeli 1993; Yalınız *et al.* 1996); (2) S-type or C_{ST}-type, syncollisional, peraluminous, two-mica granites (the Üçkapılı Pluton – Göncüoğlu 1986; the Yücebaca Pluton – Alpaslan & Boztuğ 1997; the Sarıhacılı Leucogranite – Ekici & Boztuğ 1997); (3) calcalkaline, alumino-cafemic to cafemic, collision-related plutons displaying I-type and S-type features (the Cefalıkdağ Pluton; Geven 1995); (4) an I-type or H_{LO}-type, high-K calc-alkaline, typically K-feldspar megacrystalline post-collisional monzonitic association in the composite Yozgat Batholith (Tatar & Boztuğ 1998); (5) A-type, post-collisional silica-oversaturated alkaline, syenitic–monzonitic and monzogabbroic–monzodioritic plutons (including the Hasançelebi Pluton – Yılmaz *et al.* 1993; the Kösedağ Pluton–Boztuğ *et al.* 1994; the Karaçayır Pluton–Boztuğ *et al.* 1996; the Dumluca, Murmana and Karakeban Plutons–Boztuğ *et al.* 1997; the Davulalan Pluton – Alpaslan & Boztuğ 1997; the İdişdağ Syenite–Göncüoğlu *et al.* 1997; the Baranadağ quartz Monzonite, and the Hamit and Çamsarı quartz Syenites – Otlu & Boztuğ 1998) and silica-undersaturated alkaline plutons [including some parts of the Karaçayır and Davulalan Plutons, and the Eğrialan Syenite – Boztuğ 1998; the Hayriye nepheline Syenite (Kayseri–Felahiye) – Özkan & Erkan

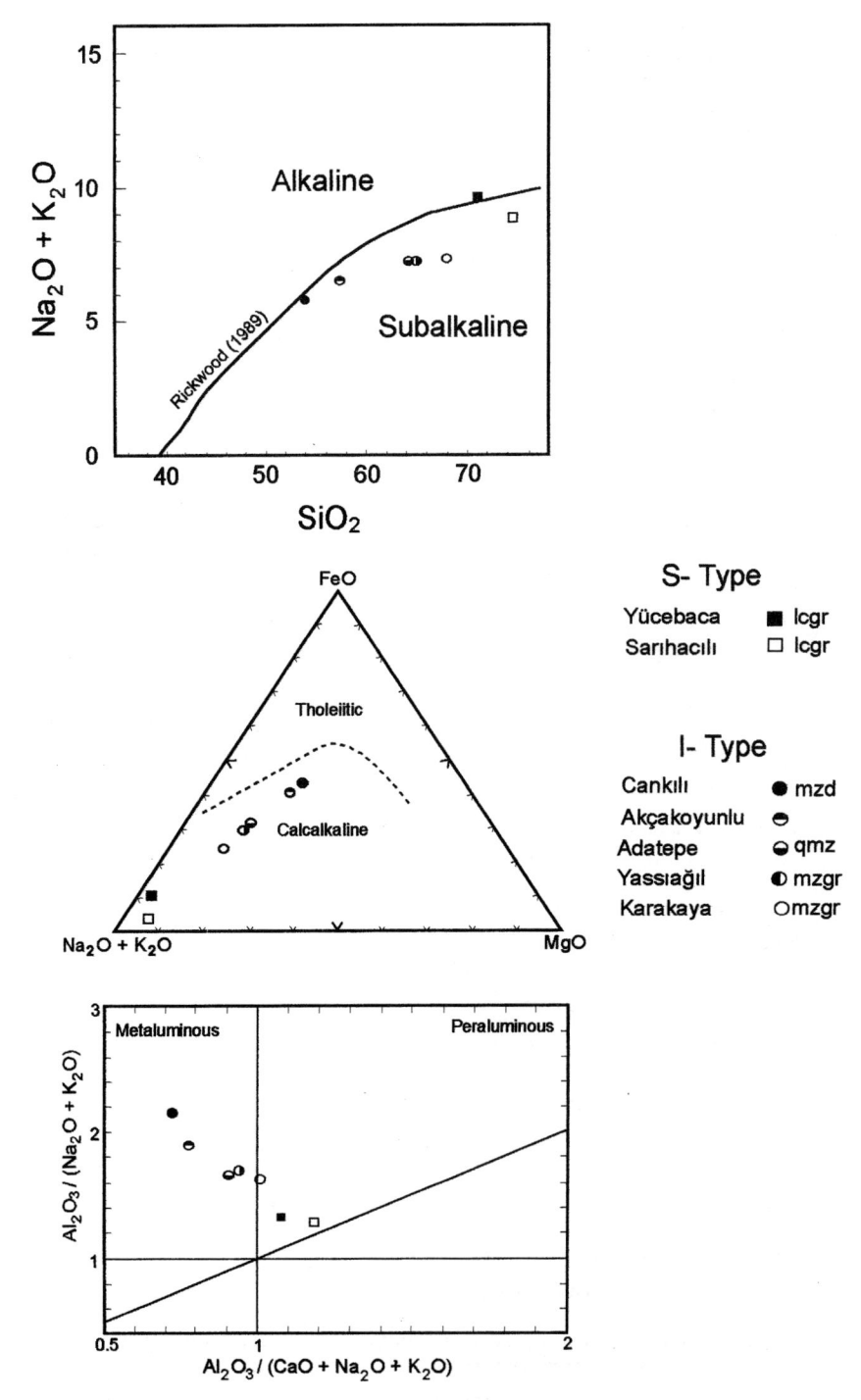

Fig. 2. Total alkali v. silica diagram [dividing line has been taken from Rickwood (1989)], AFM ternary diagram (Irvine & Baragar 1971) and $Al_2O_3/(Na_2O + K_2O)$ v. $Al_2O_3/(CaO + Na_2O + K_2O)$ diagram (Maniar & Piccoli 1989) of the S-type and I-type intrusive associations.

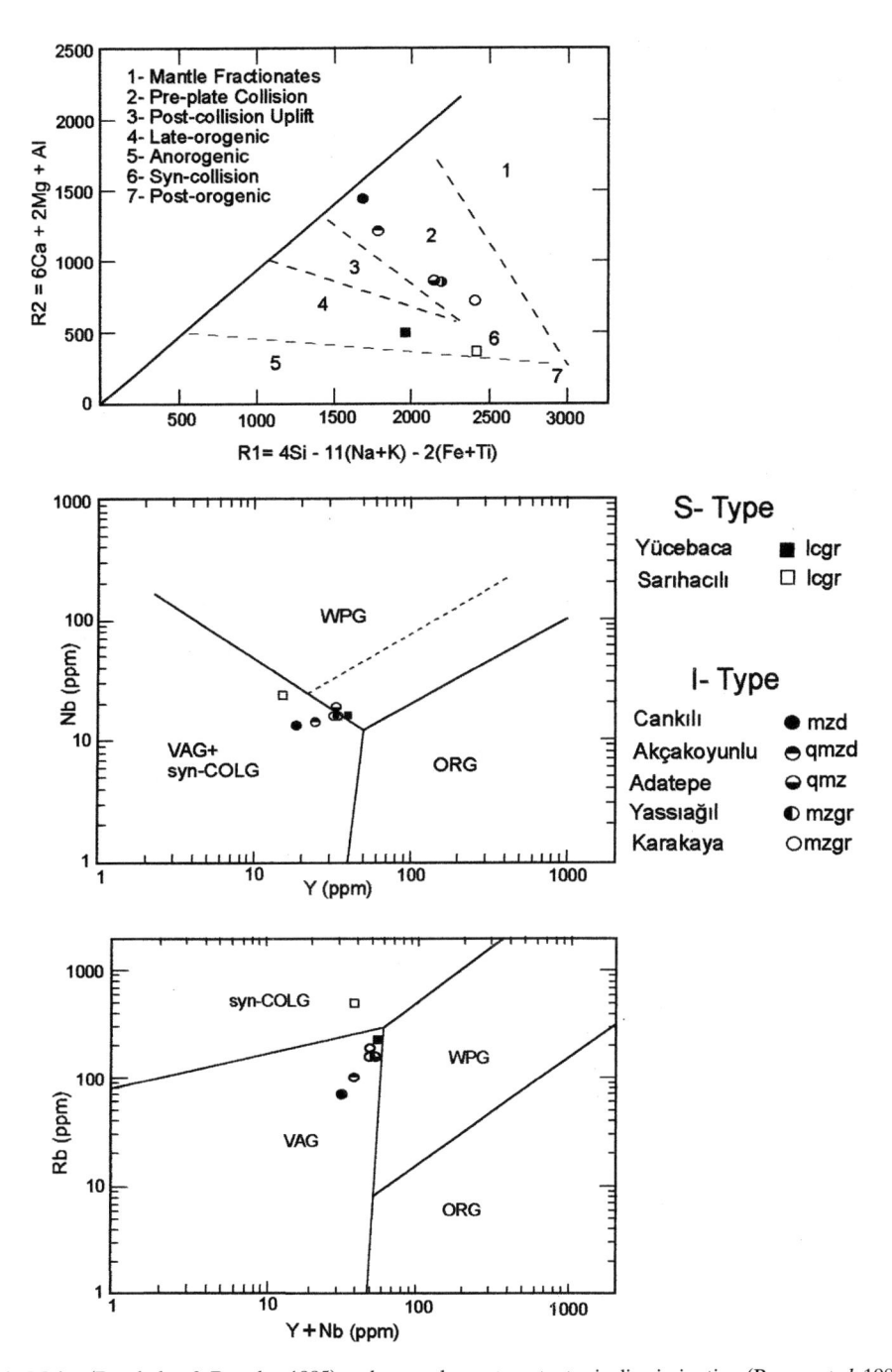

Fig. 3. Major (Batchelor & Bowden 1985) and trace element geotectonic discrimination (Pearce *et al.* 1984) diagrams for the S-type and I-type intrusive associations.

1994; the Durmuşlu nepheline–nosean–melanite Syenite porphyry and the Bayındır nepheline cancrinite Syenite – Otlu & Boztuğ 1998]; and (6) post-collisional mafic gabbroic/dioritic plutons (the Yıldızdağ Pluton, Yıldızeli–Sivas region; Boztuğ *et al.* 1998).

S-type intrusive association

The Yücebaca and Sarıhacılı Leucogranites typically represent S-type mineralogy and geochemistry (Alpaslan & Boztuğ 1997; Ekici & Boztuğ 1997) among the Central Anatolian collision-related intrusive associations. The Yücebaca Pluton is exposed within medium- to high-grade metasediments of the CACC in the Yıldızeli district of the Sivas region (Alpaslan & Boztuğ 1997). The Sarıhacılı Leucogranite constitutes part of the composite Yozgat Batholith in Central Anatolia (Ekici & Boztuğ 1997; Tatar & Boztuğ 1998). There is no stratigraphic or radiometric evidence for the ages of these leucogranites, although they are assumed to be the oldest granitic rocks in Central Anatolia by various authors (Erler & Bayhan 1995; Erler & Göncüoğlu 1996; Ekici & Boztuğ 1997; Boztuğ 1998). Both plutons consist of pinkish-grey aphanitic–granular leucogranites composed of quartz, orthoclase, albite–oligoclase and small amounts of biotite and muscovite. These rocks are also called two-mica leucogranites on the basis of their mineralogical composition (Alpaslan & Boztuğ 1997; Ekici & Boztuğ 1997). Major element chemical data (Table 2) show an apparent peraluminous, calc-alkaline (Fig. 2) aluminium saturation index (ASI) value [equivalent to molecular $Al_2O_3/(Na_2O + K_2O + CaO)$; see White & Chappell (1988)] > 1.1 and normative corundum. These are characteristics of S-type granites as described by Chappell & White (1974, 1992) and White & Chappell (1988). These types of granites are also called C_{ST}-type, an abbreviation of crustal shearing and thrusting determined by Barbarin (1990), and are attributed to shearing associated with thrusting during the crustal thickening process. The genetic classification of Barbarin (1990) seems to accord with the S-type classification of the Australian (Chappell and White) school (Chappell & White 1974, 1992; White & Chappell 1988; Chappell 1984, 1996) who propose that supracrustal-derived granitic melts are generated by ultrametamorphism during crustal thickening. The R1–R2 geotectonic classification diagram, based on major element data described by Batchelor & Bowden (1985) and trace elements (Pearce *et al.* 1984), shows a syncollisional crustal origin for the Yücebaca and Sarıhacılı Leucogranites in Central Anatolia (Fig. 3). Such an origin is also supplied in the mid-ocean ridge basalt (MORB)-normalized spider diagram which shows a considerable enrichment in the contents of Rb, Ba and K, and a depletion in Ti (Fig. 4). On the other hand, Sr shows an enrichment in the Yücebaca Pluton and a depletion in the Sarıhacılı Pluton, which could be related to secondary removal resulting from feldspar alteration or to low-temperature feldspar fractionation. The latter interpretation is more consistent with the observation that the Sarıhacılı Leucogranite contains more Rb, Ba and K than the Yücebaca Pluton, which could reflect low-temperature feldspar fractionation (Bonin 1987; Chappell 1996).

I-type intrusive association

This association constitutes part of the composite Yozgat Batholith in Central Anatolia (Fig. 1). It is composed of five mappable subunits called the Cankılı Monzodiorite, the Akçakoyunlu Quartz Monzodiorite, the Adatepe quartz Monzonite, the Yassıağıl Monzogranite and the Karakaya Monzogranite (Tatar & Boztuğ 1998). The most distinct feature of these subunits is that they all, except for the Karakaya Monzogranite, contain large K-feldspar megacrysts, visible even in hand specimens, and augite, hornblende and biotite as the major mafic. They represent the metaluminous and calc-alkaline trends of subalkaline composition (Fig. 2) with ASI values < 1.1 and include normative diopside.

These mineralogical and chemical features are considered to be evidence of I-type (Chappell & White 1974, 1992; Chappell & Stephens 1988) or H_{LO}-type (hybrid-late orogenic type; Barbarin 1990) intrusive rocks. Some detailed studies suggest that all these subunits were solidified from a single magma derived by both mixing and mingling types of interaction between coeval underplating mafic magma and crustal-derived felsic magma (Boztuğ 1998; Tatar & Boztuğ 1998). Some solidification processes, like fractional crystallization (FC) and assimilation fractional crystallization (AFC), are proposed during solidification of this hybrid magma source to yield different subunits showing a well-preserved reverse zoning in the composite Yozgat Batholith (Tatar & Boztuğ 1998). The FC process is clearly observed in the rock/MORB spider diagram, indicating that the first cooling product is the Cankılı Monzodiorite and the last one is the Karakaya Monzogranite (Fig. 4). Such derivation from both mantle and crustal source materials for the I-type or H_{LO}-type

Table 2. *Averages and standard deviations of the whole-rock major (wt%) and trace element (ppm) chemical compositions of the collision-related Central Anatolian S–I–A-type magmatic associations*

Pluton*	Rock type†	n‡	SiO$_2$	Al$_2$O$_3$	TiO$_2$	Fe$_2$O$_3$§	MnO	MgO	CaO	Na$_2$O	K$_2$O	P$_2$O$_5$	LOI¶	Total	Co	Cu	Pb	Zn	Rb	Ba	Sr	Nb	Zr	Y	Th	Ga
Yb	S/lcgr	6	70.79 (1.86)	16.56 (0.56)	0.24 (0.13)	1.31 (0.81)	0.01 (0.02)	0.33 (0.22)	1.56 (0.68)	3.88 (0.25)	5.70 (0.62)	0.07 (0.04)	0.64 (0.36)	101.34 (0.83)	na	na	na	na	229 (55)	1135 (268)	799 (171)	16 (5)	258 (69)	39 (7)	na	na
Sh	S/lcgr	3	74.49 (0.32)	14.97 (0.21)	0.03 (0.01)	0.47 (0.14)	0.02 (0.02)	0.55 (0.02)	0.50 (0.17)	4.02 (0.29)	4.75 (0.23)	0.01 (0.01)	0.48 (0.42)	100.30 (0.46)	82 (6)	12 (1)	54 (11)	36 (12)	511 (36)	nd	22 (10)	24 (6)	65 (12)	15 (11)	28 (5)	17 (1)
Cn	I/md	2	53.85 (0.35)	17.10 (0)	0.95 (0.07)	7.75 (0.78)	0.10	3.30 (0.28)	8.60 (0.14)	3.00 (0.14)	2.80 (0.28)	0.45 (0.07)	0.75 (0.07)	98.65 (0.07)	27 (2)	16 (4)	24 (3)	107 (2)	70	1137 (445)	888 (32)	13 (1)	218 (13)	18 (0)	14 (10)	20 (2)
Ak	I/qmd	7	57.31 (1.57)	16.34 (0.72)	0.79 (0.1)	7.30 (0.53)	0.13 (0.04)	3.08 (0.29)	6.81 (0.95)	2.87 (0.24)	3.60 (0.29)	0.32 (0.05)	0.76 (0.34)	99.30 (0.86)	25 (2)	19 (11)	32 (5)	100 (3)	100 (6)	1202 (203)	707 (32)	14 (3)	220 (31)	24 (4)	14 (5)	18 (2)
Ad	I/qmz	19	64.01 (2.58)	15.51 (0.57)	0.53 (0.1)	4.79 (1.29)	0.09 (0.02)	1.98 (0.46)	4.24 (1.04)	2.76 (0.14)	4.44 (0.74)	0.19 (0.04)	0.68 (0.29)	99.22 (0.95)	16 (2)	9 (7)	47 (12)	86 (3)	156 (12)	1025 (203)	602 (204)	19 (3)	229 (31)	32 (4)	27 (17)	19 (1)
Ya	I/mzgr	27	64.83 (2.03)	15.90 (0.57)	0.53 (0.06)	4.28 (0.48)	0.09 (0.01)	1.82 (0.24)	4.13 (0.56)	2.81 (0.26)	4.42 (0.28)	0.19 (0.04)	0.70 (0.41)	99.67 (0.94)	14 (1)	9 (7)	47 (12)	84 (10)	155 (37)	1117 (280)	682 (113)	16 (5)	230 (21)	31 (5)	23 (17)	19 (1)
Ka	I/mzgr	7	67.89 (0.74)	15.44 (0.34)	0.48 (0.03)	3.18 (0.17)	0.05 (0.01)	1.44 (0.11)	3.22 (0.29)	2.70 (0.06)	4.65 (0.13)	0.14 (0.02)	0.65 (0.20)	99.82 (1.12)	11 (1)	8 (3)	44 (8)	76 (1)	187 (14)	964 (166)	603 (90)	16 (2)	236 (16)	33 (1)	22 (14)	20 (2)
Dcf	A/qmz	8	65.83 (1.10)	17.18 (0.45)	0.44 (0.17)	3.26 (0.45)	0.05 (0.02)	0.92 (0.25)	1.99 (0.34)	5.02 (0.38)	5.04 (0.33)	0.16 (0.05)	0.74 (0.26)	100.63 (0.66)	11 (1)	7 (1)	23 (7)	61 (8)	172 (7)	643 (71)	211 (39)	51 (2)	280 (10)	39 (5)	24 (16)	22 (1)
Dcm	A/mgo	8	52.28 (1.88)	16.35 (1.36)	1.49 (0.69)	7.70 (1.14)	0.12 (0.01)	6.15 (1.80)	8.60 (1.06)	3.47 (0.80)	2.70 (0.53)	0.30 (0.10)	1.71 (0.70)	100.85 (0.44)	24 (3)	53 (28)	11 (2)	72 (5)	71 (18)	437 (76)	396 (92)	33 (9)	141 (52)	20 (4)	16 (14)	na
Mmf	A/qmz	8	65.00 (2.11)	16.86 (0.55)	0.49 (0.17)	3.36 (0.79)	0.04 (0.02)	1.64 (0.55)	2.84 (0.51)	4.58 (0.21)	4.69 (0.32)	0.21 (0.05)	0.72 (0.14)	100.44 (0.61)	11 (3)	5 (2)	13 (4)	51 (8)	159 (23)	959 (140)	306 (40)	24 (3)	216 (27)	33 (4)	12 (4)	–
Mmm	A/nmgo	3	46.90 (2.99)	17.08 (1.41)	1.70 (1.40)	7.30 (3.82)	0.37 (0.43)	6.60 (0.78)	13.59 (3.83)	3.18 (1.12)	2.10 (0.96)	0.46 (0.40)	2.52 (0.42)	101.70 (0.31)	23 (11)	267 (426)	14 (3)	63 (17)	78 (42)	723 (574)	488 (147)	32 (29)	140 (35)	21 (9)	16 (7)	14 (1)
Kkf	A/mz	2	63.58 (0.75)	17.42 (0.39)	0.59 (0.06)	4.10 (0.06)	0.07 (0.21)	0.73 (0.91)	2.04 (0.35)	5.87 (0.76)	4.54 (0.08)	0.19 (0.02)	0.69 (0.11)	99.81 (0.11)	na	na	na	na	na	na	na	na	na	na	na	na
	A/sy	3	62.17 (0.64)	17.40 (0.09)	0.60 (0.07)	5.16 (0.09)	0.10 (0.02)	0.75 (0.47)	1.54 (0.49)	5.94 (0.36)	5.68 (0.08)	0.12 (0.05)	0.95 (0.48)	100.42 (1.05)	17 (1)	3 (2)	20 (7)	95 (12)	179 (20)	685 (501)	118 (44)	81 (14)	265 (58)	47 (7)	12 (4)	24 (1)
	A/qs	6	67.69 (1.58)	16.21 (0.82)	0.33 (0.07)	3.29 (0.43)	0.05 (0.01)	0.60 (0.08)	1.05 (0.12)	4.84 (0.20)	5.87 (0.31)	0.11 (0.03)	0.63 (0.24)	100.65 (0.82)	10 (1)	3 (1)	22 (3)	74 (9)	213 (35)	385 (192)	92 (30)	65 (24)	363 (59)	50 (8)	17 (5)	24 (1)
Kkm	A/nmgo	3	49.10 (3.84)	17.60 (0.97)	2.00 (0.13)	10.89 (1.30)	0.15 (0.04)	3.87 (1.62)	5.34 (1.37)	5.52 (0.47)	2.74 (0.76)	0.58 (0.02)	2.20 (1.80)	99.98 (0.55)	na	na	–	–	–	–	–	na	–	–	–	–

Table 2. Continued

Pluton*	Rock type†	n‡	SiO₂	Al₂O₃	TiO₂	Fe₂O₃t§	MnO	MgO	CaO	Na₂O	K₂O	P₂O₅	LOI¶	Total	Co	Cu	Pb	Zn	Rb	Ba	Sr	Nb	Zr	Y	Th	Ga
Kd	A/sy	40	60.18 (2.10)	17.47 (0.52)	0.62 (0.07)	4.75 (0.93)	0.11 (0.03)	2.03 (0.72)	3.04 (0.67)	4.37 (0.40)	5.83 (0.54)	0.29 (0.07)	1.08 (1.04)	99.78 (1.15)	37 (22)	76 (42)	36 (26)	100 (51)	168 (42)	714 (196)	515 (162)	21 (8)	244 (78)	46 (15)	16 (10)	na
	A/qs	30	62.72 (2.26)	17.22 (0.38)	0.54 (0.07)	3.85 (0.73)	0.10 (0.03)	1.60 (0.76)	2.12 (0.57)	4.42 (0.29)	6.18 (0.52)	0.21 (0.06)	0.87 (0.49)	99.82 (1.36)	57 (31)	55 (9)	38 (10)	94 (18)	222 (43)	547 (147)	355 (133)	28 (7)	380 (159)	59 (15)	26 (10)	—
	A/mz	2	56.57 (1.51)	17.37 (0.74)	0.73 (0.00)	6.33 (0.20)	0.12 (0.01)	2.76 (0.13)	4.85 (1.31)	3.92 (0.02)	4.61 (0.58)	0.40 (0.06)	1.02 (0.26)	98.66 (0.16)	47 (2)	95 (31)	30 (2)	89 (2)	119 (47)	777 (26)	781 (72)	23 (2)	208 (14)	39 (6)	14 (7)	na
	A/qmz	7	60.11 (2.57)	17.19 (0.58)	0.61 (0.10)	4.98 (0.92)	0.11 (0.02)	2.35 (0.64)	3.29 (0.60)	4.26 (0.48)	5.12 (0.27)	0.29 (0.07)	1.71 (1.83)	100.00 (1.04)	35 (24)	78 (25)	31 (6)	94 (12)	155 (13)	658 (117)	54 (57)	18 (6)	234 (31)	39 (6)	13 (4)	—
Hç	A/sy	2	64.36 (1.18)	17.68 (0.91)	0.71 (0.00)	0.70 (0.17)	0.06 (0.02)	0.48 (0.55)	1.66 (1.20)	4.79 (0.43)	7.19 (0.22)	0.03 (0.00)	1.71 (1.39)	99.34 (0.90)	22 (8)	32 (4)	29 (2)	64 (4)	95 (17)	2804 (1133)	376 (143)	60 (14)	510 (200)	47 (3)	31 (9)	78 (2)
	A/qs	4	65.48 (1.78)	15.89 (0.72)	0.54 (0.09)	1.07 (0.19)	0.14 (0.19)	0.29 (0.18)	1.67 (1.23)	4.19 (0.64)	6.86 (0.88)	0.11 (0.06)	2.20 (1.02)	98.66 (0.49)	23 (14)	30 (4)	30 (1)	69 (4)	123 (5)	1225 (137)	92 (49)	72 (25)	484 (306)	59 (12)	47 (15)	68 (8)
Kç	A/sy	13	61.77 (2.06)	19.80 (1.52)	0.14 (0.09)	1.21 (0.88)	0.06 (0.04)	0.12 (0.08)	0.49 (0.24)	4.58 (1.48)	8.62 (1.83)	0.04 (0.02)	1.81 (0.62)	98.62 (0.50)	19 (5)	32 (6)	43 (21)	83 (21)	182 (39)	1622 (1163)	1356 (1188)	49 (22)	546 (388)	44 (8)	78 (43)	na
Dv	A/qmz	2	69.53 (0.15)	17.78 (1.24)	0.11 (0.01)	0.82 (0.85)	0.01 (0.01)	0.28 (0.01)	1.83 (0.40)	5.22 (0.34)	4.92 (0.37)	0.03 (0.00)	0.86 (0.26)	101.37 (0.05)	5 (3)	31 (2)	31 (3)	64 (2)	164 (13)	1409 (114)	1351 (234)	24 (12)	335 (101)	26 (2)	63 (28)	5 (3)
	A/qs	2	67.21 (1.21)	17.15 (0.22)	0.26 (0.11)	1.63 (0.10)	0.02 (0.02)	0.19 (0.27)	1.06 (0.76)	5.33 (0.25)	6.51 (0.33)	0.05 (0.05)	0.96 (0.42)	100.35 (0.81)	100 (80)	29 (2)	35 (3)	73 (5)	350 (3)	497 (181)	294 (103)	40 (1)	270 (181)	58 (2)	82 (9)	48 (8)
Eğ	A/sy	9	61.40 (1.31)	19.06 (0.55)	0.44 (0.10)	3.20 (0.96)	0.10 (0.04)	0.98 (0.28)	2.90 (1.30)	4.77 (0.36)	6.64 (0.59)	0.23 (0.23)	0.73 (0.50)	100.46 (1.26)	10 (3)	8 (3)	93 (26)	103 (13)	226 (48)	1089 (381)	842 (367)	36 (14)	524 (205)	46 (8)	40 (25)	na
Br	A/sy	2	60.70 (2.33)	18.93 (0.33)	0.39 (0.05)	3.74 (1.07)	0.09 (0.01)	1.17 (0.40)	3.84 (0.76)	3.48 (0.55)	7.48 (0.30)	0.19 (0.08)	0.55 (0.08)	100.52 (0.61)	13 (4)	12 (3)	38 (2)	98 (6)	206 (14)	858 (468)	111 (154)	32 (6)	265 (58)	41 (8)	24 (11)	18 (0)
	A/qmz	9	63.04 (1.43)	18.03 (0.06)	0.42 (0.04)	3.94 (0.28)	0.11 (0.01)	1.73 (0.23)	3.81 (0.54)	3.80 (0.15)	5.05 (0.20)	0.18 (0.03)	0.53 (0.12)	100.63 (0.59)	13 (1)	10 (3)	29 (16)	93 (1)	155 (17)	1062 (195)	558 (94)	18 (1)	234 (7)	34 (0)	31 (18)	19 (1)
	A/qs	6	63.82 (1.21)	17.97 (0.39)	0.40 (0.03)	3.63 (0.25)	0.10 (0.01)	1.51 (0.24)	3.33 (0.32)	3.82 (0.34)	6.12 (1.05)	0.19 (0.01)	0.47 (0.24)	100.85 (0.24)	12 (8)	15 (3)	39 (13)	91 (6)	202 (32)	1091 (250)	653 (93)	23 (3)	276 (31)	41 (5)	31 (15)	19 (2)
Hm	A/qmz	6	61.48 (0.93)	18.22 (0.45)	0.45 (0.05)	4.33 (0.47)	0.09 (0.01)	1.69 (0.27)	4.24 (0.33)	3.47 (0.17)	4.86 (0.27)	0.17 (0.02)	1.03 (0.44)	100.03 (1.12)	15 (1)	11 (3)	54 (30)	78 (2)	137 (16)	2283 (250)	741 (42)	16 (6)	256 (8)	27 (2)	25 (13)	19
	A/qs	9	64.34 (1.61)	18.00 (0.31)	0.34 (0.08)	3.13 (0.88)	0.08 (0.02)	1.26 (0.35)	2.88 (0.81)	4.05 (0.41)	6.02 (0.49)	0.16 (0.06)	0.59 (0.22)	100.84 (0.45)	10 (3)	9 (3)	53 (15)	81 (7)	249 (57)	923 (473)	615 (200)	24 (8)	294 (46)	46 (9)	65 (42)	20 (1)

Table 2. *Continued*

Pluton*	Rock type†	n‡	SiO$_2$	Al$_2$O$_3$	TiO$_2$	Fe$_2$O$_3$§	MnO	MgO	CaO	Na$_2$O	K$_2$O	P$_2$O$_5$	LOI¶	Total	Co	Cu	Pb	Zn	Rb	Ba	Sr	Nb	Zr	Y	Th	Ga
Çs	A/qs	5	67.27 (1.60)	18.24 (0.54)	0.18 (0.08)	1.54 (0.56)	0.03 (0.03)	0.47 (0.09)	1.12 (0.56)	5.06 (0.30)	6.20 (0.21)	0.06 (0.02)	0.69 (0.23)	100.86 (0.27)	4 (2)	6 (1)	84 (21)	68 (9)	384 (45)	290 (163)	235 (89)	48 (15)	440 (140)	66 (3)	94 (32)	23 (1)
	A/mzgr	2	69.99 (0.08)	16.61 (0.11)	0.11 (0.04)	1.10 (0.11)	0.03 (0)	0.47 (0.04)	0.96 (0.29)	4.78 (0.04)	5.07 (0.17)	0.04 (0.01)	0.81 (0.44)	99.96 (0.09)	3 (0)	6 (2)	59 (16)	58 (2)	375 (141)	309 (299)	191 (130)	28 (12)	217 (7)	59 (21)	61 (7)	21 (3)
Dş	A/nmsp	6	57.19 (0.57)	20.72 (0.36)	0.29 (0.05)	2.78 (0.47)	0.09 (0.01)	0.64 (0.08)	3.21 (0.82)	5.79 (0.55)	8.54 (0.27)	0.09 (0.02)	1.66 (0.27)	101.01 (0.36)	9 (1)	11 (2)	117 (23)	109 (9)	267 (40)	1436 (612)	1154 (269)	42 (11)	473 (45)	46 (4)	40 (17)	19 (1)
By	A/sy	3	63.93 (0.49)	19.70 (0.18)	0.22 (0.03)	1.77 (0.15)	0.04 (0.04)	0.37 (0.05)	1.10 (0.37)	5.56 (0.64)	7.37 (1.32)	0.05 (0)	0.62 (0.16)	100.74 (0.34)	5 (1)	5 (2)	67 (7)	78 (3)	305 (10)	256 (137)	226 (82)	46 (2)	523 (353)	54 (1)	111 (47)	23 (2)
	A/ncs	7	62.12 (1.91)	20.15 (0.70)	0.21 (0.12)	2.06 (0.96)	0.07 (0.03)	0.51 (0.20)	1.55 (0.68)	6.14 (0.97)	7.25 (0.93)	0.06 (0.05)	0.87 (0.32)	101.06 (0.46)	6 (3)	11 (10)	97 (28)	92 (8)	294 (74)	651 (767)	369 (379)	49 (9)	430 (236)	48 (10)	96 (84)	22 (4)

* The plutons are as follows: Yb, Yücebaca; Sh, Sarıhacılı; Cn, Cankılı; Ak, Akçakoyunlu; Ad, Adatepe; Ya, Yassıağıl; Ka, Karakaya; Dcf, Dumluca felsic; Dcm, Dumluca mafic; Mmf, Murmana felsic; Mmm, Murmana mafic; Kkf, Karakeban felsic; Kkm, Karakeban mafic; Kd, Kösedağ; Hç, Hasançelebi; Kç, Karaçayır; Dv, Davulalan; Eğ, Eğrialan; Br, Baranadağ; Hm, Hamit; Çs, Çamsarı; Dş, Durmuşlu; By, Bayındır. †The Rock types are as follows: (S/, I/, and A/ before the abbreviation of rock type represent S-type, I-type and A-type associations, respectively; qmz, quartz monzonite; mzgr, monzogranite; mgo, monzogabbro/monzodiorite; nmgo, nepheline monzogabbro/monzodiorite; mz, monzonite; sy, syenite; qs, quartz syenite; nmsp, nepheline–nosean–melanite syenite porphyry; ncs, nepheline–cancrinite syenite. ‡ Number of analysed rock samples. § Total iron as ferric oxide. ¶ Loss on ignition.

The numbers given in parentheses below each value correspond to the standard deviation; na, not analysed.

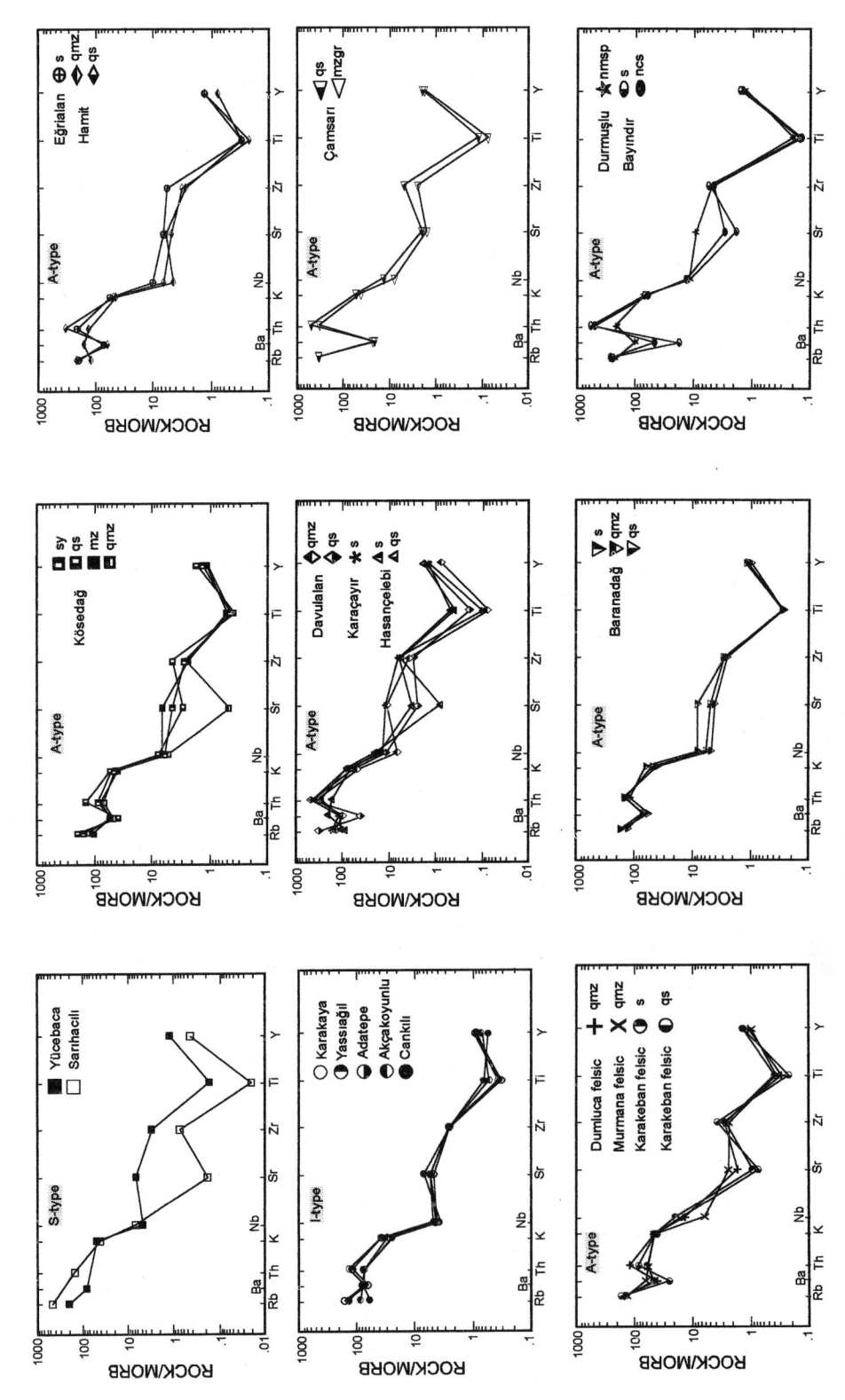

Fig. 4. MORB-normalized spider diagrams of the S-, I- and A-type intrusive associations.

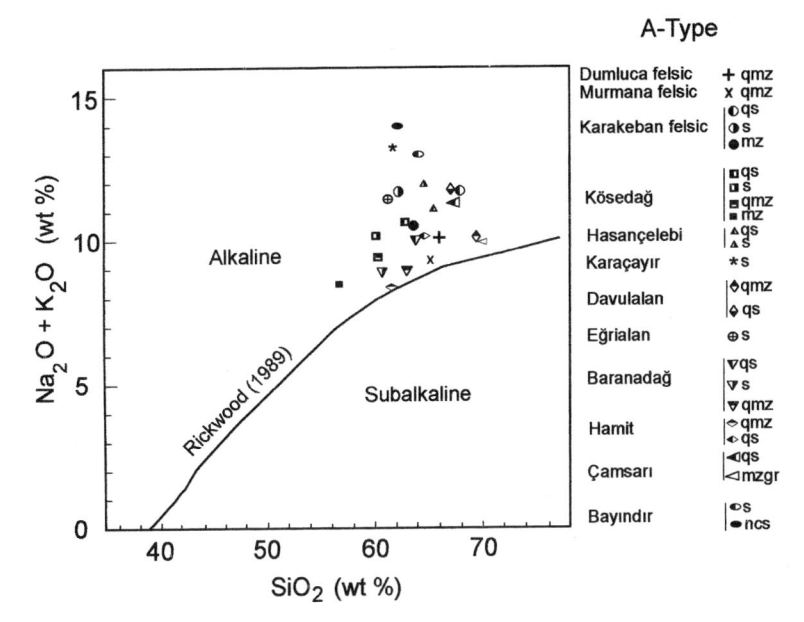

Fig. 5. Total alkali v. silica diagram of the A-type intrusive association [dividing line has been taken from Rickwood (1989)].

intrusive association in Central Anatolia seems to be a relevant signature of an arc-related origin in the geotectonic classification of this association (Fig. 3). As pointed out by Pearce *et al.* (1984) and Harris *et al.* (1986), the arc-related geochemical appearance of most post-collisional calc-alkaline granitoids is sourced in the magmatic melt from which these rocks were derived by partial melting in a post-collisional crustal thickening environment. These rock types are known to be particularly located around the triple-junction area of the VAG, syn-COLG and WPG granite types of Pearce *et al.* (1984), and are identified from trace element geotectonic discrimination diagrams as shown in Fig. 3.

A-type intrusive association

The most widespread plutonic units in the collision-related Central Anatolian intrusive suite are of alkaline composition (Fig. 1). This association comprises: (1) the silica-oversaturated alkaline and (2) the silica-undersaturated alkaline subgroups, on the basis of both mineralogical and chemical compositions (Otlu & Boztuğ 1998; Boztuğ 1998). The latter is observed to intrude the former in the field. The common mineralogical–chemical feature of these subgroups is an alkaline trend in the total alkalis v. silica diagram (Fig. 5), and they include alkaline mafic minerals such as aegirine, arfvedsonite-riebekite and hastingsite amphibole. On the other hand, the silica-oversaturated ones always contain free quartz whilst the silica undersaturated ones generally contain nepheline, and rarely nosean and cancrinite, in the modal mineralogical compositions (Özkan & Erkan 1994; Boztuğ *et al.* 1996; Göncüoğlu *et al.* 1997; Otlu & Boztuğ 1998; Boztuğ 1998). All these mineralogical and chemical characteristics identify an A-type intrusive association within the collision-related central Anatolian plutonism.

The silica-oversaturated alkaline subgroup is the most abundant one in the A-type intrusive association in Central Anatolia. It includes the Divriği region plutons (Dumluca, Murmana and Karakeban), and the Kösedağ, Hasançelebi, Karaçayır, Davulalan, Eğrialan and Kaman region plutons (Baranadağ, Hamit and Çamsarı) (Fig. 1). The silica-undersaturated alkaline subgroup is represented by the Durmuşlu nosean–nepheline–melanite Syenite porphyry and the Bayındır nepheline-cancrinite Syenite in the Kaman region (Otlu & Boztuğ 1998), the İdişdağ Pluton (Göncüoğlu *et al.* 1997) and the Hayriye Nepheline Syenite (Özkan & Erkan 1994).

Detailed geology and petrology of these units is described by Yılmaz *et al.* (1993) for Hasançelebi; Boztuğ *et al.* (1994) for Kösedağ; Boztuğ

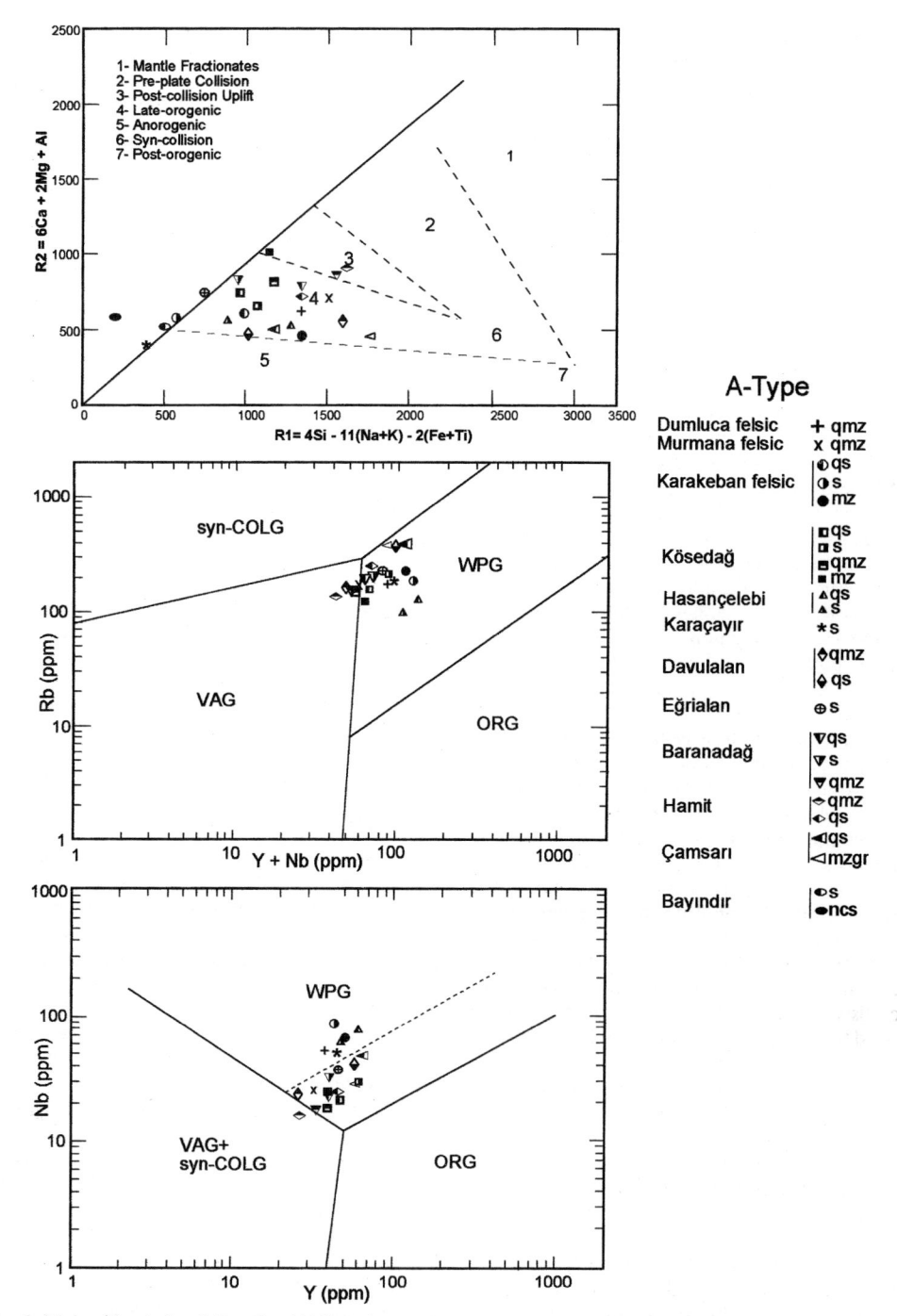

Fig. 6. Major (Batchelor & Bowden 1985) and trace element geotectonic discrimination (Pearce *et al.* 1984) diagrams for the A-type intrusive association.

et al. (1996) for Karaçayır; Alpaslan & Boztuğ (1997) for Davulalan; Boztuğ et al. (1997) for Divriği region plutons; and Otlu & Boztuğ (1998) for the Kaman region plutons. The major rock types of the silica-oversaturated alkaline plutons are monzonite, syenite and their quartz-normative variants. The mineralogical reflection of silica-oversaturated alkaline chemistry is usually represented by the presence of free quartz, which makes it possible to classify these rocks in the trace element geotectonic discrimination diagrams of Pearce et al. (1984). These rocks always include K-feldspar and plagioclase as felsic constituents. The mafic minerals are basically hastingsitic or kaersutitic amphiboles, aegirine–augite and biotites. This subgroup clearly identifies 'late orogenic' and 'WPG' settings on the Batchelor & Bowden (1985) and Pearce et al. (1984) diagrams, respectively (Fig. 6). The alkaline composition and geodynamic setting suggest that these rocks can be classified as 'post-collisional group IV magmatism' (Harris et al. 1986). The Durmuşlu nosean–nepheline–melanite syenite porphyry and Bayındır nepheline–cancrinite syenite units from the silica-undersaturated alkaline subgroup show a highly alkaline mineralogy and chemistry (Otlu & Boztuğ 1998).

The Dumluca Pluton is enriched in the most incompatible elements and yields a negative Ba anomaly. Ba readily substitutes for K in feldspar, hornblende and biotite (Wilson 1989). A negative Ba anomaly in the Dumluca, Murmana and Karakeban Plutons can be related to high temperature feldspar fractionation rather than that of hornblende and biotite, i.e. some highly alkaline rocks such as syenites and feldspathoidal syenites could have been crystallized from the same magma. The Kösedağ Batholith records a negative Ba anomaly, which can be related to feldspar fractionation in Fig. 4, similar to those of the Divriği region plutons. There are also some wide variations in the contents of Th, Nb and Sr, among which Th is remarkably high. The behaviour of Sr (Fig. 4) reflects feldspar crystallization at low temperatures. On the other hand, the behaviour of Th, Nb and Y may also indicate a fractionation or crustal contamination, or low-degree partial melting of the source material (Wilson 1989). Depletions in Rb and K in the Hasançelebi Pluton (Fig. 4) is consistent with alteration by albitization. Yılmaz et al. (1993) have already noted the existence of deuteric alteration such as scapolitization, albitization and carbonatization in this pluton. However, the high contents of Th, Nb, Zr and Y may be interpreted to show a high degree of magma fractionation or a low degree of partial melting of source material during magma genesis. The main trend of the MORB-normalized trace element spider diagram of the Karaçayır Pluton resembles that of the Hasançelebi Pluton in Fig. 4. As for the Th enrichment, Schuiling (1961) has already observed euhedral thorianite (ThO_2) among the accessory minerals in the Karaçayır Pluton. The MORB-normalized trace element distribution diagram of the Davulalan Pluton is more restricted than those of the Karaçayır and Hasançelebi Plutons, which also contain a negligible nepheline and are associated with fluorite mineralization. However, there are some distinctive features in Fig. 4, such as the Rb and K contents displaying a sharper positive anomaly than that of Th. The MORB-normalized trace element spider diagram (Fig. 4) indicates that Ba, Sr and Ti contents of the Çamsarı Quartz Syenite are lower whilst the contents of Rb, Th, Nb, Zr and Y are higher than those in the Baranadağ and Hamit subunits. This can be considered as evidence for fractionation from the Baranadağ through Hamit to the Çamsarı Units in the Kaman region plutonic complex. A similar relationship is also encountered in the Durmuşlu and Bayındır subunits (Fig. 4).

Geodynamic significance of synchronism of metamorphism and magmatism

Radiometric dates determined by various authors in different parts of Central Anatolia (Table 1) reveal that the metamorphism and magmatism occurred together in the Late Cretaceous, producing the metamorphic–magmatic rock assemblage in Central Anatolia. The regional geological setting of Central Anatolia and surroundings (Fig. 1), and the local geological features of the mapped areas, reveal that this could result from the collision between the Anatolide and the Pontide basements following consumption of the northern branch of the Neotethyan oceanic crust along the İzmir–Ankara–Erzincan Suture Zone. The crustal metasediments of the CACC could have been affected by inverted metamorphism (Le Fort 1986; Burg et al. 1994), resulting from the Anatolide–Pontide collision at the northern margin of the Anatolide–Tauride segment. This was the passive margin of the Anatolide–Pontide collisional system (Fig. 7). This new interpretation of the geodynamic evolution of Central Anatolia is also supported by the following observations: (1) initial rock compositions of the crustal metasediments in Central

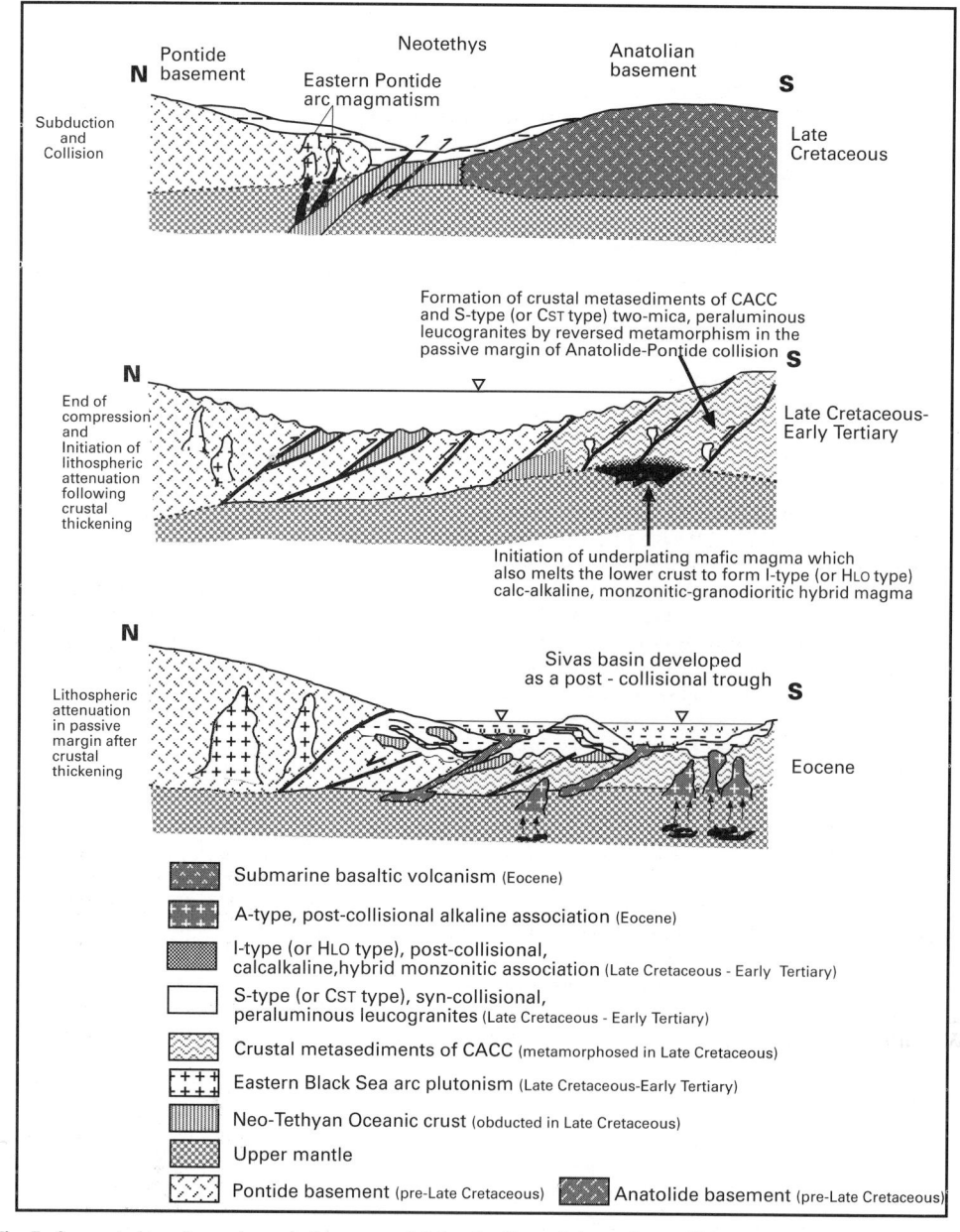

Fig. 7. Suggested geodynamic evolutionary model for the Central Anatolian collision-related plutonism.

Anatolia resemble those of Palaeozoic rocks in the Taurides, and most of the data from metamorphic petrology indicate a decrease in metamorphic grade from north (from main collision zone) to south (towards the Taurides); (2) the southerly situation of the fossiliferous Palaeozoic units in the Taurides (Özgül 1976) is supplementary evidence for the geodynamic interpretation outlined above.

Collision-related intrusives within this crystalline body may have been emplaced into the passive margin of the Anatolides by various magmatic pulses such as: (1) an S-type (or C_{ST}-type), syncollisional peraluminous episode; (2) an I-type (or H_{LO}-type), post-collisional high-K calc-alkaline hybrid episode; and (3) an A-type, post-collisional within-plate alkaline episode (Fig. 7). A syncollisional, S (or C_{ST})-type, peraluminous and two-mica leucogranitic association (the Ortaköy Pluton – Göncüoğlu 1986; the Yücebaca Pluton – Alpaslan & Boztuğ 1997; the Sarıhacılı Leucogranite – Ekici & Boztuğ 1997) is thought to be formed from minimum melt composition anatectic melts derived from supracrustal material during ultrametamorphism induced by the juxtaposition (syncollision) of the Anatolides and Pontides (Fig. 7). The I-type (or H_{LO}-type), post-collisional, metaluminous, high-K calc-alkaline, typically K-feldspar megacrystalic monzonitic association, is considered to have been derived from a hybrid magma formed by the magma mixing–mingling processes between mantle-derived underplating mafic magma and crustal-derived felsic magma. Mantle-derived underplating mafic magma can be derived from the adiabatic decompressional melting of upwelling upper-mantle material following crustal thickening (Fig. 7). After crustal thickening, uplift of the basement during core complex development (Whitney & Dilek 1997, 1998) may cause lithospheric attenuation which makes possible the adiabatic decompressional melting of upper-mantle material to yield A-type, within-plate alkaline magma. Such a melting of upper-mantle material could produce rocks of the: (1) high-K and silica-oversaturated alkaline, K-feldspar megacrystic monzonitic and syenitic subgroup; and (2) silica-undersaturated alkaline feldspathoid–sodalite syenite porphyry subgroup. The subgroup produced depends upon processes such as H_2O activity in the magma chamber, which can modify the composition of any primary magma (Bonin 1987, 1988, 1990), or upon different primary magmas derived from mantle sources by different types and degrees of partial melting under water-poor conditions (Wilson 1989; Rollinson 1993).

Conclusions

The Central Anatolian crystalline region, comprising metasediments, ophiolitic slabs and numerous intrusives, has been called the Kırşehir Massif, the Kırşehir Block or the Central Anatolian Crystalline Complex. Consideration of major and trace element chemistry, and timing of magmatic activity, suggest that metamorphism in Central Anatolia may have been generated by an inverted metamorphism induced by the Anatolide–Pontide collision. This interpretation is supported by the apparent synchronism of metamorphism and magmatism in the Late Cretaceous, and is reflected in a decrease of metamorphic grade from north (i.e. from the main suture zone) to south (i.e. towards the Taurides). However, the collision-related intrusives within the crystalline complex record some differences in geological setting, mineralogical–chemical composition and associated ore deposition. They can be classified, from oldest to youngest, as; (1) a syncollisional, S (or C_{ST})-type, peraluminous and two-mica leucogranitic association; (2) a post-collisional, I (or H_{LO})-type, metaluminous, high-K calc-alkaline, typically K-feldspar megacrystalline, hybrid monzonitic association; (3) an A-type, post-collisional and within-plate alkaline association comprising: (a) a high-K and silica-oversaturated alkaline, K-feldspar megacrystalline monzonitic and syenitic subgroup and (b) a silica-undersaturated alkaline feldspathoid–sodalite syenite porphyry subgroup.

Published data on metamorphism and magmatism in the CACC suggest that both the metamorphism and the magmatism seem to be related to the Anatolide–Pontide collision along the İzmir–Ankara–Erzincan Suture Zone in the Late Cretaceous. The metamorphism is interpreted to have been formed by inverted metamorphism during collision at the passive margin of the Anatolide Plate. High–medium-grade crustal metasediments of the CACC are thought to have been formed from the northernmost tip of the Anatolide–Tauride segment along the İzmir–Ankara–Erzincan Suture Zone due to the Anatolide–Pontide collision. This accounts for the decrease of metamorphic grade from north (i.e. from the main collision zone) to south (i.e. towards the Taurides). Subsequent magmatism may have been reinitiated at the passive margin of the Anatolides by various magmatic episodes, which included a syncollisional peraluminous pulse, a post-collisional calc-alkaline hybrid pulse and a post-collisional within-plate alkaline pulse.

This paper was written at the Department of Geological Engineering of Cumhuriyet University in Sivas. For the financial and logistic support, thanks are due to Cumhuriyet University (University Research Fund, Department of Geological Engineering). P. Kennan and S. Mittwede constructively criticized the preliminary manuscript; the author is indebted to these referees for their helpful comments and substantial improvements.

References

AKIMAN, O., ERLER, A., GÖNCÜOĞLU, M. C., GÜLEÇ, N., GEVEN, A., TÜRELİ, T. K. & KADIOĞLU, Y. K. 1993. Geochemical characteristics of granitoids along the western margin of the Central Anatolian Crystalline Complex and their tectonic implications. *Geological Journal*, **28**, 371–382.

ALPASLAN, M. & BOZTUĞ, D. 1997. The co-existence of the syn-COLG and post-COLG plutons in the Yıldızeli area (W-Sivas). *Turkish Journal of Earth Sciences*, **6**, 1–12.

——, GUEZOU, J. C., BONHOMME, M. G. & BOZTUĞ, D. 1996. Yıldızeli metasedimanter grubu içindeki Fındıcak metamorfitinin metamorfizması ve yaşı. *Geological Society of Turkey Bulletin*, **39**, 19–29 [in Turkish with English abstract].

ATAMAN, G. 1972. Ankara'nın güneydoğusundaki granitik–granodiyoritik kütlelerden Cefalıkdağ'in radyometrik yaşı hakkında ön çalışma. *Hacettepe Bulletin of Natural Sciences and Engineering*, **2**, 44–49 [in Turkish with English abstract].

AYAN, M. 1963. *Contribution a l'étude pétrographique et géologique de la région située au nordest de Kaman*. Mineral Research and Exploration Institute of Turkey (MTA) Publications, **115**.

BARBARIN, B. 1990. Granitoids: main petrogenetic classifications in relation to origin and tectonic setting. *Geological Journal*, **25**, 227–238.

BATCHELOR, B. & BOWDEN, P. 1985. Petrogenetic interpretation of granitoid rock series using multicationic parameters. *Chemical Geology*, **48**, 43–55.

BİNGÖL, E. 1989. *Geological Map of Turkey (1:2 000 000 scale)*. Mineral Research and Exploration Institute of Turkey (MTA) Publications.

BONIN, B. 1987. Reflexions a propose de la repartition des granitoides les massifs cristallins externes des alpes françaises. *Géologie Alpine*, **63**, 137–149.

—— 1988. From orogenic to anorogenic environments: evidence from associated magmatic episodes. *Schweizerische Mineralogische Petrographische Mitteilungen*, **68**, 301–311.

—— 1990. From orogenic to anorogenic settings: evolution of granitoid suites after a major orogenesis. *Geological Journal*, **25**, 261–270.

BOZTUĞ, D. 1998. Post-collisional central Anatolian alkaline plutonism, Turkey. *Turkish Journal of Earth Sciences*, **7**, 145–165.

——, YILMAZ, S. & KESKİN, Y. 1994. İç-Doğu Anadolu alkalin provensindeki Kösedağ plütonu (Suşehri–KDSivas)doğukesimininpetrografisi,petrokimyası ve petrojenezi. *Geological Society of Turkey Bulletin*, **37**, 1–14 [in Turkish with English abstract].

——, —— & ALPASLAN, M. 1996. The Karaçayır syenite, north of Sivas: an A-type, peraluminous and post-collisional alkaline pluton, Central Anatolia, Turkey. *Bulletin of the Faculty of Engineering, Cumhuriyet University, Series A – Earth Sciences*, **13**, 141–153.

——, YAĞMUR, M., OTLU, N., TATAR, S. & YEŞİLTAŞ, A. 1998. Petrology of the post-collisional, within-plate Yıldızdağ gabbroic pluton, Yıldızeli–Sivas region, CA, Turkey. *Turkish Journal of Earth Sciences*, **7**, 37–51.

——, DEBON, F., İNAN, S., TUTKUN, S. Z., AVCI, N. & KESKİN, Ö. 1997. Comparative geochemistry of four plutons from the Cretaceous–Palaeogene Central-Eastern Anatolian alkaline province (Divriği region, Sivas, Turkey). *Turkish Journal of Earth Sciences*, **6**, 95–115.

BURG, J.P., DELOR, C. P., LEYRELOUP, A. F. & ROMNEY, F. 1994. Inverted metamorphic zonation and Variscan thrust tectonics in the Reuergue area (Masif Central, France): *P*–*T*–*t* record from mineral to regional scale. *In*: DALY, J. S., CLIFF, R. A. & YARDLEY, B. W. D. (eds) *Evolution of Metamorphic Belts*. Geological Society, London, Special Publications, **43**, 423–439.

CATER, J. M. L., HANNA, S. S., RIES, A. C. & TURNER, P. 1991. Tertiary evolution of the Sivas basin, Central Turkey. *Tectonophysics*, **195**, 29–46.

CHAPPELL, B. W. 1984. Source rocks of I- and S-type granites in the Lachlan Fold Belt, Southeastern Australia. *Philosophical Transactions of the Royal Society, London, Series A*, **310**, 693–707.

—— 1996. Compositional variation within granite suites of the Lachlan Fold Belt: its causes and implications for the physical state of granite magma. *Transactions of the Royal Society of Edinburgh: Earth Sciences*, **87**, 159–170.

—— & STEPHENS, W. E. 1988. Origin of infracrustal (I-type) granite magmas. *Transactions of the Royal Society, Edinburgh: Earth Sciences*, **79**, 71–86.

—— & WHITE, A. J .R. 1974. Two contrasting granite types. *Pacific Geology*, **8**, 173–174.

—— & —— 1992. I- and S-type granites in the Lachlan Fold Belt. *Transactions of the Royal Society, Edinburgh: Earth Sciences*, **83**, 1–26.

COLLINS, W. J., BEAM, S. D., WHITE, A. J. R. & CHAPPELL, B. W. 1982. Nature and origin of A-type granites with particular reference to southeastern Australia. *Contributions to Mineralogy and Petrology*, **80**, 189–200.

COWARD, M. P. & RIES, A. C. (eds) 1986. *Collision Tectonics*. Geological Society, London, Special Publications, **19**.

EKİCİ, T. & BOZTUĞ, D. 1997. Anatolid–Pontid çarpışma sisteminin pasif kenarında yer alan Yozgat batolitinde syn-COLG ve post-COLG granitoyid birlikteliği. *Geosound*, **30**, 519–538 [in Turkish with English abstract].

ERKAN, Y. 1976. Kırşehir çevresindeki rejyonal metamorfik bölgede saptanan izogradlar ve bunların petrolojik yorumlanmaları. *Hacettepe University Earth Sciences*, **2**, 23–54 [in Turkish with English abstract].

—— 1977. Orta Anadolu Masifinin güneybatısında (Kırşehir bölgesinde) etkili rejyonal metamorfizma ile amfibol minerallerinin bileşimi arasındaki ilişkiler. *Hacettepe University Earth Sciences*, **3**, 41–46 [in Turkish with English abstract].

—— 1978. Kırşehir Masifinde granat minerallerinin kimyasal bileşimi ile rejyonal metamorfizma arasındaki ilişkiler. *Geological Society of Turkey Bulletin*, **21**, 43–50 [in Turkish with English abstract].

—— 1981. Orta Anadolu Masifinin metamorfizması üzerinde yapılmış çalışmalarda varılan sonuçlar. *Proceedings of the Geology of Central Anatolia Symposium*. Geological Society of Turkey Publications, 9–11 [in Turkish with English abstract].

—— & ATAMAN, G. 1981. Orta Anadolu Masifi (Kırşehir yöresi) metamorfizma yaşı üzerine K–Ar yöntemi ile bir inceleme. *Hacettepe University Earth Sciences*, **8**, 27–30 [in Turkish with English abstract].

ERLER, A. & BAYHAN, H. 1995. Orta Anadolu Granitoidleri' nin genel değerlendirilmesi ve sorunları. *Hacettepe University Earth Sciences*, **17**, 49–67 [in Turkish with English abstract].

—— & GÖNCÜOĞLU, M. C. 1996. Geologic and tectonic setting of the Yozgat batholith, Northern Central Anatolian Crystalline Complex, Turkey. *International Geology Review*, **38**, 714–726.

GEVEN, A. 1995. Cefalıkdağ granitoyidinin petrografi ve jeokimyası (Orta Anadolu kristalen kütlesi batısı). *Hacettepe University Earth Sciences*, **17**, 1–16 [in Turkish with English abstract].

GÖNCÜOĞLU, M. C. 1986. Geochoronological data from the southern part (Niğde area) of the Central Anatolian Massif. *Mineral Research and Exploration Institute of Turkey (MTA) Bulletin*, **105/106**, 83–96.

—— & TÜRELİ, K. T. 1993. Orta Anadolu ofiyoliti plajiyogranitlerinin petrolojisi ve jeodinamik yorumu (Aksaray – Türkiye). *Turkish Journal of Earth Sciences*, **2**, 195–203.

—— & —— 1994. Alpine collisional-type granitoids from Western Central Anatolian Crystalline Complex, Turkey. *Journal of Kocaeli University: Earth Sciences*, **1**, 39–46.

——, KÖKSAL, S. & FLOYD, P. A. 1997. Post-collisional A-type magmatism in the Central Anatolian Crystalline Complex: petrology of the İdiş dağı Intrusives (Avanos, Turkey). *Turkish Journal of Earth Sciences*, **6**, 65–76.

——, TOPRAK, V., ERLER, A. & KUŞCU, İ. 1991. *Orta Anadolu Batı Kesiminin Jeolojisi, Bölüm I, Güney Kesim*. Turkish Petroleum Corporation (TPAO) Report, No. 2909 [in Turkish].

GÖRÜR, N., OKTAY, F. Y., SEYMEN, İ. & ŞENGÖR, A. M. C. 1984. Palaeo-tectonic evolution of the Tuzgölü basin complex, Central Turkey: sedimentary record of a Neotethyan closure. In: DIXON, J. E. & ROBERTSON, A. H. F. (eds) *The Geological Evolution of the Eastern Mediterranean*. Geological Society, London, Special Publications, **17**, 467–482.

GOVINDARAJU, K. 1989. 1989 compilation of working values and sample description for 272 geostandards. *Geostandards Newsletter*, **13**, 1–113.

GÜLEÇ, N. 1994. Rb–Sr isotope data from the Ağaçören granitoid (East of Tuz Gölü): geochronological and genetical implications. *Turkish Journal of Earth Sciences*, **3**, 39–43.

GÜNDOĞDU., M. N., BROS, R., KURUÇ, A. & BAYHAN, H. 1988. Rb–Sr whole rock systematic of the Bayındır feldispathoidal syenites (Kaman-Kırşehir). *Abstracts of the Symposium for the 20th Anniversary of Earth Sciences at Hacettepe University*, 55.

HARRIS, N. B. W., PEARCE, J. A. & TINDLE, A. G. 1986. Geochemical characteristics of collision-zone magmatism. In: COWARD, M. P. & RIES, A. C. (eds) *Collision Tectonics*. Geological Society, London, Special Publications, **19**, 67–81.

İLBEYLİ, N. & PEARCE, J. A. 1997. Petrogenesis of the collision-related Anatolian Granitoids, Turkey. *European Union of Geosciences (EUG) 9*, Strasbourg – France, 502.

IRVINE, T. N. & BARAGAR, W. R. A. 1971. A guide to the chemical classification of common volcanic rocks. *Canadian Journal of Earth Sciences*, **8**, 523–548.

KALKANCI, Ş. 1974. Etude géologique et pétrochimique du sud de le région de Suşehri. Géochronologie du massif syenitique de Kösedağ (NE de Sivas – Turquie). Thèse de doctorat de 3ème cycle, L'Université de Grenoble.

LE FORT, P. 1986. Metamorphism and magmatism during the Himalayan collision. In: COWARD, M. P. & RIES, A. C. (eds) *Collision Tectonics*. Geological Society, London, Special Publications, **19**, 159–172.

MANIAR, P. D. & PICCOLI, P. M. 1989. Tectonic discrimination of granitoids. *Geological Society of America Bulletin*, **101**, 635–643.

OTLU, N. & BOZTUĞ, D. 1998. The coexistence of the silica oversaturated (ALKOS) and undersaturated alkaline (ALKUS) rocks in the Kortundağ and Baranadağ plutons from the Central Anatolian alkaline plutonism, E Kaman/NW Kırşehir, Turkey. *Turkish Journal of Earth Sciences*, **7**, 241–257.

ÖZGÜL, N. 1976. Torosların bazı temel jeoloji özellilleri. *Turkish Geological Society Bulletin*, **19**, 65–78 [in Turkish with English abstract].

ÖZKAN, H. M. & ERKAN, Y. 1994. A petrological study on a foid syenite intrusion in Central Anatolia (Kayseri, Turkey). *Turkish Journal of Earth Sciences*, **3**, 45–55.

PEARCE, J. A., HARRIS, N. B. W. & TINDLE, A. G. W. 1984. Trace element discrimination diagrams for the tectonic interpretation of granitic rocks. *Journal of Petrology*, **25**, 956–983.

POISSON, A. 1986. Anatolian micro-continents in the Eastern Mediterranean context: the neo-Tethysian oceanic troughs. *Science de la Terre, Mémoire*, **47**, 311–328.

RICKWOOD, P. C. 1989. Boundary lines within petrologic diagrams which use oxides of major and minor elements. *Lithos*, **22**, 247–263.

ROLLINSON, H. R. 1993. *Using Geochemical Data: Evaluation, Presentation, Interpretation*. Longman.

SCHUILING, R. D. 1961. Formation of pegmatitic carbonatite in a syenite–marble contact. *Nature*, **192**, 1280.

SEYMEN, İ. 1984. Kırşehir Masifi metamorfitlerinin jeolojik evrimi. *Proceedings of the Ketin Symposium*. Geological Society of Turkey Publications, 133–148 [in Turkish with English abstract].

ŞENGÖR, A. M. C. & YILMAZ, Y. 1981. Tethyan evolution of Turkey: a plate tectonic approach. *Tectonophysics*, **75**, 181–241.

TATAR, S. & BOZTUĞ, D. 1998. Fractional crystallisation and magma mingling/mixing processes in the monzonitic association in the SW part of the composite Yozgat batholith (Şefaatli–Yerköy, SW Yozgat). *Turkish Journal of Earth Sciences*, **7**, 215–230.

TOLLUOĞLU, A. Ü. 1987. Orta Anadolu Masifi Kırşehir metamorfiklerinin (Kırşehir kuzeybatısı) petrografik özellikleri. *Doğa Dergisi: Muhendislik re Gevre*, **11**, 344–361 [in Turkish with English abstract].

—— & ERKAN, Y. 1989. Regional progressive metamorphism in the central Anatolian crystalline basement, northwest Kırşehir massif, Turkey. *METU Journal of Pure and Applied Sciences*, **22**, 19–41.

WHALEN, J. B., CURRIE, K. L. & CHAPPELL, B. W. 1987. A-type granites: geochemical characteristics, discrimination and petrogenesis. *Contributions to Mineralogy and Petrology*, **95**, 407–419.

WHITE, A. J. R. & CHAPPELL, B. W. 1988. Some supracrustal (S-type) granites of the Lachlan Fold Belt. *Transactions of the Royal Society of Edinburgh: Earth Sciences*, **79**, 169–181.

WHITNEY, D. L. & DİLEK, Y. 1997. Core complex development in central Anatolia, Turkey. *Geology*, **25**, 1023–1026.

—— & —— 1998. Metamorphism during Alpine crustal thickening and extension in central Anatolia, Turkey: the Niğde metamorphic core complex. *Journal of Petrology*, **39**, 1385–1403.

WILSON, M. 1989. *Igneous Petrogenesis*. Unwin Hyman.

YALINIZ, M. K., FLOYD, P. A. & GÖNCÜOĞLU, M. C. 1996. Supra-subduction zone ophiolites of Central Anatolia: geochemical evidence from the Sarıkaraman Ophiolite, Aksaray, Turkey. *Mineralogical Magazine*, **60**, 697–710.

YILMAZ, A. 1994. An example of post-collisional trough: Sivas basin, Turkey. *Proceedings of the 10th Petroleum Congress and Exhibition of Turkey*. Turkish Association of Petroleum Geologists Publications, 21–33 [in Turkish with English abstract].

YILMAZ, S., BOZTUĞ, D. & ÖZTÜRK, A. 1993. Geological setting, petrographic and geochemical characteristics of the Cretaceous and Tertiary igneous rocks in the Hekimhan–Hasançelebi area, northwest Malatya, Turkey. *Geological Journal*, **28**, 383–398.

ZECK, H. P. & ÜNLÜ, T. 1987. Parallel whole rock isochrons from a composite, monzonitic pluton, Alpine belt, central Anatolia, Turkey. *Neues Jahrbuch für Mineralogische Monatsheft*, **1987**, 193–204.

Neogene and Quaternary volcanism of southeastern Turkey

JAN ARGER,[1] JOHN MITCHELL[2] & ROB W. C. WESTAWAY[3]

[1] Department of Geography, University of Edinburgh, Edinburgh EH8 9XP, UK
[2] Department of Physics, University of Newcastle upon Tyne, Newcastle upon Tyne NE1 7RU, UK
[3] 16 Neville Square, Durham DH1 3PY, UK
(e-mail: R.W.C. Westaway@newcastle.ac.uk)

Abstract: Potassium–argon dating indicates two episodes of basaltic magmatism in southeastern Turkey at *c.* 19–15 and *c.* 2.3–0.6 Ma. Each produced olivine–titanaugite basalts, whose chemical compositions are difficult to classify using any conventional model in both the Anatolian continental fragment and the Arabian Platform. It is proposed here that both episodes of volcanism, and the associated crustal thickening and surface uplift, result from heating of the mantle lithosphere by crustal thickening caused by inflow of plastic lower crust from adjoining regions. Thus, although this study region has remained in a plate boundary zone for tens of millions of years, its volcanism has no direct relationship to local plate motions. It is suggested instead that both episodes of volcanism are the result of loading effects caused by glacial to interglacial sea-level variations, which will cause net flow of lower crust from beneath the offshore shelf to beneath the land: the moderate glaciations of Antarctica which began in the Early–Middle Miocene, and the more intense lowland glaciations of the northern hemisphere which began around *c.* 2.5 Ma.

Eastern Turkey contains the modern boundaries between the Turkish (TR), Arabian (AR), African (AF) and Eurasian (EU) Plates, which comprise segmented strike-slip fault zones that approximate transform faults (e.g. Westaway 1994; Westaway & Arger 1996) and take up convergence between Arabia and Eurasia by allowing the Turkish Plate to move westward away from this convergent zone (Fig. 1a). Before *c.* 5 Ma, the Turkish Plate was part of Eurasia and this convergence was instead taken up directly in the AR–EU boundary zone (Fig. 1b–d).

A pilot investigation of the Neogene and Quaternary volcanism associated with this plate convergence in southeastern Turkey has been carried out, where few sites have previously been isotopically dated. The first aim of this paper is to summarize this region's tectonic evolution to indicate the significance of particular ranges of dates; the procedures for K–Ar dating and geochemical analysis are then described, and the implications of the results obtained for the development of volcanism in this region discussed.

Regional geology of eastern Turkey

The complex Mesozoic evolution of the region studied here involved the interaction of several ocean basins and continental fragments, including the Anatolian fragment which now forms much of eastern Turkey. However, the timing and geometry of these plate motions remain uncertain, many incompatible schemes having been proposed (e.g. Şengör 1979; Özkaya 1982; Michard *et al.* 1984; Dercourt *et al.* 1986; Dewey *et al.* 1986; Robertson *et al.* 1991; Robinson *et al.* 1995).

Until the Dead Sea Fault Zone (DSFZ) formed in Miocene time (e.g. Garfunkel 1981), Arabia was part of the African Plate. At *c.* 92 Ma, the AF–EU motion had adjusted from east-southeast to northeast in the vicinity of Anatolia (e.g. Dewey *et al.* 1986), requiring convergence for the first time. The oldest evidence of northward subduction of the Neotethys Ocean between Arabia and Anatolia, the calc-alkaline volcanism in eastern Turkey dated at *c.* 87 Ma by Michard *et al.* (1984), is consistent with this timing. Closure of the Neotethys involved northward subduction until *c.* 65–70 Ma, when the Kızıldağ ophiolite in southernmost Turkey and its equivalents further east were obducted onto Arabia (e.g. Y. Yılmaz *et al.* 1993). Subduction later resumed and took up the AF–EU motion until *c.* 40 Ma (e.g. Y. Yılmaz *et al.* 1993), when shortening of their continental crust began.

Oligo-Miocene plate convergence caused widespread reverse faulting in eastern Turkey (e.g. Temiz *et al.* 1993; Y. Yılmaz *et al.* 1993); and shortening continues east of the modern AR–EU–TR triple junction at Karlıova. Much

From: BOZKURT, E., WINCHESTER, J. A. & PIPER, J. D. A. (eds) *Tectonics and Magmatism in Turkey and the Surrounding Area*. Geological Society, London, Special Publications, **173**, 459–487. 1-86239-064-9/00/$15.00
© The Geological Society of London 2000.

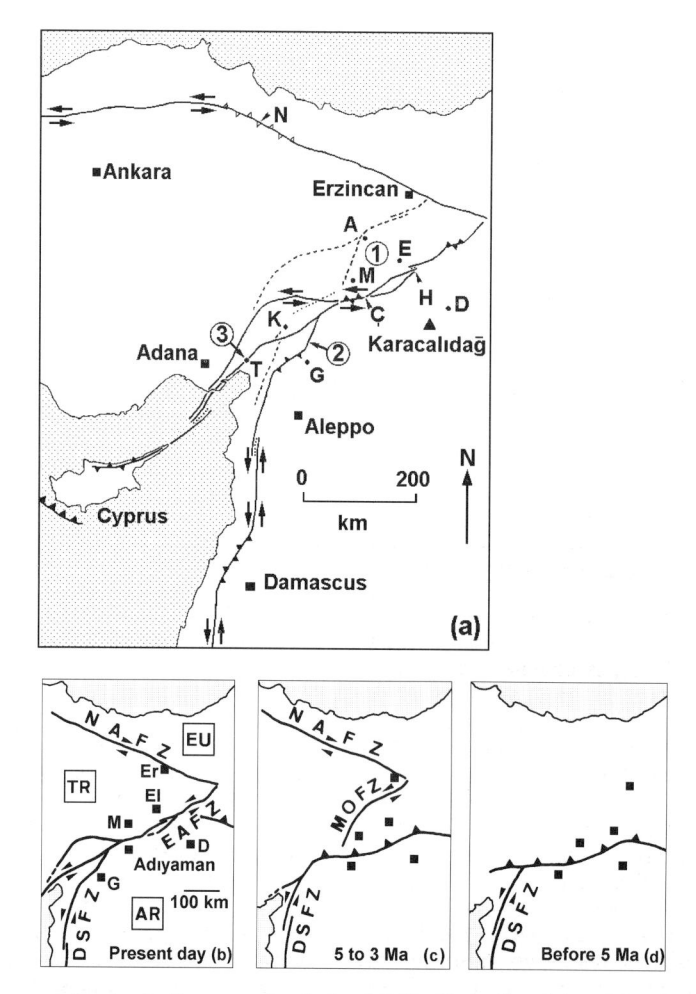

Fig. 1. (**a**) Location map showing fault zones active during 5–3 Ma (dashed) and since 3 Ma (solid), and locations 1–3 of the three field areas. Fine dot ornament marks major pull-apart basins; chevron ornament on both sides of a strike-slip fault marks stepovers, the ornament being solid if the stepover is transpressional and open if transtensional. Other chevron ornament marks the subduction zone west of Cyprus. A, Arapkir; D, Dıyarbakır; E, Elazığ; G, Gaziantep; K, Kahraman Maraş; M, Malatya; T, Toprakkale. (**b**)–(**d**) Summary maps illustrating the evolution of the plate boundary zone in eastern Turkey; from Westaway & Arger (1996, fig. 1). MOFZ is the Malatya–Ovacık Fault Zone.

of the Anatolian fragment was below sea level in Eocene time (e.g. Brinkmann 1976, p. 63) but it emerged during the Oligocene, when evaporitic sedimentation was widespread, before becoming submerged again in Early Miocene time (e.g. Gökçen & Kelling 1985). Lower Miocene marine sediment is now found at 1200–2000 m altitudes near Malatya and Elazığ (Fig. 2) (e.g. Baykal & Erentöz 1961). By Middle Miocene time, lacustrine conditions existed in much of eastern Anatolia (e.g. Leo *et al.* 1974; Asutay 1988). The adjacent Arabian Platform remained marine until Middle Miocene time (e.g. Rigo di Righi & Cortesini 1964; Tolun & Pamir 1975; Lovelock 1984; Karig & Kozlu 1990), although evidently not deeply submerged, given the widespread erosion (presumed to be at wave base) and the coastal sedimentary facies such as reefs. This area has since uplifted to typical altitudes of 500–900 m.

The modern TR–EU and TR–AR boundaries follow the North Anatolian and East Anatolian Fault Zones (NAFZ and EAFZ, respectively), which approximate transform faults (Westaway 1994) (Fig. 1a). Evidence of active shortening

Table 1. *Locations of sampling sites*

Locality	Samples	Position	Latitude	Longitude
Elazığ area				
Arapkir	A, A3, A4	DD 544 206	39°02′N	38°28′E
Örtülü	E3, E5	DD 628 065	38°54′N	38°34′E
Çipköy	E1, E2	EC 072 822	38°41′N	39°05′E
Kahraman Maraş area				
Fevzipaşa	F2	BB 903 078	37°06′N	36°39′E
Çınarlı	K1	CB 314 476	37°27′N	37°05′E
Köprüağzı	T	CB 244 364	37°21′N	37°01′E
Pazarcık	X	CB 417 467	37°27′N	37°12′E
İskenderun Gulf area				
Toprakkale 1	T1, T2	BB 456 040	37°03′N	36°09′E
Toprakkale 2	T3, T4	BB 454 043	37°03′N	36°08′E
Toprakkale 3	T5, T6, T7	BB 453 049	37°04′N	36°08′E
Gates of Issos	T8	BB 490 019	37°01′N	36°09′E
Sarımazi Beach	T9	BA 364 923	36°56′N	36°02′E
Kurtpınar	K	YF 623 912	36°56′N	35°57′E

Positions, precise to 100 m, relate to the Universal Transverse Mercator grid.

west of the Karlıova Triple Junction, reported for instance by Karig & Kozlu (1990) and Taymaz *et al.* (1991), is confined to localities where active left-lateral faulting steps to the right, where shortening is required to satisfy geometrical constraints (Westaway 1995). These localities are linked by transform faults which are subparallel to the TR–AR motion. Westaway & Arger (1996) proposed a kinematic model in which this present geometry of major faulting became active around *c.* 3 Ma (Fig. 1b). They suggested that, between *c.* 5 and 3 Ma, the relative motion between the Turkish and Arabian Plates was taken up north of the EAFZ, which did not yet exist, on the Malatya–Ovacık Fault Zone (MOFZ) (Fig. 1c). Other studies have inferred similar ages of the present phase of faulting in eastern Turkey using other local evidence. For instance, Şaroğlu (1988) estimated the age of the easternmost NAFZ as *c.* 2.5 Ma, whereas the age of the EAFZ has been estimated as Late Pliocene (i.e. *c.* 2–3 Ma, Şaroğlu *et al.* 1992b; *c.* 2 Ma, Trifonov *et al.* 1994; and 1.8 Ma, Yürür & Chorowicz 1998).

Field areas and sampling sites

Due to reported effects of alteration affecting volcanic rocks in some localities (e.g. Leo *et al.* 1974; S. Yılmaz *et al.* 1993), samples were widely taken in the fieldwork. The study region was divided into three areas, around Elazığ, Kahraman Maraş and İskenderun Gulf (Figs 1a, 2–4; Table 1). Previous isotopic dating at localities near this study's sampling sites are also summarized and all K–Ar dates published before 1977 are recalculated [using the decay constants from Steiger & Jäger (1977)].

The Elazığ area

This area of the former Anatolian continental fragment (Fig. 2) has experienced widespread and prolonged volcanism. Past studies indicate Late Cretaceous and Middle Miocene volcanism, plus a younger phase that has been variously reported as Late Miocene, Pliocene or Quaternary (see below).

The Yamadağ area experienced basaltic and andesitic volcanism in Late Cretaceous time (the Bahçedam volcanic series, 78–67 Ma; Leo *et al.* 1974), and in Middle Miocene time (the Yamadağ Series). Leo *et al.* (1974) determined K–Ar dates for the Yamadağ Series at localities 1–3 in Fig. 2 of 19.2 ± 0.5, 17.2 ± 0.5 and 14.5 ± 0.4 Ma, respectively, where eruptions occurred in a palaeolake. The rocks studied were: hornblende–augite–biotite andesite with chlorite–carbonate alteration (1); olivine basalt with calcic plagioclase feldspar and titanaugite (2); and hornblende dacite (3). Ota & Dinçel (1975) described their petrology, reporting the typical plagioclase content of the olivine–pyroxene basalts as bytownite close to labradorite (i.e. *c.* 70–75% anorthite). S. Yılmaz *et al.* (1993) analysed the geochemistry of these rocks at localities 4 and 5.

A contiguous outcrop of volcanic rocks persists eastward from Yamadağ for *c.* 40 km to Arapkir (Fig. 2); immediately west of Arapkir it is known as the Göldağ Suite. Further east, the facies of volcanic rocks interbedded with

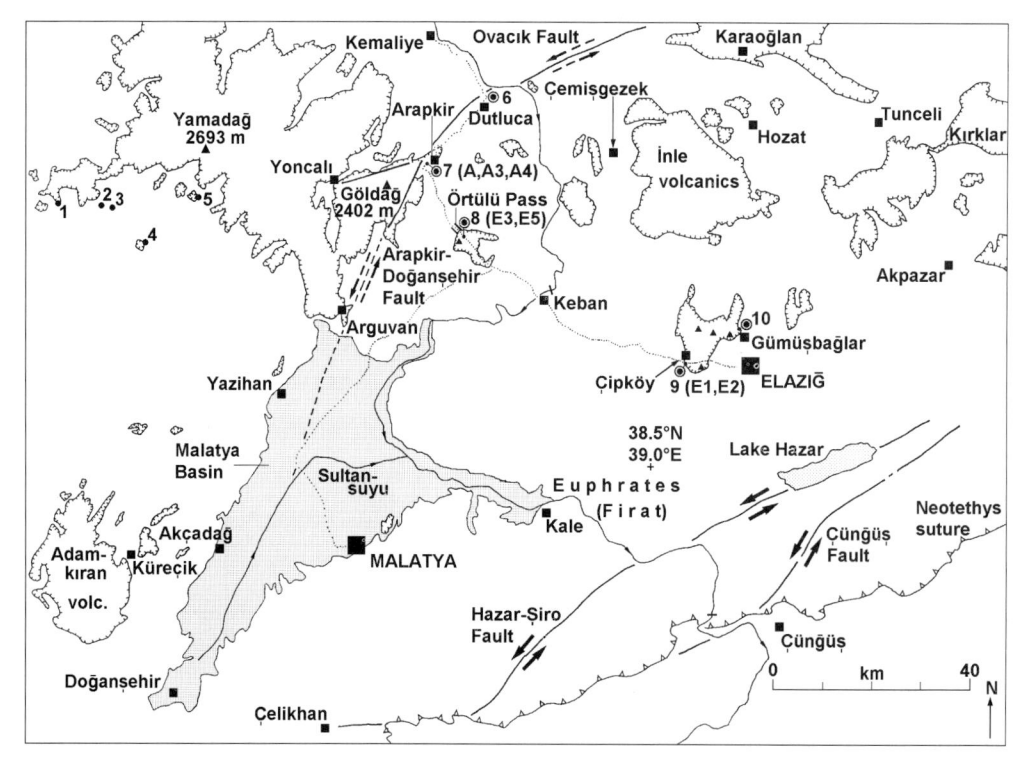

Fig. 2. Map of the Elazığ field area. Tick marks bound volcanic outcrops; line with open chevrons marks the Neotethys Suture; thick lines with bold arrows mark strands of the EAFZ (active); thick lines with dashed arrows mark strands of the MOFZ. Numbered localities are discussed in the text, 7–10 marking this study's sampling sites and Sanver's (1968) site at Gümüşbağlar. Fine dotted lines are roads providing access to the sampling sites. Triangles mark volcanic necks near these sites, from Tonbul (1987) and these investigations, and summits of other large volcanoes with altitudes in metres. Dot ornament marks the Pliocene Malatya Basin. Other information is from Altınlı & Erentöz (1961) and Baykal & Erentöz (1961).

lacustrine sediment (typically marl and limestone) is called the Karabakır Formation (e.g. Bingöl 1984). It is usually regarded as Late Miocene from its stratigraphic position above Lower Miocene marine sediments (e.g. Asutay 1988) and by deduction from geomorphological evidence. For instance, Tonbul (1987) correlated the shoreline terrace of the Karabakır palaeolake with the 'D2' erosion surface identified elsewhere in eastern Turkey by Erol (1982), which is inferred to have a Late Miocene age.

Bingöl (1984) assigned all volcanism from Arapkir to Kırklar (Fig. 2) to this Karabakır Formation. However, the evidence from Yamadağ indicates that lacustrine conditions already existed in the Middle Miocene. Furthermore, Sanver (1968) showed that samples of the 'Karabakır' volcanics near Elazığ (Fig. 2, 10) are normally magnetized and inferred their age as Brunhes (< 0.8 Ma). Innocenti *et al.* (1982) concluded that the Yamadağ volcanics are Early–Middle Miocene, İnle and Karaoğlan are Middle–Late Miocene, and the Kırklar and Gökdere volcanics are Late Miocene and Plio-Quaternary. However, sheets of the new series of geological maps (e.g. Asutay 1988) and other recent studies such as Kerey & Türkmen (1991) retain the 'official' Late Miocene age for the 'Karabakır' Formation volcanics in the vicinity of Elazığ. The lack of consensus concerning their age is none the less evident, with a potential age range for volcanism associated with lacustrine sedimentation in this area spanning from Middle Miocene to Quaternary.

Three samples (A, A3 and A4) were collected near Arapkir (Fig. 2.7). This site is *c.* 1.5 km southeast of the nearest strand of the MOFZ (the Arapkir–Doğanşehir Fault) that runs southeast of the Göldağ Volcano, and was thus on the Arabian side of this fault when it was active. The sampled flows have previously been assumed to have come from Göldağ and to be

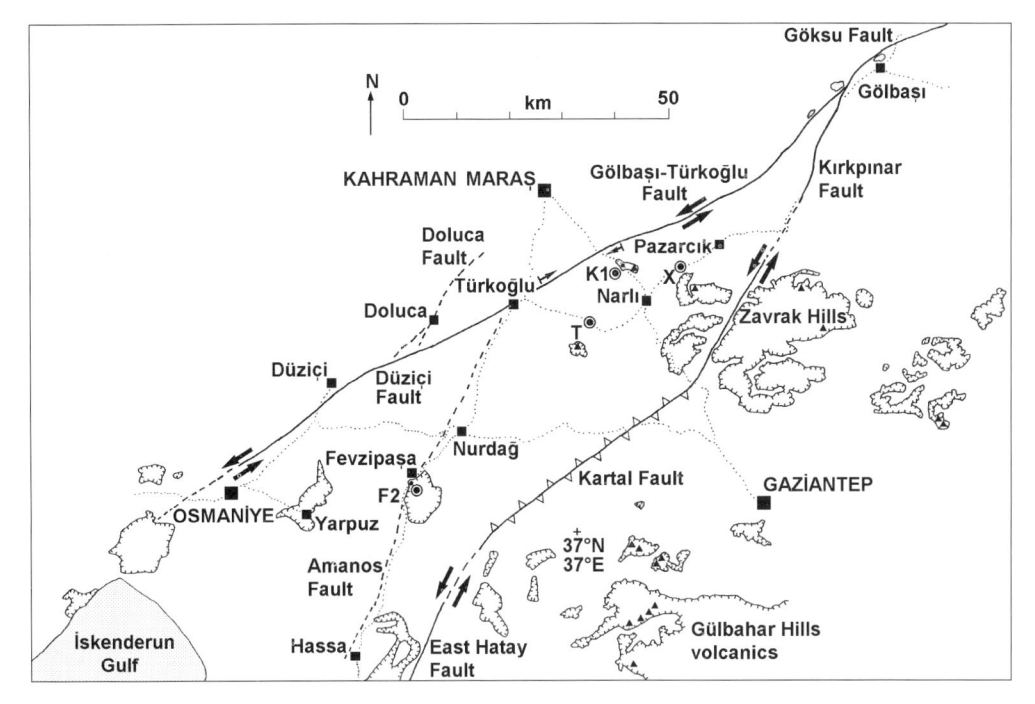

Fig. 3. Map of the Kahraman Maraş field area. Most notation is the same as in Fig. 2. Solid lines with bold arrows mark the most important active strike-slip faults, which are regarded as having been active since c. 3 Ma. Open chevrons along the Kartal Fault indicate that it is the internal fault within a stepover and not a transform fault [using Westaway's (1995) terminology]. Thick dashed lines without arrows are left-lateral faults that are regarded as having been active during 5–3 Ma, including the splay at Doluca that appears to have marked the junction of the main strands of the AF–TR and AF–AR boundaries, analogous to the modern splay near Gölbaşı. The 16 km offset of this older left-lateral faulting between Doluca and Türkoğlu matches the 16 km offset of an outcrop of ophiolite north of Narlı (double arrows), indicating the amount of slip on the Düziçi and Gölbaşı-Türkoğlu Faults since the modern fault geometry became active. Volcanic outcrop and neck positions are from Lyberis et al. (1992). Sample X came from a small outcrop in the Aksu River Valley west of Pazarcık, *not* from the much larger area of basalt which caps the Zavrak Hills to the southeast.

Late Miocene because they interbed with 'Karabakır' lacustrine sediments and cover an erosion surface that has been regarded as Middle Miocene (Tonbul, pers. comm.). However, if > 20 km of left-lateral slip on the Arapkir–Doğanşehir Fault is restored, they instead become juxtaposed with the southernmost part of the Yamadağ volcanics near Arguvan (Fig. 2).

Samples E3 and E5 were collected from a c. 10 × 10 km outcrop of 'Karabakır' Formation volcanics that overlie the lacustrine sediment near the Örtülü Pass (Fig. 2.8). This olivine basalt presumably flowed from a volcanic neck, Karakez, c. 2 km west of the site. At Çipköy (Fig. 2.9), two c. 3 m thick flows of olivine basalt are exposed. Sample E1 was from the lower flow and E2 from the darker upper flow. This site is 12 km west-southwest of Gümüşbağlar where Sanver (1968) inferred a Brunhes age of the basalt. According to mapping by Tonbul (1987), lava from necks west of Gümüşbağlar flowed southwest towards Çipköy, then down the Altınkuşak River Gorge forming the northward 'tongue' of basalt outcrop in Fig. 2.

The Kahraman Maraş area

The second field area is in the Arabian Platform south of Kahraman Maraş (Fig. 3). Most of this area remained below sea level until the Middle Miocene (e.g. Rigo di Righi & Cortesini 1964; Lovelock 1984; Karig & Kozlu 1990), and now marks the triple junction between the left-lateral faults bounding the Turkish, Arabian and African Plates. Muehlberger & Gordon (1987), Perinçek & Çemen (1990), Lyberis et al. (1992), Şaroğlu et al. (1992b), Westaway (1994), Westaway & Arger (1996) and Yürür &

Table 2. *K–Ar dating of Karasu Valley basalts by Çapan et al. (1987)*

No.	[K] (wt%)	[^{40}Ar] (wt%)	C (%)	Age (Ma)	Locality	Classification (1)	Classification (2)	Classification (3)
12	1.05	2.6 × 10^{-9}	97.6	0.35 ± 0.15	Sulumağara	Alkali olivine basalt	Subalkaline basalt	Alkali basalt (22/36)
15	0.873	2.8 × 10^{-9}	96.9	0.45 ± 0.15	Alagözbanisi	Alkali olivine basalt	Subalkaline basalt	Alkali basalt (23/34)
18	1.02	4.2 × 10^{-9}	95.1	0.60 ± 0.10	Şarklı	Quartz tholeiite	Tholeiitic basalt	Tholeiite (23/13)
22	1.14	6.2 × 10^{-9}	94.1	0.78 ∓ 0.10	Yalankoz	Olivine tholeiite	Tholeiitic basalt	Not resolved (28/28)
17	1.18	9 × 10^{-9}	94.5	1.10 ± 0.20	Söğüt	Quartz tholeiite	Tholeiitic basalt	Tholeiite (22/10)
08	1.30	1.06 × 10^{-8}	84.4	1.17 ± 0.17	Fevzipaşa	Olivine tholeiite	Tholeiitic basalt	Alkali basalt (25/33)
31	1.03	1.24 × 10^{-8}	86.4	1.73 ± 0.10	Ceylanlı	Alkali basalt	Subalkaline basalt	Alkali basalt (22/36)
53	0.723	1.03 × 10^{-8}	90.4	2.10 ± 0.20	Korogha Geri	Olivine tholeiite	Tholeiitic basalt	Not determined

Data are from table 1 of Çapan *et al.* (1987). Calculations assume decay constants and percentage abundance of ^{40}K from Steiger & Jäger (1977). [K] The percentage composition of potassium, by weight; [^{40}Ar], the percentage composition of radiogenic ^{40}Ar, by weight. Classifications are: (1) from Çapan *et al.* (1987); (2) this study, using the Le Bas *et al.* (1986) scheme; (3) this study using the Floyd & Winchester (1975) criterion of the Y/Nb ratio (with Y and Nb concentrations listed in ppm.) [This table is provided here because the Çapan *et al.* (1987) article is inaccessible.]

Chorowicz (1998) have all presented their opinions on the detailed geometry of this junction.

The abundant young volcanism of this area has been regarded as Quaternary (e.g. Tolun & Pamir 1975), and Çapan *et al.* (1987) obtained eight Plio-Quaternary K–Ar dates for sites in the Karasu Valley between Fevzipaşa and Kırıkhan (Fig. 5; Table 2). This locality is dominated by the north–south trending Amanos Mountains, whose eastern escarpment rises from the *c.* 500 m level of the valley floor to > 2200 m (Figs 3 and 5). Many previous studies have regarded this escarpment as bounded by an active left-lateral fault called the Amanos Fault, although there is no structural or geomorphological evidence for significant active strike-slip on it, nor any evidence of local historical seismicity (Westaway 1994). Instead, firstly, Westaway & Arger (1996) suggested that this fault has a minimal slip rate at present. Second, kinematic consistency between adjoining major strike-slip fault segments requires no significant slip on it since the present geometry of the EAFZ became active. Third, basalt flows that straddle this fault, dated at *c.* 1 Ma (Çapan *et al.* 1987) (Söğüt, *c.* 1.18 Ma; and Şarklı, *c.* 1.02 Ma; Table 2; Fig. 5) are offset by no more than a few hundred metres [see Şaroğlu *et al.* (1992*b*, Fig. 24)]. For instance, the left-lateral offset of the Söğüt flow has been measured as 325 m (Rojay *et al.* 1998). Using the Çapan *et al.* (1987) age for this flow, the time-averaged slip rate on this fault has since been *c.* 0.3 mm a^{-1}. As this is much smaller than the *c.* 7 mm a^{-1} relative velocity of the African and Arabian Plates (from Westaway 1994), the conclusion by Westaway & Arger (1996) that this fault does not have a significant role in the present-day regional kinematics appears sound.

Specimens from the Karasu Valley were not dated in this study. Firstly, because it appeared unnecessary to duplicate the work of Çapan *et al.* (1987) when many flows in adjoining areas have not been dated. Secondly, the specimen collected for geochemical analysis (F2, from Fevzipaşa; Fig. 4) displayed some evidence of alteration, possibly due to the widespread hydrothermal activity in this region (e.g. Tezcan 1979). Such alteration can cause anomalously young K–Ar dates by damaging the host crystal lattice and thus releasing radiogenic argon (e.g. Dalrymple & Lanphere 1969, p. 198).

Heimann *et al.* (1998) and Rojay *et al.* (1998) have carried out independent K–Ar dating of Karasu Valley basalts. Their dates range from

1.57 ± 0.08 to 0.05 ± 0.03 Ma, and are 0.08 ± 0.06 and c. 0.19 Ma for the Söğüt and Şarklı flows, respectively, making both the overall age span and the ages of individual flows systematically younger than the results of Çapan et al. (1987). Heimann et al. (1998) and Rojay et al. (1998) neither reported the checks they made for alteration of specimens nor quoted their proportions of atmospheric argon. In a specimen containing c. 1% potassium with a true age of < 100 ka, the measured argon content is > 99% non-radiogenic. The calculated radiogenic argon content thus depends critically on: (1) the ratios of argon isotopes assumed to be present in the atmosphere, which fluctuates; and (2) the concentration and isotope ratio in the mainly ^{38}Ar 'spike' which was mixed with the specimen. These parameters were not quoted by Çapan et al. (1987) either but, as they concluded that substantial proportions of radiogenic argon were present (Table 2), their precise values are not so critical. Treating these new K–Ar dating results as valid ages would require a slip rate of c. 4 mm a^{-1} on the Amanos Fault, a substantial proportion of the AF–AR Plate motion. This would require major localized deformation around Türkoğlu (Fig. 3), where this fault intersects the AF–TR Plate boundary, where none is observed (Westaway & Arger 1996). It would also require the slip rate on the Gölbaşı-Türkoğlu Fault, east of Türkoğlu, to be c. 4 mm a^{-1} greater than the slip rate on the Düziçi Fault and its in-line continuations further west (Fig. 3), whereas the available field evidence indicates that these slip rates are equal (Westaway & Arger 1996). It is also noted here, that if the same rock yields two K–Ar dates, one systematically younger than the other, it is customary to reject the younger date as a result of alteration, escape of radiogenic argon, or other possible causes considered by Dalrymple & Lanphere (1969, pp. 198–200).

Çapan et al. (1987) noted substantial compositional variations in the Karasu Valley basalts (Fig. 5; Table 2). The largest flow, at Yalankoz, was reported as olivine tholeiite, with phenocrysts of olivine, diopside and calcic plagioclase (typically 50–60% anorthite; i.e. labradorite), but some smaller flows are either more or less mafic, being classified as alkali olivine basalts and quartz tholeiites, respectively (Table 2).

Quaternary basalt flows persist for c. 300 km eastward to Karacalıdağ, a 1957 m high basaltic shield volcano which covers c. 10 000 km^2 of the Arabian Platform. Sanver (1968) reported dates of 1.5 ± 0.1 and 1.0 ± 0.1 Ma from its eastern flank, and Pearce et al. (1990) dated basalts from its northern margin at 0.9 ± 0.3 and 0.8 ± 0.9 Ma. According to Ota & Dinçel (1975) and Pearce et al. (1990), these basalts typically contain phenocrysts of olivine, titanaugite and labradorite. Apart from Karacalıdağ, the largest volcanic centres in this study area cover c. 1000 km^2 (Fig. 3) in the Zavrak and Gülbahar Hills (Tolun & Erentöz 1962). Necks and flows in the Zavrak area straddle the Kırkpınar Fault (Fig. 3), the northernmost segment of the AF–AR Plate boundary according to Westaway & Arger (1996). However, samples were not taken from this area, because to obtain a representative number of samples would have been beyond the resources available: instead, smaller areas of volcanism were concentrated on.

As well as this Quaternary volcanism, Karig & Kozlu (1990) reported basaltic volcanism around Kahraman Maraş associated with the lower part of the Kalecik sedimentary formation, of Late Burdigalian–Langhian (c. 17–15 Ma) age. They regarded this volcanism as marking regional uplift of the land surface through sea level. Further south, this basalt covers clastic marine sediments of the Selmo Formation, which are also Burdigalian (e.g. Terlemez et al. 1997). Yoldemir (1987) reported a K–Ar date of 12.1 ± 0.4 Ma for this volcanism from a site near Narlı (Fig. 3). Terlemez et al. (1997) assumed that all the volcanism in the Kahraman Maraş–Gaziantep area is Miocene, including the Zavrak Hills which were previously thought to be Pleistocene (e.g. Tolun & Pamir 1975). Without quantitative age control, the best indication of a Pleistocene age for the Zavrak volcanism is that these flat-lying basalt flows rest unconformably on an erosion surface in folded Miocene marine sediments [see, for example, Terlemez et al. (1997)]. None the less, as the regional deformation sense was different in the Miocene compared with the present day (Fig. 1), one might well anticipate that these two episodes of volcanism should be chemically distinct, thus providing a method of age control which would avoid the need to date every flow.

Sampling site K1 is at Çınarlı, near the western end of a c. 5 km long and c. 1 km wide flow mapped by Tolun & Erentöz (1962), which came from a neck c. 3.5 km east of this study site (Lyberis et al. 1992). Site X was north of Narlı in the floor of the Aksu River Valley. Sample T was collected at Köprüağzı, c. 4 km northeast of another neck mapped by Lyberis et al. (1992). These flows are all depicted in detail by Terlemez et al. (1997). Sample F2 came from a c. 60 km^2 flow from a neck situated c. 7 km east of Fevzipaşa (Fig. 5).

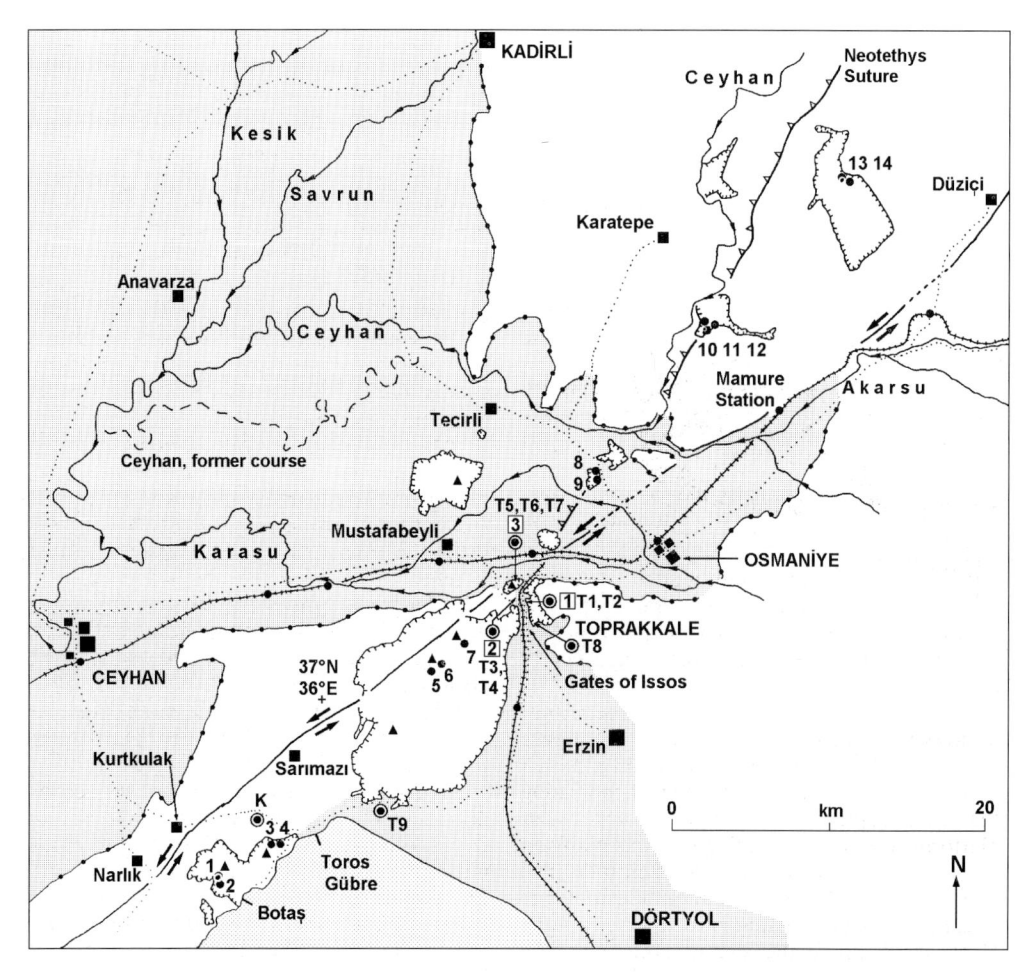

Fig. 4. Map of the İskenderun Gulf field area. Most notation is the same as in Fig. 2. Lines with double ticks are railways; dot ornament indicates the young alluvium in the Upper Cilician (Çukurova) Basin, and in the Osmaniye and Dörtyol areas. Thick line with paired arrows between Narlık, Sarımazi, Toprakkale, Mamure Station and Düziçi is the Karataş–Osmaniye Fault and its northeastward continuation. Numbers 1–3 inside boxes are Toprakkale sites 1–3; smaller markers numbered 1–14 are sampling sites for geochemical analyses by Bilgin & Ercan (1981). Dashed line west of Tecirli is a former course of the Ceyhan River. Map uses Harita Genel Müdürlüğü 1:200 000 sheets G9-Kozan [Sis], G10-Maraş and H9-Adana as a base, plus information from Tolun & Erentöz (1962), Bilgin & Ercan (1981), Gökçen et al. (1988), Karig & Kozlu (1990) and Lyberis et al. (1992). Inconsistencies between these sources are resolved using field observations from this study.

The İskenderun Gulf area

The final field area of this study was beside the S60°W trending Karataş–Osmaniye Fault (KOF) (Şaroğlu et al. 1992a) that forms the modern TR–AF Plate boundary to the north of İskenderun Gulf. Basalt flows locally cover a $c.\ 60 \pm 10$ km area (Fig. 4), oriented at $c.\ 20°$ to the KOF but subparallel to the Neotethys suture that passes offshore beneath the gulf (e.g. Karig & Kozlu 1990). Bilgin & Ercan (1981) described these rocks as typically $c.$ 45–55% plagioclase (mainly labradorite) and $c.$ 10–20% olivine, plus titanaugite. Their chemistry has been investigated further by Yurtmen et al. (2000).

The Armenian castle of Toprakkale, at the point where this suture crosses the KOF, is built of local basalt on a neck that rises to $c.$ 160 m altitude, $c.$ 100 m above the surrounding valley floors. Gottwald's (1940) archaeological report was the first study to interpret this neck, which has lost its natural shape due to quarrying and construction work. It now also adjoins a complex road and rail junction, where other excavations have made it impossible to determine the continuity of flows. About 2 km south of this site, a c. 400 m wide and c. 80 m deep dry valley, the Gates of Issos, leads south towards İskenderun Gulf (Fig. 4). It is incised into Miocene sediments, its surroundings being capped by basalt.

Sample T8 came from the flow capping the eastern escarpment of the Gates of Issos. T1 and T2 were collected c. 1.8 km further north, from a c. 3 m thick exposure of basalt east of a road junction (Fig. 4.1). T2 is more vesicular and was judged to indicate the top of an older flow beneath T1. The road surface is locally c. 8 m above a c. 2006 m wide plain where samples T3 and T4 were collected (Fig. 4.2). T5–T7 came from the north face of the Toprakkale neck, c. 5 m below the outer castle wall (Fig. 4.3). T5 and T6 are coarse-grained olivine basalt; T7 is a fine-grained, ropy vesicular basalt from c. 1 m higher. Within the castle, it was noted that thin conglomerate sheets interbedded between flows, suggesting that some time gap was involved between eruptions and that eruption occurred at the level of a surrounding river or lake.

Sample T9 came from near the southern limit of flows from the Hama Tepe Neck. This basalt surface is flush with the surrounding beach deposits, now c. 15 m above sea level, indicating that after eruption it formed part of a wave-cut platform. The landward margin of this platform is evident further inland at c. 40 m altitude. The flow which yielded sample K at Kurtpınar is presumed to have originated from Hayıtlı Tepe Neck, c. 2.5 km to the south, rather than the more distant Arnavut Tepe Neck (Fig. 4).

The KOF passes adjacent to the Toprakkale Neck on its northwest side and can be followed southwest towards Karataş (e.g. McKenzie 1976; Şaroğlu et al. 1992a). Published maps show basalt northwest of the KOF for c. 8 km distance, starting c. 4 km southwest of the northeast edge of the basalt at Toprakkale (or 3 km southwest of the Toprakkale Neck) (e.g. Bilgin & Ercan 1981; Gökçen et al. 1988). As no neck has been identified northwest of the KOF at this point, this basalt is presumed to be from Toprakkale. This offset indicates c. 11–12 km of slip on the KOF since this neck became active with c. 3–4 km of slip since its activity ceased (Westaway & Arger 1996).

Experimental work

About 500 g of material was collected at each site, for both K–Ar dating and geochemical analyses.

Isotopic dating

Potassium–argon dating at the Department of Physics, University of Newcastle upon Tyne, used the procedure of Wilkinson et al. (1986). In K–Ar dating it is important to preclude the possibility of significant alteration, as an altered sample is unlikely to have remained a closed system. This was accomplished by checking surfaces of samples (both external surfaces in the field and internal surfaces of vesicles in the laboratory) for evidence of alteration, and by petrological examinations which revealed no evidence of altered mineral grains. To anticipate results, two groups of ages were found: Early Miocene (19–15 Ma) and Plio-Quaternary (2.3–0.6 Ma), both sets of samples being petrologically equivalent. The main practical difficulty in dating the Plio-Quaternary group is the small amount of radiogenic argon present, due to the young age and the low potassium content. Mitchell & Westaway (1999) have discussed at some length how measurement errors factor into uncertainties in the radiogenic argon content and age for specimens of similar age and composition; these points are thus not repeated here. The Miocene samples contain more radiogenic argon but because they are older their potential for having become weathered is greater.

Sample T7, which was vesicular, was rejected after secondary mineralization (zeolites?) was found within it. E3, E5 and F2 were also rejected after other evidence of alteration was noted. Of 21 samples, 16 were subjected to K–Ar study. Three (T3, T5 and T9) yielded no detectable radiogenic ^{40}Ar. Sample A4 was not dated because A and A3 had already given concordant dates. Thus, 13 K–Ar dates were obtained (Table 3).

Except at Toprakkale, where some flows have evidently been buried then exhumed (see below), and other sites associated with palaeo-rivers or -lakes, all samples have apparently remained subaerial since extrusion. The Elazığ and Kahraman Maraş areas are arid, with, e.g., only 334 mm of annual rainfall at Malatya (e.g. Ionides 1937, p. 31) and 341 mm at Aleppo (Taha et al. 1981). The İskenderun Gulf area is wetter, with 625 mm of rainfall at Adana (Taha et al. 1981). Similar arid climates also typified southeast Turkey during Mio-Pliocene time (e.g. Brinkmann 1976, pp. 74–78) and are conducive to low rates of subaerial weathering. The rejected samples derive instead from volcanism associated with palaeorivers or -lakes, and were evidently immersed in water for substantial lengths of time. The freshness in outcrop of the Miocene samples here may indeed explain why

Table 3. *Potassium–argon dating*

Sample identification		[K$_2$O] (wt%)	[^{40}Ar] (mm^3 g^{-1})	C (%)	Age (Ma)
Elazığ area					
E1	3525	1.61 ± 0.02	7.66 ± 0.48 × 10^{-5}	87.5	1.47 ± 0.09
E2	3519	2.70 ± 0.05	1.63 ± 0.05 × 10^{-4}	70.0	1.87 ± 0.07
A	3516	1.30 ± 0.04	6.41 ± 0.07 × 10^{-4}	46.7	15.2 ± 0.5
A3	3517	0.99 ± 0.02	5.09 ± 0.06 × 10^{-4}	42.8	15.9 ± 0.4
Kahraman Maraş area					
K1	3536	1.29 ± 0.01	6.92 ± 0.10 × 10^{-4}	69.3	16.5 ± 0.3
X	3520	1.20 ± 0.02	6.67 ± 0.08 × 10^{-4}	53.0	17.1 ± 0.4
T	3515	0.88 ± 0.01	5.32 ± 0.08 × 10^{-4}	69.3	18.6 ± 0.4
İskenderun Gulf area					
T1	3524	1.47 ± 0.01	2.88 ± 0.45 × 10^{-5}	95.5	0.61 ± 0.10
K	3526	1.56 ± 0.02	3.17 ± 0.43 × 10^{-5}	96.7	0.63 ± 0.09
T8	3518	0.674 ± 0.015	2.60 ± 0.50 × 10^{-5}	96.7	1.20 ± 0.23
T6	3513	0.691 ± 0.008	3.11 ± 0.38 × 10^{-5}	92.4	1.39 ± 0.17
T2	3512	0.675 ± 0.001	3.71 ± 0.54 × 10^{-5}	91.9	1.70 ± 0.25
T4	3521	1.43 ± 0.04	1.04 ± 0.36 × 10^{-4}	98.7	2.25 ± 0.78

The sample identifier columns list the field number followed by the Newcastle laboratory number. [K$_2$O], The percentage composition of K$_2$O, by weight, from three analyses; [^{40}Ar], the concentration of radiogenic ^{40}Ar from two analyses; C, the higher of two atmospheric contamination values. All margins of uncertainty are ±SD. Calculations assume decay constants and abundance of ^{40}K from Steiger & Jäger (1977).

they have been misinterpreted as younger in the past (see below).

Geochemical analyses

Geochemical analyses were carried out at the Department of Geology and Geophysics, University of Edinburgh. Approximately 50 g of each specimen – free of veins, amygdales, inclusions and other evidence of alteration – was selected for crushing and grinding. Glass discs for major element analysis were prepared, using *c.* 1 g of powder, as described by Fitton & Dunlop (1985). For trace element analysis, *c.* 6 g of powder was mixed with a PVA binding agent, backed by boric acid powder, and compressed using a hydraulic press to form a 40 mm diameter pellet. Ten major elements (Table 4) and 17 trace elements (Table 5) were measured using Philips PW1450 and PW1480 X-ray fluorescence spectrometers calibrated using US Geological Survey and other standard samples. Rocks were thus classified by major element geochemistry (Table 4).

Results

Dating

In the Elazığ area, samples A and A3 from Arapkir have concordant ages whose mean is 15.5 Ma, and thus fall within the *c.* 15–19 Ma age span of the Yamadağ volcanics (Leo *et al.* 1974), which predates the slip on the MOFZ. This age supports the deduction that this site has been juxtaposed against the Göldağ Volcano by this left-lateral slip. The Çipköy site yielded ages of 1.47 ± 0.09 Ma for sample E1 and 1.87 ± 0.07 Ma for E2. As E2 overlies E1, these formal uncertainties indicate some perturbation of the K–Ar system in one or both samples. Thus, a definite 'age' cannot be assigned to this volcanism, but these dates do demonstrate that the Late Miocene age usually quoted for this 'Karabakır' Formation can no longer be accepted. The normal magnetic polarity identified by Sanver (1968) may thus indicate the Olduvai event within the Matuyama chron [age 1.79–1.95 Ma according to Hilgen (1991)], not the Brunhes as Sanver (1968) thought. If so, sample E2 yielded a valid date and E1 is perturbed.

In the Kahraman Maraş area, Çapan *et al.* (1987) constrained the Karasu Valley volcanism as *c.* 2.1–0.4 Ma (Table 2). Although no Plio-Quaternary dates were obtained in this study, sample F2 was undoubtedly Quaternary given the 1.2 Ma date for Çapan *et al.*'s (1987) sample 08 from the same flow (Fig. 5). Sites K1 (Çınarlı) X (Pazarcık), and T (Köprüağzı) instead gave dates of 16–19 Ma, the mean of both concordant ages being 16.8 Ma. They evidently belong to the Miocene volcanic event reported by Karig &

Table 4. *Major element analyses*

Sample	SiO$_2$	Al$_2$O$_3$	Fe$_2$O$_3$	MgO	CaO	Na$_2$O	K$_2$O	TiO$_2$	MnO	P$_2$O$_5$	L.O.I.	Total	Normative minerals	Classification
Elazığ area														
A	56.35	16.98	7.06	4.40	7.37	3.91	1.40	1.035	0.100	0.230	0.87	99.70	Qz (9%), Or, Al, An (43%), Di, En	Basaltic andesite
A3	54.46	17.36	7.48	5.36	7.54	3.86	1.08	1.075	0.107	0.226	0.67	99.23	Qz (6%), Or, Al, An (45%), Di, En	Basaltic andesite
A4	51.73	17.72	9.07	4.51	8.64	4.13	0.80	1.457	0.137	0.346	0.75	99.30	Qz (3%), Or, Al, An (44%), Di, En	Tholeiite*
E3	47.65	16.25	10.33	7.85	8.98	3.07	0.54	1.376	0.147	0.198	3.02	99.41	Qz (0.4%), Or, Al, An (53%), Di, En	Tholeiite*
E5	47.60	16.33	10.11	7.51	9.24	3.11	0.55	1.369	0.143	0.196	3.51	99.66	Qz (0.3%), Or, Al, An (52%), Di, En	Tholeiite*
E1	46.19	16.57	10.33	6.89	9.26	3.88	1.58	2.194	0.179	0.633	1.67	99.38	Or, Al, An (44%), Ne (2%), Di, En, Fo (9%)	Hawaiite
E2	46.41	16.24	10.00	6.84	8.79	3.95	2.83	2.121	0.171	0.617	1.34	99.31	Or, Al, An (45%), Ne (6%), Di, En, Fo (8%)	Tephrite
Kahraman Maraş area														
F2	49.81	16.76	11.24	5.69	8.52	3.63	1.28	1.994	0.163	0.364	−0.13	99.32	Qz (1%), Or, Al, An (46%), Di, En	Tholeiite†
K1	52.40	14.42	11.17	6.50	7.37	3.44	1.18	1.778	0.289	0.285	0.72	99.54	Qz (6%), Or, Al, An (41%), Di, En	Basaltic andesite
T	52.79	14.67	11.33	6.70	7.84	3.23	0.94	1.549	0.129	0.184	0.16	99.52	Qz (7%), Or, Al, An (45%), Di, En	Basaltic andesite
X	52.23	15.03	10.28	4.76	9.18	3.44	1.28	1.835	0.123	0.286	1.12	99.56	Qz (6%), Or, Al, An (43%), Di, En	Basaltic andesite
İskenderun Gulf area														
T1	47.65	15.23	13.05	8.03	9.76	3.18	0.70	1.925	0.177	0.318	−0.40	99.62	Or, Al, An (48%), Di, En, Fo (2%)	Subalkaline basalt*
T2	47.39	15.34	12.93	7.70	9.84	3.29	0.69	1.896	0.169	0.316	−0.33	99.30	Or, Al, An (47%), Di, En, Fo (3%)	Subalkaline basalt*
T3	43.90	15.36	13.30	8.35	10.40	3.71	1.43	2.880	0.176	0.850	−0.53	99.83	Or, Al, An (47%), Ne (4%), Di, En, Fo (11%)	Basanite
T4	43.87	14.94	13.24	8.78	10.25	3.68	1.37	2.831	0.178	0.842	−0.61	99.37	Or, Al, An (46%), Ne (4%), Di, En, Fo (11%)	Basanite
T5	47.51	15.03	12.70	8.43	9.81	3.07	0.75	1.862	0.166	0.310	−0.16	99.48	Or, Al, An (49%), Di, En, Fo (3%)	Subalkaline basalt*
T6	47.77	15.39	12.49	7.90	9.91	3.05	0.76	1.886	0.160	0.307	−0.14	99.49	Or, Al, An (50%), Di, En, Fo (1%)	Subalkaline basalt*
T7	48.12	15.27	12.53	7.57	9.73	3.23	0.73	1.883	0.164	0.325	0.32	99.88	Or, Al, An (48%), Di, En, Fo (0.3%)	Subalkaline basalt*
T8	47.20	15.42	13.06	7.84	9.74	3.16	0.72	1.946	0.180	0.314	−0.11	99.46	Or, Al, An (49%), Di, En, Fo (3%)	Subalkaline basalt*
T9	44.77	15.29	12.94	9.27	9.93	3.23	1.27	2.554	0.186	0.476	−0.23	99.68	Or, Al, An (47%), Ne (0.8%), Di, En, Fo (13%)	Basanite
K	47.53	15.03	12.87	8.48	9.71	3.15	0.77	1.891	0.180	0.458	−0.22	99.85	Or, Al, An (48%), Di, En, Fo (3%)	Subalkaline basalt†

Compositions of major elements are quoted in wt%. Normative minerals column indicates the presence of selected normative minerals in CIPW normative calculations determined using the algorithm of Fears (1985); Qz, quartz; Or, orthoclase; Al, albite; An, anorthite; Ne, nepheline; Di, diopside; En, enstatite; Fo, forsterite. Abundances quoted for quartz, nepheline, and forsterite are relative to total wt%. Abundance quoted for anorthite is as total of plagioclase (Al + An), as wt%. Classification field lists classification according to major element composition, after Le Bas et al. (1986). Samples which fall within the basalt (s.s.) field are subdivided according to the presence of normative minerals. For specimens which fall within the basalt field of Le Bas et al. (1986), classifications using the Y/Nb ratio are also given, according to Floyd & Winchester (1975);* and †, specimens which classify according to this method as tholeiites (Y/Nb > 1) or alkali basalts (Y/Nb < 1, respectively, using the data from Table 5.

Table 5. Trace element analyses

Element	Elazığ area							Kahraman Maraş area				İskenderun Gulf area								
	A	A3	A4	E3	E5	E1	E2	F2	K1	T	X	T1	T2	T3	T4	T5	T6	T7	T8	T9	K
Nb	10.7	10.7	12.3	7.3	7.5	68.9	69.4	2.82	21.6	12.6	20.8	17.8	18.1	51.0	51.1	17.7	17.6	17.3	18.2	32.1	27.4
Zr	166.0	164.7	174.1	117.4	117.6	251.9	250.0	174.7	141.8	113.6	155.0	119.1	120.2	199.9	195.9	117.6	118.3	117.5	118.9	172.6	149.6
Y	20.5	20.1	25.4	24.7	24.7	26.3	26.2	26.1	27.7	21.6	24.0	20.1	20.7	25.6	25.6	20.3	19.8	19.8	20.7	23.9	22.0
Sr	359.3	355.6	431.2	306.1	302.1	831.4	803.9	508.9	349.1	276.4	364.8	504.5	541.1	980.3	975.0	502.9	517.5	470.4	510.9	688.6	610.2
Rb	39.4	14.3	9.1	11.2	11.4	48.2	61.3	17.3	24.3	24.0	25.1	6.3	5.9	14.2	13.8	5.8	5.7	7.0	7.1	13.6	10.5
Th	5.7	5.7	3.5	3.9	4.7	8.1	6.4	5.7	4.9	3.4	4.2	4.2	4.2	4.0	5.0	6.0	5.7	5.0	3.6	5.7	8.1
Pb	7.3	6.1	4.5	3.9	2.3	3.7	4.3	4.4	4.5	7.0	4.3	2.4	0.6	0.7	2.3	3.7	1.7	0.6	2.1	1.6	3.8
Zn	66.6	68.7	80.2	80.0	78.6	78.0	80.2	99.1	102.5	100.8	95.4	97.9	103.2	95.3	95.4	97.8	99.5	104.3	98.3	82.9	99.0
Cu	39.0	45.8	36.3	50.7	52.1	45.8	46.2	17.4	62.2	45.7	32.8	48.6	56.1	62.1	63.2	47.5	49.6	50.7	46.1	62.0	47.9
Ni	90.5	120.6	49.5	111.8	112.9	104.3	93.8	32.2	650.0	151.1	92.8	114.5	119.6	133.9	142.5	118.2	108.4	110.3	124.1	161.2	134.7
Cr	164.3	162.5	121.1	228.3	255.1	203.1	196.7	102.0	287.6	251.4	240.4	322.3	359.9	303.4	250.0	332.1	318.7	320.6	324.4	247.8	330.1
Ce	46.2	43.9	40.7	34.9	34.4	77.6	83.4	55.1	47.1	30.4	41.3	45.4	34.4	74.9	74.6	43.8	44.1	33.9	42.6	46.9	69.8
Nd	20.2	16.0	22.3	15.6	19.1	30.1	34.7	30.2	28.6	16.0	22.3	21.6	19.0	34.2	37.0	20.7	17.8	19.7	21.8	22.3	27.3
La	22.8	22.9	19.3	16.1	17.0	38.4	39.5	27.9	27.3	9.9	17.0	19.7	17.9	36.7	32.7	16.0	19.8	13.6	13.1	18.7	38.3
V	149.8	148.4	203.5	153.3	175.1	194.0	179.3	205.4	177.0	159.1	174.3	190.3	190.1	222.7	205.4	180.5	186.7	199.0	185.9	194.7	186.4
Ba	245.1	261.0	149.4	164.2	164.0	626.2	603.0	303.2	669.3	181.0	277.3	187.4	170.0	320.6	328.5	288.3	311.9	231.6	180.8	198.3	272.0
Sc	20.3	20.3	23.1	21.9	27.6	19.9	19.3	23.5	19.0	15.3	18.5	21.0	23.0	23.7	21.8	20.7	21.5	23.6	19.5	25.1	19.8
Y/Nb	1.92	1.88	2.07	3.38	3.29	0.38	0.38	0.93	1.28	1.71	1.15	1.13	1.14	1.14	0.50	1.15	1.13	1.14	1.14	0.74	0.80
Zr/Nb	15.51	15.39	14.15	16.08	15.68	3.66	3.60	6.20	6.56	9.02	7.45	6.69	6.64	3.92	3.83	6.64	6.72	6.79	6.53	5.38	5.46

Concentrations are expressed in ppm.

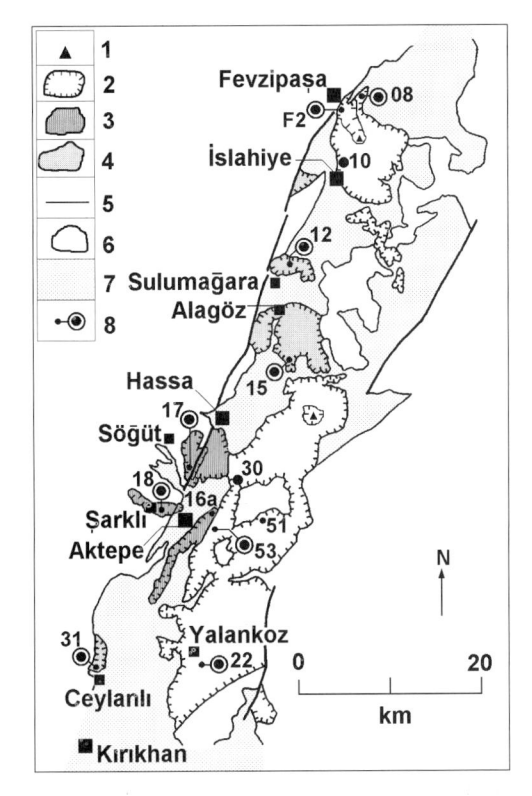

Fig. 5. Simplified geological map of the Karasu Valley, adapted from Çapan *et al.* (1987, fig. 2) with information from Şaroğlu *et al.* (1992b, fig. 24). 1, Principal volcanic necks; 2, outcrop of rock classified by Çapan *et al.* (1987) as olivine tholeiite; 3, outcrop which they classified as quartz tholeiite; 4, outcrop which they classified as alkali olivine basalt; 5, major faults where clearest in the field; 6, pre-Quaternary rocks; 7, Quaternary alluvium; 8, isotopic dating sites, comprising Çapan *et al.* (1987) sites 08, 12, 15, 17, 18, 22, 31 and 53 (Table 2), and site F2 from this study.

Kozlu (1990), the mean age here being indistinguishable from their Late Burdigalian–Langhian (*c.* 17–15 Ma) age. This event within the Arabian Platform thus occurred at the same time as the Yamadağ volcanism within the Anatolian fragment.

The İskenderun Gulf samples yielded dates of 2.3–0.6 Ma (Table 3), although given the analytical uncertainties the span may be from *c.* 1.5–0.7 to *c.* 3.0–0.5 Ma. The minimum span resembles the 1.7–0.9 Ma span for the Karacalıdağ volcanism (Sanver 1968; Pearce *et al.* 1990) and the duration of the Karasu Valley volcanism, and encompasses the dates given here for the Çipköy volcanism.

At Toprakkale site 1, the stratigraphic positions of samples T1 and T2 reflect their relative ages. The flow containing T2 erupted at 1.7 ± 0.3 Ma and either: (1) remained exposed until 0.6 ± 0.1 Ma, when it was buried beneath the flow containing T1; or (2) was buried, then exhumed, before being covered by this younger flow. At site 2, the 2.3 ± 0.8 Ma date of sample T4 means that it cannot be resolved whether it was erupted before, with or after nearby T2. Although the contact between the flows containing samples T2 and T4 has been obliterated, it seems probable that because T4 is the lowest flow exposed it is also the oldest.

Samples T6 from Toprakkale site 3 and T8 from east of the Gates of Issos Gorge, both from the highest altitudes sampled, yielded concordant dates of 1.3 Ma. Like the earlier studies, a neck east of the Gates of Issos, which could have produced the basalt there, was not observed. The flow containing sample T8 therefore most likely came from the Toprakkale Neck at roughly the same time as sample T6. This means that at *c.* 1.3 Ma, the Gates of Issos Gorge did not exist, which in turn means that interpretation (1) of samples T1 and T2 (above) is not feasible. This gorge, and the topographic low east of Toprakkale, thus formed during 1.3–0.6 Ma. The subsequent abandonment of this gorge to create the present-day dry valley is thus even younger.

Major element geochemistry

The specimens in this study have compositions ranging from basanites to basaltic andesites (Fig. 6). Petrologically, they are indistinguishable from other basalts described, e.g. Leo *et al.* (1974), Ota & Dinçel (1975), Bilgin & Ercan (1981), Pearce *et al.* (1990) and Yurtmen *et al.* (2000), from either the same outcrops or others nearby: thus, their petrology will not be discussed here. However, some specimens plot close to demarcation lines between established classifications. Therefore, specimens are classified according to major elements in this section, and immobile trace elements, following Floyd & Winchester (1975), are used later. Other possible classification schemes could also be used (e.g. Irvine & Baragar 1971).

In the Elazığ area, samples A and A3 from Arapkir are basaltic andesites, whereas A4 is a basalt (Fig. 6). This range of composition is very similar to that reported for the Yamadağ volcanics by S. Yılmaz *et al.* (1993). This evidence, that both the age and the composition of these volcanics and those of Yamadağ are equivalent, supports the suggestion that the Arapkir volcanics were originally juxtaposed against the

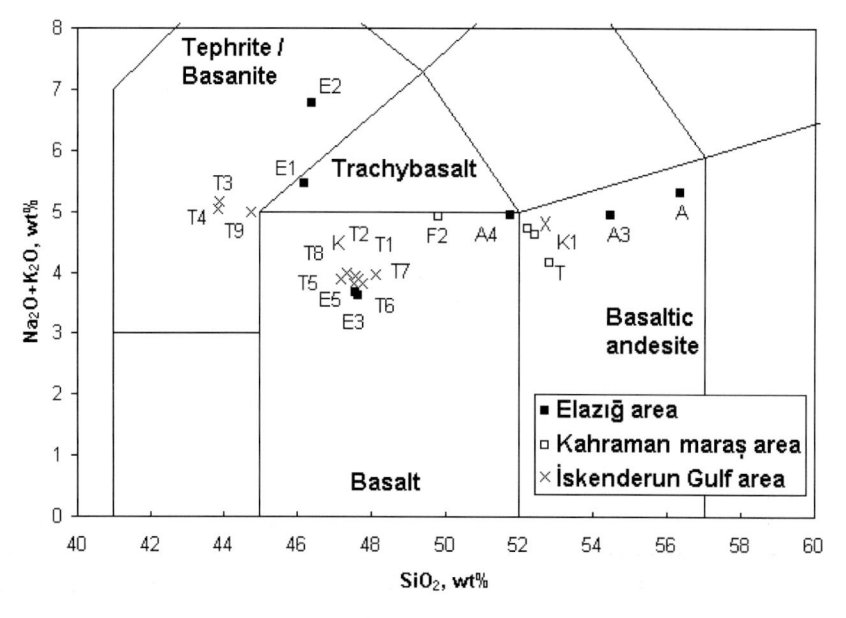

Fig. 6. Classification of specimens from this study using the Le Bas *et al.* (1986) scheme.

Yamadağ volcanics, and have since been separated by left-lateral slip on the Arapkir–Doğanşehir Fault (Fig. 2).

Samples E3 and E5 from Örtülü (Fig. 2) are marginally silica-oversaturated basalts, with compositions very similar to many samples which the present authors (Fig. 6) and others (Figs 7 and 8) have described from the

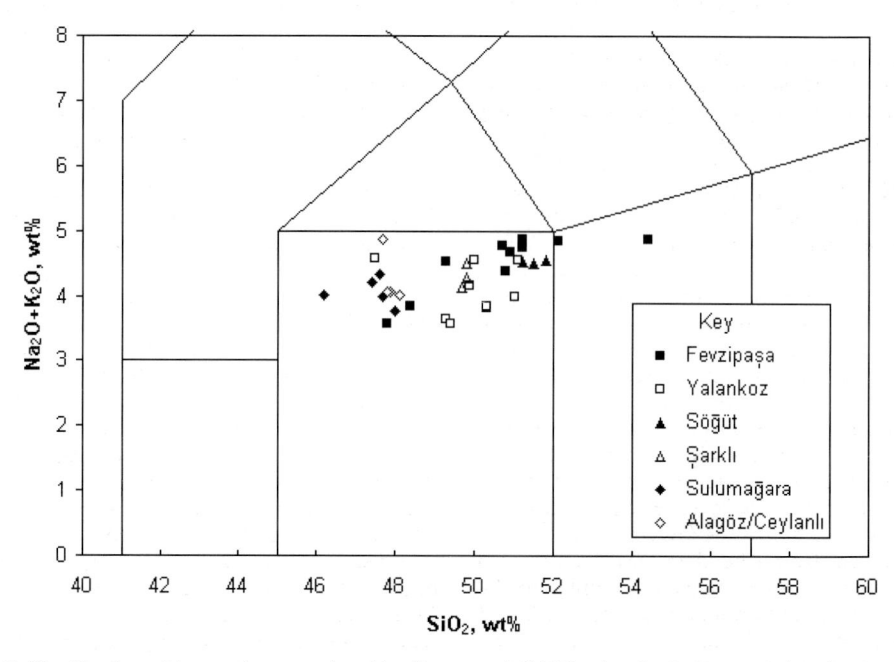

Fig. 7. Classification of the specimens analysed by Çapan *et al.* (1987) using the Le Bas *et al.* (1986) scheme.

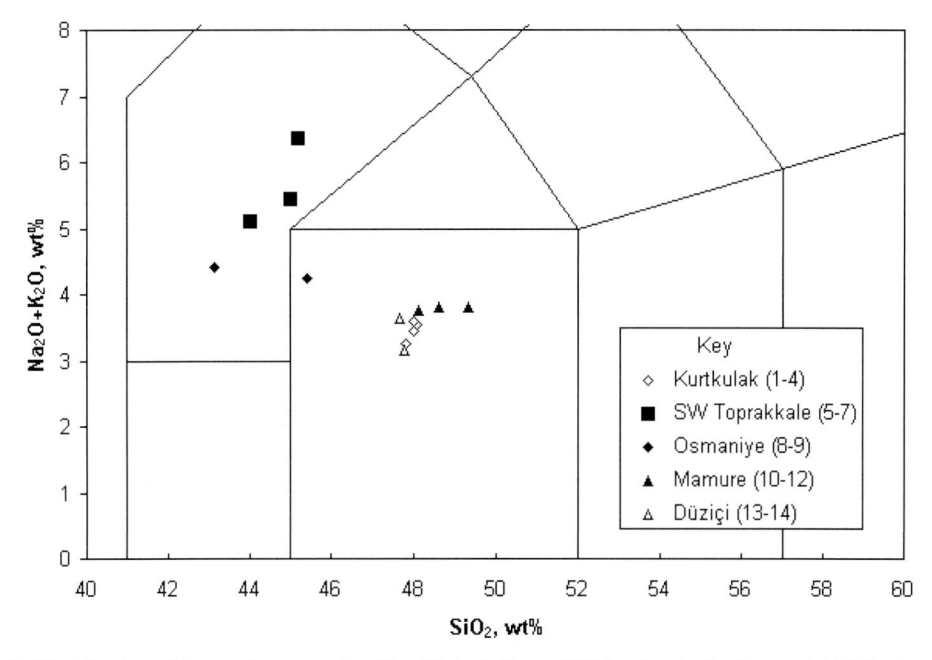

Fig. 8. Classification of the specimens analysed by Bilgin & Ercan (1981) using the Le Bas *et al.* (1986) scheme.

Kahraman Maraş and İskenderun Gulf study areas. At Çipköy, the stratigraphically younger sample, E2, is much richer in potassium and is formally a tephrite, whereas E1 is a hawaiite (Table 6).

The Miocene specimens from the Kahraman Maraş area of the Arabian Platform, K1, T and X, are all basaltic andesites, and fall within the range of composition of similar rocks of the same age from Arapkir (Fig. 6) and Yamadağ (Leo *et al.* 1974; S. Yılmaz *et al.* 1993) within the former Anatolian continental fragment. Specimen F2, another marginally silica-oversaturated basalt (Table 6), is very similar to many of the specimens analysed by Çapan *et al.* (1987), which are Quaternary from their K–Ar dating (Table 2). However, other specimens from around Fevzipaşa analysed by Çapan *et al.* (1987) are less mafic, with compositions similar to Miocene samples from the Kahraman Maraş and Arapkir areas of this study (Figs 6 and 7). Others from the Karasu Valley, also analysed by Çapan *et al.* (1987), are slightly more mafic, with some normative olivine. However, their age data in Table 2 do not reveal any clear variation in composition across the Quaternary.

Specimens from İskenderun Gulf range from subalkaline basalts to basanites, similar to the ranges in composition noted in other studies of this area (Bilgin & Ercan 1981; Yurtmen *et al.* 2000). The most mafic specimens (Table 4) are T3, T4 – which from field relationships and the K–Ar date for T4 (Table 3) marks the earliest eruptions from the Toprakkale Neck – and T9, whose age is not constrained. The less mafic younger basalts from the Toprakkale Neck are instead marginally silica-undersaturated, with compositions similar to the most mafic samples from the Karasu Valley (Figs 6–8).

Bilgin & Ercan (1981) noted that the most mafic specimens from this area are found on either side of the Toprakkale Neck (Figs 4 and 8), with less mafic rocks in the northeastern and southwestern extremities of this volcanic field. The data from this study reveal the pattern to be more complex: basanites are only found in the central part of this volcanic field but younger eruptions in this locality have produced less mafic rocks.

Given that the Toprakkale Neck marks the intersection of the active left-lateral KOF with the Neotethys suture (Fig. 4), it is suggested that the intersection of these two major discontinuities in the crust may have created a particularly easy path for magma to reach the Earth's surface. If so, then, in principle, the mafic character of flows in this area may reflect the resulting lack of crustal contamination; younger basaltic magmas may have reached the surface more slowly and thus experienced a little contamination. Similar reasoning may, in principle, explain the typical slightly more mafic character of

Table 6. *Strontium and neodymium isotope ratios*

Sample	Locality	Classification (1)	Classification (2)	SiO$_2$	^{87}Sr/^{86}Sr	^{143}Nd/^{144}Nd
12	Sulumağara	Alkali olivine basalt	Subalkaline (Fo 10.3%)	47.4	0.70335	0.51291
15	Alagöz	Alkali olivine basalt	Subalkaline (Fo 4.4%)	48.1	0.70369	0.51284
1911	Karacalıdağ		Subalkaline (Fo 0.6%)	46.7	0.70398	0.512948
1912	Karacalıdağ		Tholeiite (Qz 1.6%)	50.2	0.70431	0.512887
08	Fevzipaşa	Olivine tholeiite	Tholeiite (Qz 2.8%)	51.2	0.70441	0.51269
30	Yalankoz	Olivine tholeiite	Tholeiite (Qz 3.1%)*	50.0	0.70478	0.51280
16A	Şarklı	Quartz tholeiite	Tholeiite (Qz 4.8%)	49.8	0.70488	0.51264
17	Söğüt	Quartz tholeiite	Tholeiite (Qz 3.7%)[†]	51.2	0.70549	0.51262

Data are from Çapan *et al.* (1987), except for the two Karacalıdağ specimens from Pearce *et al.* (1990).
Classifications: (1), from Çapan *et al.* (1987); (2), from this study, based on normative compositions [Fo, forsterite; Qz, quartz; *, normative composition of Çapan *et al.* (1987) sample 22; [†], normative composition of their sample 17A].

the Quaternary basalts from the Karasu Valley, compared with the Miocene basalts from elsewhere in the Kahraman Maraş area (Figs 6 and 7). Although the DSFZ may well have already existed at the time of the Miocene volcanism (see below), none of this study's localities was near any contemporaneous active strike-slip fault. In contrast, the Quaternary magmas of the Karasu Valley may have reached the surface more easily, by rising along the line of the Amanos Fault and others en echelon to it (Figs 3 and 5), thus possibly experiencing less crustal contamination.

Trace element geochemistry

Previous investigations of the trace element geochemistry of basalts from southeastern Turkey have identified no clear reason for this magmatism. For instance, Çapan *et al.* (1987) believed that the variable composition of Karasu Valley basalts requires mixing of different primary magmas but could offer no explanation. Pearce *et al.* (1990) suggested that the basaltic volcanism at Karacalıdağ (which has a very similar composition to the Karasu Valley basalts) resulted from heating of the mantle lithosphere, but could not identify any reason for this. Yurtmen *et al.* (2000) suggested that the İskenderun Gulf volcanism has characteristics of ocean island basalts. However, its small scale, brief duration and the absence of older volcanism nearby (which could be interpreted as a plume track) makes it unlikely that a mantle plume is present beneath this region.

The most obvious variation in trace element content concerns Nb, Zr, Sr and Rb, which are concentrated in the most mafic specimens (E1, E2, T3, T4 and T9). First, ratios of the immobile trace elements Y (which is virtually constant in most specimens, at *c.* 20 ppm) to Nb were examined, in order to classify the specimens using the scheme of Floyd & Winchester (1975), where Y/Nb < 1 indicates an alkali basalt and Y/Nb > 1 indicates a tholeiite (Table 5). Using this scheme, the most mafic specimens (E1, E2, T3, T4 and T9) would all classify as alkali basalts, unless excluded by their major element compositions. Except for two, F2 and K, all specimens which classify as basalts by major elements classify as tholeiites using Y/Nb. Specimen F2 is formally an alkali basalt, although with Y/Nb *c.* 0.93 it lies near the demarcation line. However, this specimen contains a small amount of normative quartz (Table 4) and, on this basis, is marginally a tholeiite; it thus classifies differently according to the method used. Specimen K is also an alkali basalt according to Y/Nb, although subalkaline according to major elements. Similar variations in classification are evident for some Çapan *et al.* (1987) specimens from the Karasu Valley (Table 2).

The most mafic specimens contain the highest concentrations of Nb and Zr but with the lowest Zr/Nb ratios, of *c.* 4 (Table 5; Fig. 9a). In this study, marginally tholeiitic/subalkaline basalts have Zr/Nb *c.* 6–7, and least mafic specimens have Zr/Nb *c.* 15. As others (e.g. Camp & Roobol 1992) have discussed, the Zr/Nb ratio is a measure of the degree of partial melting of mantle peridotite required to produce a given rock of basaltic composition. The most mafic specimens thus involved the smallest degrees of partial melting, as can also be inferred by their relatively high concentrations of other elements which are concentrated by small-degree partial despite their separation in time. A similar effect is evident in the Elazığ area, where specimens E3 and E5 (which are not isotopically dated, but may well be Quaternary from field evidence) have the same Zr/Nb ratio of c. 15 as the Miocene specimens from Arapkir (Fig. 9a).

Whatever process has been responsible for the volcanism in this study region, it is evident that it was active in the Early–Middle Miocene and became active again around the start of the Quaternary. In an attempt to identify the cause of this volcanism, a range of empirical criteria have been applied to the data obtained in this study.

Ba/La ratio. Arculus & Powell (1986) suggested using Ba/La to distinguish basalt types. In their scheme, intraplate basalts have Ba/La of 0–2, whereas island arc basalts have Ba/La from 1 to 10 or more. The relative enrichment of island arc basalts by barium is thought to result from barium-rich fluids in subduction zones, the barium being derived from subducted sediment (e.g. Hole *et al.* 1984). All specimens here have Ba/La ≫ 2: for most it is c. 10 and for some it reaches c. 20 (Table 5). Using this criterion would lead to the rocks being classified as island arc basalts.

Zr/Rb ratio. Wilson (1989, p. 236) suggested using Zr/Rb to distinguish basalts. She noted that ocean island basalts typically have Zr/Rb of 6–15, whereas other basalts have different values: e.g. basalts that form in back-arc basins have Zr/Rb of 16 or greater. Using this criterion, most specimens here would classify as ocean island basalts. Wilson (1989, p. 138) also suggested using Sr/Rb: basalts with this ratio of c. 20–70 should classify as ocean island basalts. Using this criterion, most specimens here also classify as ocean island basalts. The only rocks with Sr/Rb > 70 are from around Toprakkale. However, if a mantle plume has somehow been responsible for the other basalts, it is not clear how these localized chemical differences can develop.

Nb content. Thompson & Fowler (1986) suggested using Nb as a criterion: basalts and basaltic andesites with < 50 ppm Nb classify as subduction related. On this basis, most specimens here classify as subduction related.

Mobile trace elements. Most specimens from this study are enriched relative to mid-ocean ridge basalts (MORB) by a factor of 3 (Sr) to 10 (Rb, Ba) in mobile trace elements (Table 5). Enrichment of this order is between that expected for

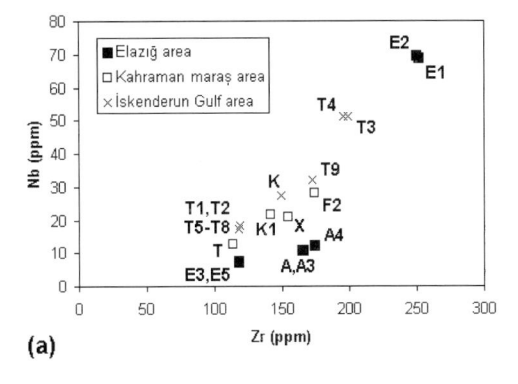

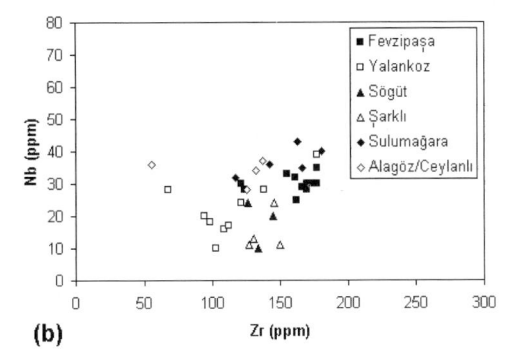

Fig. 9. Plots of Nb v. Zr concentrations for: (**a**) this study's specimens; (**b**) the specimens of Çapan *et al.* (1987). See text for discussion.

melting, such as potassium (Table 4), and thus appear unrelated to crustal contamination. This means, for instance, that around the İskenderun Gulf, rocks requiring substantially different degrees of partial melting of the mantle erupted in close proximity to each other in position and time, as is also evident using the data from Yurtmen *et al.* (2000). Another clear difference concerns the Sr/Rb ratio, which is c. 20–30 for the Elazığ and Kahraman Maraş areas but c. 60–70 or more at the İskenderun Gulf.

The Çapan *et al.* (1987) data reveal a similar pattern (Fig. 9b), including a wide range in Zr/Nb ratios at adjacent localities in equivalent groups of flows. Most relatively mafic flows yielded specimens with Zr/Nb in the range c. 4 to c. 6–7, but the tholeiitic Söğüt and Şarklı flows yielded specimens in the range c. 6–7 to c. 15. Most of their Quaternary specimens from the Karasu Valley thus yielded Zr/Nb ratios of c. 6–7 (Fig. 9b), the same as the Miocene specimens from the adjacent Kahraman Maraş area (Fig. 9a) in this study, indicating another similarity between these episodes of magmatism ocean island basalts and basalts from continental rifts, but is higher than expected for island arc basalts (e.g. Pearce 1983).

Possible causes of volcanism

It is evident that, depending on which of these (or many other) criteria are used, a case could be made that the basalts in this study region are caused by subduction, continental rifting or mantle plumes. However, there has been no subduction in this region during the Quaternary, no continental rifting and no mantle plume activity: the region has instead been dominated by strike-slip faults which behave to a good approximation as transform faults (e.g. Westaway 1994; Westaway & Arger 1996) and which thus require no regional-scale crustal extension or shortening. Most modern schemes (e.g. Aktaş & Robertson 1984; Yılmaz 1993) regard the northward subduction of the southern Neotethys Ocean has having ended in the Eocene, when continental convergence between Arabia and Anatolia began. Thus, neither subduction, extension, nor mantle plume activity is documented in the Early–Middle Miocene, either.

Previous studies (e.g. Kennett & Thunell 1975) have noted that the geological record contains evidence of significant intensifications in global volcanism in the Middle Miocene and Quaternary, irrespective of tectonic setting. These two intervals correspond to the timings of the two phases of volcanism in southeastern Turkey. However, possible local causes need to be considered before a process related to systematic global environmental change is proposed as the cause of the volcanism in southeastern Turkey.

One possibility is that the observed volcanism may be induced by strike-slip faulting. The c. 15–19 Ma age of the Miocene volcanism does suggest a possible link to the start of slip on the DSFZ, which although not well constrained is usually regarded as being at c. 15 Ma (e.g. Garfunkel 1981). The southern DSFZ has slipped left-laterally by c. 105 km (e.g. Freund $et\ al.$ 1970), of which c. 35 km has occurred since c. 5 Ma (at a c. 7 mm a^{-1} rate) and c. 70 km was earlier. Assuming the same slip rate during the Miocene yields the c. 15 Ma estimated age. The upper bound to the DSFZ age is c. 20 Ma, because basaltic dykes of this age exposed in Sinai are offset by the full 105 km distance (e.g. Garfunkel 1981). Extension in the Gulf of Suez is usually estimated to have begun at c. 24–22 Ma (e.g. Moustafa 1993), and so preceded the start of slip on the DSFZ by several million years. It is thus unclear whether the Miocene volcanism in southeastern Turkey occurred before or after the start of slip on the DSFZ. Furthermore, no mechanism is known by which pure strike-slip faulting (i.e. slip on transform faults) could initiate volcanism or volcanism could initiate strike-slip faulting.

The c. 2.3 Ma start of the younger phase of volcanism likewise suggests a possible link with the start of slip on the EAFZ, which began c. 2.8 Ma according to Westaway & Arger (1996). However, Yürür & Chorowicz (1998) instead proposed that the EAFZ became active c. 1.8 Ma. It thus appears unclear whether the volcanism or the slip on the EAFZ began first, although Westaway & Arger (1996) did note evidence that at Toprakkale (the site with the earliest Quaternary volcanism) the volcanism straddling the KOF is offset by less distance than the KOF is estimated to have slipped while the EAFZ has been active. This evidence suggests that the EAFZ became active $before$ the start of Quaternary volcanism. Furthermore, as in the Miocene, no mechanism is known by which strike-slip faulting could initiate volcanism or volcanism could initiate strike-slip faulting.

Another possibility is that the observed volcanism may be caused by crustal thickness changes resulting from strike-slip faulting which is locally oblique to the relative motion of the adjoining plates. For instance, the

---

Fig. 10. The Adıyaman region. (**a**) Geological map. 1, Metamorphic rocks of the Anatolian continental fragment (north of the Neotethys Suture): the Malatya and Pütürge metamorphics; 2, Middle Miocene and older marine sediments of the Arabian Platform; 3, Quaternary basalt; 4, Pliocene and Quaternary sediments of the Malatya and Gölbaşı Basins; 5, Late Miocene and younger lacustrine sediment of Lake Adıyaman; 6, conglomerate underlying 5. Note the rightward step in the EAFZ at Çelikhan, where it crosses the Neotethys Suture. Dashed line marks the Toraş Stream that now flows through the palaeo-outlet of Late Adıyaman. K denotes the ancient cemetery at Karakuş, where a flat 880 m hilltop marks the upper limit of lacustrine sediment. (**b**) Topographic/structural map with lacustrine sediment from (a) outlined by fine dots. Chevrons mark anticlines; solid arrow marks the outlet of the palaeolake. Topography is from Defense Mapping Agency Chart G-4A, with 1500, 2000, 2500 and 3000 ft contours labelled to the nearest metre. Other information is from Baykal & Erentöz (1961), Tolun & Erentöz (1962), Rigo di Righi & Cortesini (1964), Tolun & Pamir (1975) and Westaway & Arger (1996).

SE TURKEY VOLCANISM 477

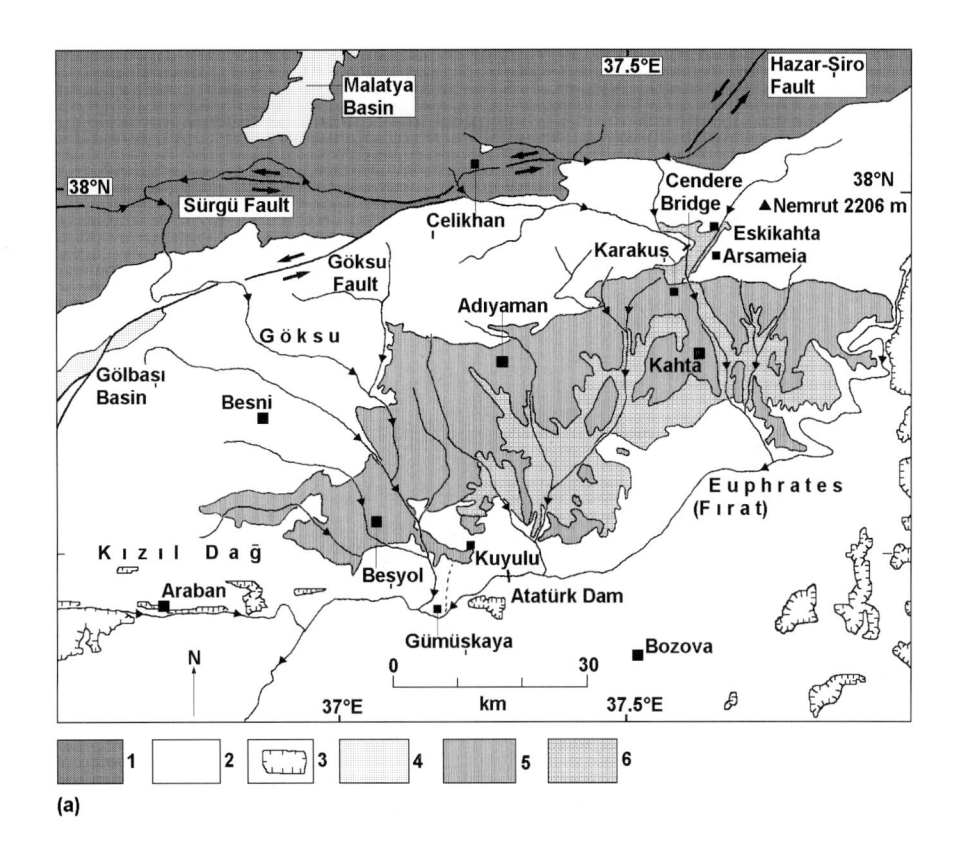

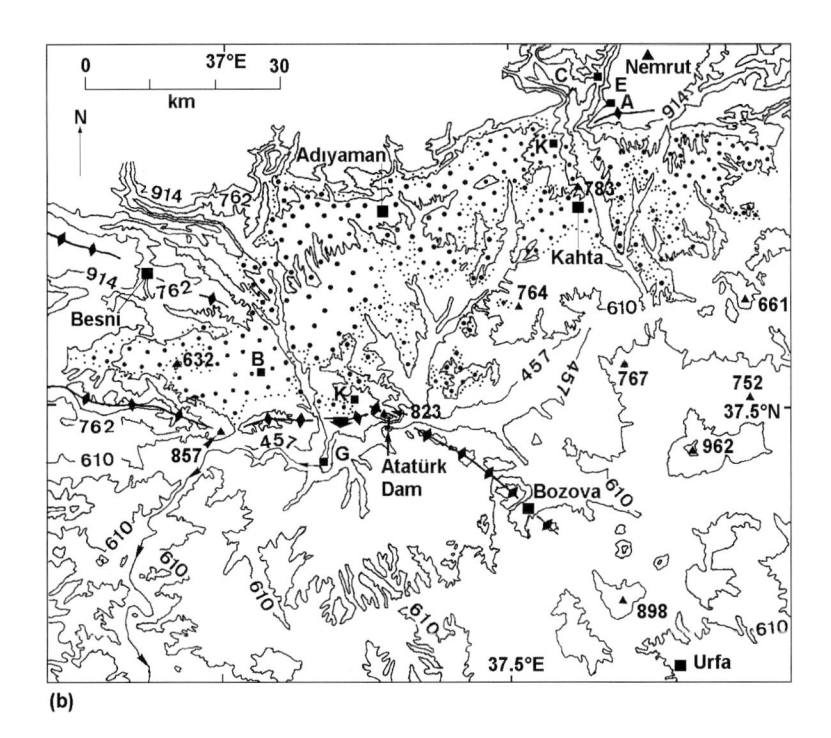

(a)

(b)

left-lateral EAFZ contains leftward steps at Gölbaşı (Fig. 10) and Lake Hazar (Fig. 1), and a rightward step where it crosses the Neotethys Suture near Çelikhan (Fig. 1). Components of localized extension are required in the vicinity of these leftward steps. In principle, the associated crustal thinning relative to surrounding localities could induce magmatism via decompression melting of the underlying mantle. However, detailed studies of both these extensional stepovers (e.g. Hempton 1984, 1985; Westaway & Arger 1996) reveal no evidence of volcanism in the vicinity. One could also argue that, in principle, the crustal thickening required to accommodate the component of distributed shortening across the Çelikhan stepover could cause heating of the underlying mantle lithosphere, potentially melting parts of it. However, detailed mapping (e.g. Tekeli & Oral 1986; Şaroğlu et al. 1992b) has documented no young volcanism within this stepover either. A third possibility is that the fracturing of the brittle upper crust caused by strike-slip faulting may provide an easy route for magma, generated by another process, to reach the Earth's surface. However, with the exception of Toprakkale (Fig. 4), volcanic necks have usually developed outside strike-slip fault zones. Thus, it is concluded that no evidence exists for a direct cause-and-effect relationship between strike-slip faulting and volcanism in this study region.

In addition to this strike-slip faulting, it has long been evident that much of the relief in southeastern Turkey is very young. Huntington (1902) inferred that much of the incision of major river gorges, such as the Euphrates, has occurred in the recent geological past. For instance, around Dutluca (Fig. 2.6), where the Euphrates Gorge crosses the MOFZ, this river has incised $c.$ 400 m into metamorphic basement beneath the young Karabakır Formation lacustrine sediments. Şaroğlu et al. (1992b) noted other geomorphological evidence of differential uplift along the EAFZ. Other clear evidence of differential uplift is provided by the tilting of a former lake basin around Adıyaman on the Arabian Platform (Fig. 10) (see also below), and by the development of marine terrace sequences along the region's coastline (e.g. Erol 1963, 1991b; Dalongeville & Sanlaville 1977; Erinç 1978). Assuming that this relief is developing in association with lower crustal roots, it requires crustal thickness changes to maintain overall isostatic equilibrium. The resulting crustal thickening will raise the Moho temperature, thus heating the mantle lithosphere and providing a potential cause of volcanism. In the absence of alternatives, this hypothesis is investigated in the next section.

Causes of Miocene and Quaternary volcanism

It is now examined whether it is feasible to regard the observed volcanism as a consequence of the crustal thickening which accompanies the observed surface uplift to maintain isostatic equilibrium. It is assumed, following Westaway (1998), that the plastic lower continental crust is highly mobile, with an effective viscosity of $c.$ 10^{19} Pa s or less, and thus readily able to flow horizontally in response to pressure gradients induced by surface processes (such as loads caused by sedimentation and negative loads caused by erosion). It is also assumed, following McKenzie (1985, 1989), that the asthenosphere is constantly experiencing small degrees of partial melting, such that incompatible elements are concentrated into the resulting metasomatic melt which percolates upward into the mantle lithosphere. Due to its low concentration, this melt will remain at each level in thermal equilibrium with its surroundings, and will thus freeze at the depth where the temperature and pressure match the melt's solidus. Frozen metasomatic melt will thus accumulate over time at a particular depth within the mantle lithosphere. Finally, it is assumed that a subsequent temperature rise can remelt this frozen melt, causing it to rise and freeze again at a shallower depth. However, if the conditions at the Moho reach the melt solidus, then the melt will escape into the crust and rise to the surface, causing volcanism.

To test this idea, one needs to know the solidus for the frozen metasomatic melt. It will be assumed that the appropriate solidus is that for 'wet' basalt, meaning for melt of basaltic composition in the presence of excess water. The experimentally determined form of this solidus has been published many times (e.g. Lambert & Wyllie 1972). At low pressures, the solidus temperature decreases with increasing pressure, reflecting the melting-point reduction caused by mixing with water; at higher pressures, it increases with pressure, due to the increased stability of hydrous minerals such as amphibole (e.g. Sood 1981, p. 116; Gill 1981, pp. 127–128). The minimum solidus temperature is 650°C at a pressure of 1.5 GPa (e.g. Lambert & Wyllie 1972).

This solidus curve has a quite complex empirical shape, but over the range of pressure, P, from 0.5 to 2.0 GPa it approximates to a parabola of the form:

$$T = A(P - P_0)^2 + B \qquad (1)$$

where A, B, and P_0 are constants, with $A = 105°C\ GPa^{-2}$, $B = 650°C$ and $P_0 = 1.5\ GPa$ (Fig. 11b). Below 1.2 GPa, the mineral assemblage produced at the solidus is a mixture of amphibole, plagioclase, clinopyroxene, olivine and water (e.g. Lambert & Wyllie 1972). On depressurization, the amphibole will decompose, releasing its water of crystallization, which along with the original excess water will escape, leaving a rock of basaltic composition.

Westaway (2000a, b) has shown that the lower continental crust is mobile enough to flow landward by distances of tens to hundreds of kilometres in response to horizontal pressure gradients caused by variations in sea level between glacials and interglacials. Because the offshore crust is thinner than the onshore crust, the offshore Moho temperature is lower and so the crust which flows from beneath the offshore shelf to beneath the land will be colder than the crust which is initially present beneath the land. This process thus affects the geothermal gradient in both onshore and offshore areas. This in turn affects the depth of the base of the brittle layer in both localities. The overall result is net flow of crust from beneath the offshore shelf to the land, causing the bathymetry of the offshore shelf to deepen and the land surface to uplift. The amplitude of the effect is proportional to a parameter DT_e, the effective temperature difference between offshore and onshore crust, which is proportional to the actual temperature difference at the depth near the Moho where the flow is concentrated, scaled by the ratio of offshore to onshore areas.

Quaternary glaciation of the northern hemisphere began at c. 2500 ka, when the first indication of lowland glaciation is revealed by evidence of ice-rafting in sediments of the North Atlantic Ocean (Shackleton et al. 1984). Variations in oxygen isotope ratios indicate that, until c. 1200 ka, the amplitude of sea-level fluctuations was about half its c. 120 m value for recent glacial cycles (e.g. Shackleton et al. 1990). In many localities, the rate of surface uplift increases from minimal values before 2500 ka, to moderate values, then to higher values after 1200 ka, evidently reflecting these two transitions (Westaway 2000a, b). During the Miocene and Pliocene, lowland glaciation was confined to Antarctica, and any small-scale sea-level fluctuations associated with the Earth's orbital cycles at this time would have induced much less intense lower crustal flow, able to cause much lower rates of surface uplift.

Increasing the crustal thickness while keeping the mantle lithosphere thickness constant, with the asthenosphere at a constant temperature, will cause the Moho temperature to rise. Crustal thickening induced by lower crustal flow can potentially raise the Moho temperature above the wet basalt solidus, thus initiating volcanism. Deformation of the continental crust as a result of lower crustal flow induced by surface processes has been designated as atectonic deformation (e.g. Kaufman & Royden 1994), in contrast with tectonic deformation which is caused by plate motions.

Testing this hypothesis, 1: predicting the timing of volcanism from uplift histories

Recent uplift of the study region is indicated by the recent incision of major gorges, like the Euphrates (Huntington 1902), and the sequences of river terraces caused by such incision (e.g. Erol 1991a, b). Much of the surrounding landscape within the Arabian Platform comprises Oligocene–Pleistocene erosion surfaces (e.g. Erol 1981, 1991a, b). Although denudation has evidently occurred along the young gorges, the much larger area of their surroundings means that the isostatic response to this denudation can be excluded as a significant cause of the recent uplift. Lower Miocene marine sediment is found above 1200 m in the Elazığ area (e.g. Baykal & Erentöz 1961) (Fig. 2). In contrast, Lower and Middle Miocene marine sediments are found only at lower levels within the Arabian Platform. The uplift since Early Miocene time has thus been greater within the Anatolian Plateau than within the Arabian Platform. However, given the many millions of years since marine conditions ended and biostratigraphic age control was lost, the timing of this uplift has previously been impossible to resolve.

South of the EAFZ, within the Arabian Platform, a 115 × 35 km palaeolake (Fig. 10), around Adıyaman, is revealed by Upper Miocene and Pliocene deposits, which locally blanket the older marine sediment, and whose facies and fossil content indicate a lacustrine environment (e.g. Tolun & Pamir 1975). Inspection of the topography reveals the southern palaeoshoreline of this Lake Adıyaman, notably its outlet at c. 470 m altitude across the Kızıldağ anticline between Kuyulu and Gümüşkaya villages, c. 50 km south of the EAFZ.

The northern limit of lacustrine sediment in Lake Adıyaman is > 800 m altitude. Southwest of Arsameia, on the eastern side of the Kâhta River Gorge, the flat-lying lacustrine silt and

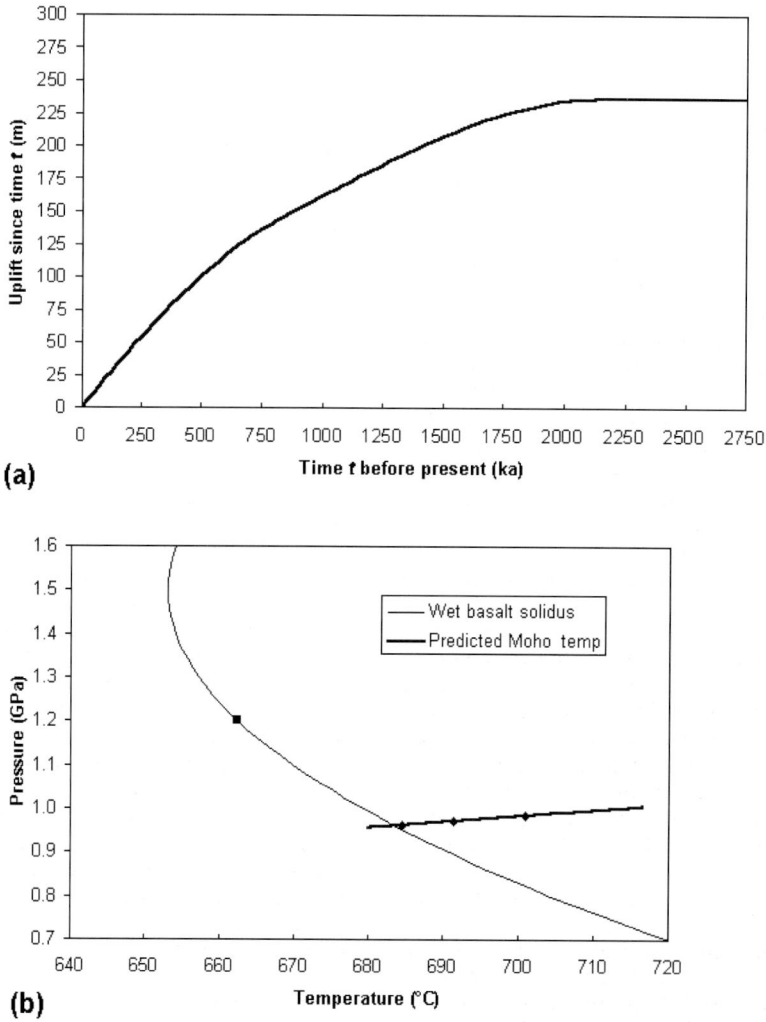

Fig. 11. (**a**) Notional Quaternary uplift history for the Arabian Platform in southeastern Turkey. The graph is calculated using the theory from Westaway (2000a,b), which quantifies the isostatic response to repeated cycles of loading and unloading of the offshore continental shelf due to repeated glacial to interglacial sea-level variations. The calculations assume the following parameter values: u (geothermal gradient in the lower crust), 20°C km^{-1}; z_b (depth of base of brittle layer), 15 km; z_i (depth where lower crustal flow is concentrated), 34 km; and k (thermal diffusivity of crust), 1.2 m^2 s^{-1}. Lower crustal flow starting at 2500 ka with an effective temperature contrast $DT_{e1} = -20$°C is assumed to be supplemented by flow with $DT_{e2} = -30$°C starting at 1200 ka. (**b**) Predicted variation in Moho temperature for the uplift profile in (a), calculated assuming that the initial Moho temperature is 680°C and the additional surface relief is isostatically compensated, with densities of 2700 and 3100 kg m^{-3} for the crust and asthenosphere, respectively. Ornament on the Moho temperature line indicates the situation at 1500, 1000 and 500 ka (i.e. 1.0, 1.5, and 2.0 Ma after the lower crustal flow is assumed to have begun). Wet basalt solidus is from Equation 1. The marker on it, at 1.2 GPa, marks the transition between melts containing normative olivine and quartz, from Lambert & Wyllie (1972). See text for discussion.

conglomerate reach c. 900 m on the southern limb of an anticline in Eocene limestone. The contact between these sediments, and the underlying lithified and folded limestone, which marks the palaeolake shoreline, can be followed at uniform altitude for many kilometres. The Kâhta and Cendere Rivers have locally incised gorges to c. 550 m altitude, exposing > 300 m thickness of the palaeolake sediment near its upper limit (Tolun & Pamir 1975).

The northern shore of this palaeolake is thus c. 400 m higher than its southern shore. Indeed, there is no land to the south high enough to contain a lake at a uniform c. 900 m altitude. The Earth's surface in this area has thus tilted south since this lake was present, with a c. 0.7° (400 m to 35 km) typical slope. The uniform dip of the lacustrine sediment, subparallel to this slope, indicates that Lake Adıyaman was only tens of metres deep, making the isostatic response to its own water unloading unimportant.

This evidence suggests that, for much of the time after marine conditions ended, conditions in this region were relatively stable with less relief than at present, and with the Anatolian continental fragment lower relative to the Arabian Platform than at present. In the relatively recent past, the uplift rate of Anatolia increased relative to the Arabian Platform, leading to the tilting of the landscape and the recent gorge incision.

As the study region's uplift history is a priori poorly constrained, a nominal amount of uplift – c. 250 m since 2500 ka – is assumed, which seems to represent a realistic lower bound for the Arabian Platform. This is modelled using two stages of intensifying lower crustal flow, using the parameter values listed in the caption to Fig. 11, to obtain the uplift history in Fig. 11a. The corresponding variations in Moho temperature and pressure are shown in Fig. 11b. Provided the initial Moho temperature was close to the solidus, volcanism starting in the Latest Pliocene or Early Pleistocene can thus be explained.

The result that typical compositions of Quaternary basalts are the same for the Arabian Platform and the Elazığ area of Anatolia has implications for the uplift histories of these regions. The present typical surface altitude of this part of Anatolia, of c. 1.5–2.0 km, requires a crustal thickness of c. 45–50 km, indicating a typical Moho pressure of c. 1.2–1.4 GPa. Under such a high pressure, the melt formed at the wet basalt solidus would produce volcanic rocks considerably less mafic than are observed, containing no normative olivine and significant normative quartz (Fig. 11b). The implication is that when local volcanism began c. 2 Ma, the crust of Anatolia was much thinner than at present, and the Earth's surface was substantially lower, possibly at a level comparable to or lower than the Arabian Platform is at present.

A final point concerns the difference in timing between volcanism in the present study region and in northeastern Turkey, where extensive Late Miocene and Pliocene volcanism has been documented (e.g. Şengör & Yılmaz 1981; Pearce et al. 1990). Along with the Caucasus, northeastern Turkey adjoins the Black and Caspian Seas, which in the Late Miocene and Pliocene were isolated from the global marine environment and experienced dramatic fluctuations in level due to cyclic variations in climate. This different timing in sea-level fluctuations may explain the different timing in volcanism. The chronologies of uplift and volcanism of the Caucasus have recently been investigated by Mitchell & Westaway (1999), who showed that crustal thickening caused by flow in the lower crust can account for this region's evolution, also.

Other instances of Miocene magmatism correlated with apparently atectonic uplift are quite numerous and may reflect the same global mechanism proposed here for southeastern Turkey. For instance, in the western USA, a pulse of magmatism occurred in the Colorado Plateau in the Middle Miocene (c. 15 Ma) (e.g. Krieger et al. 1971) while this region was uplifting (e.g. McKee & McKee 1972; Lucchitta 1979). In eastern Europe, an abrupt and widespread episode of magmatism occurred in Slovakia and western Ukraine at c. 13 Ma (e.g. Mikhaylova et al. 1974; Steininger et al. 1976). This event marked a change in this region from being submerged by a shallow shelf sea, open to the global marine environment, to a land area adjoining the landlocked Paratethys Sea (e.g. Steininger & Papp 1979). Contemporaneous brackish marine deposits of the Paratethys are now found at altitudes > 500 m across much of eastern Europe. A third example is the Middle Miocene surface uplift and volcanism of southwestern Arabia to the east of the Red Sea. It is well known that the crustal extension which led to the development of the Red Sea began c. 30 Ma and accompanied magmatism which had largely ended by c. 20 Ma (e.g. Camp & Roobol 1992). The fine-grained sediment deposited at this time indicates low relief (e.g. Jarrige et al. 1990). Subsequently, several kilometres of surface uplift has occurred in the southwestern Arabian Platform, along with several kilometres of subsidence and sedimentation within the Red Sea shelf. This dramatic change is dated to the Late Burdigalian (c. 20–16 Ma) from the start of coarse alluvial fan sedimentation (e.g. Jarrige et al. 1990), and at c. 20–14 Ma from fission track dating of the start of the associated rapid erosion (e.g. Bohannon et al. 1989). Around 12 Ma, basaltic volcanism began within the uplifting part of Arabia, whose diverse composition (e.g. Camp & Roobol 1992) resembles this study region. It is suggested that this sequence of events, which has not previously been satisfactorily explained, was initiated by the start of cyclic southern hemisphere glaciations around the end of the Early Miocene, whose transient water loading effects have caused the net landward lower crustal flow, which has led to the onshore surface uplift and offshore subsidence; and that the volcanism occurred later once the mantle lithosphere had been heated sufficiently by the onshore crustal thickening.

Testing this hypothesis, 2: predicting Nd/Sr isotopic compositions

The isotope ratios ^{143}Nd/^{144}Nd and ^{87}Sr/^{86}Sr, in specimens from the northern Arabian Platform, are predicted for comparison against the observations in Table 6 and Fig. 12, using the method of Fitton & Dunlop (1985). This method assumes that, in the distant geological past, the material now comprising the studied basalts became isolated from primitive mantle (which provides the source for MORB) as a result of small-degree partial melting. This partial melting also caused chemical fractionation, which affected the concentrations of Nd and Sr, as well as Rb and Sm whose radioactive isotopes decay to ^{87}Sr and ^{143}Nd.

MORB itself is produced by high-degree partial melting and so will not experience this chemical fractionation: as a result, the ^{143}Nd/^{144}Nd and ^{87}Sr/^{86}Sr ratios in MORB and its source material are the same. However, the chemical fractionation which occurred in the past when the metasomatic melt separated from primitive mantle by small-degree partial melting can be quantified. Concentrations of elements in this melt, C_M, can be calculated from their concentrations in the source, C_S, using the standard equation:

$$C_M = C_S/(D + F - DF) \qquad (2)$$

where D is the partition coefficient for each element and F is the degree of melting. A very small value of F, 0.0001, corresponding to 0.01% partial melting of the mantle source, is assumed. For D, 0.0075 is used for Rb, 0.039 for Sr, 0.026 for Nd and 0.033 for Sm; similar to the values estimated by Fitton & Dunlop (1985). Given the composition of primitive mantle listed in the caption to Fig. 12, the composition of the melt produced by small-degree partial melting of primitive mantle can thus be calculated.

Each successive batch of this melt is assumed to freeze within the mantle lithosphere without further chemical fractionation. The subsequent evolution of its isotopic composition can thus

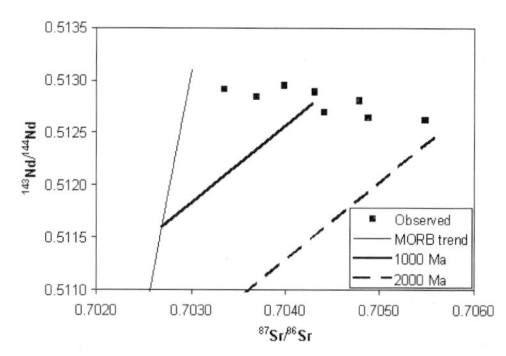

Fig. 12. Observed and predicted neodymium and strontium isotope ratios. The observed data comprise six Karasu Valley basalts (Çapan et al. 1987) and two Karacalıdağ basalts (Pearce et al. 1990) (Table 6). The line labelled MORB trend indicates the predicted variation over time in the isotopic composition of primitive mantle which forms the source of MORB. The line terminates at a point representing the typical present-day composition of MORB, which is produced from its source by high-degree partial melting which will not affect the isotope ratios. The other lines labelled 1000 and 2000 Ma indicate the predicted variation over time in the isotopic composition of material which separated from primitive mantle by small-degree partial melting at the specified times, and has since remained separate with no additional chemical differentiation. Calculations assume that the ^{143}Nd/^{144}Nd and ^{87}Sr/^{86}Sr ratios in MORB, and its source material, are 0.5131 and 0.703, respectively. Isotopic abundances (as a proportion of the total number of atoms present of each element) are assumed to be 0.2785 for ^{87}Rb, 0.099 for ^{87}Sr, 0.15 for ^{147}Sm and 0.238 for ^{143}Nd. Following Fitton & Dunlop (1985), the MORB source is assumed to comprise (in ppm): 0.15 Rb; 19.4 Sr; 1.2 Nd; and 0.44 Sm. Past ratios of these isotopes can thus be calculated using these compositions and isotope ratios, with decay constants of 1.42×10^{-5} and 6.54×10^{-6} Ma^{-1} for the decay of ^{87}Rb to ^{87}Sr and ^{147}Sm to ^{143}Nd, respectively. See text for further discussion.

be calculated using standard theory for radioactive decay given the decay constants listed in the caption to Fig. 12. It is assumed that heating of the mantle lithosphere accompanying Miocene and/or Quaternary atectonic crustal thickening will remelt all of this previously frozen material as it passes through the solidus, without additional chemical fractionation. This remelting can thus produce a mixture of batches of material which had originally separated from primitive mantle over a range of earlier times.

Using the parameter values above and in the caption to Fig. 12, compositions within the resulting basalt of 20 (Rb), 500 (Sr), 46 (Nd) and

13 (Sm) are predicted: similar to the mean compositions of Quaternary basalts in the Karasu Valley (Table 5; Çapan *et al.* 1987). The variations in the isotope ratios ^{143}Nd/^{144}Nd and ^{87}Sr/^{86}Sr can thus be calculated for MORB source back to the time of partial melting, and then forward in time for the melt to the present day.

Figure 12 compares the resulting predicted isotope ratios with observations. The different specimens have isotope ratios consistent with having initially separated from primitive mantle over a range of ages from *c.* 2000 Ma to a few hundred million years. This comparison thus indicates that melt which separated from the asthenosphere into the mantle lithosphere as a result of small-degree partial melting at various times between 2000 Ma and the present day, and then froze, before being fully remelted in the Quaternary, is predicted to match the trend of isotope ratios observed in the Quaternary basalt specimens from the Karasu Valley and Karacalıdağ.

Little is known about the age of basement in southeastern Turkey, due to the lack of exposure. However, evidence further south from Saudi Arabia indicates that the Arabian Platform developed in the Late Precambrian, between *c.* 900 and 700 Ma, as a result of the accretion of many island arcs and small continental fragments (e.g. Stoeser & Camp 1985; Unrug 1997). However, some of these continental fragments date from *c.* 2000 Ma (e.g. Stoeser & Camp 1985). The possibility that the part of the Arabian Platform in southeastern Turkey has similar composition makes it feasible for frozen melt to be preserved from this age beneath it.

The least mafic specimens listed in Table 6 have isotope ratios consistent with derivation from the earliest partial melting. This evidence, and the compositional variations in basalts from groups of flows from individual necks [evident in Figs 6–8; see also Yurtmen *et al.* (2000)] raises the possibility that many of these basalts represent complex mixtures of material derived from partial melting of primitive mantle over prolonged periods of time. The correlation of the least mafic character with the isotope ratios, indicative of the most ancient origins, suggests the possibility that typical degrees of partial melting have decreased over time, possibly reflecting gradual cooling of the asthenosphere and/or its progressive loss of relatively volatile components. This implies that the young basalts which contain a preponderance of material of relatively ancient origin classify as tholeiites; those with a greater proportion of younger material classify as alkali basalts or basanites. This, in turn, implies, e.g., that the mafic character of the earliest eruptions adjacent to the KOF at Toprakkale may indeed be caused by the relatively easy fault-controlled path to the surface, which can allow the initial eruptions to take place before significant mixing occurs. However, testing this hypothesis will require more data, notably for isotope ratios from such basanites, and is thus beyond the scope of this study.

Conclusions

It has been established that there were two episodes of magmatism in southeastern Turkey, during *c.* 19–15 and *c.* 2.3–0.6 Ma. Both had previously been reported, although the number of isotopic dates was too few to establish that either episode was of regional significance. Although some of the sampling sites in this study had been correctly dated by previous work, others had not: some Miocene sites were previously regarded as Quaternary; some Plio-Quaternary sites were instead regarded as Miocene. These results thus indicate the need for isotopic dating elsewhere in the region and demonstrate that, in the meantime, caution is warranted when using dates based on field relationships or other intuitive arguments.

It is suggested that the observed volcanism is the result of heating the mantle lithosphere by atectonic crustal thickening, which has raised frozen hydrous metasomatic melt above the solidus temperature of wet basalt. This material is regarded as having previously accumulated in the mantle lithosphere by prolonged small-degree partial melting. The lower crustal flow, which has heated the mantle lithosphere by thickening the crust and uplifting the Earth's surface, is interpreted as a consequence of moderate cyclic Miocene glaciations and enhanced glaciations since the start of the Quaternary.

We thank S. Tonbul and C. İskender for help in the field, S. Yurtmen and A. Koçyiğit for other helpful discussions, R. Ridley for his contribution to the K–Ar analyses, and G. Fitton for the geochemical analyses. We are also grateful to P. Floyd and an anonymous reviewer for their thoughtful and constructive comments.

References

AKTAŞ, G. & ROBERTSON, A. H. F. 1984. The Maden complex, SE Turkey: evolution of a Neotethyan active margin. *In*: DIXON, J. E. & ROBERTSON, A. H. F. (eds) *The Geological Evolution of the*

Eastern Mediterranean. Geological Society, London, Special Publications, **17**, 375–402.

ALTINLI, I. E. & ERENTÖZ, C. 1961. *Erzurum Sheet of the Geological Map of Turkey, 1:500 000 Scale.* Mineral Research and Exploration Institute of Turkey (MTA) Publications.

ARCULUS, R. J. & POWELL, R. 1986. Source component mixing in the regions of arc magma generation. *Journal of Geophysical Research*, **91**, 5913–5926.

ASUTAY, H. J. 1988. *Geological Map of the Malatya H27 quadrangle at 1:100 000 scale, and Accompanying Explanatory Booklet.* Mineral Research and Exploration Institute of Turkey (MTA) Publications [in Turkish with English abstract].

BAYKAL, F. & ERENTÖZ, C. 1961. *Sivas Sheet of the Geological Map of Turkey at 1:500 000 Scale.* Mineral Research and Exploration Institute of Turkey (MTA) Publications.

BİLGİN, A. Z. & ERCAN, T. 1981. Petrology of the Quaternary basalts of the Ceyhan–Osmaniye area. *Geological Society of Turkey Bulletin*, **24**, 21–30 [in Turkish with English abstract].

BİNGÖL, A. F. 1984. Geology of the Elazığ area in the eastern Taurus region. *In:* TEKELİ, O. & GÖNCÜOĞLU, M. C. (eds) *Geology of the Taurus Belt.* Proceedings of the International Tauride Symposium, Mineral Research and Exploration Institute of Turkey (MTA) Publications, 209–216 [in Turkish with English abstract].

BOHANNON, R. G., NAESER, C. W., SCHMIDT, D. L. & ZIMMERMANN, R. A. 1989. The timing of uplift, volcanism, and rifting peripheral to the Red Sea: a case for passive rifting? *Journal of Geophysical Research*, **94**, 1683–1701.

BRINKMANN, R. 1976. *Geology of Turkey.* Elsevier.

CAMP, V. E. & ROOBOL, M. J. 1992. Upwelling asthenosphere beneath western Arabia and its regional implications. *Journal of Geophysical Research*, **97**, 15 255–15 271.

ÇAPAN, U. Z., VIDAL, P. & CANTAGREL, J. M. 1987. K–Ar, Sr and Pb isotopic study of Quaternary volcanism in Karasu valley (Hatay), N-end of the Dead-Sea rift zone in SE-Turkey. *Hacettepe University Earth Sciences*, **14**, 165–178.

DALONGEVILLE, R. & SANLAVILLE, P. 1977. Témoins de lignes de rivages holocènes en Turquie méridionale. *Bulletin de l'Association Française pour l'Étude du Quaternaire*, **53**, 79–81.

DALRYMPLE, G. B. & LANPHERE, M. A. 1969. *Potassium–argon Dating: Principles, Techniques, and Applications to Geochronology.* W. H. Freeman.

DERCOURT, J., ZONENSHAIN, L. P., RICOU, L. E., ET AL. 1986. Geological evolution of the Tethys belt from the Atlantic to the Pamirs since the Lias. *Tectonophysics*, **123**, 241–315.

DEWEY, J. F., HEMPTON, M. R., KIDD, W. S. F., ŞAROĞLU, F. & ŞENGÖR, A. M. C. 1986. Shortening of continental lithosphere: the neotectonics of eastern Anatolia – a young collision zone. *In:* COWARD, M. P. & RIES, A. C. (eds) *Collision Tectonics.* Geological Society, London, Special Publications, **19**, 3–36.

ERİNÇ, S. 1978. Changes in the physical environment of Turkey since the end of the last glacial. *In:* BRICE, W. C. (ed.) *The Environmental History of the Near and Middle East since the Last Ice Age.* Academic Press, 87–110.

EROL, O. 1963. Asi Nehri Deltasının Jeomorfolojisi ve Dördüncü Zaman Denizakarsu Şekilleri (Die Geomorphologie des Orontes-deltas und der anschliessenden pleistozänen Strand- und Flussterrassen, Provinz Hatay, Türkei). *İstanbul University, Faculty of Language and History – Geography Publications*, **148**, 1–110 [in Turkish and German].

—— 1981. Neotectonic and geomorphological evolution of Turkey. *Zeitschrift für Geomorphologie, Neue Folge*, **Supplement 40**, 193–211.

—— 1982. *Geomorphological Map of Turkey at 1:2 000 000 scale.* Mineral Research and Exploration Institute of Turkey (MTA) Publications.

—— 1991*a*. Geomorphological evolution of the Taurus mountains, Turkey. *Zeitschrift für Geomorphologie, Neue Folge*, **Supplement 82**, 99–109.

—— 1991*b*. *Geomorphological Map of Turkey at 1:1 000 000 Scale.* Mineral Research and Exploration of Turkey (MTA) Publications.

FEARS, D. 1985. A corrected CIPW program for interactive use. *Computers and Geosciences*, **11**, 787–797.

FLOYD, P. A. & WINCHESTER, J. A. 1975. Magma type and tectonic setting discrimination using immobile elements. *Earth and Planetary Science Letters*, **27**, 211–218.

FITTON, J. G. & DUNLOP, H. M. 1985. The Cameroon Line, West Africa, and its bearing on the origin of oceanic and continental alkali basalts. *Earth and Planetary Science Letters*, **72**, 23–38.

FREUND, R., GARFUNKEL, Z., ZAK, I., GOLDBERG, M., WEISSBROD, T. & B. DERIN, B. 1970. The shear along the Dead Sea rift. *Philosophical Transactions of the Royal Society, London, Series A*, **267**, 107–130.

GARFUNKEL, Z. 1981. Internal structure of the Dead Sea leaky transform (rift) in relation to plate kinematics. *Tectonophysics*, **80**, 81–108.

GILL, J. B. 1981. *Orogenic Andesites and Plate Tectonics.* Springer.

GÖKÇEN, S. & KELLING, G. 1985. Oligocene deposits of the Zara–Hafik region (Sivas, central Turkey): evolution from storm-influenced shelf to evaporitic basin. *Geologische Rundschau*, **74**, 139–153.

——, ——, GÖKÇEN, N. & FLOYD, P. A. 1988. Sedimentology of a Late Cenozoic collisional sequence: the Misis complex, Adana, southern Turkey. *Sedimentary Geology*, **59**, 205–235.

GOTTWALD, J. 1940. Die Burg Til im Sudöstlichen Kilikien. *Byzantinische Zeitschrift*, **40**, 89–104.

HEIMANN, A., ROJAY, B. & TOPRAK, V. 1998. Neotectonic characteristics of the Karasu fault zone, northern continuation of the Dead Sea Transform in Anatolia (Turkey). *Proceedings of the Third International Turkish Geology Symposium, METU*, Ankara, Abstracts, 99.

HEMPTON, M. R. 1984. Results of detailed mapping

near Lake Hazar (eastern Taurus mountains). *In:* TEKELİ, O. & GÖNCÜOĞLU, M. C. (eds) *Geology of the Taurus Belt*. Proceedings of International Tauride Symposium, Mineral Research and Exploration Institute of Turkey (MTA) Publications, 223–228.

—— 1985. Structure and deformation history of the Bitlis suture near Lake Hazar, southeastern Turkey. *Geological Society of America Bulletin*, **96**, 233–243.

—— 1987. Constraints on Arabian plate motion and extensional history of the Red Sea. *Tectonics*, **6**, 687–705.

HILGEN, F. J. 1991. Astronomical calibration of Gauss to Matuyama sapropels in the Mediterranean and implications for the Geomagnetic Polarity Time Scale. *Earth and Planetary Science Letters*, **104**, 226–244.

HOLE, M. J., SAUNDERS, A. D., MARRINER, G. F. & TARNEY, J. 1984. Subduction of pelagic sediments: implications for the origin of Ce-anomalous basalts from the Mariana Islands. *Journal of the Geological Society of London*, **141**, 453–472.

HUNTINGTON, E. 1902. The valley of the upper Euphrates river and its people. *Bulletin of the American Geographical Society*, **34**, 301–318, 384–393.

INNOCENTI, F., MANETTI, P., MAZZUOLI, R., PASQUARÈ, G. & VILLARI, L. 1982. Anatolia and northwestern Iran. *In:* THORPE, R. S. (ed.) *Andesites*. Wiley, 327–349.

IONIDES, M. G. 1937. *The Régime of the Rivers Euphrates and Tigris*. E & FN Spon.

IRVINE, T. N., & BARAGAR, W. R. A. 1971. A guide to the chemical classification of the common volcanic rocks. *Canadian Journal of Earth Sciences*, **8**, 523–548.

JARRIGE, J.-J., OTT D'ESTEVOU, P., BUROLLET, P. F., MONTENAT, C., PRAT, P., RICHERT, J.-P. & THIRLET, J.-P. 1990. The multistage tectonic evolution of the Gulf of Suez and northern Red Sea continental rift from field observations. *Tectonics*, **9**, 441–465.

KARIG, D. E. & KOZLU, H. 1990. Late Palaeogene–Neogene evolution of the triple junction near Maraş, south-central Turkey. *Journal of the Geological Society, London*, **147**, 1023–1034.

KAUFMAN, P. S. & ROYDEN, L. H. 1994. Lower crustal flow in an extensional setting: constraints from the Halloran Hills region, eastern Mojave Desert, California. *Journal of Geophysical Research*, **99**, 15 723–15 739.

KENNETT, J. P. & THUNELL, R. C. 1975. Global increase in Quaternary explosive volcanism. *Science*, **187**, 497–503.

KEREY, İ. E. & TÜRKMEN, İ. 1991. Sedimentological aspects of Palu Formation (Plio-Quaternary), the east of Elazığ, Turkey. *Geological Society of Turkey Bulletin*, **34**, 21–26 [in Turkish with English abstract].

KRIEGER, M. H., CREASEY, S. C. & MARVIN, R. F. 1971. Ages of Some Tertiary Andesitic and Latitic Volcanic Rocks in the Prescott-Jerome Area, North-central Arizona. *US Geological Survey Professional Paper*, **750-B**, 157–160.

LAMBERT, I. B. & WYLLIE, P. J. 1972. Melting of gabbro (quartz eclogite) with excess water to 35 kilobars, with geological applications. *Journal of Geology*, **80**, 693–708.

LE BAS, M. J., LE MAITRE, R.W., STRECKEISEN, A. & ZANETTIN, B. 1986. A chemical classification of volcanic rocks based on the total alkali–silica diagram. *Journal of Petrology*, **27**, 745–750.

LEO, G. W., MARVIN, R. F. & MEHNERT, H. H. 1974. Geologic framework of the Kuluncak–Sofular area, east-central Turkey, and K–Ar ages of igneous rocks. *Geological Society of America Bulletin*, **85**, 1785–1788.

LOVELOCK, P. E. R. 1984. A review of the tectonics of the northern Middle East region. *Geological Magazine*, **121**, 577–587.

LUCCHITTA, I. 1979. Late Cenozoic uplift of the southwestern Colorado Plateau and adjacent lower Colorado River region. *Tectonophysics*, **61**, 63–95.

LYBERIS, N., YÜRÜR, T., CHOROWICZ, J., KASAPOĞLU, E. & GÜNDOĞDU, N. 1992. The East Anatolian fault: an oblique collisional belt. *Tectonophysics*, **204**, 1–15.

MCKEE, E. D. & MCKEE, E. H. 1972. Pliocene uplift of the Grand Canyon region – time of drainage adjustment. *Geological Society of America Bulletin*, **83**, 1923–1932.

MCKENZIE, D. P. 1976. The East Anatolian fault: a major structure in eastern Turkey. *Earth and Planetary Science Letters*, **29**, 189–193.

—— 1985. The extraction of magma from the crust and mantle. *Earth and Planetary Science Letters*, **74**, 81–91.

—— 1989. Some remarks on the movement of small melt fractions in the mantle. *Earth and Planetary Science Letters*, **95**, 53–72.

MICHARD, A., WHITECHURCH, H., RICOU, L. E., MONTIGNY, R. & YAZĞAN, E. 1984. Tauric subduction (Malatya–Elazığ provinces) and its bearing on tectonics of the Tethyan realm in Turkey. *In:* DIXON, J. E. & ROBERTSON, A. H. F. (eds) *The Geological Evolution of the Eastern Mediterranean*. Geological Society, London, Special Publications, **17**, 362–373.

MIKHAYLOVA, N. P., GLEVASSKAYA, A. M. & TSYKORA, V. N. 1974. Geomagnetic polarity epochs in the Neogene. *Doklady of the Academy of Sciences of the USSR, Earth Sciences, English Translation*, **214**, 112–114 (Russian original: *Doklady Akademii Nauk SSSR*, **214**, 1145–1148).

MITCHELL, J. & WESTAWAY, R. 1999. Chronology of Neogene and Quaternary uplift and magmatism in the Caucasus: constraints from K–Ar dating of volcanism in Armenia. *Tectonophysics*, **304**, 157–186.

MOUSTAFA, A. R. 1993. Structural characteristics and tectonic evolution of the east-margin blocks of the Suez rift. *Tectonophysics*, **223**, 381–399.

MUEHLBERGER, W. B. & GORDON, M. B. 1987. Observations on the complexity of the East Anatolian fault, Turkey. *Journal of Structural Geology*, **9**, 899–903.

OTA, R. & DİNÇEL, A. 1975. Volcanic rocks of Turkey. *Bulletin of the Geological Society of Japan*, **26**, 393–419.

ÖZKAYA, İ. 1982. Upper Cretaceous plate rupture and development of leaky transform fault ophiolites in southeast Turkey. *Tectonophysics*, **88**, 103–116.

PEARCE, J. A. 1983. The role of sub-continental lithosphere in magma genesis at destructive plate margins. *In*: HAWKESWORTH, C. J. & NORRY, M. J. (eds) *Continental Basalts and Mantle Xenoliths*. Sheva, 230–249.

——, BENDER, J. F., DELONG, S. E. *ET AL.* 1990. Genesis of post collision volcanism in Eastern Anatolia, Turkey. *Journal of Volcanology and Geothermal Research*, **44**, 190–227.

PERİNÇEK, D. & ÇEMEN, İ. 1990. The structural relationship between the East Anatolian and Dead Sea fault zones in southeastern Turkey. *Tectonophysics*, **172**, 331–340.

RIGO DI RIGHI, M. & CORTESINI, A. 1964. Gravity tectonics in the foothills structural belt of southeast Turkey. *AAPG Bulletin*, **48**, 1911–1937.

ROBINSON, A. G., BANKS, C. J., RUTHERFORD, M. M. & HIRST, J. P. P. 1995. Stratigraphic and structural development of the Eastern Pontides, Turkey. *Journal of the Geological Society, London*, **152**, 861–872.

ROBERTSON, A. H. F., CLIFT, P. D., DEGNAN, P. & JONES, G. 1991. Palaeogeographic and palaeotectonic evolution of the eastern Mediterranean region. *Palaeogeography, Palaeoclimatology, Palaeoecology*, **87**, 289–344.

ROJAY, B., TOPRAK, V. & HEIMANN, A. 1998. Neotectonic characteristics of the Karasu fault zone, northern continuation of the Dead Sea Transform in Anatolia (Turkey). *Proceedings of the Third International Turkish Geology Symposium, METU*, Ankara, Abstracts, 100.

SANVER, M. 1968. A palaeomagnetic study of Quaternary volcanic rocks from Turkey. *Physics of the Earth and Planetary Interiors*, **1**, 403–421.

ŞAROĞLU, F. 1988. Age and offset of the North Anatolian Fault. *METU Journal of Pure and Applied Sciences*, **21**, 65–79.

——, EMRE, Ö. & KUŞCU, İ. 1992a. *Active Fault Map of Turkey at 1:1 000 000 Scale*. Mineral Research and Exploration Institute of Turkey (MTA) Publications.

——, —— & —— 1992b. The East Anatolian Fault Zone of Turkey. *Annales Tectonicae*, **6**, 99–125.

ŞENGÖR, A. M. C. 1979. Mid-Mesozoic closure of Permo-Triassic Tethys and its implications. *Nature*, **279**, 590–593.

—— & YILMAZ, Y. 1981. Tethyan evolution of Turkey: a plate tectonic approach. *Tectonophysics*, **75**, 181–241.

SHACKLETON, N. J., BERGER, A. & PELTIER, W. R. 1990. An alternative astronomical calibration of the lower Pleistocene timescale based on ODP site 677. *Transactions of the Royal Society, Edinburgh: Earth Sciences*, **81**, 251–261.

—— *ET AL.* 1984. Oxygen isotope calibration of the onset of ice-rafting and history of glaciation in the North-east Atlantic. *Nature*, **307**, 620–623.

SOOD, M. 1981. *Modern Igneous Petrology*. Wiley.

STEIGER, R. H. & JÄGER, E. 1977. Convention on the use of decay constants in geo- and cosmochronology. *Earth and Planetary Science Letters*, **36**, 359–363.

STEININGER, F. F. & PAPP, A. 1979. Current biostratigraphic and radiometric correlations of the Late Miocene Central Paratethys stages (Sarmatian s. str., Pannonian s. str., and Pontian) and Mediterranean stages (Tortonian and Messinian) and the Messinian Event in the Paratethys. *Newsletters of Stratigraphy*, **8**, 100–110.

——, RÖGL, F. & MARTINI, E. 1976. Current Oligocene/Miocene biostratigraphic concept of the central Paratethys (Middle Europe). *Newsletters of Stratigraphy*, **4**, 174–202.

STOESER, D. B. & CAMP, V. E. 1985. Pan-African microplate accretion of the Arabian Shield. *Geological Society of America Bulletin*, **96**, 817–826.

TAHA, M. F., HARB, S. A., NAGIB, M. K. & TANTAWY, A. H. 1981. The climate of the Near East. *In*: TAKAHASHI, K. & ARAKAWA, H. (eds) *Climates of Southern and Western Asia, World Survey of Climatology, Volume 9*. Elsevier, 183–255.

TAYMAZ, T., EYİDOĞAN, H. & JACKSON, J. 1991. Source parameters of large earthquakes in the East Anatolian Fault Zone. *Geophysical Journal International*, **106**, 537–550.

TEKELİ, O. & ORAL, A. 1986. *Geological Map of the Malatya I-27 Quadrangle at 1:100 000 Scale*. Mineral Research and Exploration Institute of Turkey (MTA) Publications.

TEMİZ, H., GUEZOU, J. C., POISSON, A. M. & TUTKUN, Z. 1993. Tectonostratigraphy and kinematics of the eastern end of the Sivas basin (central eastern Turkey): implications for the so-called 'Anatolian block'. *Geological Journal*, **28**, 239–250.

TERLEMEZ, H. Ç. İ., ŞENTÜRK, K., ATEŞ, S., SÜMENGEN, M. & ORAL, A. 1997. *Geological Map and Accompanying Explanatory Booklet of the Gaziantep–K24 Quadrangle at 1:100 000 Scale*. Mineral Research and Exploration Institute of Turkey (MTA) Publications [in Turkish with English abstract].

TEZCAN, A. K. 1979. Geothermal studies, their present status and contribution to heat flow contouring in Turkey. *In*: CERMAK, V. & RYBACH, L. (eds) *Terrestrial Heat Flow in Europe*. Springer, 283–292.

THOMPSON, R. N. & FOWLER, M. B. 1986. Subduction-related shoshonitic and ultrapotassic magmatism: a study of Siluro-Ordovician syenites from the Scottish Caledonides. *Contributions to Mineralogy and Petrology*, **94**, 507–522.

TOLUN, N. & ERENTÖZ, C. 1962. *Hatay Sheet of the Geological Map of Turkey at 1:500 000 Scale*. Mineral Research and Exploration Institute of Turkey (MTA) Publications.

—— & PAMİR, H. N. 1975. *Explanatory Booklet for the Hatay Sheet of the Geological Map of Turkey at 1:500 000 Scale*. Mineral Research and

Exploration Institute of Turkey (MTA) Publications [in Turkish with English abstract].

TONBUL, S. 1987. Elazığ batısının genel jeomorfolojik özellikleri ve gelişimi. *Geomorphology Bulletin*, **15**, 37–52 [in Turkish with English abstract].

TRIFONOV, V. G., KARAKHANIAN, A. S. & KOZHURIN, A. I. 1994. Major active faults of the collision area between the Arabian and the Eurasian plates. *In*: BOLT, B. A. & AMIRBEKIAN, R. (eds) *Continental Collision Zone Earthquakes and Seismic Hazard Reduction*. Proceedings of the International Conference at Yerevan-Sevan, Armenia, 1–6 October 1993. IASPEI, 56–76.

UNRUG, R. 1997. Rodinia to Gondwana: the geodynamic map of Gondwana supercontinent assembly. *Geological Society of America Today*, **7**, 1–6.

WESTAWAY, R. 1994. Present-day kinematics of the Middle East and eastern Mediterranean. *Journal of Geophysical Research*, **99**, 12 071–12 090.

—— 1995. Deformation around stepovers in strike-slip fault zones. *Journal of Structural Geology*, **17**, 831–847.

—— 1998. Dependence of active normal fault dips on lower-crustal flow regimes. *Journal of the Geological Society, London*, **155**, 233–253.

—— 2000. Flow in the lower continental crust as a mechanism for the Quaternary uplift of the Rhenish Massif, north-west Europe. *In*: MADDY, D., MACKLIN, M. & WOODWARD, J. (eds) *River Basin Sediment Systems: Archives of Environmental Change*. Balkema, Rotterdam, in press.

—— 2000*a*. Quaternary uplift of small islands within the continental crust. *Tectonophysics*, in press.

—— 2000*b*. Quaternary uplift of the Rhenish Massif, north-west Europe, caused by flow in the lower continental crust. *In*: MADDY, D., MACKLIN, M. G. & WOODWARD, J. C. (eds) *River Basin Sediment Systems*. Balkema, in press.

—— & ARGER, J. 1996. The Gölbaşı basin, southeastern Turkey: a complex discontinuity in a major strike-slip fault zone. *Journal of the Geological Society, London*, **153**, 729–743.

WILKINSON, P., MITCHELL, J. G., CATTERMOLE, P. J. & DOWNIE, C. 1986. Volcanic chronology of the Meru–Kilimanjaro region, northern Tanzania. *Journal of the Geological Society, London*, **143**, 601–603.

WILSON, M. R. 1989. *Igneous Petrogenesis: A Global Tectonic Approach*. Unwin Hyman.

YILMAZ, S., BOZTUĞ, D. & ÖZTÜRK, A. 1993. Geological setting, petrographic and geochemical characteristics of the Cretaceous and Tertiary igneous rocks in the Hekimhan–Hasançelebi area, northwest of Malatya, Turkey. *Geological Journal*, **28**, 383–398.

YILMAZ, Y. 1993. New evidence and model on the evolution of the southeast Anatolian orogen. *Geological Society of America Bulletin*, **105**, 251–271.

——, YİĞİTBAŞ, E. & GENÇ, Ş. C. 1993. Ophiolitic and metamorphic assemblages of southeast Anatolia and their significance in the geological evolution of the orogenic belt. *Tectonics*, **12**, 1280–1297.

YOLDEMİR, O. 1987. *Suvarlı–Haydarlı–Narlı ve Gaziantep Arasında Kalan Alanın Jeolojisi, Yapısal Durumu ve Petrol Olanakları*. Turkish Petroleum Corporation (TPAO) Report No. 2275 [in Turkish].

YURTMEN, S., ROWBOTHAM, G., İŞLER, F. & FLOYD, P. A. (2000). Petrogenesis of basalts from southern Turkey: the Plio-Quaternary volcanism to the north of İskenderun Gulf. *This volume*.

YÜRÜR, M. T. & CHOROWICZ, J. 1998. Recent volcanism, tectonics, and plate kinematics near the junction of the African, Arabian, and Anatolian plates. *Journal of Volcanology and Geothermal Research*, **85**, 1–15.

Petrogenesis of basalts from southern Turkey: the Plio-Quaternary volcanism to the north of İskenderun Gulf

SEMA YURTMEN,[1] GEORGE ROWBOTHAM,[2] FİKRET İŞLER[1] & PETER A. FLOYD[2]

[1] *Çukurova Üniversitesi, Mühendislik-Mimarlık Fakültesi, Jeoloji Mühendisliği Bölümü, TR-01330 Balcalı, Adana, Turkey (e-mail: syurtmen@mail.cu.edu.tr)*
[2] *Department of Earth Sciences, University of Keele, Staffordshire, ST5 5BG, UK*

Abstract: The Quaternary volcanicity north of the İskenderun Gulf in the eastern Mediterranean is represented by small basaltic scoria cones and flows. Approximately 115 km² of land area is occupied by young basalts which straddle both the main (the Karataş–Osmaniye Fault Zone) strike-slip fault system which forms the Africa–Turkey Plate boundary and the suture of the southern arm of the former Neotethys Ocean. Detailed petrological and geochemical analyses of these rocks have been carried out, with the aim of trying to understand why they have erupted in this locality. The rocks consist mainly of basanites (43–46% silica; 3.9–6.5% alkalis) and some alkali olivine basalts (45% silica; 3.8–4.2% alkalis). Both the basanites and alkali olivine basalts are porphyritic, vitrophyric and highly vesiculated with euhedral and subhedral olivine (Fo_{82}–Fo_{78}) phenocrysts set in a fine-grained groundmass of olivine (Fo_{70}), plagioclase (An_{71}–An_{66}), clinopyroxene and titanomagnetite. Olivine phenocrysts contain abundant Cr-spinel and titanomagnetite inclusions. Some geochemical characteristics of these basalts indicate similarity with extension-related alkali basalts; others indicate similarity with ocean island basalt; and yet others indicate subduction-related characteristics. This complexity leads to difficulties with interpretation, especially since there is no demonstrable local extension, subduction or mantle plume activity in the vicinity.

Basaltic rocks dominate the Quaternary volcanism in the eastern Mediterranean region to the north of the İskenderun Gulf between Ceyhan and Osmaniye (Fig. 1). The eastern Mediterranean region contains three major strike-slip fault zones – the Dead Sea Fault Zone (DSFZ), and the North and East Anatolian Fault Zones (NAFZ and EAFZ, respectively) (Fig. 2) – that approximate transform faults (Westaway 1994; Westaway & Arger 1996). These fault zones form the modern plate boundaries between the Turkish (TR), Arabian (AR), African (AF) and Eurasian (EU) Plates, and accommodate northward motion of the African and Arabian Plates relative to Eurasia, and westward motion of the Turkish Plate (Şengör & Yılmaz 1981; Kelling *et al.* 1987; Yılmaz *et al.* 1988; Karig & Kozlu 1990; Perinçek & Çemen 1990; Westaway 1994; Westaway & Arger 1996; Arger *et al.* 2000). The right-lateral NAFZ forms the boundary between the Eurasian Plate and the Turkish Plate; the left-lateral EAFZ and DSFZ bound the Turkish, African and Arabian Plates (Fig. 2). Karlıova, in northeast Turkey, marks the intersection between the NAFZ and the EAFZ, and the Gölbaşı Basin on the EAFZ (Fig. 2) marks the triple junction between the African, Arabian and Turkish Plates (Karig & Kozlu 1990; Perinçek & Çemen 1990; Westaway & Arger 1996; Arger *et al.* 2000). For several workers, the triple junction between the Turkish, African and Arabian Plates is located near Kahraman Maraş (Şengör *et al.* 1985; Gülen *et al.* 1987). Active shortening continues west of the Karlıova Triple Junction where active left-lateral faulting steps to the right (Karig & Kozlu 1990; Arger *et al.* 2000). Westaway & Arger (1996) proposed a kinematic model explaining that the present-day major fault geometry developed c. 3 Ma. They suggested that between c. 5 Ma and 3 Ma, the relative motion between the Turkish and Arabian Plates was taken up north of the EAFZ, which did not yet exist, on the Malatya–Ovacık Fault Zone (MOFZ; Fig. 2). The earlier slip, which started at c. 5 Ma on other strike-slip faults nearby, marked the start of westward motion of the Turkish Plate (Westaway & Arger 1996).

The main left-lateral Karataş–Osmaniye Fault Zone (KOFZ), which is important for the structural framework of the basaltic volcanic rocks in this area, forms part of the modern Turkish–African Plate boundary. The Aslantaş Fault Zone, or Misis–Andırın Trend, follows the line of the former Neotethys Suture northeast from the study region. Westaway

From: BOZKURT, E., WINCHESTER, J. A. & PIPER, J. D. A. (eds) *Tectonics and Magmatism in Turkey and the Surrounding Area.* Geological Society, London, Special Publications, **173**, 489–512. 1-86239-064-9/00/$15.00
© The Geological Society of London 2000.

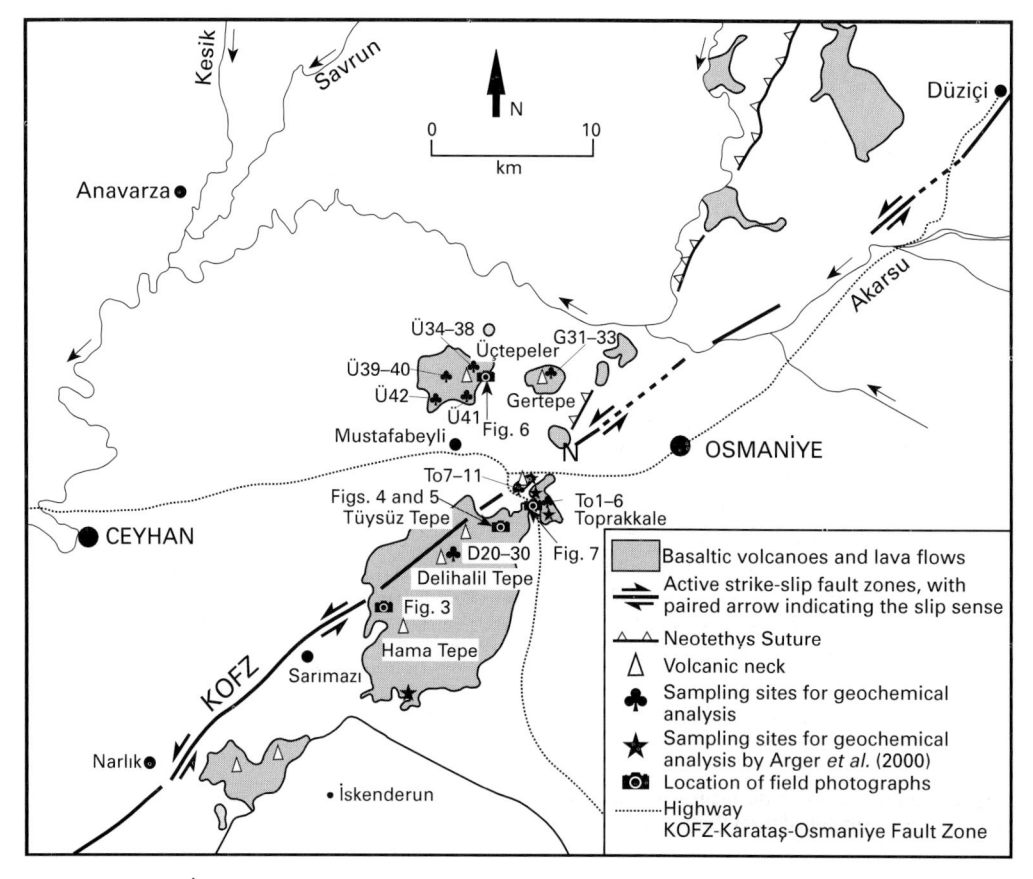

Fig. 1. Map of the İskenderun Gulf area and the distribution of basaltic volcanic rocks. Map is simplified from Bilgin & Ercan (1981) and Arger *et al.* (2000).

(pers. comm.) argued that most of the active slip continues east–northeast from the study region, linking end-on into the EAFZ. He also suggested that the active faulting steps leftward offshore, southwest of Karataş onto the former Neotethys Suture [the Cyprus–Misis–Andırın Trend of Karig & Kozlu (1990)] and that this is a pure transform fault. However, the southwest end of this transform fault bends to the right, requiring a component of transpression, hence the uplift of the Kyrenia Mountains of northern Cyprus (Westaway 1994). The active faulting thus follows the Neotethys Suture at this location, but in the study region it cuts across the Neotethys Suture in the vicinity of Toprakkale (Arger *et al.* 2000).

The Neogene evolution of these modern plate boundaries is largely recorded in the stratigraphy and structure of the basins that lie along the Turkish–Arabian and Turkish–African Plate boundaries (Kelling *et al.* 1987; Karig & Kozlu 1990). The Çukurova Basin, in the eastern Mediterranean, is made of two subsidiary troughs – the Adana and İskenderun Sub-basins – which are separated by the Misis Complex structural high (Kelling *et al.* 1987). Two main lithostratigraphic units have been recognized, namely the İsali and Karataş Formations (Schmidt 1961; Schiettecatte 1971). The İsali Formation is of Early Miocene (Aquitanian) age and is composed of an olistostromic complex. It is at least 2000 m thick and tectonically overlies the Karataş clastic rocks (Kelling *et al.* 1987). Basaltic clasts and pillow lavas in the İsali Mélange have a subduction-related chemistry and were formed in a back-arc basin setting (Floyd *et al.* 1991). The Karataş Formation is dominated by turbidite sandstones alternating with pelagic mudstones and marls. It also includes rare olistoliths of Maastrichtian and Palaeocene shelf limestone and volcanic rocks (Kelling *et al.* 1987). The thickness of the Karataş Formation varies from 1500 to 3000 m and the age ranges from Early to Late Miocene (Kelling

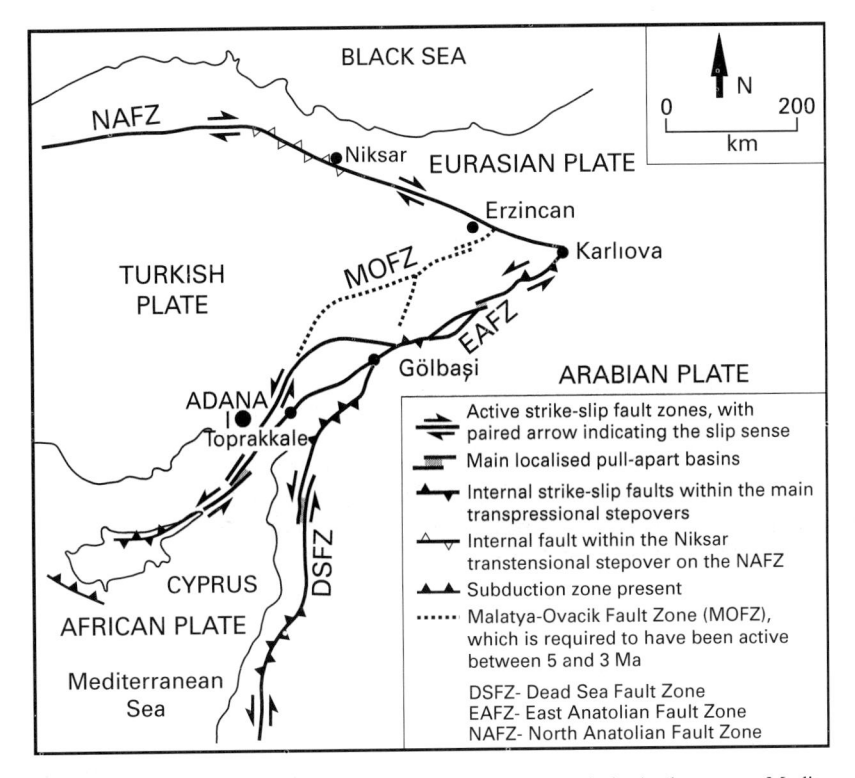

Fig. 2. Summary maps of the main active fault zones forming plate boundaries in the eastern Mediterranean region. Map is simplified from Westaway & Arger (1996).

et al. 1987). Analyses of volcaniclastic rocks in the İsali Formation and flysch-type turbidites for the Karataş Formation imply derivation from calc-alkaline arc and back-arc basin sources (Floyd et al. 1992).

The tectonic evolution of the region has been studied widely (Nur & Ben-Avraham 1978; Şengör & Yılmaz 1981; Kelling et al. 1987; Yılmaz et al. 1988; Karig & Kozlu 1990; Perinçek & Çemen 1990; Westaway 1994; Westaway & Arger 1996; Yürür & Chorowicz 1998). However, very few investigations on the petrology and geochemistry of the basalts have been published. Bilgin & Ercan (1981) described the petrology of these basalts and interpreted them as plateau basalts of tholeiitic character with a weak alkaline affinity. Parlak et al. (1997) gave a brief geochemical description of the basalts. Arger et al. (2000) did age determinations and gave some geochemical characteristics of the Toprakkale basaltic rocks.

The aim of this study is to review the petrographical and chemical composition of the basaltic rocks and to investigate the cause of the volcanism in this region. The İskenderun Gulf basin appears to basically be a sag basin (Kelling et al. 1987), but with a minor component of shortening due to a small amount of plate convergence being taken up across the Neotethys Suture where it runs along the northwestern coastline of the gulf, indicating no extension in the study area (Arger et al. 2000). However, to the north, within the Cilicia Basin, which was interpreted as being underlain by the active sinistral Yakapınar–Göksun Fault, there may be some evidence of extension. Westaway (pers. comm.) suggested that part of the interior of this basin may be underlain by a pull-apart basin where this fault zone steps leftward. There is no evidence that this is correct, although the fault geometry is not well resolved and so this possibility cannot be ruled out. Therefore, the suggestion that such a pull-apart basin may exist is the only way one can argue for a component of extension anywhere near the study area [see Arger et al. (2000) for a detailed discussion].

Distribution of the basaltic rocks

Plio-Quaternary basaltic rocks, now referred to as the İskenderun Gulf alkali volcanic rocks,

Fig. 3. Field photograph of Delihalil Tepe Volcano, viewed north-northeast from the southwest of the Delihalil Tepe Neck.

Fig. 4. Subplinian pumice-fall deposits from northeast of Tüysüz Tepe Parasitic Cone.

the main topic of this study, are exposed along the left-lateral northeast–southwest trending KOFZ (Şaroğlu *et al.* 1992) and overlie all the Neogene series in the area. These basaltic volcanic rocks have been studied at four main localities: Delihalil Tepe; Üçtepeler; Toprakkale; and Gertepe (Fig. 1).

The Delihalil Tepe Volcano presents the largest basalt exposure in the area and rises to a height of 460 m above sea level (asl). It is a large composite stratovolcano consisting of a scoria cone and several parasitic cones, pumiceous pyroclastic scoria-falls (subplinian pumice-fall characteristics) and lava flows. The scoria cone consists almost entirely of rough clinkery blocks and resembles rubbly aa lavas (Fig. 3). The subplinian pumice-fall deposits (Fig. 4) are located to the northeast of the Tüysüz Tepe Parasitic Cone. In this area, pyroclastic rocks located at the base are about *c.* 3–4 m thick and contain a variety of rock fragments, ranging in size from millimetres to 1 m (Fig. 5a). This horizon is overlain by a 50–60 cm thick soil layer. Subplinian pumice-fall deposits overlie the soil layer (Fig. 5a and b) and are 2–4 m thick, showing normal grading and good lamination. The grain size of some of the pumice fragments is up to 1.5 cm and their long axes are parallel to the deposition surface. There is a clinkery lava flow unit (20–30 cm thick)

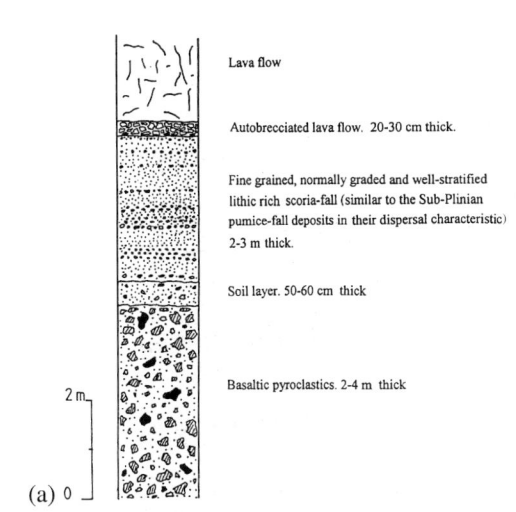

Fig. 5. (a) The columnar stratigraphical section of the pyroclastics fall deposits from the northeast of Tüysüz Tepe. (b) Field photograph of (a).

overlying this scoria-fall unit. At the topmost level, there is another lava flow which has rough clinkery and spinose-like surfaces reflecting autobrecciation of the solidified surface crust by movement below. The scoria cone is surrounded by a large lava field. Around the base of the Delihalil Tepe Volcano, within this lava field, there are many surface features such as small mounds or dome-like blisters up to 20 m in diameter. These structures are thought to be formed from local liquid pressure points inside the lava causing the plastic crust to dome up. The fluid pressure could be caused by magma, or by locally high concentrations or pockets of gas which could either be magmatic gas or vaporized groundwater over which the lava has flowed (Cas & Wright 1987). In some places, e.g. to the west of the main neck, the volcanic rocks show a bright red or pinkish coloration due to oxidative alteration caused by steam (Walker & Crosdale 1972).

Twelve samples from Delihalil Tepe have been analysed for their petrographical and geochemical characteristics. Samples numbered from D20 to D25 were collected from the northeast of the Delihalil Tepe Neck and samples D26–D30 came from the same location but from further south towards the base of the neck (Fig. 1).

The small Üçtepeler Volcano (Fig. 6) has similar features to Delihalil Tepe, with a scoria cone, including a parasitic cone, on the southern flank and a lava field fed from the southwest and northwest flanks of the volcano. The height of the scoria cone is 132 m asl. It has a small crater-like depression on its summit which opens to the southwest and northwest. Lava flows surrounding the cone are columnar jointed. Samples Ü34–Ü38 were collected from the northern part of the Üçtepeler Crater; Ü39 and Ü40 came from the base of the crater, nearer to the neck; Ü41 was collected from the lava flow surrounding the volcano to the southeast of Üçtepeler; and Ü42 was from the southwest of Üçtepeler (Fig. 1).

Toprakkale is a small monogenetic centre and is a low (151 m) irregular-shaped hill located to the northeast of Delihalil Tepe (Fig. 1). The original shape of the cone is not clear, as it lost its natural shape due to quarrying and construction work. The Armenians built a castle on it c. 900 a ago and excavated its sides to improve the local defences (Gottwald 1940). The Toprakkale basalt lava is columnar jointed at the bottom and is capped by 5–6 cm flow wrinkles similar to pahoehoe flows around the main neck. Samples To7–To11 were collected from south of the Toprakkale Neck.

Fig. 6. General areal view of the small Üçtepeler Volcano, viewed towards the west from the eastern flank of the volcano.

In the same area, c. 1 km further south of Toprakkale, lavas exhibit two different cooling phenomena; a platy jointed or flow-foliated lava flow at the base of the succession; and a columnar jointed (1–5 m) flow towards the top (Fig. 7). In the platy jointed lava, vesicularity increases from the base of the unit towards the top, where elongated (ellipsoidal) vesicles reach 10–15 cm in length. The columnar jointed lava is highly vesicular and shows slight atmospheric alteration on its surface. There is an impersistent soil layer c. 2 m thick separating these two lava flows. Where absent, the columnar jointed lava rests directly on the platy jointed lava flow (Fig. 7). Samples To1–To6 came from this location (Fig. 1), where this platy jointed lava flow is clearly seen to overlie the Karataş Formation.

Gertepe (Fig. 1) is a dome-shaped small volcano composed of aa type lavas. Basaltic rocks in this location show similar characteristics to the Delihalil Tepe Scoria Cone. At the bottom, lavas are less vesicular and show no alteration. Towards the top there is a good stratigraphical layering of lava and black ash. Samples G31–G33 were collected from the quarried section inside the cone.

Analytical techniques

Fifty samples collected from the four main centres north of the İskenderun Gulf area outlined above have been analysed on a fully automated ARL 8420 X-ray fluorescence spectrometer, calibrated against both international and internal rock standards of appropriate composition for their major and trace element concentrations. Major oxides [SiO_2, TiO_2, Al_2O_3, Fe_2O_3 (as total iron), MnO, MgO, CaO, Na_2O, K_2O and P_2O_5] analyses were performed on fused glass beads and trace element (including Cr, Cu, Ga, Nb, Ni, Pb, Rb, Sr, Th, V, Zn, Zr and light rare elements) analyses were made on pressed powder pellets. Precision was determined by pooled variance on the results obtained from replicate analyses on individual samples. Analytical details may be found in Floyd & Castillo (1992). Results of these analyses are listed in Table 1 for a representative selection of specimens; normative compositions of all samples are presented in Table 2.

Chemical analysis of rock-forming minerals in the İskenderun Gulf alkali olivine basalts and basanites were performed using electron-probe microanalysis. The analyses were made on polished sections using a Link QX 2000 electron microprobe attached to the lithium-drifted silicon detector at the Department of Earth Sciences, University of Keele. Calibration was made by using metals and silicate mineral standards. Accuracy and precision of the techniques used are given by Dunham & Wilkinson (1978), with corrections using a standard software program (ZAF4-FLS) from Link Analytical. Representative analyses of rock-forming minerals (olivine, plagioclase and clinopyroxene) are listed in Tables 3–5.

Classification and petrography of the basaltic rocks

Compositional variation and classification of the basaltic rocks are presented on a total alkali–silica diagram (Fig. 8) after Le Bas et al. (1986). It is evident (Fig. 8) that the volcanism of the İskenderun Gulf area is predominantly alkalic in nature. Arger et al. (2000) have studied the age and geochemistry of the Toprakkale basalts, determining them to be 2.3–0.6 Ma old. However, as a result of analytical uncertainties, they explained that the age span may vary between 1.5 and 0.7 Ma and 3.0–0.5 Ma. Specimens from the Toprakkale area analysed by Arger *et al.* (2000) have similar variations in compositions to the samples from this study (Fig. 9). The magma series in the İskenderun Gulf area is clearly undersaturated, with abundant normative nepheline ranging from 3.30 to 17.3 (Table 2), and is primarily composed of basanites (normative olivine ranging from 15.5 to 21.6, Table 2) and alkali olivine basalts.

Basanites

The basanites from Delihalil Tepe, Gertepe and Toprakkale all lie close to each other in the total alkali–silica diagram and show similar textural characteristics. They exhibit porphyritic, intersertal, intergranular and glomeroporphyritic textures, all of which may sometimes be seen in one thin section. Basanites are mainly composed of fresh euhedral, and occasionally subhedral, olivine phenocrysts, sometimes up to 3 mm in diameter (Fig. 10a). Olivine phenocrysts and microphenocrysts are embedded in a highly vesicular and fine-grained groundmass of olivine, plagioclase, augite, ilmenite (microprobe analysis), Ti-magnetite, and subordinate apatite and glass. Some olivine phenocrysts show no resorption whilst others are strongly embayed and resorbed. These embayed and skeletal forms are probably

Fig. 7. Columnar jointed lava flow overlying the platy jointed lava flow at a location 1 km south of Toprakkale.

Table 1. Representative X-ray fluorescence major-element (wt%) and trace element (ppm) analyses of selected İskenderun Gulf alkali basaltic rocks

	To2 Basanite	To5 Basanite	To7 Alk.Ol.B	To11 Basanite	D21 Basanite	D25 Basanite	D27 Basanite	D29-1 Basanite	D30 Basanite	G32 Basanite	G33 Basanite	Ü35 Basanite	Ü39 Basanite	Ü40 Alk.Ol.B	Ü42 Alk.Ol.B
SiO$_2$	44.24	43.55	44.41	44.44	43.24	43.35	43.70	44.3	43.46	44.31	44.97	45.86	46.41	44.43	45.43
TiO$_2$	2.99	2.91	2.94	2.93	2.97	2.96	3.11	2.93	2.96	2.67	2.62	2.63	2.65	2.65	2.55
Al$_2$O$_3$	15.78	15.17	15.66	15.42	14.44	14.57	15.22	14.91	14.46	14.36	14.46	16.06	16.67	15.48	16.14
Fe$_2$O$_3$	2.16	2.21	2.16	2.22	3.38	3.34	3.32	3.36	3.41	3.27	3.30	2.86	2.81	3.42	2.02
FeO	11.04	11.25	11.02	11.31	11.44	11.31	11.26	11.40	11.56	11.09	11.17	9.71	9.52	11.59	10.28
MnO	0.17	0.17	0.17	0.18	0.18	0.18	0.18	0.18	0.18	0.18	0.18	0.18	0.18	0.18	0.17
MgO	7.10	8.77	7.35	8.40	9.70	9.41	7.78	9.34	9.80	9.69	10.08	8.48	7.36	9.19	7.65
CaO	10.28	10.22	10.33	10.11	8.78	9.15	9.75	9.38	9.20	9.12	9.32	8.01	7.33	10.04	10.89
Na$_2$O	4.16	3.60	2.82	3.90	3.95	4.01	4.51	4.08	3.96	3.79	3.38	4.72	5.20	2.91	3.16
K$_2$O	1.45	1.42	1.16	1.50	1.83	1.87	1.99	1.73	1.88	1.59	1.53	2.36	2.72	0.96	1.00
P$_2$O$_5$	0.88	0.91	0.94	0.90	1.05	1.05	1.13	1.00	1.07	0.79	0.79	0.70	0.74	0.48	0.43
Mg-no.	49.34	54.14	50.26	52.95	54.42	53.94	49.31	53.57	54.42	55.16	55.96	55.16	52.12	52.77	52.98
K$_2$O/Na$_2$O	0.35	0.39	0.41	0.38	0.46	0.47	0.44	0.42	0.47	0.42	0.45	0.50	0.52	0.33	0.32
Cr	125	224	145	188	256	224	162	221	252	294	298	168	103	241	
Ni	90	144	98	132	182	129	170	195	191	193	142	105	165	97	146
V	199	179	198	191	189	192	195	186	193	193	186	175	153	204	
Cu	67	60	64	67	69	65	68	66	69	65	68	56	45	57	203
Pb	11	10	10	13	9	10	10	10	10	12	12	13	13	13	59
Zn	82	90	102	88	112	106	104	99	108	108	109	80	82	89	13
Ba	349	360	355	328	335	385	377	385	369	322	306	275	292	198	84
Rb	17	15	13	17	22	23	22	21	22	20	23	28	30	11	200
Sr	964	1004	1091	968	991	1077	1127	1052	1068	877	880	839	875	652	11
Ga	22	23	20	21	22	19	20	20	21	21	23	22	25	21	694
Nb	45	47	48	47	65	64	68	61	66	51	48	53	58	29	22
Zr	195	194	194	191	248	250	263	237	249	210	201	295	332	141	29
Y	30	28	30	30	29	29	31	28	30	25	24	27	27	25	146
Th	1	3	n.d	2	1	3	2	3	4	3	3	4	5	2	24
La	39	36	42	30	39	39	45	48	45	33	33	39	36	20	2
Ce	75	80	80	80	102	77	98	83	94	84	79	73	94	47	20
Nd	23	34	38	29	39	30	44	41	46	34	34	27	39	27	61
Rb/Zr	0.09	0.08	0.07	0.09	0.09	0.09	0.08	0.09	0.09	0.10	0.11	0.09	0.09	0.08	30
Zr/Nb	4	4	4	4	4	4	4	4	4	4	4	6	6	5	0.08
La/Nb	0.87	0.77	0.88	0.64	0.60	0.61	0.66	0.79	0.68	0.65	0.69	0.74	0.40	0.69	5
Y/Nb	0.66	0.59	0.62	0.64	0.45	0.45	0.46	0.46	0.45	0.49	0.50	0.51	0.47	0.86	0.69
															0.83

Alk.Ol.B, alkali olivine basalts.

Table 2. CIPW normative mineral calculations of İskenderun Gulf basalts

Sample	Or	Ab	An	Ne	Di	Ol	Mt	Il	Apt
Toprakkale									
T1	5.65	14.15	21.17	11.43	20.31	16.64	3.15	5.60	1.91
T2	8.56	11.22	20.02	12.93	21.10	15.48	3.12	5.67	1.92
T3	7.64	11.36	19.18	12.42	20.38	17.58	3.16	5.55	1.95
T4	7.82	11.43	20.02	12.60	20.58	16.89	3.13	5.53	2.00
T5	8.38	9.48	20.97	11.32	19.85	19.31	3.20	5.52	1.98
T6	9.90	11.92	18.88	12.12	19.12	16.89	3.18	5.72	1.28
T7	6.93	17.99	26.89	3.30	15.86	18.16	3.16	5.64	2.07
T8	8.11	12.05	22.45	10.34	19.92	16.46	3.13	5.57	1.97
T9	8.51	11.04	21.39	10.29	18.96	19.09	3.21	3.53	1.99
T10	8.10	11.15	21.33	10.31	18.98	19.50	3.19	5.54	1.91
T11	8.76	10.49	19.84	11.95	19.97	18.37	3.18	5.50	1.94
Delihalil									
D20	10.90	8.63	15.52	13.97	17.75	20.70	4.79	5.53	2.22
D21	10.72	9.41	16.08	12.82	17.03	21.24	4.85	5.59	2.27
D22	10.66	9.73	17.44	11.42	16.45	21.69	4.85	5.53	2.23
D23	10.90	9.24	17.03	12.01	16.90	21.39	4.83	5.48	2.23
D24	9.41	8.76	15.63	14.76	19.42	19.61	4.76	5.46	2.20
D25	10.93	8.33	16.01	13.64	18.50	20.00	4.79	5.56	2.27
D26	10.91	7.02	15.07	14.99	19.96	19.47	4.82	5.51	2.26
D27	11.55	7.19	15.08	16.37	21.11	15.76	4.72	5.80	2.42
D28	10.87	6.36	13.87	16.30	20.48	19.47	4.83	5.53	2.30
D29-1	10.00	9.33	16.83	13.21	18.63	19.68	4.76	5.44	2.13
D29-2	9.42	9.33	16.14	14.33	19.27	19.21	4.76	5.50	2.05
D30	10.91	7.74	15.79	13.60	18.54	20.76	4.85	5.52	2.29
Gertepe									
G31	9.34	12.87	18.43	9.03	17.74	21.09	4.74	5.04	1.72
G32	9.32	11.90	17.29	10.76	18.65	20.63	4.70	5.03	1.71
G33	8.89	13.46	19.38	7.92	17.44	21.64	4.70	4.89	1.69
Üçtepeler									
Ü34	15.00	10.47	13.75	17.02	15.76	17.25	4.12	5.04	1.59
Ü35	13.74	11.82	15.39	14.88	15.98	17.69	4.08	4.92	1.50
Ü36	15.61	12.18	13.73	16.59	14.97	16.38	4.04	4.96	1.54
Ü37	15.79	11.72	13.31	17.16	15.21	16.28	4.03	4.96	1.53
Ü38	15.82	11.37	13.36	17.26	15.26	16.36	4.05	4.95	1.57
Ü39	15.84	12.88	13.85	16.46	14.38	16.03	4.01	4.96	1.59
Ü40	5.60	14.86	25.96	5.10	16.64	20.94	4.89	4.97	1.03
Ü41	6.56	17.44	27.47	4.57	16.58	16.86	4.43	4.99	1.10
Ü42	5.93	15.64	26.94	6.04	20.20	16.51	2.94	4.86	0.94

Or, orthoclase; Ab, albite; An, anorthite; Ne, nepheline; Di, diopside; Ol, olivine; Mt, magnetite; Il, ilmenite; Ap, apatite.

induced by rapid cooling (Cox & Jamieson 1974). Occasionally, olivine phenocrysts in Toprakkale basanites show zoning with a homogeneous core surrounded by a continuously normally zoned mantle (Fig. 10b). Also, occasional olivine phenocrysts, from south of Toprakkale, are surrounded by a reaction rim of clear plagioclase and minute augite crystals. Large olivine phenocrysts contain some rounded green spinel inclusions as well as opaque oxide and dark greenish brown Ti-augites (Fig. 10a).

Augite is usually present as a groundmass phase and sometimes as inclusions in olivine and plagioclase. Plagioclase microphenocrysts in basanites are subhedral and show slight zoning. They also show resorption textures and enclose euhedral apatite crystals.

Magnetites are very common in the groundmass and as inclusions in the phenocrysts. The dark brown or reddish brown coloration of groundmass Fe-oxides suggests they are titanomagnetites. Microprobe analyses established that inclusions in the olivine phenocrysts are usually ferrous chrome spinels and titaniferous magnetites.

Table 3. *Representative analyses of olivines in the İskenderun Gulf alkali basaltic lavas*

	Delihalil Tepe basanite					Toprakkale basanite				Üçtepelar basanite			
	Core	Rim	Core	Rim	Core	Rim	Core	Rim	Core	Rim	Core	Rim	Core	Rim
SiO_2	38.60	38.32	38.45	37.46	38.51	37.73	38.78	36.50	38.76	36.55	37.66	34.96	38.36	36.80
TiO_2	0.00	0.00	0.00	0.00	0.00	0.00	0.00	0.00	0.00	0.00	0.00	0.00	0.00	0.00
Al_2O_3	0.00	0.00	0.00	0.00	0.00	0.35	0.00	0.00	0.00	0.00	0.00	0.00	0.00	0.00
FeO	18.02	20.60	18.56	22.57	19.05	22.13	17.69	29.80	17.15	27.04	25.52	34.62	18.57	26.86
MnO	0.24	0.18	0.00	0.48	0.00	0.40	0.00	0.34	0.00	0.30	0.28	0.59	0.00	0.49
MgO	42.30	40.18	41.76	37.87	41.39	38.49	42.22	32.38	43.13	35.28	35.18	28.76	41.77	34.74
CaO	0.23	0.34	0.27	0.47	0.30	0.48	0.24	0.39	0.25	0.43	0.48	0.53	0.32	0.56
Na_2O	0.00	0.00	0.00	0.52	0.00	0.00	0.41	0.00	0.00	0.00	0.39	0.00	0.00	0.00
Cr_2O_3	0.00	0.00	0.00	0.00	0.00	0.00	0.00	0.00	0.00	0.00	0.00	0.00	0.00	0.00
NiO	0.00	0.00	0.00	0.00	0.00	0.00	0.00	0.00	0.00	0.00	0.00	0.00	0.00	0.00
Total	99.37	99.62	99.04	99.37	99.25	99.58	99.34	99.41	99.29	99.60	99.81	99.46	99.02	99.45

Number of ions on the basis of four oxygens

Si	0.986	0.992	0.986	0.984	0.989	0.985	0.991	0.989	0.986	0.978	0.998	0.975	0.984	0.985
Ti	0.000	0.000	0.000	0.000	0.000	0.000	0.000	0.000	0.000	0.000	0.000	0.000	0.000	0.000
Al	0.000	0.000	0.000	0.000	0.000	0.000	0.000	0.000	0.000	0.000	0.000	0.000	0.000	0.000
Fe	0.385	0.446	0.398	0.496	0.409	0.483	0.378	0.675	0.365	0.605	0.565	0.807	0.807	0.398
Mn	0.005	0.004	0.000	0.011	0.000	0.009	0.000	0.008	0.000	0.007	0.006	0.014	0.000	0.011
Mg	1.611	1.550	1.596	1.482	1.585	1.498	1.608	1.308	1.636	1.408	1.401	1.195	1.597	1.386
Ca	0.006	0.009	0.007	0.013	0.008	0.013	0.006	0.011	0.007	0.012	0.014	0.016	0.009	0.016
Na	0.000	0.000	0.000	0.027	0.000	0.000	0.021	0.000	0.000	0.000	0.020	0.000	0.000	0.000
Cr	0.000	0.000	0.000	0.000	0.000	0.000	0.000	0.000	0.000	0.000	0.000	0.000	0.000	0.000
Ni	0.000	0.000	0.000	0.000	0.000	0.000	0.000	0.000	0.000	0.000	0.000	0.000	0.000	0.000
Total	2.994	3.001	2.988	3.012	2.990	3.000	3.004	2.991	2.994	3.010	3.020	3.007	2.988	2.999
Fo	81	78	80	75	79	75	81	66	82	70	71	59	80	69
Fa	19	22	20	25	21	25	19	34	18	30	29	41	20	31

Table 4. Representative analyses of clinopyroxenes in the İskenderun Gulf alkali basaltic rocks

	Delihalil Tepe basanite			Toprakkale basanite				Üçtepelar basanite			
	Core	Rim	Core	Rim	Core	Rim	Core	Rim	Core	Rim	Core
SiO$_2$	41.43	42.59	44.87	41.83	45.03	45.33	45.32	48.26	47.92	48.91	46.32
TiO$_2$	5.49	4.39	3.80	5.59	3.70	4.26	3.87	2.68	3.04	2.13	3.47
Al$_2$O$_3$	9.70	8.33	6.92	10.07	7.05	7.30	6.93	4.66	4.72	3.55	6.10
FeO	9.55	10.42	8.92	8.63	8.54	9.62	8.66	8.19	9.05	9.15	9.08
MnO	0.00	0.00	0.00	0.00	0.18	0.21	0.00	0.00	0.00	0.00	0.00
MgO	10.37	10.80	11.90	9.85	11.81	10.34	11.56	12.70	12.35	12.99	11.69
CaO	22.63	22.00	22.46	21.74	22.76	22.11	22.76	22.73	22.10	22.29	22.30
Na$_2$O	0.66	0.84	0.77	1.26	0.53	0.84	0.56	0.67	0.56	0.75	0.68
K$_2$O	0.00	0.00	0.00	0.13	0.00	0.00	0.00	0.00	0.00	0.00	0.00
P$_2$O$_5$	0.00	0.41	0.00	0.61	0.00	0.00	0.00	0.00	0.00	0.00	0.00
Total	99.82	99.78	99.63	99.71	99.60	100.01	99.66	99.87	99.74	99.77	99.64

Number of ions on the basis of six oxygens

Si	1.585	1.630	1.704	1.589	1.707	1.720	1.720	1.810	1.806	1.844	1.753
Ti	0.158	0.126	0.108	0.160	0.106	0.121	0.110	0.080	0.086	0.061	0.099
Al	0.437	0.376	0.310	0.451	0.315	0.327	0.310	0.210	0.209	0.158	0.272
Fe	0.305	0.333	0.283	0.274	0.271	0.305	0.270	0.260	0.285	0.289	0.287
Mn	0.000	0.000	0.000	0.000	0.006	0.007	0.000	0.000	0.000	0.000	0.000
Mg	0.591	0.616	0.673	0.558	0.667	0.585	0.650	0.710	0.694	0.730	0.660
Ca	0.928	0.902	0.914	0.885	0.924	0.899	0.920	0.920	0.892	0.900	0.904
Na	0.049	0.062	0.056	0.093	0.039	0.062	0.040	0.050	0.041	0.055	0.050
K	0.000	0.000	0.000	0.007	0.000	0.000	0.000	0.000	0.000	0.000	0.000
P	0.000	0.013	0.000	0.020	0.000	0.000	0.000	0.000	0.000	0.000	0.000
Total	4.054	4.063	4.048	4.036	4.034	4.026	4.030	4.030	4.013	4.036	4.024
Wo	51	49	49	52	49	50	50	48	48	47	49
En	32	33	36	32	36	33	35	38	37	38	35
Fs	17	18	15	16	15	17	15	14	15	15	16

Table 5. *Representative analyses of plagioclases in the İskenderun Gulf alkali basaltic rocks*

	Delihalil Tepe basanite						Toprakkale basanite			Üçtepelar basanite			
	Core	Rim	Core	Rim	Core	Rim	Core	Core	Rim	Core	Rim	Core	Rim
SiO$_2$	52.81	54.09	51.26	54.29	53.00	53.07	53.58	51.23	53.97	50.00	52.31	49.66	52.44
TiO$_2$	0.00	0.76	0.26	0.92	0.33	1.47	0.62	0.00	0.00	0.00	0.24	0.00	0.31
Al$_2$O$_3$	28.71	26.32	29.92	25.42	28.58	19.98	27.24	30.10	28.29	30.65	28.82	30.83	28.41
FeO	0.97	1.79	0.96	2.09	0.99	6.97	1.59	0.58	0.46	0.60	1.13	0.68	1.54
MnO	0.00	0.00	0.00	0.00	0.00	70.23	0.00	0.00	0.00	0.00	0.00	0.00	0.00
MgO	0.26	0.46	0.34	0.53	0.33	2.01	0.33	0.00	0.23	0.37	0.30	0.26	0.23
CaO	12.12	9.65	13.22	8.40	11.77	5.37	11.04	13.94	11.53	14.25	12.79	14.80	12.64
Na$_2$O	4.21	5.08	3.54	5.36	4.30	4.48	4.78	3.59	4.70	3.55	3.93	3.17	4.17
K$_2$O	0.46	1.42	0.38	2.10	0.48	5.02	0.56	0.18	0.42	0.17	0.27	0.15	0.24
P$_2$O$_5$	0.00	0.00	0.00	0.46	0.00	1.04	0.00	0.00	0.00	0.00	0.00	0.00	0.00
Total	99.54	99.69	99.88	99.57	99.78	99.76	99.74	99.62	99.58	99.58	99.78	99.56	99.98

Number of ions on the basis of eight oxygens

Si	2.409	2.476	2.343	2.493	2.414	2.520	2.450	2.340	2.450	2.290	2.390	2.280	2.400
Ti	0.000	0.026	0.009	0.032	0.011	0.050	0.020	0.000	0.000	0.000	0.010	0.000	0.010
Al	1.544	1.420	1.612	1.376	1.535	1.120	1.470	1.620	1.510	1.660	1.550	1.670	1.530
Fe	0.037	0.068	0.037	0.080	0.038	0.280	0.060	0.020	0.020	0.020	0.040	0.030	0.060
Mn	0.000	0.000	0.000	0.000	0.000	0.010	0.000	0.000	0.000	0.000	0.000	0.000	0.000
Mg	0.018	0.031	0.023	0.036	0.023	0.140	0.020	0.000	0.020	0.030	0.020	0.020	0.020
Ca	0.593	0.473	0.648	0.413	0.574	0.270	0.540	0.680	0.560	0.700	0.630	0.730	0.620
Na	0.372	0.451	0.313	0.477	0.379	0.410	0.420	0.320	0.410	0.320	0.350	0.280	0.370
K	0.027	0.083	0.022	0.123	0.028	0.300	0.030	0.010	0.020	0.010	0.020	0.010	0.010
P	0.000	0.000	0.000	0.018	0.000	0.040	0.000	0.000	0.000	0.000	0.000	0.000	0.000
Total	4.999	5.035	5.007	5.050	5.002	5.160	5.010	4.990	4.990	5.020	5.000	5.010	5.020
An	60	47	66	41	59	28	54	68	56	68	63	71	62
Ab	37	45	32	47	39	42	42	31	41	31	35	28	37
Or	3	8	2	12	2	31	3	1	3	1	2	1	1

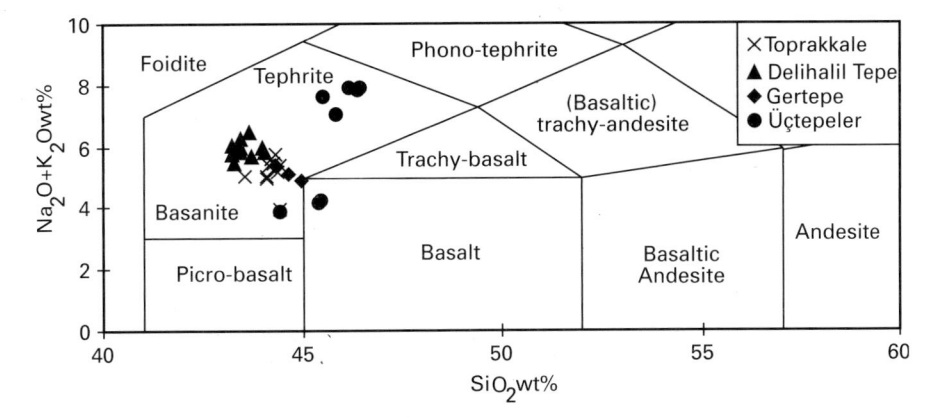

Fig. 8. Total alkali–silica diagram of the İskenderun Gulf alkali volcanics [after Le Bas *et al.* (1986)].

Alkali olivine basalts

The alkali olivine basalts are also porphyritic and, as in the basanites, olivine is the dominant phenocryst phase (Fig. 10c). Olivine phenocrysts are euhedral to subhedral, up to 1.5 mm in size and are embedded in a vesicular, fine-grained groundmass of olivine, plagioclase, augite, titanomagnetite and glass. As in the basanites, most of the olivines are resorbed and contain dark olive green spinel and titanaugite inclusions. Augite generally occurs in higher proportions in the alkali olivine basalts than in the basanites. Augites are present as euhedral to subhedral phenocrysts and microphenocrysts, as well as groundmass material. Along the margins of some altered olivine minerals magnetite and augite are present. Apatite is an accessory phase often associated with Ti-magnetite. Large gas vesicles are randomly distributed in the rock with variable carbonitization along the sides of some vesicles.

Chrome spinel, found enclosed within the olivine, was the earliest liquidus phase. The spinel was followed by olivine, clinopyroxene, plagioclase and titanomagnetite in both basanites and alkali olivine basalts, justified by petrographical observations.

Mineral chemistry

The chemical composition of the olivines from basanites and alkali olivine basalts of Delihalil Tepe are similar and relatively uniform, ranging

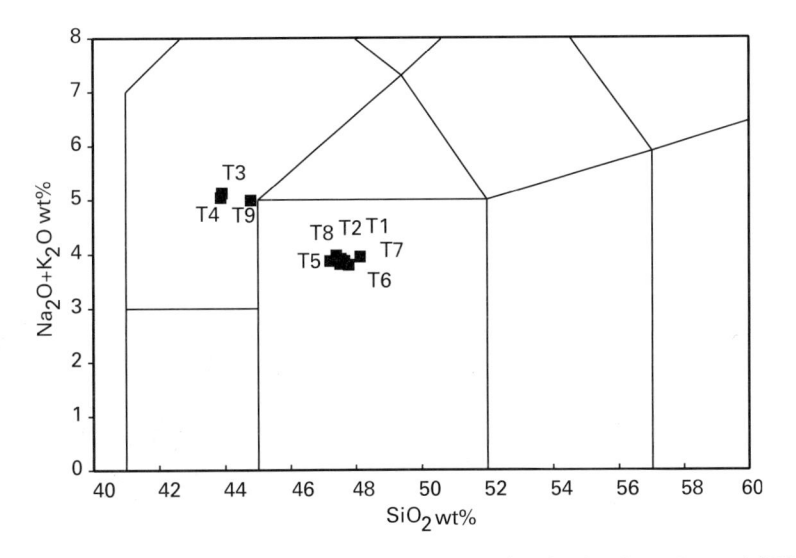

Fig. 9. Classification of Toprakkale samples analysed by Arger *et al.* (2000) using the Le Bas *et al.* (1986) scheme.

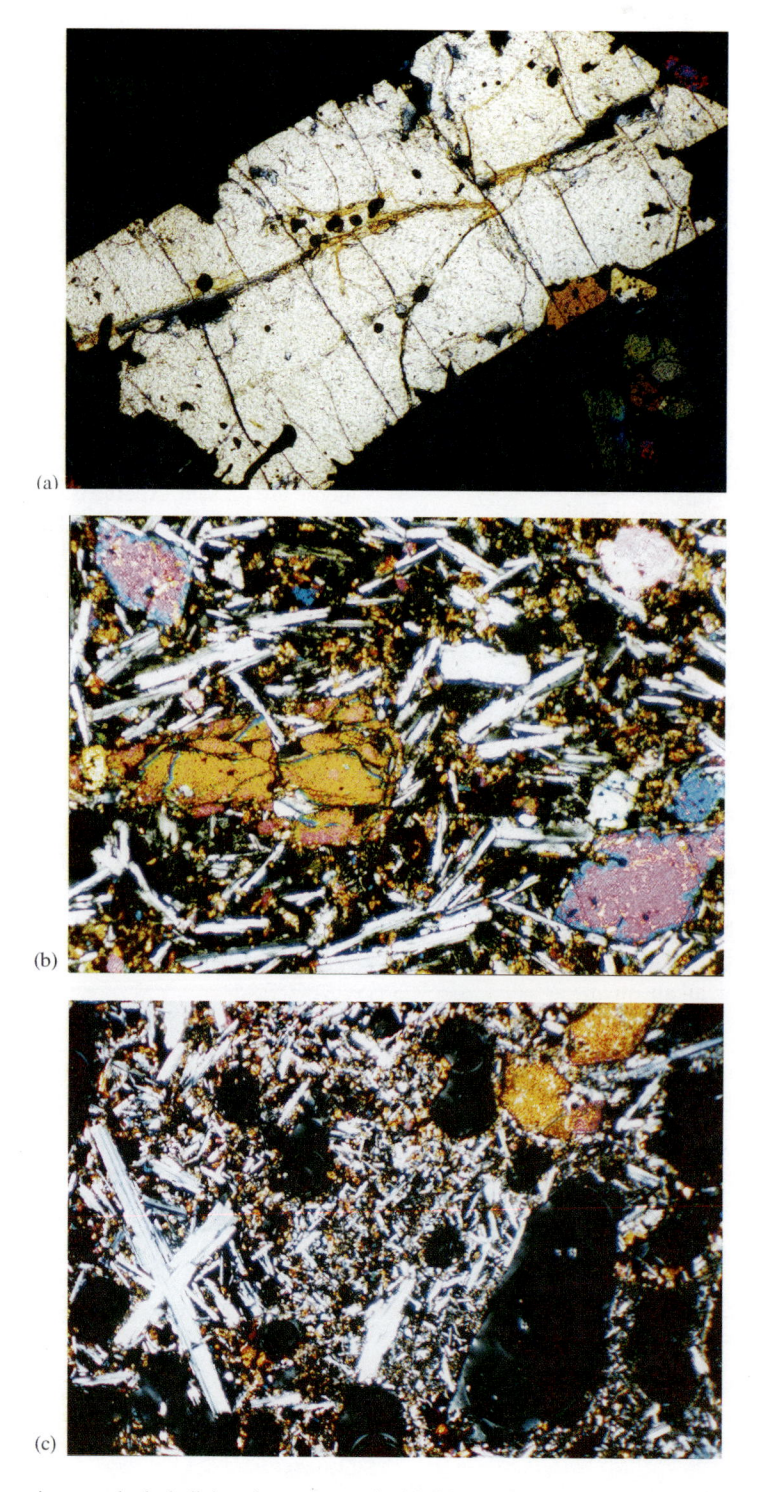

Fig. 10. (**a**) Basanite – a euhedral olivine phenocryst embedded in a highly vesicular and fine-grained groundmass. Olivine contains plenty of ferrous chrome spinel inclusions (CPL, 2.5 × 1.25 × 10). (**b**) Basanite from Toprakkale, olivine phenocrysts showing zoning with a homogeneous core surrounded by a continuously zoned mantle indicated by the variation in interference colours (CPL, 2.5 × 1.25 × 10). (**c**) Alkali olivine basalt from Üçtepeler (CPL, 1.25 × 1.25 × 10). Ol, Olivine; Plg, Plagiocalse.

from Fo_{81} to Fo_{78} in the cores (Table 3). Core compositions of olivines in the Üçtepeler alkali olivine basalts vary between Fo_{80} and Fo_{68}, and in the Toprakkale basanites vary between Fo_{82} and Fo_{61}. In the Delihalil Tepe basanites, the forsterite content of the olivine is fairly constant from the core towards the rim, but the forsterite content decreases at the rim, ranging from Fo_{78} to Fo_{75}. Olivine microphenocrysts in the groundmass are more Fe rich (Fo_{70}) than the phenocrysts. Olivines in the Üçtepeler alkali olivine basalts and the Toprakkale basanites show a slightly wider range of compositions at the olivine margins. Rim compositions in Üçtepeler range between Fo_{69} and Fo_{59}, and from Fo_{70} to Fo_{30} in Toprakkale (Table 3).

Euhedral and subhedral augites in basanites and alkali olivine basalts do not show very much variation in their core and rim compositions (Table 4). Core composition of augite in basanites from Delihalil Tepe ranges from $Wo_{51-49}En_{36-31}Fs_{18-15}$, and rim compositions range from $Wo_{52-48}En_{35-31}Fs_{18-16}$. Clinopyroxenes show similar core and rim compositions to Delihalil Tepe basanites in Üçtepeler and Toprakkale alkaline rocks (Table 4). They are all titanaugites with high CaO (22–23%) and TiO_2 (2–5%) contents. The TiO_2 content of augites in alkali olivine basalts (2–3%) is lower than the Delihalil Tepe and Toprakkale basanites (Table 4).

Plagioclase (Table 5) is only present as a groundmass phase and are microcrystalline. The chemical compositions of plagioclase in the Delihalil Tepe basanites range from $An_{66-54}Ab_{42-32}Or_{3-2}$ in the core to $An_{64-28}Ab_{47-34}Or_{31-2}$ at the rim. Plagioclase compositions in the Üçtepeler alkali olivine basalt range from $An_{71-68}Ab_{31-28}Or_{1-0}$ in the core to $An_{69-62}Ab_{37-30}Or_{2-1}$ at the rim, and in Toprakkale basanites from $An_{70-55}Ab_{33-28}Or_{2-1}$ in the core to $An_{67-26}Ab_{59-32}Or_{15-1}$ in the margins (Table 5).

The general mineralogical characteristics of the İskenderun Gulf alkaline mafic rocks indicates that the most Fo-rich composition of the olivines is Fo_{80-82} within both the basanites and the alkali olivine basalts. Clinopyroxene compositions are also, in general, similar, except that augites in alkali olivine basalts have lower TiO_2 concentrations. Plagioclases have chemical compositions in the andesine–labradorite range in both alkali olivine basalts and basanites.

Major element chemistry

Representative major element analyses for Plio-Quaternary İskenderun Gulf basanites and alkali olivine basalts are presented in Table 1. Basanites and alkali olivine basalts have SiO_2 contents in the 43–46 wt% range and MgO in the 6.04–10.08 wt% range; Mg-numbers range from 49 to 56. Total alkali ($Na_2O + K_2O$) contents of İskenderun Gulf alkali basalts range from 3.78 to 7.92%; they are all sodic with K_2O/Na_2O ratios varying between 0.32 and 0.52.

Concentrations of P_2O_5, K_2O and Na_2O are generally higher in the basanites than in alkali olivine basalts, but Al_2O_3, MgO and TiO_2 are similar (Table 1). In major oxides v. SiO_2 diagrams (Fig. 11), Al_2O_3 displays a positive correlation and $Fe_2O_3^T$ a scattered negative correlation; P_2O_5 and TiO_2 show scattered negative correlation with SiO_2; CaO and MgO also present scattered linear correlations with SiO_2.

Interpretation

Major element characteristics of the basalts in the study area suggest that the İskenderun Gulf mafic magmas are evolved (Mg-number 49–56). Also, although scattered, the linear correlations observed in major oxides v. silica diagrams may indicate the importance of different fractionating mineral assemblages during the evolution of the İskenderun Gulf magmas. They have experienced fractionation of olivine and clinopyroxene to varying degrees (Wilson 1989).

Major element concentrations of İskenderun Gulf basanites and alkali olivine basalts are closely comparable to those of average ocean island basalts (OIB) and intracontinental basalts (Table 6). The sodic ratios of the İskenderun Gulf alkali basalts (0.32–0.52) are within the range of Pliocene extension-related alkali basalts of the Pannonian Basin (Embey-Isztin et al. 1993), the Tertiary sodic mafic rocks in the Massif Central (Wilson & Downes 1991) and Early Cretaceous continental rift-type volcanic rocks of the Mescek Mountains (Harangi 1994), all with a K_2O/Na_2O rations < 1.

The Al_2O_3 concentrations of all volcanic rocks are generally high compared to average OIB (Table 6). Similar high concentrations of Al_2O_3 are a characteristic feature of Basin-and-Range extensional magmatism in the southwestern USA (Fitton et al. 1991).

İskenderun Gulf alkaline rocks have high Al_2O_3/CaO ratios, ranging between 1.5 and 2.3. It is commonly argued that the Earth's upper mantle should have an Al_2O_3/CaO ratio which is close to that of chondrites, i.e. c. 1.2, and that basalts derived by high degrees of melting, leaving mainly olivine and orthopyroxene as

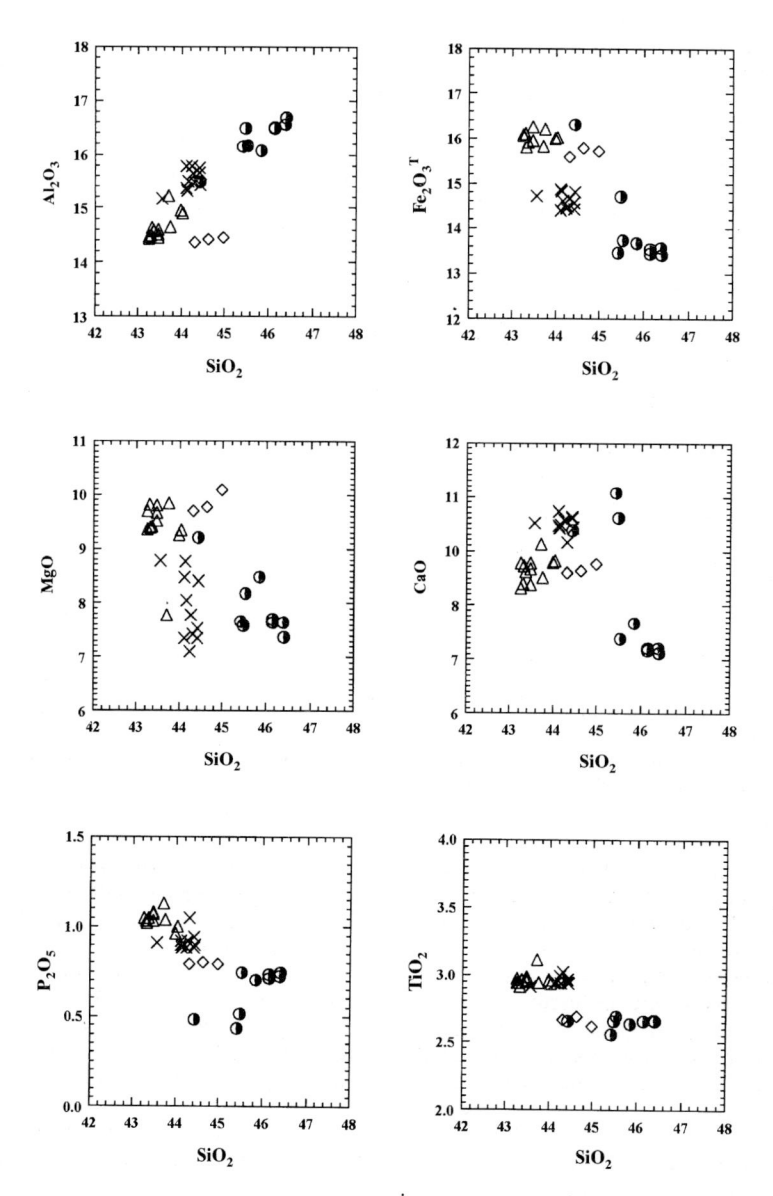

Fig. 11. Major element v. silica variation diagrams of the İskenderun Gulf alkali basalts. Open triangle, Delihalil Tepe basanites; cross, Toprakkale basanites; diamond, Gertepe basanite; half-filled circle, Üçtepeler alkali olivine basalts and basanites.

residual phases, should have an Al_2O_3/CaO ratio c. 1.2 (Frey et al. 1978). If melting leaves residual clinopyroxene, and Al_2O_3 contents of both pyroxenes are not very high, then (Al_2O_3/CaO) can be > 1.2 (Frey et al. 1978). However, it is difficult to generate Al_2O_3/CaO of 1.5 and 2.3 by partial melting leaving residual clinopyroxene, and it is more probable that the alkali olivine basalts and basanites have undergone low-pressure fractionation by settling out of olivine and Ca-rich clinopyroxene, or by crystallization of olivine, Ca-rich clinopyroxene and plagioclase.

Trace elements

The trace element compositions of İskenderun Gulf alkali olivine basalts and basanites are

Table 6. *Comparison of İskenderun Gulf alkali olivine basalts and basanites with those of Tertiary–Quaternary intracontinental alkali basalts and basanites and with average OIB*

	Cey.AB	Cey.Bas	MCAB	MCBas	Average OIB
SiO_2	45.43	43.46	46.79	41.22	44.88
TiO_2	2.55	2.98	2.36	3.22	2.94
Al_2O_3	16.14	14.50	13.99	13.32	13.48
Fe_2O_3	12.10	13.61	11.86	12.50	13.05
MnO	0.17	0.18	0.15	0.18	0.18
MgO	7.65	9.51	10.04	9.71	9.03
CaO	10.89	9.33	9.62	11.73	10.57
Na_2O	3.16	4.41	3.49	4.2	2.98
K_2O	1.00	1.88	1.34	1.88	1.18
P_2O_5	0.43	1.08	0.60	1.08	0.61
Cr	146	241	236	282	329
Ni	97	186	215	150	192
V	203	197	–	–	276
Zn	84	107	–	–	–
Rb	11	23	30	54	29
Sr	694	1077	646	1066	718
Ba	200	352	474	796	511
Zr	146	251	234	369	255
Nb	29	66	79	140	52
Th	2	3	6	12	–
Y	24	28	25	33	29
Nd	30	34	42	72	44
La	20	36	38	88	45
Ce	61	87	91	179	94

Major and trace element data for MCAB (Massif Central alkali basalt) and MCBas. (Massif Central basanite) from Wilson & Downes (1991), and average OIB data from Fitton *et al.* (1991).

presented in Table 1, and their concentrations are compared with intracontinental plate basalts and average OIB concentrations in Table 6.

The concentrations of the compatible (first transition series) elements in the basanites and alkali olivine basalts are generally low compared with average OIB and intraplate data, reflecting the more evolved nature of these rocks. Nickel is a sensitive indicator of olivine fractionation/accumulation from basaltic magmas because of its large mineral/melt partition coefficient. Nickel abundances in the İskenderun Gulf volcanic rocks vary between 97 and 165 ppm in alkali olivine basalts and 90–196 ppm in basanites. Chromium concentrations are between 146 and 241 ppm in alkali olivine basalts and 103–298 ppm in basanites. Chromium and Ni both show good positive linear correlations with MgO (Fig. 12). Vanadium concentrations in the İskenderun Gulf alkali volcanics range from 151 to 213 ppm, correlate negatively with SiO_2 and are expected to partition into clinopyroxene (Frey *et al.* 1978).

The large-ion lithophile (LIL) elements are not preferentially incorporated into early crystallizing minerals, so all become enriched in residual liquids (Wilson 1989). Figure 13a and b shows primordial mantle-normalized trace element variation diagrams of İskenderun Gulf alkaline basalts, together with typical alkali basalts from the oceanic islands of St Helena (HIMU-type) and continental Massif Central alkali basalt and basanite. The shape and the slope of the trace element patterns of the İskenderun Gulf alkaline basalts and basanites are similar to those of alkaline basalts and basanites from OIB (HIMU-type) and continental within-plate (Massif Central) areas. All İskenderun Gulf alkalic rocks showing typical humped patterns are enriched in incompatible elements (Fig. 13b), with peaks at Rb, Ba, K and Nb and with significant troughs at Th, Nd and Zr. This is also comparable to P-type mid-ocean ridge basalt (MORB) trace element patterns with peaks at Ba and Nb and troughs at Th and K. The spider diagram patterns of İskenderun Gulf alkali volcanic rocks also reveals a relative enrichment of LREE with depletion of HREE (Fig. 13b).

Incompatible element abundances of İskenderun Gulf alkaline volcanics are plotted v. Zr

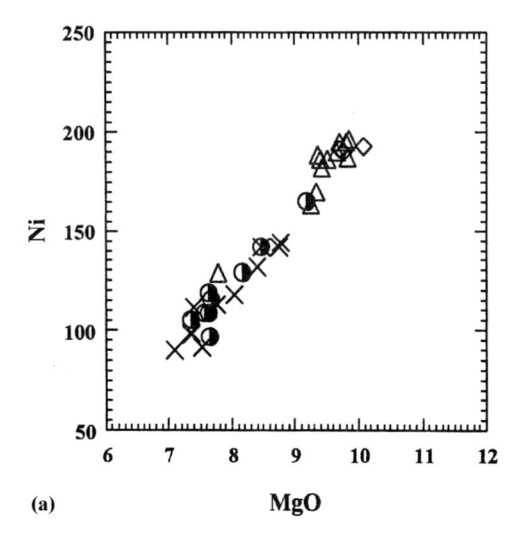

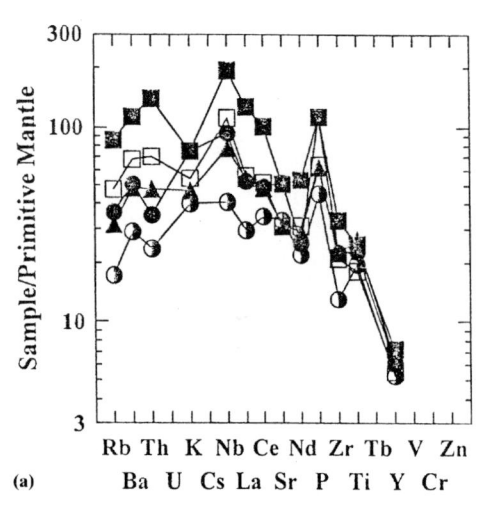

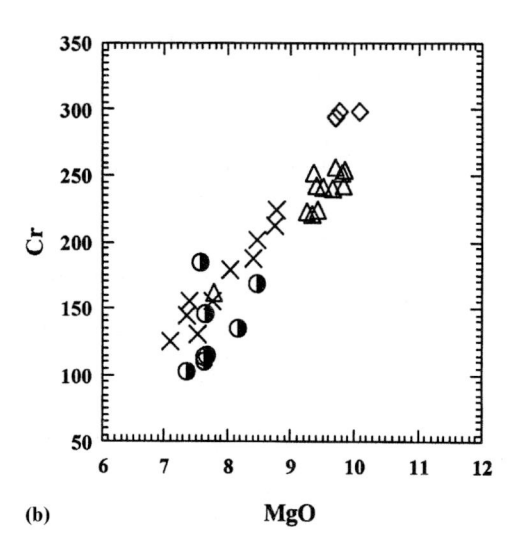

Fig. 12. Variation of Ni (**a**) and Cr (**b**) contents (ppm) v. MgO (wt%) for the İskenderun Gulf alkaline volcanic rocks. Symbols as in Fig. 11.

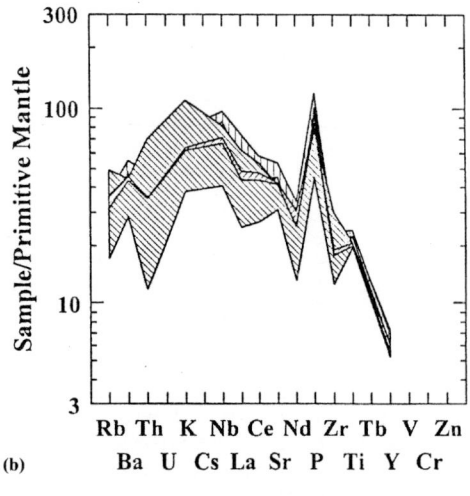

Fig. 13. Primitive mantle (Sun & McDonough 1989) normalized trace element patterns of the İskenderun Gulf alkali olivine basalts and basanites. (**a**) The patterns of the representative sample from St Helena OIB [data from Sun & McDonough (1989)], the patterns of intracontinental alkali basalts and basanites from Massif Central, France [data from Wilson & Downes (1991)] are presented along with İskenderun Gulf alkali basalt and basanite for comparison. Open triangle, St Helena; open square, Massif Central alkali basalts; filled square, Massif Central basanites; half-filled circle, İskenderun Gulf alkali olivine basalt; filled circle, İskenderun Gulf basanite. (**b**) İskenderun Gulf alkali olivine basalts and basanites envelope.

(used as a general index of fractionation) in Fig. 14. The incompatible elements tend to scatter but correlate positively with one another, and their abundances increase with alkalinity. In the Nb v. Zr diagram there is a good linear trend which intersects the Nb axis near the origin, indicating slightly more compatible behaviour of Nb than Zr during melting (Weaver *et al.* 1987). In several of the incompatible element diagrams, especially in Nb v. Zr and P_2O_5 v. Zr, volcanic rocks present two distinct linear trends. The first trend includes

Toprakkale, Delihalil Tepe and Gertepe basanites, and the second one Üçtepeler alkali basalts

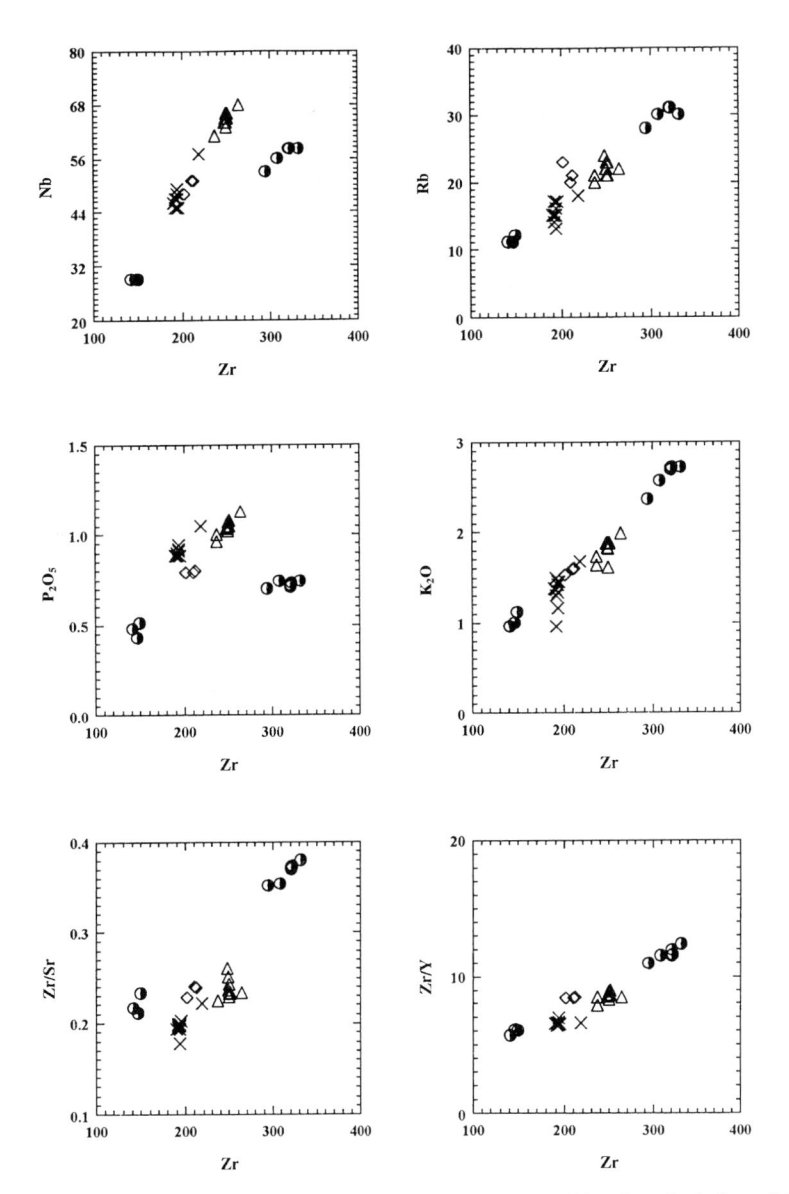

Fig. 14. Incompatible element abundances v. Zr (used as a general index of fractionation) plots of the İskenderun Gulf alkaline basalts. Symbols as in Fig. 11.

and basanites but with a large compositional gap. However, in Zr/Y v. Zr, Rb v. Zr and K_2O v. Zr diagrams all the volcanic rocks present good linear trends (Fig. 14).

Interpretations of trace element data

Fractional crystallization. The compatible trace elements support the observation made from major elements that the İskenderun Gulf mafic magmas experienced some crystal fractionation, and positive correlation of Ni with MgO suggests the presence of olivine in the fractionating assemblage. The positive correlation between Cr and MgO is possibly due to the concurrent crystallization of olivine and a Cr-rich spinel phase and/or clinopyroxene. Also, a negative correlation observed between V and SiO_2 should indicate clinopyroxene fractionation.

The incompatible element abundances in basalts are controlled by source region abundances, degree of partial melting, residual mineralogy and crystal fractionation (Weaver et al. 1987; Fitton et al. 1991). Constant ratios of certain incompatible trace elements (e.g. Nb/Zr, Ce/Zr, La/Zr and Rb/Zr) provide a useful test for the fractionation-controlled origin of the more SiO_2-rich magmas, indicating whether significant crustal contamination is likely to have occurred (Wilson 1989). The degree of partial melting from a peridotite source should be reflected in the Zr/Nb ratios of the derivative lavas. Since the distribution coefficient for Zr in clinopyroxene is about ten times larger than that for Nb (Watson & Ryerson 1986), higher Zr/Nb ratios indicate greater degrees of partial melting from mantle peridotite (Camp & Roobol 1992). The Zr/Nb ratio of the İskenderun Gulf alkali volcanic rocks is 4 and is very uniform throughout the Delihalil Tepe, Toprakkale and Gertepe basanites (Table 1); this ratio is 5 in Üçtepeler alkali basalts and 6 in Üçtepeler basanites (Table 1). These Zr/Nb ratios may indicate that the İskenderun Gulf basanites and alkali olivine basalts could be derived from a similar mantle source.

Good linear trends observed in Nb v. Zr, Zr/Y v. Zr, Rb v. Zr and K_2O v. Zr diagrams, and the constancy of incompatible element ratios in İskenderun Gulf alkaline volcanic rocks, provide strong evidence that fractional crystallization has been the dominant process in the evolution of the alkali olivine basalts and basanites. These linear trends may be interpreted as liquid lines of descent produced by low-pressure fractional crystallization of a similar source with variation in the degree of partial melting.

Crustal contamination. Basalts erupted through continental rather than oceanic crust are more likely to be modified by crustal contamination because of the lower density, lower melting temperature and the greater thickness of continental crust compared to oceanic crust. Thompson et al. (1984) suggested that the La/Nb ratio might be a useful index of crustal contamination in magmas. OIB and continental alkali basalts all have La/Nb ratios < 1, whereas continental flood basalts range from 0.5 to 7, suggesting variable degrees of contamination. La/Nb ratios of İskenderun Gulf alkali rocks range between 0.40 and 0.91; therefore, crustal contamination is not thought to be significant. The Ce/Pb ratio of average OIB is 29.7 (Fitton et al. 1991). Ratios lower than average OIB may have arisen in the oceanic environment by addition of Pb to the mantle, probably by addition of lithospheric material into the asthenosphere.

Ce/Pb ratios of lavas in the İskenderun Gulf volcanic rocks are significantly lower than the OIB average ratio, ranging from 5 to 11. In general, the trace element compositions of İskenderun Gulf alkaline mantle-derived magmas indicate that crustal contamination has been minimal.

Mantle heterogeneity. Variations in K/Ba ratios (as well as other incompatible element ratios) are sensitive indicators of mantle heterogeneity. The calculated K/Ba ratio for average unaltered MORB is 121 [Engel et al. (1965) in Basaltic Volcanism Study Project (1981)], although the observed range is from as low as c. 20, suggesting that the oceanic upper mantle is inhomogeneous with respect to those elements. K/Ba ratios in the İskenderun Gulf alkali volcanics range between 27 and 95, which is very similar to those in the Hawaii alkali series basalts with a 25–80 range (Basaltic Volcanism Study Project 1981).

Evaluation of possible origin and discussion

İskenderun Gulf alkali basalts have intraplate petrographic and geochemical characteristics. Intraplate alkali basalts from both continental and oceanic environments are known to share the same general chemical and isotopic characteristics. Major and trace element concentrations, and interelemental ratios, provide general constraints on petrochemical processes involved in magma evolution. The main focus on the determination of the mantle source characteristics is whether they are derived from the asthenosphere (including plume) or from the overlying lithosphere, or perhaps a mixture of both sources (Kay & Gast 1973; Fitton & Dunlop 1985; Fitton 1987; Weaver et al. 1987; Wilson & Downes 1991).

The incompatible element concentrations of İskenderun Gulf alkali olivine basalts and basanites are closely comparable to average continental alkali basalts and OIB of alkaline affinity, and are enriched relative to MORB. However, these basalts present some other chemical characteristics similar to extension-related alkali basalts and others indicative of subduction.

Plume origin

It is widely accepted that most continental-rift-related basalts derive from an asthenospheric mantle component which shows affinities to that of the St Helena OIB source.

Highly incompatible trace element ratios are independent of, or vary only slightly during, partial melting (if the degree of melting is not too low) and during low-grade fractionation of basaltic magmas. Therefore, they are good indicators of the mantle source of basaltic magmas (Weaver et al. 1987; Wilson 1989). Also, P-type MORB incompatible trace element patterns, with Ba and Nb peaks and Th and K troughs, may suggest that the components in the sources of P-type MORB and intraplate basalts have similar plume-related geochemical characteristics (Weaver et al. 1987; Wilson & Downes 1991).

Mantle-source characteristics of the İskenderun Gulf alkaline volcanic rocks are here compared to those of extension-related alkaline volcanic rocks in Central Europe and OIB. The geochemical characteristics of mafic magmas in Central Europe have been explained simply in terms of two-component mixing of end-members. Wilson & Downes (1991) have suggested that both asthenospheric and lithospheric mantle material has been involved in the petrogenesis of the Tertiary–Quaternary extension-related alkaline mafic magmas in Europe. The asthenospheric mantle component is regarded as a mixture of depleted mantle (DM: depleted MORB-type mantle mixed with plume material) and shows affinities to that of the St Helena OIB source. The lithospheric mantle component has enriched mantle (EM) characteristics and is isotopically similar to the source of Gough-type Dupal OIB, i.e. contamination of the OIB source region by a small amount (1–2%) of ancient pelagic sediment which was subducted into the deep mantle along with the ocean crust (Weaver et al. 1987; Wilson 1989).

Figure 15 shows the variation of Rb/Nb v. K/Nb for the İskenderun Gulf alkaline volcanic rocks compared with typical values for primitive mafic volcanic rocks from the Massif Central, France (Wilson & Downes 1991), and St Helena and Gough OIB (Sun & McDonough 1989). The alkaline volcanic rocks from north of the İskenderun Gulf define a coherent linear trend which may show that these volcanic rocks are produced by mixing of two chemically distinct mantle-source components or partial melts (Fitton et al. 1991; Wilson & Downes 1991). The majority of the alkaline volcanic rocks (Delihalil Tepe, Toprakkale and Gertepe basanites, and Üçtepeler alkali basalts) have a strong affinity to depleted MORB-type mantle mixed with plume material, the HIMU OIB-source component similar to those observed in Massif Central alkali basalts and basanites (Fig. 15). However, trace element data alone is not enough to determine whether it originates in the lithosphere or the asthenosphere. Only Üçtepeler basanites show a slight affinity to the lithospheric mantle component (EM characteristics).

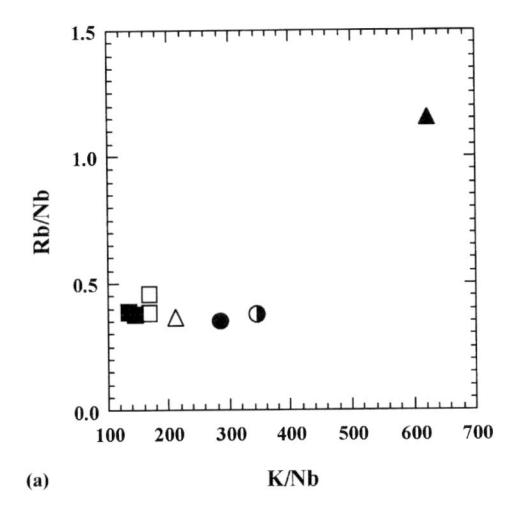

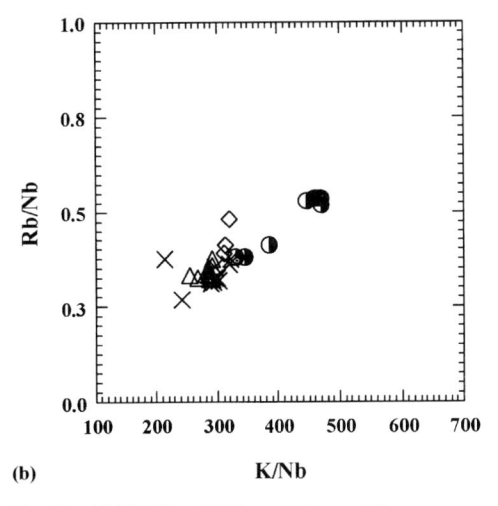

Fig. 15. (a) Rb/Nb v. K/Nb variations of the İskenderun Gulf alkali basalts plotted together with Massif Central (intracontinental) alkali basalts and basanites [data from Wilson & Downes (1991)]. The samples from St Helena and Gough are also plotted in this diagram to compare with (hypothetical) mixing end-members. Symbols as in Fig. 13. (b) All İskenderun Gulf volcanic rocks define a coherent linear trend. Symbols as in Fig. 11.

Subduction origin

Trace element distribution patterns and some incompatible trace element ratios (e.g. Rb/Nb, K/Nb) of the İskenderun Gulf alkali olivine basalts and basanites clearly indicate an incompatible element–EM source origin which was similar to that of OIB. Subduction processes may have affected the mantle source of these alkaline volcanic rocks. Subduction-related modification of the source would be reflected by enrichment in Ba, K, Rb, Pb and Sr relative to Nb (Fitton et al. 1991). The composition of the İskenderun Gulf samples does demonstrate enrichment of these elements and may indicate contribution from a mantle source which has been modified by slab-derived fluids. However, İskenderun Gulf basic magmas do not show the typical trace of a subduction signature in their incompatible element abundances such as Nb and Ti troughs (Fig. 13). Also, trace element ratios such as La/Nb and Ce/Pb, indicate that the İskenderun Gulf magmas have not been significantly contaminated with crustal material during ascent and trace element ratios therefore reflect mantle-source characteristics. Differences in their trace element patterns are due to different degrees of partial melting of a similar source (Fig. 14).

The nature of the lithospheric upper mantle is usually well constrained by the occurrence of abundant spinel lherzolite xenoliths within the alkaline volcanic rocks. In the studied alkaline volcanic rocks centres north of the İskenderun Gulf, xenoliths of spinel lherzolite have never been observed. However, the overall geochemical characteristics of these volcanic rocks indicate that they are largely from a similar source with variation in the degree of partial melting. Basanites and alkali olivine basalts in this area display nearly constant values (in the same narrow range) of mobile–immobile element ratios (La/Ce, Zr/Nb, Rb/Zr, K/Zr). Variations in these ratios (Fig. 14) reflect the diversity of magma batches feeding individual centres related to different degrees of melting (Baker 1987). The alkali basalts have low Rb (11–31 ppm) and Ba (198–385 ppm) contents compared to average OIB. Mobile–immobile trace element ratios (e.g. Ba/Nb, Rb/Nb) do not show any significant correlation with LOI and there is little petrographic evidence for sample alteration. Therefore, it can be assumed that LIL element contents of some alkali olivine basalts and basanites reflect, principally, the source characteristics and primary magmatic features. Slight K-enrichment in some samples compared to average OIB could be related to subsequent alteration.

Extension-related origin

Major and trace element chemistry of the İskenderun Gulf alkali basalts also presents an extensional magmatism signature. The sodic nature (K_2O/Na_2O ratio of 0.32–0.52) similar to extension-related alkali basalts and basanites in Europe, high Al_2O_3 content and some trace element ratios of basanites and alkali olivine basalts, and enrichment of incompatible elements, indicate extensional source characteristics.

Primary magma characteristics

A primary magma derived by partial melting of a peridotite mantle may be expected to have high Ni and Cr contents and a high Mg-number [$Mg/(Mg + Fe^{2+})$]. The Mg-number should lie between 68 and 75 if the degree of partial melting is < 30% (Green 1971; Frey et al. 1978; Green et al. 1979). Suggested compositions of Ni in the primary magmas should be within the 300–500 ppm range (Frey et al. 1978) and between 10 and 12 wt% for MgO (Sato 1977). None of the İskenderun Gulf alkaline volcanic rocks can be regarded as representing primary liquids that are in equilibrium with mantle peridotite. In general, all the samples of the İskenderun Gulf alkaline rocks are relatively evolved, with Ni < 200 ppm, Cr < 300 ppm, Mg-number < 60 and MgO in the 6–10% range. Most have compositions indicating moderate degrees of differentiation and low MgO and Ni, indicating olivine fractionation. Good positive correlation between MgO and Cr testify to clinopyroxene fractionation. This is also substantiated by major element data, as CaO and Fe_2O_3 decrease with decreasing MgO.

Conclusions

Based on geochemical characteristics, the following conclusions can be made on the Plio-Quaternary volcanic rocks of the İskenderun Gulf area.

Plio-Quaternary volcanicity produced a series of relatively small scoria cones and associated lavas that are dominantly alkaline in character. These alkaline volcanic rocks lie along the active left-lateral northeast–southwest trending KOFZ, forming part of modern the Turkish–African Plate boundary and the Neotethys Suture. This basaltic volcanism is concentrated around the point where these two structures intersect near Toprakkale and Delihalil Tepe.

Two types of basaltic groups are distinguished: dominantly basanites and alkali olivine basalts. The trace element characteristics of the alkali olivine basalts and basanites show that they could have been formed by variable degrees of partial melting of similar mantle sources, with the basanites reflecting smaller degrees of melting than the alkali olivine basalts. These low-volume, low-degree melts most likely reached the Earth's surface where the intersection created by the KOFZ and Neotethys Suture allowed their passage through the lithosphere without any detectable interaction.

Major and trace element characteristics of the İskenderun Gulf alkali basalts are similar to those of alkaline basalts of oceanic and intracontinental plate areas, and show enrichment in K, Rb, Sr, Pb, Nb and Ti. No crustal contamination effects have been recognized in the basalts. The mantle source of the İskenderun Gulf alkali basalts originated in the asthenosphere similar to the OIB source area and overall geochemical characteristics support a St Helena-type source, as proposed by Wilson & Downes (1991). This source is regarded as a mixture of depleted mantle with a plume component, has HIMU character and is classified as one of the mantle end-members for the young extension-related alkaline basalts in Europe.

The basaltic lavas are evolved in terms of Mg-number and the concentrations of compatible trace elements also suggest that they are too evolved to represent primary magmas in equilibrium with their mantle source. Low-pressure fractionation must have occurred in the mantle from a primary parent created by varying degrees of partial melting of the source. The strong depletion of the compatible elements in these evolved basalts indicated that crystallization of olivine, a Cr-rich spinel phase, and clinopyroxene controlled the composition of the residual liquids during magma evolution.

These rocks contain some characteristics of OIB (e.g. trace element patterns, and Nb/Zr, Rb/Zr and K_2O/Zr ratios), other characteristics indicative of extension-related alkali basalts (e.g. K_2O/Na_2O, high Al_2O_3 and trace element concentrations), and still some other characteristics indicative of subduction (e.g. enrichment of incompatible elements and La/Zr, Ce/Zr and Ce/Pb).

S. Y. is very grateful to G. D. Williams and G. Kelling (Department of Earth Sciences, Keele University) for providing laboratory facilities. Analytical data production was greatly helped by the expertise of D. W. Emley and M. Aikin. We are grateful to R. Westaway and A. Chambers for their encouragement and suggestions that improved the text.

References

ARGER, J., MITCHELL, J. & WESTAWAY, R. W. C. 2000. Neogene and Quaternary volcanism of southeastern Turkey. *This volume*.

BAKER B. H. 1987. Outline of the petrology of the Kenya rift alkaline province. *In*: FITTON J. G. & UPTON, G. G. L. (eds) *Alkaline Igneous Rocks*. Geological Society, London, Special Publications, **30**, 293–311.

BASALTIC VOLCANISM STUDY PROJECT. 1981. *Basaltic Volcanism on the Terrestrial Planets*. Pergamon.

BILGIN, A. Z. & ERCAN, T. 1981. Petrology of the Quaternary basalts of İskenderun Gulf–Osmaniye area. *Geological Society of Turkey Bulletin*, **24**, 21–30.

CAMP, V. E. & ROOBOL, M. J. 1992. Upwelling asthenosphere beneath western Arabia and its regional implications. *Journal of Geophysical Research*, **97**, 15 255–15 271.

CAS, R. A. F. & WRIGHT, J. V. 1987. *Volcanic Successions*. Allen & Unwin.

COX, K. G. & JAMIESON, B. G. 1974. The olivine-rich lavas of Nuanetsi: a study of polybaric magmatic evolution. *Journal of Petrology*, **15**, 269–301.

DUNHAM, A. C. & WILKONSON, F. C. F. 1978. Accuracy, precision and detection limits of energy-dispersive electron-microprobe analyses of silicates. *X-ray Spectrometry*, **27**, 50–56.

EMBEY-ISZTIN, A., DOWNES, H., JAMES, D. E. ET AL. 1993. The petrogenesis of Pliocene alkaline volcanic rocks from Pannonian Basin, eastern Central Europe. *Journal of Petrology*, **34**, 317–343.

FITTON, J. G. 1987. The Cameroon Line West Africa: a comparison between oceanic and continental alkaline volcanism. In: FITTON, J. G. & UPTON, B. G. J. (eds) *Alkaline Igneous Rocks*. Geological Society, London, Special Publications, **30**, 273–291.

—— & DUNLOP, H. M. 1985. The Cameroon Line, West Africa, and its bearing on the origin of oceanic and continental alkali basalt. *Earth and Planetary Science Letters*, **72**, 23–38.

——, JAMES, D. & LEEMAN, W. S. 1991. Basic magmatism associated with Late Cenozoic extension in the Western United States: compositional variations in space and time. *Journal of Geophysical Research*, **96**, 13 693–13 711.

FLOYD, P. A. & CASTILLO, P. R. 1992. Geochemistry and petrogenesis of Jurassic ocean crust basalts, ODP Leg 129, Site 801. In: LARSON, R., LANCELOT, Y. ET AL. (eds) *Proceedings of Ocean Drilling Project Scientific Results*, **129**, 361–388.

——, KELLING, G., GÖKÇEN, S. L. & GÖKÇEN, N. 1991. Geochemistry and tectonic environment of basaltic rocks from the Misis ophiolitic mélange, south Turkey. *Chemical Geology*, **89**, 263–280.

——, ——, —— & —— 1992. Arc-related origin of volcaniclastic sequences in the Misis Complex,

southern Turkey. *Journal of Geology*, **10**, 221–230.
FREY, F. A., GREEN, D. H. & ROY, S. D. 1978. Integrated models of basalt petrogenesis: a study of quartz tholeiites to olivine melilites from southeastern Australia utilising geochemical and experimental petrological data. *Journal of Petrology*, **19**, 463–513.
GOTTWALD, J. 1940. Die Burg Til im Sudöstlichen Kilikien. *Byzantinische Zeitschrift*, **40**, 89–104.
GREEN, D. H. 1971. Compositions of basaltic magmas as indicators of conditions of origin: applications to oceanic volcanism. *Philosophical Transactions of the Royal Society, London, Series A*, **268**, 707–725.
——, HIBBERSON, W. O. & JAQUES, A. L. 1979. Petrogenesis of mid-ocean ridge basalts. *In*: MCELHINNEY, M. W. (ed.) *The Earth: its Origin, Structure and Evolution*. Academic Press, 265–299.
GÜLEN, L., BARKA, A. & TOKSÖZ, M. N. 1987. Continental collision and related complex deformation: Maraş triple junction and surrounding structures. *Hacettepe University Earth Sciences*, **14**, 319–336.
HARANGI, S. 1994. Geochemistry and petrogenesis of the Early Cretaceous continental rift-type volcanic rocks of the Mescek Mountains, south Hungary. *Lithos*, **33**, 303–321.
KARIG, D. E. & KOZLU, H. 1990. Late Palaeogene–Neogene evolution of the triple junction region near Maraş, south-central Turkey. *Journal of the Geological Society, London*, **147**, 1023–1034.
KAY, R. W. & GAST, P. W. 1973. The rare earth content and origin of alkali rich basalts. *Journal of Geology*, **81**, 653–683.
KELLING, G., GÖKÇEN, S. L., FLOYD, P. A. & GÖKÇEN, N. 1987. Neogene Tectonics and plate convergence in the eastern Mediterranean: new data from southern Turkey. *Geology*, **15**, 425–429.
LE BAS, M. J., LE MAITRE, R. W., STRECKEISEN, A. & ZANETTIN, B. 1986. A chemical classification of volcanic rocks based on the total alkali–silica diagram. *Journal of Petrology*, **27**, 745–750.
NUR, A. & BEN-AVRAHAM, Z. 1978. The eastern Mediterranean and the Levant: tectonics of continental collision. *Tectonophysics*, **46**, 297–311.
PARLAK, O., KOZLU, H., DEMİRKOL, C. & DELALOYE, M. 1997. Intracontinental Plio-Quaternary volcanism along the African–Anatolian plate boundary, southern Turkey. *Ofioliti*, **22**, 111–117.
PERİNÇEK, D. & ÇEMEN, İ. 1990. The structural relationship between the East Anatolian and Dead Sea fault zones in southeastern Turkey. *Tectonophysics*, **172**, 331–340.
SATO, H. 1977. Nickel content of basaltic magma: identification of primary magmas and a measure of the degree of olivine fractionation. *Lithos*, **10**, 113–120.
SCHIETTECATTE, J. P. 1971. Geology of the Misis Mountains. In: CAMPBELL, J. (ed.) *The Geology and History of Turkey*. Tripoli, Petroleum Exploration Society of Libya, 305–312.
SCHMIDT, G. C. 1961. Stratigraphic nomenclature of the Adana region, petroleum district. VII. *Turkish Petroleum Administration Bulletin*, **6**, 47–62.
ŞAROĞLU, F., EMRE, Ö. & KUŞÇU, İ. 1992. *Active Fault Map of Turkey at 1:1 000 000 scale*. Mineral Research and Exploration Institute of Turkey (MTA) Publications.
ŞENGÖR, A. M. C. & YILMAZ, Y. 1981. Tethyan evolution in Turkey: a plate tectonic approach. *Tectonophysics*, **75**, 181–241.
——, GÖRÜR, N. & ŞAROĞLU, F. 1985. Strike-slip faulting and related basin formation in zones of tectonic escape: Turkey as a case study. *In*: BIDDLE, K. & CHRISTIE-BLICK, N. (eds) *Strike-slip Faulting and Basin Formation*. Society of Economic Palaeontologists and Mineralogists, Special Publications, **37**, 227–264.
SUN, S.-S. & MCDONOUGH, W. F. 1989. Chemical and isotopic systematics of oceanic basalts: implications for mantle composition and processes. *In*: SAUNDERS, A. D. J. & NORRY, M. J. (eds) *Magmatism in the Oceanic Basins*. Geological Society, London, Special Publications, **42**, 313–345.
THOMPSON, R. N., MORRISON, M. A., HENDRY, G. L. & PARRY, S. J. 1984. An assessment of the relative roles of crust and mantle in magma genesis: an elemental approach. *Philosophical Transactions of the Royal Society, London, Series A*, **310**, 549–590.
WALKER, G. P. L. & CROSDALE, R. 1971. Characteristics of some basaltic pyroclastics. *Bulletin of Volcanology*, **35**, 303–317.
WATSON, E. B. & RYERSON, F. J. 1986. Partitioning of zirconium between clinopyroxene and magmatic liquids of intermediate composition. *Geochimica et Cosmochimica Acta*, **50**, 2523–2526.
WEAVER, B. L., WOOD, D. A., TARNEY, J. & JORON, J. L. 1987. Geochemistry of ocean island basalts from the south Atlantic: Ascension, Bouvet, St. Helena, Gough and Tristan da Cunha. *In*: FITTON, J. G. & UPTON, B. G. J. (eds) *Alkaline Igneous Rocks*. Geological Society, London, Special Publications, **30**, 253–267.
WESTAWAY, R. 1994. Present-day kinematics of the Middle East and Eastern Mediterranean. *Journal of Geophysical Research*, **99**, 12 071–12 090.
—— & ARGER, J. 1996. The Gölbaşı basin, southeastern Turkey: a complex discontinuity in a major strike-slip fault zone. *Journal of the Geological Society, London*, **153**, 729–743.
WILSON, M. 1989. *Igneous Petrogenesis: A Global Tectonic Approach*. Unwin Hyman.
—— & DOWNES, H. 1991. Tertiary–Quaternary extension-related alkaline magmatism in Western and Central Europe. *Journal of Petrology*, **32**, 811–849.
YILMAZ, Y., GÜRPINAR, O. & YİĞİTBAŞ, E. 1988. Tectonic evolution of the Miocene basins at the Amanos mountains and the Maraş region. *Turkish Association of Petroleum Geologists Bulletin*, **1**, 52–72 [in Turkish with English abstract].
YÜRÜR, M. T. & CHOROWICZ, J. 1998. Recent volcanism, tectonics and plate kinematics near the junction of the African, Arabian and Anatolian plates in the eastern Mediterranean. *Journal of Volcanology and Geothermal Research*, **85**, 1–15.

Index

Page numbers in *italics* refer to Figures and page numbers in **bold** refer to Tables

A-type magmatism 451–453
Abra alba 255
Abra scythian 253
Acanthocardia andrussovia 255
Adala Limestone 367–368
Adatepe quartz monzodiorite 446
Adjara–Trialeti Unit 171–172, *173*, 174
Aegean extensional province 327
African Plate 423
Afyon Zone *26*
Akçaabat Formation 256
Akçakoyunlu quartz monzodiorite 446
Akçay Formation 374
Akliman Formation 252–253
Akşehir Fault Zone 413–416
Akşehir–Afyon Graben
  fill 407–412
  seismicity 417
  structure 412–417
  summary history 417–420
Aksu Basin 110, 111
Aktaş Dam 187, **190**
Alandere Formation *49*
Alanya Massif 107–110, 111
Alanyurt Formation 374
Alaşehir Group 365, 367
algae and algal mounds 85, 285
Almus Fault Zone 423, *424*
Alpine Orogeny 156–157
Alpine Tethys
  closure 13
  defined 1
  opening 12
Alvenius nitidus 253
amphibole chemistry 226, **228**
Anatolian Block 423–425
  Neogene volcanism palaeomagnetic study
    methods 426–429
    results 429–432
    results discussed 433–439
Anatolide–Tauride Block 25
Andana Basin 113–114, 116, 490
Andrusov Rise 237
Antalya Complex 103–107
Antalya Gulf *see* Manavgat Basin
Arabian foreland 118–120
Arabia Plate 423
Arabian Platform 461
Arapkir 462
Arapkir–Doğanşehir Fault 462
Arkhangelsky Rise 237
Armorian Terrane 5
Artvin Massif 175
Artvin–Bolnisi Unit 172, *173*, 174–176
Aslantaş Fault Zone 489
Aslantaş Formation *314*
Athlar Formation 176, *177*
Atlantic Ocean (North) 12

Avalonia 5
Aydoğu Formation 367, 368–369
Ayvalık Burhaniye Graben 355

back-arc spreading tectonic model 386, 405–406
Bağdiğin Formation *306*
Bakırköy Formation 174
Balıklıova Formation *48*, *49*, 59, 60
Baranadağ pluton 451
Barbotella intermedia 255
Barla Dağ
  map *164*
  radiolarites
    controls on sedimentation 166–169
    cycle of deposition 165–166
basalts of İskenderun Gulf
  geochemistry
    methods of analysis 494
    results 501–508
  origins 508–510
  petrography 494–501
  setting 491–493
basanite 495–501
Baskil Arc 120
Bayat Formation *304*
Bayındır syenite 451
Bayraktepe Formation 256
Bayramiç Graben 364
Bayramiç Group 363
Bella costulata 253
Bergama Graben 353, *354*, 355, *356*, *359*, *362*
Bey Dağları Platform 106
Beypazarı Group 142
Biscay Ocean 12, 13
Bitlis Massif 120
Bitlis Suture Zone 423
Bitola Basin *329*
Bittium agibelicum 253
Bittium digitatum 253
Bittium reticulatum 256
bivalves
  Eastern Paratethys 253, 254, 255, 256
  Karaburun mélange 54–55
  Manavgat Basin 279
Black Sea Palaeocene–Eocene history
  evidence 236–244
  models discussed 244–247
Blagoevgrad Basin *329*
blueschists 27, 123, 141
Borlu Formation 370
Bornova mélange *48*, *49*, 59–60
Bostancı Formation 255
Botevgrad Basin *329*
Bozçaldağ Formation 443
Bozdağ Horst *354*
Bozkır Formation *310*, *314*
Braarudosphaera bigelowi 256
brachiopods 85

bryozoa 85
   Eastern Paratethys 253, 254
   Manavgat Basin 279
Bulgaria
   tectonic regime
      Eocene–Oligocene 328–332
      Miocene–Pliocene 332–340
      Quaternary 341–343
Bunte facies *49*
Burgas Basin *329*
Büyük Menderes Graben 353, *354*, 387–388
   age 395–396
   extension timing
      evidence reviewed 396–398
      sedimentary evidence 388–389, 391–394
      structural evidence 389–391, 394–395

Çağlayan Formation 174
Çal Unit 33
calcarenite 166
Calliostema hommairei 255
Camiboğazı Formation *48*, *49*, 56, 60
Camoba Formation 89
Çamsarı pluton 451
Çandır Formation *306*, *310*, *314*
Candona sp. 255
Cankılı Monzodiorite 446
Çankırı Basin 187, **190**, 298
   fault plane analysis
      methods 300–303
      results 303–316
      results discussed 316–321
   structural setting 300
   tectonostratigraphy 299
carbonate blocks
   in Karakaya Orogen 87–90
   see also limestones
carbonate platform environment 92–93
carbonate ramp 167
Carboniferous
   limestones 36
   metamorphic rocks 34
   molasse 35
   plate settings *8*
Cardium edulis 256
Cardium fittoni 255
Cardium sp. 253
Carpathian–Balkan Chain 326
Caspiolla sp. 255
Catapsyndrax dissimilis 281
Central Anatolia
   tectonic regime 406
   see also Akşehir–Afyon Graben
Central Anatolian Crystalline Complex (CACC) 203
   intrusive events 443–446
      A-type 451–453
      I-type 446–451
      S-type 446
   metamorphic events 442–443
   ophiolites 187, 188, 203, 204
      comparison with Sarıkaraman ophiolite 214
      emplacement model 215–217
      geochemistry **187**, *193*, **207**, **208**, **209**, 210–214
      petrography 206
      stratigraphy 204–206
   ophiolites and metabasites of Kırşehir Block 185
      alteration effects 188
      lithological units 185
      magmatic groups 185–186
         geochemistry 188–198
         petrography 187
      tectonic modelling 199–200
   plate setting *454*
Central Black Sea Rise 237
Central Sakarya ophiolite complex 148
Central Sakarya Terrane
   evolution 152–157
   tectonostratigraphy 142–150
characteristic remnant magnetism (ChRM) 426
chert
   Chios mélange 65
   Karaburun mélange 52–53
Chios (Greece) 45–46
   stratigraphy
      correlation with Karaburun Peninsula 70–71
      Mesozoic 67–70
      Palaeozoic 61–67
   tectonic model 71–78
Chios mélange *49*
   description 61–66
   interpretation 66–67
Chlamys digitalina domgeri 253
Chlamys opercularis domgeri 253
Çiçekdağ ophiolite 187, 188, 204
   comparison with Sarıkaraman ophiolite 214
   emplacement model 215–217
   geochemistry **187**, *193*, **207**, **208**, **209**, 210–214
   petrography 206
   stratigraphy 204–206
Cilicia–Andana Basin 116
Cimmerian Block 2
Cimmeride deformation 25
Cimmeride Orogen 27, 35, 36–37
   Central Sakarya 155–156
Çipköy 463
CIPW norm **497**
Cladocoropsis Limestone *49*
clinopyroxene
   İskenderun Gulf volcanics **497**, **499**
   Pozantı–Karsantı ophiolite 222–224
Coccolithus pelagicus 256
Congeria sp. 255
conglomerate 255, 256
Conkbayırı Formation 256
coral of Manavgat Basin 279, 281–282
coralline algae of Manavgat Basin 279
cover rocks of east Pontian–Transcaucasus 179
Crassostrea gryphoides 253
Cretaceous
   Central Sakarya 147–148
   Karaburun Peninsula 59
   plate settings *14*
   Taurus Mts area 125–127
Çukurova Basin 490
cumulates, gabbroic
   locations 219–220
   Pozantı–Karsantı study 220–221
      geochemistry 222–226

cumulates, gabbroic–*cont.*
    petrogenesis 227
    petrography 221–222
    tectonic significance 227–231
    types 220

Dağküplü Mélange 150
Dead Sea Fault Zone 423, 489
Demirci Graben 359, *369*, 370
Denizcik Formation 375
Denizgiren Group 47
Derbent Limestone 87, 89
Deyim Formation *304*, *306*, *310*, *314*
Dikili Group 360
Discoaster brouweri 256
Ditrupa sp. 253
Dizilitaşlar Formation *304*
Djerman Basin *329*
Doğancık Formation 410
Drama Basin *329*
Dreissena inequivalis 256
Dreissena sp. 255
Dumluca pluton 451, 453
Dures Basin *329*
Durmuşlu syenite 451
Dursunlu Formation 412

earthquakes 417, *418*
East Anatolian Fault Zone 424, 489
Eastern Black Sea Basin 235
    evolution 237–239, 244, 246, 247
Eastern Paratethys
    palaeogeography *257*, *259*, *260*, *261*, *263*, *264*, *265*
    sedimentary succession
        Black Sea 252–256
        Marmara 256
Ecemiş Fault Zone 424
    Neogene palaeomagnetic study
        methods 426–429
        results 429–432
        results discussed 433–439
Ecemiş–Sivas Fault Zone *296*
echinoderms
    Eastern Paratethys 253, 255
    Karaburun mélange 54–55
    Manavgat Basin 279
eclogite 27
Edremit Graben 353, 359, 367
Eğrigöl Basalts 360
Elazığ volcanics 461–463
Elphidium sp. 255
Emremsultan olistrostrome 150
Eocene tectonic setting
    Bulgaria–Greece 328–332
    Taurus Mts area 127–129
Eocimmerian event 2, 7–9
epi-Baikalian (Pan-African) domain 2
erosion surfaces, sub-Quaternary 340–341
Esenkaya Formation 374
Etili Graben 361, 364
Eupatorina littoralis 255
Euxinocythere sp. 255
Evrenli Group 366, 367

extension modelling 386, 405–406
    phases identified 394–397
Ezine Group 363, 364
Ezine Horst 361

fanglomerate 256
fault plane analysis
    method 300–303
    results 303–316
    results discussed 316–321
feldspar
    İskenderun Gulf volcanics **497**, **500**
    Pozantı–Karsantı ophiolite 226, **229**
foraminifera
    Eastern Paratethys 253, 254, 256
    Manavgat Basin 279
    Northern Biofacies Belt 85–87
    Southern Biofacies Belt 83–85
fusulines 85

gabbro cumulates *see* cumulates
gastropods
    Eastern Paratethys 256
    Manavgat Basin 279
Gates of Issos escarpment 467
Geceleme Formation 273, **276**, 285–289
Gediz Graben 353, *354*, *359*, 364–365, 386
geochemistry
    Central Anatolia intrusives *445*, *447*, *448*, *449*, *452*
    Çiçekdağ ophiolite **207**, **208**, **209**, 210–214
    Dağküplü Mélange 150
    İskenderun Gulf alkali volcanics
        methods of analysis 494
        results 501–508
    Pozantı–Karsantı ophiolite 222–226
    volcanics of SE Turkey
        methods of analysis 468
        results 471–474
Georgia *see* Transcaucasus (southern)
Gephyrocapsa caribbeanica 256
Gerence Formation 47, *48*, *49*, 54, 60
Gibbula bajarunasi 253
Gibbula chokrakensis 253
Girdapdere meta-olistrostrome 152–153
Globigerina spp. 279, 281, 287
Globigerinoides spp. 281
Globivalvulina spp. 86
Gödene Zone 106
Gökova Graben 353, 372, 375
Göktepe Formation 374
Göktepe Metamorphics 151–152
Gölbaşı Basin 489
Gölbaşı–Türkoğlu Fault 465
Göldağ Suite 461
Gölyaka Formation 409
Gördes graben 370
Gotse Delchev Basin *329*
Gotse Delchev Graben 335
Govedartsi Basin *329*
Gözpınarı Formation 410–411
Göztepe Formation 256
granodiorite of Sakarya zone 34–35
Greece
    tectonic regime

Greece–*cont.*
    Eocene–Oligocene 328–332
    Miocene–Pliocene 332–340
    Quaternary 343–345
Guleman ophiolite 120
Gülpınar Graben 361
Güvercinlik Formation *48*, *49*, 56, 57

Hadım Nappe 103
Halimeda 279
Hallstatt facies *49*
Hallstatt Meliata Ocean 7, 9
Hamit pluton 451
Hancılı Formation *314*
Hastacondana spp. 255
Hayriye pluton 451
Heliastraea 279
Helicosphaera kaptneri 256
Hellenic Trench *26*
Hellenides 326
hematite 426
Hemigordiopsis 86
Heterostegina 279
Hıdırlık Formation 254–255
Hodul Unit 33–34, 36
Hoyran–Beyşehir–Hadım Nappes 103
Hun Superterrane 3, 6, 17

I-type magmatism 446–451
İcikler Horst *369*, 370
İdişdağ pluton 451
igneous activity *see* plutonism *also* volcanism
Ihtiman Basin *329*
Ihtiman Graben 340
ilmenite 426
Imbricated Bajburt–Karabakh Unit 172, *173*, 176–178
İncik Formation *304*, *306*, *310*
Inner Tauride Belt 185
Inner Tauride Suture *26*
Intra-Pontide Ocean 203
Intra-Pontide Suture Zone *26*, 139, 140
intrusive events
    Central Anatolia 446–453
        timing 453–455
    Taurus Mts area 120
    Western Anatolia 379
İsalı Formation 490
İskenderun Gulf 466
    basalts
        geochemistry 494, 501–508
        origins 508–510
        petrography 494–501
        setting 491–493
İskenderun sub-basin 490
isotopic dating
    Neogene–Quaternary volcanics
        method 467
        results 468
Isparta Angle 110, 271
    see also Akşehir–Afyon Graben
İspendere ophiolite 120
İstanbul nappe 139
İstanbul Terrane 140

İstanbul zone 27
İzmir–Ankara ophiolite suture *354*
İzmir–Ankara Suture 14, *26*, 27
İzmir–Ankara–Erzincan Ocean 184, 185
İzmir–Ankara–Erzincan Suture Zone (İAESZ) 45–46

Jurassic
    Central Sakarya 147
    Chios 70
    Karaburun Peninsula 56–59
    plate settings *12*
    radiolarites
        controls on sedimentation 166–169
        cycle of deposition 165–166
        stratigraphy of Barla Dağ *165*
        tectonic setting of Taurus Mts area 125

K/Ar dating
    Black Sea 243
    Neogene–Quaternary volcanics
        method 467–468
        results 468–471
Kahraman Maraş 463–466
Kalkanlıdağ Formation 443
Kale–Tavas basin *354*, *359*, *371*, *374*
Kaleboynu Formation 185–186, 187 **191**
Kali Formation 374
Kamartsi Basin *329*
Kamenitsi Basin *329*
Karabakır Formation 462, 463
Karabalçık Formation *304*
Karaburun mélange
    description *48*, 49–53
    interpretation 53–54
Karaburun Peninsula 45–46
    history of research 46–47
    stratigraphy 47–48
        correlation with Chios 70–71
        Mesozoic 54–61
        Palaeozoic 49–54
    tectonic model 71–78
Karagöl Formation 374
Karakaya Complex 27, 46
Karakaya Formation 27, 360
    basal features 30–33
Karakaya monzogranite 446
Karakaya Orogen 27
    fate of Upper Permian 87–90
Karakeban pluton 451, 453
Karaköy–Evciler pluton 363
Karareis Formation 47
Karataş Formation 490
Karataş–Osmaniye Fault Zone 489
Karçal Intrusive Suite 175
Kargöztepe Fault Zone 413
Karhova Triple Junction 489
Karkinit Basin 244
Karlıovo Basin *329*
Karlıovo Graben 336
Karnobat–Aitos Basin *329*
Karnobat–Itos Graben 337
Karpuzçay Formation 273, **276**, 289–291
Kastamonu granitoids 33

Kavak Formation *306*
Kavardarci Basin 329
Kayapınar metacarbonates 152
Kayıkbaşıburnu Formation 255
Kazdağ Horst 361
Kemer Zone 106
Kestanbolu pluton 363
Khrami massif 175
Kırıkkale–Erbaa Fault Zone 423, *424*
Kırklar 462
Kırşehir Block 185
　alteration effects 188
　lithological units 185
　magmatic groups 185–186
　　geochemistry 188–198
　　petrography 187
　tectonic modelling 199–200
Kırşehir Massif 25
Kızılçay Group 142
Kızıcık Formation 175
Kızılırmak Fault Zone 298
Kızıldağ Group 365, 366, 367
Kızıldağ ophiolite 459
Kilçak Formation *314*
Kirazlı Formation 256
Kocaçay Formation *304*
Komotnin Basin 329
Köprü Basin 110
Korce Basin *329*
Kösedağ batholith 453
Kostenets Basin *329*
Kostenets Graben 340
Köstere Formation 407–409
Kozak Granite 360
Kozak Horst 355, *356, 357*, 360
Kraguleva Basin *329*
Kraljevo Basin *329*
Küçükkuyu Group 363
Kultak Formation 375
Kumluca Zone 106
Kura Basin *26*
Küre Complex 33
Kurtkuyusu Formation 253–254
Kustendil Basin *329*
Kütahya–Bolkardağ Belt 141
Kyrenia–Misis Lineament 116

Lamemmaptychus 166
Larisa Basin *329*
limestones
　Chios mélange 61–63
　Eastern Paratethys 253, 254, 255, 256
　Karaburun mélange 50–52
　Manavgat Basin 282–285
　Permian–Carboniferous 36
　see also carbonate blocks
Lithophyllium 279
Lithothamnium 279
Lomonosov Massif *242*, 243
Louisettita 86
Loxoconcha sp. 255
Lycian Bozkır Nappes 141
Lycian Nappe Front *354*
Lycian Nappes 101–103, *354*, 374

Mactra spp. 255, 256
Maden Complex 120–121
magnetite 426
major element analysis
　Central Anatolia intrusives *445, 447, 448, 449, 452*
　Chios mélange **57**
　gabbro cumulates 222, **223**
　İskenderun Gulf volcanics **496**, 503–504
　Karaburun Peninsula **56**
　volcanics of SE Turkey
　　methods of analysis 467
　　results **469**, 471–474
Manavgat Basin 111, 271, 272
　facies analysis 277–291
　stratigraphy 273–274
　structural setting 275–277
　summary of evolution 291–292
Mauretanian Ocean 3
Mediterranean Sea
　formation of 251
　tectonic setting 43–44, 97–99
Meliata Ocean 12
Meliata Ridge 7
Meliata Rift 8
Menderes Massif *26*, 46, *354*, 365, 367
Menderes–Central Anatolian Unit 141
Mesta Graben 335
metamorphism and metamorphic rocks
　Central Anatolia 442–443
　　timing 453–455
　Central Sakarya 141–142, 142–144, 146–147, 151–153
　massifs of S Turkey 120
Mihhikaya metacarbonates 152
miliolids, Eastern Paratethys 253
mineral analysis and mineralogy
　Göktepe Metamorphics 151
　Pozantı–Karsantı ophiolite 222–224, 226, **228, 229**
Miocene
　palaeogeography 258–266
　stratigraphy 277–291
　tectonic regime
　　Bulgaria–Greece 332–340
　　Taurus Mts area 129–130
Miogypsina 279
Mirkovo Basin *329*
Misis Complex 490
Misis–Andırın Complex 115–116
Misis–Andırın Trend 489
Moesian Platform *26*
molasse, Carboniferous 35
mudstone, Eastern Paratethys 255
Murmana pluton 451, 453
Mut Basin 114–115
Mytilaster volhynicus 255

Naldökendağ Formation 443
nannofossils 256
Nassarius limatus 253
Neogene basins of S Turkey 111–113
Neogene volcanism
　Anatolia 425–426
　Ecemiş Fault Zone palaeomagnetic study
　　methods 426–429

Neogene volcanism–*cont.*
  results 429–432
  results discussed 433–439
  S Turkey 120–121
  SE Turkey
    dating 467–468, 468–471
    field work 461
    geochemistry 467, 471
    interpretations 475–483
Neotethys
  closure 459
  defined 1
  modelling evolution 44
  opening 2, 10–11, 38, 99
Neotethys Suture 489, 490
Nestos Basin *329*
Nilüfer Unit 30–33
Nis Basin *329*
Nohutalan Formation *48*, 59
North Anatolian Fault Zone 361, 423, 489
North Anatolian Ophiolite Mélange 300
North Anatolian–Lesser Caucasus ophiolite belt 178–179
Northern Biofacies Belt
  foraminiferal assembly 85–87
  lateral continuity 90–91
  origins 87–90
Nuculana fragilis 253

oceanic plateau collision model 30–34, 36–37
Ohrid Basin *329*
Oligocene tectonic settings
  Bulgaria–Greece 328–332
  Taurus Mts area 129
olistostromes 36, 150
olivine analyses **497**, **498**
Olympus, Mt, tectonic setting 332
Operculina 279, 282
ophiolite
  belts in Turkey 183–185
  Central Sakarya ophiolite complex 148
  Çiçekdağ ophiolite 187, 188, 204
  Guleman ophiolite 120
  İspendere ophiolite 120
  Kırşehir Block study
    magmatic groups 185–186
      alteration effects 188
      geochemistry 188–198
      petrography 187
      tectonic modelling 199–200
  occurrence 219–220
outcrops *16*
Pontian–southern Transcaucasus suture 178–179
Pozantı–Karsantı ophiolite 220–221
  geochemistry 222–226
  petrogenesis 227
  petrography 221–222
  tectonic significance 227–231
Sarıkaraman ophiolite 187, 188
Taştepe ophiolite 148, 150
Orbulina universa 287
Ordovician plate settings *4*
Ören Formation 374
Ören Graben *372*, 374–375

Örenli–Eğiller Graben 355
Orhanlar Greywacke 33
orogenic collapse tectonic model 386, 405–406
orthopyroxene chemistry 224, 226, **228**
Osmankahya Formation *304*
ostracods 253, 255, 256
Ostrea spp. 253, 256
Otluk metaclastites 152
Ova Regine 406
Oymapınar Limestone 273, 282–285, **285**
Özbek Formation 256

Padesh Basin *329*
Padesh Graben 335
Palaeocene
  sediments
    Black Sea 236–244
    Karaburun Peninsula 59
  tectonic setting 127
palaeogeography
  Antalya Complex *108*
  Eastern Paratethys 257, 259, 260, 261, 263, 264, 265
  Permian 92–93
palaeomagnetic analysis
  methods 426–429
  results 429–432
  results discussed 433–439
palaeostress measurement 295
  case study in Çankırı Basin 298, 300
  fault plane analysis 300–321
Palaeotethyan orogen 27
Palaeotethys
  closure 2, 7, 25
  defined 1
  modelling evolution 44
  opening 3–6
Palakaria Basin *329*
Palakaria Graben 340
Pamphylian Suture *26*
Pan-African (epi-Baikalian) domain 2
Pangaea, formation of 7
Paradacna abichi 255, 256
Paradagmarita 85, 86
Paratethys, formation of 251–252
  see also Eastern Paratethys
Pec Basin *329*
Pechenega–Camena Fault 238
Pelagonian Zone *26*
Peri-Arabic Belt 203
Permian
  foraminifera
    Northern Biofacies Belt 85–87
    Southern Biofacies Belt 83–85
  sediments 36
  palaeogeography 92–93
  plate settings *8*, *9*
  tectonic evolution 124–125
Pernik Basin *329*
Persink–Bobov Basin *329*
petrogenetic studies 227
petrography
  Central Anatolian Crystalline Complex ophiolites 187
  Çiçekdağ ophiolite 206

petrography–*cont.*
 Dağküplü Mélange 150
 İskenderun Gulf volcanics 494–502
 Pozantı–Karsantı ophiolite 221–222
plagioclase
 İskenderun Gulf volcanics **497, 500**
 Pozantı–Karsantı ophiolite 226, **229**
plagiogranite 175, 188
plant fossils 253
plate modelling of ocean closure 199–200
Pleistocene *see* Quaternary
Pliocene
 basalts *see* İskenderun Gulf volcanics
 sediments 266
 tectonic regime
  Bulgaria–Greece 332–340
  Taurus Mts area 130
Plovdiv Basin *329*
plutonism
 Central Anatolia 446–453
  timing 453–455
 western Anatolia 379
Pontides 25
Pontides, Eastern
 evolution 179–180
 lithotectonic units 174–179
Pontoniella sp. 255
Porites 279
Pozantı–Karsantı ophiolite 220–221
 geochemistry 222–226
 petrogenesis 227
 petrography 221–222
 tectonic significance 227–231
Prespansko Basin *329*
Pristina Basin *329*
Prosenik Basin *329*
Prosenik Graben 337
Prosodacno sp. 255
Protopeneroplis striata 166
Prototethys 3
pseudobrookite 426
Pseudoemiliana lacunosa 256
Pütürge Massif 121
pyroxene
 İskenderun Gulf volcanics **497, 499**
 Pozantı–Karsantı ophiolite 222–224, 226, **228**

Quaternary
 tectonic regime 130–131, 341–345
 volcanism
  İskenderun Gulf
   geochemistry 494, 501–508
   origins 508–510
   petrography 494–501
   setting 491–493
  SE Turkey
   dating 467–468, 468–471
   field work 461
   geochemistry 467, 471
   interpretations 475–483

radiolarian chert 55–56
radiolarites 55, 123

Barla Dağ
 controls on sedimentation 166–169
 lithological cycle 165–166
 history of interpretation 164
rare earth element analysis 222, **224**, 507–508
Razlog Basin *329*
Razlog Graben 335
Reticulocondana sp. 255
Rheic Ocean 3
Rhodope–Strandja Zone *26*
Rioni Basin *26*

S-type magmatism 175, 446
Saccocoma 166
St George Fault 239
Sakarya Composite Terrane 142
Sakarya Microcontinent 139
Sakarya Zone *26*, 27
 clastic sequences 33–34
 collision complexes 27–33
 Hercynian units 34–35
 orogeny timing 35
 pre-Jurassic outcrops *28, 29*
 subduction history 35–36
 tectonic setting 27
Sandanski Basin *329*, 332–334
sandstones
 Eastern Paratethys 256
 Karaburun mélange 53
 geochemistry **187**, *193*
Sarantsi Basin *329*
Sarantsi Graben 335
Sarıhacılı leucogranite 446
Sarıkaraman ophiolite 187, 188, 214
Sarıkum Formation 255
Sart Group 365, 366, 367
Scythian Platform *26*
seismicity in Akşehir–Afyon Graben 417, *418*
Sekköy Formation 374
Selendi Graben 370
Semail Ocean 14
Sevan–Akera Suture *26*
shale
 Eastern Paratethys 256
 Karaburun mélange 53
Shantia 85, 86
Sheinovo Basin *329*
Sheinovo Graben 336
siltstone in Eastern Paratethys 255
Silurian plate settings 6
Simitli Basin *329*
Skadarsko Basin *329*
Skopje Basin *329*
slickensides *415*
 in fault plane analysis
  method 300–303
  results 303–316
  results discussed 316–321
Sliven Basin *329*
Sliven Graben 337
Sofia Basin *329*, 336
Soğukkuyu Metamorphics 146–147
Soğüt Metamorphics 141–142, 142–144
Sömdiken Metamorphics 142, 151

Southern Biofacies Belt
  foraminiferal assembly 83–85
  lateral continuity 90–91
  origins 87–90
Southern Neotethys Ocean 184, 203
Southern Turkey *see* Taurus Mts area
spider diagrams *450*
Spiratella tarchanensis 253
Srednagorie Basin *329*
Straldja Basin *329*
Straldja Graben 337
stress inversion study *see* fault plane analysis
Strumeshnitsa Basin *329*
Strumitza Basin *329*
Strymon Basin *329*
Stylophora 279
Sub-Balkan Graben System 336–340
subduction, Palaeotethyan 35–36, 36–37
subduction–accretion complexes 27–30
subsidence curves *5, 11, 13, 15*
Süleymanlı Formation *304, 306, 310, 314*
Sungurlare Basin *329*
Sungurlare Graben 337
Sungurlu Fault Zone 298
suture zones *16*

Tarbellastraea 279
Taşköprü Formation 411–412
Taştepe ophiolite 148, 150
Tauride–Anatolide Platform 139, 141
Tauride Belt 141
  biofacies belts 90–91
Taurus Mts area
  divisions 99–100
    Central segment 113–116
    Eastern segment 118–124
    Western segment 101–113
  tectonic evolution
    Cenozoic 127–131
    Mesozoic 124–127
tectonic modelling 122–124, 386, 405–406
tectonostratigraphy
  Çankırı Basin 299
  Central Sakarya 142–153
tempestite 166
Tepekli Conglomerate 273, **275**, 277–282
Tepeköy Metamorphics 141–142, 144–146
Tethyan cycle 7
Tethyan domain defined 1
  Alpine Tethys 12
  Palaeotethys 2, 3–6, 7, 25, 44
  Prototethys 3
  Neotethys 2, 10–11, 38, 99, 44, 459
Tethyan ocean basin, south *see* Taurus Mts area
Tethyan sutures 14–17
Tetovo Basin *329*
Texularia 279
Thermaikos Basin 329
thermomagnetic analysis
  methods 428–429
  results 429–432
  results discussed 433–439
Thrace Basin *26, 329*, 338–340
Tirana Basin *329*

Toprakkale 467
Toygar Volcanics 367, 368
trace element analysis
  Central Anatolia intrusives *445, 447, 448, 449, 452*
  gabbro cumulates 222, **223**
  İskenderun Gulf volcanics **496**, 505–508
  Karaburun Peninsula **56**
  volcanics of SE Turkey
    methods of analysis 467
    results **470**, 471, 474–476
Transcaucasus (southern)
  evolution 179–180
  tectonic units 174–179
Triassic
  olistostromes 36
  plate settings *9, 10*
  sediments
    Chios 67–69
    Karaburun Peninsula 54–56
  tectonic evolution 124–125
Trikola Basin *329*
Trocholina spp. 166
tubiphytes 166
Tuğlu Formation *304, 306, 310*
Tundja Basin *329*
Tundja Graben 337
Turkey, plate tectonic setting *26*, 459–461
Tvarditsa basin *329*
two-stage graben tectonic model 405–406
Typhlocyprela sp. 255

Uşak–Ulubey Graben 370

Valais Ocean 12, 13
Valenciennius sp. 255
Vardar Ocean 12
Vardar Suture *26*
Vardar–İzmir Ankara Ocean 141
Vardar–İzmir–Ankara Suture Zone 139, 140
Vardar–İzmir–Ankara–Erzincan (VİAE) Ocean 203
Variscan 2, 7–9
  Central Sakarya 153–155
Varlık Group 175
Veneripus aurea 256
Vetren Basin *329*
Vetren Graben 336, 337
viscous remnant magnetism (VRM) 426
Viviparus bifarcinatus 256
volcanics and volcanism
  Anatolia 425–426
  Bulgaria–Greece 328–329
  Chios mélange **57**, 63–65
  Ecemiş Fault Zone palaeomagnetic study
    methods 426–429
    results 429–432
    results discussed 433–439
  eastern Pontides 174
  İskenderun Gulf basalts 466
    geochemistry **494**, 501–508
    origins 508–510
    petrography 494–501
    setting 491–493
  Karaburun mélange 53, **56**
  SE Turkey

Karaburun mélange–*cont.*
　interpretations 475–483
　observations
　　dating 467, 468
　　field work 461
　　geochemistry 467, 471

West Black Sea Fault *26*
West Crimean Fault *26*
Western Anatolia, structural evolution 375–380
Western Black Sea Basin 235
　evolution 239, 244, 246–247, 2237
Western Taurides *see* Manavgat Basin

Yakasinek Fault 412–413
Yamadağ volcanics 461
Yapraklı Formation *304, 306*
Yarkındağ Formation 374
Yassıağıl monzogranite 446

Yatağan Basin 374–375
Yatağan Formation 375
Yaykıl Formation 255
Yaylaçayı Formation *304*
Yenidere Formation 374
Yeşilyurt Formation 367, 369
Yoncalı Formation *304*
Yozgat batholith 446
Yücebaca leucogranite 446
Yükesekova Arc 120

Zagore Basin *329*
Zajecar Basin *329*
Zeytinçay Formation 365–366
Zeytindağ Graben *358*, 361
Zeytindağ Group 360
Zeytindağ–Maruflar Horst *358*, 361
Ziyarettepe Formation 175